78.00
100K

NUCLEI FAR FROM STABILITY
FIFTH INTERNATIONAL CONFERENCE

AIP CONFERENCE PROCEEDINGS 164

RITA G. LERNER
SERIES EDITOR

NUCLEI FAR FROM STABILITY

FIFTH INTERNATIONAL CONFERENCE

ROSSEAU LAKE, ONTARIO, CANADA 1987

EDITOR:
IAN S. TOWNER
CHALK RIVER NUCLEAR
LABORATORIES

AMERICAN INSTITUTE OF PHYSICS   NEW YORK 1988

Authorization to photocopy items for internal or personal use, beyond the free copying permitted under the 1978 US Copyright Law (see statement below), is granted by the American Institute of Physics for users registered with the Copyright Clearance Center (CCC) Transactional Reporting Service, provided that the base fee of $3.00 per copy is paid directly to CCC, 27 Congress St., Salem, MA 01970. For those organizations that have been granted a photocopy license by CCC, a separate system of payment has been arranged. The fee code for users of the Transactional Reporting Service is: 0094-243X/87 $3.00.

Copyright 1988 American Institute of Physics.

Individual readers of this volume and non-profit libraries, acting for them, are permitted to make fair use of the material in it, such as copying an article for use in teaching or research. Permission is granted to quote from this volume in scientific work with the customary acknowledgment of the source. To reprint a figure, table or other excerpt requires the consent of one of the original authors and notification to AIP. Republication or systematic or multiple reproduction of any material in this volume is permitted only under license from AIP. Address inquiries to Series Editor, AIP Conference Proceedings, AIP, 335 E. 45th St., New York, NY 10017.

L.C. Catalog Card No. 87-73214
ISBN 0-88318-364-1
DOE CONF-870970

Printed in the United States of America.

Contents

Preface xx
    J. C. Hardy and I. S. Towner
International Advisory and Canadian Organizing Committees xx
Photographs xx

Mass Measurements, Mass Formulae

Direct Mass Measurements of Light Neutron-Rich Nuclei
Using Fast Recoil Spectrometers 1
    D. J. Vieira, J. M. Wouters, and the TOFI Collaboration
Mass-Measurements Far from Stability of Neutron-Rich Light Nuclei 11
    W. Mittig et al.
Mass Measurements of Short-Lived Isotopes in a Penning Trap 22
    F. Kern et al.
Beta-Decay Energies and Systematics of Nuclear Structure Effects 30
    M. Graefenstedt et al.
Mass Measurements of ^{162,163}Ta and Nuclides in their α Decay Chains 41
    E. Hagberg et al.
Experimental Mass Excess of ^{49,50}K and ^{40,42}Cl 51
    Ch. Miéhé et al.
The 1986–87 Atomic Mass Predictions 53
    P. E. Haustein
Mass Predictions, Partial Difference Equations
and Higher-Order Isospin Effects 62
    P. J. Masson and J. Jänecke
Nuclei Far from Stability via the "ETFSI" Mass Formula 66
    J. M. Pearson, A. K. Dutta, and F. Tondeur
Infinite Nuclear Matter Model for Masses of Atomic Nuclei 76
    L. Satpathy
Masses of Nuclei in the Infinite Nuclear Matter Model 80
    L. Satpathy and R. C. Nayak
A New, Simplified, Interpolation Method for Estimation of Microscopic
Nuclear Masses Based on P-Factor, $P = N_p N_n / (N_p + N_n)$ 84
    P. E. Haustein, D. S. Brenner, and R. F. Casten
The Fermion Dynamical Symmetry Model of Nuclear Collective Motion 89
    D. H. Feng
Empirical Mass Formula with Proton-Neutron Interaction 97
    T. Tachibana et al.
Occupation Number Dependent Effective Interactions
and the Odd-Even Mass Difference 101
    A. S. Jensen and A. Miranda

Moments and Radii

Static Magnetic and Electric Moments Near Closed Shells
from Laser Spectroscopy 105
    G. Huber

Investigation of Nuclear Charge Radii and Electromagnetic Moments of Rare
Earth Elements by Laser Spectroscopy 115
    G. D. Alkhazov et al.

New Techniques and Results of Collinear Laser Spectroscopy:
Radon, Xenon, and Holmium 126
    R. Neugart et al.

Laser Measurements of Radii and Moments of Barium Nuclei
Near the Proton Drip Line 136
    D. A. Eastham et al.

Studies of Oriented Nuclei at the KOOL Facility 140
    D. Vandeplassche et al.

Low Temperature Nuclear Orientation of Nuclei Far from Beta Stability.
Plans of Investigations at ISOL-Facility YASNAPP-2 157
    T. I. Kracikova et al.

Nuclear Spins and Moments of ^{9}Li and ^{11}Li 161
    E. Arnold et al.

Production of Spin Polarized ^{15}C in Heavy-Ion Reaction
and Measurement of g-Factor for the $1/2^+$ Ground State 165
    K. Asahi et al.

TRISTAN Research on Neutron-Rich Nuclei: Magnetic Moments
and Nuclear Structure 174
    A. Wolf et al.

Deformation of Light Br Isotopes 184
    A. G. Griffiths et al.

Magnetic Moments of Neutron Deficient Yttrium Nuclei 193
    I. Berkes et al.

Nuclear Ground State Properties of Strontium Isotopes ($78 \leqslant A \leqslant 100$)
by Laser Spectroscopy 197
    F. Buchinger et al.

Laser Spectroscopy in the $Z<82$ Region 201
    St. Becker et al.

Resonant Ionization Spectroscopy of Laser-Desorbed Gold Isotopes 205
    J. K. P. Lee et al.

Nuclear Moments and Change in the Charge Radii of Neutron-Deficient
Silver and Lead Isotopes 209
    U. Dinger et al.

Nuclear Structure Studies Using Beams of Radioactive Nuclei 213
    I. Tanihata

The Neutron Halo of Nuclei at the Neutron Drip Line 223
    B. Jonson et al.

Interaction Radii of Neutron-Rich Nuclei 233
    A. Gillibert et al.

Evidence for EMSO Effects in Radii of the Tin Isotopes 237
    D. Berdichevsky, R. Fleming, and D. W. L. Sprung

Shell Corrections to the Proton Mean Square Radius 241
    A. S. Jensen and A. Miranda

Shape Coexistence, Intruders

Nuclear Shape Coexistence and the Study of Nuclei Far from Stability 245
   J. L. Wood

A Shell-Model Description of Intruder States 255
   K. Heyde

Oblate-Prolate Shape Competition in Z=34–38 Nuclei 268
   J. H. Hamilton et al.

A Microscopic View of Large Deformation Nuclei with A=60–80 278
   D. P. Ahalpara and S. P. Pandya

Spectroscopy of the Transitional A=96–98 Nuclei;
New Region of Fast First Forbidden Transitions 282
   H. Mach et al.

Evidence for Shape Coexistence in Neutron-Rich Rh Ag Isotopes 286
   N. Kaffrell et al.

Transition Probabilities Between Intruder States in Heavy Ag Isotopes 296
   B. Fogelberg et al.

Shape Coexistence in the Region Around Z=82 305
   M. Huyse et al.

Low Energy E0 Transitions in Odd-Mass Nuclei of the Neutron Deficient
$180 < A < 200$ Region 313
   E. F. Zganjar et al.

Spectroscopy

The Light Exotics 324
   K. K. Seth

A Shell-Model Study of Light Exotic Nuclei 334
   N. A. F. M. Poppelier et al.

Spectroscopic Measurements with a New Method:
The Projectile-Fragments Isotopic Separation 344
   J. P. Dufour et al.

The Spectroscopy of N=Z Nuclei from $^{64}_{32}$Ge towards $^{100}_{50}$Sn 354
   C. J. Lister et al.

Neutron-Rich Nuclei Produced in Multinucleon-Transfer Reaction 365
   W.-D. Schmidt-Ott et al.

Is the Region Above ^{78}Ni Doubly Magic? 375
   J. C. Hill et al.

Lifetime Measurements in ^{72}Br 385
   S. Ulbig et al.

New Neutron Deficient Nuclei Near Mass Number A=80 389
   U. Lenz et al.

The $\pi g_{9/2} - \nu g_{7/2}$ Interaction in Neutron-Rich Nuclei with $A \approx 100$ 393
   G. Lhersonneau et al.

Neutron-Rich $A \simeq 100$ Nuclei: Nilsson Configurations, Deformation,
Pairing Reduction 403
   B. Pfeiffer et al.

Nilsson Orbitals and Rotational Bands of Deformed $A \simeq 100$ Nuclei 407
   F. K. Wohn et al.

Nuclear Spectroscopy of On-Line Mass Separated Fission Products
with Mass Numbers from A=110–120 411
    J. Äystö et al.
Low-Spin States in ^{122}Cs 415
    B. Weiss et al.
Decay of New Mass-Separated Neutron-Deficient La and Ce Isotopes 419
    J. Genevey et al.
The Daresbury Recoil Separator and γ-Ray Spectroscopy Near ^{130}Nd:
A ΔJ=1 Superdeformed Band 425
    A. N. James and D. C. B. Watson
Structure of the Odd–Odd and Odd-Mass Sb Nuclides 429
    C. A. Stone et al.
Energy Levels and Structure of Light Rare-Earth Nuclei,
^{136,138,140}Sm and ^{132,134,136}Nd, via Beta Decay 441
    B. D. Kern et al.
Identification and Structure of p-Rich Rare-Earth Nuclei Investigated Using a
He-Jet Fed On-Line Mass-Separator 445
    R. Béraud et al.
Level Structure of Neutron-Rich Odd-Mass Pr Nuclides 455
    P. F. Mantica et al.
The Structure of ^{143}Ba and ^{147}Ce 459
    J. D. Robertson et al.
Decay Studies of Neutron Deficient Rare Earth Isotopes with OASIS 463
    J. Gilat et al.
Decays of Neutron Deficient Gadolinium Isotopes 473
    R. Turcotte et al.
Q-Values and Isomer Energies from High-Resolution Alpha-, Proton-,
and Gamma-Ray Spectroscopy Above ^{146}Gd 477
    D. Schardt et al.
Proton-Neutron Interaction and the Relative Isomer Masses in ^{148}Tb 489
    J. Styczen et al.
Beta-Decay of $27/2^-$ Isomers in N=83 Nuclei 494
    P. Kleinheinz et al.
Superdeformed Nuclei at High Spin 499
    P. J. Twin
Spectroscopy of Neutron-Deficient Tantalum-to-Rhenium Isotopes 509
    F. Meissner et al.
Study of Transitional Doubly-Odd ^{186}Ir and ^{184}Ir 513
    A. Ben Braham et al.
Detection of Low-Energy Conversion Electrons and Location of Isomeric
States 517
    P. Kilcher et al.
The Alpha-Decay of Mass-Separated ^{225}Th 521
    C. F. Liang, P. Paris, and Ch. Briançon
Role of Higher-Multipolarity Deformations in the Properties of "Octupolly"
Deformed Nuclei 529
    P. Rozmej et al.
Search for Stable Octupole Deformation in the Nucleus ^{225}Fr 533
    D. G. Burke et al.

Astrophysics

Nuclear Physics Requirements for Studies of Nuclear Energy Generation
and Nucleosynthesis 543
    J. W. Truran
Nuclear Structure Effects far from Stability and their Consequences
on Rapid-Neutron-Capture Processes 558
    K.-L. Kratz et al.
Neutron-Rich Nuclei and Nucleosynthesis 568
    Yu. S. Lyutostansky et al.
The ^{22}Na(p,γ) ^{23}Mg Reaction: Preparation of a ^{22}Na Target 572
    L. Buchmann et al.
Identification and Half-Life Determination of Very Neutron-Rich Copper
Isotopes 578
    E. Lund et al.
Strangelet Mass Formula and Its Implications to the Solar-Neutrino Problem .. 581
    K. Takahashi and R. N. Boyd
Sudbury Neutrino Observatory 585
    G. T. Ewan

Gamow-Teller Strength

Quenching of Gamow-Teller Strength 593
    I. S. Towner
Gamow-Teller Strength Functions from Nucleon Scattering Experiments 604
    O. Häusser
Improvement of the Gross Theory of β-Decay 614
    T. Tachibana et al.
Gamow-Teller Beta Decay of $^{29-31}$Na, Comparison
with Shell-Model Estimates 624
    P. Baumann et al.
The Beta Strength in the Proton-Rich Argon Isotopes 634
    K. Riisager et al.
Gamow-Teller Strength in the Beta Decay of Mirror Nuclides 640
    J. Honkanen et al.
Precision Half-Life Measurements of the Mirror Nuclei
in the $f_{7/2}$-Shell Using an IGISOL 650
    H. Hama et al.
The Beta Decay of ^{48}Mn: Gamow-Teller Quenching in fp-Shell Nuclei 654
    T. Sekine et al.
Gamow-Teller Beta Decay of Even Nuclei Near ^{100}Sn 656
    K. Rykaczewski et al.

Particle Decay

Nuclear Structure Effects on α Reduced Widths 665
    K. S. Toth et al.
Alpha-Decay of Light Protactinium Isotopes 675
    T. Faestermann et al.

Anisotropic α-Emission of On-Line Separated Isotopes 679
    J. Wouters et al.
Beta-Delayed Charged Particles Decay from Light Nuclei Near the Drip Line .. 690
    J. C. Jacmart et al.
The β^+ and EC Decay of ^{69}Se, Possible Shape-Coexistence
and Superdeformation Effects in ^{69}As 695
    Ph. Dessagne et al.
Beta-Delayed Proton Decay in the Lanthanide Region 697
    J. M. Nitschke et al.
Study of Beta Decay Half-Lives and Beta-Delayed Multi-Particle Final States
at Michigan State University 708
    D. Mikolas et al.
Radioactivities with $146 \leqslant A \leqslant 152$ Investigated at the OASIS Facility;
Evidence for ^{147}Tm β-Decay 718
    K. S. Toth et al.
On the Production Mechanisms of Nuclei Far from Stability at GANIL
Energies and Their Application to New β-Delayed Neutron Emitters 722
    D. Bazin et al.
New Region of Deformation for Neutron-Rich Nuclei
and β-Delayed Neutron Emission 727
    Yu. S. Lyutostansky, M. V. Zverev, and I. V. Panov
Structure Effects in the Calculation of Beta Strength Functions
and Half Lives of Rb and Br Isotopes 731
    S. Rab and A. Shihab-Eldin
Decay of Delayed-Neutron Precursor ^{93}Rb and the Simple Statistical Model 735
    G. D. Alkhazov et al.
Proton Radioactivity of Medium Heavy Nuclei 739
    T. Faestermann et al.
Beta-Delayed Two-Proton Emission as a Nuclear Probe 749
    D. M. Moltz et al.

New Nuclei, Heavy-Element Synthesis, Fission

Production and Identification of New Nuclei at the Neutron
and Proton Drip-Lines for Light Isotopes 757
    D. Guillemaud-Mueller et al.
Identification of the New Neutron-Rich Isotopes $^{70-74}$Ni and $^{74-77}$Cu
in Thermal Neutron Fission of ^{235}U 768
    M. Bernas et al.
Identification of New Neutron-Rich Rare-Earth Isotopes Produced in
^{252}Cf Spontaneous Fission 782
    R. C. Greenwood et al.
Production of the Heaviest Elements on the Borderline of Nuclear Stability
by Cold Fusion Reactions 786
    F. P. Hessberger et al.
Study of Secondary Reaction Experiments in the Superheavy
and Actinide Regions 796
    A. Marinov, S. Eshhar, and D. Kolb

Heavy Fragment Radioactivities 800
    P. B. Price
Evidence for Bimodal Fission in the Heaviest Elements 810
    E. K. Hulet
On the Fission of the Heaviest Fermium Isotopes 821
    S. Ćwiok, P. Rozmej, and A. Sobiczewski
From Symmetric Cold Fission Fragment Mass Distributions to Extremely
Asymmetric Alpha Decay 827
    D. N. Poenaru et al.

INSTRUMENTATION

A Residue Implantation Detection System on the Daresbury Recoil Separator   831
    P. J. Woods et al.
Proposed Recoil Mass Spectrometer for Heavy Ion Reactions 835
    J. D. Cole et al.
The Projectile-Fragment Separator at the Darmstadt SIS/ESR Facility 839
    P. Armbruster et al.
Tests of a Large Air-Core Superconducting Solenoid
as a Nuclear-Reaction-Product Spectrometer 845
    R. L. Stern et al.
UNISOR On-Line Nuclear Orientation Facility 849
    I. C. Girit et al.
ISOL-Facility YASNAPP-2 853
    K. Ya. Gromov, V. G. Kalinnikov, and V. M. Tsupko-Sitnikov
Pulsed Laser Spectroscopy of Bunched Beams 857
    V. Egorov

CLOSING REMARKS

A View of the Future 860
    P. Armbruster
Program 865
List of Participants 870
Author Index 881

Preface

The series of conferences on Nuclei far from Stability began in 1966 at Lysekil, Sweden and has proceeded at about five-year intervals since: at Leysin, Switzerland (1970); Cargèse, France (1976); and Helsingor, Denmark (1981). The fifth in the series, and the first to be held outside of Europe, took place at Rosseau Lake, Ontario, Canada in September 1987.

The scientific program gave prominence to the advances in the study of exotic nuclei that have occurred since the last conference in 1981, with emphasis on the most recent results in experiment and theory. But as with all evolving fields the volume of data is increasing and a greater focus is needed on the broad understanding of classes of phenomena and regions of nuclei. Thus the conference program contained a mix of review talks and presentations of important new results.

This book represents volume two of the conference proceedings and contains most of the invited presentations and the contributed papers of the attendees. Volume one contains the one-page abstracts submitted to the conference and is available from the conference organizers. The organizing and advisory committees wrestled long over the following question: should the proceedings contain only the texts of the oral presentations or should they also contain contributions from attendees that were submitted and, in many cases, displayed as posters? In the end it was decided to follow the tradition of this conference series and allocate space in the proceedings to all accepted contributions, even though the number of pages made available to each author was quite small. We trust this decision meets with the reader's approval. The punctuality of essentially all authors in preparing finished manuscripts of a high technical quality has allowed this volume to go to press promptly.

In preparing the conference program the organizers were helped by the International Advisory Committee whose membership is listed below. We should like to thank the members for their time and their advice.

The conference was sponsored by Atomic Energy of Canada Limited, the International Union of Pure and Applied Physics, and the Natural Sciences and Engineering Research Council of Canada. In addition we would like to thank the Site Services branches of the Chalk River Nuclear Laboratories for their technical assistance.

The success of the conference reflects the very long hours put in by our colleagues Erik Hagberg, Vernon Koslowsky, Hermann Schmeing, Wayne Perry, and Mike Watson; the efficiency and industry of the conference secretariat: Suzanne Sheridan-Cole and June Elliott; the management of the Clevelands House resort hotel under J. Richard Lees; the hospitality of Jim Blake of Camp Queen Elizabeth; and the ingenuity of Dieter Wendel of Honey Harbour in transporting 200 delegates to beautiful Beausoleil Island in a flotilla of small boats for the conference excursion.

John C. Hardy
Chairman of the Organizing Committee

Ian S. Towner
Proceedings Editor

Chalk River Nuclear Laboratories
October, 1987

International Advisory Committee:

P. Armbruster (Germany)
J. H. Äystö (Finland)
C. Détraz (France)
G. T. Ewan (Canada)
P. G. Hansen (Denmark)
J. C. Hardy (Canada) **Chairman**
P. E. Haustein (USA)
B. Johnson (Sweden)

V. A. Karnaukhov (USSR)
H. J. Kluge (Switzerland)
E. Roeckl (Germany)
N. J. Stone (England)
J. L. Wood (USA)
N. Zeldes (Israel)
Z. Y. Zhou (China)
J. Żylicz (Poland)

Canadian Organizing Committee:

R. C. Barber (Manitoba)
J. M. D'Auria (Simon Fraser)
H. C. Evans (Queen's)
E. Hagberg (Chalk River)
J. C. Hardy (Chalk River) **Chairman**
V. T. Koslowsky (Chalk River)
S. K. Mark (McGill)
H. Schmeing (Chalk River)
I. S. Towner (Chalk River)

The local organizing committee comprised the five Chalk River members noted above.

Photos by
Wayne Perry, Mike Watson, and Nico Poppelier

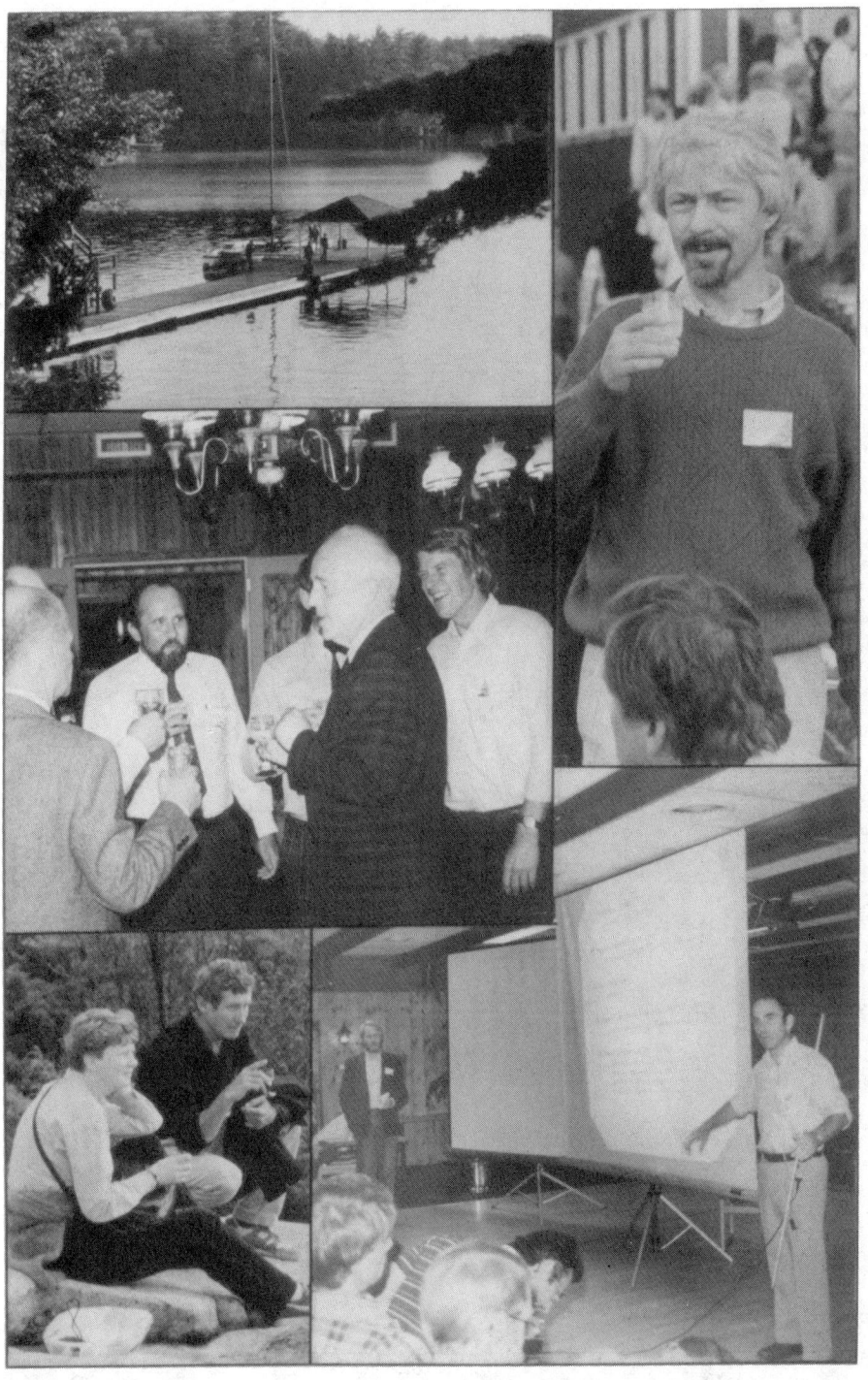

DIRECT MASS MEASUREMENTS OF LIGHT NEUTRON–RICH NUCLEI USING FAST RECOIL SPECTROMETERS

David J. Vieira, Jan M. Wouters, and the TOFI Collaboration
Los Alamos National Laboratory, Los Alamos, NM 87545

ABSTRACT

Extensive new mass measurement capabilities have evolved with the development of recoil spectrometers. In the Z=3–9 neutron–rich region alone, 12 neutron–rich nuclei have been determined for the first time by the fast–recoil direct mass measurement method. A recent experiment using the TOFI spectrometer illustrates this technique. A systematic investigation of nuclei that lie along or near the neutron–drip line has provided a valuable first glimpse into the nuclear structure of such nuclei. No evidence for a large single–particle energy gap at N=14 is observed; however, a change in the two–neutron separation energy trend is found at N=15. This change is correctly predicted by shell model calculations, and is interpreted in terms of the smaller $1s_{1/2}$–$1s_{1/2}$ interaction compared to that of the $0d_{5/2}$–$0d_{5/2}$ neutron–neutron interaction.

INTRODUCTION

In the study of light exotic nuclei, the ground state mass is often one of the first pieces of quantitative information to be learned about a "new" or previously unobserved nucleus. Because the mass of a nucleus reflects the interplay of nuclear and Coulomb forces, valuable insight into the nuclear structure of such nuclei can be obtained from their measurement. In the present case, we seek to test and refine our understanding of nuclei that lie far from the valley of β-stability. Do these exotic nuclei have the same type of shell structure as nuclei that lie close to stability? Can the new regions of deformation being observed in exotic nuclei be understood in terms of conventional models? And does the nucleon–nucleon pairing energy decrease with increasing isospin, as recently suggested by Vogel *et al.*?[1] We can take the first steps toward answering these and other interesting questions by making a systematic investigation of the masses of exotic nuclei.

In this paper, we concentrate on the masses of light neutron–rich nuclei. The current status of the mass surface in this region is shown in Fig. 1, where the shading of each square shows the accuracy with which the mass of each isotope has been measured. Since the compilation of Wapstra and Audi,[2] several new measurements have been performed (indicated by the darker shaded squares in Fig. 1). Of the 27 isotopes whose masses have been determined for the first time, 21 were measured with fast recoil spectrometers. This new technique relies on the direct mass measurement of fast reaction products. The method is not only fast (transit time $\sim$1 μsec) and accurate (σ_m/m = 10–100 ppm; σ_m = 100–1000 keV, depending on production rates), but it also permits the simultaneous mass determination of many nuclei throughout an entire A and Z region. The first examples of such measurements have been carried out in the light mass region; these examples are highlighted in this and the following paper.[3]

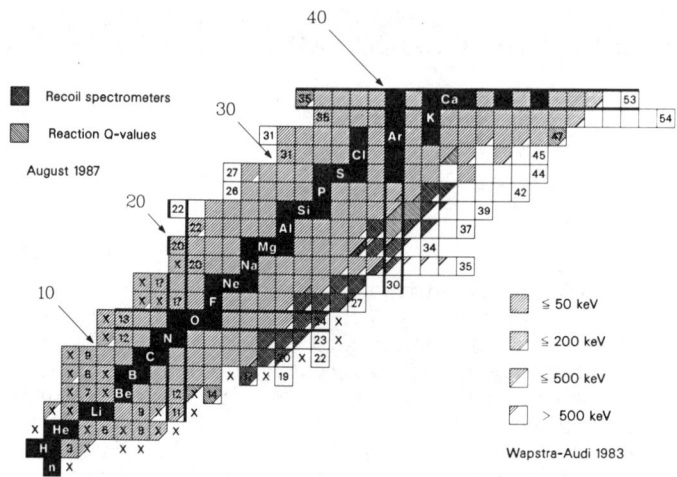

FIG. 1. A section of the chart of the nuclides showing the light mass region. Stable nuclei are indicated by black squares; nuclei whose masses have been determined are shaded according to their quoted accuracy (see Ref. 2); open squares indicate nuclei which have been identified to be stable with respect to prompt nucleon (two-nucleon) emission, but for which no mass has been reported; and X indicates nuclei that are nucleon unstable. Recent mass measurements are indicated by darker shaded squares (see legend).

The recoil spectrometer developments have been pioneered by two groups, one working with the energy-loss spectrometer (SPEG) at the Grand Accélérateur National d'Ions Lourds (GANIL) and the other using the Time-of-Flight Isochronous (TOFI) spectrometer at the Los Alamos Meson Physics Facility (LAMPF). At GANIL, projectile fragmentation reactions are used to produce exotic nuclei, and a combined velocity and magnetic rigidity measurement is employed to produce mass determinations. In contrast, LAMPF uses proton-induced target fragmentation reactions as a source of neutron-rich nuclei and employs a new type of time-of-flight recoil spectrometer to extract masses from a high-resolution, mass-to-charge determination. Both groups have made extensive measurements throughout the light-mass neutron-rich region. We will restrict our discussion to the Z=3-9 neutron-rich region, whereas the following paper will concentrate on the Z=10-15 region.

TOFI MEASUREMENTS IN THE Z=3-9 REGION

To illustrate these measurements, we will briefly describe a recent experiment using the TOFI spectrometer (see Ref. 4 for more details).

A schematic of the TOFI spectrometer and its associated transport line is shown in Fig. 2. Large yields of exotic nuclei are produced in the scattering chamber when LAMPF's 1-mA, 800-MeV proton beam strikes a 1-mg/cm^2

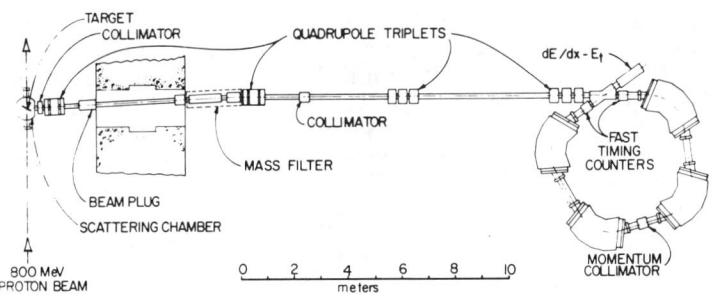

FIG. 2. Schematic of the TOFI spectrometer and its associated transport line.

^{nat}Th target. Some of the reaction products recoiling out of the target are captured by a secondary beam transport line.[5] Because proton–induced target fragmentation is not strongly angle dependent at this energy, the transport line has been conveniently located at ~90 degrees to the primary beam.

The transport line consists of four electromagnetic quadrupole triplets and a small, separated–sector mass–to–charge filter. A collimator is located at the intermediate focus position (approximate halfway through the line), where a mass–to–charge dispersed image of the illuminated target spot is produced. Here the high rate of uninteresting ions with mass–to–charge ratios (M/Q) of < 2.0 are greatly reduced to avoid count rate problems in the spectrometer. In the second half of the transport line, the velocities of the recoils are measured over a path length of ~10 m by thin–foil, secondary–electron, microchannel–plate (MCP) intensified, fast–timing detectors[6] located both just downstream from the M/Q collimator and at the entrance of the spectrometer.

The spectrometer has been designed to be isochronous so that the flight time of ions passing through the system provides a precise measure of the mass–to–charge ratio.[7] Four identical dipole magnets are arranged so that ions of a particular mass–to–charge ratio, but of lower (or higher) velocity than the selected mean velocity, take a shorter (or longer) path length through the system in such a way that the same overall flight time results. In addition, the spectrometer is one–to–one imaging and nondispersive overall so that small–area, fast–timing detectors can be used while a reasonably large solid angle and momentum–to–charge acceptance ($\Omega = 2.5$ msr and $\delta(p/Q)/(p/Q) = 4\%$) are maintained. The fourfold unit–cell symmetry of the system leads to small timing aberrations that result in high mass–to–charge resolution for a wide range of M/Q species without using ray–tracing techniques. In this experiment where the spectrometer was set for a momentum–to–charge of 210 MeV/c/Q, a timing resolution of ~180 ps (FWHM) was obtained between two MCP fast–timing detectors (one located at the entrance and the other at the exit of the spectrometer). For a typical flight time of ~500 ns, a relative time resolution—and consequently mass–to–charge resolution—of $\delta T/T = \delta(M/Q)/(M/Q) = 3.6 \times 10^{-4}$ was obtained. This performance was limited by the fast–timing detectors and their associated electronics and *not* by the intrinsic resolution of the spectrometer.

Masses are extracted from the measured mass-to-charge (time-of-flight) spectra that have been gated according to atomic number and charge state (see Fig. 3). The atomic number was determined from measurements of the ion's velocity and stopping power; the charge state was obtained from the measured velocity, total energy, and mass-to-charge ratio. In the present experiment, the stopping power and total energy were measured in a ΔE (20-μm) - E (720-μm) silicon detector telescope positioned immediately downstream from the MCP detector at the exit of TOFI. To further reduce the effects of isobaric cross-contamination, the mass resolution of TOFI was used to directly resolve isobars.

Using the direct mass measurement approach, the centroid of each M/Q line (determined by moments analysis) is related to the mass of that particular species. The centroids of lines with known masses were used to calibrate the system so that the masses of "unknown" lines could be determined. Small corrections to each centroid were made to account for time-to-amplitude walk and nonlinearities of the electronics. A typical calibration contained ~80 known

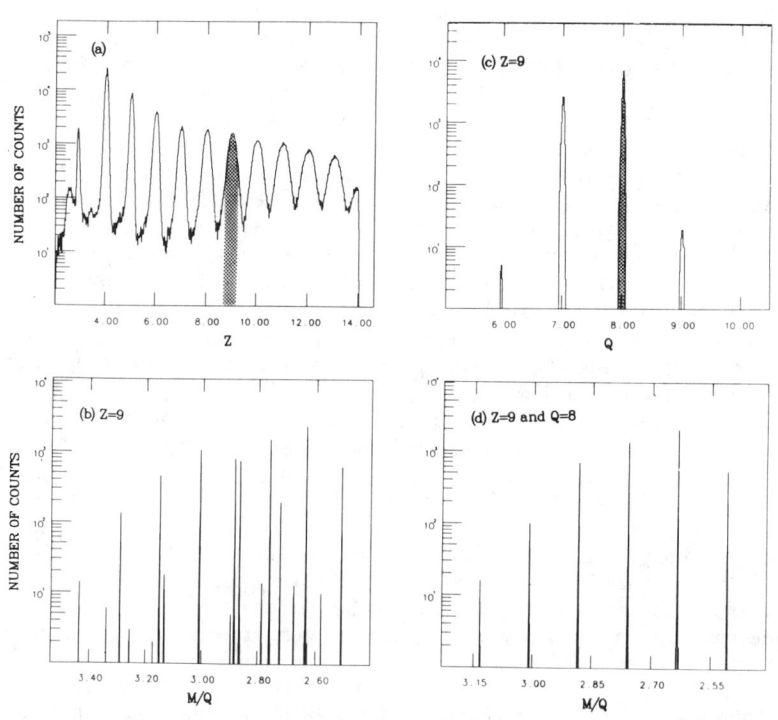

FIG. 3. (a) An ungated Z spectrum collected in a 24-h period; (b) and (c) the corresponding M/Q and Q spectra obtained for Z=9 selected events; (d) the resulting Z=9 and Q=8 gated M/Q spectrum. The measured Z, Q, and M/Q resolutions are 3.5%, 1%, and 3.6×10^{-4} (FWHM), respectively.

mass-to-charge lines. Table I shows final masses determined from a weighted average measurement of several runs (and different charge states where possible) taken over a 4-week data collection period. The quoted uncertainties result from a sum of statistical, calibration, and systematic uncertainties; the latter equaled ~2% of the line (~17 keV/Q).

DISCUSSION

Figure 4 shows a comparison of all recent mass measurements in the Z=3-9 neutron-rich region. There is excellent agreement in all cases except ^{14}Be and ^{22}N, were the measurements disagree at slightly beyond the one standard deviation level. Refinements in the fast-recoil direct mass measurement technique are indicated by both the reductions in errors and the determination of additional neutron-rich nuclei with each successive experiment. Of the nuclei that have been shown to be stable with respect to prompt neutron (or two-neutron) emission,[12] only the masses of ^{19}B, ^{22}C, and ^{23}N remain to be measured.

In Fig. 5, the masses are plotted in terms of the two-neutron separation energy, S_{2n}, vs neutron number. This plot provides a convenient way of removing odd-even neutron-pairing energy effects so that other nuclear structure features, such as energy gaps in the single-particle levels or changes in nuclear deformation, stand out more clearly. Of particular interest here is the question of a subshell closure at N=14. If a large energy gap between the neutron $0d_{5/2}$ and $1s_{1/2}$ energy levels did exist, a significant decrease in S_{2n} values after N=14 would be expected. However, for the neutron-rich isotopes of oxygen, fluorine,

TABLE I. Total number of observed events and determined mass excesses (errors given in parentheses).

^{A}Z	No. of Events	Mass Excess (μu)[a]	(MeV)
^{11}Li	168	43780 (130)	40.78 (.12)
^{14}Be	95	42660 (150)	39.74 (.14)
^{17}B	106	46830 (180)	43.62 (.17)
^{19}C	700	35180 (130)	32.77 (.12)
^{20}C	82	40360 (240)	37.60 (.22)
^{20}N	13,015	23380 (130)	21.78 (.12)
^{21}N	2,733	26930 (210)	25.09 (.20)
^{22}N	110	34340 (250)	31.99 (.23)
^{23}O	949	15700 (150)	14.62 (.14)
^{24}O	61	20000 (500)	18.6 (.5)
^{25}F	3,312	12210 (150)	11.37 (.14)
^{26}F	363	19820 (210)	18.46 (.20)
^{27}F	15	27500 (700)	25.6 (.7)

[a] Although nuclei with long-lived ($\tau > 150$ ns) isomeric states are rare in this region, the population of an, as yet unknown, isomeric state in one of these isotopes cannot be excluded on the basis of this data.

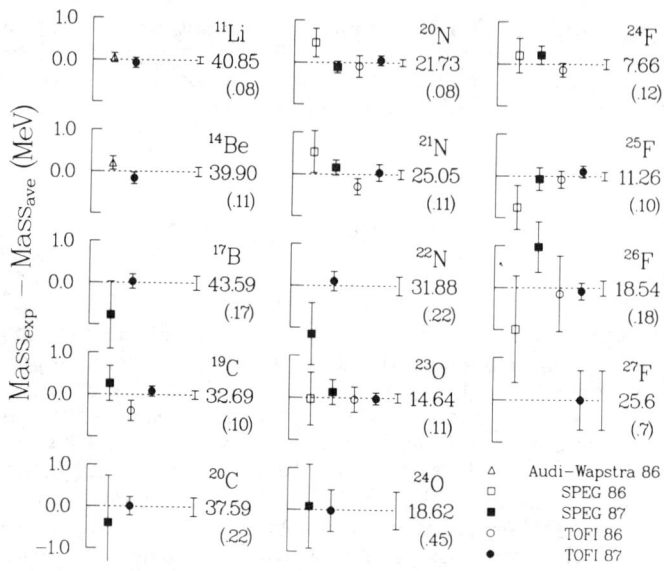

FIG. 4. A comparison of recent mass measurements for Z=3–9 neutron–rich nuclei. The weighted average mass excess is given in MeV for each isotope; errors are given in parentheses. The legend references are: Audi–Wapstra 86 – Ref. 8; SPEG 86 – Ref. 9; SPEG 87 – Ref. 10; TOFI 86 – Ref. 11; and TOFI 87 – Ref. 4 and this work.

and neon, no change in the S_{2n} trend is found at N=14, but rather at N=15! What is happening in this region?

As noted in Ref. 11, the spherical shell model calculations of Wildenthal et al.[13] reproduce the S_{2n} values throughout the sd shell region very well, with the notable exception of the deformed N=20 region. Based on the good agreement found in the N=14–16 region (see Fig. 6) and by the fact that the ground state wave functions are calculated to be relatively pure a simple shell model interpretation of the observed S_{2n} trend can be advanced. First, we point to the fact that no large decrease in S_{2n} values is observed in going from N=14 to N=15. This indicates that no large energy gap exists between the $0d_{5/2}$ and $1s_{1/2}$ neutron levels. (In Ref. 13, ~1-MeV energy gap between these levels is given, which is a typical ground state level spacing in these light nuclei.) Second, the more rapid decrease in S_{2n} values between N=15 and N=16 can be explained by the difference in the $0d_{5/2}$–$0d_{5/2}$ and the $1s_{1/2}$–$1s_{1/2}$ neutron-neutron interaction energies (the former is calculated to be nearly twice as large as that of the latter), for it is not until N=16 that no $d_{5/2}$–$d_{5/2}$ interaction is involved in the removal of two-neutrons. This provides a qualitative interpretation of the change in the S_{2n} trend observed at N=15.

When these mass measurements are compared to the psd shell model calculations of Brown et al.[14] (see Fig. 6), good agreement is found for the carbon and nitrogen isotopes, but increasing deviations are noted as one moves to lighter nuclei. The theoretical under-estimation of binding found for ^{26,27}F

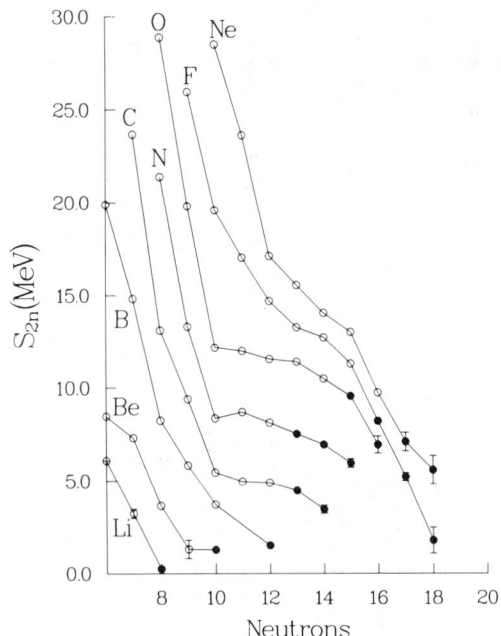

FIG. 5. Two–neutron separation energy vs neutron number for isotopes of lithium to neon. Solid circles indicate nuclei that have been measured by fast recoil spectrometers, and the open circles show data taken from Ref. 8. Error bars are indicated where they are larger than the symbol size.

could be related to the onset of prolate deformation, as observed for the N=19–20 isotopes of sodium and magnesium, but the measurements of ^{29,30}Ne are needed to confirm this interpretation.

In Fig. 7, comparisons to three other approaches are seen. The macroscopic-microscopic model of Tachibana et al.[15] gives reliable masses, except in the N=15–17 region. This likely reflects a deficiency in the microscopic part of the calculation because for nuclei that lie close to stability (e.g., ^{28}Si), a rapid decrease in binding occurs after N=14; we do not observe this feature in the case of these neutron–rich nuclei. Thus, calculations that rely on the masses of nuclei lying close to stability would be expected to underestimate the binding energy of these exotic nuclei. Similar trends are also noted for the macroscopic-microscopic models of Möller and Nix[16] and Möller et al.[17] (not shown in Fig. 7) and, to a much smaller extent, for the latest Garvey-Kelson mass relationship predictions of Jänecke and Masson[18] and the updated modified shell model calculations of Wouters et al.[4] Finally, a significant overestimation of binding is evident for the Garvey-Kelson predictions for ^{20}C and ^{21,22}N. This can be explained in a manner analogous to the arguments above, but in this case the nuclei used by the mass relationship are at or near the deformed ^{20}Ne region. The increased binding afforded by this deformation then leads to a more bound

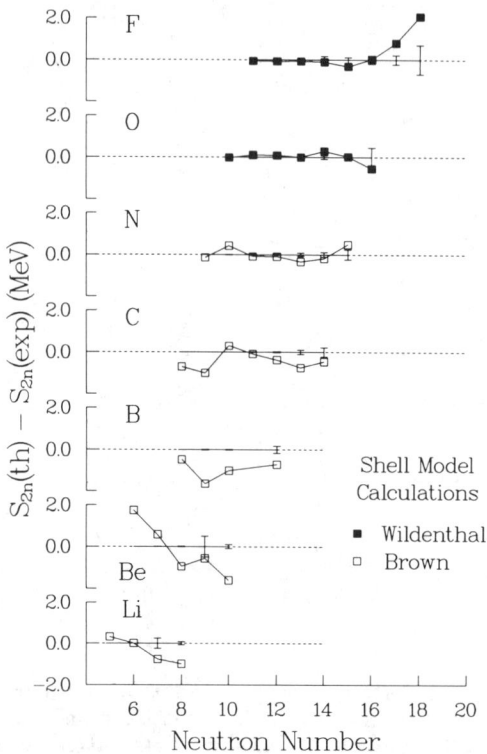

FIG. 6. Two–neutron separation energy difference between shell model predictions and those experimentally determined *vs* neutron number. The solid squares indicate calculations of Ref. 13 and the open squares indicate Ref. 14. Error bars represent the experimental uncertainties.

mass prediction for N=14, 15 isotones of carbon and nitrogen, which are expected to be spherical in shape.

CONCLUSION AND OUTLOOK

In summary, the development of fast recoil spectrometers has significantly advanced our mass measurement capabilities for exotic light nuclei. Some 12 isotopes have been determined for the first time in the Z=3–9 neutron–rich region alone. This progress has lead to an improved understanding of the light mass surface. There is no evidence for a subshell closure at N=14 in these measurements; however, the indication of a weaker $s_{1/2}$–$s_{1/2}$ interaction compared to that of the $d_{5/2}$–$d_{5/2}$ neutron–neutron interaction is evident in

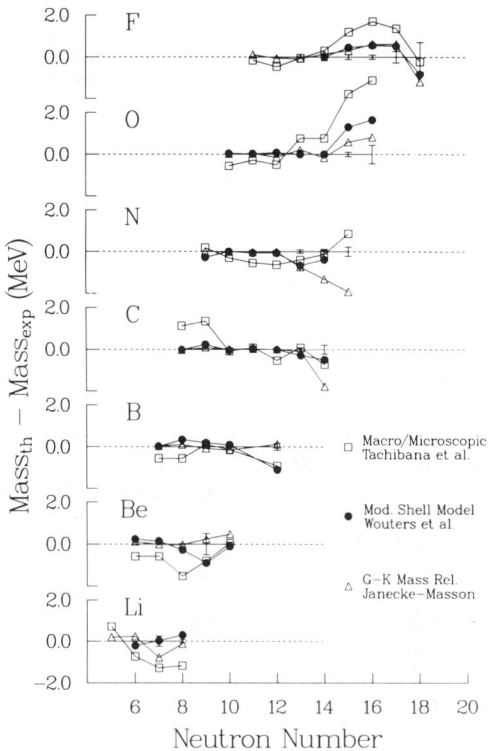

FIG. 7. Mass difference between theoretical predictions of Ref. 15 (open squares), Ref. 4 (solid circles), and Ref. 16 (open triangles), and the measured experimental results *vs* neutron number, as in Fig. 6.

S_{2n} trend and is supported by shell model calculations. In the future, further progress can be expected as fast recoil spectrometers are applied to the mass measurement of heavier nuclei.

ACKNOWLEDGMENTS

We wish to acknowledge G. T. Garvey, P. Möller, J. R. Nix, and B. H. Wildenthal for valuable discussions and comments concerning this work. We thank B. A. Brown for providing us with his shell model calculations before publication. Special thanks are extended to C. E. Lowe, J. H. Heiken, and G. L. Tietjen for help in preparing this manuscript. The following members of the TOFI collaboration have also made significant contributions to this work and are gratefully acknowledged: G. W. Butler, (Los Alamos National Laboratory); K. Vaziri and V. G. Lind (Utah State University); R. H. Kraus, Jr. and

D. S. Brenner (Clark University); F. H. Wohn (Iowa State University); K. E. G. Löbner (University of Munich); A. H. Wapstra (Nuclear Institute for Nuclear and High-Energy Physics, Amsterdam). This work was performed under the auspices of the U.S. Department of Energy.

REFERENCES

1. P. Vogel, B. Jonson, and P. G. Hansen, Phys. Lett. **139B**, 227 (1984).
2. A. H. Wapstra and G. Audi, Nucl. Phys. **A432**, 1 (1985).
3. W. Mittig, et al., these proceedings.
4. J. M. Wouters, et al., to be published.
5. K. Vaziri, et al., Nucl. Instr. Meth. **B26**, 280 (1987).
6. R. H. Kraus, Jr., et al., Nucl. Instr. Meth. **A** (in press).
7. J. M. Wouters, et al., Nucl. Instr. Meth. **A240**, 77 (1985); J. M. Wouters, et al., Nucl. Instr. Meth. **B26**, 286 (1987).
8. G. Audi and A. H. Wapstra, Interim Mass Adjustment, April 1986, At. Data Nucl. Data Tables (in press).
9. A. Gillibert, et al., Phys. Lett. **B176**, 317 (1986).
10. A. Gillibert, et al., Phys. Lett. **B192**, 39 (1987).
11. D. J. Vieira, et al., Phys. Rev. Lett. **57**, 3253 (1986).
12. F. Pougheon, et al., Europhys. Lett. **2**, 505 (1986).
13. B. H. Wildenthal, Prog. Part. Nucl. Phys. **11**, 5 (1984); B. H. Wildenthal, M. S. Curtin, and B. A. Brown, Phys. Rev. C **28**, 1343 (1983).
14. B. A. Brown, private communication; M. S. Curtin, et al., Phys. Rev. Lett. **56**, 34 (1986).
15. Tachibana, et al. At. Data Nucl. Data Tables (in press); and these proceedings.
16. P. Möller and J. R. Nix, At. Data Nucl. Data Tables (in press).
17. P. Möller, et al., *ibid.*
18. J. Jänecke and P. J. Masson, *ibid.*

MASS-MEASUREMENTS FAR FROM STABILITY OF
NEUTRON-RICH LIGHT NUCLEI*

W. Mittig, C. Grégoire, Y. Schutz, Zhan Wen Long[+]
GANIL BP 5027, 14021 Caen-Cedex, France
A. Gillibert, L. Bianchi, H. Dumont, B. Fernandez, J. Gastebois
CEN Saclay, DPhN/BE, 91191 Gif-sur-Yvette, France
A. Cunsolo, A. Foti
INFN, Sezione di Catania and Dipartimento di Fisica
Corso Italia 57, F-95129 Catania, Italy
C. Stephan
IPN, BP.1, 91406 Orsay, France
M. Morjean, Y. Pranal
CEA/DAM, BP.12, 91680 Bruyères-le-Châtel, France

I. INTRODUCTION

The study of nuclei far from stability is a verification of nuclear models that generally have been established using the properties of stable nuclei. One of the critical variables that should be reproduced by these models are the nuclear binding energies, that constitute a global test of these models.

For nuclei near stability, the determination of the Q-value of transfer-reactions will give results of a very good precision[1]. However, this method fails for nuclei very far from stability due to the very low cross-sections for multi-nucleon transfer reactions. The determination of the Q-value of β-decay is problematic due to the non-discrete β-spectrum and the need of knowledge of the decay scheme.

Thus if high enough production rates are obtained, the direct measurement of the mass has considerable advantages for nuclei very far from stability. However, the desired precision of mc^2 of less than 0.5-1 MeV, necessary to obtain crucial tests of the models, signifies a precision of better than $2 \cdot 10^{-5}$ for a mass 50 nucleus. This implies a high resolution measurement device, reasonable production rates of the nuclei of interest, and very low systematic errors. This will be discussed in the next two chapters. Some of the results have been published recently[2,3].

II. THE EXPERIMENTAL DEVICE

The magnetic rigidity Bρ of a particle of velocity v having the charge Q and the relativistic mass m is given by

$$B\rho = \frac{mv}{Qe} \qquad (1)$$

* Experiment performed at the GANIL National Laboratory.

Once the atomic number and the charge state of the particle are known, the mass may be obtained from equation (1) by the simultaneous measurement of its velocity and its magnetic rigidity. After some evident algebra , the atomic mass or, equivalently, the mass excess and the binding energy are obtained.

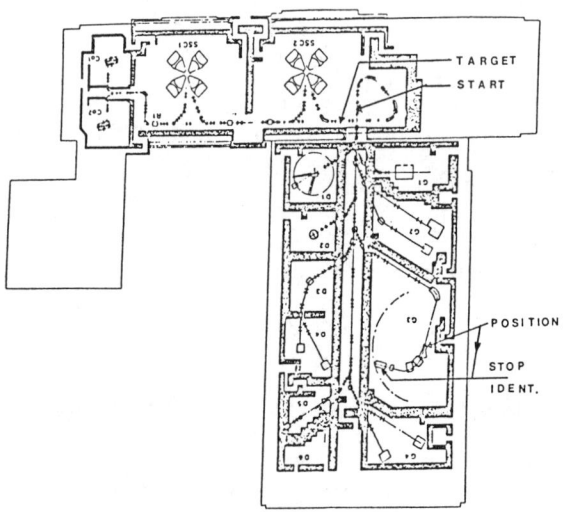

The figure 1 shows schematically the experimental device that permitted to obtain the necessary precision. It uses the standard beam transport line of GANIL and the magnetic high resolution spectrograph SPEG[4]. The production target is placed near the exit of

Fig. 1 : General schematic lay-out of the experimental device used for the mass-measure-ments.

the second cyclotron CSS2. The secondary beam is selected by the α-shaped spectrograph. The primary beam may be eliminated by a powerful steerer situated just before the target. At the exit of the α-shaped spectrograph we have a μ-channel device that detects the secondary electrons produced by the secondary beam passing through a thin, CsI coated mylar foil. It delivers a fast start signal for the time of flight measurement. The intrinsic time resolution of this detector is 150 ps.

The secondary beam is transported to the magnetic spectrograph SPEG[4]. The beam line is tuned in such a way that it is achromatic at the focal plane of the spectrograph. Therefore, all particles are focalised in a 2 cm diameter spot, independent of their magnetic rigidity. Thus, they may be detected by high quality solide state detectors, consisting of a 50 μ, a 300 μ and a 6000 μ thick cooled Si-detector. This telescope permits an unambiguous identification of the atomic number and the charge state.

A second μ-channel device just in front of the solid state detector telescope delivers the stop signal for the time of flight measurement. Due to the achromatic tuning, the length of the path of the particles is independent of the angle and position on the target, thus the velocity is obtained from the time of flight t and the length L of the path by $v = L/t$. In the present case, L was 86 meters, and a typical flight time is 1 μs.

The horizontal coordinate perpendicular to the trajectory is measured at two places (figure 1) giving for each particle (using TRANSPORT[5] notations)

$$X_1 = (\frac{X}{\delta})_1 \delta + (\frac{X}{X})_1 X + (\frac{X}{\Theta})_1 \Theta$$
$$X_2 = (\frac{X}{\delta})_2 \delta + (\frac{X}{X})_2 X + (\frac{X}{\Theta})_2 \Theta$$
(2)

where (X/δ) is the dispersion (X/X) is the magnification, and (X/Θ) is the dependence of X on the angle Θ. The magnetic rigidity is given by Bρ = Bρ$_0$ (1+δ), where Bρ$_0$ is the magnetic rigidity of the central trajectory.

The achromatic tuning results in $(\frac{X}{\delta})_2 = 0$, and placing the detectors at a focal point gives $(\frac{X}{\Theta})_1 = (\frac{X}{\Theta})_2 = 0$. Having two equations for two unknown variables, the magnetic rigidity may be obtained from (2) without influence of the object size. The position resolution being 1 mm, the dispersion $(\frac{X}{\delta})$ of the order of 10 cm/%, the resolution of $\frac{\Delta B}{B}$ is about 10^{-4}.

There are two position detectors in the focal plane, each measuring the X, Y coordinates. Thus the horizontal angle Θ and the vertical angle φ are determined. This allows to check higher order aberrations. They were found to be small.

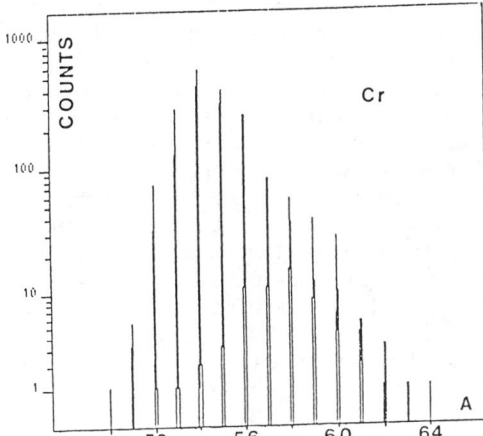

The final mass-resolution is thus determined by the following contributions : Bρ-measurement, 1.10^{-4} ; time measurement, 2×10^{-4} ; energy and angular straggling in the position detectors, 2.10^{-4}. Thus we expect a final resolution

Fig. 2 : Mass-spectrum obtained for Cr isotopes with a ^{86}Kr beam at 44 A.MeV on a 150 mg/cm^2 Ti target.

of $\frac{\Delta m}{m}$ FWHM of $3.4 \ 10^{-4}$. The experimental resolution of 5×10^{-4} is near this value. The quality of the spectra is illustrated by figure 2. The final precision of the mass-measurement is given by the precision with which we may determine the position of a pic in the mass-spectrum. This is given for a gaussian line shape[6] by

$$\frac{\Delta m}{m} \text{ final} = \frac{\Delta m}{m} \text{ FWHM} \cdot \frac{1}{2.35} \cdot \frac{1}{\sqrt{N}}$$
(3)

where $\frac{\Delta m}{m}$ FWHM is the resolution, and N are the counts in this pic. Thus we need at least 100 counts to get the desired precision. The transmission of the experimental device being of the order of 10^{-3} of the total yield, this corresponds to a cross-section of the order 10^{-30} cm^2, for a standard experiment with a 1 μ Ae 40 Ar beam.

III. SYSTEMATIC ERRORS

When a pic position in the mass-spectrum is determined with a precision given by equation (3), an important experimental problem is the conversion of this position into an absolute value of mass or binding energy. Fortunately, together with nuclei of unknown mass, an important number of nuclei with known mass are transmitted for a given setting of the beam line. Thus the calibration is obtained from these nuclei. We optimized the target-thickness in order to get broad isotope distributions. Nonetheless, this supposes that the calibration depends not, or only smoothly, on the mass and the atomic number of the particle, and that the calibration is not affected by differential non linearities.

Two types of measurements enter in our mass determination. One is position measurements. With a dispersion of 10 cm/%, a 10^{-5} error corresponds to 0.1 mm. All the particles covering the some spacial domain of the detectors, this is not very critical. In our device, we had to correct for the energy loss in the first position sensitive detector. This was done using the ΔE-Si detector. The second type is the time measurement for the determination of the velocity. If one uses a time to amplitude converter (Tac) and an analog to digital converter of 8192 channels, the time is converted to a channel number. The precision of 1.10^{-5} signifies an absence of differential non linearities below 1/10 of a channel. Even taking higher order terms in the calibration curve, it is difficult or impossible to achieve such a precision.

In order to avoid this problem, instead of measuring directly the time difference of the start-stop signals T_1, T_2, we measured the time differences t_{1r} and t_{2r} between T_1 and T_2 and a quartz oscillator with a period Δt_{osc} of 80 ns. The oscillator having no causal relation with the particles, each of the time spectra is random. The time of flight is than obtained by the relation

$$t = t_{2r} - t_{1r} + n \Delta t_{osc} + \text{constant} \qquad (4)$$

where n is an integer number that can be determined by comparison with time difference between the start-stop signal. Equation (4) depends therefore in the mean only on the mean calibrations and this method thus avoids completely the influence of differential non-linearities.

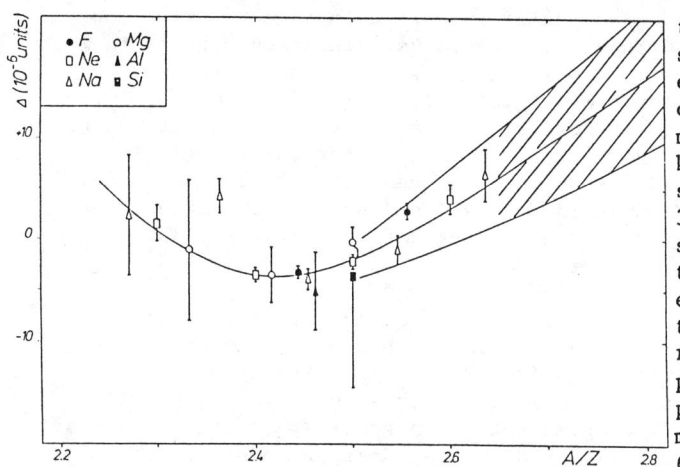

As test of the absence of systematic errors, the ratio of the measured mass and the known mass is shown on figure 3. Besides a smooth parabolic trend that can easily be corrected, the scattering of the points corresponds to a systematic error of $6 \cdot 10^{-6}$.

Fig. 3 : Ratio of the measured mass over masses from tables[8] minus one, in 10^{-6} units.

IV. ISOMERS AND DECAY PROPERTIES

In the present measurement, the aim is to measure the ground state mass. Therefore, the question arises if there is a contribution from longlived isomers. Isomers with a lifetime of less than about 100 ns will not contribute because they will decay before reaching the start detector. Lifetimes of the order of 1μsec or bigger however may influence the results. The experimental evidence of such states is Doppler broadening of the lines or delayed γ or β emission. What is the best experimental evidence depends on the lifetime of the state, and difficulty is increasing with increasing lifetime. We surrounded our Si-telescope by a NaI(Tl)-detection system. It subtended essentially 4Π for γ's and 7% of 4Π for β.

We looked thus systematically for contributions by isomers. Only in the case of ^{32}Al we found an indication of an isomer having a period of 1μs, with about 2% of the total yield. However, such a small contribution is near our experimental limit and it must be confirmed. It will not modify significantly the mass-measurement. We may conclude that there is no strong contribution of isomers in the 1μs to 10 ms range.

As a byproduct of these measurements, we obtained total reaction cross-sections of exotic nuclei[9] and some β-lifetimes. The anticoincidence with this 4Π detector cleans the final spectra obtained because it eliminates wrong identification due to reactions in the Si-detectors.

V. RESULTS

We performed experiments using ^{40}Ar and ^{86}Kr beams of $4 \cdot 10^{11}$

particles/s respectively. Optimizing target thickness to get best
yields for neutron rich nuclei and broad isotope distributions
for calibration purposes, in the actual conditions the Ar beam is
more favourable up to $Z \sim 14$, whereas for higher Z-values the
Kr-beam gives better yields. The potentiality of the Kr beam is
illustrated by fig. 2. For Cr, a total of 8 isotopes with unknown
mass are observed ($^{57-64}$Cr), two of them for the first time
(^{63}Cr, ^{64}Cr). Similar results have been obtained for neighbouring
Z values. With the planned future increase of beam intensity at
GANIL by a factor of 10 and an improved transmission, this and
other heavy beams will clearly open a wide field of
investigation.

We will present here the data from measurements with Ar
beam, the results for Kr not being completely analysed. The numerical results may be found in ref. 2, 3. Vieira et al. (ref. 9
and this conference), have recently obtained new results with a
different method. The agreement is in general good in the cases
of overlap. The most important differences are observed for
^{27,28}Ne, where we obtain a mass excess of (6.96±0.28) MeV for
^{27}Ne and (11.43±0.52) MeV for ^{28}Ne to be compared to the values
of Vieira et al. of (5.6±0.6) MeV and (10.7±0.4) MeV respectively. As a general trend, the mass excesses of Vieira et al. are
systematically lower than ours by 0.48 MeV. This difference
should be clarified in future.

The aim of the present measurements is the test of nuclear
models. We will compare our results to different classes of
models. We will concentrate here on the region Z=9-15. The numerical values of the mass excesses may be found in ref. 2, 3.

EMPIRICAL MODELS

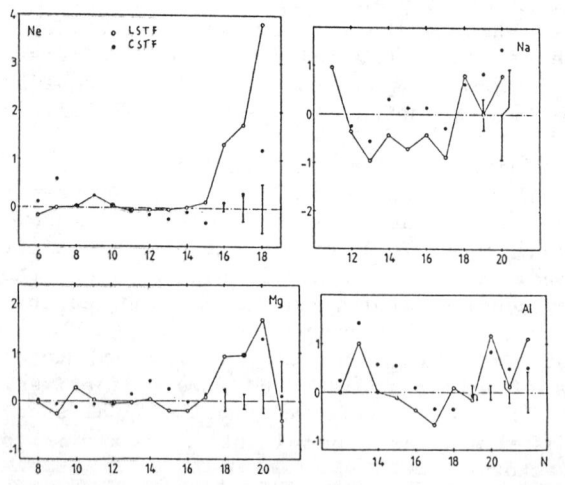

In empirical models one tries, using some general ideas of nuclear structure, to obtain a convenient parametrization of the
nuclear masses. One of the principal aims of such a formulation is
a good numerical reproduction and pre-
diction of masses. It may be used to obtain interesting informations such as the mean pairing energy[10]. These models

Fig. 4 : Experimental mass excess minus the mass excess predicted
by ref. 11 in MeV, as a function of neutron number.

contain however very often an important number of free parameters, which makes it difficult to extract information of such fits. The model of Uno and Yamada[11] has up to 500 free parameters. In figure 4, the results are compared to this model. On this figure and on the following, we present the difference of experimental mass excess and the mass excess of the model. Negative values for example signify that the nucleus is experimentally stronger bound than predicted. Experimental errors are shown as errors bars around the zero line. The calculation of Uno and Yamada reproduced reasonably well the existence or non-existence of nuclei in this mass-region (ref. 12 and this conference). It uses a macroscopic contribution described by a liquid drop model plus a shell correction. The shell correction separates the contributio of neutrons and protons. They used two versions with 250 (CSTF) and 500 (LSTF) free parameters respectively, for fitting the whole mass table. The danger of extrapolation with a big number of free variables is illustrated by Ne isotopes. The LSTF version is fastly diverging from experiment, whereas the CSTF-version is better behaved. Quite generally, in this mass region, the LSTF shows no significant improvement with respect to the CSTF which has half the free parameters. The transverse mass-relations of Comay-Kelson[13] give a similar quality of the fit to the data.

SEMI-EMPIRICAL MODELS

In semi-empirical models, one tries to separate as far as possible the smooth macroscopic part in an empirical parametrization, such as the liquid drop model, and to describe the fluctuating microscoic part by a more fundamental formulation. A model of such type is the one of Möller et al.[14]. They describe the well known mean behaviour by a liquid drop model plus a pairing correction and they try to reproduce smaller variations by microscopic shell corrections of the Strutinsky type[15].

The number of free parameters adjusted to reproduce the binding energies is reduced to five. The binding energy is maximised as a function of deformation, which is thus an interesting prediction of this model. The comparison of experiment with this model is shown on figure 5. The general agreement is reasonable. One may note that in the

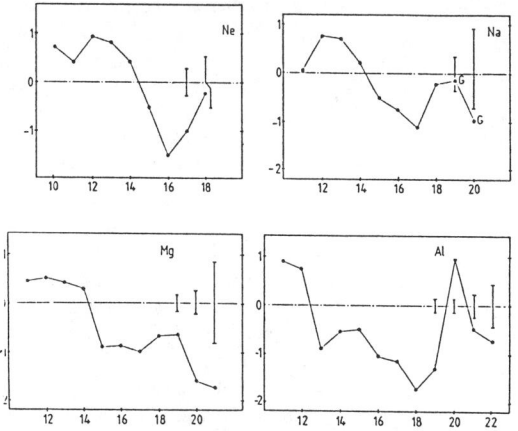

Fig. 5 : As figure 4, for ref. 14

mean, the model understimates the binding energy as a function of neutron excess. This tendancy corresponds to about 200 keV per excess neutron. There is some indication that this trend diminues going from O to Al. It will be interesting to understand which effects provides this extra binding energy with respect to the model.

MORE BASIC MODELS

The best would be to start as far as possible from first principles, this is to start from some nucleon-nucleon interaction and to try to solve the nuclear many body problem. This is actually not possible, and therefore important simplifications are necessary. Only some calculations exist for nuclei of relatively simple structure such a ^{18}O using a realistic nucleon-nucleon interaction. In most calculations, matrix elements of an effective interaction are obtained directly from experiment and they are used to solve the problem.

The shell-model calculation of Wildenthal[17] is an exemple for such a procedure. Considering only particles in the 2s-1d shell, the single particle energies and the two body matrix elements are obtained from experimental data, such as spectra of ^{17}O, ^{39}Ca and from ground and first excited state properties. Once these matrix elements are known, the hamiltonien of the system may be solved by diagonalisation. Even if in this type of calculation, as in the preceeding ones, a big number of parameters are determined from experiment, it has the advantage that wave functions of ground states and excited states are obtained, which results in rich possibilities of calculation of decay properties.

The result of this calculation is compared to the experiment on figure 6. The general agreement is excellent. However, strong discrepancies are observed for ^{28}Ne, ^{31}Na, $^{31,32}Mg$. For ^{31}Na and $^{31,32}Mg$, this way may be due to the influence of a deformation. Ref. 14 predicts the onset of a strong deformation ($\beta\sim0.3-0.4$) near the magic number N=20 for neutron rich nuclei, mainly having Z=11,12. It is interesting to note that for ^{33}Al, $^{33,34}Si$ and ^{35}P the model prediction is again in good agreement with experiment. We measured the β-lifetime for ^{33}Al t1/2= (54 ± 12) ms which is in agreement

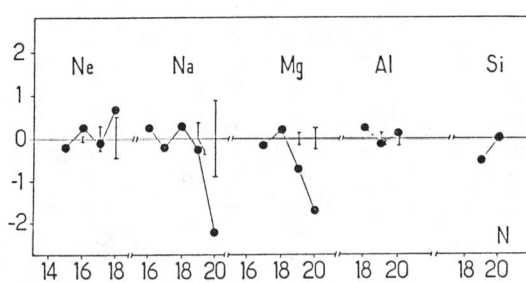

Fig. 6 : as fig. 4 for ref. 17

with the predicted value of 34 ms. This, too, is an indication that the wave function is correct.

A deformation implies that the shell closure N=20 disappears and configurations of the 1f-2p still have to be taken into account. The influence of the 1f1/2 shell changes mainly the binding energy for N>20, showing that it is essential to include this shell in the calculation.

A recent calculation of Poves et al.[19] has included the 1f7/2-2p3/2 shells and it predicts strong admixtures of 2 particle-2 hole configurations $(sd)^{-2} (fp)^2$ for N>20 for Ne, Na and Mg. For F, Al and Si, this admixture is less pronounced. However, the mass-prediction of this calculations is not very satisfying, mainly for Al and Si isotopes.

Recently, self-consistant Hartree-Fock + BCS calculations have been performed for even-even nuclei by Naulin[20], using the Skyrme III force[21]. The results of the mass calculation are shown on figure 7. One notes that the binding energies are approximately correct for N=Z, due to the fact that the interaction has been adjusted for these nuclei. The discrepancy increases with increasing (N=Z). This shows that the Hartree-Fock calculations do not reproduce such a Wigne like term. This was already observed by Flocard[22]. Note that the differences observed are in the opposite sense as the one of ref.14.

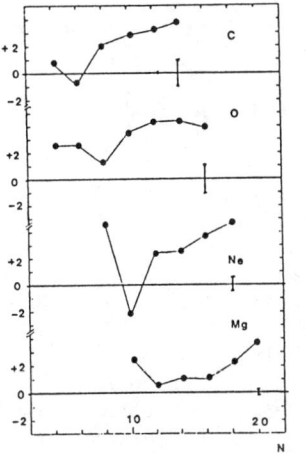

Fig. 7 : as fig. 8 for ref. 20.

VI. CONCLUSION

We developped experimental methods at GANIL that give precise mass-measurements of very neutron rich nuclei. The knowledge of nuclear masses has been extended by several isotopes for atomic numbers from Z = 4 to 15. The agreement with recent measurements of Vieira et al.[9] is satisfactory, even if a systematic difference of about 0.5 MeV exists.

These results were compared to theoretical results. Empirical relations were seen to produce important extrapolation errors, mainly if a big number of free parameters is involved. Best agreement with the data is obtained by the shell model calculation of Wildenthal, except for nuclei near N=20 and Z=11. In this region, other calculations predict a strong deformation. Our results favour such an interpretation, however it would be very interesting to measure ^{32}Na and ^{33}Mg, which we hope will be possible in near future.

The extension of mass measurements prolongating the known isotope chains showed that the model of Möller et al. does not give the correct isospin dependence of the binding energy, underestimating it for large neutron excess. The selfconsistant Hartree-Fock calculation of Naulin reproduces reasonably the experimental data for Z = 6 to 12. The comparison with the long series of isotopes makes appear however a discrepancy which is similar to a Wigner term $|N-Z|$. This model thus overestimates the binding energy for neutron rich nuclei. It is essential to clearly understand the origin of this discrepancy and study the consequent change of the wave functions.

REFERENCES

1) See for example : C. Détraz, F. Naulin, M. Langevin, P. Roussel, M. Bernas, F. Pougheon and J. Vernotte, Phys. Rev. C 15 (1977) 1738
   C.A. Gagliardi, D.C. Semon, R.E. Tribble, L.A. Van Audelin, Phys. Rev. C 34 (1986) 1663
2) A. Gillibert, L. Bianchi, A. Cunsolo, B. Fernandez, A. Foti, J. Gastebois, Ch. Grégoire, W. Mittig, A. Péghaire, Y. Schutz and C. Stephan Physics Letters B 176 (1986) 317
3) A. Gillibert, W. Mittig, L. Bianchi, A. Cunsolo, B. Fernandez, A. Foti, J. Gastebois, Ch. Grégoire, Y. Schutz and C. Stephan Phys. Lett. 192 (1987) 39
4) P. Birien and S. Valero, Note CEA n° 2215
5) K.L. Brown, note CERN 80-04 (1980)
6) J.B. Kennedy and A.M. Neville, Basic Statistical Methods (Harper and Row, New York, 1984)
7) W. Mittig, J.M. Chouvel, Zhan Wen Long, L. Bianchi, A. Cunsolo, B. Fernandez, A. Foti, J. Gastebois, A. Gillibert, C. Grégoire, Y. Schutz, C. Stephan. Preprint GANIL P.87-07, and contribution to this conference by A. Gillibert et al.
8) G. Audi, A.H. Wapstra, the 1986 Audi Wapstra Midsteam Mass Evaluation
9) D.J. Vieira, J.M. Wouters, K. Vaziri, R.H. Krans, H. Wollnik, G.W. Butler, F.K. WOh, and A.H. Wapstra Phys. Rev. Letters 57 (1986) 3253
9a) P.J. Woods, R. Chapman, J.L. Durell, J.N. Mo, R.J. Smith, B.R. Fulton, R.A. Cunningham, P.V. Drumm, L.K. Fifield Phys. Lett. B 182 (1986) 297
10) A.S. Jensen, P.G. Hansen, B. Jonson, Nucl. Phys. A431 (1984) 393
11) M. Uno, M. Yamada, INS-NUMA-40 (1982)
12) F. Pougheon, D. Guillemaud-Mueller, E. Quiniou, M.G. Saint-Laurent, R. Anne, D. Bazin, M. Bernas, D. Guerreau, J.C. Jacmart, S.D. Hoath, A.C. Mueller, C. Détraz, Europh. Letters 2 (1986) 505
13) E. Comay, I. Kelson, At. Dat. and Nucl. Data Tab. 17 (1976) 463

14) P. Möller and J.R. Nix, to appear in At. Data and Nucl. Dat. Tab. preprint LA-UR-863983 ; see too P. Möller, W.D. Myers, W. J. Swiatecki, J. Treiner, preprint LBL-22686
15) V.M. Strutinski, Nucl. Phys. A95 (1967) 420
16) J. Shupin, T.T.S. Kuo, D. Strottmann, Nucl. Phys. A408 (1983) 310
17) B.H. Wildenthal, S.M. Curtin, B.A. Brown, Phys. Rev. C28 (1983) 1343.
18) A. Watt, R.P. Singhal, M.H. Storm, R.R. Whitehead, J. Phys. G7 (1981) L145
19) A. Poves, J. Retamos, Phys. Lett. 184 (1987) 311
20) F. Naulin, thesis, Orsay 1987, and to be published
21) P. Bonche, H. Flocard, P.H. Heenen, S.J. Krieger, M.S. Weiss, Nucl. Phys. A443 (1985) 39
22) H. Flocard, thesis, Orsay 1975

† Permanent address: IMP, Lanzhou, China

MASS MEASUREMENTS OF SHORT-LIVED ISOTOPES IN A PENNING TRAP

F. Kern, P. Egelhof, T. Hilberath, H. Kalinowsky, H.-J. Kluge, K. Kunz,
L. Schweikhard and H. Stolzenberg,
Institut für Physik, Universität Mainz, D-6500 Mainz, FRG

R.B. Moore
Foster Radiation Laboratory, McGill University, Montreal, Canada H3A 2B2

G. Audi
Laboratoire René Bernas du CSNSM, Bât. 108, 91 406 Orsay, France

G. Bollen and the ISOLDE Collaboration
CERN, EP Division, CH-1211 Geneva, Switzerland

ABSTRACT

A mass spectrometer has been set up at the on-line isotope separator ISOLDE at CERN/Geneva. Mass-separated radioactive ions are stored in a Penning trap. Their mass is determined by a measurement of the cyclotron frequency in the magnetic field of a superconducting magnet. A resolving power of up to 300.000 and a precision of some 10 keV were determined in case of mass measurements of neutron-deficient Rb and Cs isotopes. The resonance of the isobars ^{88}Sr and ^{88}Rb were clearly resolved and evidence was obtained for an isomer in ^{122}Cs.

INTRODUCTION

Masses of nuclei far from stability are key input parameters for nuclear models and serve as test for nuclear-mass predictions needed in astrophysical calculations. Until now, precise direct mass determinations are restricted to the valley of stability or its direct neighbourhood[1]. Far away from this valley nuclear binding energies are measured indirectly by means of mass differences obtained as Q values from nuclear decays or reactions. The only exceptions up to now are a series of direct mass measurements on alkali isotopes performed by the Orsay group at ISOLDE by means of a Mattauch-Herzog spectrometer[2,3] and very recently performed time-of-flight experiments at Los Alamos[4] and GANIL[5,6] on light isotopes.

Mass measurements of heavier isotopes far away from stability by a determination of Q values suffer from errors introduced by summing up the errors of the many mass differences which link the mass of the isotope under investigation to that of a well known mass and, in addition, from uncertainties due to an insufficient knowledge of the nuclear level schemes of nuclei far away from stability. Hence, a method is very desirable where the masses are directly determined, i.e. independently of the knowledge of level schemes and masses of neighbouring isotopes. This can be achieved by measuring the cyclotron frequency of ions stored in a Penning trap. Such a system which can be loaded with ions delivered from an on-line isotope separator has been developed at Mainz and recently

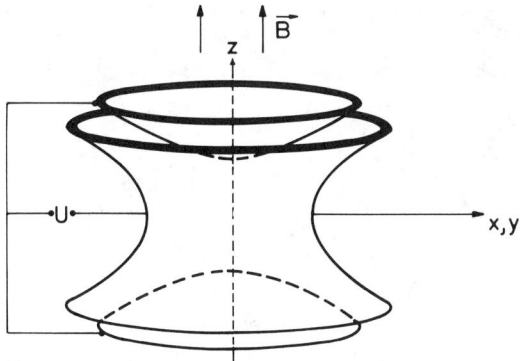

Figure 1: Scheme of a Penning trap. The shapes of the electrodes are hyperbolical with rotational symmetry around the z axis which is also the direction of the magnetic field.

installed at the ISOLDE facility at CERN[7]. First test measurements were carried out on Rb isotopes in December 1986 at the ISOLDE facility at CERN/Geneva[8,9], where the isobars ^{88}Sr and ^{88}Rb were resolved. In June 1987 further tests were done with neutron-deficient Cs isotopes. The results are reported in this contribution. Evidence was obtained for the first resolution of an isomeric state in a direct mass measurement.

PRINCIPLE OF THE MASS MEASUREMENTS AND EXPERIMENTAL SET-UP

Particles with charge e and mass m perform a cyclotron motion in a homogeneous magnetic field B given by

$$\omega_c = (e/m)\, B. \qquad (1)$$

In order to achieve a high resolving power $R = \nu_c/\Delta\nu_c$, the cyclotron frequency ν_c should be as large and the line width $\Delta\nu_c$ as small as possible. According to the uncertainty priciple the ultimate line width depends on the coherent interaction time T_{RF} with the applied radio frequency field ($\Delta\nu_c \simeq 1/T_{RF}$). Hence, it is straightforward to use a Penning trap (Fig. 1) in the magnetic field of a superconducting magnet because it combines long confinement time with maximum ν_c.

A combination of static, magnetic and electric fields is used to establish a three-dimensional trapping potential[10,11]. Due to the presence of an electrostatic field, the cyclotron frequency is modified. Nevertheless the undisturbed cyclotron frequency ν_c can be measured by inducing a two-quantum transition[12]. In resonance the trapped ions gain energy out of the applied radio frequency (RF) field. This is detected by a time-of-flight (TOF) method[13]: The charged particles are ejected out of the trap, drift to a channel plate detector (MCP) and their TOF is determined. The mean TOF as a function of the RF shows a resonance which gives the mass of the stored ions (Fig. 2).

The ion storage technique has not yet been used in nuclear physics for the study of short-lived nuclei. The reason is an experimental difficulty: The ions investigated so

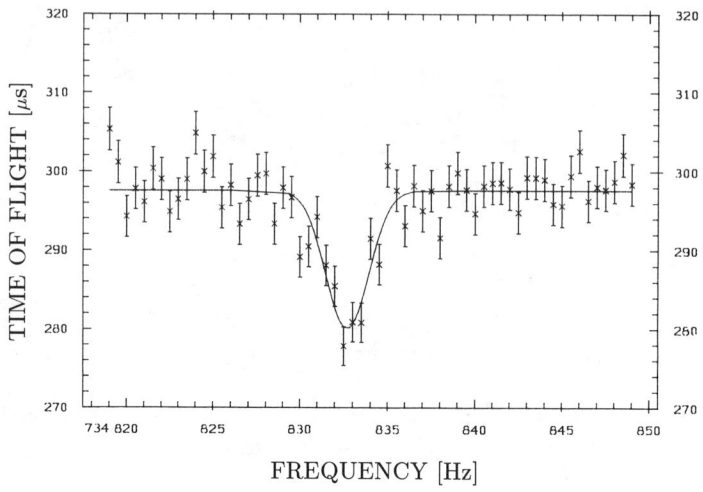

Figure 2: Cyclotron resonance of ^{124}Cs ($T_{1/2} = 30.8\,s$) as obtained by the mean time of flight as a function of the frequency of the applied RF field.

far in an ion trap were generally created inside the trap. However, in case of the study of short-lived isotopes these rare species have to be produced outside the trap and then guided with high efficiency into the trap. We have developed and tested a technique which allows to capture in-flight a bunch of ions in a Penning trap. Details can be found in Ref. 14.

The experimental set-up for the mass measurements is shown in Fig. 3. It is connected with a beam-line at the on-line separator ISOLDE/CERN and has a tandem configuration. A Penning trap in an electromagnet is used to collect the ions after re-ionization from a Re foil in which the ISOLDE ion beam is implanted. The ions are transfered into a second Penning trap installed in the 5.7 T field of a superconducting magnet. Here, the measurement of the cyclotron frequency takes place. Details are given in Ref. 14.

MASS MEASUREMENTS AND PERFORMANCE

The first test experiments with radioactive isotopes have only be performed recently and concerned neutron-deficient Rb and Cs isotopes, which are produced at ISOLDE with high yields up to 10^{11} mass-separated ions per second and mass number[7]. The isotopes with the shortest half life investigated were ^{124}Cs ($T_{1/2} = 30.8\,s$) and ^{122}Cs with $T_{1/2} = 4.2\,min$ and $21.0\,s$. The aim of these experiments was to check the performance of the apparatus under the realistic conditions of a beamtime at an on-line isotope

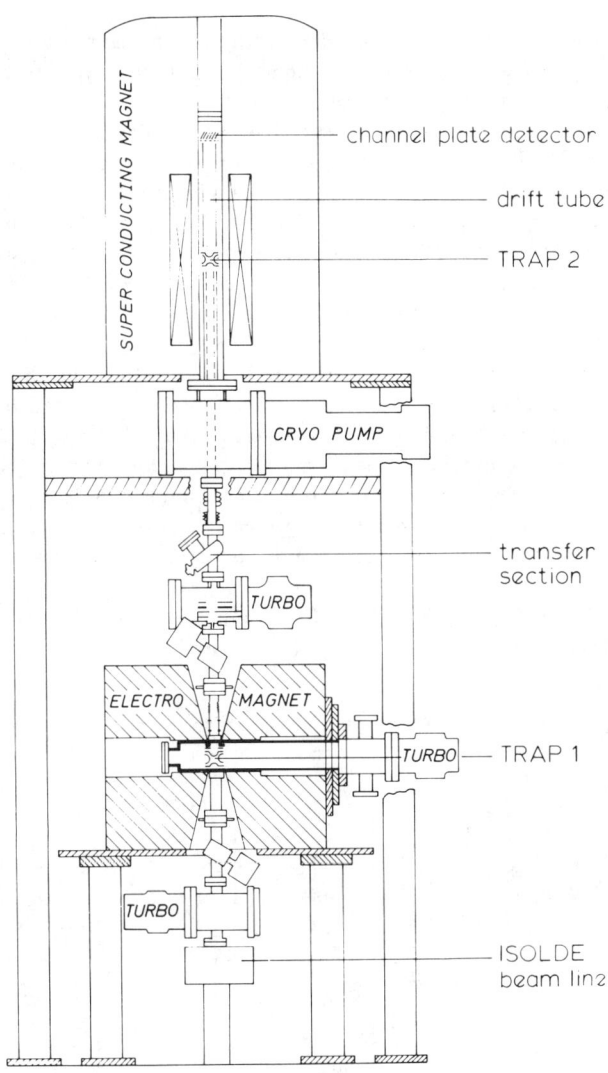

Figure 3: Experimental set-up used for direct mass measurement of radioactive isotopes at the on-line isotope separator ISOLDE at CERN. Trap 1 acts as a bunching device for the continuous 60 keV ion beam delivered by ISOLDE. The bunch of 1 keV ions extracted out of trap 1 is transfered to and captured in-flight in trap 2 where the mass measurement takes place. After inducing the radiofrequency the ions are ejected out of trap 2 and time-differentially counted by the channel plate detector.

separator. The results given below were obtained during a 2-days run in December 1986 ($^{77,78,85,86,88}Rb$ and ^{88}Sr) and a 8-hour run in June 1987 ($^{122,124,125,127,135,136}Cs$). A further test run is scheduled for September 1987.

Up to now the following performance has been achieved:

Trapping Efficiency: Up to 70 % of an ion bunch ejected out of trap 1 can be retarded and captured in-flight in trap 2. More details are given in Ref. 14.

Sensitivity: Typically 50 ions were stored in trap 2 at a time for the measurement of the cyclotron frequency. Since the detection scheme for the resonance by TOF is destructive, the trap has to be refilled after each cycle. The cycle time is about 1 s. Minimum 10^4 to 10^5 stored and detected ions are required for a resonance curve with sufficient statistics. Hence one mass measurement typically takes $30\,min$ to $1\,hour$. The overall efficiency was determined to be $\varepsilon = 5 \cdot 10^{-6}$ defined as the ratio between the number of ions implanted into the foil of trap 1 to those detected by the MCP. Since the trapping efficiency for a bunched ion beam in trap 2 is up to 70 %, the main losses are clearly due to evaporation, ionization, trapping and ejection in the bunching trap 1. Much better efficiencies will be obtained if a laser ion source is installed at the on-line mass separator which directly delivers a pulsed ion beam[15] or the continuous beam of the on-line separator is bunched in a Paul trap as proposed and tested by one of us (R.B.M.).

Storage Time: The storage time of trap 2 has been determined to be $\tau \simeq 15\,min$ for Cs ions at a vacuum pressure of $p \leq 10^{-9}\,mbar$.

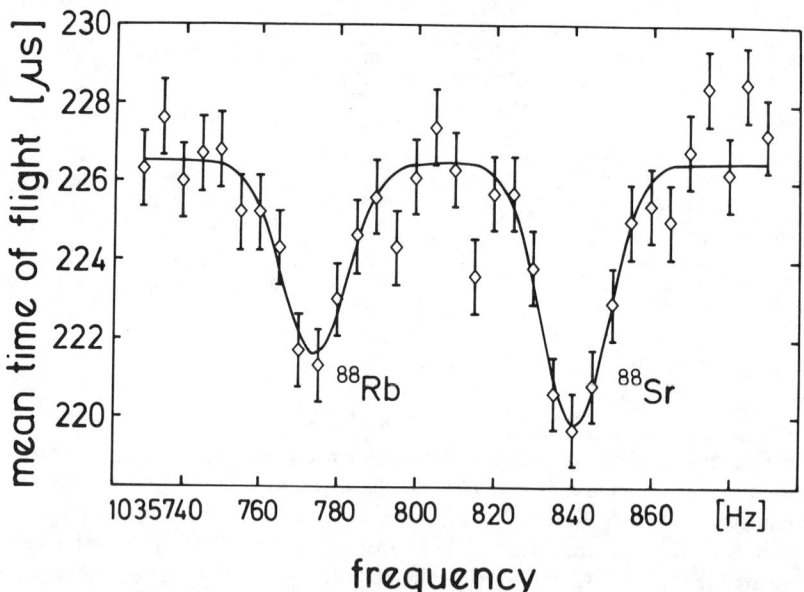

Figure 4: Simultaneously recorded cyclotron resonance of the isobars ^{88}Rb and ^{88}Sr which have a mass difference of 5.3 MeV. The power-broadened line width is due to higher RF power used during this measurement.

Resolution: A resolving power of $3 \cdot 10^5$ for K and Cs has been achieved in the on-line experiments. This corresponds to $130\,keV$ and $440\,keV$ and allows to resolve the cyclotron frequencies of isobars and in many cases also of isomers. Very recently, test experiments with stable ^{133}Cs yielded a resolving power of 760 000 corresponding to a $FWHM = 180\,keV$. Fig. 4 shows the signal obtained for the isobars $^{88}Rb/^{88}Sr$ which have a mass difference of $5.3\,MeV$. Both were produced at ISOLDE with comparable yields and cannot be separated by the ISOLDE mass separator. Although the resonances are broadened to $FWHM = 20\,Hz$ by RF power broadening they are clearly resolved owing to the still high resolving power of the Penning trap technique.

The Orsay group has investigated a number of isomeric states in neutron-deficient Cs isotopes by laser spectroscopy[16]. The same group could not resolve in their mass experiments[2,3] the ground and isomeric states. We have observed in a measurement at $A = 122$ a broadened structure (Fig. 5), which can be fitted by two Gaussians with center frequencies differing by $3\,Hz$ ($500\,keV$), line widths of $2.7\,Hz$ and heights of 3 and 6 %. Two half-lives are known for ^{122}Cs ($T_{1/2} = 4.2\,min$ and $T_{1/2} = 21.0\,s$). The signals of Fig. 5 were obtained by starting the experiment immediately after collection of the $A = 122$ ions on the foil. If the start of the experiment was delayed by some minutes, the resonance at lower frequency disappeared and we observed only one resonance with a line width of about $2\,Hz$ and a height of the typical magnitude of 7 %. Hence, there is strong evidence that we have resolved the resonances of the ground and isomeric states for the first time in a mass experiment. Due to lack of beam time, the measurement could not be repeated in June 1987.

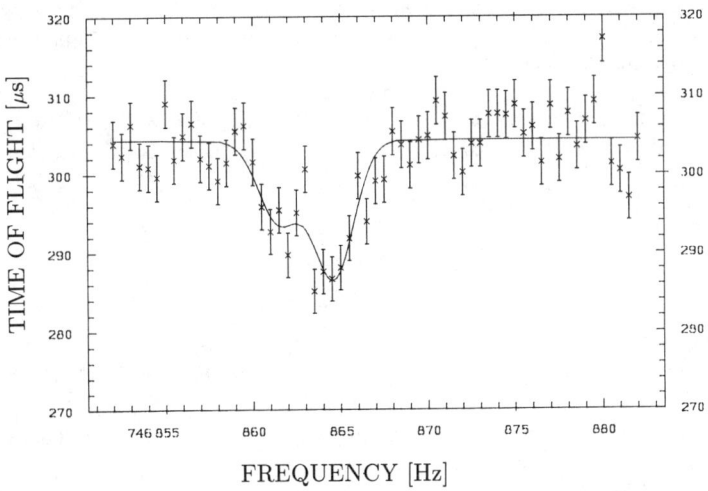

Figure 5: Mean time of flight of ^{122}Cs ions as a function of applied RF frequency. The resonance at lower frequency (higher mass) is attributed to the $T_{1/2} = 21\,s$ isomer.

Accuracy: The statistical uncertainty of the centroid of the resonances is about 5 % of the line width corresponding to $\Delta\nu_c/\nu_c \leq 2\cdot 10^{-7}$ or $20\,keV$ for ions with mass number $A = 100$. The magnetic field at the place of trap 2 is repeatedly measured by the determination of the cyclotron frequency of a stable alkali isotope. In test experiments with those reference isotopes it was found that the ratios of the cyclotron frequencies coincide with the tabulated mass ratios within the statistical uncertainties. However, sometimes there are jumps by 0.3 to 0.5 Hz (60 to 80 keV) in the resonance frequencies between groups of consistent data. The reason is not known until now but presumably due to changing magnetic fields in the environment at the ISOLDE/synchro-cyclotron site. Similar results were obtained in case of the test experiments with short-lived isotopes. In almost all cases the masses determined by the Penning trap technique coincide with the Orsay results within one standard deviation. It should be noted, however, that the accuracy of the Penning-trap results are more or less the same for all isotopes independent of the distance from the valley of stability. The errors of the Orsay experiments[3] are a factor of 2 smaller near stability ($A \geq 127$), comparable at $A \simeq 125$ and a factor of 2 to 3 larger at $A \leq 124$.

CONCLUSION

We were able to demonstrate that the Penning-trap technique is well suited to mass measurements far away from the valley of stability. After calibration of the magnetic field the masses are measured in an absolute way without the necessity of knowing the masses of neighbouring isotopes. The high resolving power of the technique enables to avoid problems and uncertainties by the isobaric impurities of on-line beams and even allows to separate resonances of isomers and groundstates. The accuracy of the technique is better than $100\,keV$ for $A \simeq 100$ and sufficient for tests of nuclear models by masses of nuclei far from stability.

This work has been funded by the German Federal Minister for Research and Technology (Bundesminister für Forschung und Technologie) under Contract No. Mz-458-I.

REFERENCES

1. A.H. Wapstra and G. Audi, Nucl. Phys. A432, 1 (1985)
2. G. Audi, M. Epherre, C. Thibault, A.H. Wapstra and K. Bos, Nucl. Phys. A378, 443 (1982)
3. G. Audi, A. Coc, M. Epherre-Rey-Campagnolle, G. Le Scornet, C. Thibault and F. Touchard, Nucl. Phys A449, 491 (1986)
4. D.J. Vieira, J.M. Wouters, K. Vaziri, R.H. Kraus Jr., H. Wollnik, G.W. Butler, F.K. Wohn and A.H. Wapstra, Phys. Rev. Lett. 57, 3253 (1986)
5. A. Gillibert, L. Bianchi, A. Cunsolo, B. Fernandez, A. Foti, J. Gastebois, C. Gregoire, W. Mittig, A. Peghaire, Y. Schutz and C. Stephan, Phys. Lett. B176, 317 (1986)

6. A. Gillibert, W. Mittig, L. Bianchi, A. Cunsolo, B. Fernandez, A. Foti, J. Gastebois, C. Gregoire, Y. Schutz and C. Stephan, Phys. Lett. B192, 39 (1987)
7. ISOLDE Users Guide, ed. by H.-J. Kluge, CERN, Geneva, Yellow Report, CERN 86-05 (1986)
8. G. Bollen, P. Dabkiewicz, P. Egelhof, T. Hilberath, H. Kalinowsky, F. Kern, H. Schnatz, L. Schweikhard, H. Stolzenberg, R.B. Moore, H.-J. Kluge, G.M. Temmer, G. Ulm and the ISOLDE Collaboration, Hyperfine Interactions, Vol. 37-38, in print
9. H. Stolzenberg, diploma work, Mainz, 1987, unpublished
10. F.M. Penning, Physica (Utrecht) 3, 873 (1936)
11. L.S. Brown and G. Gabrielsen, Rev. Mod. Phys. 58, 233 (1986)
12. H.-J. Kluge, H. Schnatz and L. Schweikhard, Z. Phys. D3, 189 (1986)
13. G. Gräff, H. Kalinowsky and J. Traut, Z. Phys. 297, 35 (1980)
14. H. Schnatz, G. Bollen, P. Dabkiewicz, P. Egelhof, F. Kern, H. Kalinowsky, L. Schweikhard, H. Stolzenberg, H.-J. Kluge and The ISOLDE Collaboration, Nucl. Instr. and Meth. A251, 17 (1986)
15. H.-J. Kluge, F. Ames, W. Ruster and K. Wallmeroth, Proc. Accelerated Radiaoctive Beam Workshop, Vancouver Island, 1985, eds. L. Buchmann and J.M. D'Auria, TRIUMF Proceedings TRI-85-1, 119 (1985)
16. C. Thibault, F. Touchard, S. Büttgenbach, R. Klapisch, M. De Saint Simon, H.T. Duong, P. Jacquinot, P. Juncar, S. Liberman, P. Pillet, J. Pinard, J.L. Vialle, A. Pesnelle, The ISOLDE Collaboration and G. Huber, Nucl. Phys. A367, 1 (1981)

BETA-DECAY ENERGIES AND SYSTEMATICS OF NUCLEAR STRUCTURE EFFECTS*

M. Graefenstedt, U. Keyser, F. Münnich and F. Schreiber
Institut für Metallphysik und Nukleare Festkörperphysik
Technische Universität Braunschweig
D-3300 Braunschweig, Mendelssohnstr.3, Germany

ABSTRACT

A systematic study of the beta-decay energies of very neutron-rich fission products in the mass regions $85 \leq A \leq 109$ and $130 \leq A \leq 150$ has been performed in recent years at the mass separators LOHENGRIN and OSTIS at Grenoble, France. From the experimental values of these decay energies - or of binding energies derived from them - it is possible to investigate systematics of nuclear structure effects as a function of mass number A, proton number Z or neutron number N. Several such effects have been observed in the mass regions around $A=100$ and $A=145$, namely subshell closures at $Z=40$ (38) and $N=56$, and deformations for nuclei with $N \geq 60$ and $N \geq 90$. Their existence is clearly demonstrated in the plots of the two-neutron and two-proton separation energies calculated from the Q_β-values measured.

INTRODUCTION

In this report, we want to present results of beta-decay energy measurements, which were started several years ago at the mass separator LOHENGRIN and OSTIS of the Institute Laue-Langevin at Grenoble. All these Q_β-values have been derived from experimental β-endpoint energies, which have been measured with a data acquisition and detector assembly consisting of a plastic scintillator telescope [1,2] and a large Ge(Li) or Ge(HP) γ-detector. In this way, β-transitions populating different levels in the same nucleus or those belonging to other isobars present in the sources could be identified through their known coincident γ-transitions. The coincidence system used has been described in a recent publication [3], including a discussion of the techniques applied for the calibration of the β-detector and the computer evaluation of the experimental data.
With very few exceptions, the Q_β-values presented in Tables I and II are based on several (up to 15) different β-endpoint energies measured in coincidence with different γ-transitions in the same nucleus.

EXPERIMENTAL RESULTS

In our latest experiments at the LOHENGRIN separator, neutron-rich nuclei - mainly with mass numbers $107 \leq A \leq 109$ - have been measured, using targets of ^{235}U and ^{239}Pu with high independent yields. The β-endpoint energies have been obtained from βγ-coincidence matrices (256 β- × 2048 γ-channels) measured on-line with an improved data

---

* This work has been funded by the German Federal Minister of Research and Technology (BMFT) under the contract number 06BS452I.

Table I: Q_β-values of light fission products; recent results (preliminary)

Nuclide	$T_{1/2}$	γ-gate	Present value	Other authors	Wapstra [5]
$^{87}_{34}Se_{53}$	5.8s	243 573 1168 3687	7250±150	–	7170±420*
$^{107}_{42}Mo_{65}$	2.9s	384 400 484 1237	6160± 80	–	6450±660*
$^{107}_{43}Tc_{64}$	21 s	103 177 1264 1573	4820± 90	–	4750±420*
$^{107}_{44}Ru_{63}$	3.8m	463 848 1272 1306	2890±160	3150±300 [6]	3150±300
$^{108}_{43}Tc_{65}$	5.0s	242 851 1548 2217	7710± 50	–	7710±730*
$^{108}_{44}Ru_{64}$	4.5m	165	1420±185	1320±100 [6]	1390±610
$^{108}_{45}Rh_{63}$	16.8s	434 498 619 932	4525±120	4500±600 [6]	4430± 50
$^{109}_{43}Tc_{66}$	1.4s	195 490 1159 1268	6240±200	–	5900±580*
$^{109}_{44}Ru_{65}$	34.5s	359 1305 1928	4160±100	–	4200±300*

* estimated from systematics

acquisition system [4]. Preliminary Q_β-values for the nuclei studied are given in Table I, where only those β-transitions have been taken into account, which are coincident with the strongest γ-transitions in the daughter nucleus. Also shown in this table are the Q_β-values taken from the compilation of Wapstra et al. [5], and values reported by other authors [6].

The large error of the Q_β-value of ^{108}Ru in [5] is due to the existence of isomerism in ^{108}Rh. In the present coincidence experiments it was possible (cf. ref. [4]) to determine the decay of ^{108}Ru to the ground state of ^{108}Rh and so to reduce the error of the Q_β-value (the error given in [6] is only the statistical one and does not take into account the isomerism).

In Table II, revised Q_β-values of all our earlier work are presented (cf. ref. [3] and additional literature therein). These revised values have been calculated with an improved data handling procedure to decompose the complex β-spectra and to substract the background [7]. The error in the β-energy calibration could be reduced by using standards with high accuracy [3], which were not known at the time of the measurement. In the forth column of this table, results of other authors are presented. The experimental errors of those Q_β-values in this column, which were measured with Ge(HP)-detectors, have been multiplied by a factor of 2-3 in all cases, where the original errors are below 50 keV, because important systematic errors, especially those due to the detector response function, have not been taken into account in general in the calculation of the error [12]. The adopted value in the fifth column of Table II is the weighted mean of the values of column 3 and 4. In the last column of this table, the Q_β-values from "The 1983 Mass Table" [5] are presented. Those values which are estimated from systematic trends are marked by an asterisk.

Finally it should be noted that for the nuclides $^{96-98}$Y, ^{98}Rb, ^{100,102,104}Nb, ^{131}Sn, ^{134}Sb and ^{136}I, also the β-decay energies of isomeric states could be determined; no other measurements of these Q_β-values have been reported in literature.

NUCLEAR STRUCTURE EFFECTS IN THE MASS REGION A ≅ 100

Shell effects manifest themselves as sudden discontinuities of the slope of the nuclear mass surface along lines parallel to the axes in the N,Z-plane and are easily recognized in plots of binding or decay energies as a function of N,Z or A. At main magic numbers, these discontinuities are observed throughout the whole known region of the nuclear chart. At semi-magic numbers, however, where the discontinuities are considerably smaller, the effect is local, i.e. it is confined within a rather small interval of N or Z.

The graph of the two-neutron separation energies S(2n) as a function of N is shown in Fig. 1 for all known nuclei with 48≤N≤65. Those S(2n)-values, which have been calculated from our own experiments only, are marked by heavy dots in this figure. The steep decrease of the S(2n)-values at N=51 is due to the main neutron shell closure; it occurs for all nuclei with N=51 or N=52. A second, but smaller change of the slope of the curves is observed at N=56, which is due to the closure of the $2d_{5/2}$ neutron subshell. It is evident from Fig. 1 that this subshell closure is a local effect: it occurs only for nuclei with 36≤Z≤42. The question, if this subshell effect is present also for Z=35, cannot be answered at the moment, because the Q_β-values of the very neutron-rich Br isotopes with N≥56 are not yet known. We intend to measure these decay energies in the near future at ISOLDE.

Table II: Q_β-values of neutron-rich nuclei

Nuclide	$T_{1/2}$	Present result	Other authors	Adopted values	Wapstra [5]	Refs.
$^{85}_{34}Se_{51}$	33 s	6185±90	-	6185±90	6190±100	-
$^{86}_{34}Se_{52}$	14.1s	5095±100	-	5095±100	5120±120	-
$^{90}_{36}Kr_{54}$	32.2s	4380±25	-	4380±25	4390±30	-
$^{92}_{36}Kr_{56}$	1.8s	6000±60	6160±80	6055±50	6160±80	8
$^{92}_{37}Rb_{55}$	4.5s	8095±25	8080±160	8095±20	8120±12	8
			8111±50			9
			8080±50			10
$^{93}_{36}Kr_{57}$	1.3s	8600±100	-	8600±100	8530±120	-
$^{93}_{37}Rb_{56}$	5.8s	7455±35	7560±120	7465±25	7443±13	8
			7485±50			9
			7440±50			10
$^{94}_{37}Rb_{57}$	2.7s	10335±45	10325±60	10335±35	10307±27	9
			10353±100			11
$^{95}_{37}Rb_{58}$	0.38s	9280±45	9300±50	9290±35	9280±60	12
$^{96}_{37}Rb_{59}$	0.19s	11590±80	11547±100	11565±50	11750±50	11
			11553±90			12
$^{96}_{38}Sr_{58}$	1.0s	5345±50	5330±60	5365±25	5416±20	13
			5413±50			9
			5362±50			12
$^{96}_{39}Y_{57}$	6 s	7030±70	7120±70	7075±35	7140±40	9
			7080±50			12
	10 s	8030±150	-	8030±150	-	-
$^{97}_{37}Rb_{60}$	0.17s	10440±60	10450±60	10445±40	10520±70	12
$^{97}_{38}Sr_{59}$	0.44s	7420±80	7450±60	7440±45	7470±70	12
$^{97}_{39}Y_{58}$	3.7s	6645±70	6702±50	6670±40	6680±60	12
	1.2s	7280±150	-	7280±150	-	-
$^{98}_{37}Rb_{61}$	114ms	12440±75	12343±150	12450±55	12430±80	11
			12520±90			12
	96ms	12710±120	-	12710±120	-	-
$^{98}_{38}Sr_{60}$	0.65s	5815±40	5820±40	5820±30	5880±60	12
$^{98}_{39}Y_{59}$	0.65s	8840±55	8980±100	8885±40	8910±60	13
			8900±60			12
	2 s	9780±200	-	9780±200	-	-

Table II: continued

Nuclide	$T_{1/2}$					
$^{99}_{37}Rb_{62}$	59ms	10960±130	–	10960±130	11310±90	–
$^{99}_{38}Sr_{61}$	0.29s	8360±75	–	8360±75	7950±120	–
$^{99}_{39}Y_{60}$	1.5s	7605±60	–	7605±60	7610±80	–
$^{99}_{40}Zr_{59}$	2.1s	4550±35	4510±80	4540±30	4590±70	14
$^{100}_{38}Sr_{62}$	0.17s	7075±100	–	7075±100	6700±720*	–
$^{100}_{39}Y_{61}$	0.94s	9310±70	–	9310±70	9900±400*	–
$^{100}_{40}Zr_{60}$	7.1s	3335±25	3350±35	3340±20	3330±70	14
$^{100}_{41}Nb_{59}$	1.5s	6245±25	6330±70	6265±25	6240±30	14
	3.1s	6745±75	–	6745±75	–	–
$^{101}_{40}Zr_{61}$	2.4s	5520±45	–	5520±45	5780±120	–
$^{101}_{41}Nb_{60}$	7.1s	4590±30	4500±60	4570±25	4630±70	14
$^{102}_{40}Zr_{62}$	2.1s	4605±30	–	4605±30	4590±410*	–
$^{102}_{41}Nb_{61}$	1.3s	7210±35	–	7210±35	7210±70	–
	4.5s	7335±40	–	7335±40	–	–
$^{103}_{40}Zr_{63}$	1.3s	6945±85	–	6945±85	7500±510*	–
$^{103}_{41}Nb_{62}$	1.5s	5530±30	–	5530±30	5500±120	–
$^{103}_{42}Mo_{61}$	67.5s	3750±60	–	3750±60	4000±200*	–
$^{103}_{43}Tc_{60}$	54.2s	2615±45	–	2615±45	2654±11	–
$^{104}_{41}Nb_{63}$	4.8s	8105±90	–	8105±90	8700±500*	–
	0.99s	8320±80	–	8320±80	–	–
$^{104}_{42}Mo_{62}$	1.0m	2155±40	–	2155±40	2000±300*	–
$^{104}_{43}Tc_{61}$	18.2m	5590±60	5620±70	5605±45	5620±70	15
$^{105}_{41}Nb_{64}$	2.8s	6485±70	–	6485±70	7000±400*	–
$^{105}_{42}Mo_{63}$	36.7s	4950±45	–	4950±45	5000±300*	–
$^{105}_{43}Tc_{62}$	7.6m	3640±55	–	3640±55	3800±200*	–
$^{106}_{42}Mo_{64}$	8.2s	3510±45	–	3510±45	3200±500*	–
$^{106}_{43}Tc_{63}$	36 s	6545±45	–	6545±45	6700±300*	–

Table II: continued

Nuclide	$T_{1/2}$					Ref
$^{131}_{50}Sn_{81}$	50 s	4600±110	4595±200	4595±95	4650±120	16
	39 s	4680±120	-	4680±120	-	-
$^{134}_{51}Sb_{83}$	0.85s	8420±120	8400±300	8370±100	8410±110	17
			8240±240			16
	10.3s	8510±110	-	8510±110	-	-
$^{135}_{52}Te_{83}$	18 s	5960±100	5950±240	5960±90	5960±100	16
$^{136}_{52}Te_{84}$	17.5s	5095±100	5100±150	5095±85	5090±90	18
$^{136}_{53}I_{83}$	85 s	6925±70	-	6925±70	6930±50	-
	44.8s	7705±120	-	7705±120	-	-
$^{137}_{52}Te_{85}$	8.5s	6925±130	-	6925±130	7020±310*	-
$^{137}_{53}I_{84}$	24.2s	5880±60	-	5880±60	5880±80	-
$^{138}_{53}I_{85}$	6.4s	7820±70	-	7820±70	7820±70	-
$^{139}_{53}I_{86}$	2.3s	6815±100	-	6815±100	6820±100	-
$^{142}_{55}Cs_{87}$	1.7s	7280±40	7230±70	7285±30	7317±20	8
			7330±50			19
$^{143}_{55}Cs_{88}$	1.8s	6240±70	6287^{+60}_{-250}	6265±45	6280±40	19
$^{143}_{56}Ba_{87}$	14.5s	4210±70	4260±60	4240±35	4250±40	19
			4240±70			20
$^{144}_{55}Cs_{89}$	1.0s	8560±80	8450±70	8500±55	8470±60	19
$^{144}_{56}Ba_{88}$	11.9s	3055±70	-	3055±70	2970±90	-
$^{144}_{57}La_{87}$	40.9s	5435±90	-	5435±90	5600±110	-
$^{145}_{55}Cs_{90}$	0.6s	7930±75	-	7930±75	7790±130	-
$^{145}_{56}Ba_{89}$	4.3s	4925±80	-	4925±80	4950±110	-
$^{145}_{57}La_{88}$	24.8s	4110±80	-	4110±80	4120±90	-
$^{145}_{58}Ce_{87}$	2.98m	2530±50	-	2530±50	2530±50	-
$^{146}_{55}Cs_{91}$	0.3s	9310±60	-	9310±60	9410±90	-
$^{146}_{56}Ba_{90}$	2.2s	4030±50	-	4030±50	4270±110	-
$^{146}_{57}La_{89}$	6.2s	6620±70	6640±50	6630±40	6386±30	11
$^{147}_{56}Ba_{91}$	0.72s	5750±50	-	5750±50	5710±590*	-

Table II: continued

$^{147}_{57}La_{90}$	2.2s	4945±55	–	4945±55	5190±310*	–
$^{147}_{58}Ce_{89}$	57 s	3290±40	–	3290±40	3310±70	–
$^{148}_{58}Ce_{90}$	48 s	2060±75	–	2060±75	2050±100	–
$^{148}_{59}Pr_{89}$	2.0m	4965±100	–	4965±100	4960±100	–
$^{149}_{58}Ce_{91}$	5.0s	4190±75	–	4190±75	3700±300*	–
$^{150}_{58}Ce_{92}$	4.8s	3010±90	–	3010±90	3080±100	–
$^{150}_{59}Pr_{91}$	6.1s	5690±80	–	5690±80	5100±200*	–

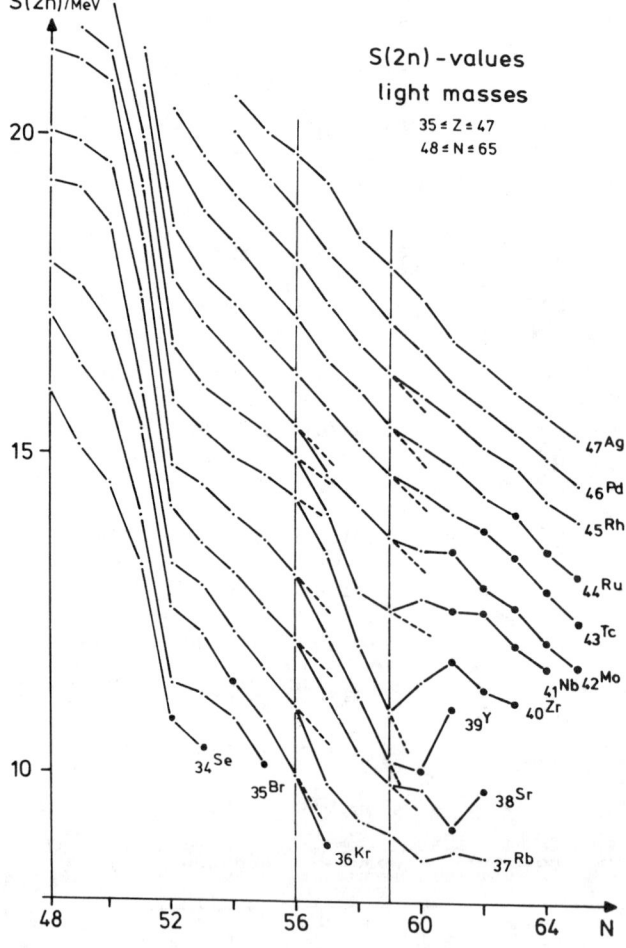

Fig. 1. Two-neutron separation energies S(2n) for nuclei in the mass range 83≤A≤111. Heavy dots: values calculated from our Q_β-values only.

The magnitude of the change in slope at N=56 attains its maximum value for the Zr isotopes and decreases, when |Z-40| increases. This is an example of the "mutual support of magicity" and is due to the fact that Z=40 is a semi-magic proton number. The change in the slope of the S(2n)-curve for the neutron-rich Zr isotopes shows clearly the sudden phase transition from spherical to deformed shape in the ground state of these nuclei at N=60. This onset of deformation is the reason for the sudden rise of the S(2n)-curves at N=60, which is observed for all nuclei with proton number Z≤45.

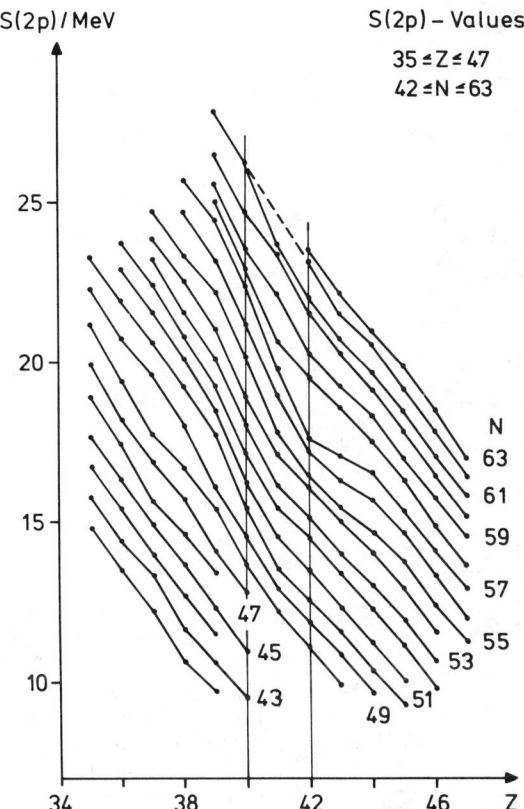

Fig. 2. Two-proton separation energies S(2p) for nuclei in the mass range 77≤A≤110.

In Fig. 2, the S(2p) isotonic lines in this mass region are shown as a function of Z. One observes a discontinuity around Z=40 which, however, is smaller than the discontinuity in the S(2n)-curves at N=56. This change in slope is limited to neutron numbers 48≤N≤60 and depends sensitively upon N. The mutual support of magicity is also present in this figure and explains the peculiar change in the slope of the S(2p)-curve for the isotones with N=57.

Contrary to Fig. 1, where the change in the S(2n)-curves associated with the semi-magic N=56 extends from N=56 to N=58, the discontinuity in Fig. 2 is observed between Z=39 and Z=42, depending upon neutron number N. According to Zeldes [21], this is due to a considerably smaller pairing energy of the last $2p_{1/2}$ proton pair in Zr. So it seems that for neutron numbers N<56, also Z=38 becomes a semi-magic proton number due to the closure of the $1f_{5/2}$ proton subshell.

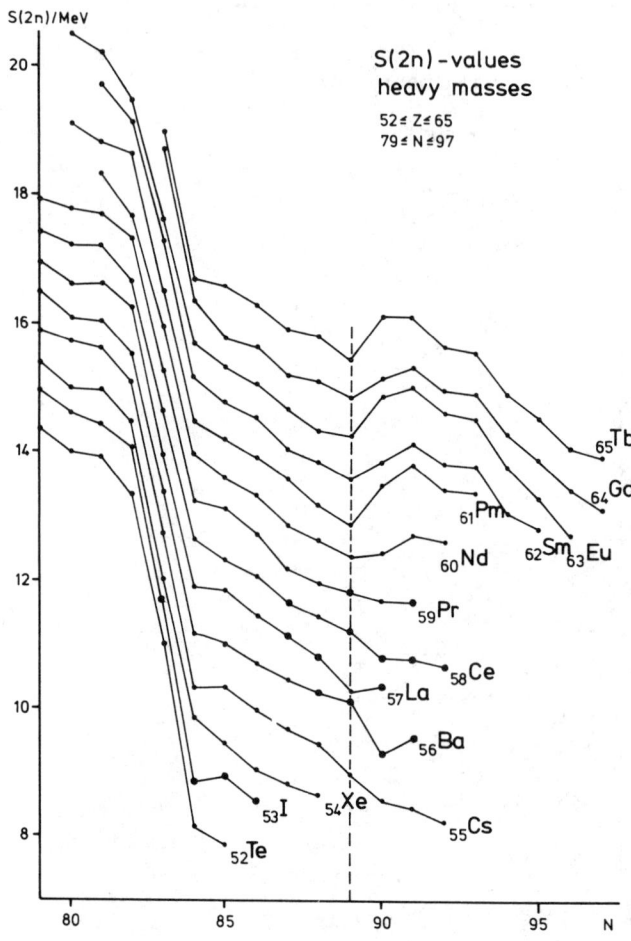

Fig. 3. Two-neutron separation energies S(2n) for nuclei in the mass range $131 \leq A \leq 162$.

NUCLEAR STRUCTURE EFFECTS IN THE MASS REGION A ≈ 145

The S(2n)-curves for all known nuclei in this mass region are shown in Fig. 3. The main shell closure at N=82 manifests itself in the steep decrease of the S(2n)-values for all nuclei with N=83 and N=84. The slope of the curves increases at neutron number N=90 for all nuclei with Z≥60, which is again caused by the onset of deformation. An irregular behaviour of the S(2n)-curves is observed for the neutron-rich nuclei with 55≤Z≤58, especially for ^{146}Ba (N=90), where the S(2n)-value sharply decreases. This irregularity is a "local effect" and might possibly be due to an octupol deformation, which

has been predicted to occur for the ground state of this nucleus [22].
One more structure effect seems to be present in Fig. 3: the decrease of the S(2n)-curve for Cs at N=88. No subshell, however, exists at N=88 according to the shell model. An explanation for this change in slope can possibly be derived from the S(2p)-curves, which are shown in Fig. 4. As can be seen, there is a slight change in the slopes of the isotonic lines with N≥85 for 56≤Z≤58, which might be due to a rather weak subshell closure effect at Z=56. If this interpretation is correct, then the situation in both mass regions considered here is quite similar:
mass region A≅100: Subshell closure N=56, deformation N≥60;
mass region A≅145: Subshell closure Z=56, deformation Z≥60.
But it is certainly necessary to extend the experiments to still heavier isotopes in both mass regions in order to get a more solid basis for the interpretation of the nuclear structure effects discussed above.

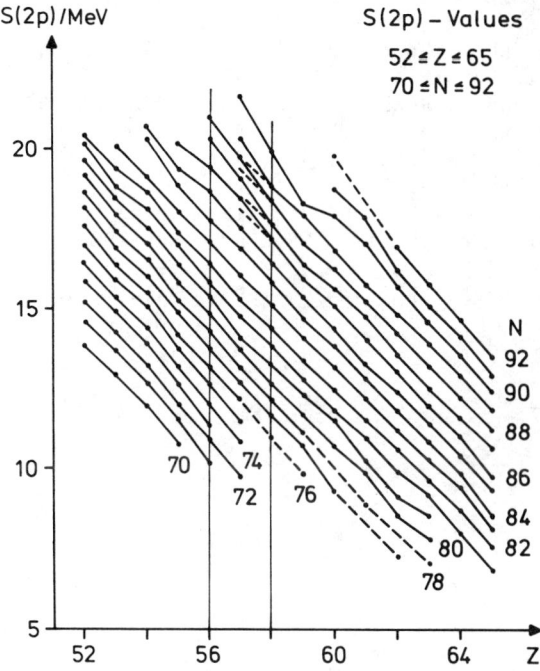

Fig. 4. Two-proton separation energies S(2p) for nuclei in the mass range 122≤A≤157.

We are indepted to Mrs. Ch. Laupheimer for her help in the presentation of the data and the preparation of the manuscript.
The financial support of the Bundesministerium für Forschung und Technologie is gratefully acknowledged.

REFERENCES

1. M. Graefenstedt, Diplomarbeit Braunschweig 1982.
2. F. Schreiber, Diplomarbeit Braunschweig 1986.
3. M. Graefenstedt et al., Z.Physik A, Atomic Nuclei 324(1986).
4. M. Graefenstedt, Thesis Braunschweig 1987.
5. A. H. Wapstra and G. Audi, The 1983 Atomic Mass Table, Nucl. Phys. A432, No.1 (1985).
6. W. R. Pierson, H. C. Griffin and C. D. Coryell, Phys. Rev. 127, 1708 (1962).
7. B. Pahlmann, Thesis Braunschweig 1982.
8. F. Wohn and W. L. Talbert, Jr., Phys. Rev. C18, 2328 (1978).
9. R. Decker et al., Z. Physik A294, 35 (1980).
10. R. Iafigliola et al., Can. J. Chem. 61, 694 (1983).
11. D. S. Brenner et al., Phys. Rev. C26, 2166 (1982).
12. F. Bloennigen et al., Proc. AMCO-7; THD Schriftenreihe Wissenschaft und Technik 26, 134 (1984).
13. P. Peuser et al., Nucl. Phys. A332, 95 (1979).
14. D. Vogel et al., Jahresbericht Kernchemie Mainz 1981, p. 43.
15. K. Sümmerer et al., Z. Physik A287, 287 (1978).
16. E. Lund, K. Aleklett and G. Rudstam, Nucl.Phys.A286,403(1977).
17. A. Kerek et al., Nucl Phys. A195, 177 (1972).
18. F. Schussler et al., Z. Physik A283, 43 (1977).
19. R. Decker et al., Z. Physik A301, 165 (1981).
20. F. Schussler et al., Z. Physik A290, 359 (1979).
21. N. Zeldes, Institut für Kernphysik, T.H. Darmstadt 1984, IKDA 84/20.
22. W. Nazarewicz et al., Nucl. Phys. A429, 269 (1984).

MASS MEASUREMENTS OF ^{162,163}Ta AND NUCLIDES IN THEIR α DECAY CHAINS

E. Hagberg, X.J. Sun, V.T. Koslowsky, H. Schmeing and J.C. Hardy
Atomic Energy of Canada Ltd., Chalk River Nuclear Laboratories,
Chalk River, Ontario, Canada K0J 1J0

ABSTRACT

Two weak α groups have been observed in ^{127}I-induced reactions on ^{40}Ca. The energies, apparent half-lives and excitation functions of these α groups have been measured. We confirm and make unambiguous a previous tentative assignment of these α groups to ^{162,163}Ta. Our assignment is based on their measured half-lives as well as on the good agreement between their excitation functions and those exhibited by well-known γ rays from the decays of ^{162,163}Ta, identified in earlier studies with isotope-separated samples. The mass excesses of ^{162,163}Ta, as well as those of seven other nuclides in their α-decay chains, have been deduced from the α-decay data obtained in the present work and in previously published studies. The three heaviest members of the ^{163}Ta chain are shown to have proton separation energies that are less than zero.

INTRODUCTION

The region of α-decaying nuclides just below the Z=82 closed shell is of special interest for systematic studies of atomic masses. Because α decay provides a characteristic signature for identification, the limit of known nuclides in this region has been pushed to species with half-lives much shorter than a second, so a large number of isotopes have been identified for each element. In addition, the α-decay mode also offers a simple, but precise method for determining parent-daughter mass differences, which extends even to the remotest nuclides. Thus the large body of α-decaying nuclides below the Z=82 closed shell form a region where we have a good knowledge of relative atomic masses. Furthermore, the nuclides in this region are frequently connected by several, sequential α decays and they can therefore be organized in long decay chains. Because the relative masses of the members of such a chain are known, their absolute masses can all be deduced if the absolute mass of any one member be established.

Most of the α-decay chains do not contain a nuclide with a known mass. A large number of potentially interesting mass data can therefore not be used to their fullest extent. Considerable effort has been directed recently towards establishing mass connections between members of such decay chains and nuclides with known masses[1-4]. The missing link is usually either a very weak α-decay branch or a connection that can be obtained through reaction or β-decay Q-values.

The formerly unknown α-decay branches of ^{162,163}Ta were once two such missing links. Their non-observation meant that the masses of these two nuclides, as well as seven others in their α-decay chains, were unknown. Last year the first report on the α decay of ^{162,163}Ta was published[4]. However, the data obtained in this work were rather sparse and, furthermore, the weak α group assigned to ^{163}Ta had previously been assigned to ^{164}Ta in a study by a different group[5]. Such discrepancies in assignments of α groups have occurred before in this region and they can be satisfactorily resolved only by further investigations (see, for example, the case of rhenium isotopes[6]).

We report here a new and independent study of the α decays of ^{162,163}Ta. The main emphasis of our work was to obtain conclusive proof of the identity of the two α groups assigned to these nuclides by Runte et.al.[4]. Our major tool to achieve this goal was to compare decay data for these α groups with simultaneous data obtained on the γ decay of ^{162,163}Ta. The γ rays from these two isotopes have been unambiguously assigned through studies with isotope-separated samples from the ISOCELE-2 facility[7].

EXPERIMENTAL METHOD

The nuclides investigated in this study were produced with a beam of ^{127}I from the Chalk River tandem accelerator superconducting cyclotron (TASCC) facility. The target was 1.1 mg/cm^2 thick natural calcium. A 711 MeV beam was delivered by the cyclotron but the energy of the incident beam on target was selected by the insertion of a series of molybdenum beam degraders immediately before the target.

A He-jet transport system was used to bring the reaction products to a counting location in a shielded room. The energetic reaction products were first slowed down in a 7.4 mg/cm^2 thick molybdenum recoil degrader and then thermalized in the helium-filled interior of a 120 mm long target chamber. The helium, containing NaCl aerosol, was swept out of the target chamber and transported through a 14 m long teflon capillary to the next room in 370 ms. Samples of the transported activities were collected on the tape of a small tape-transport system and periodically moved to the counting location. A 300 mm^2, 100 μm thick surface barrier detector and a Ge(Li) detector were positioned in close geometry at this location. Eight sequential spectra of both α particles and γ rays were measured with the two detectors.

The measured energies of the α groups are sensitive to the energy loss sustained in the accumulated aerosol material of each sample. In the present experiment the energies of all visible α groups were therefore established relative to that of the prominent and precisely known 5,037 keV α group from $^{154}Tm^{8}$). The energy dispersion was determined from the well-known α groups of ^{154}Tm, ^{157}Yb and $^{150}Dy^{8,9}$).

Possible background radiation from activities produced in reactions of the beam with materials other than the calcium target were investigated in a separate experiment. In that experiment the target was removed but all molybdenum degraders were present and all other parameters were the same as in the subsequent experiments with the target in place. No α groups were observed under these conditions and no γ rays of an energy similar to those reported from ^{162,163}Ta were found.

The appropriate thickness of the recoil degrader was investigated in a series of tests. It was found that a 7.4 mg/cm^2 thick molybdenum degrader caused all target reaction products to stop well inside the He-filled target chamber for all incident beam energies used in this work.

RESULTS

Six α spectra obtained at different incident beam energies are shown in fig.1. These spectra are quite complex with many unresolved α groups. Our analysis of all spectra was therefore based on the following procedure. A common peak width was used in fitting all α groups found in the same spectrum. Minor variations in the common width were found from spectrum to spectrum and were accounted for by small changes in the experimental conditions. Initially, all α spectra were fitted with the minimum number of peaks required by a visual inspection of the data. Additional peaks were then added if the agreement was poor.

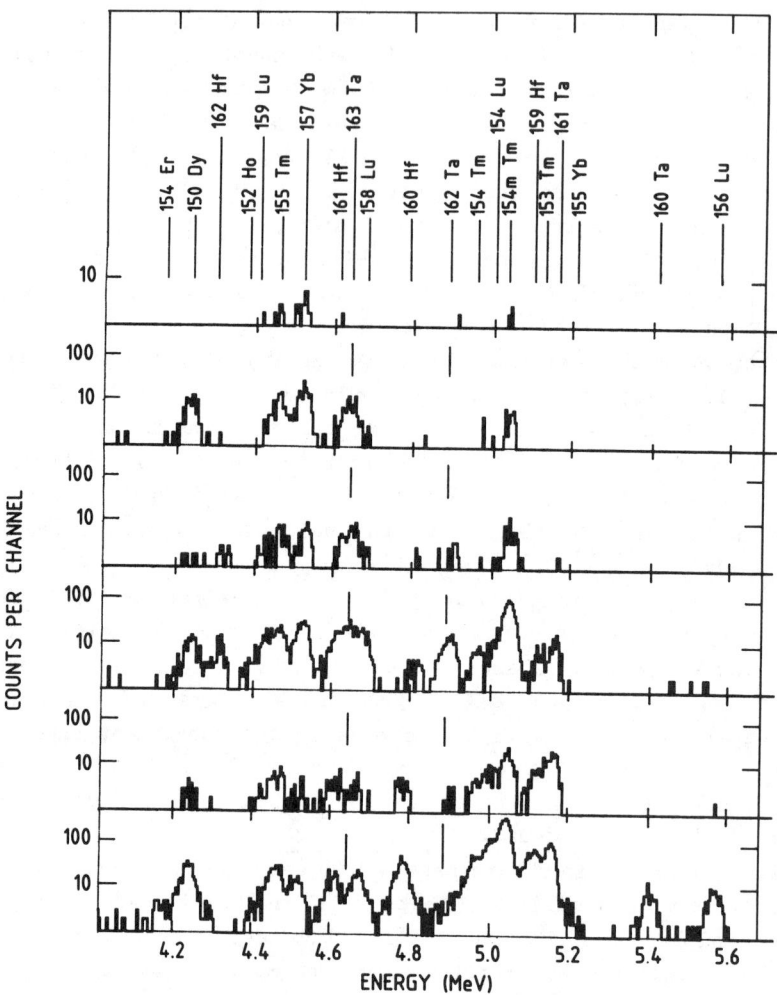

Fig.1 Alpha spectra from activities produced in ^{127}I on ^{40}Ca reactions. From top to bottom the six spectra were obtained with bombarding energies of 540 MeV, 558 MeV, 598 MeV, 628 MeV, 670 MeV and 711 MeV respectively.

The α groups observed in our spectra were assigned to previously known activities if they fulfilled three conditions. First, the energy determined for the α group had to be consistent in all spectra where this group was found and it also had to agree with the literature value. This requirement was especially important for the proper decomposition of multiplets. Second, the half-lives determined for a particular α peak at all bombarding energies had to be consistent with one another and in agreement with the literature value. Finally, the deduced yields of a proposed activity had to be consistent for all spectra obtained with the same incident energy, and the excitation function had to be in reasonable agreement with that computed by the ALICE code[10].

All α groups are assigned in figure 1 according to these criteria. The two α groups with measured energies of 4,892 keV and 4,635 keV agree with those assigned by Runte et.al.[4] to ^{162}Ta and ^{163}Ta, respectively, and they are labelled accordingly in the figure. Our data on these two α groups are not consistent with any other known α emitters.

The excitation functions for the 4,892 keV and 4,635 keV α groups are shown in fig.2 together with those of γ rays from ^{162,163}Ta and ^{162,163}Hf. It is evident from fig.2a that a mass-162 assignment to the 4,892 keV α group is appropriate especially when the good agreement between the excitation function for this α group and that of the 284 keV γ ray, unambiguously assigned to ^{162}Ta by the ISOCELE separator group[7], is taken into

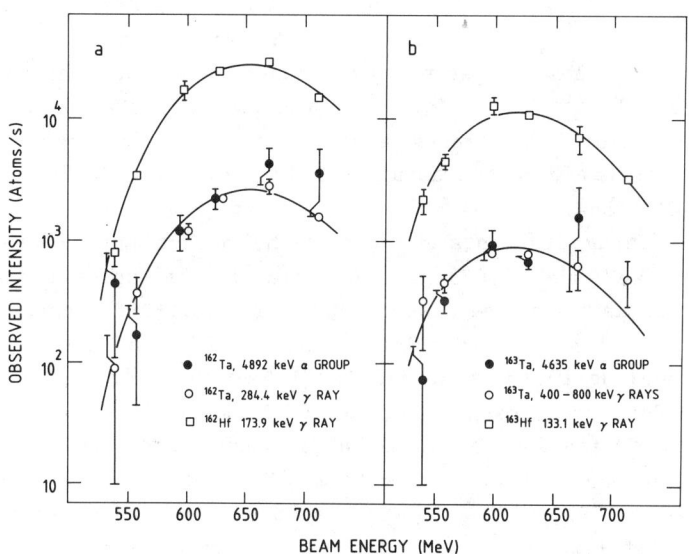

Fig. 2 Yields of ^{162,163}Ta and ^{162,163}Hf deduced from α and γ-ray intensities. The latter have been corrected for known branching ratios and detector efficiencies; the former are normalized to yield the branching ratios in table 1. In each part of the figure the two smooth curves, intended to guide the eye, are identical in shape.

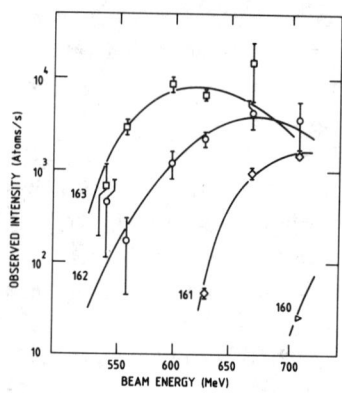

Fig. 3 Yields of $^{160-163}$Ta deduced from their α-group intensities, corrected for branching ratios and detector efficiency.

account. Similarly, it is shown in fig.2b that the 4,635 keV α group originates from mass 163. The lower energy γ rays attributed to ^{163}Ta by the ISOCELE group[7]) were found in our work to be partially obscured by γ rays of nearly the same energy from the decays of ^{164}Ta and ^{160,161}Lu. Our excitation function corresponding to γ rays from ^{163}Ta was therefore taken to be the weighted average of the 449 keV, 451 keV, 628 keV and 713 keV γ rays.

Further support for our mass assignments of the 4,892 keV and 4,635 keV α groups is shown in fig.3 where their excitation functions are shown together with those from the well-known α groups of ^{160,161}Ta. It is evident that these excitation functions exhibit the trends expected for the four masses. In addition, the experimental data shown in fig.3 also agree well with ALICE predictions.

The preceding arguments determine the parent mass numbers for the 4,892 keV and 4,635 keV α groups; the specific element responsible can now be established by half-life comparisons. Possible elements are restricted to Ta, Hf, and Lu since the compound nucleus in our reaction is ^{167}Ta. Some of our measured half-life data are shown in fig.4. It is seen in fig.4a that our measured half-life for the 4,892 keV α group agrees well with that measured for the 284 keV γ rays from ^{162}Ta, but disagrees strongly with those of ^{162}Hf and ^{162}Lu, 37.6 s and 1.4 m, respectively. Our measured half-life for the 4,635 keV α group, shown in fig.4b, agrees well with that of ^{163}Ta, 10.5 ± 1.8 s[7]), but disagrees strongly with those of ^{163}Hf and ^{163}Lu, which are longer than 40 s.

We assign the 4,892 keV α group to ^{162}Ta and the 4,635 keV α group to ^{163}Ta based on our excitation function and half-life data. Our assignments confirm those of Runte et.al.[4]), and thus contradict the previous ^{164}Ta assignment of Schrewe et.al.[5]). Our measured decay data for ^{162,163}Ta are given in table 1

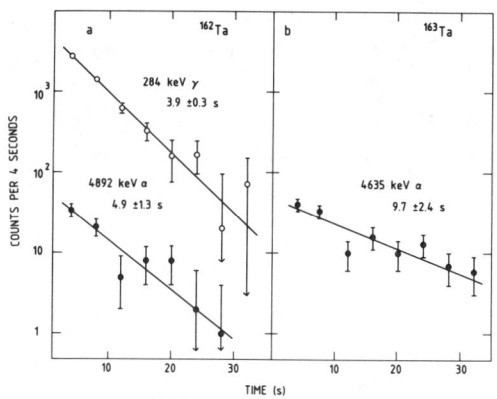

Fig. 4 Half-life curves obtained for ^{162,163}Ta radiations in one experiment accounting for about 30% of all our decay data.

together with those that have been obtained elsewhere. The agreement is generally good and the uncertainties have been reduced in many cases.

The α-particle branching ratios given in table 1 were deduced with the assumption that the strongest γ rays from the ^{162,163}Ta isotopes carried 100% of the decay intensity. Since we do not have a detailed knowledge of the decay schemes of these isotopes, the branching ratios should be taken as upper limits. In the case of ^{163}Ta the actual branching ratio must be much smaller than the

Table 1: Decay properties of neutron-deficient Ta isotopes

	^{162}Ta		^{163}Ta	
	present work	literature	present work	literature
E_α (keV)	4892 (6)	4880 (10)[a]	4635 (8)	4630 (10)[a]
$T_{1/2}^\alpha$ (s)	5.1 (11)	5 (3)[a]	10 (2)	10.9 (14)[a]
$T_{1/2}^\gamma$ (s)	3.52 (12)	3.5 (2)[b]	--	10.5 (18)[b]
b_α (%)	≤0.081 (13)	≤0.065 (14)[a]	≤0.28 (4)	--

a) Ref. 4 b) Ref. 7

limit in table 1 since that value results in a reduced α width that is almost an order of magnitude higher than that expected from systematics in this region. The β-decay strength of ^{163}Ta is therefore probably spread out over many states in ^{163}Hf and no single γ ray carries a large intensity. Such a conclusion also explains our observation of intense γ rays from the $4^+ \rightarrow 2^+ \rightarrow 0^+$ cascade in ^{162}Hf but only very weak γ rays in the even-odd nucleus ^{163}Hf.

DISCUSSION

The masses of the α-decay daughters of ^{162,163}Ta are known. The mass excesses of all nuclides in their α-decay chains can therefore be deduced from the measured ^{162,163}Ta α-decay energies and they are shown in table 2. So far, including the two α-decay chains discussed here, four chains have been linked to nuclides with a known mass. They are indicated in fig.5 and they form a region of nuclides with known masses that extends all the way out to the limits of known nuclei.

This provides a good opportunity for the exploration of systematic trends. As an example, we have deduced the proton separation energies of nuclides in the ^{163}Ta α-decay chain. Three of them, ^{179m}Tℓ, ^{175}Au and ^{171}Ir, were found to have negative proton separation energies of -1,663 ± 131 keV, -877 ± 119 keV and -279 ± 107 keV, respectively. ^{175}Au and ^{171}Ir therefore join a small group of 12 nuclides, which have been shown conclusively to

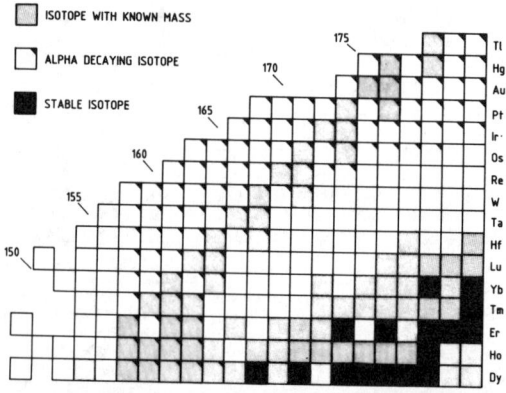

Fig. 5 Section of the chart of nuclides showing known α emitters and nuclides with known absolute masses just below the Z=82 closed shell.

be unstable towards ground state proton emission. It is also evident that other known nuclei, which differ from these by the removal of two neutrons, such as ^{173}Au, ^{169}Ir and ^{167}Ir, also have negative proton binding energies, although the actual values are not yet known.

The Q_p value for ^{179m}Tℓ is quite large. If the proton decay mode of this nucleus were not hindered, the partial half-life would be less than a microsecond and the proton branch close to 100%. However, only the α-decay mode of this nucleus has been observed[12] and the measured half-life of approximately 7 ms indicates that the proton decay of ^{179m}Tℓ is strongly hindered.

Further studies will undoubtedly link other α-decay chains to nuclides with a known mass, thus providing additional information on the masses of nuclides very far from the stability line. Consequently, the exploration of systematic trends in this region will be of particular interest.

Table 2: Deduced mass excesses

decay	α-decay energy (keV)	nuclide	mass excess (keV)
		^{158}Lu	-47,485 (175)[a]
^{162}Ta → ^{158}Lu	4,892 (6)	^{162}Ta	-40,045 (175)
^{166}Re → ^{162}Ta	5,372 (10)[b]	^{166}Re	-32,115 (175)
^{170}Ir → ^{166}Re	6,027 (5)[c]	^{170}Ir	-23,520 (175)
^{174}Au → ^{170}Ir	6,530 (20)[d]	^{174}Au	-14,410 (175)
		^{159}Lu	-47,717 (60)[a]
^{163}Ta → ^{159}Lu	4,635 (8)	^{163}Ta	-42,540 (79)
^{167}Re → ^{163}Ta	5,136 (9)[b]	^{167}Re	-34,854 (94)
^{171}Ir → ^{167}Re	5,925 (3)[c]	^{171}Ir	-26,362 (106)
^{175}Au → ^{171m}Ir	6,440 (20)[e]	^{175}Au	-17,156 (118)[f]
^{179m}Tℓ → ^{175}Au	7,200 (20)[d]	^{179m}Tℓ	- 7,367 (130)

a) Ref.11 b) Ref.6 c) Ref.8 d) Ref.12 e) Ref.13
f) Isomeric transition in ^{171}Ir is 190 ± 0.5 keV[11]

References

1. U.J. Schrewe, E. Hagberg, H. Schmeing, J.C. Hardy, V.T. Koslowsky, K.S. Sharma, and E.T.H. Clifford, Phys. Rev. C25 (1982) 3091.
2. L. Spanier, S.Z. Gui, H. Hick, and E. Nolte, Z. Phys. A299 (1981) 113.
3. W.-D. Schmidt-Ott, R. Kantus, E. Runte, U.J. Schrewe and R. Michaelsen, Phys. Rev. C24 (1981) 2695.
4. E. Runte, T. Hild, W.-D. Schmidt-Ott, U.J. Schrewe, P. Tidemand-Petersson and R. Michaelsen, Z. Phys. A324 (1986) 119.
5. U.J. Schrewe, E. Hagberg, H. Schmeing, J.C. Hardy, V.T. Koslowsky, and K.S. Sharma, Z. Phys. A310 (1983) 295.
6. U.J. Schrewe, E. Hagberg, H. Schmeing, J.C. Hardy, V.T. Koslowsky, and K.S. Sharma, Z. Phys. A315 (1984) 49.
7. C.F. Liang, P. Paris, M.G. Porquet, J. Obert, and J.C. Putaux, Z. Phys. A321 (1985) 695.
8. S. Della Negra, C. Deprun, D. Jacquet, Y. Le Beyec, Ann. Phys. Fr 7 (1982) 149.
9. J.D. Bowman, R.E. Eppley, and E.K. Hyde, Phys. Rev. C25 (1982) 941.
10. M. Blann, University of Rochester report UR NSRL-181 (1978).
11. G. Audi, private communication 1987.
12. J.R.H. Schneider, S. Hofman, F.P. Hessberger, G. Münzenberg, W. Reisdorf, and P. Armbruster, Z. Phys. A312 (1983) 21.
13. C. Cabot, C. Deprun, H. Gauvin, B. Lagarde, Y. Le Beyec, and M. Lefort, Nucl. Phys. A241 (1975) 341.

EXPERIMENTAL MASS EXCESS OF ^{49,50}K AND ^{40,42}Cl

Ch. Miehé, Ph. Dessagne, P. Baumann, A. Huck, G. Klotz,
A. Knipper and G. Walter
Centre de Recherches Nucléaires, 67037 Strasbourg Cedex France
C. Richard-Serre
CERN 1211 Geneva 23 Switzerland
G. Marguier
Institut de Physique Nucléaire 69622 Villeurbanne, France

The Q_β energy of the radioactive decay of the neutron rich ^{49,50}K and ^{40,42}Cl have been measured. The isotopes, produced by the fragmentation of uranium carbide target with 2µA proton beam from the CERN 600 MeV synchrocyclotron, are mass separated in the connected ISOLDE facility and carried to a fast tape transport system. The β decay energy spectra are registered by means of a scintillation telescope in coincidence either with the delayed neutron time of flight spectra of ^{49,50}K or with the γ rays following the decay of ^{40,42}Cl. The β distributions are analysed by the shape fitting technique[1]. The calibration of the β telescope is achieved in beam and by means of radioactive sources. According to the decay schemes[2] several independent Q_β measurements could be performed for each nucleus (fig.1,2,3).

From this work, the first experimental value of the mass excess of ^{49,50}K[3] and ^{42}Cl has been obtained and a new determination has been achieved for ^{40}Cl[4,5]. For each nucleus (table I), the reported Q_β value results from the weighted mean of the different independent determinations. The experimental mass excesses deduced from these values and from the mass excesses of the daughter nuclei[6] are in rather good agreement with the various theoretical estimates[7,8] for the four isotopes under study.

Table I. Q_β values and experimental mass-excesses

Isotope	Q_β (MeV)	Mass-Excess (MeV)
^{49}K	10.97 ± 0.07	− 30.33 ± 0.07
^{50}K	14.05 ± 0.30	− 25.50 ± 0.30
^{40}Cl	7.32 ± 0.08	− 27.77 ± 0.08
^{42}Cl	8.63 ± 0.20	− 25.79 ± 0.20

REFERENCES

1. L.A. Parks, C.N. Davids and R.C. Pardo, Phys. Rev. C 15, 730 (1977).
2. A. Huck et al. in Proceedings of the 4th International Conference on Nuclei far from Stability, Helsingør, 1981, CERN Report 81.09, 378, (1981).

3. Ch. Miehé et al., Phys. Rev. 33, 1736 (1986).
4. Kh. Gurach et al., Sov. J. Nucl. Phys. 19, 596 (1974).
5. L.K. Fifield et al., Nucl. Phys. A417, 534 (1984).
6. A.H. Wapstra and G. Audi, Nucl. Phys. A432, 1 (1985).
7. A.H. Wapstra and K. Bos, At. Data Nucl. Data Tables 17, 474 (1976).
8. K. Takahashi et al., Report IKDA 76/26, (1976).

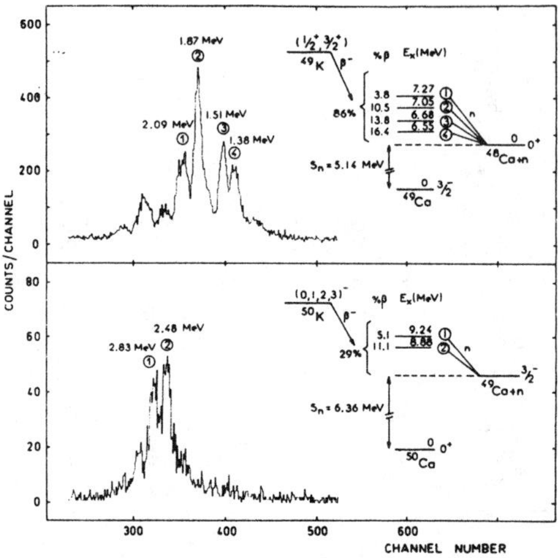

Fig. 1. Neutron time of flight spectra for ^{49,50}K.

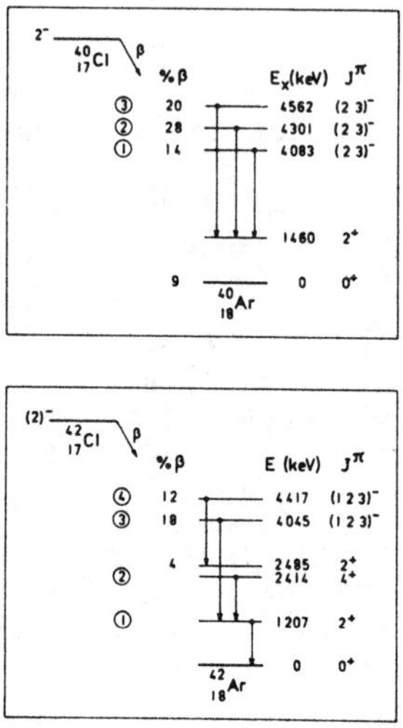

Fig. 2. Partial decay schemes of ^{40,42}Cl.

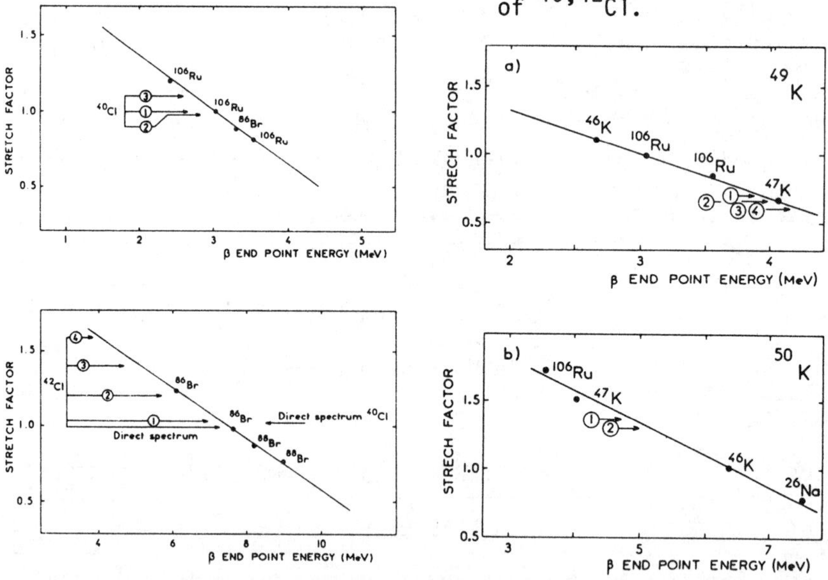

Fig. 3. Beta end-point calibration lines.

THE 1986 – 87 ATOMIC MASS PREDICTIONS

P. E. Haustein
Chemistry Department
Brookhaven National Laboratory
Upton, NY 11973 USA

ABSTRACT

A project to perform a comprehensive update of the atomic mass predictions has recently been concluded and will be published shortly in <u>Atomic Data and Nuclear Data Tables</u>. The project evolved from an ongoing comparison between available mass predictions and reports of newly measured masses of isotopes throughout the mass surface. These comparisons have highlighted a variety of features in current mass models which are responsible for predictions that diverge from masses determined experimentally. The need for a comprehensive update of the atomic mass predictions was therefore apparent and the project was organized and began at the last mass conference (AMCO-VII). Project participants included: Pape and Anthony; Dussel, Caurier and Zuker; Möller and Nix; Möller, Myers, Swiatecki and Treiner; Comay, Kelson and Zidon; Satpathy and Nayak; Tachibana, Uno, Yamada and Yamada; Spanier and Johansson; Jänecke and Masson; and Wapstra, Audi and Hoekstra. An overview of the new atomic mass predictions is presented. Magnetic tape copies of the new predictions may be obtained by written request.

INTRODUCTION

A project to produce a comprehensive update of atomic mass predictions began in 1984 at the Seventh International Conference on Atomic Masses and Fundamental Constants (AMCO-7), held in Seeheim, Federal Republic of Germany.[1] At that time nearly a decade had passed since the publication of the collection of mass tables in the 1975 Mass Predictions.[2] During this interval several additional sets of mass predictions were published individually.[3-5] Experimentalists meanwhile had produced many new nuclear species and the masses of many of these had been determined and tabulated.[6] In many instances, these measurements provided mass values in regions quite far removed from the line of beta stability.

ANALYSES OF THE OLDER PREDICTIONS

These new data proved to be very sensitive indicators of the predictive qualities of the older mass models. Analyses[1,7,8] of the 1975 Mass Predictions and later ones, as made by comparisons of the predictions with several hundred mass measurements of the new nuclides, revealed several interesting features. The correlation between a particular model's predictive quality for new nuclides and that model's goodness-of-fit to the known mass surface was generally a complicated one. Some models employed large numbers of adjustable parameters, and thereby achieved good fits to the known mass surface. A portion of this group of models did not however necessarily provide comparably accurate predictions beyond it. In

some cases, large deviations developed far off the line of beta
stability. In contrast to this, some other mass models employed more
fundamental approaches (using smaller numbers of parameters) and,
consequently, were not able to reproduce as well the fine details of
known mass surface. Some of this group of models nevertheless gave
more reliable predictions. In addition to these global comparisons,
a number of systematic features in the individual models that
resulted in poor predictions in certain mass regions were identified
when differences between predicted and measured masses were plotted
against a variety of relevant physical variables.

THE MASS PREDICTIONS PROJECT AND ITS GOALS

In light of these trends, a comprehensive update of the atomic
mass predictions was indicated. It was clear that such a project
would benefit significantly from the newer and larger body of mass
data now available. It would also be aided by the understanding that
had been acquired in the analysis of the models and from additional
theoretical insights that have been gained in the last decade.
Additionally, a goal of the project would be the use of a common
database of experimental masses, so that the adjustment of model
parameters to known masses would be done with the same database for
all models. This facilitates comparison among different calculations
and it removes questions concerning the sensitivity of the older
predictions to the different databases used in the 1975 predictions
and in later ones.

Potential contributors to this project were contacted at AMCO-7
and shortly thereafter. Interest in the project was judged to be
sufficient to proceed. Coordination of the project from Brookhaven
began in late 1984 and continued through the summer of 1987. In the
spring of 1986, Wapstra, Audi, and Hoekstra[9] provided a mid-stream
atomic mass evaluation that served as the common database for the
mass models. Tables of the new atomic mass predictions were received
at Brookhaven in the late 1986 and early 1987. Once the new master
table of mass predictions had been assembled, numerical and graphical
methods were employed to make several types of comparisons among the
models. The aim of this effort is to give the users of these new
predictions more information about the predictions than that which
has been traditionally supplied before. This allows one to gauge the
reliability of the predictions in regions that border the known mass
surface, i.e., in the regions where new mass measurements are most
likely to be performed in the coming years.

CONTRIBUTORS TO THE 1986 - 87 ATOMIC MASS PREDICTIONS

Ten sets of new atomic mass predictions were prepared. They
were combined with the latest evaluation[9] of experimental masses as
tabulated by Wapstra, Audi, and Hoekstra. Several of the new sets of
predictions came from revised and updated versions of models that
were also presented in the 1975 Mass Predictions. Predictions from
other models appear for the first time. The table of experimental
masses also contains some predictions based on systematics.

Contributors to the mass predictions project

Pape and Antony: Masses of Proton-Rich $T_z < 0$ Nuclei via the Isobaric Mass Equation

Dussel, Caurier, and Zuker: Mass Predictions Based on α-Systematics

Möller and Nix: Nuclear Masses from a Unified Macroscopic-Microscopic Model

Möller, Myers, Swaitecki, and Treiner: Nuclear Mass Formula with a Finite-Range Droplet Model and a Folded-Yukawa Single-Particle Potential

Comay, Kelson, and Zidon: Mass Predictions by Modified Ensemble Averaging

Satpathy and Nayak: Masses of Atomic Nuclei in the Infinite Nuclear Matter Model

Tachibana, Uno, Yamada, and Yamada: Empirical Mass Formula with Proton-Neutron Interactions

Spanier and Johansson: A Mass Formula with Few Free Parameters

Jänecke and Masson: Mass Predictions from the Garvey-Kelson Mass Relations

Masson and Jänecke: Masses from an Inhomogeneous Partial Difference Equation with Higher-Order Isospin Contributions

Wapstra, Audi and Hoekstra: Atomic Masses from (Mainly) Experimental Data

NUMERICAL AND GRAPHICAL ANALYSES OF THE NEW PREDICTIONS

This section presents some numerical and graphical results that give the users of the new mass predictions a means of comparing the different sets of predictions. In Table I the number of nuclei whose masses were used for refinement of adjustable parameters and the number of these adjustable parameters is given for each model. The average and root-mean-square deviations (in keV) of the calculated masses to these experimental data are also shown. Average deviations are typically within a few keV of zero. Non-zero ones signal, in a global sense, whether predicted masses for database nuclei reflect slight underbinding or overbinding in the calculation. The rms-deviations, the minimization of which yields "best values" for adjustable parameters, exhibit a trend noted by Tondeur.[10] There is typically a direct inverse correlation between the rms-deviations of the models and the number of adjustable parameters used in the models. Models which achieve the smallest rms-deviation with a minimum number of adjustable parameters are thought to have most successfully and proficiently incorporated the relevant physical features of the mass surface.

It is useful to explore additional aspects of the mass models beyond those global ones just discussed. It is very informative to examine differences between the calculated mass surface and the experimental one in regions of the N, Z plane in a format like that

Table I Comparison of the Mass Models

Model	Parameters Used	Database Nuclei Used	RMS Deviation (keV)	Average Deviation (keV)	Nuclei[a] Predicted
Pape and Antony	--	85	271	123	381
Dussel, Caurier, & Zuker	45	1328	287	-13	1984
Möller & Nix	26	1593	849	13	4635
Möller, Myers, Swiatecki, & Treiner	29	1593	777	14	4635
Comay, Kelson, and Zidon	b	1632	424	13	6537
Satpathy & Nayak	238	1593	456	1	3481
Tachibana, Uno, Yamada & Yamada	281	1657	538	22	7204
Spanier & Johnson	28	886	699	-111	4162
Jänecke & Masson	928	1633	339	19	5860
Masson & Jänecke	471	1582	344	14	4383

[a] Includes the database nuclei used.
[b] Subsets of the known masses.

of the chart of the nuclides. For a few representative models these differences are shown in Figures 1 - 3. The location of the major shell closures are shown along the proton or neutron axis. Differences (Delta = calculated mass - experimental mass) are plotted using filled (delta > 0) or open (delta < 0) symbols whose size signals the magnitude of the difference. Ranges of delta values (in keV) are represented by the following symbols: $-\infty <$ □ < -800; $-800 <$ O < -400; $-400 <$ ◇ < -200; $-200 < - < -100$; $-100 < \bullet < 100$; $100 < + < 200$; $200 <$ ◆ < 400; $400 <$ ● < 800; $800 <$ ■ $< +\infty$. Several features are apparent from such plots. A model such as that of Jänecke and Masson displays relatively few isotopes with large delta values, a reflection of the small rms-deviations which that model achieved using numerous parameters. In contrast, the model of Möller and Nix exhibits more isotopes with a wider spread of delta values. Here this is a reflection of the larger rms-deviations of this model which operates with small numbers of parameters. In addition one notes that significant trends in the delta values are associated with shell closures. A model that displays intermediate characteristics between these extremes is that of Satpathy and Nayak.

It is useful to recall that in almost all situations, it is the mass difference between two isotopes that is of interest. A mass model may place the calculated surface slightly above or below the measured one in a particular region. This is, of itself, unimportant provided that the mass difference between two isotopes is correctly predicted. Another point to keep in mind is that the reliability of extrapolations of the models for mass predictions into <u>unmeasured</u>

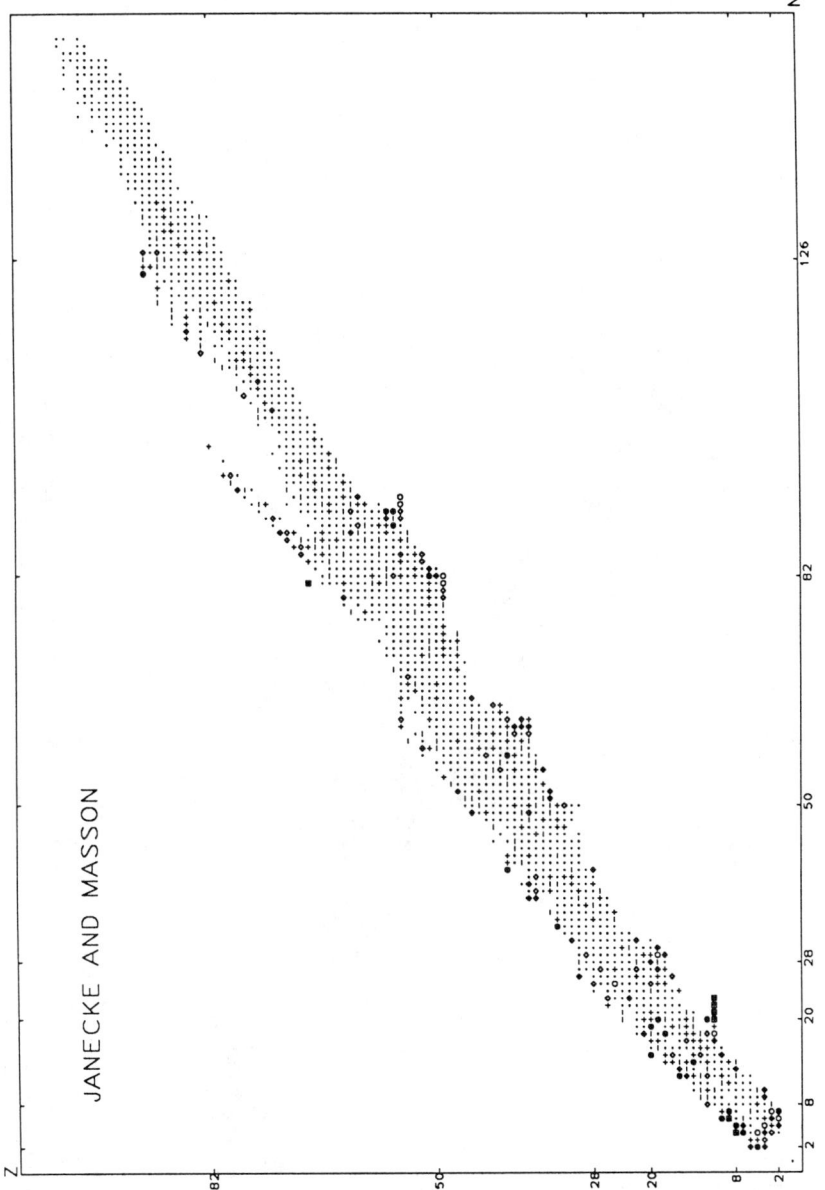

Figure 1. Delta values for the model of Jänecke and Masson plotted in the format of the chart of the nuclides. Symbol definitions are given in the text.

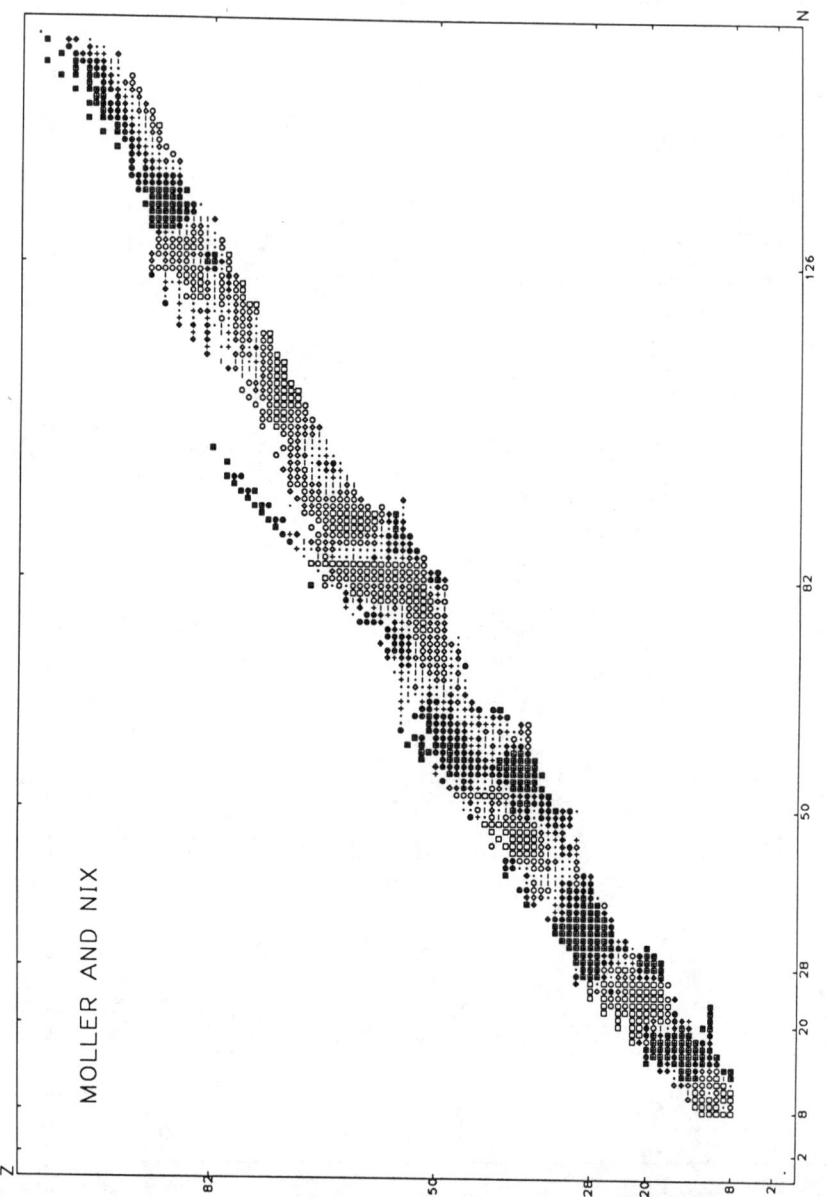

Figure 2. Delta values for the model of Möller and Nix plotted in the format of the chart of the nuclides. Symbol definitions are given in the text.

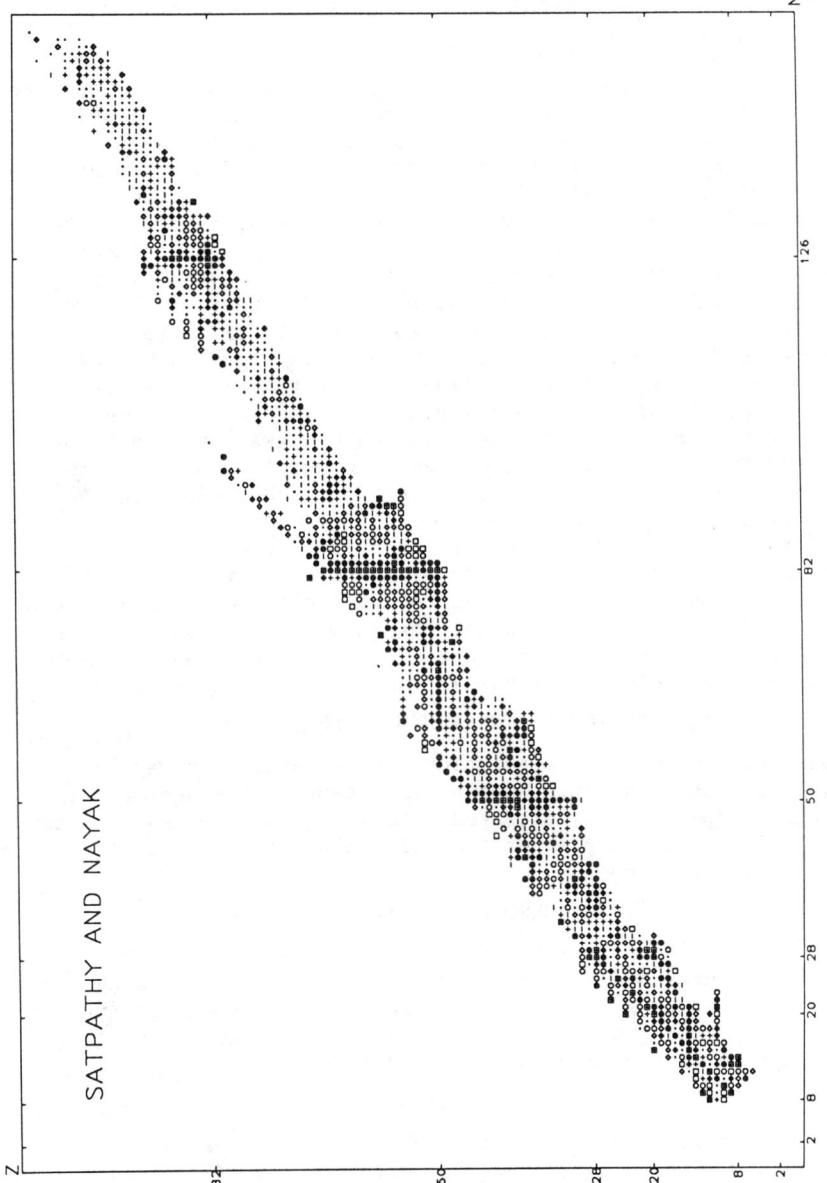

Figure 3. Delta values for the model of Satpathy and Nayak plotted in the format of the chart of the nuclides. Symbol definitions are given in the text.

regions (beyond the database used in fitting is not a function of how well the model reproduced the <u>measured</u> surface, although it is generally assumed that this latter capability manifests a crucial understanding of the underlying physical features which should aid accurate extrapolation.

Plots of delta values like those in Figures 1 - 3 may be used to gauge the reliability of predictions from the various mass models in regions that border the experimentally known surface. This is done by first locating the isotope(s) of interest in the Z and N plane. One then examines the delta values of those measured isotopes nearby. This is done to identify the extent of displacement of the predicted surface relative to the measured one. This is indicated by the sign and magnitude of the delta values and is denoted by one of the several plotting symbols. A constant displacement of the two surfaces is signaled by clusters of symbols of the same type. Convergence or divergence of the two surfaces is signaled by changes in types of plotted symbols along lines of constant Z, N, A, or isospin. Since the extent of displacement of the two surfaces generally varies smoothly and slowly in regions away from shell closures and beyond the lightest nuclei, the position of the actual mass surface relative to the predicted one in such regions can be estimated by this procedure. Examination of trends of this type across different model types can also identify one or a group of models that may be particularly appropriate for a given mass region. In the lightest nuclei or those near to shell closures, the extent of displacement of the two surfaces may change much more rapidly with nucleon number, mass number, or neutron excess. Consequently closer examination of the relevant trends is called for in these cases.

DISTRIBUTION OF THE NEW MASS PREDICTIONS

Copies of the new mass predictions on magnetic tape may be obtained by written request. Send a blank tape to the address above. Unless specifically directed otherwise, tape copies will be prepared at 6250 bpi. Copies of the delta plots for each of the models will also be provided.

ACKNOWLEDGMENTS

Research carried out at BNL under contract with the U. S. Department of Energy and supported by its Office of High Energy and Nuclear Physics. The author wishes to acknowledge the assistance of Professor J. Jänecke and Dr. E. Comay who supplied the computer software that permitted production of the figures.

REFERENCES

1. P. E. Haustein, Proc. 7th Int. Conf. on Atomic Masses and Fundamental Constants (AMCO-7) O. Klepper, Ed., Technische Hochschule Darmstadt, Darmstadt, Fed. Rep. Germany, 1984, p. 413.
2. S. Maripuu, At. Data Nuc. Data Tables, 17, 411 (1976).
3. J. E. Monahan and F. J. D. Serduke, Phys. Rev. <u>C17</u>, 1196 (1978).
4. P. Möller and J. R. Nix, Jr., At. Data Nuc. Data Tables, <u>26</u>, 165 (1981).

5. M. Uno and M. Yamada, Report INS-NUMA-40, Science and Engineering Research Laboratory, Waseda University, Japan, 1982.
6. A. H. Wapstra and G. Audi, Nucl. Phys. A432, 1 (1984).
7. P. E. Haustein, Am. Chem. Soc. Symp. Ser. 324, Nuclei Off the Line of Stability, R. A. Meyer and D. S. Brenner, Eds., pp. 126-131, 1986.
8. P. E. Haustein, "Predictive Properties of Atomic Mass Models: Their Relationship to Nuclear Decay Modes by Spontaneous Charged Particle Emission", book Chapter in Charged Particle Emission from Nuclei, Vol. 1, CRC Press, M. Ivascu and D. N. Poenaru, Eds. (in press).
9. A. H. Wapstra, G. Audi, and R. Hoekstra, Midstream (1986) Atomic Mass Evaluation, available from the National Nuclear Data Center, Brookhaven National Laboratory.
10. F. Tondeur, CERN Report 81-09, p. 81 (1981).

MASS PREDICTIONS, PARTIAL DIFFERENCE EQUATIONS AND HIGHER-ORDER ISOSPIN EFFECTS

P. J. Masson and J. Jänecke
The University of Michigan, Ann Arbor, Michigan 48109, USA

ABSTRACT

The Garvey-Kelson mass relation has been extended by introducing inhomogeneous source terms to improve problems with long-range extrapolations. Such mass relations are third-order partial difference equations with solutions representing mass equations. It was found that inhomogeneous source terms based on shell-dependent Coulomb and symmetry energy terms are not sufficient to improve upon extrapolations. However, contributions from higher-order perturbations in isospin (mostly cubic) have a significant effect. A many-parameter mass equation was constructed as the solution of an inhomogeneous difference equation with properly adjusted shell-dependent source terms. The standard deviation for reproducing the experimental mass values is σ_m = 194 keV. Nuclear contributions were subjected to the constraint of charge symmetry, and Coulomb displacement energies are reproduced with σ_c = 41 keV. Mass predictions for over 4000 nuclei with A > 16 and both N ⩾ Z and N < Z (except N = Z = odd for A < 40) are reported.

Mass relations are recursion relations which can be used to predict unknown masses. When viewed as partial difference equations they can also be used to construct mass equations. The Garvey-Kelson relation[1] has been very successful to predict masses and binding energies of nuclei close to the known nuclei. However, long-range extrapolations display systematic effects. Figs. 1 and 2 show systematic deviations when a band of nuclei along the line of β-stability is used as data base to predict the masses of the known nuclei outside this band. Fig. 1 displays the average deviations for bands of different widths. Cubic contributions in isospin are suggested. Fig. 2 displays the individual residuals for one specific example. Coulomb and symmetry energy contributions to nuclear binding energies energies are known to generate inhomogeneous source terms.[2,3] Detailed shell-model expressions were therefore derived and included in the mass equation. Fig. 3 shows again the comparison between the experimental and predicted mass values outside the data base along the line of β-stability. Only minor improvements are observed. It was therefore concluded that shell-dependent terms in isospin of the type $f(A)(T - T_{stab})^3$ must be included in the description of the symmetry energy. Such terms arise from higher-order perturbations in isospin due to subshell mixing, core excitations or departures from simple coupling schemes and also indirectly from deformation effect. Fig. 4 shows a graphical representation of these higher-order shell-dependent contributions which were constructed from the data taking

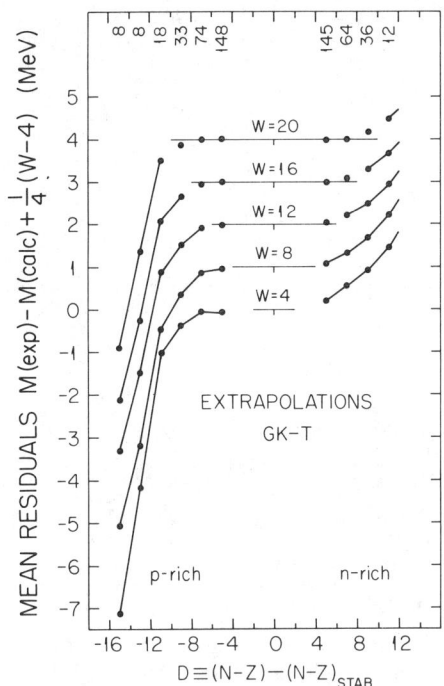

Fig. 1. Differences between experimental and extrapolated masses as function of the distance from the line of β-stability averaged over all mass numbers A. The horizontal bars (displaced) represent the widths of the data bases.

into consideration the combined effects of the homogeneous and inhomogeneous parts of the partial difference equation. Fig. 5 shows greatly improved extrapolations when these higher-order terms are included. The solutions were subjected to the constraint of charge symmetry of nuclear forces. They are therefore valid for nuclei with both N > Z and Z > N, and the accuracy of the Coulomb energies (σ_C = 41 keV for the Coulomb displacement energies[4]) is preserved. The final many-parameter mass equation (see Ref. 5 for details) was derived using all available data as input. The standard deviation for reproducing the data is σ_m = 194 keV, and mass predictions for over 4000 nuclei with A > 16 were made.

Supported in part by the U. S. National Science Foundation.

1. G. T. Garvey, W. J. Gerace, R. L. Jaffe, I. Talmi and I. Kelson, Rev. Mod. Phys. 41, S1 (1969).
2. J. Jänecke, in Atomic Masses and Fundamental Constants 7, ed. O. Klepper (THD Schriffenreihe 26, Darmstadt 1984), p. 420.
3. J. Jänecke and E. Comay, Nucl. Phys. A346, 108 (1985).
4. E. Comay and J. Jänecke, Nucl. Phys. A410, 103 (1983); J. Jänecke and E. Comay, Phys. Lett. 140B, 1 (1984).
5. J. Jänecke and P. J. Masson, Atomic Data Nucl. Data Tables (1987); P. J. Masson and Jänecke, ibidem.

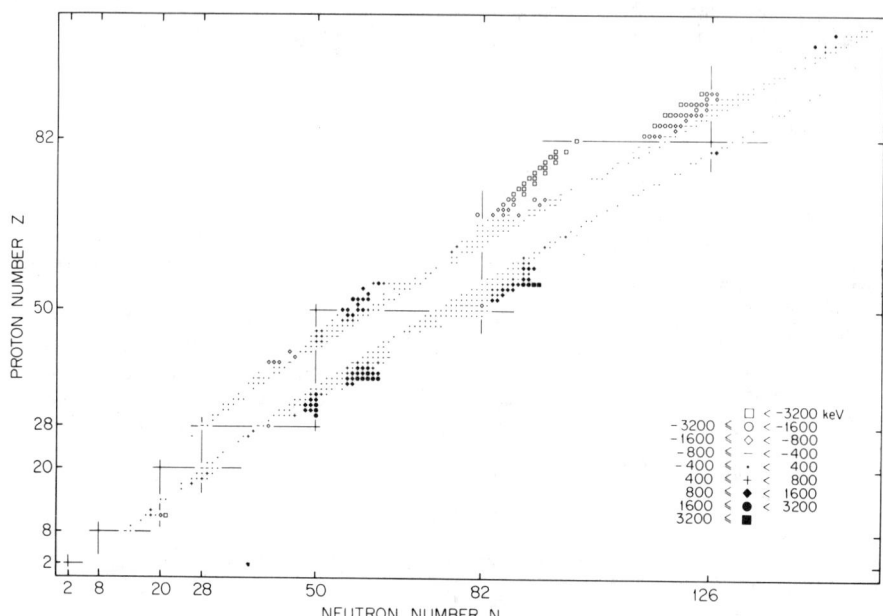

Fig. 2. Differences between experimental and calculated (extrapolated) masses for the transverse Garvey-Kelson relation. Only the empty band of nuclei along the line of β-stability is used as data base in the fitting procedure. (See also Figs. 3 and 5.)

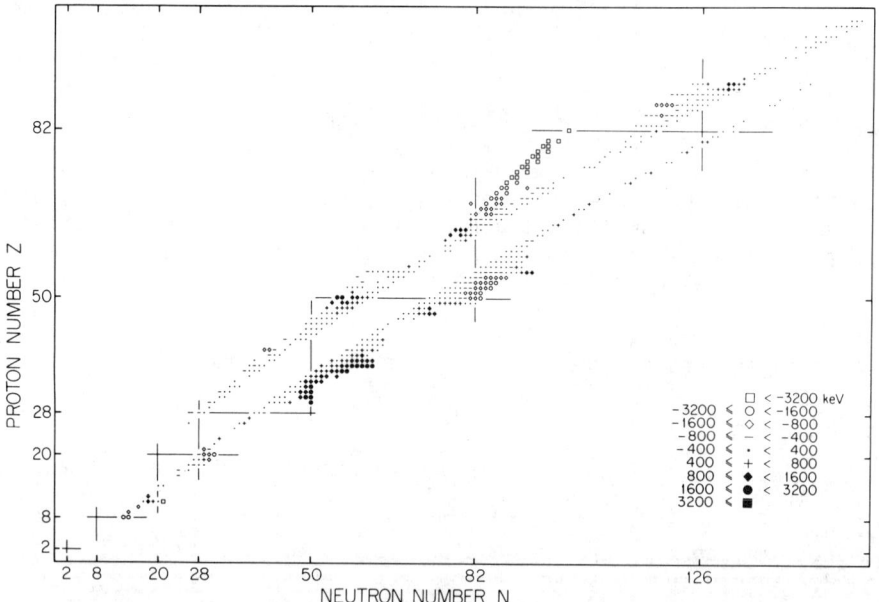

Fig. 3. Differences between experimental and calculated (extrapolated) masses. Shell-dependent Coulomb and symmetry energy terms (quadratic in isospin) are included as inhomogeneous source terms.

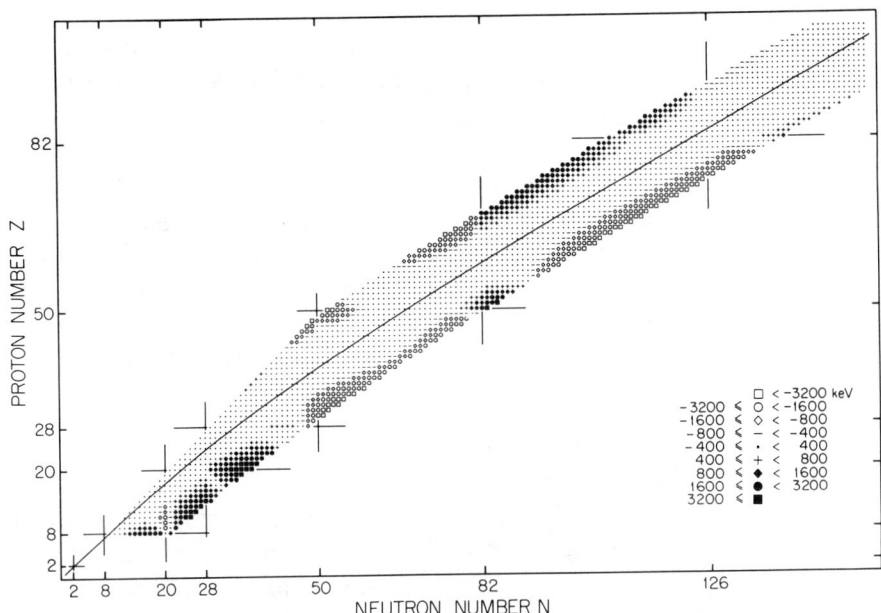

Fig. 4. Shell-dependent symmetry-energy contributions of higher order in isospin (T > 2).

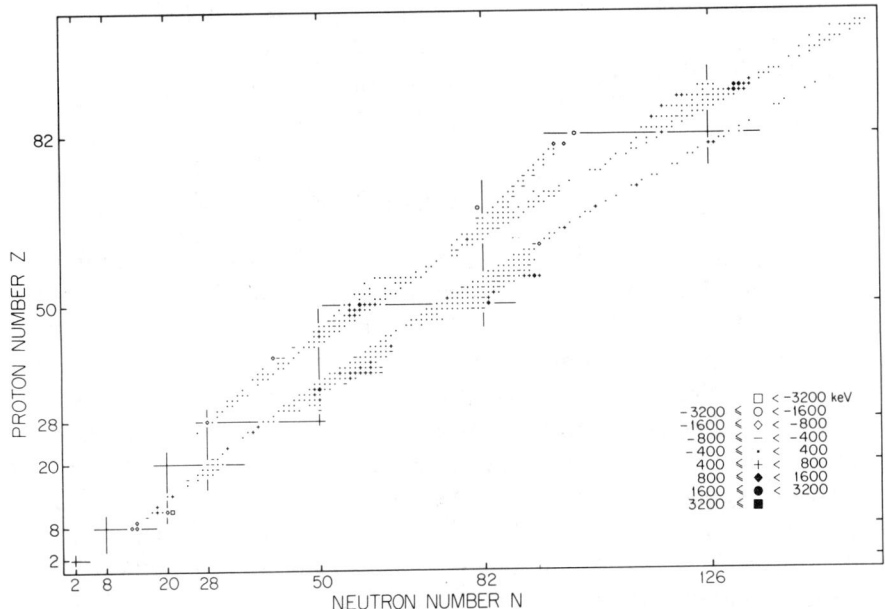

Fig. 5. Differences between experimental and calculated (extrapolated) masses. The inhomogeneous source terms include those used in Fig. 3 and the higher-order terms of Fig. 4.

NUCLEI FAR FROM STABILITY VIA THE "ETFSI" MASS FORMULA

J. M. Pearson, A. K. Dutta
Laboratoire de Physique Nucléaire, Université de Montréal,
Montréal, Qué. H3C 3J7, Canada
and
F. Tondeur
Institut d'Astrophysique, Université Libre de Bruxelles,
B1050 Bruxelles, Belgium.

We have developed an approach to the mass formula based on the extended Thomas-Fermi method with shell corrections calculated by a variant of the Strutinsky method. For extrapolation from known nuclei to unknown nuclei close to the neutron-drip line this method is essentially equivalent to the Hartree-Fock method, for a given form of force, but is so much faster that the construction of a complete mass table becomes feasible. Preliminary partial fits to the data (491 spherical and 25 deformed nuclei) give rms errors significantly smaller than is the case with mass formulas based on the droplet model.

1. INTRODUCTION.

A considerable effort continues to be devoted to the construction of nuclear mass formulas of ever greater sophistication. Much of the motivation for this lies in the fact that the evolution of stellar nucleosynthesis, and in particular of the so-called r-process, depends crucially on the binding energies of nuclei that lie so close to the neutron-drip line that there is no possibility of being able to measure them in the laboratory. It thus becomes of the greatest importance to be able to make reliable extrapolations of nuclear binding energies away from the known region, relatively close to the stability line, out towards the n-drip line. Also of interest is extrapolation in another direction: extension of the stability line to explore the possible existence of superheavy nuclei.

If one is to have any confidence in the ability of a mass formula to extrapolate reliably it must not only give a good fit to the available data but also have a sound theoretical basis. Clearly, if two different mass formulas give comparable fits to the data but extrapolate differently, one would prefer the one with the better theoretical foundation. We believe, therefore, that it is vitally important to develop mass formulas that are as rigorously based as possible. The ideal would be to be able to derive all nuclear properties from "real" nuclear forces, but it will be impossible to do this with anything like the required precision in the forseeable future.

Mass formulas in use at the present time[1-5] are based rather on the so-called "macroscopic-microscopic" approach[6], in which the smoothly varying part of the binding energy is represented by one form or another of the drop(-let) model (DM), to which must be added microscopic corrections to take account of the fluctuations associated with shell-model and pairing effects. Now while this approach gives acceptable fits to the data (ref.3, for example, has rms errors of 0.835 Mev for 1323 masses, and 1.331 Mev for 28 fission barriers), it is open to theoretical criticism on two different counts:

a) In the usual forms of the DM the characteristic "leptodermous" expansion in powers of $A^{-1/3}$ is truncated prematurely[7-9]. This difficulty appears to be rectified at least partially in the so-called "finite-range" droplet model[10,11], but there remains the problem of a premature truncation of the expansion in powers of $I=(N-Z)/A$. Our group showed[11] that when this model was fitted to the data errors as large as 15 Mev could result on extrapolating out to the n-drip line.

b) It is difficult to relate the macroscopic and microscopic parts coherently. To be specific, the calculation of the microscopic corrections involves the use of a single-particle (s.p.) potential, and since this must be generated by the distribution of nucleons in the nucleus it constitutes a link between the microscopic and macroscopic parts of the mass formula. Now the actual way in which the s.p. potential is generated from the nucleon distribution is by folding some two-body force over the latter, but in no form of the DM is there an unambiguous prescription for choosing this force. Furthermore, the density distribution itself is determined only in a very crude way in the DM.

Both these classes of difficulty are avoided when the binding energy is calculated by the Hartree-Fock (HF) method. In the first place there is no approximation based on a power-series expansion. Secondly, there is no separation of the total energy into macroscopic and microscopic parts, so complete consistency between the two is automatically guaranteed. A mass formula based on the HF method (or rather on the HF + BCS method, since pairing always has to be taken into account) will, therefore, be much more theoretically secure than one based on a DM. This method represents, in fact, the most fundamental approach to a mass formula that has any chance of succeeding, even though it is much less rigorous than an approach based on the "real" nuclear forces.

The ideal procedure to be followed with this method would consist in taking some suitable form of effective interaction and fitting its parameters, along with those of the pairing force, to all the data on masses, fission barriers, radii, etc., as in the present DM fits. Unfortunately, the method suffers from the defect of requiring a very large amount of computer time, especially for deformed nuclei, with the result that its systematic application has been somewhat limited. In particular, the available HF effective forces have been fitted to relatively few of the available data, thereby detracting from the reliability of the method as a means of extrapolating.

Here we present an approach of intermediate complexity that we have developed: it is based on the extended Thomas-Fermi (ETF) method for the macroscopic part, with microscopic corrections calculated by the so-called Strutinsky-integral (SI) method.[12,13] When the underlying Skyrme-type force is fitted to the data the extrapolations to the n- drip line are very close to those given by the HF method, but being much more rapid computationally the method offers a practical approach to the ultimate task of constructing a mass table. We give a brief review of the method in Section 2, summarizing refs. 12 and 13, while in Section 3 we discuss our preliminary fits.

2. SKYRME-ETFSI METHOD.

The basis of our method is a generalized Skyrme force:

$$v_{ij} = t_0(1+x_0 P_\sigma)\delta(\vec{r}_{ij})$$

$$+ t_1(1+x_1 P_\sigma)[p_{ij}^2 \delta(\vec{r}_{ij}) + \text{h.a}]/2\hbar^2 + t_2(1+x_2 P_\sigma)\vec{p}_{ij}\cdot\delta(\vec{r}_{ij})\vec{p}_{ij}/\hbar^2$$

$$+t_3(1+x_3P_\sigma)[\rho_{qi}(\vec{r}_i)+\rho_{qj}(\vec{r}_j)]\gamma\delta(\vec{r}_{ij})/6$$

$$+(i/\hbar^2)W_0(\vec{\sigma}_i+\vec{\sigma}_j).\vec{p}_{ij}\times\delta(\vec{r}_{ij})\vec{p}_{ij} \qquad (1)$$

To this we add the constraints $x_1=-(4+5x_2)/(5+4x_2)$ and $t_1 = -t_2(5+4x_2)/3$ in order that the effective and real nucleon masses be equal, $M^*=M$, a choice which allows a good fit to the s.p. energies near the Fermi surface without having to take particle-vibration coupling into account.

Adopting now the ETF semi-classical approximation to the kinetic-energy and spin-current densities, τ_q and $\vec{J}_q$, respectively, we express the total energy as a function of the nucleon densities, ρ_q, and their gradients: we use the full fourth-order (in powers of $\hbar$) expansions of Grammaticos and Voros [14,15]. The resulting expressions are given in refs. 12 and 13; ref. 13 also discusses our adoption of the (u,v) coordinate system of Brack et al (BGH)[9] for axially symmetric nuclei (we cannot handle triaxial nuclei, as such, but this is of no great significance as far as masses are concerned). The density distributions, ρ_q, are then parametrized according to the Fermi form, and we minimize the corresponding energy with respect to its parameters. (For deformed nuclei we have to make the additional *ansatz* that ρ_q depends only on the u-variable, as in BGH.)

It is inevitable with the ETF method that the energy will vary smoothly from one nucleus to another: the shell effects are effectively lost in truncating the semi-classical expansions. Thus we are forced back to a macroscopic-microscopic approach again, and have to add shell corrections. However, compared to the DM mass formulas there is an important difference, since we can now determine quite unambiguously the s.p. fields $U_q(\vec{r})$ and $\vec{W}_q(\vec{r})$ appearing in the s.p. Schrodinger equation,

$$[(-\hbar^2/2M)\nabla^2 + U_q(\vec{r}) + \vec{W}_q(\vec{r}).(-i\vec{\nabla}\times\vec{\sigma})]\phi(\vec{r}) = \epsilon\phi(\vec{r}) \qquad (2)$$

the solutions to which are required for the determination of the shell (and pairing) corrections. One simply folds the same Skyrme-type force involved in the ETF functional over the density distribution obtained in the macroscopic part of the calculation. There is thus a high degree of coherence between the macroscopic and microscopic parts, the unifying factor being the Skyrme-type force that underlies both. With Eq.(2) solved the calculation of the shell and pairing corrections can proceed.

The shell correction is written in the usual way as

$$\delta = \Sigma' \epsilon_\mu - \widetilde{\Sigma}' \epsilon_\mu \qquad (3)$$

To calculate the smoothed term here we could use the conventional method of Strutinsky smoothing (see ref. 16 for a review), as in refs. 3-5, for example. However, it is well known that this method contains some ambiguities, mainly because of the continuum s.p. states, which means that they may be expected to become particularly troublesome towards the drip lines. Fortunately, in the present case, where we know the density distributions from which the s.p. potential is generated, these problems can be avoided, since another method, due to Chu et al[17], is available. This method, which we call the Strutinsky-integral method, is based on a direct application of the Strutinsky theorem (see, for example, Eq. (2.131) of ref. 18), from which it is easy to show that

$$\widetilde{\Sigma}' \epsilon_\mu \approx \Sigma \int d^3\vec{r} (\tau_q + \rho_q U_q + \vec{J}_q \cdot \vec{W}_q) \qquad (4)$$

where τ_q, ρ_q, and $\vec{J}_q$ are the smoothed densities that emerge from the minimization in the macroscopic part of the calculation, while U_q and $\vec{W}_q$ are the corresponding fields, appearing in Eq. (2). This prescription for shell corrections is very simple to apply and quite unambiguous, even at the drip lines (note that it cannot be used with drop (-let) models, since the distributions ρ_q and $\vec{J}_q$ giving rise to the s.p. fields are not known there).

Executing this method requires that the s.p. field resulting from the macroscopic part of the calculation be diagonalized, in order to get the s.p. energies appearing in Eq. (3). Thus this method of calculating shell corrections may be regarded as doing one iteration of HF on top of ETF. Since this is by far the most time-consuming part of the whole calculation we see that the ETFSI method is an order of magnitude faster than HF.

We handle pairing by doing BCS with a δ-function force:

$$v_{ij} = V_p \delta(\vec{r}_{ij}) \qquad (5)$$

Although this increases the computer time as compared to the "constant-G" approximation, it allows an excellent fit to the data with a constant strength, which in turn leads to a more reliable extrapolation.

Comparison with HF. For a given set of force parameters this method overbinds nuclei by between 3 and 7 Mev, as compared to HF. Now while errors this large would be inacceptable in modern mass formulas it is not inevitable that this will pose a problem in practice, since whatever method is being used the force parameters are always fitted to the data. Rather, the crucial question is: when the ETFSI method is extrapolated out to the n-drip line, will it agree with the HF extrapolation? In ref.12 we showed that the discrepancy between the two methods at the n-drip line is less than 1 Mev for the total energy, and less than 0.2 Mev for the n-separation energy, S_n, and beta-decay energy, Q_β, the quantities of ultimate importance for the r-process. Likewise we showed in ref.13 that a similar level of agreement with HF can be found for fission barriers at the n-drip line. We thus conclude that while the ETFSI method is an order of magnitude faster than HF it gives essentially the same extrapolations from known nuclei out to the n-drip line, for a given *form* of force.

Mass Table. Once an effective interaction has been fitted to the data a further advantage of the ETFSI method emerges. All quantities involved in the total energy, including the s.p. energies (but not their sum, of course) vary smoothly with N, Z, and deformation, so it will be possible to construct a mass table by interpolation. This is not possible with the HF method itself, and it should be possible to gain two more orders of magnitude in computer time.

3. PRELIMINARY FITS.

We are fitting our force parameters to the 1986 nuclear-mass compilation of Audi and Wapstra[19]. We begin with an automatic least-squares fit to a sample of 491 spherical nuclei with $Z \geq 20$. This sample is selected on the basis of Z or N not being more than two units away from a magic number, excluding the few such nuclei known or suspected to be deformed; because of the Wigner effect we likewise exclude nuclei with $Z \approx N$. So far we have kept γ fixed at the value 1/3, which guarantees a nuclear-matter incompressibility K of close to 235 Mev. (Always we find a_v to be very close to -15.9 Mev.) Fits are made with different fixed values of k_F, J (Fermi momentum and volume-symmetry coefficient, respectively, of nuclear matter), and x_2.

	$x_2 = -0.5$	$x_2 = 1.0$	expt.
ϵ_{sph}	0.655	0.689	
$\epsilon_{def(A \leq 240)}$	0.782	1.16	
^{246}Pu	1845.0	1846.4	1845.9
^{246}Fm	1835.3	1836.4	1836.3
^{250}Cm	1868.1	1869.5	1869.0
^{250}Fm	1863.1	1864.2	1864.7
^{252}Cf	1878.7	1880.0	1880.5
^{256}Fm	1900.0	1901.2	1901.7
^{256}Rf	1887.7	1889.4	1889.7
^{260}Nh	1906.5	1907.3	1908.1

Table I. Fits to binding energies (Mev) for two values of x_2.

We find that $k_F = 1.335$ fm^{-1} leads to the best fit to charge radii, and keep this value in this report. Excellent fits to the mass data are found with J in the range 27 to 29 Mev, and we take here the mean value of 28 Mev, even though our preliminary fission barrier calculations favour J=27 Mev. We tried several values of x_2 and found -0.5 to give the best fit to the masses of the 491 spherical nuclei, as shown in the first column of Table I, in which the first line gives the rms error of this fit (the rms error in the fit to the 68 most neutron-rich spherical nuclei given in the 1986 mass compilation is even better: 0.525 Mev). The force thus determined by this fit is then used to calculate the energies of 25 deformed nuclei, the rms error for the 17 of these nuclei with A≤240 being shown in the second line of Table I. We believe that if these deformed nuclei had been included in the fit that defines the force they would have agreed with experiment even better.

However, it is clear from Table I that the fit for $x_2 = -0.5$ is definitely deteriorating beyond A=240 (note that the heaviest spherical nucleus in the fit has A=218). In this respect we see from the second column of Table I that there is a definite improvement with a force having $x_2=1.0$. Nevertheless, this second force, as it stands, has an inacceptable agreement for the deformed nuclei with A≤240. Conceivably, this could be remedied by including these nuclei in the fit, but it seems fairly clear that the overall fit can never be as good as for the force with $x_2=-0.5$.

	$x_2 = -0.5$	$x_2 = 1.0$	Myers
B	1795.8	1798.6	1817.7
S_n	1.8	1.8	5.8
Q_β	13.3	13.3	12.2

Table II. Extrapolation to ^{266}Pb.

In any case, in deciding which value of x_2 to persist with in improving the fit to the data, it is worthwhile to check on the implications for extrapolation to highly n - rich nuclei. Table II shows the case of doubly-magic ^{266}Pb, the last column being the DM prediction of ref. 1.While the total binding energy obviously depends on the choice of force, it will be seen that S_n and Q_β, the quantities of ultimate astrophysical interest, are quite independent of this choice. This is something that one should bear in mind, given the large amounts of computer time that fits of this sort entail. Nevertheless, we intend to investigate the effect on the fits of varying K (by varying γ) before passing to a final force and beginning the construction of the mass table.

We wish to thank the Centre de Calcul at the Université de Montréal for its extremely generous allocation of computer time, and P.Haustein for sending us the 1986 mass compilation ahead of publication. This work was supported in part by NSERC of Canada.

REFERENCES.

1. W.D. Myers, Atomic Data and Nuclear Data Tables 17 411 (1976)

2. H. von Groote, E.R. Hilf and K. Takahashi, Atomic Data and Nuclear Data Tables, 17 418 (1976)

3. P. Möller and J.R. Nix, Nucl. Phys. A361 117 (1981)

4. P. Möller and J.R. Nix, Atomic Data and Nuclear Data Tables 26 165 (1981)

5. P. Möller and J.R. Nix, Atomic Data and Nuclear Data Tables (submitted 1986)

6. W.D. Myers and W. J. Swiatecki, Nucl. Phys. 81 1 (1966)

7. J.M. Pearson, Nucl. Phys. A376 501 (1982)

8. F. Tondeur, J.M. Pearson and M. Farine, Nucl. Phys. A394 462 (1983)

9. M. Brack, C. Guet and H.-B. Hakansson, Physics Reports 123 277 (1985)

10. P. Möller, W.D. Myers, W. J. Swiatecki and J. Treiner, *Proc. 7th Int. Conf. on Atomic Masses and Fundamental Constants, AMCO-7*, ed. O. Klepper (THD, Darmstadt 1985) p. 457; Atomic Data and Nuclear Data Tables (to appear 1987)

11. A.K. Dutta, J.-P. Arcoragi, J.M. Pearson, R. Behrman and M. Farine, Nucl. Phys. A454 374 (1986)

12. A.K. Dutta, J.-P. Arcoragi, J.M. Pearson, R. Behrman and F. Tondeur, Nucl. Phys. A458, 77 (1986)

13. F. Tondeur, A.K. Dutta, J.M. Pearson and R. Behrman, Nucl. Phys. A470 93 (1987)

14. B. Grammaticos and A. Voros, Ann. Phys. <u>123</u> 359 (1979)

15. B. Grammaticos and A. Voros, Ann. Phys. <u>129</u> 153 (1980)

16. M. Brack and P. Quentin, *Proc. Int. Conf. on Nuclear Self-Consistent Fields*, Trieste (1975), ed. G. Ripka and M. Porneuf (North-Holland, Amsterdam, (1975) p. 353

17. Y.H.Chu, B.K. Jennings, and M. Brack, Phys. Lett. <u>68B</u> 407 (1977)

18. P. Ring and P. Schuck, *The Nuclear Many-Body Problem* (Springer-Verlag, New York, 1980)

19. G. Audi and A.H. Wapstra, Atomic Data and Nuclear Data Tables (to appear 1987)

INFINITE NUCLEAR MATTER MODEL FOR MASSES OF ATOMIC NUCLEI

L. Satpathy
Institute of Physics, Bhubaneswar-751005, India.

ABSTRACT

The ground-state energy of an atomic nucleus with asymmetry β is considered to be equivalent to the energy of a perfect sphere made up of the infinite nuclear matter of the same asymmetry plus a residual energy η called the local energy. η represents the energy due to shell, deformation diffuseness and exchange Coulomb effect etc. Using this picture and the generalised Hugenholtz-Van Hove theorem of many-body theory a new mass formula has been developed.

INTRODUCTION

The nuclei posses two categories of properties namely the universal and the individualistic. The shell, deformation and diffuseness etc are the individualistic properties of the nuclei. The liquid like behaviour of the nuclei can be termed as their universal property. At a fundamental level these two types of properties have a common origin in the microscopic many-body dynamics involving nucleon-nucleon interaction. A model based on many-body theoretic foundation is proposed which is called the infinite nuclear matter [INM] model. It uses the generalised Hugenholtz-Van Hove theorem[1] of many-body theory. It is naturally hoped that through this model these two categories of properties can be well accounted for and a successful mass formula can be obtained.

THE INM MODEL

The details of the model can be seen elsewhere[2]. In this model the ground-state energy of a nucleous is considered equivalent to the energy of a perfect sphere made up of the infinite nuclear matter pluse the residual characteristic energy η called the local energy. Consider a nucleus with neutron number N, proton number Z, mass number A= N+Z and the symmetry parameter β =(N-Z)/(N+Z). We take an infinite nuclear matter with the same asymmetry parameter β and consider a perfectly spherical volume of radius R inside it which contains N neutrons and Z protons. R is related to A through $R = r_0 A^{1/3}$ with r_0 being a constant. The density of the nuclear matter is ρ_0. Let the energy contained in this volume be E(A,Z). Then cut out this perfect sphere from the sea of nuclear matter by switching on the surface force.
Subsequently switch on the Coulomb force and the pairing force. While the surface force will tend to contract the sphere and increase its density the Coulomb force will tend to expand and decrease the density. It has been shown by Brandow[3] and also generally believed that these two effects nearly cancel. Thus the density of our isolated sphere will have the same density ρ_0 as infinite nuclear matter. Let $E^S(A,Z)$ be the energy of this cut out sphere hereafter referred to as INM sphere. Then $E^S(A,Z)$ will be given by

$$E^S(A,Z) = E(A,Z) + a_s A^{2/3} + a_c Z^2/A^{1/3} - \delta(A,Z) \quad (1)$$

where a_s and a_c are the surface and Coulomb coefficients and $\delta(A,Z)$ is the usual pairing term given by

$$\delta(A,Z) = \begin{cases} \Delta A^{-0.5} & \text{for even-even nuclei} \\ 0 & \text{for odd-A nuclei} \\ -\Delta A^{-0.5} & \text{for odd-odd nuclei} \end{cases} \quad (2)$$

Here Δ is a parameter. The Coulomb coefficient is given by $3e^2/5r_0$ and the density of nuclear matter is related to r_0 through $\rho_0 = 3/4\pi r_0^3$. The groundstate energy $E^F(A,Z)$ of a real finite nucleus with mass number A and charge number Z can be considered equal to the energy $E^S(A,Z)$ of the INM sphere plus the characteristic residual energy η. Then the energy of a finite nucleus splits into global and local parts:

$$E^F(A,Z) = E^S(A,Z) + \eta(A,Z)$$
$$= E(A,Z) + a_s A^{2/3} + a_c Z^2/A^{1/3} - \delta(A,Z) + \eta(A,Z) \quad (3)$$

Putting
$$f(A,Z) = a_s A^{2/3} + a_c Z^2/A^{1/3} - \delta(A,Z) \quad (4)$$

Eq(3) can be rewritten as

$$E^F(A,Z) = E(A,Z) + f(A,Z) + \eta(A,Z) \quad (5)$$

Then Eq(5) represents a mass formula with two unknown function $E(A,Z)$ and $\eta(A,Z)$. $E(A,Z)$ being the property of infinite nuclear matter will satisfy the extended HVH theorem which assumes the following form for the ground-state

$$\frac{E}{A} = \frac{1}{2}\left[(1+\beta)\epsilon_n + (1-\beta)\epsilon_p\right] \quad (6)$$

where β, ϵ_n and ϵ_p are the asymmetry parameter, neutron and proton Fermi energies respectively. With the help of Eq(5) the relation (6) reduces to

$$\frac{E^F(A,Z)}{A} = \frac{1}{2}\left[(1+\beta)\epsilon_n^F(A,Z) + (1-\beta)\epsilon_p^F(A,Z)\right] + S(A,Z)$$
$$- (N/A)\left[\eta(A,Z) - \eta(A-1,Z)\right]$$
$$- (Z/A)\left[\eta(A,Z) - \eta(A-1,Z-1)\right] \quad (7)$$

where ϵ_n^F and ϵ_p^F are the neutron, and proton Fermi energies in the finite nuclei and

$$S(A,Z) = \left(\frac{1-A}{A}\right) f(A,Z) + \left(\frac{N}{A}\right) f(A-1,Z) + \left(\frac{Z}{A}\right) f(A-1,Z-1) \quad (8)$$

$\eta(A,Z)$ is expected to be much smaller than the corresponding total energy $E(A,Z)$. Hence terms like $\{\eta(A,Z) - \eta(A-1,Z)\}$ and $\eta(N,Z)/A$ in Eq(7) can be neglected. More appropriately we assume that the sum of the terms involving η on the r.h.s. is equal to the similar terms on the l.h.s. So we get

$$\frac{\eta(A,Z)}{A} = \left(\frac{N}{A}\right)\left[\eta(A,Z) - \eta(A-1,Z)\right] + \left(\frac{Z}{A}\right)\left[\eta(A,Z) - \eta(A-1,Z-1)\right]$$

or eqjuivalently

$$\eta(N,Z) = \left(\frac{N}{A-1}\right)\eta(N-1,Z) + \left(\frac{Z}{A-1}\right)\eta(N,Z-1) \quad (9)$$

Hence Eq(7) reduces to

$$\frac{E^F(N,Z)}{A} = \frac{1}{2}\left[(1+\beta)\epsilon_n^F(N,Z) + (1-\beta)\epsilon_p^F(N,Z)\right] + S(A,Z) \quad (10)$$

Now expressing ϵ_n^F and ϵ_p^F in terms of the binding energies, we obtain from Eq(10)

$$\frac{E^F(N,Z)}{A} - \frac{1}{2}\left[(1+\beta)\{E^F(N,Z) - E^F(N-1,Z)\} + (1-\beta)\{E^F(N,Z) - E^F(N-1,Z-1)\}\right] = S(A,Z) \quad (11)$$

This mass relation was derived[1] before and its success has been demonstrated. S is a function of the three universal parameters a_s, a_c and Δ which are determined by fitting Eq(10) with all known masses. Now solution of Eq(6) is of the form

$$E = -a_v A + a_a \beta^2 A \quad (12)$$

where a_v and a_a are two constants which can be identified as the volume and asymmetry coefficients. Using Eqs(12), (6) and (5) we obtain

$$-a_v + a_a \beta^2 = \frac{1}{2}\left[(1+\beta)\epsilon_n^F + (1-\beta)\epsilon_p^F\right]$$
$$- f(A,Z) + (N/A) f(A-1,Z) + (Z/A) f(A-1,Z-1) \quad (13)$$

a_v and a_a are the only two remaining unknown universal parameters in the above equation, which are determined again by fitting (13) to known masses. Now all the terms in the mass formula (5) are known except η. The values of η can be determined from the experimental binding energies of nuclei through Eq(5). These values can be called the experimental values of η. Eq(9) is then used as a recurrence relation to find η's of nuclei whose masses are not yet measured. This relation has been found to have excellent extrapolation properties. The $\eta's$ of nuclei in a given region can be expressed in terms of few parameters called the local parameters.

CONCLUSION

The mass formula has been found to be quite successful in predicting masses of nuclei far from stability. In fact the challenging problem of predicting the masses of Na isotopes has been satisfactorily resolved in this mass formula. Different saturation properties of nuclear matter have been obtained^2 in this mass formula which agree with the prediction of Day^4 based on recent many-body calculation using modern two-body forces. This is quite intriguing and deserves attention. Elaborate discussion on this point can be seen in Ref.^{2}.

1. L.Satpathy and R.Nayak, Phys. Rev. Lett.$\underline{51}$, 1243 (1983).

2. L.Satpathy J. Phys. C;Nucl. Phys. $\underline{13}$, 761 (1987).

3. B.H.Brandow, Ph.D.thesis, Cornel Universite 1964.

4. B.Day, Comments Nucl.Part. Phys. $\underline{11}$, 115 (1983), Phys.Rev.Lett.$\underline{47}$, 226 (1981).

MASSES OF NUCLEI IN THE INFINITE NUCLEAR MATTER MODEL

L.Satpathy
Institute of Physics,Bhubaneswar-751005,India.

R.C.Nayak
Khallikote College,Physics Department,
Berhampur(Gm)-760001,India.

ABSTRACT

The ground-state masses of 3481 nuclei in the range $18 \leq A \leq 267$ have been calculated using the infinite nuclear matter model based on the generalised Hugenholtz-Van Hove theorem. In this model there are two kinds of parameters: Global and local. The five global parameters which characterise the properties of the sphere made up of infinite nuclear matter are determined once for all by fitting the masses of all nuclei (756) in the recent mass table with error bar less than 30 keV The local parameters are determined for 25 regions defined by $\Delta A=8$ or 10. The r.m.s. deviation for the calculated masses from the experiment is 397 keV for the 1572 nuclei used in the least square. fit. Sample results on Na isotopies and other recently measured masses have been given. The derived saturation properties of nuclear matter has been discussed.

INTRODUCTION

In the infinite nuclear matter [INM] model[1] the ground-state energy of a nucleus is considered to be equivalent to the energy of a perfect sphere made up of infinite nuclear matter plus the residual characteristic energy η called the local energy. This sphere will be referred to as INM sphere. η represents the energy due to shell, deformation and diffuseness etc. Thus the global properties of a nucleus are contained in the INM sphere and the local or characteristic properties in the η . It has been possible in this model, by the use of generalized Hugenholtz-Van Hove theorem to separate these two categories of properties and relate the global properties to a combination of the binding energies of three neighbouring nuclei. Similarly the local properties have been separately related. Since this model is exclusively built in terms of the properties of infinite nuclear matter, it is most suitable to extract saturation properties from the ground-state energies of nuclei.

CALCULATION AND RESULTS

We have made prediction for the mass excess of 3481 nuclei in the range $18 \leq A \leq 267$ using the INM model. The details can be found in Ref.[1,2]. In this model there are five global parameters which characterise the properties of the INM sphere. The values of these parameters have been determined by fitting the masses of all nuclei (756) with error bar less than 30 keV in the recent mass table of Audi and

Wapstra.[3] The values of the parameters so obtained are a_V =18.333 MeV, a_S =25.846 MeV, a_c =0.841 MeV, a_a=36.211 MeV and Δ =11.709 MeV. The entire periodic table was derived into 25 regions defined by Δ A=8 or 10. For each region the values of the local parameters were determined by taking known masses in that region. There are altogether 233 local parameters. The root mean square deviation for the calculated masses from the experiment is 397 keV for the 1572 nuclei used in the least square fit. We have presented here only some sample results. It is well known that the Na isotopes pose a challange to most of the mass formulae. As a measure of the success of this mass formula we present in Table I a comparison of our predictions on Na isotopes with those of the recent predictions of Kelson group[4] and Janecke group[5]. Similarly in Table II a comparison is given for a few recently measured masses kby Vieira et al.[6] Our predictions agree quite well with experiment.

Table I. Comparision of mass excess in MeV. Error given in parenthesis.

Isotopes	Expt.	Present work	Camay Kelson & Zidon	Janecke and Masson
Na^{28}	-.1.14(0.14)	-0.91(0.54)	-1.05(0.42)	-0.87
Na^{29}	2.65(0.15)	2.95(0.28)	2.26(0.64)	2.12
Na^{30}	8.21(0.25)	7.6(0.22)	8.81(0.80)	8.36
Na^{31}	11.83(0.58)	10.93(0.22)	13.58(1.07)	12.74
Na^{32}	16.55(0.74)	18.58(0.26)	22.42(1.2)	21.37
Na^{33}	21.47(1.14)	21.87(0.32)	28.62(1.41)	27.69
Na^{34}	26.65(3.57)	25.99(0.54)	37.24(1.55)	36.48
Na^{35}		29.28(0.73)		

Table.II. Same as Table I

A_Z	Expt.	Present work	Camay Kelson	Janecke Masson
^{19}C	32.30(0.24)	32.55(0.98)	32.47(0.31)	32.57
^{27}Ne	5.6(0.6)	6.19(0.64)	6.6(0.45)	6.7
^{28}Ne	10.7(0.4)	8.59(1.05)	10.4(0.64)	10.41
^{32}Al	-11.33(0.2)	-11.37(0.16)	-10.81(0.35)	-11.29

Table.II(Continued)

A_Z	Expt.	Present work	Comay Kelson	Janecki Masson
^{33}Al	-8.84(0.23)	-9.42(0.19)	-8.66(0.5)	-9.14
^{34}Al	-3.5(0.4)	-3.51(0.22)	-2.22(0.59)	-2.66
^{36}Si	-12.9(0.6)	-13.63(0.26)	-11.82(0.51)	-11.87
^{37}P	-19.31(0.4)	-18.61(0.28)	-18.64(0.32)	-18.53

GLOBAL PARAMETERS AND THE SATURATION PROPERTIES OF NUCLEAR MATTER

The values of the parameters a_v, a_s, a_c, a_a and Δ are significantly different compared to those of the usual liquid drop or droplet model. We would like to stress here that our values refer to the properties of a sphere made up of infinite nuclear matter which is a hypothetical system. Thus our value a_v =18.333 MeV is the binding energy per nucleon in nuclear matter w. The value of the saturation Fermi momentum k_F can be obtained from the Coulomb coefficient a_c which is the characteristic of INM sphere. Our value a_c=0.841 MeV gives k_F =1.48 fm^{-1}. These saturation values are significantly different from the traditionally accepted values of 15.96 MeV and 1.36 fm^{-1}. Recently Day has made exhaustive calculations for the saturation properties of nuclear matter using several modern two-body forces and has concluded that if the nuclear forces is of two-body type then the saturation point would lie inside on oval area (see Fig.1) in $w \propto k_F$ plane. Suprisingly the saturation point corresponding to our above value of w and k_F falls (shown as a triangle) inside this oval. It is quite intriguing and we feel this result should be viewed with seriousness. The empirical values of w and k_F are deduced from two different sources. Traditionally w is identified with the volume coefficient of liquid drop or droplet model. Similarly the empirical value of k_F is obtained from the electron scattering data on nuclei. However in the present model these two data are obtained from the same source namely the binding energies of nuclei through a model which is built in terms of the properties of infinite nuclear matter. Hence the determination of the saturation properties of nuclear matter in the present model is more consistent. Extensive discussion on this point is reported elsewhere[1].

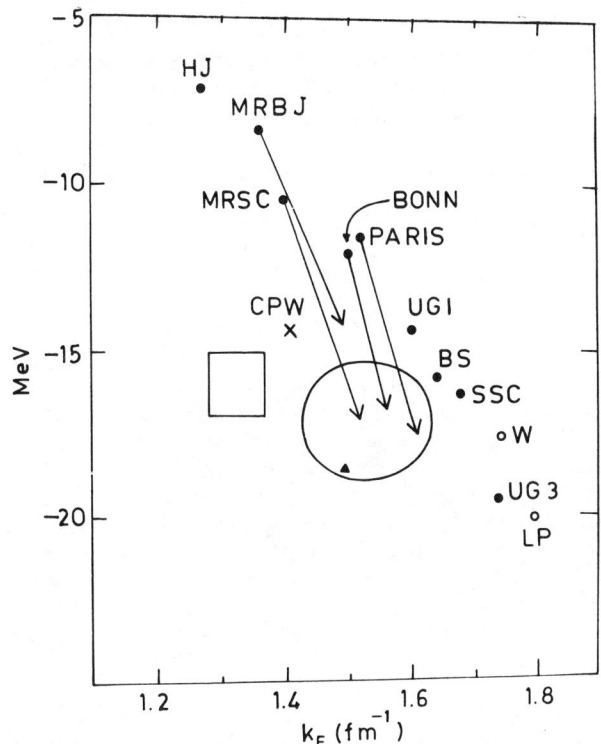

Fig.1. Nuclear matter saturation point Diagram taken from Ref.[7]. Oval represents the theoretical calculation of Day[7]. The triangle is the new saturation point obtained in the present work. The square is the usual saturation point. for symbols see Ref.[7].

1. L.Satpathy J. Phys. G; Nucl. Phys. 13,761 (1977).

2. L.Satpathy and R.Nayak, submitted to Atomic Data and Nuclear Data Table. preprint IP/BBSR/87-27,(1986-87). Mass prediction project.

3. A.H.Wapstra and G.Audi Nucl. Phys. A432, 1 (1985).

4. E.Comay, I.Kelson and Zidan, private circulation by P.E.Haustein, Brookhaven National Laboratory. The 1986-87 Mass prediction project.

5. P.J.Masson and J.Jonecke private circulation by P.E.Haustein, Brokhaven National Laboratory 1986-87 Mass prediction project.

6. D.J. Vieira et al, Phys. Rev. Lett. 57, 3253 (1986).

7. B.D.Day Cooments Nucl. Part. Phys. 11, 115 (1983), Phys. Rev. Lett. 47, 226 (1981).

A NEW, SIMPLIFIED, INTERPOLATION METHOD FOR ESTIMATION OF MICROSCOPIC NUCLEAR MASSES BASED ON THE P-FACTOR, $P = N_p N_n/(N_p + N_n)$

P. E. Haustein[+], D. S. Brenner[*], and R. F. Casten[+]

[+]Brookhaven National Laboratory, Upton, NY 11973 USA
[*]Clark University, Worcester, MA 01610 USA

ABSTRACT

A new semi-empirical method, based on the use of the P-factor ($P = N_p N_n/(N_p + N_n)$), is shown to simplify significantly the systematics of atomic masses. Its use is illustrated for actinide nuclei where complicated patterns of mass systematics seen in traditional plots versus Z, N, or isospin are consolidated and transformed into linear ones extending over long isotopic and isotonic sequences. The linearization of the systematics by this procedure provides a simple basis for mass prediction. For many unmeasured nuclei beyond the known mass surface, the P-factor method operates by <u>interpolation</u> among data for known nuclei rather than by extrapolation, as is common in other mass models.

INTRODUCTION

Many observables related to nuclear excitations have been shown[1] to behave in a smooth and simple manner when plotted against the valence nucleon product $N_p N_n$ which is a measure of the integrated valence proton-neutron interaction. More recently a normalized quantity, the P-factor, $P = N_p N_n/(N_p + N_n)$, has been shown[2] to provide even greater uniformity in such systematics than that afforded by $N_p N_n$ alone. It is of interest to investigate whether these formalisms can now be applied successfully to atomic mass systematics. However, it is essential to delineate two important components of atomic masses. The first (a macroscopic one), not inherent in other observables that have been systematized using the $N_p N_n$ model, is a strong, basically structure independent, secular scaling of the mass with A, Z, or N. The second (a microscopic one) contains the structure and deformation dependence. We have used the model of Möller and Nix[3] to isolate the second component empirically by subtracting their calculated spherical macroscopic masses from the experimental masses.[4] The resultant quantity is termed a "semi-empirical macroscopic (SEM) mass" and is a measure of the structure dependent component of the atomic mass.

THE P-FACTOR AND THE ACTINIDES

Figure 1 shows plots of actinide SEM-masses versus N_n and against the P-factor for Z = 84-104 with N ⩾ 128. Data are considered only up to midshell (assuming that the next shell closures occur at Z = 126 and N = 184). Ignoring the distinction for the moment between open and closed points in the P-factor plot, one sees a general consolidation of the data. On closer examination, when individual sequences, N_i = constant, i = p or n, are isolated one finds a remarkable linearization of the data. Data points with the same N_p or N_n fall on the same nearly linear curves (Figure 2) whose slopes vary smoothly and systematically from large positive values at the beginning of the shell to values near-zero for N_p = N_n = 9 and then to negative values (panel 2f). The special regularity of these features is signaling that the P-factor, which measures the number of proton-neutron interactions per valence nucleon, quantifies an important physical feature that governs the evolution of the microscopic mass as the nuclear valence space is occupied. The regularity is also the essential ingredient leading to a new method of mass prediction discussed below.

BEHAVIOR OF THE HEAVY ACTINIDES (WITH N $\geq$ 146)

The open points in Figure 1b show a different feature which is illustrated in more detail for Cm and Cf in the inset. Starting from N = 146, the data display a leveling off relative to the descending line seen for lower P values, followed by an up-bend. C. L. Wu et al.[5] have noted this feature and proposed an interesting explanation in terms of the Fermion Dynamical Symmetry Model. The up-bend occurs when certain SU(3) representations become forbidden due to the large number of valence nucleons which are present in the normal parity orbitals. (See also the contribution of Feng et al., this conference.)

THE P-FACTOR AND MASS PREDICTIONS

The P-factor plot of the actinides with its linearization of the SEM-sequences is a novel feature that can be used to advantage for mass predictions when enough members of the isotopic and isotonic sequences are known to establish the common line shared by other unmeasured nuclides. A special aspect of these plots is that many unmeasured actinide nuclei have P-factors that fall between the ranges of points that correspond to the groupings of isotopic and isotonic species. Consequently, microscopic masses for these unmeasured nuclei (that generally lie beyond the known mass surface) are obtained by an <u>interpolation</u> procedure, rather than by extrapolation as has been commonly used before in mass models. For a given set of isotopes (N_p fixed) the interpolation is performed along the line defined by data points with $N_p = N_n$. A complementary procedure is to interpolate along other lines using the N_n value of the nuclide in question instead of the N_p of its element. Very similar results are obtained by the two methods. Addition of the Möller-Nix spherical macroscopic mass to the interpolated microscopic mass yields the mass excess. Some examples of atomic mass predictions in the actinides are given in Table 1. Comparisons of these predictions to other mass models[6] reveals that they track very closely the model of Liran and Zeldes, whose

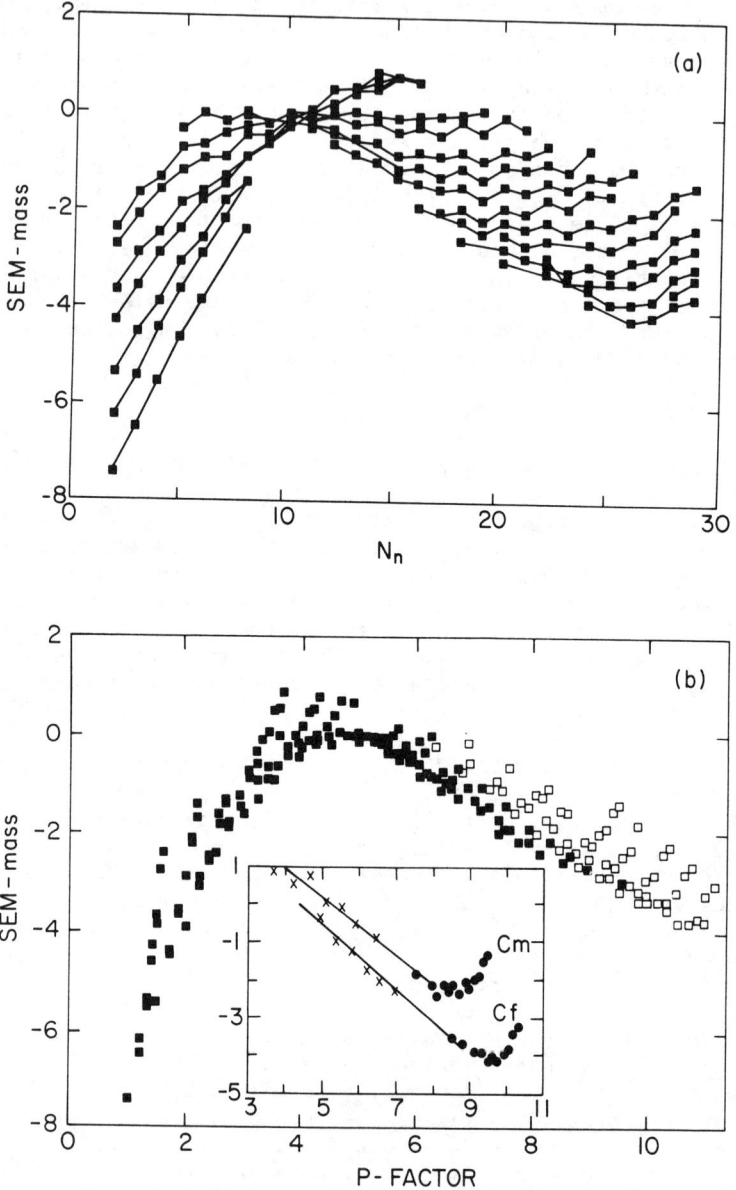

Figure 1. N_n and P-factor plots of SEM-masses (in MeV) for the actinides. In the N_n plot (a) solid lines connect isotopic chains. In (b) the SEM masses for $N \geq 146$ are plotted by open squares (see text). The inset illustrates the upturn for $N \geq 146$ for two high Z sequences.

TABLE I

Actinide Mass Predictions from the P-factor

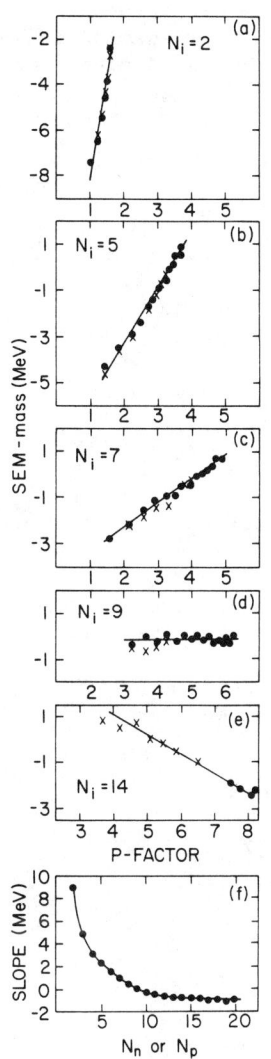

Figure 2. P-factor plots for selected N_p and N_n values in the actinides. Slopes (per unit P) of the fitted straight lines are shown in (b).

Nuclide	SEM-Mass (MeV)[1]	Predicted Mass Excess (MeV)
^{225}U	-0.50	27.08
^{226}Np	0.03	32.24
^{227}Pu	0.21	36.54
^{228}Am	0.31	42.36
^{229}Cm	0.43	47.77
^{230}Bk	0.44	54.65
^{232}Cm	-0.53	46.29
^{233}Cm	-0.77	47.23
^{234}Cm	-0.98	46.76
^{235}Cm	-1.25	47.94
^{236}Cm	-1.41	47.81
^{237}Cm	-1.60	49.35
^{239}Cm	-2.08	51.18

[1]Results were averaged from interpolations along $N_p = N_n$ lines for N_p and N_n of each nuclide.

prescription for the microscopic term is very elaborate compared to the simplified procedure using the P-factor. Further examination of the relationship of the P-factor to atomic masses will indicate if this apparent simplification is general and widely applicable. A comprehensive set of predictions (for A > 100) is in preparation and will be published shortly.

ACKNOWLEDGEMENTS

We are grateful for illuminating discussions with A. E. L. Dieperink, I. Talmi, D.-H. Feng, C. L. Wu, I. Kelson, and O. Scholten. Research has been performed under contracts DE-AC02-76CH00016 and DE-AC02-79ER10493 with the U.S. Department of Energy and by NATO grant no. RG 85/0036.

REFERENCES

1. R. F. Casten, Phys. Rev. Lett. $\underline{54}$, 1991 (1985).

2. R. F. Casten, D. S. Brenner, and P. E. Haustein, Phys. Rev. Lett. $\underline{58}$, 658 (1987).

3. P. Möller and J. R. Nix, Los Alamos National Laboratory Report LA-UR-90-1966; At. Data and Nuc. Data Tables $\underline{26}$, 165 (1981).

4. A. H. Wapstra, G. Audi, and R. Hoekstra, Midstream (1986) Atomic Mass Evaluation, available from the National Nuclear Data Center, BNL.

5. C. L. Wu et al., Phys. Lett. $\underline{194B}$, 447 (1987).

6. S. Maripuu, At. Data and Nuc. Data Tables $\underline{17}$, 411 (1976).

The Fermion Dynamical Symmetry Model of Nuclear Collective Motion

Da Hsuan Feng
Department of Physics and Atmospheric Science, Drexel University, Philadelphia, Pennsylvania 19104

Abstract : A microscopic fermion model of nuclear collective motion is discussed. The model has a SU_3 rotational limit, a SO_6 γ-soft limit and three vibrational limits (SU_2, SO_7 and $SO_5 \times SU_2$), with *fully microscopic* connections between these dynamical symmetries and the underlying shell structure. It is shown that under certain approximations both the phenomenological Interacting Boson Model (IBM) and the Geometrical Model (GM) can be recovered from this model.

1. INTRODUCTION

As I can see from the list of speakers of this conference, most of the speakers in this conference are experimentalists. Thus, it is indeed an honor to be the first of the very few theorist to present our work on the recently proposed *Fermion Dynamical Symmetry Model* (FDSM)[1]. This model was initially proposed by **Cheng-Li Wu** (*Jilin University*), **Mike W. Guidry** (*University of Tennessee*), **Jin-Quan Chen** (*Nanjing University*), **Xuan-Gen Chen** (*Drexel University*) and myself. Many subsequent works are in collaborations (since the FDSM was proposed) with *Joe Ginocchio, Rick Casten, Hua Wu, Ke-Xia Wang, Zeng-Ping Li, Wei-Min Zhang* and *Xiao-Ling Han*.

The awkward number of interacting particles(~10^2) renders the microscopic description of nuclear collective motion a very difficult subject. The shell model is, of course, one of the most fundamental microscopic model of nuclear structure and was successful in providing an understanding of the properties of nuclei manifested in the low energy structure in regions where the number of valence nucleons is not large [2]. Unfortunately for the medium-heavy to heavy weight nuclei, especially for those far away from closed shells, the configuration space for such nuclei becomes enormous and therefore is *completely unreachable* within the shell model even for the most advanced computers. The *Fermion Dynamical Symmetry Model* is in fact *a prescription for the many-body solutions of the spherical shell model* in the heavy or medium-heavy mass regions.

2. WHAT IS THE FDSM?

The basic idea of the FDSM is illustrated in Fig.1. The original shell model Hamiltonian H_{sh} is usually an extremely large and full matrix. This means that the off-diagonal matrix elements cannot generally be neglected due to strong collectivity. Hence, even if we were only interesting in the structure at low energy (schematically indicated in the small box on the lower right hand corner of the matrix in Fig.1a), we must still diagonalize the entire matrix, which is clearly a hopeless (and probably unrewarding) task. This is why the traditional shell model method is not applicable for heavy nuclei. The key to the FDSM is to transform the shell model Hamiltonian into a basis called the k-i basis . In this basis the Hamiltonian has the form as shown in Fig.1b.

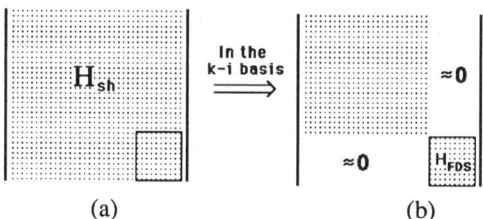

Fig.1 *The shell model Hamiltonian in the k-i basis*

Thus, for the low energy structure we need only to diagonalize a much smaller matrix H_{FDS} which is what we shall refer to as the Hamiltonian of the FDSM. The Hilbert space of the H_{FDS} is, obviously, severely truncated. This makes the calculations tractable. In this sense the FDSM can be viewed as the shell model truncation in the k-i basis.

Let me now explain what is the k-i basis. *The k-i basis is a single-particle basis with the angular momentum j decomposed into an active and an inert parts* instead of the usual orbital and spin parts.

$$j = j_{active} + j_{inert} \qquad (1)$$

The active j_{active} is defined as the part of the single-particle angular momentum for which two identical nucleons can *only* couple to angular momentum $J = 0$ and 2, while the inert part j_{inert} is the remaining single-particle angular momentum. The dominance of the S-D pairs in the low-lying collective modes suggests that the inert parts of the angular momenta for each identical nucleon pair tend to couple to zero and become inactive at low-energy. This is why we call this part of the single-particle angular momentum "inert". In this way, the part with integer value is called the "pseudo orbital angular momentum" (denoted as **k**) while the part with half-integer the "pseudo-spin" (denoted as **i**). Hence, Eq.(1) can be rewritten as

$$j = k+i. \qquad (2)$$

Correspondingly, a single-particle creation operator $b^{\dagger}_{km_k \, im_i}$ in the k-i basis can be defined as

$$a^{\dagger}_{jm} = \sum_{m_k m_i} <k \, m_k \, i \, m_i | j \, m > b^{\dagger}_{km_k \, im_i} \qquad (3)$$

where $a^{\dagger}_{jm}$ is the creation operator in the original shell model single-particle basis. Hence, by definition, there are only two alternatives for the active angular momentum in the k-i decomposition. For the active **k**, *k must be 1*; for the active **i**, *i must be 3/2*.

The k-i basis has several very important properties. First of all, simply by freezing the inert part of the angular momenta (couple to zero), a coherent S-D subspace, which we believe to be the major portion of the space responsible for the low lying nuclear collective motion can be easily "carved" out from the entire shell model space. In the k-i basis the S and D pairs are defined as

$$S^{\dagger}(k) = \sqrt{\Omega_{ki}/2} \; [b^{\dagger}_{ki} b^{\dagger}_{ki}]^{00}_{00} \; , \; D^{\dagger}_{\mu} = \sqrt{\Omega_{1i}/2} \; [b^{\dagger}_{1i} b^{\dagger}_{1i}]^{20}_{\mu 0} \; , \; D^{\dagger}_{\mu} = \sqrt{\Omega_{k3/2}/2} \; [b^{\dagger}_{k3/2} b^{\dagger}_{k3/2}]^{02}_{0\mu}$$

(for any k and i) (for k-active) (for i-active)

$$(4)$$

where $\Omega_{ki}=(2k+1)(2i+1)/2$. These S and D pairs have very simple structure in the k-i basis and yet they are highly coherent in the original shell model basis. In particularly, the S-pair is just the Cooper pair in the pairing condensate. The coherent nature inherent in these pairs are precisely the desired property for describing nuclear collective motion. Secondly, the space built up by such S and D pairs has SO_8 symmetry for the i-active scheme and Sp_6 symmetry for the k-active scheme. Each of these symmetries has also a multi-chain dynamical symmetries as shown in Fig.2 .

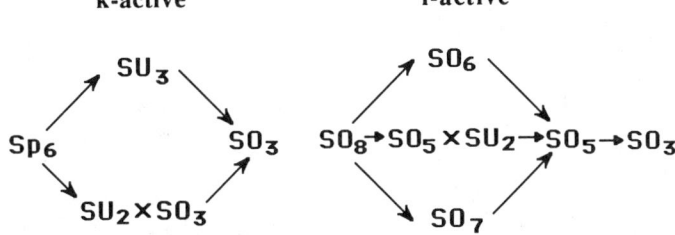

Fig. 2 *The group chains of the SO_6 and SO_8 symmetries*

In this subspace, all the IBM dynamical symmetries[3] can be recaptured from the fermionic level: The spectra of the dynamical symmetries $SO_8 \supset SO_5 \times SU_2$ and $Sp_6 \supset SU_2 \times SO_3$ are identical to those of the IBM U_5 (vibrational) limit, the $SO_8 \supset SO_6$ is identical to the IBM O_6 γ-soft limit, and the $Sp_6 \supset SU_3$ is identical to the IBM SU_3 (rotational) limit. Some of the transition and transfer matrix elements, however, because of the fermionic nature of the system, differ from the IBM ones by a Pauli factor. In addition there is a SO_7 limit *which has no counterpart* in the IBM dynamical symmetry of U_6. These fermion dynamical symmetries were first found by Ginocchio [4]. However, it was unclear at that time that such a k-basis can be utilized as a physical basis to describe realistic nuclei since neither the physical meaning of the k-i decomposition nor the correspondence between the k-i basis and the physical shell were understood. Furthermore, there was some difficulties associated with the $Sp_6 \supset SU_3$ chain[4,5]: namely when the nucleon pair number $N>\Omega/3$, the representation $(2N,0)$, which describes the ground band of rotational nuclei, is Pauli disallowed. This is, of course, in contradiction with the fact that rotational nuclei often occur near mid-shell ($N \approx \Omega/2$). This difficulty renders the Ginocchio model to be viewed as a "non-realistic" one.

A significant step made by the FDSM is the establishment of the one-to-one correspondence between the k-i basis and the shell model basis. This is shown in Table 1. By including the abnormal single particle level in a major shell, the symmetry of the S-D subspace in a major shell becomes either $SO_8 \times SU_2$ or $Sp_6 \times SU_2$ (SU_2 denotes the quasi spin group associated with the S pairs (i.e. $S^\dagger(k=0)$ of eq.(4)) in the abnormal parity level) depending on which valence physical shell one is considering.. In other words, the k-i values and their associated symmetries are entirely determined by the underlying shell structure of nuclei. Also, we can show that the presence of the SU_2 symmetry for the abnormal level will "resurrect" the $Sp_6 \supset SU_3$ symmetry and therefore has no difficulty in having an axially-symmetric rotational limit. Details of this discussion can be found in reference [1]. It will not be discussed here due to space limitation.

Of course to build a complete basis for a many-body system, one must activate the inert parts of the angular momenta of pairs other than S and D. To succinctly display this fact, the model defines a new quantum number u,

$u \equiv$ *the number of nucleons which do not form coherent S and D pairs*

which we shall call the "heritage", (the coherent S and D pairs are defined in eq.(4)). The full fermion shell model space can therefore be "stratified" according to this number: u=0 for the S-D subspace; u=1 for S and D pairs plus an odd particle; u=2 for S and D pairs plus one broken pair and so on. For light nuclei or nuclei very close to the doubly closed shells, one is generally able to include all possible u's, in which case the entire shell model space is recaptured. On the other hand, for low lying low spin states of heavy nuclei, we may truncate the shell model space to S-D subspace. For high spin states, the physics dictates us to consider the broken pair (or pairs), thereby enlarging the space somewhat. Such an enlargement renders the calculations still within reach if u<4. Thus in heavy mass regions, the k-i basis provides a very convenient and physically reasonable truncation scheme to carry out detailed numerical computations within the spherical shell model.

3. THE TEST OF THE FDSM

Since the FDSM is microscopic, it is a testable theory. The theory may be tested at two levels. The first concerns the global predictions of the nuclear systematics by the theory in the analytical symmetry limits, that is to ask 'do the collective modes implied by the FDSM occur in a systematic fashion in the low energy nuclear structure?'. The second concerns microscopic calculations including numerical diagonalization of symmetry breaking terms, that is to ask 'do the collective modes of the FDSM have the correct microscopic structure to describe the detailed behavior of individual nuclei?'. In the remaining discussion presented here, it is the first aspect which we shall concentrate on. Indeed, the results, as we shall see, are very encouraging. As a first check of this model, we have shown that the IBM is an approximation of the FDSM when there is no broken pair (u=0) and the Pauli effects are negligible [1]. We have also found empirical evidence for the SO_7 fermion dynamical symmetry in nuclei, which was predicted by the FDSM[6].

Table 1
Reclassification of Shell Model Single-Particle Basis

No.	1	2	3	4	5	6		7		8	
n	0	1	2	3	3	4	4	5	5	6	6	7
k	0	1	1	0	1	0	2	0	1	1	0
i	$\frac{1}{2}$	$\frac{1}{2}$	$\frac{3}{2}$	$\frac{7}{2}$	$\frac{3}{2}$ $\frac{9}{2}$	$\frac{3}{2}$	$\frac{11}{2}$	$\frac{1}{2}$	$\frac{7}{2}$ $\frac{13}{2}$	$\frac{3}{2}$	$\frac{9}{2}$ $\frac{15}{2}$
SYM. CONFIGURATION	$s_{1/2}$	$p_{1/2}$ $p_{3/2}$	$s_{1/2}$ $d_{3/2}$ $d_{5/2}$	$f_{7/2}$	$p_{1/2}$ $p_{3/2}$ $f_{5/2}$ $g_{9/2}$	$s_{1/2}$ $d_{3/2}$ $d_{5/2}$ $g_{7/2}$	$h_{11/2}$	$p_{1/2}$ $p_{3/2}$ $f_{7/2}$ $f_{5/2}$ $h_{9/2}$ $i_{13/2}$		$s_{1/2}$ $d_{3/2}$ $d_{5/2}$ $g_{7/2}$ $g_{9/2}$ $i_{11/2}$ $j_{15/2}$	
			$G_6 G_8 G_3$		$G_6 G_8 G_3$	G_8		G_6		G_6	
Ω_0	0	0	0	0	5	6		7		8	
Ω_1	1	3	6	4	6	10		15		21	
n	2	8	20	28	50	82		126		184	

No. labels the shell ordering, n, k, i label the principle, pseudo-orbit and pseudo-spin quantum numbers, and Ω_0 and Ω_1 are the pair degeneracies of the abnormal-parity and normal-parity levels for each shell. The number n means the maximum allowable nucleon number up to and including that particular major shell. The symbols G_6, G_8 and G_3 are short-hand notation for the symmetries:

$G_6 = (Sp_6^k \times SO_3^i) \times (SU_2 \times SO_3)$ (k-active)

$G_8 = (SO_8^i \times SO_3^k) \times (SU_2 \times SO_3)$ (i-active)

$G_3 = (SU_3^k \times SO_6^i) \times (SU_2 \times SO_3)$ (k/i active)

where SO_3^k, SO_3^i, and SO_3, denote the rotational groups associated with angular momenta k, i, and j_0 (abnormal). For s-d shell there is no $SU_2 \times SO_3$ factor due to the absence of an abnormal-parity level in the shell.

Furthermore we have shown that by taking into account the broken pairs (u≠0) the FDSM can describe all the basic features of high-spin phenomena (Band crossing, rotational alignment, back bending, loss of collectivity and band termination, etc.) [7]. Since these results are already published, I will not discuss here. Interested readers may consult our papers. Here I shall only sketch some of our more recent results. Once again, details can be found from the references or the soon to be published preprints.

3-1. Evidence for nuclear shell symmetries [8]

One of the striking characteristics of the FDSM is the relationship between the filling of valence shells and the allowed collective degrees of freedom. According to the FDSM, once a valence shell is decided, the highest shell symmetry automatically follows. It is therefore desirable to find a physical observable which is sensitive only to the highest symmetries of the valence shells (Sp_6 or SO_8 symmetries for neutrons and protons), but not sensitive to the details of the sub-groups. Then, this observable could be used as a highest shell symmetry indicator. To this end, it was found that E2 electromagnetic transition branching ratio for the even-even nuclei,

$$R_{22} = \frac{B(E2, 2_2^+ \to 0_g^+)}{B(E2, 2_1^+ \to 2_g^+)} \quad (5)$$

has precisely this property, where $B(E2, 2_2^+ \to 0_g^+)$ is the B(E2) value from the second 2^+ state to the ground state, while $B(E2, 2_2^+ \to 2_g^+)$ is the B(E2) from the second 2^+ state to the first 2^+ state. According to FDSM, it can be shown that for nuclei which have both valence neutrons and protons in the Sp_6 shells, R_{22} ranges from zero to a large value around 0.7 (the so-called Alaga value); for SO_8 shells, R_{22} must always be zero. This is why R_{22} is such an excellent detector of the fermion dynamical symmetry for each valence shell. To this purpose, we have tabulated all the available data on R_{22} for 173 nuclei with $36 \leq Z \leq 100$. The results of our tabulation are displayed in Fig.3. *The most remarkable feature of this figure is that for nuclei with both neutron and proton numbers less then 82, all the data are condensed at the bottom of the figure. Only nuclei with nucleons (either neutrons or protons or both) with Sp_6 symmetry will there be a chance to reach the Alaga value.* Therefore Fig. 3 convincingly shows that, as predicted by the FDSM, the shells 7 and 8 possess Sp_6 symmetry and shell 6 possesses SO_8 symmetry. For nuclei with $Z \leq 50$ (shown on the left hand side of Fig. 3c) where the valence proton holes are in shell 5, the highest symmetry of valence protons could either be Sp_6, SO_8 or $SO_6 \times SU_3$ (k-i both active). The small R_{22} values seem to indicate that due to the n-p interactions, proton SO_8 symmetry is "picked up" to form the $SO_8^\nu \times SO_8^\pi$ symmetry.

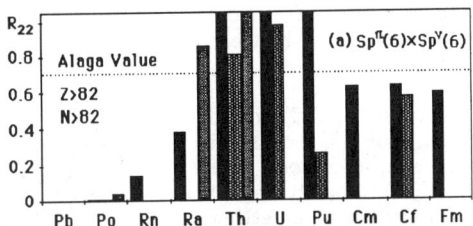
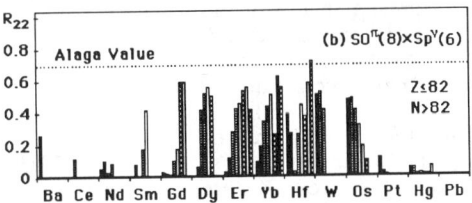
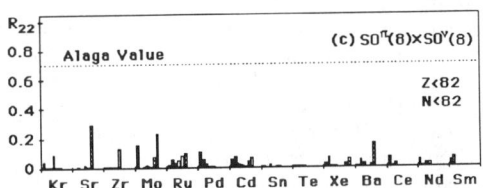

Fig.3 The B(E2) ratios R_{22} for 173 even-even nuclei.

3-2. The microscopic mass formula of actinides [9]

A mass formula for actinides is derived from the FDSM Hamiltonian microscopically and has only a few parameters of the effective interaction. These parmeters are determined by fitting the existing data. The derivation is based on the symmetry limits of SU_3 or SU_2

plus a first order perturbation, namely the symmetric wavefunctions (with respect to the interchange of neutron and proton pairs) of $SU_2^\pi \times SU_2^\nu$ and $SU_3^{\pi+\nu}$ are utilized for the description of the vibrational and rotational regions respectively but the Hamiltonian is kept the same by including the pairing and the n-p quadrupole-quadrupole interactions in the form of perturbations. The agreement between the theory and the data is excellent (see Fig.4) with a r.m.s. deviation for the ground state mass of 322 nuclei (including even-even odd and odd-odd nuclei) to be about 0.34 MeV. It is extremely interesting to note that when the proton number is greater than 8 or the neutron number greater than 13, or both, the total number of pairs in the normal levels can be estimated to be larger then $\Omega_1/3$. Therefore the $(2N_1,0)$ representation is forbidden as discussed, and the ground state masses should exhibit some sort of "jump" from the $(2N_1,0)$ representation to the $(2N_1-4,2)$, $(2N_1-8, 4)$, etc. with each pair added to the system beyond the $\Omega_1/3$ rule. In Fig. 5 this effect is clearly displayed. It is seen that without considering the change of representation, the deviation will increase as a step function when the nucleon pair number is larger than 8 (for protons) or 13 (for neutrons). If we take the change of the SU_3 representations into account, the "jump" as we observed in Fig. 4 vanishes. Hence, this can be taken as an evidence of the $\Omega_1/3$ Pauli rule.

3-3 *The microscopic Particle-Rotor Hamiltonian* [10]

A microscopic Particle-Rotor Hamiltonian is derived from the FDSM Hamiltonian with haritage quantum number $u \neq 0$ in the SU_3 limit. A comparison with the phenomenological Particle-

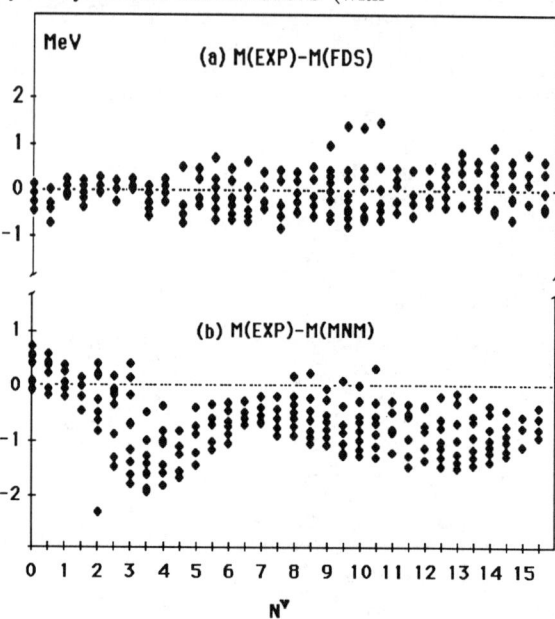

Fig.4 Discrepancy between calculation and experimental mass of actinides. The quantities M(FDS) and M(EXP) are the masses calculated by the FDSM formula and the experimental data respectively. The r.m.s. deviation of 322 nuclei is 0.34 MeV. For comparison, the results calculated by the Moller-Nix mass formula (M(MNM)) are also presented (Atom Data and Nucl. Data Table **26** (1981) 165).

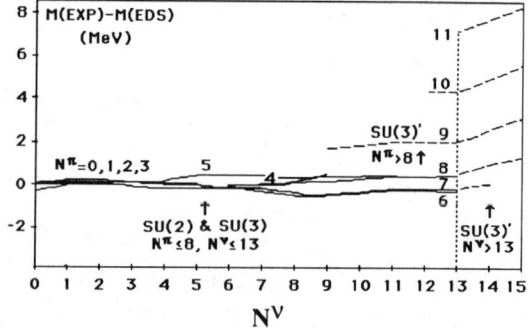

Fig.5 The FDSM calculations in this figure assume that the ground state belongs to the $(2N_1,0)$ SU_3 representation when $N^\pi \geq 4$ ($N^\nu \geq 5$), and SU_2 representation otherwise. The figure clearly indicates that when $N^\nu > 13$ or $N^\pi > 8$, the discrepancy increases linearly as $N^\pi(N^\nu)$ increases which implies a representation change from $SU(3) \to SU(3)'$. In this figure, $SU(3)$ denotes $(2N_1,0)$ representation and $SU(3)'$ denotes $(2N_1 - 4\Delta N_1, 2\Delta N_1)$ representation where ΔN_1 is $N_1 - \Omega_1/3$ (the number of nucleon pairs beyond $\Omega_1/3$).

Rotor Hamiltonian (i.e. the Geometrical Model, GM for short) [11] using the Nilsson scheme is shown in Fig. 6. Although it appears that these two approaches are completely different: the phenomenological one assumes a deformed core with an extra particle moving in the Nilsson potential, while the FDSM starts from the spherical shell model without introducing any deformation and Nilsson scheme, they essentially arrived at the same results. Indeed, the well known three coupling limits in the GM (strong coupling, rotational alignment and the weak coupling limits) are individually well reproduced.

4. SUMMARY

The study of nuclear structure has traditionally been carried out under the auspices of two models: the *shell model* and the *geometrical model*. The shell model is useful for determining the properties of nuclei near the doubly closed shells where the predominant nuclear modes are single-particle like; the geometrical model is supposed to work for nuclei far away from the doubly closed shells where collective modes dominate. It was later on suggested that the collective motion in low energy can also be described by a U_6 boson model (IBM) without the necessity of introducing the concept of deformation. The FDSM now seems to be able to

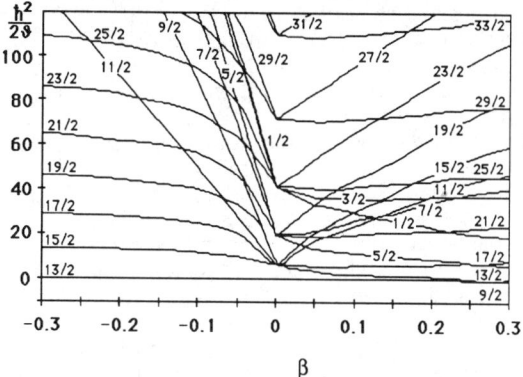

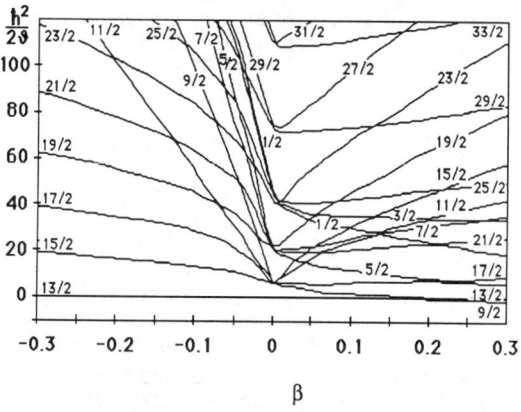

Fig.6 The model calculation of the $i_{13/2}$ band for an odd system. The Spectrum is plotted as a function of $[E(J)-E(13/2)]/(\hbar^2/2\vartheta)$. (a) The Particle-Rotor Mode (PRM) calculations. In this calculation, the fermi surface is chosen to be below all the $i_{13/2}$ level. (b) The FDSM calculations. The strength of $P^2(k) \cdot P^2(i)$ is assumed to be $\kappa = 60.45\beta$. The relationship is adjusted to give roughly the same scale as the results of the PRM calculations. In both calculations, only one particle in the abnormal $i_{13/2}$ shell is assumed. For details, see ref.[10].

unify all three previous models. I have mentioned from the very beginning of this talk that the FDSM can be regarded as the shell model in the truncated k-i basis. For the lower shells (e.g. the s-d shell) where the Hilbert space can be fully taken into consideration without truncation, by including all possible heritage quantum number u's, *the FDSM is identical to the traditional shell model*. For the larger shells where the traditional shell model is not possible, the FDSM can be employed as a truncation scheme for shell model calculations. As I have mentioned earlier that the *IBM is in fact an approximation of the FDSM under the conditions that u=0 (no broken pairs) and the Pauli effects are negligible* (i.e. $\Omega_1 \to \infty$). It is further demonstrated in section 3 that : (1) a *microscopic Particle-Rotor Model Hamiltonian* can be derived from the FDSM which produces essentially the same results as the

Geometrical Model (GM) ; (2) all the important *coupling schemes* (weak coupling, strong coupling and rotational-aligned coupling) in high-spin physics of odd mass nuclei do naturally occur for the odd-mass FDSM; (3) The FDSM provides an *accurate microscopic mass formula* for the actinides. Very recently, we have obtained many additional results. Due to the lack of space, I will not be able to present them here. May I refer the interested readers to consult our forthcoming paper [12]. For example, it turns out that the equivalence of the FDSM model and the particle-rotor model at $N_1 \rightarrow \infty$ limit can be proven analytically; The abnormal Nilsson levels in one major shell can be derived from the FDSM; the well known K-selection rule of the GM can in fact be derived microscopically from the FDSM when $N_1 \rightarrow \infty$. Also, the CAP effect, the stretching effect and the Coriolis attenuation can all emerge naturally. Besides, not only the ground state masses but also the first 2^+ excitation energies for the whole actinide region, from vibrational to rotational nuclei, can be well reproduced by the FDSM, etc. The culmination of all the above mentioned results is certainly very suggestive to the notion that the Geometrical Model (GM) of the Copenhagen School may be regard as another limit of the FDSM when $N_1 \rightarrow \infty$ (classical limit). The "bottom line" of this discussion, i.e. the relationships among these models, is schematically shown in the "cartoon" of Fig.7.

This work is supported by the Physics and International Divisions of the United States National Science Foundation.

REFERENCES

[1]. C.-L. Wu, D. H. Feng, X.-G. Chen, J.-Q. Chen and M. W. Guidry, Phys. Lett. *168B*,313 (1986); Phys. Rev. *C36*, 1157 (1987); J.-Q. Chen, D. H. Feng and C.-L. Wu, Phys. Rev. *C34*, 2269 (1986); R. F. Casten, D. H. Feng, J. N. Ginocchio and C.-L. Wu, Mod. Phys. Lett. *A1*, 61 (1986);

[2] M. Mayer, Phys. Rev. *74*, 235(1948); O. Haxel, J. H. D. Jensen and H. E. Suess, Phys. Rev. *75*, 1766(1949); B. H. Wildenthal in *Proceedings of the Conference on Nuclear Shell Model*, M.Vallieres and B.H.Wildenthal (eds.) (World Scientific Publications, Singapore, 1985) and refs. therein.

[3] A. Arima and F. Iachello, Ann. Rev. of Nucl. Part. Sci. *31*, 75(1981).

[4] J. N. Ginocchio, Phys. Rev. B79, 173(1978); Ann. of Phys. *126*, 234(1980).

[5] K. T. Hecht, Notas de Fisica VIII (1985).

[6] R. F. Casten, C.-L. Wu, D. H. Feng, J. N. Ginocchio and X.-L. Han, Phys. Rev. Lett. *56*, 2578 (1986);

[7] M. W. Guidry, C.-L. Wu, D. H. Feng, J. N. Ginocchio, X.-G. Chen and J.-Q. Chen, Phys. Lett. *176B*, 1 (1986). M. W. Guidry, C.-L. Wu, Z.-P. Li, D. H. Feng and J. N. Ginocchio, Phys. Lett. *187B*, 210 (1986).

[8] X.-L. Han, M. W. Guidry, D. H. Feng, K.-X. Wang and C.-L. Wu, Phys. Lett. *192B*, 253 (1987).

[9] C.-L. Wu, X.-L. Han, Z.-P. Li, M. W. Guidry and D. H. Feng, Phys. Lett. **194B**, 447 (1987).

[10] H. Wu, D. H. Feng, C.-L. Wu, Z.-P. Li and M. W. Guidry, Phys. Lett. **193B**,163 (1987);

[11].See A. Bohr and B. R. Mottelson, Nuclear Structure Vol I and II, (Benjamin, New York, 1975).

[12] C.-L. Wu et al.(to be published).

EMPIRICAL MASS FORMULA WITH PROTON-NEUTRON INTERACTION

Takahiro Tachibana, Masahiro Uno,* So Yamada and Masami Yamada
Science and Engineering Research Laboratory, Waseda University
3-4-1 Okubo, Shinjuku-ku, Tokyo 160, Japan

ABSTRACT

An atomic mass formula consisting of a gross part, an average even-odd part and an empirical shell part is studied. The gross part is, apart from a small atomic term, taken to be the sum of nucleon rest masses, Coulomb energies and a polynomial in $A^{1/3}$ and $|N-Z|/A$. The shell part includes, in addition to proton and neutron shell terms, a proton-neutron interaction term which represents the mutual support of nuclear magicities and the cooperative deformation effect. After the first construction of such a formula, refinements have been made in two respects. One is a separate treatment of $Z=N$ odd-odd nuclei suggested by a quartet model, and the other is an improvement of the proton neutron interaction term. By these refinements the root-mean-square deviation of calculated masses from the 1986 Audi-Wapstra masses has been reduced from 538 keV to 460 keV.

1. INTRODUCTION

Previously, Uno et al.[1] constructed mass formulas in which shell effects are represented by two terms, i.e. a proton shell term and a neutron shell term. These terms have many parameters and their values were determined from comparison with experimental data rather than from nuclear models. Recently, we have advanced this kind of study by including a proton-neutron interaction term. The first formula along this line was constructed last year, and will be published elsewhere.[2] In Sec.2 we will give a briefly sketch of it. After completing this formula, we have made some refinements, which are described in Sec.3.

2. FIRST FORMULA

We assume that the atomic mass of the nuclide with Z protons and N neutrons are expressed as

$$M(Z,N)=M_g(Z,N)+M_{eo}(Z,N)+M_s(Z,N) \quad (1)$$

Here, $M_g(Z,N)$ is the gross part and is given, in MeV units, as

$$M_g(Z,N)c^2=M_H c^2 Z+M_n c^2 N+a(A)A+b(A)|N-Z|+c(A)(N-Z)^2/A \\ +E_C(Z,N)-14.33\times 10^{-6} Z^{2.39} \quad (2)$$

where M_H is the mass of a ^{1}H atom, M_n is the neutron mass, $E_C(Z,N)$ is the Coulomb energy and the last term is the binding energy of atomic electrons. The coefficients $a(A)$, $b(A)$ and $c(A)$ depending on the mass number A are assumed to be polynomials in $A^{-1/3}$ as

$$a(A)=a_1+a_2 A^{-1/3}+a_3 A^{-2/3}+a_4 A^{-1} \\ b(A)=b_1+b_2 A^{-1/3}+b_3 A^{-2/3}+b_4 A^{-1} \quad (3) \\ c(A)=c_1+c_2 A^{-1/3}+c_3 A^{-2/3}+c_4 A^{-1}$$

---

*Now at Elementary and Secondary Education Bureau, Ministry of Education, Science and Culture, 3-2-2 Kasumigaseki, Chiyoda-ku, Tokyo 100, Japan

The parameters a_i, b_i and c_i are either determined from comparison with experimental data[3] or fixed to certain values to avoid their drift to unreasonable values during optimization.

The Coulomb energy is taken as

$$E_C(Z,N) = \frac{3e^2 Z^2}{5 r_{0\text{eff}} A^{1/3}} [1 - \frac{5}{4}(\frac{9}{4\pi^2})^{1/3} C_x Z^{-2/3}]$$

$$\approx 0.864 \frac{Z^2}{r_{0\text{eff}}(\text{in fm}) A^{1/3}} (1 - 0.764 C_x Z^{-2/3}) \text{ MeV} \quad (4)$$

with

$$r_{0\text{eff}}(\text{in fm}) = 1.175[1 + 0.1(\frac{N-Z}{A})^2 - 0.3 A^{-1/3}] + 1.4[1 - 0.85(\frac{N-Z}{A})^2] A^{-2/3}$$

$$- 0.27(1 - A^{-1/3}) \frac{N(N-Z)}{A^2} + 0.432 \frac{Z^2}{150 A^{4/3}(1 - A^{-1/3})} (1 - 0.764 C_x Z^{-2/3})$$

$$+ 0.432 \frac{Z^2 N^2}{20 A^{11/3}} (1 - 0.764 C_x Z^{-2/3}), \qquad C_x = 1.3 \quad (5)$$

This expression has been obtained from considerations of the compression by surface tension, the surface thickness, the proton-neutron displacement, and the isoscalar and isovector expansions by Coulomb forces, and the parameter values have been determined from comparison with Coulomb displacement energies and isotope shift data.

The average even-odd part is assumed to be

$$M_{eo}(Z,N) = 13[2Z+10]^{-1/2} \delta_{o\text{dd}Z} + 13[2N+10]^{-1/2} \delta_{o\text{dd}N}$$

$$- (50 A^{-1} - 110 A^{-4/3} + 60 A^{-5/3}) \delta_{o\text{dd}Z} \delta_{o\text{dd}N} \quad (6)$$

with

$$\delta_{o\text{dd}Z} = \begin{cases} 1 & \text{for odd } Z \\ 0 & \text{for even } Z \end{cases} \qquad \delta_{o\text{dd}N} = \begin{cases} 1 & \text{for odd } N \\ 0 & \text{for even } N \end{cases}$$

The shell part is taken as

$$M_s(Z,N) = P_Z(A) + Q_N(A) - h_1 A^{2/3} [X - h_3 + (X^2 - 2h_2 X + h_3^2)^{1/2}]$$

$$- \alpha \theta(P_Z(A) + Q_N(A)) [P_Z(A) + Q_N(A)]^2 Z^2 / A \quad (7)$$

with

$$P_Z(A) = (2Z/A)^{2/3} P_Z^0 \qquad Q_N(A) = (2N/A)^{2/3} Q_N^0 \quad (8)$$

$$X = P_Z(A) Q_N(A) \quad (9)$$

$$\theta(P_Z(A) + Q_N(A)) = \begin{cases} 0 & \text{for } P_Z(A) + Q_N(A) \leq 0 \\ 1 & \text{for } P_Z(A) + Q_N(A) > 0 \end{cases} \quad (10)$$

In Eq.(7), the third term in the right-hand side is the proton-neutron interaction term with three parameters h_i. This term represents the mutual support of nuclear magicities and the cooperative deformation effect. The last term of Eq.(7) is to represent the effect of deformation on the Coulomb energy, and we take $\alpha = 0.003$ MeV^{-1}. The values of the shell parameters P_Z^0 and Q_N^0 have been determined from considerations of charge symmetry and fit of calculated masses to experimental data.[3]

The degree of agreement of this formula with 1986 Audi-Wapstra masses[3] is shown for some nuclidic regions in Fig.1. The root-mean-square deviation between them is 538keV. For more details about this formula, see Ref.2.

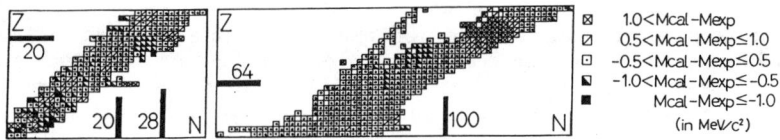

Fig.1. Comparison between the masses calculated from the first formula and the experimental masses in selected nuclidic regions.

3. REFINEMENTS OF THE FORMULA

We refine the above formula in two respects.

First, the quartet model[4] suggests that the odd-odd nucleus with $Z=N$ has a large even-odd energy than given by Eq.(6). Therefore, by analyzing the data we add to the even-odd part a correction term

$$\Delta M_{eo}(Z,N(=Z))=120A^{-4/3}-150A^{-5/3} \quad \text{for odd } Z=N \quad (11)$$

The second refinement is made on the proton-neutron interaction energies, in particular, on those in the region of deformed nuclei. It seems that the reduction of energy due to cooperative deformation of proton and neutron distributions is large in the last half of a shell than in the first half. To take account of it we modify Eq.(9) as

Table 1. The gross-part parameters in MeV and the parameters for the proton-neutron interaction term in the modified formula. The values in parentheses are fixed in the optimization.

i	1	2	3	4
a_i	−15.7883	16.71	(15.0)	(−30.0)
b_i	(0.0)	(0.0)	(0.0)	35.22
c_i	26.13	−21.73	−22.47	(−30.0)
h_i	0.0011 MeV^{-1}	3.0 MeV^2	4.0 MeV^2	

$$X=[P_Z(A)-\beta P'_Z(A)][Q_N(A)-\beta Q'_N(A)] \quad \text{if } P_Z(A)-\beta P'_Z(A)\geq 0,\ Q_N(A)-\beta Q'_N(A)\geq 0 \quad (12)$$

where $P'_Z(A)$ and $Q'_N(A)$ are "derivatives" of $P_Z(A)$ and $Q_N(A)$, and actually obtained as

$$P'_Z(A_1)=\frac{1}{[Z/20]}\sum_{z=1}^{[Z/20]}\frac{1}{2z}[P_{Z+z}(A=A_1+z)-P_{Z-z}(A=A_1-z)] \quad (13)$$

and with a similar equation for $Q'_N(A)$. In Eq.(13), [$Z/20$] means the integral part of $Z/20$, and $P'_Z(A)=0$ for $Z<20$ and $Q'_N(A)=0$ for $N<20$. For $Z>103$, Eq.(13) is somewhat modified, and for $N>149$, the corresponding fomula for Q'_N is also

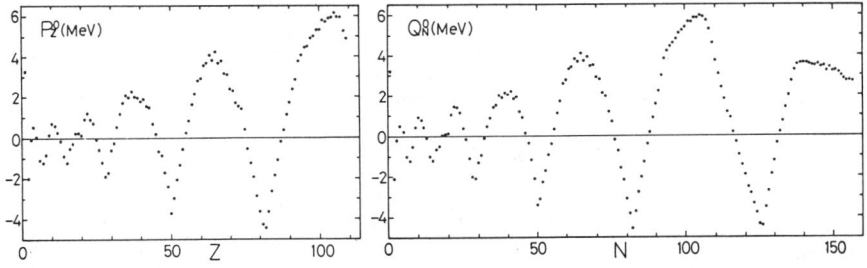

Fig.2. The proton shell parameters P_Z^0 in the modified formula.

Fig.3. The neutron shell parameters Q_N^0 in the modified formula.

modified. We found that β=6 is suitable.

The values of the parameters for this modified formula are given in Table 1 and in Figs. 2 and 3, and the general degree of agreement of this formula with 1986 Audi-Wapstra masses[3] is shown in Fig.4. The root-mean-square deviation between them is 460 keV, which is considerably smaller than that of the first formula.

The table of masses calculated from this refined formula will be sent upon request.

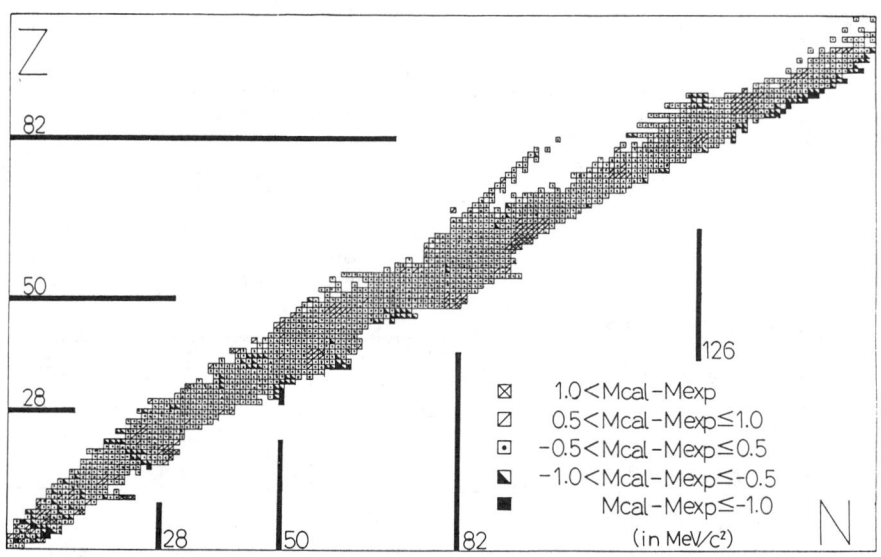

Fig.4. Comparison between the masses calculated from the modified formula and the experimental masses.

REFERENCES

1. M.Uno and M.Yamada, Prog. Theor. Phys. **53**, 987 (1975); **65**,1322 (1981); M.Uno and M.Yamada, in Atomic Masses and Fundamental Constants 6, edited by J.A.Nolen,Jr. and W.Benenson (Plenum Pub.Co., New York, 1980), p.141; M.Uno, M.Yamada, Y.Ando and T.Tachibana, Bull. Sci. Eng.Res. Lab. Waseda Univ. No.97, 19 (1981); M.Uno and M.Yamada,"Atomic Mass Prediction from the Mass formula with Empirical Shell Terms", Institute for Nuclear Study, University of Tokyo, INS-NUMA-40 (1982).
2. T.Tachibana, M.Uno, M.Yamada and S.Yamada, Atomic Data and Nuclear Data Tables, to be published.
3. A.H.Wapstra, G.Audi, and R.Hoekstra, Atomic Data and Nuclear Data Tables, to be published.
4. M.Cauvin, V.Gillet, F.Soulmagnon and M.Danos, Nucl. Phys. **A361**, 192 (1981); M.Danos and V.Gillet, Z. Phys. **249**, 294 (1972); M.Uno, R.Sugaya and M.Yamada, in Proc. 7th International Conference on Atomic Masses and Fundamental Constants, edited by O.Klepper (Darmstadt, 1984), p.443.

OCCUPATION NUMBER DEPENDENT EFFECTIVE INTERACTIONS AND THE ODD-EVEN MASS DIFFERENCE

A.S. Jensen and A. Miranda
Institute of Physics, University of Aarhus
DK - 8000 Aarhus C, Denmark

ABSTRACT

The pairing matrix elements are parametrized in terms of the occupation numbers in analogy to the density dependence of the Skyrme interaction. Generalized gap-equations are derived and the neutron excess dependence of the odd-even mass difference is calculated and shown to reproduce global experimental values for a suitable parameter set. It is pointed out that moments of inertia and odd-even radius differences should provide tests of the model.

- oOo -

Analysis of the measured nuclear odd-even mass differences reveals a global dependence on the neutron excess.[1] Attempts to understand this global trend using the standard BCS-theory of pairing in nuclei were not successful unless unusual types of neutron-proton pairings were explicitly added to the usual Skyrme + Pairing Hamiltonian.[2]

We assume here that the trial wave function for the nuclear ground state is again the product BCS wave function, i.e. without explicit build-in neutron-proton correlations. However, the effective residual coupling is complicated. We assume that the effects we are looking for are tied up with proton and neutron pairing states in the vicinities of their respective Fermi levels. For these states, the bare Brueckner's G-matrix in the particle-particle and hole-hole *pairing* channels get appreciable renormalisations. These are important for reproducing the overall phenomenological strengths of the nuclear pairing matrix elements for neutrons and for protons. We follow the Skyrme procedure that works so well for the particle-hole channels. The Skyrme effective interaction, to be used in conjunction with Hartree-Fock wave functions, is a kind of a phenomenological G-matrix. As one recalls, it was originally assumed to be a sum of two- and three-body interactions, which together with Hartree-Fock wave functions produced a specific effective momentum and density dependent two-body interaction.

We likewise first assume that the structure of a phenomenological pairing G-matrix could be obtained, in conjunction with product BCS wave functions, by considering a further four-body interaction. By virtue of Wick's theorem, it yields an effective two-body pairing interaction. This leads to not only further density dependence but also introduces specific sums of terms involving contractions of type $(u_p v_p)^{k_1} (u_n v_n)^{k_2}$, where k_1 and k_2 are small integers. We therefore parametrize our effective pairing matrix elements with products of the above type. However, the Skyrme interaction actually used in practice does not necessarily depend linearly on the density, as it would in the original model with three-body potentials. Likewise, we do not insist that our effective interaction has necessarily something to do with a definite four-body interaction.

The Hamiltonian consists of a Hartree-Fock part and a *residual* pairing interaction:

$$H = \sum_{qv} \varepsilon_{vq}^{HF} a_{vq}^+ a_{vq} + \frac{1}{4} \sum_{qv_1 v_2} \langle v_1 \bar{v}_1 | V_{pair} | v_2 \bar{v}_2 \rangle_q \quad (1)$$

$$a_{v_1 q}^+ a_{\bar{v}_1 q}^+ a_{\bar{v}_2 q} a_{v_2 q} ,$$

where q=n,p and the remaining notation is obvious. We assume the parametrization

$$V_{v_1 v_2}^q \equiv \langle v_1 \bar{v}_1 | V_{pair} | v_2 \bar{v}_2 \rangle_q = - [G + a(\tilde{\Delta}_p + \tilde{\Delta}_n)^\beta + b\tilde{\Delta}_{np}^\gamma \quad (2)$$

$$+ c(\tilde{\Delta}_p^\delta + \tilde{\Delta}_n^\delta)] x_{v_1 q} x_{v_2 q}$$

$$\tilde{\Delta}_q = \sum_v x_{vq} u_{vq} v_{vq} \quad (3)$$

$$\tilde{\Delta}_{np} = \sum_{v_1 v_2} y_{v_1 v_2} u_{v_1 n} v_{v_1 n} u_{v_2 p} v_{v_2 n} , \qquad (4)$$

where v_{vq} are the variational parameters entering the BCS wave function for an even-even nucleus (i.e. a product of n and p wave functions with $u^2_{vq} + v^2_{vq} = 1$). Introducing the chemical potentials as usual, we minimize the expectation value of the resulting Hamiltonian using this BCS wave function. The BCS expressions for u and v emerge with $E_{vq} \equiv \sqrt{\varepsilon^2_{vq} + \Delta^2_{vq}}$ and

$$\Delta_{vq} = -\frac{1}{2} \sum_{v_1} V^q_{v_1 v} u_{v_1 q} v_{v_1 q} - \frac{1}{4} (\tilde{\Delta}^2_q + 2 \sum_{v_1} x^2_{v_1 q} v^4_{v_1 q})$$

$$\cdot \frac{\partial V^q_{vv}}{\partial v_{vq}} u_{vq} / (u^2_{vq} - v^2_{vq}) \qquad (5)$$

$$\varepsilon_{vq} = \varepsilon^{HF}_{vq} - \mu_q + V^q_{vv} \cdot v^2_{vq} . \qquad (6)$$

The odd-even mass difference δU_q is now computed as the expectation value of H in a single quasi-particle state relative to that in the even-even BCS state (in principle, a new minimization ought to be performed). The result is

$$\delta U_q = \frac{1}{2} \Delta_{v_F q} (-V^q_{v_F v_F}) / x_{v_F q} , \qquad (7)$$

where v_F is the index of the Fermi level with $\varepsilon_{v_F q} = 0$.

For given parameters $(G, a, b, c, \beta, \gamma, \delta, x_{vq} y_{v_1 v_2})$, the equations (3), (4) and (5) should be solved to give Δq and $\tilde{\Delta} q$. Then δU_q can be calculated and compared to the phenomenological expression

$$\delta U_q = 7.36 \text{MeV} (1 - 7.0 (N-Z)^2 / A^2) / A^{1/3} . \qquad (8)$$

To understand the main point, let us simplify drastically: x_{vq} is non-vanishing (and constant) only within the interval S_q around $\varepsilon_{v_F q}$ and $y_{v_1 v_2}$ is non-vanishing (and constant) only when both energies are within the respective intervals S_q. With $\beta=\gamma=\delta=1$ and the uniform model assumption we find (c=0 without loss of generality)

$$\tilde{\Delta}_q = 2\Delta_q g_q \ln(2S_q/\tilde{\Delta}_q) \qquad (9)$$

$$\Delta_q = \delta U_q + \frac{1}{4}(a+b\tilde{\Delta}_{-q})(\tilde{\Delta}_q^2 + 4S_q g_q) \qquad (10)$$

$$\delta U_q = \frac{1}{2}\tilde{\Delta}_q(G + a(\tilde{\Delta}_n + \tilde{\Delta}_p) + b\tilde{\Delta}_n \cdot \tilde{\Delta}_p), \qquad (11)$$

where g_q is the doubly degenerate single particle level density. Using the $A^{-1/3}$-dependence of S_q and δU_q and the A-dependence of g_q, eqs. (9)-(11) imply $GA \equiv G_0$, $aA^{5/3} \equiv a_0$, $b \cdot A^{7/3} \equiv b_0$ and $A^{-1/3}$ and $A^{2/3}$-dependencies of Δ_q and $\tilde{\Delta}_q$, respectively. Expanding all quantities in eqs. (8)-(10) to second order in (N-Z)/A, the expansion coefficients are related by a set of equations. Here we use g_n=N/90MeV and g_p=Z/90MeV in fair agreement with Skyrme single-particle energies. Choosing the N-Z dependence of S_q such that $\Delta_q = \delta U_q$, the traditional result of a=b=0 is obtained. Other assumptions for S_q lead to non-vanishing values of the effective pairing force parameters. Calculating other pairing sensitive quantities like moments of inertia as function of N-Z could presumably determine S_q. The real test would then come by computing, for example, the odd-even root mean square radius differences.

1. A.S. Jensen, P.G. Hansen and B. Jonson, Nucl.Phys. A **431** (1984) 393.
2. A.S. Jensen and A. Miranda, Nucl.Phys. A **449** (1986) 331.

STATIC MAGNETIC AND ELECTRIC MOMENTS NEAR CLOSED SHELLS FROM LASER SPECTROSCOPY

G. Huber
Institut für Physik, Universität Mainz
6500 Mainz, Fed. Rep. of Germany

ABSTRACT

Magnetic moments, electric quadrupole moments and isotope shifts of elements close to $Z = 50$ and $Z = 82$ are discussed. Single particle character and coupling rules are valid. Parabolic slope of the charge radii at $Z = 50$ is discussed. At $Z = 82$ intruder states dominate the charge radii.

INTRODUCTION

The advances in laser spectroscopy on nuclei far from stability since the Helsingör conference 1981 are considerable in the amount of data and also in experimental techniques. Then the long isotope chains of alkali elements and of Ca, Cd, Ba, Yb and Hg /1/ yielded a systematic information on nuclear structure and shape via isotope shift and quadrupole moments mainly. The change in the radius increase at a neutron shell closure was established as a general feature and transitions to new regions of deformation had been confirmed (N=20, 40, 60, 64, 88, 105).

Today we have already a nuclear chart studied systematically with neighbouring elements as shown in Fig. 1 taken from a recent review article of E.W. Otten /2/. All simple atomic spectra, alkali-like and earth alkaline like, have been studied. At this conference the latest data are presented on the very heavy elements, the bulk of rare earth elements and elements near closed proton shells.

The novel detection techniques proliferate and some of them have already proofed their superior sensitivity as those presented at this conference. At the same time the isotope production methods go beyond the ISOLDE type spallation and H.E. fission especially with low energy reactions with light /3/ and heavy ions /4,5,6/.

NUCLEI NEAR CLOSED SHELLS

It has thus become possible to study very specific questions using different production techniques and also complicated atomic transitions. One of those questions is the behaviour of transitional nuclei near closed shells: charge radii increase rapidly on both sides of $N = 50$ in the Rb /7/ and Sr /8/ chains whereas calculations of different types cannot reproduce this trend /9,10/. Instead of studying the behaviour of nuclei at neutron shells it is easier to study isotope chains close to proton shells in order to observe a large number of different nuclei.

The observation of long chains in Sn /11,12/, In /13,14/, Cd /15/ and Ag /16/ shows a particularly regular behaviour of spins, moments and charge radii at $Z = 50$ whereas a similar trend at $Z = 82$ in the isotope chains of Pb /17,18/, Tl /19,20/, Hg /21/ and Au

/22/ is masked by the well-known shape isomerism and flipping in Hg and intruder states in Au.

EXPERIMENTAL TECHNIQUES

The recent experiments use quite different techniques like thermal atomic beams and frequency doubled cw lasers /11,17/, collinear spectroscopy with metastable state depopulation /12,13,18,19,20/ partly using isotope storage in an ion source with bunched release /23/. Whereas these experiments are based on optical fluorescence, a three step resonance ionization mass spectroscopy (RIMS) has been performed on the Au isotopes /22/ with pulsed lasers and recent development includes a pulsed laser-induced desorption (PLID) to increase the sensitivity /24/.

DIPOLE AND QUADRUPOLE MOMENTS AT Z = 50

These are obtained from the hyperfine structure (hfs) of the observed spectra. The centre of gravity shift of these patterns between isotopes is the IS and a measure of $\delta\langle r^2 \rangle$. Since $Z \leq 50$ is a region of spin isomerism many of these moments have been studied as a function of selected spin states along isotope series.

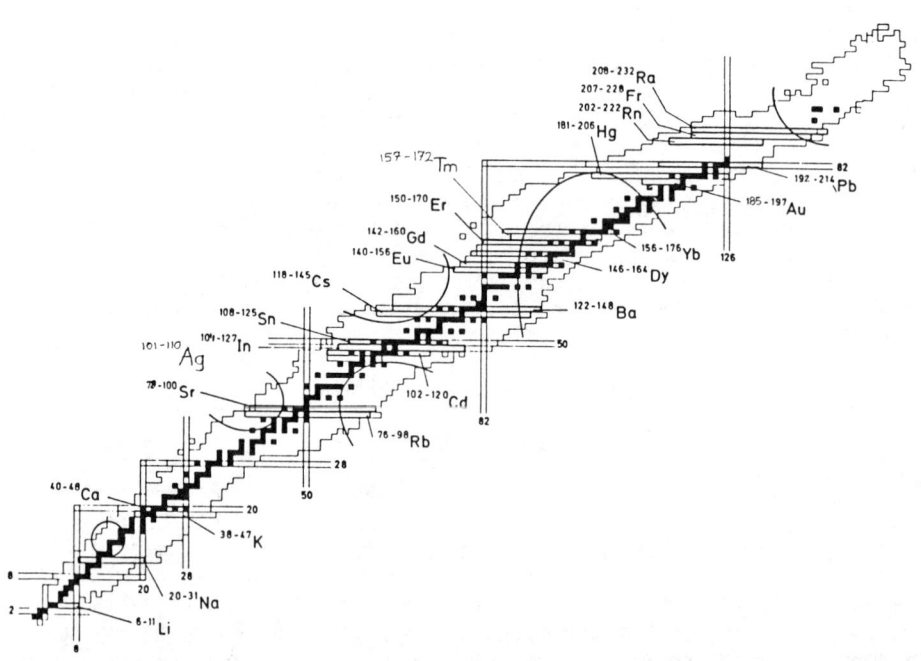

Fig. 1: Chart of nuclides with extended isotope chains studied by laser spectroscopy.

The atomic parameters for magnetic dipole interaction have been obtained by comparison with the hfs of stable isotopes. The quadrupole interaction could be analyzed similarly in the case of indium but the tin and silver isotopes where analyzed with calculated field gradients. The nuclear spin I has been directly determined or corrected in many low spin states ($I \leq 3$). Assignments for higher spin states could be given.

The moments of the odd proton states can be well described by the particle plus rotor model /25/. For the $I = 9/2$ ground states in $^{105-127}$In the magnetic moments are almost constant with $\mu_I \simeq 5.5$ n.m. The experimental quadrupole moments for the $I = 9/2$ states are constant with $Q_s \simeq 0.8$ b within in the range of $^{105-121}$In and decrease gradually to 0.6 b for ^{127}In. This reflects the single particle character near the closed proton shell and the reduced effect of neutron core polarization. The experimental results are reproduced in the above model using $g_s = 0.5\ g_{s,free}$ and $g_r = Z/A$ and an effective charge $e_p = 1.5$ e for the spectroscopic quadrupole moments /26/. Here, the hole plus vibrating core calculations yield a constant value of $Q_s \simeq 0.7$b.

In contrast to the magnetic moments of the $I = 9/2$ ground states the moments of the $I = 1/2$ isomeric states exhibit negative moments close to the Schmidt line $\mu_s = -0.263$ n.m. for $A \leq 117$ and go beyond this value to reach finally $\mu = -0.433$ n.m. at $A = 125$, that is 1.6 times the Schmidt value. For a $p_{1/2}$ moment core polarization should not contribute so that $g_s = g_{s,free}$ has to be chosen /27/. The hole plus vibrating core model then predicts a constant negative moment $\mu = -0.27$ n.m., close to the Schmidt value, independent of the neutron number. In order to reproduce the observed trend a substantial change in the admixture of the $[|p_{3/2}> \otimes |2^+>]_{I=1/2}$ configuration is needed since its negative magnetic moment yields $\mu = -0.85$ n.m. for $g_R = Z/A$. Such a change in core polarization does not appear in the magnetic moment of the $I = 9/2$ ground state.

The moments of the odd Ag isotopes have been described by three different types of calculations, namely three hole cluster-vibrational core coupling calculations, the interacting boson-fermion model calculations, and calculations within the Nilsson model, as has been shown recently /28,29/. The calculation of quadrupole moments of the $I = 7/2$ and $I = 9/2$ states within the three hole cluster coupled to a vibrating core model /30/ agrees very well with the experimental data in ^{101}Ag and $^{103-109}$Ag.

The moments of the odd-odd indium isotopes, can be reproduced by the additivity rule of magnetic moments using the values of the neighbouring isotones and isotopes. A similar additivity rule is valid for the spectroscopic quadrupole moments of the odd-odd indium isotopes. This is shown in Fig. 2 where a good agreement for magnetic moments and spectroscopic quadrupole moments is seen except for the states with $I = 8$. This has been attributed in ref. /13/ to a change from rotational aligned to strong coupling of a $1h_{11/2}$ neutron. Using the strong coupling scheme a deformation parameter ß can be extracted which varies slowly over 30 isotopes and isomers with $I \geq 1$.

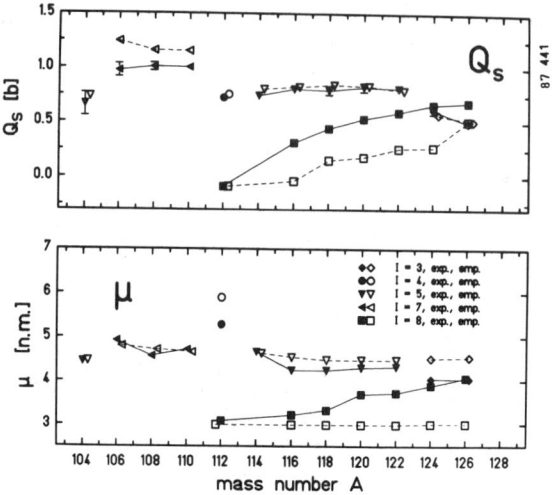

Fig. 2: Nuclear moments of the odd-odd In isotopes. Full symbols for experimental and open symbols for semi-empirical values from coupling rules.

ISOTOPE SHIFT AT Z = 50

Usually the mean square charge radius is discussed on the basis of the two parameter formula

$$<r^2> = <r^2>_{spherical} (1 + \frac{5}{4\pi} <\beta^2>) \qquad (1)$$

where β is the quadrupole deformation parameter of the nuclear field $r(\theta) = r_0/(1 + \beta Y_{20}(\theta) - (1/4\pi)\beta^2)$. This parameter β is commonly a measure of the static intrinsic deformation. But in the present investigation of almost spherical and transitional nuclei the role of the dynamical contribution to $<\beta^2>$ is predominant and adds to be static deformation to give a very simple functional slope in the isotope shift.

The change in the charge radii has been extracted from the isotope shift using calculated electron densities as discussed in /13/. The mass shift in the medium mass range is of the same order as the volume shift and has been fixed by a comparison with K_α x-rays in tin and cadmium. By this uniform procedure the systematic uncertainties in IS are reduced to a common scaling error of less than 10 % and the errors in the stable reference isotopes in the order of 3 %.

In the region between the two major neutron shell closures N = 50 and N = 82 the experimental slope can be well fitted by a parabolic slope added to a constant odd-even staggering:

$$\delta \langle r^2 \rangle_{66,N} = a(N-66) + b(N-66)^2 + c(1-1^N) \qquad (2)$$

The linear term can be identified with a pure spherical volume increase as given by the droplet model or some spherical Hartree-Fock calculations. The quadratic term is symmetric around the mid-shell and can be related to a quadrupole polarization of the core by the

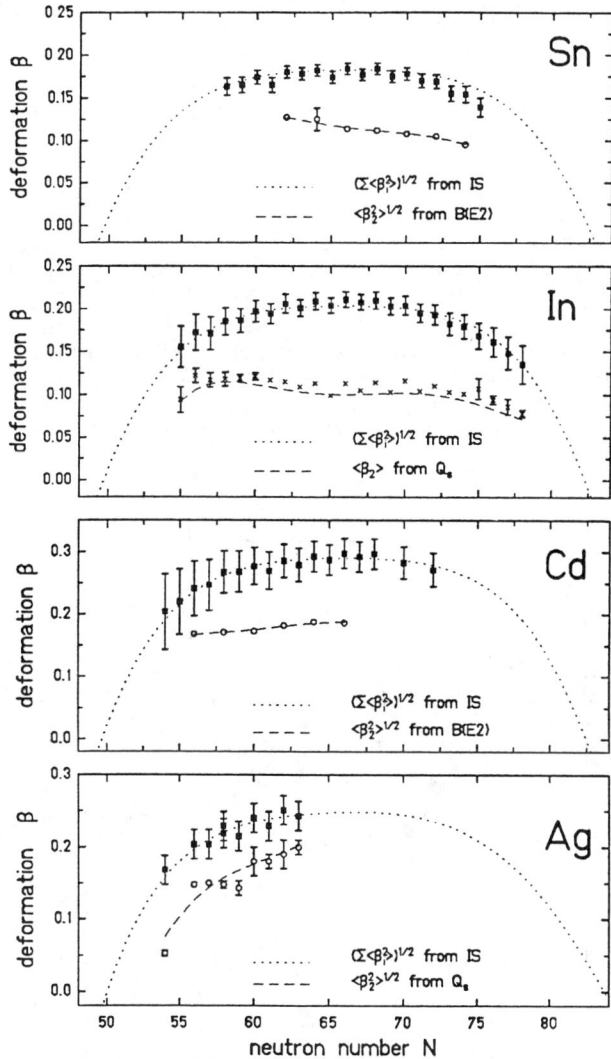

Fig. 3: Deformation parameters for isotope chains with Z = 47 - 50 deduced from the isotope shift according to the formula in the text in full symbols. The open symbols represent B(E2) values in Sn and Cd and quadrupole moments in strong coupling in In and Ag.

neutrons similar to the polarization responsible for the odd-even staggering. This mechanism has been proposed and applied successfully to the calcium and also lead isotopes /31/. Subtracting the linear part in the above formula the reminding structures can be related to the static and dynamical deformations $<\beta^2>$ in the two parameter formula. If one sets $<\beta^2> = 0$ at closed neutron shells an absolute measure can be given for the isotope chains as presented in Fig. 3. The four elements exhibit parabolic slopes.

A comparison to other deformation parameters like B(E2) values in tin and cadmium or intrinsic quadrupole moments in indium reveals a substantial discrepancy: the deformation measure in the isotope shift is up to a factor 1.5 larger than the generally accepted one. In the case of silver the data are still incomplete but here the discrepancy with respect to the intrinsic quadrupole moments is obviously reduced. The difference has to be attributed to quadrupole (and higher-order) surface vibrations induced by the valence particle polarization of the core. For even-even nuclei the higher moments are taken into account by the Hartree-Fock field but β_2, β_3, β_4 have to be introduced separately /32/. The tin isotope chain is partly reproduced by such calculations but disagreement persists for the outermost isotopes which support well the parabolic slope.

The parameters of the parabolic fit of all 4 elements are shown in Fig. 4 and are compared with the droplet model predictions /33/

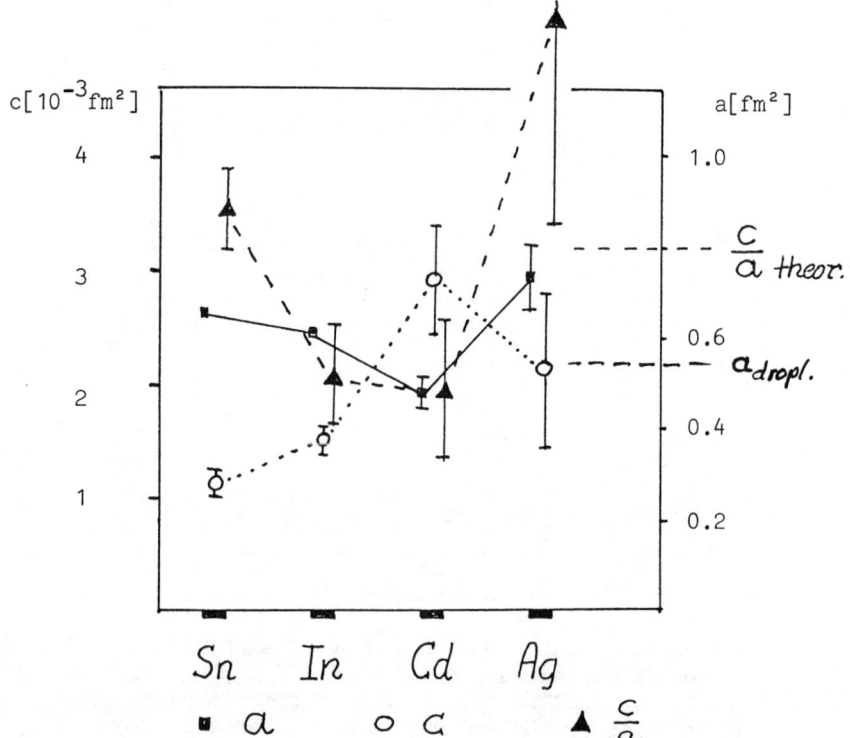

Fig. 4: The parameter a, b, c/a in formula (2) are shown for Z = 47 - 50. $a_{Droplet}$ and $(c/a)_{theor.}$ are given for comparison.

and the core polarization model /31/. Deviations with the latter might be partly due to the quality of the Cd data and the incomplete data in Ag.

ISOTOPE SHIFTS AT Z = 82

The nuclei close to N = 104 are known to exhibit shape coexistence in Hg and charge radii with large odd-even staggering and huge isomer shifts. Now we know $^{192-214}$Pb /17,18/ with a marked kink at the N = 126 shell closure consistent with the heavier elements Fr /34/ and Ra /35/. Before shell closure the isotope shift coincides with the droplet model prediction /33/ not leaving any space for deformation contribution in the two parameter formula. A discussion of steeper slopes and the need of improved droplet parameters has recently been given by /36/. Introducing giant monopole and quadrupole perturbations in a HF-RPA calculation with Skyrme forces seem to reproduce the above mentioned kink and - mainly due to the monopole resonance - the odd-even staggering as well /37/. The most neutron-deficient isotopes as shown in Fig. 5 exhibit a reduced slope in the region of a low lying intruder state /37/ giving rise to the isomer shift in the neighbouring Tl isotopes as discussed in /20/. Beyond the shape coexistence in Hg the recent measurements at ISOLDE with ionization spectroscopy (RIMS) reveal a drastic shape change attributed to a $\pi h_{9/2}$ intruder state /22/. In contrast to the Hg and Tl isotopes no shape isomerism of flipping has been observed in Au although the neighbouring light Pt isotopes are known to show a large odd-even staggering in the spacing of the ground state rotational bands /38/.

CONCLUSION

The elements near closed proton shells studied by laser spectroscopy reveal many facettes of nuclear structure. Single particle character is seen in the magnetic dipole and electric quadrupole moments at Z = 50. The influence of surface vibrations on the slope and staggering of mean square charge radii is rather unperturbed in medium masses with a dominance of quadrupole and octupole modes. The polarization models can be tested on an already important set of data. It is evident that this pure structure seen in the parabolic slope is present all over the nuclear chart except at spherical or deformed shell closures and gives substantial contributions in the two parameter formula for transitional nuclei. The heavy elements can be described by monopole and quadrupole modes mainly for the smooth part. Near the neutron midshell the nuclear shape is affected by intruder states which seem to dominate at least in Au the nearby proton shell closure.

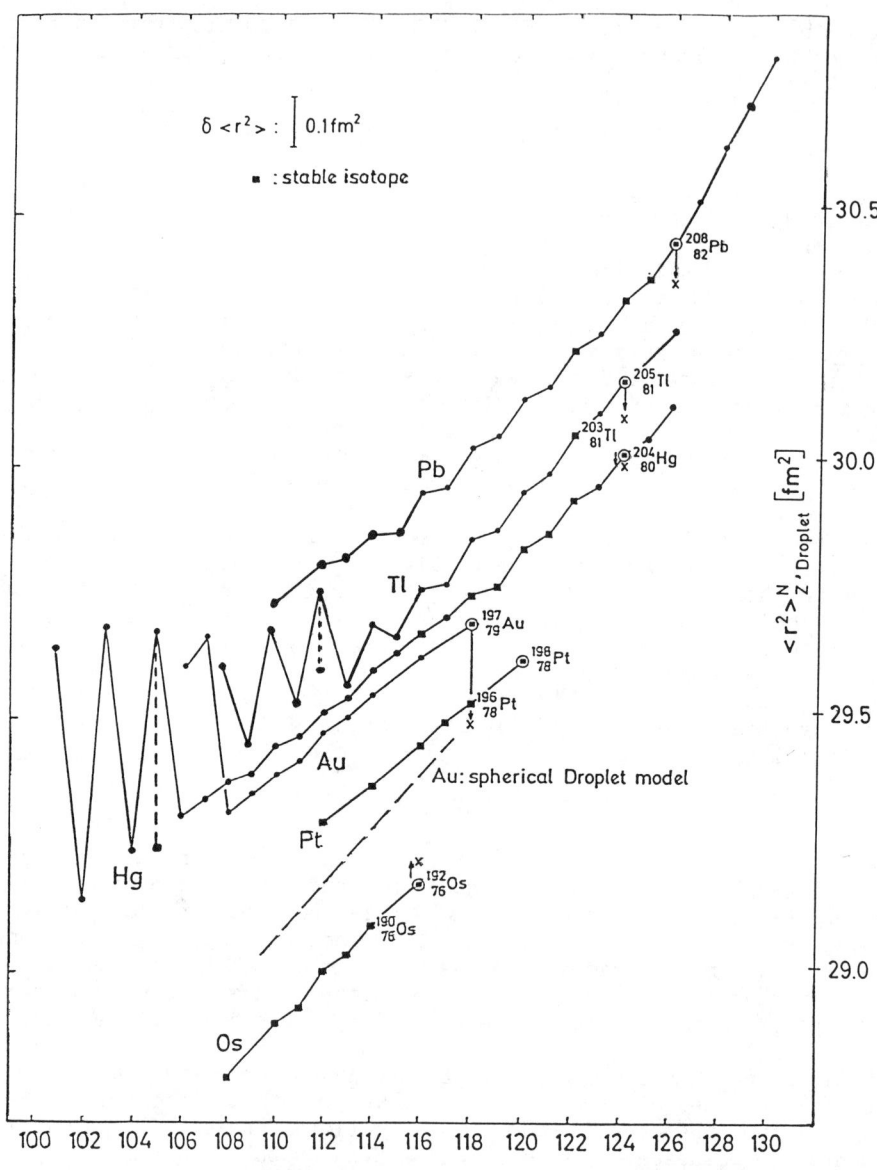

Fig. 5: Charge radii at Z = 82 on an absolute scale by combining optical isotope shift data and radii from x-ray shift. The slope of the Tl isotopes is too steep and not a result of a complete analysis of the isotope shift. The influence of intruder states in Pb, Tl and Au start at N = 113, N = 112 and N = 107, respectively.

OUTLOOK

The data of the heavy elements are challenge for theoretical models describing simultaneously the smooth and shape instable region of the $Z \leq 82$ elements. The experimental effort will certainly extend the data for Au, Pt, Pb but it would be interesting to go beyond $Z = 82$ and study Bi and Po in spite the technical difficulties. This point is true as well for the $Z = 50$ region where the study of Sb and Te would not be easy though highly interesting since $Z = 50$ is the most pure shell model case without intruder states.

The data need to be refined and extended for Cd and Ag. The extension of the Sn chain is extremely helpful for the test of HF-calculations. There is an increasing need for confident atomic field parameters used in laser spectroscopy. This discussion is most important for rare earth and very heavy elements. With the perspectives of experiments in heavy ion storage rings some fundamental questions might be settled in atomic screening of nuclear moments and polarization of the nuclear field.

REFERENCES

/1/ Proceedings of 4th International Conference on Nuclei far from Stability, Helsingør 1981, ed. P.G. Hansen, O.B Nielsen, CERN 81-09
/2/ E.W. Otten, Nuclear Radii and Moments of Unstable Isotopes, in: Treatise on Heavy Ion Physics, Vol. 8, ed: D.A. Bromley, to be published
/3/ G. Novicki et al., Phys. Rev. Lett. $\underline{39}$, 332 (1977)
/4/ D.A. Eastham et al., J. Phys. $\underline{G12}$, $\overline{205}$ (1986)
/5/ G. Ulm et al., Z. Phys. $\underline{A321}$, $\overline{395}$ (1985)
/6/ H.K. Carter et al., Proc. SPIE $\underline{426}$, 60 (1983)
/7/ C. Thibault et al., Phys. Rev. $\overline{C23}$, 2720 (1981)
/8/ M. Anselment et al., Phys. Rev. $\overline{D3}$, 421 (1986)
   F. Buchinger et al., Phys. Rev. $\overline{C32}$, 2058 (1985)
   D.A. Eastham et al., submitted to Phys. Rev. Lett.
/9/ X. Campi et al., Phys. Rev. $\underline{C22}$, 2605 (1980)
/10/ W. Myers et al. Nucl. Phys. $\overline{A470}$, 107 (1987)
/11/ M. Anselment et al. Phys. Rev. $\overline{C34}$, 1052 (1986)
/12/ J. Eberz et al., Z. Phys. $\underline{A326}$, 121 (1987)
/13/ J. Eberz et al., Nucl. Phys. $\overline{A464}$, 9 (1987)
/14/ T. Kühl et al., Nucl. Instr. Meth. $\underline{B26}$, 419 (1987)
/15/ F. Buchinger et al., Nucl. Phys. $\underline{462}$, 305 (1987)
/16/ U. Dinger et al., to be published
/17/ M. Anselment et al., Nucl. Phys. $\underline{A451}$, 471 (1986)
/18/ U. Dinger et al., submitted to Z. Phys.
/19/ R. Neugart et al., Phys. Rev. Lett. $\underline{55}$, 1559 (1985)
/20/ J. Bounds et al., Phys. Rev. Lett. $\underline{55}$, 2269 (1985)
/21/ G. Ulm et al., Z. Phys. $\underline{A325}$, 247 ($\overline{1986}$)
/22/ K. Wallmeroth et al., Phys. Rev. Lett. $\underline{58}$, 1516 (1987)
/23/ R. Kirchner, Nucl. Instr. Meth. $\underline{B26}$, 235 (1987)
/24/ U. Krönert et al., Appl. Phys. B in press

/25/ K. Heyde et al., Phys. Rep. 102, 291 (1983)
/26/ K. Heyde, private communication (1986)
/27/ A. Arima et al., Prog. Theor. Phys. 12, 623 (1954)
/28/ K. Heyde et al., Phys. Lett. B179, 1 (1986)
/29/ E. Hagn, Phys. Lett. B184, 309 (1987)
/30/ V. Paar, Nucl. Phys. A211, 29 (1973)
/31/ I. Talmi, Nucl. Phys. A423, 189 (1984)
/32/ J. Dobaczewski et al., Phys. Rev. C29, 1540 (1984)
/33/ W. Myers et al., Nucl. Phys. A410, 61 (1983)
/34/ A. Coc et al., Phys. Lett. B163, 66 (1985)
/35/ K. Wendt et al., Z. Phys. D4, 227 (1987)
/36/ D. Berdichevski et al., Z. Phys. A322, 141 (1985)
/37/ H. Sagawa et al., submitted to Nucl. Phys. A
/38/ K. Heyde et al., Nucl. Phys. A466, 189 (1987)
/39/ E. Hagberg et al., Phys. Lett. 78B, 44 (1978)

INVESTIGATION OF NUCLEAR CHARGE RADII AND ELECTROMAGNETIC MOMENTS OF RARE EARTH ELEMENTS BY LASER SPECTROSCOPY

G.D.Alkhazov, A.E.Barzakh, V.P.Denisov, V.S.Ivanov, I.Ya.Chubukov

Leningrad Nuclear Physics Institute, USSR Academy of Sciences, Gatchina, Leningrad district, 188350

N.B.Buyanov, V.S.Letokhov, V.I.Mishin, S.K.Sekatsky, V.N.Fedoseev

Institute of Spectroscopy, USSR Academy of Sciences Troitzk, Moscow district, 142092

ABSTRACT

Isotope shifts and hyperfine structure of optical lines of neodymium isotopes with the mass numbers A=132, 134-142, samarium isotopes with A=138-145, 147, 149, 150, 152, 154, europium isotopes with A=138-151, holmium isotopes with A=152-165, and thulium isotopes with A=156-172 have been studied at the IRIS mass separator laser nuclear facility. For these isotopes, nuclear electromagnetic moments and isotopic changes in mean square charge radii are determined. Characteristic features of the charge radii N-dependence in the regions at $N < 82$ and $88 < N < 94$ are revealed.

INTRODUCTION

The ground-state study of atomic nuclei is an important task of nuclear physics. A description of these states is an attribute of any consistent nuclear theory aspiring to universality. Of special value is experimental information on long isotopic chains.

One of the most effective means of obtaining such information is laser optical spectroscopy technique. It allows determining spins, magnetic dipole and electric quadrupole moments, and isotopic changes in mean square charge radii (MSCR) [1,2]. These nuclear characteristics are derived from the isotope shifts (IS) and hyperfine structure (HFS) of optical lines.

Under the joint research programme of the Leningrad Nuclear Physics Institute of the USSR Academy of Sciences, and the Institute of Spectroscopy of the USSR Academy of Sciences, the IS and HFS for radioactive atoms were measured at the IRIS mass separator facility in Gatchina (near Leningrad) by the highly sensitive laser spectroscopy technique of resonance atomic photoionization [3]. Rare earth element isotopes were the subject of the investigation. For nuclei in this region, of interest are the following points:

(1) neutron shell effect in charge radii in the neighbourhood of the magic neutron number N=82;
(2) transition from spherical to deformed nuclei at $N < 82$;
(3) characteristic features of the transition from spherical to strongly deformed nuclei in the vicinity of the "critical" number N=90 for nuclei with even and odd proton numbers Z;
(4) influence of the proton shell closure at Z=64 on nuclear properties.

By now we have studied long chains of Eu [3,4,5], Tm [6,7,8], Sm [9], Ho [10] and Nd [10] isotopes. The present contribution is a summary of the experimental **data** obtained and is aimed at offering a brief outline of our studies.

EXPERIMENTAL METHOD

The resonance atomic photoionization technique and experimental setup have been described elsewhere[3,4,5,7]. Measurements were carried out as follows. The isotopes under study were produced in the tantalum target of the mass separator by irradiation with 1-GeV protons. The ion beam of a required isotope was directed from the mass separator into a hot tantalum crucible. A collimated atomic beam emerged in the opposite direction from the crucible and was crossed at right angles by three dye laser beams. To increase the efficiency of photoionization, the laser beams crossed the atomic beam several times using successive reflections from two plane mirrors. The laser beam frequencies were tuned to the frequencies of three consecutive atomic transitions. As a result of this three-step excitation, the atoms were raised to an autoionization state, and the photoions resulting from the autoionization of this state were detected with an electron multiplier. The optical spectrum was represented in terms of the number of photoions as a function of the first-step excitation laser frequency. Some of the laser radiation was split off and directed into a separate vacuum chamber (reference chamber) to obtain the HFS photoionization spectra of stable isotopes. These spectra were utilized for frequency scale calibration. The dye lasers were pumped by three pulsed copper-vapor lasers, the pulse repetition rate being 10 kHz. For the first excitation step, a CW single-frequency laser (Spectra Physics Mod.380) was used. The output from this continious laser was amplified in a pulsed laser amplifier pumped by one of the Cu-vapor lasers. The bandwidth of the amplified radiation was 30 - 50 MHz.

The observed widths (100 - 300 MHz) of the optical

lines under study were determined primarily by the residual Doppler broadening due to the angular divergence of the atomic beam and by the saturation of the atomic transitions involved. The minimum mass-separated ion fluxes still sufficient for the measurement to be possible were $10^4 - 10^5$ ions/s.

Nuclear spins I, dipole magnetic moments μ, quadrupole electric moments Q_s and isotopic changes in charge mean square radii $\Delta \langle r^2 \rangle^{AA'}$ were derived from the measured IS and HFS. The required atomic constants were estimated with the standard semiempirical methods or were found from the known values of electromagnetic moments and isotopic changes in charge radii for some isotopes. Details of the treatment of the data and numerical values of the measured quantities may be found in ref.[11].

RESULTS AND DISCUSSION

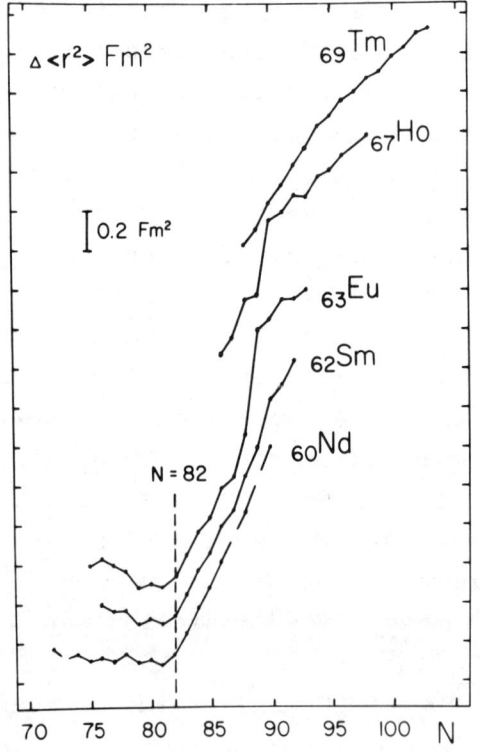

Fig.1. Values of $\Delta \langle r^2 \rangle$ as a function of N for Nd, Sm, Eu, Ho and Tm isotopes.

Figure 1 shows $\Delta \langle r^2 \rangle$ curves for the isotopes of the elements under study, plotted taking account of the data reported in [12-15] for stable and long-lived isotopes and also those given in [16] for $^{140-153}$Eu. The neutron shell effect, that is different rates

of charge radius changes on both sides of the number
N=82, is clearly seen in the isotopic chains of Nd, Sm,
and Eu. The effect is apparently mainly due to an increase in nuclear zero-point vibrations both at $N<82$
and at $N>82$. At $88>N>82$ the above three isotopic
dependences of charge radii coincide practically with
each other. At $N<82$ the isotopic behaviour of charge
radii features a strong Z-dependence.

Figure 2 presents isotopic changes in MSCR at $N<82$
for the europium, samarium, and neodymium isotopes studied by us and also for the barium, cesium and xenon isotopes investigated earlier using other experimental methods [1]. The charge radii decrease as neutron number N
becomes less for Xe (Z=54), Cs (Z=55) and Ba (Z=56)
isotopes, remain practically constant down to N=72
for Nd (Z=60) isotopes and start to increase for Sm
(Z=62) and Eu (Z=63) isotopes in the region at $N<76$.
This strong Z-dependence in the isotopic trends of charge radii is evidently connected with different deformations of the nuclei considered. We have estimated
rms deformations of these nuclei using the data on
$\Delta \langle r^2 \rangle$ and the droplet model. The values obtained
agree fairly well with the quadrupole deformation parameters deduced from the

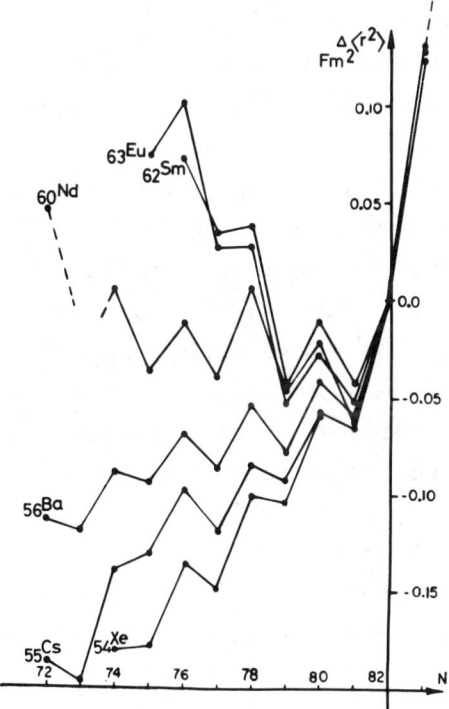

Fig.2. Values of $\Delta \langle r^2 \rangle$ as a function of N at $N<82$ for Sm, Eu, and Nd (this work) and Ba, Cs, and Xe [1].

lifetimes and with those obtained from the Grodzin's phenomenological formula.

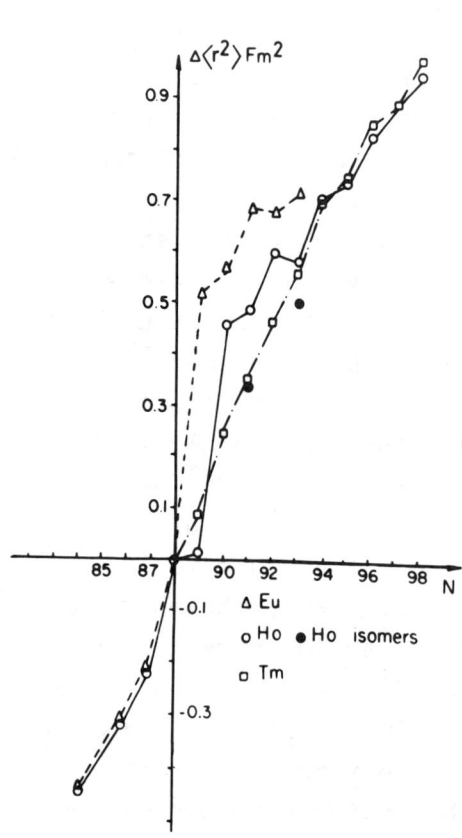

Fig.3. Values of $\Delta \langle r^2 \rangle$ as a function of N at $88 < N < 94$ for Eu, Ho, and Tm.

Figure 3 presents $\Delta \langle r^2 \rangle$ curves for studied in the present work proton odd isotopes of Eu, Ho and Tm in the nuclide region near "critical" neutron number N=90. Similar data for proton even isotopes were obtained previously in other works [1]. The isotopic dependences of MSCR for nuclei with the proton number close to the magic Z=64 (Sm, Z=62; Gd, Z=64; and Dy, Z=66) were found to undegro a sudden change when passing from the nuclei with N=88 to those with N=90 associated with a sharp change in deformation. This change in the isotopic dependences decreases as one moves away from Z=64 and vanishes entirely for Yb (Z=70) and Ba (Z=56) [1].

Figure 4 shows differential changes in MSCR for even-even [1] and for the odd-even nuclei studied by us. For odd-nuclei, the general trend of transition from a smooth to a stepwise change in MSCR at N= 88 - 90, as one approaches the "magic" number Z=64, remains, but as distinct from even-Z nuclei, it is not smooth in Z.

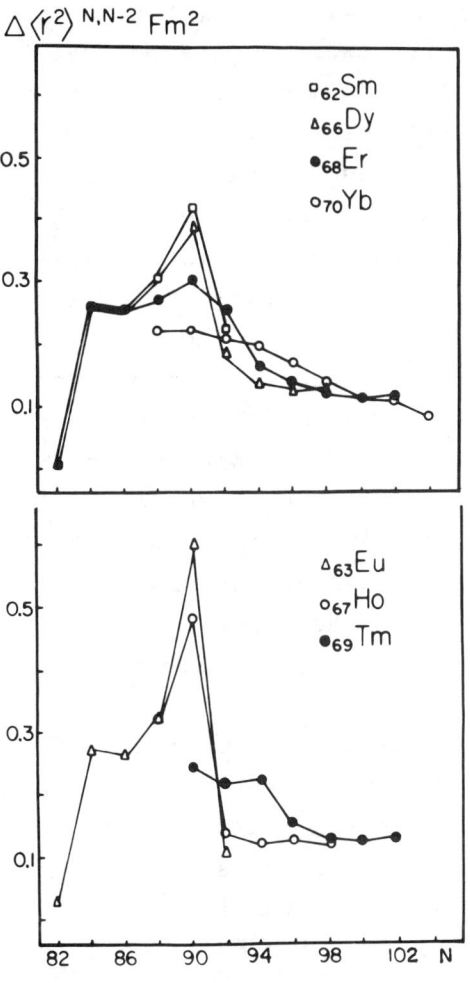

Fig.4. Differential dependence of $\Delta \langle r^2 \rangle^{N,N-2}$ versus neutron number for Dy, Er, Yb[1], and Sm[11] and similar data (our results) for odd-proton nuclei Eu, Ho, and Tm.

While there is no sudden change in MSCR for thulium isotopes (Z=69), the next odd-Z isotopic chain (Ho, Z=67) exhibits a clearly manifest jump exceeding the maximum jump observed for even-Z nuclei (samarium isotopes, Z=62). The isotopic MSCR dependences for odd-Z nuclei are less smooth than for their even-Z neighbours. For thulium isotopes charge radius dependence features a kink at N=94 marking an acceleration of the decrease of MSCR with decreasing N. This irregularity correlates with the changes of the spins and magnetic moments of the respective odd-even nuclei in the thulium isotopic sequence.

The comparison between the results of macroscopic-microscopic calculations and the measured values of I, μ, and Q_s for odd-odd thulium isotopes with $N < 94$ points to an irregularity in the sequence of filling of proton states [7,8]. We have also calculated ground-state properties of thulium isotopes in the framework of the Hartree-Fock method with the Skyrme effective interaction (with forces S3 and SKM'). The calculated Q_s-values are compared with the experimental ones in fig.5. It is seen that the theory describes the A-dependence of the quadrupole moments fairly well when using both types of the effective interactions. We note that unfortunately the experimental uncertainties in the Q_0-values are relatively large. For this reason the comparison of the calculated and experimental Q_0-values is not decisive. However, as follows from fig.6, the theory with S3-force does not reproduce well enough the isotopic behaviour of MSCR. The force SKM' gives better results at N>94. It also allows to reproduce qualitatively the irregularity in the MSCR decrease with N at N<94, though the calculated values of MSCR decrease sharper than the experimental ones. The discrepancy between the theory and the experiment is likely to be due to the gradual enhancement of the zero-point vibrations as the perma-

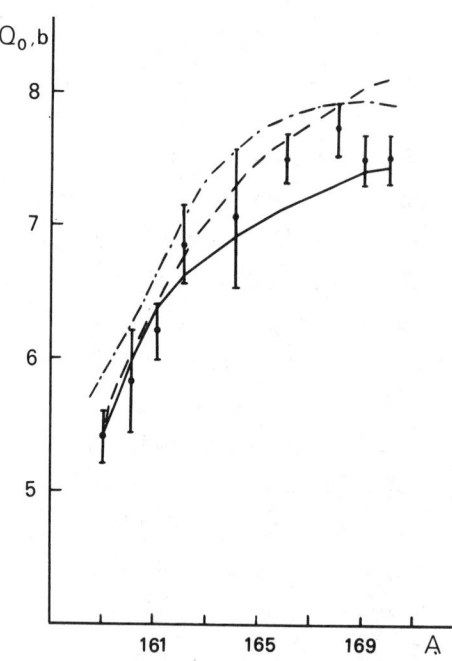

Fig.5. Comparison of the experimental intrinsic quadrupole moments for Tm isotopes with predictions of the macroscopic-microscopic calculations (solid line) and of the deformed HF-method with the S3 force (dashed line) and the SKM' force (dash-dotted line).

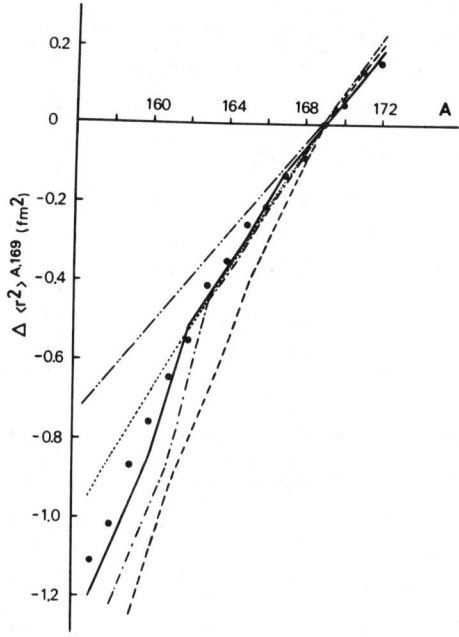

Fig.6. Comparison of the experimental changes in MSCR in Tm isotopes (circles) with predictions of the macroscopic-microscopic calculations (solid line) and of the deformed HF-method with the S3 force (dashed line) and the SKM' force (dash-dotted line). Predictions of the spherical HF-method (dotted line) and the droplet model (double dot-dashed line) are also presented.

nent deformation decreases. The marked difference between the isotopic changes in MSCR calculated with the forces S3 and SKM' is caused by their different contributions to the deformation. Fig.6 shows also the results of calculations of the nuclear charge radii using the spherical HF approach and that of the spherical droplet model. A considerable disagreement between the results of these calculations is seen. However, the reason of this discrepancy is not clear.

A distinctive peculiarity in MSCR isotopic dependences in the region of deformed nuclei at $88 < N < 94$ is seen in Fig.3. At $82 < N < 88$, as well as at $N > 94$, the isotopic dependences of MSCR for various isotopic chains (Eu, Ho, and Tm) are close to one another. In the region at $88 < N < 94$, the behaviour of MSCR as a function of N exhibits a strong Z-dependence.

Among other interesting features of odd-Z nuclei revealed by us are the opposite signs of the odd-even staggering in the charge radii of Eu and Ho isotopes and a marked weakening of the odd-even staggering for Tm isotopes with the same number of neutrons. It may be also noted that the sudden change in the isotopic dependence of the holmium MSCR occurs in the transition from N=89 to N=90 and not from N=88 to N=89 as is the case with Eu.

CONCLUSION

A high-efficient high-resolution laser experimental setup is being operated at the mass-separator of IRIS facility making it possible to investigate the ground and long-lived isomeric states of atomic nuclei by measuring the IS and HFS of atomic lines using the resonance atomic photoionization technique. In studying rare earth element isotopes the technique proved to be more effective than some other methods used for the purpose previously. During the past three years we have measured isotopic changes in MSCR, spins, and electromagnetic moments for some seventy radioactive isotopes and isomers. Our results together with those obtained by other authors [1] allowed new interesting effects to be revealed: (1) a strong Z-dependence of the isotopic behaviour of MSCR at $N < 82$ and in the region at $88 < N < 94$, (2) correlation between irregularities in the isotopic dependence of MSCR and in the behaviour of magnetic moments for thulium isotopes, and (3) transitions from a smooth to a stepwise change in MSCR of odd-Z nuclides in the region at N= =88 - 90 when approaching the magic number Z = 64, the character of this transition with Z being sharper than in even - Z nuclides. Our calculations for thulium isotopes show that the data on $\Delta \langle r^2 \rangle$ are rather useful in selecting the type of effective interaction within the framework of the Hartree-Fock method [17]. Further experimental and theoretical ground-state studies may provide a deeper insight into the nuclear structure of rare earth elements.

REFERENCES

1. E.W.Otten, International school-seminar on heavy ion physics. Dubna 1983, p.158.
2. V.I.Balykin et al., UFN (Sov.) 132, 293 (1980).
3. A.N.Zherikhin et al., ZhETF (Sov.) 86, 1249 (1984).
4. V.N.Fedoseev et al., Optics Communications 52,24(1984).
5. G.D.Alkhazov et al., Preprint LNPI N 1161, L-d, 1986.
6. A.E.Barzakh et al., Preprint LNPI N 1079, L-d, 1985.
7. G.D.Alkhazov et al., Preprint LNPI N 1145, L-d, 1985.
8. G.D.Alkhazov et al., Izv.AN SSSR,ser.fiz.50,2366(1986).
9. G.D.Alkhazov et al., Preprint N 1213, L-d, 1986.
10. G.D.Alkhazov et al., Preprint N 1283, L-d, 1987.
11. G.D.Alkhazov et al., Preprint LNPI N 1309,L-d, 1987.
12. H.Brand, B.Seibert, A.Steudel, Z.Phys.A296,281(1980).
13. W.H.King, A.Steudel, M.Wilson, Z.Phys. 265,207(1973).
14. G.D.Alkhazov et al., Atomic masses and fundamental constants 7 (O.Klepper, Darmstadt, 1984), p.327.
15. D.A.Eastham et al., Atomic masses and fundamental constants 7 (O.Klepper, Darmstadt, 1984), p.322.
16. S.A.Ahmad et al., Z.Phys.321,p.35,(1985).
17. A.E.Barzakh, V.E.Starodubsky, Yad.Fiz.(Sov.) 45, 45 (1987).

NEW TECHNIQUES AND RESULTS OF COLLINEAR LASER SPECTROSCOPY RADON, XENON AND HOLMIUM

R. Neugart, E. Arnold, W. Borchers, W. Neu,
G. Ulm and K. Wendt
Institut für Physik, Universität Mainz, D-6500 Mainz
and the ISOLDE Collaboration, CERN, CH-1211 Geneva

ABSTRACT

The moments and radii of extremely neutron-rich radon and xenon isotopes have been investigated by collinear fast-beam laser spectroscopy. The experiment has become possible owing to the breakthrough in sensitivity achieved by the detection of optical pumping via state-selective collisional ionization. The results are dicussed within the systematics of nuclear shapes around N=136 and N=90. In the latter context measurements on the neutron-deficient holmium isotopes give new details about the development of nuclear deformation.

INTRODUCTION

The success of laser spectroscopy in providing moments and radii of nuclei far from stability has provoked new efforts to gain significant factors in sensitivity. A few promising approaches are based on the collinear fast-beam technique which - owing to its conceptual simplicity - can be applied to nearly any beam from a conventional on-line isotope separator.

The principle is to measure hyperfine structures and isotope shifts in a laser-induced optical transition from a properly prepared state of the ion or neutral atom. Narrow absorption linewidth and Doppler-tuning ensure the desirable resolution and precision. Concerning sensitivity a crucial point is the detection. The universal concept of counting fluorescence photons suffers from scattered laser light, collisional excitation and radioactivity. This is where particle detection schemes come into play: The problems will be solved, if one succeeds to selectively count the ions that have interacted with the laser light. An obvious procedure would be stepwise laser ionization[1,2] which is efficiently performed by use of pulsed high-power lasers. For continuous-beam conditions, this involves serious duty-cycle losses. On the other hand, cw lasers give low efficiency and optical pumping into levels outside the ionization ladder. Therefore, alternatives have been sought in exploiting the effects of optical pumping:

(i) The RADOP technique - i.e. radioactive detection of optically induced nuclear polarization - has been adapted to the fast-beam conditions and applied to the lithium isotopes of which the exotic ^{11}Li deserves special interest[3].

(ii) The state selectivity of charge-transfer neutralization has been used to detect very sensitively the laser excitation of Sr$^+$ and Ba$^+$ beams[4,5].

(iii) For neutral beams the collisional ionization can discriminate between atoms in different states[6], provided that the cross-sections are sufficiently different.

We report the application of such a scheme to the very neutron-rich isotopes of radon and xenon, whereby we have gained several orders of magnitude in sensitivity compared to fluorescence detection.

ION-DETECTED SPECTROSCOPY ON RARE GASES

It is a common feature of the rare-gas spectra that the ionization energy is more than 10 eV for the ground state and only about 4 eV for the metastable first excited state in which one electron is promoted from the closed np^6 valence shell to the next higher $(n+1)s$ shell. This involves a considerable difference in the cross-sections for electron stripping. Fig.1 shows the relevant part of the energy level diagram for the example of radon. A fast beam of metastable atoms in the J=2 state of the $6p^57s$ configuration - designated $7s[3/2]_2$ - is prepared in the charge-transfer neutralization of the original ion beam with caesium vapour. Laser excitation to $7p[3/2]_2$ at 705.5 nm depopulates the metastable level and pumps the

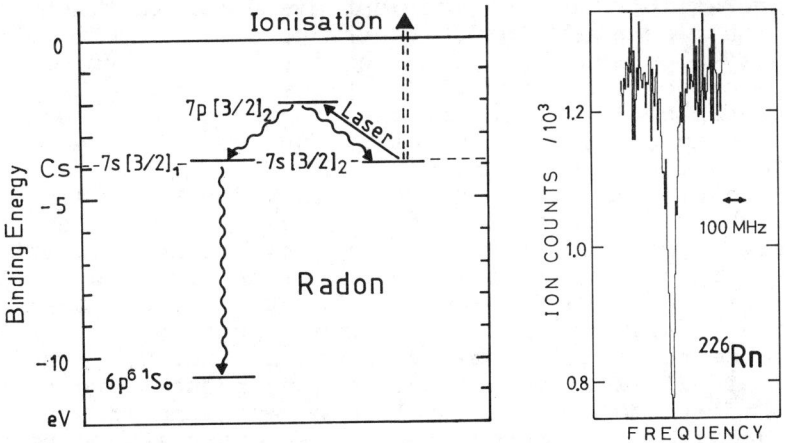

Fig. 1. a) Partial energy-level diagram of radon. The caesium ground-state energy is close to the metastable $7s[3/2]_2$ level. b) Resonance signal of ^{226}Rn.

atom via $7s[3/2]_1$ into the low-lying $^1S_0$ ground state. In passing the beam through a gas target, one can ionize predominantly the metastable atoms and thus detect the optical pumping as a flop-out signal on the ion current.

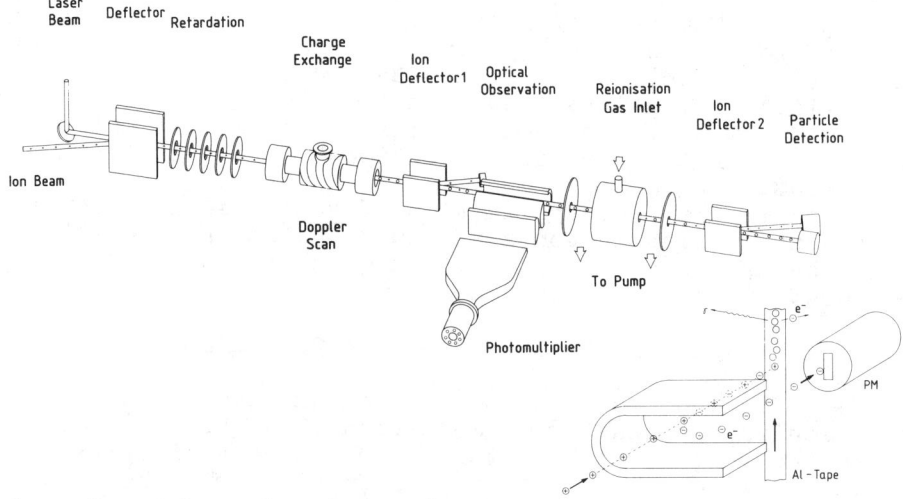

Fig. 2. Schematic view of the experimental setup. The ion detection system is shown in detail.

Fig.2 gives a schematic view of the experimental setup used at ISOLDE. The front part including fluorescence detection is essentially identical with the standard apparatus[7] that has been used extensively to study nuclear moments and radii. The differentially pumped stripping target has an effective thickness of 10 cm at a pressure of 10^{-3} to 10^{-2} mbar optimized for the individual gas. Ions created in this target are detected by a secondary-electron multiplier whose cathode is formed by a remote-controlled moveable metallic tape. This is indispensable for removing the considerable background from the radioactive decay of the nuclei collected on the detector during the experiment.

Another important source of background consists in the isobaric contaminations of the very weak beams of mass-separated radioactive atoms: Only to some extent the neutralization and ionization reactions are specific for the particular chemical element. For the "cold" plasma ion source of ISOLDE we have found that after careful outgassing the beams of heavy Rn and Xe isotopes are very clean, if all less volatile reaction products are trapped in a water-cooled transfer line between the ThC_2 or UC_2 target and the ion source. Before we discuss some results of the initial experiments on $^{223-226}$Rn and $^{141-146}$Xe, we shall demonstrate the sensitivity for the example of the discovered new isotope, ^{146}Xe.

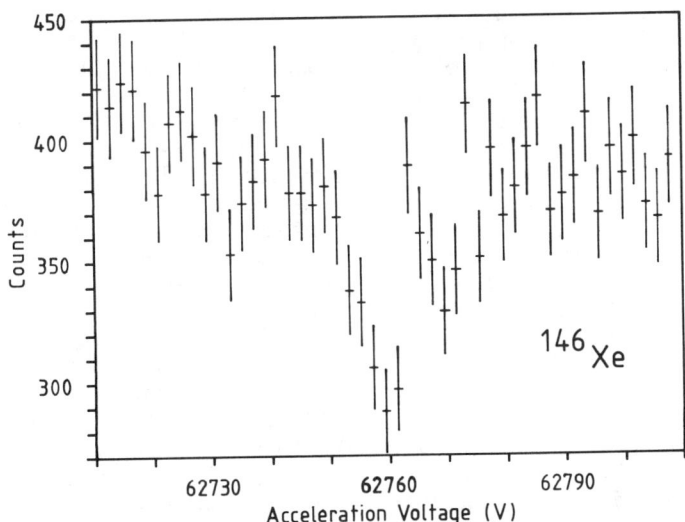

Fig. 3. Resonance signal of ^{146}Xe from a beam of originally 400 atoms/s.

The resonance of Fig.3 represents the transition $6s[3/2]_2 - 6p[3/2]_2$ for the I=0 isotope ^{146}Xe, measured relative to ^{136}Xe thus yielding the isotope shift. The total measuring time over a 2.5 times larger scanning range was 35 min with a beam of about 400 atoms/s from ISOLDE. The signal efficiency is given by a 50% total transmission of the beam line and our apparatus, 50% charge exchange and 10% ionization efficiencies and a 50% flop-out signal on the ion current. These numbers refer to the 2×10^{-3} mbar Cl_2 target used in most of our experiments. A constant background of about 10 counts/s is ascribed to the general radioactivity level of the experimental area and should be removed by more careful shielding. It is obvious that even without significant improvements mass-separated beam intensities well below 100 atoms/s are sufficient for a measurement within reasonable time. This should be compared to the fluorescence detection limit for rare gases which is about 10^6 atoms/s in the far-red transitions that are accessible from the metastable state.

RESULTS ON RADON AND XENON

For radon the present measurements on $^{223-226}$Rn continue the earlier work on the abundantly produced isotopes $^{202-212, 218-222}$Rn for which preliminary results have been reported[8]. The particular interest in this continuation arises from the theoretical and empirical evidence for strong octupole correlation effects in the Ra-Th region around neutron number N=136. The Rn nuclei are probably marking the boundary of this region. Indeed the spins, moments, and the odd-even staggering of the isotope shifts indicate unusual ground-state properties for ^{219}Rn and ^{221}Rn. We shall briefly discuss how this trend develops towards the heavier isotopes. Table 1 gives the spins, the hyperfine interaction parameters A and B in the $7s[3/2]_2$ state, and the deduced magnetic dipole and electric quadrupole moments for the odd-A isotopes between ^{218}Rn and ^{226}Rn. The magnetic moments are slightly recalibrated compared to ref.8, as a recent direct g_I-factor measurement[9] on ^{209}Rn gives a reference value for the evaluation according to

$$A = \mu\, H_e(0)/IJ \qquad (1)$$

where $H_e(0)$ is the magnetic hyperfine field.

Table 1: Spins and moments of the odd-A Rn isotopes. The magnetic moments include a 1% error for the hfs anomaly, and the quadrupole moments are not corrected for the Sternheimer effect.

	I	A (MHz)	B (MHz)	$\mu(\mu_K)$	Q_s(b)
219	5/2	-564.9(1.2)	-5316(10)	-0.442(5)	0.93
221	7/2	-18.4(6)	2184(5)	-0.020(1)	-0.38
223	7/2	-708.5(3)	-4564(3)	-0.776(8)	0.80
225	7/2	-635.0(8)	-4777(9)	-0.696(8)	0.84

There has been no rigorous theoretical interpretation of the ^{219}Rn and the ^{221}Rn moments, but it seems that the transitional nature of these nuclei, together with a weakened tendency of parity-mixing is responsible for changes in the neutron states from radium[10,11] to radon. In contrast to ^{221}Ra, the isotone ^{219}Rn with the same spin I=5/2 has a magnetic moment close to the prediction of reflection symmetric particle-rotor calculations[12]. However, the value of the I=7/2 for ^{221}Rn, different from I=3/2 for ^{223}Ra, is difficult to interpret in combination with the almost zero magnetic moment and the small negative quadrupole moment.

For the isotopes ^{223}Rn and ^{225}Rn it is surprising to find again the spin I=7/2 with moments that are completely different from ^{221}Rn. They can be accounted for by assuming a 7/2- orbital based on the $j_{15/2}$ configuration. The isotones ^{225}Ra and ^{227}Ra, on the contrary, have I=1/2 and I=3/2 and are interpreted within the reflection-asymmetric scheme.

This qualitative difference is further manifested in the odd-even staggering of the mean square charge radii. An inverted staggering - i.e. larger $<r^2>$ of the odd-N isotopes compared to their even-N neighbours - is observed for $^{220-228}$Ra, and we have found arguments to interpret this behaviour as an effect of the octupole deformation: The reflection-asymmetric shape is stabilized by polarization due to the odd neutron. It is thus an effect in $<\beta_3^2>$ counteracting the normal staggering which is for instance found in the light Ra and Rn isotopes below N=126. In Fig.4 this is plotted in the form of differences

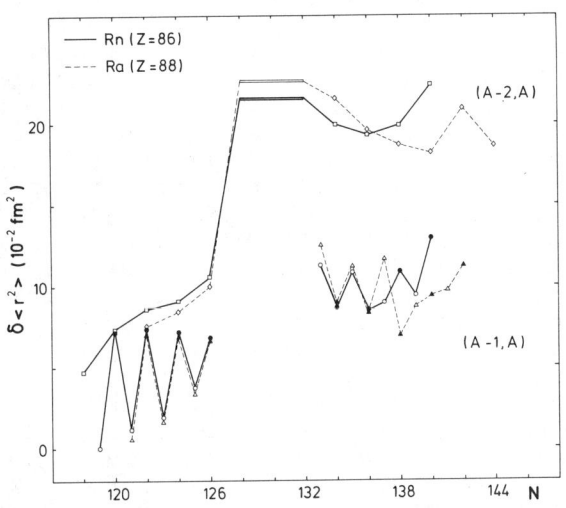

Fig. 4. Brix-Kopfermann plot of $\delta<r^2>$ for Rn (solid lines) and Ra (dashed lines). The upper curves correspond to the addition of neutron pairs and the lower ones for the addition of single neutrons (for symbols see text).

$\delta<r^2>^{A-1,A}$ between odd A and even A-1 (open symbols) and between even A and odd A-1 (full symbols). The striking similarity of both zig-zag curves disappears at N=135 (^{223}Rn and ^{225}Ra), where for radon also the nuclear moments indicate a change in the ground-state structure. As the normal staggering is restored for ^{223}Rn and ^{225}Rn, it appears natural to interpret them by reflection-symmetric configurations and to assume $\beta_3 \neq 0$ only for ^{221}Rn and ^{219}Rn.

The recent xenon measurements include a few neutron-deficient test candidates and the systematically studied isotopes $^{136-146}$Xe ($^{136-140}$Xe from standard fluorescence spectroscopy). These neutron-rich nuclei with $82 \leq N \leq 92$ belong to the classical region of transition from spherical to deformed nuclear shapes which is crossed by the stability line along the lighter rare-earth elements.

Xe (Z=54) is the element closest to the magic proton number Z=50 for which isotopes around N=90 can be reached, and it remains to be analyzed to what extent these nuclei still develop deformed shapes like the neighbouring Ba (Z=56)[7].

SYSTEMATICS IN THE RARE-EARTH REGION: HOLMIUM

For the rare-earth elements between Nd and Dy - i.e. around Z=64 - the onset of deformation occurs in a sharp transition between N=88 and 90. Here the systematics of moments and radii mainly covers the even-Z elements[13,14] and europium[15-17] of which the two stable isotopes happen to fall on these two neutron numbers (see also the discussion by G.D. Alkhazov[18] about recent results of the Leningrad-Moscow collaboration). The measurements including neutron-deficient isotopes show that the shape transition for Z=63 is more pronounced than for all even-Z neighbours. This is likely to be connected with the nearby Z=64 subshell closure. From the results on even-Z elements one also concludes that holmium lies at the boundary between the sharp and the smooth onset of deformation. Therefore, measurements on holmium should help to clarify the influence of the odd proton which for most of the isotopes can be found in a high-spin ground state or a low-spin isomer.

We have studied $^{151-165}$Ho for which an adequate technique is the conventional photon-detected collinear laser spectroscopy. The experiment was performed in the transition $4f^{11}6s^2$ $^2F_{15/2}$ -> $4f^{11}6s6p(15/2,1)_{17/2}$ of neutral holmium at 410.4 nm. The yields from spallation in a thick Ta-foil target were between 10^6 and 10^{10} atoms/s.

Fig.5 gives the change in the mean square charge radii $\delta<r^2>$ for $^{151-165}$Ho in comparision with the curve for $^{140-156}$Eu (ref. 16,17), both evaluated from isotope shift measurements. Despite the striking similarity the two curves indicate a remarkable difference in the structure of the N=89 isotones: ^{152}Eu is strongly deformed in the I=3 ground state, whereas in ^{156}Ho both the I=4 ground state and the I=1 isomer are on the near-spherical branch. Beyond N=90 the initial deformation is smaller and continues to increase towards the heavier isotopes. A particularly interesting feature is the isomerism which gives rise to large shape differences between the I=5 ground states and the I=2 isomers in ^{158}Ho and ^{160}Ho. The isomer shifts below N=90 are small. While the onset of deformation for the holmium ground states occurs suddenly between N=88 and 90, the deformation for the isomeric states increases gradually with the neutron number.

All measured spins are compiled in Table 2. They are readily deduced from the spectra, because the number of

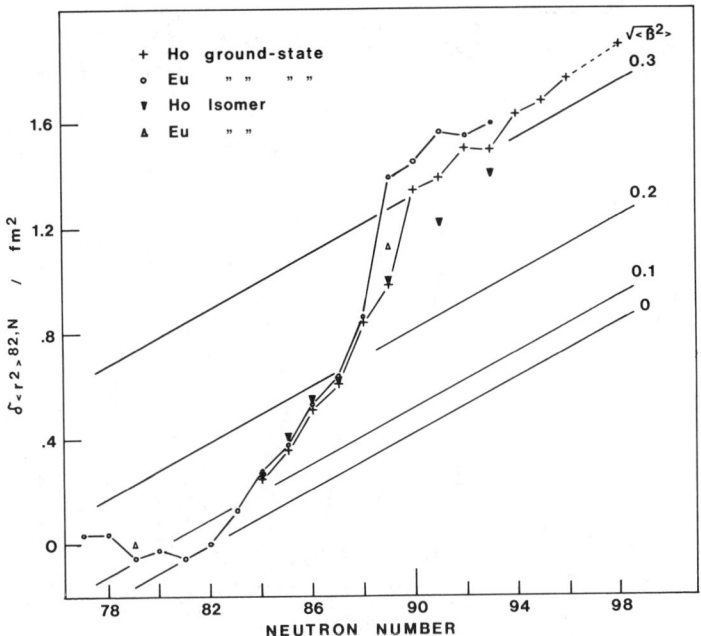

Fig. 5. Plot of $\delta\langle r^2\rangle^{82,N}$ for Eu and Ho. Lines of constant deformation are calculated from the droplet model.

Table 2: Spins of the ground states (I) and isomeric states (I^m) of Ho.

A	151	152	153	154	155	156	157
I	11/2	9	11/2	8	5/2	4	7/2
I^m	1/2	2	1/2	2	-	1	-

A	158	159	160	161	162	163	165
I	5	7/2	5	7/2	1	7/2	7/2
I^m	2	-	2	-	-	-	-

strong hyperfine structure components ($\Delta F=+1$), observed in the $J=15/2 \rightarrow J=17/2$ transition, is equal to $(2I+1)$ for $I<8$. In a few cases the determined spin differs from earlier assignments. As an example we give the hyperfine structure of ^{156}Ho (N=89) in Fig.5 where the ground state spin is obviously I=4 and the three small components of the I=1 isomer are marked by arrows.

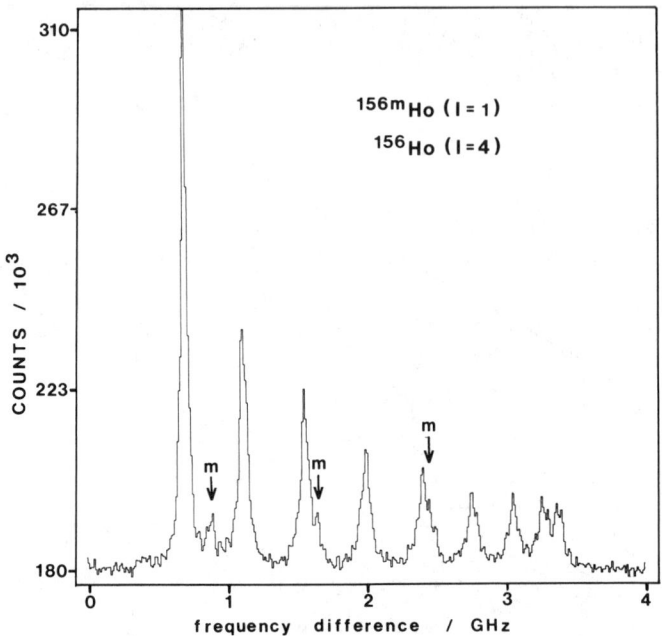

Fig. 6. Hyperfine spectrum for ^{156}Ho and ^{156m}Ho.

The interpretation of the proton configurations is straightforward for the odd-A isotopes whose moments are consistent with spherical shell model assignments of $\pi h_{11/2}$ and $\pi s_{1/2}$ for ^{151}Ho and ^{153}Ho, $\pi d_{5/2}$ for ^{155}Ho and the rather pure $\pi[532\ 7/2]$ Nilsson configuration[19] for the deformed $^{157-165}$Ho.

This work has been funded by the German Federal Minister for Research and Technology (BMFT) under the contract number 06 MZ 458 I.

REFERENCES

1. G.S. Hurst, M.G. Payne, S.D. Kramer, and J.P. Young, Rev. Mod. Phys. 51, 767 (1979).
2. V.I. Balykin, G.I. Bekov, V.S. Letokhov, and V.I. Mishin, Sov. Phys. Usp. 23, 651 (1980).
3. E. Arnold, J. Bonn, R. Gegenwart, W. Neu, R. Neugart, E.W. Otten, G. Ulm, and K. Wendt, contribution to this conference.
4. R.E. Silverans, G. Borghs, P. De Bisschop, and M. Van Hove, Hyperfine Interactions 24-26, 181 (1985).
5. F. Buchinger, E.B. Ramsay, R.E. Silverans, P. Lievens, E. Arnold, W. Neu, R. Neugart, K. Wendt, and G. Ulm, contribution to this conference.

6. R. Neugart, W. Klempt, and K. Wendt, Nucl. Instr. & Meth. B17, 354 (1986).
7. A.C. Mueller, F. Buchinger, W. Klempt, E.W. Otten, R. Neugart, C. Ekström, and J. Heinemeier, Nucl. Phys. A403, 234 (1983).
8. W. Borchers, R. Neugart, E.W. Otten, H.T. Duong, G. Ulm, and K. Wendt, Hyperfine Interactions 34, 25 (1987).
9. M. Kitano, F.P. Calaprice, M.L. Pitt, J. Clayhold, W. Happer, M. Kadar-Kallen, M. Musolf, G. Ulm, K. Wendt, T. Chupp, J. Bonn, R. Neugart, E.W. Otten, and H.T. Duong, submitted to Phys. Rev. Letters.
10. G.A. Leander and R.K. Sheline, Nucl. Phys. A413, 375 (1984).
11. W. Nazarewicz, P. Olanders, I. Ragnarsson, J. Dudek, G.A. Leander, P. Möller, and E. Ruchowska, Nucl. Phys. A429, 269 (1984).
12. S.A. Ahmad, W. Klempt, R. Neugart, E.W. Otten, K. Wendt, and C. Ekström, Phys. Lett. 133B, 47 (1983).
13. R. Neugart, in Lasers in Nuclear Physics (C.E. Bemis, jr., and H.K. Carter), Nuclear Science Research Conference Series, Vol. 3 (Harwood, Chur-London-New York, 1982), p. 231.
14. R. Neugart, K. Wendt, S.A. Ahmad, W. Klempt and C. Ekström, Hyperfine Interactions 15/16 181 (1983).
15. V.N. Fedoseev, V.S. Letokhov, V.I. Mishin, G.D. Alkhazov, A.E. Barzakh, V.P. Denisov, A.G. Dernyatin and V.S. Ivanov, Optics Commun. 52, 24 (1984).
16. K. Dörschel, W. Heddrich, H. Hühnermann, E.W. Peau, W. Wagner, G.D. Alkhazov, E.Ye. Berlovich, V.P. Denisov, V.N. Panteleev and A.G. Polyakov, Z. Phys. A317, 233 (1984).
17. S.A. Ahmad, W. Klempt, C. Ekström, R. Neugart, and K. Wendt, Z. Phys. A321, 35 (1985).
18. G.D. Alkhazov et al., contribution to this conference.
19. C. Ekström and I.-L. Lamm, Physica Scripta 7, 31 (1973).

LASER MEASUREMENTS OF RADII AND MOMENTS OF BARIUM NUCLEI NEAR THE PROTON DRIP LINE

D.A. Eastham, J.R.H. Smith, J. Groves, D.D. Warner and D.W.L. Tolfree
Science and Engineering Research Council, Daresbury Laboratory,
Warrington WA4 4AD, United Kingdom

S.A. Wells, J.A.R. Griffith and D.E. Evans
Department of Physics, University of Birmingham, Birmingham B15 2TT,
United Kingdom

M.J. Fawcett, I.S. Grant and J. Billowes
Department of Physics, University of Manchester, Manchester M13 9PL,
United Kingdom

P.M. Walker
Department of Physics, University of Guildford, Guildford GU2 5XH,
United Kingdom

ABSTRACT

A new technique of laser spectroscopy has been used to measure the magnetic dipole and electric quadrupole moment of ^{121}Ba, and the r.m.s. charge radii of ^{120,121}Ba. The results are discussed in terms of the unified model.

EXPERIMENTAL

A new technique[1] of laser spectroscopy, FACS (fluorescent atom coincidence spectroscopy), has been developed at the NSF, Daresbury Laboratory, to measure moments and r.m.s. charge radii of radioactive nuclei. Using this technique it is now possible to make measurements with as little as 50 atoms/ions s^{-1} intersecting the laser beam and this has considerably extended the range of accessible nuclei produced in heavy ion reactions. The basic experimental arrangement is shown in fig. 1. A laser beam is collinear with the mass separated beam from the Daresbury On-line Isotope Separator DOLIS. Laser light is scattered from the fast moving atoms or ions and detected with a series of photomultipliers. If only fluorescent light is detected then a limiting sensitivity in the arrangement is about 10^4-10^5 atoms/ions s^{-1}. A considerable improvement in sensitivity is obtained by associating each scattered photon with a particular atom or ion (photon tagging) which is achieved by detecting the atoms/ions further downstream from the scattering region. When only photon-atom/ion concidence events (after allowing for the atom/ion transit time) are recorded, most of the scattered light background is eliminated and the sensitivity is improved by 2 to 3 orders of magnitude. The technique has been applied to strontium ions[2] and atoms[3] and more recently to barium ions using the strong λ = 460.7 nm ($^2S_{1/2}$ to $^2P_{3/2}$) transition in BaII.

Isotopes of barium near the proton drip line were produced by bombarding ^{92,94}Mo with 160-190 MeV ^{32}S beams from the NSF

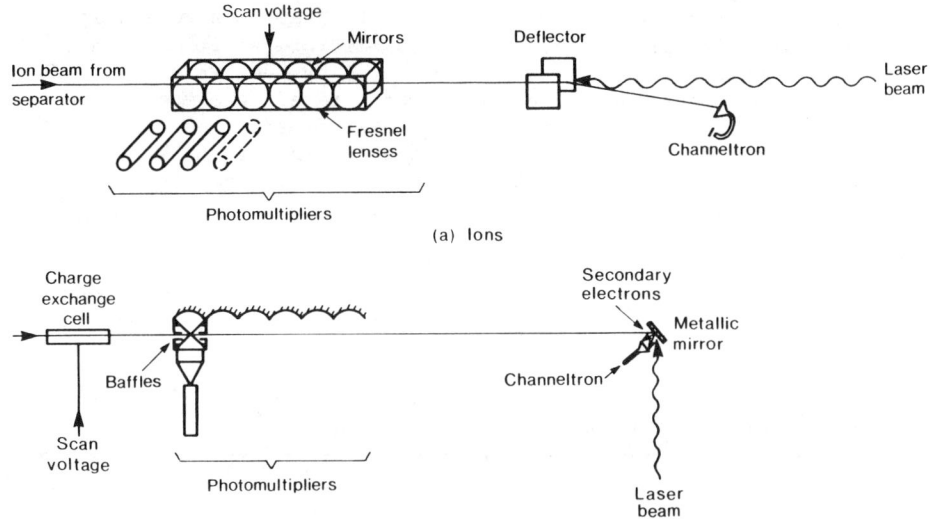

Fig.1. Arrangement for coincidence spectroscopy.

accelerator. Figure 2 shows resonance spectrum for the two barium isotopes studied so far. The resonance condition is obtained by varying the voltage on the light collector whilst keeping the laser frequency fixed at a wavelength close to the barium transition. In this way the velocity of the ions was scanned across the region where the Doppler shifted frequency matched the ionic transition frequency.

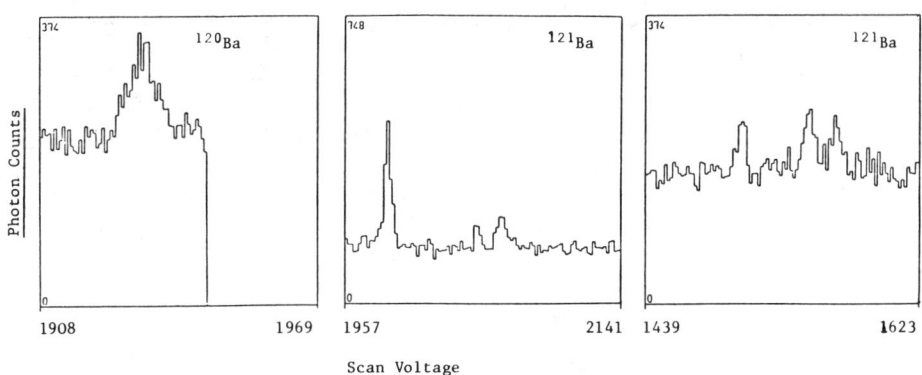

Fig.2. Barium coincidence spectra.

RESULTS

From the peak positions the isotope shifts of ^{120,121}Ba relative to ^{124}Ba and the hyperfine splittings for ^{121}Ba were used to calculate the changes in mean square radius and the ground state moments of ^{121}Ba. To deduce the changes in mean square radius relative to ^{124}Ba the mass and field shift parameters were calculated by a King plot procedure with optical isotope shift data in the λ = 533.6 nm line in atomic barium measured by Mueller et al[4]. In this way our measurements have been incorporated onto the proton-rich end of the overall chain of barium isotopes[5]. The moments of ^{121}Ba were calculated by measuring the splitting of the stable isotopes ^{135,137}Ba and using the known quadrupole and dipole moments for these isotopes.

Table I. Results of Calculations.

^{120}Ba $\delta_v^{124,120}$ = 390(6) MHz, $\langle r^2\rangle_{120} - \langle r^2\rangle_{124}$ = -0.087(2) fm^2
(λ = 455.4 nm Ba II)

^{121}Ba $\delta_v^{124,121}$ = 21(10) MHz, $\langle r^2\rangle_{121} - \langle r^2\rangle_{124}$ = -0.009(2) fm^2
μ = 0.659(2) nm   Q = 1.78(21)b

DISCUSSION

Changes in the mean square radius of barium are plotted in fig. 3 and for comparison the caesium isotopes have also been included. The general trend away from the closed shell N = 82 is of increasing deformation with a regular odd-even staggering up to N = 65 where there is a sudden distinct jump. A deformation of

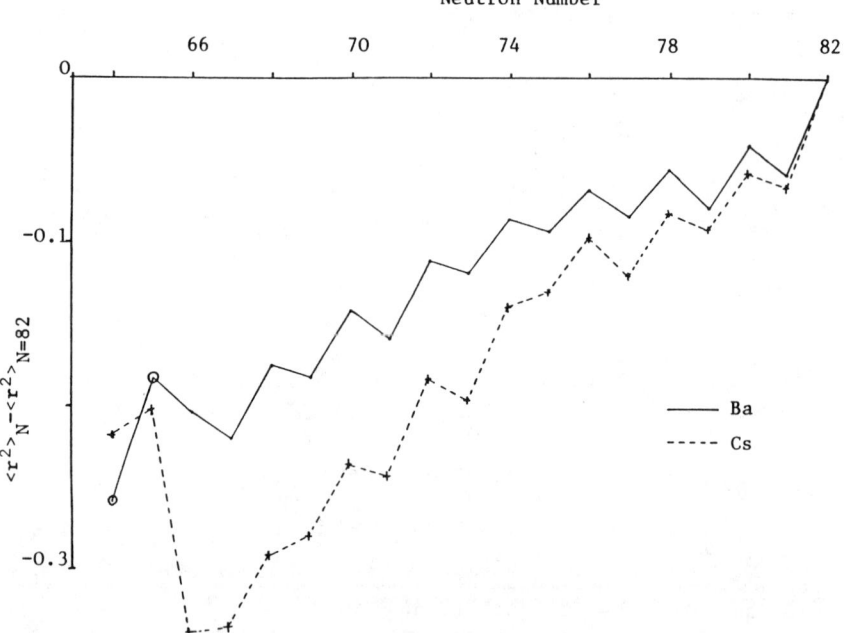

Fig. 3. Changes in mean square radii for barium and caesium.

$\beta = 0.33$ for ^{121}Ba is calculated from the measured quadrupole moment and this is similar to the value obtained from the change in the r.m.s. radius (assuming only quadrupole distortions and a monopole expansion as predicted by the Droplet model). The observed h.f. splitting defines the spin of the ground state as 5/2 and yields a positive dipole moment consistent with the assignment of the 5/2 [413] Nilsson orbital. A Nilsson calculation with $\mu = 0.475$, $\kappa = 0.064$, $\beta = 0.33$ indeed yields this orbital as the ground state and gives a magnetic moment in reasonable agreement with the empirical result. The 5/2 [413] orbital is reasonably flat as a function of distortion whereas the 5/2 [402] orbital which constitutes the g.s. in ^{123}Ba is strongly oblate driving. This difference may play a role in producing the unexpected change in radius, particularly since a similar, but even more pronounced, anomaly has been observed in the Cs nuclei at the same neutron number, $N = 65$ (fig. 3).

REFERENCES

1. D.A. Eastham et al, Opt. Commun. <u>60</u>, 293 (1986).
2. D.A. Eastham et al, J. Phys. <u>G12</u>, L205 (1986).
3. D.A. Eastham et al, Phys. Rev. (in press).
4. A.C. Mueller et al, Nucl. Phys. <u>A403</u>, 234 (1983).
5. C. Thibault et al, Nucl. Phys. <u>A367</u>, 1 (1981).

STUDIES OF ORIENTED NUCLEI AT THE KOOL FACILITY

D. Vandeplassche, J. Wouters, E. van Walle, N. Severijns,
J. Van Haverbeke and L. Vanneste,
Instituut voor Kern- en Stralingsfysika, Leuven University,
3030 Leuven, Belgium

ABSTRACT.

The method of on line low temperature nuclear orientation is presented and illustrated with typical examples taken from the measurements performed at KOOL. A special emphasis is put on the combination of recent particle detection techniques with the standard γ detection. Some unexpected results with a repercussion on the method itself are discussed.

In this contribution we present a general view of the experiments performed at the KOOL facility, which is our on line low temperature nuclear orientation set-up, installed at the on line mass separator LISOL since 1981.

In general terms, hyperfine techniques are applied to on line separated isotopes in order to obtain complementary information to the purely spectroscopic studies. Nuclear orientation can, from this point of view, be characterized as follows. (i) The method has a high sensitivity, since it relies upon single radioactive counting. This feature is still enhanced by the recent particle detection techniques (section 2.3). (ii) high precision resonance techniques are available for moment and/or spin determination (section 2.4). (iii) It is especially important for on line applications that several observables may be obtained simultaneously, during a single run. (iv) It is a fundamental property that the observations are sensitive to the magnitude and the sign of radiation multipolarity mixing ratios (section 2.2).

Some of these aspects are unique to the method, whereby nuclear orientation deserves its place at on line facilities. However, for its realisation up to the present status, many technical difficulties had to be overcome.

1. INTRODUCTION: STATIC ORIENTATION OF NUCLEI.

Consider an ensemble of nuclei under such conditions, that the m-sublevels are unequally populated: the ensemble is said to be oriented. If the nuclei are radioactive, the radiation is emitted anisotropically, and its angular distribution depends on the relative populations and on the characteristics of the radiation.

In our case the interaction which is responsible for the orientation is of the magnetic dipole type, and is assumed to have axial symmetry. Then the general expression of the normalized angular distribution can be simplified to[1] :

$$W(\Theta) = 1 + f\sum_k B_k A_k U_k Q_k P_k (\cos\Theta) \qquad (1)$$

In this expression the population distribution over the m-sublevels determines the orientation parameters B_k, the characteristics of the observed radiation (i.e. initial spin, final spin, multipolarity) are entered into the angular distribution coefficients A_k, the deorienting effects of unobserved preceding transitions into the U_k parameters. The geometry of the distribution is described by the Legendre polynomials $P_k(\cos\Theta)$; the effects of finite solid angle detection are taken into account by the Q_k coefficients. The correction factor f is introduced to allow for a fraction of the nuclear ensemble not being oriented.

In order to obtain a statically oriented ensemble, one needs to fulfill the Boltzmann condition, which in our case reads: $\mu B/k_B T \lesssim 1$. In spite of magnetic fields of several tens or even hundreds of T being obtainable at the site of the nuclei (see next section), the constraint to the temperature remains heavy: base temperatures below 15 mK for common cases. The importance of the cryogenic equipment is thus obvious.

In standard anisotropy measurements data points are taken at $\Theta = 0$ and at $\Theta = \pi/2$, both as a function of temperature. On line applications request a separate normalization of each data point, due to short lifetimes and production instabilities. Therefore our analyses are based on the quantity $W(0)/W(\pi/2)$:

$$\frac{W(0)}{W(\pi/2)} = \frac{[N(0)/N(\pi/2)]_{cold}}{[N(0)/N(\pi/2)]_{warm}}, \qquad (2)$$

where $N(\Theta)$ is the number of counts detected at angle Θ.
Sets of data points $\frac{W(0)}{W(\pi/2)}(T_i)$ are then fitted to the theoretical expression based on (1). Temperatures are obtained from calibrated nuclear orientation thermometers, mostly ^{60}CoCo s.c. and ^{57}CoFe.

Our original one line nuclear orientation set-up is described in [2]. Its basic concepts are unchanged, but many improvements have been made to that design [3]. A ^{3}He/^{4}He dilution refrigerator with a horizontal beam access tube at 4K of 1m long, and terminated with a movable diaphragm at 77K, permits a base temperature with open beam access of ~11mK. A top loading facility allows fast removal or loading of 14mm diameter sample holders. The main improvements concern enhanced beam control and monitoring, and better thermal contact between sample holder and mixing chamber. At the same time the evolution towards ion sources for cyclotron and separator of longer lifetime and higher stability proves to be an important factor in the overall improvement of the system.

As mentioned earlier, in order to obtain sufficiently high magnetic fields at the site of the nuclei to be oriented, we make use of the hyperfine fields experienced by impurities on substitutional places in a magnetized iron host. Using separated ion beams of 50 keV a proper substitutionality may be obtained by ion implantation, i.c. at low temperature and low dose. This aspect of on line nuclear orientation, and the connected problem of the relaxation towards a Boltzmann equilibrium after the implantation, are of fundamental importance for the technique itself. It will be treated separately in section 3.

2. SURVEY.

The large variety of measurements that may be performed by nuclear orientation can be catalogued according to the object of their study, or according to the detection method used. It is a recent acquirement that, besides the standard γ-detection techniques, we are able to detect α-particles with high resolution. Therefore solid state detectors have been installed at 0 to $\pi/2$ directions, without any window between source and detector. Combined with the use of implanted sources, this set-up guarantees that high resolutions may be obtained. The absence of background in the recorded spectra increases the sensitivity of the detection by about 2 orders of magnitude as compared to normal γ-detection, making $\sim 10^2$ particles/s feasible. Clearly the same techniques may be applied for β-, X-ray, conversion electron detection. Some technical details will be given in subsection 2.3.

In order to distinguish the measurements according to the object of their study, expression (1) is to be used as a guide.

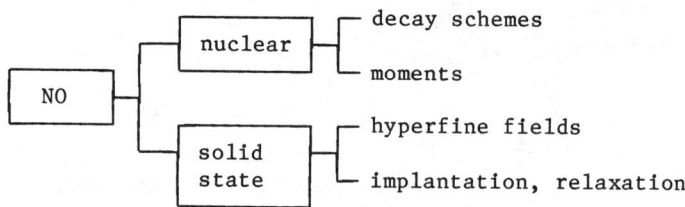

In many practical cases, however, these distinctions cannot be clearly made, thus manifesting the variety of results that can be obtained simultaneously.

In the following subsections we will try to present the facets of the method separately, using recent experiments as examples.

2.1. INTEGRAL MAGNETIC MOMENT DETERMINATION IN γ-DETECTION: 186-185Au.

Whereas the slightly oblate shapes of the ground states in 187,188Au as well as their configurations are well established, rather different configurations and shapes are expected in 185,186Au [4,5]. Calculations show that the magnetic moment is sensitive to an interplay between the prolate $\pi 1/2^-[541]$ and $\pi 3/2^-[532]$ Nilsson orbitals. The pure configurations yield calculated moments smaller than 1.7 μ_N.

In the decay schemes of both nuclei [6,7] well established E2
transitions are suited for a direct fit with the magnetic moment
as the only free parameter. The obtained anisotropies are displayed in fig. 1. The fits result in:

$\mu(^{185}Au) = 2.22(14)\mu_N$
$\mu(^{186}Au) = 1.07(13)\mu_N$.

The magnetic moments are in agreement with the proposed prolate
proton configurations in the ground states. The result is fully
supported by recent laser spectroscopy work [8]. The observed
behaviour is related to the ground state in ^{185,186}Au being
built upon the intruder state across the Z = 82 shell closure of
h9/2⁻ origin.

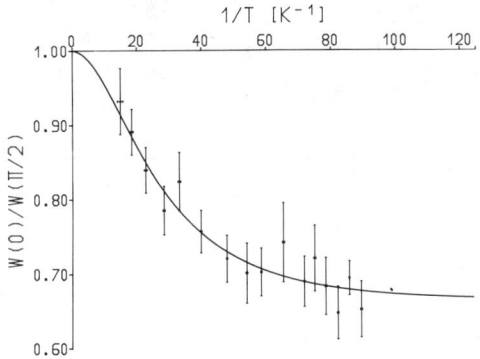

Fig. 1. ^{185}AuFe, 418 keV: anisotropy data points as a function of inverse temperature and their best fit.

2.2. MIXING RATIOS OF γ-TRANSITIONS IN Pd NUCLEI.

It is a great advantage of nuclear orientation that the angular
distribution coefficients A_k are strongly dependent on the
magnitude and the sign of the multipole mixing ratio in the
observed transition. Confining to δ(E2/M1) mixing ratios [1]:

$$A_k = \frac{F_k(11)+2\delta F_k(12)+\delta^2 F_k(22)}{1+\delta^2} \qquad (2)$$

These ratios of transition matrix elements are frequently a
stringent test of model descriptions, with mostly poor agreement.
Especially the sign changes of δ's are remarkable features, and in
the tabulations of ref. 9 such sign changes are found in a number
of nuclei having a closed (sub)shell. If subshell closures are
indeed reflected in sudden sign changes of δ-values, it may be
interesting to investigate the case N = 56 — cfr. the doubly
semi-magic ^{96}Zr [10]. It can be reached in the region right below
Z=50 in the Pd nuclei (^{102}Pd), where there is no immediate
spectroscopic indication (from $E(2_1^+)$) of subshell closure.

^{102,104}Ag have been produced and oriented, yielding 15 excited state spin values and 25 δ mixing ratios in the Pd daughters. Among these the most remarkable ones, in view of a sign change at N=56, are displayed in table 1. Especially in the $\delta(3_1^+ \to 2_1^+)$ and $\delta(2_2^+ \to 2_1^+)$ the effect is undoubted. There is at present no theoretical explanation.

Table I. Comparison of δ-values in the even-even Pd nuclei.

Transition	A = 102		A = 104		A = 106 [a]	
	E_γ (keV)	δ(E2/M1)	E_γ (keV)	δ(E2/M1)	E_γ (keV)	δ(E2/M1)
$2_2^+ \to 2_1^+$	978	2.8(2)	786	−25.0(5)	616	−14.1(15)
$3_1^+ \to 2_1^+$	1556	[4.4, 9.0][b]	1265	[−14.0,−11.0][b]	1046	−3.8(3)
$3_1^+ \to 2_2^+$	577	−	480	[−29.0,−22.0][b]	430	−9.0(1)
$4_2^+ \to 4_1^+$	862	−0.31(5), 1.8(2)[c]	759		703	−1.7(13)
$4_3^+ \to 4_1^+$	1025	0.01(8)	858			

a) Ref. 11
b) Limits of δ values for extreme U_k coefficients
c) A distinction between the two possible mixing ratios is not possible for this transition.

2.3. PARTICLE DETECTION.

If one wants to combine high resolution α-detection with an adequate cooling for nuclear orientation, only one solution is available: the installation of particle detectors inside the 4K radiation shield surrounding the sample position. Together with our sample preparation by implantation, the absence of any window between source and detector is a fundamental prerequisite for high resolution. For the characteristics of the detectors themselves the reader is referred to a separate contribution to this conference [12]. The tails of the cryostat with the special 4K shield and the detectors in place are presented in fig. 2. The sample holder is tilted 20° in order to allow for the full opening angle towards the detector. The realization of the helium-tight electrical feedthroughs at 4K was a cryogenic difficulty, but it was required in order to minimize the distance from detector to preamplifier. The energy resolutions we obtain presently are ∼20 keV at 6 MeV. As an illustration fig. 3 shows an accumulation of α spectra taken at different masses of At and Po isotopes.

The odd astatines decay by α emission, from $\pi h9/2^-$ to the $\pi h9/2^-$ groundstates in Bi. The recorded anisotropies (fig. 4) show a strong dependence on the neutron number, and a sign change

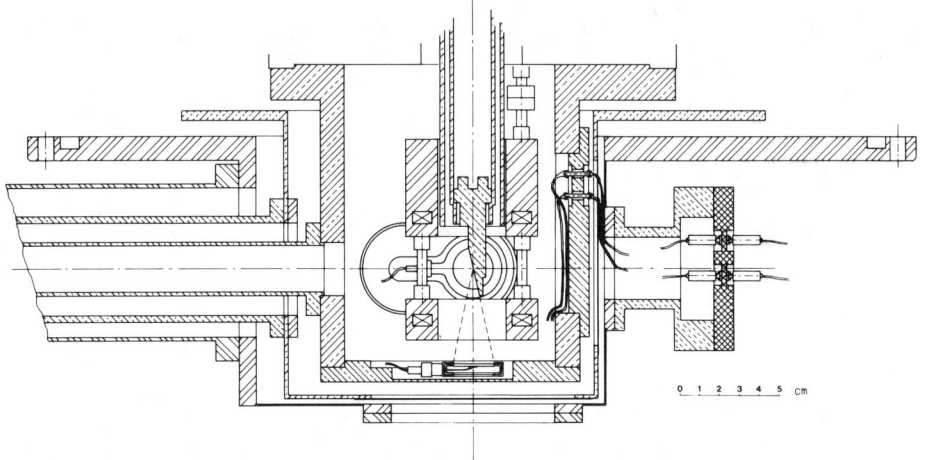

Fig. 2. Cryostat tails with particle detectors installed.

occurs: the emission changes from preferentially parallel to preferentially perpendicular to the direction of the nuclear spin when decreasing N. The change in α anisotropy originates in the variation of the mixing ratios of the α waves with L=2 and L=0 [13], and in fact this δ is the only measurable α-decay parameter really sensitive to L ≠ 0. Here again the nuclear orientation technique is particularly powerful, since it is sensitive to very small δ-values (in dominant L=0 α transitions, as e.g. in At, the L≠0 component contributes less than half a percent), and it provides the knowledge of both magnitude and sign of the relative amplitudes. In order to make the link towards nuclear structures and/or shapes, both are needed [14,15]. A full description and interpretation of our results is given in [12].

The present set-up for particle detection has also been used for the observation of β anisotropies. As a test and calibration case we measured ^{70}As. Its value of μB being known, the fraction f could be obtained from E2 γ transitions. The other γ-anisotropies have yielded several E2/M1 mixing ratios. The analysis of the β anisotropies was performed in 2 energy regions. As is shown by the following numerical results, both A_1-values are in good agreement with the calculations:

energy region [MeV]	calc. A_1	exp. A_1
1.72 - 2.15	-.41(4)	-.40(3)
2.39 - 2.75	-.62(3)	-.59(5)

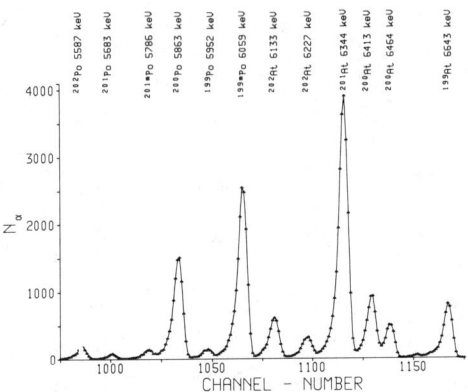

Fig. 3. Accumulation of α spectra (At and Po isotopes).

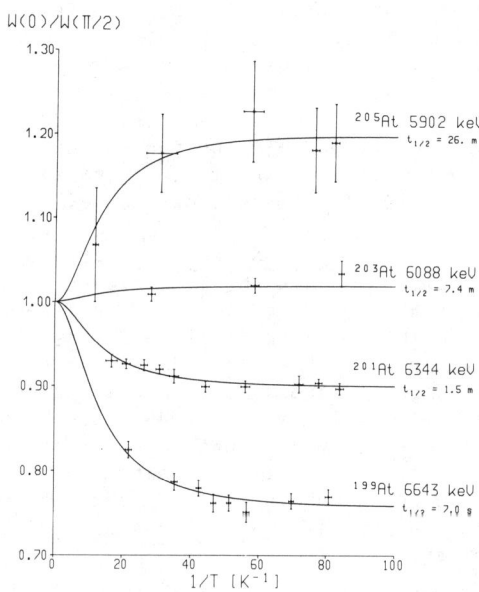

Fig. 4. Anisotropy data points and best fits of odd $^{199-205}$AtFe.

We conclude that the set-up is suitable to obtain β anisotropy results, even at this energy around 3 MeV, at an accuracy level of better than 10%.

A very recent measurement concerns ^{69}As. The β anisotropy data result in a fit value of $\mu = +1.58(16)\mu_N$ for pure Gamow-Teller, and covering practical limits of Fermi admixture. Our result is significantly larger than the previously reported one ($\mu = 1.2(2)\mu_N$ [16]), and it removes the need to postulate a proton configuration that is different from ^{71,73}As.

2.4. HIGH PRECISION MAGNETIC MOMENT AND SPIN: NMR/ON ON ^{106}In.

^{106}In has been the first isotope on which the well known method of "nuclear magnetic resonance on oriented nuclei" [17] was applied on line. The method consists of a resonant destruction of the orientation by a radio frequency (RF) field, corresponding to the Zeeman energy splitting. This partial destruction is observed by a diminished anisotropy of the radiation. The obtainable accuracy is typically one order of magnitude better than in "integral" measurements, and also therefore these results are extremely useful for calibration purposes.

We adapted our on line apparatus for this method by installing an RF-coil around the continuously implanted sample, and by optimizing the overall stability of its operation. To give each data point its own normalization (cfr. section 1), we use the ratio of the anisotropies recorded with and without frequency modulation. Moreover, the temperature is constantly monitored for the presence of RF-power induced pseudo-resonances.

In order to register the resonance curve of ^{106}In we monitored the anisotropy of the 633 keV γ-transition, while scanning the frequency region corresponding to our integral result [18]. A coarse indication around 153 MHz was refined, leading to the curve in fig. 5, obtained with a modulation width of 375 kHz and a frequency step of 500 kHz. A Gaussian was fitted to the data points, yielding a center frequency of 153.1(3)MHz and a fwhm of 2.4 MHz. We then deduce g(^{106}In) to be 0.7030(19).

The spin value of the oriented state in ^{106}In remaining unknown (6 or 7), we inserted the resulting μB-values for both spins in the analysis of the anisotropy of the 998 keV $6^+ \to 4^+$ E2 γ transition in ^{106}Cd. The comparison is presented in fig. 6: for spin 6 the observed anisotropy can never be reproduced, even for a fully substitutional implantation, whereas for spin 7 an excellent fit is obtained. A moment value of $4.921(13)\mu_N$ is then derived, and it is in perfect agreement with the recent value from laser spectroscopy for spin 7 [19].

This spin now being fixed (also in ^{108}In), it finally puts an end to a long controversy. The spin of ^{104}In remains unresolved up-to-now, but should be accessible with the same method: using the laser spectroscopy result [18] we obtain center frequencies of 141.5, 163.5 and 193.2 MHz for spins 7, 6 and 5, respectively.

148

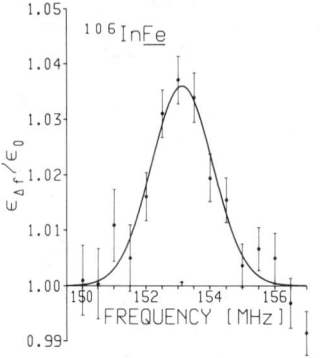
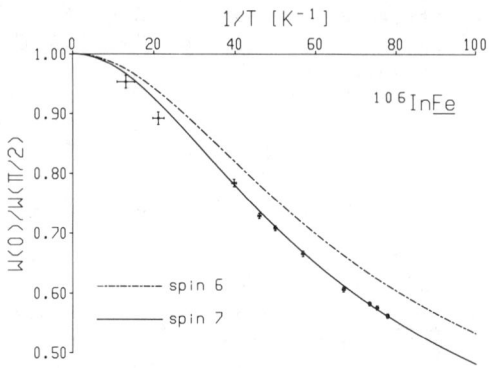

Fig. 5. NMR/ON on ^{106}In<u>Fe</u>: data points and Gaussian fit.

Fig. 6. Anisotropy data points of ^{106}In<u>Fe</u>, 998 keV, with 2 fit curves: full line for spin 7, dashed line for spin 6.

3. SOLID STATE ASPECTS.

The way to obtain orientation in our method is clearly based on solid state effects: internal magnetic fields and thermal equilibrium with the lattice. Therefore the results obtained in these domains are of fundamental importance for the technique itself.

3.1. IMPLANTATION.

The quality of the implantation is usually expressed as the substitutional fraction f (see (1)). This 2-site approximation has proven to be adequate in all cases of low temperature implantation we studied up-to-now. The value of f is sensitive to the fact that the implantation occurs at low temperatures [20], and to the surface treatment of the Fe foil.

Another approximation, that is valid in most cases at a 1% accuracy level, is the neglection of hyperfine anomalies. This allows to perform relative measurements: in a long isotopic chain it suffices to determine f for one isotope, which can then be used as a calibrator for the others (implanted under identical conditions).

To obtain f-values one mainly has 3 ways: (i) the anisotropy of a well known transition from an isotope for which NMR/ON data are available (cfr. ^{106}In); (ii) a saturated anisotropy, where temperature dependent and independent parameters become uncorrelated; (iii) a multiparameter fit in a combined analysis of several anisotropies (cfr. infra). In table 2 we present the f-values for the several elements we "cold-implanted" up-to-now — they should be taken as lower limits of what is achievable.

Table II. List of substitutional fractions obtained in low temperature, low dose implantation. The values are to be taken as lower limits. The value for F is given for immediate relaxation (see section 3.2).

element	F	As	Ag	In	Au	At	Po	Fr
substitutional fraction	(.96)	.90	.97	.95	.95	.76	.78	.76

3.2. RELAXATION.

The problem of relaxation arises whenever the timescales of the radioactive decay and of the search for thermal equilibrium become comparable. This situation results in incompletely oriented ensembles, and thus in diminished anisotropies. We adopt the relaxation formalism of Klein [21] and obtain corrected values of the orientation parameters B_k (see (1)), in our specific situation of continuous cold implantation: a secular equilibrium. The relaxation constants C_K are used as a fixed or as a free parameter. ^{106}In is now a well suited case for the investigation of relaxation effects, again monitoring the 998 keV line. The lifetime of 6.2 min and a C_K of 2058 s·mK (calculated from the measured relaxation time in ^{110m}In [22]) makes one expect an anisotropy reduction of ~14%. From the fit in fig. 6 we obtain a reduction of 6.5% only (the fit to the fraction yields f = 0.944(12)). In fact, if the same data points are fitted to C_K and assuming 100% substitutionality, we obtain an upper limit for the relaxation: $C_K \lesssim 600$ s·mK. Since ^{110m}In and ^{108}In, with lifetimes of 4.2h and 58 min, resp., both implant with a fraction around .95, we conclude there is no evidence for relaxation effects at all for the system ^{106}In(Fe), <u>cold</u> implanted into a <u>magnetized</u> iron foil.

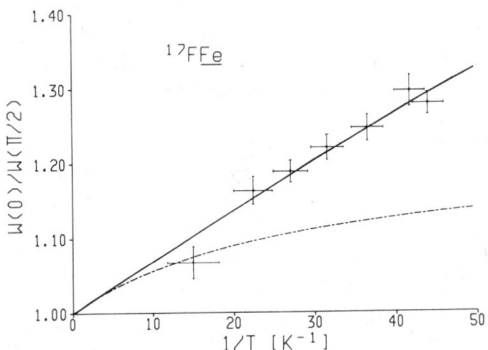

Fig. 7. Anisotropy data points of the β-emission of ^{17}FFe, with 2 fit curves: the full line indicates the best fit with immediate relaxation, the dashed line shows the maximal effect for C_K = 2900 s·mK.

Such a surprising result has also been obtained in light nuclei, like β-decaying ^{17}F ($t_{1/2}$ = 64 s). The theoretical anisotropy of its β emission can be calculated: the angular distribution coefficient A_1 is determined from the log ft value [23]. On the other hand one can estimate the relaxation constant from the knowledge of the interaction [24,25], resulting in C_K ≈ 2900 s.mK. In fig. 7 we present, together with the experimental data points, a comparison between the full anisotropy curve (theoretical, immediate relaxation) and the one obtained for C_K = 2900 s.mK, where the temperature dependence of the relaxation behaviour is well visible. Here again, the result clearly suggests all relaxation effects to be absent: not only can the size of the anisotropy not be reproduced, but moreover is the shape of the temperature dependence significantly different. More generally, up-to-now we do not have any measurement where relaxation effects after cold implantation into magnetized iron are demonstrated. Before making any further conclusions, however, the matter must be investigated in far more detail. It seems doubtless, that such a fast evolution towards an oriented ensemble must originate in some presently neglected interaction, which might arise <u>during</u> the implantation process in a cold magnetized host material. A transient field-like mechanism could perhaps be invoked.

3.3. HYPERFINE FIELDS.

Obviously the anisotropy measurement of an element with known angular distribution coefficients and known magnetic moment yields a value for the hyperfine field. Recently our α detection technique has permitted the determination of the hyperfine field of Fr in Fe [26]. A full discussion is given in a separate contribution to this conference [12].

4. ^{199m}Po.

This case is chosen for a more complete treatment in this overview of on line nuclear orientation, because it shows the typical present situation of the method. Experiments are performed with γ detectors outside and particle detectors inside the cryostat, and the recorded anisotropies are analyzed in a combined way.

The schematics of the ^{199}Po decay are represented in fig. 8. For this "multi-branch" decay structure we obtained anisotropy data in several daughter nuclei, down to ^{195}Tl and ^{199}Pb. In this treatment we will limit ourselves to the magnetic interaction of the $13/2^+$ isomeric state in ^{199}Po, and the excited level structure of ^{199}Bi. This level scheme has been carefully studied by standard γ spectroscopic methods on line [27] and in beam [28], but most of the spin assignments remain between parentheses.

The magnetic interaction frequency of ^{207}PoFe has been determined with high accuracy by NMR/ON [29] as 575.08($\overline{20}$)MHz.

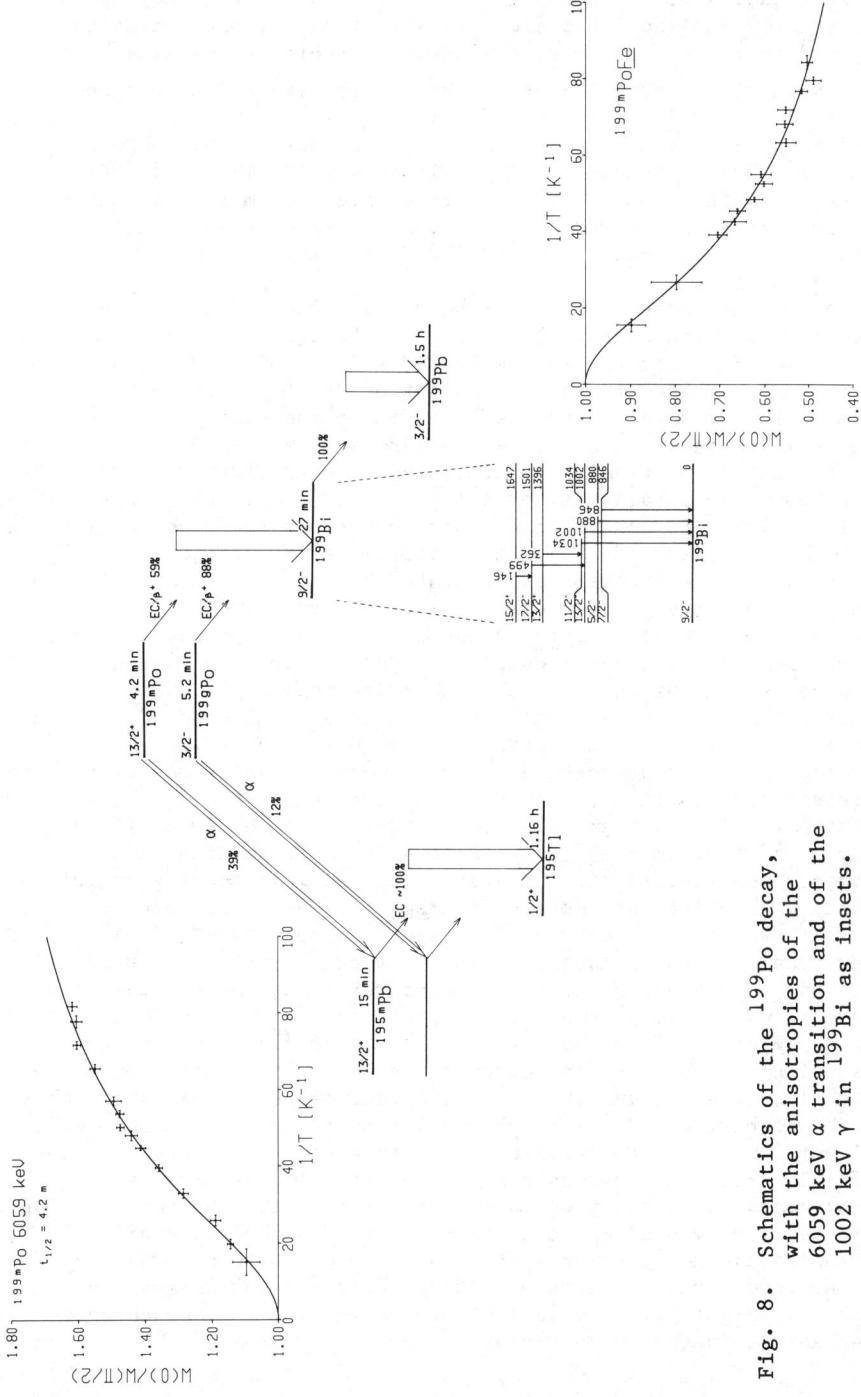

Fig. 8. Schematics of the ^{199}Po decay, with the anisotropies of the 6059 keV α transition and of the 1002 keV γ in ^{199}Bi as insets.

Unfortunately, the dependence of the resonance frequency on the externally applied field did not allow for a precise value of the hyperfine field: $B_{hf}(\vec{\text{PoFe}}) = +238(16)$T. Therefore we will quote the magnetic interaction µB, to be interpreted relatively to the value for ^{207}PoFe, which is $188.61(7)\mu_N\cdot$T.
The Po nuclei have been produced by a 160 MeV Ne beam onto a combined Ir and Re target. Approximately $1\cdot10^4$ ions/s ^{199m}Po were extracted from the FEBIAD ion source and implanted into the Fe foil at temperatures down to 13 mK. The recorded α anisotropy and the γ anisotropy of the 1002 keV line in ^{199}Bi are shown as insets of fig. 8.

Among the parameters that govern the anisotropies, µB and f are "general", in the sense that they must be equal for several radiations issuing from the same oriented state. This permitted us to perform a combined fit on the anisotropies of 5 different transitions (1α and 8γ data sets) with 5 free parameters, yielding the "general" parameters as: $\mu B = 235(7)\mu_N\cdot$T, $f = .78(2)$.
With these being fixed, the remaining anisotropies could be analyzed. Our results allow to obtain a consistency check on several spin assignments. An important discrepancy is found concerning the levels at 846 and 880 keV, with proposed $I^\pi = (5/2^-)$ and $(7/2^-)$ respectively. With these assignments the 846 keV γ transition to the $9/2^-$ groundstate would have E2 character. However, this transition shows a <u>positive</u> anisotropy, totally precluding E2. Since both a simple core coupling model and the systematics of the heavier uneven Bi indicate the occurrence of a $5/2^-$ and a $7/2^-$ state around 1 MeV, we conclude the levels at 846 and 880 keV to have $I^\pi = 5/2^-$ and $7/2^-$, respectively (cfr. fig. 8).
There is no further contradiction between our anisotropies and the present spin assignments of refs. 27 and 28.
From our anisotropy analyses on ^{201m}Po we obtain, in a similar way to ^{199m}Po: $\mu B = 237(8)\mu_N\cdot$T. It may be interesting to compare our results with $vi13/2^+$ states in neighbouring nuclei by placing them into an isotone diagram. Fig. 9 shows these magnetic moment systematics for Pb (laser spectroscopy [30]), Po (DPAD [31] and our results) and Rn isotopes (laser spectroscopy [32] and DPAD [33]). The Rn results are given for fixed a reference value, whereas our results use an adopted hyperfine field of 238 T. There is, in such a plot, a clear suggestion that the absolute values of the magnetic moments tend to decrease regularly by adding protons to the closed shell, and that the different isotopic chains all show a comparable, rising slope when removing neutrons from the N=126 closure. Our results certainly confirm a trend that was already visible from the previously known values. The feature is not reproduced by calculations up-to-now — in fact, the reproduction of the experimental magnetic moments in this region is a long standing problem. The present data, together with trends from other recent measurements on $vf5/2$, $vp3/2$ and $\pi h9/2$ magnetic moments, might give new indications concerning the quenching mechanisms in the nuclei around ^{208}Pb.

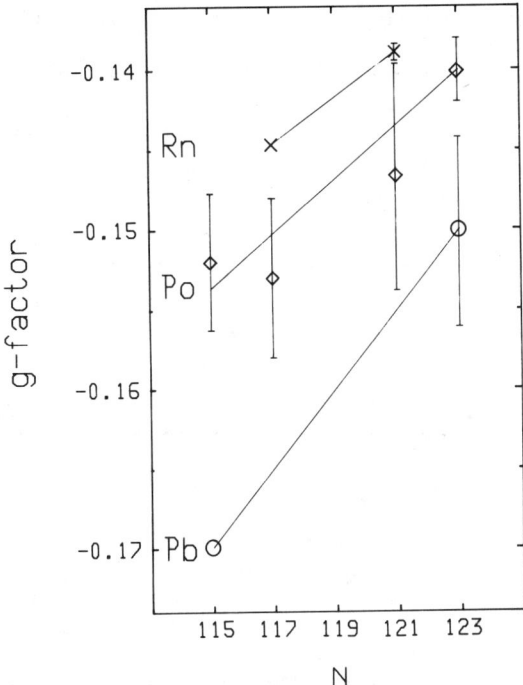

Fig. 9. $\nu i13/2^+$ magnetic moment systematics in Pb, Po, Rn.

5. CONCLUSION.

Since the last conference in this series the technique of on-line low temperature nuclear orientation has developed from its first steps to a now well established method far from stability — this overview has been intended to illustrate it. Its present day status has been shown by recent experimental results in γ, α and β detection, and the strength of their combination indicated.

In the near future the particle detection techniques have to be refined in order to perform e^- and X-ray anisotropy measurements. A more extensive use must be made of NMR/ON, hopefully extending its applicability towards lower production rates by taking advantage of the high sensitivity of the particle detection. Recent cryogenic tests with a shortened (25 cm) 4K-cooled side access tube have been very promising and provide a way to enhance the beam handling quality. Finally, the fundamental advantage of nuclear orientation, that quadrupole moments may be determined by sign and by magnitude still has not been applied on line.

To terminate this overview we present in table 3 an up-to-date survey of on-line nuclear orientation results.

Table III. Survey of on line nuclear orientation results.

Anisotropies	Oriented isotopes
γ integral	$^{104-110}$In a)
	$^{101-104}$Ag a)
	^{185,186,189m}Au, ^{191}Hg, ^{185}Ir a)
	$^{118-121}$Cs, $^{117-122}$I, ^{74m,75}Br b)
NMR/ON	^{106,108}In, ^{104}Ag, ^{189m}Au a), ^{116}Sb b)
α integral	$^{197-203}$Po, $^{198-205}$At, $^{205-209}$Fr
	+ daughters a)
β integral	^{17}F, ^{15}O, ^{19}Ne, ^{69,70}As a)

a) KOOL (ref. 12, 17, 34, 35).
b) DOLIS-COLD (Daresbury) (ref. 36-38).

We wish to thank P. Schoovaerts for technical assistance, M. Huyse and the separator crew, and the operators of the CYCLONE cyclotron at Louvain-la-Neuve. We gratefully acknowledge discussions with E. Hagn, P. Herzog and K. Heyde. This work is financially supported by the "Nationaal Fonds voor Wetenschappelijk Onderzoek" and by the "Interuniversitair Instituut voor Kernwetenschappen".

REFERENCES.

1. K.S. Krane in 'Nuclear Orientation', N.J. Stone and H. Postma, eds., North-Holland Publishing Company, Amsterdam (1985), ch.2
2. D. Vandeplassche, L. Vanneste, H. Pattyn, J. Geenen, C. Nuytten and E. van Walle, Nucl. Instr. & Meth. 186 (1981) 211
3. D. Vandeplassche, E. van Walle, J. Wouters, N. Severijns and L. Vanneste, Hyp. Int. 22 (1985) 483
4. C. Ekström, L. Robertsson, S. Ingelman, G. Wannberg and I. Ragnarsson, Nucl. Phys. A348 (1980) 25
5. M.G. Porquet, C. Bourgeois, P. Kilcher and J. Sauvage-Letessier, Nucl. Phys. A411 (1983) 65
6. M. Finger, R. Foucher, J.P. Husson, J. Jastrzebski, A. Johnson, G. Astner, B.R. Erdal, A. Kjelberg, P. Patzelt, A. Hoglund, S.G. Malmskog and R. Henck, Nucl. Phys. A188 (1972) 369
7. B. Roussière, C. Bourgeois, P. Kilcher, J. Sauvage and M.G. Porquet, Nucl. Phys. B34 (1986) 2014
8. K. Wallmeroth, G. Bollen, A. Dohn, P. Egelhof, J. Grüner, F. Lindenlauf, U. Krönert, J. Campos, A. Rodriguez Yunta, M.J.G. Borge, A. Venugopalan, J.L. Wood, R.B. Moore and H.J. Kluge, Phys. Rev. Lett. 58 (1987) 1516

9. J. Lange, K. Kumar and J.H. Hamilton, Rev. Mod. Phys. $\underline{54}$, 1 (1982) 119
10. M.L. Stolzenwald, G. Lhersonneau, S. Brant, G. Menzen and K. Sistemich, Z. Phys. $\underline{A327}$ (1987) 359
11. R. Eder, E. Hagn and E. Zech, Phys. Rev. C $\underline{30}$ (1984) 676
12. J. Wouters, D. Vandeplassche, E. van Walle, N. Severijns, J. Vanhaverbeke and L. Vanneste, contribution to this conference
13. J. Wouters, D. Vandeplassche, E. van Walle, N. Severijns and L. Vanneste, Phys. Rev. Lett. $\underline{56}$ (1986) 1901
14. P.J. Brussaard and H.A. Tolhoek, Physica $\underline{24}$ (1958) 233
15. H.J. Mang and J.O. Rasmussen, Mat. Fys. Skr. Dan. Vid. Selsk. $\underline{2}$, 3 (1962)
16. W. Hogervorst, H.A. Helms, G.J. Zaal, J. Bouma and J. Blok, Z. Phys. $\underline{A294}$ (1980) 1
17. N.J. Stone in 'Nuclear Orientation', N.J. Stone and H. Postma, eds., North-Holland Publishing Company, Amsterdam (1985), ch. 13
18. D. Vandeplassche, E. van Walle, J. Wouters, N. Severijns and L. Vanneste, Phys. Rev. Lett. $\underline{57}$ (1986) 2641
19. J. Eberz, U. Dinger, G. Huber, H. Lochmann, R. Menges, R. Neugart, R. Kirchner, O. Klepper, T. Kühl, D. Marx, G. Ulm and K. Wendt, Nucl. Phys. $\underline{A464}$ (1987) 9
20. C. Nuytten, D. Vandeplassche, E. van Walle and L. Vanneste, Phys. Lett. $\underline{A92}$ (1982) 139
21. E. Klein in 'Nuclear Orientation', N.J. Stone and H. Postma, eds., North-Holland Publishing Company, Amsterdam (1985), ch. 12
22. E. Hagn, E. Zech and G. Eska, Z. Phys. $\underline{A300}$ (1981) 339
23. S. Raman, T.A. Walkiewicz and H. Behrens, Atomic Data and Nuclear Data Tables $\underline{16}$ (1975) 451
24. K. Sugimoto, A. Mizobuchi, K. Nakai and K. Matuda, J. Phys. Soc. Jap. $\underline{21}$ (1966) 213
25. C. Fahlander, K. Johansson, B. Lindgren and G. Possnert, Hyp. Int. $\underline{7}$ (1979) 299
26. J. Wouters, D. Vandeplassche, E. van Walle, N. Severijns and L. Vanneste, to be published in Phys. Lett.
27. R.E. Stone, C.R. Bingham, L.L. Riedinger, R.W. Lide, H.K. Carter, R.L. Mlekodaj and E.H. Spejewski, Phys. Rev. $\underline{C31}$ (1985) 582
28. W.F. Piel, T. Chapuran, K. Dybdal, D.B. Fossan, T. Lönnroth, D. Horn and E.K. Warbuton, Phys. Rev. C $\underline{31}$ (1985) 2087
29. P. Herzog, H. Walitzki, K. Freitag, H. Hildebrand and K. Schlösser, Z. Phys. $\underline{A311}$ (1983) 351
30. M. Anselment, W. Faubel, S. Göring, A. Hanser, G. Meisel, H. Rebel and G. Schatz, Nucl. Phys. $\underline{A451}$ (1986) 471
31. 'Table of Isotopes', C.M. Lederer and V.S. Shirley, eds., John Wiley & Sons, New York (1978)
32. W. Borchers, R. Neugart, E.W. Otten, H.T. Duong, G. Ulm and K. Wendt, Hyp. Int. $\underline{34}$ (1987) 25

33. O. Häusser, J.R. Beene, T. Faestermann, T.K. Alexander, D. Horn, A.B. McDonald and A.J. Ferguson, Hyp. Int. 4 (1978) 219
34. E. van Walle, D. Vandeplassche, J. Wouters, N. Severijns and L. Vanneste, Phys. Rev. B 34 (1986) 2014
35. D. Vandeplassche, E. van Walle, C. Nuytten and L. Vanneste, Phys. Rev. Lett. 49 (1982) 1390
36. V.R. Green, C.J. Ashworth, J. Rikovska, T.L. Shaw, N.J. Stone, P.M. Walker, C. Leyland and I.S. Grant, Nucl. Inst. & Meth. B26 (1987) 482
37. V.R. Green, N.J. Stone, T.L. Shaw, J. Rikovska, K.S. Krane, P.M. Walker and I.S. Grant, Phys. Lett. B173 (1986) 115
38. T.L. Shaw, V.R. Green, N.J. Stone, J. Rikovska, P.M. Walker, S. Collins, S.A. Hamada, W.D. Hamilton and I.S. Grant, Phys. Lett. B153 (1985) 221

LOW TEMPERATURE NUCLEAR ORIENTATION OF NUCLEI FAR FROM BETA STABILITY. PLANS OF INVESTIGATIONS AT ISOL-FACILITY YASNAPP-2

T.I.Kracikova, M.Finger, V.N.Pavlov, V.M.Tsupko-Sitnikov
Joint Institute for Nuclear Research, Dubna, USSR   141980

ABSTRACT

Low temperature nuclear orientation of radioactive nuclei of the Eu, Gd, Tb, Yb, Lu, Bi, At and Np isotopes was performed. Some results on corresponding daughter nuclei are presented which are compared with theoretical predictions. Recent results obtained by the NMR/ON experiments are also given. Progress in realization of nuclear orientation system on-line to mass-separator and the new 680 MeV JINR proton phasotron in described.

The first off-line investigations of the radioactive decay of nuclei oriented at low temperatures have been started in JINR in 1975. This was facilitated (i) by the successful operation of the ^{3}He-^{4}He dilution refrigerator built by the group of Neganov in 1965, (ii) by the unique capability of the JINR synchrocyclotron to produce radionuclides far from beta stability and (iii) by the experience gained in nuclear spectroscopic investigations in JINR, particularly, a well-developed radiochemical technique for separation of radionuclide fractions from the targets as well as an efficient mass-separation technique to prepare the monoisotopic sources. A combination of the above factors ensured a high standard of investigations and created the unique possibilities in the range of radionuclides available for investigations, in production of the samples and in efficiency of the refrigerator and rate of the sample cooling. All this allowed to obtain ample information on nuclear, atomic and solid state physics.

As it was reported in Ref./1/, our main concern during past several years was to set up the system (the SPIN facility) for nuclear orientation (NO) study of short lived nuclei produced by spallation reaction with protons using mass-separator both on-line (ISOL system) and off-line to the JINR 660 MeV synchrocyclotron. The SPIN facility is realized in two steps: the off-line SPIN-1 facility was built (see /1/) and the on-line SPIN-2 system is being set up at the new JINR phasotron.

Present version of the SPIN-1 facility is designed for the study of angular distribution and temperature dependence of the anisotropy of gamma rays and NMR/ON studies. Some recent experimental results and prospects for the SPIN-2 facility are given.

Over 40 nuclides (isotopes of Nd,Pm,Eu,Gd,Tb,Dy,Tm,Yb,Lu,Ta,Re, Bi,At and Np) with half-lives from about 50 min to several years (^{167}Lu,$T_{\frac{1}{2}}$=51.5 min; ^{148}Tb,$T_{\frac{1}{2}}$=60 min and ^{154}Eu,$T_{\frac{1}{2}}$=8.5 y, see /2,3,4/) were investigated at the off-line facility /1/. On studying such a wide range of nuclides, much attention was paid to work up the methods of the source and sample preparation using the gadolinium and iron matrices (see /1,2,5-7/). As a result, we produce the sources which practically contain no admixtures and the samples in gadolinium of superior quality (with approximately complete substitution in

© American Institute of Physics 1988

the lattice sites).

Our results show that for the purpose of nuclear structure studies iron is a good matrix for the majority of nuclei (gamma-ray anisotropies up to 50%). Moreover, it was shown for ^{146}Eu that saturation of the iron matrix is the same at implantation voltages of 35 and 65 kV (see /6/).These are the important results for the on-line studies when the samples are prepared by implantation.

Absolute values of the magnetic dipole moments (μ) of the $^{145-149}$Eu and ^{147,149}Gd ground states /8,9/ have been determined. Magnetic moments of the odd-A nuclei in this region have absolute values decreasing with the increase of collective effects. However, in vicinity of Z=64 and N=82 ($^{145}_{63}$Eu$_{82}$, $^{147}_{64}$Gd$_{83}$) the values of /μ/ slightly increase approaching to those accounted for by the single-particle model. This confirms a shell closure at Z=64.

The NO study of the decay of ^{169}Yb polarized in iron and gadolinium was performed /9/. Gamma rays from ^{169}YbGd, obtained by the decay of implanted and melted sample of ^{169}LuGd (gamma ray anisotropies up to 40%), were observed to have small anisotropies indicating that electric quadrupole and magnetic dipole interactions are comparable for ytterbium in gadolinium.

Systematic NO studies of the decay of ^{146,148,152,154}EuFe, ^{148,150,152,154,156}TbGd and ^{172}LuGd and ^{172}LuFe allowed to obtain the considerable amount of data on the level spin-parities and E0/E2, E2/M1 and M2/E1 mixing ratios of transitions in corresponding daughter Sm, Gd and Yb nuclei /3,4,6,10-12/. These data were found to be very helpful in new interpretation of some states in terms of quasi-ground, -beta and -gamma bands. The experimental data indicate a shape transition upon crossing from N=88 to N=90. This is most apparent in the mass region A=150 where the E2/M1 mixing ratios of 2_2- 2_1, 4_2- 4_1, 2_3- 2_1 and 3_1- 2_1 transitions all change a sign and, according to data available, this remains till N=102 (^{172}Yb). The properties of Sm, Gd and Yb nuclei were considered within the framework of the IBA-1 model. Roughly viewed, the model is good for the quasi-band level energies, fair for the E2 transitions and poor for E0 transitions. Difficulties can be explained by the extra states (particle or mixed) which represent a short coming of the simple IBA.

The low-lying levels of ^{167,169,171,173}Yb nuclei were studied using the decays of ^{167,169,171,173}LuGd (see /2,14-16/). The main aim of these studies was to obtain the data which could serve as a sensitive test of the applicability of the quasiparticle-phonon model including Coriolis interaction to the deformed odd-mass nuclei. The experimental results show that Coriolis coupling is very important for odd-A Yb nuclei /14-18/, particularly, for the positive-parity bands. For instance, in the 7/2$^+$ 633 band it grows rapidly with increasing spin value, however, when going from the lighter to heavier isotopes, the role of Coriolis coupling is quickly decreasing. The band levels are depopulating by competing E2 and M1+E2 transitions. All experimental E2/M1 mixing ratios are negative and their magnitudes smoothly decrease with increase of spin. Indeed, with the growth of spin value, the amplitude of main component 7/2 633 and consequently the contribution of collective effects to the E2 transition probability become smaller.

The irregularity of level spacing in the $1/2^-$ 521 band is clearly reflected in the E2/M1 mixing ratios: transitions connecting the levels lying furthest apart have large E2 admixture (15%), while the close-lying levels are connected by nearly pure M1 transitions (E2 $\leq$ 1%). An agreement between the experiment and theory is particularly good for this band. The most interesting seems to be that upon crossing the N=99 to N=101, there is a sign change of the (E2/M1) values, and this may be explained by filling of the 1/2 521 level at N=101 (the ground state of ^{171}Yb).

Comparison of the experimental results on odd A Yb nuclei with those calculated using different nuclear models showed that the level energies and probably the transfer-reaction cross sections do not represent sensitive tests of nuclear models. It has to be noted that the collective admixtures in the quasiparticle states, though they are small in the low-lying levels of ^{167,169,171,173}Yb, are responsible for renormalization of the Coriolis matrix elements requiring no "artificial" attenuation as it was done in all previous calculations.

Recently, we have started the systematic NO studies of radioactive nuclei in the trans-lead and trans-uranium region. Preliminary results[19,20] on ^{204,206}Bi, ^{208,209,210}At and ^{239}Np contain much information on the levels and transitions in ^{204,206}Pb, ^{208,209,210}Po and ^{239}Pu. The experimental data are analysed now in order to check in details the shell-model predictions for the nuclei close to double-magic nucleus ^{208}Pb.

Time of nuclear spin-lattice relaxation on ^{60}Co in the temperature range from 10 to 40 mK was investigated /21/ in Pd and Pt based alloys with giant magnetic momenta. The NMR/ON studies were performed using modified version of eddy current heating method. Abrupt decrease of spin-lattice relaxation time was observed with increasing Pt content in $(Pd_{1-x} Pt_x)_{99} Co_1$ alloy, while the relaxation rate in pure Pd and Pt is practically the same. This effect may be explained by specific mechanism of interaction of the Co atoms with their crystalline neighbourhood. Decreasing of the spin-lattice relaxation time of ^{60}Co in Pd in comparison with Pt disagree with an assumption of the non-frozen orbital momentum of Co in Pd as a source of positive hyperfine field, since the orbital momentum should also contribute to the relaxation mechanism.

The samples of PdCo were artificially saturated by hydrogen and influence of the hydrogen admixture on the magnetic hyperfine field of ^{60}Co in the $Pd_{99}Co_1$ alloy was studied. It was shown that the hydrogen admixture leads to significant decrease of hyperfine field from 21 to 16 T. At the same time, the saturation by hydrogen speeds up the spin-lattice relaxation.

The SPIN-2 facility will be the nuclear orientation system connected to the mass-separator of the YASNAPP-2 facility on-line to the new 680 MeV JINR proton phasotron. The intensity of the proton beam on the target and ion-source unit of the mass-separator is 1-2 μA.

The side-access ^{3}He-^{4}He dilution refrigerator capable of maintaining temperature of 10 mK and less in continuous operational mode was built for the SPIN-2 facility. This will allow direct implantation of desired ions into a target at orientation temperature.

With such a new and fast technique of sample preparation one can expect that cooling down time will be limited by the spin-lattice relaxation time. In the case when ferromagnetic matrices are used, the limit of half-lives of nuclei accessible to nuclear orientation can be extended to about 1-100 s. If one takes into account experimental results on the yields of neutron deficient nuclei in spallation reaction quite high counting rates can be expected. Detection of gamma, beta and alpha radiation in single and coincidence modes and NMR/ON can be utilized. The laser-induced nuclear orientation can be also employed in order to polarize nuclei with very short half-lives ($T_{\frac{1}{2}} > 1$ ns).

With this new powerful on-line NO technique wide range of investigations with nuclei far from the beta-stability line can be performed.

REFERENCES

1. M.Finger et al. Hyp.Int. 22, 461 (1985).
2. T.I.Kracikova et al. Czech.J.Phys. B31, 527 (1981).
3. J.Dupak et al. Proc. 30d Conf.on Nuclear Spectroscopy and Structure of Atomic Nuclei,Leningrad (Nauka,Leningrad,1980),p.240.
4. T.I.Kracikova et al. Hyp.Int. 34, 127 (1987).
5. V.A.Deryuga et al. Izv.Acad.Nauk USSR (Ser.Fiz.), 46, 867 (1982).
6. T.I.Kracikova et al. J.Phys. G10, 571 (1984).
7. S.Davaa et al. Nucl.Phys. (USSR) 45, 635 (1987).
8. T.I.Kracikova et al. Hyp.Int. 15/16, 73 (1983).
9. T.I.Kracikova, S.Davaa, M.Finger, Hyp.Int. 34, 69 (1987).
10. T.I.Kracikova et al. J.Phys. G10, 667 (1984).
11. T.I.Kracikova et al. Proc. 35th Conf. on Nuclear Spectroscopy and Structure of Atomic Nuclei,Leningrad (Nauka,Leningrad,1985),p.98.
12. P.O.Lipas et al. Physica Scripta 27, 8 (1983).
13. T.I.Kracikova et al. J.Phys. G10, 1115 (1984).
14. S.Davaa et al. J.Phys. G8, 1585 (1982).
15. T.I.Kracikova et al. Nucl.Phys. A440, 203 (1985).
16. S.Davaa et al. JINR, R6-84-556 (Dubna, 1984).
17. J.Kvasil et al. Czech.J.Phys. B31, 1376 (1981); B33, 626 (1983).
18. T.I.Kracikova et al. Czech.J.Phys. B35, 1084 (1985); B36, 581 (1986).
19. P.Malinsky et al. Book of Abstracts VIIth Int.Conf. on Hyperfine Interactions (Bangalore, India, 1986).
20. P.Šimeček et al. Hyp.Int. 34, 131 (1987).
21. V.Bartoš et al. JINR, R6-86-357 (Dubna, 1986).

NUCLEAR SPINS AND MOMENTS OF ^{9}Li AND ^{11}Li

E. Arnold, J. Bonn, W. Neu, R. Neugart,
E.W. Otten, K. Wendt

Institut für Physik, Universität Mainz
D-6500 Mainz, Fed. Rep. of Germany

G. Ulm* and The ISOLDE Collaboration
CERN, CH-1211 Geneva 23, Switzerland

ABSTRACT

Nuclear spins and moments of ^{9}Li and ^{11}Li have been measured by optical pumping of a fast atomic beam. The angular asymmetry of the ß-radiation from the polarized nuclei was used to detect the hfs of the 2s $^2S_{1/2}$ - 2p $^2P_{1/2}$ resonance line and the NMR signal in a LiF or LiNbO$_3$ crystal. The results I = 3/2 and μ_I = 3.6673(25) n.m. indicate a pure 1p$_{3/2}$ state of the valence proton of ^{11}Li. A first preliminary result for the ratio of quadrupole moments is extracted from the NMR to be Q(^{9}Li) / Q(^{8}Li) = 0.88(3).

The interest in the exotic nuclei ^{9}Li and ^{11}Li has been renewed recently by the observation of a strong increase in the matter radius between ^{9}Li and ^{11}Li /1/. In order to achive a broader experimental basis to decide upon the structure of this weakly bound nuclei /2,3,4/ we have measured spin and magnetic moment of ^{11}Li /5/ and the quadrupole moment of ^{9}Li. The high sensitivity necessary to work with an ISOLDE yield of as few as 2·10^3 ions/s of ^{11}Li was reached by combining collinear fast-beam laser spectroscopy and ß-RADOP, i.e. optical pumping of the fast atomic beam, its implantation into a crystal and subsequent detection of the ß-decay asymmetry. The set-up used is shown in Fig. 1.

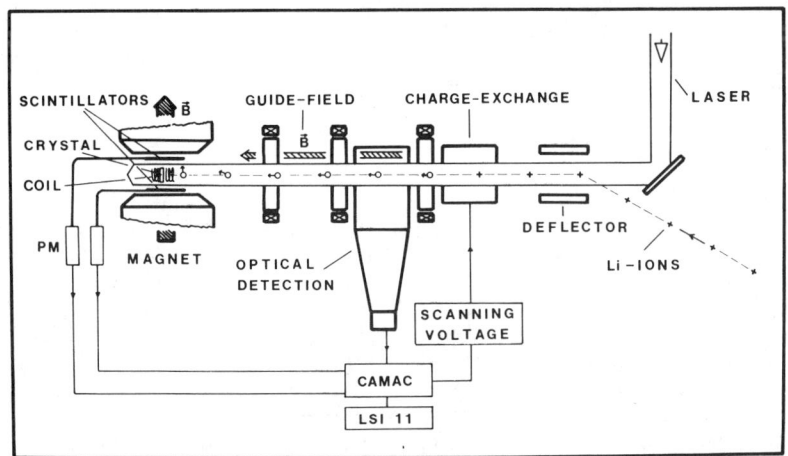

* Present address: Physikalisch-Technische Bundesanstalt, Institut Berlin, D-1000 Berlin 10, FRG

The Li ion beam from ISOLDE /6/ is superimposed with a circularly polarized laser beam. After neutralization by charge exchange optical pumping induces nuclear polarization due to hfs interaction of the nucleus and the valence electron. In front of the magnet nucleus and shell are decoupled in the increasing magnetic field and the nuclear polarization is rotated adiabatically into the axis of the magnet. The asymmetry signal of the implanted nuclei is recorded by two pairs of plastic scintillators each covering a solid angle of about 11 % of 4π. A LSI-11 microcomputer controls the Doppler-tuning potential at the charge exchange cell, the rf scanning and the data taking process via CAMAC.

The method has been tested on ^{8}Li (I = 2; $T_{1/2}$ = 838 ms). Nuclear moments are given relative to this reference isotope.

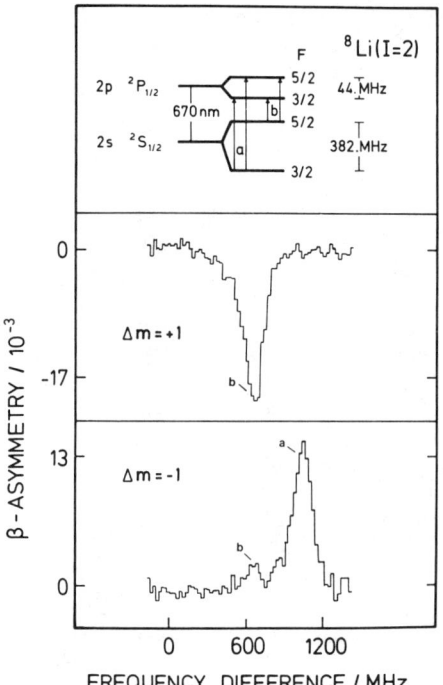

Fig. 2 shows the ß-asymmetry signal of ^{8}Li as a function of laser frequency. The complex line shape is caused by i) the unresolved hfs of the excited state (ΔW_{HFS} ($^2P_{1/2}$ = 44 MHz)), ii) energy loss by multiple excitation in the charge exchange cell and iii) optical pumping in the region where the polarization is rotated. They cause a destruction of the polarization affecting the F = I + 1/2 component when pumping with σ^--light and the F = I - 1/2 for σ^+-light. Therefore, two independent measurements were performed and combined to give the hfs pattern of the atomic ground state.

Fig. 3 shows the optical spectrum of ^{11}Li obtained in four hours of running time collecting a total of 10^6 events. The centre of gravity for ^{11}Li hfs is shifted by $\delta\nu^{9,11}$ = 8968(48) MHz in agreement with the calculated mass shift of $\delta\nu_m$ = 8991 MHz. The latter is estimated from the known value of $\delta\nu^{6,7}$ = 10534.8(2.0) MHz /7/ neglecting the estimated field shift of only a few MHz.

The assumption of I = 3/2 leads to a result of μ_I = 3.74(15) n.m. obtained from the hfs splitting of ΔW_{HFS} = 920(39) MHz. This number is in good agreement with the value of $\mu_I(^{11}Li)$ = 3.6673(25) n.m. obtained from ß-NMR in LiF. Spins and magnetic moments are compiled in Table 1.

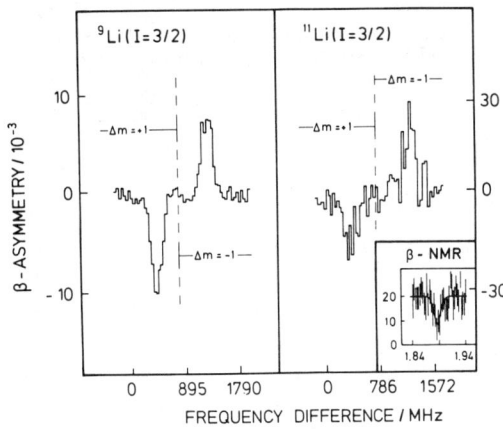

Fig. 3: β- asymmetry signals of ^{9}Li and ^{11}Li as a function of laser frequency
insert: β-NMR signal of ^{11}Li

Table 1: Spins, hfs-splittings and magnetic moments of radioactive Li-isotopes

	I	ΔW$_g$ [MHz]	μ$_I$(hfs) [n.m.]	μ$_I$(NMR) [n.m.]
^{8}Li	2	387(17)[a] 382.543(7)[b]	1.673(73)[a]	1.65335(35)[c]
^{9}Li	3/2	856(16)[a]	3.469(64)[a]	3.4391(6)[d]
^{11}Li	3/2[a]	920(39)[a]	3.74(15)[a]	3.6673(25)[a]

a) ref. /5/; b) ref. /8/; c) ref. /9/; d) ref. /11/

Fig. 4 shows β-NMR signals of ^{8}Li and ^{9}Li to determine the quadrupole moment of ^{9}Li. For this experiment the crystal was changed to LiNbO$_3$ which provides a reasonable electrical field gradient.

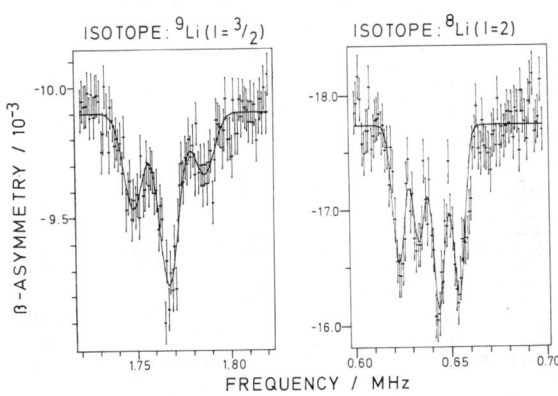

Fig. 4: β-NMR of ^{8}Li and ^{9}Li in LiNbO$_3$.

NMR resonances were taken under identical conditions for ^{8}Li and ^{9}Li. The extremely low concentration of radioactive nuclei in the crystal excludes perturbations due to radiation damage. For single quantum transitions the line splits into an ecquidistant multiplet. The centre of gravity is given by the magnetic interaction

$$\Delta E_M = g_I \mu_N <H_o>. \qquad (1)$$

The energy differences for the triplet in the case of ^{9}Li are

$$(I = 3/2) \qquad \Delta E_Q (I = 3/2) = 1/2 \; e^2 \; q \; Q_S \qquad (2)$$

and

$$(I = 2) \qquad \Delta E_Q (I = 2) = 1/4 \; e^2 \; q \; Q_S \qquad (3)$$

in the case of ^{8}Li. Neither in ^{9}Li nor in ^{8}Li indications for multiple quantum transitions could be observed. From the resonances shown in Fig. 4 we extract a ratio of quadrupole moments of $Q_S(^9Li)/Q_S(^8Li) = 0.88(3)$ the error including statistical uncertainties only. The quadrupole splitting found for ^{8}Li in LiNbO$_3$ ($\nu_Q(^8$Li in LiNbO$_3$)= 42.5(6) kHz) is in perfect agreement with the value given by Ackermann et al. /10/.

To compare our value for $Q_S(^9Li)$ with the result quoted by Correll et al. /11/ we have to refer it to $Q_S(^7Li)$ using the ratio

$$|Q(^8Li) / Q(^7Li)| = 0.78(1) \qquad /12/$$

Our result is

$$|Q(^9Li) / Q(^7Li)| = 0.69(3)$$

in contrast to the value of $|Q(^9Li)/Q(^7Li)| = 0.88(18)$ quoted by Correll et al. A more detailed discussion of the result on ^{9}Li will be given in a forthcoming paper /13/.

This work has been funded by the German Federal Minister for Research and Technology (BMFT) under the contracts number 06 MZ 458 I and 03 B21A29 7.

REFERENCES

/1/ I. Tanihata et al., Phys. Rev. Lett. 55, 2676 (1985)
/2/ F.C. Barker and G.T. Hickey, J. Phys. G3, L23 (1977)
/3/ Y. Chang and M.R. Meder, Phys. Rev. C30, 1320 (1984)
/4/ P.G. Hansen and B. Jonson, (submitted to Europhysics Letters)
/5/ E. Arnold et al., Phys. Lett. B (in print)
/6/ H.L. Ravn, Phys. Rep. 54, 201 (1979)
/7/ M. Fuchs and H.-G. Rubahn, Z. Phys. D2, 253 (1986)
/8/ R. Neugart, Z. Phys. 261, 237 (1973)
/9/ R.C. Haskell and L. Madansky, Phys. Rev. C7, 1277 (1973)
/10/ H. Ackermann et al., Phys. Lett. 52B, 54 (1974)
/11/ F.D. Correll et al., Phys. Rev. C28, 862 (1983)
/12/ D. Dubbers et al., Z. Phys. A282, 243 (1977)
/13/ E. Arnold et al., to be published

PRODUCTION OF SPIN POLARIZED ^{15}C IN HEAVY-ION REACTION AND MEASUREMENT OF g-FACTOR FOR THE 1/2$^+$ GROUND STATE

K. Asahi and M. Ishihara[*]
The Institute of Physical and Chemical Research (RIKEN)
Saitama 351-01, Japan

T. Shimoda, T. Fukuda[**], N. Takahashi, K. Katori,
K. Hanakawa, T. Itahashi, Y. Nojiri, and T. Minamisono
Osaka University, Osaka 560, Japan

N. Ikeda[**], S. Shimoura, and A. Nakamura
Department of Physics, Kyoto University, Kyoto 606, Japan

ABSTRACT

Spin-polarized ^{15}C nuclei produced in ^{15}N-induced reaction were transmitted through an achromatic spectrometer, and implanted in a high-purity graphite stopper in the presence of a static magnetic field. An rf field was applied on the stopper, and the spin inversion induced via nuclear magnetic resonance was observed through a change in the measured up/down asymmetry of β rays from ^{15}C. From the observed resonance frequency the g-factor for the ground state of ^{15}C has been determined to be $|g| = 2.63 \pm 0.14$.

INTRODUCTION

Production of wide variety of unstable isotopes in heavy-ion reactions, when combined with a certain means for isotope separation, constitutes a useful tool for the studies of β decays and nuclear structures in the regions far from stability. The usefulness would be still further augmented if these product nuclei are spin polarized. Recently the tilted foil technique has been successfully applied by Rogers et al.[1)] to polarize a short-lived isotope ^{33}Cl produced in the d(^{32}S,^{33}Cl) reaction in their g-factor measurement. For products ejected at higher velocities where ions are predominantly in a fully stripped state, however, this method is not applicable unless some additional technique is incorporated.

A systematic study of the spin polarization in (^{14}N,^{12}B)[2)] reactions on various target nuclei has revealed that the projectile-like fragments produced in heavy-ion reactions at incident energies around 10 MeV/u are substantially polarized. They found a systematic behavior of the polarization as a function of reaction Q-value and target mass, which is compatible with a simple macroscopic model of heavy-ion collisions. This result

---

[*] Now also at Department of Physics, The University of Tokyo, Tokyo 113, Japan.
[**] Present address: The Institute of Nuclear Study, University of Tokyo, Tokyo 188, Japan.

suggests that the spin polarization in this type of reaction is a rather general phenomenon which is applicable also for product nuclei other than ^{12}B.

We report on a measurement of the magnetic moment for the ground state of a neutron-rich nucleus ^{15}C ($I^{\pi}=1/2^+$, $T_{1/2}=2.449$ s, $Q_\beta=9.772$ MeV).[3] Spin polarized ^{15}C nuclei were produced using (^{15}N, ^{15}C) reaction. The nuclear magnetic resonance (NMR) has been observed by detecting the change in asymmetry in the angular distribution of β rays from Gamow-Teller transition to the $1/2^+$ excited state of ^{15}N.

Measurement of the magnetic moment provides useful information on nuclear structure and interaction. In a single-particle shell model where the ground state of ^{15}C is represented by a single neutron in the $s_{1/2}$ orbit outside the ^{14}C(0^+) core, the g-factor is predicted to be given by the neutron spin g-factor. The prediction can be improved by considering the first-order correction arising from a small admixture of the wave function corresponding to the 1^+ excitation of proton from the $p_{3/2}$ orbit in the core to the $p_{1/2}$ orbits. In fact the observed g-factor for the first excited state($5/2^+$) are well explained by this picture.[4] Viewing from the opposite direction, it is interesting to note that the measurement of the ground state g-factor, if interpreted as the g-factor of a nucleon residing in the $s_{1/2}$ orbit, represents a direct observation of the intrinsic moment of nucleon embedded in the nucleus. Possible effects on the nucleon magnetic moment due to modification of nucleon properties in a nuclear medium have been discussed recently.[5]

Experimentally, on the other hand, it is only for a few nuclei that the $I = 1/2$ magnetic moment is already known. This is because the $I = 1/2$ states do not bear the rank-2 spin orientation (alignment) which would facilitate observation of spin precession through the angular distribution of the de-excitation γ rays. Under these circumstances the observation of asymmetry in β ray

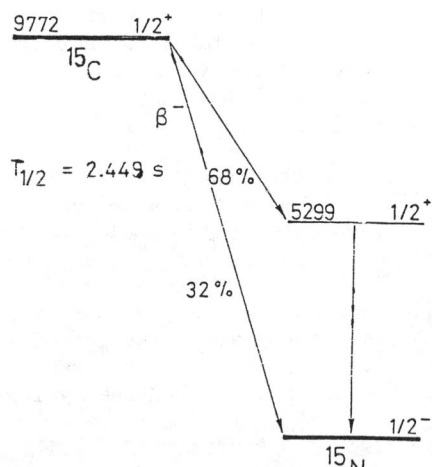

Fig. 1. Main decay branches in the β decay of ^{15}C. Values presented are taken from ref. 3)

angular distribution from polarized nucleus is the only sensitive means to determine the I = 1/2 moment.

In addition to the Gamow-Teller transition to the $1/2^+$ excited state of ^{12}N to be utilized for the g-factor measurement, the β decay of ^{15}C also has a branch to the $1/2^-$ ground state.[6] The measurement of β asymmetry for this $1/2^+ \to 1/2^-$ first-forbidden transition is by itself interesting since the role played by virtual pion in the weak axial current is predicted to enhance in this class of transition.[7] This interest affords an additional motivation to the present study of ^{15}C magnetic moment, since the spin control technique by means of NMR is indispensable for the precise measurement of β asymmetry free from systematic error.

EXPERIMENTAL METHOD

A 158 MeV ^{15}N beam from the AVF cyclotron at the Research Center for Nuclear Physics (RCNP), Osaka University, was used to produce ^{15}C isotope via the (^{15}N,^{15}C) reaction. The beam was pulsed with the beam-on and -off periods of 3 s and 7 s, respectively. A spectrometer DUMAS[8] provided a means for isotope separation of the projectile-like fragments. DUMAS, which has been constructed at RCNP mainly for use in the polarization transfer measurement for light ion reactions, is an achromatic spectrometer consisting of two dipole magnet and the associated focussing elements, as shown in Fig. 2. It has two focal planes, the first (F1) being the momentum-dispersive one on which particles with

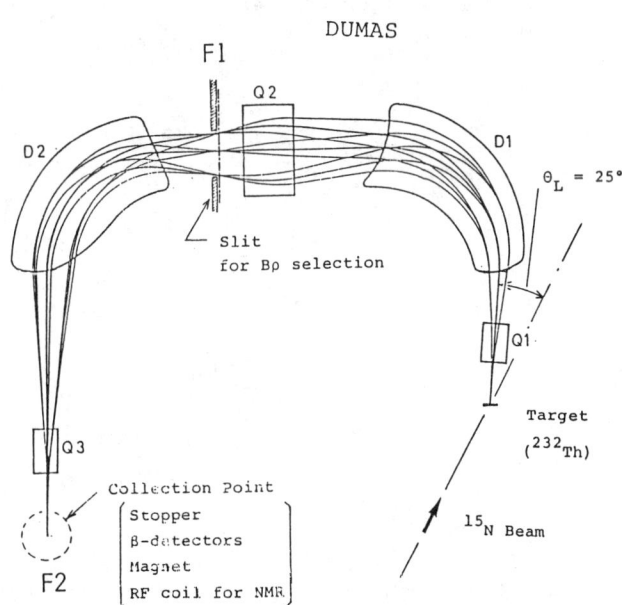

Fig. 2. Schematic view of the setup used for the collection and separation of projectile-like fragments from the heavy-ion reaction.

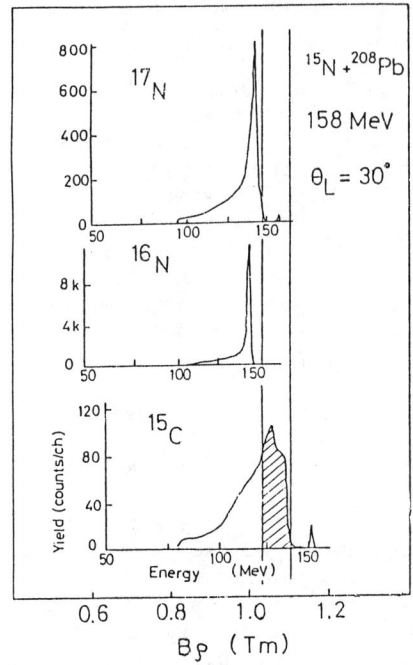

Fig. 3. Schematic illustration of isotope separation through the Bρ value.

different Bρ-values focus at different positions, and the second (F2) the achromatic one where all the particles are collected into the same position. In the present experiment the reaction products emitted from the target at $\theta_L = 25°$ were transmitted to DUMAS. They were selected by the slit placed on F1 which allowed those with Bρ = 0.99 - 1.07 Tm to pass through, and then were collected at F2. Fig. 3 schematically illustrates the isotope separation of neutron-rich products through the Bρ value.

A stopper was placed at the collection point F2. For the stopper substance we employed high-purity graphite, for the reason discussed later. The experimental arrangement used for the β ray detection and the NMR is shown in Fig. 4. Static magnetic field B_0

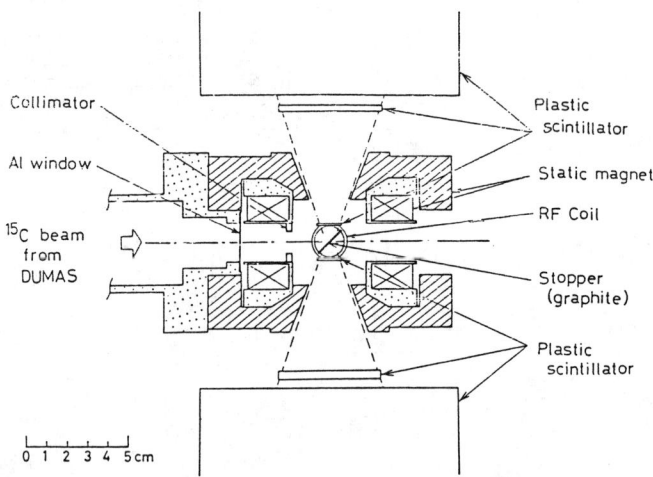

Fig. 4. Arrangement of the β detectors, stopper, static magnet, and rf coils.

Fig. 5. Time chart for the beam bombardment, application of the rf field, and the β detection.

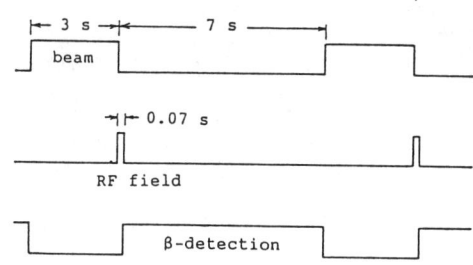

= 74.5 mT perpendicular to the reaction plane was applied on the stopper. A pair of coils for the rf field were installed around the stopper, for the purpose to be described later. β Rays emitted from the stopper were detected by using counter telescopes placed above and below the stopper. The β rays were detected during the beam-off period of the pulsed beam, as shown in Fig. 5. The observed energy and time spectra indicated that the main part of the detected β rays were those from ^{15}C.

The angular distribution of β rays[9] in the decay of polarized ^{15}C to the first excited state $(1/2^+)$ of ^{15}N is given by

$$W(\theta) \simeq 1 - (2/3)P\cos\theta, \qquad (1)$$

where P denotes the spin polarization of ^{15}C and θ the angle of β emission. Here the relevant β transition is assumed to be mainly due to the Gamow-Teller matrix element. Then the up/down ratio of the β ray yield is expressed in terms of P as

$$R = W(0°)/W(180°) \simeq 1 - (4/3)P. \qquad (2)$$

Several points have to be clarified to assure the preservation of polarization: Firstly the depolarization due to hyperfine interaction in a free ion is negligible, since in the present experiment the kinetic energies of the products are so high that they are predominantly in a fully stripped state during the flight to the stopper. Secondly the spin precession due to the fields inside the spectrometer is calculated to be small, using the well defined relation between the spin precession and the orbital deflection in an external field. Thirdly, spin relaxation in the stopper can be made negligible by choosing the material for the stopper. In fact the spin-lattice relaxation time T_1 for ^{15}C in graphite is estimated to be as long as 11 s at the room temperature, based on the T_1 value reported[10] for ^{13}C.

The adiabatic fast passage method of the NMR technique was employed. An rf field perpendicular to B_0 was applied for 70 ms duration preceding the β detection period, with the frequency being swept in the specified range ν to ν + Δν. When the frequency was swept across the resonance frequency $ν_0$ orientation of ^{15}C spins was inverted. The spin inversion was detecteded as a change in the measured up/down ratio R of the β ray yield. In addition to the resonance frequency, the polarization P can be deduced from the observed size of the change in R, by using expression

$$P = (3/8)[R_{off}/R_{on} - 1], \qquad (3)$$

where R_{on} and R_{off} denote the up/down ratio measured with the spin orientation inverted and not inverted, respectively.

RESULTS AND DISCUSSION

Change in β asymmetry due to the NMR was observed in the run with the rf field sweeping over the wide frequency window 1.167 - 2.331 MHz, indicating that the resonance ν_0 lies in this frequency region. Using Eq.(3), the spin polarization around 4 % was obtained for ^{15}C produced in ^{232}Th(^{15}N,^{15}C) reaction at E_{lab} = 158 MeV. P values changing from positive to negative values were observed when Bρ of ^{15}C products were varied. Fig. 6 shows the observed spin polarization as a function of Bρ value.

The resonance scans with narrower frequency windows were then performed. The results obtained for several different values of Δν

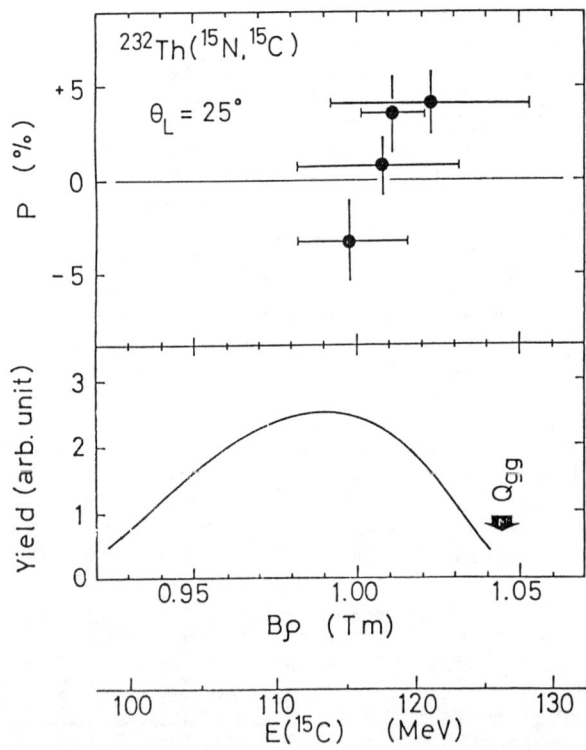

Fig. 6. Observed spin polarization of ^{15}C in the ^{232}Th(^{15}N,^{15}C) reaction, as a function of the ^{15}C kinetic energy.

are displayed in Fig. 7. From the spectrum obtained with the narrowest window ($\Delta\nu = 0.075$ MHz), the g factor for the ^{15}C was determined to be $|g| = 2.63 \pm 0.14$. The overall uncertainty associated with this measurement is determined solely by the rf window width $\Delta\nu$ in which the resonance is located. The Knight shift and other chemical shifts are assumed to be negligible compared to the experimental error asigned. The ^{15}C ground state wave function can be written[3,9] as ^{14}C(0^+) × $\nu 1s_{1/2}$ with small admixtures of other configurations such as ^{14}C(2^+) × $\nu 0d_{5/2}$ and ^{14}C(2^+) × $\nu 0d_{3/2}$. In fact an experimental spectroscopic factor close to unity has been reported for ^{14}C(d,p)^{15}C($1/2^+$) reaction.[11] The observed g factor, however, deviates substantially from the Schmidt value $g_{Schmidt} = g_n = -3.83$ for the single $s_{1/2}$ neutron configuration, where g_n denotes the free neutron g factor. A simple estimate of correction for the first-order core polarization,[12] $\delta g^{1st} = 0.23$, explains only a small part of the observed deviation $\delta g^{exp} = +1.20 \pm 0.14$ from the Schmidt value (where the negative sign for the observed value is assumed). This is in contrast to the case for the $5/2^+$ excited state of ^{15}C, for which a $\nu d_{5/2}$ single particle model including the first-order correction well explains the observed g factor.[4] Clearly, a more

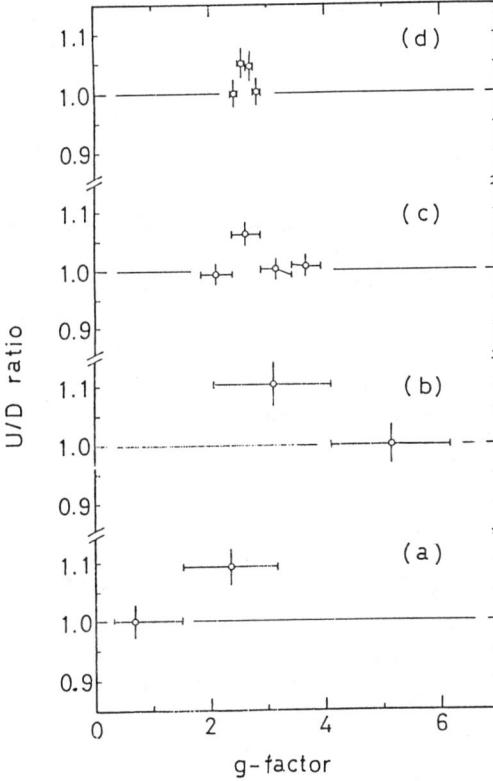

Fig. 7. Up/down ratio of β-ray yield, measured after the application of an rf field with the frequency swept over the specified ranges (represented by horizontal bars).

elaborate theoretical treatment including higher order configurations is needed to clarify the situation for the ^{15}C ground state.

CONCLUSION

^{15}C isotope produced in ^{15}N-induced reaction was separated from other β activities using the achromatic spectrometer DUMAS. Use of the DUMAS largely eliminated interference due to the background β activities and enabled the detection of a rather small effect of spin polarization. Spin polarization of $P \simeq 4$ % was observed via the up/down asymmetry of β rays from ^{15}C. The ^{15}C products were implanted in a high-purity graphite stopper, and the spin inversion induced by NMR was detected through the observation of the change in the asymmetry. From the observed resonance frequency the g-factor for the ground state of ^{15}C has been determined.

The present result has shown that the in-flight isotope separation applied to the projectile-like fragments from the heavy-ion reactions indeed provides a new tool for the detailed studies of nuclei far from stability. This method makes it possible to utilize the reaction-induced polarization, which previously has been only possible in light-ion reactions at low energies where limitted number of reaction channels are open. Spin polarization $P \simeq 4$ % obtained in the present experiment was not very large but was still enough to be used in determining g-factor. It would be very interesting to explore how the spin polarization in products behaves as the beam energy goes up towards the domain of the projectile fragmentation.[13]

REFERENCES

1) W.F. Rogers, D.L. Clark, S.B. Dutta and A.G. Martin, Phys. Lett. 77B, 293 (1986).
2) K.H. Tanaka, Y. Nojiri, T. Minamisono, K. Asahi, and N. Takahashi, Phys. Rev. C34, 580 (1986).
3) D.E. Alburger and D.J. Millener, Phys. Rev. C20, 1891 (1979).
4) J. Asher, D.W. Bennett, B.A. Brown, H.A. Doubt, and M.A. Grace, J. Phys. G6, 251 (1980).
5) G. Karl, G.A. Miller, and J. Rafelski, Phys. Lett. 143B, 326 (1984).
6) E.K. Warburton, D.E. Alburger, and D.J. Millener, Phys. Rev. C29, 2281 (1984).
7) K. Kubodera, J. Delorme, and M. Rho, Phys. Rev. Lett. 40, 755 (1978).
8) T. Noro, T. Takayama, H. Ikegami, M. Nakamura, H. Sakaguchi, H. Sakamoto, H. Ogawa, M. Yosoi, T. Ichihara, N. Isshiki, M. Ieiri, Y. Takeuchi, H. Togawa, T. Tsutsumi, and S. Kobayashi, J. Phys. Soc. Japan 55 Suppl., 470 (1986).
9) M. Morita, "Beta Decay and Muon Capture" (W.A. Benjamin Inc.,

     Reading, Massachusetts, 1973), p.317.
10) G.P. Carver, Phys. Rev. $\underline{B2}$, 2284 (1970).
11) F.E. Cecil, J.R. Shepard, R.E. Anderson, R.J. Peterson and P. Kaczkowski, Nucl. Phys. $\underline{A255}$, 243 (1975).
12) H. Morinaga and T. Yamazaki, "In-Beam Gamma-Ray Spectroscopy" (North-Holland Publ. Co., Amsterdam, 1976), Chap. 3; H. Noya, A. Arima, and H. Horie, Prog. Theor. Phys. Suppl. $\underline{8}$, 33 (1958).
13) D. Guerreau, J. de Phys. $\underline{C4}$, 207 (1986).

TRISTAN RESEARCH ON NEUTRON RICH-NUCLEI:
MAGNETIC MOMENTS AND NUCLEAR STRUCTURE

A. Wolf[*], R. F. Casten, and R. L. Gill
Brookhaven National Laboratory, Upton, New York, 11973, USA

D. S. Brenner and H. Mach
Clark University, Worcester, Massachusetts, 01610, USA

John C. Hill, F. K. Wohn, and J. A. Winger
Ames Laboratory, Iowa State University, Ames, Iowa, 50011, USA

ABSTRACT

The TRISTAN research program is briefly outlined. The discussion then turns to the measurements of magnetic moments of excited states in neutron-rich fission product nuclei using the perturbed angular correlation method. The results for 2^+_1 states are analyzed in light of the systematics of magnetic moments in the A=70-200 region, and in relation to collective nuclear models.

INTRODUCTION

TRISTAN, an isotope separator on-line to the Brookhaven National Laboratory High Flux Beam Reactor, is used to study neutron-rich fission product nuclei far from stability. With an array of excellent ion sources, it provides access to about 400 isotopes far from stability. The facility currently operates as a joint BNL-Ames Laboratory-Clark University collaboration. The four main (interrelated) areas of research (and recent results) are:

• Study of phase transitions, collectivity, and the onset of deformation in nuclei, with special emphasis on the crucial role of the p-n interaction.

Results: a) Development of the $N_p N_n$ scheme and the related idea of the P-factor[1]. This approach to nuclear systematics simplifies the interpretation of each transitional region, unifies the interpretation of different regions, provides quantitative "rules" for the onset of deformation and clearly distinguishes the monopole and quadrupole components of the p-n interaction. Recent work on nuclear masses in this scheme is also promising and seems to have predictive power. b) $g(2^+_1)$-factor measurements near A=100, 150 providing quantitative empirically-based information on the effective numbers of valence protons and neutrons and on the evolution of shell gap structure[2-3]. c) Spectroscopy near A=100, 150 leading to further information on the nature, mixing and coexistence of spherical and deformed states[4], and on the possibility of pairing free rotations[5]. d) New techniques for lifetime measurements in the 100 psec range[6].

[*]On leave from Nuclear Research Center Negev, Beer Sheva, Israel

- Symmetries in Nuclei

    Results: Discovery of the first good example (30 years after its proposal) of a vibrational nucleus, namely ^{118}Cd, in which Aprahamian and co-workers[7] discovered the complete 3-phonon quintuplet and possible evidence for higher phonon excitations.

- New Magic Regions

    Results: Spectroscopy and g-factor studies near ^{78}Ni (^{82}Ge and ^{83}As) and ^{132}Sn (^{136}Xe, ^{138}Ba, and ^{132}Te).[8-9] Recent shell model calculations indicate ^{132}Sn to be a valid doubly magic core, but are not as successful for the N=50 isotones above ^{78}Ni.

- Astrophysics Related to r-process Nucleosynthesis

    Results: Half-life, Q_β value and decay scheme for ^{80}Zn,[10] one of the three crucial "waiting point" nuclei whose properties provide key constraints on the stellar supernova "exposure time" for r-process nucleosynthesis. New isotopes near the r-process path at A≈130 are currently under study.

In this report we will primarily concentrate on the presentation of the results of magnetic moment measurements which were undertaken five years ago at TRISTAN, and have so far provided data for neutron-rich nuclei far from stability. These data could not be obtained by techniques other than mass separation of fission products. The results will be discussed in light of a systematic analysis of g-factors of 2^+_1 excited states in 65 nuclei in the range A=70-200, and the existing B(E2) data for these nuclei. Fine structure effects in the $g(2^+_1)$ vs. A dependence were observed in the A=100, 150, and 170 regions. These effects can be related to the underlying shell structure, and, in the case of the neutron-rich Er and Yb isotopes, to a "saturation" in the proton-neutron interaction which takes place towards the middle of the N = 82-126 shell.

EXPERIMENTAL TECHNIQUES AND RESULTS

The TRISTAN ISOL facility was described in some detail in previous publications[11]. It provides intense beams of short-lived neutron-rich nuclei produced by thermal neutron fission induced by a $3 \times 10^{10} n_{th}/cm^2$-sec neutron beam incident on a 5g ^{235}U target. During the past six years, several ion sources have been developed, each of which is best suited to producing certain elements: a) the surface ionization source[12] mainly produces high activites of Rb and Cs isotopes; b) the thermal source[13] was designed to optimize the yield of rare-earth elements; c) the high-temperature plasma source[11] produces a large variety of elements (e.g., Zn, Ga, Ge, Ag, In, I, Xe) with relatively high yields. Typically, most yields at the output of the separator are of the order of 10^3-10^8 atoms/sec depending on mass number, half-life, and type of source used. The radioactive beam is deposited on an aluminized plastic tape, and the accumulated activity is subsequently transported to the desired counting position. For the magnetic moment measurements, the counting position is in the center of a superconducting magnet capable of providing a field of up to 6.25 Tesla with inhomogeneity <1% in a volume of about 1 cm^3.

Four large (80-120 cm^3) hyperpure Ge detectors are set at about 10 cm from the center of the magnet to detect gamma radiations from the source. The magnet was specially designed so that the absorption between the source and the detector is minimal. This is particularly important for very deformed nuclei, where the transitions between low-lying levels have relatively low energy.

The electronics and data acquisition system associated with the four detector system has been described previously[14]. It is capable of recording coincidence events between any two detectors. Thus, six angles are measured simultaneously. To determine a magnetic moment, we use the integral perturbed angular correlation method (IPAC) and calculate the double ratio:

$$R^2(\theta) = \frac{I(\theta,B)}{I(\theta,-B)} \bigg/ \frac{I(-\theta,B)}{I(-\theta,-B)} \tag{1}$$

where $I(\theta,\pm B)$ is the actual number of coincidence events in the specific γ-γ cascade, at angle θ, with magnetic field up or down. In general, it is convenient to use $0^+ \to 2^+ \to 0^+$ cascades, which have a very large anisotropy and give large effects. For these cascades, $R(\theta)$ has a maximum at about 150°. Consequently, three of the six angles were set at 150°, two at 120°, and one at 90°. The effect is also strongly dependent on the g-factor, the applied magnetic field, and the half-life of the intermediate state. The maximum available field (6.25T) imposes the limitation $T_{1/2} > 0.05$ nsec for excited states that can be investigated using the IPAC technique. In Fig. 1 we

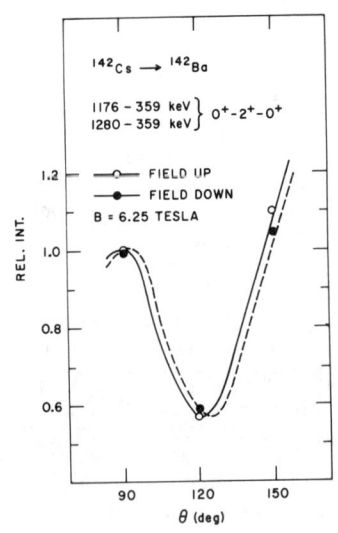

Fig. 1. Results of PAC measurement for ^{142}Ba.

Table I Experimental g-factors measured at TRISTAN

Isotope	Level Energy (keV)	J^π	g_{exp}
^{97}Zr	1264	$7/2^+$	0.39(4)
^{98}Sr	144	2^+	0.38(7)
^{132}Te	1775	6^+	0.79(9)
^{136}Xe	1694	4^+	0.80(15)
^{138}Ba	1899	4^+	0.80(14)
^{142}Ba	359	2^+	0.426(48)
^{144}Ba	199	2^+	0.33(5)
^{146}Ba	181	2^+	0.27(7)
^{146}Ce	259	2^+	0.24(5)
^{148}Ce	158	2^+	0.37(6)

present the perturbed angular correlation recently measured with field up and field down for two $0^+ \to 2^+ \to 0^+$ cascades in ^{142}Ba. The results for the two cascades were summed to improve statistics. The solid (dashed) lines were calculated for field up (down) using $g(2^+_1)$ = 0.42. The experimental errors are smaller than the size of the points. In this case, $T_{1/2}(2^+_1)$ = 0.066 nsec and the effect at 150° was quite small: R(150°) = 1.053(6). A much larger effect, R(150°) = 1.47(9), was obtained, for example, for the 71-144 keV, $0^+ \to 2^+ \to 0^+$ cascade in ^{98}Sr, with a field of only 1.7T, mainly because in that case $T_{1/2}(2^+_1)$ = 2.7 nsec.

In Table I we present g-factor values that have been measured at TRISTAN. In the following discussion we will concentrate only on the $g(2^+_1)$ data and present an analysis in the context of systematics of magnetic moments across the periodic table.

SYSTEMATICS OF g-FACTORS OF 2^+_1 STATES IN THE RANGE A=70-200

The simplest prediction for the g-factor of the lowest-lying 2^+ state in a collective (vibrational or rotational) nucleus is just Z/A. This is the hydrodynamical approach, which assumes that all, and only, the protons contribute to the magnetic moment. A correction to this model was suggested by Greiner[15], who took into account the different pairing forces between protons and neutrons. This effect causes a reduction of the g-factors by about 15-20%. The experimental data does indeed show such a reduction. However, in many cases, deviations from the simple Z/A dependence have been observed[16]. An alternative approach, which gives a much more accurate prediction of $g(2^+_1)$ in many nuclei, is provided by the neutron-proton version of the Interacting Boson Approximation (IBA-2). In this model, $g(2^+_1)$ is given by the simple relation[16]:

$$g(2^+_1) = g_\pi N_\pi/N_t + g_\nu N_\nu/N_t, \qquad (2)$$

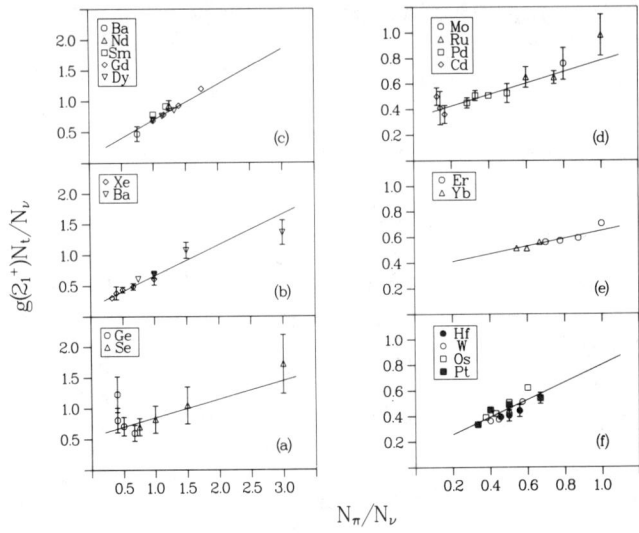

Fig. 2. Systematics of $g(2^+_1)$ data for six groups of nuclei in A=70-200 mass range. The lines are linear fits of $g(2^+_1) \times N_t/N_\nu$ vs N_π/N_ν.

for totally symmetric states in the proton, neutron degrees of freedom. $g_\pi(g_\nu)$ are the proton (neutron) boson g-factors, $N_\pi(N_\nu)$ are the respective numbers of bosons, and $N_t = N_\pi + N_\nu$. The essential difference between this IBA expression (2) and the hydrodynamic model does not depend specifically on the boson assumption or on the detailed IBA Hamiltonian but rather resides in the fact that the former assumes that the g-factor, and indeed all collective properties, arise from the <u>valence</u> nucleons whereas the hydrodynamic approach assumes that the entire nucleus participates equally. The difference is thus that between geometric and core plus valence particle pictures. In order to test to what extent the experimental data indeed follows the above linear relation, we analyzed[17] $g(2^+_1)$ values for 65 nuclei in the range A=70-200, using equation 2. In order not to obscure gross variations of g_π, g_ν with mass number, the data was divided into six groups, following the major proton and neutron shells. Nuclei in the immediate vicinity of closed shells were not included in the analysis. For each group, a linear fit was made, assuming g_π, g_ν are constant for all nuclei in the group. The linear fits (Fig. 2) clearly show that equation 2 gives a good description of the data. Quantitatively, values of g_π, g_ν were obtained for each group. In general, a smooth behavior of g_π, g_ν vs A is observed[17], with g_π slightly increasing and g_ν decreasing for the heavier nuclei. For nuclei with A>130, we obtain g_π = 0.5-0.7, g_ν = 0.05-0.15. These results are in contradiction with the simple expectation g_π = 1.0, g_ν = 0.0, based on the assumption that there are only orbital contributions to the collective g-factors.

A significant deviation from the smooth dependence of g_π, g_ν vs A occurs for the neutron-rich Er,Yb isotopes. This is evident from Fig. 2e, where the slope of the line is much smaller, and the intercept much larger (g_π = 0.30, g_ν = 0.35), than those in 2c (Ba-Dy) and 2f (Hf-Pt), which contain deformed nuclei with similar structure. This anomalous behavior will be discussed in a following section.

Now we proceed to discuss the TRISTAN results in light of the systematics presented above.

$g(2^+_1)$ DATA IN THE A=100 REGION (in collaboration with K. Sistemich)

The latest result from TRISTAN in this mass region is $g(2^+_1)$ = 0.38(7) for ^{98}Sr. Two other g-factor values for N=60 isotones are known to date: $g(2^+_1)$ = 0.22(5) for ^{100}Zr,[18] and $g(2^+_1)$ = 0.42(7) for ^{102}Mo.[19] The values for Sr,Mo fit well in the systematics of Fig. 2d. In fact, if we use $g_\pi = g_\nu \approx 0.4$ for A=100 nuclei[17], we obtain $g(2^+_1)$ = 0.4, in excellent agreement with the experimental data. The reason for these particular values of g_π, g_ν is unclear at present. They may be due to the underlying shell structure in this region.

The strongly reduced value for ^{100}Zr can not be explained with the above values of g_π, g_ν. It has been suggested[19] that the Z=38 subshell is still active in this nucleus, thus reducing the effective number of valence protons, and consequently the g-factor. However, when $g_\pi = g_\nu$, equation (2) reduces to $g(2^+_1) = g_\pi$, and thus the low value for ^{100}Zr is difficult to reconcile with the new result for ^{98}Sr and the systematics in the A=100 region. It might be possible to resolve this problem with a detailed shell model calculation in which the contributions of the different proton and neutron configurations are taken into account, but, at present it remains a puzzling experimental result.

DISSIPATION OF THE Z=64 SUBSHELL IN THE A=150 REGION

The existence of a proton subshell closure at Z=64 has been well established[20]. It was found[21] that this subshell is active for N≤88, but "disappears" when N≥90. Since in IBA-2, $g(2^+_1)$ depends on N_π, N_ν (equation (2)), we expect g-factor data in the A=150 region to be affected by the existence of the Z=64 subshell, due to a change in N_π of a given element when N changes around 90. Such an effect was observed experimentally[22] for Ce, Nd, and Sm isotopes. This suggests that we can in fact use experimental $g(2^+_1)$ data to extract the number of valence protons using equation (2), and thus follow the dissipation of the Z=64 subshell. In order to do this we need: a) to obtain a set of g_π, g_ν values; b) to make an assumption about the values of N_ν, or use a second equation so as to calculate both N_π and N_ν simultaneously.

g_π, g_ν can be deduced from our systematic analysis (Fig. 2). For the A=150 region we have: g_π = 0.63, g_ν = 0.05. These values were obtained from a linear fit (Fig. 2c) in which isotopes with N≤88 have not been included. Although it is reasonable to assume that N_ν can be simply counted in the major N=82-126 shell,

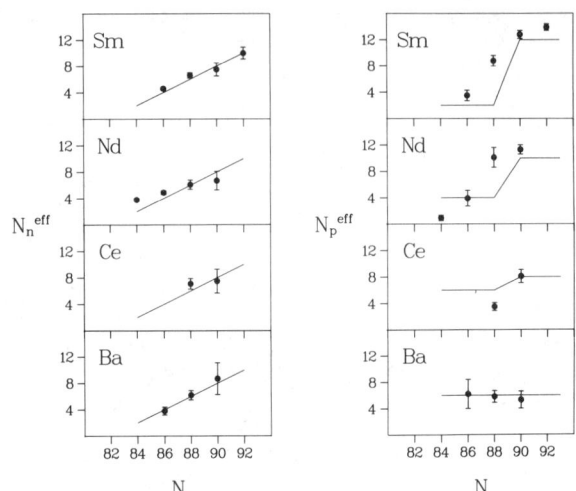

Fig. 3. Effective numbers of valence protons and neutrons in A=150 region, extracted from $g(2^+_1)$ and B(E2) data.

we will take here a different approach[23]. Namely, we will use B(E2) data for the 2^+_1 states involved, in addition to $g(2^+_1)$ data, so that both N_π and N_ν can be obtained. This procedure involves the use of analytical formulas for the B(E2) transition probabilities. We used the expression:

$$B(E2:0^+ \to 2^+) = 5/N_t \, (e_\pi N_\pi + e_\nu N_\nu)^2 \qquad (3)$$

for the vibrational nuclei[24] in the A=150 region, and:

$$B(E2:0^+ \to 2^+) = 0.05 \, (1+1/N_t)^2 \, (e_\pi N_\pi + e_\nu N_\nu)^2 \qquad (4)$$

for the deformed nuclei[25]. Both formulas are based on the IBA. The effective charges e_π, e_ν were deduced by linear fits to experimental data, similarly to the calculation of g_π, g_ν. Then, we used the $g(2^+_1)$ data from Table I for Ba, Ce isotopes, in addition to $g(2^+_1)$ values for Nd, Sm,[26] and B(E2) data from the recent compilation of Raman et al.[27], and extracted the effective numbers of valence protons and neutrons, N_p^{eff}, N_n^{eff}, by solving equations (2) and (3) or (2) and (4). The results are given in Fig. 3. We see that while N_n^{eff} follows very closely the values we would obtain by "normal" counting in the 82-126 shell, N_p^{eff} for Ce, Nd, and Sm exhibits a significant drop for N≤88. This behavior is consistent with the existence of a Z=64 subshell, and shows that the dissipation is gradual in the region N=86-92. Moreover, the Ba isotopes do not show any change in N_p^{eff}, as expected since Z=56 is below midshell for both 50-64 and 50-82 shells.

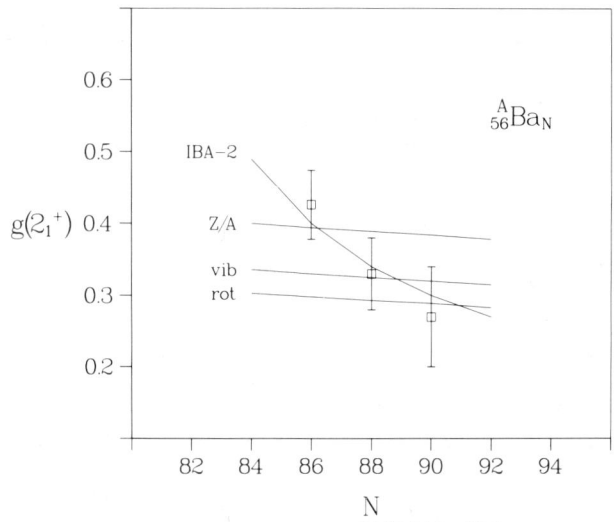

Fig. 4. Experimental $g(2^+_1)$ data for ^{142,144,146}Ba isotopes.

COMPARISON BETWEEN IBA AND HYDRODYNAMICAL MODELS FOR ^{142,144,146}Ba ISOTOPES

As we mentioned in a previous section, both IBA-2 and the hydrodynamical models give reasonable predictions for the gross mass dependence of $g(2^+_1)$. Therefore, in order to differentiate between these models, fine structure effects should be investigated. One example is the N-dependence of $g(2^+_1)$ for the transitional nuclei in the A=150 region which was discussed above. We saw that the valence particles play a major role in determining the g-factors, in good agreement with the IBA. Another example is the mass dependence of $g(2^+_1)$ for the neutron-rich ^{142,144,146}Ba isotopes. These g-factors were measured at TRISTAN (see Table I), and the results are plotted in Fig. 4 vs neutron number. Also shown in the figure are the Z/A line, the curves that include Greiner's correction[15] for vibrational and rotational nuclei, and the IBA-2 curve based on equation 2. We used $g_\pi = 0.63$, $g_\nu = 0.05$ in equation 2. From Fig. 4 we see that: a) the mass dependence of $g(2^+_1)$ is very different from Z/A, but in good agreement with IBA-2; b) for ^{142}Ba, $g(2^+_1)$ is about two standard deviations larger than the hydrodynamical value which includes Greiner's correction. These results illustrate clearly that accurate g-factor measurements can help to distinguish between different nuclear models.

ANOMALOUS g_π, g_ν IN NEUTRON-RICH Er,Yb ISOTOPES

In the systematic analysis presented earlier we noticed that in the Er,Yb region $g_\pi = 0.30$, $g_\nu = 0.35$, i.e., very different from $g_\pi \approx 0.6-0.7$ and $g_\nu \approx 0.05-0.10$ which were observed for the deformed nuclei in the A=150-200 region (see Figs. 2c, 2e, 2f). This is rather surprising in view of the fact that the neutron-rich Er,Yb

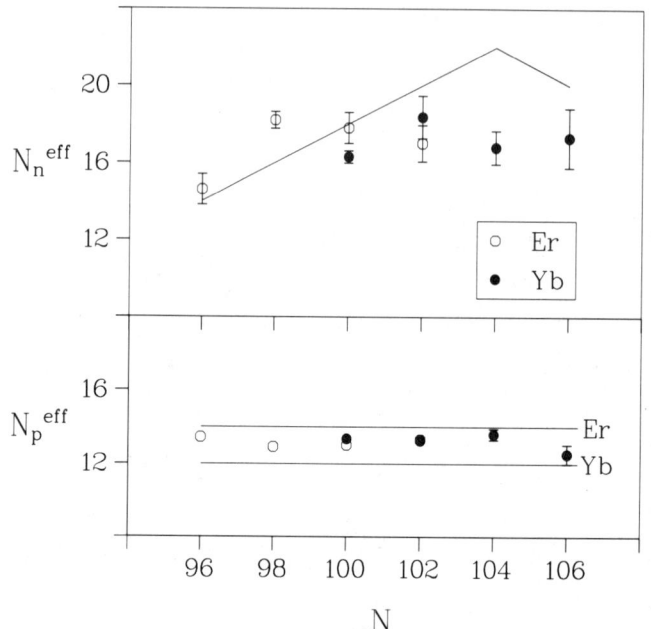

Fig. 5.

Effective numbers of valence particles in neutron-rich Er, Yb isotopes, extracted from $g(2^+_1)$ and $B(E2)$ data.

isotopes considered are not very different in structure from the neighboring Gd, Dy, Hf, and W isotopes. A possible interpretation of this anomaly is that, like the A=150 region discussed earlier, there is a change here in the effective N_π, N_ν, with respect to "normal" counting. In order to follow this assumption quantitatively, we will assume "normal" g_π, g_ν values ($g_\pi = 0.65$, $g_\nu = 0.08$) and calculate N_p^{eff}, N_n^{eff} from the experimental data. Again, we use B(E2) data and relation (4) as a second equation in addition to $g(2^+_1)$ data and relation (2). The results are shown in Fig. 5. We see that while N_p^{eff} for both Er and Yb are quite close to the "normal" values of 14 and 12, respectively, the N_n^{eff} values tend to "saturate" for N>100, and are about 20-30% below "normal". This reduction can be explained in terms of a "saturation" in the integrated proton-neutron interaction near midshell as the protons and neutrons fill orbits with varying spatial overlap. That is, the p-n interaction summed over all occupied orbits no longer scales with $N_p N_n$. This can be expressed in terms of lower "effective" $N_p N_n$ values and is reflected in the present approach to the data in terms of reduced effective N_n values. A quantitative calculation[28] of the proton-neutron interaction in a deformed field supports this argument and clearly shows a "saturation" of $(N_p N_n)^{eff}$ for $N \geq 100$.

CONCLUSIONS

The results presented in this work show that magnetic moments of excited states in neutron-rich nuclei provide valuable nuclear structure information. In this paper we concentrated only on the significance of $g(2^+_1)$ data and its relation to collective nuclear models. The results for other than 2^+_1 states (Table I) were found to be of importance for understanding the shell structure of the respective nuclei. In the future we plan to do more experiments in odd-mass nuclei and also to study higher spin states in even-even nuclei, particularly around the double-magic ^{132}Sn nucleus.

ACKNOWLEDGEMENTS

The authors are indebted to A. Piotrowski for his help in developing the ion sources. Thanks are also due to K. Sistemich for collaborating on the ^{98}Sr experiment and to D. D. Warner for collaboration on the systematic analysis of g-factors. Research has been performed under contract DE-AC02-76CH00016 with the United States Department of Energy.

REFERENCES

1. R. F. Casten, D. S. Brenner and P. E. Haustein, Phys. Rev. Lett. 58, 658 (1987).
2. A. Wolf et al., Phys. Lett. 123B, 165 (1983).
3. R. L. Gill et al., Phys. Rev. C33, 1030 (1986).
4. F. K. Wohn et al., Phys. Rev. C33, 677 (1986); F. K. Wohn et al., Phys. Rev. Lett. 51, 1873 (1983).
5. L. K. Peker et al., Phys. Lett. 169B, 323 (1986).
6. H. Mach et al., to be published.
7. A. Aprahamian et al., Phys. Rev. Lett. 59, 535 (1987).
8. Z. Berant et al., Phys. Rev. C31, 570 (1985).
9. B. Fogelberg et al., Nucl. Phys. A451, 104 (1986).
10. R. L. Gill et al., Phys. Rev. Lett. 56, 1874 (1986).
11. A. Piotrowski, R. L. Gill and D. C. McDonald, Nucl. Instrum. & Methods B26, 249 (1987).
12. M. Shmid et al., Nucl. Instrum. & Methods 211, 287 (1983).
13. A. Piotrowski et al., Nucl. Instrum. & Methods 224, 1 (1984).
14. A. Wolf et al., Nucl. Instrum. & Methods 206, 397 (1983).
15. W. Greiner, Nucl. Phys. 80, 417 (1966).
16. M. Sambataro et al., Nucl. Phys. A423, 333 (1984).
17. A. Wolf, R. F. Casten and D. D. Warner, Phys. Lett. 190B, 19 (1987).
18. A. Wolf et al., Phys. Lett. 97B, 195 (1980).
19. G. Menzen et al., Z. Physik A321, 593 (1985).
20. M. Ogawa et al., Phys. Rev. Lett. 41, 289 (1978).
21. R. F. Casten et al., Phys. Rev. Lett. 47, 1433 (1981).
22. A. Wolf, D. D. Warner and N. Benczer-Koller, Phys. Lett. 158B, 7 (1985).
23. A. Wolf and R. F. Casten, Phys. Rev. C, in press.
24. J. N. Ginocchio and P. Van Isacker, Phys. Rev. C33, 365 (1986).
25. R. F. Casten and A. Wolf, Phys. Rev. C35, 1156 (1987).
26. N. Benczer-Koller et al., Ann. Israel Phys. Soc. 7, 133 (1984).
27. S. Raman et al., At. Data and Nucl. Data Tables 36, 1 (1987).
28. R. F. Casten, K. Heyde and A. Wolf, to be published.

DEFORMATION OF THE LIGHT Br ISOTOPES.

A.G. Griffiths, C.J. Ashworth, M.J. Reynolds, J. Rikovska, N.J. Stone,
W.B. Walters and J.P. White.
Clarendon Laboratory, Oxford University, Oxford OX1 3PU.

P.M. Walker.
Daresbury Laboratory and University of Surrey, Guildford.

P. Herzog and U. Dammrich.
University of Bonn, D-5300, Bonn.

K. Schloesser and I.S. Grant.
The ISOLDE Collaboration, CERN, CH-1211, Geneva

ABSTRACT.

The nuclear magnetic moments of ^{74m,75,77g}Br have been measured by Low Temperature Nuclear Orientation, using the reaction of a 150MeV ^{28}Si beam on ^{54}Fe. Comparison with particle plus triaxial rotor (PTR) calculations assign the [312]3/2 orbital as the ground state for both ^{75}Br and ^{77}Br, and gives evidence for a considerable increase in the nuclear deformations of ^{75,77}Br as compared to ^{79}Br.

INTRODUCTION.

Among the nuclei off the line of stability, those in the N $\sim$ Z $\approx$ 40 region are of particular interest, exhibiting large axially symmetric quadrupole deformations, triaxiality and shape co-existence [[1,2,3,4]]. In the present work we report on recent nuclear orientation measurements at Daresbury, Bonn and Oxford which yield the nuclear magnetic dipole moments of ^{74m,75,77g}Br. These moments, by identifying the single particle orbitals present give direct evidence of the equilibrium ground state deformation of these isotopes.

EXPERIMENTAL DETAILS.

Sources of ^{77}Br (3/2$^-$, 57h) implanted into iron foils were prepared at the ISOLDE facility, CERN and studied both in Bonn and Oxford at temperatures down to 2mK. The isotopes ^{75}Br (3/2$^-$, 98m) and ^{74m}Br (4$^-$, 41.5m) were produced using 150 Mev ^{28}Si on a ^{54}Fe target which formed part of the FEBIAD ion source of the Daresbury isotope separator DOLIS. After acceleration to 60keV the selected Br ions were implanted into an iron foil on the copper cold finger of the on-line dilution refrigerator. The presence of an inactive contaminant beam at mass 75 required that the ^{75}Br source was built up by implanting for 2.5h, followed by cooling to temperatures down to 10mK with the separator closed off, this cycle being repeated several times. ^{74m}Br was implanted with the iron foil at temperatures down to 15mK.

RESULTS.

The implanted Br nuclei are oriented at low temperatures through the large magnetic hyperfine interaction that they experience in an iron lattice. Ge(Li) detectors, placed parallel and perpendicular to the axis of orientation, measure the angular distribution of radiation from the polarised nuclei, given by

$$W(\theta) = 1 + f \sum_{k\ even} B_k U_k A_k P_k(\cos\theta)$$

where the symbols have their usual meaning [5]. The strength of the hyperfine interaction, μB_{hf}, determines the temperature dependence of the anisotropy of angular distribution, defined as $(W(0)/W(90)-1)\%$. Experimental data on the 728keV, 377keV and 575keV transitions from the decays of ^{74m}Br, ^{75}Br and ^{77}Br respectively are shown in fig.1. Taking the measured hyperfine field of BrFe (B_{hf}=83.31(8)T [6]) and using a model in which a fraction f of nuclei experience the full hyperfine field while the remainder undergo no interaction, the nuclear magnetic moments given in table 1 are deduced. It has been shown [6] that this simple model fails to completely describe the true distribution of available lattice sites, but at a level which will not influence the present results.

Table 1.
Spins And Magnetic Moments In The Light Br Isotopes.

Nuclide	Spin	Moment (expt.) μ_N	Moment (calc.) [a] Conf.	μ	Conf.	μ
^{79}Br	$3/2^-$	+2.1064 [7]	$\pi[301]3/2$	+2.0	$\pi[312]3/2$	+0.75
^{77}Br	$3/2^-$	±0.845(16)	$\pi[301]3/2$	+2.0	$\pi[312]3/2$	+0.75
^{76}Br	1^-	±0.5482 [7]	$\pi[301]3/2$ $\nu[422]5/2$	−1.5	$\pi[312]3/2$ $\nu[422]5/2$	−0.57
^{75}Br	$3/2^-$	±0.73(9)	$\pi[301]3/2$	+2.0	$\pi[312]3/2$	+0.75
^{74m}Br	4^-	±1.45(13)	$\pi[301]3/2$ $\nu[422]5/2$	+1.7	$\pi[312]3/2$ $\nu[422]5/2$	+0.24

[a] theoretical odd-odd moments adapted from Rb calculations [1]
theoretical odd-even moments from present work.

P.T.R. CALCULATIONS FOR LIGHT ODD-A Br ISOTOPES.

The experimental levels in ^{75,77}Br were analysed within the framework of a generalised particle-asymmetric rigid rotor model [9], in order to obtain a consistent set of parameters with which to calculate the nuclear magnetic moments. The total Hamiltonian is the sum of two terms, H_{sp} and H_r. The first term, H_{sp} describing the odd particle motion with respect to the core in the intrinsic system, is represented by the Nilsson anisotropic oscillator potential expressed in the doubly stretched co-ordinate system

$$V_{osc} = \frac{1}{2}\hbar\omega_o(\epsilon,\gamma)\rho^2\left\{1 - \frac{2}{3}\epsilon\sqrt{\frac{4\pi}{5}}\left(Y_{20}\cos\gamma - \frac{1}{\sqrt{2}}\sin\gamma(Y_{22} + Y_{2\,-2})\right)\right\}$$
$$- 2\kappa\hbar\omega_o(\epsilon,\gamma)\left(\vec{l}\cdot\vec{s} - \mu(\rho^4 - \langle\rho^4\rangle_N)\right)$$

where κ and μ are adjustable coupling parameters. The multipole moments of the core field, represented by ϵ and γ define the nuclear deformation via the components of the oscillator frequency

$$\omega_k = \omega_o(\epsilon,\gamma)\left\{1 - 2\frac{\epsilon}{3}\cos(\gamma + \frac{2k\pi}{3})\right\}$$

the factor ω_o being scaled to conserve the nuclear volume according to the requirement $\omega_x\omega_y\omega_z$ =constant.

The second term, H_r, is the particle rotor Hamiltonian

$$H_r = \sum_k \frac{\hbar^2}{2J_k}(I_k^2 + j_k^2 - 2\xi\vec{I}_k\cdot\vec{j}_k)$$

where $\vec{I}_k$ and $\vec{j}_k$ are the total angular momentum of the nucleus and the odd nucleon respectively and ξ is the Coriolis attenuation factor. The constant moments of inertia, J_k, are those predicted by the hydrodynamical model

$$J_k = \frac{4\hbar^2}{E_{2+}}\sin^2(\gamma + \frac{2k\pi}{3})$$

with the moment of inertia parameter E_{2+} approximated by the empirical relation of Grodzins [10]

$$E_{2+} \approx \frac{1100}{A^{7/3}\epsilon^2} \text{ MeV}$$

Pairing is introduced via a standard BCS calculation on the adiabatic single particle levels. The parameters varied were κ and μ, associated with the $\vec{l}\cdot\vec{s}$ and $\vec{l}\cdot\vec{l}$ terms, the moment of inertia parameter E_{2+} and the deformation parameters ϵ and γ.

The single particle orbitals near to the Z=35 Fermi level are shown in fig.2. The coupling parameters used, $\kappa = 0.07$ $\mu = 0.34$, are consistent with the values appearing in [11]. Figure 3 compares the calculated and experimental negative parity levels as a function of quadrupole deformation.

No γ deformation was found to be required, a level built on the [310]1/2$^-$ orbital becoming the ground state for $\gamma > 25°$. The moment of inertia parameter was set to 350keV, consistent with Grodzin's relation, with the Coriolis attenuation parameter ξ fixed at unity. This produced the ground state band [12] (marked by *) as far as spin 9/2$^-$ but predicted higher spins far too high in energy. The experimental level scheme has a 5/2$^-$ level at 162keV and a (5/2,7/2$^-$) level at 114keV. Calculations cannot produce a second prolate low lying 5/2$^-$ state. If this level is identified with the [303]7/2$^-$ orbital then the deformation is accurately determined as $\epsilon = 0.345$. With these parameters the ground state consists of 87% [312]3/2$^-$ and 10% [310]1/2$^-$ with static magnetic dipole and electric quadrupole moments of $0.77\mu_N$ and 0.60eb respectively. The measured moment of $\pm 0.845(16)\mu_N$ identifies the [312]3/2$^-$ deformationally aligned ($\langle K \rangle = 1.47$) level as the ground state. Of the positive parity states, the possibility of assigning the [431]3/2$^+$ orbital, originating from the spherical $g_{9/2}$ shell, to the ground state is precluded by its large calculated moment of $1.9\mu_N$.

DISCUSSION.

Nilsson orbital calculations (fig.2 [2]) give the 35th proton ground state as the [301]3/2$^-$ orbital at zero deformation and for oblate and prolate deformations of $|\epsilon| < 0.2$. For oblate deformations greater than -0.2 the ground state is [404]9/2$^+$, whilst for increasing prolate deformation the 35th orbital is successively [440]1/2$^+$ ($0.2 < \epsilon < 0.3$), [312]3/2$^-$ ($0.3 < \epsilon < 0.4$), [431]3/2$^+$ ($\epsilon \sim 0.4$) and [303]7/2$^-$ ($\epsilon > 0.4$).

Summarising the results of table 1, we can assign the following ground state configurations and deformations

Nuclide	Configuration	Deformation
^{79}Br	π[301]3/2	$\epsilon < 0.2$
^{77}Br	π[312]3/2	$0.3 < \epsilon < 0.4$
^{76}Br	π[312]3/2 ν[422]5/2	$0.3 < \epsilon < 0.4$
^{75}Br	π[312]3/2	$0.3 < \epsilon < 0.4$
^{74m}Br	π[301]3/2 ν[422]5/2	$\epsilon < 0.2$

Between ^{79}Br and ^{77}Br the measured reduction by a factor of two in the magnetic moment of the 3/2$^-$ ground state establishes clearly a trend of increasing deformation leading to occupation of the [312]3/2$^-$ orbital in ^{77}Br in place of the [301]3/2$^-$ orbital in ^{79}Br. For ^{76}Br, in addition to the 1$^-$ ground state with magnetic moment well understood in terms of a π[312]3/2$^-\nu$[422]5/2$^+$ configuration, there is a short lived 4$^+$ isomer, possibly the π[431]3/2$^+\nu$[422]5/2$^+$ configuration with a larger prolate deformation of $\epsilon \sim 0.4$. In ^{75}Br the moment again identifies the [312]3/2$^-$ state. However the measured moment for the high spin isomer in ^{74}Br, proposed as 4$^-$, is not compatible with the stretched combination of the $\pi\nu$ configuration found in ^{76}Br.

Equilibrium shape calculations by Bengtsson [8] mirror these conclusions, predicting increasing deformation with falling N, passing through $\epsilon=0.20$ at ^{78}Br. The discrepancy in ^{74m}Br may be resolved by a measurement of the ground state moment (0,1$^-$, 25m).

In many A$\sim$80 even-even nuclei, co-existing 0$^+$ and 2$^+$ states corresponding to very different deformations have been observed [14,15,16]. In particular, the total energy surface of ^{76}Kr exhibits two minima [3], one prolate with $\epsilon \sim 0.35$, $\gamma \sim 0°$ and the other, 550keV above, oblate with $\epsilon \sim 0.35$ and $\gamma \sim -60°$. By contrast, in ^{74}Se it is the oblate state with $\epsilon \sim 0.25$, $\gamma \sim -60°$ which lies 200keV below the prolate deformation with $\epsilon = 0.31$, $\gamma \sim 0°$. The deformations of several nuclei in this region, calculated from transitional quadrupole moments [17] show a consistent quadrupole deformation of order 0.35, indicating that the predicted prolate to oblate transition [8] in this region occurs via the γ degree of freedom, the prolate potential having approximately zero γ deformation as confirmed in the present calculations.

Both the present calculations and those of [3] predict that the yrast negative parity prolate states of both ^{77}Br and ^{75}Br have wavefunctions which strongly overlap those of corresponding states on the oblate side allowing a large degree of prolate-oblate band mixing which might account for the poor fit to the ground state band. The $g_{9/2}$ low-K positive parity band, arising from the [440]1/2 orbital, has no such analogue on the oblate side where high-K $g_{9/2}$ levels are present, and so will remain unmixed resulting in a rotational band which is well fitted by the calculations.

In an alternative approach using a particle quadrupole-phonon model [13], it is shown that bands exhibiting rotation-like features in ^{75}Se and ^{75,77}Kr do not give unique evidence for the deformation of their band heads. Rather, they can be better described as only a subset of a more complete excitation spectra of a particle coupled to a vibrational core, thereby accounting for many of the low lying states. This phenomenon cannot be invoked to explain the relatively complicated pattern of ground state bands in ^{75,77}Br. The measured bromine moments give unequivocal evidence for ground state deformation, so that we are justified in assuming a rotational nature for bands in these isotopes, probably with a complex Coriolis interaction.

The systematics of the odd-A $_{35}$Br and $_{37}$Rb isotopes show a striking similarity in magnetic moments between nuclei with the same neutron number, both showing sudden changes at N=42 associated with a sharp increase in deformation leading to the transition from occupation of the $\pi[312]3/2$ to the $\pi[301]3/2$ orbital. A survey of the known Br and Rb ground state spins, moments and single particle configurations is given in table 2.

Table 2.

Experimental Nuclear Magnetic Dipole Moments In Odd-A Br And Rb Isotopes.

N	$_{35}$Br Spin	$_{35}$Br Moment μ_N	Configuration	$_{37}$Rb Spin	$_{37}$Rb Moment μ_N	Configuration
38	(3/2$^-$)					
40	3/2$^-$	0.73(9)	[312]3/2	3/2$^-$	0.652(7) [1]	[312]3/2
42	3/2$^-$	0.845(16)	[312]3/2	5/2$^+$	3.36(4) [1]	[422]5/2
44	3/2$^-$	2.1064 [7]	[301]3/2	3/2$^-$	2.05(2) [7]	[301]3/2
46	3/2$^-$	2.2706 [7]	[301]3/2	5/2$^-$	1.43(2) [7]	[303]5/2
48	(3/2$^-$)			5/2$^-$	1.3534 [7]	[303]5/2
50	3/2$^-$			3/2$^-$	2.7518 [7]	

ACKNOWLEDGMENTS.

The authors wish to express their gratitude to the members of the ISOLDE collaboration for their kind help in the production of the ^{77}Br sample.

REFERENCES.

[1] C. Ekstrom, et al., Nuc. Phys. **A311** (1978)269
    Phys. Scripta **22** (1980)344
[2] W. Nazarewicz, et al., Nuc. Phys. **A435** (1985)397
[3] L. Luhmann, et al., Phys. Rev. **C31** (1985)828
[4] J.H. Hamilton, et al., Rep. Prog. Phys. (GB) **48** (1985)631
[5] N.J. Stone and H. Postma, Low Temperature Nuclear Orientation, (North Holland, Amsterdam)1986
[6] P. Herzog, et al., Z. Phys. **B64** (1986)853
[7] C.M. Lederer and V.S. Shirley, Table Of Isotopes (Wiley)1978
[8] R. Bengtsson, et al., Phys. Scripta **29** (1984)402
[9] S. Larsson, et al., Nuc. Phys. **A307** (1978)189
[10] L. Grodzins, Phys. Lett. **2** (1962)88
[11] S.E. Larsson, et al., Nuc. Phys. **A261** (1976)77
[12] H. Schäfer, et al., Z. Phys. **A293** (1979)293
[13] R.A. Meyer, et al., To Be Published.
[14] J.H. Hamilton, et al., Phys. Rev. Lett. **32** (1974)239
    Phys. Rev. Lett. **36** (1976)340
    Phys. Rev. Lett. **49** (1982)308
[15] R.B. Piercey, et al., Phys. Rev. **C25** (1982)1941
    Phys. Rev. Lett. **47** (1981)1514
[16] K.P. Lieb, et al., Phys. Rev. **C15** (1977)939
[17] K.P. Lieb, et al., In the proceedings of the symposium on Recent Advances In The Study Of Nuclei Off The Line Of Stability, Chicago, 1985

Fig 1. Angular Distribution Of Gamma Radiation From Oriented Nuclei As A Function Of Temperature.

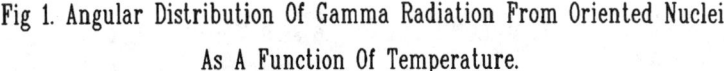

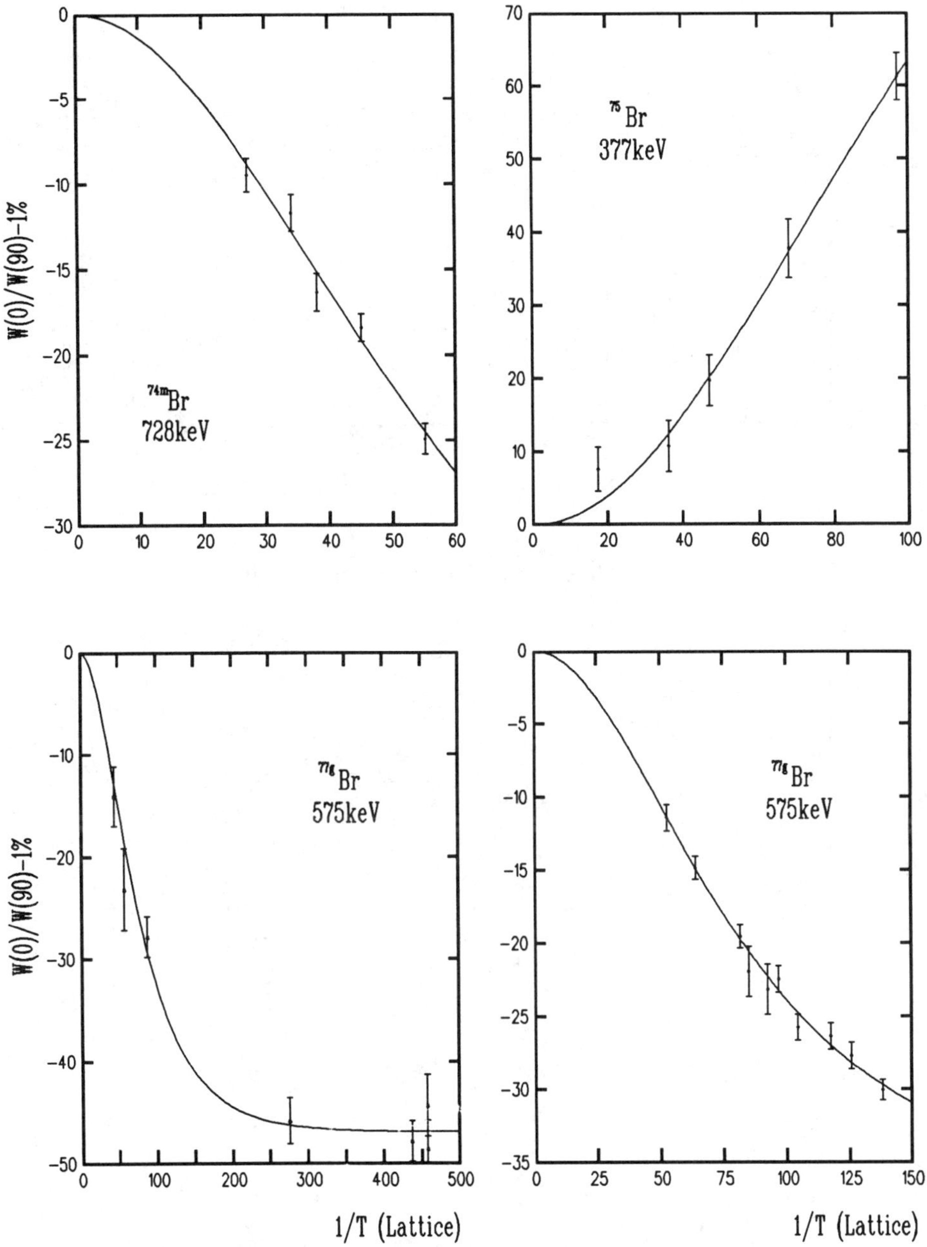

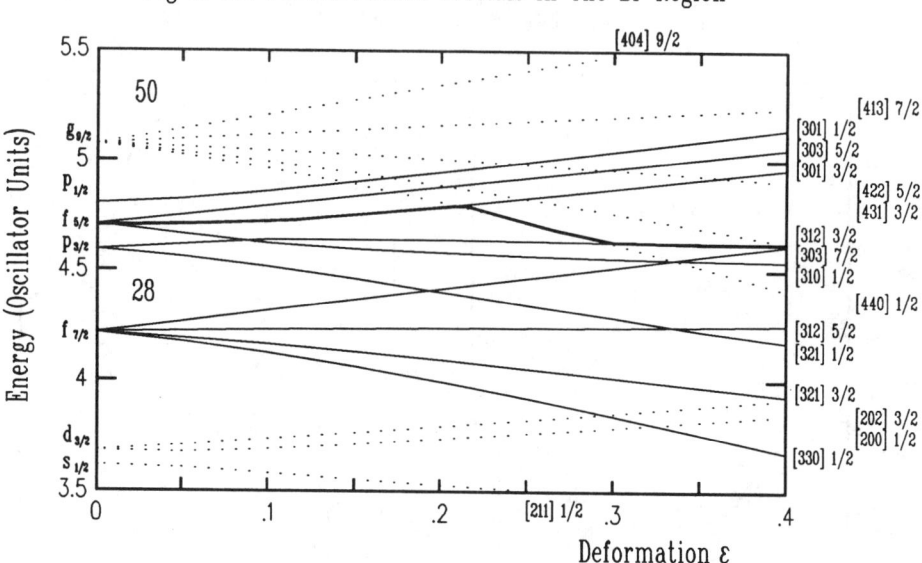

Fig 2. Odd Proton Nilsson Orbitals In The Br Region

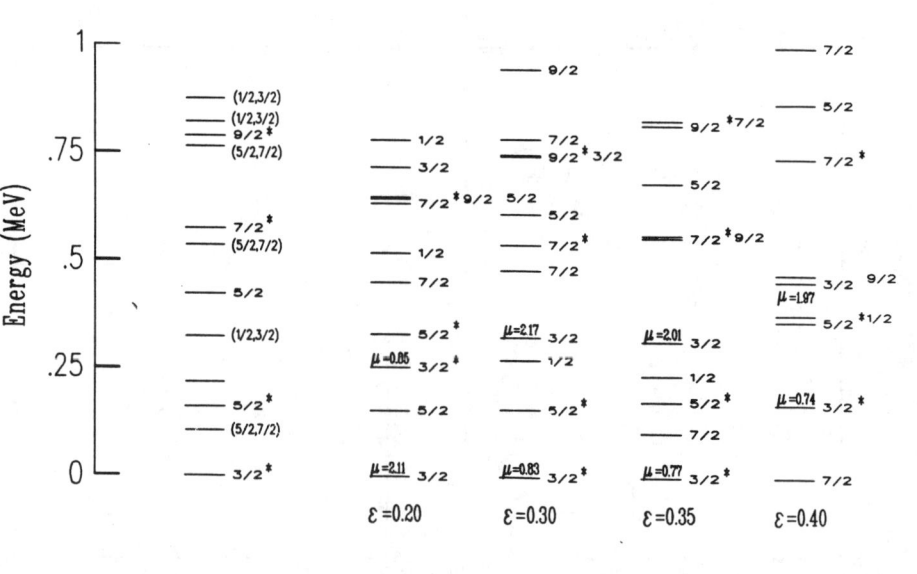

Fig 3. Negative Parity Energy Levels In ^{77}Br.

MAGNETIC MOMENTS OF NEUTRON DEFICIENT YTTRIUM NUCLEI

I. Berkes*, O. El Hajjaji*, M. Fahad*, R. Hassani*, J. Giroux*,
G. Marest*, G. Marguier*, N.J. Stone+, J.Rikovska+ and V.R. Green+,

*Institut de Physique Nucléaire (and IN2P3), Université Lyon-1
43, Bd du 11 Novembre 1918, F-69622 Villeurbanne Cedex - France

+Clarendon Laboratory, Parks Road, OXFORD, OX1 3PU, U.K

ABSTRACT

This paper describes recent low temperature nuclear orientation (LTNO) work on neutron deficient $^{85m,86,86m}Y$ nuclei. Results are compared with experimental systematics of neighbouring nuclei and particle core coupling calculations.

INTRODUCTION

Nuclei with $40 \leqslant N, Z \leqslant 50$, despite the vicinity of $Z = 40$ and $N = Z = 50$ shell closures exhibit very complex coexistence of spherical and collective features. Systematics of measured ground and isomeric $9/2^+$ states for nuclei with $38 \leqslant N \leqslant 54$ and $33 \leqslant Z \leqslant 43$ shows regularity of the $9/2^+$ excitation energy and its correlation with the value of magnetic dipole moment. An effect of proton-neutron correlation is seen with filling up the neutron $g_{9/2}$ orbital below $N = 50$. This correlation influences magnetic moments of odd-A nuclei and the simple additivity of magnetic moments for odd-odd nuclei in this region. Moreover, a rotation aligned $\Delta J = 2$ band built on both proton and neutron $9/2^+$ states has been observed in a number of nuclei in this region, corresponding to moderate deformation of about 0.10 - 0.15, with about 50% due to collective effects. Thus, magnetic moment measurement can reveal a significant nuclear structure information.

In yttrium nuclei the low lying proton states are $(p_{1/2})^1$ and $(g_{9/2})^1$, with, at $A < 90$, neutron configurations $(g_{9/2})^n$. LTNO measurements have been made on $^{85}_{39}Y^m_{46}$ $(9/2^+ [\pi (g_{9/2})^1 \nu (g_{9/2})^6])$, $^{86}_{39}Y^g_{47}$ $(4^- \pi (p_{1/2})^1 \nu (g_{9/2})^7])$, $^{86}_{39}Y^m_{47}$ $(8^+ [\pi (g_{9/2})^1 \nu (g_{9/2})^7])$ and $^{87}_{39}Y^m_{49}$ $(9/2^+ [\pi (g_{9/2})^1 \nu (g_{9/2})^9])$. The magnetic moment $|\mu(^{87}Y^m)| = 6.10(23) \mu_N$ [1] is known.

Yttrium isotopes can be expected to polarise strongly when present as a trace impurity in iron, utilising the large internal hyperfine field. This field has been determined by spin-echo NMR to be $B_{hf}(\underline{Y}Fe) = -28.6(6)T$ [2,3].

NUCLEAR ORIENTATION MEASUREMENTS

Samples of ^{86g,86m,87m}Y were prepared by Rb(α, xn) reactions followed by recoil implantation. The 48 MeV α particle beam from the University of Lyon synchrocyclotron was incident on a sandwich target consisting of a stack of thin Al foils on which a layer of RbI was deposited followed by 1 micron Fe foils. Several samples were made and cooled to temperatures close to 5 mK. Gamma-ray angular distribution measurements were made simultaneously on the 381 keV M4 transition in the decay of ^{87}Y^m, the 208 keV mixed M1/E2 transition in the decay of ^{86}Y^m, and on many transitions in the decay of ^{86}Y, using a single detector placed at an angle of 10° to the orientation axis.

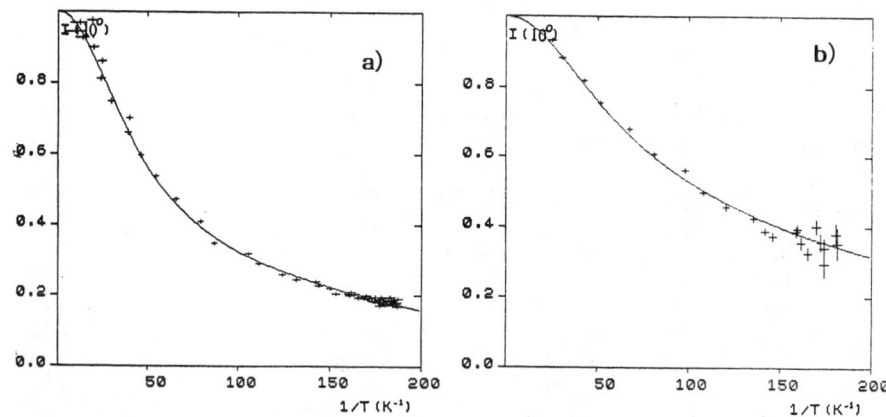

Figs 1 : Normalised intensity at $\theta = 10°$ versus inverse temperature in a typical sample for a) the 381 keV M4 transition in ^{87}Y^m (fit parameters given in text) b) for the 208 keV M1/E2 transition in ^{86}Y^m : fitted $|\mu(8^+)| = 4.8(3)\ \mu_N$.

Sources of ^{85m,86g}Y were prepared in situ in iron foils by fusion-evaporation reactions of 100-110 MeV ^{32}S from the Oxford FN Tandem [4]. They were cooled to temperatures close to 5 mK in an applied field of 0.7T. Two gamma detectors were used, at 0° and 90° to the orientation axis. Results for the strong 1220 keV transition in ^{85}Y^m are shown in fig. 2.

Uncertainty in the mixing ratio [5] δ (E2/M1) in the 208 keV transition in ^{86}Y^m contributes to the error in the moment extracted from orientation data. In an auxiliary experiment, the K-electron conversion coefficient α_K(208) was measured to be 0.044(4), hence δ (E2/M1) = + 1.07(22), (sign determined from the orientation data).

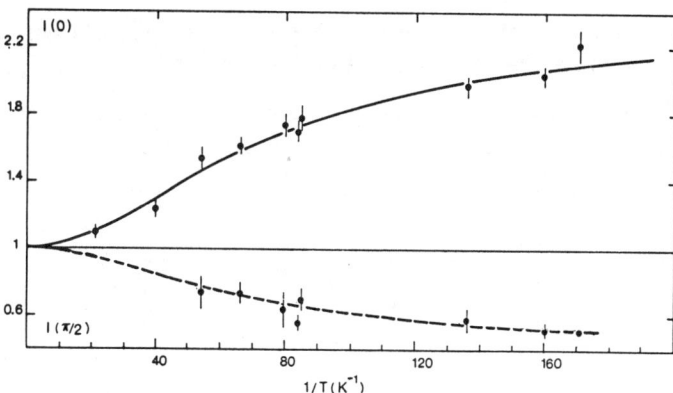

Fig.2: W(0°) and W (90°) versus inverse temperature for the 1220 keV M1/E2 transition in the decay of $^{85}Y^m$. Fitted : $|\mu (9/2^+)| = 6.2(5)\mu_N$.

DATA ANALYSIS

Analysis of the data was made using a model in which a fraction f of the Y ions experiences the full hyperfine field $B_{hf}(1)$ associated with undisturbed substitutional lattice sites in iron and the remainder 1-f of non-substitutional nuclei experiences a second field $B_{hf}(2)$. Thus :

$$W(\theta, T) = 1 + f \sum_{\lambda=even} B_\lambda (B_{eff}(1), T) U_\lambda A_\lambda Q_\lambda P_\lambda (\cos \theta)$$
$$+ (1-f) \sum_{\lambda=even} B_\lambda (B_{eff}(2), T) U_\lambda A_\lambda Q_\lambda P_\lambda (\cos \theta)$$

where $B_{eff}(i) = B_{hf}(i) - B_{pol}$ and the other symbols have their usual meaning [6]. Details of the model will be published elsewhere [7].

Taking the known magnetic dipole moment of $^{87}Y^m$, the best values for the parameters $B_{hf}(1)$ (averaged over six sources) were : $B_{hf}(1)$ = - 28.0(11)T and $B_{hf}(2)$ = - 7.8(11), including error in the moment of $^{86}Y^m$. With these fields, (but with the factor f as a free parameter for the data taken on Oxford sources), data on $^{85m,86m}Y$ were fitted for their magnetic moments, with results given in Table 1. The fits to both W (0°) and W (90°) data for $^{85}Y^m$ and $^{87}Y^m$ show that attenuation due to the strong electric quadrupole interaction observed in melted or annealed alloys [3] is small for the dominant site 1 in non heat-treated ferromagnetic iron. For $^{86}Y^g$ the anisotropies observed were small, with only weak transitions showing barely statistically significant effects which were not consistent in all experiments. Temperature close to 5 mK was maintained for approximately 15h, sufficiently long as compared to the spin lattice relaxation time of $^{86}Y^g$ nuclei in iron. We conclude that the moment of this state is less than 0.6 μ_N.

Table 1 : Deduced nuclear magnetic dipole moments of Y isotopes.

Nucleus	Spin/Parity	Half Life	Transition Analysed (keV)	Multi-polarity	Deduced Moment $|\mu/\mu_N|$
$^{85}Y^m$	$9/2^+$	4.9 h	1220 2124	M1/E2 M1/E2	6.2(5)
$^{86}Y^g$	4^-	14.5 h	see text	-	< 0.6
$^{86}Y^m$	8^+	46 m	208	M1/E2	4.8(3)

DISCUSSION

Energy levels in $^{85,87}Y$ have been used to fit parameters of the particle-asymmetric rotor model and to calculate magnetic moments of the $9/2^+$ states. The energyy levels of ^{85}Y are well reproduced by the calculation, which gives $\mu(^{85}Y^m)$ = + 5.7 μ_N. The fit was less satisfactory for the levels of the almost vibrational-like ^{87}Y, giving $\mu(^{87}Y^m)$ = + 5.8 μ_N. Parameters of the model used are essentialy in line with those published previously in this mass region. Details of the calculation will be discussed elsewhere.

Moments of $^{86}Y^{g,m}$ have been calculated using $g_\pi(p_{1/2})$ = - 0.274, $g_\pi(g_{9/2})$ = + 1.36(5), $g_\nu(g_{9/2})$ = - 0.216 taken from $^{89}Y_{50}$, $^{85}Y^m_{46}$ and $^{83}Kr_{47}$, resp., giving $\mu \approx$ + 4.6 μ_N for the metastable, and $\approx$ - 0.84 μ_N for the ground state. Other experimental 8^+ state moments in this region are also somewhat higher than the calculated values, indicating a possible neutron proton interaction in the $g_{9/2}$ shells.

REFERENCES
1) M.Kontani, K.Asayama and J. Itoh, J. Phys. Soc. Japan **20**, 1737(1965).
2) G.Marest, R.Haroutunian and I.Berkes, Hyp. Int. **4**, 425(1978).
3) G.Marest, R.Haroutunian and I.Berkes, Phys. Rev. **C17**, 287(1978).
4) A.L.Allsop, B.I.Bradley, V.R.Green and N.J.Stone, Hyp. Int. **15/16**, 313 (1983).
5) Nuclear Data Sheets **25**, 553 (1978).
6) K.Krane in Low Temperature Nuclear Orientation, eds. N.J.Stone and H.Postma (North Holland 1986) ch.2.
7) In preparation, to be submitted to Hyperfine Interactions.
8) S.E.Larsson, G.Leander and G.Ragnarsson, Nucl.Phys., **A307**, 189(1978).

NUCLEAR GROUND STATE PROPERTIES OF STRONTIUM ISOTOPES (78≤A≤100) BY LASER SPECTROSCOPY

F. Buchinger, E.B. Ramsay
McGill University/FRL, Montréal, Canada H3A 2B2

R.E. Silverans, P. Lievens
Katholieke Universiteit, Leuven, Belgium

E. Arnold, W. Neu, R. Neugart, K. Wendt
Universität Mainz, Mainz, FRG

D. Berdichevsky, R. Fleming, D.W.L. Sprung
McMaster University, Hamilton, Canada L8S 4M1

G. Ulm[*], the ISOLDE Collaboration
CERN, Geneva, Switzerland

ABSTRACT

We report on the measurements of nuclear ground state spins, moments and changes in mean square charge radii in a series of Strontium isotopes between A=78 and A=100 by collinear fast-beam laser spectroscopy. The experiment was carried out on-line at the ISOLDE mass separator at CERN using the traditional photon counting as well as a new particle detection method of fast beam spectroscopy. The results are presented and discussed in the light of the rapid change of nuclear shape within the investigated isotopic series.

INTRODUCTION

The series of Strontium (Sr) isotopes reaches from the valley of stability (at N=50) with isotopes showing spherical shape to strongly deformed isotopes on both sides of the stability line. Since the variation of the nuclear shape happens over a remarkably short interval of approximately 20 neutrons, this even Z element has found considerable interest from the experimental, as well as from the theoretical physics side. The predicted strong ground state deformation at N≈40 and N≈60[1] has been established through measurements of BE2 values[2]. Laser spectroscopy experiments of the ground state spins, moments and charge radii mainly cover the transitional region of the neutron deficient isotopes[3,4]. We have extended those studies to Sr isotopes in the mass region 78 ≤ A ≤ 100. This coherent information on nuclear ground state properties at both sides of the neutron shell closure, allows a systematic study of the development of the strong deformations under various aspects.

[*] Present address: PTB, D-1000 Berlin 1

EXPERIMENT

Nuclear spins (I), magnetic moments (μ_I) and spectroscopic quadrupole moments (Q_s) were obtained from optical hyperfine structure (hfs) measurements in the 5s $^2S_{1/2}$ - 5p $^2P_{3/2}$ transition (λ = 407.8 nm) of the Sr ion. Information about changes in mean square charge radii ($\delta<r^2>$) were derived from the isotopic shifts (IS) in the same line. The experiment was carried out on-line at the ISOLDE mass separator at CERN. We used the technique of collinear fast-beam laser spectroscopy and an experimental procedure similar to that one described in Ref. 5 for the measurements of the isotopes between A=78 and A=98. The measurements were extended to ^{100}Sr by the first on-line application of a new variant of collinear fast-beam laser spectroscopy proposed and developed by Silverans et al.[6]. A detailed description of the apparatus and its application in the Sr measurements is given in Ref. 7 and we restrict ourselves to a schematic description of the technique. In an optical pumping region the Sr ions in the fast beam interact in collinear geometry with the laser light. When the frequency tuned laser light is resonant with the Doppler shifted 5s-5p transition, multiple excitation at resonance leads to a strong depopulation of the ground state and correspondingly to a high occupation of the low-lying 4d metastable states. The ion beam is then sent through a Na-vapour charge transfer cell. Since the neutralization cross-section (σ) for Sr ions in Na is state dependent (e.g. $\sigma(4d)/\sigma(5s)$ = 1.5 at 60 keV ion energy), the optical pumping process at resonance can be conveniently detected by particle counting. For this purpose the ions are subsequently separated from the neutral particles by an electrostatic deflector and the atoms are counted via secondary electron emission after impact on an aluminum tape. Frequency calibration of the laser scan in the ^{100}Sr measurements was achieved by recording the signal of a reference isotope (e.g. ^{94}Sr) by optical detection at two different ion energies and unchanged laser settings.

RESULTS and DISCUSSIONS

The magnetic moments and spectroscopic quadrupole moments of the odd A isotopes and isomers in the investigated Sr series are shown in Tab. I. For the calculation of μ_I and Q_s from the ratio of the hfs constants of the respective isotopes, we have used μ_I = 1.093602 (1) nm[8] and Q_s = 0.335 (20) b[9] for ^{87}Sr, where Q_s is corrected for Sternheimer type polarization effects. In Tab. 1 we have also listed the nuclear ground state spins of the respective isotopes. Our spin measurements confirm the most recent assignments from nuclear spectroscopy studies[8,10,11] except for ^{93}Sr (I=5/2) and ^{97}Sr (I=1/2). For this isotope, which is situated on the borderline between spherical and strongly deformed nuclear shape in the narrow range of neutron numbers where shape coexistence is indicated[1,12], a spin assignement of I=3/2 was recently suggested by several authors[12,13]. Our results characterize the ^{97}Sr ground state as a $s_{1/2}$ shell model state, a conclusion also supported by the nearly identical magnetic moments of ^{95}Sr and ^{97}Sr (see Tab. I).

A	I	μ_I (nm)		Q_s (b)	
79	3/2	-0.474	(2)	0.744	(54)
81	1/2	0.5440	(4)		
83	7/2	-0.830	(2)	0.766	(58)
83m	1/2	0.582	(1)		
85	9/2	-1.0011	(9)	0.283	(22)
85m	1/2	0.6008	(4)		
87	9/2	-1.093602	(1)	0.335	(20)
87m	1/2	0.6282	(6)		
89	5/2	-1.1488	(7)	-0.274	(19)
91	5/2	-0.8868	(6)	0.044	(5)
93	5/2	-0.7942	(5)	0.265	(19)
95	1/2	-0.5379	(4)		
97	1/2	-0.500	(1)		

Table I: Spins (I), magnetic moments (μ_I) and spectroscopic quadrupole moments (Q_s) in the investigated Sr series.

Changes of mean square charge radii were obtained from the optical data following the standard procedure of Heilig and Steudel[14]. We use a calibratration factor (notation as in Ref. 14) of F=1579(47) MHz/fm^2 and a specific mass shift (SMS) of SMS=-0.19(15) NMS with NMS being the normal mass shift in the Sr-D2 transition. The resulting $\delta<r^2>$ values are plotted relative to ^{88}Sr as reference isotope in Fig. 1. Similar to the Sr neighbour Rb[15], a decrease in $<r^2>$ is observed for the neutron deficient isotopes when N is increased. For the neutron rich isotopes a steep increase of $<r^2>$ between N=59 and N=60 is found after a more regular variation for the isotopes between the closed neutron shell at N=50 (^{88}Sr) and N=59, respectively.

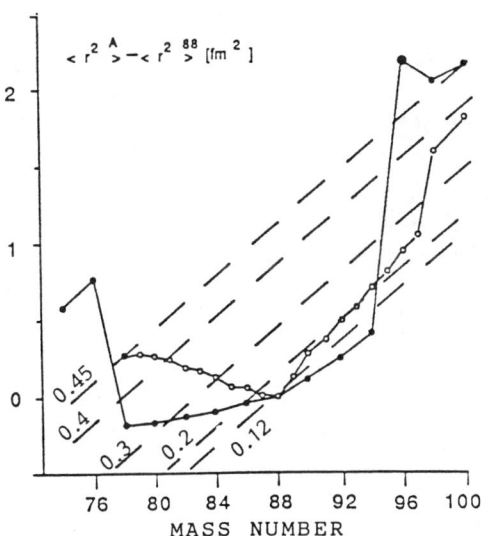

Fig. 1: Changes of mean square charge radii of Sr isotopes (open circles). The predictions from the droplet model for deformations $<\beta>^{1/2}$ ranging between 0.12 and 0.45 are shown as broken lines. HF+BCS results (force SKa) are given as full dots.

The experimental values can be interpreted in terms of changes in ground state deformation, by comparing them to the predictions from the droplet model[16]. The droplet-isodeformation curves calculated with the model parameter set of Ref. 17 for deformations between β = 0.12 and β = 0.45 are included in Fig. 1. The predicted sudden onset of deformation on both sides of the line of stability[1] is observed only for the neutron rich isotopes, whereas on the neutron deficient side the deformation develops more smoothly. The deformations deduced for the strongly deformed isotopes at the extrema of the investigated isotopic series agree with those derived from BE2 values (see Tab. 2). An indication for static prolate deformation of the lightest isotopes comes from the intrinsic quadrupole moment of ^{79}Sr, Q_0=+3.72 (27) b, derived from Q_s under the assumption of strong coupling. The Q_0 of ^{79}Sr is similar in magnitude to that one of its even neighbour ^{78}Sr, where Q_0=3.28 (21) b is found from its BE2 value.

Table II. Deformation parameters derived from BE2 values[2]

A	78	80	82	84	86	88	98	100
$<\beta^2>^{1/2}$	0.43(3)	0.38(2)	0.29(1)	0.21(2)	0.13(1)	0.117(3)	0.36(2)	0.37(1)

For the neutron deficient transitional isotopes the puzzling discrepancy between deformations obtained from the IS in the frame of the droplet model and from BE2 values, already discussed for the stable isotopes[18], persists towards lower neutron number.
For a more thorough interpretation of the trend of Sr charge radii, we have extended our calculations of mean square charge radii in the HF+BCS approach[18] to isotopes between A=74 and A=100. The results are included in Fig. 1. Similar to the findings of other authors[19], an abrupt onset of static ground state deformation is predicted for the heavy as well as for the light Sr isotopes. A detailed discussion of these calculations will be presented in a forth-coming more extended presentation of the results.

1. J.H. Hamilton et al., Rep. Prog. Phys. **48,** 631 (1985) and references therein.
2. S. Raman et al., At. Data Nucl. Data Tables **36,** 1 (1987) and references therein.
3. D.A. Eastham et al., Daresbury Lab. Preprint DL/NUC/P243E.
4. M. Anselment et al., Z. Phys. **D3,** 421 (1986).
5. K. Wendt et al., Z. Phys. **A318,** 125 (1984).
6. R.E. Silverans et al., Nucl. Instr. Meth., **B26,** 591 (1987).
7. R.E. Silverans et al., to be published.
8. Table of Isotopes, ed. by C.M. Lederer and V.S. Shirley (Wiley, N.Y. 1978).
9. S.M. Heider et al., Phys. Rev. **A16,** 1371 (1977).
10. B. Pfeiffer et al., Proc. 4th Int. Conf. on Nuclei Far From Stability, CERN 81-09, 423 (1981).
11. C.J. Lister et al., Phys. Rev. Lett. **49,** 308 (1982).
12. R.A. Meyer, Hyp. Int. **22,** 385 (1985) and references therein.
13. K.L. Kratz et al., in: Nuclei off the Line of Stability, 190th Nat'l. Meeting Am. Chem. Soc., Chicago (1985), Eds. R.A. Meyer and D.S. Brenner, ACS Symp. Series **324,** 159 (1986).
14. K. Heilig et al., At. Data Nucl. Data Tables **14,** 613 (1974).
15. C. Thibault et al., Nucl. Phys. **A367,** 1 (1981).
16. W.D. Meyers et al., Nucl. Phys. **A410,** 61 (1983).
17. D. Berdichevsky et al., Z. Phys. **A322,** 14 (1985).
18. F. Buchinger et al., Phys. Rev. **C32,** 2058 (1985).
19. P. Bonche et al., Nucl. Phys. **A443,** 39 (1985) and references therein.

LASER SPECTROSCOPY IN THE Z < 82 REGION

St. Becker, M. Gerber, Th. Hilberath, H.-J. Kluge,
U. Krönert and K. Wallmeroth

Institut für Physik, Universität Mainz,
D-6500 Mainz, FRG

and

The ISOLDE Collaboration, CERN,
CH-1211 Geneva 23, Switzerland

The nuclei with Z < 82 and N ≃ 104 have attracted considerable attention during the last years following the observation of a sudden change of the nuclear deformation in the light Hg isotopes by optical spectroscopy[1,2]. In the meantime systematic measurements have been performed in the mass range 181 ≲ A ≲ 206. An update of the existing information has been published recently[3].

Similar data on properties of nuclear ground and isomeric states of neighbouring elements are of interest in order to study the influence of pairing and shell effects. Hence we started a program to investigate neutron-deficient Au isotopes at the on-line isotope separator ISOLDE/CERN.

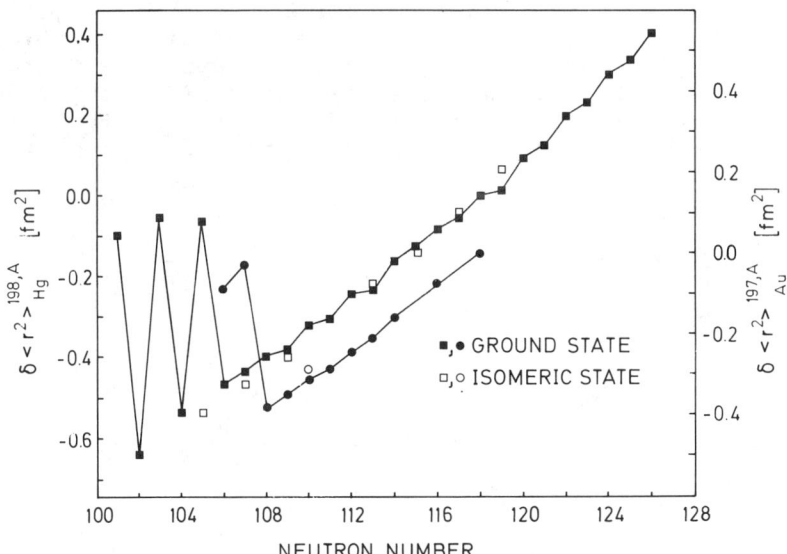

Fig.1: Changes of mean-square charge radii of Hg (left scale, squares) and Au isotopes (right scale, dots).

Several techniques were tested and their detection efficiencies determined. These are polarization spectroscopy[4], fluorescence spectroscopy[5], two-photon spectroscopy[6], resonance ionization mass spectroscopy (RIMS)[7] and a combination of RIMS with pulsed-laser induced desorption[8]. Data on the hyperfine structure and isotope shift could be obtained[4,5,7] for the Au isotopes in the mass range $185 \leq A \leq 197$. The resulting changes in the nuclear charge radii are shown in Fig. 1 together with the data of Hg[3]. An onset of strong deformation is observed at ^{186}Au which continues for ^{185}Au in contrast to the situation in Hg. The stabilization of strongly deformed (prolate) nuclear shape is attributed to the down-sloping $\pi 1h_{9/2}$ intruder state.

An extension of the investigations to still more neutron-deficient Au isotopes requires a drastic increase in detection efficiency at low background. In the case of the RIMS experiments[7], the ISOLDE Hg beam was implanted into an oven. A continuous thermal atomic beam was formed by heating the oven after a suitable delay in order to let the Hg nuclei decay to the Au daughter isotope. The Au atoms were ionized by stepwise, resonant excitation with pulsed dye laser beams via intermediate atomic levels into the continuum. The ions resulting were detected mass-selectively by time-of-flight (TOF) spectroscopy. In resonance one Au ion was detected per 10^8 atoms implanted into the oven at a background event rate of 1 per 1000 laser shots.

The main reason for the low detection efficiency of $\epsilon = 10^{-8}$ is the mismatch between the continuous evaporation of the Au atoms and low repetition rate of a high-power, pulsed laser system: A duty cycle of only 10^{-4} is calculated for the interaction of a continuous

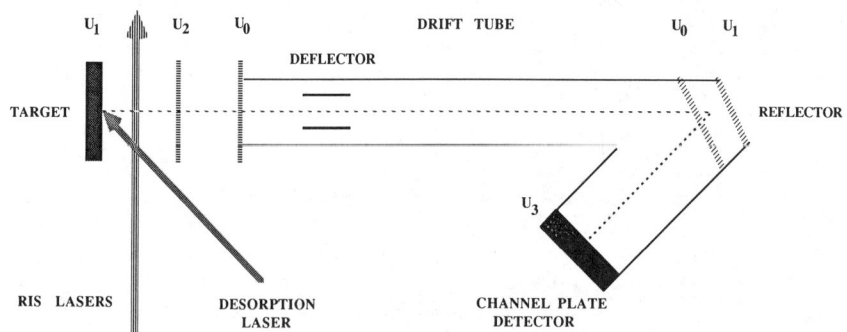

Fig.2: Experimental setup for pulsed-laser induced desorption and resonance ionization mass spectroscopy. The potentials are $U_0 = 0$ V, $U_1 = 2.6$ kV, $U_2 = 2.4$ kV and $U_3 = -2$ kV.

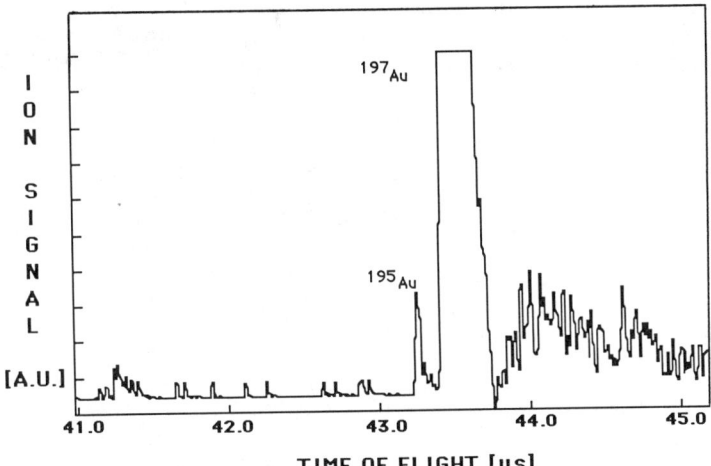

Fig.3: Part of the time-of-flight signal in case of resonance ionization mass spectroscopy of $5 \cdot 10^9$ atoms of ^{195}Au ($T_{1/2}$ = 183 d) implanted at 60 keV into graphite.

thermal atomic beam with the laser beams (diameter d = 1 cm) of a pulsed dye laser system, pumped by a Nd:Yag laser (repetition rate v_{rep} = 10 Hz).

We could increase the detection efficiency to $\epsilon = 10^{-5}$ by producing a pulsed thermal atomic beam[8,9]. A second Nd:Yag laser was used in order to desorb ^{195}Au ($T_{1/2}$ = 183 d) implanted at ISOLDE as ^{195}Hg into graphite (Fig. 2). When the desorbed atoms reached the region of interaction with the laser beams used for resonance ionization, the bunch of atoms was probed by these lasers, and the resulting ions were detected in the same way as before[7]. Fig. 3 shows part of a TOF spectrum in case of a sample with $5 \cdot 10^9$ implanted atoms of ^{195}Au. The large peak at mass A = 197 is due to stable ^{197}Au, and the small one at 41.3 μs is most probably due to ^{181}Ta. Both elements are present as impurities in the target material. About 60% of the implanted activity could be **released** by pulsed-laser induced desorption. As in the RIMS experiment the background was as low as 1 event per 1000 laser shots. Hence, optical investigation of isotopes with A < 185 and other refractive elements seems to be in reach.

This work has been funded by the German Federal Minister for Research and Technology (Bundesminister für Forschung und Technologie) under Contract No. Mz-458-I.

REFERENCES

1. J. Bonn et al., Phys. Lett. 8B (1972) 308
2. J. Bonn et al., Z. Phys. A276 (1976) 203
3. G. Ulm et al., Z. Phys. A325 (1986) 247
4. H.-J. Kluge et al., Z. Phys. A309 (1983) 187
5. J. Streib et al., Z. Phys. A521 (1985) 537
6. G. Bollen et al., J. Opt. Soc. Am. B4 (1987) 329
7. K. Wallmeroth et al., Phy. Rev. Lett. 58 (1987) 1516
8. U. Krönert et al., submitted to Appl. Phys. B
9. Similar experiments are performed by a Montreal-Orsay Collaboration: J. Lee, private communication (1987)

RESONANT IONIZATION SPECTROSCOPY OF LASER-DESORBED GOLD ISOTOPES

J.K.P. Lee, G. Savard, J.E. Crawford and G. Thekkadath
Foster Radiation Laboratory, McGill University
Montréal, Canada

H.T. Duong, J. Pinard and S. Liberman
Laboratoire Aimé Cotton, CNRS II, Orsay, France

F. Le Blanc, P. Kilcher, J. Obert, J. Oms, J.C. Putaux, B. Roussière and
J. Sauvage
Institut de Physique Nucléaire, Orsay France

and the ISOCELE Collaboration

ABSTRACT

A system for laser spectroscopic studies on implanted radioactive samples has been developed and installed at the ISOCELE isotope separator. Mass separated gold isotopes collected in a graphite substrate are desorbed by pulses from a Nd/YAG laser. Three synchronized dye laser pulses produce ions in a resonant ionization process; these are detected and identified in a time-of-flight mass spectrometer. Results of isotope shift and hyperfine structure measurements on gold isotopes with A=186, 187, 190, 192, 194, 195 and 196 are presented.

INTRODUCTION

The neutron-deficient heavy nuclei just below the Z=82 proton shell closure are characterized by the co-existence of nuclear shape isomers and sudden change of nuclear shape between ground states of neighbouring isotopes. This phenomenon was first discovered in mercury isotopes[1]. Since then, the nuclear structure of nuclei in this region was extensively studied via various nuclear reactions and nuclear spectroscopic studies, and similar nuclear shape changes were predicted for the neighbouring elements. Recently, these predictions for Tl and Au were verified via laser spectroscopic studies[2,3]. However, for platinum and lighter elements, no experimental data were yet available.

From an experimental point of view, laser spectroscopic studies of Au, Pt, Ir and Os present some major difficulties. These elements all have very high evaporation temperature and do not easily diffuse out of host material. As a result, radioactive ion beams of these elements are very difficult to produce. While gold isotopes are available at ISOCELE, ions of the other elements are not yet produced anywhere in sufficient quantity. Also, for these elements, the allowed atomic transitions from their ground states lie in the UV region - a major inconvenience particularly for CW laser studies. Therefore, despite the strong interest in the ground state properties of these nuclei, few laser spectroscopic measurements have been attempted.

We have recently developed a novel laser spectroscopic method for studies of these elements. It utilizes two distinct capabilities of pulsed lasers: to momentarily heat up a surface to a very high temperature; and to efficiently and selectively ionize atoms of a particular element, permitting their masses to be measured via resonant ionization mass spectrometry (RIMS). The technique was first tested using atoms evaporated onto a substrate surface[4]; later an on-line setup was installed at ISOCELE. Here, we describe the experimental method used and the initial results obtained with radioactive gold ion beams.

EXPERIMENTAL METHOD

The schematic diagram of the experimental setup is shown in Fig. 1. The mass separated gold ions from ISOCELE are implanted onto a graphite substrate. The rotation of the cylinder brings the implanted sample to the atom-laser interaction region. A laser pulse from a Nd-YAG laser, operated at 532 nm wavelength, is used to heat up the substrate surface and re-evaporate the implanted atoms. The choice of the wavelength is based on a previous study[4] which shows the frequency-doubled 532 nm output to be preferable to 1064 nm for desorption purposes. A few microseconds later, three pulsed laser beams, all produced via dye lasers pumped by the same excimer laser, pass in front of the heated spot and selectively ionize the gold atoms via the resonant ionization spectroscopic (RIS) process. The ions created were accelerated by the electrodes and identified by the time-of-flight mass spectrometer.

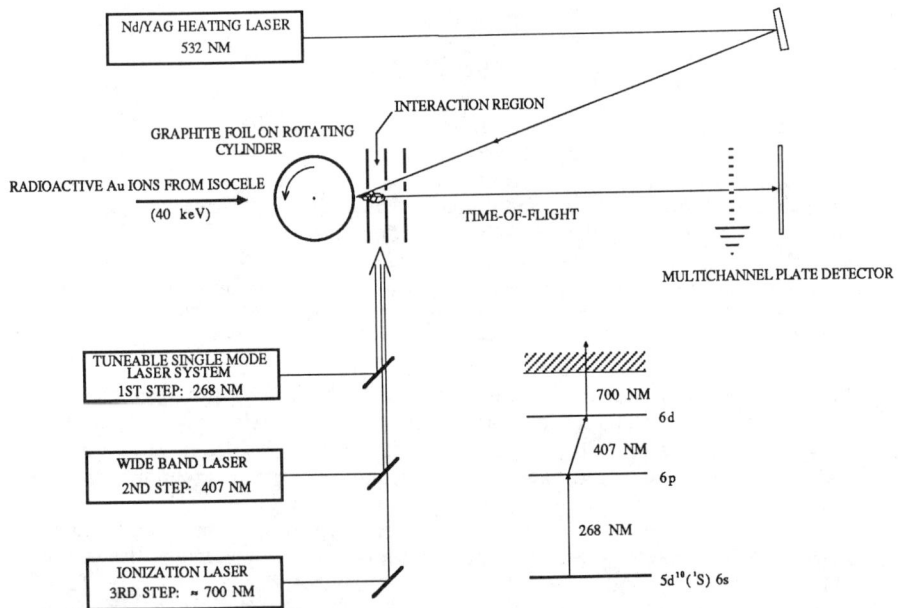

Fig. 1 Schematic Diagram of the Experimental Setup

The RIS scheme used was a three-step two-resonance process. The first resonant absorption corresponds to the transition from $(6s)S_{1/2}$ to $(6p)P_{1/2}$ levels (268 nm) in gold atoms. A high resolution pulsed laser beam of this wavelength was obtained by inserting an extra dye cell, pumped by the excimer laser, inside a commercial single-mode CW dye laser cavity operating at 536 nm, with its output subsequently amplified and frequency doubled[5]. With this system, it is possible to control the steady scan of the laser frequency while maintaining an excellent linewidth, which is essentially Fourier limited at 65 MHz for 536 nm wavelength. The second resonant excitation (407 nm) corresponds to the $(6p)P_{1/2}$ - $(6d)D_{3/2}$ transition, and is provided by a commercial pulsed dye laser with about 5 GHz linewidth. The last step of photoionization is the broad-band output from the spontaneous emission of pyridin dye centred at 700 nm. In the actual experimental setup, the first and third laser beams were collinear to each other but perpendicular to the 407 nm beam. This arrangement reduces the Doppler broadening effect and permits an overall resolution of 450 MHz.

In a typical run, suitable quantities of radioactive atoms (about 10^{10} atoms or more) were accumulated over a 2 mm x 8 mm spot. The heating laser spot was focussed to about 3 mm x 50 μm size and successive laser shots were slowly moved across this area while the laser frequency was varied. It is estimated that each laser shot would desorb about 10^6 atoms. At resonance, about 100 ions reach the detector.

EXPERIMENTAL RESULTS

The data obtained on hyperfine structure (HFS) and isotope shifts (IS) of several gold isotopes are shown in Fig. 2 and listed in columns 3 and 6 of Table 1. The deduced magnetic moments μ and nuclear deformation parameters $<\beta^2>^{1/2}$ together with those from the literature are also shown. To facilitate comparison, analysis procedures similar to those outlined in Ref. 3 were followed, where a liquid droplet model was used to calculate $\delta<r^2>_{sph}$, and to deduce $\delta<r^2>$. The $<\beta^2>^{1/2}$ of ^{197}Au was assumed to be 0.11.

For the isotopes ^{194,196}Au, the IS results presented here are new. As expected, the deduced deformation parameter is similar to those of the neighbouring isotopes. With the present resolution, the hyperfine splitting of the ground level for ^{192}Au can be resolved and a negative sign for this magnetic moment was obtained. For ^{187}Au, the magnetic moment deduced from laser spectroscopy experiments[3] is different from earlier measurements[6]. The present result is in agreement with the earlier laser spectroscopy work and the cause of the discrepancy with other measurements is still not clear. The sudden change of nuclear deformation between ^{187}Au and ^{186}Au observed previously was also clearly reproduced.

We have also attempted to extend this method to platinum. Data for HFS and IS of the stable isotopes were obtained. However, the overall efficiency reached was at least three orders of magnitude lower than that for gold and no data for radioactive Pt isotopes were obtained. The cause for this drastic drop in efficiency is currently being investigated.

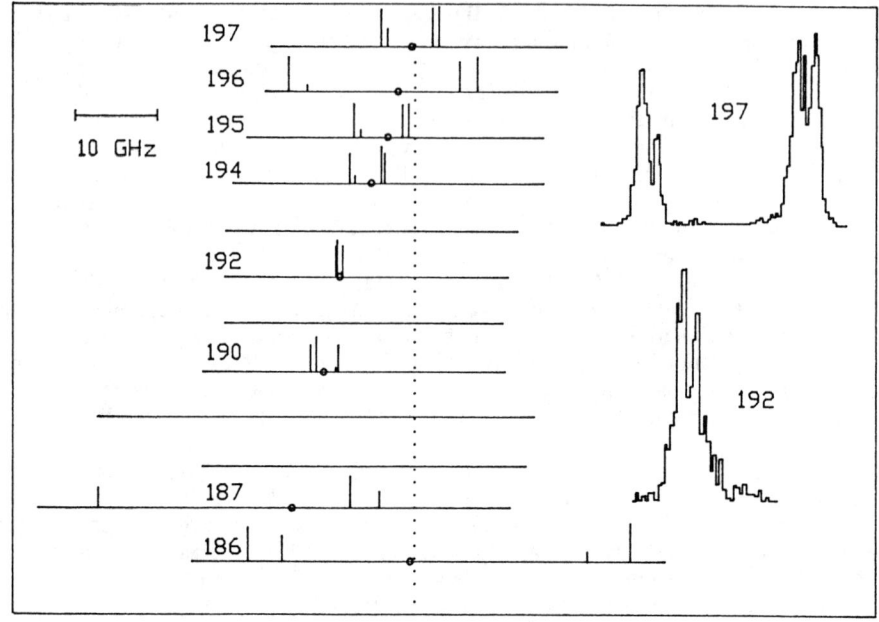

Fig. 2 HFS and IS of Radioactive Gold Isotopes

Table 1. The experimental results and deduced magnetic moments and deformation parameters.

Au	I	$A_{hfs}(gs)$ (GHz)	μ_I (n.m.)	μ_I(ref.3) (n.m.)	$\delta\nu^{197,A}$ (GHz)	$\delta\langle r^2\rangle^{197,A}$ (GHz)	$\langle\beta^2\rangle^{1/2}$ calc.	ref.3
197	3/2	3.04(5)	0.148	0.148158(8)	0	0	0.11	0.11
196	2	8.41(22)	0.580(15)	0.5914(14)	1.72(15)	-0.043(4)	0.114	
195	3/2	3.04(9)	0.157(5)	0.151(7)	3.21(23)	-0.080(6)	0.120	0.121
194	1	2.28(8)	0.079(3)	0.076(4)	5.19(13)	-0.130(4)	0.121	
192	1	-0.22(6)	-0.0076(21)[a]	±0.0081(11)	8.17(16)	-0.204(4)	0.131	0.130
190	1	-1.87(18)	-0.065(7)	-0.068(26)	10.93(31)	-0.282(8)	0.139	0.139
187	1/2	30.78(72)	0.531(12)	0.535(15) 0.72(7)[6]	14.61(39)	-0.365(10)	0.160	0.155
186	3	-12.41(32)	-1.284(33)	-1.263(29)	0.05(84)	0.021(15)	0.252	0.246

a) sign determined in this work

1. J. Bonn et al., Phys. Lett. **38B**, 308 (1972) and Z. Phys. **A276**, 203 (1976).
2. J.A. Bounds et al., Phys. Rev. Lett. **95**, 2269 (1985).
3. K. Wallmeroth et al., Phys. Rev. Lett. **58**, 1516 (1987) and J. Streets et al., Z. Phys. **A321**, 537 (1985).
4. J.K.P. Lee et al., Nucl. Inst. & Meth. **B26**, 444 (1987).
5. J. Liberman et al., Rev. Cethedec-ondes **NS83-2**, 37 (1983) and J. Pinard et al., Op. Com. **20**, 344 (1977).
6. C. Ekström et al., Nucl. Phys. **A348**, 25 (1980).

NUCLEAR MOMENTS AND CHANGE IN THE CHARGE RADII OF NEUTRON-DEFICIENT SILVER AND LEAD ISOTOPES

U. Dinger, J. Eberz, G. Huber, R. Menges, S. Schröder
Institut für Physik, Universität Mainz, 6500 Mainz, FRG

R. Kirchner, O. Klepper, T. Kühl, D. Marx, G. D. Sprouse
Gesellschaft für Schwerionenforschung, 6100 Darmstadt, FRG

ABSTRACT

The nuclear moments and the change in the charge radii of neutron deficient isotopes of silver (^{101,103,104,105,105m,106m}Ag) and lead (^{192,194,195,196}Pb) have been determined by collinear laser spectroscopy.

INTRODUCTION

Collinear laser spectroscopy at efficient on-line mass-separators has proven to be a powerful tool in the investigation of optical hyperfine structures and isotope shifts in long isotopic chains[1]. The analysis of the spectra yields the nuclear moments and the change in the charge radii of the studied nuclei. Systematic studies in dependence of the neutron and proton numbers in the vicinity of closed shells may reveal details of nuclear structure possibly hidden by deformation effects in nuclei far from closed shells. However high demands on targets and ion-sources have to be made and more complicated excitation and detection schemes have to be applied when leaving the domain of the alkali and earth alkali-elements. Continuing our studies around Z=50, we have added data of neutron-deficient silver isotopes (Z=47)[2] to the existing data of cadmium (Z=48)[3], indium (Z=49)[4] and tin (Z=50)[5]. In the Z=82 region the systematics of the lead-isotopes[7] could be extended down to ^{192}Pb[6].

EXPERIMENTAL

The neutron deficient nuclei have been produced as fusion evaporation products at the mass-separator on-line to the UNILAC at GSI[8]. ^{16}O- and ^{12}C-beams with intensities of up to 600 pnA and energies of up to 10 MeV/u served as projectiles striking internal molybdenum and tungsten foil targets of a discharge ion-source. The silver experiments were performed on mass-separated beams of typically 1×10^7 particles per second, whereas 4×10^5 pps were sufficient for the spectroscopy of the lead isotopes. To achieve chemical selectivity and to increase the signal to noise ratio the ion-source could be operated in a bunched mode[9]. After extraction from the ion-source and mass-separation the 54keV-ionic beam was neutralized in sodium vapour. The hyperfine components of the studied atomic transition were induced by the light of a collinearly over-laid cw-dye laser beam. Finally the resonance fluorescence was detected by a photon counting system. A detailed description of the set-up has been given by Kühl et al.[10]

RESULTS

Silver: The ground state transitions of neutral silver are in the ultra-violet part of the spectrum, which is not within reach of our laser system. Fortunately the charge exchange also populates metastable atomic states. Hence we have studied the 547 nm line connecting the long lived $4d^9 5s^2\ ^2D_{5/2}$-state with the alkali-like $4d^{10}6p\ ^2P_{3/2}$-state. Depending on the nuclear spin up to twelve hyperfine components have been observed. The nuclear moments have been

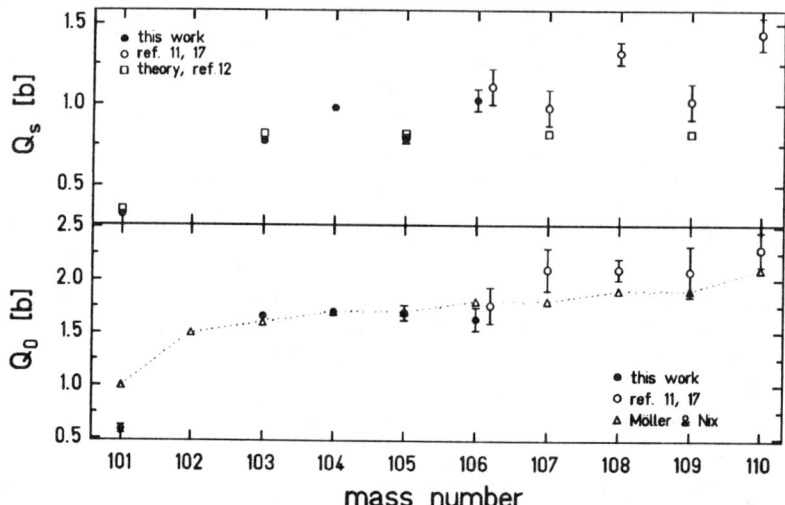

Figure 1: Quadrupole moments of the silver isotopes. The errors of our data include only the experimental contributions. A systematic shift of about ±10 percent due to uncertainties in the Sternheimer corrections and the electric field gradient is still possible.

---

determined from the hyperfine splitting. From the shift of the whole pattern relative to a stable reference isotope the change in the charge radii has been obtained. The magnetic moment of ^{105}Ag(I=7/2) has been determined to be $\mu = 4.414(13)$ n.m., fitting well in the systematics of the magnetic moments of this anomalously coupled state in $^{103-109}$Ag. The accuracy of some other magnetic moments could be increased. Figure 1 shows preliminary results for the spectroscopic quadrupole moments, corrected for Sternheimer shielding (R=-0.15)[20] as obtained from the $4d^9 5s^2\ ^2D_{5/2}$-hyperfine structure. Our result for ^{106}Ag is in agreement with the value given by Berkes et al.[11] The moments of the odd isotopes, especially the rather small value of Q_s in ^{101}Ag are in excellent agreement with the predictions of the Alaga-Paar model[12], treating the odd silver nuclei in terms of a three quasiparticle cluster coupled to a vibrating core. Except of ^{101}Ag the intrinsic quadrupole moments Q_0 as obtained from Q_s in the strong coupling limit are rather constant and in agreement with the calculations of Möller and Nix[13].

The isotope shift was calibrated using muonic x-ray data[14] for the stable isotopes and both Hartree-Fock[15] and Dirac-Fock[16] calculations for the electronic factor. Figure 2 shows preliminary results of the analysis, including former measurements[17] which have been recalibrated with the muonic value, which in turn has been justified by an analysis of the isotope shift in several transitions of the stable isotopes, using the HF- and DF-calculations. The prominent features of the radii are a rather strong odd-even-staggering and the small isomer shift in ^{105}Ag. The mean slope of the radii as predicted by the spherical droplet model[18] appears to be somewhat too small. According to the two parameter ansatz for the radii

$$\delta <r^2> = \delta <r^2>_{sph.} + \overline{<r^2>_{sph.}} \times \frac{5}{4\pi} \times \delta <\beta_2^2>$$

the I=1/2 isomer of ^{105}Ag is more deformed than the I=7/2 ground state. A parabolic behaviour of the radii, as observed in the chains of cadmium[3], indium[4] and tin[5] is indicated, further measurements to follow this trend to the neutron-rich side are in preparation. The deformations extracted from the quadrupole moments in the strong coupling limit are again as in the case of

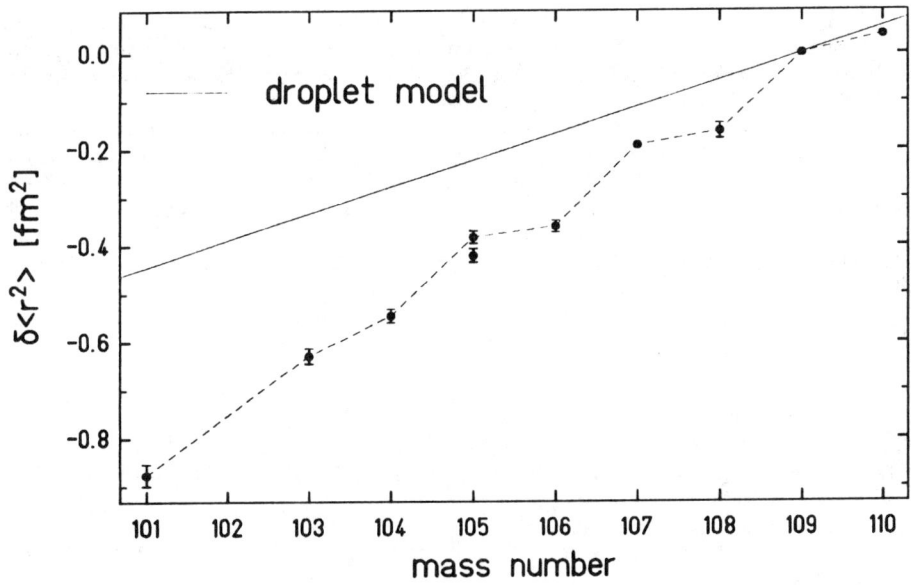

Figure 2: Change in the charge radii of the silver isotopes

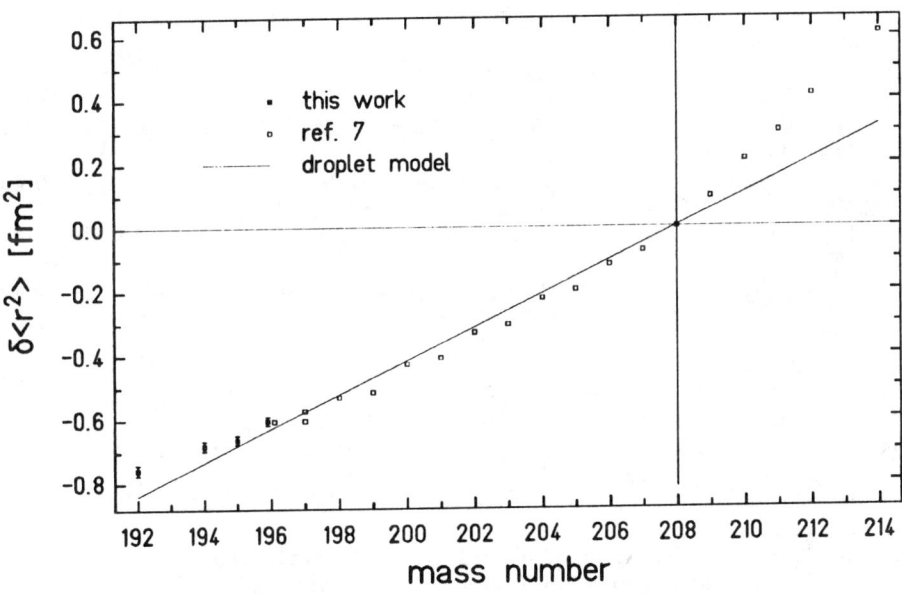

Figure 3: Change in the charge radii of the lead isotopes

cadmium, indium and tin smaller than the ones obtained from the isotope shift using the two parameter ansatz for the radii. Since the isotope shift in contrast to the hyperfine structure is also sensitive to dynamic deformations, this systematic behaviour indicates that zero point vibrations may be important in this region of the nuclear chart.

Lead: The $6p^2\,^1D_2 \to 6p7s\,^3P_1$ ($\lambda = 723\ nm$) transition has been studied. Using the conversion factors given in ref.7 the nuclear moments of ^{195}Pb have been determined from the $^3P_1$-hyperfine structure to $\mu = -1.1318(13)\ n.m.$, $Q_s = 0.286(95)eb$. The isotope shift has been calibrated via a King-plot with the data of ref.7. Figure 3 shows the change of the radii including the results of ref.7. Below the kink at the shell closure the radii follow a straight line. At the lower end however the slope is somewhat decreased, indicating a change in deformation, which may be induced by the 0^+-intruder state approaching the ground state in these nuclei[19]. Further measurements reaching to the middle of the neutron shell are planned for the near future.

REFERENCES

1. E. W. Otten, to be published in: Treatise on Heavy-Ion Physics Vol. 8 (Plenum Press, New York, 1987)
2. U. Dinger et al., to be published
3. F. Buchinger, P. Dabkiewicz, H. J. Kluge, A. C. Mueller, E. W. Otten, Nucl. Phys. A462, 305 (1987)
4. J. Eberz, U. Dinger, G. Huber, H. Lochmann, R. Menges, R. Neugart, R. Kirchner, O. Klepper, T. Kühl, D. Marx, G. Ulm, K. Wendt, Nucl. Phys. A464, 9 (1987)
5. J. Eberz, U. Dinger, G. Huber, H. Lochmann, R. Menges, G. Ulm, R. Kirchner, O. Klepper, T. Kühl, D. Marx, Z. Phys. A326, 121 (1987)
   M. Anselment, K. Bekk, A. Hanser, H. Hoeffgen, G. Meisel, S. Göring, H. Rebel, G. Schatz, Phys. Rev. C34, 1052 (1986)
6. U. Dinger, J. Eberz, G.Huber, R. Menges, S. Schröder, R. Kirchner, O. Klepper, T. Kühl, D. Marx, G. D. Sprouse, submitted to Z. Phys. A
7. M. Anselment, W. Faubel, S. Göring, A. Hanser, G. Meisel, H. Rebel, G. Schatz, Nucl. Phys. A451, 471 (1986)
8. C. Bruske, K. H. Burkard, W. Huller, R. Kirchner, O. Klepper, E. Roeckl, Nucl. Instr. Methods 186, 61 (1981)
9. R.Kirchner, O. Klepper, D. Marx, G.-E. Rathke, B. Sherrill, Nucl. Instr. Methods A247, 265 (1986)
10. T. Kühl, R. Kirchner, O. Klepper, D. Marx, U. Dinger, J. Eberz, G. Huber, H. Lochmann, R. Menges, G. Ulm, Nucl. Instr. and Meth. B26, 419 (1987)
11. I. Berkes, B. Hlimi, G. Marest, E. H. Sayouty, R. Coussement, F. Hardeman, P. Put, G. Scheveneels, Phys. Rev. C30, 2026 (1984)
12. V. Paar, Nucl. Phys. A211, 29 (1973)
13. P. Möller, J. R. Nix, At. Data Nucl. Data Tables 26, 165 (1981)
14. R. Engfer, H. Scheuwly, J. L. Vuiellheumer, H. K. Walter, A. Zehnder, At. Data Nucl. Data Tables 14, 509 (1974)
15. P. Aufmuth, private communication
16. B. Fricke, private communication
17. W. Fischer, H. Hühnermann, T. Meier, Z.Phys. A274, 79 (1975)
18. W. D. Myers, K.-H. Schmidt, Nucl. Phys. A410, 61 (1983)
19. P. van Duppen, E. Coenen, K. Deneffe, M. Huyse, J. L. Wood, Phys. Lett. 154B, 354 (1985)
20. R. Sternheimer, private communication; Phys. Rev. 164, 10 (1967); Phys. Rev. 130, 1423 (1963)

Nuclear Structure Studies Using Beams of Radioactive Nuclei

Isao Tanihata

RIKEN, 2-1 Hirosawa, Wako, Saitama 351-01, JAPAN

1. Introduction

Recent developments of high-energy beams of radioactive nuclei enabled us studies of nuclear structure far from stability. One of the primary interest which has been opened by this new method, is the size of light unstable nuclei which has been determined by the measurements of interaction cross sections.[1,2] The isospin dependence of the matter radii of isobars become possible to be studied. The matter radius of ^{11}Li nucleus is found to be extremely large compared to the neighboring nuclei. The large size together with the momentum distribution of the projectile fragment of ^{11}Li suggests an existence of a giant neutron halo in ^{11}Li.[3] Other measurement in which the magnetic moments of short-lived partner of mirror nuclei in f-shell are also in progress.[4] Here I would like to review these studies which use radioactive-nuclear beams.

Production of Secondary Radioactive Beams

Two important properties of the projectile fragmentation at high-energy opened a new possibility of using beams of radioactive nuclei. One is large production cross sections of wide variety of nuclei including extremely neutron- and proton-rich species. The other is the persistence of velocity; i.e. the fragments are emitted into a very narrow cone with a velocity nearly equal to that of the projectile. Because of these properties, high-energy beams of radioactive nuclei can be produced by simple separation technique.

Two beam lines for secondary radioactive nuclei exist at BEVALAC in Berkeley. One is B42 or HISS beam line which delivers beams of energy up to 2.1 GeV/nucleon (for $A/Z=2$ nuclei).[5] It has been used for measurements of the interaction and the fragmentation cross sections. The other, which is called as B44 and delivers lower energy beams up to 200 MeV/nucleon, is being used for measurements of the magnetic moments and the lifetimes of radioactive nuclei.[4] At GANIL, also two beam lines delivers radioactive beams; one is the LISE spectrometer[6] and the other is the beam line connected to the SPEG spectrometer.[7] It is extremely impressive how the combination of high-intensity intermediate-energy heavy ions and these separators provided many discoveries of new isotopes and enabled measurements of mass and lifetimes of exotic nuclei.

The Interaction Cross Sections and the Nuclear Radii

The interaction cross sections were measured for isotopes of He, Li, Be, B, and C at 790 MeV/nucleon at the BEVALAC. The interaction cross section (σ_I) discussed here is defined as the total cross section for the process of nucleon(s) (proton and/or neutron) removal from the incident nucleus. A transmission type measurement provided good accuracies (~ 1%) for most of σ_I. Details of experimental system are seen in refs. 1 and 5. The σ_I was then calculated by the equation,

$$\sigma_I = (1/N_t) \log(\gamma_0/\gamma), \quad (1)$$

where γ is the ratio of the number of non-interacting nuclei to that of incoming nuclei for a target-in run, and γ_0 is the same ratio for an empty-target run. The number of the target nuclei per cm² is written as N_t.

At high energy, the interaction cross sections are known to reflect the geometrical size of nucleus. An interesting and important finding from the interaction cross section is the separability of the interaction nuclear radii between a target and a projectile. The interaction nuclear radius is defined by the equation

$$\sigma_I(p,t) = \pi [R_I(p) + R_I(t)]^2, \quad (2)$$

where $R_I(p)$ and $R_I(t)$ are the interaction nuclear radii of the projectile and the target, respectively. The equation looks as if black sphere nucleus is assumed to define R_I. It is not however necessary to assume the black sphere nuclei but only requires a separability of R_I. This separability can be tested by the experimental data with different combinations of a projectile and a target. **Figure 1** shows the R_I for Li and Be isotopes determined by the interaction cross sections with three different target nuclei. Agreement of radius values determined from different targets shows that the separability of R_I hold very well. Theoretically Sato and Okuhara calculated the interaction cross sections using Hartree-Fock wave function and Glauber model, and confirmed also the separability for nuclei of all mass range.[8] **Figure 2** shows R_I of light p-shell nuclei so far determined. In general R_I follows $1.18A^{1/3}$ dependence.

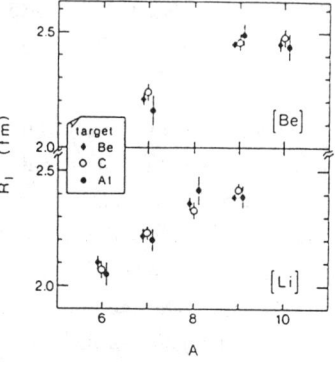

Fig. 1 Interaction radius determined by different targets.

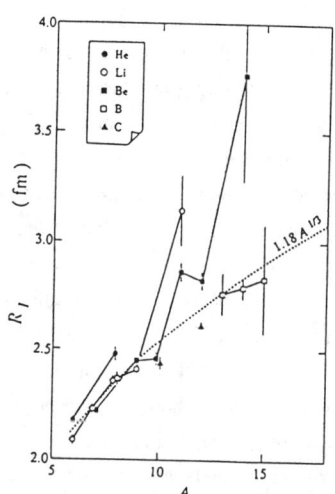

Fig. 2 Interaction nuclear radii in p-shell.

From the experimental data above 800 MeV/nucleon, it is known that the total reaction cross section is independent of the beam energy within 1 %.[9] The interaction nuclear radius, therefore, is well defined quantity above this energy. However, the relation to the nuclear distribution is not clear without a help of model calculation. In addition, the value may be energy dependent at lower energy because of the change in nucleon-nucleon cross sections. To obtain a nuclear size which is directly related to the nucleon distributions, a Glauber type calculation of the interaction cross sections were performed. The interaction cross sections were calculated using three different model distributions, i.e. Gaussian, harmonic oscillator,[10] and the droplet model.[11] With all three distribution functions, equal values of root-mean square (*rms*) radius were obtained when the parameter of the distributions were adjusted to reproduce an observed σ_I value. Although it could be an accidental luck for light nuclei, this important finding enable us to compare the *rms* radii directly to that of theoretical predictions.

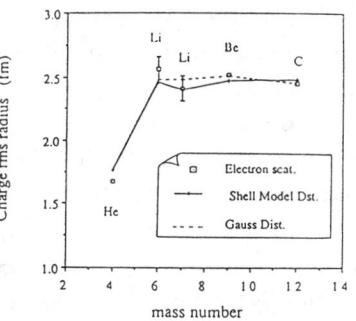

Fig. 3 Comparison of charge *rms* radii.

Figure 3 shows a comparison of charge *rms* radii of stable nuclei determined from the σ_I to those determined by electromagnetic probes. The charge *rms* radii were calculated from the proton distribution by folding the charge distribution in proton itself. In the Gaussian distribution, the proton and the neutron *rms* radii were assumed to be equal, and in harmonic oscillator the proton *rms* radii were different from that of neutron because of the occupation number difference. An excellent agreement of *rms* radii determined by two methods shows an validity of the present method. The *rms* radii of nucleon distribution for unstable nuclei as well as those of stable nuclei thus determined are shown in **Fig. 4**. The discussion on the data will be given in the later section.

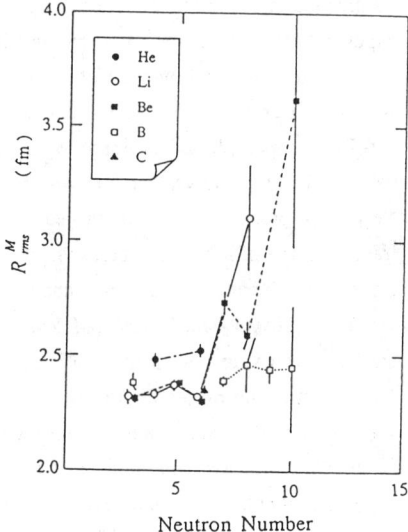

Fig. 4 Root-mean square radii of nucleon distribution.

Fragmentation of radioactive nuclei

Projectile fragmentation was extensively studied using beams of stable nuclei. Two important findings were made from the experimental data. One is the Q_{gg} rule similar to that obtained in low energy collisions in which the production cross sections follow a simple rule when they are plotted against the ground state Q-value.[12] The abrasion-ablation (collision and evaporation) model has then been developed to explain the cross sections. The other is the regularity of the momentum distribution of the projectile fragments. It was found that a momentum distribution has a Gaussian shape in the projectile frame, and the width of the Gaussian σ is independent of target mass and beam energy but depend only on the mass numbers of a projectile (A_B) and a fragment (A_F). The dependence of σ on A_B and A_F can be expressed as,

$$\sigma(P) = \sigma_0 \sqrt{A_F(A_B - A_F)/(A_B - 1)} \quad , \tag{3}$$

where $\sigma_0 \approx$ 90 MeV/c.[13] It has been shown that σ_0 is the constant related to the Fermi momentum (P_F) of the nucleons in the projectile as,

$$P_F = \sigma_0 \sqrt{5} \tag{4}$$

and therefore related to the internal motion of the nucleons. This idea was further developed by Fujita and Hüfner[14,15] and the momentum distributions of the one-nucleon removal channel in projectile fragmentation were studied in detail. They showed that the momentum distribution of a nucleon could be determined up to momentum around 500 MeV/c.

Recently we have measured a projectile fragmentation of ^{11}Li at 790 MeV/nucleon.[16] The experimental setup is the same one used for the interaction cross section measurements. **Figure 5** shows the fragmentation cross sections of ^{11}Li to Li and He isotopes plotted against the minimum Q value. The data shows an interesting behavior for He isotope production. In the projectile fragmentation of a stable nucleus, a production cross section is smaller for a nucleus with larger Q value. In present data Li isotopes follow this rule

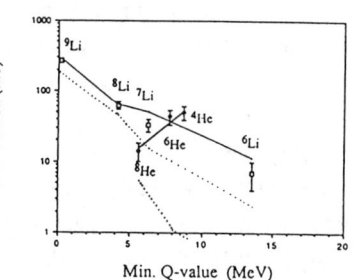

Fig. 5 Fragmentation cross section of ^{11}Li.

but He isotopes do not. In the collision stage of the fragmentation, nuclei which have similar A/Z to that of the projectile nuclei are produced with larger cross sections. In the evaporation stage, on the other hand, nuclei near the stability line tend to accumulate. For the fragmentation of a stable nucleus, unfortunately, nuclei near the stable line are produced copiously in both stages because a nucleus with larger Q value is also farther from the stability line. However in ^{11}Li, A/Z is much larger than that of stable nuclei, therefore neutron-rich nuclei are produced more abundantly in the collision stage of the reaction. The larger cross section of ^{4}He indicates, therefore, an importance of the evaporation process for the production of He isotopes. On the other hand, the collision stage takes an important role for production of Li isotopes because larger production cross sections are observed for neutron rich isotopes.

A cascade type calculation of nucleus-nucleus collisions was made to verify the idea. Dotted lines in **Fig. 5** show the production cross sections calculated only with the collision stage using a Wood-Saxon type density distribution with the central density of proton and neutron to be equal, thus having a neutron skin of thickness about 0.4 fm. The production cross section for nuclei near the stability line are much smaller than the observed values. Solid lines in the figure show the result of the calculation including a evaporation processes. An excellent agreement is obtained. It thus shows the importance of the evaporation processes for production of He isotopes. Further studies are in progress to study a sensitivity of the production cross sections to the neutron skin.

The transverse momentum distributions of fragments of ^{11}Li by the carbon target were also measured at 790 MeV/nucleon.[16] **Figure 6** presents a transverse momentum distribution of ^{9}Li, which shows two Gaussian structure. One component has a wide width σ= 95 ± 12 MeV/c and the other has a narrow width σ= 23 ± 5 MeV/c. The two-component structure had not been observed in fragmentation of stable nuclei and therefore is a new feature so far only seen in ^{11}Li fragmentation. In the Fermi gas model, it is difficult to consider two different Fermi momenta in a nucleus.

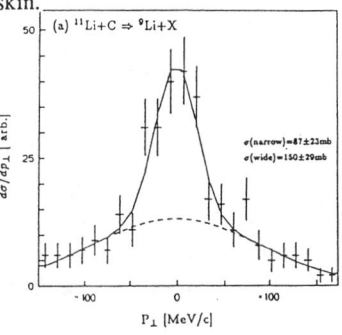

Fig. 6 Transverse momentum distribution of ^{9}Li produced by the fragmentation of ^{11}Li.

Therefore a simple view based on Fermi gas model fails to explain the data. If we follow the Hüfner's argument, the narrow peak indicates an existence of nucleons with extremely small momentum fluctuation inside ^{11}Li. The wide width is consistent with the momentum fluctuations of usual nucleons. If we extend the method used for stripping reactions to many nucleon removal. The momentum width of the projectile fragment is related to the separation energy of last neutrons and then written as,

$$\sigma^2 = u <\varepsilon> \frac{A_F (A_B - A_F)}{A_B} \quad (5)$$

where u is the atomic mass unit and $<\varepsilon>$ is an average separation energy of the removed nucleons. The observed narrow width gives $<\varepsilon>$ = 0.34 ± 0.16 MeV which can be compared with separation energy of last neutrons; $\Delta E[(^{10}\text{Li}+n) - (^{11}\text{Li})]$ = 0.96 MeV and $\Delta E[(^9\text{Li} + 2n) - (^{11}\text{Li})]$ = 0.19 MeV. The broad width, on the other hand, gives $<\varepsilon>$ = 6.0 ± 1.5 MeV which is consistent with the normal nucleon separation energy. Therefore it is considered that the narrow component is produced by reactions in which weakly bound last neutrons are removed from ^{11}Li. The broader component is due to the removal of normally bound neutrons. The small momentum fluctuation of last two neutron also indicates the existence of a long neutron tail in the neutron density distribution. This neutron tail or halo in ^{11}Li will be discussed more later.

2. Matter *rms* Radii of Light p-shell Nuclei

Now we discuss the radii obtained from the interaction cross section. **Figure 4** shows the rms radii of nucleon distributions, as a function of neutron number (N), determined by assuming the harmonic-oscillator model distribution. The radii of Li and Be isotopes depend similarly on N. It is seen that the majority of *rms* radii of p-shell nuclei are almost constant (2.3 -2.4 fm) except ones close to the neutron drip line. Large increases of the radius from ^{9}Li to ^{11}Li and from ^{10}Be to ^{11}Be are observed, and rates of increases are similar to each other. It is interesting to note that the spin-parity of ^{11}Be ground state is 1/2+ indicating the major contribution of $2s_{1/2}$ shell instead of $p_{1/2}$. In fact, it is seen that the 1/2+ state of nuclei with $N=7$, which locates 5.18 MeV higher than the 1/2- state in ^{15}O (Z=8), becomes lower as Z decreases i.e., 3.09 MeV for ^{13}C (Z=6) and 1/2+ state is lower by 0.32 MeV for ^{11}Be (Z=4).[18] Therefore, the large radii of ^{11}Be is considered to be due to the strong mixture of sd-shell component with long tail in the neutron wave function. The similar large radii of ^{11}Li may also due to the mixture of sd-shell component. Now let us see the data from several view points.

Isospin Dependence of The Matter rms Radii

The *rms* radii of nucleon distribution are plotted for isobars of mass number $A=$ 6,7,8,9,11, and 12 in **Fig. 7**. A pair of nuclei with same isospin (T) but different T_z, ^{7}Li - ^{7}Be or ^{8}Li - ^{8}B , shows equal radii. It suggests that the Coulomb effect on the radii is negligible for these light nuclei. On the other hand, a nucleus with larger isospin shows a larger radius except for $A=9$ isobars. The isospin dependence of isobar radii are compared with theoretical predictions as follows.

Firstly as one of the gross theory, the droplet model[11] was used and results of the model calculations are shown by the dotted lines in **Fig. 7**. Since the original parameters used in ref. 11, determined from the data of heavier nuclei, gave *rms* radii systematically ~0.3 fm larger than the experimental ones, the nuclear diffuseness parameter was arbitrary set to 0.7 fm (instead of 0.9 fm used in ref. 11) to obtain an over all fit. The predicted values are almost independent of isospin and do not reproduce the experimental trend, i.e. the strong isospin dependence. The droplet model allows proton and neutron density distribution to vary so as to minimize the total nuclear energy but assume the constant surface diffuseness. The disagreement to the data suggests that those changes are not enough to explain the observed isospin dependence.

Secondly, comparisons with predictions by a Hartree-Fock(HF) calculations using the Skyrme potential[8] were also made and shown in **Fig. 7**. Two different sets of potential parameters SIII and SV were used.[18] Two sets of parameters are basically different in the density dependence of effective interactions; SIII includes a strong density-dependent interaction, while SV includes no density-dependent interaction.

Both of the calculations give fair agreements to the data within 0.2 fm for most of the nuclei. The radii calculated with SV potential, however, gives much weaker isospin dependence than observed. On the other hand, the radii obtained with SIII potential gives good agreement to the observed isospin dependence, except for $A=9$ isobars. The comparison of the data with the HF calculation, therefore, indicates an importance of the density dependent interaction in understanding the radii of light exotic nuclei.

Isobars of $A=9$ show abnormal behavior in the data, namely the isobar with smaller isospin (^{9}Be) has larger radius than the isobar with larger isospin (^{9}Li). It has been known that a specific structure of ^{9}Be nucleus; i.e. a strong configuration of two-α clusters weakly bound by a neutron, gives a large radius for the charge distribution. This anomalous structure was not taken into account for the present HF calculation and therefore it is understandable that the calculation gave different isospin dependence of nuclear radii. In light nuclei the clustering aspect is therefore considered to be important for understanding the radii.

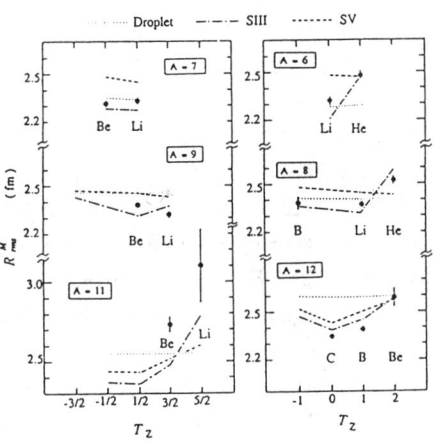

Fig. 7 Isospin dependence of the *rms* radii.

Giant Neutron Halo in ^{11}Li ?

As already shown ^{11}Li, which has neutron number of p-shell closure in naive shell model, shows an unusually large radius. An existence of a large deformation and/or a long tail in the matter distribution may be considered. A recent determination of spin value (3/2), however, indicates that ^{11}Li is not largely deformed. Also the magnetic moment is close to the Schmidt value of the $p_{3/2}$ proton state and therefore consistent with not too large deformation. On the other hand, the small momentum fluctuation observed in the last two neutron, which is seen in the momentum spectrum of ^{9}Li for ^{11}Li fragmentation, indicates the long neutron tail.

The mass difference between ^{11}Li and ^{9}Li+2n is only 190 ± 100 keV. When the binding energy of last two neutron is small, an existence of dineutron near the nuclear surface is predicted by Migdal.[19] Recently, Hansen and Jonson applied this idea to ^{11}Li.[20] They used a model in which a dineutron is orbiting around the well bound ^{9}Li core. They assumed that the potential between ^{9}Li core and the dineutron to be a square well. Because of the small binding energy of the dineutron, the wave function has long exponential tail outside the potential. The decay constant ρ of the exponential is given by,

$$\rho = \frac{\hbar}{\sqrt{2\mu B}} \quad (6)$$

where μ is a reduced mass of the system and B is the binding energy. Using the experimentally determined ^{9}Li rms radii, they found a relation between binding energy and the rms radius of ^{11}Li as shown in **Fig. 8**. The experimental values of binding energy and radius are right on the curve and thus consistent with the model.

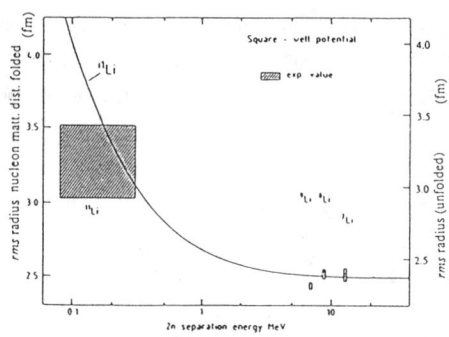

Fig. 8 Relation between binding energy of 2n system and the rms radii of ^{11}Li.

From the discussion above, (i) a large size of ^{11}Li, (ii) a small momentum fluctuation of last two neutron, and (iii) a small binding energy, are consistent with an existence of long neutron tail or halo. The decay constant of this tail density is considered about 4 times larger than that of normal nucleus, because the spread of momentum σ is about a quarter of that of normal Fermi gas.

A large increase of nuclear size for neutron rich nuclei were also suggested for heavier nuclei (Z= 5 - 12) by a group at GANIL.[21] They measured total reaction cross sections of neutron rich nuclei by a detection of γ rays at 60 MeV/nucleon. They observed that the size of nucleus monotonically increases as neutron excess increase. Because this experiment used a thick target, the cross section is integrated over the beam energy. The understanding of the data, therefore, is much more complicated due to the low energy character of the reaction. However, it provides an interesting suggestion that the inflation of nuclei is common thing for neutron rich nuclei. Experiments on nuclei near neutron drip line are surely interesting for studying low density neutron matter.

3. Measurement of the Magnetic Moments of Mirror Nuclei in f-shell

A new experimental method has recently been developed for studying magnetic moments of mirror nuclei in the $f_{7/2}$ shell. As at least one partner of mirror pairs in this region is short-lived (0.1 - 1 s) β emitter, preparation of spin polarized isotopes for NMR studies presents quite a challenge. In this experiment a nuclear polarization is produced by the tilted foil method. Then the polarization is detected by the asymmetric emission of β ray.

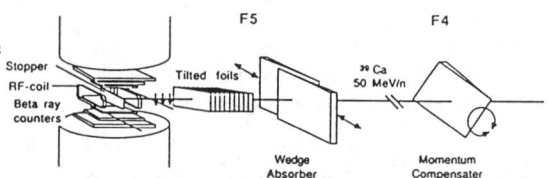

Fig. 9 Experimental set up for the measurement of magnetic moments.

Figure 9 shows the experimental set up for the NMR study of ^{39}Ca at BEVALAC. A secondary beam of ^{39}Ca ion was produced from the 220 MeV/nucleon ^{40}Ca beam and separated by the beam line. Separated beam is then guided to the momentum dispersive focus shown as F4 in **Fig.9**. A thin wedge is used to compensate the energy spread. This compensation of energy is most important for application of the tilted foil method because the polarization is produced only for ions with sufficient number of electrons. If energy spread of the ^{39}Ca ion beam is large, some of the ^{39}Ca pass through the tilted foils at energy where only few electrons are bound. With this wedge compensator the energy spread at the final stopping point of the ^{39}Ca ions was reduced to 3 MeV/nucleon (HWHM). A stack of ten foils, tilted at 60° relative to the beam axis, were used to polarize the ions just upstream of a CaF$_2$ catcher.

Figure 10 shows a time spectrum of the β rays emitted from the catcher. It represents a clean decay spectrum of ^{39}Ca. From the asymmetry change in the β-ray emission, NMR was successfully detected. The β-ray asymmetry change of (0.65 ± 0.19) % was observed. This method can be applied not only for mirror nuclei but also for other short-lived nuclei. One possible limitation of the method is the production of a polarization by the tilted foil method because, in general, produced polarization is less than 5 %. If one can apply other method, by which a larger polarization can be obtained, the region of study will widely be extended. The selection of scattering angle for lower energy fragmentation (≤ 200 MeV/nucleon) or use of polarized Na vapor would be such a possibility.

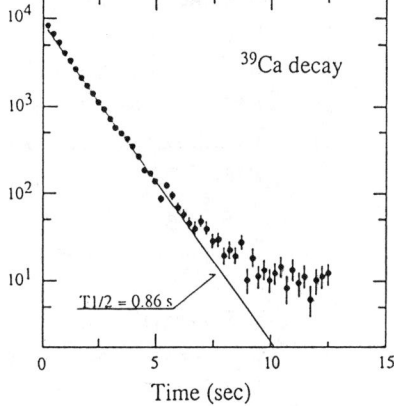

Fig. 10 The time spectrum of β rays emitted from CaF stopper.

Acknowledgement

Author would like to express his great thanks to the colleague who contributed to the experiments (E690H and E732H) at Berkeley. Fruitful discussions with Drs. M. Sano, H. Sato, and K. Ogawa are gratefully acknowledged. The present work was supported in part by the Grant in Aid for Scientific Research from the Ministry of Education, Culture and Science, Japan, by the U. S. DOE under contract No. DE-AC03-76SF0098, by the INS-LBL Collaboration, and by the Japan-US cooperative Research Program from the Japan Society for the promotion of Science. Author also thanks to the Yamada Science Foundation for their kindfull support.

References

1. I. Tanihata et al., Phys. Lett. **160B**, 380 (1985).
2. I. Tanihata et al., Phys. Rev. Letters **55**, 2679 (1985).
3. I. Tanihata, Proceedings of the Particles and NucleiI nternational Conference (PANIC) Kyoto, Japan, April 1987, to be published. RIKEN report RIKEN-AF-NP-59. Also I. Tanihata et al., RIKEN-AF-NP-60, 1987, submitted for the publication.
4. Y. Nojiri, et al., Hyperfine Interactions **35**, 1019 (1987).
5. I. Tanihata, Hyperfine Interaction **21**, 251 (1985).
6. J. P. Dufour et al., Nucl. Inst. Methods in Phys. Res. **A248**, 267 (1986).
7. A. Gillibert et al., Phys. Lett. **B176**, 317 (1986)
8. H. Sato and Y. Okuhara, Phys. Rev. C **34**, 2171 (1986).
9. A. S. Goldhaber and H. H. Heckmann, Ann. Rev. Nucl. Sci. **28**, 161 (1978).
10. R. B. Elton, Nuclear Sizes (Oxford Univ. Press, London 1961) pp. 21 -22.
11. W. D. Myers and K-H Schmidt, Nucl. Phys. **A410**, 61 (1983).
12. H. Sato, Private communications. INS Univ. of Tokyo report, INS-NUMA-60.
13. A. S. Goldhaber, Phys. Lett. **53B**, 306 (1974).
14. T. Fujita and J. Hüfner, Nucl. Phys. **A343**, 493 (1980).
15. J. Hüfner and M. C. Nemes Phys. Rev. C **23**, 2538 (1981).
16. T. Kobayashi et al., private communication, to be published.
17. C. M. Lederer and V. S. Shirley ed., Table of Isotopes, Seventh edition (Wiley-Interscience).
18. M. Beiner, H. Flocard, Nguyen Van Giai, and P. Quentin, Nucl. Phys. **A238**, 29 (1975).
19. A. B. Migdal, Soviet J. Nucl. Phys. 16, 238 (1973).
20. P. G. Hansen and B. Jonson, Private Communication.
21. A. Gillibert et al., Contribution to this conference. Also W. Mittig et al., GANIL report 0.87-07.

THE NEUTRON HALO OF NUCLEI AT THE NEUTRON DRIP LINE

B.Jonson, S.Mattsson, G.Nyman, O.Tengblad
Dept. of Physics, Chalmers University of Technology, Göteborg,
Sweden

M.J.G.Borge
Insto. de Quimica-Fisica "Rocasolano", CSIC, Madrid, Spain

P.G.Hansen
Institute of Physics, University of Aarhus, Aarhus, Denmark

K.Riisager
EP-Division, CERN, Geneva, Switzerland

ABSTRACT

Empirical evidence suggests that neutron pairing plays an important role for the stability of nuclei near the neutron drip line. The low binding of these nuclei is shown to lead to the formation of a neutron halo surrounding the nucleus. The halo may lead to large cross sections for Coulomb dissociation.

1. INTRODUCTION

We have recently pointed out[1] that the very low binding energies of the last neutrons near the neutron drip line will lead to a swelling of their wave functions so that a neutron halo is formed around the nucleus. We also note that experimental data suggest that the n-n force in the T=1, S=0 state is an essential ingredient for stabilizing nuclei near the drip line. The over-all effect is large : a crude estimate suggests that the neutron "gas" surrounding ^{11}Li occupies a 25 times larger volume than the ^{9}Li "core", so that this phenomenon should be observable in a number of ways, to be discussed below. One particularly interesting consequence is that the halo may be associated with a soft E1 mode leading to large cross sections for Coulomb dissociation in collisions with heavy nuclei.

2. EXPERIMENTAL BACKGROUND

It is very striking that not only do most of the predicted particle-stable, neutron-rich nuclei remain undiscovered, we do not even have any good idea as to their production in the laboratory. Only for the lightest elements has there within the last few years been remarkable progress, of which we give some examples in this section. More details and many references can be found in a recent review[2].

2.1 Nuclear stability.

New experiments at GANIL, ref.[3] and discussed also at this conference, seem to have detected all predicted particle-stable isotopes up to nitrogen, Z=7, see Figure 1. It is striking that for

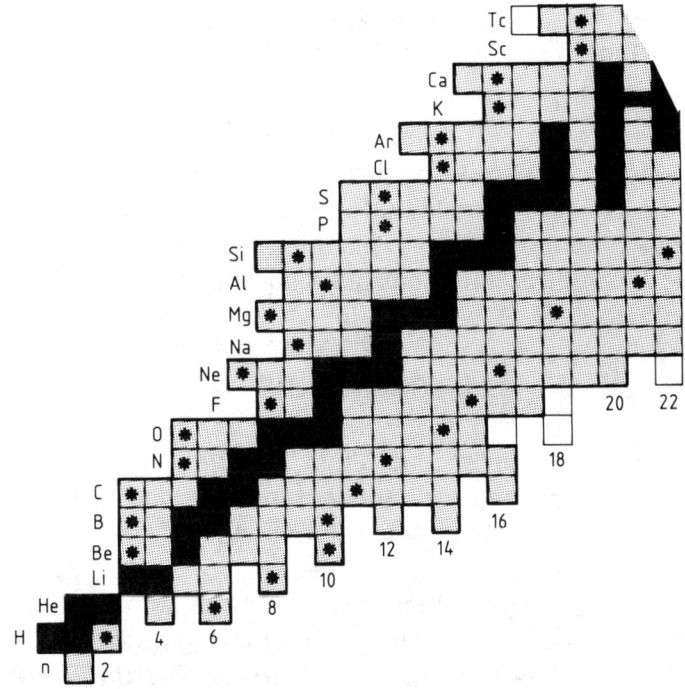

Figure 1. The nuclear chart: nuclides detected experimentally and theoretical estimates of the limits to stability. The black squares represent the stable isotopes in nature. The heavy line surrounds the (shaded) area of isotopes detected experimentally with half-lives longer than 10^{-7}s, and the external squares in thin lines represent predicted but as yet unobserved isotopes. The stars mark the heaviest isotope for which the half life is known experimentally. All isotopes of all elements up to nitrogen now appear to have been detected. Note also that the proton drip line has been reached for all light elements up to Z=21, scandium.

each element the heaviest isotope has even neutron number N (=2n), and in all but two cases the isotope with N = 2n - 1 is unbound. For helium and boron also the isotope with N = 2n - 3 is unbound. This pattern seems to indicate that pairing is essential for nuclear stability at the drip-line.

2.2 Nuclear masses

The heaviest n-drip-line isotope for which the mass is known[4] is ^{11}Li. The corresponding two-neutron binding energy B_{2n} is only 190 ± 110 keV. The isotope ^{10}Li is unbound with respect to neutron emission. Other interesting information comes from Seth et al.[5], who have exploited the (π^-, π^+) reaction to observe resonant states, even beyond the drip line such as the ground state of ^{9}He, unbound by 1.13 MeV with respect to neutron emission. They point out that the neutron-rich helium isotopes are considerably more bound than extrapolations from near stability would suggest. This could be an indication that the nucleus responds to the large neutron excess by a change in structure : the halo, which is our subject here, is one model of such a change.

2.3 Nuclear radii.

The nuclear volume shifts in atomic spectra are so small in light nuclei that it is difficult to deduce charge radii. Total reaction cross sections have been used to infer the mass radii of p-shell nuclei in a series of elegant experiments by Tanihata et al.[6]. While most of the p-shell nuclei turn out to have matter radii around 2.50 fm, similar to the measured charge radii, the neutron-rich nuclei ^{6,8}He and ^{11}Li have root-mean square (rms) matter radii of 2.73 ±0.04, 2.70 ±0.03, and 3.27 ± 0.24 fm, respectively. We return to our interpretation of these results in Sect. 3. Similar experiments at GANIL [7] also find strong increases in the radii of heavier (C to Mg) nuclei with increasing neutron excess.

2.4 The pairing energy

The strong role played by pairing (Sect. 2.1) would at first sight seem evident. It is only very recently that it has been found [8], see Fig. 2, that the pairing is most important at the

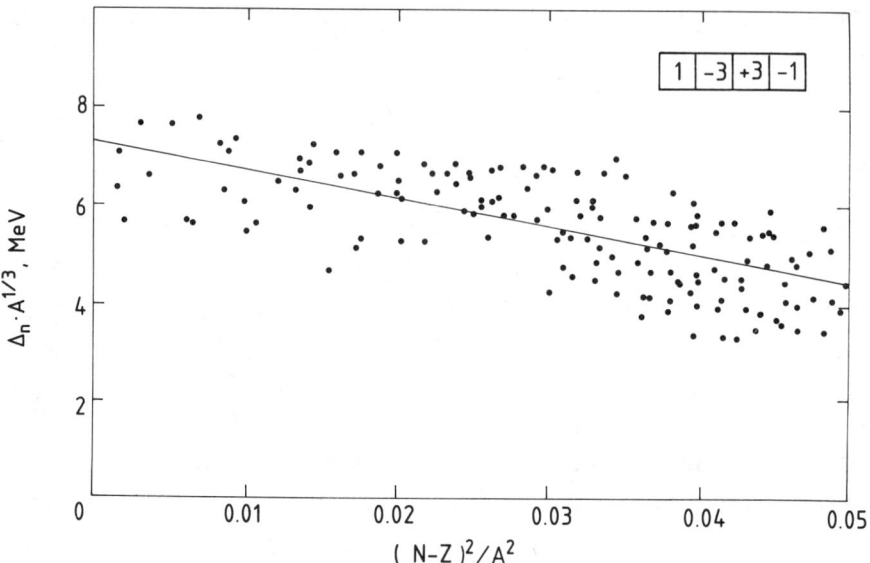

Figure.2 Experimental neutron pairing energies Δ_n represented as $A^{1/3}\Delta_n$ versus the isospin symmetry parameter $I^2 = (N-Z)^2 A^{-2}$, excluding points with magic Z or N. The pairing energies were calculated from the diagram shown in the inset. The straight line is the least-squares fit
$$A^{1/3}\Delta_n = 7.36 \, (1 - 8.15 \, I^2) \text{ MeV}.$$

N = Z line and decreases strongly with increasing neutron excess. This is accounted for by semiempirical expressions[8,9] incorporating an *ad hoc* isospin-symmetry term[*]. When these expressions are extrapolated to the neutron drip line they predict that pairing essentially vanishes there, which is definitely not the case. We conclude that neither the old treatment of pairing nor the new one have much predictive power near the neutron drip-line, and that new and precise experiments will have the last word.

2.5 Exotic decay modes of ^{8}He and ^{11}Li.

These isotopes with half-lives of only 117 ms and 8.5 ms, respectively, are the heaviest drip-line nuclei for which one has detailed experimental information, see the review [2]. Of

---

[*] The origin of this term is an interesting open question, which neither of refs. [8,9] attempt to answer. One suggestion has been given by ref. [10].

particular interest are a number of decay modes involving particle emission, see Table 1. New experiments[11], now in progress, plan to search for beta-delayed deuteron emission, which could be the signature of the beta decay of a neutron pair in the halo. The beta-delayed tritons, see ref.[12] and earlier work cited therein, could have a similar origin.

Table 1.
Beta-delayed particle decay modes of ^{8}He and ^{11}Li.

	Particle	p_x (%)
^{8}He	n	16 ± 1
	^{3}H	0.9 ± 0.1
^{11}Li	n	84.9 ± 0.8
	2n	4.1 ± 0.4
	3n	1.9 ± 0.2
	^{3}H	0.010 ± 0.004

3. THE NEUTRON HALO

In the absence of any detailed theory of drip-line nuclei we consider in this section for simplicity a model consisting of two neutrons coupled to a core (^{6}He or ^{9}Li). This is suggested by the experimental data (Fig. 1), which show that the additional attractive contribution from the n-n interaction usually is essential for binding the last neutron(s) to the nucleus. This special case of the three-body problem was considered by Migdal [13], who conjectured that several nuclei should exhibit a bound state of this kind, which he interpreted as a dineutron near the nuclear surface. He also pointed to the possibility that states more complex than dineutrons may exist.

It is interesting to note in passing that the possibility of loosely bound states of three particles interacting via short-range forces also has been discussed in molecular physics, and that it has been suggested that the trimer He_3 held together by van der Waals forces may be stable at low temperatures. The problem has been discussed in a recent paper by Macek [14], who gives references to earlier work.

For semi-quantitative estimate of this effect one may approximate the nucleus by a quasi-deuteron consisting of a core with mass M coupled to a dineutron ^2n with mass m. Letting μ denote the reduced mass of this system and B its binding energy, one has immediately that the wave function in the force-free region must decay exponentially with a decay length ρ given by

(1) $$\rho = \hbar/(2\mu B)^{1/2},$$

which for the ^{11}Li parameters takes the value 8.2 fm. It is clear that a neutron halo inevitably arises and that the pairing effect, in part, reflects the force between free neutrons. The external wave function of the system can for small B be approximated by

(2) $$\psi(r) = (2\pi\rho)^{-1/2} \frac{\exp[-r/\rho]}{r} \left[\frac{\exp[R/\rho]}{(1+R/\rho)^{1/2}} \right],$$

for a square-well potential with radius R and with r denoting the distance between ^9Li and the dineutron. For surface phenomena it is a good approximation to use the wave function as written here over the entire range of r. This wave function with the corresponding inner part was used[1] to calculate nuclear matter radii as shown in Fig. 3. The increase in nuclear size is well reproduced by this simple estimate.

Another, even, simpler estimate of the halo effect was given by Bang et al.[15], who calculated the rms radius for the most loosely bound neutron in a central potential, the depth of which

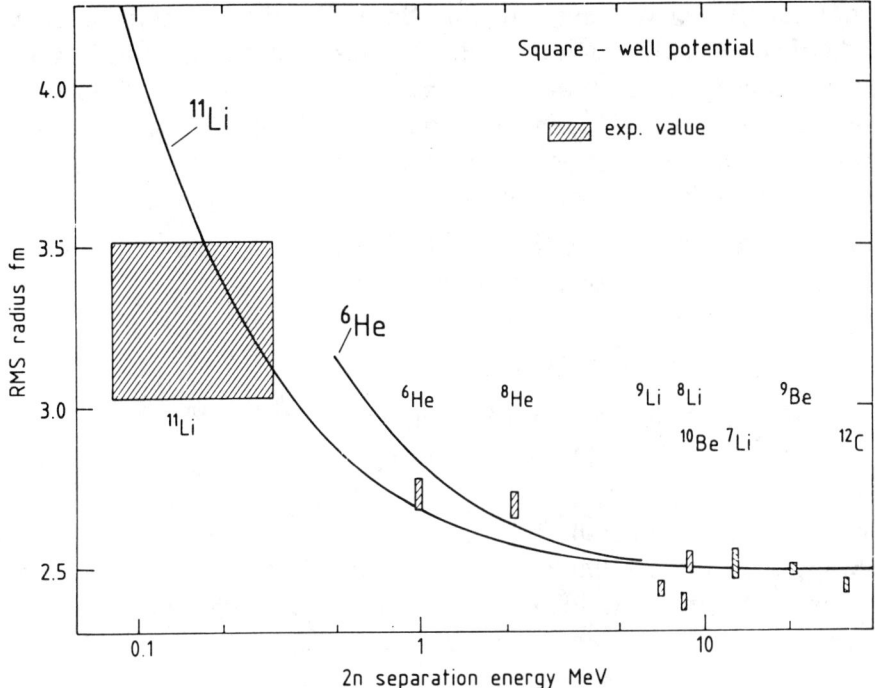

Figure 3. The matter rms radius calculated in the dineutron cluster model as a function of the 2n separation energy and for mass parameters corresponding to ^{6}He and ^{11}Li. The points are experimental values. The theory was normalized at large binding energies and has no predictive power there.

was adjusted to give the experimental binding energy. For ^{6,8}He and ^{11}Li the calculation was for two uncorrelated neutrons, each with one half of the 2n separation energy. The results are in good agreement with experiment and with the cluster model estimates given in Fig. 3. This shows that the qualitative features of the halo effect are insensitive to the details of the model : one model assumes no n-n correlation and the other approximates the dineutron by a point particle.

4. COULOMB DISSOCIATION

The analogy developed in Sect. 3 of the drip-line nucleus with the deuteron, the most loosely bound object in classical nuclear physics, suggests that it would be interesting to calculate the cross section for Coulomb dissociation in the field of a struck nucleus. The deutron problem has been considered by Oppenheimer[16] and more recently by Fäldt[17]. The wave function (2) can be used to calculate the dissociation probability for a momentum transfer q. From the Rutherford cross section for collision of two ions with charges Z_1 and Z_2 and relative velocity v one obtains[1] after integration over q the approximate dissociation cross section

$$(3) \qquad \sigma_d \cong \frac{2\pi Z_1^2 Z_2^2 e^4 m}{3v^2(M+m)MB} \left[\frac{e^{2x}}{1+x} \right] \ln(q_{max}/q_{min}),$$

valid for small values of B (and consequently small $x = R/\rho$). Taking the upper limits approximately as the momentum transfer at grazing incidence and in the adiabatic limit, respectively, we find the order of magnitude

$$(4) \qquad q_{max}/q_{min} \cong (\mu v^2/B)^{1/2},$$

where a number of terms of order unity have been left out. The cross sections (3) can be very large, especially for high Z_2, low binding energies B and not too large energies. For 100 MeV/A ^{11}Li incident on uranium one calculates a cross section of 6-22 barns for the process ^{11}Li $\longrightarrow$ ^{9}Li + 2n where the interval corresponds to the experimental error on the 2n separation energy. Such reactions might offer a new way of studying clusters and their separation energies.

It is probably useful to underline at this point that the - still hypotetical - Coulomb dissociation mode discussed here is a <u>soft</u> E1 mode. With the reservation that the Hamiltonian underlying eq. (2) is ill suited for a discussion of the spectrum of excited states, we may use the energy-weighted sum rule and find for ^{11}Li an average excitation energy of the order of 1 MeV only. Clearly, this process has little to do with the - currently much discussed - Coulomb disintegration caused by ultrarelativistic ions and driven largely by a relativistic factor of γ, left out in eq. (4). (At the energies discussed in ref.[1]) γ is in the range 1 - 2.)

5. CONCLUDING REMARK

Naturally enough, theories of neutron drip-line nuclei have until now consisted in an extrapolation of the structure observed at stability, shells, warts and all. These extrapolations have been held in high esteem, which may be linked to the fact that they could not be verified experimentally (the psychological law behind this seems a general one, and is probably not the least of the paradoxes of astrophysics). The striking experimental progress in the study of drip-line nuclei within the last years now is forcing us to think more seriously about their structure. We suspect that this subject is one that will play an important role also at the next conference on nuclei far from stability.

REFERENCES

1. P.G.Hansen and B.Jonson, Europhys. Lett., 4, 409 (1987).

2. P.G.Hansen and B.Jonson, Beta-delayed particle emission from neutron-rich nuclei, contribution to the book Particle Emission From Nuclei, to be published by CRC Publ.Co.

3. M.Langevin, E.Quiniou, M.Bernas, J.Galin, J.C.Jacmart, F.Naulin, F.Pougheon, R.Anne, C.Détraz, D.Guerreau, D.Guillemaud-Mueller, and A.C.Mueller, Phys.Lett. 150B, 71 (1985).
F.Pougheon, D.Guillemaud-Mueller, E.Quiniou, M.G. Saint-Laurent, R.Anne, M.Bernas, D.Guerreau, J.C.Jacmart, S.D.Hoath A.C.Mueller, and C.Détraz, Europhys. Lett. 2 ,505 (1986).

4. C.Thibault, R.Klapisch, C.Rigaud, A.M.Poskanzer, E.Perils, L.Lessard, and W.Reisdorf, Phys. Rev. C12 , 644 (1975).

5. K.K.Seth, M.Artuso, D.Barlow, S.Iversen, M.Kaletka, H.Nann, B.Parker,and R.Soundranayagam, Phys.Rev.Lett. 58 , 1930 (1987).

6. I.Tanihata, H.Hamagaki, O.Hashimoto, Y.Shida, K.Yoshikawa, K.Sugimoto, T.Yamakawa, T.Kobayashi, and N.Takahashi, Phys.Rev.Lett. 55, 2676 (1985).
I.Tanihata, H.Hamagaki, O.Hashimoto, S.Nagamiya, Y.Shida, O.Yamakawa, K.Sugimoto, T.Kobayashi, D.E.Greiner, N.Takahashi, and Y.Nojiri, Phys.Lett. 160B, 380 (1985).

7. W.Mittig, J.M.Chouvel, Z.W.Long, L.Bianchi, A.Cuusolo, B.Fernandez, A.Foti, J.Gastebois, A.Gillibert, C.Grégoire, Y.Schutz, and C.Stephan, to be published.

8. P.Vogel, B.Jonson, and P.G.Hansen, Phys.Lett. 139B, 227 (1984). A.S.Jensen, P.G.Hansen, and B.Jonson, Nucl.Phys. A431, 393 (1984).

9. D.G.Madland, and J.R.Nix, to be published in Nucl. Phys.

10. A.S.Jensen, and A.Miranda, Nucl.Phys. A449, 331(1986).

11. M.J.G.Borge et al, proposal to the ISOLDE Committee CERN IP-46 (April 1987), unpublished.

12. M.J.G.Borge, M.Epherre-Rey-Campagnolle, D.Guillemaud-Mueller, B.Jonson, M.Langevin, G.Nyman, and C.Thibault, Nucl.Phys. A460, 373 (1986).

13. A.Migdal, Yad. Fiz. 16, 427 (1972); English translation Soviet J. Nucl. Phys. 16, 238 (1973).

14. J.Macek, Z.Phys. D3, 31 (1986).

15. J.M.Bang, F.A.Gareev, and G.S.Kazacha, Radii of p-shell Nuclei, private communicartion and to be published (1987).

16. R.J.Oppenheimer, Phys. Rev. 47, 845 (1935).

17. G.Fäldt, Phys. Rev. D2, 846 (1970).

INTERACTION RADII OF NEUTRON-RICH NUCLEI[*]

A. Gillibert, L. Bianchi, H. Dumont, B. Fernandez, J. Gastebois,
CEN Saclay, DQhN/BE, 91191 Gif-sur-Yvette, France

W. Mittig, J.M. Chouvel, C. Grégoire, Y. Schutz, W.L. Zhan,
GANIL, BP 5027, 14021 Caen-cedex, France

C. Stephan, IPN Orsay, BP 1, 91406 Orsay, France

M. Morjean, Y. Pranal,
CEA/DAM, BQ.12, 91680 Bruyères-le-Châtel, France

A. Cunsolo, A. Foti,
INFN and Dipartimento di Fisica, Corso Italia 57, I-95129
Catania, Italy

ABSTRACT

We present total cross-section measurements corresponding to nuclear reactions of neutron rich fragments of a ^{40}Ar beam on a silicon target. The dependence on the neutron excess of the fragments is discussed.

Doing a mass measurement at GANIL[1], we have to measure the total energy of the incoming ions. This is done with the use of a thick silicon solid state detector. The reaction probability in the detector is high enough to measure a total reaction cross section.

With a high intensity primary beam ($\sim 10^{11}$ particles per second), and a thick tantalum target ($\sim$500 mg/cm^2), we are able to transport a secondary beam up to the focal plane of the SPEG spectrometer[2]. With an adequate choice of the magnetic rigidity Bρ, we shall get rid of the residual primary beam and only fragmentation products will be transported if they satisfy to the phase space constraints of the line (in momentum, angle and position). The intense primary beam is stopped far upstream, so that no background is present on the spectra measured in the focal plane.

The path of the ions is equal to 120 meters from the target up to the focal plane. At the end of the path, the ions are stopped in a silicon solid state detector telescope ΔE_1 (300 μm)- ΔE_2 (300 μm) - E (600 μm). The Z identification may be achieved with the relation $\sqrt{\Delta E}/t$ where t is a time of flight corresponding to a given path. The ΔE detector gives a convenient time reference for these events where a reaction occurs in the E detector.

[*]Experiment performed at the GANIL National Laboratory.

To measure the total reaction cross-section we used the so called associated radiation method[3] where one assumes that a reaction gives rise at least to one photon or a detectable light particle. The Si-telescope is surrounded by a 4π NaI detector device (figure 1) which allows to give the signature that a nuclear interaction has taken place during the slowing down process of the ions. The NaI detectors on the side are mainly triggered by γ's and, with less efficiency, by neutrons. The smaller central NaI detector will in addition detect, all light charged particles in the forward angle cone. A detection efficiency equal to 70 % for a single γ was measured with a calibrated ^{60}Co source. Using the measured multiplicities a combined detection efficiency of 95 % was obtained.

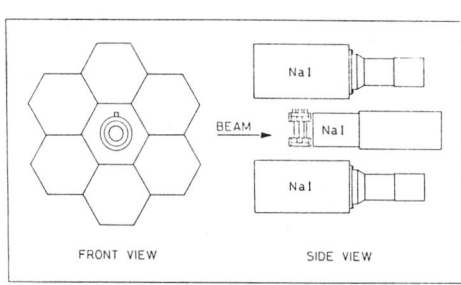

Figure 1 : Experimental device. The ions are stopped in a ΔE, ΔE telescope, constituted successively by $\Delta E1$ and $\Delta E2$ (300 µm) and E (6000µm).

The total reaction cross section for a thick target may be obtained by the ratio of coincidences between the telescope and the NaI detectors and the number of incident ions corrected for the range of the ions in the silicon and for the small efficiency loss.

Due to the magnetic selection, the velocity of the light fragments (A≤40) depends only on their ratio mass over charge. In order to compare different isotopes, it is convenient to define another parameter in which the energy dependence will be removed. The so called reduced geometrical strong absorption radius r_o defined by

$$\sigma_R(E) = \pi\ r_o^2\ f(E_p,\ A_p,\ Z_p,\ A_t,\ Z_t)$$

satisfies this condition. Here, f is a functional depending on the energy E_p of the projectile, the mass and charge of both projectile and target. If we know how to parametrize exactly $f(E, A, Z)$, r_o should be a constant independent of the nucleus. We have on one hand the experimental reaction probability Σ_R integrated over all the energies between the incoming energy E_{max} of the fragment before any interaction up to the energy of the Coulomb barrier E_{cb} between the fragment and ^{28}Si; on the other hand a variation law $\sigma_R(E)$ depending on the energy. Thus, it is necessary to perform the integration :

$$\Sigma_R = \int_0^{R_{max}} \sigma_R(E)\, dR \quad \alpha \quad \int_{E_{max}}^{E_{cb}} f(E)\, \frac{dR}{dE}\, dE \qquad (1)$$

where we took the commonly used stopping powers and ranges from tables of Hubert et al[4].

In the simplest strong absorption model, the variation of the transmission coefficient T_l defined for each partial waves by

$$\sigma_R = \frac{\pi}{k^2} \sum_l (2l + 1)\, T_l \qquad (2)$$

is a step-function and $T_l = 0$ when target and projectile have no overlap at the absidal distance. We obtain then the well known parametrization :

$$\sigma_R = \pi\, r_0^2\, (A_t^{1/3} + A_p^{1/3})\, (1 - \frac{E_{cb}}{E_{cm}}) \qquad (3)$$

where E_{cm} is the total energy in the center of mass frame. Here we chose a more refined parametrization, taken from Kox et al[5]

$$\sigma_R = \pi r_0^2\, (A_t^{1/3} + A_p^{1/3} + \frac{a A_t^{1/3} A_p^{1/3}}{A_t^{1/3} + A_p^{1/3}} - C)\, (1 - \frac{E_{cb}}{E_{cm}}) \qquad (4)$$

with an asymmetry term a = 1.9. We assumed a linear dependence for the transparency term C(E)

$$C(E) = 0.14 + 0.015\, E/A \qquad (5)$$

in agreement with the measurements on stable nuclei reported in ref.5.

The function $\sigma_R(E)$ to integrate is displayed in a linear

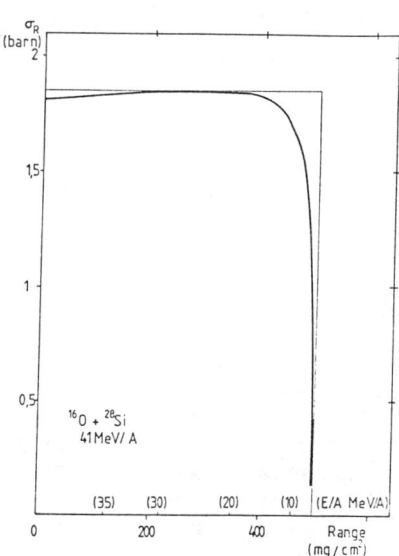

Figure 2 : Variation of $\overline{\sigma_R}(E)$ (eq.4) as a function of the range in the silicon for the reaction $^{16}O + ^{28}Si$. The incoming ^{16}O ions have after the fragmentation process and the magnetic selection an energy equal to 41 MeV/A.

scale on figure 2. We see that an uncertainty on the parametrization of eq.(4), if any, at low incident energy, has few consequences on the result of integration. That integration is a critical step in the calculation of r_o^2 and must be controlled. We checked the consistency of the method since we performed the measurements for two values of $B\rho$, that is two different values of E_{max} and R_{max} for each nucleus. The two deduced values of r_o^2 are in agreement within the error bars : this is a necessary condition to use the energy dependance of eq.(4).

We present here the results corresponding to the fragmentation of the ^{40}Ar beam at 60 MeV/A on a thick tantalum target (500 mg/cm^2). Figure 3 shows the strong dependence of the calculated values of r_o^2 against the neutron excess N − Z. That trend is observed whatever the atomic number may be. That effect then cannot be assigned to the difference of the nucleon-nucleon interaction in the states T=0 and T1. The open question is the influence of a neutron excess on the matter distribution, its half density radius and diffuseness.

The matter distribution may change from a stable nucleus, say ^{23}Na, up to a nucleus much farer from stability like ^{32}Na, as it was earlier predicted with Hartree-Fock calculations[6]. However, the amplitude of the observed dependance is much stronger than expected.

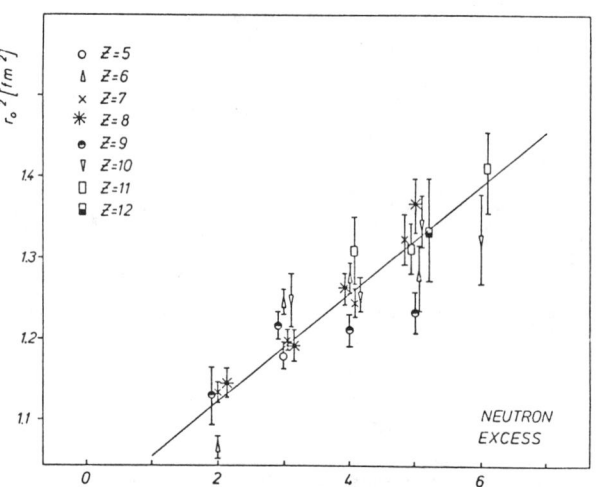

Figure 3 : Variation of the model-dependent r_o^2 parameter against the neutron excess N − Z

References

1- W. Mittig et al, this conference and therein references
2- P. Birien, S. Valero, Note CEA 2215 (1981)
3- J.F. Bruandet, J. de Phys. C4 (1986) 125
4- F. Hubert et al, Annales de Physique 5 (1980) 1
5- S. Kox et al, Phys. Rev. C35 (1987) 1678
6- X. Campi et al, Nucl. Phys. A251 (1975) 193

EVIDENCE FOR EMSO EFFECTS IN RADII OF THE TIN ISOTOPES

D. Berdichevsky, R. Fleming and D.W.L. Sprung
Physics Department, McMaster University
Hamilton, Ontario Canada L8S 4M1.

ABSTRACT

A study of measured charge density differences of the tin nuclei provides evidence for the electromagnetic spin-orbital contribution to the charge density.

INTRODUCTION

Recent measurements[1-4] of the mean square (m.s.) charge radii of the tin isotopes cover a wide range of neutron number: $58 \leq N \leq 74$ for $Z = 50$. These neutrons are filling the s,d,g shell and the $1h_{11/2}$ intruder orbital. The data show a steady, nearly linear increase of m.s. charge radius as N increases: see figs. 1 and 4 of ref. 4. At the same time, BE(2) and BE(3) values[5] imply an almost constant value for the contribution to the m.s. radius from "zero-point" motion of the the nuclear shape. These data now allow a systematic study of the validity of theoretical calculations of $\delta \langle r^2 \rangle$ in this closed proton shell element.

In this contribution we build on the isotope shift study of the tin isotopes of ref. 6. In evaluating the nuclear charge density we pay attention to a number of small corrections: proton finite size, neutron finite size, centre of mass correction and electromagnetic spin-orbit interaction (EMSO)[7]. The main contribution comes of course from the proton matter distribution. It has been checked by Hasse et al. that incoherent (non-collective) particle-hole correlations[8] when added consistently to a HF calculation using the filling prescription affect only weakly the isotope shift derived from the HF calculation alone. Also, there are indications from experiment that the "zero-point" oscillations of the nuclear surface are nearly constant over the series of tin isotopes. We have therefore not included RPA correlations in this analysis.

In this note we compare data on the differences of mean square (ms) charge radii $\langle r^2 \rangle_{ch}^A - \langle r^2 \rangle_{ch}^{A-2} = \Delta_2 \langle r^2 \rangle_{ch}$ to results of HF+BCS calculations using the effective interaction Ska of Köhler[9]. We have assumed spherical symmetry, as in the earlier work of Campi et al.[6]. Pairing has been included with a fixed strength BCS interaction rather than by fixing the value of the pairing gap. Spurious centre of mass (cm) motion was accounted for in the HF self consistency as discussed by Butler et. al.[10] for the G0-P force, and using the $(1 - 1/A)$ approximation for the Skyrme-type force Ska.

RELATIVISTIC CONTRIBUTIONS

It was pointed out by Bertozzi et al.[7] that relativistic effects of order $1/m^2$ make small but important contributions to the nuclear charge density as observed in electron scattering, and that these can form an essential part of the isotope shift. The main contribution comes from folding the point proton density with the proton electric form factor $G_{Ep}(q^2)$. To this is added a similar contribution for neutrons [denoted NFS for neutron finite size] and the electromagnetic spin orbit interaction[11,12], [denoted EMSO]. We can write (* denotes convolution)

$$\rho_{ch}(r) = \sum_{\tau=n,p} \{\tilde{G}_{E\tau}*\rho_\tau(r) + \tilde{\mu}_\tau*\rho_\tau^{eq}(r)\} \tag{1}$$

where all the nucleon form factors include the Darwin-Foldy and cm motion effects as follows:

$$\tilde{G}_E(q^2) = G_E(q^2)\, e^{+b^2 q^2/4A} / [1 + (\tfrac{\hbar q}{mc})^2]^{1/2} \tag{2}$$

$$\tilde{\mu}(q^2) = \tilde{G}_M(q^2) - \tfrac{1}{2}\tilde{G}_E(q^2) \tag{3}$$

For the EMSO contribution,

$$\rho_\tau^{eq} = -\tfrac{1}{2}(\tfrac{\hbar}{mc})^2 \tfrac{1}{r^2}\tfrac{\partial}{\partial r}(r^2 \rho_{s\ell}(r)) \tag{4}$$

with a spin-orbit weighted density

$$\rho_{s\ell}(r) = \sum_k v_k^2\, (2j+1)_k\, (\vec{\sigma}\cdot\vec{\ell})_k\, |u_k(r)|^2 , \tag{5}$$

v_k^2 being the pairing probability of orbital k. Following ref. 10, the oscillator length used in the cm correction is

$$b^2 = \tfrac{2}{3}\langle r^2\rangle\, f(A) \approx A^{1/3}\, \text{fm}^2 \tag{6}$$

For the proton electric form factor we used the three-gaussian model fitted by Chandra and Sauer[13]. The magnetic form factors are taken to be proportional to G_{Ep}. For G_{En}, we used the difference of two gaussians as in ref 6, adjusted to $r_n^2 = -0.1192(18)\,\text{fm}^2$.

In so far as we are primarily interested in the nuclear charge radius, we can write[12]:

$$\langle r^2\rangle_{ch} = \langle r^2\rangle_p + \{r_p^2 + \tfrac{3}{4}(\tfrac{\hbar}{mc})^2 - \tfrac{3}{2}\tfrac{b^2}{A}\}$$
$$+ \tfrac{N}{Z} r_n^2 + \tfrac{1}{Z}(\tfrac{\hbar}{mc})^2 (\hat{\mu}_p \Sigma_p + \hat{\mu}_n \Sigma_n) \tag{7}$$

where the terms on the rhs are respectively the point proton ms radius from a HF calculation, the proton ms electric radius, the Darwin-Foldy and c.m. motion corrections, the NFS term and finally the proton and neutron EMSO contributions. The latter involve, in view of (4), (5) the sums

$$\Sigma_\tau = \sum_{k \in \tau} (2j+1)_k \, v_k^2 \, (\vec{\sigma} \cdot \vec{\ell})_k \qquad (8)$$

and the $q^2 = 0$ values $\hat{\mu}_p = 2.29278$, $\hat{\mu}_n = -1.9135$. Notice that the EMSO effect on the ms radius depends on the occupations of the orbits and not on the details of the magnetic form factor. Also, LS coupling closed shells where both members of the spin-orbit doublet are completely full or empty, make no contribution to eq.(8). On the other hand, ^{48}Ca (ref. 7), ^{88}Sr (ref. 14) and ^{124}Sn do show significant EMSO effects. In ^{208}Pb, the opposite sign of $\hat{\mu}_p$, $\hat{\mu}_n$ leads to a cancellation between proton and neutron contributions, but ^{207}Pb shows a particularly strong EMSO effect[11] at small radii.

In the case of the tin nuclei, one finds that the ms charge radii $\langle r^2 \rangle_{ch}$ have a close to linear dependence on the mass number. Taking differences of eq.(7) gives:

$$\Delta_2 \langle r^2 \rangle_{ch} = \Delta_2 \langle r^2 \rangle_p + \frac{2}{Z} r_n^2 + \frac{1}{Z} \left| \frac{\hbar}{mc} \right|^2 \hat{\mu}_n (\Sigma_n^A - \Sigma_n^{A-2}) \qquad (9)$$

These are plotted in the figure and show the deviation from strictly linear behaviour on a very expanded scale. In the HF calculation, up to mass 116 the neutrons are filling primarily the "spin-down" orbitals $g_{7/2}$ and $d_{3/2}$; in this case the neutron finite size (NFS) term has opposite sign to the EMSO contribution and the two essentially cancel. Beyond 116, one is filling the $1h_{11/2}$ shell; then the EMSO effect reinforces the NFS term and both act to decrease the observed charge radius. This effect is seen in the nearly constant value of the HF differences from $A = 126$ to 132. As can be seen in the figure, the calculation including the EMSO and NFS contributions (black dots) is in much better accord with the data [ref.1, circles and ref.4, triangles] for these neutron-rich isotopes than is the calculation omitting them (barred circles).

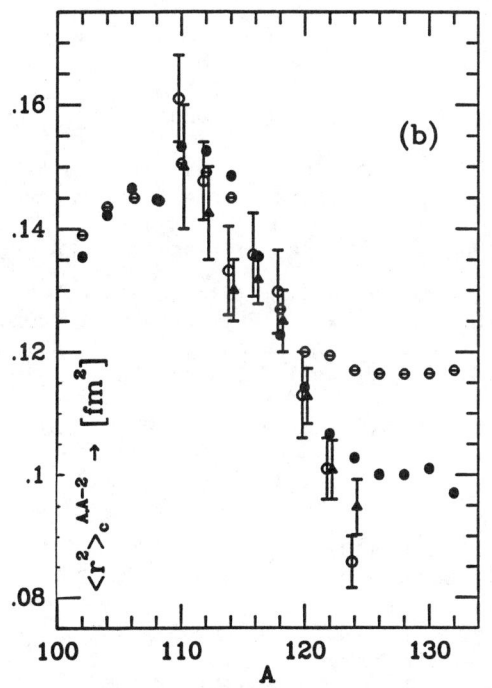

In addition to the terms given in eq.(9), there should be a

contribution from "zero-point" motion of quadrupole or higher multipole oscillations of the nuclear density. The amplitude of this motion can be related to the observed $B(E,\lambda)$ values. The evidence from $B(E,2)$ and $B(E,3)$ values[5] is that this contribution to the tin radii is constant for $110 < A < 124$, and therefore makes no contribution to $\Delta_2<r^2>_{ch}$. It would be very interesting to see measurements extended to $A > 128$, where theory predicts a constant value, in sharp contrast to the steady decrease of the experimental values of $\Delta_2<r^2>_{ch}$ in the measured region. This could provide striking confirmation of this small but distinctive contribution to measured charge radii.

ACKNOWLEDGMENTS

We are grateful to NSERC (Canada) for continued support under operating grant A-3198.

REFERENCES

1. Anselment M., Bekk K., Hanser A., Hoeffgen H., Meisel G., Goering S., Rebel H., Schatz G.; Phys. Rev. C 34, 1052 (1986)
2. Baird P.E.G., Blundell S.A., Burrows G., Foot C.J., Meisel G., Stacey D.N., Woodgate G.K.; J. Phys. B (Atom. Mol.) 16, 2485 (1983)
3. Silver J.D. and Stacey D.N.; Proc. Roy. Soc. (London) A332, 139 (1973)
4. Eberz J., Dinger U., Huber G., Lochmann H., Menges R., Ulm G., Kirchner R., Klepper O., Kuehl T.U., Marx D.; Zeit. f. Phys. A 326, 121 (1987)
5. Endt P.M.; At. Data & Nucl. Data Tables 26, 47 (1981): Christi A., Häusser O.; Nucl. Data Tables 11, 281 (1972): Stelson P.H., McGowan F.K., Robinson R.L., Milner W.T.; Phys. Rev. C 2, 2015 (1970): Stelson P.H., Grodzins L.; Nucl. Data A 1, 21 (1965)
6. Campi X., Sprung D.W.L., Martorell J.; Nucl. Phys. A223, 541 (1974)
7. Bertozzi W., Friar J., Heisenberg J., Negele J.W.; Phys. Lett. 41B, 408 (1972)
8. Hasse R.W. and Schuck P.; Nucl. Phys. A438, 157 (1985), ibid. A445, 205 (1985)
9. Köhler H.S.; Nucl. Phys. A258, 301 (1976)
10. Butler M.N., Sprung D.W.L., Martorell J.; Nucl. Phys. A422, 157 (1984)
11. Martorell J. and Sprung D.W.L.; Zeit. f. Phys. A 298, 153-7 (1980)
12. Nishimura M. and Sprung D.W.L.; Progr. Theor. Phys. (Lett.) 77, 781-6 (1987)
13. Chandra H. and Sauer G.; Phys. Rev. C 13, 245 (1976)
14. Buchinger F., Corriveau R., Ramsay E., Berdichevsky D., Sprung D.W.L.; Phys. Rev. C 32, 2058 (1985)

SHELL CORRECTIONS TO THE PROTON MEAN SQUARE RADIUS

A.S. Jensen and A. Miranda
Institute of Physics, University of Aarhus
DK - 8000 Aarhus C, Denmark

ABSTRACT

The shell correction to the charge mean square radius is calculated in a simple model. For a spherical double magic nucleus we find about -0.1 fm^2 for any nucleon number. The deformation dependence is investigated and it is concluded that any model correct to a better accuracy must include pairing effects in some approximation.

- oOo -

The mean square charge radii of series of isotopes have been measured[1] with very high precision. On passing a magic neutron number, the behaviour changes discontinuously. An explanation in terms of shell structure is clearly required.

Let us calculate the shell correction $\delta \langle r^2 \rangle$ to the proton mean square radius $\langle r^2 \rangle$ in a deformed harmonic oscillator shell model. For this we need to compute both $\langle r^2 \rangle$ and its smooth part $\langle \tilde{r}^2 \rangle$. For a total number Z of protons, the proton mean square radius is given by

$$Z \langle (\frac{r}{b})^2 \rangle = \omega \left[\frac{\Sigma_1}{\omega_1} + \frac{\Sigma_2}{\omega_2} + \frac{\Sigma_3}{\omega_3} \right] \quad (1)$$

$$\Sigma_\varkappa = \sum_{k\,occ} (n_\varkappa(k) + \frac{1}{2}) \quad (2)$$

where $\omega_\varkappa$ is the oscillator frequency in the $\varkappa$-direction, b is the length parameter given by $b^2 = \hbar/m\omega$, and $\omega^3 = \omega_1 \omega_2 \omega_3$.

The corresponding smooth part is calculated using the Strutinsky averaging procedure.[2] This rather lengthy but straight-forward calculation leads to the result

$$Z \langle (\frac{\tilde{r}}{b})^2 \rangle = \frac{1}{4}(3Z)^{4/3} \cdot (s+(3Z)^{-2/3} \cdot (1 - \frac{1}{2} sr) \quad (3)$$

$$r = \frac{1}{3}(\omega_1^2 + \omega_2^2 + \omega_3^2)/\omega^2 \quad (4)$$

$$S = \frac{\omega^2}{3}(\omega_1^{-2} + \omega_2^{-2} + \omega_3^{-2}) \tag{5}$$

Specializing to axial symmetry, we define the deformation parameter q by

$$\omega_1 = \omega_2 = \omega q^{1/3} \; ; \; \omega_3 = \omega q^{-2/3}. \tag{6}$$

Then Σ_κ can easily be calculated analytically for values of q smaller than the value for which the last occupied level crosses an unoccupied level. When the quantum numbers of the last occupied level are $(n_\perp, n_z) = (n, N_f - n)$, we get by direct summation for prolate nuclei

$$Z\langle(\tfrac{r}{b})^2\rangle_{\text{prol.}} =$$

$$= \tfrac{1}{4}(3Z_0)^{4/3}(\tfrac{2}{3} q^{-1/3} + \tfrac{1}{3} q^{2/3})(1 + \tfrac{1}{3}(3Z_0)^{-2/3})$$

$$+ \tfrac{2}{3}(Z-Z_0)(Z-Z_0 + \tfrac{1}{4})^{1/2}(q^{-1/3} - q^{-2/3} \tag{7}$$

$$+ (Z-Z_0)q^{2/3} \cdot ((3Z_0)^{1/3} + \tfrac{1}{2} + \tfrac{1}{3}(3Z_0)^{-1/3})$$

where Z_0 is the number of protons in the fully occupied lower oscillator shells. Similarly, eq. (3) can be rewritten as

$$Z\langle(\tfrac{\tilde{r}}{b})^2\rangle = \tfrac{1}{4}(3Z)^{4/3} \cdot (\tfrac{2}{3} q^{-2/3} + \tfrac{1}{3} q^{4/3})$$

$$\tag{8}$$

$$+ \tfrac{1}{4}(3Z)^{2/3} \cdot \tfrac{1}{18}(13 - 2q^2 - 2q^{-2}).$$

For an oblate deformation, the proton mean square radius is obtained analogously. The shell correction $\delta\langle r^2\rangle = \langle r^2\rangle - \langle\tilde{r}^2\rangle$ to the proton mean square radius is now computed. For a closed proton shell we find assuming $b^2 = A^{1/3}$ fm$^2 = (2Z_0)^{1/3}$ fm^2 that

$$\delta\langle r^2\rangle = -\tfrac{1}{8} b^2 (3Z_0)^{-1/3} = -0.11 \text{fm}^2 \tag{9}$$

for any nucleon number.

The radius of a spherical (doubly magic) nucleus can therefore be expected to be smaller than that of other nuclei by this amount. In the extreme shell model neutrons and protons are coupled only through the mean field, which usually is assumed to be the same for both kinds of particles. Therefore, the charge distribution is expected to be spherical only in the neighbourhood of doubly magic nuclei. When the neutron number is changed for a given isotope the deformation of the charge distribution is also expected to vary. The interesting quantity then becomes the shell correction to the proton mean square radius as function of deformation. The dependence of $\delta\langle r^2\rangle$ on q is shown in Fig. 1 for various relative occupancies of the last shell. For magic proton number $\delta\langle r^2\rangle$ decreases as q is changed away from the spherical value. Averaging for q=1, the various shell corrections lead to the expected value of zero. It may therefore seem surprising that $\delta\langle r^2\rangle$ is an overall decreasing function of q when oscillations around zero would be expected for such fluctuating quantities.

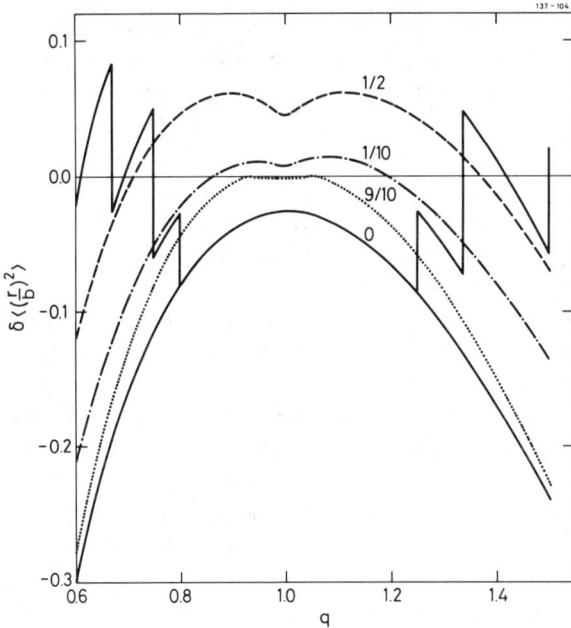

Figure 1: The shell correction $\delta\langle(r/b)^2\rangle$ to the proton mean square radius as function of q. The 4 curves are labelled with their corresponding fraction of levels that are occupied in the last shell. The step functions are the result of level crossings for the closed shell Z=70 in the harmonic oscillator model.

The explanation is simple and actually already illustrated in the figure. Level crossings occurring as q moves away from unity change the occupancies of the single particle states. In turn this changes $\langle r^2\rangle$, but not $\langle \tilde{r}^2\rangle$. Since protons are lifted into higher shells of larger radius, the shell correction for a spherical closed proton shell configuration will increase in the steps shown in Fig. 1.

The variations in $\delta\langle r^2\rangle$ due to level crossings are quite large compared to its size at q=1. If the level of accuracy is better than the size of this shell correction, the picture is clearly not satisfactory. The same conclusion holds for Hartree-Fock calculations which schematically are similar to those of the non-selfconsistent shell model. However, including also pairing in some approximation leads to an additional smearing which significantly will modify the results. A pairing gap Δ gives a diffuse occupation probability in intervals about Δ above and below the Fermi energy. For q=1 the distance between shells is much larger than the empirical Δ-values and the smearing can be expected to be insignificant for magic numbers. As q increases, the levels around the Fermi energy become closer and significant smearing due to pairing can be expected. This effect increases the radius, since the levels from the higher lying shell get a finite occupation probability while that of the lower lying levels gets less. A small amount of smearing of the step functions in Fig. 1 would for the magic proton number give a minimum $\delta\langle r^2\rangle$ at a q-value larger than 1. It is conceivable, however, that the pairing effect for the weaker shells of more realistic level schemes is large enough to move this minimum to the spherical configuration.

In conclusion, if the mean square radius is obtained to an accuracy better than 0.1 fm^2, then the correct interpretation must include effects of shell structure. This could for instance be accomplished by a Strutinsky calculation or a full microscopic Hartree-Fock computation. However, this will not be sufficient as pairing effects then will also be significant and must be included in either the BCS-approximation or in a full-scale Hartree-Fock-Boguliubov calculation.

1. K. Heilig, Hyp.Int. 24 (1985) 349. 2. M. Brack and H.C. Pauli, Nucl.Phys. A207 (1973) 401.

NUCLEAR SHAPE COEXISTENCE AND THE STUDY OF NUCLEI FAR FROM STABILITY

J.L. Wood
School of Physics, Ga. Tech., Atlanta, Ga. 30332, USA

ABSTRACT

The systematic features of shape coexistence are briefly outlined. The most useful spectroscopic fingerprints for identifying shape coexistence far from stability are presented. Directions for future work are discussed.

INTRODUCTON

Shape coexistence or shape isomerism is emerging[1] as a widely-occurring low-energy degree-of-freedom of the nucleus. The single most extensive mass region where it is found is the "far from stability" region which is approximately bounded by $78 \leq Z \leq 83$, $98 \leq N \leq 120$ (see, e.g., ref. 1,2,3,4,5). The factors controlling shape coexistence are rather subtle and not always simple. Two basic theoretical approaches have been taken (see, e.g., ref. 1): the first is essentially the Strutinsky shell correction method; the second incorporates an explicit dependence on proton and neutron number through a proton-neutron force[6]. The latter approach predicts shape coexistence will be found near closed shells when the other type of nucleon is near mid shell, e.g., ^{116}Sn. Subshells also play a role[7]: they can suppress coexistence if they occur at or near mid shell for a closed shell for the other type of nucleon, e.g., ^{90}Zr; and they can give rise to shape coexistence by playing the role of major closed shells, e.g., ^{100}Zr.

Studies of nuclei far from stability have made major contributions to this developing picture. Most notably, the neutron-deficient region (mentioned above) centered on $Z=82$ and $N=104$ and the neutron-rich region near $Z=40$, $N=60$ are only widely accessible to study by far-from-stability techniques. Further, studies can be made in a <u>systematic</u> way to reveal the dependence on changing proton and neutron number. Sudden changes in

ground-state properties such as charge radii (see,
e.g., ref. 8,9,10,11) are among the most dramatic
changes known for changing particle number, and are a
consequence of these coexisting structures.

Some of the major new experimental results are
summarized in this review, with emphasis on the
physical significance of the results. Most
importantly, directions for future work are suggested.

THE PHENOMENOLOGY OF SHAPE COEXISTENCE

A variety of classes of systematic behavior for
shape coexistence are expected on the basis of models[6]
with explicit proton and neutron number. These are
shown in Fig. 1. Examples of all of these classes are
known experimentally. Fig. 1a illustrates the
behavior[1] of e.g, the even-mass Sn and odd-mass Tl
isotopes, respectively. Fig. 1b describes the behavior
of, e.g., the even-mass Pt isotopes[12] and odd-mass Cs
isotopes[13], respectively. In Fig. 1c, behavior
characteristic of the odd-mass Au isotopes[3] is
depicted. Fig. 1d depicts the type of behavior[1] seen
in the N=50 isotones, where a mid-shell maximum is seen
in the closed subshell nucleus ^{90}Zr, for the neutron
2p-2h states. Finally, subshells in turn give rise to
coexistence as shown schematically in Fig. 1e, and
manifested in the behavior of the neutron-rich Zr
isotopes (see, e.g., ref. 14).

In Fig. 2, the regions of the nuclear mass surface
where shape coexistence is expected are depicted. This
view reveals the important role that far-from-stability
studies have and will continue to play in the
investigation of nuclear shape coexistence. Thus, the
mid-shell regions centered on: N=20, Z=14; N=82, Z=66;
Z=8, N=14; Z=28, N=39; Z=82, N=104 all depend heavily
or totally on nuclear spectroscopy far from the line of
beta stability. Even regions centered on stable nuclei
are critically dependent on studies of very unstable
nuclei to provide essential systematic details: e.g.,
the neutron-rich N≈50 isotones and the neutron-rich Cd
(Z=48) isotopes.

Often, existing spectroscopic information is
limited, and sometimes circumstantial or even
contradictory. An example is the region centered on
N=20, Z=14. Excited-state systematics of the N=20

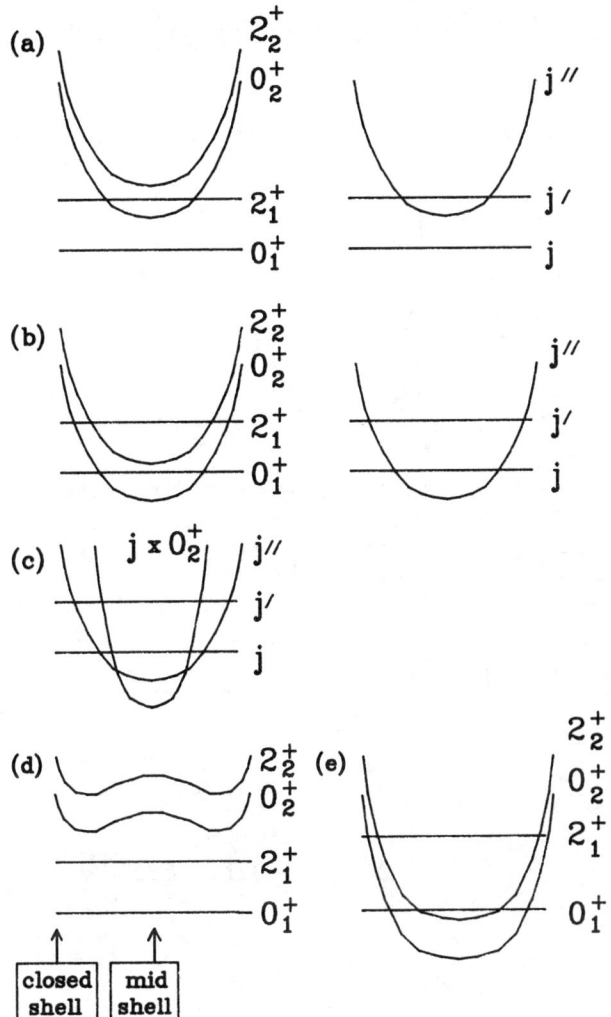

Fig. 1. Different types of intruder state systematics:
(a) Even- and odd-A major shell, excited-state intruder;
(b) Even- and odd-A major shell, ground-state intruder;
(c) Odd-A major shell with core (0_2^+) intruder;
(d) Even-A major shell with intersecting subshell at mid shell; (e) Even-A subshell (subshell disappears with rapid onset of deformation where the intruder state becomes the ground state).

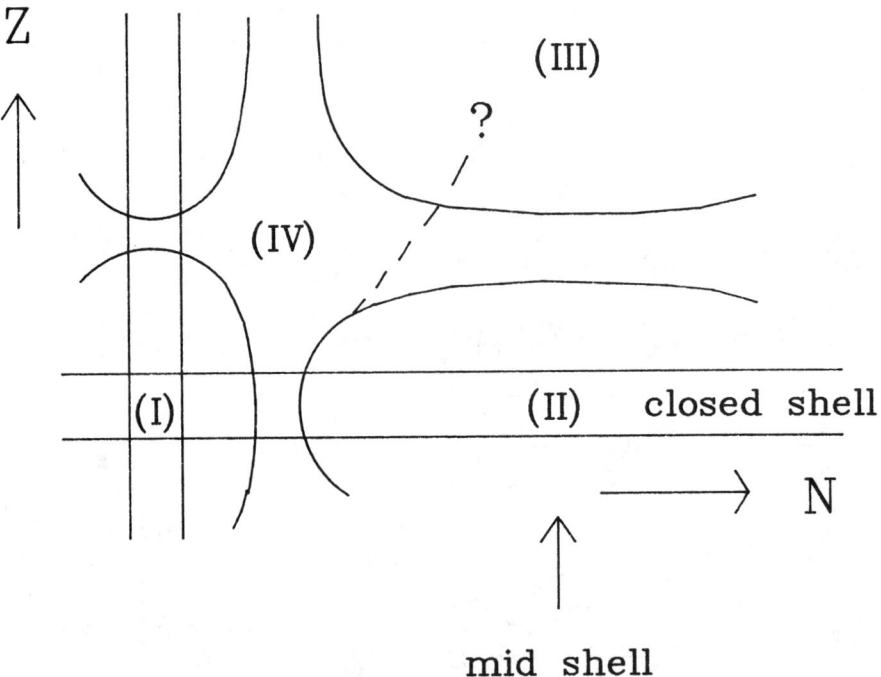

Fig. 2. A schematic view of regions of the nuclear mass surface that exhibit different major types of structure: (I) shell model, (II) pairing model, (III) deformation, (IV) transitional behavior. Shape coexistence can be expected in region (II).

isotones are shown in Fig. 3. Ground-state properties of the Na isotopes in the N=20 region are shown in Fig. 4. Contrary to the expected closed shell behavior of ^{31}Na, namely a decrease in $\delta<r^2>$ and a decrease in S_{2n}, ground-state deformation is implied. Evidently, studies of excited states in ^{34}Si and further information on excited states of ^{32}Mg are highly desirable (cf. Fig. 3). The resolution of the contradictory masses of ^{32}Mg is crucial. While the first mass results[19] in this region were a surprise, it would be now difficult to understand the absence of extra binding in ^{32}Mg (cf. refs. 16,17).

Accumulating spectroscopic proof of shape coexistence in nuclei remains a major challenge. In Fig. 5 a summary is given of the basic types of shape coexistence fingerprints that have been used. Some of these were already discussed in ref. 1. Isotope shift data remain a dramatic indicator of shape coexistence. This has recently been reemphasized[10] in the neutron-deficient odd-mass Au isotopes. However, if the intruding deformed states remain excited states, electromagnetic transition rates may be the only fingerprint. This is the case[20,21] in the neutron-rich Ag isotopes.

A particularly promising new fingerprint appears to be electric monopole transitions. These are clearly seen in even-mass nuclei (see, e.g. refs. 22,23). There is now accumulating evidence[3,4] for electric monopole transitions in odd-mass nuclei also. However, until lifetime data are available, this signature remains qualitative. Currently, there are no available data on absolute E0 transition rates in odd-mass nuclei.

A novel fingerprint is the use of alpha decay across closed shells to identify intruder states. Thus, the ground states of the neutron-deficient Bi isotopes decay[5,24] to the intruder states in the neutron-deficient Tl isotopes by unhindered alpha decay.

FUTURE AND CONCLUSIONS

The above sketch of the systematics of shape

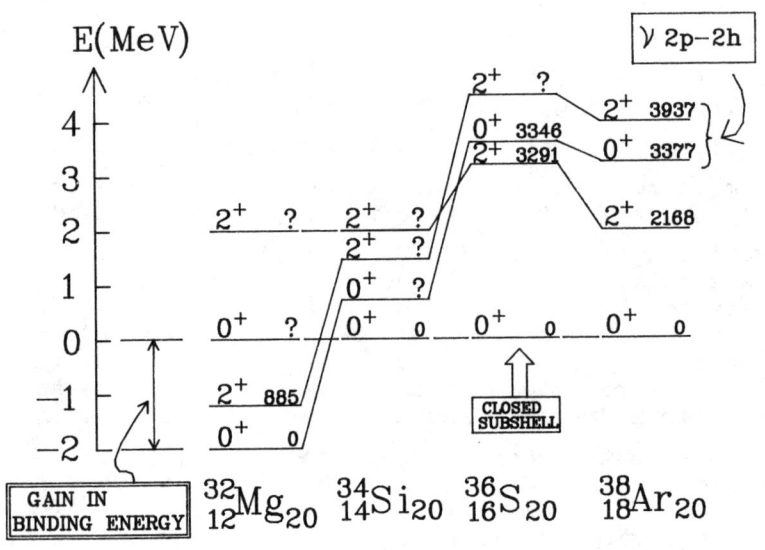

Fig. 3. Excited state systematics in the even-mass N=20 isotones. The low-lying 2^+_1 state in ^{32}Mg is interpreted as resulting from a ground state intruder configuration. The ground state of ^{32}Mg should have an anomalously large mean-square charge radius. The ground-state binding energy of ^{32}Mg has been reported variously as anomalous[13] and normal[14].

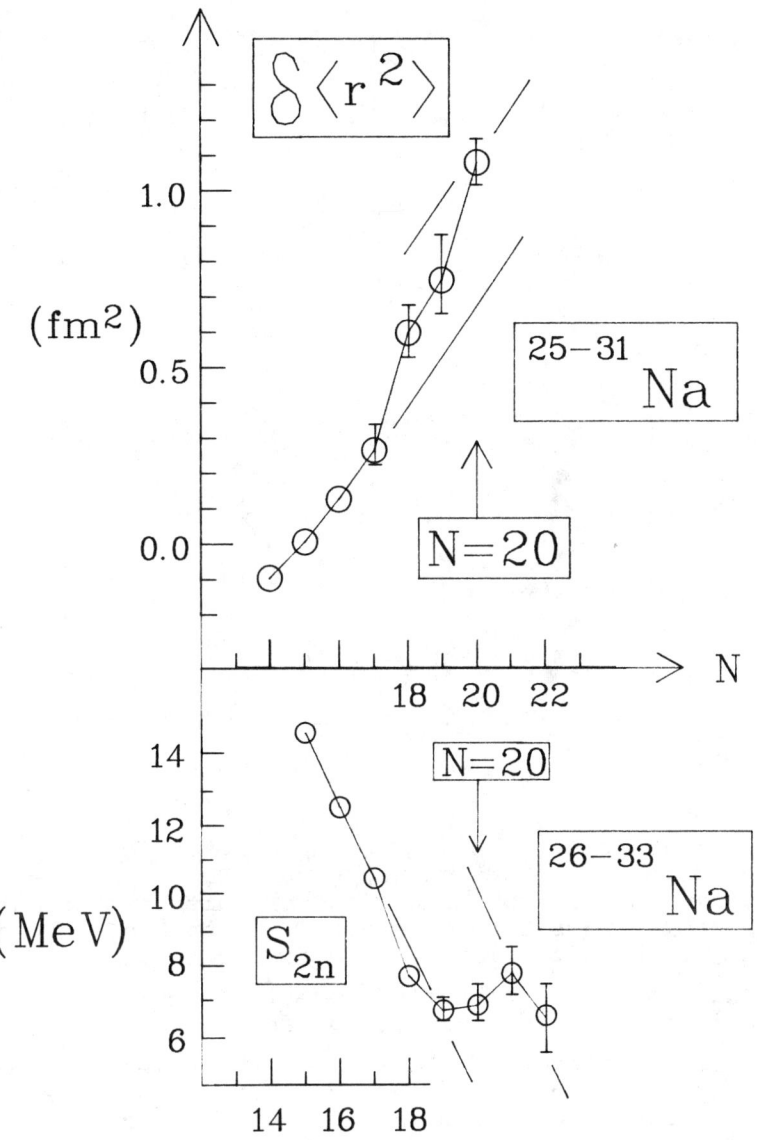

Fig. 4. Two-neutron separation energy[15], S_{2n} and isotope shift[16], $\delta\langle r^2\rangle$ systematics for the neutron-rich Na isotopes. The discontinuity at N=20 indicates an increased ground-state mean-square charge radius and increased ground-state binding energy.

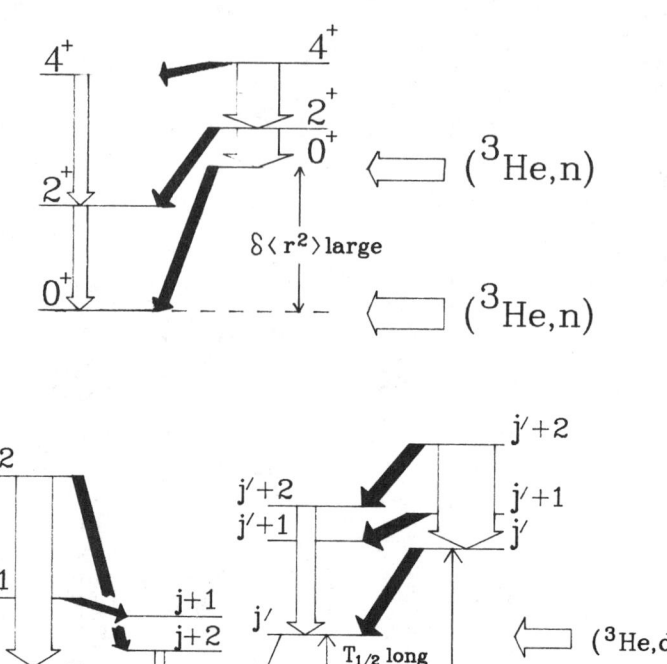

Fig. 5. A schematic view of spectroscopic fingerprints of shape coexistence. The most direct proof is coexisting bands with very different electric quadrupole transition strengths. These are shown as white arrows with widths reflecting strengths. In odd-mass nuclei retardation[1] of electromagnetic transitions between coexisting bands occurs. Large isotope and isomer shifts, $\delta\langle r^2\rangle$,[2] are indicators of underlying shape coexistence. Electric monopole transitions appear to be a signature of coexisting states. These are shown as black arrows, widths again reflecting strengths. (Near the stability line, enhanced one- and two-nucleon transfer cross sections (e.g., (^{3}He,n), (^{3}He,d) (d,^{3}He) for Z<50,82) can reveal intruder configurations which underlie[1] coexisting states.)

coexistence suggests that many more nuclei should exhibit the phenomenon. Major regions where little or no data exist include those centered on: N=86, Z=72; Z=84, N=112; N=130, Z=90 (see ref. 25). The "fate" of intruder states and shape coexistence further away from closed shells is also an open question. Studies in the region centered on Z=76, N=104 would be of great value. This will require mass separation of refractory elements. More detailed spectroscopy in established regions of shape coexistence is needed, e.g., isotope shift measurements of the neutron-deficient Pt isotopes should^{11} reveal a sudden change between ^{187}Pt and ^{186}Pt. Finally, E0 transitions need to be quantified as a spectroscopic fingerprint of coexisting bands. This will necessitate lifetime measurements down to the 50 ps range. (It should be noted that E0 transitions and isotope (and isomer) shifts are dynamic and static aspects of the same operator: the square of the electric charge radius. This unity is unexplored theoretically and experimentally.)

ACKNOWLEDGEMENTS

This work is based on collaborations with K. Heyde, E.F. Zganjar, and the UNISOR consortium; and was supported by U.S. Dept. of Energy grant DE-FG05-87ER40330, U.S. DOE contract DE-AC05-76OR00033 (UNISOR), and a NATO research grant # RG-86/0452.

REFERENCES

1. K. Heyde et al., Phys. Repts. <u>102</u>, 291 (1983); K. Heyde, invited contribution to this conference.
2. J.H. Hamilton et al., Rep. Prog. Phys. <u>48</u>, 631 (1985).
3. E.F. Zganjar et al., contribution to this conference.
4. C.D. Papanicolopulos et al., ibid.
5. M. Huyse et al., ibid.
6. K. Heyde et al., Nucl. Phys. <u>A466</u>, 189 (1987).
7. J.L. Wood in "Proc. Int. Symposium on In-Beam Nuclear Spectroscopy", Debrecen, Hungary, May 14-18, (1984), ed. Z. Dombradi and T. Fenyes, p. 31.
8. J. Bonn et al., Phys. Lett. <u>38B</u>, 308 (1972).
9. C. Thibault et al., Phys. Rev. <u>C23</u>, 2720 (1981).
10. K. Wallmeroth et al., Phys. Rev. Lett. <u>58</u>, 1516 (1987).
11. J.L. Wood, in "Lasers in Nulcear Physics", ed. C.E. Bemis and H.K. Carter, Nuclear Science Research

Conference Series (Harwood Academic Publ., NY., 1982), p. 481.
12. G.D. Dracoulis et al., J. Phys. G12, L97 (1986).
13. U. Garg et al., Phys. Rev. C19, 207 and 217 (1979).
14. R.A. Meyer et al., Phys. Lett. 177B, 271 (1986).
15. C. Detraz et al., Phys. Rev. C19, 164 (1979).
16. C. Detraz et al., Nucl. Phys. A394, 378 (1983).
17. D.J. Vieira et al., Phys. Rev. Lett. 57, 3253 (1986).
18. G. Huber et al., Phys. Rev. C18, 2342 (1978).
19. C. Thibault et al., Phys. Rev. C12, 644 (1975).
20. B. Fogelberg et al., contribution to this conference.
21. N. Kaffrell et al., ibid.
22. P. van Duppen et al., Phys. Rev. C35, 1861 (1987).
23. W.B. Walters et al., contribution to this conference.
24. E. Coenen et al., Phys. Rev. Lett. 54, 1783 (1985).
25. L.K. Peker and J.H. Hamilton, contribution to this conference.

A SHELL-MODEL DESCRIPTION OF INTRUDER STATES

K.Heyde,

Institute for Nuclear Physics, Proeftuinstraat,86 B-9000 Gent(Belgium)

ABSTRACT

Starting from a shell-model approach for the description of intruder states where particle-hole excitations across a major shell (or a sub-shell) form the basic excitation mechanism, a simple expression for the mass dependence of the excitation energy for intruder states in even-even, odd-mass and odd-odd nuclei can be derived. We also indicate the equivalence with an alternative approach starting from a deformed field where residual pairing correlations are handled subsequently.

A number of new applications are discussed e.g. the existence of a scaling law for the intruder exciation energy in the Pb region, the possibility for intruder excitations in the N = 20 region of the sd-shell nuclei and the special role played by the E0 operator in observing possible evidence for intruder states via the measurement of isotope shifts over large series of isotopes of isotones.

INTRODUCTION

In nuclei, the major part of the nucleonic motion is approximated by the average nuclear field. The energy separation between the last filled single particle orbitals and the lowest unfilled orbitals at magic numbers is found empirically to vary from $\simeq 10$ MeV (light nuclei) to $\simeq 3.5$ MeV (Z = 82 gap). It is precisely the existence of closed shells that makes the study of nuclear structure approximately tractable. Most nucleons contribute to the "core" of the nucleus and thus to the average field, leaving only a small number of nucleons (particles or holes) outside closed shells. With reference to Fig.1, this is the situation that applies to region I where present day state-of-the art shell model techniques[1] permit a detailed description of low-lying nuclear excited state properties. In region II, the residual two-body force between the like nucleons, the pairing force, plays the dominant role in determining low energy nuclear structure. Here,

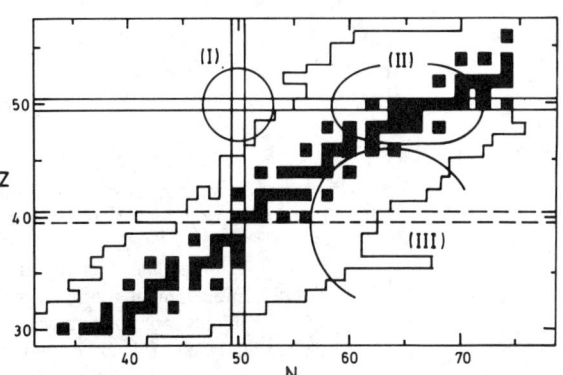

Fig.1. Schematic division in the Z ≅ 40,50 and 50 ≤ N ≤ 82 mass table into three major regions i.e. (I) region near doubly-closed schells ; (II) region of possible intruder states ; (III) region of strongly deformed nuclei.

techniques such as generalized seniority[2] (or the BCS and broken-pair methods[3]) become useful in describing the systematic behaviour of nuclear excited states. In region III, in addition to the pairing force, the residual proton-neutron interaction starts playing a major role at low energy. The most widely used descriptions are phenomenological in nature.Most notably, the geometric description in terms of shape variables[4] and algebraic descriptions in terms of interacting bosons[5] have been used.

© American Institute of Physics 1988

Concentrating on regions far from stability which can be of type I,II or III, depending on the particular A value considered, it has become evident[6] in the last few years, that shell-model intruder states occur at low energy in many nuclei of type II region. It has been shown, studying the many experimental data[6], that such states probably result, in zeroth order, from particle-hole (p-h) excitations across the major closed shells. In odd-mass (and eventually also in odd-odd nuclei) they appear as 1p-2h (or 1h-2p) states amongst the regular 1h (1p) states in e.g. the In,Tl (the Sb,Bi) nuclei. In even-even nuclei, intruder states then appear as 2p-2h excitations[7] in the Z=50 and the very neutron deficient Z=82 nuclei. The systematics suggests that these states appearing at very low excitation energy do form a widely occuring low energy degree of freedom of the nucleus.

SHELL-MODEL DESCRIPTION

1. Intruder states in even-even nuclei

Using the generalized seniority[2] (or broken-pair[3]) description for the proton and neutron 0^+ and 2^+ coherent pair operators S_π^+, D_π^+ (and S_ν^+, D_ν^+) and excited pair operators $S_\pi^{+'}; S_\nu^{+'}$... , one can, in principle, (and in practice in region I and partly in II) construct a truncated shell-model basis and calculate e.g. the lowest 0^+ levels in even-even nuclei by diagonalizing the shell-model energy matrix. This is carried out in the space spanned by the configurations

$$(S_\pi^+)^{N_\pi}(S_\nu^+)^{N_\nu}, \ (S_\pi^+)^{N_\pi+1}(S_\pi^{+'})(S_\nu^+)^{N_\nu}, \ (S_\pi^+)^{N_\pi}(S_\nu^+)^{N_\nu+1}(S_\nu^{+'}), \ \ldots \tag{1}$$

where N_π and N_ν describe the number of valence proton and neutron pairs.

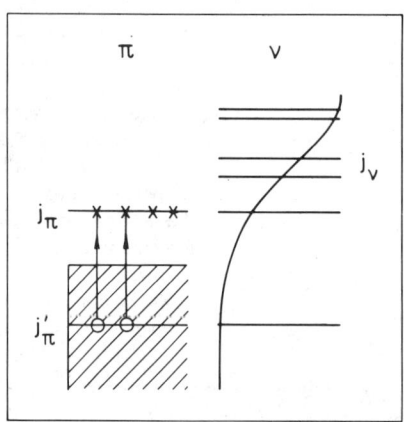

Fig.2. Schematic representation of a proton intruder 2p-2h 0^+ configuration where j_π denotes the regular orbital; $j_{\pi'}$ the intruder orbital, and j_ν the neutron orbitals.

This becomes in many cases, a rather cumbersome numerical problem so we shall only look to the lowest extra pair excited states (where excitations across a major or subshell are considered) and study how the residual interactions affect is unperturbed energy. In particular, near (or at) closed shells, the particle-hole excitations result in two extra pairs, denoted by the extra S_π^+ and $S_\pi^{+'}$ pair operators, where the primed operator creates a hole pair in the closed shell (see Fig.2). In such cases, the proton 2p-2h or neutron 2p-2h configuration (depending on the particular shell closure) will occur lowest in energy. Eventually, α-pair excitations can also be considered. In the remaining, we consider the proton 2p-2h configuration as the lowest intruder state where excitations across the closed shell occur and calculate the different energy contributions for such states (see Fig.2).

Calling

$$|0^+_{GS}> = (s^+_\pi)^{N_\pi} (s^+_\nu)^{N_\nu} |\tilde{0}> ,$$

$$|0^+_I> = (s^+_\pi)^{N_\pi+1} (s^{+'}_\pi)(s^+_\nu)^{N_\nu} |\tilde{0}> ,$$
(2)

respectively the 0^+ ground-state and the lowest 0^+ intruder 2p-2h configuration, one derives for the total energy difference, the expression (with $\hat{H}$ the total Hamiltonian operator)

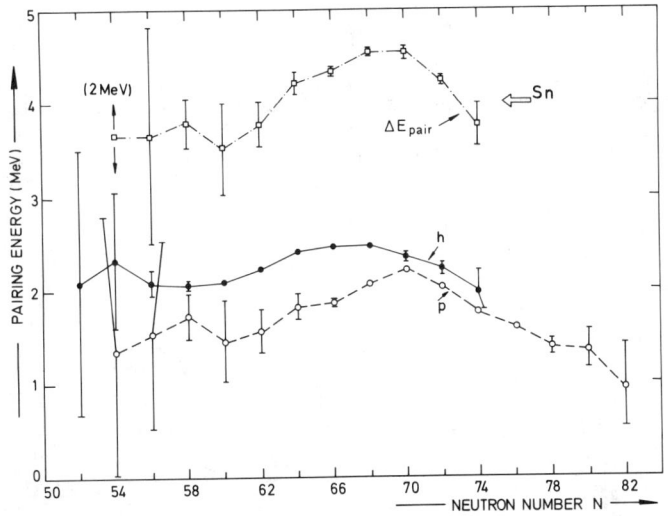

Fig.3. The pairing energy contribution to the 2p-2h configuration with $Z=50$ region, according to eq. (4). The separate hole pairing (h) and particle pairing (p) energies are also given.

$$E_{intr} \equiv <0^+_I|\hat{H}|0^+_I> - <0^+_{GS}|\hat{H}|0^+_{GS}>$$

$$= 2(\epsilon_{j\pi} - \epsilon_{j\pi'}) - \Delta E_{pair} + \Delta E_M + \Delta E_Q ,$$
(3)

where the first term describes the unperturbed 2p-2h energy, ΔE_{pair} the extra pairing correlation energy amongst the particle and the hole pair, e.g.

$$\Delta E_{pair} \equiv \langle 0_I^+ | \hat{V}_{\pi\pi} | 0_I^+ \rangle - \langle 0_{GS}^+ | \hat{V}_{\pi\pi} | 0_{GS}^+ \rangle ,$$

$$\simeq 2 \cdot S_p(Z,N) - S_{2p}(Z,N) , \qquad (4)$$

where S_p and S_{2p} denote proton and two-proton separation energies. Typical pairing correlation energies are shown in Fig.3 for the $Z = 50$ mass region.

Furthermore,

$$\Delta E_M \equiv 2 \sum_{j_\nu} (2j_\nu + 1) v_{j_\nu}^2 \left[\bar{E}(j_\pi j_\nu) - \bar{E}(j_\pi' j_\nu) \right] , \qquad (5)$$

where ΔE_M describes the self-energy shift for the 2p-2h unperturbed energy due to the filling of

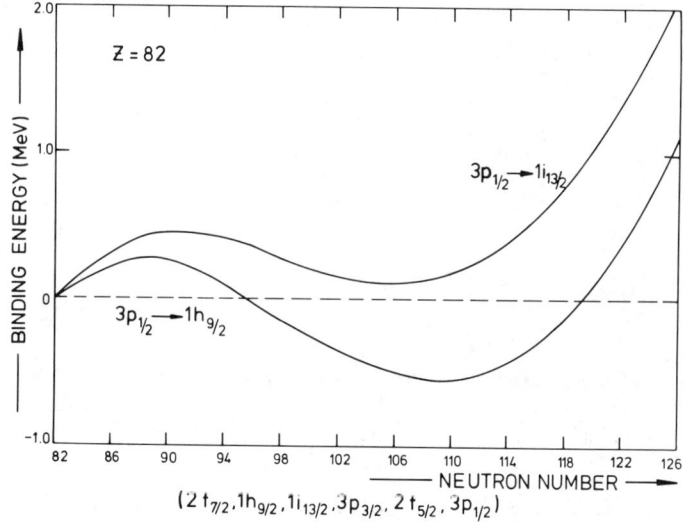

Fig.4. The monopole energy ΔE_M (see eq.5) for the $Z = 82$ region. The type of excitation $j_{\pi'} \to j_\pi$ is indicated.

the neutron orbitals j_ν (with occupation probabiliteis $v_{j_\nu}^2$ and $\bar{E}(j_\pi j_\nu)$ the spin-averaged proton-neutron matrix elements), and which is equal to the energy difference $\langle 0_I^+ | \hat{V}_{\pi\nu} | 0_I^+ \rangle - \langle 0_{GS}^+ | \hat{V}_{\pi\nu} | 0_{GS}^+ \rangle$. In all calculations to be discussed, we have used a zero-range delta-δ interaction since only the monopole part gives a contribution to the shift in the single-particle energies. This term accounts for a typical shell-model variation in the excitation energy for the intruder 2p-

2h state and can, when the gap between filled and unfilled orbitals is rather small (the $Z=40$ and $Z=64$ subshells), completely eradicate the subshell gap. An illustration for the $Z = 82$ mass region is given in Fig.4.

Even though the remaining multipole components of the proton-neutron interaction do not give a direct contribution to the binding energy in the 0^+ ground-state and intruder state, the quadrupole proton-neutron interaction can polarize these 0^+ states by admixing 2^+ core excited components[8,9]. Thereby, the 0^+ state modifies into

$$\overline{|0^+_\pi \otimes 0^+_\nu\rangle} \to |0^+_\pi \otimes 0^+_\nu\rangle + \alpha |2^+_\pi \otimes 2^+_\nu\rangle + \ldots , \qquad (6)$$

and using second order perturbation theory, the relative energy gain becomes[7]

$$\Delta E_Q \equiv \langle \tilde{0}^+_I | \hat{V}_{\pi\nu} | \tilde{0}^+_I \rangle - \langle \tilde{0}^+_{GS} | \hat{V}_{\pi\nu} | \tilde{0}^+_{GS} \rangle ,$$

$$= \sum_{j_\nu} \frac{\bar{\kappa}^2}{\Delta E} \cdot \frac{2}{5} \frac{(2j_\nu+1)^2 u^2_{j_\nu} v'^2_{j_\nu}}{2(2j_\nu-1)} \cdot F(j_\pi, j_\nu, j'_\pi, n_\pi) , \qquad (7)$$

when, for $\hat{V}_{\pi\nu}$ a quadrupole proton-neutron $\kappa \hat{Q}_\pi \cdot \hat{Q}_\nu$ force is used.

When, on the other hand, more 2^+ components get admixed into both the ground state and in particular in the intruder state, leading to more complicated polarization effects and inducing deformation, the SU(3) collective wave functions can be used to calculate the above energy difference, with as a result[7] (for large N_ν)

$$\Delta E_Q \equiv \langle 0^+_I | \kappa \hat{Q}_\pi \cdot \hat{Q}_\nu | 0^+_I \rangle - \langle 0^+_{GS} | \kappa \hat{Q}_\pi \cdot \hat{Q}_\nu | 0^+_{GS} \rangle ,$$

$$= 2 \kappa N_\pi N_\nu . \qquad (8)$$

This particular term is a very important contribution maximizing the relative gain in binding energy of the intruder 2p-2h configuration at or very near to mid-shell configurations in region II.

Collecting now all the contributions to the intruder energy, as given in Eq.3, one observes that for a given mass region, using as starting values, the constant unperturbed 2p-2h energy and the almost stable pairing interaction energy at the doubly-closed shell nucleus, the number dependence of the lowest 0^+ intruder excitation energy is governed by the proton-neutron interaction with

— the monopole force giving specific shell-effects modulating the unperturbed energy,

— the quadrupole component, giving a smooth N-dependence, for the intruder energy, minimizing the excitation energy at or near mid-shell.

In Figs.5 and 6, some specific results for both the Pb ($Z=82$ region) and for the subshell $Z=40$ region are shown[7)10] in order to present the general trends implied by the above method.

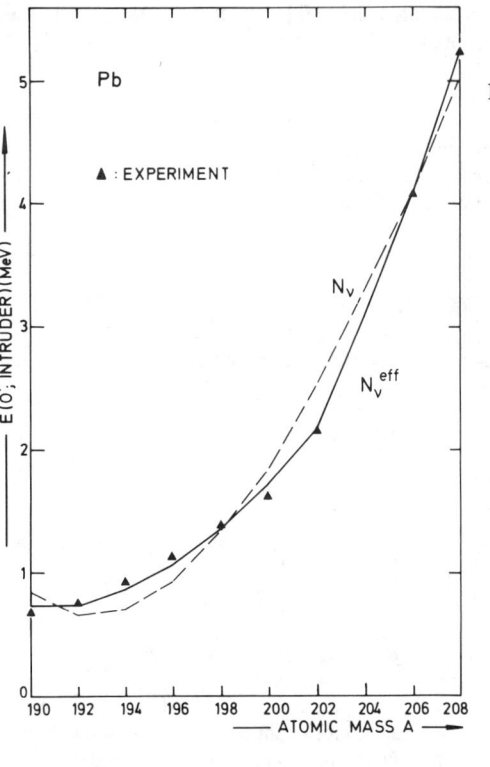

Fig.6. The excitation energy for the lowest 2p-2h 0^+ intruder state for the $Z = 40$ region. The hatched region gives a possible range for the intruder 0^+ state according to whether the quadrupole binding energy ΔE_Q is calculated for the (56,82) or from the (50,82) shells.

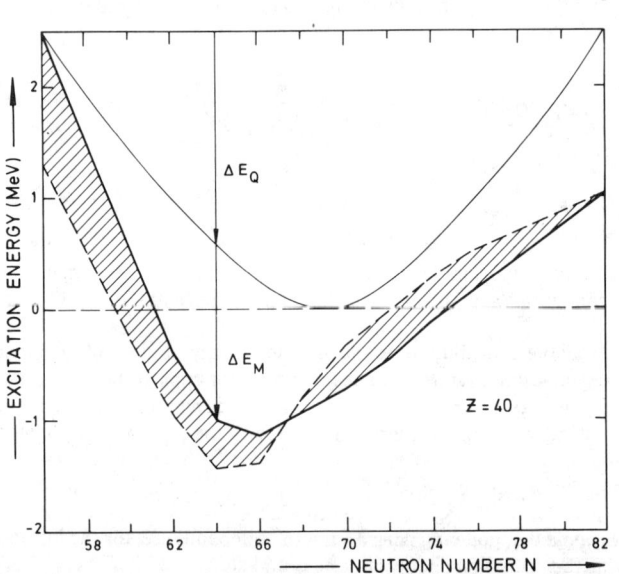

Fig.5. The results from a fit using the SU(3) expression for ΔE_Q (eq.8) and using the full expression (3), to the 0^+ intruder levels in the Pb nuclei. The full line is obtained including a subshell closure at $N = 114$.

2. Extension to odd-mass and doubly-odd nuclei.

Using the same methods outlined before with now as the regular state a 1h and the 1p-2h excitation as intruder configuration (see Fig.7), the relative excitation energy becomes

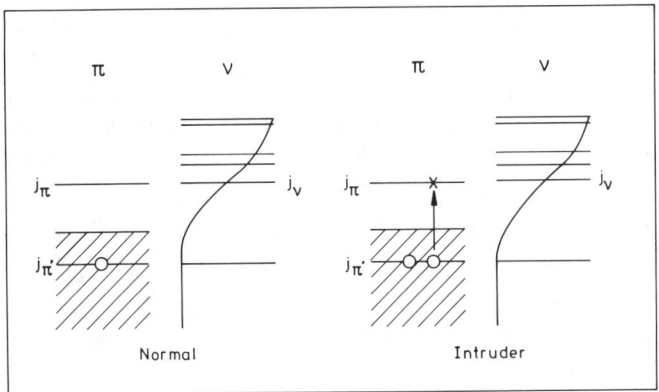

Fig.7. Schematic representation of a proton intruder 1p-2h j_π configuration, where $j_{\pi'}$ denotes the normal, low-lying 1h orbital. The neutron orbitals filling, are denoted by j_ν.

$$E_{intr}(j_\pi) \equiv E_{j_\pi} - E_{j_{\pi'}} = (\epsilon_{j_\pi} - \epsilon_{j_{\pi'}}) - \Delta E_{pair}(odd)$$
$$+ \Delta E_M(odd) + \Delta E_Q(odd) + \overline{\Delta E_Q}(odd), \qquad (9)$$

where $\Delta E_{pair}(odd)$ denotes the pairing interaction amongst the holes only and $\Delta E_M(odd)$ is the monopole shift as given before, but now for a 1p-1h excitation i.e. $\Delta E_M(odd) \simeq 1/2 \cdot \Delta E_M$.

The quadrupole binding energy is somewhat more complicated and contains:

— that part coming from the interaction of the hole pair with the neutron pair distribution, and, using the SU(3) wave functions from the even-even nucleus, results into

$$\Delta E_Q(odd) = 2\kappa N_\pi N_\nu , \qquad (10)$$

which is half of the contribution obtained for the 2p-2h intruder state at closed shells.

— extra terms result because of the quadrupole interactions of the odd particle with the core, given by

$$\Delta E_Q(\text{odd}) = \langle j_\pi \otimes 0^+_I | \kappa \hat{Q} \cdot \hat{Q} | j_\pi \otimes 0^+_I \rangle$$

$$- \langle j_{\pi'} \otimes 0^+_{GS} | \kappa \hat{Q} \cdot \hat{Q} | j_{\pi'} \otimes 0^+_{GS} \rangle$$

$$+ \langle j_\pi (j_{\pi'})^{-2} 0^+ | \kappa \hat{Q} \cdot \hat{Q} | j_\pi (j_{\pi'})^{-2} 0^+ \rangle , \quad (11)$$

where the third term turns out to become zero. The first two terms do not contribute in lowest order. In second order perturbation theory, when the quadrupole proton-neutron force admixes 2^+ configurations into the 0^+ states, one gets

$$\overline{\Delta E_Q}(\text{odd}) = 2\alpha^2/\Delta E_\alpha - 2\beta^2/\Delta E_\beta ,$$

with

$$\alpha = \langle j_{\pi'} \otimes 0^+_{GS} | \kappa \hat{Q} \cdot \hat{Q} | j_{\pi'} \otimes 2^+_{GS} \rangle ,$$

and

$$\beta = \langle j_\pi \otimes 0^+_I | \kappa \hat{Q} \cdot \hat{Q} | j_\pi \otimes 2^+_I \rangle \quad (12)$$

In the vibrational limit of particle-core coupling, the difference becomes[11]

$$\overline{\Delta E_Q}(\text{odd}) \simeq \frac{4\pi \cdot a}{\Delta E} \left[\frac{1}{2j_\pi + 1} \overline{\langle j_\pi || \hat{Q} || j_\pi \rangle}^2 - \frac{1}{2j_{\pi'} + 1} \overline{\langle j_{\pi'} || \hat{Q} || j_{\pi'} \rangle}^2 \right] , \quad (13)$$

with a the particle-core coupling strength and $\Delta E \simeq \hbar\omega_2$; the phonon excitation energy. For typical particle-core coupling strengths near closed shells in the 50-82 and 82-126 mass regions, this quantity never exceeds $\simeq 100$ keV. Taking into account these simplifying assumptions, one obtains again a rather simple expression for the intruder state excitation energy in the odd-mass nuclei[12].

For the odd-odd nuclei, completely similar arguments[13] now lead to the expression for the intruder excitation energy considering the full proton-neutron intruder multiplet of states $(j_\pi j_\nu)I$

$$E_{intr}(j_\pi j_\nu;I) = E(j_\pi j_\nu;I) - E(j_{\pi'}j_\nu;I)$$

$$= (\epsilon_{j_\pi} - \epsilon_{j_{\pi'}}) - \Delta E_{pair}(odd) + \Delta E_M(odd)$$

$$+ \Delta E_Q(odd) + \overline{\Delta E_Q(odd)} + \langle j_\pi j_\nu,I|\hat{V}_{\pi\nu}|j_\pi j_\nu,I\rangle$$

$$- \langle j_{\pi'}j_\nu,I|\hat{V}_{\pi\nu}|j_{\pi'}j_\nu,I\rangle \quad , \tag{14}$$

and only the specific proton-neutron residual two-body matrix elements remain for modifying the intruder state energy referring to odd-mass nuclei.

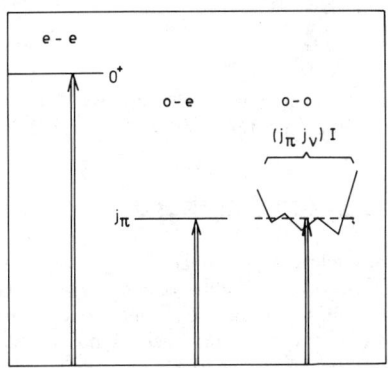

Fig.8. Illustration of the lowest 0^+ intruder state (in even-even nuclei), the lowest j_π intruder state (in odd-mass nuclei) and of the $(j_\pi j_\nu)I$ intruder multiplet (in odd-odd nuclei), according to eqs. 3,9 and 14.

The three different situations are now schematically compared in Fig.8. Comparing the results of Eqs.3, 9 and 14, one observes that the lowest intruder states in even-even, odd-mass and odd-odd nuclei fulfill the simple relationship of scaling[14] with the number of pairs at the closed shell configuration i.e. :

$$\frac{1}{2}E_{0^+}(intr) \simeq E_{intr}(j_\pi) \simeq \overline{E_{intr}(j_\pi j_\nu;I)} \quad . \tag{15}$$

This expression (15) is illustrated most dramatically in Fig.9 comparing the lowest intruder states in even-even Pb nuclei (0^+ intruder state), the $9/2^-$ intruder state in odd-A Tl nuclei, the $1/2^+$ intruder state in odd-A Bi and the 10^- intruder state in the odd-odd Tl nuclei.

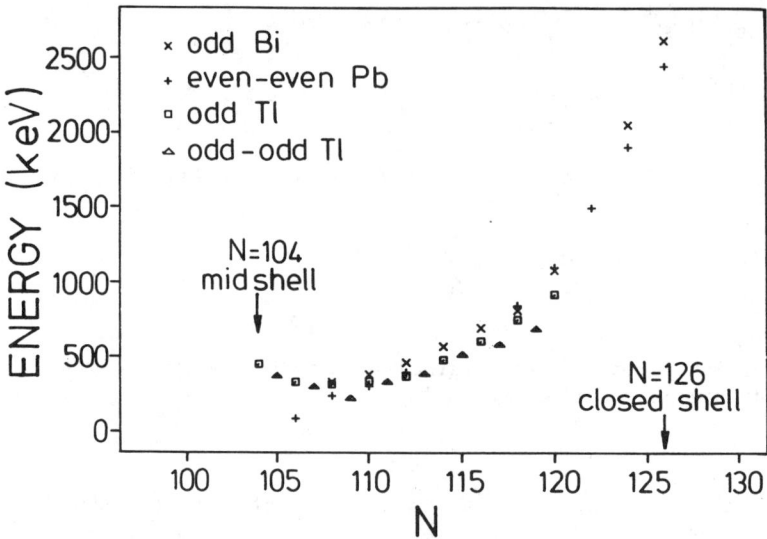

Fig.9. Systematics of the intruder state excitation energies in the odd-mass Tl, Bi nuclei, the odd-odd Tl nuclei (the 10⁻ level) and half of the excitation energy in the even-even Pb nuclei.

DEFORMED SHELL-MODEL APPROACH

Near closed shells, when the average single-particle field is allowed to exhibit quadrupole deformation, steep up- and down sloping orbitals occur. Thereby, the energy needed for exciting particle-hole configurations or more general, 2 quasi-particle excitations across the otherwise spherical closed shell, is lowered in a major way. The unperturbed single-particle energy difference even becomes zero at a number of crossing points (see the Nilsson single-particle orbitals). The pair scatter amongst the deformed orbitals, added to the liquid drop energy of the deformed nucleus results in a total potential energy surface. Thereby, the excitations that we called intruder particle-hole configurations in sect.2, now correspond to deformed 2 qp configurations having a quadrupole equilibrium deformation rather different from the 0 qp ground state equilibrium value. Moreover, due to the quadrupole deformed average field, a large deformation energy is gained with respect to the low-lying spherical 0 qp configurations. Such calculations have been carried out by now in many mass regions by the Lund group[15] and more recently, by Bengtsson and Nazarewicz[16] in the Pb region. Here, we show some of their pertinent results concerning the ^{186,196,202}Pb nuclei as well as the total excitation energy for the lowest 0^+ intruder and the high-spin intruder 11^- level(see Fig.10).

The relation with the discussion before, starting from the spherical shell-model can now be elucidated . When the number of valence nucleons (neutrons in the Pb case) is increasing out of the closed shell, together with a steady increase in the equilibrium value for the lowest 2qp excitations (corresponding with the spherical 2p-2h intruder states) , a lowering in excitation energy up to about mid-shell results due to the quadrupole deformation energy. Thereby an effective valence nucleon number dependence of the excitation energy for the deformed 2 qp state results,

265

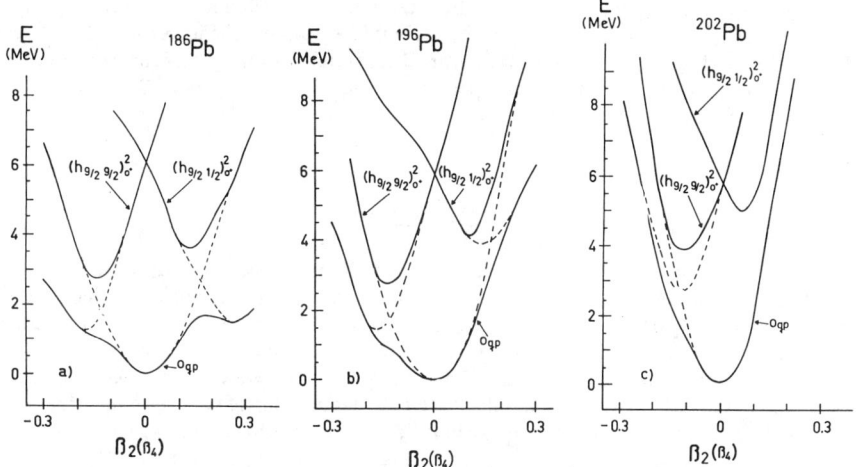

Fig.10.a. Results of the total potential energy surfaces for ^{186,196,202}Pb. Solid lines correspond with the adiabatic configurations while the dashed lines have been used for the diabatic configurations. (from Bengtsson and Nazarewicz, ref.16).

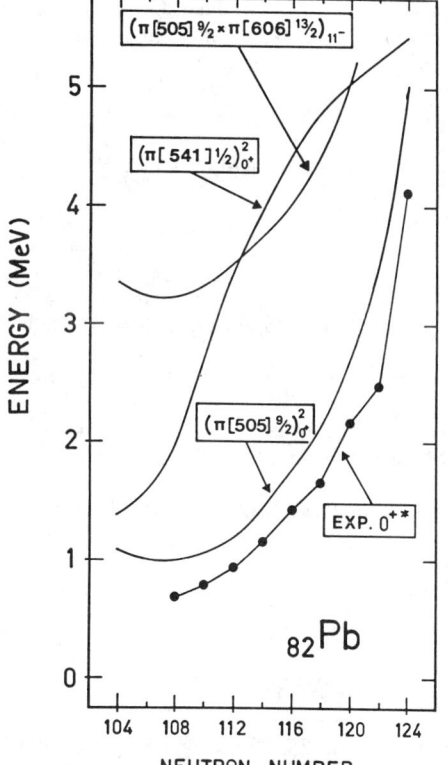

Fig.10.b. The excitation energies for the 2qp configurations as indicated on the figure (from Bengtsson and Nazarewicz, ref.16).

very similar to the dependence when starting from the spherical shell-model. The advantage, however, of the calculation starting from the spherical shell-model, where the unperturbed 2p-2h energy is first lowered by the pairing correlations and later by the residual proton-neutron monopole and quadrupole interactions, is that a very specific analytic expression (see Eqs.3, 9 and 14) for the lowest intruder states results.

CONCLUSION

In the present contribution, we have extensively discussed the possibility of describing intruder states near closed shells in both even-even odd-mass and odd-odd nuclei starting from the basis assumption that particle-hole excitations across major shells, interacting with a large number of valence nucleons of the other type through the proton-neutron force, are respossible for the very low excitation energy and mass dependence. This also explains the presence or absence of intruder states related to the difference in magnitudo of the expectation value of the proton-neutron interaction energy $<\hat{Q}_\pi.\hat{Q}_\nu>$ compared to the size of the shell-gaps in different mass regions.

More detailed applications relating to light sd-shell nuclei ($N \simeq 20$), relating to the importance of measurements of nuclear radii and E0 transitions in identifying intruder states and concentrating on subshell and double sub-shell closures will be discussed by J.L.Wood.

The author is most gratefull to R.Casten, P.Federman, W.Hesselink, M.Huyse, J.Kantele, R.A.Meyer, J.L.Wood for stimulating discussions and to M.Waroquier, J.Ryckebusch, J.Moreau and J.Van Maldeghem in Gent for contributing to some of the ideas and results. This work is parthly supported by NATO RG-86/0452 and the IIKW and the NFWO.

REFERENCES

1. B.H.Wildenthal in : Proc. of the Int. Symp. on Nuclear Shell Models, eds. W.Vallières and B.H.Wildenthal (World Scientific, Singapore, 1985), 346 ;
   M.Waroquier et al., Phys. Reports 148, 249 (1987)

2. I.Talmi, Nucl. Phys. A172, 1 (1971)

3. P.Ring and P.Schuck, The nuclear many-body problem (Springer, New-York, 1980), Ch.6

4. A.Bohr and B.Mottelson, Nuclear Structure, Vol.2 (Benjamin, Reading, 1975), ch.4

5. A.Arima and F.Iachello, in : Advances in Nuclear Physics, vol.13, 139 (1984) and F.Iachello and A.Arima, The Interacting Boson Model (Cambridge Univ. Press, 1987),

6. K.Heyde, P.Van Isacker, M.Waroquier, J.L.Wood and R.A.Meyer, Phys. Reports 102, 291 (1983)

7. K.Heyde et al., Nucl. Phys. A466, 189 (1987)

8. I.Talmi, Nucl. Phys. A423, 189 (1984)

9. F.Barranco and R.A.Broglia, Phys. Lett. 151B, 90 (1985)

10. P.Van Duppen, E.Coenen, K.Deneffe, M.Huyse and J.L.Wood, Phys. Rev. C35, 1861 (1987)

11. V.Paar, Nucl. Phys. A211, 29 (1973)

12. K.Heyde et al., to be publ.

13. M.Huyse, E.Coenen, K.Deneffe, P.Van Duppen, K.Heyde and J.Van Maldeghem, to be publ.

14. J.Van Maldeghem and K.Heyde, to be publ.

15. S.G.Nilsson et al., Nucl. Phys. A131, 1 (1969)

16. R.Bengtsson et al. Phys. Lett. 183B, 1 (1987) ; R.Bengtsson and W.Nazarewicz, to be publ.

OBLATE-PROLATE SHAPE COMPETITION IN Z = 34-38 NUCLEI

J. Eberth, M. Wiosna, T. Mylaeus, N. Schmal, S. Skoda,
J. Busch, W. Teichert, P. v. Brentano
University of Köln, Köln, W. Germany

J. H. Hamilton, A. V. Ramayya, X. Zhao, C. F. Maguire, W.-C. Ma,
J. Kormicki, S. Wen, L. Chaturvedi, Z.-M. Chen
Vanderbilt University, Nashville, TN. 37235

T. M. Cormier, M. Satteson
University of Rochester, Rochester, NY. 14627

J. D. Cole
INEL, Idaho Falls, ID. 83415

R. B. Piercey, M. A. Herath-Banda
Space Astronomy Lab., Univ. of Florida, Gainesville, FL. 32601

E. F. Zganjar, M. O. Kortelahti
Louisiana State University, Baton Rouge, LA. 70803

G. A. Leander, W. Nazarewicz
UNISOR and Joint Institute, Oak Ridge, TN. 37831

ABSTRACT

Now magic numbers for deformed shapes are established at N and Z of 38 and N of 60. These shell gaps at large deformation ($\beta \sim 0.4$) are magic when the proton and neutron shell gaps reinforce each other. Other shell gaps for 34 and 36 are predicted to be important for oblate deformation. The competition between these oblate and the 38 prolate gaps as well as the 40 spherical shell gap are considered. In ^{71}Se coexistence of oblate ($\beta \sim -0.24$) and large prolate ($\beta \sim 0.4$) shapes are observed. In ^{72}Se the excited prolate band with large deformation which coexists with the near-spherical ground state is found to dominate the yrast spectrum to 28^+. The moment of inertia of this band is essentially that of a rigid body. This moment of inertia supports the suggestion that the "super deformation", $\beta \sim 0.4$, being observed in this region may be associated with a collapse of pairing.

INTRODUCTION

A new region of very large prolate ground state deformation ($\beta \sim 0.4$) associated with shell gaps at N and Z of 38 for such large prolate β was first reported in ^{74,76}Kr at the last conference on nuclei far from stability[1,2]. Based on those results and new data for ^{80}Sr (ref. 3), it was proposed[1,4] that this unusually large deformation for nuclear ground states arose from the reinforcing of the proton and neutron shape driving forces associated with the shell gaps at β ~ 0.4 for N and Z of 38 as seen in the then available calculations of the Nilsson single particle levels. Simultaneously, new calculations of Möller and Nix[5] in a folded Yukawa plus single particle potential predicted this new region of very strong deformation centered around N = Z = 38. The single particle levels in this model likewise have gaps for 38 at β ~ 0.4, as shown in Fig. 1, which is the same for both protons and neutrons.[6]

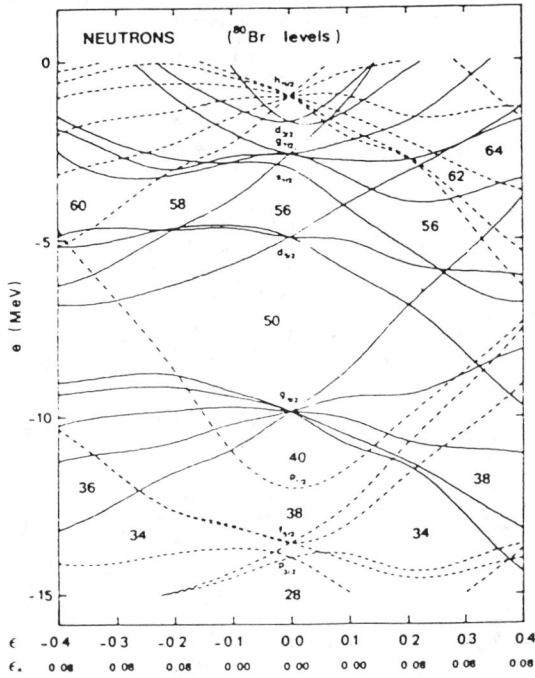

Fig. 1. The single-particle levels calculated with a folded Yukawa single-particle potential (ref. 6).

DEFORMED SHELL GAPS AND SHELL GAP REINFORCEMENT

Subsequently, we have shown that the shell gaps at β ~ 0.4 for N and Z of 38 (refs. 7, 8, 9) as well as N = 60 (refs. 7, 9) are new magic numbers but now for deformed shapes in addition to the well-known spherical magic numbers. These "deformed" magic numbers confirm the long-standing theoretical predictions of Brack et al.[10] that there should be "deformed" shell gaps which stabilize a deformed nuclear shape just as the spherical magic numbers (shell gaps at 2, 8, 20, ...) stabilize a spherical nuclear shape. The importance of the Z = 38 shell gap at large prolate deformation reinforcing the N = 38 and N = 60 shell gaps at the same large (β ~ 0.4) deformation to drive the nuclei in these regions to what may be called "super deformation" for nuclear ground states is illustrated in Fig. 2. This figure also illustrates the importance of spherical shell gap reinforcement as well. Both $^{88}_{38}Sr_{50}$ and $^{90}_{40}Zr_{50}$ have very large first excited 2^+_1 energies which are characteristic of spherical double magic (double closed shell) nuclei like $^{208}_{82}Pb_{126}$. Here the 38, 40 weaker spherical subshell gaps are reinforced by the strong N = 50 spherical shell gap to produce spherical double magic nuclei. Even the weaker N = 56 spherical subshell gap can reinforce Z = 38 and 40 to keep the Sr and Zr nuclei strongly spherical out to N = 58. Then, at N = 60, there

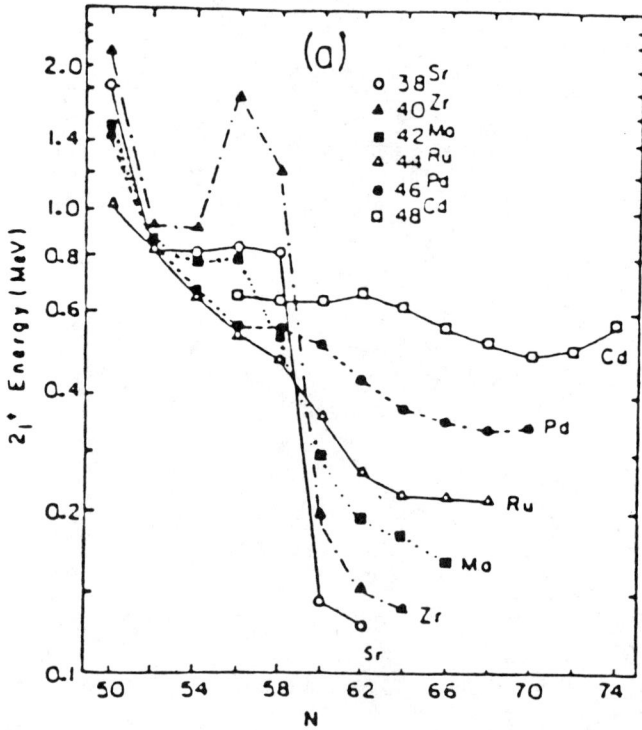

Fig. 2. The 2^+_1 energies of heavy Sr to Cd nuclei which show the sudden onset of large deformation for N = 60 in Sr and Zr nuclei with Z = 38 and 40.

is a sudden onset in $^{98}_{38}Sr_{60}$ and $^{100}_{40}Zr_{60}$ of unusually large ground state deformation. However, moving away from Z = 38 to Z $\geq$ 42 (Fig. 2) this sudden onset of deformation quickly disappears to show clearly the importance of the reinforcement of the Z = 38 and N = 60 shell gaps. Likewise, the reinforcement of the spherical shell gaps at Z = 40 and N = 56 quickly disappears as N or Z changes. Note it is the switch in reinforcement of two spherical to two different deformed shell gaps that leads to the suddenness of the onset of deformation in ^{98}Sr and ^{100}Zr. The switch in importance from the 40 spherical to the 38 deformed shell gaps is further seen in the recent discovery[11] of $^{80}_{40}Zr_{40}$ which was found to be strongly deformed and not spherical double magic. The possibility of this switch was one of the arguments presented in the proposal to develop the UNISOR facility[12]. Reinforcing shell gaps indeed explain a wide range of changing deformation including, for examples, the sudden onset of deformation in the rare earths around Z = 64 and the disappearance of this suddenness as Z goes away from 64 (refs. 7, 9 for examples) and spherical double magic character of $^{146}_{64}Gd_{82}$.

Note these two new regions with ground state deformation ($\beta \sim$ 0.4) may be called "super deformed" ground states by comparison with $\beta \lesssim 0.25$ for the well-known deformed rare earth and actinide ground states. This "super deformation" may be a signal that these nuclei are approaching the long-sought, pairing free structures but now at or near the ground state rather than at high spins as earlier predicted[13]. This possibility is discussed elsewhere in these proceedings[14].

OBLATE SHELL GAPS IN THE A = 70 REGION

In Fig. 1 there are also shell gaps at other deformations such as N = Z = 34 for oblate deformation $\beta \sim -0.25$ and for N = Z = 36 large oblate deformation of $\beta = -0.4$. The large oblate shell gaps at N = Z = 36 could be theoretically expected to be as important as the N = Z = 38 shell gaps for the prolate shape with $\beta \sim 0.4$. However, based on the strong evidence that the prolate shell gaps can stabilize a deformed shape only when the proton and neutron shell gaps reinforce each other, one would expect that similar reinforcement would be necessary for the oblate shapes to be observed. Already the prolate shapes of $^{74,75,76}_{36}Kr_{38,39,40}$ show that the N = 38 prolate gap at $\beta \sim 0.4$ is stronger than the Z = 36 shell gap at $\beta \sim -0.4$. Nevertheless, it is important from the theoretical standpoint to establish if these oblate gaps at 34 and 36 can stabilize oblate shapes and in which nuclei.

Searches to identify a prolate to oblate phase transition in the ground states of nuclei in the A = 70 region was begun at the Recoil Mass Spectrometer (RMS) at the University of Rochester. One may also observe prolate-oblate transitions as functions of spin and energy in a given nucleus. One possible sequence to observe such a ground state phase transition is the Br nuclei going from $^{75}_{35}Br_{40}$ to $^{73}_{35}Br_{38}$ to $^{71}_{35}Br_{36}$. In our first RMS studies[15] we identified levels in $^{73,75}Br$ which have strong prolate ground state deformations with ^{73}Br with N = 38 having somewhat larger deformation than ^{75}Br (N = 40) as expected.

OBLATE-PROLATE COMPETITION IN $^{71}_{34}Se_{37}$

Further experiments were carried out at the University of Köln tandem and with the RMS at Rochester on the mass 71 region. Here we report our studies of $^{71}_{34}Se_{37}$. In Köln an 8-fold neutron multiplicity filter in combination with the 10-fold OSIRIS anticompton spectrometer[16] was used to carry out n-γ, n-γ(θ), and n-γ-γ coincidence studies. A new five sector neutron detector (built by Vanderbilt and LSU) and four new NaI detectors (built at Vanderbilt) which were calibrated

Fig. 3. Gamma rays in coincidence with recoil mass 71 only, RM-71-protons and RM-71 neutrons from bottom to top, respectively.

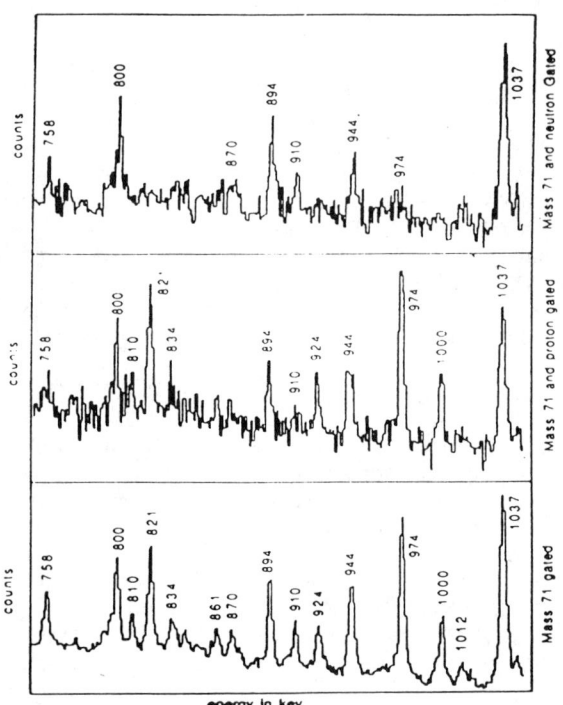

for fast protons were used to select coincidence gates along with a recoil mass gate from the Rochester RMS, to do recoil mass-γ, RM-n-γ, and RM-p-γ double and triple coincidences. The reaction $^{16}O + ^{58}Ni$ was studied with 41-65 MeV ^{16}O ions.

In Fig. 3 are shown partial γ-ray spectra with only a single recoil mass 71 gate, and with double RM-71-neutron and RM-71-proton gates. For example, the 894 keV transition is clearly enhanced in the RM-neutron gated spectrum compared to the RM-p gated one, and RM only gated one as are other transitions in ^{71}Se from levels populated in the (2p,n channel) to mass 71. Transitions in ^{71}As populated in the 3p channel to mass 71 are enhanced in the RM-proton gated spectrum. Combining the data from the Köln and Rochester experiments yielded the level scheme shown in Fig. 4. Of particular importance is the measured E2/M1 mixing ratio $\delta = 1.5 \pm 0.3$ for the 894 keV transition.

A theoretical analysis of the levels of ^{71}Se in which a deformed shell model of the nucleonic orbits, BCS pairing, the Strutinsky method to determine self-consistent shapes, and cranking or coupling

Fig. 4. Level structure of ^{71}Se.

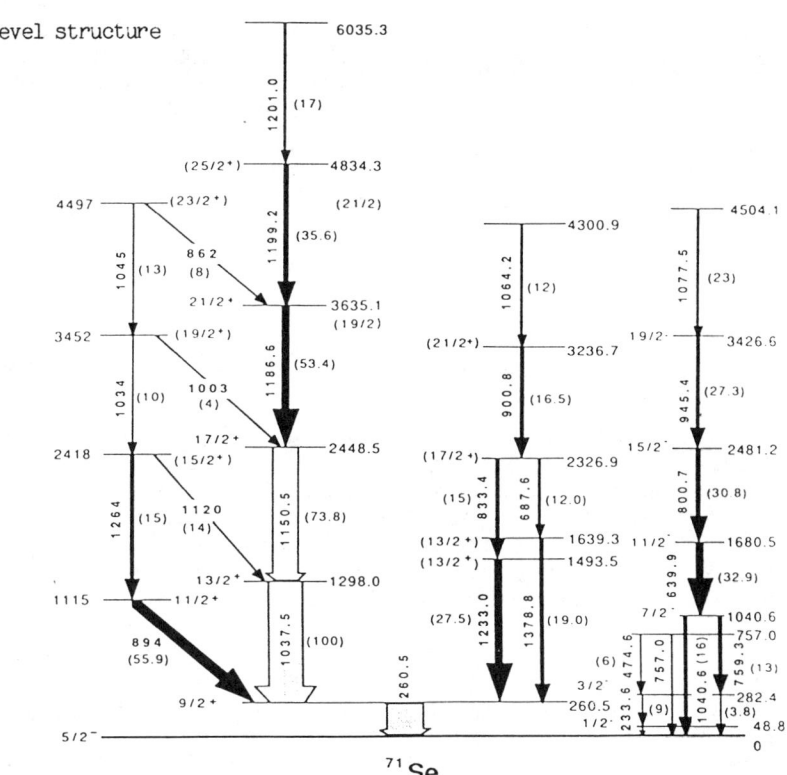

of the quasiparticles to a triaxial rotor core in order to include non-adiabatic effects of rotation were used. A shell model potential of the Woods-Saxon type with the "universal" parameter set[17] was used with its shape specified by quadrupole (β), hexadecapole (β_4) and triaxial (γ) deformation coordinates[17]. The pairing interaction is given by Dudek et al.[18], the implementation of the Strutinsky and cranking procedures are described in ref. 17 and references therein, and the core-quasiparticle coupling is carried out as in ref. 18b.

The potential energy of deformation for both positive and negative parity exhibits two coexisting equilibrium shapes with prolate and oblate axial deformations, respectively. This shape coexistence persists when the potential is cranked to modest rotational frequencies, however, the equilibrium deformations become slightly triaxial (in the collective sector).

The calculated axial deformations and relative energies of the band heads are given in Table I. The corresponding bands nicely account for the main features of the observed band structures in Fig. 4.

Table I. Calculated properties of the lowest prolate and oblate quasiparticle states of each parity. The Nilsson label is the conventional one (see ref. 20). The excitation energy E* is given relative to the lowest state for oblate and prolate shape separately.

	π	β	$β_4$	Nilsson Label	E* (keV)
oblate	-	0.24	-0.003	1/2[321]	(0)
oblate	+	0.24	-0.006	9/2[404]	8
prolate	-	0.38	0.004	3/2[312]	(0)
prolate	+	0.30	-0.006	3/2[431]	529

The lowest negative-parity levels and the positive parity $\Delta I = 1$ sequence to the left in Fig. 4, correspond to the oblate solutions, while the positive and negative parity high spin sequences to the right in Fig. 4 are identified with the prolate solutions. The moment of inertia for the high-spin sequences in Fig. 4 increases from left to right (taking into account rotational alignment), in qualitative agreement with the calculated deformations of Table I.

The $5/2^-$, $1/2^-$, $3/2^-$ level sequence of the ground state band is reproduced by a rotor-quasiparticle calculation for oblate or any oblatish triaxial shape at $|β| = 0.24$. This is essentially a decoupled $f_{5/2}$ structure, with wave functions that are roughly equal admixtures of the adiabatic 1/2[301], 3/2[321] and 1/2[321] orbitals. A test of this interpretation is provided by the γ-decay of the $3/2^-$ level. The calculated B(M1) rate is more than 3 times larger to the $1/2^-$ than the $5/2^-$ level, in agreement with the experimental branching ratio.[19]

The large moment of inertia of the negative parity states above $7/2^-$ suggests that these states belong to the strongly deformed prolate band built on the 3/2[312] configuration. This configuration was predicted to be the ground state of ^{71}Se, some 450 keV below the lowest oblate state, but in light of the spectroscopic evidence discussed above and below it appears that an oblate shape is more favored in nature than in the model. It is surprising that only the signature corresponding to the spin states ...$7/2^-$, $11/2^-$,... is observed, since the other ($f_{5/2}$-like) signature is slightly favored in the model calculations. The $\Delta I = 1$ transitions that connect the two signatures are, however, expected to be an order of magnitude weaker than the observed $\Delta I = 2$ transitions.

For positive parity there are two disconnected sets of levels at high spin. The set to the right is connected by relatively low-energy E2 transitions and appears to belong to the favored signature of a decoupled $g_{9/2}$ band. Such a band would be based on the low-Ω $g_{9/2}$ suborbitals that lie near the Fermi level for prolate shape. The levels to the left in Fig. 4 form a $\Delta I = 1$ sequence on the $9/2^+$

level, and would thus appear to be associated with the oblate 9/2[404] orbital. There is a sizable signature splitting in this $\Delta I = 1$ sequence, which can be reproduced by assuming a triaxial (γ) deformation in the non-adiabatic model calculations. If the core moment of inertia and the γ deformation of the particle-rotor model are fitted to the observed spacings of the $9/2^+$, $11/2^+$ and $13/2^+$ levels, the resulting value of γ is 32°, closer to oblate than prolate but just barely. Such large deviations from axial symmetry are often needed to describe moderately deformed bands in odd-A nuclei[21,22]. The γ deformation of the rotor model may then be simulating the effect of dynamical shape fluctuations[23]. A predominantly oblate deformation is also implied by the measured mixing ratio of $\delta = +1.5 \pm 0.3$ for the $11/2^+$ to $9/2^+$ transition.

For high Ω $g_{9/2}$ orbitals and a deformation range $0.2 < \beta_2 < 0.3$, $-0.01 < \beta_4 < 0.0$ $g_K - g_R$ is negative. The relation sign (δ) = sign ($g_K - g_R/Q_0$) gives a negative value of the intrinsic quadrupole moment Q_0 from the positive sign of the experimental mixing ratio. Though the mixing ratio indicates an oblate deformation the discussed structure appears to be predominantly triaxial. Thus, it is seen that in ^{71}Se we observe the transition from a prolate structure in the heavier selenium isotopes to an oblate deformation of the $g_{9/2}$ bands in the very neutron deficient selenium isotopes. This becomes very clear from the recent results[26] for ^{69}Se where an oblate $g_{9/2}$ band with low signature splitting was discovered.

SHAPE COEXISTENCE AND "SUPER DEFORMATION" IN ^{70,72}Se

The coexistence of bands of levels built on near-spherical (probably small oblate) ground states and strongly deformed excited bands in $^{70,72}_{34}\text{Se}_{36,38}$ was established earlier[27,28]. The ground states of these nuclei are likely associated with the shell gaps for Z = 34 at oblate deformation $\beta \sim -0.25$ and the excited deformed band with the prolate N = 38 shell gap at $\beta \sim 0.4$. Recently the high spin levels of ^{70,72}Se have been extended to 14^+ and 28^+, respectively[29]. Plots of the angular momenta as a function of gamma-ray energy for these nuclei are shown in Fig. 5. The deformed bands totally dominate the yrast spectra in these nuclei in sharp contrast to $^{68,70}_{32}\text{Ge}_{36,38}$ (with two less protons) where the ground state bands fork at low spins. What is most striking is the constant moment of inertia seen in ^{72}Se at high spins with the moment of inertia essentially the rigid body value. This constant rigid body moment of inertia gives additional support to the proposal[14] that these "super deformed" ($\beta \sim 0.4$) structures in this region may be arising from a strong reduction in pairing, even approaching a pairing free state.

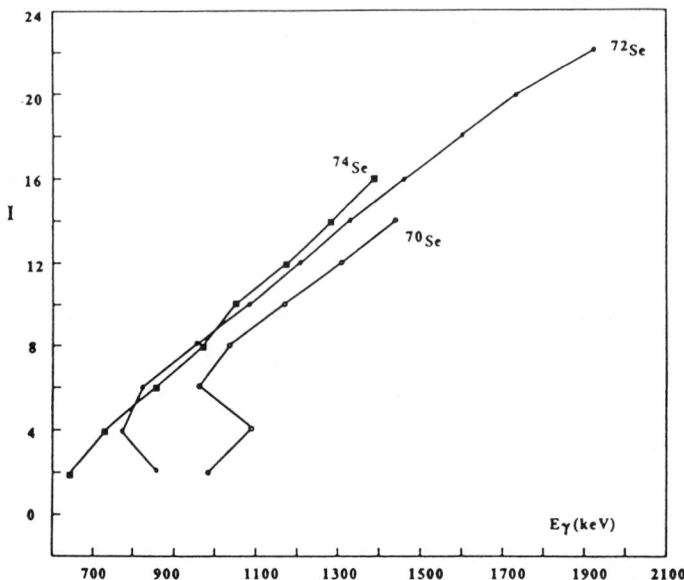

Fig. 5. Plot of the angular momentum as a function of the γ-ray energies for ^{70,72,74}Se.

ACKNOWLEDGEMENTS

The work at Vanderbilt University, Louisiana State University and UNISOR is supported by the U.S. Department of Energy under Contract Nos. DE-AS05-76ER05034, DE-AS05-76ER04935, and DE-AC05-76OR00033, respectively. The University of Rochester Facility is supported by the National Science Foundation. The work at University of Köln is supported by the German Federal Ministry of Research and Technology under Contract No. 060K272.

1. J. H. Hamilton, R. B. Piercey, R. Soundranayagam, A. V. Ramayya, C. F. Maguire, X.-J. Sun, Z. Z. Zhao, J. Roth, L. Cleemann, J. Eberth, T. Heck, W. Neumann, M. Nolte, R. L. Robinson, H. J. Kim, S. Frauendorf, J. Döring, L. Funke, G. Winter, J. C. Wells, J. Lin, A. C. Rester and H. K. Carter, in Proc. IV Int. Conf. on Nuclei Far From Stability, CERN 81-09 (1981), pp. 391-396.
2. R. B. Piercey, Phys. Rev. Lett. 47, 1514 (1981).
3. M. Barclay, L. Cleemann, A. V. Ramayya, J. H. Hamilton, J. D. Cole, A. de Lima, W.-C. Ma, S. C. Pancholi, R. Soundranayagam, H. Yamada, W. Ellerby, R. L. Robinson, and W. Milner, private communication (1981) as reported in ref. 4.
4. J. H. Hamilton, Proc. Int. Symp. on Nuclear Collectivity, A. Arima and T. Marumori, eds., Tokyo: INS (1981), p. 87.
5. P. Möller and J. R. Nix, At. Data and Nucl. Data Tables 26, 165 (1981); Am. Chem. Soc. 183rd Nat. Conf., Abs. Nucl. 16 (1982).

6. R. Bengtsson, P. Möller, J. R. Nix, and J.-Y. Zhang, Physica Scripta 29, 402 (1984).
7. J. H. Hamilton, Proc. Int. Symp. on Nuclear Shell Models, M. Vallieres and B. H. Wildenthal, eds., Singapore: World Scientific (1985), p. 31.
8. J. H. Hamilton, A. V. Ramayya, C. F. Maguire, R. B. Piercey, R. Bengtsson, P. Möller, J. R. Nix, J.-Y. Zhang, R. L. Robinson, and S. Frauendorf, J. Phys. G: Nucl. Phys. 10, L87 (1984).
9. J. H. Hamilton, Progress in Particle and Nuclear Physics, Vol. 15, A. Faessler, ed., New York: Pergamon Press (1985), p. 107.
10. M. Brack, J. Damgaard, A. S. Jensen, H. C. Pauli, V. M. Strutinsky, and C. Y. Wong, Rev. of Mod. Phys. 44, 329 (1972).
11. C. J. Lister et al., elsewhere these Proceedings.
12. J. H. Hamilton, Izv. Akad. Nauk, USSR Ser., Fiz 36, 143 (1972).
13. B. R. Mottelson and J. G. Valatin, Phys. Rev. Lett. 5, 511 (1960).
14. L. K. Peker and J. H. Hamilton, elsewhere in these proceedings.
15. S. Wen, A. V. Ramayya, S. J. Robinson, C. F. Maguire, W.-C. Ma, X. Zhao, T. M. Cormier, P. M. Stwertka, J. D. Cole, E. F. Zganjar, R. B. Piercey, M. A. Herath-Banda, J. Eberth, and M. Wiosna, J. Phys. G: Nucl. Phys. 11, L173 (1985).
16. R. M. Lieder, et al., Inst. Meth. 220, 363 (1984).
17. J. Dudek, Z. Szymanski, and T. Werner, Phys. Rev. C23, 920 (1981).
17. W. Nazarewicz, J. Dudek, R. Bengtsson, T. Bengtsson, and I. Ragnarsson, Nucl. Phys. A435, 397 (1985).
18. J. Dudek, A. Majhofer, and J. Skalski, J. Phys. G6, 447 (1980).
18b. S. E. Larsson, G. Leander, and I. Ragnarsson, Nucl. Phys. A307, 189 (1978).
19. J. Eberth, L. Cleemann, N. Schmal, Int. Symp. on In-Beam Nuclear Spectroscopy, (Debrecen, Hungary, 1984)Z. S. Dombradi and T. Fenyes, Eds., Budapest:Akademiai Kiado, p. 23.
20. S. E. Larsson, et al., Nucl. Phys. A261, 77 (1976).
21. J. Meyer-ter-Vehn, Nucl. Phys. A249, 111 and 141 (1975).
22. H. Toki and A. Faessler, Phys. Lett. 63B, 121 (1976).
23. G. Leander, Nucl. Phys. A273, 286 (1976).
24. A. Bohr and B. R. Mottelson, Nuclear Structure, Vol. 2, New York: Benjamin (1975).
25. A. R. Mokhtar, et al., Proc. Int. Conf. on Nuclear Shapes, Vol. 1 (Crete, 1987), G. S. Anagnostatos, et al., eds., p. 62 and references therein.
26. M. Wiosna, J. Busch, J. Eberth, M. Liebchen, T. Mylaeus, N. Schmal, R. Sefzig, S. Skoda, and W. Teichert, Phys. Lett. (in press).
27. J. H. Hamilton, et al., Phys. Rev. Lett. 32, 239 (1974).
28. A. Ahmed, et al., Phys. Rev. C24, 1486 (1981).
29. T. Mylaeus, J. Busch, J. Eberth, M. Liebchen, R. Sefzig, S. Skoda, W. Teichert, M. Wiosna, P. v. Brentano, K. Schiffer, K. O. Zell, A. V. Ramayya, K. H. Maier, H. Grawe, and A. Kluge, (to be publilshed).

A Microscopic View of Large Deformation Nuclei with A=60-80

D.P. Ahalpara and S.P. Pandya

Physical Research Laboratory
Ahmedabad 380009, India

Abstract

Nuclei in the mass region A=60-80 are characterized by sudden onset of large deformations or shape changes with changes of N or Z, as well as by shape coexistence in the same nucleus. An attempt is made to understand these properties in terms of deformed shell model calculations.[1] The configuration space includes $p_{3/2}$, $f_{5/2}$, $p_{1/2}$ and $g_{9/2}$ orbits, and a modified Kuo interaction[2] is used. As examples, we consider even-even N=Z nuclei ^{68}Ge, ^{68}Se, ^{72}Kr and ^{76}Sr.

Text

It has now been known for more than a decade that nuclei with masses in the range A=60-80 show curious collective properties, such as large deformations, sudden onset of collectivity with small changes in neutron or proton numbers, and shape coexistence in the same nucleus. To understand these features in terms of a microscopic model (e.g. shell model) is highly desirable, since one can then gain some insight into the nature of the effective interactions and their role, and also the influence of the configurations involved in such nuclei. Since large-scale spherical shell model calculations are very difficult in this region, we have resorted to shell-model calculations in deformed Hartree-Fock basis.[1]

Active single particle orbits spanning the configuration space are $p_{3/2}$, $f_{5/2}$, $p_{1/2}$, $g_{9/2}$ and in this space we consider an effective interaction generated by Kuo, and partially modified by Bhatt et al.[2] For a given number of active neutrons and protons we carry out Hartree-Fock calculations (preserving axial symmetry) giving the lowest energy prolate and oblate solutions. These now provide a basis for deformed shell model calculations. A variety of configurations can be generated, and for each of these we redo a variational constrained calculation, obtaining the best mean field and deformation for this configuration. Standard projection techniques give a band of states with good angular momentum J for each of these intrinsic states, and finally the Hamiltonian operator is diagonalized for all states with the same J.

In figures 1 and 2 we show the HF spectra for the lowest prolate and oblate shapes of nuclei with N = Z = 32, 34, 36 and 38 respectively. For each solution we also give the deformation parameter β calculated from the intrinsic quadrupole moment. It is found that for Ge and Se, the HF energies of prolate and oblate shapes are nearly degenerate, while for Kr and Sr, the oblate state is lower in energy by 5-6 MeV. The obvious reason for this is the use of harmonic oscillator wave functions, and the limited configuration space. We have not included $2s_{1/2}$, $1d_{5/2}$ and $1d_{3/2}$ orbits which would have resulted in lowering the energies of the K=1/2, 3/2 and 5/2 components of the $g_{9/2}$ orbit considerably. This would not only give larger energy for the prolate shape, but also would increase substantially the calculated deformations for the prolate solutions. This is clear from the two prolate solutions we have shown for Sr. If the higher 1s and 2d orbits had been included the more deformed solution would have been very much lower in energy.

It is interesting that for oblate solutions, gaps appear in the single particle HF spectra for N,Z = 32, 34 at small deformations (β~0.2) and for N,Z = 36 at large deformations (β~0.4). Kr is the most deformed nucleus on oblate side, and the reason is clear considering the orbits--and their

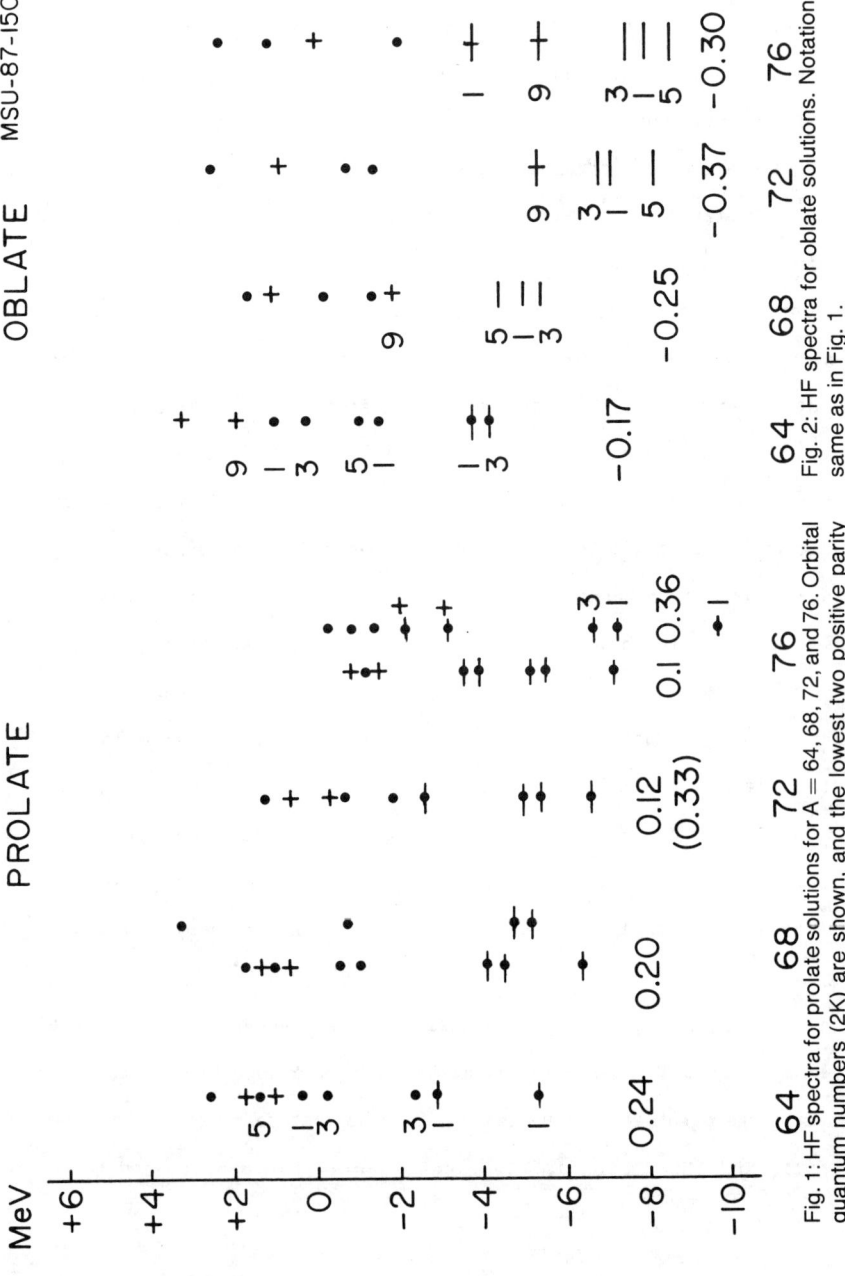

Fig. 1: HF spectra for prolate solutions for A = 64, 68, 72, and 76. Orbital quantum numbers (2K) are shown, and the lowest two positive parity orbits are marked with +. The values of B are shown below the HF spectra. The orbits occupied in the lowest HF solution are shown as solid bars.

Fig. 2: HF spectra for oblate solutions. Notations are the same as in Fig. 1.

intrinsic quadrupole moments--that are being fitted. Also for Se, one can see that a highly deformed band of states can be obtained at quite low energies by exciting nucleons to the first unoccupied orbit $k = 9/2^+$. The coexistence phenomena is thus easily explained.

On the prolate side, as we explained earlier, although a gap appears at $N,Z = 34$ for small deformations, the expected gap at $N,Z = 38$ fails to appear. For Se, a spherical state is almost degenerate with the intrinsic state with $\beta = 0.2$. For Kr and Sr highly deformed states with $\beta = 0.33$ and 0.36 can be obtained at low excitation energies, and would be ground states if the configuration space were larger. Thus one can again expect a coexistence of large and small deformations states in these nuclei.

Thus HF calculations explain a great deal of systematics and other features of nuclei in A=60-80 range. More detailed calculations of projection of good J states, band mixing and other properties are in progress.

References

1. A.K. Dhar, D.R. Kulkarni and K.H. Bhatt, Nucl. Phys. A238, 340 (1975).
2. D.P. Ahalpara, A. Abzouzi and K.H. Bhatt, Nucl. Phys. A445, 1 (1985).

Spectroscopy of the Transitional A=96-98 Nuclei; New Region of fast First Forbidden Transitions *

H. Mach[a], R. L. Gill[b], G. Molnár[b,c], R. F. Casten[b], A. Wolf[b,d], and J. A. Winger[e]

[a] Clark University, Worcester, Massachusetts, 01610
[b] Brookhaven National Laboratory, Upton, New York, 11973
[c] Institute of Isotopes, Budapest, H-1525, Hungary
[d] Nuclear Research Center Negev, Beer Sheva, Israel
[e] Ames Laboratory, Iowa State University, Ames, IA 50011

ABSTRACT

Absolute γ-intensities have been measured for the strong transitions in the decay chain ^{98}Rb $\to$ ^{98}Sr $\to$ ^{98}Y $\to$ ^{98}Zr and ^{98}Nb $\to$ ^{98}Mo and found ≈ 4 times higher than those currently adopted. Shape coexistance is established in ^{98}Y with a $J^\pi=1^-$ "spherical" ground state and with a previously reported deformed band with band-head at 495.7 keV. A fast first-forbidden β-transition is observed between ^{98}Y^g and ^{98}Zr^g, which alongside with the previously reported ^{96}Y^g $\to$ ^{96}Zr^g decay forms a new region, other than ^{208}Pb and ^{16}O, of such fast transitions.

INTRODUCTION

The transitional nuclei in the A=100 region provide a unique opportunity to study the deformation process taking place in close vicinity to stable nuclei and with structures amenable to shell model calculations. Of particular interest is the role of the highly overlapping $\pi(1g_{9/2})$ and $\nu(1g_{7/2})$ spin-orbit partners in the onset of deformation[1] and on the formation and dynamics of the intruder bands.[2,3] Crucial to the understanding of this region is a recent study[2] of the ^{96}Zr levels populated in the β-decay of the 0^- ^{96}Y ground state (gs) which has affirmed almost complete subshell closures (Z=40 and N=56) in ^{96}Zr in which there is very little ($\approx 4\%$) mixing of the spherical gs and the "deformed" first excited 0^+ states. More importantly it established the $0^- \to 0^+$ ^{96}Y^g to ^{96}Zr^g β-transition is one of the fastest first forbidden transitions known.[4] Such transitions are observed only in the regions of double shell closures in ^{16}O and ^{208}Pb.[5]

In this study we have measured the absolute γ- and E0- intensities and deduced β-branching ratios for the low-spin β decays in the A=98 nuclei. Of particular interest is the structure of the odd-odd ^{98}Y (N=59,Z=39) which is positioned in the center of the transitional region, neighbouring the nearly spherical[2] ^{96}Y and fully deformed[6] ^{100}Y isotopes. The coexistence of spherical

---

* The work has been performed under contracts DE-AC02-76CH00016 and DE-AC02-79ER10493 with the United States Department of Energy.

and collective structures is well established[7] in neighbouring even-even nuclei, and thus also expected at low excitation energy in ^{98}Y. A fully deformed band with a band-head at 495.4 keV has been already reported[8-9], and yet spherical structures arising from coupling of a valence neutron particle and a proton hole to the almost spherical ^{98}Zr core could not be probed without the crucial spin-parity assignment to the ground state, which remained uncertain. In particular, positive parity assignments based on strong β-feeding to both the ^{98}Y gs and the state at 600 keV conflicte with the E1 and M1 character of the 481 - 119 keV γ-ray cascade connecting these levels, which requires them to have opposite parities. Our new absolute γ-intensities in the A=98 nuclei are found $\approx$4 times higher than presently adopted[8] and resolve the above-mentiond ambiguity.

ABSOLUTE γ-INTENSITIES IN A=98 CHAIN

The measurement was performed at the TRISTAN mass separator of the fission products on line to the High-Flux-Beam-Reactor at Brookhaven National Laboratory. Details of those measurements are presented elsewhere.[10] Pure mass separated ^{98}Rb activity was continuously deposited for a period of time long in comparison to the half-lives of the key nuclei. γ-ray singles spectra were accumulated at the end of the irradiation process and γ-γ coincidences were measured with two Ge detectors.

We have identified lines from individual decays in the A=98 mass chain[8] and from A=97 nuclei[11] populated in the β-delayed neutron decay of ^{98}Rb. Relative γ-intensities compare closely to the values adopted for each decay scheme.[8,11] We obtained the absolute* γ-intensities for the strongest transitions in the A=98 chain by a direct comparison to the standard A=97 lines and assuming intensity balance within A=97 and 98 chains. The key result is that the β-branchings into excited levels are found to be $\approx$4 times higher than the adopted values[8] resulting in a substantial reduction in the gs β-feedings. The differences between our results and the adopted values can be traced to one number, namely the absolute intensity for the "standard" 787.6 keV ^{98}Mo line to which lines in the A=98 chain have been normalized[8] before. The new values have a strong impact on the resultant gs β-feedings discussed next.

Rb → Sr: ~50% of the beta intensity remains unaccounted[8] for by the level scheme of ^{98}Sr, and thus, we assume that a large fraction of it feeds the gs with logft $\approx$5 suggesting J^{π}=1$^+$ for the low spin isomer of ^{98}Rb.

Sr → Y: We have deduced the decay scheme of ^{98}Sr to ^{98}Y from our coincidence data and complemented by published results.[8,9,12] The transition multipolarities[9,12] are consistent with the requirements of the decay scheme and provide evidence that the parities of the gs and the above-mentioned 600 keV levels must be opposite. It is now in agreement with the new limit of $\leq$ 10% of β-branching into the gs of ^{98}Y deduced from the absolute total intensity for the known γ-transitions feeding the gs.

---

* The absolute intensity is expressed as the number of γ- or E0-transitions per 100 β-decays of the parent.

The 600 keV level in ^{98}Y is strongly populated in the β-decay of even-even ^{98}Sr with $\log ft$=4.3 which implies J^π=1$^+$. Thus on the basis of γ-ray transition multipolarities, the gs must have negative parity. Furthermore, it has spin J=1, since it decays by fast first-forbidden transitions, with $\log ft \approx 6$, to a number of 0$^+$ and 2$^+$ levels in ^{98}Zr. Moreover, the transition to the 0$^+$ "spherical" gs of ^{98}Zr is particularly fast. The $\log ft$ of 5.8 (based[8] on $t_{1/2}$=0.65 s) or 5.7 (if $t_{1/2}$=0.55 s, see ref. 13) for this first forbidden transition is lower than the accepted[4] lower limit of 5.9. Such fast first forbidden transitions occur in the vicinity of doubly magic nuclei where there is a strong overlap between pure spherical configurations.[5] Indeed there is substantial evidence[2] to consider ^{96}Zr and perhaps even[3] ^{98}Zr, as almost doubly magic nuclei; in particular, one of the fastest $0^- \to 0^+$ transitions with a $\log ft$=5.6, is observed[2] in the ^{96}Y$^g \to$ ^{96}Zr^g decay. The fast β-decay transition indicates that the ^{98}Y gs arises from a simple coupling of one neutron particle and one proton hole to the "spherical" even-even ^{98}Zr core, and thus, can be also interpreted as "spherical". The 39-th proton in the $\pi 2p_{1/2}$ orbit[1] can only couple with an odd neutron in either the $3s_{1/2}$ or $2d_{3/2}$ orbit to form a J^π=1$^-$ state. However, the $3s_{1/2}$ neutron must be excluded as the main component of the β-decay since it leads to an empty $3s_{1/2}$ subshell contrary to the expectation[3] that the almost magic character of the ^{98}Zr g.s. structure is due to the $3s_{1/2}$ subshell closure.

Y $\to$ Zr: The absolute intensity for the E0 854.0 keV line in ^{98}Zr can be now revised from the previously adopted[8] value of 4.9(18) to the new value of 15(3), which essentially removes the inconsistency of the earlier result with the value of 26(5) deduced[14] from the E0 conversion electrons measured in the ^{235}U(n,f) reaction. The new value derived here was obtained from the ratio[8] of intensities for the E0 854.0 keV and 1222.8 keV γ-lines in ^{98}Zr and the new absolute intensity of the latter line.

The new γ- and E0-intensities alter the important relative β-feeding to the 0$^+$ gs and the 0$^+$ first excited state in ^{98}Zr as well as the associated $\log ft$ values. The new values are $\log ft$=5.8 (^{98}Zr^g) and $\log ft$=6.2 (^{98}Zr*). The gs is a spherical configuration dominated by proton $(p_{1/2})^2$ component relative to Z=38 while the excited 0$^+$ state is a more collective state with a large $\pi(g_{9/2})^2$ amplitude. Since the β-decay matrix element connecting the odd $d_{3/2}$ neutron in ^{98}Y with a $p_{1/2}$ proton is large while transition to a $g_{9/2}$ proton is highly hindered, the β-feeding of the 0_2^+ state is due to an admixture of the gs configuration. Consequently, from the ratio of the comparative half-lives, ft, approximately equal to the square of the amount of mixing, and one deduces a 30% admixture of the spherical configuration in the deformed 0_2^+ state.

Nb $\to$ Mo: The absolute intensity of the "standard" 787.4 keV γ-line in ^{98}Mo were found to be 13(1) in contrast to earlier adopted[8] value of 3.2(5). Furthermore, the ratio of E0 intensities[8] for the 854.0 keV (^{98}Zr) and 734.6 keV (^{98}Mo) lines, 0.56(6), leads to the new absolute intensity of 26(6) for the latter line in agreement with $I_{ABS} \approx 30$ deduced in Ref. 14, but different than the adopted[8] value of 5.5(11). We have made a direct measurement of the absolute intensity of the E0 734.6 keV line using mass separated ^{98}Zr source. From a combination of β-γ coincidences and singles β-spectra we have deduced the value of 29(2) consistent with the results of our γ studies and ref. 14. The revised β-feedings to the 0_1^+ and 0_2^+ levels in ^{98}Mo give new logft values of 4.7 and 4.8, respectively. By analogous arguments as before, this suggests an almost equal mixing of those 0^+ levels provided the 1^+ gs of ^{98}Nb is a pure "deformed" configuration. The configuration mixing calculations of Sambataro and Molnar[15], who predicted a 40/60% mixture of the involved 0^+ configurations, are consistent with those results.

CONCLUSIONS

New absolute γ-intensities for the A=98 mass chain have been found higher than the adopted values by a factor of $\approx$4. The difference can be traced to the absolute intensity for 787.6 keV ^{98}Mo line. The new results alter the β-feeding branches to gs and excited states for A=98 nuclei and thereby the experimentally deduced mixing of spherical and deformed states. These results provide solutions to a number of outstanding discrepancies in these key transitional nuclei and suggest that the gs of ^{98}Y is the "spherical" $(\pi 2p_{1/2}\nu 2d_{3/2})1^-$ configuration and that it coexists with the previously reported[8-9] deformed band at 495.7-keV. The fast first-forbidden β-transition observed in the decay of ^{98}Y^g → ^{98}Zr^g, alongside the previously reported[2] ^{96}Y^g → ^{96}Zr^g decay, reinforces the view of the $^{96-98}$Zr region as a new region, besides ^{208}Pb and ^{16}O, where such fast transitions occur.

REFERENCES

1. P. Federman and S. Pittel, *Phys. Rev.* **C20**, 820 (1979).
2. H. Mach et al., preprint, to be published.
3. R.A. Meyer et al., *Phys. Lett.* **B177**, 271 (1986).
4. S. Raman and N.B. Grove, *Phys. Rev.* **C7**, 1995 (1973).
5. J. Damgaard, R.Broglia, and C. Riedel, *Nucl. Phys.* **A135**, 310 (1969).
6. F.K. Wohn et al., preprint to be published.
7. see for example: F.K. Wohn et al., *Phys.Rev.* **C33**, 677 (1986).
8. H.-W.Muller, *Nucl.Data Sheets* **39**, 467 (1983).
9. G. Lhersonneau et al., in *Recent Advances in the Study of Nuclei off the Line of Stability*, Eds. R.A.Meyer and D.S.Brenner (ACS Symposium Series 324, Washington, 1986) p. 202.
10. H. Mach and R. L. Gill, preprint, to be published.
11. B. Haesner and P. Luksch, *Nucl. Data Sheets* **46**, 607 (1985).
12. K. Kawade et al., *Z.Phys.* **A304**, 293 (1982).
13. P. L. Reeder et al., in *Recent Advances in the Study of Nuclei off the Line of Stability*, Eds. R.A.Meyer and D.S.Brenner (ACS Symposium Series 324, Washington, 1986) p. 171.
14. B.Fogelberg, *Phys.Lett.* **37B**, 372 (1971).
15. M. Sambataro and G. Molnár, *Nucl. Phys.* **A376**, 201 (1982).

EVIDENCE FOR SHAPE COEXISTENCE IN NEUTRON-RICH Rh AND Ag ISOTOPES

N. Kaffrell, J. Rogowski, H. Tetzlaff, N. Trautmann
Institut für Kernchemie, Universität Mainz, D-6500 Mainz, Germany

D. De Frenne, K. Heyde, E. Jacobs
Laboratorium voor Kernfysica, Proeftuinstraat 86,
B-9000 Gent, Belgium

G. Skarnemark
Department of Nuclear Chemistry, Chalmers University of Technology,
S-41296 Göteborg, Sweden

J. Alstad
Department of Chemistry, University of Oslo, N-0315 Oslo, Norway

M.N. Harakeh, J.M. Schippers, S.Y. van der Werf
Kernfysisch Versneller Instituut, Rijksuniversiteit,
NL-9747 AA Groningen, The Netherlands

W.R. Daniels, K. Wolfsberg
Los Alamos National Laboratory, Los Alamos, New Mexico 87545, USA

ABSTRACT

Neutron-rich nuclei of Rh and Ag have been investigated by γ-ray single and $\gamma\gamma(t)$-coincidence measurements after the β^--decay of their precursors. The Ru and Pd activities for the decay studies were produced in the ^{249}Cf(n_{th},f) reaction and separated from the fission product mixture using on-line chemical procedures. In addition, single-particle transfer reactions like (d,^{3}He) and (^{3}He,d) were performed as far as stable target material has been available.
The level schemes of ^{105,107,109}Rh and ^{113,115}Ag that were obtained clearly show two different types of excitation: (I) proton-hole states, strongly excited in the pick-up reaction, related to a more or less spherical shape of the nucleus, and (II) proton-particle states with a rotational band-like structure pointing towards deformation of the nucleus. In the case of ^{109}Rh a deformation parameter $\beta=0.32$ could be determined. The latter excitations, observed with high cross sections in the stripping reaction, may be interpreted as a rotational band built upon the $1/2^+$[431] Nilsson configuration. The systematics of this shape coexistence in the investigated mass region are discussed.

INTRODUCTION

The phenomenon of shape coexistence in medium-heavy and heavy odd-mass nuclei has recently been reviewed in detail by Heyde et al.[1]. For the odd-mass In (Z=49) and Ag (Z=47) nuclei, it was shown that the $2d_{5/2}$ and/or $1g_{7/2}$ shell-model states intrude across the

Z=50 shell closure giving rise to a rotational-like positive-parity band with $J^\pi=1/2^+, 3/2^+...$(intruder band), coexisting with spherical hole states ($1g_{9/2}$, $2p_{1/2}$, $2p_{3/2}$, $1f_{5/2}$) and $1g_{9/2}$ and $2p_{1/2}$ core coupled configurations. In the In nuclei, where the most extensive spectroscopic information is available, an interpretation of these intruder bands as decoupled bands built on the $2d_{5/2}$ and $1g_{7/2}$ orbitals with some degree of mixing is favoured[1,2] over a description as a single deformed band[3] built on the $1/2^+$[431] Nilsson orbital.

For the Ag isotopes, the data are less complete and especially nothing is known on the more neutron-rich nuclei including the mid-shell region at neutron number N≈66 where a maximum quadrupole deformation, i.e. lowest energy for these intruder bands, is expected on the basis of the residual proton-neutron interaction. In odd-mass Rh (Z=45) isotopes these intruder bands are not known yet at all. Here, the more deformed underlying cores may probably favour an identification and the interpretation as rotational states built on a single-Nilsson configuration. Hence, we have started a systematic investigation of neutron-rich Rh and Ag isotopes by post β^--decay γ-ray spectroscopic measurements and by single-particle transfer reactions like (d,^{3}He) and (^{3}He,d) with stable target material, where available.

EXPERIMENTAL PROCEDURES

For the study of ^{105}Rh, the 4.4 h ^{105}Ru was produced by the ^{104}Ru(n,γ) reaction. In all other cases the Ru and Pd activities have been produced by thermal neutron-induced fission of ^{249}Cf (≈300 μg) in the TRIGA Mainz reactor at a flux density of $6 \cdot 10^{11}$ neutrons/cm$^2 \cdot$s. A N$_2$/KCl gas jet system[4] was used to transport the fission products out of the reactor into the continuously working on-line chemical separation system[5] SISAK 2. The SISAK system is based on several consecutive liquid-liquid extraction steps and provides a continuous flux of activity of the separated element. A detailed description of the chemical separation procedures used for Ru and Pd has been given elsewhere[6,7]. The activities leaving the SISAK-system were pumped through a 30-ml polyethylene cell placed between the detectors. The average delay times from the target site to the counting position are now 4 s for Ru and 6 s for Pd activities.

The γ-rays following the β^--decay were studied by γ singles (including multispectrum analysis) and $\gamma\gamma$(t) coincidence measurements using different Ge X- and γ-ray detectors with active volumes of 2 - 120 cm^3 (energy resolution: 250 eV FWHM at 5.9 keV for the X-ray detectors, 1.8 keV FWHM at 1332 keV for the γ-ray detectors). More experimental details are given elsewhere[8-10]. As an example two γ-ray spectra taken in the study of the β^--decay of ^{109}Ru are presented in Fig. 1.

The single-particle pick-up and stripping reactions summarized in Table I were studied with 50 MeV deuteron or ^{3}He beams from the KVI cyclotron at Groningen. The targets consisted of 70-300 μg/cm^2 ^{102,104}Ru, ^{108,110}Pd or ^{114,116}Cd layers (enriched to 93-99% in the respective isotope) produced by sputtering the material on thin carbon backings of 20-35 μg/cm^2. The outgoing ^{3}He particles or deuterons

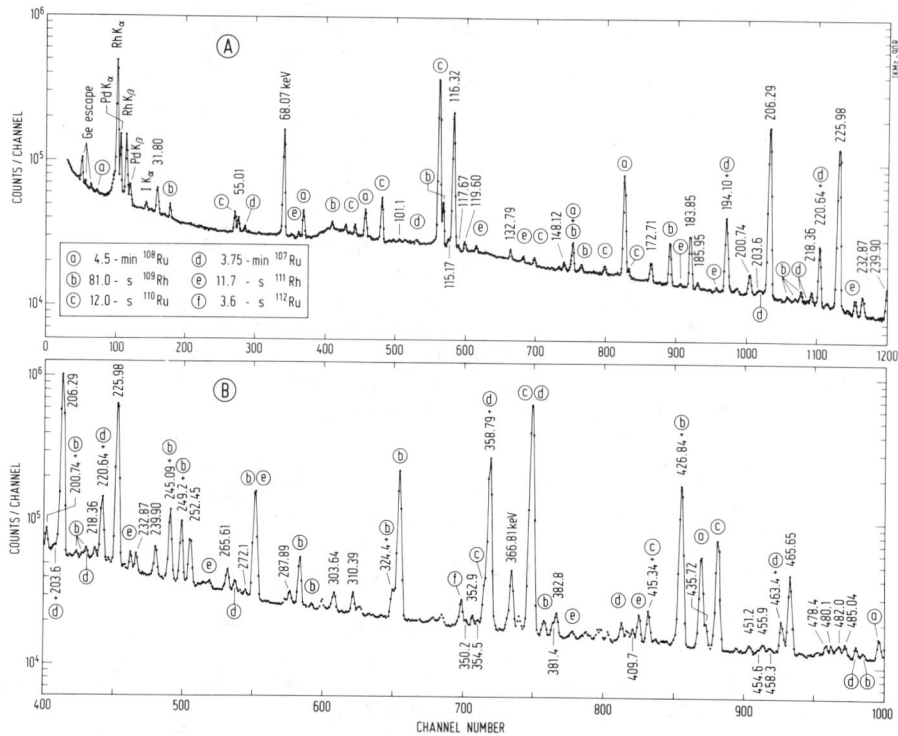

Fig. 1. Low-energy part up to 240 keV of the γ-ray spectrum of the ruthenium fraction measured with an X-ray detector (A), and the γ-ray spectrum in the energy region from 200 to 500 keV taken with a 32 cm^3 Ge detector (B). The energies of the γ-rays assigned to the decay of ^{109}Ru are given in keV. Other γ-rays are labelled in both spectra according to the explanation given in the insert.

Table I. Production mode of the activities and the nuclear reactions applied in the studies of levels in ^{103,105,107,109}Rh and ^{113,115}Ag.

Nucleus	Decay studies	Reaction studies
^{103}Rh	—	^{102}Ru(^{3}He,d)
^{105}Rh	^{104}Ru(n,γ)^{105}Ru(β$^-$)	^{104}Ru(^{3}He,d)
^{107}Rh	^{249}Cf(n,f)^{107}Ru(β$^-$)	^{108}Pd(d,^{3}He)
^{109}Rh	^{249}Cf(n,f)^{109}Ru(β$^-$)	^{110}Pd(d,^{3}He)
^{113}Ag	^{249}Cf(n,f)^{113}Pd(β$^-$)	^{114}Cd(d,^{3}He)
^{115}Ag	^{249}Cf(n,f)^{115}Pd(β$^-$)	^{116}Cd(d,^{3}He)

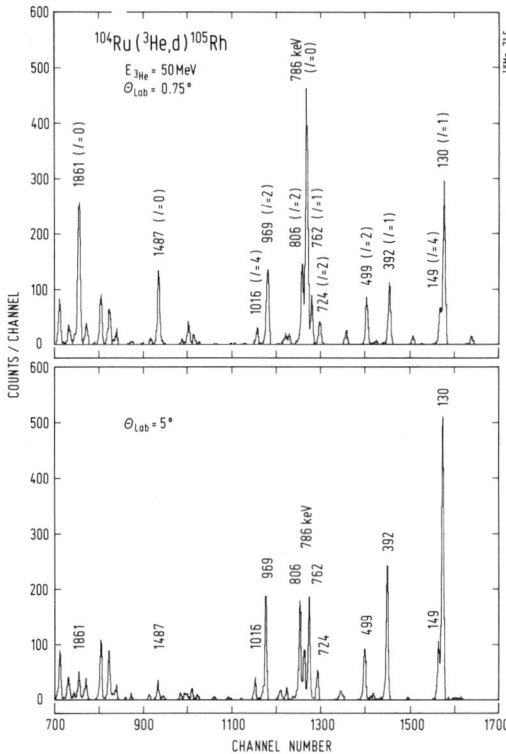

Fig. 2. Deuteron spectra taken at Θ_{Lab}=0.75 and 5° in the ^{104}Ru(^{3}He,d)^{105}Rh reaction at $E_{3_{He}}$=50 MeV. Excitation energies of levels in ^{105}Rh are given in keV.

were momentum analyzed with the QMG/2 spectrograph and detected using position sensitive detection systems in the focal plane of the spectrograph. Typical energy resolutions obtained were ≃30 keV for the (d,^{3}He) and ≃12 keV FWHM for the (^{3}He,d) reactions. As an example the deuteron spectra taken at 0.75 and 5° in the ^{104}Ru(^{3}He,d)^{105}Rh reaction are shown in Fig. 2. As the angular distributions for ℓ=0 transfers are strongly peaked in forward direction and as we were able to measure at very low angles, a straightforward identification of candidates for the J^π=1/2$^+$ head of an intruder band is already possible from these two spectra.

For the study of angular distributions, particle spectra were taken at several laboratory scattering angles ranging from 0.75-32.5° with $\Delta\Theta$=3.4-6.0°. The analysis of the spectra and the DWBA calculations were performed as described earlier[8-12].

RESULTS AND DISCUSSION

The complete level schemes of ^{105,107,109}Rh and ^{113,115}Ag deduced from our γγ coincidence and nuclear reaction studies can be found elsewhere[8-11]. For the spin and parity assignments, the ℓ-

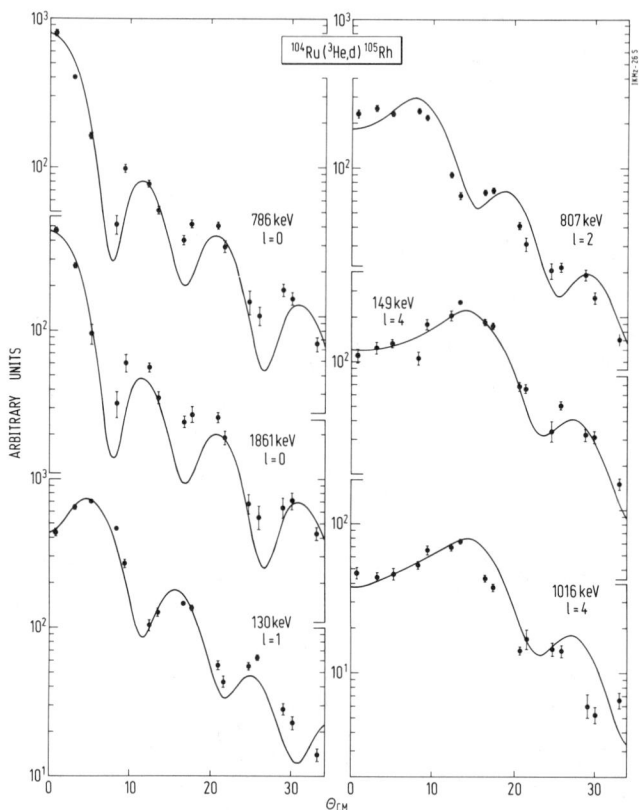

Fig. 3. Angular distribution of deuterons from the ^{104}Ru(^{3}He,d)^{105}Rh reaction at E_{3He}=50 MeV.

transfer results from the (d,^{3}He) and (^{3}He,d) reactions have been taken into consideration together with the analyzing power results of Flynn et al.[13], where available. As examples some selected angular distributions of deuterons from the ^{104}Ru(^{3}He,d)^{105}Rh reaction together with the one-step DWBA predictions are presented in Fig. 3. In addition, results from the β$^-$-decay studies, like selection rules for β- and γ-decay and a few conversion coefficients as well as the systematic trends of the levels in well-known neighbouring nuclei have been used.

In Figs. 4 and 5 the low-energy level systematics in ^{109,111}Ag (Refs.[1,13-15]) and ^{113,115}Ag and ^{103,105,107,109}Rh (present results) are shown. In Fig. 4, where the negative-parity states are presented, only a smooth change in the level energies occurs with increasing neutron number. This behaviour is also observed for the spectroscopic factors in the pick-up reactions. In all cases, the largest $2p_{3/2}$ and $1f_{5/2}$ single-particle strength is not carried by the first but second excited $3/2^-$ and $5/2^-$ levels, respectively. This favours an interpre-

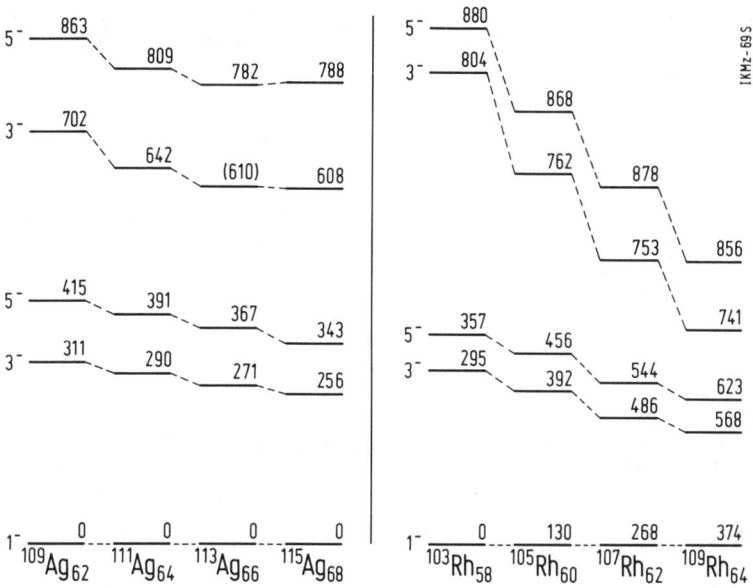

Fig. 4. Low-lying negative-parity states in odd-mass neutron-rich Ag and Rh isotopes. Level energies are given in keV; the indicated spin values are twice the actual values. Data are taken from Refs.[1,8-11,13-18].

tation of the first 3/2⁻ and 5/2⁻ levels as members of a multiplet based on the coupling of the first 2⁺ state in the even-even core to the $2p_{1/2}$ state. The theoretical description of the negative-parity states in the Rh isotopes in the framework of the interacting boson-fermion model[19] (IBFM-1) reveals a general agreement with the experimental results, although the calculated energies are higher and the energy spacing of the 3/2⁻, 5/2⁻ doublets is slightly underestimated.

For the positive-parity states, shown in Fig. 5, a similar smooth behaviour concerning excitation energies is observed for the first 7/2⁺ and 9/2⁺ levels in all nuclei. While the 9/2⁺ level contains about 30-50% of the $1g_{9/2}$ sum rule limit, the 7/2⁺ state is only very weakly excited in the pick-up reactions and can be considered as a core coupled configuration.

The main aim of the present study, however, is concerned with the positive-parity states (1/2⁺, 3/2⁺, 5/2⁺ and 7/2⁺) drawn as thick lines in Fig. 5, which represent possible candidates for an intruder band. They are getting lower in energy with increasing neutron number and reach an energy minimum in the Ag nuclei at exactly N=66, i.e. the middle of the N=50 and 82 shell closures.

From the 'fingerprints' for intruder states connected with shape coexistence outlined in Ref.[1] we will first discuss the band-like structure and the expected strong E2 transitions between the band members. The energies of the possible members of a collective rotational-like band with K=1/2 should fulfil the rotational formula

$$E_J = E_0 + A[J(J+1)+a(-1)^{J+1/2}(J+1/2)]. \qquad (1)$$

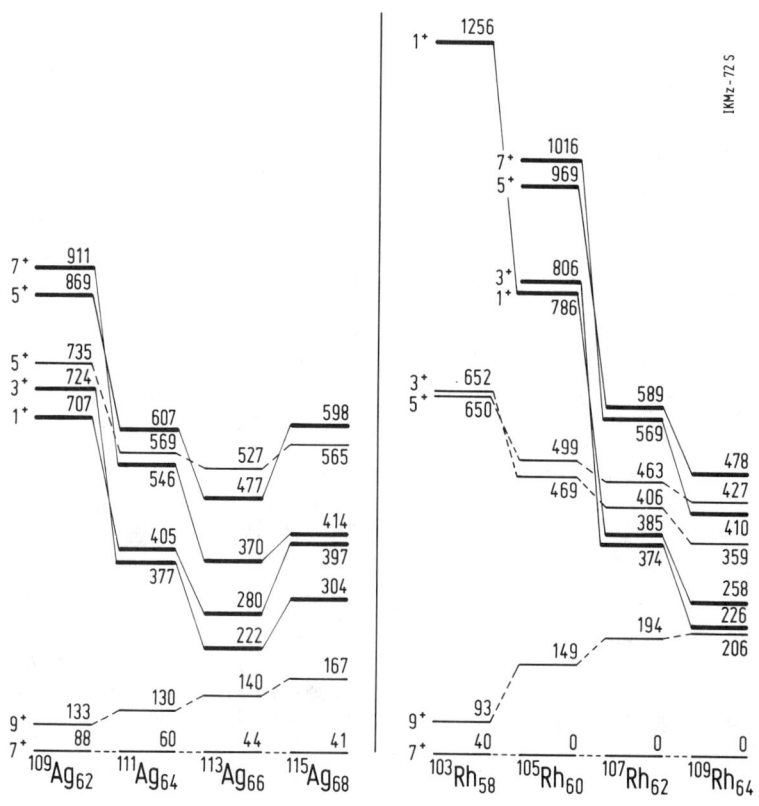

Fig. 5. Same as Fig. 4 for positive-parity states.

Table II. The parameter sets used to fit the rotational formula (1).

Nucleus	A/keV	a	E_0/keV
^{105}Rh	19.70	-0.66	758.1
^{107}Rh	19.92	-1.153	346.1
^{109}Rh	20.16	-1.498	211.9
^{113}Ag	17.08	-1.92	230.2
^{115}Ag	15.57	-2.76	336.7

Using the parameter sets given in Table II best agreement ($\Delta E \leq 1$ keV for the Rh and $\Delta E \simeq 4$ keV for the Ag nuclei) with all experimental energy values is obtained. Also the strong E2 intra-band transitions expected from this rotational band assumption are not in conflict with the observed decay pattern of these levels. This can clearly be demonstrated for the 31.8 keV transition (73%M1+27±4%E2) connecting the $1/2^+$ and $3/2^+$ members in ^{109}Rh. With the measured lifetime of 28.7 ns for the 258 keV level in ^{109}Rh an experimental B(E2) value of 5400 e^2fm^4 is obtained, which shows that the 31.8 keV E2 transition is strongly enhanced (174 Weisskopf units). Similar results for ^{113,115}Ag are reported by Fogelberg et al.[20].

From the experimental B(E2) value for the 31.8 keV transition in ^{109}Rh the intrinsic quadrupole moment Q_0 can be deduced and from this a deformation parameter $\beta=0.32\pm0.03$ results.

The second fingerprint is related to the expected retarded electromagnetic transitions between the coexisting states. The lifetime of the 226 keV level in ^{109}Rh has been determined to be 1660 ns and multipolarity E2 has been obtained via the α_K value for the transition from this level to the $7/2^+$ ground state. Thus, compared to the Weisskopf estimate a hindrance factor of $F_W=58$ is obtained for this E2 transition.

As expected, the possible members of the intruder bands are not or only very weakly excited in pick-up reactions. Unfortunately, single-particle stripping reactions are restricted to the lighter Rh and Ag nuclei, due to the lack of stable target material. However, as we could demonstrate for ^{105}Rh (see Fig. 2), the corresponding levels in this nucleus (see Fig. 5) are strongly excited in the ^{104}Ru(^{3}He,d) stripping reaction.

CONCLUSION

Combining the results from the decay studies and the single-particle transfer reactions we have obtained evidence for coexisting states in ^{105,107,109}Rh and ^{113,115}Ag. In addition to the more or less pure spherical shell-model states like the $1g_{9/2}$, $2p_{1/2}$, $2p_{3/2}$, and $1f_{5/2}$ configurations, we have observed in the same energy region levels with a rotational-like structure pointing towards a deformed nuclear shape. In the case of ^{109}Rh a deformation parameter $\beta=0.32$ related to the $1/2^+ \rightarrow 3/2^+$ transition was found. From the Nilsson diagram for protons in this region it is quite obvious that for such a deformation the $1/2^+[431]$ configuration is very close to the Fermi level for Z=45,47 nuclei.

ACKNOWLEDGMENTS

The authors are grateful to H. Folger, GSI Darmstadt, for preparation of the targets used in the nuclear reaction studies and to K.H. Gläsel for his help during the experiments at the Mainz reactor. We would like to thank the US Transplutonium Research Committee for making the ^{249}Cf available. Profs. A.J. Deruytter, G. Herrmann, A.C. Pappas and J. Rydberg are acknowledged for their interest in this work. This research was supported by the Bundesministerium für Forschung und Technologie (Germany), the Interuniversity Institute for Nuclear Sciences (Belgium), the Natural Science Research Council (Sweden), the Research Council for Science and Humanities (Norway), the Stichting voor Fundamenteel Onderzoek der Materie (FOM) with financial support from the Nederlandse Organisatie voor Zuiver Wetenschappelijk Onderzoek (ZWO) and the Department of Energy (USA).

REFERENCES

1. K. Heyde, P. Van Isacker, M. Waroquier, J.L. Wood and R.A. Meyer, Phys. Reports 102, 291 (1983).
2. K. Heyde, M. Waroquier and R.A. Meyer, Phys.Rev. C17, 1219 (1978).
3. W. Dietrich, A. Bäcklin, C.O. Lannegård and I. Ragnarsson, Nucl. Phys. A253, 429 (1975).
4. E. Stender, N. Trautmann and G. Herrmann, Radiochem. Radioanal. Lett. 42, 291 (1980).
5. G. Skarnemark, P.O. Aronsson, K. Brodén, J. Rydberg, T. Björnstad, N. Kaffrell, E. Stender and N. Trautmann, Nucl. Instr. Meth. 171, 323 (1980).
6. G. Skarnemark, K. Brodén, Mao Yun, N. Kaffrell and N. Trautmann, Radiochim. Acta 33, 97 (1983).
7. P.O. Aronsson, E. Ehn and J. Rydberg, Phys. Rev. Lett. 25, 590 (1970).
8. N. Kaffrell, P. Hill, J. Rogowski, H. Tetzlaff, N. Trautmann, E. Jacobs, P. De Gelder, D. De Frenne, K. Heyde, G. Skarnemark, J. Alstad, N. Blasi, M.N. Harakeh, W.A. Sterrenburg and K. Wolfsberg, Nucl. Phys. A460, 437 (1986).
9. N. Kaffrell, P. Hill, J. Rogowski, H. Tetzlaff, N. Trautmann, E. Jacobs, P. De Gelder, D. De Frenne, K. Heyde, S. Börjesson, G. Skarnemark, J. Alstad, N. Blasi, M.N. Harakeh, W.A. Sterrenburg and K. Wolfsberg, Nucl. Phys. A470, 141 (1987).
10. J. Rogowski, Diploma thesis, Universität Mainz (1985).
11. J. Rogowski, N. Kaffrell, D. De Frenne, E. Jacobs, K. Heyde, M.N. Harakeh, J.M. Schippers, S.Y. van der Werf and H. Folger, Institut für Kernchemie, Universität Mainz, Ann. Report 1986, p. 8 (1987).
12. P. De Gelder, D. De Frenne, K. Heyde, N. Kaffrell, A.M. van den Berg, N. Blasi, M.N. Harakeh and W.A. Sterrenburg, Nucl. Phys. A401, 397 (1983).
13. E.R. Flynn, F. Ajzenberg-Selove, R.E. Brown, J.A. Cizewski and J.W. Sunier, Phys. Rev. C24, 902 (1981); C25, 2851 (1982); C27, 2587 (1983).
14. R.E. Anderson, J.J. Kraushaar, I.C. Oelrich, R.M. DelVecchio, R.A. Naumann, E.R. Flynn and C.E. Moss, Phys. Rev. C15, 123 (1977).
15. S.Y. van der Werf, B. Fryszczyn, L.W. Put and R.H. Siemssen, Nucl. Phys. A273, 15 (1976).
16. J. Blachot, Nucl. Data Sheets 41, 111 (1984).
17. D. De Frenne, E. Jacobs and M. Verboven, Nucl. Data Sheets 45, 368 (1985).

18. N.K. Aras and W.B. Walters, Phys. Rev. $\underline{C11}$, 927 (1975).
19. J. Jolie, P. Van Isacker. K. Heyde, J. Moreau, G. Van Landeghem, M. Waroquier and O. Scholten, Nucl. Phys. $\underline{A438}$, 15 (1985).
20. B. Fogelberg, E. Lund, Y. Zongyuan and B. Ekstöm, Contribution to this Conference.

TRANSITION PROBABILITIES BETWEEN INTRUDER STATES IN HEAVY Ag ISOTOPES

B. Fogelberg, E. Lund, Ye. Zongyuan[1], and B. Ekström
The Studsvik Neutron Research Laboratory
S-61182 Nyköping, Sweden

ABSTRACT

The level schemes of ^{113,115}Ag as observed in the decays of Pd isotopes have been established using γ-ray and electron spectroscopy on mass separated samples of Pd. A set of low lying levels of intruder type was observed for both nuclei. The E2 transitions between intruder states were found to be enhanced by a factor of about 10^2 which clearly indicates shape coexistence in these nuclei. In addition we have studied the systematics of the normal states and the B(E3) of the isomeric transitions in ^{113,115,117}Ag. These latter data suggest that collective effects become increasingly more important also for the normal states of the heavy odd mass Ag nuclei.

INTRODUCTION

The properties of the low energy levels of the neutron rich Ag isotopes are of relevance for several facets of nuclear structure, such as the properties of the proton three quasi-particle states including the much discussed J-1 anomaly (see e.g. Ref.[1]), the occurrence and properties of intruder states, and the strength of the p-n interaction in comparison to pairing[2,3]. The phenomenon of intruder states have been extensively discussed[4] by Heyde and coworkers, who predict that such levels of proton nature should be found at particularly low energies near the mid-point of a neutron shell, which for the isotopes of Ag corresponds to A=113. The expected low excitation energies are naturally greatly facilitating a detailed experimental study of the intruder levels.

The level structure of ^{113,115}Ag was practically unknown when the current experiments were initiated. During the course of the work we have become aware of a similar study performed by radio-chemical methods by the Mainz group. The results[5], with regard to the level structure of ^{113,115}Ag, are in good agreement with our data.

EXPERIMENTS

All measurements were made using mass separated sources from the OSIRIS[6] on-line facility for fission products. The recently developed[7] high temperature plasma ion source has a good efficiency for Pd and also for the isobaric activities of Ag. A rather compre-

---

[1] Present address: Institute of Atomic Energy, P.O. Box 275-60 Beijing, China

TABLE 1

Conversion coefficients for transitions in ^{113,115}Ag

E_γ (keV)	ICC exp.			Multi-polarity
^{113}Ag 43.6	K	90	± 40	E3
	L	700	± 300	
95.7	K	0.46	± .04	M1
147.7	K	0.38	± .15	E2
222.1	K	<0.03		E1
^{115}Ag 41.1	K	>40		E3
	L	>300		
87.2	K	0.63	± .07	M1
92.7	K	2.1	± 1.1	E2
110.4	K	0.6	± .3	M1 or E2
125.5	K	0.21	± .02	M1
255.5	K	0.035	± .004	M1
303.9	K	<0.009		E1
342.7	K	0.017	± .003	M1 or E2

hensive series of photon measurements were made using Ge spectrometers of both coaxial and LEP types. Some examples of the data obtained are shown here in Figs. 1 and 2, but a full account of the β-decay properties of the Pd isotopes, including tables of γ-ray data will be given in a forthcoming publication.

The conversion coefficients of a number of low energy transitions were determined by simultaneous measurements of electrons and γ-rays using respectively a small ion implanted Si detector and a LEP spectrometer. A brief summary of the results regarding conversion coefficients is given in Table 1, and an example of an electron spectrum is shown in Fig. 3. The conversion electron data proved to be of vital importance for the proper characterization of the low lying levels in the Ag nuclei and also

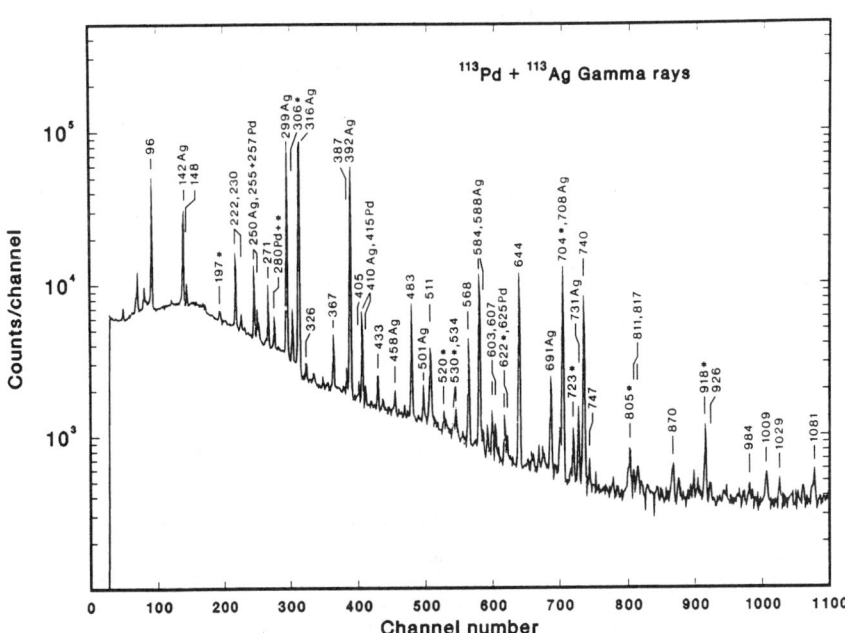

Fig. 1. A portion of a γ-ray spectrum obtained for the A=113 isobars.

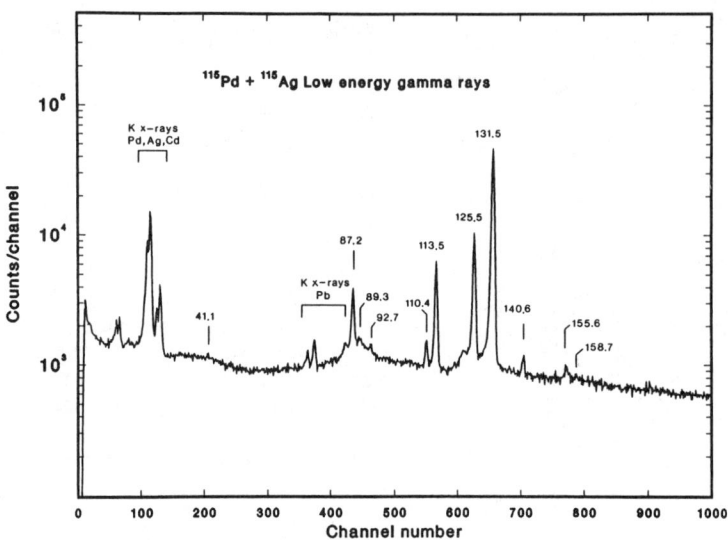

Fig. 2. A LEP detector spectrum of the A=115 isobars. The 113.5, 131.5 and 155.6 keV γ-rays follow the decay of ^{115}Ag.

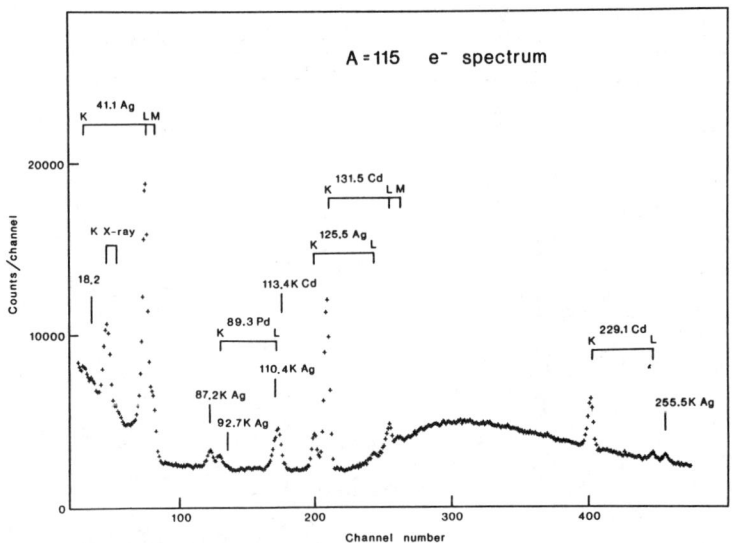

Fig. 3. An electron spectrum of the A=115 isobars recorded with a room temperature Si detector. Transitions are labelled according to the nucleus in which they proceed. The complex line near ch. 180 was resolved by comparing the current data with a spectrum taken using a low temperature ion source. The latter spectrum is more simple since the contribution from Pd activity is negligible.

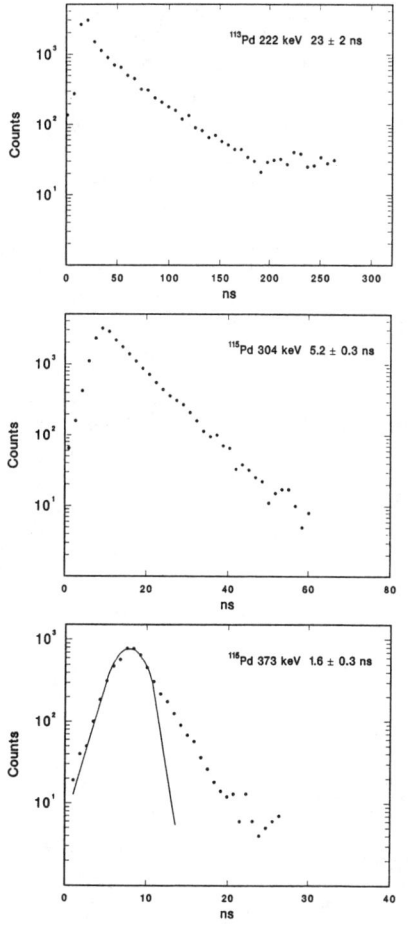

Fig. 4. Examples of delayed coincidence curves. The shape of the prompt distribution is indicated in one case.

for the study of isomeric transitions and their branching ratios. This latter quantity is necessary in order to derive the transition probabilities of the isomeric transitions. The rates of the $7/2^+ \rightarrow 1/2^-$ E3 transitions in the Ag isotopes are of special interest as discussed later. A separate measurement was therefore made also for the A=117 isobars in order to search for the previously unknown isomeric transition in ^{117}Ag. The transition was found at an energy of 28.6 keV with a branching of (6±1.5)%/d.

The half lives of the intruder states were measured with the βγ delayed coincidence method using a plastic scintillator in the start channel and a planar HPGe detector to provide stop signals. Half lives or significant limits of half lives were obtained for most of the intruder states by a straight forward analysis of the delayed coincidence time distributions, of which some are shown in Fig. 4. The very weak population of the 273.8 keV level in ^{113}Ag precluded an accurate determination of its half life. A study of

TABLE 2

Level half lives measured in ^{113,115}Ag

Level energy (keV)		half life (ns)	Level energy (keV)		half life (ns)
^{113}Ag	222.1	23 ± 2	^{115}Ag	303.9	5.2 ± 0.3
	273.8	≈ 30[a]		396.6	0.8 ± 0.3
	369.8	< 0.8		414.2	1.6 ± 0.3
	477.0	< 0.5		597.7	< 0.8

[a] The uncertainty is of the order of a factor of two.

the γ-ray intensities in various delayed gates indicates a value of the order of 30 ns, but the half life can be a factor of two different from this. The level half lives are given in Table 2.

RESULTS AND DISCUSSION

The decay schemes. Only one β-decaying state was found for ^{113}Pd. A unique $J^\pi = 5/2^+$ could be assigned to this state from its β-decay scheme. In the case of ^{115}Pd we observed two β-decaying levels connected with an isomeric E3 transition. The decay scheme allows $J^\pi = 3/2^+$ or $5/2^+$ for the ground state and $9/2^-$ or $11/2^-$ for the isomer of ^{115}Pd, with the latter of these alternatives being strongly favoured by empirical systematics and by the shell model. Turning to the J^π assignments of the levels of ^{113,115}Ag (see Figs 5,6) it is to some extent necessary to rely on the existing level

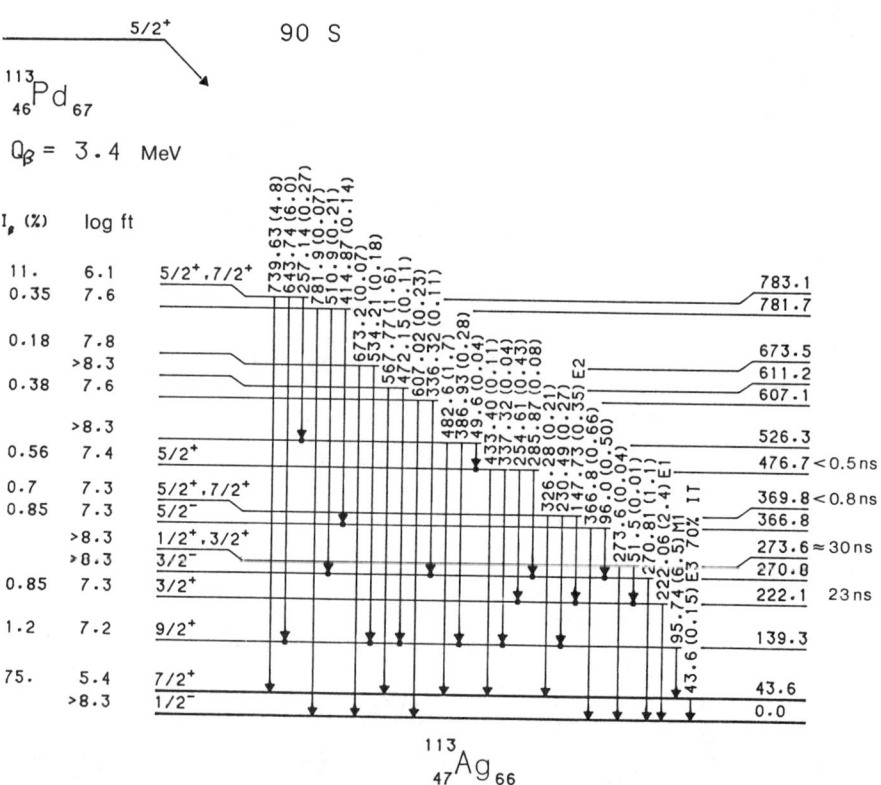

Fig. 5. Low energy levels of ^{113}Ag. Transition intensities are in units of %/d.

systematics of the lighter odd mass Ag isotopes, where the ground state has $J^\pi=1/2^-$ and the first few excited states are assigned $7/2^+$, $9/2^+$, $3/2^-$ and $5/2^-$. A set of analogous levels can readily be identified in ^{113,115}Ag, and the values of J^π are supported by the transition multipolarities deduced from our conversion electron data. The remaining assignments given in Figs 5,6 follow from the selection rules of the β- and γ-transitions.

The intruder levels. The systematics of levels of intruder type in $^{105-111}$Ag has been compiled by Heyde and co-workers[4], and shows four levels with $J^\pi=1/2^+$, $3/2^+$, $5/2^+$ and $7/2^+$. Our data suggests strongly that the corresponding levels in ^{113}Ag are found at 222, 274, 370 and 477 keV and in ^{115}Ag at 304, 397, 414 and 598 keV. It is, however, not possible to uniquely assign values of J^π which agree with the expectations for an intruder band. In ^{115}Ag, two of the levels appear to have $J^\pi=5/2^+$. The lower lying one of these, at

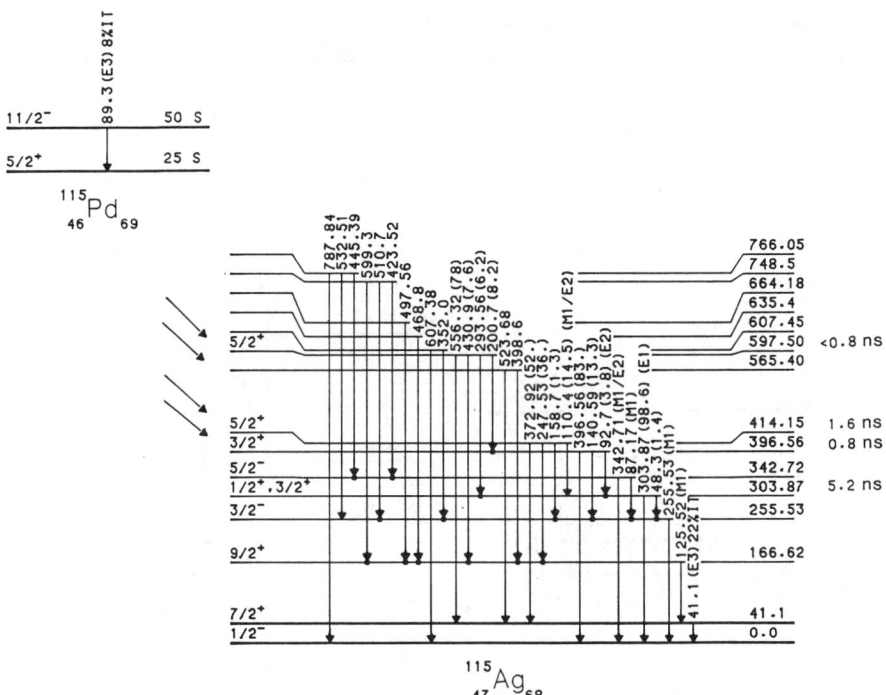

Fig. 6. Low energy levels of ^{115}Ag. The feeding of the levels is complex due to the two Pd isomers. Some strong β-transitions from the Pd ground state are shown as arrows. Relative intensities are given for the transitions depopulating some levels.

414 keV is uniquely assigned $5/2^+$ due to the presence of the 158.7 keV transition to the $3/2^-$ state. This transition is quite weak, so there is a finite probability that the observed coincidence with the 255.5 keV γ-ray is indirect and that the 158.7 keV transition consequently should be placed elsewhere in the level scheme. This would allow $7/2^+$ for the 414 keV level, which actually does not improve the situation, since then the strong 110.4 keV E2 transition to the 304 keV level requires this latter state to have $J^\pi=3/2^+$, giving us an intruder band with two $3/2^+$ levels. Apart from this possible deviation from the expected properties of an intruder band, the levels discussed above have properties which are typical for the coexistence of deformed states in spherical odd-mass nuclei. Our measurements of conversion electrons and level half lives also provide more accurate and more detailed information on the intruder bands in ^{113,115}Ag than have been gotten for such bands in other nuclei away from the stability line. The most important feature of our data is that the transition probabilities have been measured for several transitions with a well established E2 multipolarity. Some of the data on transition probabilities are given in Fig. 7, showing that the intra band transitions typically are enhanced by a factor of about 10^2, and that at least one intruder state in each nucleus decays to the normal $9/2^+$ level with a strength of a few SPU. The situation is thus very similar to what is found[4] in ^{115}In which

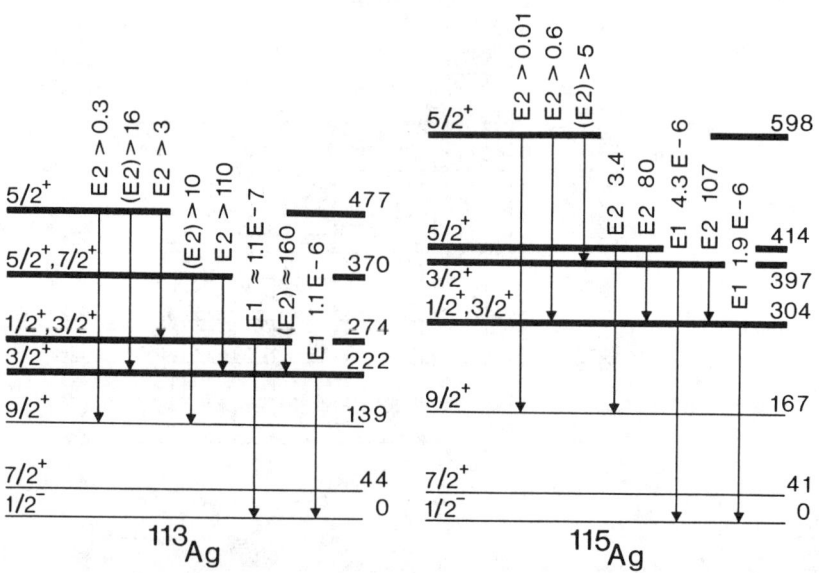

Fig. 7. A summary of the transition probabilities between and from the intruder states in ^{113,115}Ag. Transition multipolarities which are well established from the electron measurements or the level schemes are given without parenthesis.

otherwise is the best studied case in this mass region, and one may conclude that the properties of the coexisting intruder bands as opposed to the normal states, are as sharply defined in Ag as in In despite the higher complexity due to two additional proton holes.

It should be remarked that the suggestion[4] that also the ground states of the heavy Ag isotopes could be intruder configurations is not supported by the data on level systematics which is now at hand. The minimum excitation energy of the intruder band head is found in ^{113}Ag, precisely at the middle of the neutron shell, and the systematic trend of the excitation energies appears to be nearly parabolic around this minimum.

The normal states. The currently obtained information on the normal states of the Ag isotopes is rather scanty and consists largely of a determination of the positions of the $1/2^-$, $7/2^+$, $9/2^+$, $3/2^-$ and $5/2^-$ levels in ^{113,115}Ag and of the $1/2^-$ and $7/2^+$ levels in ^{117}Ag. We were also able to determine the hindrance factors of the $7/2^+ \to 1/2^-$ E3 transitions in $^{113-117}$Ag by measurements of the isomeric branching ratios. It is quite interesting to compare the empirical data to the detailed shell model calculations by Paar[1]. His treatment, which well reproduces several features including the J-1 anomaly, is based on a model in which a cluster of three valence nucleon holes in a shell model potential are coupled to the quadrupole vibrational field of the nucleus. It is noteworthy that an increased depression of the J-1 $7/2^+$ state below the $9/2^+$ level (see Fig. 2 of Ref.[1]) is obtained in the calculations as a function of the strength of the coupling to the vibrational field. Empirically, see Fig. 8, we find the same effect as a function of neutron number.

Fig. 8. (Top) Systematics of levels in the odd mass Ag isotopes. The lowest lying intruder level is indicated with a short heavy bar. (Bottom) Hindrance factors of the $7/2^+ \to 1/2^-$ transitions in Ag. Error bars are given for the values derived from the current experiments.

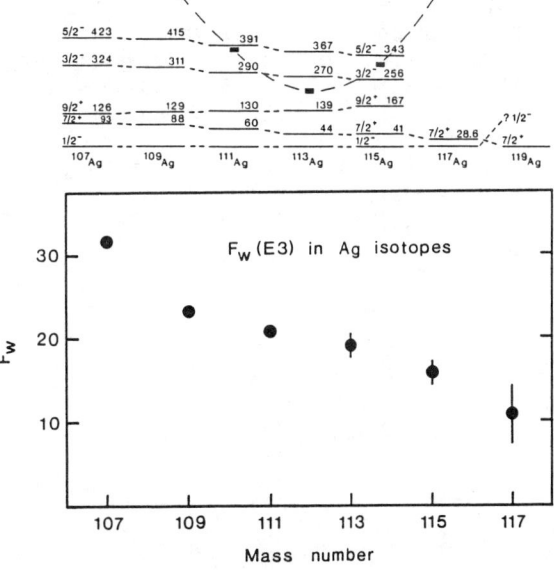

With the assumption that the Hamiltonian used by Paar at lest approximates the true situation, one would expect that the stronger coupling suggested by the lowering of the $7/2^+$ level also leads to an increased collectivity of the normal Ag levels. This would tend to relax the forbiddenness of the isomeric E3 transitions, as is also observed experimentally. It is most interesting to note that both the splitting of the $7/2^+$ and $9/2^+$ levels and the B(E3) continue to increase beyond the middle of the neutron shell at A=113. There are thus reasons to believe that the presumed increased coupling to collective degrees of freedom is not simply related to a softness near the middle of the shell. The depression of the $7/2^+$ level actually makes it the ground state in ^{119,121}Ag which are the heaviest isotopes for which reliable data exists, suggesting a significant quadrupole deformation[3] in the region near these nuclei.

An alternative explanation of the very low excitation energies of the $7/2^+$ states is based on the assumption[2] of a cancellation of the pairing energy, partly due to a strong coupling of the $\pi g_{9/2}$ and $\nu g_{7/2}$ levels. This point of view appears less likely with regard to the systematics of Fig. 8. It can also be remarked that a level likely to represent the 1^+ state formed by these orbitals state has been observed[8] by us as an excited state in ^{116}Ag, which implies that the attractive coupling of these orbitals is actually weaker in the Ag nuclei than in the doubly odd isotopes of In.

Further clues to the mechanism behind the increasing depression of the $7/2^+$ states can possibly be gotten from the transition probabilities of the $9/2^+ \rightarrow 7/2^+$ M1+E2 transitions. Studies of these transitions are now being planned by us.

1. V. Paar, Nucl. Phys. A211, 29(1973)
2. L.K. Peker, J. H. Hamilton and P.G. Hansen, Phys. Lett. 167B, 283(1986)
3. K. Heyde and V. Paar, Phys. Lett. 179B, 1(1986)
4. K. Heyde, P. van Isacker, M. Waroquier, J.L. Wood and R.A. Meyer, Physics Reports 102, 291(1983)
5. N. Kaffrell, private communication and contribution to this conference.
6. G. Rudstam, Nucl. Instr. and Meth. 139, 239(1976)
7. L. Jacobsson, B. Fogelberg, B. Ekström and G. Rudstam, Nucl. Instr. and Meth. B26, 223(1986)
8. Ye Zongyuan et al. (to be published)

SHAPE COEXISTENCE IN THE REGION AROUND Z=82

M. Huyse, E. Coenen, K. Deneffe, P. Van Duppen
LISOL, University of Leuven, B-3000 Leuven, Belgium

ABSTRACT

A survey of the experimental work done at the Leuven Isotope Separator on Line facility on intruder-state systematics and shape coexistence in the region around Z=82 is given. The excitation energy of the intruder based states in the odd Bi, odd Tl, odd-odd Tl and even-even Pb nuclei show, in function of the neutron number, a remarkable coinciding behavior. This indicates the general applicability of the scaling law near the Z=82 proton closed shell.

1. INTRODUCTION

In many odd-mass nuclei with one or three valence particles (holes) outside a single closed shell and with a maximal number of valence nucleons of the other type, intruder states occur at very low excitation energy, compared with the regular single-particle or single-hole excitations. Detailed studies on odd-mass nuclei have been carried out throughout the nuclear mass table, especially in the Z=50 (Ag, In, Sb, I) and Z=82 (Au, Tl, Bi) regions[1]. It was shown both experimentally and theoretically that such intruder states are mainly due to particle hole (p-h) excitations across the closed shell.

Also, in even-even nuclei in the Z=50 region, low-lying intruder states have been observed, with the $J^\pi=0^+$ rotationallike bands in $^{112-118}$Sn as the most conspicious examples[2].

It is at LISOL (Leuven Isotope Separator On Line) that similar intruder states in neutron-deficient even-even Pb nuclei, were observed for the first time[3] and studied in great-detail[4,5,6]. Also further information on odd-mass nuclei around Z=82 was obtained: the observation of hindered and unhindered α decay of the odd-mass Bi nuclei through the Z=82 shell closure towards states in Tl has proven to be an extremely powerful method to identify intruder states[7]. Finally the same method has also been used to study intruder states in odd-odd Tl nuclei[8]. A large set of experimental data on intruder-state energies in these nuclei is now available from the neutron-shell closure at N=126 down to the neutron mid-shell at N=104. This has led to potential energy surface calculations[3,9] within the Nilsson model, including different coexisting shapes, and to calculations within the shell model, incorporating explicitly the proton-neutron residual interaction[10]. More details about the different theoretical approaches can be found in a contribution of K. Heyde to this conference[11]. We will concentrate here on the experimental evidence on intruder states in the Pb region that comes out of the work at LISOL.

2. EXPERIMENTAL PROCEDURES

The experimental results which will be reviewed here, are all coming from β^+/EC and α-decay studies of mass-separated neutron-deficient Bi and Po nuclei produced in heavy ion fusion reactions. Experimental details on the LISOL separator can be found in ref. 12. Table I lists the used reactions.

Table I. List of the used reactions to produce neutron-deficient Bi and Po isotopes

target	beam	mass-separated nuclei
5mg/cm^2 ^{nat}Ir	≤ 127 MeV ^{14}N	^{198,200}Po , ^{198,200}Bi
16mg/cm^2 ^{nat}Re	≤ 210 MeV ^{16}O	^{193,194,195,196,197}Bi
12mg/cm^2 ^{nat}W	≤ 225 MeV ^{20}Ne	^{94,196}Po
8mg/cm^2 ^{181}Ta	≤ 230 MeV ^{20}Ne	^{191,192}Bi
5mg/cm^2 ^{165}Ho	≤ 280 MeV ^{32}S	^{189,190}Bi

In order to obtain an optimum production rate for the different nuclei the beam energy was degraded by putting Ta-degrader foils (2.5 μm) in the beam just in front of the target.
The activities so produced were mass separated and implanted into an aluminized Mylar tape which periodically moved the activity.
Singles γ spectra were taken with two Ge-detectors with a resolution of 2.0 keV and a relative efficiency of 20% at the 1332.5 keV line of ^{60}Co. Singles x-ray spectra were accumulated with a Lege-type (Low Energy Germanium) detector. This detector had an active surface of 1500 mm^2 and a resolution of 580 eV for the 122 keV line of ^{57}Co. Calibration sources were used for energy and intensity calibration of these detectors. A Si(Li) detector (KEVEX-type, thickness=5mm) was used to detect conversion electrons. For the 624 keV e$^-$ line of ^{137}Cs a resolution of 2.5 keV was obtained. This e$^-$ detector faced the radioactive source directly at a variable distance between 3 and 6 cm. Energy and intensity calibrations were performed using strongly converted transitions with known conversion coefficients. Multi-scaled α-singles spectra were taken with a Si surface-barrier detector (450 mm^2 - 500 μm, resolution 19 keV for the 5.486 MeV α-line in ^{241}Am).

3. RESULTS

3.1. The even-even Pb nuclei

Due to the existence of a low spin β^+/EC-decaying state in ^{196}Bi, fed up to 40% by isomeric decay of higher spin states, it was easy to observe an E0 transition, giving evidence for an anomalous low-lying

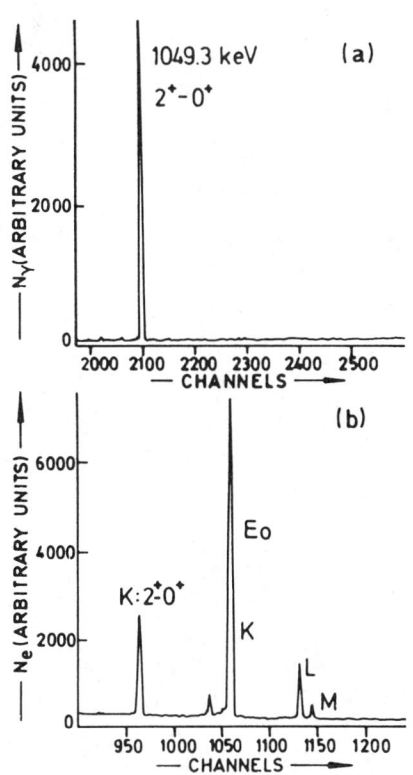

Fig. 1. *Comparison between (a) the gamma, and (b) the electron spectra at mass A=196.*

0^+ state in ^{196}Pb only 94 keV abouve the first excited 2^+ state[3].
The surprisingly low lying 0^+ state deexcites probably completely to the ground state, giving a strong electron signal in an almost background-free region (see Fig. 1).
It was also possible to observe similar E0 transitions in the nuclei ^{192,194,198,200}Pb[3,4]. The corresponding 0^+ states are at 1625.5 ± 0.5 keV (^{200}Pb), 1392.0 ± 0.2 keV (^{198}Pb), 930.1 ± 0.2 keV (^{194}Pb); and 768.5 ± 0.2 keV (^{192}Pb) excitation energy, becoming for ^{194}Pb and ^{192}Pb the first excited state. By using e^--γ and γ-γ coincides it was possible to construct level schemes for ^{192,194,196}Pb[6]. As an example, a part of the ^{196}Pb level scheme is shown in figure 2. A cascade of two strong γ-rays (enhanced stretched E2's) gives strong evidence for the beginning of a $\Delta I=2$ band built upon the 0_2^+ state. We could observe also such structure in ^{192,194}Pb.

The above mentioned method, to observe the 0^+ states by conversion electron spectroscopy following the β^+/EC decay of the odd-odd Bi-isotopes cannot be used easily for A ≤ 190. This is because of increasingly strong α-decay branches in the neutron-deficient Bi isotopes and the decreasing production yields. Another method has been developed: it consists in studying the α decay of the even-even Po nuclei. It has been possible to confirm the 769 keV 0^+ level in ^{192}Pb and to locate an excited 0^+ state in ^{190}Pb at 669 keV[5].

In Fig. 3 all information on 0^+ states is collected. The work from LISOL spans from ^{200}Pb down to ^{190}Pb. The candidates for intruder states are connected with a dotted line (for more details see ref. 6).

3.2 The odd-mass Bi and Tl nuclei

By studying the β/EC and α-decay of the $\pi h9/2$ groundstates in ^{189,191,193,195}Bi it has been possible to prove the suggestions

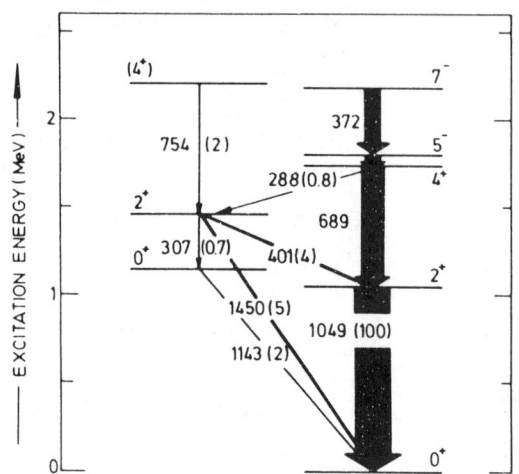

Fig. 2. A part of the ^{196}Pb level scheme. The numbers in parantheses are relative γ-ray intensities.

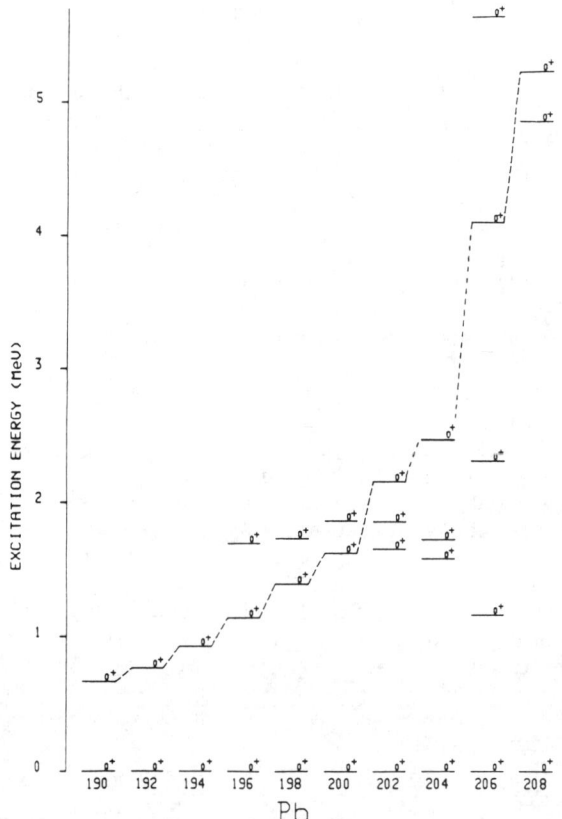

Fig. 3. The excitation energy of the 0^+ intruder states in the $^{190-208}$Pb nuclei. The candidates for intruder states are connected with a dotted line. References to other work ($A \geq 202$) can be found in ref. 6 and 13.

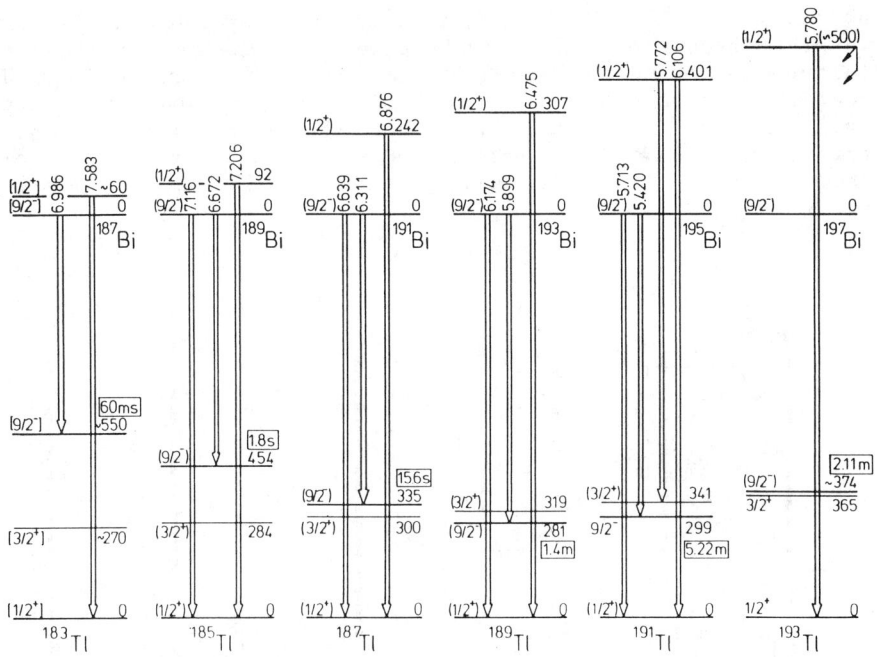

Fig. 4. Low energy systematics in the odd-mass Tl isotopes and the α-decay schemes of the odd-mass Bi isotopes. For further details see ref. 7.

made by Vakthel et al[14] concerning the placement of the different α lines[7]. The decays are typically characterized by unhindered α decay for the ^{A}Biπs1/2 intruder → $^{A-4}$Tlπs1/2 groundstate and ^{A}Biπh9/2 ground- → $^{A-4}$Tlπh9/2 intruder state transitions and by strongly hindered α decay for the ^{A}Biπh9/2 ground- → $^{A-4}$Tlπs1/2 groundstate transitions. These decay branches are given in fig. 4.

The systematics of the Tl and Bi intruder state energies are shown in fig. 7.

An important conclusion from this work[7] is that α-decay studies and the determination of absolute α-hindrance factors provide strong spectroscopic fingerprint for the intruder states in the odd-mass Tl and Bi isotopes. In many of the cases here, it is the only way to determine the intruder-state excitation energies. This is because the $1/2^+$ → $9/2^-$ M4 transitions are observed[15] to be the most strongly hindered M4 transitions known in any odd-mass nuclei.

3.3. The odd-odd Tl nuclei

The same method (hindered vs unhindered α-decay) is used to characterize shell-model intruder states in odd-odd Tl nuclei. In the decay of ^{190,192,194}Bi four α branches are observed[8]. The two most intense ones are nearly degenerate in energy and have a hindrance factor near unity. They are coming from the unhindered α decay of two isomers in Bi to completely identical and thus intruder levels in Tl. The two weak α lines at higher energy are the hindered decay towards lower-lying normal states in Tl. The decay schemes based on energy, intensity and coincidence relations are given in fig. 5.

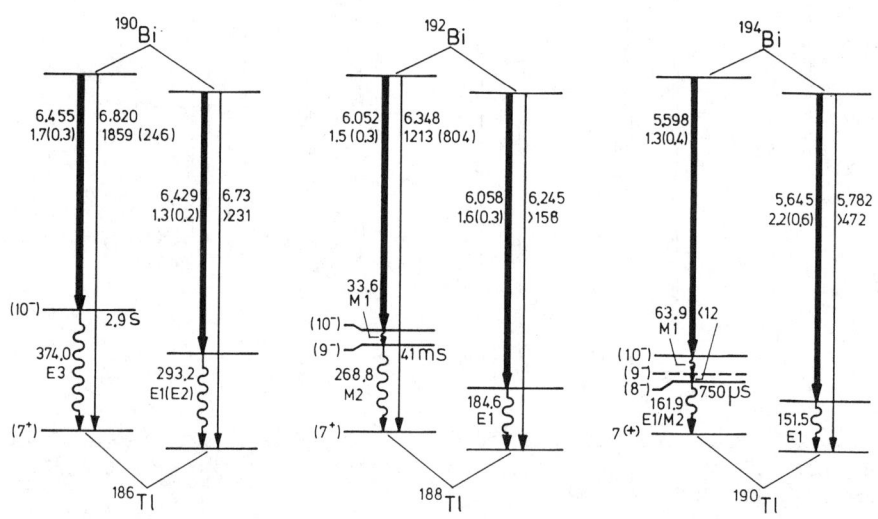

Fig. 5. Decay schemes of ^{190,192,194}Bi. For the α decay the energy (in MeV) and the hindrance factor are given. For the γ decay the energy (in keV) and the multipolarity are given.

Our level scheme work on ^{186,188,190}Tl is completing the in-beam work of Kreiner et al.[16] who was not able to observe the low-energy transitions within the intruder-based multiplets. In the higher mass Tl nuclei he does observe such transitions with energies down to 61 keV. A summary of Kreiner's results together with our work is given in fig. 6. It is plotted in such a way that the bandstructure built upon the $\pi h9/2 \times \nu i13/2$ intruder multiplet can be compared with the bandstructure on the $\pi h9/2$ intruder state in the odd-mass Tl isotopes (the data is taken from the review work of K. Heyde et al.[1]). In ref. 8 the procedure is given to extract from the experimental position the $\pi h9/2 \times \nu i13/2$ intruder states the "unperturbed" position, i.e. the position of the intruder multiplet without the disturbance of the residual π-ν interaction. It is this energy that is set out in fig. 7.

4. DISCUSSION

The excitation energy of all known intruder states in the Bi, Pb and Tl nuclei are given in fig.7. A remarkable coinciding systematic behavior of the excitation energy of the intruder based states in function of the neutron number is now present for the odd Bi, odd Tl, odd-odd Tl and even-even Pb nuclei. This shows the general applicability of the scaling law near the Z≈82 proton closed shell[10,11].

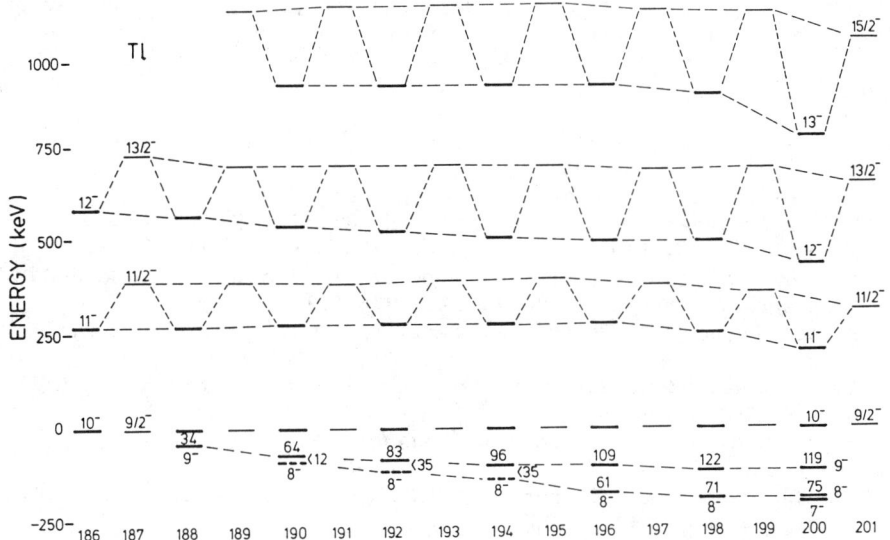

Fig. 6. A comparison between the intruder based bandstructure in odd-odd and odd mass Tl nuclei. An unobserved transition is assumed in ^{190,192,194}Tl. The indicated spins are tentative. The energy spacing in the $\pi h9/2-\nu i13/2$ multiplet is given in keV.

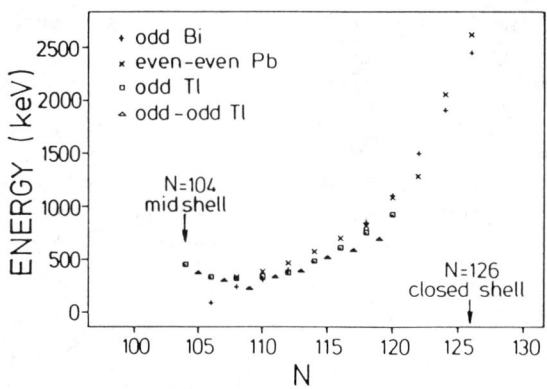

Fig. 7. Systematics of the intruder-state excitation energies. The excitation energies of the 0^+ intruder states in the even-even Pb nuclei is divided by two.

ACKNOWLEDGEMENTS

We express our special thanks to J.L. Wood and K. Heyde for the many stimulating discussions and help during the course of this work. The authors wish to thank J. Gentens and D. Wouters for their technical assistance and the Inter-Universitary Institute for Nuclear Sciences and the National Fund for Scientific Research (Belgium) for financial support. We thank Dr. H. Folger and the target lab crew of GSI, Darmstadt for the tungsten and irridium targets.

REFERENCES

1. K. Heyde, P. Van Isacker, M. Waroquier, J.L. Wood and R.A. Meyer, Phys. Rep. 102, 291 (1983).
2. J. Bron, W.J.A. Hesselink, A. Van Poelgeest, J.J.A. Zalmstra, M.J. Uitzinger, H. Verheul, K. Heyde, M. Waroquier, P. Van Isacker, and H. Vincx, Nucl. Phys. A318, 335 (1979).
3. P. Van Duppen, E. Coenen, K. Deneffe, M. Huyse, K. Heyde and P. Van Isacker, Phys. Rev. Lett. 52, 1974 (1984).
4. P. Van Duppen, E. Coenen, K. Deneffe and M. Huyse in Proc. Int. Symp. on In-Beam Nuclear Spectroscopy (Debrecen, Hungary, May 1984) eds. Zs. Dambradi and T. Feynes (Akadémia Kiadó, Budapest, 1984) p. 439.
5. P. Van Duppen, E. Coenen, K. Deneffe, M. Huyse and J.L. Wood, Phys. Lett. 154B, 354 (1985).
6. P. Van Duppen, E. Coenen, K. Deneffe, M. Huyse and J.L. Wood, Phys. Rev. C35, 1861 (1987).
7. E. Coenen, K. Deneffe, M. Huyse, P. Van Duppen and J.L. Wood, Phys. Rev. Lett. 54, 1783 (1985).
8. M. Huyse, E. Coenen, K. Deneffe, P. Van Duppen, K. Heyde and J. Van Maldeghem, to be published.
9. R. Bengtsson and W. Nazarewicz, to be published
10. K. Heyde, J. Jolie, J. Moreau, J. Ryckebusch, M. Waroquier, P. Van Duppen, M. Huyse and J.L. Wood, Nucl. Phys. A466, 189 (1987).
11. K. Heyde, this conference.
12. M. Huyse, K. Deneffe, J. Gentens, P. Van Duppen and D. Wouters, Nucl. Instr. and Meth. B26, 105 (1987).
13. J. Kantele, M. Luontama, W. Trzaska, R. Julin, A. Passoja and K. Heyde, Phys. Lett. 171B, 151 (1986).
14. V.M. Vakhtel, S.G. Kadmenskii, A.A. Martynov and V.I. Furman, Yad. Fiz. 28, 1241 (1978) [Sov. J. Nucl. Phys. 28, 639 (1978)].
15. R.A. Braga, W.R. Western, J.L. Wood, R.W. Fink, R. Stone, C.R. Bingham and L.L. Riedinger, Nucl. Phys. A349, 61 (1980).
16. A.J. Kreiner, C. Baktash and G. Garcia Bermudez, Phys. Rev. Lett. 47, 1709 (1981).

LOW ENERGY E0 TRANSITIONS IN ODD-MASS NUCLEI OF THE NEUTRON DEFICIENT 180<A<200 REGION*

E. F. Zganjar and M. O. Kortelahti
Department of Physics and Astronomy
Louisiana State University, Baton Rouge, LA 70803, USA

J. L. Wood and C. D. Papanicolopulos
School of Physics
Georgia Institute of Technology, Atlanta, GA 30332, USA

ABSTRACT

The region of neutron-deficient nuclei near Z=82 and N=104 provides the most extensive example of low-energy shape coexistence anywhere on the mass surface. It is shown that E0 and E0 admixed transitions may be used as a fingerprint to identify shape coexistence in odd-mass nuclei. It is also shown that all the known cases of low energy E0 and E0 admixed transitions in odd-mass nuclei occur where equally low-lying 0^+ states occur in neighboring even-even nuclei. A discussion of these and other relevant data as well as suggestions for new studies which may help to clarify and, more importantly, quantify the connection between E0 transitions and shape coexistence are presented.

INTRODUCTION

Nuclei in the far from stability region between $78 \leq Z \leq 83$ and $104 \leq N \leq 109$ have been discussed by many authors (see e.g. refs. 1, 2) as the most extensive region known exhibiting coexisting nuclear shapes. We reported earlier on the observation in ^{185,187}Au (Z=79; N=106, 108) of a systematic pattern of bands interconnected with low-energy very converted transitions[3,4]. In contrast to other reports (e.g. ref. 5) which indicated that the very converted cases were anomalous M1 transitions, we have clearly shown that these very converted transitions are entirely consistent with an E0 multipole component and that they closely match low-energy E0 transitions in the neighboring doubly-even isotopes. This is shown in fig. 1 for the case of ^{185}Au. The systematic pattern of transitions with E0 components in ^{185}Au (and ^{187}Au, see ref. 4) bears a close resemblance to E0 and E0 admixed transitions in the neighboring doubly-even Pt and Hg isotopes which are considered cores for certain configurations in the odd-mass Au isotopes. The relevant Hg and Pt cores are shown in fig. 2; and the corresponding conversion coefficients are given in Table I.

The $9/2^-$ state at 9 keV (see fig. 1) is the proton $h_{9/2}$ particle (intruder) state. The particle character has the consequence that the appropriate core for this configuration in ^{185}Au is ^{184}Pt. The $11/2^-$ state at 220 keV is the proton $h_{11/2}$ hole state. The appropriate core for proton hole configurations in ^{185}Au is ^{186}Hg.

---

*Work supported in part by DOE grant DE-FG05-84ER40159 (LSU), DE-FG05-87ER40330 (Ga. Tech.), and DOE contract DE-AC05-76OR00033 (UNISOR).

See refs. 6, 7 for a discussion of the coupling of these configurations to Pt and Hg cores for the case of ^{189}Au. The association between ^{185}Au states and ^{184}Pt, ^{186}Hg states, shown in fig. 1 by dashed lines, is suggested as reflecting the major components of the particle- and hole-coupled wave functions. Such a pattern of bands connected by low-energy transitions with E0 components is suggested to reflect a new type of particle-core coupling in nuclei.

Early in-beam γ-ray spectroscopy on ^{184,186}Hg showed that their yrast bands become deformed above the 2^+ state[8,9] and studies of the decay of ^{188,186,184}Tℓ → ^{188,186,184}Hg at UNISOR and ISOCELE clearly revealed two coexisting bands which approach each other with decreasing neutron number[10-14]. Not shown in fig. 2 is ^{184}Hg where the 0^+ is at 375 keV[11]. In-beam work[15] on ^{182}Hg(N=102) indicates that the deformed configuration in that nucleus is nearing a minimum at the mid neutron shell (N=104) as one would expect based on a microscopic model[16-18] which incorporates an explicit dependence on proton and neutron number through a proton-neutron force. These Hg data provide a classic example of the coexistence of levels built on different shapes in even-even nuclei. More recently excited 0^+ deformed states have been observed[19] in the neutron deficient, closed shell, Z=82, Pb isotopes $^{190-198}$Pb. These 0^+ configurations also exhibit a continuous drop in energy with decreasing mass as one approaches N=104 such that the deformed 0^+ even becomes the first excited state in $^{190-194}$Pb.[19]

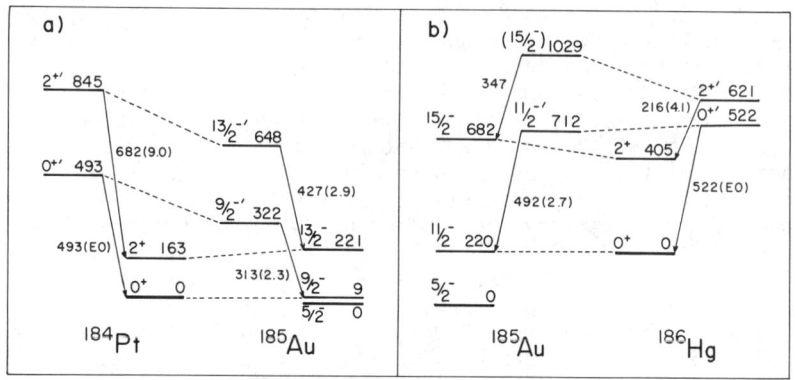

Fig. 1. Part a): Portions of the $h_{9/2}$ and $h'_{9/2}$ bands in ^{185}Au compared to the 0^+ and $0^{+'}$ bands in ^{184}Pt. Part b): Portions of the $h_{11/2}$ and $h'_{11/2}$ bands in ^{185}Au compared to the 0^+ and $0^{+'}$ bands in ^{186}Hg. Only transitions with $\alpha_K > \alpha_K(M1)$ are shown. All energies are in KeV. The numbers in parentheses following the transition energy are the ratios $\alpha_K(\text{expt})/\alpha_K(M1 \text{ theory})$ for transitions where $I \neq 0$. The data are from refs. 3, 4, and 20. The ^{185m,g}Hg decay contains three 347 keV transitions and thus a multipolarity for the 1029 keV → 682 keV 347 keV transition (part b) could not be determined.

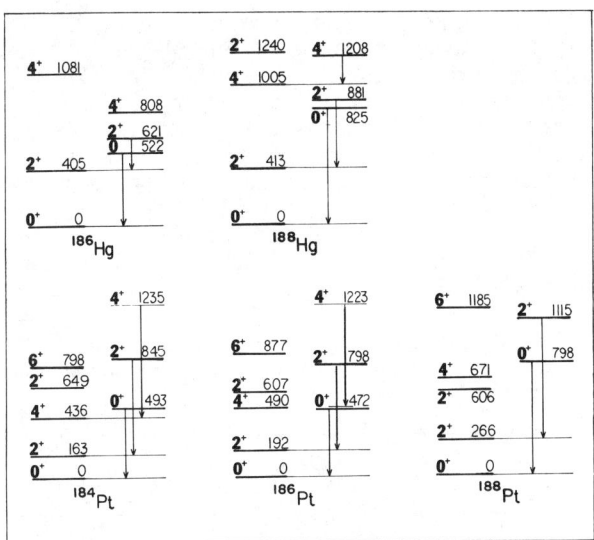

Fig. 2. Systematics of the low-lying levels in ^{186,188}Hg, ^{184,186,188}Pt. Transitions which are pure E0 or with E0 admixtures are indicated. Data are from refs. 21(^{184}Pt), 13(^{186}Hg), 22(^{186,188}Pt), and 14(^{188}Hg).

Table I. K-conversion coefficients for the $I_i^\pi \to I_f^\pi$ ($I\neq 0$) transitions in the Hg and Pt isotopes shown in fig. 2.

Isotope	E_γ(keV)	E_i(keV)	I_i^π	Experiment α_K	ref.	Theory[23] α_K(M1)	α_K(E2)
^{186}Hg	216	621	2^+	3.1 (7)	13	0.76	0.14
^{188}Hg	468	881	2^+	0.087 (6)	14	0.093	0.022
	203	1208	4^+	0.99 (11)	14	0.89	0.16
^{184}Pt	682	845	2^+	0.28 (6)	21	0.031	0.010
	799	1235	4^+	0.049 (10)	21	0.020	0.0074
^{186}Pt	606	798	2^+	>0.10*	22	0.041	0.013
	733	1223	4^+	0.057 (22)	22	0.025	0.0091
^{188}Pt	849	1115	2^+	0.22 (1)	22	0.016	0.0067

*Doublet. Lower limit computed from I_K/I_γ given in ref. 22. The other component is probably E2.

DISCUSSION

Low-energy transitions with E0 components between states with $I^\pi \neq 0^+$ are fairly common in doubly-even nuclei. However, in odd-mass nuclei, low-energy transitions with E0 components are very rare; and even relaxing the low-energy criterion, they are still uncommon. In Table II we list (in increasing order by energy) all the cases known to us in odd-mass nuclei of low-energy transitions (E < 430 keV) with E0 components. While it is surprising that only 16 cases below 430 keV are known, one immediately notes that 13 of the 16 cases are odd-mass neighbors of the even-even nuclei listed in Table I. Low-lying (< 600 keV) excited 0^+ states are not commonly known in nuclei. We list all the cases known to us in Table III. Note that all the isotopes listed in Table II are odd-mass neighbors of even-even nuclei listed in Table III. Table III can be used to suggests other odd-mass candidates for low-energy E0 admixed transitions.

Table II. The lowest-energy transitions with E0 components known between low-lying levels in odd-mass nuclei. All known cases with $E_\gamma \leq 430$ keV are shown.

Isotope	E_γ (keV)	E_i (keV)	I_i^π	ref.
185Au	205	497	$5/2^+$	24,25
187Pt	260	260	$3/2^-$	26,27
187Pt	262	288	$5/2^-$	26,27
185Au	281	515	$5/2^+$	24
183Au	284	----not certain----		28
185Au	289	289	$5/2^-$	24
183Au	297	----not certain----		28
185Au	313	322	$9/2^-$	3,4,24
195Pb	318	1126	$11/2^+$	29
187Au	323	444	$9/2^-$	30,33
185Au	330	371	$3/2^+$	24,25
185Pt	340	521	$3/2^-$	31
187Au	388	742	$13/2^-$	30,33
195Pb	401	1177	$9/2^+$	29
185Au	427	648	$13/2^-$	3,4,24
231Th	429	634	$7/2^-$	32

The unique occurrence of E0 and E0 admixed low-energy transitions in both even- and odd-mass nuclei for $78 \leq Z \leq 80$ and $104 \leq N \leq 109$, together with the systematic features[20] for 185,187Au observed at UNISOR, leads us to propose a new type of multiple band structure in odd-mass nuclei (fig. 1 for 185Au). The essential features are: two bands interconnected by transitions with strong E0 components, a small (< 500 keV) energy separation between the bands, and the absence of well-defined strong coupling in the bands.

Pairs of bands interconnected by transitions with intense E0 admixtures are known in strongly-deformed odd-mass nuclei. Some examples of note are 231Th (ref. 32), 155Eu (ref. 41), and 173Lu (ref. 42). The systematic occurrence of such pairs has also been reported[43] in the actinides. An example for 235U is presented in fig. 3. To our knowledge, however, the present work reports the first systematic occurrence of pairs of bands interconnected by

Table III. The lowest first excited 0^+ states known in doubly-even nuclei. All known cases with $E_x \leq 635$ keV are shown.

Isotope	E_x	ref.	comments
^{98}Sr	215	34	
^{100}Zr	331	35	
^{184}Hg	375	11,36	
^{178}Pt	(422)	39	0^+ spin inferred
^{176}Pt	(433)	39	0^+ spin inferred
^{186}Pt	472	22	
^{184}Pt	492	21	
^{182}Pt	500	40	
^{180}Pt	(500)	-	estimate from systematics
^{186}Hg	522	13	
^{152}Gd	615	37	
^{230}Th	635	38	

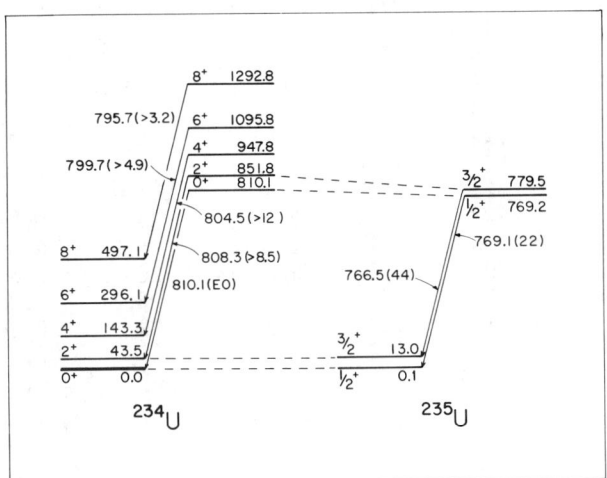

Fig. 3. Comparison of the ground state and β-bands in ^{234}U and ^{235}U. All energies are in keV. Only transitions with $\alpha_K > \alpha_K$(M1) are shown. The numbers in parentheses following the transition energy are the ratios α_K(expt)/α_K(M1 theory) for transitions where I ≠ 0. The ^{235}U ground state is 7/2⁻ and the basic configuration of the bands is |631↓|. The data are from refs. 44(^{234}U) and 45(^{235}U).

transitions with intense E0 admixtures outside of the traditional strongly-deformed regions, and it is significantly lower in energy.

K-conversion coefficients for $I_i^\pi \to I_f^\pi$ transitions with E0 components for the cases given in Table II and, for completeness, the additional transitions with energy > 430 keV, which will be needed for the remaining figures, are presented in Table IV. Obvious missing entries include Hg and Tℓ isotopes. Data for [189,193]Hg (ref. 46) [195]Hg (ref. 47) and [197]Hg (ref. 48) are either preliminary or the transitions are much higher in energy. Studies on [193,195]Tℓ are underway (ref. 49).

Table IV. K-conversion coefficients for $I_i^\pi \to I_f^\pi$ transitions with E0 components for the cases given in Table II and, for completeness, the additional transitions with energy >430 keV.

Isotope	E_γ(keV)	E_i(keV)	I_i^π	Experiment		Theory[23]	
				α_K	ref.	α_K(M1)	α_K(E2)
[183]Au	284	-not certain-		1.6	28	0.33	0.069
	297	-not certain-		3.9	28	0.28	0.064
[185]Au	205	497	$5/2^+$	2.5 (5)	24,25	0.82	0.16
	281	515	$5/2^+$	0.77 (41)	24	0.34	0.073
	289	289	$5/2^-$	0.52 (20)	24	0.32	0.068
	313	322	$9/2^-$	0.58 (15)	3,4,24	0.26	0.056
	330	371	$3/2^+$	2.8 (8)	24,25	0.22	0.049
	427	648	$13/2^-$	0.33 (3)	3,4,24	0.11	0.027
	492	712	$11/2^-$	0.21 (2)	3,4,24	0.077	0.020
[187]Au	323	444	$9/2^-$	0.66 (17)	30,33	0.12	0.035
	388	742	$13/2^-$	0.64 (9)	30,33	0.23	0.054
	657	881	$11/2^-$	0.10 (2)	33	0.036	0.011
[185]Pt	340	521	$3/2^-$	0.35	31	0.19	0.042
	542	723	$3/2^-$	0.19	31	0.055	0.015
[187]Pt	260	260	$3/2^-$	3.2 (6)	26,27	0.39	0.086
	262	288	$5/2^-$	5.7 (14)	26,27	0.38	0.084
	499	508	$3/2^-$	0.8 (3)	26,27	0.068	0.018
[195]Pb	318	1126	$11/2^+$	4.9 (32)	29	0.34	0.056
	401	1177	$9/2^+$	5.4 (27)	29	0.17	0.033
[231]Th	429*	634	$7/2^-$	0.66 (4)	32	0.28	0.037

*Additonal cases for [231]Th are not shown.

The Pt isotopes are particularly interesting because they undergo a pronounced change[22] in their excitation spectra from higher to lower mass numbers, for example, between [188]Pt and [186]Pt. With regard to odd-A Pt isotopes, a similarly drastic change occurs[31,50] between [187]Pt and [185]Pt. The $\alpha_K > \alpha_K$(M1) transitions for these two cases are shown in fig. 4. In [187]Pt (fig. 4a.), the states connected by the 260.5 and 262.7 keV transitions appear[27] to be related to the 0^+ ground states and 0^+ excited states at 472 and 798 keV respectively in [186,188]Pt (fig. 2). This is particularly interesting since the [186]Pt and [188]Pt cores are so different. In [185]Pt (fig. 4b), a rotational band (not shown) is built on the $9/2^+$ ground state identified as the $9/2^+$[624]

configuration due to strong coupling between the particle and the prolate core. The states connected by the 542.2 and 340.1 keV transitions are most likely related to the coupling of the 1/2⁻[521] configuration to the 0^+ and 0^+ excited states in ^{184,186}Pt. Previous work[31] on ^{185}Pt indicates that the very converted transitions may be anomalous M1, like those so assigned[5] in ^{185}Au. The evidence presented for anomalous M1 transitions in ^{185}Au comes from the electron intensities assigned by Bourgeois et al.[5] to the 321K and 330K conversion lines. They claimed to observe $\alpha_K > \alpha_K(M1)$ for transitions between levels with different spins. We have clearly shown[20], however,

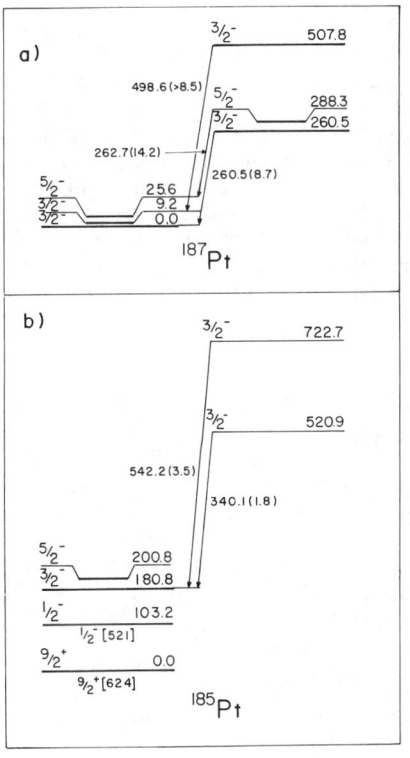

Fig. 4. Transitions in ^{187,185}Pt with $\alpha_K > \alpha_K(M1)$. All energies are in keV. The numbers in parentheses following the transition energy are the ratios $\alpha_K(\text{expt})/\alpha_K(M1 \text{ theory})$. Data are from refs. 26,27 (^{187}Pt) and 31,51 (^{185}Pt).

that multiplet structure is indicated for these lines and that the very converted components occur between levels of like spin and, consequently, can carry an E0 component. Indeed, we find all transitions with $\alpha_K > \alpha_K(M1)$ in ^{185}Au and ^{187}Au (ref. 20) and ^{185}Pt (ref. 51) to be consistent with $\Delta I = 0$.

Transitions with $\alpha_K > \alpha_K(M1)$ between positive parity states in ^{185}Au are also observed as shown in fig. 5. The significant E0 admixture in the 330.0 and 205.2 keV transitions suggests coexisting bands for the positive parity proton-hole states ($s_{1/2}$, $d_{3/2}$, $d_{5/2}$) in the ^{186}Hg core similar to that observed for the $h'_{11/2}$ band. However, a strongly-coupled sequence of levels was observed[52] in-beam to feed the $h^{-1}_{11/2}$ decoupled band (no spin assignments were made in ref. 52). This is shown in fig. 6. The cascade and crossover intraband transitions from the in-beam data clearly indicate strong-coupling. The absence of transitions indicated by dashed lines is a puzzle, but it may well be that the head of the band lies below the decay sequence of the in-beam data. Support for this interpretation comes from work[53,54] on ^{187}Hg, but one still remains puzzled by the absence of the 13/2$^-$ level in the decay work. In this regard it should be noted that the $h_{11/2}$ 13/2$^-$ member at 681.1 keV lies unusually low in energy (by systematics). This may indicate mixing of the two 13/2$^-$ states with the elusive 13/2$^-$ being repelled to a higher energy location. The lack of a clearly observable 346.3 keV transition is also part of the remaining difficulty.

We have observed[24] a similar strongly-coupled → decoupled sequence (via decay spectroscopy alone) for the positive parity proton-hole bands shown in fig. 7 (a continuation of fig. 5 to higher energies). Note the cascade and crossover interband transitions at higher energy. Note also the missing, but otherwise expected, transitions indicated by dashed lines. It is, therefore, not entirely certain that the 370.7, 496.3 keV levels and the 559.2, 683.0, 861.3, 1014.1 keV levels belong to the same band. This data on odd-mass core shape isomerism, represented by figs. 6 and 7, imply that the situation at low-spin is complex, probably due to mixing. This represents a new type of degree of freedom in odd-mass nuclei and needs to be clarified in ^{185}Au and searched for in other nuclei.

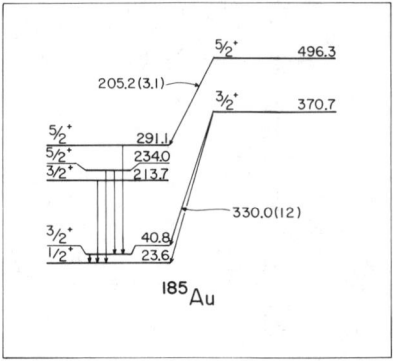

Fig. 5. Transitions with $\alpha_K > \alpha_K(M1)$ between positive parity states in ^{185}Au. All energies are in keV. The numbers in parentheses following the transition energy are the ratios $\alpha_K(\text{expt})/\alpha_K(M1\ \text{theory})$. Data are from ref. 24.

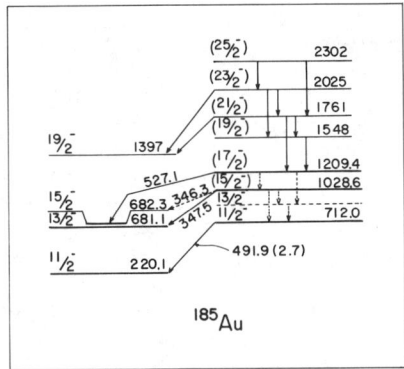

Fig. 6. Portions of the $h_{11/2}$ structure in ^{185}Au as seen in decay and in-beam spectroscopy. The 11/2⁻ [220 keV], 15/2⁻ [682 keV] and 11/2⁻ [712 keV], (15/2⁻) [1029 keV] levels and the 491.9 keV transition were shown earlier in fig. 1b. All energies are in keV. The levels and transitions with energies to 0.1 keV are seen in decay spectroscopy (ref. 24). The upper levels with energies to 1 keV are seen only in in-beam spectroscopy (ref. 52). The 1209.4, 682.3 and 220.1 keV levels and the 527.1 keV transition are seen both in-beam[52] and in the ^{185m,g}Hg decay[24].

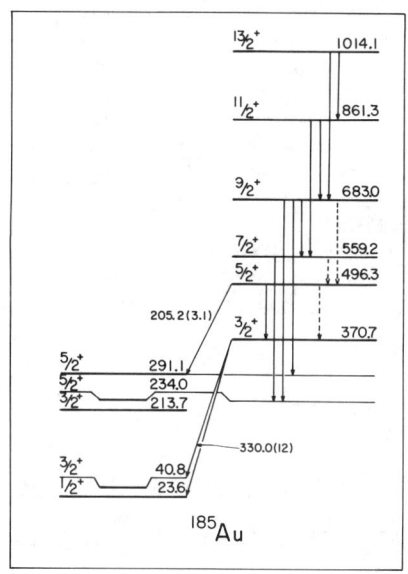

Fig. 7. Possible continuation of the upper positive parity band in ^{185}Au shown in fig. 5 (370.7 and 496.3 keV). All energies are in keV and other notations are as indicated for fig. 5. Data are from ref. 20.

CONCLUSIONS AND FUTURE

Recent work bearing on the data presented here include the observation[55] of a very large isotope shift, $\delta\langle r^2\rangle$, between ^{187}Au and ^{186}Au ($\beta \approx 0.25$ for ^{185,186}Au); the extension of the Hg isotope shift data to ^{182}Hg and the observation of the ^{185m}Hg/^{185g}Hg isomer shift (the largest isotope/isomer shifts known anywhere -- see ref. 56);

evidence for coexisting bands in ^{187}Hg from a study[57] of high-spin states; and the observation[58] of sudden changes in the yrast structure of ^{176,178}Pt.

The most outstanding example[59] of an electric monopole transition between states of very different shapes is probably the decay of the fission isomer to the ground state in ^{238}U. The B(E0) for the decay of ^{238m}U is the smallest[59] ever measured. While E0 and E0 admixed transitions appear to be a promising fingerprint for shape coexistence (see ref. 60 for a review of E0-E2-M1 multipole admixtures), shape change alone would not seem to be the solution to understanding the strong low-energy E0 transitions in the neutron deficient Pb, Au and Hg isotopes, even though these shape changes appear to be large in some cases. E0 transitions need to be quantified if they are to be used as a fingerprint for coexisting bands. This will, of course require lifetime measurements in the 50 ps range.

It should also be noted that E0 transitions and isotope shifts are dynamic and static aspects of the same operator -- the square of the electric charge radius[18]. A quantitative determination of the E0 components in the $I^\pi \to I^\pi$ transitions is also needed. This is essentially impossible by conversion-electron measurements alone, but can be accomplished in combination with γ anisotropies from nuclei oriented at low temperature. This is vitally important when an E0+E2 admixture masquerades as M1 in the internal-conversion process.

While such longer-term developments are underway, important short-term studies which need to be done to extend the data include: searches for E0 admixtures in the neutron deficient odd-mass Hg and Tℓ isotopes; extensions of the studies in Pt, Au and Pb to both lighter and heavier isotopes; isotope shift measurements on the neutron deficient Pt nuclei to verify the inferred large $\delta \langle r^2 \rangle$ change between ^{187}Pt and ^{186}Pt; and, finally, a search for the other odd-mass candidates suggested by the even-even nuclei with low-lying 0^+ states presented in Table III.

ACKNOWLEDGEMENTS

We gratefully acknowledge the contribution to this work by the other UNISOR scientists, the UNISOR staff, and the HHIRF staff.

REFERENCES

1. J. H. Hamilton et al., Rep. Prog. Phys. 48, 631 (1985).
2. K. Heyde et al., Phys. Rep. 102, 291 (1983).
3. E. F. Zganjar et al., in Nuclei off the Line of Stability, ed. R. A. Meyer and D. S. Brenner, ACS Symp. Series 324, 1986, p. 245.
4. E. F. Zganjar, et al., in Nuclear Structure Reactions and Symmetries, Vol. 2, ed. R. A. Meyer and V. Paar (World Scientific Publ. Co., Singapore, 1986), p. 716.
5. C. Bourgeois et al., Nucl. Phys. A386, 308 (1982).
6. V. Berg et al., Nucl. Phys. A244, 462 (1975).
7. J. L. Wood et al., Phys. Rev. C14, 682 (1976).
8. D. Proetel et al., Phys. Rev. Lett. 31, 896 (1973).
9. N. Rud et al., Phys. Rev. Lett. 31, 1421 (1973).
10. J. H. Hamilton et al., Phys. Rev. Lett. 35, 562 (1975).
11. J. D. Cole et al., Phys. Rev. Lett. 37, 1185 (1976).
12. R. Beraud et al., Nucl. Phys. A284, 221 (1977).

13. J. D. Cole et al., Phys. Rev. C16, 2010 (1977).
14. J. D. Cole et al., Phys. Rev. C30, 1267 (1984).
15. W. C. Ma et al., Phys. Lett. B139, 276 (1984).
16. K. Heyde et al., Nucl. Phys. A466, 189 (1987).
17. K. Heyde, contribution to this conference.
18. J. L. Wood, contribution to this conference.
19. P. Van Duppen et al., Phys. Rev. C35, 1861 (1987).
20. C. D. Papanicolopulos et al., Phys. Lett. B, preprint.
21. M. Cailliau et al., J. de Phys. 35, 469 (1974).
22. M. Finger et al., Nucl. Phys. A188, 369 (1972).
23. F. Rosel et al., At. Data Nucl. Data Tables 21, 291 (1978).
24. C. D. Papanicolopulos, Ph.D. thesis, Ga. Tech., 1987.
25. C. D. Papanicolopulos et al., Contribution to this Conference.
26. A. Ben Braham et al., Nucl. Phys. A332, 397 (1979).
27. B. E. Gnade et al., Nucl. Phys. A406, 29 (1983).
28. M. I. Macias-Marques et al., Nucl. Phys. A427, 205 (1984).
29. J. C. Griffin, Ph.D. thesis, Ga. Tech., 1987.
30. E. F. Zganjar et al, in Proc. 4th Int. Conf. on Nuclei Far From Stability, Helsingør, Denmark, 1981, ed. P. G. Hansen and O. B. Nielsen, CERN Report 81-09, p. 630.
31. B. Roussiere et al., Nucl. Phys. A438, 93 (1985).
32. D. H. White et al., Phys. Rev. C35, 81 (1987).
33. M. A. Grimm, Ph.D. thesis, Ga. Tech., 1978.
34. F. Schussler et al., Nucl. Phys. A339, 415 (1980).
35. T. A. Khan et al., Nucl. Phys. A283, 105 (1977).
36. J. D. Cole, Ph.D. thesis, Vanderbilt, 1978.
37. C. M. Baglin, Nucl. Data Sheets 30, 1 (1980).
38. Y. A. Ellis-Akovali, Nucl. Data Sheets 40, 385 (1983).
39. E. Hagberg et al., Nucl. Phys. A318, 29 (1979).
40. J. P. Husson et al., in Proc. Third Int. Conf. on Nuclei Far From Stability, Cargese, 1976, ed. R. Klapisch, CERN 76-13, p. 460.
41. P. T. Prokofjev et al., Nucl. Phys. A455, 1 (1986).
42. E. G. Funk et al., Phys. Rev. C10, 2015 (1974).
43. T. von Egidy et al., Phys. Lett. 81B, 281 (1979).
44. W. Z. Venema et al., Phys. Lett. 156B, 163 (1985).
45. J. Almeida et al., Nucl. Phys. A315, 71 (1979).
46. G. M. Gowdy, Ph.D. thesis, Ga. Tech., 1976.
47. G. M. Gowdy et al., Nucl. Phys. A312, 56 (1978).
48. R. A. Braga et al., Phys. Rev. C19, 2305 (1979).
49. C. R. Bingham, private communication.
50. J. L. Wood, in proc. 4th Int. Conf. on Nuclei Far From Stability, Helsingør, Denmark, 1981, ed. P. G. Hansen and O. B. Nielsen, CERN Report 81-09, p. 612.
51. J. Schwarzenberg, Ga. Tech., private communication.
52. A. J. Larabee et al., Phys. Lett. 169B, 21 (1986).
53. J. L. Wood et al., Bull. Am. Phys. Soc. 25, 739 (1980).
54. F. Hannachi et al., Z. Phys. A325, 371 (1986).
55. K. Wallmeroth et al., Phys. Rev. Lett. 58, 1516 (1987).
56. G. Ulm et al., Z. Phys. A325, 247 (1986).
57. F. Hannachi et al., Z. Phys. A325, 371 (1986).
58. G. D. Dracoulis et al., J. Phys. G12, L97 (1986).
59. J. Kantele et al., Phys. Rev. Lett. 51, 91 (1983).
60. J. Lange et al., Rev. Mod. Phys. 54, 119 (1982).

THE LIGHT EXOTICS*

Kamal K. Seth[†]

Northwestern University, Evanston, IL 60201, USA

1. INTRODUCTION

I have recently adopted a custom originated by Dennys Wilkinson. According to this custom, all talks are best begun by talking about things which need not be talked about. So I first want to talk about why one should be interested in the study of exotic nuclei, especially the very light ones – the isotopes of hydrogen and helium which are the subject of my talk. Obviously, to this audience such an introduction is hardly necessary, but here I go, anyway.

There are three different reasons for studying the light exotics. The first reason is a general and philosophical one. In order to best study what holds a system together, one needs to find out what extreme conditions break it down. For a nuclear system the extreme conditions may be: a) very high spins, b) extremes of isospin, c) very high energy and/or matter density. As we all know there are many afficianados of high spins, and many aspirants of high energy/matter density. We, at this conference are the self-avowed lovers of the extremes of isospin.

The second reason for the interest in light exotics doesn't quite come down to earth either, – it is astrophysical. To put it very succinctly, the stuff of light exotics is the stuff neutron stars are made of. Hopefully we can learn about the highly neutron-rich matter in bulk by studying the same matter in such exotic nuclei.

With the third reason we finally do get down to earth. Calculations of the structure and stability of few-nucleon systems can now be done almost from first principles, and therefore in such nuclei one can test subtle aspects of the nuclear interactions, e.g., the need for three-body and many-body forces, better than anywhere else.[1]

Having preached to the converted, I will now briefly review what we know about Z = 0, 1 and 2 nuclei.

2. WHAT WE HAVE KNOWN FOR SOMETIME

2.1 *Polyneutron Nuclei*

We know that the dineutron (n^2) system is unbound by only ~70 keV. This gives one some hope that perhaps polyneutron systems may be bound or near-bound. Since it is believed that if tetraneutron (n^4) is stable, heavier

---

*This work was supported in part by the U.S. Department of Energy.

[†]Work done in collaboration with Drs. B. Parker and R. Soundranayagam.

polyneutron systems (neutron droplets) have a high probability of being stable too, most of the experimental efforts have concentrated on searching for stable tetraneutrons. Three different approaches have been tried. In the first approach a reaction is made in which a tetraneutron may be expected to be produced. The tetraneutron is then expected to be absorbed by the target to produce an observable radioactivity. The latest of a long series of this type of experiments was done by de Boer et al.[2] No evidence for ^{4}n was found. The second type of experiment involves the search for an enhancement in the phase space for the pion double charge exchange reaction ^{4}He(π^-, π^+). Three earlier experiments had reached a negative conclusion about the existence of ^{4}n. In the latest experiment done at LAMPF[4] a broad structure was indeed observed in the spectra of outgoing π^+ but it could be positively identified as being not due to any resonant state of the four neutron system. Despite early unsuccessful attempts to search for tetraneutrons by means of heavy-ion reactions,[5] Belozyorov et al.[6] at Dubna have recently searched for tetraneutron in three different reactions: ^{7}Li$(^{11}$B, ^{11}O$)^4$n, ^{7}Li$(^9$Be, ^{12}N$)^4$n, and ^{9}Be$(^9$Be, ^{14}O$)^4$n. No statistically significant departures from phase-space were found in any of the reactions.

The chapter on tetraneutrons seems to be closed, at least for now.

2.2 The Hydrogen Isotopes

^{1,2,3}H are known to be particle stable. Early phase-shift analyses of the n–t scattering data indicated two resonant states of ^{4}H, unbound with respect to ^{3}H+n decay by 3.4 MeV (2^-) and 5.5 MeV (1^-) respectively.[7] A study of the pion capture reaction $\pi^- + {}^7$Li $\to {}^4$H + t, ^{4}H $\to$ t + π confirms the 2^- ground state at 2.7 $\pm$ 0.6 MeV, but finds no evidence for the second state.[8]

^{5}H has been the subject of many experimental searches. I will not remind you of the infamous story of the reputed β-activity of ^{5}H. Suffice it to say that the latest experiments were all negative. So were the searches in ^{9}Be$(\alpha, {}^8$Be$)$, ^{3}H(t, p), ^{7}Li(π^-, d), ^{7}Li(π^-, pn) reactions. At Helsingor I presented some results from our study of the ^{7}Li(π^-, π^+)X reaction.[9] No narrow peak corresponding to ^{7}H was found in this experiment, but I argued that the shape of the π^+ spectrum was that corresponding to the phase space for ^{5}H + n + n + (π^+) break-up. Admittedly, this could not be considered as evidence for the existence of ^{5}H. I also presented some inconclusive preliminary results from our ^{6}Li(π^-, p)X experiment, and stated that there was no evidence for a bound, or unbound but narrow, peak due to ^{5}H.

To summarize the situation, all searches for $A \geq 5$ isotopes of hydrogen appear to have been unsuccessful so far.

2.3 The Helium Isotopes

Helium has been traditionally a gold-mine for exotic isotope prospectors.

We know that ^{3,4}He are stable, the even-isotopes ^{6}He, and ^{8}He are particle-stable, and the odd-isotopes ^{5}H and ^{7}H are particle-unstable. We note however that the odd-isotopes ^{5}He and ^{7}He are much less unstable than one would infer on the basis of the transverse Garvey-Kelson mass formula, in which the experimental masses of the even isotopes ^{6}He and ^{8}He are used. The GK relations give -1.26 MeV and -2.29 MeV for single neutron separation energies of ^{5}He and ^{7}He, respectively. Experimentally ^{5}He and ^{7}He are found to be unbound by only 0.89 MeV and 0.45 MeV respectively, and have rather narrow widths, $\Gamma(^5\text{He}) = 0.60$ MeV, $\Gamma(^7\text{He}) = 0.16$ MeV. The GK relation also predicts ^{9}He to be unbound with respect to ^{8}He+n by 3.55 MeV. It appears that if the extra binding trend of ^{5,7}He continues, ^{9}He may be unbound by a considerably smaller amount. It may therefore also have quite narrow width. What remains is to find a suitable reaction to populate it. Of course, if we find ^{9}He, what is to stop us from dreaming about the doubly-magic darling of exotica lovers, ^{10}He!

I now want to talk about some of the experiments we have recently done in order to search for and study the heavy isotopes of hydrogen and helium.

3. THE EXPERIMENTAL SET-UP

The experiments reported here were done with pion beams available at the EPICS facility at LAMPF. The apparatus has been described in detail elsewhere,[9,10] and is schematically illustrated in Fig. 1. The important points to note are the following.

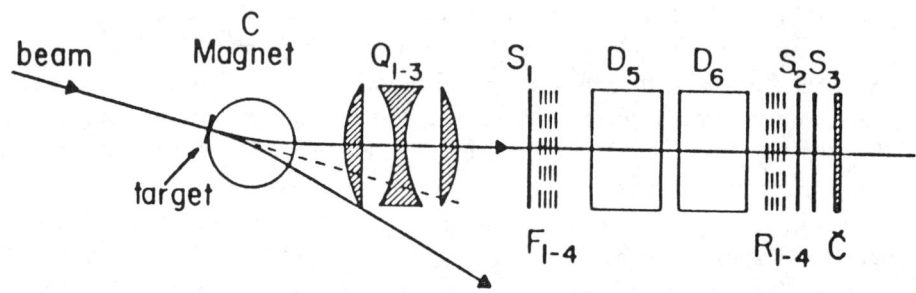

Fig. 1. Schematic of the spectrograph system at the EPICS facility at LAMPF as used in the present measurements.

Despite the fact that LAMPF holds the world-record for the intensity of its 800 MeV primary proton beam ($\sim$1 mA), the π^- intensity at the EPICS channel is still only $\sim$2 - 5 $\times$ 10$^7\pi^-$/sec between $T(\pi^-) = 100$ - 300 MeV. This translates to 3 to 8 picoamps of 'beam'. This 'beam' is spread over an area of $\sim$8cm $\times$ 20cm at the target, forcing one to use huge targets, often exhausting the world-supply of an enriched isotope (from 5 – 150 grams).

The second point to keep in mind is that we are always measuring double charge exchange, DCX (π^-, π^+) cross sections in the range of 1 to 100 nb/sr. At forward angles the rare π^+ are accompanied by several orders of magnitude more prolific yield of $e^\pm$, $\mu^\pm$, p, d, etc. Thus a very high level of particle identification is required. This is accomplished by several means. First, the target is placed inside a C-magnet so that the primary beam, the elastic and inelastic π^-, e^-, and μ^- are swept away from the direction of the magnetic spectrometer. Second, the electrons are identified with great efficiency by a threshold gas Čerenkov counter. Third, time of flight is measured in the ~9 meters path through the spectrometer. Finally, at the exit of the focal plane we identify muons by ranging out pions in a stack of carbon absorbers (not shown in Fig. 1).

The net result of these elaborate means of particle identification is that we have essentially no contaminant particles in our π^+ spectra and almost no particles are ever seen in the disallowed regions of missing mass. The energy resolution of the EPICS spectrometer system is FWHM ~150keV. For the DCX experiments, however, it is usually larger and is determined by the thickness of the target used. In each experiment the energy resolution is determined by making an elastic scattering measurement with the same target. Absolute cross sections are obtained by making measurements of π-p elastic scattering under identical conditions and normalizing to the known π-p cross sections ($\pm 5\%$).

4. HELIUM-9 AND ITS EXCITED STATES

At the Helsingor conference we reported[9] our preliminary results for the reaction $^9\text{Be}(\pi^-, \pi^+)^9\text{He}$ done at $T(\pi) = 194$ MeV and $\theta(\text{lab}) = 15°$. In this experiment an energy resolution of FWHM $\approx$ 1 MeV was obtained with a 2.40 gm/cm^2 thick ^{9}Be target. ^{9}He(g.s.) was clearly oberserved at the end of the phase-space dominated by ^{8}He+n break-up. The atomic mass excess of ^{9}He(g.s.) was determined to be 40.80 $\pm$ 0.10 MeV. There were indication of excited states at ~1.5 MeV and 4 MeV, but nothing definitive could be said about them.

There is one interesting consequence of this ^{9}He mass measurement. If it is used in the local GK transverse relationship, it predicts that the doubly magic nucleus ^{10}He is unstable against one neutron decay by 0.31 $\pm$ 0.14 MeV and against two neutron decay by 1.44 $\pm$ 0.14 MeV. These numbers are much smaller than any of the earlier predictions. They give us the hope that if ^{10}He can be searched for in the missing-mass spectrum of a two-body reaction, it is quite likely that it will be found that its ground state is either bound, or is barely unbound and has a small width. Unfortunately, one can not think of too many such reactions; $^{10}\text{Be}(\pi^-, \pi^+)$ appears to be a rather unique but remote possibility! Perhaps projectile fragmentation techniques can help.

The measured mass of ^{9}He corresponds to the g.s. being unstable against single-neutron decay by 1.13 $\pm$ 0.10 MeV only. This is much smaller than the estimates, ranging from 3.55 MeV to 2.36 MeV, obtained by the use of the GK relation and other mass-systematics. Recently shell-model calculations for $A \leq 16$ nuclei have been reported in which no 'inert core' is assumed.[12,13] These calculations use a translationally invariant interaction which allows them to avoid the problem of spurious states due to center of mass motion. Results of calculations using two different interactions have been reported. In the first calculation (A) in the $(0+1)\hbar\omega$ space Van Hees and Glaudemans[12] determined the two-body matrix elements empirically from a fit to the energies of 136 levels in $A = 4$–16 nuclei. In the second calculation by Poppelier, Wood and Glaudemans[13] both the $(0+1)\hbar\omega$ space (B) and the $(0+2)\hbar\omega$ space (C) were used. The two-body matrix elements were derived from the realistic Reid soft-core potential and the mass-independent differences between the effective and the free nucleon-nucleon interactions were obtained by a fit to 146 energy levels. As noted in Fig. 2, which illustrates the results of these calculations, the $(0+1)\hbar\omega$ calculation of Poppelier et al. (calculation B) predicts that ^{9}He is unbound against single neutron decay by 1.21 MeV. This agrees with our experimental result almost exactly, while the other two calculations, A and C, predict a much more unbound ^{9}He. Of course, the excellent result of calculation B may be fortuitous, It is therefore necessary to examine other predictions of these calculations in order to determine if any of them is indeed markedly superior to the others.

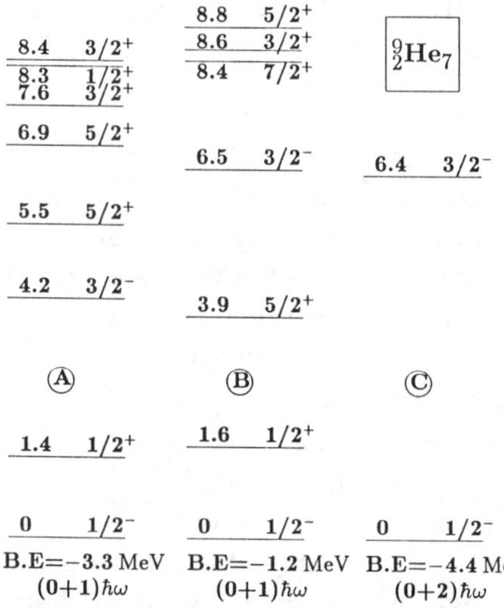

Fig. 2. Predicted excited state spectra for ^{9}He.
(A) is from ref. 12. (B) and (C) are from ref. 13.

The excited state spectra predicted by the three calculations are shown in Fig. 2. We note that all calculations predict the $1/2^-$ ground state and a $1/2^+$

state at ~ 1.5 MeV excitation. However, there are notable differences beyond that point. For example, between the two $(0+1)\hbar\omega$ calculations, A and B, there is a dramatic inversion of the $3/2^-$ and $5/2^+$ states. The agreement between the results of calculation B and the predictions of the $(0+2)\hbar\omega$ calculation C (which does not contain positive parity levels) again suggests that calculation B is to be preferred over A.

In order to investigate the excited states of ^{9}He we made[11] a new measurement of the ^{9}Be$(\pi^-, \pi^+)^9$He reaction with a thinner target (0.93 g/cm^2). The missing mass spectrum observed is shown in Fig. 3.

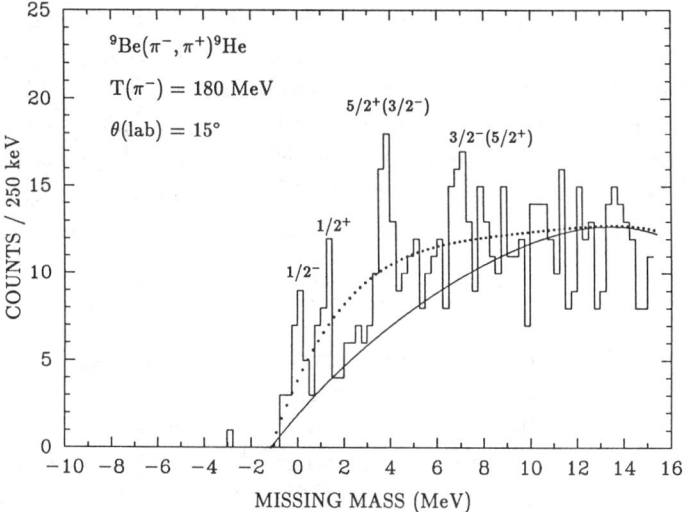

Fig. 3. Missing mass spectrum for the reaction ^{9}Be$(\pi^-, \pi^+)^9$He. The curves are discussed in the text. (From ref. 11.)

With the better energy resolution achieved in this experiment (FWHM = 420 ± 10 keV), we identify in this spectrum at least three peaks with widths ≥ 420 KeV over and above the continuum. In order to determine the level of statistical significance for this conclusion, we have made χ^2-based likelihood analysis of the data. As the shape for the non-resonant continuum spectrum we have used both a 4th-order polynomial and a mixture of 3-body (^{8}He + n + π^+) and 4-body (^{7}He + 2n + π^+) phase spaces. We find that for the different constraints on the shape of the continuum spectrum, the hypothesis of no peaks in the data has a maximum likelihood of ~ 40-50 % (dotted curve in Fig. 3), whereas introducing four Lorentzian peaks at 0 MeV (FWHM = 420 ± 100 KeV), 1.2 MeV (FWHM = 420 ± 100 KeV), 3.8 MeV (FWHM = 500 ± 100 KeV) and 7.0 MeV (FWHM = 550 ± 100 KeV) with a polynomial background (shown by solid curve), leads to a very significant increase in the maximum likelihood, to ~ 95-

98%. We therefore conclude that the states at 0 MeV, 1.2 MeV and 3.8 MeV are certain, and the 7.0 MeV state is highly probable. The correspondence of the experimental level spectrum with the predictions of both calculations A and B is very good (perhaps fortuitously), and we are tempted to identify the first three states with the predicted sequence of the $1/2^-$, $1/2^+$, and $(5/2^+, 3/2^-)$ states. The broad structure at ~ 7 MeV excitation appears to corresponds to the $(3/2^-, 5/2^+)$ state predicted at ~ 6.7 MeV. If our identifications are correct it is indeed remarkable that shell model calculations whose parameters are optimized in the valley of stability, should work so well so far from the valley.

In order to put our conclusions on a firmer basis it is necessary to measure the J^π of the various states of ^{9}He. In a DCX experiment this is hardly possible, except in one case. One can attempt to determine whether any of these states have identifiable L=0 components in their angular distributions and therefore $J^\pi = 3/2^-$. This may allow a choice to be made between the different interactions of calculations A and B. In order to investigate this possibility and to investigate higher-lying narrow states it is planned to make a high-resolution, high-statistics investigation of the ^{9}Be$(\pi^-, \pi^+)^9$He reaction at several angles in the near future.

5. ON THE EXISTENCE OF ^{6}H

As mentioned earlier, several experiments have been done in search of ^{5}H and ^{7}H. These include our own pion experiments[9] and a very recent series of heavy-ion experiments [^{9}Be + ^{11}B → ^{16,15,14}O + ^{4,5,6}H] done by Belozyorov et al. at Dubna[14] All experiments reported so far reach a negative conclusion about the existence of ^{5}H and ^{7}H. The prevailing wisdom has been that if ^{5}H and ^{7}H do not exist there is even smaller probability for the existence of isotopes containing odd numbers of neutrons. Thus ^{6}H is considered unlikely to exist. What are the expectations on the basis of theoretical considerations? No predictions of ^{6}H mass using the transverse GK relation are possible. The calculations of Poppelier et al.,[13] referred to earlier, predict a 2^- g.s. for ^{6}H with a total B.E.=+2.5 MeV in the $(0+1)\hbar\omega$ calculation, and a 1^- g.s. with total B.E.= -3.2 MeV (unbound) in the $(0+2)\hbar\omega$ calculation. These numbers correspond to ^{6}H being unbound with respect to ^{3}H+3n by 6 to 12 MeV. To summarize, it seems highly improbable that ^{6}H exists. However, in two recent Russian papers it is claimed that ^{6}H does indeed exist.

Aleksandrov et al.[15] at the Khurchatov Institute have claimed the "observation of unstable superheavy hydrogen isotope ^{6}H in the reaction ^{7}Li$(^7$Li,^{8}B$)^6$H" at T(^{7}Li) = 82 MeV (see Fig. 4). Over and above the phase-space for ^{3}H + 3n break-up, a continuous background of pile-up events, and peaks due to reactions on impurity contents of ^{16}O and ^{12}C in the target, these authors identify an enhancement with a width of 1.8 ± 0.5 MeV which they attribute to ^{6}H ground-

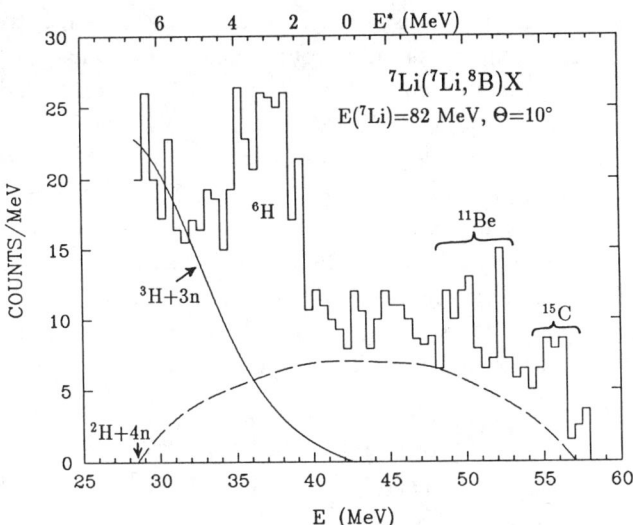

Fig. 4. ^{8}B spectrum obtained in the ^{7}Li(^{7}Li, ^{8}B)X reaction. The energy scale on the top is based on ^{3}H + 3n break-up illustrated by the solid curve. The dashed curve indicates the pile-up contribution. (From ref. 15.)

state unbound with respect to ^{3}H + 3n by 2.7 ± 0.4 MeV. A production cross section of 60 nb/sr is inferred. This claim has been 'verified' by another heavy-ion experiment done at Dubna. Belozyorov et al.[14] claim to have observed the same enhancement in the ^{9}Be(^{11}B,^{14}O)^{6}H reaction done at T(^{11}B) = 88 MeV. The ^{6}H production cross sections was claimed to be is 16 ± 8 nb/sr (see Fig. 5).

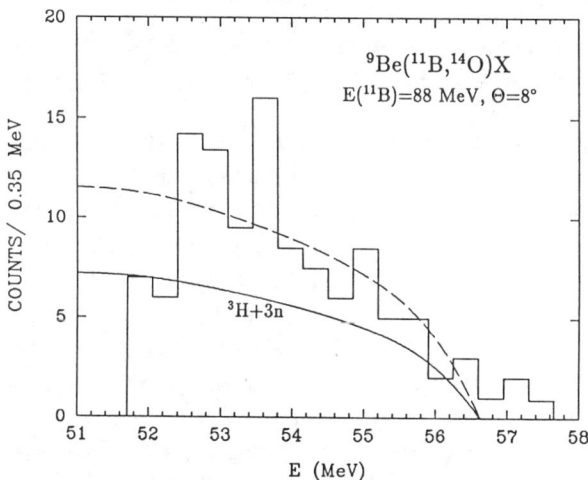

Fig. 5. ^{14}O spectrum obtained in the ^{9}Be(^{11}B, ^{14}O)X reaction. The dashed curve corresponds to 1.6 × the solid curve used by the authors for phase-space contribution (Based on ref. 14.)

Since the existence of ^{6}H has great theoretical significance, we have attempted to verify it by means of pion double charge exchange on ^{6}Li. The ^{6}Li(π^-,π^+)X reaction was studied with a 8" × 4" × 0.700 g/cm^2 solid target of lithium hydride (enriched to 95.5% ^{6}Li) which had a total impurity (elements other than lithium and hydrogen) content of < 0.5% by weight. A 220 MeV beam of ∼ 5 × 10^7 π^-/sec was used. Since $l = 1$ or 3 is expected[7] for the (^{6}Li)1$^+$ → 2$^-$ or 1$^-$(^{6}H) transition, the outgoing π^+ were detected at $\theta = 30°$. From auxiliary elastic scattering runs it was determined that the energy resolution, FWHM ≈ 0.6 MeV. The excitation energy spectrum (with E* = 0 MeV corresponding to the break-up reaction with X = ^{3}H + 3n) obtained in a 36 hour exploratory run is shown in Fig. 6. The curves in the figure show least-squares

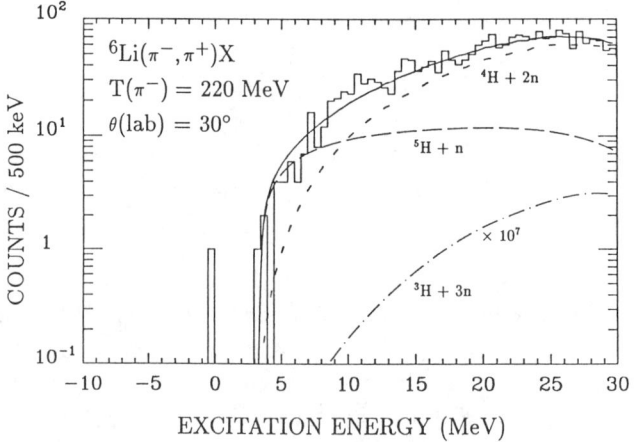

Figure 6. Excitation energy spectrum (E* = 0 for X = ^{3}H + 3n) for the reaction ^{6}Li(π^-,π^+)X. The curves are phase-space fits described in the text.

fitted phase-space contributions for the indicated break-up channels, and their sum (solid curve). Despite relatively poor statistics it is obvious from Fig. 6 that there is no enhancement at E* ≈ 2.7 MeV and that the data are fit quite satisfactorily by a sum of phase space contributions for ^{4}H+2n and ^{5}H+n break-up. We therefore put an upper limit of 1.6 nb/sr (corresponding to 1 count in the 3 MeV region, 2.5 ± 1.5 MeV) for the production of ^{6}H, if it exists.

It is difficult to compare our reaction with the heavy-ion reactions and to refute existence of ^{6}H claimed in the two Russian papers. We can, however, examine the experiments carefully and isolate their weak points.

The experiment of Aleksandrov et al.[18] was done with a telescope of three solid state detctors at θ(lab) = 10°. No magnetic analysis was done. The authors complain about large pile-up background (presumably due to elastic scattering) and call it the "most dangerous source of background". However, they also claim that they know the shape of their pile-up spectrum accurately and can reliably subtract it. We find this difficult to believe, and would like to see the experiment repeated under better conditions, preferably with a magnetic spectrometer.

The experiment of Belozyorev et al.[17] severely lacks in statistics. We do not know why they normalize their phase-space curve as they do. However, if the curve is normalized 60% higher (shown by solid line in Fig. 5) we find that the data are fit with the pure phase-space prediction quite reasonably. The fit without any resonance has a value of $\chi^2 = 13$ for 16 degrees of freedom. We believe that this experiment must be repeated with much better statistics before any reliable conclusions can be drawn.

REFERENCES

1. V. R. Pandharipande and R. B. Wiringa, Rev. Mod. Phys. **51**, 821(1979); also V. R. Pandharipande, priv. comm.
2. F. W. N. de Boer, Nucl. Phys. **A350**, 149(1980).
3. L. Gilly et al., Phys. Lett. **19**, 335(1965); A. Stetz et al., Phys. Rev. Lett. **47**, 782(1981); J. E. Ungar et al., Phys. Lett. **144B**, 333(1984).
4. E. R. Kinney et al., Phys. Rev. Lett. **57**, 3152(1986).
5. O. D. Brill et al., Phys. Lett. **12**, 51(1964); J. Cerny et al., Phys. Lett. **53B**, 247(1974).
6. A. V. Belozyorov et al., Dubna preprint E7-87-140.
7. S. Fiarman and W. E. Meyerhof, Nucl. Phys. **A206**, 1(1973).
8. U. Sennhauser et al., Phys. Lett. **103B**, 409(1981).
9. K. K. Seth, Proc. 4th Internat. Conf. on Nuclei Far from Stability, Helsingor, P. G. Hansen and O. B. Nielsen editors, CERN Report No. CERN 81-09(1981), p. 655.
10. K. K. Seth et al., Phys. Lett. **173**, 397(1986).
11. K. K. Seth et al., Phys. Rev. Lett. **58**, 1930(1987).
12. A. G. M. Van Hees and P. W. M. Glaudemans, Z. Phys. **A314**, 323(1983), and **315**, 223(1984).
13. N. A. F. M. Poppelier, L. D. Wood, and P. W. M. Glaudemans, Phys. Lett. **157B**, 120(1985).
14. A. V. Belozyorov et al., Nucl. Phys. **A460**, 352(1986).
15. D. V. Aleksandrov et al., Sov. J. Nucl. Phys. **39**, 323(1984).

A SHELL-MODEL STUDY OF LIGHT EXOTIC NUCLEI

N.A.F.M. Poppelier, J.H. de Vries,
A.A. Wolters, and P.W.M. Glaudemans
Department of Physics and Astronomy, Rijksuniversiteit Utrecht
P.O. Box 80.000, 3508 TA Utrecht, The Netherlands

ABSTRACT

We present results of shell-model calculations on exotic p-shell and sd-shell nuclei. We have determined the effective interaction from an empirical procedure using both energies and electromagnetic moments of $T \leq 2$ nuclei as input. Binding energies, spectra, dipole moments, and radii of ground states are compared with experimental data.

INTRODUCTION

During the past few years it has become possible to produce and identify exotic nuclei and at the same time to determine a number of their physical properties, using sophisticated experimental techniques. This makes it interesting to compare the experimental results with the predictions of nuclear structure models. The region of the light exotic nuclei is an excellent testing ground for a shell-model approach to the description of light nuclei that we have developed over the past few years. This approach has been applied to the normal light nuclei with good results. The main question we hope to answer with the present work is: can non-exotic and exotic nuclei be described in one model?

In an earlier paper[1] we already presented results of calculations for exotic p-shell nuclei. In that calculation we used a $(0+1)\hbar\omega$ model space, which was constructed without employing an inert core, and the effective interaction was written as $V_{\text{eff}} = V + (V_{\text{eff}} - V)$. For the realistic interaction V we used the Reid soft-core interaction[2]. The difference between the realistic interaction and the effective interaction was parametrized in terms of nine Talmi integrals. We determined these Talmi integrals empirically from binding energies of ground states and excited states of both normal and non-normal parity.

The present calculations are also performed in a $(0+1)\hbar\omega$ model space without employing an inert core but using a different effective interaction. Instead of decomposing the effective interaction as in the previous paper[1], we now parametrize the effective interaction directly in terms of Talmi integrals. The required number of Talmi integrals depends on the model space concerned, and is 29 in the present case. These Talmi integrals are determined from energies as well as electromagnetic multipole moments, in a way similar to that used in recent studies[3,4]. The reason for applying this approach to the region of exotic nuclei is the following. We have found that several Talmi integrals are not determined well enough from binding energies alone. If we determine the effective interaction as described above we obtain better results for energies and dipole moments of normal p-shell nuclei[3]. We expect to see the same improvement for exotic p-shell nuclei.

In the rest of this paper we will concentrate on the results obtained in the $0\hbar\omega$ model space, since we are interested in ground-state properties only.

Calculation of the Hamiltonian (and other operators) and the eigenfunctions has been performed on a Cyber 205 computer with RITSSCHIL[5] and related programs, all developed in Utrecht.

METHOD

If one considers a nucleus as a system of A particles with mass m and momenta p_k experiencing a two-body interaction V_{kl} then the Hamiltonian for such a nucleus can be written as[6,7]:

$$H = T + V = \sum_{k=1}^{A} \frac{p_k^2}{2m} + \sum_{k<l}^{A} V_{kl}. \qquad (1)$$

Because V_{kl} depends only on the relative coordinates, this Hamiltonian is translationally invariant.

The right-hand side of Eq. (1) can be separated into a part which depends on the coordinates of the nuclear center-of-mass and a part which depends on intrinsic coordinates[8]. The nucleus is localized near the origin of the laboratory system by adding a center-of-mass potential of harmonic-oscillator form to the Hamiltonian, an operation that does not affect the intrinsic structure of the Hamiltonian. The modified Hamiltonian can be written as

$$H = \sum_{k=1}^{A} \left[\frac{p_k^2}{2m} + \tfrac{1}{2} m\omega^2 r_k^2 \right] + \sum_{k<l} \left[V_{kl} - \frac{m\omega^2}{2A}(r_k - r_l)^2 \right], \qquad (2)$$

and can be separated into a part depending on the center-of-mass coordinates only and a part depending on the intrinsic coordinates. The fact that this Hamiltonian is still separable ensures that the non-physical (spurious) states due to the excitation of the center of mass can be removed from the spectrum following the treatment of Van Hees et al[7].

In order to find an expression for the two-body matrix elements of a translationally invariant interaction one applies transformations between jj-coupling and ls-coupling, and a transformation from laboratory coordinates to relative and center-of-mass coordinates[7]:

$$\langle \rho_a \rho_b \| V \| \rho_c \rho_d \rangle =$$
$$\sum_{nln'l'NLSj} C(\rho_a\rho_b; nlNLSj)\, C(\rho_c\rho_d; n'l'NLSj) \langle n(lS)j \| V \| n'(l'S)j \rangle, \quad (3)$$

where the coefficients $C(\rho_a\rho_b; nlNLSj)$ are transformation coefficients[7]. The radial part of the relative matrix elements $\langle n(lS)j \| V \| n'(l'S)j \rangle$ appearing in Eq. (3) can be expanded in Talmi integrals[4,9]

$$\langle nl | V | n'l' \rangle = \sum_{p=\lambda}^{\lambda+n+n'} B(nl, n'l'; p)\, I_p, \quad (4)$$

where $\lambda = (l+l')/2$, and the Talmi integrals I_p are defined by

$$I_p = \frac{\sqrt{2}}{b\Gamma(p+\frac{3}{2})} \int_0^\infty e^{-r^2/2b^2} \left(\frac{r}{b\sqrt{2}}\right)^{2p+2} V(r)\, dr. \quad (5)$$

Since the calculations are performed in a finite model space, the interaction V should be replaced by an effective interaction V_{eff}. In our approach the following assumptions are made with respect to the effective interaction:

- the effective interaction is translationally invariant, rotationally invariant, and parity conserving, which implies that the effective interaction is the sum of a central term, a tensor term, a spin-orbit term, and a quadratic spin-orbit term which is negligible in our case;

- isospin (T) is a good quantum number;

- the Talmi integrals of the effective interaction can be determined accurately enough from properties of states of normal and non-normal parity in $T \leq 2$ nuclei; and

- the Talmi integrals of the effective interaction are mass-independent in the region A=4–40.

The values of the Talmi integrals are determined empirically by means of a least-squares fitting procedure using binding energies of ground states and excited states as well as magnetic dipole moments and electric quadrupole moments as input[3].

We can parametrize the Hamiltonian as

$$H = \sum_{k=1}^{n_\lambda} H_k \lambda_k, \quad (6)$$

and the electromagnetic multipole moment operator as:

$$M = \sum_{k=1}^{n_\mu} M_k \mu_k, \qquad (7)$$

where the λ_k are the parameters of the Hamiltonian, in our case $\hbar\omega$ and the Talmi integrals, and the μ_k are the parameters of the moment operators, in our case the gyromagnetic ratios and the nucleon charges.

Using Rayleigh-Schrödinger perturbation theory one can show[3] that the first-order corrected energies $E_k + \delta E_k$ and moments $m_l + \delta m_l$ are then given by:

$$E_k + \delta E_k = \sum_{r=1}^{n_\lambda} \langle \psi_k | H_r | \psi_k \rangle (\lambda_r + \delta\lambda_r), \qquad (8)$$

and

$$m_l + \delta m_l = \sum_{s=1}^{n_\mu} \langle \psi_l | M_s | \psi_l \rangle (\mu_s + \delta\mu_s) + 2 \sum_{r=1}^{n_\lambda} \sum_{k \neq l} \frac{\langle \psi_k | H_r | \psi_l \rangle}{E_k - E_l} \langle \psi_k | M | \psi_l \rangle \, \delta\lambda_r \quad (9)$$

respectively. The term to be minimized in a least-squares fitting routine, including energy levels as well as moments, is given by[3]:

$$Q^2 = \sum_{k=1}^{n_L} v_k^2 \{ E_k + \delta E_k - E_k^{exp} \}^2 + \sum_{l=1}^{n_M} w_l^2 \{ m_l + \delta m_l - m_l^{exp} \}^2, \qquad (10)$$

where n_L, n_M, v_k, and w_l are the number of experimental energy levels, the number of experimental moments, the weight factors of the energies, and the weight factors of the moments, respectively.

MODEL SPACE AND EFFECTIVE INTERACTION

The present shell-model calculations are performed on exotic isotopes of nuclei with Z=2–9 in a complete $0\hbar\omega$ model space. This means that all excitations of nucleons within major shells are taken into account, whereas excitations from one major shell into another are excluded. In this way we obtain only normal-parity states. The effective interaction in this model space is parametrized in terms of $\hbar\omega$ and 29 Talmi integrals.

The set of experimental data used in the fitting procedure consisted of 205 binding energies of ground states and excited states, and all experimentally known magnetic dipole moments (47) and electric quadrupole moments (21) of $T \leq 2$ states of both normal and non-normal parity in A=4–21 nuclei. In principle, this also enables us to predict properties of excited states, although we will concentrate on ground-state properties only.

In preliminary calculations we found that a few parameters of the effective interaction are not determined accurately enough from data on nuclei in the mass region A=4–21. Therefore we included the binding energies of the ground state of ^{40}Ca and the lowest two states of ^{39}K in the fit procedure, assuming that these energies could help to determine parameters of the effective interaction which otherwise would be determined only weakly.

Table 1: Parameters of the effective interaction (MeV)

Channel	Order		Value	Channel	Order		Value
singlet	$T=0$	$p=1$	+0.32	singlet	$T=1$	$p=0$	−6.13
singlet	$T=0$	$p=2$	−3.52	singlet	$T=1$	$p=1$	+0.29
singlet	$T=0$	$p=3$	+2.61	singlet	$T=1$	$p=2$	+0.85
triplet	$T=0$	$p=0$	−9.75	singlet	$T=1$	$p=3$	−1.08
triplet	$T=0$	$p=1$	−3.49	singlet	$T=1$	$p=4$	−3.15
triplet	$T=0$	$p=2$	−1.38	triplet	$T=1$	$p=1$	+1.18
triplet	$T=0$	$p=3$	−0.76	triplet	$T=1$	$p=2$	+1.39
triplet	$T=0$	$p=4$	+0.54	triplet	$T=1$	$p=3$	+1.10
tensor	$T=0$	$p=1$	+0.09	tensor	$T=1$	$p=1$	+0.51
tensor	$T=0$	$p=2$	+0.60	tensor	$T=1$	$p=2$	+0.85
tensor	$T=0$	$p=3$	+1.97	tensor	$T=1$	$p=3$	+0.98
tensor	$T=0$	$p=4$	+1.22				
LS	$T=0$	$p=2$	−0.45	LS	$T=1$	$p=1$	−0.89
LS	$T=0$	$p=3$	+0.77	LS	$T=1$	$p=2$	−0.13
LS	$T=0$	$p=4$	+0.57	LS	$T=1$	$p=3$	+0.02
$\hbar\omega$			+8.99				
b			+2.15				
σ_{rms}			+0.76				

The parameters of the effective interaction obtained in the procedure described above are given in Table 1. This table gives the values of $\hbar\omega$, the Talmi integrals, and the obtained average energy deviation for the present interaction. The value of the harmonic-oscillator size parameter b given in Table 1 is derived from $\hbar\omega$ using the formula $b = \sqrt{(\hbar/M\omega)}$.

Since we are working within the isospin formalism, the difference between protons and neutrons is not taken into account. As a consequence, the energies calculated by diagonalization of the Hamiltonian do not contain a Coulomb part.

The values of the Coulomb energy E_C have been obtained by a least-squares fitting procedure using measured ground-state binding energies in the region Z=2–12 and A=3–25 as input with the assumptions that

- the Coulomb energies are state independent and depend only on Z,

- the Coulomb energies are given by the difference in experimental energy of states belonging to the same isobaric multiplet.

Table 2 gives the Coulomb energies obtained using this procedure. These are added to the calculated binding energies to obtain the total binding energies, as discussed below.

Table 2: Empirically determined Coulomb energies (MeV)

Z	E_C	Z	E_C
2	0.76	6	7.62
3	1.57	7	10.37
4	3.14	8	13.67
5	5.02	9	17.10

ENERGIES

Using the interaction described above in a $0\hbar\omega$ model space, we have calculated binding energies of exotic nuclei with $Z=2-9$. In Figure 1 we compare theoretical binding energies of the ground states of exotic nuclei with available experimental data[10,11]. In Figures 1a and 1b the experimental binding energies are also compared with the predictions from the Garvey-Kelson mass formula. One should keep in mind that the latter approach is only capable of giving ground-state binding energies and contains considerably more parameters than the effective interaction used in our work.

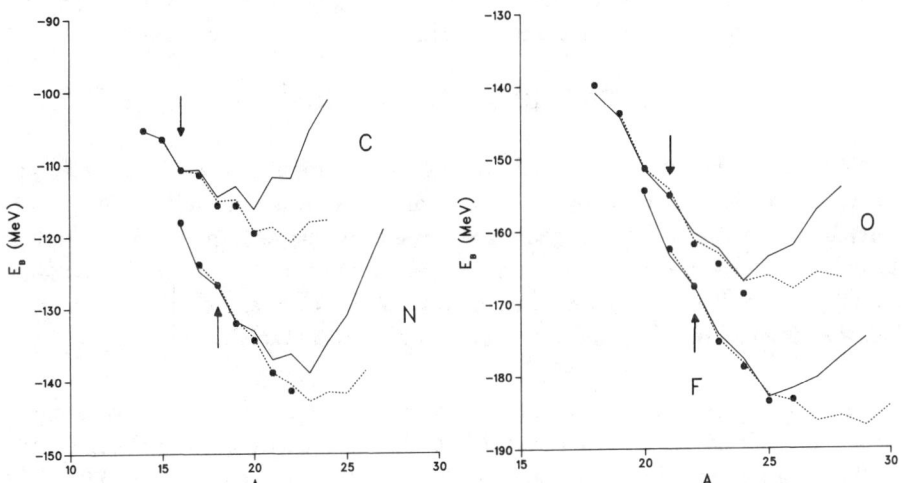

Figure 1: Binding energies (MeV) of ground states of a) C and N isotopes, and b) O and F isotopes (solid line: present work; dashed line: Garvey-Kelson mass formula) The arrows mark the isotopes with $T_z = 2$. Data of isotopes beyond the marked ones are not included in the fit.

The figures show that for large neutron/proton ratios the binding energies obtained in our calculation deviate from the experimental values and from the predictions obtained from the Garvey-Kelson formula. The explanation for this could lie in the fact that only experimental data of light nuclei with $T \leq 2$ have been used in the determination of the effective interaction. In other words: exotic nuclei are truly exotic in the sense that an effective interaction appropriate for a unified description of normal and exotic nuclei can not be determined uniquely from data on normal nuclei only.

Table 3: Stability of carbon isotopes with respect to neutron emission (+: stable; −: unstable)

Nucleus	Theory	Experiment	Nucleus	Theory	Experiment
^{16}C	+	+	^{21}C	−	−
^{17}C	+	+	^{22}C	+	+
^{18}C	+	+	^{23}C	−	
^{19}C	−	+	^{24}C	−	
^{20}C	+	+			

From Figures 1a and 1b we can immediately derive which isotopes are stable with respect to neutron emission. The predictions for particle stability, for the case of carbon, are given in Table 3. This table clearly shows that, despite the deviation between theoretical and experimental binding energies, the predictions for particle stability agree with the experimental results, except for ^{19}C.

MAGNETIC DIPOLE MOMENTS

Mass excesses, dipole moments, and lifetimes for many exotic nuclei have already been determined experimentally. Since there is still quite some uncertainty in the theoretical binding energies—see previous sections—the theoretical lifetime has considerable uncertainty. However, we expect that dipole moments can be quite accurately calculated, especially since we use an effective interaction that was determined using dipole moments of $T \leq 2$ states as input.

In Table 4 we present theoretical dipole moments of Li isotopes as an example. These dipole moments are calculated with renormalized gyromagnetic ratios, since these are also determined in the fit procedure. We obtained the following values for these parameters: $g^l_\pi = 1.02$, $g^s_\pi = 5.51$, $g^l_\nu = 0.02$, and $g^s_\nu = -3.84$.

Our predictions should be compared with the results of future experiments to establish whether or not our approach, which reproduces the observed dipole moments of normal light nuclei very well[3,4], is still valid in the region of exotic nuclei.

Table 4: Magnetic dipole moments of $^{6-18}$Li

Nucleus	J^π	Simple pred.[a]	Present work	Experiment
^{6}Li	1^+	0.63	0.82	0.82
^{7}Li	$3/2^-$	3.79	3.25	3.26
^{8}Li	2^+	1.25	1.63	1.65
^{9}Li	$3/2^-$	3.79	3.28	3.44
^{10}Li	1^+	2.84	2.88	—
^{11}Li	$3/2^-$	3.79	3.78	3.73
^{12}Li	2^-	1.88	1.60	—
^{13}Li	$3/2^-$	3.79	3.44	—
^{14}Li	4^-	1.88	1.77	—
^{15}Li	$3/2^-$	3.79	3.32	—
^{16}Li	4^-	1.87	2.05	—
^{17}Li	$3/2^-$	3.79	3.07	—
^{18}Li	3^-	0.58	1.09	—

[a]Schmidt formula or additivity formula[6]

MATTER RADII OF LITHIUM ISOTOPES

It has been argued[12] that nuclei with a large neutron/proton ratio show an effect which could be described as a neutron halo. Due to the low binding energy of some of the neutrons with respect to the rest of the nucleus, these neutrons tend to be found on the outside of the nucleus. A signature of this phenomenon would be matter radii which are considerably larger than the corresponding charge radii. In the case of the lithium isotope ^{11}Li a matter radius of 3.10(23) fm has been measured[14].

It is interesting to see if such an effect shows up in our calculations. In the $0\hbar\omega$ model space it is almost trivial to calculate charge radii and matter radii, if we make use of the fact that in our model the single-particle wave functions are of harmonic-oscillator type. If we assume that all active nucleons are in the p shell, there are simple formulae for these radii[7]:

$$\text{charge radius:} \quad \langle r^2 \rangle = \left(\frac{5}{2} - \frac{3}{2A} - \frac{2}{Z}\right) b^2 + a^2 \qquad (11)$$

$$\text{matter radius:} \quad \langle r^2 \rangle = \left(\frac{5}{2} - \frac{11}{2A}\right) b^2, \qquad (12)$$

where charge radii are calculated for a proton size of $a=0.85$ fm, and matter radii are calculated for point nucleons.

In a $(0+2)\hbar\omega$ model space, radii are calculated as expectation values of the corresponding operators, again using finite-size (point) nucleons for the charge (matter) radii. For this latter calculation we used an interaction for p-shell nuclei in a $(0+2)\hbar\omega$ model space, also determined empirically from energies and

Table 5: Charge and matter radii (fm) of Li isotopes

Nucleus	Charge radius			Matter radius		
	$0\hbar\omega$	$(0+2)\hbar\omega$	Experiment [a]	$0\hbar\omega$	$(0+2)\hbar\omega$	Experiment [b]
^{6}Li	2.48	2.41	2.57(10)	2.33	2.26	2.32(03)
^{7}Li	2.50	2.55	2.41(10)	2.42	2.50	2.33(02)
^{8}Li	2.52	2.60	–	2.49	2.62	2.37(02)
^{9}Li	2.53	2.64	–	2.54	2.72	2.32(02)
^{10}Li	2.55	2.64	–	2.58	2.77	–
^{11}Li	2.55	2.64	–	2.62	2.81	3.10(23)

[a]Ref. [13], model-independent analysis
[b]Ref. [14], point nucleons

static electromagnetic moments[4].

Although it would seem more consistent in shell-model calculations of radii and quadrupole moments to use a b parameter derived from the parameter $\hbar\omega$, we use an effective b value in Table 5. We have found that the parameter $\hbar\omega$ is not independent from the other parameters of the effective interaction, but that, instead, only a linear combination of $\hbar\omega$ and some Talmi integrals of order $p = 0$ is an independent parameter. This observation implies that we are free to modify $\hbar\omega$, and thus also to modify b. We have therefore used an effective value b_{eff} which reproduces the observed charge radii of ^{6}Li and ^{7}Li as well as possible. For the calculations in the $0\hbar\omega$ and $(0+2)\hbar\omega$ model spaces this results in $b_{\text{eff}} = 1.85$ fm, and $b_{\text{eff}} = 1.93$ fm, respectively.

The results of the $0\hbar\omega$ and the $(0+2)\hbar\omega$ calculations are given in Table 5. This table shows that only the predicted matter radius of ^{11}Li agrees fairly well with the measured value. Furthermore, our calculations predict a gradual increase of the matter radii with increasing mass number A, in contrast to experiment, and therefore do not support the idea of a neutron halo in ^{11}Li.

CONCLUSIONS

We have presented the results of shell-model calculations for light exotic nuclei, that is for exotic isotopes in the region $Z=2-9$. In previous papers[3,4] it was shown that a shell-model description using an empirically determined, mass-independent effective interaction gives good results in the case of ordinary, i.e. non-exotic, p-shell nuclei. From the present work it appears that such an approach can be extended to exotic p-shell nuclei, provided we also use data on nuclei with larger isospin values, $T \geq 5/2$, to determine the effective interaction. The theoretical binding energies provide an indication that these data are necessary in order to determine the $T = 1$ part of the effective interaction more accurately.

Therefore we hope that the present work will stimulate experimentalists to determine mass excesses and other properties of exotic nuclei in the near future.

The first author should like to express his gratitude to Shell Nederland B.V. for providing a travel grant.

This work was performed as part of the research program of the "Stichting voor Fundamenteel Onderzoek der Materie" (FOM) with financial support from the "Nederlandse Organisatie voor Zuiver Wetenschappelijk Onderzoek" (ZWO).

REFERENCES

1. N.A.F.M. Poppelier, L.D. Wood, and P.W.M. Glaudemans: Phys. Lett. **B157**, 120 (1985)

2. R.V. Reid: Ann. Phys. (NY) **50**, 411 (1968)

3. A.G.M. van Hees, A.A. Wolters, and P.W.M Glaudemans: Phys. Lett., to be published (1987)

4. A.A. Wolters, A.G.M. van Hees, and P.W.M. Glaudemans: to be published

5. D. Zwarts: Comp. Phys. Comm. **38**, 365 (1985)

6. P.J. Brussaard and P.W.M. Glaudemans: Shell-model applications in nuclear spectroscopy (North-Holland, Amsterdam 1977)

7. A.G.M. van Hees and P.W.M. Glaudemans: Z. Phys. **A314**, 323 (1983)

8. J.B. Aviles: Ann. Phys. **50**, 393 (1968)

9. I. Talmi: Helv. Phys. Acta **25**, 185 (1952)

10. A.H. Wapstra and G. Audi: Nucl. Phys. **A432**, 1 (1985)

11. A. Gillibert, W. Mittig, L. Bianchi, A. Cunsolo, B. Fernandez, A. Foti, J. Gastebois, C. Grégoire, Y. Schutz, and C. Stephan: GANIL, internal report 1987

12. M.J.G. Borge, P.G. Hansen, B. Jonson, S. Mattson, G. Nyman, K. Riisager, and O. Tengblad: contribution to this conference

13. H. de Vries, C.W. de Jager, and C. de Vries: At. Data Nucl. Data Tables **36**, 495 (1987)

14. I. Tanihata: contribution to this conference

SPECTROSCOPIC MEASUREMENTS WITH A NEW METHOD : THE PROJECTILE-FRAGMENTS ISOTOPIC SEPARATION

J.P. Dufour, R. Del Moral, F. Hubert, D. Jean, M.S. Pravikoff,
A. Fleury
C.E.N. Bordeaux, Le Haut Vigneau, F-33170 Gradignan, France

H. Delagrange, A.C. Mueller
GANIL, B.P. 5027, F-14021 Caen Cedex, France

K.-H. Schmidt, E. Hanelt, K. Summerer
G.S.I., D-6100 Darmstadt, Fed. Rep. of Germany

J. Frehaut, M. Beau, G. Giraudet
Commissariat à l'Energie Atomique,
Centre d'Etudes de Bruyères-le-Chatel
B.P. n° 12, F-91680 Bruyères-le-Chatel, France

ABSTRACT

A new method for the isotopic separation of projectile fragments is exposed. The application to the LISE spectrometer at GANIL provided radioactivity data on light neutron-rich nuclei (mass range 14-40). The observation on ^{17}B of the beta-delayed four neutron decay mode is reported for the first time.

INTRODUCTION

The ISOL technique has been very successful at exploiting the fragmentation of heavy nuclei by light energetic projectiles. One could however still hope to improve the Z-universality and shorten the transport times through the alternative technique of the so-called recoil spectrometers. For that purpose the inverse kinematics is highly desirable, although not strictly necessary[1]. This became conceivable first at the Bevalac and then at GANIL when heavy ion beams reached the proper energy regime for projectile fragmentation. It was first shown by Symons[2] that the relativistic energies allowed the use of a very simple spectrometer to identify exotic nuclei. The simplicity stemmed from the kinematics, for the fragments were forward focused in a narrow cone and had velocities very close to that of the beam. A few years ago the prospects for GANIL were more uncertain since the characteristics or even the existence of the projectile fragmentation was not clearly established below 100 MeV/n. However the beam intensity was much higher than at the Bevalac and that prompted the decision of building the LISE spectrometer. The design of the spectrometer allowed both a Time-of-Flight spectrometer mode and a new momentum-loss mode. This paper is focused on the development of this second mode and on the spectroscopic results it allowed to obtain. Other experimental results from LISE are exposed in these proceedings by D. Guillemaud-Mueller and J.C. Jacmart.

THE LISE SPECTROMETER

The originality of the device lies in its refocusing capability. As shown in fig. 1, a first section allows a momentum analysis at an intermediate focal plane and a second refocusing section provides an exit beam spot about 6 mm in diameter. The two sections also correspond to two different background environments : the target and the beam dump are separated from the experimental area by a thick concrete wall. Additional concrete has been recently added (where the dotted lines are drawn in fig. 1) to reach the very low background conditions required for neutron spectroscopy. An additional original feature of the spectrometer is the possibility to lower the magnetic rigidity of the second section so that a solid material (called here the "degrader") placed at the intermediate focal plane can slow down the ions. The purpose of this procedure is to slow down the high Z ions more than the low Z. The selection by the first dipole can thus be complemented with a further "momentum-loss" selection by the second dipole. This method, already applied to separate mono-charged elementary particles[3], finds a new extension in nuclear physics which requires a charge separation as well as a mass separation. The association of the reaction mechanism properties with the momentum-loss method results in a Projectile-Fragments Isotopic Separation (PFIS). Only a summary of the properties of the PFIS method will be presented here. The very detailed characteristics of the LISE spectrometer and of the PFIS method can be found in references 4 and 5. The study of a given nucleus requires the tuning of four parameters : the target thickness, the degrader thickness, the dipole 1 magnetic rigidity ($B\rho_1$) and the dipole 2 magnetic rigidity ($B\rho_2$).

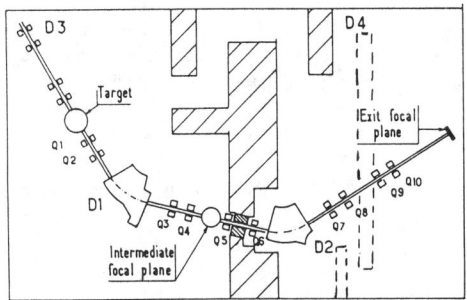

Fig. 1 : Schematic drawing of the LISE spectrometer at GANIL

The selection by the first dipole would allow a strict A/Z separation if the velocity of all the produced fragments was constant and if the target was thin. Even at intermediate energies (40-100 MeV/n), the reaction mechanism preserves the beam velocity rather well in average but a significant distribution width (close to the prediction from Goldhaber) is observed which causes a band A/Z ± Δ(A/Z) of nuclei to be simultaneously transmitted as shown in fig. 2 In addition to this unavoidable velocity spread the target also induces a spread since the fragments formed in the first or the last matter layers do not exit the target with identical velocities. The target thickness needs to be optimized to fill the Δp/p acceptance of the spectrometer. The increase of the target thickness results in wider and wider projectile-fragments exit energy distributions : the total number of ions inside the accepted Δp/p window levels off and then decreases. Typical optimum target thicknesses are 25% that of the beam range. Too large target thicknesses not only do not increase the transmitted rate but also widen the A/Z selected band through the widening of the energy distribution. The calculation of the optimum target must be made for each tuning according to the A, Z of the fragment and the opening of the slits defining the momentum acceptance.

The second selection is independent of the reaction mechanism. All the ions entering the degrader are slowed down differently according to their A, Z and also their velocity, which in turn depends on A and Z, once the Bρ is fixed (v_1 (A, Z) is proportionnal to $B\rho_1 * A/Z$). The $B\rho_2$ tuning can be analytically expressed as a function of $B\rho_1$, A, Z and d the degrader's thickness :

$$B\rho_2 = B\rho_1 (1 - d/R_1)^{1/2\lambda} \quad (1)$$

with $R_1 (A, Z, B\rho_1) = KB\rho_1^{2\lambda} (Z^{2\lambda-2}/A^{2\lambda-1})$ (2)

K and λ being constants depending on the stopping material only.

These formula can be easily derived with the assumption (excellent in the intermediate energy domain) that the range R of the heavy ions is proportional to $(\beta\gamma)^\lambda A/Z^2$, β and γ being the usual relativistic factors. Equation (1) shows that the ions selected at the same $B\rho_2$ are those having the same ranges R_1 (A, Z, $B\rho_1$) at the exit of the first dipole. This point illustrates for example the fact that the method can separate the A and A + 1 isotopes having the same Z and dE/dx but different ranges. The ions not discriminated by the second selection are thus those having the same $Z^{2\lambda-2}/A^{2\lambda-1}$ ratios as shown in fig. 2.

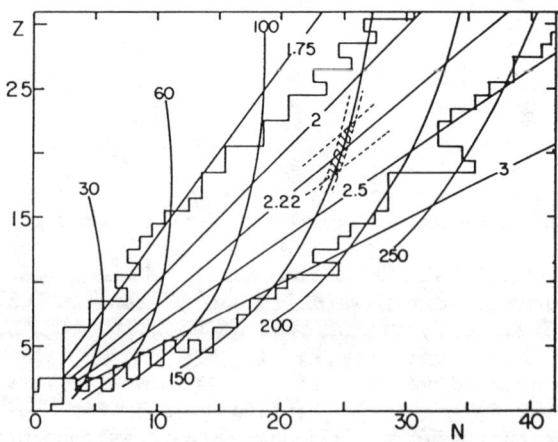

Fig. 2 : The selection operated by the degrader is shown here by a band in the N,Z plane. The lightly hatched area represents a typical selected domain which often includes only one isotope, while in less favorable cases two or three are included.

Equation (1) also leads to the determination of the shape that the degrader must have to conserve the achromatism of the line. At given distance x perpendicular to the beam axis, the magnetic rigidity is $B\rho_1(x) = B\rho_1(1+x/D_1)$ where D_1 is the dispersion of the first dipole. In order to preserve the linear dispersion, the ratio $d(x)/R_1(x)$, and also $d(x)/B\rho(x)^{2\lambda}$ must be maintained constant. One then finds that the degrader's shape must be such that :

$$d(x) = d_0 \ (1 + x/D_1)^{2\lambda} \qquad (3)$$

It is noteworthy that this equation is independent of $B\rho_1$, $B\rho_2$, A, and Z. This "accidental" property is very useful since it allows the use of a unique degrader for several tunings on different ions. The only reasons to change the degrader are the mass and charge resolutions of the second selection which can be shown[4] to linearly scale with d/R_1. In practice this just implied that in one experiment with a ^{40}Ar beam at 60 MeV/n, a set of three aluminium degraders (600μ, 900μ, 1200μ) allowed to select a variety of ions ranging from ^{15}C to ^{40}S. The achieved resolution is shown in fig. 2 by a curved band crossing the A/Z ± Δ(A/Z) band. The width of this second selection band can be estimated along a constant-Z cut, yielding a $(A/\Delta A)_d$ resolution. It must be noted that this momentum-loss resolution is very different from the usual momentum resolution of a magnet since it bears on ions all having the same initial magnetic rigidity $B\rho_1$. For degrader thicknesses going to zero, the $(A/\Delta A)_d$ also goes to zero and the width of the selection 2 in fig. 2 increases to infinity. In the case of LISE tuned on ^{37}P, the measured $(A/\Delta A)_d$ resolution reached 100 and was obtained from the ratio of the ^{37}P image diameter to the measured interval between the ^{37}P and ^{36}P images in the exit focal plane.

BETA-DELAYED GAMMA SPECTROSCOPY

Two experiments have been carried on with a ^{40}Ar beam at 60 MeV/n. The experimental detection set-up placed at the exit of LISE is shown in fig. 3.

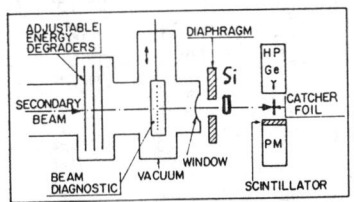

Fig. 3 : Schematic drawing of the experimental set-up placed at the exit of LISE

The detection of gammas was made by a single Ge detector (40% efficiency) in the first experiment and two NaI detectors plus a Silicon detector were added for the second experiment. The average beam intensities were 500 nA and 1 μA respectively. For the study of a given isotope the LISE tuning required less than 15 minutes and the typical data-taking time was six hours (with some exceptions !). The production rates and the nature of the measurements are listed in table 1 for the 22 studied isotopes .

Table 1 : Summary of the spectroscopic data obtained by fragmentation of a ^{40}Ar beam at GANIL. A star notes a first study, open triangles and circles respectively note improvements and confirmations of previous measurements

AZ	^{17}C	^{19}N	^{20}N	^{22}O	^{23}F	^{24}F	^{25}Ne	^{26}Ne	^{27}Na	^{30}Mg
$T_{1/2}$ (s)	o 0,20(6)	* 0,30(8)	* 0,07(4)	Δ 2,25(15)	*	* 0,34(8)	o 0,62(3)	* 0,23(6)	o 0,295(20)	Δ 0,342(18)
γ	*	*	*	*	Δ	*	o	*	o	o
N/s	20	>40	8	23	>200	41	510	50	>10^3	860

^{32}Al	^{33}Al	^{34}Al	^{35}Al	^{34}Si	^{35}Si	^{36}Si	^{37}Si	^{36}P	^{37}P	^{38}P	^{40}S
Δ 0,031(6)	*	* 0,05(2)	o 0,03(1)	* 3,9(11)	* 0,78(12)	* 0,45(6)	*	o 5,35(53)	* 2,31(12)	* 0,61(14)	* 8,6(21)
Δ	* ·	*	*	*	Δ	*	*	*	Δ	*	*	*
1700	260	40	5	>1600	430	100	6	22000	3300	225	58

The bulk of information gathered on these decay is very large and only part of it can be presented here. A first list of gammas, intensities and half-lives has already been published[6]. The half-lives were all measured with beam pulsations uncorrelated to the nuclei implantation. The isotopes listed above lie in three different shells : p, sd and fp. The comparison of the experimental data with shell-model calculations is most complete in the sd shell where Wildenthal's predictions[7] are detailed and include the β decay. The accuracy of these predictions is very impressive as can be seen in fig. 4 where the experimental over calculated half-life ratios are shown. The mean deviation X is the one defined by Klapdor which averages the r_{max} = SUP (T_{exp}/T_{calc} ; T_{calc}/T_{exp}) values. In the case of ^{24}F, the ambiguity (2$^+$, 3$^+$) in the calculation for the fundamental explains the two points placed in fig. 4.

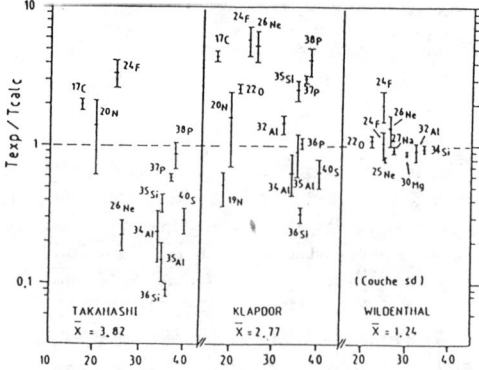

Fig. 4 : The ratios of the experimental over theoretical half-life determinations are plotted for different predictions.

Nuclei in the sd shell : an interesting property of the PFIS
selection is a very good rejection rate of the daughter of the separated isotope. The beta branching of ^{23}F and ^{34}Si to the ground
states of ^{23}Ne and ^{34}P have thus been obtained through the daughter's
total activity determination. The experimental values (30 ± 8% and
65.5 ± 2.3% respectively) compare well with the theoretical
predictions : 33.6% and 56%. Another specificity of the apparatus is
its tuning versatility and very short transport time. Those
properties were especially useful in the study of ^{26}Ne. When LISE was
tuned on ^{26}Ne very little gamma activity was observed in a 600 μs
beta-gamma coincidence window. The spectrometer was then tuned on
^{26}Na and it was possible to observe the direct decay of a 82.5 ± 0.5
keV excited state not in coincidence with betas. A multispectrum
time-analysis starting after each implantation of a ^{26}Na isotope
allowed the determination of the half-life, 9 ± 2 μs, of this excited
state. Assuming no direct feeding of the ^{26}Na ground state ($3^+ \rightarrow 0^+$
transition) leads to a decay scheme, shown in fig. 5, in good
agréement with the theory.

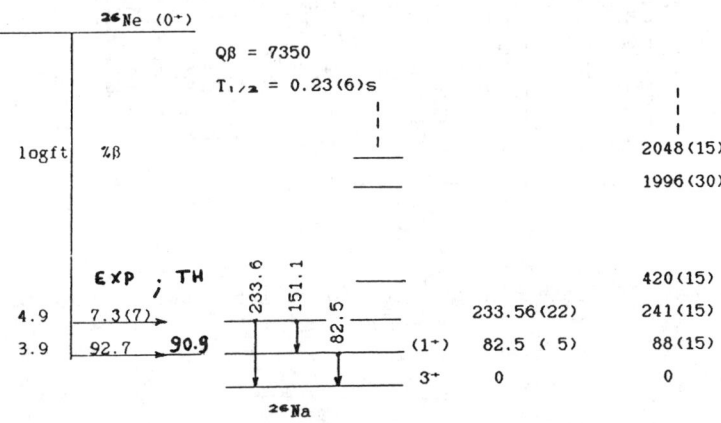

Fig. 5 : Partial decay scheme of ^{26}Ne. The half-life of
the 82.5 keV level was measured $T_{1/2}$ = 9 ± 2 μs

The case of ^{22}O revealed unexpected experimental difficulties. A
first measurement of the half-life made by Murphy and collaborators[8]
in 1982 yielded a value equal to 0.93 ± 0.35 s. When the beam was
pulsed with 2s beam-off and 2s beam-on periods, we observed an
astonishingly high beam-off activity on the 637 keV main gamma-ray
observed when LISE was tuned on ^{22}O. The second experiment allowed to
measure the half-life of different gammas for several pulsation
rates. The independent determinations all agree and converge to give
an average value $T_{1/2}$ = 2.25 ± 0.15 s, more than three standard
deviations away from the 1982 value. It thus appears that the high
beam-off activity first observed was due to a statistical fluctuation
and not to a possible isomer. The spectroscopic data also lead to
some difficulties with the observation of a strong 70.9 keV gamma ray
in coincidence with the 637 keV transition. If a low energy 3^+ state
(possibly at 70.9 keV) exists in ^{22}F as suggested by shell-model

calculations, then the measurements of excited states by Stokes and collaborators[9], or Orr[10] and collaborators, are likely to have treated the 3^+, 4^+ doublet as one single state and that could displace their reported energies by about half the doublet width (35 keV).

<u>Nuclei in the fp shell</u> : the extension of the shell-model description from the sd to the fp shell, further away from the ^{16}O inert core, has been recently improved by Warburton and Becker[11]. The comparison

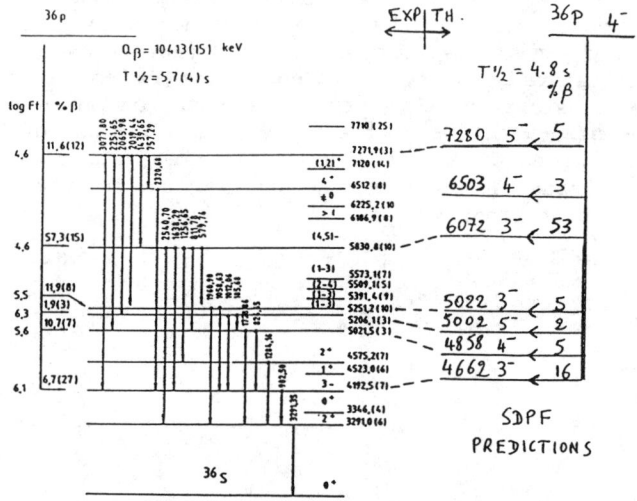

Fig. 6 : ^{36}P decay scheme. The theoretical SDPF predictions are from ref. 11

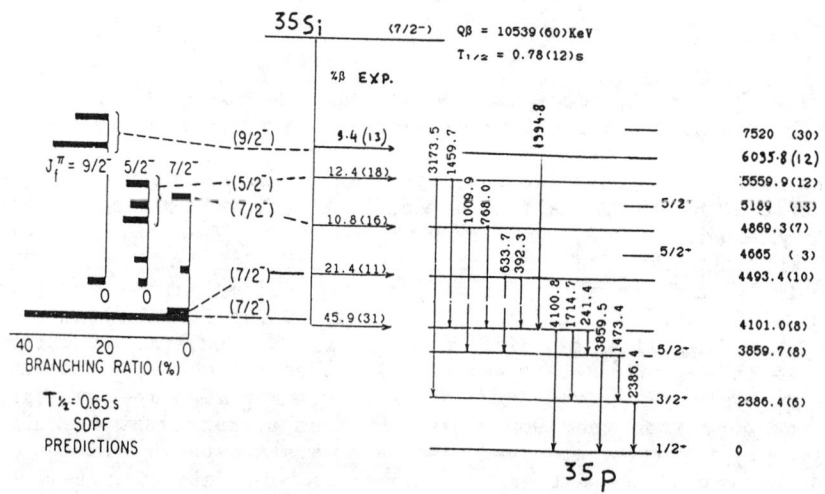

Fig. 7 : ^{35}Si decay scheme

of the experimental data with the calculation is very encouraging as can be seen in fig. 6 and 7 where the decay schemes of ^{36}P and ^{35}Si are represented. The SDPF interaction thus proves its predictive power for the N = 21 isotones from ^{38}Cl to ^{35}Si. Our data on ^{34}Al are still too scarce for a fruitful comparison. The experimental data need to be very extensive in this transition region where the decay from negative parity ground-states feeds very high energy excited states in the daughter. A good example is the 12% feeding from ^{36}P of the 7271 keV state in ^{36}S not seen in the first study[12] of this nucleus. Extension of these calculations is in progress for ^{37}P and ^{36}Si for which good data are available.

FIRST OBSERVATION OF THE BETA-DELAYED FOUR NEUTRON DECAY MODE

A very recent experiment was carried on with a ^{22}Ne beam at 60 MeV/n. The nuclei observed in this run are $^{17-18-19}$C, ^{17}B, ^{14}Be. The experimental set-up was designed for a maximum detection efficiency on the rare β,xn channels. The beta efficiency was improved by implanting the ions at a 2.5 mm depth in a 5 mm plastic scintillator. The photomultiplier gain was adjusted for normal beta detection while the implantation signal saturated the tube for less than 1 μs. This detector was placed in the hollow central part of a 4 π neutron detector described in ref. 13. The efficiency was measured to be 74% for neutrons at energies in the 0-3 MeV range. The identification of the nuclei separated by LISE was made with a single 300μ thick silicon detector which provided both a ΔE and a start signal for a time of flight identification using the HF machine as a stop signal. For these light neutron-rich nuclei the PFIS separation includes essentially two isotopes of the type $^{A+2}$Z+1, ^{A}Z. For the study of ^{17}B, ^{19}C could thus be considered as a contaminant, but due to the low counting rates and the short half-lives of these nuclei, the observed decays can easily be time-correlated to the last ^{17}B or ^{19}C implantation. The beam was stopped after any ^{17}B or ^{19}C implantation for a period of 100 ms. A target effect was clearly observed : the Be target allowed a better production of ^{18}C and ^{19}C than the Ta target whereas the opposite was true for ^{17}B. The case of ^{19}C, which is not a "pure fragmentation" product since it bears one more neutron than ^{22}Ne, seems to show that the mechanism leading to this nucleus is more of the incomplete fusion type (favored by Be) than of the deep-inelastic type (favored by Ta).

The observation of high neutron multiplicities must be accompanied by a full account of pile-up abundances, especially when using detectors with long thermalization times. With the detector used here, this average time was 11 μs and a beta-started 50 μs window allowed a full neutron collection. Another 50 μs time window was open 100 μs after the beta signal to measure the exact background conditions for each event. This procedure has been used in the metrology of neutron multiplicities in the fission process[13]. The multiplicity spectra obtained for ^{17}B are shown in fig. 8.

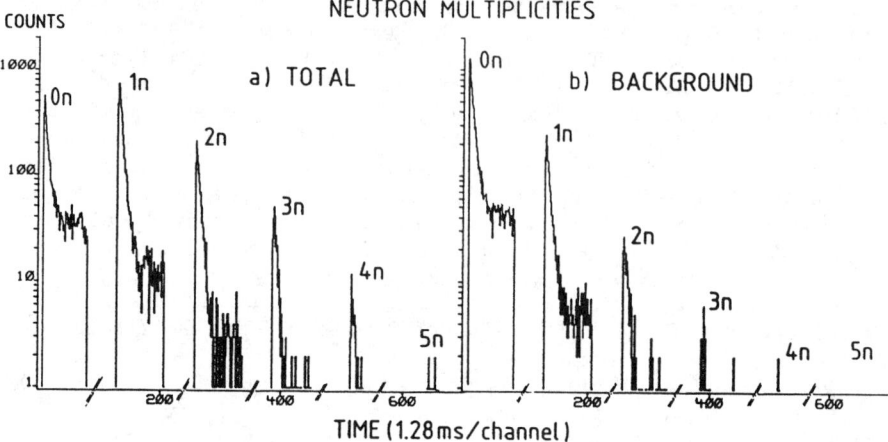

Fig. 9 : The neutron multiplicity is displayed as a function of the time difference between implantation and decay. Spectrum a) shows the data taken in a beta-coincident time window while, b) shows the data taken in a slightly decorrelated time window testing random coincidences

From these raw data two sets of corrections are made : i) the pile-up of true correlated neutrons with background events is corrected using the second coincidence window data, ii) the detection efficiency is taken into account. The measured half-lives and β, xn branchings for ^{14}Be, ^{17}B and ^{19}C are given in table 2. The observation in the case of ^{17}B of a beta-delayed four neutrons radioactivity is thus reported for the first time.

Table 2 : Preliminary beta-delayed multi-neutron branching ratios and half-lives of ^{14}Be, ^{17}B, ^{19}C

	$T_{1/2}$ (ms)	0n	1n	2n	3n	4n	Nb. events
^{14}Be	4.8(10)	0.27	0.59	0.15	$<10^{-2}$	-	2000
^{17}B	5.0(5)	0.24	0.62	0.11	$3.3 \, 10^{-2}$	$(4\pm2) \, 10^{-3}$	20000
^{19}C	30 (10)	-	-	-	-	-	1600

References

1 - J.M. WOUTERS et al. NIM A240 (1985) 77
2 - T.J.M. SYMONS. Phys. Rev. Lett. 42 (1979) 40
3 - A.P. BANFORD. "The transport of charged particle beams".
    E. & F.M. Spon Ltd 1966 London p. 127
4 - J.P. DUFOUR et al. NIM A248 (1986) 267
5 - R. ANNE et al. NIM A257 (1987) 233
6 - J.P. DUFOUR et al. Z. Phys. A324 (1986) 487
7 - B.H. WILDENTHAL et al. Phys. Rev. C28 (1983) 1343
8 - M.J. MURPHY et al. Phys. Rev. Lett. 49 (1982) 455
9 - R.H. STOKES et al. Phys. Rev. 178 (1969) 1789
10 - N. ORR. Private communication. CANBERRA, AUSTRALIA
11 - E.K. WARBURTON et al. Phys. Rev. C34 (1986) 1031
                           Phys. Rev. C35 (1987) 1851
12 - J.C. HILL. Phys. Rev. C25 (1982) 3104
13 - J. FREHAUT. NIM 135 (1976) 511

THE SPECTROSCOPY OF N=Z NUCLEI FROM $^{64}_{32}$Ge TOWARDS $^{100}_{50}$Sn

C.J. Lister, B.J. Varley, W. Gelletly, A.A. Chishti
Department of Physics, Schuster Laboratory
The University of Manchester
Manchester M13 9PL, United Kingdom

A.N. James and T. Morrison
Department of Physics, Oliver Lodge University
The University of Liverpool
Liverpool L69 3BX, United Kingdom

H.G. Price and J. Simpson
Nuclear Structure Facility, S.E.R.C. Daresbury
Warrington, WA4 4AD, United Kingdom

O. Skeppstedt
Department of Physics, Chalmers University of Technology
S-412 96 Goteborg, Sweden

ABSTRACT

The gamma decay of low-lying states in the N=Z nuclei $^{64}_{32}$Ge, $^{68}_{34}$Se, $^{72}_{36}$Kr, $^{76}_{38}$Sr and $^{80}_{40}$Zr have been studied using the Daresbury Recoil Separator operated in coincidence with an array of Compton Suppressed Ge detectors. The nuclei were produced in inverse heavy ion reactions using ^{54}Fe and ^{58}Ni beams and were populated via 2n decay channels. Production cross sections decreased from 500 μb to 10 μb across the region. Abrupt shape changes were found to occur along the N=Z line. The sensitivity of the method is discussed and the implications of the results on nuclear structure calculations. The prospects of advancing towards doubly magic ^{100}Sn are considered.

INTRODUCTION

The domain of nuclides accessible for spectroscopic study has increased drastically during the last two decades. New accelerators and sensitive detector systems have permitted access to regions of isotopes which lie far from the valley of stability. A broader picture of low energy nuclear behavior has emerged which has stimulated a host of theoretical attempts to generalize nuclear models to encompass all bound systems.

The study of nuclei with N≈Z and A≈80 is representative of this progress. Pioneering work with heavy ion beams[1,2] indicated both shape coexistence and substantial deformation while more recent experiments have shown the region to contain some of the most deformed nuclei known[3,4,5]. Rapid changes of shape have been observed, both with varying neutron or proton number, and with rotational frequency[2,6]. These changes have promoted much theoretical interest and calculations[7-15] have revealed a delicate interplay between classical collective effects and quantum shell corrections. The shell effects are especially distinct in nuclei with N=Z where both neutrons and protons act together to deform the nucleus to a common shape which is most bound. Thus, an experimental determination of low-lying nuclear shapes can be directly related to gaps in the single particle level sequence predicted from theory.

In this contribution we report on a series of experiments designed to increase the information on excited states in nuclei with N=Z. We have advanced the frontier of this type of measurement from N=Z=32 to N=Z=40 using the recently commissioned Daresbury Recoil Separator. In studying these nuclei, which have been sought for many years[1], we have increased the sensitivity for observing the gamma decay of excited states by about a factor of fifty, from $\sigma \approx 500$ μb to $\sigma \approx 10$ μb. This has been essential for the present studies and has great potential for exploring weakly populated states in nuclei nearer stability. The advance was made possible by detecting gamma radiation in an array of 10-20 Compton suppressed Ge detectors operated in coincidence with the Recoil Separator.

In the following sections we present some details of the technique, the results we have obtained so far, a comparison with some theoretical predictions and a consideration of forthcoming difficulties in searching for nuclei near the spherical double shell closure at N=Z=50.

EXPERIMENTAL TECHNIQUE

The recoil separator and its performance are described in ref. 18 and a forthcoming paper. It is a 0° electromagnetic separator with a solid angle of 10 msr. It consists of a double Wien filter to reject non-interacting beam particles to a level of $\leq 10^{-7}$ and allows the selection of the velocities of reaction residues to be studied. Residues of suitable velocity were deflected by a 50° dipole magnet onto a focal plane which was dispersed in mass but not in energy. The ions were detected at the focal plane by a carbon foil and position sensitive channel plate detector before being stopped in a split anode ion chamber similar to the one described by James et al.[19]. The transport and focussing of the beam of residues was determined by three quadrupole triplets and two sextupoles. A schematic layout of the device is shown in fig. 1.

Surrounding the target position was a ball support frame consisting of rings of 5 ports at angles of 143°, 117°, 101° and 79° to the beam line. Several combinations of detectors were used, the most fruitful being a combination of 10 Compton suppressed Gamma-X detectors in the 143° and 117° rings and 10 conical NE213 neutron detectors in the 101° and 79° ports. This arrangement was not ideal for the detection of neutrons, as the majority of the neutron flux passes nearer 0°. However, the data were useful in evaluating the additional sensitivity obtained by

Fig. 1 A block diagram of the components in the Daresbury Recoil Separator.

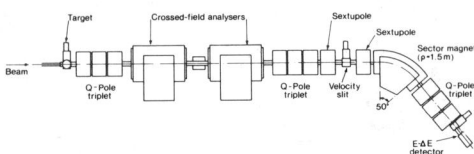

detecting the evaporated neutrons and a modified arrangement is being finalized.

The N=Z nuclei under study were populated using inverse heavy ion fusion reactions followed by pure neutron cooling for several reasons. The use of a heavy beam on a light target causes a high degree of kinematic focussing of recoil products into the separator. Further, the high recoil velocities produced (up to v/c = 6.5%) improve the Z-separation measured in the ion chamber. Pure neutron evaporation involving relatively low energy neutrons keeps the recoil cone small (and the transport efficiency high) while it also means that the isotope being sought has the highest Z of the recoil products, and hence the largest $\partial E/\partial X$ signal in the ion chamber. For example, the inverse compound nuclear reaction $^{24}Mg(^{58}Ni,2n)^{80}Zr$ reaction at 190 MeV was used to populate ^{80}Zr. At this energy 10% of the fusion reaction cross section proceeds via two nucleon evaporation which is dominated by two proton emission forming ^{80}Sr with a cross section of 44 ± 4 mb and proton plus neutron evaporation forming ^{80}Y with a cross section of 2 ± 1 mb. The inverse reaction has a center of mass recoil velocity, v/c = 5.5%. A target of 500 μg/cm$_2$ was found to produce the maximum yield of residues within the velocity acceptance (±3%) of the separator. To compensate for the thin target, beams of up to twenty particle nanoamps of ^{58}Ni were accelerated to 190 MeV using the Nuclear Structure Facility at Daresbury. The optimum beam energy was selected by directly measuring absolute gamma ray yield curves and through use of compound nuclear evaporation codes. Mass resolution was measured to be better than 1 part in 250 (fwhm) for each charge state so clean selection of mass was straightforward. In general two masses were observed on the focal plane corresponding to two and three nucleon emission. In the Strontium and Zirconium studies the three nucleon reaction channels (mainly

3p and 2pn evaporation) were about 5 times stronger than the two nucleon channels and a mechanical mask was placed between the position detector and ion chamber to suppress three nucleon evaporation events. In the case of Germanium, Selenium and Krypton production this procedure was less important and careful selection of beam energy allowed a reduction of three nucleon evaporation. Again, by way of example we consider the cse of ^{80}Zr. Here, two masses were dominant on the focal plane; A=79 and 80, both with charge state q=24$^+$. Of these, the A=79 recoils were 5.5±0.1 times stronger than those with A=80. The mask was positioned to admit ions with A/q of 3.333 ± 0.015. Most events involved ions with A=80 q=24$^+$, but a few A=77 q=23$^+$ ions were recorded. The latter events were from ^{77}Rb(αp evaporation) and ^{77}Sr(αn evaporation) and could be easily separated as they had lower energies and energy losses in the ion chamber. Less than 0.1% of the ion chamber events were associated with Zr ions and so Z selection was critical for the success of the experiment. The highly inverse reactions producing ^{64}Ge, ^{68}Se and ^{72}Kr achieved distinct Z-resolution. The case for the ^{16}O(^{58}Ni,2n)^{72}Kr is shown in fig. 2 where the energy losses of ions in the ion chamber are shown.

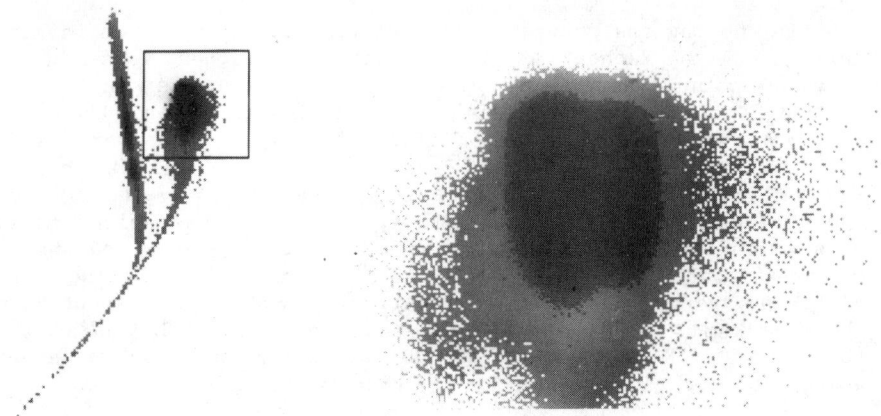

Fig. 2 Data from the split anode ion chamber. To the left is the full EΔE map, while to the right the recoil products are shown.

After a small numerical manipulation of the energy loss data to make an effective energy loss signal, ΔE_1, which was independent of energy, E, a two dimensional spectrum of energy loss against coincidence gamma ray energy was created. Events were all filtered for time-of-flight, position, total energy etc. and time randoms were subtracted. A series of gamma ray spectra were then produced corresponding to increased energy loss ΔE_1. The intensity of photopeaks in these spectra was then extracted.

Typical data for the A=72 and A=80 chains are shown in figs. 3 and 4.

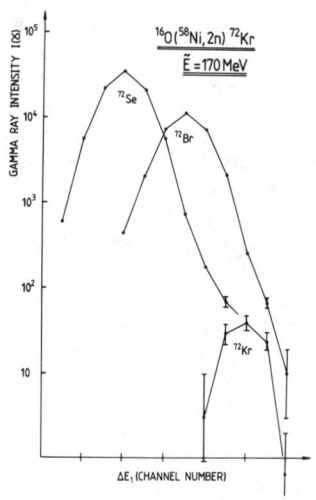
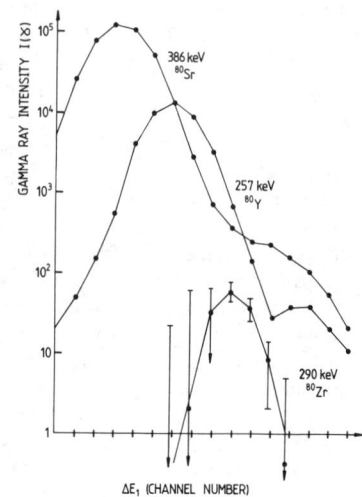

In the regions where the 2n reaction products were anticipated tails due to Rutherford scattering from carbon ions in the isobutane ion chamber gas were found. Again, in the lighter systems the Z-separation was sufficiently high, and the cross sections for the 2n channels sufficiently large, that very little data analysis was required. However, for ^{76}Sr and ^{80}Zr careful subtractions of low from high energy loss spectra were required. Again by example for the A=80 study, an energy loss selection was made which produced the optimum spectrum of Zr events and spectra of transitions known to belong to ^{80}Sr and ^{80}Y (obtained from different energy loss cuts) were carefully subtracted. About 4% of the optimum ^{80}Y and 0.1% of the optimum ^{80}Sr data were subtracted. Figs. 5, 6 show the resulting spectrum of events associated with ^{80}Zr and with ^{72}Kr respectively.

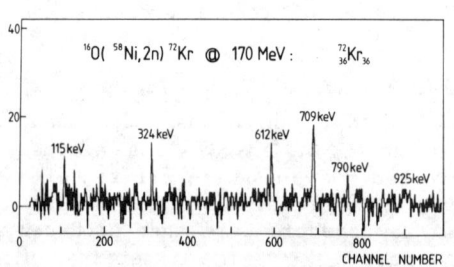
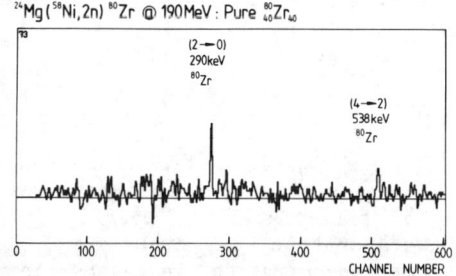

The data shown in figs. 3 and 4 reflect the relative reaction production rates of 2p, pn and 2n isotopes convoluted with gamma ray detection efficiency, nuclear decay scheme sequence, and recoil separator transport efficiency. For the 2p:2n relative cross section we assume that we are observing the normal $0^+ \to 2^+ \to 4^+$ ground state cascade of gamma rays in an even-even nucleus and the strongest transition, when corrected for internal conversion reflects 100% of the production cross section. The ion transport efficiency requires great care as it differs by a factor of about 5 between the 2p and 2n channels. It can be measured directly for the strong pn and 2p channels by comparing γ-ray singles to recoil-γ coincident spectra. However, this is not generally possible for the 2n channel due to its small cross section. A computer simulation of particle evaporation, scattering in the target and charge state fractionation was used together with the experimental 2p and pn data to predict the transport of ions produced in the 2n channel. However, this uncertainty of efficiency is the major contributor to the overall uncertainty on absolute production cross sections.

RESULTS

A considerable body of data have been amassed which have interest not only in their implications for spectroscopy of N=Z nuclei but also in the fields of high spin physics, fusion-evaporation mechanism studies and stopping power measurements. A few complete events have been collected in which γ-rays, evaporated particles and recoiling nuclei are all sampled. This type of data are entirely new and will require much analysis to benefit from their complexity. Most of these data were collected in July and only preliminary results are given here, mostly resulting from relatively straightforward scans of the data. However, the power of the method has enabled us to make considerable advances which are discussed below.

Table I contains details of the beams, targets and production cross sections for the N=Z isotopes we have studied.

$^{64}_{32}$Ge has been observed in three laboratories during the last year. At Oxford the ^{54}Fe(^{12}C,2n)^{64}Ge reaction was examined[16] using a neutron wall detector. The experiment identified the first excited state in ^{64}Ge, but indicated that 500 μb is approaching the limits of sensitivity for neutron wall detectors. The Pennsylvania group[17] used the ^{40}Ca(^{27}Al,p2n)^{64}Ge reaction and identified several candidate transitions, mostly at the limit of sensitivity. Our experiments used the ^{54}Fe+^{12}C reaction as a test for measuring the distribution of ^{64}Ge ions over a variety of velocities and charge states and their efficiency of transport through the separator. All the candidate transitions suggested to belong to ^{64}Ge by the Pennsylvania group were confirmed in our study.

Table I. A Summary of Measurements Made on N=Z Nuclei

Nucleus	Reaction	Beam Energy	Production Cross Section	Strongest Identified Transition (keV)
$^{64}_{32}$Ge	^{12}C(^{54}Fe,2n)^{64}Ge	155 MeV	500±200 μb[a]	902
$^{68}_{34}$Se	^{12}C(^{58}Ni,2n)^{68}Se	175 MeV	---	854[3]
$^{72}_{36}$Kr	^{16}O(^{58}Ni,2n)^{72}Kr	170 MeV[b]	60±15 μb	709
$^{76}_{38}$Sr	^{24}Mg(^{54}Fe,2n)^{76}Sr	177 MeV	10±5 μb[c]	261
$^{80}_{40}$Zr	^{24}Mg(^{58}Ni,2n)$^{80}_{40}$Zr	190 MeV	10±5 μb	289

[a]Measured by Ooi et al. ref. 16.
[b]A WO$_3$ target on Ta was used; this is the calculated mean beam energy.
[c]Preliminary results.

$^{68}_{34}$Se has not been previously observed neither in beam nor decay, although an erroneous decay was claimed then refuted[26]. In our ^{12}C(^{58}Ni,2n)^{68}Se study at a beam energy of 175 MeV on a 200 μg/cm^2 ^{12}C foil the recoiling nuclei were sufficiently fast (v/c=6.4%) to be well over the peak of the Bragg curve and Z-resolution was the best that we achieved. The highly irregular behavior of ^{70,72}Se appears to be replaced by a smoother sequence which indicates that a single shape plays a decisive role, though whether oblate or prolate is not yet clear.

^{72}Kr was formed in the ^{58}Ni + ^{16}O reaction. A WO$_3$ oxide target produced recoils with a wide velocity distribution which made optimization of the separator less critical. Again the Z-resolution was extremely good, an a cut on the first anode signal was sufficient to isolate ^{72}Kr without the more sophisticated analyses required for Sr and Zr.

The spectrum of transitions associated with ^{72}Kr differ from that of the heavier ^{74,76,78}Kr isotopes as the strongest transition is not lowest in energy. The spectrum closely resembles that of light ^{70,72}Se isotopes where coexistence of near spherical and prolate shapes is known. To understand this pattern of gamma rays, more data are required to permit a recoil-γ-γ coincidence analysis. However, by placing the transitions in order of intensity a close analogy to the decay scheme of ^{72}Se is found. Under the (extreme) assumption that the ^{72}Se-^{72}Kr analogy is complete, a two level mixing analysis suggested by Piercey[3] produces several systematic insights. The interaction (and mixing) between shapes is strong and increasing for lighter nuclei. The relative position of oblate (?) and prolate shapes vary regularly with neutron number; with the oblate configuration becoming the ground state in ^{72}Kr. The most deformed prolate configuration appears to be in the N=38 isotope ^{74}Kr which is consistent with our new data on $^{76}_{38}$Sr and $^{80}_{40}$Zr.

A band of states has been found in ^{76}Sr following the ^{24}Mg(^{54}Fe,2n)^{76}Sr reaction at 177 MeV. The three strongest transitions were also found in ^{40}Ca(^{40}Ca,2p2n)^{76}Sr data two years ago in a neutron wall experiment, although the assignment was then considered tentative due to oxygen contamination of the target. The first excited state was in a clean part of the gamma ray spectrum and an energy loss curve was easily extracted which was distinctly different from Kr and Rb transitions. The transitions indicate that the N=Z=38 nucleus is the most deformed in this region and the comparison with the ^{80}Zr data provide a sensitive test for many models.

The data on ^{80}Zr has been most carefully analyzed and has been published[25]. Two transitions are clearly identified (fig. 5) and a candidate for a third. The 290 keV line has an angular distribution similar to that of ^{80}Sr with 60±10% of the events occurring in the first ring of detectors and observation which is consistent with an E2 decay expected from the deexcitation of the first excited state in a deformed even-even nucleus. Although poor in statistical quality, this represents a first step towards spectroscopy of these exotic nuclei.

COMPARISON OF RESULTS WITH PREDICTIONS

The A=80 region exhibits a delicate interplay of classical collective behavior with quantum shell effects and has been the subject of considerable interest over the last two decades. The studies fall into three broad groups: Potential energy surface (PES) calculations to predict the most bound configurations and their binding energies; Hartree-Fock calculations of shape, mass and density distributions, and IBA studies of the development of various types of collectivity.

At present, our data on the new N=Z nuclei are primitive and require refinement before a detailed understanding of these nuclei is possible and quantitative model tests be made. We must assume the sequences we have observed are the normal $J^\pi = 0^+, 2^+, 4^+$ cascades which are strongly populated in even-even nuclei formed in heavy ion reactions, and for now we must estimate quadrupole deformation from the $J^\pi = 2^+$ excitation energy following the phenomenological estimate of Grodzins[20]. We have tested this estimate by comparing its predictions to deformations extracted from lifetime data. The test was made for Strontium isotopes with A=78-86. We assumed all the nuclei were axially symmetric-ellipsoids with mean radii $R_0=1.2A^{1/3}$ fm. We find that the Grodzins estimate closely follows the lifetime data[24] and that ^{76}Sr probably has a deformation close to $\beta_2=0.44$, the largest known in this region.

The data may be compared to several potential energy surface (PES) calculations. These theoretical estimates combine the classical collective nuclear properties of the liquid drop model with quantum corrections due to shell effects using the Strutinsky formalism[21]. Several early PES calculations[7,8] predicted soft, oblate shapes, but more recent estimates[9-12,15] indicate that there is great sensitivity of the most bound shape to the choice of model parameters. Again turning to ^{80}Zr for example, for which we estimate $\beta_2 = 0.4$, Moller and Nix[9] used a folded Yukawa potential for the quantum correction and considered axially symmetric shapes. They predicted $\varepsilon_2=0.39$ $\varepsilon_4=0.09$ ($\varepsilon_2=0.95\beta_2$). Heyde et al.[11] predicted a near spherical shape with a harmonic oscillator potential, while Nazarewicz et al.[12], using a Woods-Saxon potential, predicted $\beta_2=0.38$ and an axially symmetric, but γ-soft shape. The great sensitivity of the predicted shape to model parameters has been discussed by Heyde[11] and Galeriu[15] who both underline the close link between experiment and theory in this region.

Within the framework of the interacting boson model and related $N_\pi N_\nu$ scheme[22,23], ^{80}Zr and ^{76,78}Sr lie equally far from the major shell closures at nucleon numbers 28 and 50. Consequently the spectrum of excited states should be identical. The ratios of $E(4)/E(2)$ are indeed similar, being 2.83, 2.81 and 2.86 for ^{76,78}Sr and ^{80}Zr respectively. However, the model implies that the restricted valence space which limits the number of active protons or neutrons to less than 11 is not sufficient to allow quadrupole residual interactions to polarize the nuclei into axially symmetric shapes.

Two groups have attempted fully microscopic Hartree-Fock calculations in this region[13,14]. Large prolate deformations are predicted in both cases. Bonche et al.[13] predict that ^{78}Sr and ^{80}Zr should be γ-soft but that ^{76}Sr should be an axially symmetric rotor, the only one in this region. Clearly, the similicity of the $E(4)/E(2)$ ratios does not support this suggestion, nor the low value of $E(4)/E(2)=2.83$ which should be 3.33 for an axially symmetric shape.

PROGRESS TOWARDS $^{100}_{50}$Sn

It should be possible to advance the study of N=Z nuclei to $^{96}_{48}$Cd via the ^{28}Si(^{58}Ni,2n)$^{84}_{42}$Mo, ^{32}S(^{58}Ni,2n)$^{88}_{44}$Ru, ^{40}Ca(^{54}Fe,2n)$^{92}_{46}$Pd and ^{40}Ca(^{58}Ni,2n)$^{96}_{48}$Cd reactions before suitable beam and target combinations run out and 3 or 4 neutron evaporation channels must be sought. However, most experimental factors make progress towards heavier systems more difficult. The increasing Coulomb barriers for fusion imply hotter compound nuclei. Even near the barrier three and four nucleon evaporation dominate compound nuclear cooling and two nucleon evaporation becomes very weak ($\approx$ 10 mb). The compound systems lie further and further from the valley of stability and so competition between proton and neutron emission increasingly favor charged particle cooling. The reactions are less inverse which reduces the kinematic focussing into the separator and lowers the separator efficiency. The problem of Z-resolution, the limiting factor in the present studies, get worse. Slower ions means less separation between adjacent Z's. Further the fractional change between Z=49 and 50 is more difficult to resolve than between Z=31 and 32. Finally, the approach to ^{100}Sn will involve near spherical nuclei emitting high energy gamma rays which will be detected less efficiently.

The situation is not entirely desperate however, and the history of these conferences on nuclei far from stability reflect our ability to surmount many difficulties. New recoil separators are being constructed with 3-5 times higher efficiency, ion chambers and Bragg chambers are improving and larger more efficient gamma ray arrays being constructed. The use of coincidences with evaporated particles can help to resolve the problem of Z-measurement.

During the last two years we have made excellent progress in this field. The spectroscopy of intermediate mass N=Z nuclei is in its infancy and many measurements need to be made, especially in order to study the competition between oblate and prolate shapes of states in ^{68}Se and ^{72}Kr. However, our initial measurements have already provided tests of a variety of theoretical models and will hopefully continue to stimulate theoretical interest. Cross section measurements indicate that evaporation codes at present only provide a qualitative guide to exotic nuclear production and considerable effort is required in studying the prediction of exotic isotope yields.

Increase of sensitivity allows us to move further from stability and push back the frontiers of our ignorance. It also allows us to investigate nuclei nearer stability in far greater detail and a combination of these studies can unify our pictures of nuclear behavior in general.

REFERENCES

1. E. Nolte, W. Kutschera, Y. Shida, H. Morinaga, Phys. Lett. **33B** (1970) 294; E. Nolte, Y. Shida, W. Kutschera, R. Prestele and H. Morinaga, Z. Phys. **A268** 267 (1974).
2. J.H. Hamilton et al., Phys. Rev. Lett. **32** 239 (1974).
3. R.B. Piercey et al., Phys. Rev. Lett. **47** 1514 (1981).
4. C.J. Lister, B.J. Varley, H.G. Price and J.W. Olness, Phys. Rev. Lett. **49** 308 (1982).
5. L. Luhmann, K.P. Lieb, C.J. Lister, J.W. Olness, H.G. Price and B.J. Varley, Europhys. Lett. **1** 623 (1986).
6. H.G. Price, C.J. Lister, B.J. Varley, W. Gelletly and J.W. Olness, Phys. Rev. Lett. **57** 1842 (1983).
7. D. Bucurescu et al., Rev. Roum. Phys. **24** 971 (1979).
8. S. Aberg, Phys. Scr. **23** 23 (1982).
9. P. Moller and J.R. Nix, Nucl. Phys. **A361** (1981) 117 and At. Data and Nucl. Data Tables **26** 165 (1981).
10. I. Ragnarsson and R.K. Sheline, Phys. Scr. **29** 385 (1984).
11. K. Heyde, J. Moreau and M. Waroquier, Phys. Rev. **C29** 1859 (1984).
12. W. Nazarewicz et al., Nucl. Phys. **A435** 397 (1985).
13. P. Bonche, H. Flocard, P.H. Heenen, S.J. Krieger and M.S. Weiss, Nucl. Phys. **A443** 39 (1985).
14. D.P. Ahalpara, K.H. Bhatt and A. Abzouzi, Nucl. Phys. **A445** 1 (1985) and refs. therein.
15. D. Galeriu, D. Bucurescu and M. Ivascu, J. Phys. **G12** 329 (1986).
16. S.S.L Ooi et al., Phys. Rev. **C34** 1153 (1986).
17. J. Gorres et al., Phys. Rev. Lett. **58** 662 (1987).
18. A.N. James, Daresbury Study Weekend, DL/NUC/R20 publ. SERC. Daresbury 84 (1979).
19. A.N. James et al., Nucl. Inst. Meth. **212** 545 (1983).
20. $\beta_2 = 1228/[A^{7/3} E(2_1^+)]$; L. Grodzins, Phys. Lett. **2** 88 (1962); F.S. Stephens et al., Phys. Rev. Lett. **29** 438 (1972).
21. V.M. Strutinsky, Nucl. Phys. **A55** 420 (1967).
22. R.F. Casten, Phys. Rev. Lett. **58** 658 (1987) and reference therein.
23. S.L. Tabor, Phys. Rev. **C34** 311 (1986).
24. C.J. Lister et al., Phys. Rev. Lett. **49** 308 (1982); **Table of Isotopes**, Ed. C.M. Lederer and V.S. Shirley Publ. Wiley and Sons, New York 1978.
25. C.J. Lister et al., Phys. Rev. Lett. (1987) In press.
26. F. Kearns, At. Nucl. Data Tables **33** 481 (1981).

NEUTRON-RICH NUCLEI PRODUCED IN MULTINUCLEON-TRANSFER REACTION

W.-D. Schmidt-Ott, P. Koschel, F. Meissner, U. Bosch, E. Runte
University of Göttingen, 3400 Göttingen, Fed. Rep. of Germany

R. Kirchner, H. Folger, O. Klepper, E. Roeckl, D. Schardt
GSI Darmstadt, 6100 Darmstadt, Fed. Rep. of Germany

K. Rykaczewski
Uniwersytet Warszawski, 00681 Warszawa, Poland

ABSTRACT

A progress report on neutron-rich chromium-to-nickel isotopes is presented, including the new half-lives of ^{60}Cr, ^{60g}Mn, and ^{68}Ni of 0.57(6), 51(6), and 19^{+3}_{-6} s, respectively. The chromium and nickel isotopes show again larger β-decay rates than predicted. The measurement of continuous β-spectra has been used for the element assignment. For ^{66}Co, from the β-endpoint energy a Q_β=9.7(5) MeV was derived. The ^{68}Ni experiment was achieved by effective copper suppression with a bunched ion-source of the GSI mass-separator. The production of light isotopes in the reaction ^{50}Ti+^{238}U close to the Coulomb barrier was measured, however, the expected mass-transfer in quasi-fission could not be traced.

INTRODUCTION

In a series of investigations multinucleon-transfer in deeply inelastic. and quasielastic heavy-ion reactions has been used for the production of neutron-rich isotopes. About thirty new nuclei were found outside the region of fission fragments, and the research was performed for chromium-to-nickel [1-4], the lanthanides [5-8], actinium and radium [9]. We are presently concerned with the isotopes given in fig. 1. The nuclei are presented with their production yields. For both reactions a steep decrease of the yields was observed when going to more neutron-rich isotopes.

For a number of nuclei larger β-decay rates than predicted by calculations were observed [3], and this result included in synthesis calculations of the natural occuring isotopes. Compared with earlier equilibrium conditions [12] the inclusion of shorter half-lives shifted the isotope distribution to higher A-values in the region of the abundance hump at A=80 [3], however, this was not sufficient to obtain agreement with the natural abundances. Consequently, modified r-process surroundings with higher neutron densities are proposed [13, 14].

* supported by Fed. Min. of Res. and Tech. (BMFT) under contract 06GO456

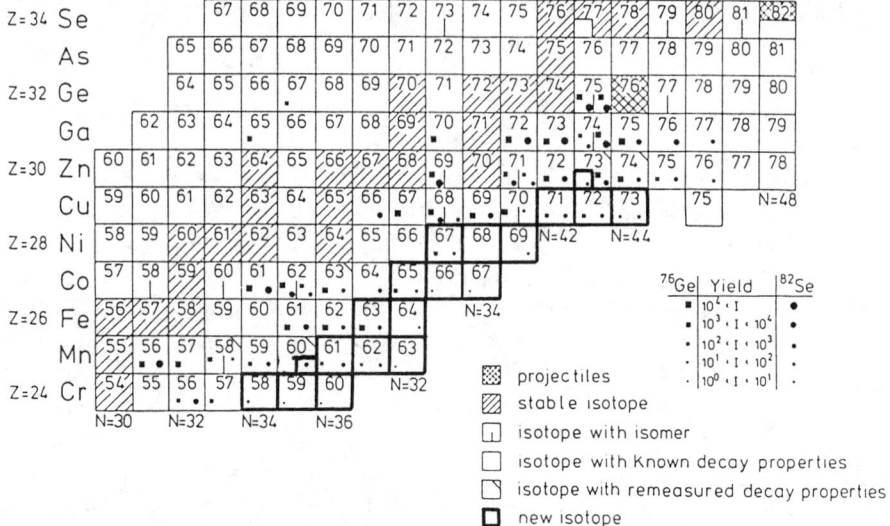

fig. 1. Neutron-rich nuclei in the light-isotope region, the new isotopes from this work are indicated, the yields normalized to 10 particle*nA, for 9 MeV/u-^{76}Ge, and 11.4 MeV/u-^{82}Se on ~40 mg/cm^2 ^{nat}W are given.

The present spectroscopic investigations may serve as a test for calculations of ground-state β-half lives. For refined r-process calculations also the inclusion of excited states in the neutron-rich nuclei are of interest [14].

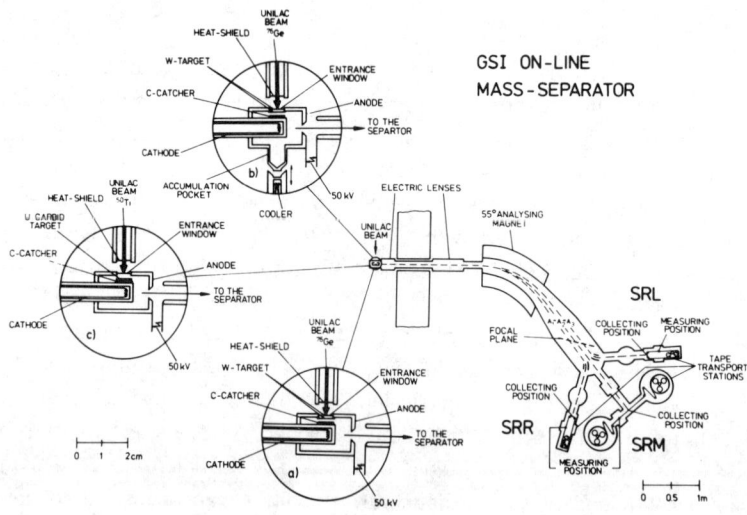

fig. 2. Mass-separator and three ion-sources used in present experiments, a) FEBIAD-F, b) bunched ion-source, and c) uranium-carbide source

EXPERIMENTAL METHOD

Our experiments were performed at the GSI mass-separator on line with the UNILAC. The apparatus is sketched in fig. 2. Three types of ion-sources were used, a high-temperature FEBIAD-F source [15, 16], a source with bunched release [17] applied for the chemical separation of copper and nickel, and one with a heat-resistant uranium-carbide target. The mass-separated beams were collected at three tape stations and cyclically transported to the measuring positions. One station was equipped with a β-ray telescope of 20% efficiency for the measurement of high-energy β-rays, and a facing Ge detector, and two of the stations had a 0.5 mm-thick β-detector surrounding the sample with 70% efficiency, and two Ge detectors used for the decay-scheme work.

NEW NUCLEI ^{60}Cr, and ^{60g}Mn

These two nuclei were identified in a recent measurement using for production 11.5 MeV/u-^{76}Ge and a 36 mg/cm^2-^{nat}W target in the ion-source of fig. 2a. The UNILAC beam current was 15-30 particle*nA. The effective background reduction in the γ-ray spectrum measured at mass A=60 in coincidence with β-rays is seen in fig. 3. The transition at 272 keV in the singles spectrum is the isomeric transition in ^{60}Mn [2]. The decay of 60m Mn was remeasured and a detailed decay scheme derived from the γγ-coincidences and using known excitation states of ^{60}Fe [18]. This is included in the level scheme of the A=60 isobars in fig. 4.

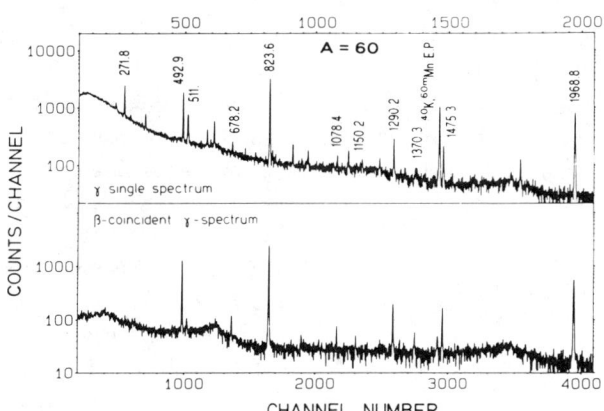

fig. 3. γ-spectrum measured at A=60; singles, and coincident with β-rays.

For ^{60}Mn the half-life was improved. The new value of 1.77(2) s was used for the analysis of the decay of β-rays. the decay was measured in two intervals as shown in fig. 5.
The subtraction of the 1.77 s component from the total decay curve measured with 4 s cycle-time, yielded a β-ray activity with 0.57(6) s half-life. This short-lived component was also seen in the decay of γ-coincident β-rays. A 51(6) s-activity was found with 320 s collection-measuring cycle.
The assignment of 0.57 s β-rays to the decay of ^{60}Cr was done on grounds of the systematic behaviour of the β-half lives and production yields decreasing towards the

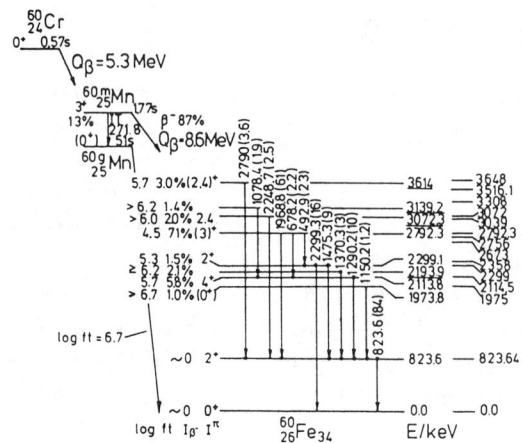

fig. 4.
The decay of A=60 isobars. New transitions in ^{60}Mn-decay are underlined, relative intensities are given in brackets, levels to the right are from reaction work.

neutron-rich side. In the odd-odd ^{60}Mn, besides the 3^+-isomeric state a 0^+-ground state with up to now unknown half-life was expected [2]. In accordance with our observation only β-ray singles and no coincidences with γ-rays can be expected for the transition to the ^{60}Fe-ground state, therefore, the assignment of the 51 s-activity to the decay of ^{60g}Mn seems to be plausible.

We have compared the ^{60}Cr half-life with the two predictions [10, 11], and have included the ratios of theoretical to experimental half-lives in the compilation. In fig. 6 our new half-life values are given, including the ^{68}Ni half-life. The four light isotopes in fig. 6 were produced by medium-energy fragmentation, the rest by multinucleon transfer. The earlier observed enhancement of β-decays in a local region is also present for ^{60}Cr and ^{68}Ni. The much lighter and heavy isotopes stay in agreement with expectation.

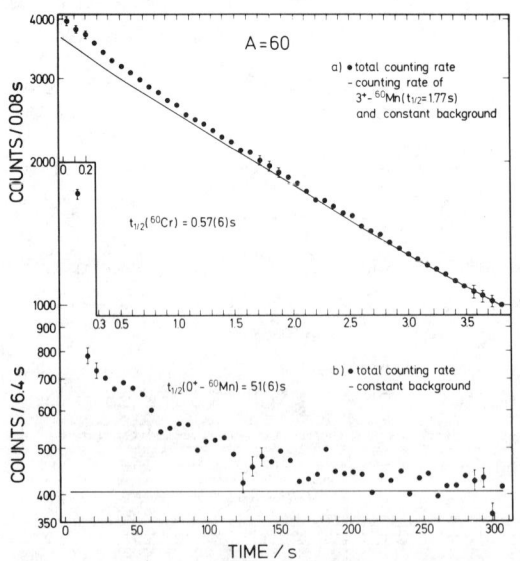

Very recently we have started the calculation of excited states with the Warsaw-Lund program to apply in the region of neutron-rich nuclei, and in a later stage to extend to half-life determination.

fig. 5. Decay of β-rays measured at A=60 with a) 4s-cycle for 4.9 h, b) 320 s-cycle for 1.8 h

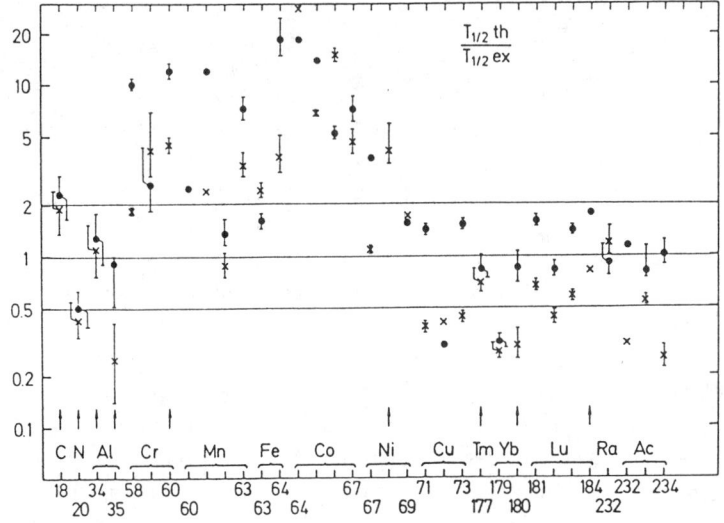

fig. 6. Ratios of predicted and experimental half-lives. Experiments, refs. 1-9, 19, and present work; theory, microscopic model [10] given by asterisks, gross theory [11] by dots.

CONTINUOUS β-SPECTRA AND ENDPOINT ENERGIES

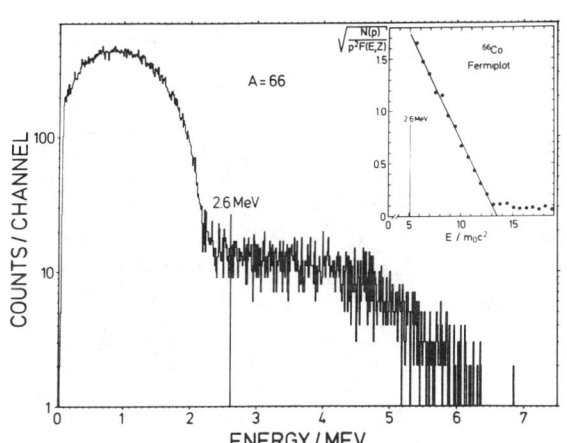

fig. 7. Beta-rays at A=66 measured with 1.6 s cycle-time in 6.7 h; inset, Fermi-analysis

In the region of the light isotopes the element assignment after mass-separation can be difficult. The measurement of β-ray spectra may be used for identification, if the increasing Q_β-values in an isobaric chain allow to distinguish between the isotopes. We have used the β-ray telescope for the measurement of the ^{66}Co, and ^{67}Co spectrum. The result at A=66 is given in fig. 7 [4]. In the energy region 0 to 2.6 MeV mainly the ^{66}Cu is present, above are the β-rays of ^{66}Co with a half-life of 0.23 s. Using a threshold in the spectrum the copper activity is suppressed. A Fermi-analysis of the β-spectrum was performed and is shown in fig. 7. The endpoint energy is 7.0(5) MeV. In ^{66}Co-decay the main transition is leading to the 2.67 MeV excitation state [3,4]. Thus the experimental Q_β-value is 9.7(5) MeV, which can be compared with Q_β-values from the atomic mass tabulations, 10.2 MeV [20], 9.66 MeV [21], and 9.1 MeV [22].

HALF-LIFE OF ^{68}Ni

The magic proton number and semi-magic neutron-number of ^{68}Ni attract particular interest for this configuration. It certainly is one of the links in the r-process, and therefore its half-life is relevant for realistic calculations of this process and resulting abundances [3]. Some of the ^{68}Ni-properties like mass excess and excited states are known [23-25], but not its half-life. An earlier search for ^{68}Ni produced in multinucleon transfer was unsuccessful [2], mostly because of the disturbance of isobaric copper which in this reaction is produced several hundred times more frequent than nickel [2,4].

In the present measurement an effective suppression of copper was achieved by making use of the difference in adsorption enthalpies at a tantalum surface at different temperatures [26]. For isotope production an 11.5 MeV/u ^{76}Ge-beam and an internal 38 mg/cm^2-^{nat}W target plus 11.5 mg/cm^2-graphite catcher were used in the ion-source with a cooled pocket, fig. 2b. The β-branches and γ-rays characteristic for identification of ^{68}Ni are given in fig. 8.

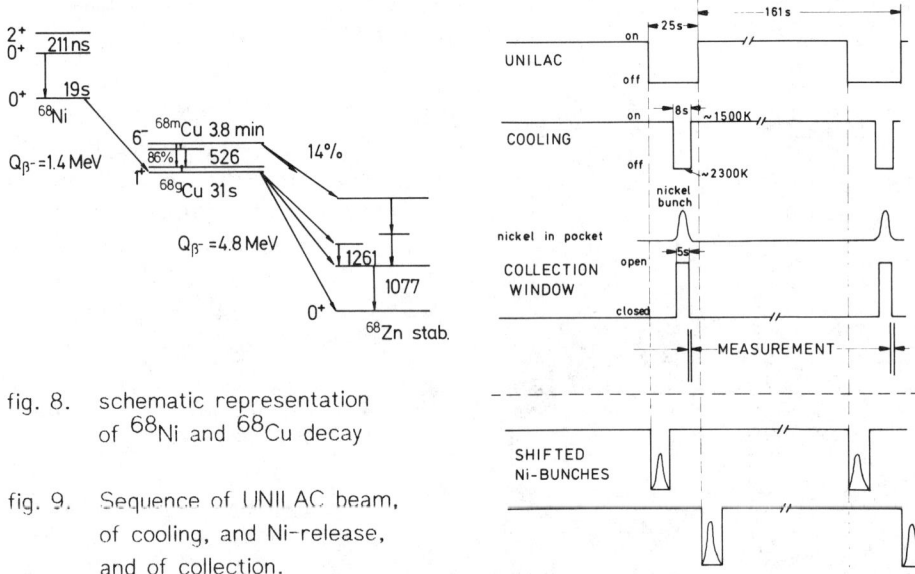

fig. 8. schematic representation of ^{68}Ni and ^{68}Cu decay

fig. 9. Sequence of UNILAC beam, of cooling, and Ni-release, and of collection.

It is expected that in ^{68}Ni decay the low-spin ^{68}Cu-ground state and not the high-spin isomeric state is populated. The UNILAC beam was periodically interrupted, as shown in fig. 9. The nickel is collected in the cooled pocket of the ion-source, while the copper is effusing out continuously. A Ni-bunch is released by a heating-pulse, i.e. cooling off, and is collected by opening a flag in the separator-beam line, and subsequently measured. The activity is nickel and some isobaric copper. Only ^{68}Cu was collected by shifting the Ni-bunch in time as compared to the collection window, comp. fig. 9. The decay of γ-rays from the Ni-bunch and the ^{68}Cu-measurement are given in fig. 10. The 526 keV γ-ray is characteristic for directly produced copper in the reaction, decaying with 3.8 min half-life. The 1077 keV radiation in the "copper-only" measurement comes from directly produced ^{68g}Cu, the growth and decay following the ^{68m}Cu decay and a small contribution from ^{68m}Cu decay.

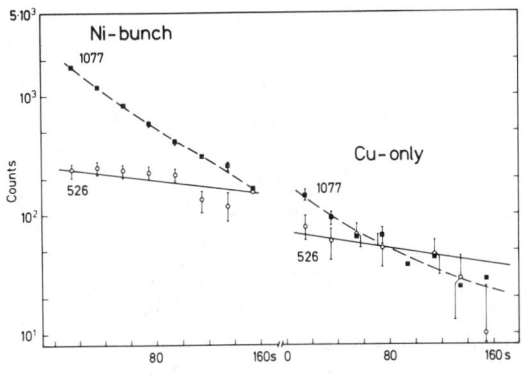

fig. 10.
Decay curves of 526 keV and 1077 keV γ-rays, to the left with Ni-bunch, measured for 2.6 h, right "Cu-only", measured for 1.2 h.

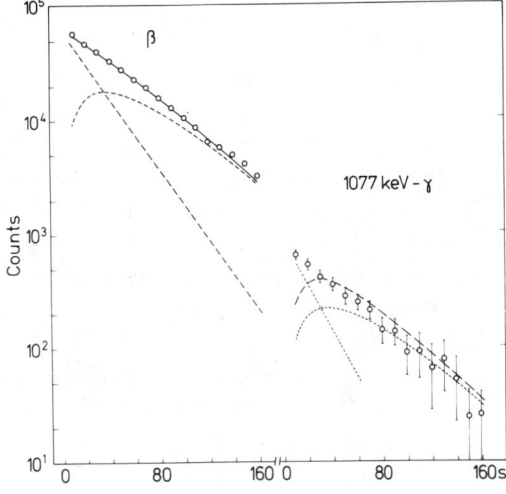

fig. 11.
Left: β-decay following the decay of ^{68}Ni after subtraction of directly produced ^{68}Cu; dashed curve is fit to experimental data with condition: number of parent ^{68}Ni-decays equals number of daughter ^{68}Cu-decays, and half-life of ^{68}Cu being 31 s.
Right: γ-rays following the decay of ^{68}Ni after subtraction of directly produced ^{68}Cu; dashed curve fitted in interval 100-160 s leaving a short-lived component, dotted curve is growth and decay calculated with half-lives of 31 s for ^{68g}Cu, and 13 s for ^{68}Ni.

The increase in the relative intensity of 1077 keV radiation with Ni-bunch is due to the feeding from ^{68}Ni decay. Considering the more than one hundred times intense copper production [2, 4], and about three times larger nickel detection, a suppression factor >400 in this experiment was estimated. Similarily, the decay of the β-rays was measured. The contribution of the copper activity in the β-, and γ-ray spectrum was subtracted using for normalization the 526 keV γ-ray intensity and assuming that the ratio of ^{g}Cu to ^{m}Cu is the same in Cu-only and Ni+Cu measurements. The difference curves are shown in fig. 11, the β-curve exhibiting much better statistics. Therefrom the ^{68}Ni-half life was derived in two ways.

Firstly, assuming a short ^{68}Ni-half life, a fit of the growth-and-decay curve was done from 100-160 s, and extrapolated to smaller times. It was confirmed, that the such derived ^{68}Ni-half life corresponded to the one obtained from the difference curve. Secondly, since the decaying ^{68}Ni is feeding the ^{68g}Cu, the total number of both decays is equal. This information and using $T_{1/2}(^{68g}Cu) = 31$ s, the other half-life was varied in a calculation to obtain an optimal fit to the experimental curve. This is shown for the β-curve in fig. 11, yielding the ^{68}Ni half-life of 19 s. In the γ-curve, the 19 s half-life is describing the growth-and-decay curve in the region 100-160 s, however, contrary to expectation, a short-lived direct component is observed. One explanation for this may be an error in the subtraction of the short-lived copper.

We have also tried to fit a growth-and-decay curve with half-lives of 13 s and 31 s, respectively. With the exception of the first two data points, such a curve is explaining the measurement without Cu-correction within the limited statistics. Since the decay in the 100-160 s interval is sensitive for ^{68}Ni half-life values >20 s, we obtain an unsymmetric error interval, namely the ^{68}Ni-half life of 19^{+3}_{-6} s. This value was compared with the prediction of the microscopic model [10] in fig. 6, the corresponding ratio derived with gross theory [11] being > 100 and out of scale in fig. 6.

SEARCH FOR LIGHT FRAGMENTS IN THE ^{50}Ti + ^{238}U REACTION

On the basis of yield extrapolations, some more isotopes are within reach using multinucleon transfer in quasielastic and deeply inelastic reactions, especially if higher detection efficiencies for decay-radiation are reached. In this respect the neutron detection is an interesting alternative [19]. However, the decreasing yields make future experiments more difficult, and alternative reactions like high-energy fragmentation [19,27] or very asymmetric fission [28] are of interest.

For this reason we have presently investigated another reaction for isotope production, the so called "cold" mass-transfer in a quasi-fission reaction. A rearrangement of masses was observed between ^{48}Ca and ^{248}Cm [29] at bombarding energies close to the Coulomb barrier. This effect cannot be explained in terms of deeply inelastic reactions or compound-nucleus formation followed by fission [30].

Specifically, production cross-sections up to 0.2 mb were found in the Ca+Cm experiment for fragments such as ^{222}Ra, ^{225}Ac, and ^{227}Th with broad isotopic distributions for each element [29]. Assuming a di-nuclear process, the complementary light fragments ^{74}Ni, ^{71}Co, and ^{69}Fe should be produced with similar cross-section.

Our present search for the light fragments of this reaction type made use of a ^{50}Ti beam and a ^{238}U-carbide target of thickness 5 mg/cm^2, comp. fig. 2c. Nearly all reaction fragments came to a stop in 12 mg/cm^2 graphite, therefore the experiment was not affected by the unknown angular dependence of this reaction. By measuring the target thickness of the uranium carbide of known composition

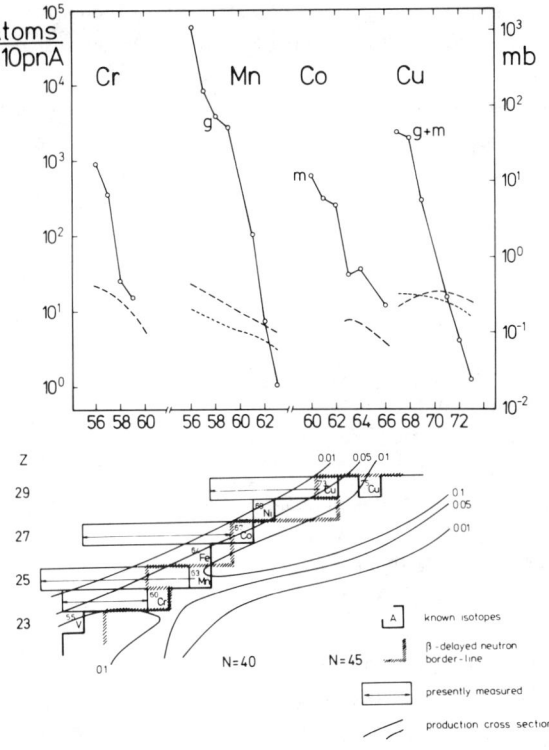

fig. 12.
Comparison of experimental yields, full lines, with calculated cumulative cross sections. Dashed curves derived with isotope distribution in fig. 13, dotted curves from a shifted primary distribution.

fig. 13.
Isotope production of light fragments as expected for binary quasi-fission reaction [29].

before and after the 21 h-run, it has been confirmed that the target material withstood the experiment and that the ^{50}Ti-energy in the uranium was 1.1 - 0.88 times the Coulomb barrier in the beginning, and 1.1 - 0.96 at the end. The present yields can be normalized to the number of ^{238}U atoms in the target and compared with isotopic yields in the earlier experiments [2,4] giving very similar normalized yields, thus suggesting the same production process. Our results are given in fig. 12, showing the steep decrease of the multinucleon transfer reaction [2,4]. The cross section distribution of the heavy fragments [29] was converted in the contour plot of production rates of the light partners in a di-nuclear reaction as given in fig. 13. The expected half-lives of these very neutron-rich and unknown isotopes are small [10,11]. Therefore, during collection the distribution is shifted by β-decays towards the valley of stability. Cumulative cross sections were thus derived and are compared with the experimental yields in fig. 12. The effect of nuclear excitation [31] was also respected by shifting the primary distribution of fig. 13 to the left; the dotted curves in fig. 12 correspond to an evaporation of 3 neutrons prior to β-decay. In both calculations we have respected the effect of β-delayed neutron emission [32]. By respecting the ion-source efficiency of 10% [16], the experimental yields of manganese and copper isotopes in fig. 12 can be converted in production cross-sections, as given by the ordinate to the right and compared with the calculated dashed and dotted curves. Since the calculated values for ^{63}Mn, and ^{72,73}Cu are by an order of magnitude larger than the observed ones, the quasi-fission production of these isotopes seems to be very unlikely in our experiment. It was also tested that the reaction took place in uranium and not in the tantalum-entrance window of the ion-source. We have also searched for ^{239}U and ^{242}Np. Assuming 10% ion-source efficiency for these elements, their production was less than 1/5 of the expected cross section [29].

REFERENCES

1) E. Runte, et al., Nucl. Phys. **A399** (1983) 163
2) E. Runte, et al., Nucl. Phys. **A441** (1985) 237
3) U. Bosch, et al., Phys. Lett. **164B** (1985) 22
4) U. Bosch, et al., Nucl. Phys. submitted for publication
5) R. Kirchner, et al., Nucl. Phys. **A378** (1982) 549
6) K. Rykaczewski, et al., Z. Phys. **A309** (1983) 273
7) K. Rykaczewski, et al., Z. Phys. A to be published
8) E. Runte, et al., Z. Phys. **A328** (1987), in press
9) K.-L. Gippert, et al., Nucl. Phys. **A453** (1986) 1
10) H.V. Klapdor, et al., At. Data Nucl. Data Tables **31** (1984) 81
11) K. Takahashi, et al., At. Data Nucl. Data Tables **12** (1973) 101
12) H.V. Klapdor, et al., Z. Phys. **A299** (1981) 213
13) W. Hillebrandt, Progress in Particle and Nuclear Physics **17**, A. Faessler (ed.), Pergamon Press, New York (1986) 215
14) G.J. Mathews, et al., ACS Symposium Series **324**, Nuclei off the line of stability, R.A. Meyer, D.S. Brenner (eds.), Am. Chem. Soc., Washington DC (1986) 134
15) R. Kirchner, et al., Nucl. Instr. Methods **186** (1987) 295
16) R. Kirchner, et al., Nucl. Instr. Methods in Physics Research **B26** (1987) 235
17) R. Kirchner, et al., Nucl. Instr. Methods in Physics Research **B26** (1987) 204
18) P. Andersson, et al., Nucl. Data Sheets **48** (1986) 251
19) R. Anne, et al., Nouvelle du GANIL **20** (1987) 14
20) P. Möller, and R. Nix, At. Data Nucl. Data Tables **56** (1981) 165
21) M. Uno, and M. Yamada, INS-NUMA-40 (1982)
22) P. Möller, W.D. Myers, W.J. Swiatecki, and J. Treiner, LBL-Report 22686 (1986)
23) T.S. Bhatia, et al., Z. Phys. **A281** (1976) 65
24) M. Bernas, et al., Phys. Rev. **C24** (1981) 756
25) M. Bernas, et al., J. de Phys. **45** (1984) L-851
26) R. Kirchner, GSI Scient. Report 1986 (1987) 259
27) D. Guerreau, J. de Phys. **47** (1986) C4-207
28) P. Armbruster, et al., IPNO-Report DRE 87-145 (1987)
29) H. Gäggeler, et al., Phys. Rev. **C33** (1986) 1983
30) G. Guardino, et al., Nucl. Phys. **424** (1984) 157
31) H. Keller, et al., GSI Scient. Report 1986 (1987) 38
32) K.L. Kratz, et al., Z. Phys. **A263** (1973) 435

IS THE REGION ABOVE ^{78}Ni DOUBLY MAGIC?

John C. Hill, J. A. Winger, and F. K. Wohn
Ames Laboratory and Iowa State University, Ames, Ia. 50011

R. L. Gill and A. Piotrowski[a]
Brookhaven National Laboratory, Upton, N.Y. 11973

X. Ji and B. H. Wildenthal[b]
Drexel University, Philadelphia, Pa. 19104

ABSTRACT

The region above ^{78}Ni would be doubly magic if the numbers Z=28 and N=50 retain their magic character far to the neutron-rich side of stability. We report results of a study of the states in ^{82}Ge (^{78}Ni core + 4 protons) and ^{83}As (^{78}Ni core + 5 protons) populated in the β⁻ decay of ^{82}Ga and ^{83}Ge, respectively. Mass separated sources were produced at the TRISTAN separator facility. The measured level schemes were compared to shell-model calculations in which we assume an inert ^{78}Ni core and extra-core protons that are allowed to fill with minor restrictions all orbits between Z=28 and 50. The calculations are generally in good agreement with the measurements except for a level in ^{83}Ge at 711 keV which cannot be easily explained. Finally the degree to which ^{78}Ni can be considered a valid doubly magic core is discussed.

INTRODUCTION

A central problem in nuclear physics is to determine the extent to which the proton and neutron numbers which are magic near stability retain their magic character far from the stable valley. On the neutron-rich side of stability the regions centered around ^{78}Ni and ^{132}Sn will be doubly magic if the numbers Z=28,50 and N=50,82 retained their magic character. Both of these regions are primarily studied by observing the decay of short-lived neutron-rich nuclei produced in fission.

In order to determine the character of the region around ^{132}Sn extensive studies of the N=82 isotones and the neutron-rich Sn(Z=50) isotopes have been carried out. Studies[1] have shown ^{132}Sn to be a very tightly bound nucleus with the lowest-lying excited states above 4 MeV. This good shell closure was also

---

[a]Permanent address: Institute for Nuclear Studies, Warsaw, Poland.
[b]Permanent address: Department of Physics and Astronomy, University of New Mexico, Albuquerque, N.M. 87131.

confirmed in studies[2,3] of ^{131}Sn (^{132}Sn core + 1 neutron hole) and ^{133}Sb (^{132}Sn core + 1 proton).

It is more difficult to study the character of the region near ^{78}Ni since it is further to the neutron-rich side of stability than ^{132}Sn, and the relevant fission yields are lower. We know of no information on excited states in Ni(Z=28) nuclei heavier than ^{68}Ni. The most practical method to explore the doubly magic character of the ^{78}Ni region is to study the structure of the neutron-rich N=50 isotones above ^{78}Ni, which are shown in Fig. 1. Although the structure of the N=50 isotones has been well studied near the valley of stability down to ^{86}Kr and some information is available on ^{85}Br and ^{84}Se from decay and reaction studies, no information is available on the structure of the lighter N=50 isotones except for a study[4] of ^{82}Ge (^{78}Ni + 4 protons) populated in the decay of ^{82}Ga.

In this work we present results of studies of the decays of ^{82}Ga and ^{83}Ge to the N=50 isotones ^{82}Ge and ^{83}As, respectively. The level schemes are compared with shell-model calculations in which an inert ^{78}Ni core is assumed and the extra-core protons are allowed to fill the $f_{5/2}$, $p_{3/2}$, $p_{1/2}$, and $g_{9/2}$ orbits between Z=28 and 50. Finally, we discuss the degree to which the concept of a doubly magic core at ^{78}Ni is valid.

Fig. 1. The N=50 isotones above ^{78}Ni.

EXPERIMENTAL METHODS

Sources of ^{82}Ga and ^{83}Ge were obtained using the TRISTAN mass separator on-line to the High Flux Beam Reactor at Brookhaven National Laboratory. A layout of the TRISTAN system is shown in Fig. 2. The experiments described here were carried out in the "γ-ray spectroscopy" beam line. (A more complete description of the TRISTAN program will be presented in this conference.) The sources were produced by a high-temperature plasma ion source[5] containing a target of 5g of enriched ^{235}U in a neutron flux of

3×10^{10} n/cm^2·s. The beams were mass separated and collected on a movable Aℓ-coated mylar tape.

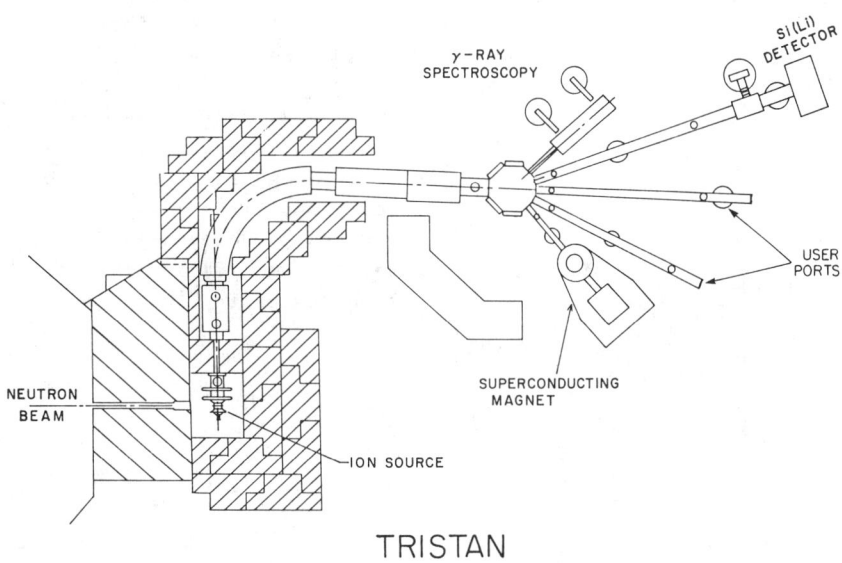

Fig. 2. Layout for the TRISTAN separator system.

Singles and coincidence measurements were carried out with two Ge γ-ray detectors in 180° geometry viewing the point of deposit. The singles spectra were β gated with a signal from a thin plastic scintillator. Enhancement of the activities of interest were made by moving the tape after a short buildup and decay cycle. In each case time-sequential γ-spectra were collected in order to measure half-lives of the nuclides of interest and to facilitate isobar identification. γ-γ concidence events were recorded on magnetic tape as address triplets representing γ-ray energies and their time separation. Angular correlation measurements were carried out for γ-ray cascades in ^{82}Ge using four Ge detectors. The angular correlation results were of poor statistical quality due to weakness of the ^{82}Ga source and run time limitations, but some restrictions on J$^\pi$ values for ^{82}Ge excited states were obtained and are discussed below.

STRUCTURE OF ^{82}Ge

The decay scheme for ^{82}Ga which populates levels in ^{82}Ge (^{78}Ni + 4 protons), is shown in Fig. 3. It is based on our γ singles, γ-γ coincidence, and γγ(θ) measurements. The results are consistent with an earlier level scheme of Hoff and Fogelberg[4] but with three additional γ transitions and an added level at 2713 keV. The ^{82}Ga half-life was measured to be 0.62±0.02s. The J^π values shown in Fig. 3 are based on γ branching, systematics, and results from angular correlations. In addition, six γ rays were attributed to delayed-neutron emission from ^{82}Ga leading to excited states in ^{81}Ge.

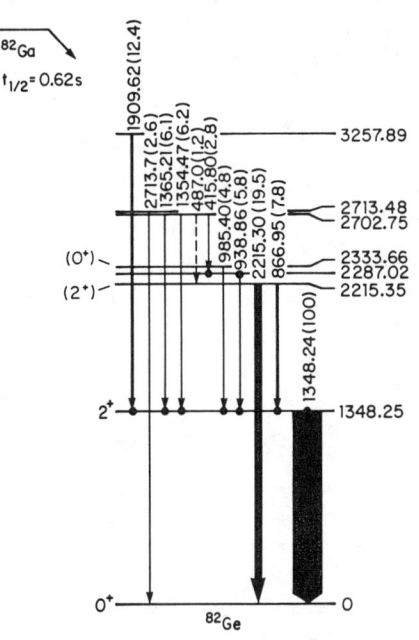

Fig. 3. Decay scheme for ^{82}Ga.

STRUCTURE OF ^{83}As

The decay scheme for ^{83}Ge which populates levels in ^{83}As (^{78}Ni + 5 protons), is shown in Fig. 4. It is based on our γ singles and coincidence measurements. Previous to this work no information was available on excited states in ^{83}As although ^{83}Ge had been identified by milking ^{83}Se from a ^{83}Ge source and its half-life determined[6] to be 1.9±0.4s. This is in good agreement with our measured value of 1.85±0.06s. No evidence for delayed neutron emission was observed for ^{83}Ge, but a 1348-keV γ ray with a short half-life attributable to delayed-neutron emission from ^{83}Ga was seen. We discuss the level structure of ^{83}As in some detail below.

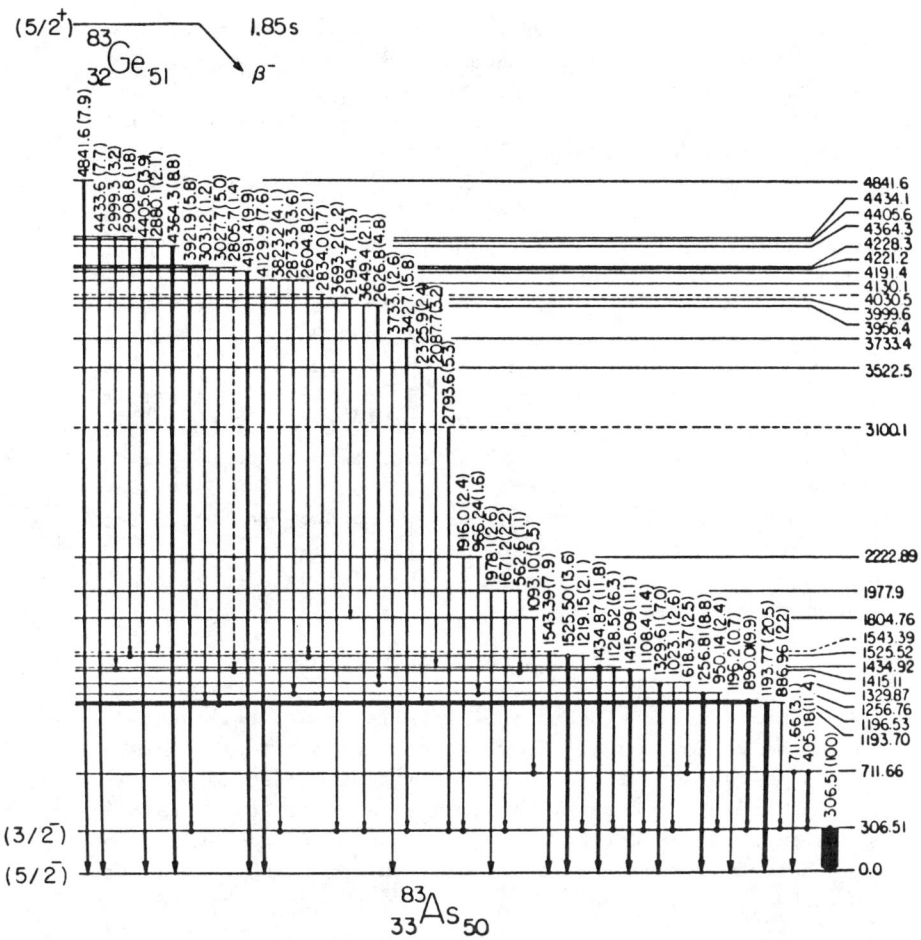

Fig. 4. Decay scheme for ^{83}Ge.

We assume J^{π} for the ground and 306-keV levels in ^{83}As to be $5/2^-$ and $3/2^-$, respectively, since it is well established that the first two orbitals in the proton shell between 28 and 50 are $f_{5/2}$ and $p_{3/2}$. The $3/2^-$ assignment is consistent with the spacings between the $3/2^-$ and $5/2^-$ levels in the N=50 isotones ^{85}Br and ^{87}Rb of 345 and 402 keV, respectively. A second excited state at 711 keV lies lower than in ^{85}Br and ^{87}Rb. Its status is not clear but is discussed below.

A major feature of the ^{83}As level scheme is a group of eight levels between 1.1 and 1.6 MeV. The β feedings and energies of these levels are consistent with their interpretation as being mostly negative-parity three-quasiparticle states outside the ^{78}Ni core that are fed by first-forbidden β transitions. Between 1.6 and 3.5 MeV only four levels are observed. A similar "gap" is observed for ^{85}Br but disappears with the addition of a proton pair to form ^{87}Rb.

A second important feature of the ^{83}As level scheme is that almost half (42%) of the β intensity to excited states in ^{83}As feeds a set of 13 levels between 3.5 and 4.9 MeV. A similar concentration of β strength is observed for ^{85}Br but ^{87}Rb has a much more uniform level density between 2.0 and 6.0 MeV. In ^{83}As 28% of the β intensity to excited states feeds seven levels lying in the narrow energy range from 4.13 to 4.43 MeV. A similar pattern was observed[7] in the decay of ^{136}I to the N=82 isotone ^{136}Xe. These levels can be thought of as neutron particle-hole states formed when a $g_{9/2}$ or $p_{1/2}$ neutron in ^{83}Ge decays by an allowed β transition to a $g_{9/2}$ or $p_{1/2}$ proton in ^{83}As. The resulting neutron particle-hole configurations $d_{5/2}(g_{9/2})^{-1}$ or $d_{5/2}(p_{1/2})^{-1}$ could then couple to the five valence protons to give a number of states.

SHELL-MODEL CALCULATIONS

A major motivation of this work is to determine the degree to which the region above ^{78}Ni is doubly magic. Thus we calculate levels in the N=50 isotones by assuming a closed core of 28 protons and 50 neutrons (^{78}Ni). The calculations are similar in style to those of Wildenthal and Larson[8] for the N=82 isotones.

Recently, a series of calculations[9] for the N=50 isotones above the $f_{5/2}p_{3/2}$ shell closure at ^{88}Sr were carried out up to A=99. A ^{88}Sr closed core was used and the extra-core protons were allowed to fill only the $p_{1/2}$ and $g_{9/2}$ orbits. A conclusion from these studies was that the subshell closure at ^{88}Sr is valid, seniority breaking in the $p_{1/2}g_{9/2}$ subspace is minimal, and a good description of the N=50 isotones between ^{88}Sr and ^{100}Sn is obtained by considering only the $p_{1/2}g_{9/2}$ subspace. For A<88 calculations[10] have been performed for ^{87}Rb, where a single quasiproton was coupled to the collective motions of a ^{86}Kr core. Also the lowest 2^+, 4^+, and 3^- states in ^{86}Kr and ^{88}Sr have been modeled using two-quasiparticle configurations for the open shell and particle-hole configurations for the closed core.[11] Calculations are lacking for N=50 isotones with A<86.

Elementary shell model considerations suggest that the proton orbits which dominate the low-lying wave functions of the light (Z=28-38) N=50 nuclei are $f_{5/2}$ and $p_{3/2}$ the $p_{1/2}$ and $g_{9/2}$ orbits lying 1 to 2 MeV higher. We assume an inert core (^{78}Ni). The resulting assumption that the $f_{7/2}$ orbit is completely filled is consistent with information from proton-transfer reactions[12] on ^{86}Kr and ^{88}Sr which show the lowest fragments of the $f_{7/2}$ hole state to occur above 3 MeV in excitation energy. These reactions also show that the $f_{5/2}$ and $p_{3/2}$ orbits are effectively filled at ^{88}Sr.

The above considerations suggest that a model space restricted to the $f_{5/2}$, $p_{3/2}$, $p_{1/2}$ and $g_{9/2}$ orbits might be sufficient to explain the low-lying features of the N=50 isotones. The extra core particles are allowed to fill the above orbits with some minor restrictions on the filling of the $g_{9/2}$ orbit. Also the seniority is limited to a maximum of 4. Two separate calculations were carried out which are discussed in the following.

In the first calculation the two-body matrix elements (TBME) were parameterized in terms of a single scaling parameter using the surface-delta interaction (SDI). Other parameters required were the 3 single-quasiparticle energy (SQPE) separations. Initial values for the 3 SQPE parameters were obtained from pertinent level separations in ^{83}As and ^{85}Br. The strength of the SDI was estimated from the first two 2^+ states in both ^{82}Ge and ^{86}Kr. Levels for the N=50 isotones with A=82 to 86 were calculated iteratively in order to find a best set of values for the four parameters described above.

In the second calculation, all 65 TBME were used as adjustable parameters in an iterative least squares search to find the optimum effective model interaction which best reproduces the observed level energies. In this free parameter (F.P.) calculation 138 level energies in the A=82-96 range of N=50 nuclei were used as the data set to which the 35 best-determined linear combinations of the 65 TBME and 4 SQPE were fitted.

COMPARISON OF CALCULATIONS AND EXPERIMENT

A comparison of the experimentally determined level scheme for ^{82}Ge with the above calculations is shown in Fig. 5. Since ^{82}Ga appears to decay directly to both 0^+ and 2^+ states in ^{82}Ge, a reasonable J^π for the ^{82}Ga ground state is 1^- from coupling a $f_{5/2}$ proton to a $d_{5/2}$ neutron. For both calculations the 2_1^+, 2_2^+, and 0_2^+ states are accurately modeled. Also the number of low-spin (0,1,2) states between 0 and 3 MeV is well reproduced. For all states below 3 MeV, the major proton component of the wave function consists of combinations of the $f_{5/2}$ and $p_{3/2}$ orbitals.

The experimental information on ^{83}As is more complete than for ^{82}Ge and a detailed comparison with the calculations (see Fig. 6) is more illuminating. The 306-keV state is well reproduced and the calculations indicate that most of the $f_{5/2}$ and $p_{3/2}$ single quasiparticle strength resides in the ground and 306-keV states, respectively. The SDI calculation shows most of the $p_{1/2}$ single quasiparticle strength to be split between the two $1/2^-$ levels between 1 and 2 MeV with the majority of the strength in the higher-lying level. According to the SDI calculation, a rather pure $g_{9/2}$ state lies at about 2.5 MeV. This state would not be strongly populated by β decay from the ^{83}Ge ground state assumed to be $5/2^+$.

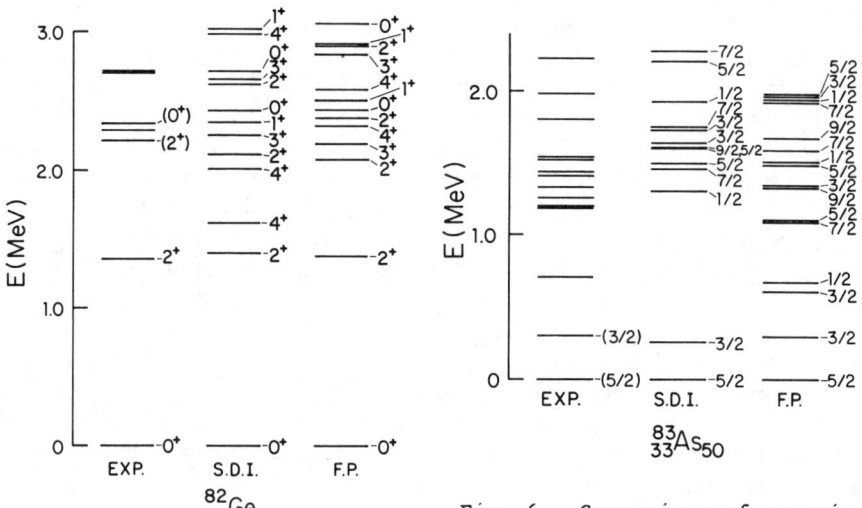

Fig. 5. Comparison of experimentally determined levels in ^{82}Ge with results from two shell-model calculations.

Fig. 6. Comparison of experimentally determined levels in ^{83}As with results from two shell-model calculations in which only negative-parity levels are shown.

An important feature of the measured level scheme for ^{83}As is a group of 10 levels between 1.1 and 2.0 MeV which are fed by first-forbidden β⁻ decay. These levels correspond nicely to 9 levels between 1.3 and 2.0 MeV with J^π values ranging from $1/2^-$ to $9/2^-$ given by the SDI calculation. In all cases, except for the $1/2^-$ state at 1.924 MeV, the states are three-quasiparticle with $f_{5/2}$ and $p_{3/2}$ protons dominating the makeup of the wave functions. A significant failure of the SDI calculation was its inability to reproduce the 711-keV level in ^{83}As or to lower it from the group of levels between 1.3 and 2.0 MeV. This provided the initial motivation for the calculation in which all the strengths of the two-body matrix elements were adjustable. As can be seen from

Fig. 6, the calculation now produces 2 states between 600 and 700 keV, either of which could represent the state observed at 711 keV. Neither of these states contain a significant portion of the $p_{1/2}$ proton strength. The level density is somewhat higher than for the SDI calculation in that 12 states are produced between 1 and 2 MeV, but both calculations get about the right level density for the region. Furthermore, since states with $J^{\pi} = 1/2^-$ or $9/2^-$ would be populated by first-forbidden unique β^- transitions and our experimental β^- feedings do not rule out any such transitions below 3 MeV, it is difficult to choose between the two calculations.

CONCLUSIONS

In this work we have experimental results on states in the N=50 isotones ^{82}Ge and ^{83}As populated in the β^- decay of ^{82}Ga and ^{83}Ge. In addition two different shell-model calculations have been carried out to determine to what extent ^{78}Ni can be considered a valid doubly magic core. The calculations are in good agreement with the limited data available on ^{82}Ge. For both ^{82}Ge and ^{83}As the dominant configurations of almost all of the states below 3 MeV consist of only $f_{5/2}$ and $p_{3/2}$ protons.

The structure of ^{83}As provides a more detailed test of the model. Most of the $f_{5/2}$ and $p_{3/2}$ single-quasiparticle strength resides in the ground and 306-keV levels, respectively, but the $p_{1/2}$ strength is split between two levels below 2 MeV. Ten levels observed between 1.1 and 2.0 MeV are characterized by the calculations as three-quasiparticle configurations dominated by $f_{5/2}$ and $p_{3/2}$ orbitals. The level density is well approximated by either calculation. The level observed at 711 keV is something of an enigma. It is not produced by the SDI calculation but two levels are obtained from the other calculation.

The low-lying structures of the N=50 isotones ^{84}Se, ^{85}Br, and ^{86}Kr are also reasonably well reproduced by the calculations. Therefore in most cases the approximation that ^{78}Ni can be considered a good doubly magic core is valid, but this conclusion is tentative due to the limited data on the N=50 isotones with A<86. There is no convincing evidence for particle-hole or intruder states below 2 MeV of excitation although the level at 711 keV in ^{83}As is not understood. Thus the region around ^{78}Ni appears to be reasonably doubly magic. It should be noted though that recently a study of states in the N=50 isotone ^{84}Se populated in the (t,p) reaction[13] showed strong population of 0^+ states at 2.24 and 2.65 MeV. This suggests that neutron particle-hole excitations occur at relatively low energies thus the ^{78}Ni core may not be as magic as the ^{132}Sn core where such low energy excitations are not observed.

ACKNOWLEDGEMENTS

This work was supported by the U.S. Department of Energy under contracts W-7405-eng-82 and DE-AC02-7600016.

REFERENCES

1. T. Bjornstad, M.J.G. Borge, J. Blomqvist, R. D. Von Dincklage, G. T. Ewan, P. Hoff, B. Jonson, K. Kawade, A. Kerek, O. Klepper, G. Lovhoiden, S. Mattason, G. Nyman, H. L. Ravn, G. Rudstam, K. Sistemich, O. Tengblad, and Isolde Collaboration, Nucl. Phys. A453, 463 (1986).
2. B. Fogelberg and J. Blomqvist, Nucl. Phys. A429, 205 (1984).
3. J. Blomqvist, A. Kerek, and B. Fogelberg, Z. Phys. A314, 199 (1983).
4. P. Hoff and B. Fogelberg, Nucl. Phys. A368, 210 (1981).
5. A. Piotrowski, R. L. Gill, and D. C. McDonald, Nucl. Instr. and Meth. in Phys. Res. B26, 249 (1987).
6. P. Del Marmol and P. Fettweis, Nucl. Phys. A194, 140 (1972).
7. W. R. Western, J. C. Hill, W. L. Talbert, Jr., and W. C. Schick, Jr., Phys. Rev. C 15, 1822 (1977).
8. B. H. Wildenthal and D. Larson, Phys. Lett. 37B, 266 (1971).
9. J. Blomqvist and L. Rydstrom, Physica Scripta 31, 31 (1985).
10. P. Hoffman-Pinther, Z. Physik A283, 85 (1977).
11. V. Gillet, B. Giraud, and M. Rho, J. Phys. (Paris) 37, 189 (1976).
12. L. R. Medsker, H. T. Fortune, S. C. Headley, and J. N. Bishop, Phys. Rev. C 12, 1516 (1975; J. F. Harrison and J. C. Hiebert, Nucl. Phys. A185, 385 (1972).
13. H. T. Fortune, private communication.

LIFETIME MEASUREMENTS IN ^{72}Br*

S. Ulbig, F. Cristancho, J. Heese, K. P. Lieb,
Th. Osipowicz and B. Wörmann,
II. Physikalisches Institut, Universität Göttingen,
D-3400 Göttingen, Fed. Rep. Germany

J. Eberth, Th. Mylaeus and E. Wiosna,
Institut für Kernphysik der Universität zu Köln
D-5000 Köln 41, Fed. Rep. Germany

ABSTRACT

The lifetimes of 12 yrast and yrare states in the odd-odd nucleus ^{72}Br were measured via the recoil distance Doppler shift technique and the reaction ^{58}Ni(^{16}O,pn) at 53 MeV beam energy. The negative parity states are grouped into two rotational bands with in-band E2 strengths of 50 - 100 Wu. Although having similar collective E2 strengths, the positive parity yrast spectrum up to spin 9^+ is dominated by the $\pi g_{9/2} \times \nu g_{9/2}$ residual interaction, but rotational bands evolve above spin $I = 9$.

INTRODUCTION

Gamma ray spectroscopy of the neutron deficient selenium, bromine, krypton and rubidium isotopes following heavy ion fusion reactions has given evidence for some of the most quadrupole deformed nuclei of the Periodic Table [1,2]. In particular the odd-Z even-N nuclei feature very regular yrast bands with prolate deformation $\beta_2 = 0.33$-0.42 and nearly constant moment of inertia parameters $\mathcal{J}/\hbar^2$, close to the rigid body values. The results of our systematic investigations on ^{73,75}Br and ^{77,79}Rb (Refs. 3-6) suggest that the proton in the low-Ω $g_{9/2}$ intruder orbits polarizes the adjacent even-even core, removes its shape coexistence and weakens the pairing correlations[5,7].

The present study of the odd-odd nucleus ^{72}Br, the lightest known bromine isotope, was motivated by two "band like" structures previously assigned by Garcia-Bermudez and collaborators [8]: while the negative parity yrast states with K = 1$^-$ follow a rotational pattern, the presumed positive parity yrast states below spin 9^+ do not, although several stretched E2 transitions among these states are collective, too ($\sim$100 Wu). We have recently established the yrast bands up to spins 15^+ resp. 14^- (Ref. 9); they follow rather well a I(I+1) dependence with the same moment of inertia in all bands. In order to better disentangle the interplay of collective and (2qp) degrees of freedom, we now have performed recoil distance lifetime measurements.

---

*Supported by Deutsches BMFT under contract 06 Gö 456

EXPERIMENT

The 53 MeV ^{16}O beam of the Cologne FN tandem accelerator was directed onto a selfsupporting 660 μg/cm^2 thin ^{58}Ni foil of 99.8 % isotopic enrichment. The beam and the recoil nuclei were stopped in a stretched 20 μm thick Ta stopper foil, at a variable distance of between 2 μm and 9 mm to the target. At the measured mean velocity of the ^{72}Br recoils, v = 4.2(1) μm/ps, lifetimes up to 3 ns could be determined. The gamma radiation was measured in four Ge detectors of 16 - 30 % efficiency, positioned at 0°, 53°, 138° and 223° to the beam. Two large NE213 neutron detectors were mounted horizontally above and below the plunger target chamber, subtending about 2π solid angle. By gating the γ-ray spectra with evaporated neutrons, the predominant charged particle evaporation channels ^{72}Se + 2p and ^{69}As + αp were removed and lines in ^{72}Br in the 700 - 1300 keV range were accessible. Fig. 1 illustrates portions of the neutron gated spectra at 0° and 138° displaying the Doppler shifted and stopped components of the 353 keV ($8^+ \to 6^+$) transition.

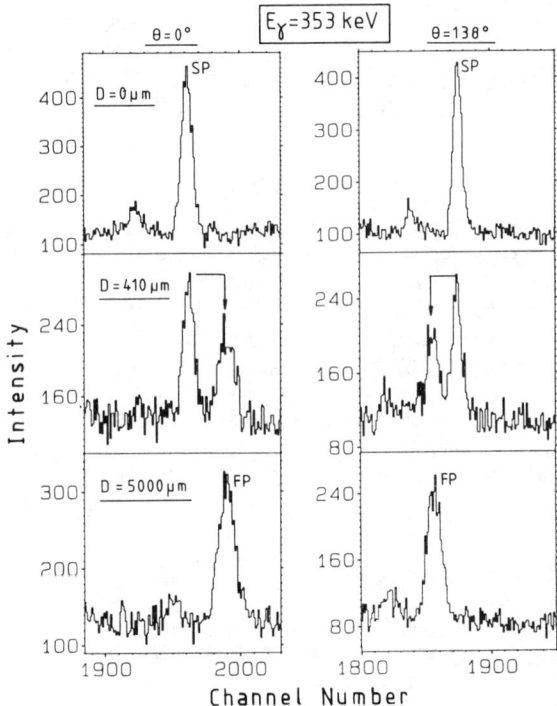

Fig. 1: Doppler shifted and unshifted components of the 353 keV line in ^{72}Br measured at 0° and 138° in coincidence with neutrons

Table I: Lifetimes and E2 and M1 transition strengths in ^{72}Br

State E_x (keV)	I^π	τ(ps)	Transition E_γ (keV)	Multipol.	B(E2) (e^2fm^4) B(M1) (μ_N^2)
334	(4^-)	740(180)	202	E2	$1645(^{505}_{310})$
			115	M1	$22(6)\ 10^{-3}$
371	(4^-)	3.1(6) ns	240	E2	~ 225
			153	M1	$\sim 0.7\ 10^{-3}$
469	(5^-)	530(235)	250	E2	$995(^{790}_{305})$
			135	M1	$8(^7_3)\ 10^{-3}$
			98	M1	$\sim 18\ 10^{-3}$
660	(6^-)	224(23)	326	E2	975(100)
718	(6^-)	2.5(3) ns	249	M1	$< 2\ 10^{-3}$
960	(7^-)	13(3)	491	E2	$1520(^{455}_{285})$
			242	M1	$97(^{29}_{18})\ 10^{-3}$
1190	(8^-)	23(3)	472	E2	1140(150)
			530	E2	213(33)
1320	(8^-)	< 5	661	E2	>1300
1614	(9^-)	< 6	654	E2	>1130
398	(2^+)	146(29)			
669	(4^+)	153(20)	270	E2	1990(270)
			290	E2	670(100)
992	(6^+)	123(11)	324	E2	1780(160)
1346	(8^+)	103(6)	353	E2	1425(85)
1449	(9^+)	85(9)	104	M1	0.56(6)
2189	(10^+)	< 3	843	E2	> 540

All stretched $\Delta I = -1$ transitions assumed to be pure M1

The γ-ray flux at the chosen beam energy was carefully evaluated from singles and neutron gated spectra taken at 55° in order to provide for the correct feeding intensities and times. The analysis was done with the program CHRONOS [10]. Continuum feeding times were neglected as they are estimated to be much shorter than the lifetimes in consideration. This is the result of Doppler shift attenuation data and Monte Carlo simulations of the γ-ray flux after heavy ion fusion reactions in this mass region [4,11].

RESULTS

Table I summarizes the measured mean lives and lifetime limits of states between 0.33 and 2.19 MeV excitation and the B(E2) and B(M1) values of stretched quadrupole resp. dipole transitions, on the basis of the partial level scheme displayed in Fig. 2. Most E2 strengths are in the range 0.1-0.2 e^2b^2 pointing to a deformation parameter $|\beta_2| = 0.34$ (6). The M1 strengths of negative parity transitions are generally weak (<0.1 μ_N^2), however, the $9^+ \to 8^+$ M1 transition is very large in size (0.54 μ_N^2) suggesting that both states have the same (2qp) $\pi g_{9/2}$ $\nu g_{9/2}$ structure. One may also note in Fig. 2 and Table I a doubling of the 4^-, 6^- and 8^- states involving weaker transition probabilities (E2 $\sim$200 e^2fm^4; M1 <2 10^{-3} μ_N^2). The intrinsic structures of these bands and their mixing are presently under investigation.

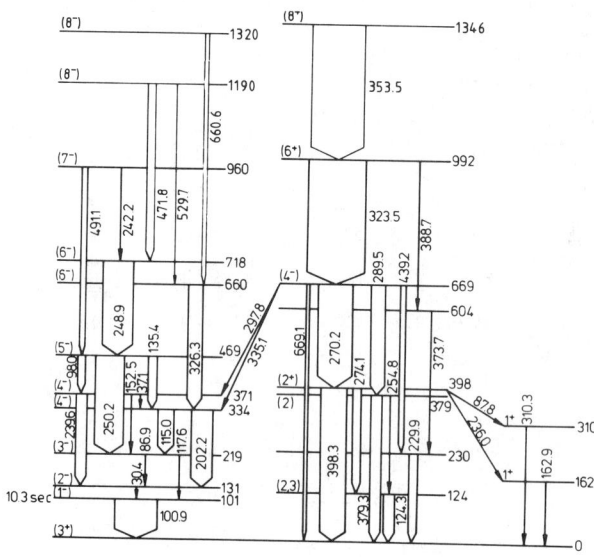

Fig. 2: Partial level scheme of ^{72}Br as reported in Refs. 8,9 and used in the analysis of the recoil distance measurements.

REFERENCES

1. J.H.Hamilton, P.G.Hansen, E.F.Zganjar, Rep. Progr. Phys. 48, 631 (1985)
2. K.P.Lieb, in "Nuclear Reactor Research for Developing Countries and Nuclear Spectroscopy Research" (World Sci. Publ. Comp., 1986) p.619
3. B.Wörmann, et al., Z. Phys. A322, 171 (1985); J.Heese, et al., to be publ.
4. L.Lühmann, et al., Phys. Rev. C31, 828 (1985)
5. L.Lühmann, et al., Europhys. Lett. 1, 623 (1986)
6. J.Panqueva, et al., Nucl. Phys. A389, 424 (1982)
7. W.Nazarewicz, et al., Nucl. Phys. A435, 397 (1985)
8. G.Garcia-Bermudez, et al., Phys. Rev. C25, 1396 (1982); C26, 1748 (1982)
9. S.Ulbig, et al., to be publ.
10. H.P.Hellmeister, L.Lühmann, program CHRONOS, Göttingen, 1981, unpubl.
11. F.Christancho, K.P.Lieb, to be publ.

NEW NEUTRON DEFICIENT NUCLEI NEAR MASS NUMBER A = 80[*]

U. Lenz, K.E.G. Löbner, U. Quade, K. Rudolph, W. Schomburg,
S.J. Skorka, M. Steinmayer
Sektion Physik, University of Munich, D-8046 Garching, FRG

ABSTRACT

Gamma-ray spectra of the new neutron deficient nuclei ^{80}Zr, ^{79}Y, and ^{80}Y have been measured in coincidence with the evaporation residues identified according to their Z- and A-values with the Munich heavy ion recoil spectrometer in the reaction ^{24}Mg + ^{58}Ni. The relative cross sections can be explained with the assumption that the protons and alpha-particles evaporate from strongly deformed compound states. A comparison of the signature splitting of the tentative yrast band of ^{79}Y with the rotational bands with K = 5/2 in neighbouring nuclei, suggests a low γ-deformation for ^{79}Y and ^{77}Sr relative to ^{81}Y, ^{79}Sr, ^{75}Kr and ^{77}Kr.

EXPERIMENT

In the reaction of 177 MeV ^{58}Ni on ^{24}Mg gamma-ray spectra have been measured of the very weakly populated reaction channels ^{80}Zr, ^{79}Y and ^{80}Y with relative cross sections of 0.01, 0.15 and 1.3 %, respectively. For many other isotopes the gamma-ray assignments have been confirmed.

These measurements have been performed with the Munich recoil spectrometer[1] by measuring γ-ray spectra at the target position in coincidence with evaporation residues identified at the end of the recoil spectrometer according to their Z- and A-values. A unique assignment of the measured γ-rays to a certain isotope is possible in this way[2].

A schematic view of the experimental set-up is shown in fig. 1

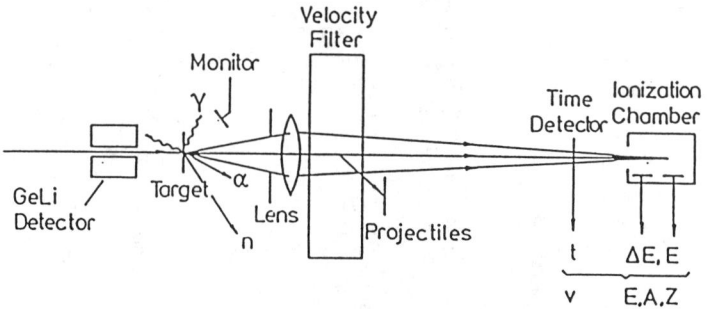

Fig. 1: Schematic view of the experimental set-up to measure gamma rays in coincidence with evaporation residues identified according to their Z- and A-values.

[*] This project is supported by the Bundesministerium für Forschung und Technologie.

The Z-values of the evaporation residues are derived from the energy loss and total energy measurement in the ionization chamber. The mass numbers A are identified from the total energy and the time of flight measurement relative to the pulsed beam (FWHM ≈ 400 ps). The γ-ray spectra measured in coincidence with evaporation residues are corrected for Doppler shift.

For all strong reaction channels, for which at least one gamma line could be observed in the single-spectrum the detection efficiency of the recoil spectrometer for this evaporation channel can be determined from the ratio: coincidence of gamma-ray intensity with identified recoil to single gamma-ray rate. Those experimentally determined detection efficiencies agree well with Monte Carlo calculations[1,3]. Thus the results of those Monte Carlo calculations have been used to determine the efficiencies of the recoil spectrometer for other (weaker) evaporation channels.

INFORMATION ABOUT THE NUCLEAR REACTION MECHANISM

Table I shows the measured relative cross sections for the different evaporation residues corrected for the detection efficiencies. These values can be approximately reproduced with the evaporation program GROGI2 of Grover and Gilat[4] by introducing deformation dependent optical-model transmission coefficients suggested earlier for the nuclear de-excitation at high angular momenta (see e.g. Blann[5] and Nicolis et al.[6] This corresponds to a reduction of the effective threshold for proton- and alpha-emission yielding higher relative cross sections for charged-particle emission than for neutron emission (see table I).

Evaporation residue	Evaporated light particles	Experiment (abs.errors)		Theory deformed	Theory spherical
80 Zr	2n	0.01	(.006)	0.01	0.025
80 Y	p + n	1.3	(0.4)	2.8	5.3
80 Sr	2p	12.1	(1.6)	8.7	14.4
79 Y	p + 2n	0.15	(.05)	0.20	0.25
79 Sr	2p + n	19.6	(2.0)	21.7	24.3
79 Rb	3p	47.0	(3.3)	40.0	36.4
78 Kr	4p	1.5	(0.3)	2.5	0.63
77 Sr	α + n	1.05	(0.20)	0.25	0.44
77 Rb	α + p	3.90	(.63)	4.3	11.9
76 Rb	α + p + n	3.10	(.42)	1.5	1.0
76 Kr	α + 2p	10.2	(1.2)	16.3	4.2

Table I: Reaction yields in % of the total cross section (400 mb) of the reaction 177 MeV ^{58}Ni+^{24}Mg → ^{82}Zr*. The experimental values corrected for the different efficiencies are compared to calculations from deformed (with deformation parameter ß = 0.55) and spherical compound nuclei.

SOME NUCLEAR STRUCTURE INFORMATION FOR THE NEW NEUTRON DEFICIENT NUCLEI

⁸⁰Zr: ^{80}Zr is the heaviest nucleus with N = Z identified until now. We could confirm the assignment of the 289 keV γ-ray, to the $2^+ \to 0^+$ transition in ^{80}Zr of Lister[7]. The energy of the 2^+ level in ^{80}Zr is similar to the 2^+ levels in neighbouring even-even nuclei, which are known to be strongly deformed (for details see ref. 7).

⁷⁹Y: From the measured γ-rays in ^{79}Y a preliminary level scheme has been constructed with a rotational band up to spin 25/2 with a rather constant moment of inertia (see fig. 2).

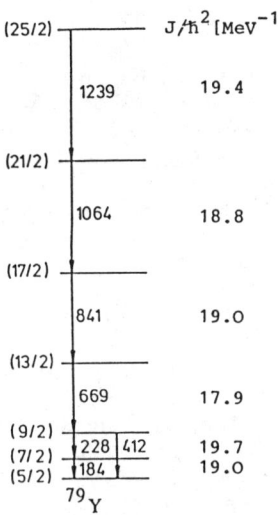

Fig.2:

Tentative level scheme of the new nucleus ^{79}Y. The numbers on the right of the figure are effective moments of inertia.

In Table II the signature splitting of the $K^\pi = 5/2^+$ [422] band in several odd-mass nuclei has been investigated. Tabulated are the energy ratios f of the $K^\pi = 5/2^+$ bands for the given nuclei with Z = 39, N = 39 or N = 41.

$$f = \{[E(7/2) - E(5/2)] - 7/16[E(9/2) - E(5/2)]\} / [E(9/2)-E(5/2)]$$

A strict I(I+1) dependence of the energy levels within the band would yield f = 0. Deviations from f = 0 can probably be interpreted as indication of γ-deformation.

A first order perturbation theory neglecting Coriolis coupling and changes of deformation yields: $\tan^2\gamma = 0.34$ f. The second order Coriolis K-mixing, which could also produce this staggering is expected to be higher for ^{77}Sr than for ^{79}Sr and higher for ^{75}Kr than for ^{77}Kr, since for ^{77}Sr and ^{75}Kr the quasiparticle energies of the $1/2^+$[440], $3/2^+$[431] and 5/2[422] bands should be closer than for the nuclei with 2 neutrons more ^{79}Sr and ^{77}Kr. The ground states of ^{79}Sr and ^{77}Kr correspond in first approximation probably to the Nilsson level 3/2$^-$[301] yielding a Fermi level above the 5/2$^+$[422] Nilsson level. (See e.g. the Nilsson diagram in ref. 8).

Nuclei	ratio f	References for data
^{79}Y	0.009	present data
^{81}Y	0.120	Lister et al.[10]
^{77}Sr	0.006	Lister et al.[11]
^{79}Sr	0.036	Lister et al.[11]
^{75}Kr	0.058	Herath-Banda et al.[12]
^{77}Kr	0.100	Nolte et al.[13]

Table II: Signature splitting. The energy ratio f (see text) is given for the energy spacings of the $K^\pi = 5/2^+$ band in odd-mass nuclei with Z = 39, N = 39 or N=41.

These data thus suggest a relatively small triaxiality for the nuclei ^{79}Y and ^{77}Sr in comparison with the other neighbouring nuclei. This is in agreement with theoretical calculations of Bonche et al.[9], which yield small triaxiality for the cores ^{76}Sr and ^{78}Sr.

REFERENCES

1) K. Rudolph, D. Evers, P. Konrad, K.E.G. Löbner, U. Quade, S.J. Skorka, and I. Weidl, Nucl. Instr. 204, 407 (1983)
2) K.E.G. Löbner U. Lenz, U. Quade, K. Rudolph, W. Schomburg, S.J. Skorka, M. Steinmayer, Nucl. Instr. B26, 302 (1987)
3) R. Pengo, D. Evers, K.E.G. Löbner, U. Quade, K. Rudolph, S.J. Skorka, I. Weidl, Nucl. Phys. A411, 255 (1983)
4) J. Gilat, BNL 50246, 1970, Phys. Rev. C1, 1432 (1970)
5) M. Blann, Phys. Lett. 88B, 5 (1979)
6) N.G. Nicolis, T.M. Cormier, P.M. Stwertka, and M.G. Herman, Phys. Lett. 178 B, 339 (1986)
7) C.J. Lister, Proceedings of the Internat. Nucl. Phys. Conference, Harrogate, U.K., August 1986, Vol. 2, p. 471
8) W. Nazarewicz, J. Dudek, R. Bengtsson, T. Bengtsson, I. Ragnarsson, Nucl. Phys. A 435, 397 (1985)
9) P. Bonche, H. Flocard, P.H. Heenen, S.J. Krieger, and M.S. Weiss, Nucl. Phys. A 443, 39 (1985)
10) C.J. Lister, R. Moscrop, B.J. Varley, H.G. Price, E.K. Warbuton, J.W. Olness, and J.A. Becker, J. Phys. C: Nucl. Phys. 11, 969 (1985)
11) C.J. Lister, B.J. Varley, H.G. Price, and J.W. Olness, Phys. Rev. Lett. 49, 308 (1982)
12) M.A. Herath-Banda, A.V. Ramayya, L. Cleemann, J. Ebert, J. Roth, T. Heck, N. Schmal, T. Mylaens, V. Koenig, B. Martin, K. Bethge and G.A. Leander, J. Phys. C: Nucl. Phys. 13, 43 (1987)
13) E. Nolte, and P. Vogt, Z. Physik A 275, 33 (1975)

THE $\pi g_{9/2} - \nu g_{7/2}$ INTERACTION IN NEUTRON-RICH NUCLEI WITH $A \sim 100$

G. Lhersonneau, S. Brant*, H. Ohm, K. Sistemich
Institut für Kernphysik, Kernforschungsanlage Jülich,
Postfach 1913, D-5170 Jülich, F.R. Germany

ABSTRACT

The investigation of members of the $[\pi g_{9/2}, \nu g_{7/2}]$ multiplet in the spherical odd-odd nuclei near ^{96}Zr provides information on the strength of the $\pi g_{9/2} - \nu g_{7/2}$ spin-orbit partner interaction which is important for nuclear-structure studies at $A \sim 100$. A value of $\Delta_1 + \sim -1.5$ MeV is deduced for the interaction matrix element from the energies of the 1^+ members. This does not exhibit exceptional strength compared to that of similar configurations in other nuclear regions.

INTRODUCTION

The interaction between protons and neutrons in the spin-orbit partner configurations $\pi g_{9/2}$ and $\nu g_{7/2}$ with large spatial overlap is supposed to have a decisive influence on the structure of the neutron-rich nuclei with $A \sim 100$. It is held responsible for the rapid transition at $N \sim 60$ from spherical to deformed nuclear shapes since it seems[1] to cause a promotion of protons from the $p_{1/2}$ and $p_{3/2}$ into the $g_{9/2}$ orbitals as soon as the $\nu g_{7/2}$ configuration starts to be filled beyond $N = 56$.

Therefore it is important for the study of the $A \sim 100$ region to get quantitative information on the strength of this interaction. This can be obtained through a comparison of the properties of the deformed nuclei of this region with calculations which use the interaction strength as a paramater. A more clear-cut way is, however, the analysis of the properties of selected states in the non-deformed odd-odd nuclei ($N \leq 59$) which may have a rather pure $[\pi g_{9/2}, \nu g_{7/2}]$ configuration, and the energies of which reflect directly the p-n interaction strength.

* Alexander-von-Humboldt fellow, on leave of absence from University of Zagreb, Yugoslavia

The identification of states with the corresponding configuration is a difficult task because of the experimental limitations in the investigation of nuclei far from stability. It is shown in the following that the β^- decay properties allow nevertheless the identification of some of these levels. The determination of the $\pi g_{9/2} - \nu g_{7/2}$ interaction strength from the energies of these states requires knowledge on the quasiparticle energies in the neighbours of the odd-odd nuclei and the calculation of the splitting of the $[\pi g_{9/2}, \nu g_{7/2}]$ multiplet, which can be done either with Paar's parabolic rule[2] or in a more sophisticated way, e.g. with IBFFM[3]. Here the results are presented which have been obtained for ^{96}Y. It is then discussed which values of the interaction-matrix elements can be deduced from the masses and the quasiparticle energies, and the results are compared to information for other mass regions.

MEMBERS OF THE $[\pi g_{9/2}, \nu g_{7/2}]$ MULTIPLET

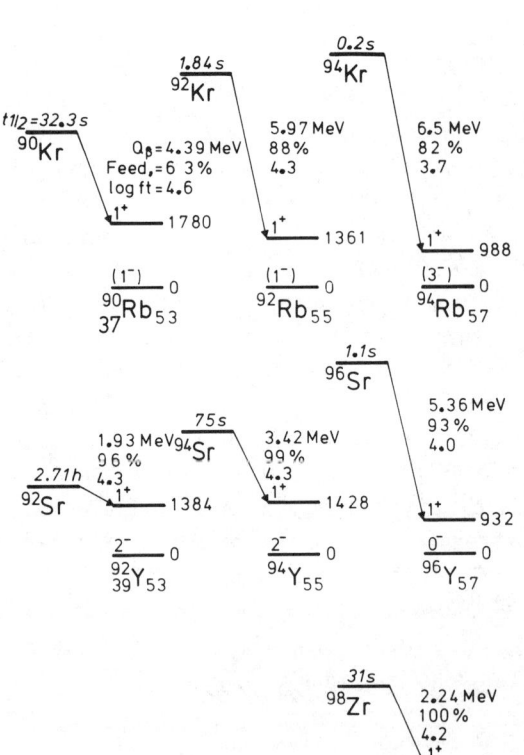

A characteristic property[4,5] of the β^- decays of 0^+ ground states of the even-even nuclei in the $A \sim 100$ region is the strong population of individual levels with log ft ~ 4, see Fig. 1. These β^- decays are most probably $\nu g_{7/2} \to \pi g_{9/2}$ – GT transitions to the 1^+ members of the $[\pi g_{9/2}, \nu g_{7/2}]$ multiplets.

Fig. 1. Characteristics of $0^+ \to 1^+$ β^- decays at $A \sim 100$ (data from Refs. 5 - 7).

This interpretation of the β^- decays is supported by the fact that their strength corresponds to the usual quenching of the GT transitions. The following values have been deduced from a comparison of the measured transition probabilities with the ones calculated with the formalism of Ref. 8:

β⁻ decay	logft	B(GT)$_{exp}$	B(GT)$_{th}$	Quench.
^{88}Kr(0⁺) → ^{88}Rb(1⁺)	4.4	0.15	0.23	0.66(16)
^{96}Sr(0⁺) → ^{96}Y (1⁺)	4.0	0.39	0.78	0.50(12)
^{94}Sr(0⁺) → ^{94}Y (1⁺)	4.3	0.19	0.61	0.31(8)
^{92}Sr(0⁺) → ^{92}Y (1⁺)	4.3	0.19	0.44	0.43(11)

Further support results from the fact that the excitation energies of the 1⁺ levels follow closely the trend of the energies of the $\pi g_{9/2}$ and $\nu g_{7/2}$ quasiparticle configurations as is shown for the Y isotopes in Fig. 2.

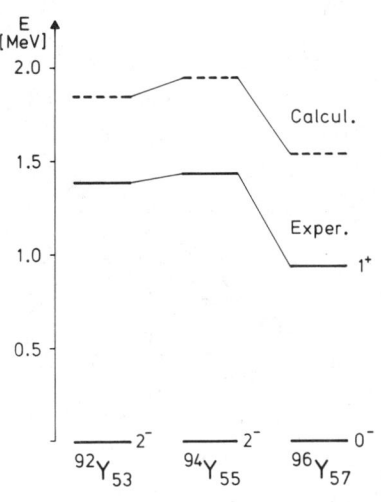

Thus the energies of the 1⁺ states are known for several odd-odd nuclei. In order to study the p-n interaction in detail it is important to identify also other members of the multiplet. Several members have been identified in neighbours of the doubly magic ^{90}Zr but not near ^{96}Zr, a region which is of more interest in connection with the study of the A ~ 100 nuclei. One exception is ^{96}Y where a β⁻ decaying isomer has been observed[9] which is probably the 8⁺ −

Fig. 2. Comparison of the energies of supposed $[\pi g_{9/2}, \nu g_{7/2}]$ 1⁺ levels with those calculated using the parabolic rule.

member of the $[\pi g_{9/2}, \nu g_{7/2}]$ multiplet[10]. Unfortunately the excitation energy of this isomer is not known but Q$_\beta$ measurements show[11] that it should lie at about 1.2 MeV.

QUASIPARTICLE ENERGIES AT A ~ 100

The knowledge on quasiparticle configurations in the odd-mass neighbours of the odd-odd nuclei near ^{96}Zr is still scarce. Although extended level schemes have been established for many of these nuclei, the nature of the states is still unknown. The spins and parities and the underlying structure have been determined only for a few of them.

Thus the quasiparticle energies have to be deduced to a large extent under the assumption that the lowest-lying levels with the corresponding spin and parity basically have single-particle nature. This assumption should be reasonable especially for the important 9/2$^+$ and 7/2$^+$ levels in the odd-proton and odd-neutron nuclei, respectively, since there are fewer candidates for strong admixtures than for levels with lower spin. The resulting quasiparticle pattern is shown in Fig. 3.

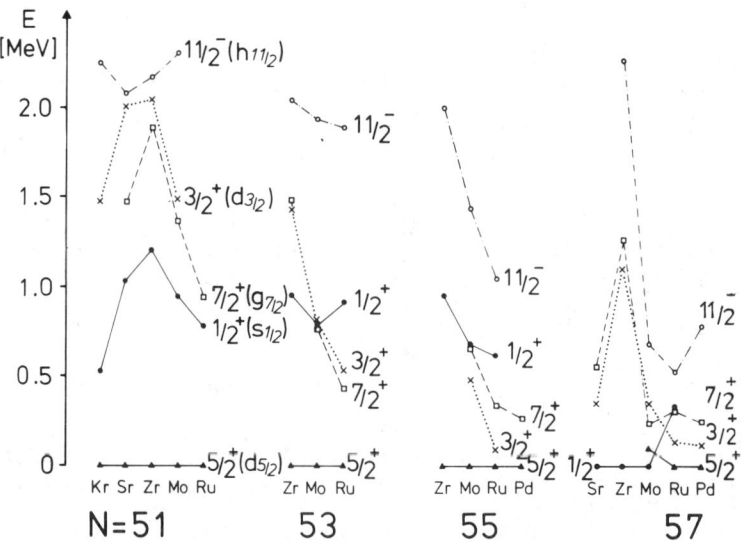

Fig. 3. Trends of the quasiparticle energies at A ~ 100.

It is apparent that there is a lack of information especially for Z ≤ 40 and that the pronounced influence of the Z = 40 subshell closure on the neutron configurations renders a reliable extrapolation difficult.

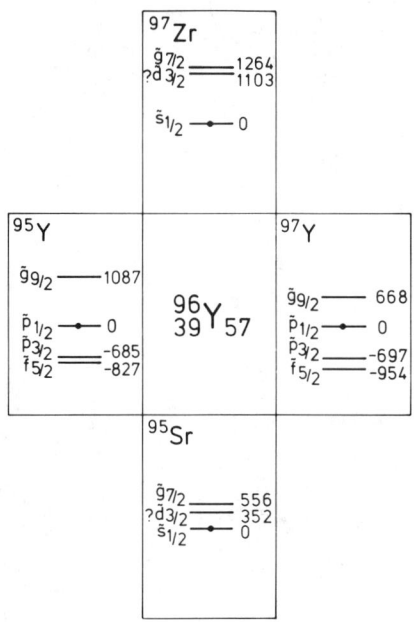

Fig. 4 shows the situation for ^{96}Y which is an especially favourable case for the study of the $\pi g_{9/2} - \nu g_{7/2}$ interaction because of the knowledge about its levels[7] and the quasiparticle energies in its neighbours. The proton-quasiparticle energies change smoothly across N = 57 while the neutron energies vary more drastically across Z = 39. The interpretation of the 352 keV level in ^{95}Sr as $d_{3/2}$-quasiparticle state is doubtful since its occurrence at low excitation energy is at variance[3] with the experimental data on ^{96}Y.

Fig. 4. Quasiparticle energies used in the calculations for ^{96}Y.

THE $[\pi g_{9/2}, \nu g_{7/2}]$ 1^+ AND 8^+ STATES IN ^{96}Y

The p-n interaction can be estimated from a comparison of the excitation energy of the 1^+ level at 932 keV in ^{96}Y (cf. Fig. 1) with the sum of the $\pi\tilde{g}_{9/2}$ and $\nu\tilde{g}_{7/2}$ quasiparticle energies in the odd-mass neighbours if the splitting of the $[\pi g_{9/2}, \nu g_{7/2}]$ multiplet is taken into account. This can be done with the parabolic rule[2]. The calculations have been performed[12] after fitting the parameters of Paar's expression for the splitting $\delta E(j_1, j_2, J)$ to the $[\pi g_{9/2}, \nu d_{5/2}]$ multiplets in ^{90}Y, ^{92}Nb and ^{96}Nb, cf. Fig. 5a.

The results of the calculations with the fitted parameters $\alpha_1^0 = 0.3$ MeV, $\alpha_2^0 = 3.6$ MeV and the quasiparticle energies interpolated between ^{95}Y and ^{97}Y as well as between ^{95}Sr and ^{97}Zr are shown in Fig. 5b. The calculated 1^+ member lies considerably higher than the experimental one. The lowering of the 1^+ level with respect to the unperturbed position of the $g_{9/2} - g_{7/2}$ configuration amounts to about 1.2 MeV.

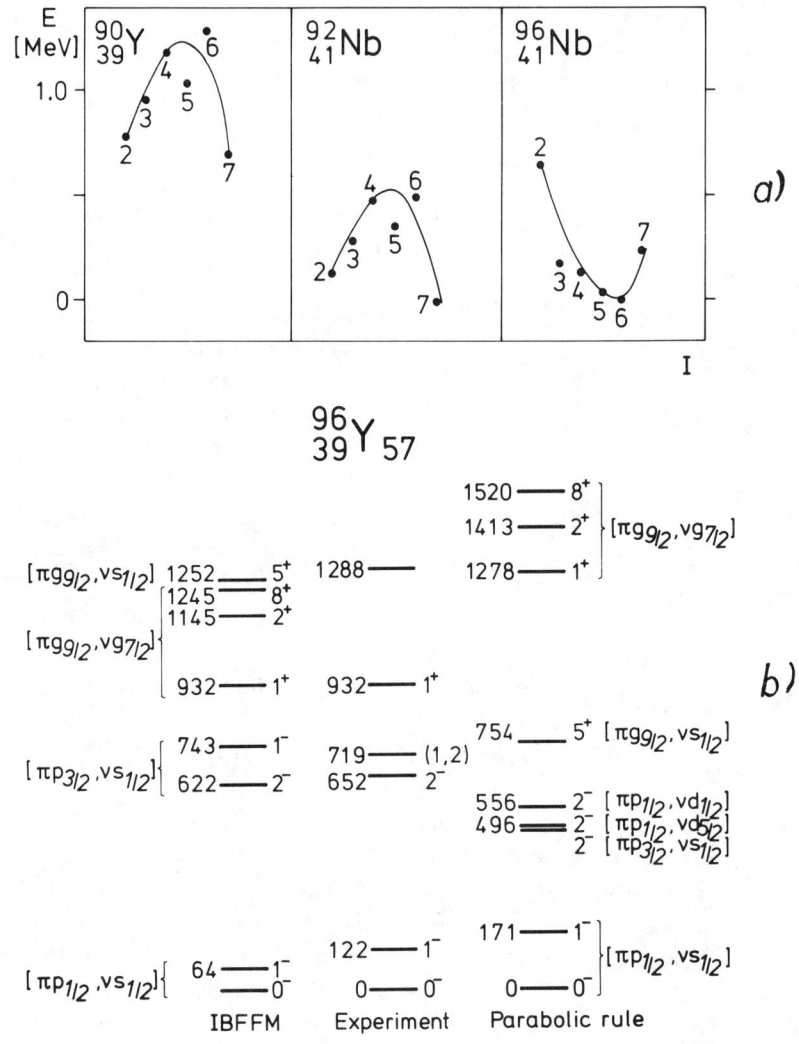

Fig. 5. a) Fits to the $[\pi g_{9/2}, \nu d_{5/2}]$ multiplets near ^{96}Y.
b) Comparison of the ^{96}Y data with the results of calculations.

Although it is expected from the parabolic rule that the 8^+ member of the $[\pi g_{9/2}, \nu g_{7/2}]$ multiplet is lowered by the same amount, the results of the calculations with the parabolic rule cannot account for the isomeric character ($t_{1/2}$ = 10 s) of this state since the $[\pi g_{9/2}, \nu s_{1/2}]$ 4^+, 5^+ doublet is predicted to lie considerably below the 8^+ state which would enable a γ deexcitation.

The isomerism can be explained, however, through more sophisticated calculations[3] in terms of IBFFM which predict the 4^+ and 5^+ states to lie above the 8^+ level. The strength of the residual force (surface delta interaction) amounts to $V_\delta = -0.16$ MeV which is not exceptionally large. If, however, the interpretation[7] of the 1288 keV level in ^{96}Y as the 2^+ member of the $[\pi g_{9/2}, \nu g_{7/2}]$ multiplet is correct then the interaction must be stronger than usual. Hence, in order to get a definite answer on the residual force from the calculations on ^{96}Y it is indispensable to determine the nature of the 1288 keV level or the excitation energy of the 8^+ member of the p-n multiplet.

THE INTERACTION MATRIX ELEMENTS

The p-n two-body residual interaction matrix element can be deduced from a balance of the nuclear masses, the excitation energy of the p-n coupled state and the single particle excitation energies in the neighbouring odd-mass nuclei in the way outlined in Ref. 13. The necessary input information is available in a few cases in the $A \sim 100$ region. One of them is again ^{96}Y, where the energy of the $[\pi g_{9/2}, \nu g_{7/2}]$ multiplet without a residual interaction can be calculated as

$$E_0 = B(^{95}Y^*) + B(^{95}Sr^*) - B(^{96}Y) - B(^{94}Sr)$$

The binding energies are corrected for the excitation energies of the $9/2^+$ (1087 keV) and $7/2^+$ (556 keV) states of ^{95}Y and ^{95}Sr, respectively, and ^{94}Sr in its ground state is taken as core. With the masses from Ref. 14 a value of 2475 keV is deduced. Then the residual-interaction matrix element in the 1^+ state is

$$\Delta([\pi g_{9/2}, \nu g_{7/2}] \, 1^+) = (-2475 + 932) \text{ keV} = -1543 \text{ keV}$$

in reasonable agreement with the lowering of the 1^+ state compared to the unperturbed position of the $[\pi g_{9/2}, \pi g_{7/2}]$ configuration deduced from the estimates with the parabolic rule discussed in the previous section.

Z=43	^{93}Tc $E_{9/2^+}=0$	^{94}Tc $E_{1^+}=442$ $\Delta_{1^+}=-1478$	95	96	97	98	^{99}Tc $E_{9/2^+}=0$	^{100}Tc $E_{1^+}=0$ $\Delta_{1^+}=-1075$	^{101}Tc $E_{9/2^+}=0$	^{102}Tc $E_{1^+}=0$ $\Delta_{1^+}=-1159$
42	^{92}Mo	^{93}Mo $E_{7/2^+}=1363$	94	95	96	97	^{98}Mo	^{99}Mo $E_{7/2^+}=236$	^{100}Mo	^{101}Mo $E_{7/2^+}=238?$
41	^{91}Nb $E_{9/2^+}=0$	^{92}Nb $E_{1^+}=1089$ $\Delta_{1^+}=-1483$	93	94	95	96	^{97}Nb $E_{9/2^+}=0$	^{98}Nb $E_{1^+}=0$ $\Delta_{1^+}=-1676$	99	100
40	^{90}Zr	^{91}Zr $E_{7/2^+}=1882$	92	93	94	95	^{96}Zr	^{97}Zr $E_{7/2^+}=1264$	98	99
39	89	90	91	92	93	94	^{95}Y $E_{9/2^+}=1087$	^{96}Y $E_{1^+}=932$ $\Delta_{1^+}=-1543$	^{97}Y $E_{9/2^+}=667$	^{98}Y $E_{1^+}=600$ $\Delta_{1^+}=-565$
38	88	89	90	91	92	93	^{94}Sr	^{95}Sr $E_{7/2^+}=556$	^{96}Sr	^{97}Sr $E_{7/2^+}=308?$
	N=50	51	52	53	54	55	56	57	58	59

Fig. 6. Input data and results of the determination of the p-n interaction-matrix elements for the $[\pi g_{9/2}, \nu g_{7/2}]$ 1^+ states.

The procedure can be applied to some other odd-odd nuclei in the A ~ 100 region, see Fig. 6. The results for those cases where a relatively inert spherical core can be assumed, namely ^{96}Y, ^{92}Nb, ^{98}Nb, ^{94}Tc agree reasonably well. In contrast, the values for the transitional nuclei ^{98}Y (with the assumption that the isomeric level[15] at 308 keV in ^{97}Sr is the $g_{7/2}$ quasiparticle state), ^{100}Tc and ^{102}Tc deviate considerably from the average of the former. Here the procedure to project the lowering of the excitation energy of the 1^+ state with respect to the unperturbed position into a residual two-body interaction is less justified because of strong core polarization effects. Even in the case of ^{96}Y the results of the IBFFM calculations show a contribution from core polarization.

CONCLUSIONS

The available data on the spherical nuclei (with N ≤ 59) of the A ~ 100 region allow a study of the strength of the important interaction between protons and neutrons in the spin-orbit partner configurations $\pi g_{9/2}$ and $\nu g_{7/2}$. In the odd-odd nuclei, the 1^+ members of the $[\pi g_{9/2}, \nu g_{7/2}]$ multiplet can be identified through the selective feeding through the β^- decay of the 0^+ ground states of the even-even parents. The 1^+ excitation energies show that the interaction-matrix element for this configuration amounts to about -1.5 MeV.

In order to put this value into the correct perspective it should be compared to the matrix elements for analogous stretched configurations in other mass regions. There is, however, little information available about the corresponding 1^+ members of spin-orbit partner multiplets. In fact, the matrix element for the $\pi h_{11/2} - \nu h_{9/2}$ interaction is known[16] from studies in the ^{146}Gd region to amount to about -1.4 MeV. A value of -1.1 MeV has been calculated[17] for the $\pi i_{13/2} - \nu i_{11/2}$ interaction in ^{210}Bi. Since the interaction matrix elements scale[18] with $\frac{1}{A}$, it turns out that the deduced strength of the $\pi g_{9/2} - \nu g_{7/2}$ interaction is smaller than in the case of the $h_{11/2}$ protons and the $h_{9/2}$ neutrons.

Thus, it is rather the position just above the Fermi surface than an exceptional p-n interaction strength which seems to lead to the important influence of the $\pi g_{9/2}$ and $\nu g_{7/2}$ orbitals on the structure of the $A \sim 100$ nuclei.

Further studies on the spherical nuclei of the $A \sim 100$ region should be performed in order to obtain more insight into the p-n interaction. It is necessary to identify additional members of the $[\pi g_{9/2}, \nu g_{7/2}]$ multiplets since only thereby the interaction potential can be deduced reliably. Equally necessary is the identification of quasiparticle configurations in the odd-mass nuclei in order to put calculations for the odd-odd nuclei on solid footing.

The autors acknowledge with pleasure fruitful discussions with Dr. P. KLeinheinz, Prof. V. Paar, Prof. O.W.B. Schult, Prof. I.S. Towner. They thank Prof. K. Heyde for a code for BCS calculations.

REFERENCES

1. P. Federman and S. Pittel, Phys. Rev. C20, 820 (1979)
2. V. Paar, Nucl. Phys. A331, 16 (1979)
3. S. Brant, G. Lhersonneau, K. Sistemich, M.L. Stolzenwald and V. Paar, to be published
4. K. Sistemich, G. Lhersonneau, G. Menzen, H. Ohm and M.L. Stolzenwald, Proc. Int. Conf. on Nuclear Structure, Reactions and Symmetries, Dubrovnik, Yugoslavia, 5-14 June 1986, World Scientific Publishing Co., Singapore, 1986, R.A. Meyer, V. Paar, eds., Vol. 2, p. 706
5. Table of Isotopes, C.M. Lederer and V.S. Shirley (eds.), 7th Edn. (John Wiley and Sons Inc. 1978)
6. G. Lhersonneau, D. Weiler and H. Ohm, Annual Rept. 1985, Institut für Kernphysik, Kernforschungsanlage Jülich, F.R. Germany, Report Jül-Spez 344 (ISSN 0170-8937), p. 26 (1986)
7. G. Jung, B. Pfeiffer, P. Hungerford, S.M. Scott, F. Schussler, E. Monnand, J.A. Pinston, L.J. Alquist, H. Wollnik and W.D. Hamilton, Nucl. Phys. A352, 1 (1981)
8. I.S. Towner, Nucl. Phys. A444, 402 (1985)
9. K. Sistemich, G. Sadler, T.A. Khan, H. Lawin, W.D. Lauppe, H.A. Selič, F. Schussler, J. Blachot, E. Monnand, J.P. Bocquet and B. Pfeiffer, Z. Physik A-Atoms and Nuclei 281, 169 (1977)
10. M.L. Stolzenwald, G. Lhersonneau, S. Brant, G. Menzen and K. Sistemich, Z. Physik A-Atoms and Nuclei 327, 359 (1987)
11. R. Stippler, F. Münnich, H. Schrader, J.P Bocquet, M. Asghar, G. Siegert, R. Decker, B. Pfeiffer, H. Wollnik, E. Monnand and F. Schussler, Z. Physik A284, 95 (1978)
12. G. Lhersonneau, M.L. Stolzenwald and K. Sistemich, Annual Rept. 1986 Institut für Kernphysik, Kernforschungsanlage Jülich, F.R. Germany, Report Jül-Spez-403 (ISSN 0170-8937) p. 19 (1987)
13. J.P. Schiffer and W.W. True, Rev. Mod. Phys. 48, 191 (1976)
14. A.H. Wapstra and G. Audi, Nucl. Phys. A432, 1 (1985)
15. K.-L. Kratz, H. Ohm, A. Schröder, H. Gabelmann, W. Ziegert, B. Pfeiffer, G. Jung, E. Monnand, J.A. Pinston, F. Schussler, G.I. Crawford, S.G. Prussin and Z.M. de Oliveira, Z. Physik A-Atoms and Nuclei 312, 43 (1983)
16. P. Kleinheinz, Kernforschungsanlage Jülich, private comm. and separate contribution to this Book
17. T.T.S. Kuo and G.H. Herling, NRL Memorandum Report 2258 (Naval Research Laboratory, Washington, 1971)
18. G.E. Brown, Many-Body Problems (North-Holland, Amstemdam, 1972)

NEUTRON-RICH A ≅ 100 NUCLEI:
NILSSON CONFIGURATIONS, DEFORMATION, PAIRING REDUCTION

B. Pfeiffer[1], V. Harms[1], E. Monnand[2], P. Möller[3]
and K.-L. Kratz[1]

[1] Institut für Kernchemie, Universität Mainz, FR- Germany
[2] Departement de Recherche Fondamentale, Centre d'Etudes Nucleaires, Grenoble, France
[3] Department of Mathematical Physics, Lund Institute of Technology, Sweden

ABSTRACT

Appropiate Nilsson parameters and deformations for RPA shell model calculations of A≅100 neutron-rich nuclei are derived. They allow Nilsson orbital assignments even on the basis of gross β-decay properties. Matching of the A≅100 and A≅150 mass regions yields evidenve for a local N=64 subshell. A study of the influence of the pairing strength on quasi-particle configurations indicates a pairing reduction of only 50-60% at A≅100, in contrast to a recent proposition of total pairing collapse.

DETERMINATION OF APPROPIATE NILSSON PARAMETER SETS

To gain a more quantitative picture of the influence of the rapidly changing nuclear structure at A≅100 on gross β-decay properties as well as on Nilsson orbital assignments, we have compared experimental data with shell model expectations for different structures to extract information on Nilsson parameters (κ,μ), single particle states and deformation. The calculations were performed with the RPA shell-model code of Krumlinde and Möller[1], which treats pairing in the BCS approximation and the Gamow-Teller (GT) residual interaction in the RPA. From detailed γ-spectroscopic work at OSTIS(ILL, Grenoble) several rotational bands could be evidenced in A=99 isotopes. By systematically varying the parameters for the RPA-model -such as (κ,μ), deformation and pairing strength- a consistent description of the β-strength distribution S_β as well as the band structure of ^{99}Sr and ^{99}Y was obtained. In the case of ^{99}Y (see Fig. 1), our Nilsson orbital assignments agree with results on "K^π" bands of IBFM/PTQM calculations[2].

It should be pointed out that an appropiate choice of Nilsson parameters and deformation has to be made. Parameters fitting experimental data of spherical nuclei near stability, like the standard A=100 set[3], tend to become inappropriate for far-unstable deformed nuclei. This, together with the assumption of unrealistically strong deformation and Coriolis mixing, are the reasons for incorrect Nilsson assignments to excited bands in ^{99}Y by Wohn et al.[4]. For the very neutron-rich region, rather the N=60 or the A=110 sets[5,6] have to be used, which in consequence do not require extreme quadrupole deformation and band mixing[7].

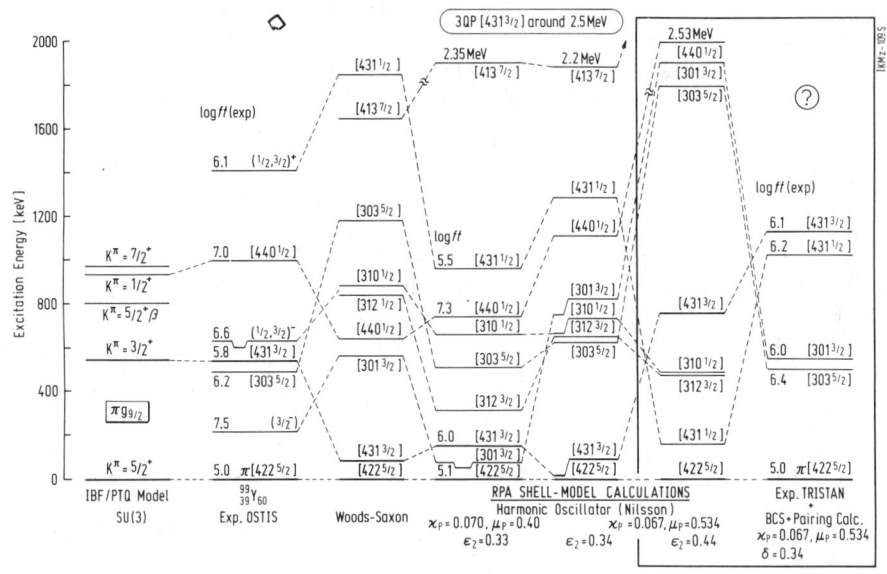

Fig. 1: Comparison of experimental band heads in ^{99}Y with predictions from the IBFM/PTQ- and RPA shell model with different parameter sets.

With the optimized parameter sets reliable estimates for isotopes with -at present- very low production yields excluding sophisticated measurements can be obtained. Already from comparing measured gross β-decay properties ($T_{1/2}$, P_n, main features of S_β) to RPA-calculations Nilsson orbitals can be tentatively assigned to ground states, e.g. π[431 3/2] for ^{101}Rb, ν[411 3/2] for ^{101}Sr and π[422 5/2] for ^{101}Y, respectively.

THE N=64 SUB-SHELL AND THE RANGE OF DEFORMED NUCLEI

Alternatively we have correlated Nilsson configurations in odd-N A≅100 nuclei with corresponding ones observed in odd-Z A≅150 rare-earth isotopes (for details see Ref.[9]). This procedure yields reliable estimates for

rotational bands built on the ν-Nilsson orbitals
[411 3/2], [413 5/2] and [422 3/2] in neutron-rich Sr, Zr
and Mo isotopes. By comparing extrapolations for the
ground state bands in ^{101,102}Sr with recent measurements
performed by us at CERN-ISOLDE, slight increases in the
energies of the first excited states seem to indicate a
local N=64 subshell effect for Z≅38, which limits the region of highly deformed nuclei to N=60-64. An application
of the quartet model of Daley et al.^{10} to the A≅100 region yielded similar results^9. It is important to note
that this model treats all two-nucleon correlations on an
equal basis and consequently includes no specific
$\nu g_{7/2} \pi g_{9/2}$ interaction^{11} which is believed to determine
the deformation at A≅100.

PAIRING COLLAPSE IN DEFORMED A ≅ 100 NUCLEI?

Recently Peker et al.^{12} have postulated the existence of (nearly) pairing-free $K^\pi=1^+$ rotational bands with
$\Theta \gtrsim 0.9\ \Theta_{rig}$ in odd-odd A≅100 nuclei due to a particularly
strong $\nu g_{7/2} \pi g_{9/2}$ interaction. To obtain more insight
into the question of a correspondence between moment of
inertia and pairing reduction we have performed shell model calculations in the RPA for comparison with experimental data on $T_{1/2}$, logft values and energies of 2QP-
states^{13}. As an example of our results, in Fig. 2 the
partial level scheme of ^{100}Y is shown, together with RPA
predictions of GT β-decay to the lowest-energy 2QP-states
for different assumptions on the pairing strength parameter Δ. It is evident that certain 2QP-energies and RPA
logft change drastically as a consequence of pairing reduction. As was suggested in Ref.^{12}, the 11 keV level
should have the 2QP configuration ν[411 3/2] π[422 5/2].
The 1^+ state at 975 keV could correspond to the
ν[411 3/2] π[431 1/2] configuration predicted around 1.1
MeV by our model. Assuming ^{100}Y to be pairing-free,
however, the 975 keV level would have a ν[411 1/2]
π[431 3/2] configuration which appears to be blocked for
$\Delta \gtrsim 3/\sqrt{A}$. The ν[422 3/2] π[422 5/2] configuration assigned
in Ref.^{12} to the 975 keV level lies high in energy (1.4
and 3.0 MeV for $\Delta=12/\sqrt{A}$ and $\Delta \cong 0$, respectively) and it is
even blocked (logft(RPA)≅8-11). Furthermore, we found no
evidence for a 2QPS around 2.5 MeV with a logft of about
4, as predicted for $\Delta \cong 0$.

With regard to the role of a specific $\nu g_{7/2} \pi g_{9/2}$ interaction in the A≅100 region, first postulated by Federman and Pittel^{11}, our RPA calculations show (in agreement
with other authors^{14,15}) that besides the above-mentioned
ν-π interaction, also other orbitals are important for
the onset of deformation around N=60, i.e. the occupancy
of the $\pi d_{5/2}$ and the $\nu d_{5/2}$ and $\nu h_{11/2}$ orbitals.

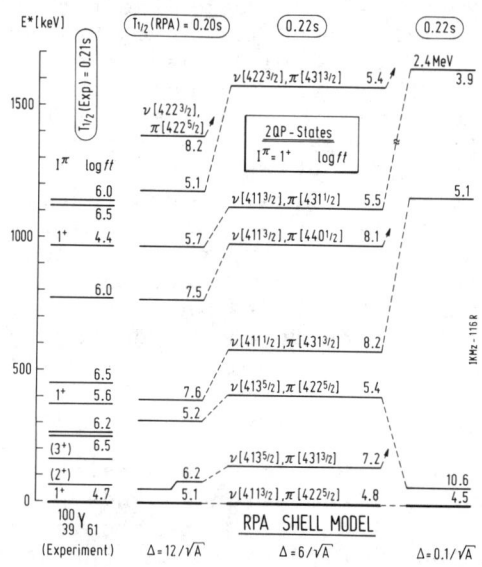

Fig. 2: Comparison of the partial level scheme and logft values for ^{100}Y with predictions for low-energy 2QPS from RPA calculations for different assumptions on pairing. For discussion, see text.

We conclude from our results that the low-lying rotational bands with $K^\pi = 1^+$ are not pairing-free. Furthermore, our calculations do not substantiate the conjecture of a selective $\nu g_{7/2} \pi g_{9/2}$ interaction in leading to a total collapse of pairing.

This research was supported by the BMFT(06MZ 552)

REFERENCES

1. J. Krumlinde, P. Möller, Nucl.Phys. A427,419(1984)
2. B. Pfeiffer et al., Z.Phys. A325,487(1986)
3. S.E. Larsson et al., Nucl.Phys. A261,77(1976)
4. F.K. Wohn et al., Phys.Rev. C31,634(1985)
5. R.A. Meyer, Hyperfine Interactions 22,385(1985)
6. J. Krumlinde et al., Nucl.Phys. A417,419(1984)
7. V. Paar et al., to be published
8. B. Pfeiffer et al., Spec. Meeting on Delayed Neutron Properties, Birmingham(1986), to appear as BRC-Report
9. B. Pfeiffer, K.-L. Kratz, Z.Phys. A327,163(1987)
10. H.J. Daley et al., Phys.Rev.Lett. 57,198(1986)
11. P. Federman, S. Pittel, Phys.Rev. C20,820(1979)
12. L. Peker et al., Phys.Lett. 169B,323(1986)
13. K.-L. Kratz et al., Phys.Lett. 187B,215(1987)
14. S.K. Khosa et al., Phys.Lett. 119B,257(1982)
15. A. Kumar, M.R. Gunye, Phys.Rev. C32,2116(1985)

NILSSON ORBITALS AND ROTATIONAL BANDS OF DEFORMED A≃100 NUCLEI

F.K. Wohn, John C. Hill and J.A. Winger
Ames Laboratory, Iowa State University, Ames, IA 50011

R.F. Petry and J.D. Goulden
University of Oklahoma, Norman, OK 73019

R.L. Gill, A. Piotrowski[a] and H. Mach[b]
Brookhaven National Laboratory, Upton, NY 11973

ABSTRACT

Highly deformed Rb, Sr, Y, Zr and Nb isotopes are being studied using the TRISTAN facility at BNL. γ-spectroscopic studies of the β decays of $^{99-102}$Sr to their odd-Z Y daughters have been completed and studies of the β decays of ^{99}Rb and ^{101}Y are nearing completion. This report summarizes our results on $^{99-102}$Sr decays, then focuses on experimental or theoretical results for certain Nilsson orbital assignments. In particular, the Nilsson assignments for the ground state of ^{99}Rb and the 536-keV K=3/2 band in ^{99}Y are discussed.

INTRODUCTION

Nuclei with A≃100 exhibit three unusual characteristrics: 1) an extremely abrupt onset of deformation at N≃60, 2) the coexistence of nearly spherical and highly deformed shapes near N≃60, and 3) rotational moments of inertia for deformed bands that are unusually large. Experimental results over the last decade on several nuclei have clearly established these characteristics. Some recent reviews of these results are given in Ref. 1. The underlying cause of these abrupt changes in the structure of nuclei with N≃60 and Z≃40 has yet to be established. This motivates further studies, both experimental and theoretical, of nuclei in the A≃100 region.

For the deformed odd-A nuclei in this region, the determination of the Nilsson orbitals near the Fermi surface is crucial to the eventual understanding, on a microscopic level, of the interactions that give rise to the unusual characteristics of these A≃100 nuclei. The microscopic explanation of Federman and Pittel[2] involves strong attractive n-p interactions between "spin-orbit partners" like the $\nu g_{7/2}$ and $\pi g_{9/2}$ spherical orbitals or the $\nu 3/2[411]$ and $\pi 5/2[422]$ deformed orbitals. Studies[3-6] at TRISTAN of levels in deformed Y nuclei (99≤A≤102) indicate that the orbitals $\pi 5/2[422]$ and $\nu 3/2[411]$ are ground-state or low-lying orbitals in these nuclei.

SUMMARY OF BANDS IN $^{99-102}$Y

Five rotational bands have been identified in ^{99}Y and ^{101}Y.[3,5] The three lowest bands can be identified with the Nilsson orbitals $\pi 5/2[422]$, $\pi 5/2[303]$ and $\pi 3/2[301]$. Our particle-rotor calculations

---

[a] Permanent address: Institute for Nuclear Studies, Warsaw, Poland
[b] Present affiliation: Clark University, Worcester, MA 01610

© American Institute of Physics 1988

reproduce both the energy-level spacings and the γ-decay patterns (for both intraband M1+E2 and interband M1 or E1 transitions) of all observed γ rays in these three bands. Higher-lying bands (above ~1 MeV) that can be associated with the 1/2[431] and 3/2[431] orbitals, are found to have similar features in ^{99}Y and ^{101}Y. However, unlike the three lower-energy bands, these two higher-energy bands deviate from particle-rotor model predictions for a single quasiparticle coupled to an axially symmetric (prolate) core.[3,5]

Nearly "pairing-free" $K^{\pi}=1^+$ bands in the odd-odd nuclei ^{100}Y and ^{102}Y, with bandheads established via low logft values, have been proposed.[4,6] (Our observed γ intensities and deduced levels differ significantly from those of a previous ^{100}Y study.[7]) Low-lying two-quasiparticle 1^+ states (predicted from neighboring odd-even and even-odd nuclei) are expected to be strongly fed in β decay of the even-even Sr parent nuclei. These $K^{\pi}=1^+$ bands (and similar bands in ^{102,104}Nb) have moments of inertia nearly equal to the rigid moment of inertia. The apparent "pairing-free" character of such bands may be caused by an unusually strong n-p interaction.[2] However, the extent to which this may instead be caused by Coriolis band mixing is at present unknown. (As was shown in our calculations[3] for ^{99}Y, Coriolis mixing can cause an increase of ~15% in the empirically deduced moment of inertia of a low-lying band in an odd-A nucleus.)

THE ^{99}Rb GROUND-STATE ASSIGNMENT

For the ground state of ^{99}Rb (Z=37), the orbital 3/2[431] is predicted by the Nilsson scheme for a wide choice of Nilsson model parameters. This assignment was used in the report of Pfeiffer et al.[8] on the decay of ^{99}Rb. However, in the latest revision of the Nuclear Data Sheets for A=99,[9] the ^{99}Rb ground state was reevaluated to be $(5/2^+)$ and this $(5/2^+)$ assignment resulted from logft values deduced[9] from an unpublished decay scheme provided by Pfeiffer.[10] Our detailed study[11] at TRISTAN of ^{99}Rb decay gives a decay scheme with 72 γ rays placed among 27 levels in ^{99}Sr. Although the K=3/2 ground band[8,12] and a 422-keV K=5/2 band[8] are still present in our decay scheme,[11] the intensities of two important γ rays (at 91 and 125 keV) differ significantly from those in Ref. 9.

A disagreement with our earlier results[12] for the 91- and 125-keV γ intensities was noted in Ref. 8. Our results[11,12] show that a 125-keV γ ray occurs in both ^{99}Rb and ^{99}Sr decays but the 91-keV γ ray occurs only in ^{99}Rb decay, whereas a 91-keV γ ray is also attributed in Ref. 8 to ^{99}Sr decay. (Although we cannot rule out a very weak 91-keV γ ray in ^{99}Sr decay, we find no evidence for its existence in our data.) Our γ-ray placements and deduced intensities are based upon time-sequential γ spectra and γ-γ coincidence gates set on established γ rays in the decays of both ^{99}Rb and ^{99}Sr. For example, our value for the γ intensity of the ^{99}Rb member of the 125-keV doublet is the weighted average of eight separate ways of deducing this intensity.[11] With the intensity of the 91-keV γ ray set to 1000, we find a ^{99}Rb 125-keV γ-ray intensity of 162±12. In comparison, Ref.9 gives an intensity of 400 (with no uncertainty given) and our preliminary report[12] used the value 240±40. These γ rays are the lowest two ΔI=1 transitions in the K=3/2 ground band and the large descrepancy in their relative intensities gives rise to a crucial difference in the deduced logft values to the 5/2 and 7/2 members of the ground band, as is shown in Table I.

Table I. β-decay of ^{99}Rb to 3/2[411] band members in ^{99}Sr

Level in ^{99}Sr I$^\pi$	E(keV)	deduced from Ref. 9 β feeding	logft	deduced from Ref. 11 β feeding	logft
3/2$^+$	0	38.5%	4.9	35.0%	5.0
5/2$^+$	90.7	−0.9%	−	12.7%	5.4
7/2$^+$	215.7	7.6%	5.6	<0.9%	>6.6
9/2$^+$	377.6	1.8%	6.2	<0.2%	>7.2

With our ^{99}Rb decay scheme,[11] the logft values shown in Table I are quite consistent with the 3/2[431] prediction of the Nilsson scheme for the ground state of ^{99}Rb. (The weak β feedings to the 7/2$^+$ and 9/2$^+$ levels are at the "noise" level of our data.) However, use of preliminary β-decay results[10] in Ref. 9 led to the tentative value of (5/2$^+$) since both the 3/2$^+$ and 7/2$^+$ levels appeared to be fed by allowed β decay. A more detailed discussion of the intensity differences for the γ rays will be given in Ref. 11.

THE 536-keV K=3/2 BAND IN ^{99}Y

In the study[3] of the β decay of ^{99}Y, the K=3/2 rotational band (with bandhead at 536 keV, 5/2 level at 656 keV and 7/2 level at 817 keV) was assigned as 3/2[301] based on β- and γ-decay results. A new assignment of 3/2[431] was recently proposed.[13] In Ref. 14 the 3/2[431] assignment is shown to be inconsistent with the observed[3] γ-decay pattern from the levels of the 536-keV band. The argument is summarized here. Both the 536-keV K=3/2 and the 487-keV K=5/2 bands in ^{99}Y decay only by dipole γ rays.[3] If Coriolis mixing is negligible, then the dipole γ-decay pattern for a given level of one band to levels of another band is given by the Alaga rule, i.e. the reduced intensity, B(M1) or B(E1), is proportional to the square of a Clebsch-Gordon coefficient. The 536-keV band has low-energy intraband M1 γ rays as well as γ rays to both a 487-keV 5/2[303] band and the 5/2[422] ground band. Three different γ-decay modes severely constrain acceptable Nilsson orbital assignments.

The key to an acceptable 536-keV assignment is provided by the ratios of intensities of intraband M1 γ rays to the ground-band γ rays. With a 3/2[431] assignment[13], the γ rays to the ground band would also be M1 and the energy-cubed factor would make intraband M1 γ intensities weaker than interband M1 γ intensities by a factor of ~100. Since the 5/2[422] and 3/2[431] bands have unique parity, the factor of ~100 is nearly independent of the Nilsson parameters.[14] For the 3/2[301] assignment,[3] however, the interband γ rays to the 487-keV band are M1 and the γ rays to the ground band are E1. As is shown in Ref. 14 (and earlier in Ref. 3) all γ rays from the 536-keV band can be accurately reproduced if the band is 3/2[301].

The Coriolis mixing of the closely spaced 487-keV 5/2[303] and 536-keV 3/2[301] bands in ^{99}Y, which is well determined by the level energies of the observed (mixed) bands,[3,14] is seen to have little influence on the E1 γ-decay pattern from the 487-keV band but has a significant influence on the E1 γ-decay pattern calculated for the 536-keV band.[14] In summary, Coriolis mixing in ^{99}Y has significant effects only for the 3/2[301] band, as both the 5/2[303] 487-keV band and the 5/2[422] ground band are in excellent agreement with the Alaga rules. [For the 5/2[422] ground band, a single parameter, $(g_K-g_R)/Q_0$, accurately determines the E2/M1 mixing for all levels of the band.] Similar arguments could also be made for the analogous bands in ^{101}Y, for which the lowest K=3/2 band is also 3/2[301].[5]

THE N=61 ISOTONES ^{99}Sr AND ^{101}Zr

Our studies of the decays of ^{99}Rb and ^{101}Y show that in both cases, the ground state of the N=61 daughter is 3/2[411] and is fed by allowed β decay with a logft of 5.0. When levels in these two K=3/2 bands were first published,[12] there was no β decay evidence to support the 3/2[411] assignment. We observe a K^π=3/2$^+$ band at 1072 keV in ^{99}Sr that is fed by allowed β decay. The most likely Nilsson assignment for this band is 3/2[422].[8] Candidates for 5/2[413], 3/2[541], and 5/2[532] bands in these two N=61 isotones will be discussed in forthcoming publications.

This work was supported by the U.S. Department of Energy under contracts W-7405-eng-82, DE-AS05-79ER10495 and DE-AC07-76CH00016.

REFERENCES

1. Nuclei Off the Line of Stability, ed. by R.A. Meyer and D.S. Brenner, Symposium Series No. 324, Amer. Chem. Society (1986).
2. P. Federman and S. Pittel, Phys. Lett. 69B, 385 (1977) and 77B, 29 (1978); Phys. Rev. C 20, 820 (1979).
3. R.F. Petry, H. Dejbakhsh, J.C. Hill, F.K. Wohn, M. Shmid and R.L. Gill, Phys. Rev. C 31, 621 (1985); F.K. Wohn, J.C. Hill and R.F. Petry, Phys. Rev. C 31, 634 (1985).
4. F.K. Wohn, J.C. Hill, J.A. Winger, R.F. Petry, J.D. Goulden, R.L. Gill, A. Piotrowski and H. Mach, Phys. Rev. C (in press).
5. R.F. Petry, J.D. Goulden, F.K. Wohn, J.C. Hill, R.L. Gill and A. Piotrowski, Phys. Rev. C (to be published).
6. J.C. Hill, J.A. Winger, F.K. Wohn, R.F. Petry, J.D. Goulden, H. Mach, R.L. Gill and A. Piotrowski, Phys. Rev. C 33, 1727 (1986).
7. J. Münzel, B. Pfeiffer, U. Stohlker, G. Bewersdorf, H. Wollnik, H. Lawin and E. Monnand, Z. Phys. A321, 515 (1985).
8. B. Pfeiffer, E. Monnand, J.A. Pinston, J. Münzel, P. Möller, J. Krumlinde, W. Ziegert, K.-L. Kratz, Z. Phys. A317, 123 (1984).
9. H.-W. Müller and D. Chmielewska, Nucl. Data Sheets 48, 663 (1986).
10. B. Pfeiffer - Priv. Comm. to Nucl. Data Sheets (1985).
11. J.D. Goulden, R.F. Petry, F.K. Wohn, J.C. Hill, J.A. Winger, R.L. Gill, H. Mach and A. Piotrowski (to be published).
12. F.K. Wohn, J.C. Hill, R.F. Petry, H. Dejbakhsh, Z. Berant and R.L. Gill, Phys. Rev. Lett. 51, 873 (1983).
13. B. Pfeiffer, S. Brant, K.-L. Kratz, R.A. Meyer and V. Paar, Z. Phys. A325, 487 (1986).
14. F.K. Wohn, Phys. Rev. C (in press).

NUCLEAR SPECTROSCOPY OF ON-LINE MASS SEPARATED FISSION PRODUCTS WITH MASS NUMBERS FROM A = 110 TO 120 *

J. Äystö, J. Honkanen, P. Jauho, V. Koponen, H. Penttilä,
P. Taskinen and M. Yoshii†
Department of Physics, University of Jyväskylä,
SF-40100 Jyväskylä, Finland

C.N. Davids
Argonne National Laboratory, Argonne, Illinois 60439, USA

J. Żylicz
Department of Physics, Warsaw University, PL-00681 Warsaw, Poland

ABSTRACT

First studies of new neutron rich nuclei below Z=46 have been performed at the on-line separator facility, IGISOL. Beta decays of ^{111}Tc, ^{115}Rh, ^{116}Rh and ^{118}Pd have been observed for the first time. The total $0^+ \rightarrow 1^+$ beta strength for the ^{114,116,118}Pd decays suggests a large hindrance factor. The level properties of neutron rich even Pd isotopes are also discussed.

INTRODUCTION

Study of neutron rich nuclides in the A = 110-120 region is of particular interest because of many reasons, including the understanding of the interplay between collective and single particle degrees of freedom and the role of the $(\pi g_{9/2}, \nu g_{7/2}; 1^+)$ configuration in their structure and particularly in their beta decay. Also extension of the half-life and binding energy systematics to this poorly known region is important. By utilizing recently the ion guide fed on-line isotope separator IGISOL in combination with the 20 MeV proton induced fission of ^{238}U this group of nuclei has become available for the first time as on-line mass separated beams[1,2]. The first results of the experiments on fission products at the IGISOL facility include the mapping of the limit of known neutron rich nuclei, the study of the strength of the $0^+ \rightarrow 1^+$ -type Gamow-Teller (GT) decays and the study of the level structure of even-even nuclei, all in the A = 110 - 120 region.

NEW ISOTOPES

The limit of known neutron rich isotopes in the A = 110-120 region lies close to the line of the most probable Z produced in the p + ^{238}U -fission. Several new β-activities have been observed and up to now three new isotopes ^{111}Tc, ^{115}Rh and ^{116}Rh have been identified. In general, the nuclidic identifi-

---

* Supported by the Research Council for Natural Sciences of the Academy of Finland.
† Present address:National Laboratory for High Energy Physics KEK, Tsukuba, Japan

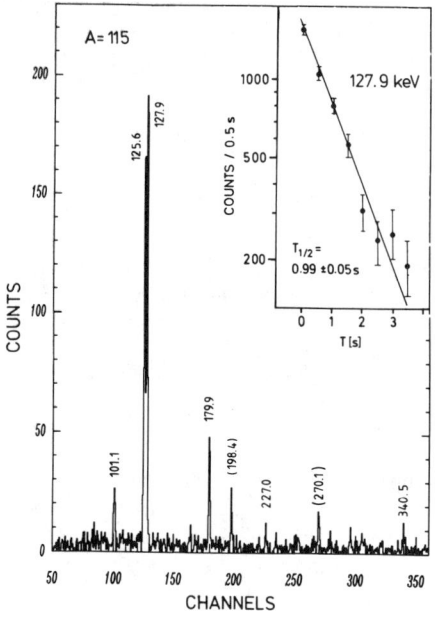

cation is based on the selection of the mass number and on the X-γ and β-γ coincidence relations. Fig. 1 shows an example of such a measurement performed to identify the γ-lines following the β-decay of ^{115}Rh. The decay of the 127 keV transition in coincidence with β-rays is used to obtain the half-life of ^{115}Rh.

The half-lives of new isotopes observed in this work are given in Table 1 together with the predictions of Klapdor et al.[3] and Takahashi et al[4].

The observed half-lives for Rh isotopes agree excellently with the predictions of Klapdor et al. Largest deviations from this theory are expected and indeed have been seen to occur across the shell closures. For this reason it is rather surprising to

Fig.1 Gamma spectrum at mass 115 in coincidence with Pd X-rays.

Table 1. Observed and predicted half-lives of isotopes discovered by the IGISOL.

Nuclide	Obs.$T_{1/2}$ [s] (this work)	Theor.$T_{1/2}$ [s] (ref. 3)	Theor.$T_{1/2}$ [s] (ref. 4)
^{111}Tc	0.30(3)	1.47	2
^{115}Rh	0.99(5)	0.98	4
^{116}Rh	0.68(6) 0.9(4)	0.61	2

see the large discrepancy between this theory and our experimental value for ^{111}Tc. Part of the explanation may lie beyond the large difference in deformation of these two groups of nuclei. Further experimental and theoretical work is clearly needed to clarify the problem.

$0^+ \rightarrow 1^+$ GAMOW-TELLER STRENGTH

Near the closed shells even-even nuclei decay typically by very fast $0^+ \rightarrow 1^+$ GT transitions. Good examples of such decays can be found near the ^{100}Sn region, where an extensive analysis of the strength of these decays is being performed[5]. We have extended such measurements to a slightly more deformed neutron rich Pd-isotopes, where an additional hindrance in the decay strength is anticipated. The $0^+ \rightarrow 1^+$ GT transitions of neutron rich Pd isotopes result mainly from the $\nu g_{7/2} \rightarrow \pi g_{9/2}$ transformation feeding the 1^+ final states in

Ag isotopes containing the $(\pi g_{9/2}, \nu g_{7/2}^{-1}; 1^+)$ component in their wave function. The total strength is then obtained by summing over all the observed 1^+ states, which can readily be compared with model calculations.

By employing the concept of the extreme single particle shell model together with pairing, the beta strength involving the $\nu g_{7/2} \to \pi g_{9/2}$ transformation can be written as[5]

$$S_{\beta_-}(GT) = \frac{160}{9} \cdot v_n^2 \cdot u_p^2,$$

where v_n^2 and u_p^2 are pairing functions for the neutrons in the $g_{7/2}$ orbital and for the proton holes in the $g_{9/2}$ orbital. In the simplest approximation of the shell model v_n^2 and u_p^2 have values of 1 and 4/10, respectively. This results in the log ft-value of 2.74.

Fig.2. shows the summary of our preliminary experimental results for the decays of ^{114}Pd, ^{116}Pd and ^{118}Pd. Except for our improved half-life the data on ^{114}Pd is from ref.(6). Many new transitions were observed for ^{116}Pd as compared with those in ref.(7). The decay of ^{118}Pd was observed for the first time; its existence has been inferred earlier in a radiochemical study[8].

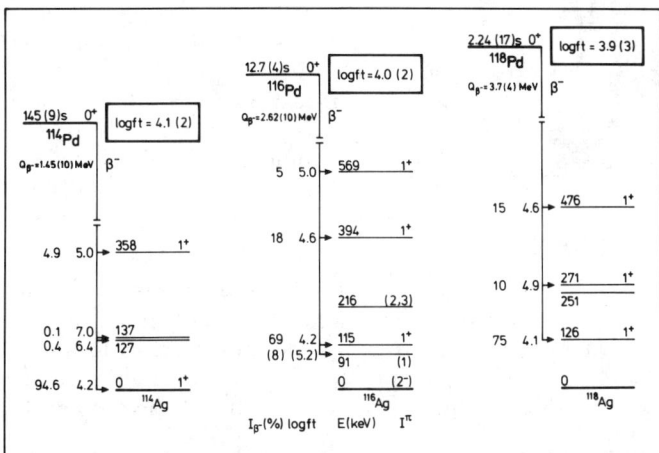

Fig.2. The observed beta strength of ^{114,116,118}Pd.

The total observed beta strength is shown inside rectangles in fig.2 for ^{114,116,118}Pd. This summed strength can now be compared with the above calculation for the pure $\nu g_{7/2} \to \pi g_{9/2}$ transition. The experimental strength is considerably hindered being only about 4, 5.5 and 7 % of the theoretical one in cases of ^{114}Pd, ^{116}Pd and ^{118}Pd, respectively. More elaborate calculations, including the collective effects, are clearly needed.

LEVEL STRUCTURE OF EVEN Pd ISOTOPES

High energy beta decays of low- and high-spin isomers of Rh isotopes should cover a wide range of spin values and levels in neutron rich even Pd isotopes. In this work we have used these β-decays to construct the level schemes for the $^{110-116}$Pd isotopes[9,10]. Fig.3. shows the existence of at least two β-decaying states in ^{116}Rh. One is associated with a spin 1^+ feeding the low-lying 0^+ and 2^+ states and the other(s) with a spin value of 5 or more. The spin assignments of the daughter levels are mainly based on their gamma-decay pattern and on the systematics of $^{110-116}$Pd isotopes.

The 1^+ isomer could be the member of the $\pi g_{9/2} \nu g_{7/2}$ multiplet, which decays to the low-lying low-spin states of even Pd with a log ft of about 5.5, typical to allowed transitions to collective states. The Rh isomers with $I \geq 5$ have their origin in the multiplets formed by coupling a $g_{9/2}$ proton to $d_{5/2}$, $d_{3/2}$, $s_{1/2}$ on $h_{11/2}$ neutron. Since the allowed β-decay is mainly mediated via the decay of a $g_{7/2}$ neutron to a $g_{9/2}$ proton the remaining odd $g_{7/2}$ neutron can couple with the spectator neutrons to form two-quasiparticle states in the daughter, lying above the pairing gap at ~ 2.5 MeV[10]. The strong β-branch to the 2450 keV level could represent such a transition.

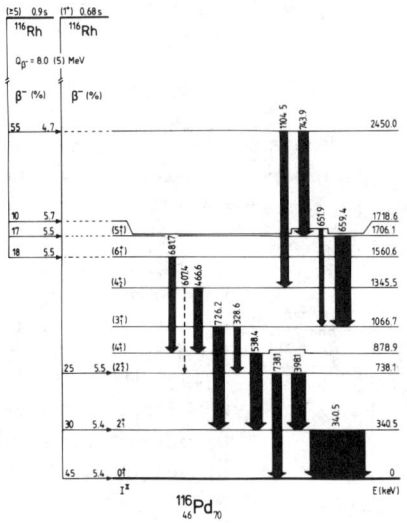

Fig.3. The decay scheme of ^{116}Rh.

The idea that the proton-neutron interaction is primarily responsible for the development of collectivity and deformation in nuclei was recently demonstrated by a simple $N_p N_n$ scheme, where the $N_p N_n$ is the product of the number of valence nucleons[11]. In this scheme a similar collective structure to ^{116}Pd should be found in the isotone ^{124}Xe. Indeed, the level energies of the ground state bands coincide almost perfectly and the two lowest levels of γ-band (2_1^+, 3_1^+) are at slightly higher energy in ^{124}Xe. More detailed description of the collectivity of the studied Pd isotopes is given in ref. 10.

We would like to thank Professor J. Kantele for stimulating discussions.

REFERENCES

1. J. Äystö, P. Taskinen, M. Yoshii, J. Honkanen, P. Jauho, J. Ärje and K. Valli, Nucl. Instr. Meth._B26_ (1987) 394
2. J.Ärje, J. Äystö, H. Hyvönen, P. Taskinen, V. Koponen, J. Honkanen, A. Hautojärvi and K. Vierinen, Phys. Rev. Lett. 54 (1985) 99
3. H.V. Klapdor et al., At. Data Nucl. Data Tables 31 (1984) 81
4. K. Takahashi et al., At. Data Nucl. Data Tables 12 (1973) 101
5. J. Dabaczewski et al., in Weak and Electromagnetic Interactions in Nuclei, H.V. Klapdor (ed.), Springer Verlag, Heidelberg, 1986, p.248
6. D. Meikrantz et al., Radiochimica Acta 29 (1981) 93
7. J. Blachot et al., Nucl. Data Sheets 32 (1981) 287
8. H. Weiss et al., Phys. Rev. 188 (1969) 1893
9. J. Äystö et al., JYFL Preprint 7/87, Jyväskylä, Finland, to be published
10. J. Äystö et al., JYFL Preprint 8/87, Jyväskylä, Finland, to be published
11. R. Casten et al., Phys. Rev. Lett. 58 (1987) 658

LOW-SPIN STATES IN ^{122}Cs

B. Weiss[+], A. Gizon
Institut des Sciences Nucléaires, 38026 Grenoble, France

C.F. Liang, P. Paris and the Isocele Collaboration
Centre de Spectrométrie Nucléaire et de Spectrométrie de Masse,
F-91406, Campus Orsay, France

ABSTRACT

The decay of ^{122}Ba to the odd-odd ^{122}Cs levels has been studied with on-line mass-separated samples at the Isocele facility. A complex level scheme including 42 new low-spin excited states has been established using gamma and conversion electron spectroscopy techniques. From a series of new measurements, the $T_{1/2} = 0.36$s ^{122}Cs isomeric state is proposed at 127 keV with 5⁻ spin and parity assignments.

INTRODUCTION

Prior to this study, no level scheme was available for ^{122}Cs. The ground-state was known to be a 1^+, $T_{1/2} = 21$s state with a mixing of two $\{\pi,\nu\}$ configurations[1,2] and the excitation energies of the two isomeric states ($T_{1/2} = 4.2$ min, $I = 8$ and $T_{1/2} = 0.36$s[3]) are still unknown. From in-beam measurements, a $\Delta I = 1$ collective structure has been identified on a $I = 6^-$ or 7^- band-head having probably a $\{\nu g_{7/2}^{-1}, \pi h_{11/2}\}$ configuration at unknown excitation energy[4].

The present study on ^{122}Cs has been performed mainly in the ^{122}Ba → ^{122}Cs decay. To elucidate the lower part of the deduced level scheme, a new series of precise measurements has been undertaken on the 0.36s isomeric state and its decay mode.

EXPERIMENTS AND RESULTS

Thick metallic targets of cerium or lanthanum were bombarded with a 280 MeV ^{3}He beam from the Orsay synchrocyclotron. Activities were extracted and mass-separated at the on line Isocele-2 facility. Cesium samples were classically collected while the barium ones were extracted as BaF$^+$ ions. Gamma-ray measurements in the singles and γ-γ-t modes were performed using coaxial detectors and/or intrinsic Ge planar with a thin Be window. Conversion electron spectra were recorded by means of a Si(Li) detector associated with a magnetic filter. Three-parameter e-γ-t and e-x-t coincidences were performed.

<u>Decay of the 0.36s isomeric state</u>. An example of conversion electron spectrum obtained with 3s collection period is displayed in fig. 1. Setting the magnetic field of the filter at 225 gauss, electrons have been recorded in the 20 - 300 keV energy range. Experimental and theoretical conversion coefficient values and multipolarities are given in Table I, the α_i being normalized to the 331 keV E2 ($2^+ \to 0^+$)

[+]Permanent address : Laboratoire de Radiochimie, Faculté des Sciences F-06034 Nice, France.

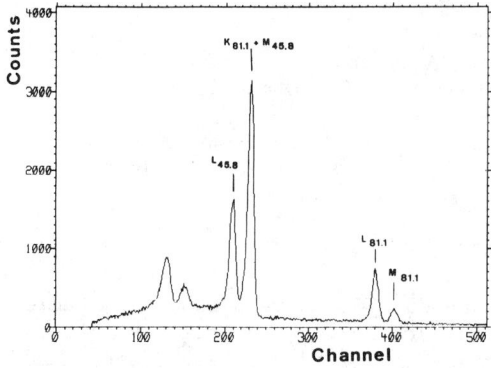

Fig. 1. Part of the low-energy conversion electron spectrum of the 0.36 s ^{122}Cs isomeric state.

transition in ^{122}Xe. The 45.8 keV line was in prompt coincidence neither with the 81.2 keV γ-ray nor with a 35.4(81.2-45.8) keV line. Indeed a careful analysis of time spectra gated with Cs X-rays or low-energy lines in 3-parameter coincidence data shows that the 45.8 keV is associated with a long half-life ($T_{1/2}$ > 1 μs). Then, the 0.36 s ^{122}Cs isomeric state could decay simply via the 81.2 (M2) - 45.8 (E2) cascade to the $T_{1/2}$ = 21 s, 1$^+$ ground-state. The long-lived state at 45.8 keV could be a 3$^+$ level whereas spin and parity assignments 5$^-$ could be proposed for the 0.3 s isomer located at 127.0 keV. This simple cascade decay mode is also supported by the intensity balance which is correct within 25 % incertitude.

^{122}Cs level scheme - Based upon γ-γ coincidence results, multipolarities of low-energy transitions, energy sum relationships and transition intensity balance, a ^{122}Cs level scheme including 42 excited states has been proposed. Its low-energy part is presented in fig. 2. The transitions shown in dotted lines have been tentatively placed, based on energy differences. The 1$^+$ assignments have been deduced from log ft values.

DISCUSSION

From the present work, a new low-spin low-energy level structure has been proposed for the $^{122}_{55}$Cs$_{67}$ nucleus. No similar characteristics are observed in the $^{124}_{55}$Cs$_{69}$ and $^{120}_{53}$I$_{67}$ neighbours when looking at the strongly fed 1$^+$ states. As the 1$^+$ ^{122}Cs ground state is complex in itself, it is difficult to propose {π ν} configurations for the 3$^+$ and 5$^-$ excited states. Due to their low excitation energy in odd-A Ba the ν 5/2$^+$ [413] and ν 7/2$^-$ [523] configurations could be considered to account for these 3$^+$ and 5$^-$ excited states, respectively. More investigation in this nucleus is necessary to establish its properties.

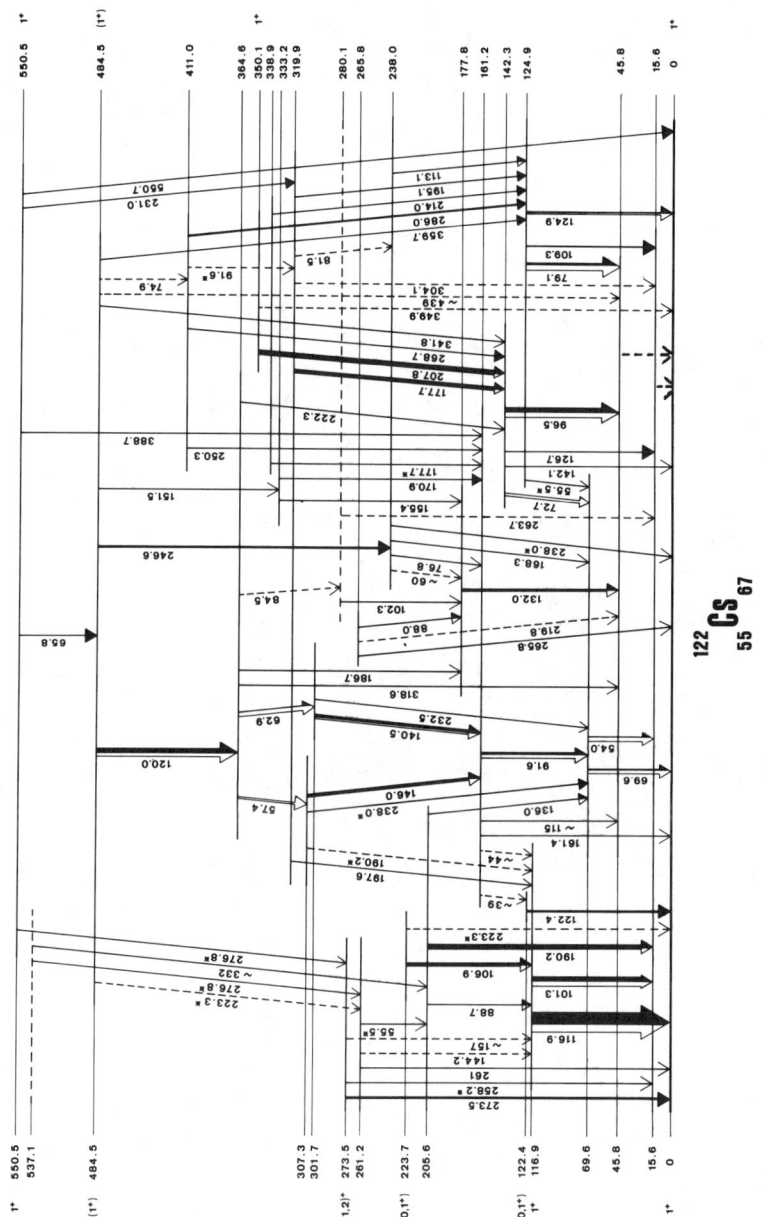

Fig. 2. Partial low-energy level scheme of ^{122}Cs

Table I. Characteristics of transitions observed in the 0.3s ^{122m}Cs

E_γ (keV)	I_γ	Line	I_e-	Exp. α_i or K/L	Theory E1	E2	E3	M1	M2	M3	Mult.
81.20 (10)	2.26 (11)	K	500 (10)	α_k = 11.6 (8)	0.335	2.26	12.1	1.41	15.6	114	
		L	154 (4)	K/L = 3.25(15)	7.32	1.70	0.29	7.5	4.4	1.7	M2
		M	37 (2)	α_M = 0.86(9)	0.0092	0.286	9.47	0.038	0.76	15.5	
45.85 (15)	0.80 (10)	L	253 (8)	α_L = 16.6$\{^{19.6}_{14.3}$	0.235	19.5	1,100	1.0	39	1.150	E2
		M	60 (5)								

REFERENCES

1. C. Ekström et al., Nucl. Phys. A292, 144 (1977) and CERN Rep. 81-09, 12 (1981)
2. C. Thibault et al., Nucl. Phys. A367, 1, (1981).
3. B. Weiss et al., Z. Phys. A313, 173 (1983).
4. M.A. Quader et al., Phys. Rev. C33, 1109 (1986)

DECAY OF NEW MASS-SEPARATED NEUTRON-DEFICIENT La AND Ce ISOTOPES

J. Genevey, A. Gizon, N. Idrissi, B. Weiss
Institut des Sciences Nucléaires, 38026 Grenoble, France

R. Béraud, A. Charvet, R. Duffait, A. Emsallem,
M. Meyer[+], T. Ollivier, N. Redon
Institut de Physique Nucléaire, 69622 Villeurbanne, France

ABSTRACT

By use of a He jet system coupled to a Bernas-Nier ion-source, several new mass-separated A = 122 - 127 isotopes reached in heavy ion fusion reactions at SARA have been identified and studied. From experimental decay properties of La isotopes, systematics of low-lying energy levels have been extended for even-even and odd-A barium. New informations on Ce decay schemes are briefly reported.

INTRODUCTION

Systematic experimental studies of various nuclear properties over a transitional region are of fundamental importance to our understanding of the nuclei. While new results on high-spin states properties and theoretical approaches are developed to describe deformations, shape coexistence, softness or non-axial properties in nuclei, low-energy low-spin states in weakly deformed A = 120 - 130 Ba, La and Ce isotopes appear experimentally poorly known.

The availability of various heavy ion beams at SARA has allowed the production of neutron-deficient nuclei of this transitional region. The present work concerns spectroscopy of low-lying levels of some of these isotopes extracted with a He-jet system coupled to a medium-current separator ion source.

EXPERIMENTAL

In the present work, our first goal was to produce and identify the neutron-deficient isotopes in this mass region and to verify or establish their main characteristics of decay. Most of the results given below have been obtained on light isotopes of Ce, La and Ba (A = 122-127) by use of a He-jet system working alone or coupled to a Bernas-Nier medium current ion-source of the SARA on-line isotope mass separator [1]. With several proton-rich combinations of the target and the projectile nuclides, heavy-ion fusion reactions at 5-6 MeV/nucleon incident energy provide us with means for studying very neutron-deficient nuclei in the light rare earth region. Due to the large angular momentum transfers involved, the production of long-lived high-spin isomeric states of residual products is favored in these reactions.

The experiments have been carried out with thin ($\sim$ 2mg/cm^2) ^{92}Mo ^{94}Mo or ^{96}Mo targets bombarded with ^{32}S, ^{35}Cl, ^{36}Ar beams. Typical

[+]Present address : ISN 38026 Grenoble, France

intensities (1 to 4.10^{11} part./s) were delivered at SARA with the ECR source and the first variable energy cyclotron working alone.

The details concerning the coupling of the He-jet to the ion-source mass separator were recently reported at the 11th EMIS Conference[1]. The performances were given for the ^{92}Mo + ^{35}Cl reaction at 191 MeV beam energy because, as expected, appreciable productions of A = 124 La, Ba and Cs were observable. With diffusion pump oil as aerosol at the He-jet system, suitable chlorination and catcher temperatures (1250°C for La) at the ion-source, typical coupling efficiencies reach 1-2 % for Ba and La isotopes. The low background detection station and its tape driver system have been designed to receive activity coming either directly from the He-jet system via a 1 mm diameter 6 m long capillary or from the mass separator. In such a way, γ-ray multianalysis, γ-X (Z-identification) and γ-γ coincidences are performed with the He-jet working alone whereas the mass identification is made with the coupled set up.

RESULTS AND DECAY SCHEMES DEDUCED

Results observed from our experiments, concerning mainly La and Ce decays with half-lives in the 1s-2min range are discussed by mass number.

A = 122 : In measurements performed with the ^{92}Mo + ^{36}Ar system, several γ-rays at 196.1, 372.7, 742.8, 939.5 and 971.7 keV have been easily observed with $T_{1/2}$ = (8.5 ± 0.6)s and in coincidence with Ba X-rays. These results which give evidence of ^{122}La production are in agreement with those reported previously[2] from β-delayed proton emission. From γ-γ coincidences, only the 2^+ and 4^+ levels of the ground band [2,3] have been established; the next 513 keV ($6^+ \rightarrow 4^+$) transition has not been observed in this experiment but two new levels at 939.5 and 1167.8 keV respectively have been placed in the ^{122}Ba level scheme. From their decay and by analogy with heavier even-even Ba, they correspond very likely to the two first members (2^+_2, 3^+) of the gamma band; from a weak 703-373 coincidence the next member (4^+_2) could be located at 1272 keV. The absence of other ^{122}Ba excited states is certainly associated to a weak ^{122}La production but the preferential feedings to low-spin states could also indicate that the 8s ^{122}La state has a spin value of 5 or 4.

The ^{122}Ba → ^{122}Cs decay, studied in details at Isocele (Orsay) is reported at this Conference. From present measurements, no new information had been extracted for the ^{122}Cs isomeric states. The 93, 104 or 132 keV lines proposed below the $\Delta I = 1$ collective band and its probable $(6,7^-)$ $(\nu g^-_{7/2}, \pi h_{11/2})$ base state[4] have not been observed so, their location above the 1^+ ^{122}Cs g.s. remains unknown.

A = 123 : In the ^{92}Mo + ^{36}Ar reaction, only weak γ-rays at 66, 113 and 178 keV could be assigned to the ^{123}Ce → ^{123}La decay. Up to now, this weak ^{123}Ce production has not been observed on separated samples. In the He-jet γ-γ coincidence measurements a part of the 66 keV complex line was in coincidence with the La K X-rays and with both the 113 and 178 keV lines. The previously reported $T_{1/2}$ = 3.8s.[2] has been checked on the 113 keV transition alone (the 66 and 178 lines being complex in He-jet spectra). From the 1.2s ^{124}Pr β-delayed proton activity, lines at 70, 113 and 166 keV were also proposed in ^{123}La [2]. Up to now, even if there is an agreement on the 113 keV transition, there is no chance to connect the base states of the three collective bands recently identified in ^{123}La [3] and the 5/2+ g.s. assignment proposed for ^{123}Ce cannot be confirmed.

In ^{92}Mo + ^{35}Cl and ^{94}Mo + ^{36}Ar, a (16 ± 1)s activity has been assigned to the ^{123}La → ^{123}Ba decay. The 5/2+ g.s. spin [15], the strong lines at 92.7 (E1), 110.0, 133.3 and 169 keV and their in beam angular distributions [5] have been used to establish the relative position of the lower lying levels in ^{123}Ba (fig. 2). Several other lines have been identified in the same decay; low-energy conversion electrons are needed to place the first members of the ($s_{1/2}$, $d_{3/2}$) positive rotational sequence.

A = 124 : The ^{124}Ce production has been followed with a special attention. It was missing in the A = 124 mass-separated samples ; nevertheless, in the ^{92}Mo + ^{36}Ar reaction where the La x-rays were relativily strong, a coincident 134 keV line (T1/2 ∿ 8 s) [6] could be assigned to the ^{124}Ce decay, in agreement with previous results. Moreover, this line is one of the three stronger ones assigned to ^{124}La at the Daresbury recoil separator [7]. Other lines have been observed in coincidence with La x-rays but a ^{124}La isomeric state could be established only via further very low-energy conversion electron measurements.

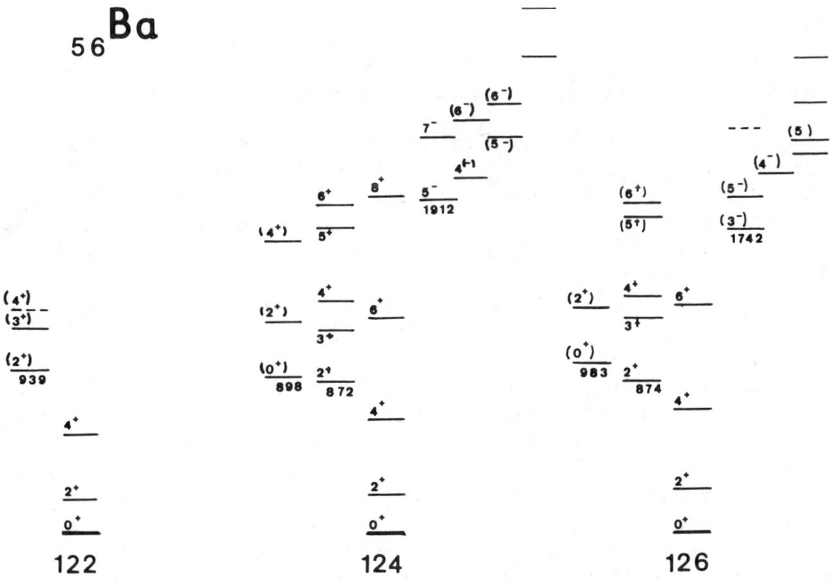

Fig. 1. Excited levels in even-even barium observed from La → Ba decays.

The ^{124}La → ^{124}Ba (T1/2 = 29s) decay has been extensively studied in the present work. The ^{124}La mass separated samples have been extracted with the ^{92}Mo + ^{35}Cl reaction at 191 MeV. In the ^{124}Ba level scheme deduced [8], several spin assignments previously established by log ft values and by systematics have been confirmed by in-beam angular distribution measurements [9]. The ^{124}Ba levels observed are presented in fig.1. The main ^{124}La precursor produced by H.I. reaction and which feeds the barium levels has very likely a spin 7 or 8 but observation of low-spin direct feedings requires the existence of a low-spin ^{124}La in our experiments. A precise search for two different half-lives has been undertaken but the low-spin component production was too weak to solve the problem.

Table I Gamma lines assigned to Ce decays

Isotope	γ-rays (keV) coincident with La KX rays	Coincidence cascades
^{127}Ce	(58.6); 114.8; (120); 177; 253; 398	58 - 177 58 - 115 - 253
^{126}Ce	(61.4); 82.0; (116.4); (120); 136 ; 188	61 - 136 - 188
^{125}Ce	(56); (491)	56 - 491
^{124}Ce	134	
^{123}Ce	(66); 113; (178)	66 - 113 66 - 178

A = 125 : In the ^{96}Mo + ^{35}Cl reaction, a cascade of two lines (56-491 keV) has been related to a ∿ 10s ^{125}Ce activity. In addition, the 56 keV line is well observed in the A = 125 separated samples. Unfortunately, these new lines have no connections with previous informations and are unable to precise the relative position of 11/2⁻ and 9/2⁺ states in ^{125}La.

The ^{125}La → ^{125}Ba (T$_{1/2}$ = 76s) decay has been studied with both ^{96}Mo and ^{94}Mo targets. From in-beam measurements [11] $h_{11/2}$ and $g_{7/2}$ neutron-hole collective bands were identified in ^{125}Ba. The decay scheme established in the present work is very similar to the ^{127}La → ^{127}Ba one [12]. A new rotational sequence built on the 1/2⁺ g.s. has been observed with a 3/2⁺ state at 43.7 keV. From low-energy conversion electron spectra measured at Isocele (Orsay), an E1 multipolarity has been established for the strong 67 keV transition which deexcites the 7/2⁻ state so, a 7/2⁻ → 5/2⁺ placement is proposed. The same 5/2⁺ (d 5/2) is fed by the g7/2⁺ band via the 169 keV M1 transition [11]. A careful analysis of both low-energy electron and gamma spectra has been made to extract the 7/2⁻ → 3/2⁺ and/or 5/2⁺ (d5/2) → 1/2⁺ g.s. expected transitions. Up to now, no clear evidence can be retained ; the d5/2 energy is probably very close of 20 keV as proposed in the systematics (fig. 2). From feedings and log ft values deduced to the 11/2 and 13/2 ^{125}Ba states, the ^{125}La g.s. is very likely 11/2⁻.

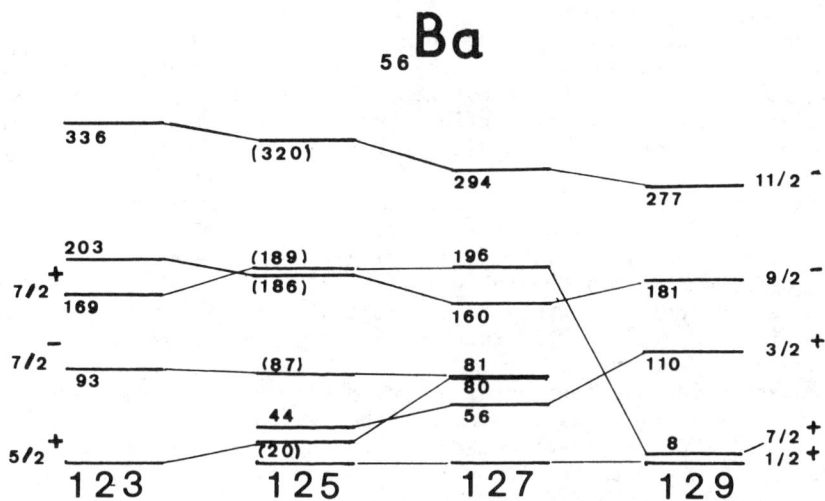

Fig. 2. Systematics of low-lying states in odd-A barium isotopes

A = 126 : The γ-rays assigned to a ^{126}Ce production, as in the ^{94}Mo + ^{35}Cl system, are reported in table I. Except for the weak 61 keV line, all were present in the A = 126 mass-separated spectra and the stronger ones agree with the known $T_{1/2} = (50 \pm 4)$s [6]. The results are partly in agreement with in-beam experiments [4] and the 137 and 116 keV transitions are very likely near the ^{126}La g.s.. Nevertheless, conversion electrons and transition multipolarities are needed to built the level scheme and to identify the ^{126}La isomeric states.

The ^{126}La → ^{126}Ba decay has been well studied. Many of the observed excited states (fig. 1) agree with in-beam measurements [13]. Two new levels at 983 and 1296 keV have been established and their decay modes suggest (0⁺) and (2⁺) labels, respectively. Obviously, conversion electrons are needed to identify E0 transitions. Direct feedings observed to the various excited states in ^{126}Ba requires two parallel decay processes, one from a low-spin ^{126}La (I = 1 or 2) and another one from a medium spin ^{126}La (I = 5 or 6). As the low-spin part represents approximately 10 % of the total activity only one half life (64 ± 3)s has been detected in our measurements.

A = 127 : A relatively strong (32s)^{127}Ce activity has been produced with the ^{96}Mo + ^{35}Cl system. In addition to the 58 keV transition already identified [6], five new ones have been detected (Table I). Direct identification of ground states is missing in both ^{127}Ce and ^{127}La. From in-beam measurements [14] ^{127}La g.s. can be considered as 11/2⁻ but a 3/2⁺ state is expected in its vicinity. So, lines observed in the present work very likely follow β-decay modes between Ce and La positive states. Further precise experiments are needed to clarify the ^{127}La low-energy scheme.

DISCUSSION

These results show that the He-jet system coupled to the on-line SARA mass separator provide us with a good tool for nuclear structure studies in the light rare-earth region. The measurements concerning Ce decays require larger productions but those obtained with La samples are already remarkable and able to give new informations on Ba nuclei as indicated below.

The results on ^{122,124,126}La extend the even-even barium systematics (fig. 1) of several states poorly or not fed by in-beam experiments. From comparisons with neighbouring xenons one observes significant differences : i) the energy of the 2_2^+ level appears more stable in Ba isotopes ; ii) this 2_2^+ level remains always above the 4^+g.s. band ; iii) new second (0^+) states found in ^{124,126}Ba are located at relatively low energy.

An extension has been obtained for the low-energy level systematics in odd-A Ba (fig. 2) combining both previous ground-state spin assignments [15] and the present measurements. A small γ-deformation is needed to reproduce these energy-level systematics in $^{123-127}$Ba, in agreement with spins and magnetic moments.

REFERENCES

1. A. Plantier et al., Nucl. Instr. Meth. B26, 314 (1987)
2. J.M. Nitschke et al., Z. Phys. A312, 265 (1983) ; A316, 249 (1984) and A.325, 485 (1986)
3. R.A. Wyss et al., Contribution Int. Conf. on Nuclear Shapes, Crete Greece, June 1987.
4. M.A. Quader et al., Phys. Rev. C33, 1109 (1986)
5. N. Yoshikawa et al., J. Phys. 40, 209 (1979)
6. D.D. Bogdanov et al., Nucl. Phys. A307, 421 (1978).
7. A.N. James et al., Daresbury, Ann. Rep. 1985/1986, p. 103.
8. A. Gizon et al., Proc. 4th Colloque Franco-Japonais, Seillac, France, 1986, P. 189.
9. J. Ph. Martin et al. Z. Phys. A326, 337,(1987).
10. J. Tamura et al., NDS 32, 497 (1981).
11. J. Gizon et al., Z. Phys A285, 259 (1978)
12. C.F. Liang et al., Proc. 4th Int. Conf. on Nuclei far from Stability, Helsingør 1981, p. 487.
13. K. Schiffer et al. Z. Phys. A327, 251 (1987)
14. P.J. Smith et al. J. Phys. G : Nucl. Phys. 11, 1271 (1985)
15. A.C. Mueller et al. Nucl. Phys. A403, 234 (1983)

THE DARESBURY RECOIL SEPARATOR AND γ-RAY SPECTROSCOPY NEAR ^{130}Nd
A ΔJ = 1 SUPERDEFORMED BAND

A N James* and D C B Watson
Oliver Lodge Laboratory, University of Liverpool, U.K.

ABSTRACT

The Recoil Separator at the Daresbury NSF heavy ion tandem is designed to produce mass separated beams of the products of compound nucleus reactions. Evaporation product nuclei far from stability can be isolated in the focal plane by mass and there, in low background, detectors may be used to study their decay modes. An alternative use of the instrument is to take signals from a split anode ion chamber in the focal plane which provides a prompt A and Z signature for events detected in-beam where the compound nucleus was formed. Results from an experiment using the POLYTESSA array of 20 escape suppressed germanium detectors in this way to study residues from the compound nucleus ^{132}Sm are given. This experiment forms part of a survey studying particle configurations and nuclear shapes for the most neutron deficient nuclei in the $h_{11/2}$ region.

THE RECOIL SEPARATOR

The main components of the Daresbury Recoil Separator[1] are indicated in figure 1. The vertical magnetic field and horizontal electric field in the crossed field devices (Table I) produce a velocity dispersed beam of residues with the centre of mass velocity undeflected. The accelerator beam is deflected away from

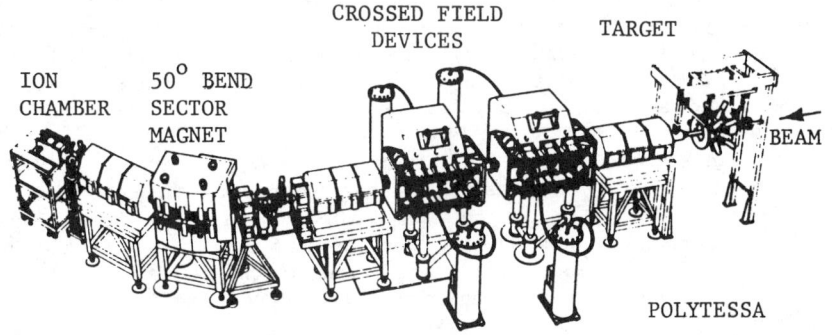

Figure 1   The Daresbury Recoil Separator arranged for tagging γ-rays observed in-beam with atomic mass A and atomic number Z.

*The willing collaboration of many colleagues on this project is acknowledged.

Table I   Properties of the Crossed Field Devices
             Wien filters 100mm diameter

Magnetic		Electrostatic	
Maximum field	0.5T	Operating field	$<3\text{MVm}^{-1}$
Pole tip length	1000mm	Electrode length	1280mm
Magnet gap	350mm	Separation	100mm
Effective length	1240mm	Depth	200mm
Field Uniformity	1/1000	Field Uniformity	1/500

Table II   Properties of the Daresbury Recoil Separator

Dimensions		Performance	Horiz	Vert
Overall length	13000mm	Target beam size	±0.5mm	±3.5mm
Target to Quadrupole	350mm	Acceptance angle	±45mr	±45mr
Quadrupole aperture	150mm	Final image size	±1.5mm	±10mm
Sextupole aperture	150mm			
Wien filter aperture	100mm	Velocity acceptance	±2%	
Sector magnet angle	50°	A/Q acceptance	±1.2%	
Radius of curvature	1500mm	A/Q resolution FWHM	0.35%	

the entrance to the sector magnet. The sector magnet has its dispersion matched to the dispersion of the crossed field devices in such a way as to focus ions of the same A/Q into one image. There is dispersion in A/Q at the final image. Quadrupole and sextupole lenses ensure high efficiency and good image quality (Table II).

Secondary electrons from a carbon foil 20µg/cm² are detected with a position sensitive microchannel plate multiplier to give A/Q readout[1,2]. An isobutane filled ion chamber[3] with two anodes gives energy loss information in which Z = 59(Pr) is separated from Z = 60(Nd) by about half the FWHM resolution in Z. This is sufficient for subsequent analysis to yield γ-ray spectra from each detected nuclear species without contamination by any other species[2].

Compound nucleus residues from the bombardment of a ^{58}Ni target by a 300MeV ^{74}Se beam have been detected in coincidence with γ-rays detected in the POLYTESSA array of 20 escape suppressed germanium detectors. The overall detection efficiency through the Recoil Separator was measured to be 7% for ^{130}Nd nuclei. Four rotational bands have been identified in ^{129}Pr and three in ^{129}Nd.

A ΔJ = 1 SUPERDEFORMED BAND

The γ-ray spectrum for ^{129}Pr shown in figure 2a is dominated

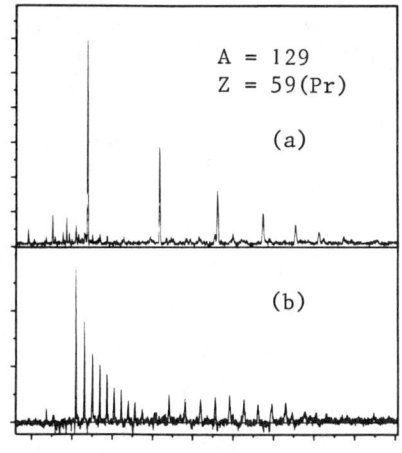
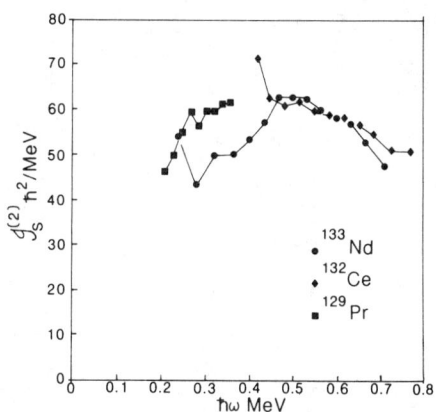

Figure 2. Gamma-ray spectra in ^{129}Pr, a) Recoil gated, b) γ gated on the 185keV transition.

Figure 3. Moments of inertia $\mathcal{J}^{(2)}$ scaled using $A^{5/3}$ to mass 132. The data for the ^{132}Ce and ^{133}Nd superdeformed bands are taken from reference 5.

by the rotationally aligned decoupled $h_{11/2}$ band common in this mass region[4]. A γ-γ coincidence spectrum gating on a weak line (in figure 2a) at 185keV is shown in figure 2b and corresponds to a rotational band of 12 levels in which all ΔJ = 2 and ΔJ = 1 transitions are observed, signature splitting is less than 1keV. The experiment gives no evidence either for the feeding mechanism or for the way the band is depopulated. The energy increase between the successive transitions is nearly constant and the data have been used to calculate the moment of inertia $\mathcal{J}^{(2)}$ = dI/dω with the result shown in figure 3. Comparison with ΔJ = 2 superdeformed bands in this mass region[5] clearly shows that this band in ^{129}Pr is also "superdeformed" and will have ε_2 ~ 0.4. (Because $\mathcal{J}^{(2)}$ ~ constant at ℏω ~ 0.3MeV it is not unreasonable to assume $\mathcal{J}^{(1)} = \mathcal{J}^{(2)}$ in which case the spins of the levels run from ~ 10 to ~ 20.)

Using mass gated recoil-γ coincidences the ratio of intensities of γ-rays at 101° (79°) to 143° (37°) have been measured for γ-ray peaks not contaminated by other mass 129 γ-rays. The results shown in figure 4 are consistent with the $h_{11/2}$ band being stretched E2 transitions and the ΔJ = 1 transitions in the 185keV band being mixed, with the ratio of E2/M1 strengths increasing up the band. By contrast a band of transitions in ^{129}Pr ending in a 104keV transition is consistent with pure M1 transitions, no cross over transitions are detected in this

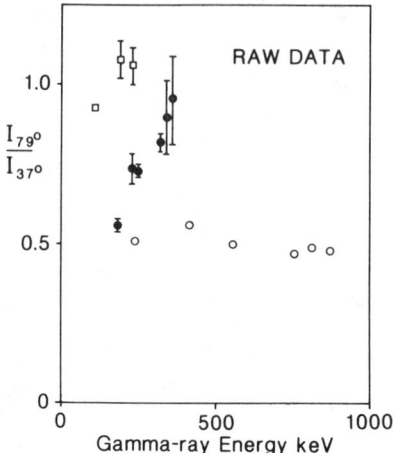

Figure 4. Ratio of intensities at two angles for selected clean γ-rays. Open circles: stretched E2 transitions in the $h_{11/2}$ band. Open squares: stretched M1 transitions in the 104keV band. Filled circles: ΔJ = 1 transitions in the 185keV band.

band. The 104keV band also has zero signature splitting but normal deformation, it is thought to be due to the odd proton being in the $g_{9/2}[404]9/2$ Nilsson orbital. This orbital is the only one near the Fermi surface at $\varepsilon_2 \sim 0.3$ which gives zero signature splitting. With its large magnetic moment perpendicular to the rotational axis this orbital should produce large M1 strengths.

The superdeformed shape of rapidly rotating Ce and Nd isotopes is thought to be due to action of $\nu i_{13/2}$ orbits. A possible configuration to account for the 185keV band in ^{129}Pr would combine the 4th oscillator shell $g_{9/2}[404]9/2$ orbit with a core driven to high deformation by the action of the 6th oscillator shell $\nu i_{13/2}$ quasiparticles[6].

REFERENCES

1. A N James et al, Submitted to Nuclear Instruments and Methods.
2. K L Ying et al, J. Phys. G: Nucl. Phys. 12 (1986) L211-L215.
3. A N James et al, Nucl. Instr. and Meth. 212 (1983) 545-553.
4. J R Leigh et al, Nucl. Phys. A213 (1973) 1-21.
5. R Wadsworth et al, J. Phys. G: 13 (1987) L207-L212.
6. T Bengtsson, preprint, NORDITA-87/14 N.

Structure of the Odd-Odd and Odd-Mass Sb Nuclides

C.A. Stone, S.H. Faller, J.D. Robertson and W.B. Walters
University of Maryland, College Park, MD 20742 U.S.A.

Abstract

We have performed experiments on the β^- decay of several Sn nuclides near ^{132}Sn. The multiplet structure of ^{132,130}Sb was studied and an extensive level scheme was constructed for the decay of the 7⁻ isomer in ^{130}Sn. These investigations were extended to ^{131,129,127}Sb where the properties of the more complex multiplets were determined. Extensive high-spin level schemes were constructed for the decay of ^{131,129}Sn and a 19/2⁻, β^--decaying isomer was placed in the ^{129}Sb level scheme. A partial level is presented for the decay of the 19/2⁻ isomer into levels of ^{129}Te.

Introduction

The properties of nuclides near shell closures have been extensively studied in order to characterize the effective nucleon-nucleon interaction. Much of our information on the effective interaction has come from mass regions where a double shell closure lies near the line of stability, such as at ^{16}O, ^{40,48}Ca or ^{208}Pb. Nuclides in these mass regions are easily populated by a variety of different reactions and they provide a wide range of information. Recent improvements in mass separation yields have made it possible to extend these studies to the ^{132}Sn mass region which lies away from the line of stability.

The nuclide ^{132}Sn has 50 protons and 82 neutrons and is the strongest doubly magic nuclide beyond ^{16}O. An important feature of the nuclides with Z>50 and N<82 is that they form the last mass region in which the neutron and proton single particle orbitals are the same. Figure 1 shows the single proton levels of ^{133}Sb and the single neutron-hole levels of ^{131}Sn. Orbitals which are observed in this region are $1g_{7/2}$, $2d_{5/2}$, $2d_{3/2}$, $3s_{1/2}$ and $1h_{11/2}$, commonly referred to as the gddsh model space. These orbitals cover a wide range of angular momenta, giving a wide range of interactions which can be sampled in a particular nuclide.

Nuclear structure within the ^{132}Sn region is marked by the presence of the

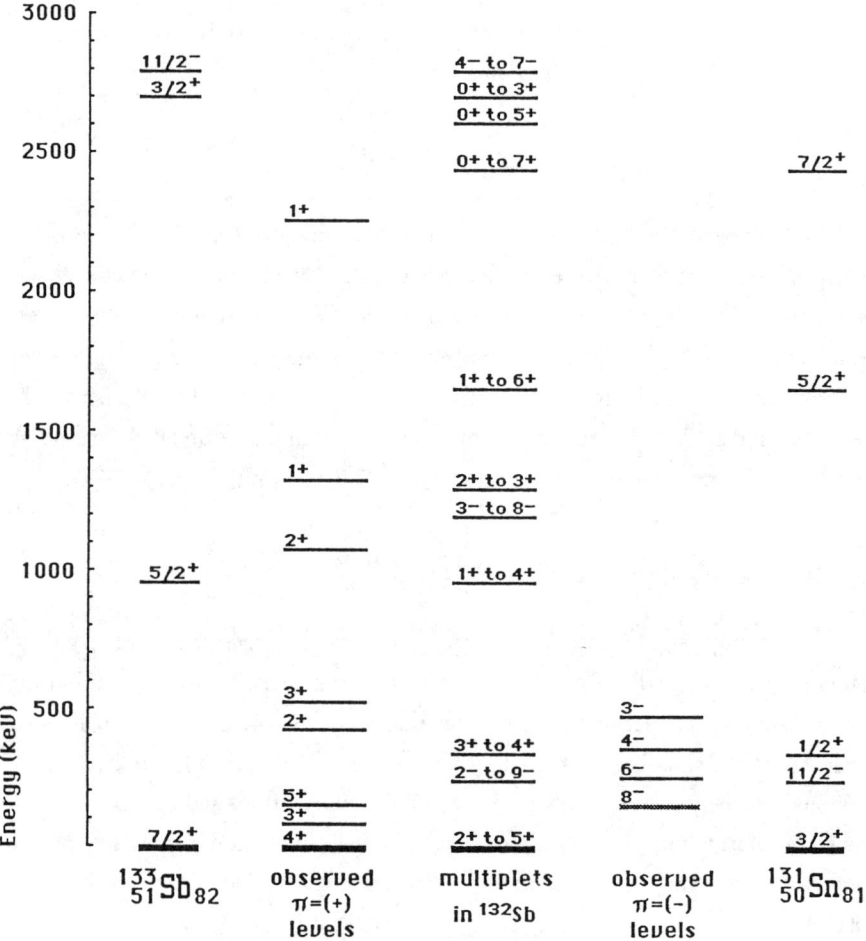

Figure 1. Levels observed in the decay of ^{132}Sn. The center set of levels represents degenerate multiplets in ^{132}Sb. Energies for these multiplets were taken as a sum of the single proton and single neutron-hole level energies in ^{133}Sb and ^{131}Sn, respectively.

unnatural-parity $1h_{11/2}$ orbitals. Work by Fogelberg et al.[1] on ^{131}Sn showed that the 11/2⁻ neutron hole level lies very close in energy to the 3/2⁺ and 1/2⁺ neutron hole levels. A result of this is that most of the odd-N Sn, Te and Xe nuclides and several of the odd-odd Sb and I nuclides have two long-lived β⁻-decaying isomers. Besides producing low-lying high-spin levels, the $h_{11/2}$ orbitals give rise to a set of high-spin levels at higher excitation energies. High-spin isomers, with half lives ranging from nsec to several min are known in this region above excitation energies of 1.5 MeV. Studying the properties of the high-spin levels can reveal some interesting features of nuclear structure that can not be observed in other mass regions.

We have performed experiments on nuclides in the ^{132}Sn mass region in order to characterize properties of the effective interaction. One goal of these experiments was to identify the full $\pi g_{7/2} \otimes \nu d_{3/2}$ multiplet in odd-odd ^{132}Sb and ^{130}Sb. A second goal of the studies on ^{130}Sb was to investigate the high-spin levels populated in the decay of 7⁻ ^{130}Sn. These investigations were extended to the more complex multiplet structures in odd-mass ^{131,129,127}Sb. We focused on identifying levels belonging to the $\pi g_{7/2} \otimes 2^+$ and $\pi d_{5/2} \otimes 2^+$ multiplets and those belonging to the high-spin negative-parity multiplets. One of the high-spin levels identified in ^{129}Sb is a 19/2⁻ β⁻-decaying isomer. We have studied the decay of this isomer into high-spin levels of ^{129}Te.

The Structure of Odd-Odd ^{132}Sb

Figure 1 shows the ^{132}Sb levels observed in the decay of ^{132}Sn. Single proton and single neutron-hole levels are represented by levels in ^{133}Sb and ^{131}Sn, respectively, and the proton-neutron multiplets are represented by the center set of levels. The energy of a degenerate multiplet is taken as a sum of the single proton and the single neutron-hole level energies.

There are four levels in the ground state, $\pi g_{7/2} \otimes \nu d_{3/2}$ multiplet, ranging in spin from 2⁺ to 5⁺. Since it is a particle-hole multiplet, the levels are arranged with a concave-up ordering, the 3⁺ and 4⁺ levels are below the 2⁺ and 5⁺ levels. Somewhat higher in energy is a second 3⁺ level. It is one of two levels in the $\pi g_{7/2} \otimes \nu s_{1/2}$ multiplet. We were unable to identify a level near this 3⁺ level that could be a candidate for the 4⁺ level. The 1⁺ and 2⁺ levels near 1200 keV are part of the $\pi d_{5/2} \otimes \nu d_{3/2}$ multiplet. Much of the β⁻ feeding to ^{132}Sb is through the allowed $\nu d_{3/2} \to \pi d_{5/2}$ Gamow-Teller β⁻ decay that populates this 1⁺ level. Levels in the $\pi d_{5/2} \otimes \nu d_{3/2}$ multiplet probably have a concave-up ordering with the 1⁺, 2⁺ and 4⁺ levels pushed up in energy. The splitting of levels in the $\pi d_{5/2} \otimes \nu d_{3/2}$ multiplet is similar to that in the $\pi g_{7/2} \otimes \nu d_{3/2}$ multiplet. It is much smaller than the apparent splitting of the levels in the $\pi g_{7/2} \otimes \nu d_{5/2}$ multiplet. Only the 1⁺ level (at 2268 keV) has been identified in this multiplet and it has moved up ca. 500 keV from its degenerate position.

Four levels have been identified which belong to the $\pi g_{7/2} \otimes \nu h_{11/2}$ multiplet. Spins and parities of 3⁻, 4⁻ and 6⁻ are proposed for the levels at 483, 348, and 254 keV, respectively. These levels were placed in the ¹³²Sb level scheme but the position of the 8⁻ level is not known. This 8⁻ level is a 4.2-min isomer. No transitions have been identified which either feed the isomer from the 5⁺ or 4⁻ levels or which decay to the 4⁺ ground state. We can say that the 8⁻ level is within about 200 keV of the 4⁺ ground state, else we would observed an M4 transition between the two isomers. Identification of these levels gives an indication of the magnitude of the splitting in the $\pi g_{7/2} \otimes \nu h_{11/2}$ multiplet. Most calculations on the negative-parity levels in ¹³²Sb do not predict such a small splitting for these levels.

The Structure of Odd-Odd ¹³⁰Sb

The nuclide ¹³⁰Sn is the only known even-even nuclide with a second β⁻-decaying isomer. Decay of this 7⁻ isomer populates a different set of levels in ¹³⁰Sb than does the 0⁺ isomer. Figure 2 shows levels observed in the decay of the ¹³⁰Sn isomers. Levels to the left are populated in the decay of the 0⁺ isomer and those to the right are populated in the decay of the 7⁻ isomer. The 5⁺ level was identified in the decay of the 7⁻ isomer but is shown for clarity with the low-spin levels.

The low-spin structure of ¹³⁰Sb is similar to that of ¹³²Sb. The 4⁺ level is the lowest level in the $\pi g_{7/2} \otimes \nu d_{3/2}$ multiplet and the other levels in this multiplet are ordered in a concave-up fashion. This indicates that the multiplet has a particle-hole configuration. For the multiplet to have a particle-hole configuration, the change in occupancy, from ¹³²Sb to ¹³⁰Sb, has to be predominantly in the $h_{11/2}$ orbitals. Otherwise the multiplet would flip giving a 2⁺ or 5⁺ ground state. The splitting of these levels in ¹³⁰Sb is smaller than in ¹³²Sb suggesting that the $d_{3/2}$ occupancy has decreased somewhat. Only the 3⁺ member of the $\pi g_{7/2} \otimes \nu s_{1/2}$ multiplet has been identified. Although our data are quite extensive, we could not identify other levels near 300 keV. The small difference between the energy of the 3⁺ level and the energy of the degenerate multiplet probably indicates that the 3⁺-4⁺ splitting is not large.

Decay of the 7⁻, ¹³⁰Sn isomer has allowed us to identify many high-spin negative

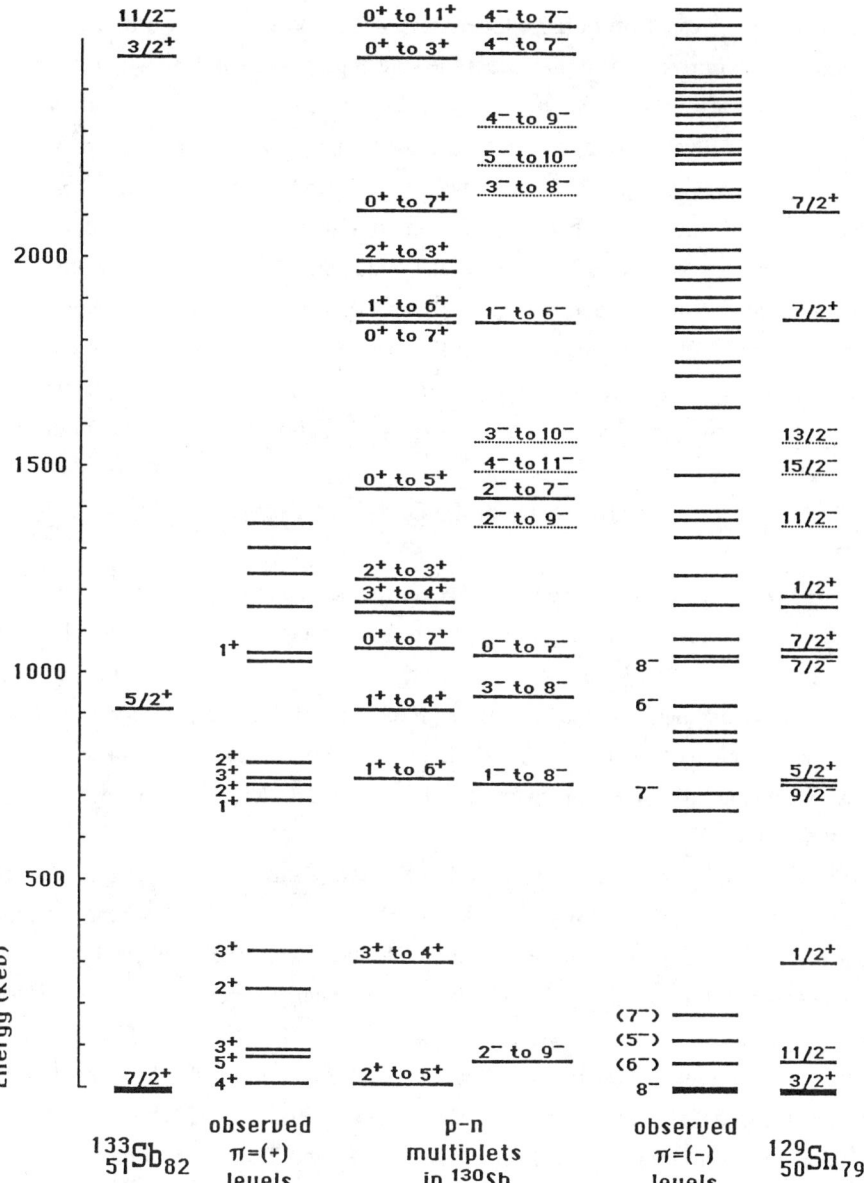

Figure 2. Levels observed in the decay of the 0^+ and 7^- isomers of ^{130}Sn. The negative-parity levels are populated in the decay of the 7^- isomer and the positive-parity levels are populated in the decay of the 0^+ isomer.

parity levels in ^{130}Sb. Most of these levels have a spin in the range of 6-8. Removing two neutrons from ^{132}Sb has not affected the splitting of levels in the $\pi g_{7/2} \otimes \nu h_{11/2}$ multiplet. Four levels belonging to the $\pi g_{7/2} \otimes \nu h_{11/2}$ multiplet have been observed. Using these four levels, along with the four observed $\pi g_{7/2} \otimes \nu h_{11/2}$ levels in ^{132}Sb, it appears that the $\pi g_{7/2} \otimes \nu h_{11/2}$ level splitting is not quenched significantly in ^{130}Sb. This may be expected since, as the $h_{11/2}$ orbitals can hold more neutrons than the $d_{3/2}$ orbitals, the occupancy term is less sensitive to the removal of the neutron pair. Particularly strong β^- decay is found to levels at 733, 938, and 1044 keV, whose proposed spins and parities are 7^-, 6^- and 8^-, respectively. These levels may be identified as members of the $\pi d_{5/2} \otimes \nu h_{11/2}$ multiplet. They are fed strongly by the allowed β^- decay of the $(\nu d_{3/2} \nu h_{11/2})_{7^-}$ isomer in ^{130}Sn through the $\nu d_{3/2} \rightarrow \pi d_{5/2}$ allowed Gamow-Teller decay.

The Structure of the Odd-A Sb Isotopes

The study of multiplets to characterize nuclear structure can be extended to odd-A nuclides. Their odd nucleon can be considered as a particle (hole) coupled the the adjacent even-even nuclide. When the even-even core is in its ground state, the levels that are formed are just the single particle levels. Excitation of the core will produce multiplets of levels due to the coupling of the odd particle to the excited core. Since the first excited state in an even-even nuclide is 2^+, the lowest multiplet is the odd particle coupled to the 2^+ state.

The low-spin systematics of the odd-A Sb isotopes are shown in Fig. 3. Levels for ^{131,129,127}Sb are from our work[3] while those for ^{123,125,133}Sb are from the various Nuclear Data Sheets. The five levels belonging to the $g_{7/2} \otimes 2^+$ multiplet can be seen in ^{131}Sb between 1100 and 1500 keV. These levels are nearly degenerate except for the $7/2^+$ level which lies ca. 300 keV above the other levels. The behavior of the $g_{7/2} \otimes 2^+$ multiplet is very regular moving to the lighter odd-A Sb isotopes. The $3/2^+$ level drops at almost the same rate as the single particle $5/2^+$ level. Both the $9/2^+$ and $11/2^+$ levels remain almost constant in energy from ^{131}Sb to ^{123}Sb and the $7/2^+$ level moves upward. A $1/2^+$ level and a second $9/2^+$ level are seen around 1800 keV in ^{131}Sb. These levels arise from the $d_{5/2} \otimes 2^+$ multiplet, a multiplet built on the $5/2^+$ single particle state. The energy of the upper $9/2^+$ level and that of the $1/2^+$ level drops in the lighter Sb isotopes but repulsion from the $9/2^+$ member of the $g_{7/2} \otimes 2^+$ multiplet seems to slow the drop of the $9/2^+$ level energy.

Figure 4 shows the high-spin levels of ^{131}Sb up to about 2 MeV. These levels

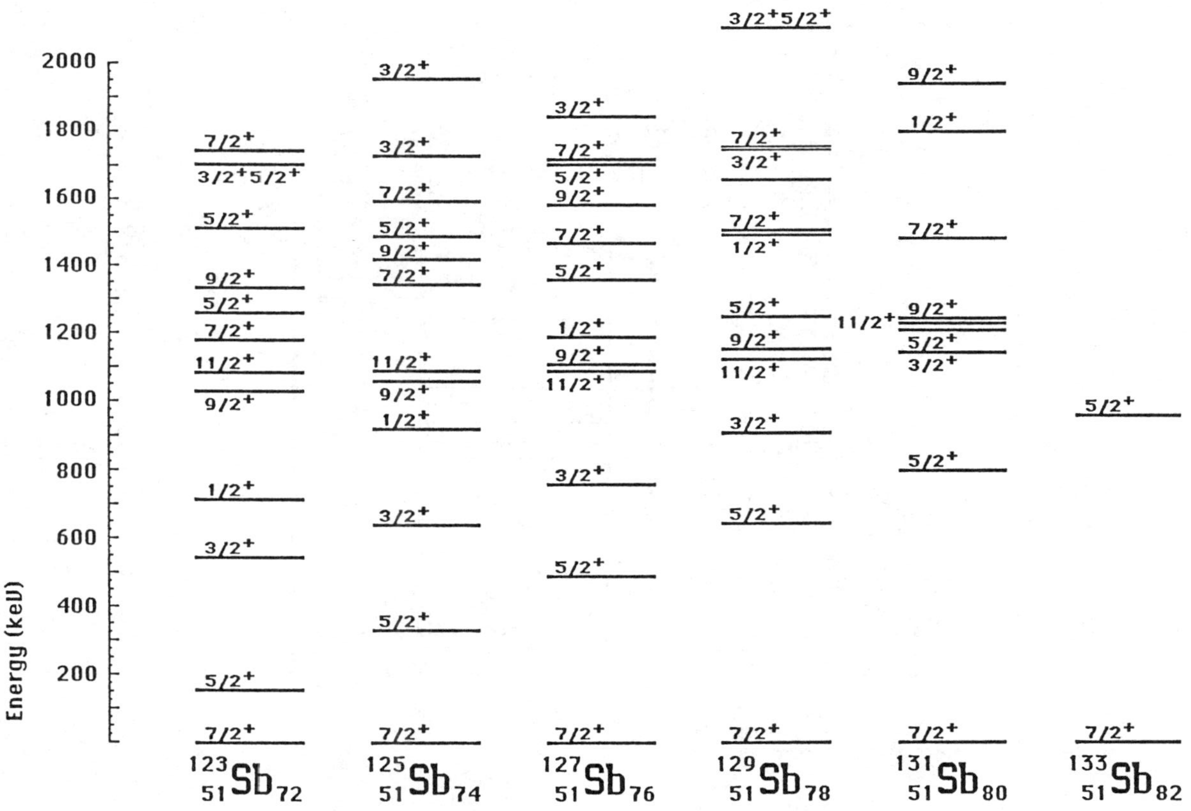

Figure 3. Systematics of the positive-parity levels in the odd-mass Sb isotopes.

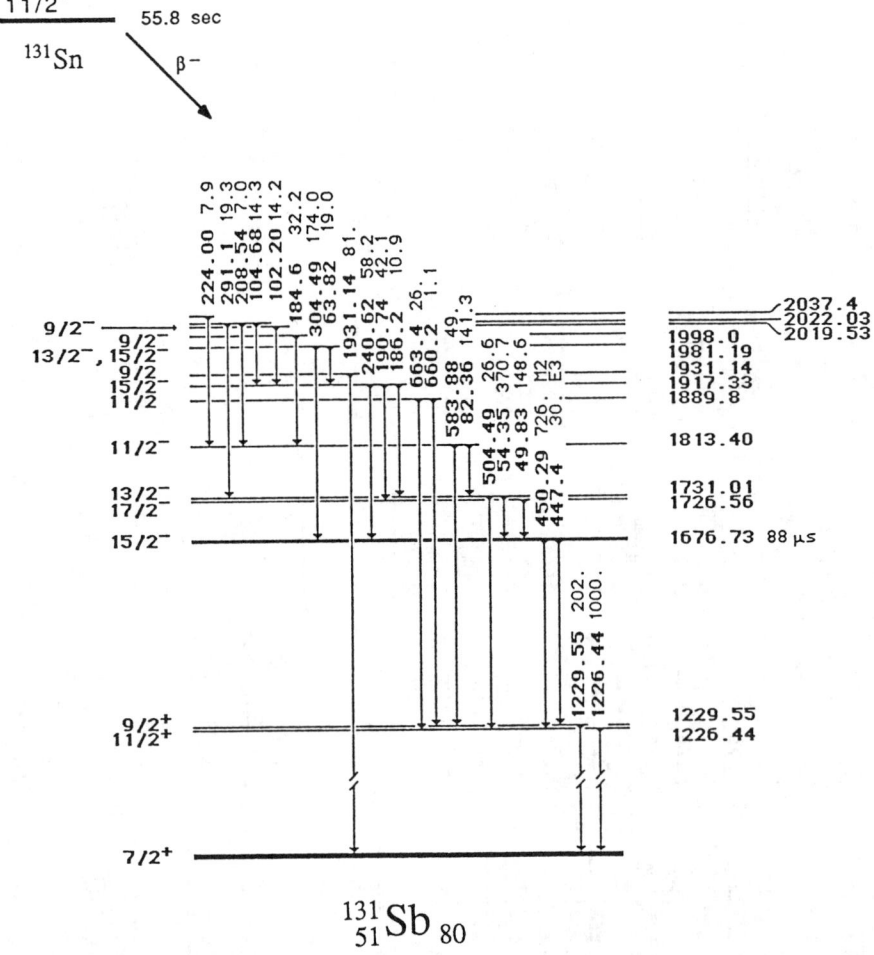

Figure 4. A part of the ^{131}Sb level scheme populated in the decay of the 11/2$^-$ ^{131}Sn isomer.

are from multiplets in which there is an odd neutron (hole) in the $h_{11/2}$ orbitals. The $\pi g_{7/2} \otimes (\nu h_{11/2} \otimes \nu d_{3/2})_{7^-}$ multiplet has the lowest energy of the high-spin multiplets. One of its levels lies below the other high-spin levels giving rise to an isomeric level whose half life has been measured to be 88 μs.[4] We have established that the 450-keV transition from this isomer has an M2 multipolarity and this limits the spin of the isomer to 15/2⁻. Levels at 1726, 1731 and 1813 keV are also likely to be members of the $\pi g_{7/2} \otimes (\nu h_{11/2} \otimes \nu d_{3/2})_{7^-}$ multiplet.

A partial level scheme for the high-spin system of ^{129}Sb is shown in Fig. 5. There are a larger number of levels near 2 MeV with several low energy transitions between them. We have performed conversion-electron measurements on the 723-keV transition from the 1851-keV and found that it is an M4 transition. This would suggest that the spin of the level is 19/2⁻. Gamma multiscaling data show that the half life of the 723-keV transition is 17.7-min and that the 1129-keV transition has a 17.7-min component to its half life. Huck et al.[5] identified the presence of this 17.7-min isomer but did not place it in the ^{129}Sb level scheme. Coincidence data indicate that the 1861-keV level is also an isomeric state but we can only say that the half life is considerably longer than the coincidence resolving time in our experiments, which was ca. 2 μs. An intense, unobserved 9.8-keV E2 transition feeds the 1851-keV level. Two other transitions depopulate the isomer feeding both the 11/2⁺ and 9/2⁺ levels which, with the 9.8-keV E2 transition, limits its spin to 15/2⁻.

Presence of the 19/2⁻ isomer is not found in ^{127}Sb or ^{131}Sb. In these nuclides the 15/2⁻ member of the $\pi g_{7/2} \otimes (\nu h_{11/2} \otimes \nu d_{3/2})_{7^-}$ multiplet is the lowest level and an isomer with a half life in the μs range. The levels in this multiplet move up in energy from ^{131}Sb to ^{129}Sb. In ^{131}Sb the 15/2⁻ isomer lies at 1677 keV while in ^{129}Sb it is 200 keV higher at 1861 keV. The 15/2⁻ isomer, though moves up in energy at a faster rate than the 19/2⁻ level causing the two levels to switch positions and resulting in a 19/2⁻ isomer. This shifting of the levels to higher energies continues to ^{127}Sb but the 19/2⁻ and 15/2⁻ levels switch again and only a 15/2⁻ isomer is observed.

Decay of the 19/2⁻ Isomer in ^{129}Sb

The 19/2⁻ isomer in ^{129}Sb decays by two modes, the 723-keV M4 internal transition and an 85% β⁻-decay branch to levels of ^{129}Te. Figure 6 shows a preliminary level scheme for the β⁻ decay of this isomer. The 5/2⁺ level at 544 keV was identified in previous studies on the structure of ^{129}Te along with the 1228-keV level and the (7/2⁻,9/2⁻)

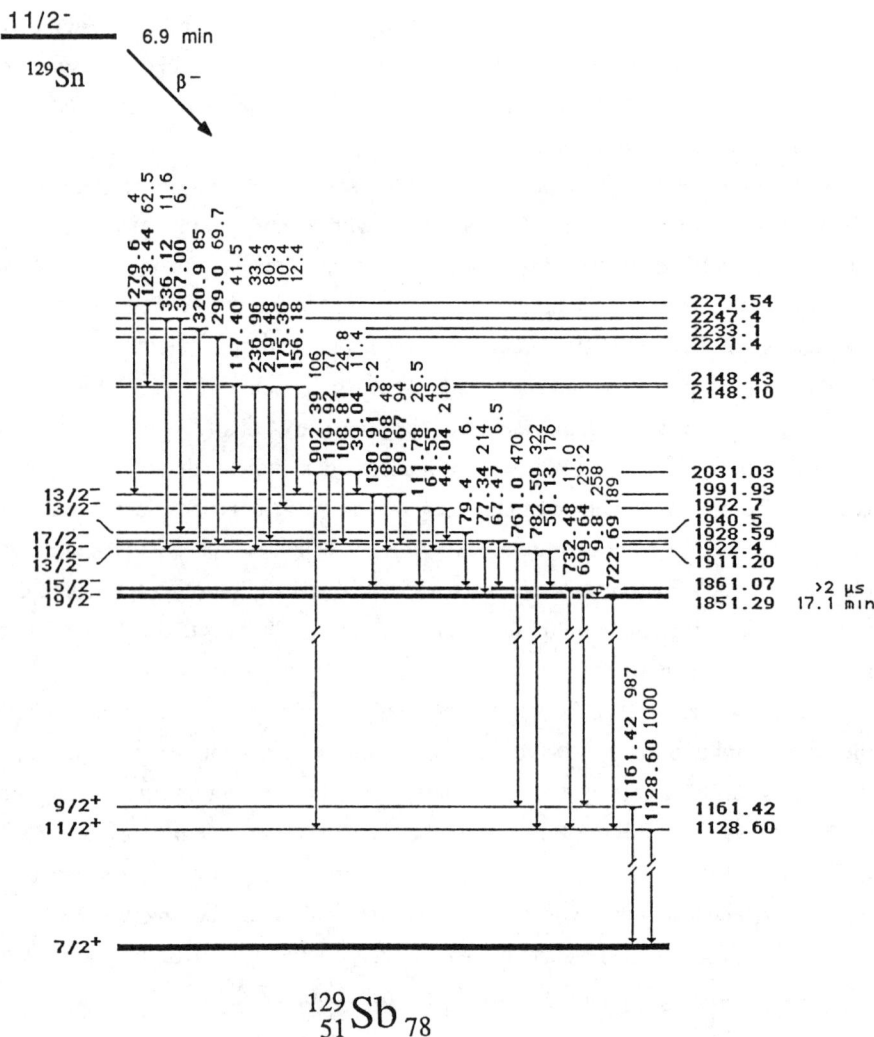

Figure 5. A part of the ^{129}Sb level scheme populated in the decay of the $11/2^-$ ^{129}Sn isomer.

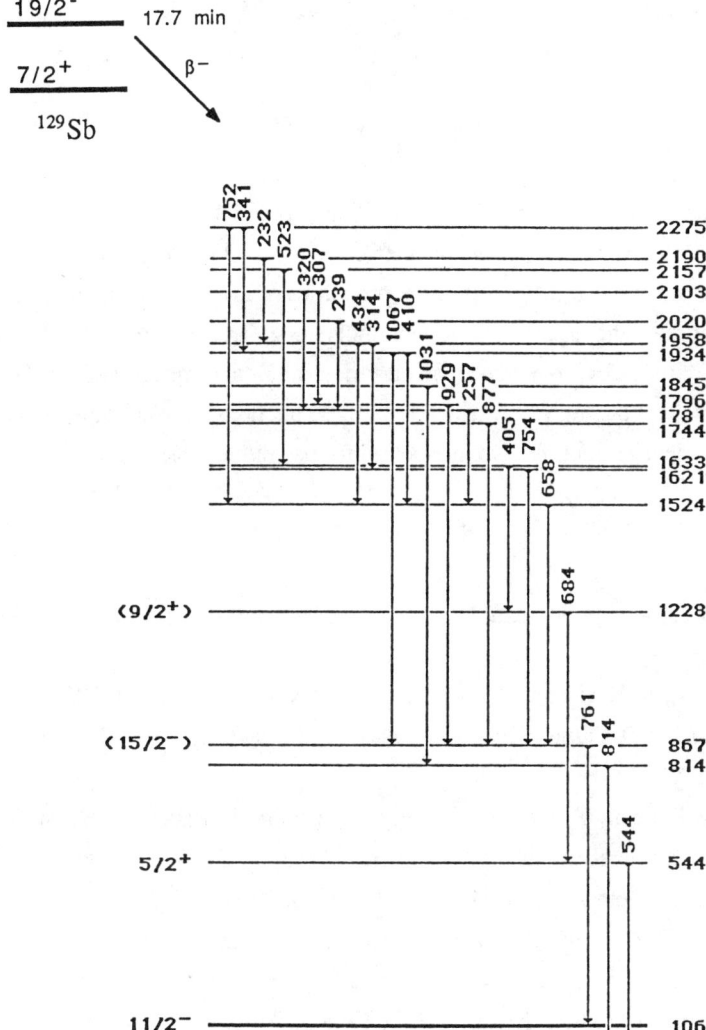

Figure 6. A preliminary level scheme for ^{129}Te which is populated in the decay of the 19/2⁻ ^{129}Sb isomer.

negative-parity levels at 760 and 876 keV. The latter levels are probably a part of the $(vh_{11/2} \otimes 2^+)_{7/2^- - 15/2^-}$ m multiplet. As the 761-keV transition carries most of the intensity from the decay of this isomer, we have identified it as the $15/2^-$ to $11/2^-$ transition. Note that the 1958-keV level branches to both the positive- and negative-parity systems.

Conclusions

Recent studies on the performance of two-body potentials showed that the commonly used potentials do not properly characterize the interactions involving $h_{11/2}$ nucleons.[6] At the time of these calculations, the properties of levels in the nuclides near ^{132}Sn were either not yet observed or not well understood. Our experiments on the structure of the Sb nuclides have yielded an enormous amount of information on the multiplet structure and the high-spin structure of these nuclides. We are now in a much better position to develop an effective interaction for use in the ^{132}Sn region.

References

1. B. Fogelberg, A. Aprahamian, R.L. Gill, H. Mach and D. Rehfield, Phys. Rev. **C31**, 1026 (1985).
2. J. Blomqvist, A. Kerek and B. Fogelberg, Z. Physik, **A314**, 199 (1983).
3. C.A. Stone, Ph.D. Dissertation, University of Maryland, College Park, MD 20742 USA.
4. D. Weiler, P. Kohl, G. Menzen, G. Lhersonneau and K. Sistemich, Annual Report, Insitut fur Kernphysik, 5170 Julich, 1985, pp. 31.
5. H. Huck, M.L. Perez, J.J. Rossi, Phys. Rev. **C26**, 621 (1982).
6. C.A. Stone, W.B. Walters, S.D. Bloom and G.J. Mathews, Shell-Model Calculations near ^{132}Sn using a Realistic, Effective Interaction, in R.A. Meyer and D.S. Brenner (eds.), "Nuclei Off the Line of Stability", ACS Conf. Ser. 324, (1986), pp. 70-77.

This work has been supported by the United States Department of Energy.

ENERGY LEVELS AND STRUCTURE OF LIGHT RARE-EARTH NUCLEI, ^{136,138,140}Sm and ^{132,134,136}Nd, via BETA DECAY

B.D. Kern
University of Kentucky, Lexington, KY 40506
G.A. Leander, R.L. Mlekodaj, and H.K. Carter
UNISOR, Oak Ridge Associated Universities, Oak Ridge, TN 37831
M.O. Kortelahti and E.F. Zganjar
Lousiana State University, Baton Rouge, LA 70803
R.A. Braga, R.W. Fink, and C.P. Perez
School of Chemistry, Georgia Institute of Technology
Atlanta, GA 30332
P.B. Semmes[+] and W. Nazarewicz[#]
Joint Institute for Heavy-Ion Research,
Oak Ridge National Laboratory, Oak Ridge, TN 37831.

ABSTRACT

Levels in ^{136,138,140}Sm were populated by the beta decay of Eu, following (HI,pxn) reactions and on-line mass separation. Members of the γ band were observed in all three daughter nuclei. Spectroscopic calculations were made using the triaxial rotor model, with all parameters derived microsopically from a Woods-Saxon deformed shell model. Comparison with the data supports the characterization of these nuclei in terms of a triaxial intrinsic shape. Improved decay schemes for ^{132,134,136}Nd are given.

INTRODUCTION

A transition from spherical to deformed shape in increasingly neutron-deficient Sm and Nd isotopes with N<82 has long been predicted on the basis of elementary shell structure considerations and was borne out by recent systematic measurements of yrast level energies in these nuclei. The detailed nature of this shape transition is of special interest for the study of nuclear structure in view of the unusual nature of the analogous shape transition in the Sm isotopes at N>82. Deformed shell model calculations agree with the observation that experimentally available nuclear species with 50<N<82 are softer with respect to triaxial gamma-deformation than their N>82 counterparts.

The tandem accelerator at the Holifield Heavy Ion Research Facility provided beams that were used to produce radioactive ions through the ^{92}Mo(^{46,48}Ti, ypxn) and ^{112}Sn(^{28}Si, ypxn) reactions at energies from 170 to 250 MeV. The radioactive ions were introduced into a high-temperature ion source and passed through the UNISOR mass separator. Also, in order to enhance the yields, He-jet measurements were carried out without mass separation. In both experimental

[+]On leave from TN Technological University, Cookeville, TN 38505.
[#]On leave from Technical University, Pl-00-662, Warsaw, Poland.

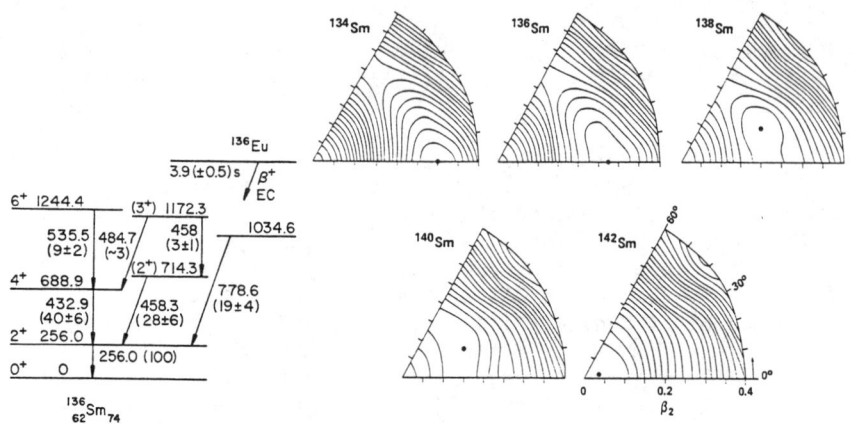

Fig. 1. Partial level diagram of ^{136}Sm.

Fig. 2. Potential-energy surfaces in the β,γ quadrupole deformation plane.

arrangements, radioactive ions were transported on a continuous plastic tape from a collection point to a counting location which was situated between two Ge detectors. The counting time intervals were equal to the collection time intervals; these time intervals were varied from 5 s to 75 s.

In this work, level schemes for the N = 74-78 isotopes of Sm were established by observing γ rays following the β decay of ^{136,138,140}Eu, and for the N = 72-76 isotopes of Nd, following the β decay of ^{132,134,136}Pm. Gamma bands were observed in each nucleus. The level diagram of ^{136}Sm is shown in Fig. 1.

The potential energy of the β_2, γ and β_4 deformation was calculated using the triaxially deformed "Warsaw" Woods-Saxon potential[1]. The potential-energy surfaces shown in Fig. 2 were calculated with a square lattice of 0.05 and at steps of 0.04 in β_4. The contour line separation is 250 keV. Spectroscopic properties were calculated at the interpolated minima in β_2, γ, and β_4. Quadrupole transition moments, B(E2: $2^+ \rightarrow 0^+$) reduced transition rates, rotational moments of inertia with respect to the three principal axes, and the energy levels of ^{138,140}Sm were calculated. Details of these calculations will be available in a future publication[2].

RESULTS

The prediction of triaxiality in this region was made by several workers[2]. Here, in Fig. 2, the transition from a well-deformed prolate shape in ^{134}Sm to near-spherical shape in ^{142}Sm is seen to proceed via triaxial shapes in ^{138,140}Sm.

The triaxial rotor Hamiltonian, without adjustment of parameters, reproduces the energies of levels of ^{138,140}Sm rather well (Fig. 3).

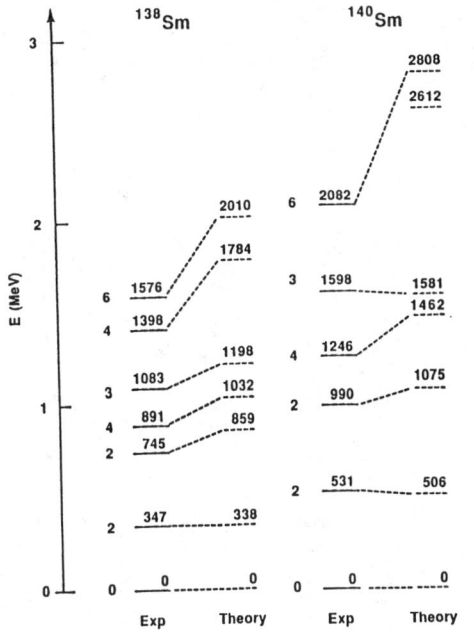

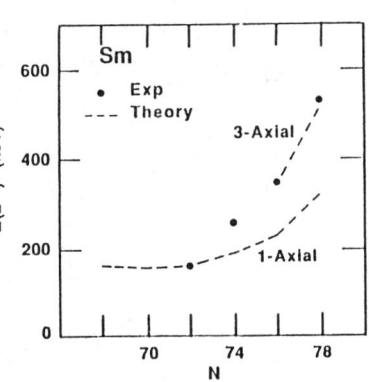

Fig. 3. Comparison between the experimental and calculated levels of the triaxial nuclei.

Fig. 4. Comparison between the experimental and calculated energies of the first 2^+ levels. The "1-axial" energies are those of a semiclassical rotor.

The experimental energies of the lowest 2^+ states are compared in Fig. 4 with the "1-axial" values which are derived by cranking the nucleus around one axis, as is appropriate semiclassically or for an axial shape, and with the "3-axial" values which were obtained from quantal triaxial rotation. The triaxial theory fits the ^{138,140}Sm energies and the axial symmetry theory fits the ^{134}Sm energy. The energies of the 2_2^+ and 3_1^+ members of the γ band are reasonable. The states of spin 4 and higher are systematically lower in experiment than in theory due to the variable moment of inertia effect.

Finally, it is noted that the large change in the energy of the 2_1^+ levels between ^{134}Sm (163 keV) and ^{140}Sm (531 keV) can be understood since the triaxial rotation tends to increase the 2_1^+ level energy of ^{140}Sm.

^{132,134,136}Nd

Recently acquired data has enabled improved decay schemes of ^{132,134,136}Nd[3] (Fig. 5-7). Not shown, but available elsewhere[2], are the $β_2$, γ and $β_4$ calculated values for the Nd isotopes. They show that at ^{132}Nd there is a trend toward a stiffly prolate shape; experimentally this is indicated by the increase in energy of the $2_γ^+$ level as it moves above that of the 4_1^+ level.

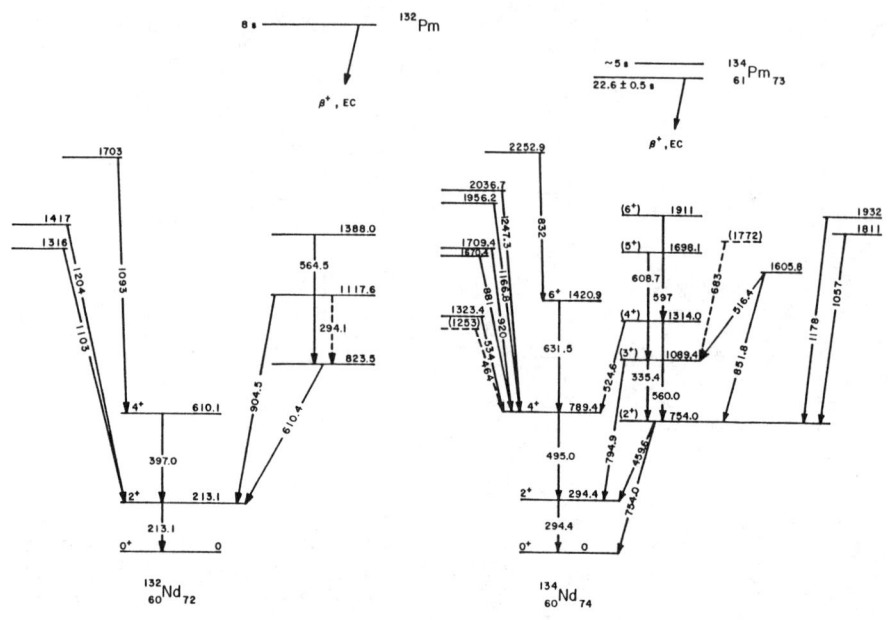

Fig. 5. Partial decay scheme of ^{132}Nd. The level at 823.5 keV is assumed to be the 2^+ of the γ band.

Fig. 6. Partial decay scheme of ^{134}Nd.

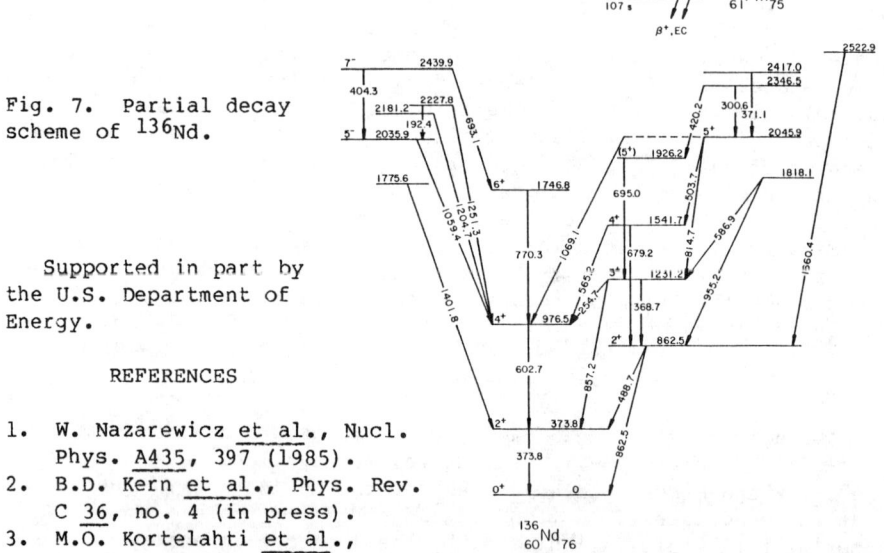

Fig. 7. Partial decay scheme of ^{136}Nd.

Supported in part by the U.S. Department of Energy.

REFERENCES

1. W. Nazarewicz et al., Nucl. Phys. A435, 397 (1985).
2. B.D. Kern et al., Phys. Rev. C 36, no. 4 (in press).
3. M.O. Kortelahti et al., Z. Phys. A 327, 231 (1987).

IDENTIFICATION AND STRUCTURE OF p-RICH RARE-EARTH NUCLEI INVESTIGATED USING A HE-JET FED ON-LINE MASS-SEPARATOR

R. Béraud[1], A. Charvet[1,*], R. Duffait[1], A. Emsallem[1],

J. Genevey[2], A. Gizon[2], M. Meyer[2,+], N. Redon[1], and D. Rolando-Eugio[1]

1 Institut de Physique Nucléaire (et IN2P3), Université Cl. Bernard Lyon-1, 43, Bd du 11 Novembre 1918, 69622 Villeurbanne Cedex, France.

2 Institut des Sciences Nucléaires (IN2P3 et USTMG), 53, avenue des Martyrs, 38026 Grenoble Cedex, France.

ABSTRACT

Activities produced via fusion-evaporation reactions with ^{32}S, ^{35}Cl and ^{36}Ar beams in thin ^{112}Sn and ^{106}Cd targets have been transported using a He-jet coupled to a mass separator. Gamma-ray and X-ray techniques have been used to study the excited levels of ^{138,136}Sm and ^{134,132}Nd fed in the β-decay of ^{138}Eu(12 s), ^{136}Eu(3.7 s), ^{134}Pm (21 s) and ^{132}Pm (5 s). Results are presented through systematics and discussed in the framework of recent self-consistent calculations including the γ degree of freedom.

INTRODUCTION

The region of p-rich rare-earth nuclei has been, since a couple of years very actively investigated by many experimental groups and theoreticians as well. The interest in these studies was mainly due to the occurence of strongly deformed isotopes on both sides of doubly magic $^{146}_{64}$Gd$_{82}$ nucleus. Axial macroscopic-microscopic calculations by Leander and Möller[1] have predicted large deformation near the proton drip-line and Ragnarsson et al.[2] have shown that the γ-degree of freedom could give major effects around N=76. More recently self-consistent calculations[3] of triaxial deformations have shown that some N=76 isotones exhibit a stable triaxial shape.

A lot of experimental work has been devoted to the $132 \leqslant A \leqslant 138$ mass region both from β-decays[4,5,6] or in-beam spectroscopy[7,8] giving new information on the high spin state structure of these neutron deficient isotopes. Fusion-evaporation reactions using both n-deficient target and projectile are still a powerfull way to produce p-rich nuclei with substantial yields and forward peaking kinematics. They were used in this work in connection with ISOL techniques in order to extend the systematics of nuclear properties as far as possible from the stability valley.

---

* deceased on May 28th, 1987.
+ permanent address : IPN Lyon

2. EXPERIMENTAL

2.1 SARA and ISOL facilities

The present work has been possible thanks to the development of an ECR ion source on the SARA accelerator at Grenoble and simultaneous construction of a He-jet fed on-line mass-separator. The principle of the connection has been described in details and performances of the system already reported[9]. The experiments described below were carried out with low energy (5-6 MeV/u), high intensity (> 3.10^{11} particles/s) beams accelerated by the first K=90 cyclotron of the SARA facility. The low consumption of the ECR ion source allowed to use enriched gas such as ^{36}Ar as projectile.

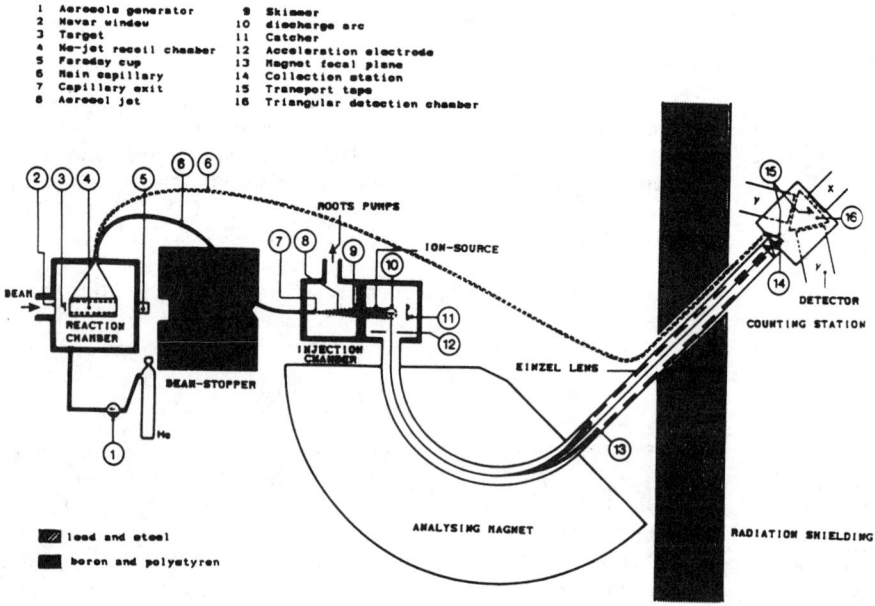

Figure 1 : Draft of the He-jet + separator coupling facility.

2.2 Layout of the separator and principle of operation

A schematic plan view of the He-jet fed on-line separator is given in **figure 1**. This naive representation shows how the system can be used either with the He-jet alone or with the coupled system. In the former mode of operation γ-γ , X-γ ray coincidence measurements and γ-ray multianalysis spectra for decay half-life determination are carried out. From the K_α X-ray gated γ spectra the Z identification of the daughter nucleus is obtained allowing an unambiguous identification when associated with spectra recorded after mass separation in the later mode of operation.

The radioactive products recoil from the target with mean energy corresponding to full linear momentum transfer in the fusion-evaporation process and are then thermalized in the 1-2 bars He pressurized sheet-type recoil chamber placed in the middle of the reaction chamber. He gas is fed through a temperature controled oven (400°C) in order to give optimum transfer yield with $PbCl_2$ aerosols.

A high flow pumping system composed of three roots (8000 m^3/h - 3000 m^3/h - 350 m^3/h) and a primary pump (120 m^3/h) is used to skim off the He from the injection chamber where the pressure may be maintained in the range 10^{-1} - 10^{-2} torr depending on the conductance of the main capillary.

The ion-source is a modified version[9] of the Bernas-Nier type medium current source also successfully applied by Schmeing et al.[10] who developped a high temperature version. It is worth to mention two advantages of the method when compared to more conventional systems :
i)   the delay time is mainly due to the mean transit time in the capillary (100 ms is our lower limit) and
ii) the target is located far away from the hot environment of the source so that it was possible to use low melting point element targets such as Tin or Cadmium.

The measured coupling efficiencies (skimmer + source + magnet + lens transport) are in the range 1-2 % and allowed us to get a number of original results on the p-rich rare-earth isotopes near N=82.

3. RESULTS

3.1 General

Figure 2 gives the new isotopes studied in this work. Most of experiments have been carried out using 191 MeV ^{35}Cl, 170 MeV ^{32}S and 234 MeV ^{36}Ar beams on 1-3 mg/cm^2 self-supporting enriched ^{112}Sn and ^{106}Cd targets.

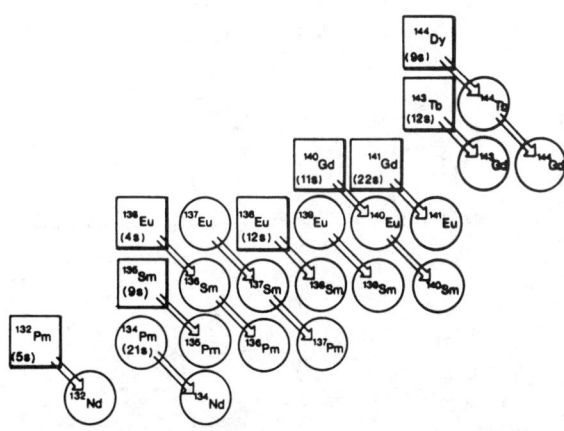

Figure 2 : Part of the nuclear chart related to the mass region investigated. Near N=82 β-decays investigated experimentally are denoted by arrows. New isotopes are indicated by solid squares.

The largest productions were generally associated with (2p,n), (2p), (2p,2n) reaction channels but rather important yields were also observed with (3p,xn) channels in good agreement with ALICE code predictions[11].

Half-lives, gamma-ray energies and intensities have been measured in the various β-decays and the decay scheme constructed on the basis of γ-γ coincidences. When possible, the direct feeding of the ground state has been estimated and thus log ft values calculated in order to get spin-parity assignments.

3.2 ^{138}Eu → ^{138}Sm

Among the data obtained in the light rare-earth region a very interesting case is that of ^{138}Sm fed by the β-decay of ^{138}Eu ($T_{1/2}$=12 s) produced via the ^{106}Cd + ^{35}Cl reaction. In **figure 6** are represented two γ spectra showing the same energy range, one after mass separation and the second in coincidence with K_α X-rays of Sm.

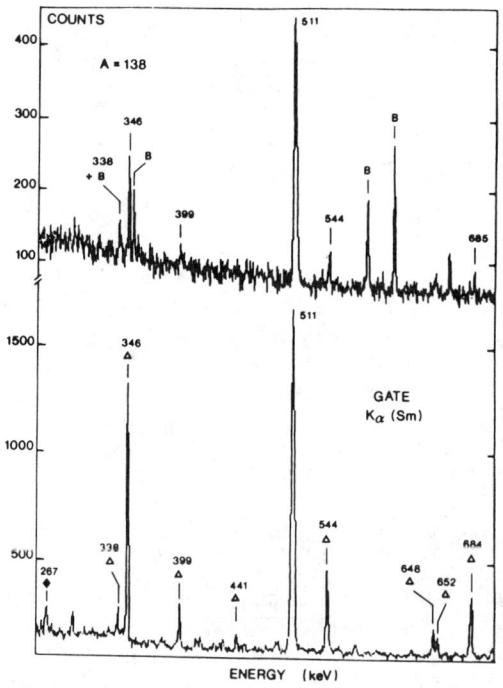
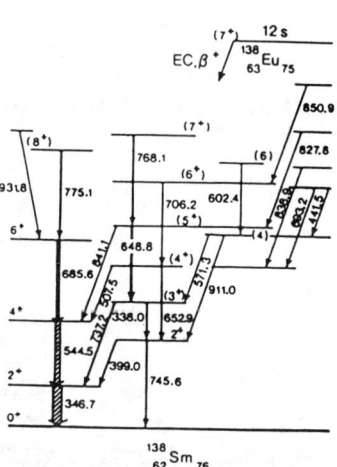

Figure 3 : Identification of 12 s ^{138}Eu. γ-rays found in coincidence with Sm-K_α gate (lower spectrum) are also observed in the mass-separated spectrum (upper part) at A=138.

Figure 4 : Decay scheme of 12s ^{138}Eu.

From γ-γ coincidence data we have constructed the level scheme represented in **figure 4**. Both ground-state and quasi-gamma bands are populated up to the (8^+) and (7^+) levels respectively giving (7^+) as probable parent state of ^{138}Eu. The assignment of $2'^+$ below the first 4^+ state will be discussed in details later.

3.3 ^{136}Eu → ^{136}Sm

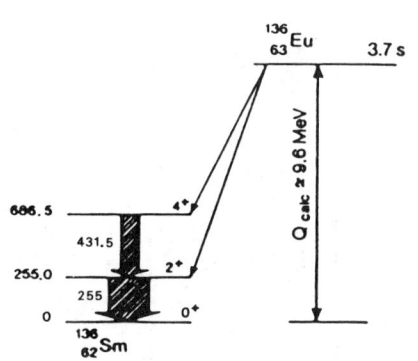

Figure 5: Partial decay scheme of 3.7 s ^{136}Eu → ^{136}Sm.

A 234 MeV ^{36}Ar beam has been used to bombard a ^{106}Cd target and by means of the set-up described previously a new activity with a half-life of (3.7 ± 0.5)s was found and unambiguously assigned to ^{136}Eu → ^{136}Sm decay. A partial decay scheme is shown in **figure 5** and it is corroborated by in beam experiments on ^{136}Sm[7,12].

3.4 ^{134}Pm → ^{134}Nd and ^{132}Pm → ^{132}Nd

Using the same reaction ^{36}Ar + ^{106}Cd as in previous section it was possible to identify the (5 ± 1)s ^{132}Pm from the decay curves of the 213.0 keV and 398.5 keV γ-rays and the (21 ± 1)s ^{134}Pm from the 294.4, 494.8 and 631.3 keV γ-rays as shown in **figures 6 and 7**.

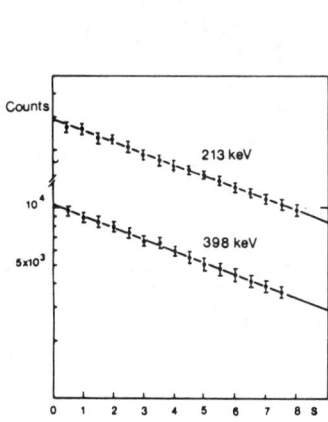

Figure 6: Decay curves of the two γ-rays ascribed to (5 ± 1)s ^{132}Pm.

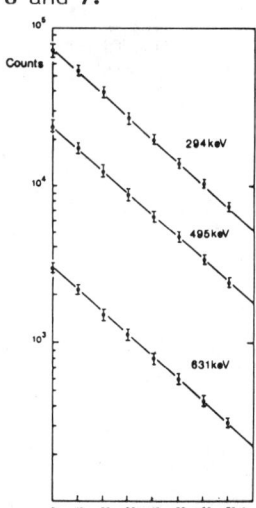

Figure 7: Decay curves of the main γ-lines associated to the (21 ± 1)s ^{134}Pm.

From the data obtained in this experiment it has been possible to construct a partial level scheme of ^{132}Nd and ^{134}Nd shown in **figures 8** and **9** respectively. It is worth mentionning the good agreement of these results with a very recent note by M. Kortelahti et al.[13].

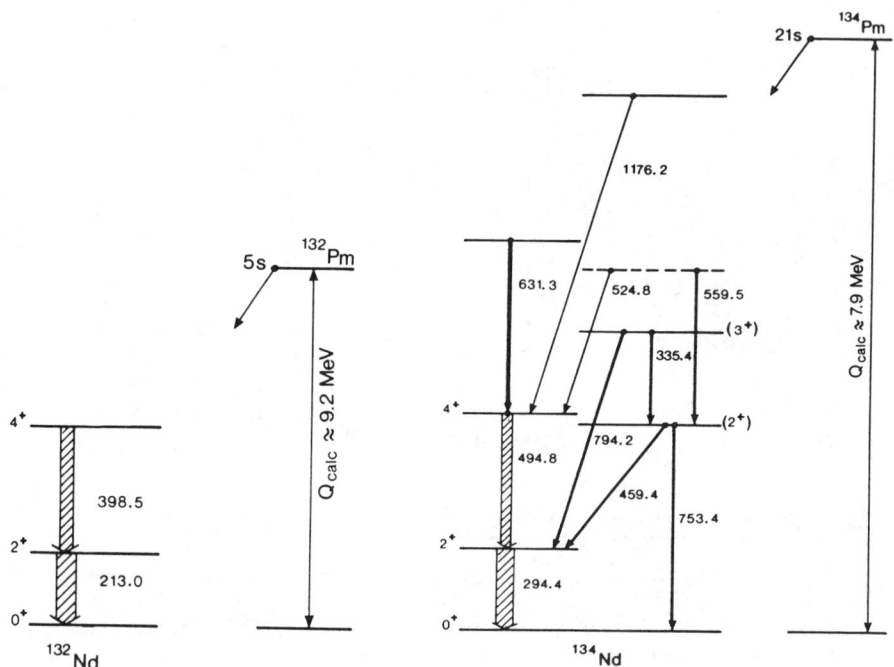

Figure 8 : Preliminary decay scheme of 5s ^{132}Pm.

Figure 9 : Partial decay scheme of 21s ^{134}Pm.

ANALYSIS OF THE RESULTS

Macroscopic-microscopic calculations have predicted from N=72,70 in Sm, Nd isotopes the emergence of a large axial prolate deformation [1] with $\varepsilon_2 > 0.28$. Experimentally, if we consider the systematics of first excited level 2^+ energies of even-even nuclei in this region (**figure 10**), we recognize the energy gap characteristic of shell closure for N=82 indicating a spherical shape. On the side N < 82, the energy of the first excited 2^+ state goes down smoothly exhibiting the onset of a deformation. The energy ratio $E(4+)/E(2+)$ of the first two excited g.s. band states of proton-rich nuclei (**table 1**) shows, near N=76, a possible γ-unstability of these nuclei with a value 2.5 whereas for the more neutron-deficient isotopes, this ratio reaches values close to 3.33 consistent with a pure axially-symmetric rotor.

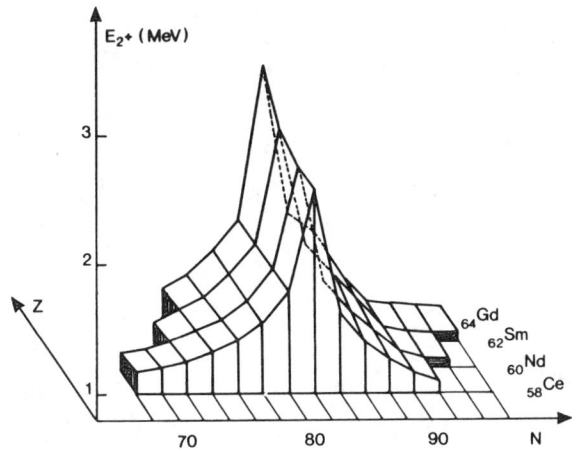

Figure 10 : Systematics of first excited level 2^+ energies of e-e nuclei around N=82 for $58 \leqslant Z \leqslant 64$.

			^{138}Gd	^{140}Gd	^{142}Gd	^{144}Gd	Z = 64
			2.74	2.54	2.40	2.34	
		^{134}Sm	^{136}Sm	^{138}Sm	^{140}Sm	^{142}Sm	Z = 62
		2.94	2.69	2.57	2.35	2.33	
^{128}Nd	^{130}Nd	^{132}Nd	^{134}Nd	^{136}Nd	^{138}Nd	^{140}Nd	Z = 60
3.18	3.06	2.86	2.68	2.62	2.40	2.33	
^{126}Ce	^{128}Ce	^{130}Ce	^{132}Ce	^{134}Ce	^{136}Ce	^{138}Ce	Z = 58
3.06	2.93	2.80	2.64	2.56	2.38	2.30	
N = 68	N = 70	N = 72	N = 74	N = 76	N = 78	N = 80	

Table 1 : Ratio of E_{4^+}/E_{2^+} energies in Gd, Sm, Nd, Ce isotopes.

In the case of Nd and Sm isotopes (**figure 11**), the level spacing of the ground state band is decreasing both side of N=82 but these two regions differ in the behavior of the second 2^+ level : on the neutron rich nuclei side, this state remains markedly above the first 4^+ level characteristic of a pure axially deformation whereas, on the proton rich side, it lies under the 4^+ level (^{138,140}Sm, ^{134,136,138,140}Nd) and the existence in ^{138}Sm, ^{134,136,138}Nd of an additionnal sequence of 3^+, 4^+, 5^+, ... levels based on this 2^+ state is characteristic of triaxial deformation with γ closed to 30° in the simple picture of a pure triaxial rotor[14]. Asymmetric equilibrium solutions have been already suggested in this region by macroscopic-microscopic calculations of ref. 2.

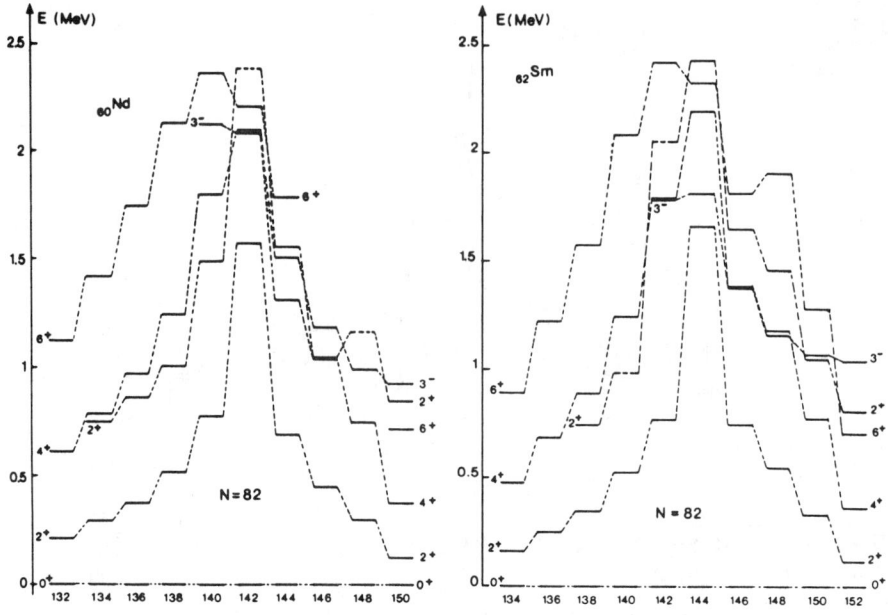

Figure 11 : Evolution of the first excited states in the even-even isotopes of Nd and Sm with $72 \leqslant N \leqslant 90$.

In order to take into account in microscopic theories possible triaxial deformations, lattice Hartree-Fock + BCS calculations for axially asymmetrical solutions[15] have been performed nowadays for the Sm e-e isotopes[3,16,17]. They have predicted the emergence of a strong axial deformation with prolate shapes for ^{134,132}Sm via triaxial shapes for ^{136,138,140}Sm, the ^{142}Sm remaining spherical as the ^{144}Sm semi-magic nucleus. For instance in agreement with our experimental data, the potential energy surface of ^{138}Sm obtained by these microscopic calculations presents an absolute minimum at $Q_0 \sim 8b$ and $\gamma \sim 25°$ located ~ 0.6 MeV below two axial local minima almost degenerated in energy. The existence of such stable triaxial shapes has been also predicted for the intrinsic states in neighboring N=76 isotones of Gd and Nd and interpreted in terms of neutron shell-effects.

The He-jet coupled system used after (HI,xn) reactions has proved to be an efficient tool for the study of heavy exotic nuclei. In-beam studies of odd isotopes are highly desirable to confirm these structure effects. Therefore complementary experiments have been initiated with the "Château de Cristal", the analysis of which are underway.

ACKNOWLEDGEMENTS

The authors are highly indebted to A. Plantier, J.L. Vieux-Rochaz, S. Vanzetto, V. Boninchi, G. Margotton and the SARA staff for technical assistance during the experiments, J.P. Richaud and L. Vidal for target preparation. We are grateful to J. Meyer, P. Quentin, M.S. Weiss, P. Bonche, H. Flocard and P.-H. Heenen for enlightening theoretical discussions and fruitful collaboration. We also wish to thank Dr. Juha Äystö and Dr. J. Honkanen (University of Jyväskylä) and Dr. K. Deneffe (I.K.S. Leuven) for intensive collaboration in this work.

REFERENCES

1 G.A. Leander and P. Möller,
   Phys. Lett. **B110** (1982) 17.

2 I. Ragnarsson, A. Sobiczewski, R.K. Sheline, S.E. Larsson and B. Nerlo-Pomorska,
   Nucl. Phys. **A233** (1974) 329.

3 N. Redon, J. Meyer, M. Meyer, P. Quentin, M.S. Weiss, P. Bonche, H. Flocard and P.-H. Heenen,
   Phys. Lett. **B181** (1986) 223.

4 M. Novicki, D.D. Bogdanov, A.A. Demyanov, Z. Stachura,
   Acta Phys. Polon. **B13** (1982) 879.

5 P.A. Wilmarth, J.M. Nitschke, P.K. Lemmertz and R.B. Firestone,
   Z. Phys. **A321** (1985) 179.

6 N. Redon, T. Ollivier, R. Béraud, A. Charvet, R. Duffait, A. Emsallem, J. Honkanen, M. Meyer, J. Genevey, A. Gizon and N. Idrissi
   Z. Phys. **A325** (1986) 127.

7 C.J. Lister, B.J. Varley, R. Moscrop, W. Gelletly, P.J. Nolan, D.J.G. Love, B.J. Bishop, A. Kirwan, D.J. Thornley, L. Ying, R. Wadsworth, J.M. O'Donnell, H.G. Price and A.H. Nelson,
   Phys. Rev. Lett. **55** (1985) 810.

8 S. Lunardi, F. Scarlassara, F. Soramel, S. Beghini, M. Morando and S. Signorini
   Z. Phys. **A321** (1985) 177.

9   A. Plantier, R. Béraud, T. Ollivier, A. Charvet, R. Duffait, A. Emsallem, N. Redon, V. Boninchi, S. Vanzetto, A. Gizon, J. Tréherne, N. Idrissi, J. Genevey and J.L. Vieux-Rochaz,
    Nucl. Instr. & Meth. in Phys. Res. **B26** (1987) 314.

10 H. Schmeing, J.S. Wills, E. Hagberg, J.C. Hardy, V.T. Kolowsky and W.L. Perry,
    Nucl. Instr. & Meth. in Phys. Res. **B26** (1987) 321.

11 M. Blann,
    Code Alice 85/300, Report UCID 20169 (1984).

12 A. Makishima, M. Adachi, H. Taketani and M. Ishii,
    Phys. Rev. **C34** (1986) 576.

13 M. Kortelahti, E. F. Zganjar, R.L. Mlekodaj, B.D. Kern, R.A. Braga, R. W. Fink and C.P. Perez,
    Z. Phys. **A327** (1987) 231.

14 A.S. Davydov and B.F. Filippov,
    Nucl. Phys. **8** (1958) 237.

15 P. Bonche, H. Flocard, P.-H. Heenen, S.J. Krieger and M.S. Weiss,
    Nucl. Phys. **A443** (1985) 39.

16 N. Redon, A. Béraud, A. Charvet, R. Duffait, A. Emsallem, J. Genevey, A. Gizon, M. Meyer, J. Meyer, P. Quentin, M.S. Weiss, P. Bonche, H. Flocard and P.-H. Heenen,
    9ème Session d'Etudes Biennale de Physique Nucléaire,
    Aussois, 9-13 Mars 1987, LYCEN/8702 (1987) S11.

17 N. Redon, J. Meyer, M. Meyer, P. Quentin, P. Bonche, H. Flocard, P.-H. Heenen,
    Abstract in this conference and to be published.

LEVEL STRUCTURE OF NEUTRON-RICH ODD-MASS Pr NUCLIDES

P. F. Mantica, Jr., E. F. Baum, J. D. Robertson, C. A. Stone, and W. B. Walters
University of Maryland, College Park, MD 20742

D. F. Kusnezov and R. A. Meyer
Lawrence Livermore National Laboratory, Livermore, CA 94550

ABSTRACT

Results of recent studies of the structures of odd mass neutron-rich Pr nuclides, ^{143}Pr, ^{145}Pr, and ^{147}Pr populated in the decay of ^{143}Ce, ^{145}Ce, and ^{147}Ce, respectively, will be described and discussed. PTQM calculations for the structure of these nuclides will be presented and compared with the experimentally observed levels.

INTRODUCTION

The odd-mass Pr nuclides lie in a critical position among the nuclides with $50 < Z < 64$, as they can be treated as having either 9 protons beyond the closed shell at $Z = 50$ or as having 5 holes in the closed subshell at $Z = 64$. The nuclides with smaller Z: La, Cs, and I are clearly viewed as particles beyond $Z = 50$, while considerable success has been obtained for the levels of Pm and Eu by considering them as holes in the $Z = 64$ subshell closure.[1,2] As neutrons are added beyond $N = 82$, considerable evidence has been presented to indicate the breakdown of the $Z = 64$ subshell closure so that in the region beyond $N = 90$ even Eu and Pm must be treated as having particles beyond $Z = 50$.[3] This work was initiated to provide additional experimental information for the odd-mass Pr nuclides to serve as tests for calculations to extend the understanding of the breakdown of the $Z = 64$ subshell closure.

The decay of 3-min ^{145}Ce has been studied at LLNL where it was produced by the ^{148}Nd(n,a)^{145}Ce reaction. At the on-line mass separator, TRISTAN, the decay of ^{145}Ce was studied in gamma-ray singles, gamma-ray coincidence, gamma-ray angular correlation, time-dependent singles, and conversion-electron experiments. The decay of 56-s ^{147}Ce has also been studied at TRISTAN in gamma-ray singles, gamma-ray coincidence, and time dependent singles experiments. The results for the lower-lying levels are shown in Fig. 1.

These new data have established the position of the 3/2$^+$ level at 188 keV in ^{145}Pr and determined spin and parity assignments of 3/2$^-$ and 5/2$^-$, respectively, for the strongly beta-fed levels at 786 and 1210 keV as shown in Fig. 2 along with the systematics of the $N = 86$ odd-Z isotones. As is seen, the 3/2$^+$ level moves to even lower energies in the lighter $N = 86$ isotones. In ^{147}Pr an additional low-lying level has been established at 28 keV that decays by a 25.4-keV transition to the 2.6-keV level. These data indicate that the ground and 28-keV levels differ in spin by 2 units of angular momentum and when examined in light of the beta decay properties of ^{147}Pr, and can be used to suggest that the ground state of ^{147}Pr may be 3/2$^+$.

The results for the PTQM calculations for ^{143}Pr and ^{145}Pr are shown in Fig. 3. The PTQM calculations were carried out assuming that Pr has 5 holes in the $Z = 64$ subshell closure and are seen to provide reasonable fits for the experimental data. The lowered position of the lowest 3/2$^+$ in ^{145}Pr is accounted for by these calculations.

The calculations have not yet been attempted for ^{147}Pr. The density of low-lying levels indicates a much more complicated structure. It is not possible to identify rotational structure from among these levels. If the 93-keV level is a member of a band build on either the ground state or 2.6 keV level, the rotational constant would be quite high compared to

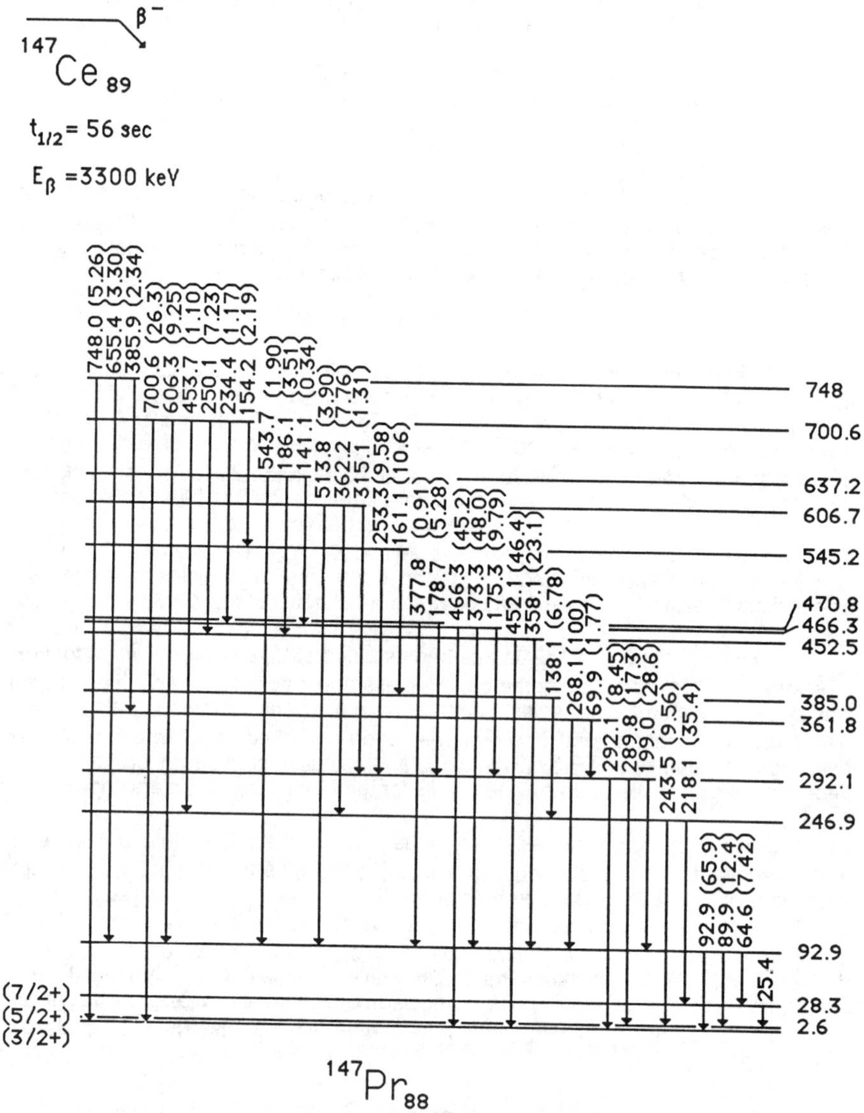

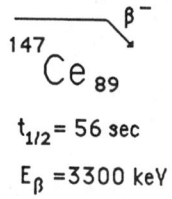

Figure 1. The low lying levels of ^{147}Pr populated in the decay of 56-s ^{147}Ce.

457

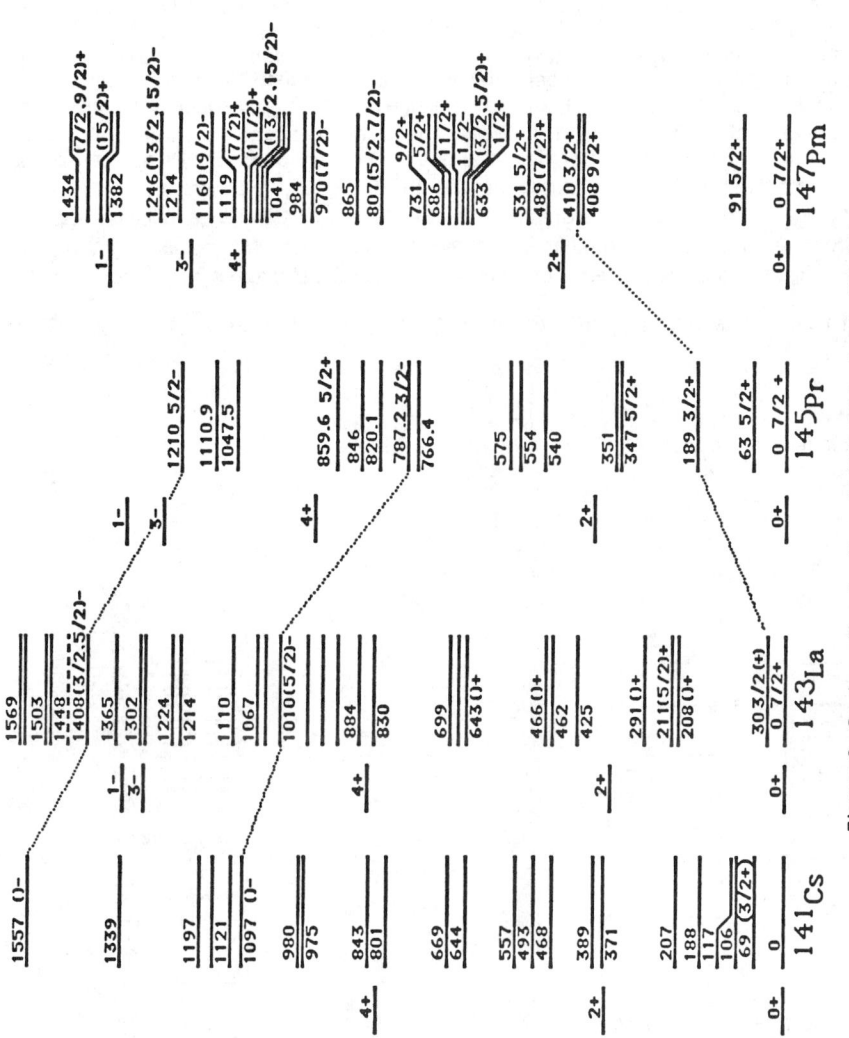

Figure 2. Systematics for the odd-Z N=86 isotones, 55≤Z≤61.

those in well deformed lanthanides such as ^{151}Pm and indicate considerably less deformation. On the other hand, considerably stronger coupling will be required in the PTQM calculations to produce 4 levels below 100 keV. What data are available for ^{149}Pr, do suggest that it is well deformed.[4]

These data suggest that the transition from spherical to deformed structure in the Pr nuclides is quite spread out. In the higher Z Eu and Pm nuclides, the transition is quite clear between N = 88 and N = 90. The N = 88 Eu and Pm nuclides can be readily described in intermediate coupling or IBFA calculations without large static deformation, while the N = 90 Eu and Pm nuclides are quite clearly deformed with well established Nilsson level structures. This sharp transition is atrributed to the rapid breakdown of the Z = 64 subshell closure. The complex structue of ^{147}Pr, however, indicates that either the breakdown of the subshell closure is more advanced at N = 88 with smaller numbers of neutrons in the Pr nuclides, or the possible reflection asymmetric structures identified[5,6,7] around ^{145}Ba are also maintaining some influence in ^{147}Pr. As the lowered 3/2+ level in the N = 86 Cs and La nuclides appears to arise from the presence of the reflection asymmetric structures, the possible 3/2+ ground state of ^{147}Pr supports the presence of reflection asymmetric structures in ^{147}Pr.

This work has been supported by the U. S. Department of Energy.

REFERENCES

1. O. Scholten and T. Ozzello, Nucl. Phys. **A424**, 221 (1984).
2. O. Scholten and N. Blasi, Nucl. Phys. **A380**, 509 (1982).
3 . A. Wolf et al., Phys. Lett. **123B**, 165 (1985).
4 . B. Pfeiffer et al., J. Phys. (Paris) **38**, 9 (1977).
5. J. D. Robertson et al., Phys. Rev. C **34**, 1012 (1986).
6. G. A. Leander et al., Phys. Lett. **152B**, 284 (1985).
7. W. R. Phillips, et al. , Phys. Rev. Lett. **57**, 3257 (1986).

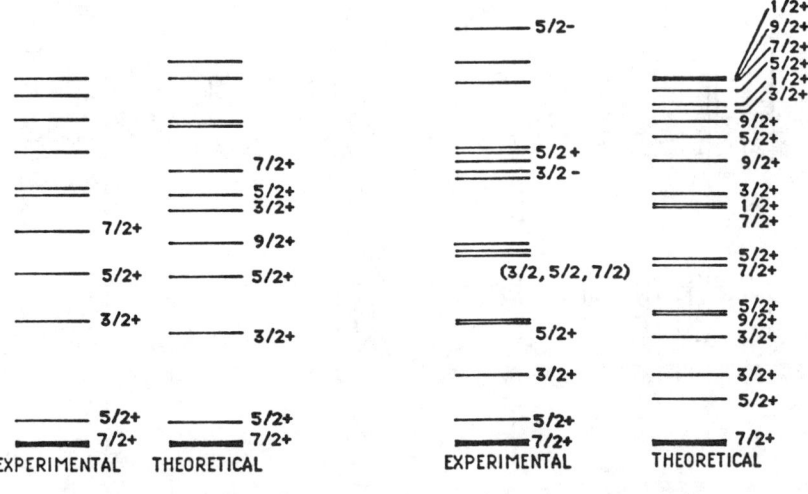

Figure 3. PTQM calculation results for the low lying levels in ^{143}Pr and ^{145}Pr.

The Structure of ^{143}Ba and ^{147}Ce

J.D. Robertson*, W.B. Walters, S.H. Faller, C.A. Stone, E.M. Baum, P.F. Mantica,
Dept. of Chemistry, University of Maryland, College Park, MD 20742
R.L. Gill and A. Piotrowski†
Physics Dept., Brookhaven National Laboratory, Upton, NY 11973

ABSTRACT

An investigation of the β^- decay of ^{143}Cs and ^{147}La was undertaken at the on–line mass separator TRISTAN. E_γ, I_γ, γ–γ coincidences, and γ–γ angular correlations were measured. Level schemes for ^{143}Ba and ^{147}Ce were constructed with proposed spin and parity assignments based upon measured transition multipolarities and γ–γ angular correlation coefficients.

INTRODUCTION

In a 1985 Physics Letter it was proposed that nuclei with N=88-90 in the near vicinity of ^{145}Ba may form a new region of octupole deformation similar to that which is observed in the Z=88–90 light actinide region.[1] Our subsequent investigation of the decay of ^{145}Cs indicated, however, that ^{145}Ba does not break reflection symmetry in the intrinsic frame.[2] The parity doublets which characterize the octupole deformed light odd–A actinides were not observed in the structure of ^{145}Ba. On the other hand, the results of our study did suggest that octupole correlations may play a role in determining the low–lying structure in this region.[2] No firm conclusions about the strength of such correlations could be made, however, by examining a single isolated case. For this reason, we have continued our study of reflection asymmetry in medium mass nuclei by investigating the structure of ^{143}Ba and ^{147}Ce in an effort to complete the systematics of the Z=56 isotopes from the N=82 neutron shell and the systematics of the N=89 isotones from the Z=64 proton subshell.

RESULTS

The decay scheme proposed for ^{147}Ce is shown in Figure 1. It resolves many of the discrepancies between the two previous works by Schussler et al.[3] and Shmid et al.[4] A spin and parity of 5/2$^-$ is assigned to the ground state (g.s.) based upon the systematics of the region. The g.s. J^π of the N=89 isotones is : ^{151}Sm=5/2$^-$, ^{149}Nd=5/2$^-$, and ^{145}Ba=5/2$^{(-)}$. The g.s. J^π of the odd–A Ce isotopes is : ^{141}Ce=7/2$^-$, ^{143}Ce=3/2$^-$, and ^{145}Ce=5/2$^-$. If the spin of the g.s. is 5/2$^-$, then the M1/E2 multipolarity of the 117–keV[3] transition limits J^π of the 117–keV level to 3/2$^-$, 5/2$^-$, or 7/2$^-$. A spin of 7/2$^-$ is, however, unlikely because of the large beta branch to the 117-keV level from a 3/2$^+$ parent. And finally, the spin and parity of the 438-keV level is limited to 3/2$^+$, 5/2$^+$, or 7/2$^+$ by the E1 multipolarity of the 438-keV gamma.[3]

The lower portion of the decay scheme proposed for ^{143}Ba is shown in Figure 2. The g.s. spin of ^{143}Ba has been measured to be 5/2.[5] Negative parity is assigned to the g.s. based upon the systematics of the region. The g.s. J^π of the N=87 isotones is : ^{151}Gd=7/2$^-$, ^{149}Sm=7/2$^-$, ^{147}Nd=5/2$^-$, and ^{145}Ce=(5/2)$^-$.[6] The g.s. J^π of the odd-A Ba isotopes is : ^{139}Ba=7/2$^-$ and ^{141}Ba=3/2$^-$. The E2 multipolarity reported for the 33- and 117-keV transitions[7] limits the J^π of the 33- and 117-keV levels to 1/2$^-$

*Present address: Lawrence Berkeley Lab, One Cyclotron Rd., Berkeley CA 94720
†Present address: Institute for Nuclear Studies, 05-400 Swierk, Poland

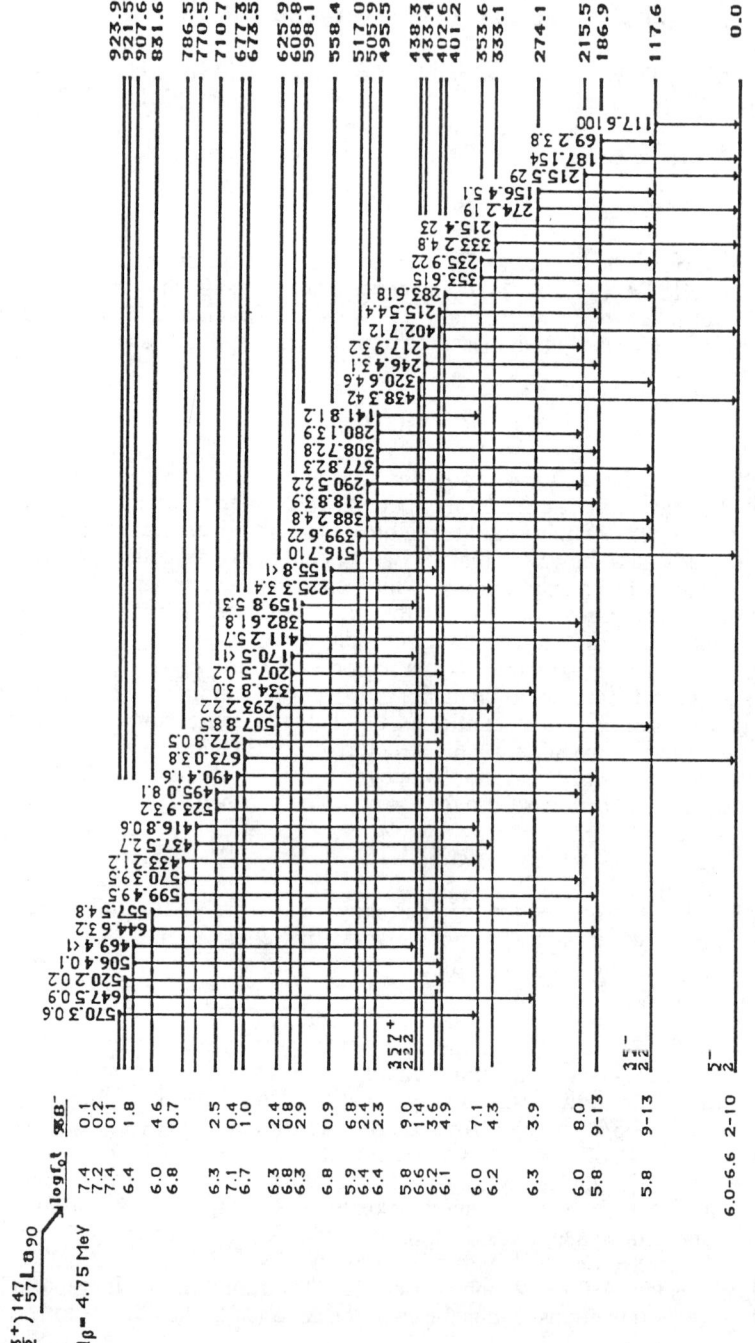

Figure 1. The decay scheme of ^{147}Ce. The beta branching was calculated assuming an absolute intensity of 0.23±0.02 for the 117-keV gamma.6

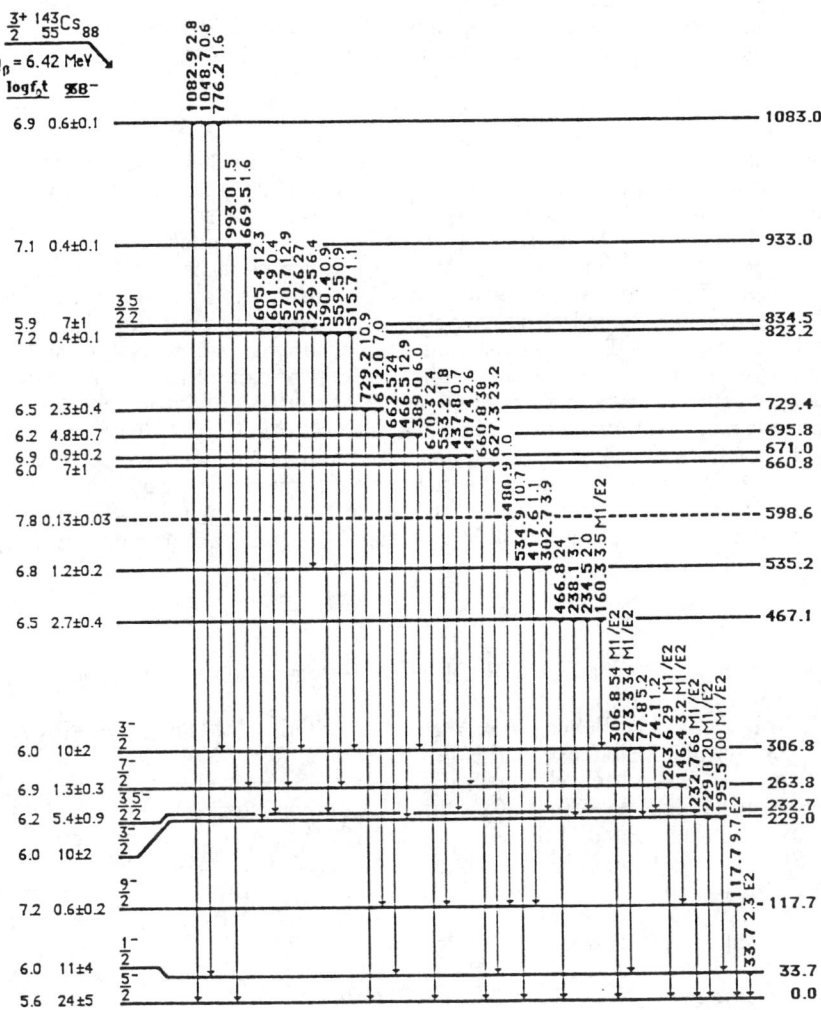

Figure 2. The lower portion of the decay scheme of ^{143}Ba. The beta branching was calculated assuming an absolute intensity of 0.126±0.02 for the 195-keV gamma.[9]

or 9/2⁻. A spin of 1/2⁻ is assigned to the 33-keV level based upon the strong beta branch from a 3/2⁺ parent. A spin of 9/2⁻ is assigned to the 117-keV level as the observed angular correlation for the 146-117 cascade rules out a spin 1/2 assignment (A_{22}=-0.122±0.032). The J^π of both the 229- and 306-keV levels is limited to 3/2⁻ by the two M1/E2 transitions from these levels to the 1/2⁻ 33-keV level and the 5/2⁻ ground state. Likewise, the M1/E2 multipolarity of both the 263-keV g.s. transition and the 146-keV transition[7] limits the J^π of the 263-keV level to 7/2⁻.

CONCLUSION

Little can be said about the structure of ^{147}Ce without further experimental work and definite spin and parity assignments. The great similarity between the low-lying structure of ^{145}Ba and ^{147}Ce does, however, suggest that the same group of nuclear states are responsible for the structure observed in these two isotopes.

Meyer et al. have proposed that the low-lying structure of the N=87 isotopes can be accounted for in terms of a coexisting $(f_{7/2})^{-3}$ cluster and a quasi-deformed $h_{9/2}$ band.[8] The downward trend of the 9/2⁻ state in the N=87 isotopes from 380 keV in ^{151}Gd to 117 keV in ^{143}Ba, however, runs counter to what is observed in the N=83 and N=85 isotopes. The energy of the $h_{9/2}$ orbital remains flat across the N=83 isotopes and actually drops towards Z=64, not Z=56, in the N=85 isotopes. A second alternative is that the structure of ^{143}Ba could arise from an $(f_{7/2})^{-3}$ cluster coupled to the quadrupole phonon. At the extreme coupling limit, the 9/2⁻ level would be pushed down and would track the drop in the quadrupole phonon energy across the N=86 isotopes towards Z=56. This approach, however, cannot account for the low-lying 1/2⁻ state at 33 keV as it is impossible to construct a spin 1/2 state from an $(f_{7/2})^{-3}$ cluster coupled to a 2⁺ phonon.

REFERENCES

1. G.A. Leander, W. Nazarewicz, P. Olanders, I. Ragnarsson, and J. Dudek, Phys. Lett. 152B, 284 (1985).
2. J.D. Robertson, S.H. Faller, W.B. Walters, R.L. Gill, H. Mach, A. Piotrowski, E.F. Zganjar, H. Dejbakhsh, and R.F. Petry, Phys. Rev. C 34, 1012 (1986).
3. F. Schussler, B. Pfeiffer, H. Lawin, E. Monnand, J. Munzel, J.A. Pinston, and K. Sistemisch, in Proceedings of the 4th Int. Conf. on Nuclei Far From Stability, ed. by P.G. Hansen and O.B. Nielsen (Skolen, Helsingor, 1981), pg. 589.
4. M. Shmid, Y.Y. Chu, G.M. Gowdy, R.L. Gill, H.I. Liou, M.L. Stelts, R.E. Chrien, R.F. Petry, H. Dejbakhsh, C. Chung, and D.S. Brenner, in Proceedings of the 4th Int. Conf. on Nuclei Far From Stability, ed. by P.G. Hansen and O.B. Nielscn (Skolen, Helsingor, 1981), pg. 576.
5. A.C. Mueller. F. Buchinger, W. Klempt, E.W. Otten, R. Neugart, C. Ekstrom, and J. Heinemeier, Nucl. Phys. A403, 234 (1983).
6. J.D. Robertson, Ph.D. thesis, University of Maryland, 1986 (unpublished).
7. M.S. Rapaport and A. Gayer, Int. Jour. Appl. Radiat. and Isotopes 36, 689 (1985).
8. R.A. Meyer, J.T. Meadows, and E.S. Macias, J. Phys. G8, 1413 (1982).
9. B. Sohnius, M. Brugger, H.O. Denschlag, and B. Pfeiffer, Radiochimica Acta 37, 125 (1984).

DECAY STUDIES OF NEUTRON DEFICIENT RARE EARTH ISOTOPES WITH OASIS*

J. Gilat,** J.M. Nitschke, P.A. Wilmarth, K. Vierinen*** and R.B. Firestone
Lawrence Berkeley Laboratory, University of California, Berkeley, CA 94720

ABSTRACT

We report results on the decay of ^{124}Pr, ^{124,125}Ce, ^{124,125}La, $^{134-136}$Eu, $^{134-136}$Sm, $^{134-136}$Pm, ^{144}Ho, ^{141,142,144}Dy, ^{140,141,142,144}Tb, $^{140-142}$Gd, and $^{140-142}$Eu, produced by ^{92}Mo(H.I.,xpyn) reactions at the Berkeley SuperHILAC, and studied with the OASIS on-line mass separator facility. Half-lives, delayed proton branching ratios, γ-ray energies and intensities, partial decay schemes and several J^π assignments are presented. Level systematics of the even mass Nd and Sm isotopes and of the $\nu h_{11/2}$ - $\nu s_{1/2}$ isomers for N=77 are discussed.

INTRODUCTION AND EXPERIMENTAL

The OASIS on-line mass separator facility at the Berkeley SuperHILAC[1] is imminently suited for the study of highly neutron deficient, short-lived nuclides. The salient features of the system are shown in Fig.1. The ready availability of a wide variety of heavy ion beams from the SuperHILAC, including sufficiently intense beams (about 0.1 particle microamperes) of light isotopes of the first transition metal group (e.g. ^{58}Ni, ^{54}Fe, ^{52}Cr, ^{46}Ti, etc.) provides the means of producing, via the compound nucleus mechanism, adequate quantities of highly neutron deficient species, even though their production cross-sections are often less than 1mb. The high temperature (>2800 °C) surface ionization ion source, high ionoptical transmission and computer stabilized H.V. and magnet power supplies of the mass separator permit a rapid and interference-free separation of the desired mass chain. Unknown masses are extrapolated from known, stable isotopes via an NMR mass meter. Finally, the electrostatic ion deflector and fast programmable tape transport system enable time resolved studies of the desired species by an array of particle, x-ray and γ-ray detectors, in singles and coincidence modes, under relatively low background conditions in a shielded spectroscopy laboratory.

In the framework of a systematic investigation of neutron deficient rare earth isotopes, we have observed and characterized about 60 isotopes and isomers, ranging from Z=57 (La) to Z=71 (Lu) and with 124≤A≤152. About half of these species were previously unknown. For the others, the body of available decay data has been substantially extended.

In this brief report we present results for about 30 of these nuclides, produced by the interaction of a variety of beams with enriched ^{92}Mo targets. The assignment of different decay modes and half-lives observed in a given mass chain to the relevant species is based, in all cases, on the observation of appropriate characteristic x-rays. The results include half-lives, γ-ray energies and relative intensities and, for some cases, partial decay schemes and delayed proton branching ratios.

---

* This work was supported by the Director, Office of Energy Research, Division of Nuclear Physics of the Office of High Energy and Nuclear Physics of the U.S. Department of Energy under Contract DE-AC03-76SF00098.
** On leave from Soreq Nuclear Research Center, Yavne 70600, Israel.
*** On leave from University of Helsinki, SF 00170, Finland.

© American Institute of Physics 1988

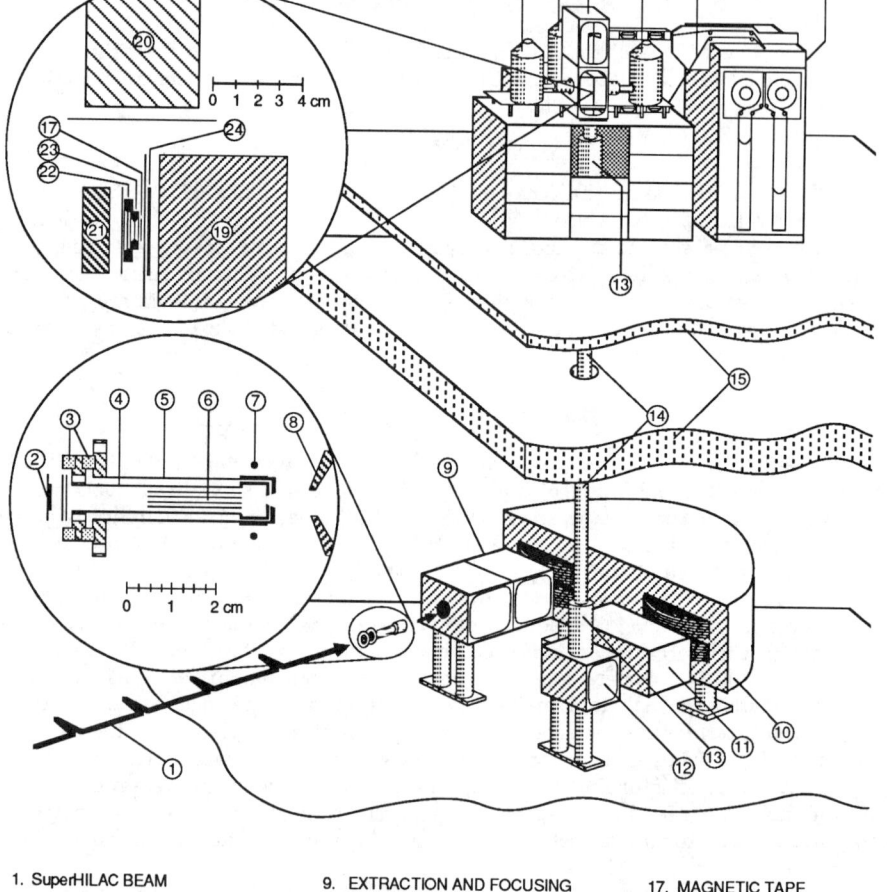

1. SuperHILAC BEAM
2. TARGET
3. INSULATORS (BeO)
4. ION SOURCE ANODE (Ta)
5. ION SOURCE CATHODE (Ta)
6. CAPILLARY TUBES (Ta)
7. EB FILAMENT (Ta)
8. EXTRACTION ELECTRODE
9. EXTRACTION AND FOCUSING
10. ANALYZING MAGNET
11. FOCAL PLANE DETECTOR BOX
12. ELECTROSTATIC MIRROR
13. ELECTROSTATIC QUADRUPOLE
14. TRANSFER LINE
15. CONCRETE SHIELDING
16. TAPE DRIVE (IBM 729)
17. MAGNETIC TAPE
18. DETECTOR BOX
19. N-TYPE Ge DETECTOR (52%)
20. N-TYPE Ge DETECTOR (24%)
21. HPGe DETECTOR
22. 718 µm Si DETECTOR
23. 13.8 µm Si DETECTOR
24. 1mm PILOT F SCINTILLATOR

Fig. 1. A simplified representation of the OASIS mass separator online at the Lawrence Berkeley Laboratory SuperHILAC. The separator and tape system are approximately to scale. The major components are highlighted above.

RESULTS

1. Mass 124

^{124}Pr, ^{124}Ce and ^{124}La were produced via the ^{92}Mo (^{36}Ar, xpyn) reaction at 174 MeV, with x = 1,2, & 3 and y = 3, 2, & 1 respectively.

^{124}Pr: This new delayed proton precursor is produced with a very low cross section, so we could only measure its proton decay half-life of 1.2(2) s and assign one γ-ray of 142 keV to its decay[2]. Our interpretation of this γ-ray as the $2^+ \rightarrow 0^+$ transition in ^{124}Ce is confirmed by recent in-beam data[3].

^{124}Ce: In ref. 4, a 6(4) s decay of La x-rays was attributed to ^{124}Ce. We assign γ-rays of 120, 253, 544 and 560 keV to this decay. The intensity of these γ-rays is quite low with respect to the ~6 s component of the 511 keV annihilation peak or the La K$_\alpha$ and K$_\beta$ x-rays. This indicates that the even-even ^{124}Ce decays predominantly to a low spin (1$^+$?) lanthanum isomer, not observed in our experiments.

^{124}La: Our measured half-life of 30(2) s agrees with that quoted by ref. 4. The following γ-rays (energies in keV, relative intensities in brackets) are associated with this decay: 229.8 [100], 421.2 [95], 576.3 [63], 694.5 [11], 1033 [~10] and 1262 [~10]. The first four lines correspond, respectively, to the $2^+ \rightarrow 0^+$, $4^+ \rightarrow 2^+$, $6^+ \rightarrow 4^+$ and $8^+ \rightarrow 6^+$ transitions in the ground state rotational band of ^{124}Ba, while the other two establish a second 2^+ level at 1262 keV. This decay pattern, shown in Fig. 2, is a clear indication of the high spin nature (7$^+$?) of the parent ^{124}La. No direct evidence for the decay of the low spin (1$^+$?) isomer produced in the decay of ^{124}Ce could be found.

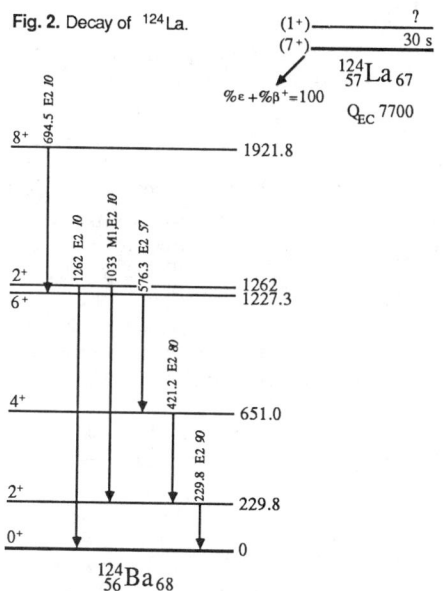

Fig. 2. Decay of ^{124}La.

2. Mass 125

The reactions ^{92}Mo(^{36}Ar,2pn) and ^{92}Mo(^{36}Ar,3p) at 152 MeV were used to produce ^{125}Ce and ^{125}La. The analysis of the data is incomplete, and only preliminary results are presented.

^{125}Ce: Bogdanov et al.[4] attributed a La x-ray activity decaying with a half-life of 11(4) s to ^{125}Ce. In ref. 2 we reported a delayed proton branch in ^{125}Ce and measured its half-life as 10(1) s. From the proton branching ratios to the 2^+, 4^+ and 6^+ levels in ^{124}Ba we assigned a spin of $5/2^+$ to the ground state of ^{125}Ce. The γ-ray spectrum attributed to this decay is very complex: over 70 γ-rays between 50 and 2550 keV were tentatively identified by half-life analysis and coincidence with La x-rays. The strongest γ-rays at 194, 325, 370 and 379 keV decay with a half-life of 14(1) s, rather than 10(1) s obtained from delayed proton and La x-ray decays. This difference of half-lives is probably due to an isomer feeding the same set of levels in ^{125}La, but its exact nature has not yet been elucidated.

^{125}La: Our half-life of 70(3) s agrees with that of ref. 4. In addition to the 43.7 and 67.2 keV γ-rays assigned in ref. 4, we have identified 30 γ-rays between 70 and 1270 keV. The transitions at 98.8, 134.0, 168.6, 216.7, 232.7 and 384.4 keV were also observed in ref. 5 and attributed to the $h_{11/2}$ and $g_{7/2}$ rotational bands.

3. Mass 134

The reactions ^{92}Mo(^{46}Ti,xpyn) with x = 1, 2 & 3 and y = 3, 2 & 1 at 212 MeV were used to produce ^{134}Eu, ^{134}Sm and ^{134}Pm. Only a 4s tape cycle, optimized for the study of ^{134}Eu, was used, so that only an estimate of the half-life of ^{134}Sm could be obtained, and the half-life of ^{134}Pm could not be measured.

^{134}Eu: This new delayed proton precursor was produced with a very low cross section, so we could only measure its half-life of 0.5(2) s. Based on production cross section systematics, this is probably the high spin ($\pi h_{11/2}$) member of the isomeric pair typical of most even mass Eu isotopes.

^{134}Sm: Our estimated half-life of 10(2) s is consistent with the value of 12(3) s of ref. 6. The following nineteen γ-rays were assigned to this decay (energies in keV, relative intensities in brackets): 51 [1.6], 105 [3], 107 [1], 110 [11], 112 [7], 117 [3], 119 [100], 130 [2], 141 [3], 162 [11], 186 [7], 219 [28], 224 [2], 229 [12], 257 [6], 280 [14], 299 [18], 380 [16], 419 [9], and 15 of them placed in a decay scheme with levels in ^{134}Pm at 112.4, 118.9, 229.0, 280.0, 304.9, 409.5, 419.0, 537.5 keV. Since they are populated in the decay of the even-even ^{134}Sm these are presumably low (J=1,2) spin states.

^{134}Pm: Twenty one γ-rays between 290 and 1750 keV were identified with this decay; eighteen were placed in a decay scheme comprising levels in ^{134}Nd at 294.4(2$^+$), 735.9(2$^+$), 789.3(4$^+$), 1089.1(3$^+$), 1318.8(4$^+$), 1384.0, 1420.5(6$^+$), 1541.8, 1605.9, 1669.4, 1697.7, 1956.4 and 2231.9 keV. The levels at 294, 789 and 1421 keV are members of the ground state rotational band; the levels at 736, 1089 and 1319 keV appear to belong to a γ-vibrational band. This part of the decay scheme is in agreement with that recently proposed by Kortelahti et al.[7], but our placement of some of the higher lying levels differs from theirs. The population of the 6$^+$ level in ^{134}Nd suggests a J$^\pi$ = 5$^+$ for the 25s ^{134}Pm parent. No direct evidence for the decay of the low spin ^{134}Pm species produced in the decay of ^{134}Sm was observed.

4. Mass 135

^{135}Eu, ^{135}Sm and ^{135}Pm were produced by ^{92}Mo(^{46}Ti,xpyn) with x = 3, 2 & 1 and y = 0, 1, & 2 respectively, at 192 MeV.

^{135}Eu: A half-life of 1.5(2) s for this new isotope was determined from the decay of the Sm K x-rays, and a weak γ-ray at 120.8 keV was assigned to its decay. Delayed proton emission, though energetically possible, was not observed, consistent with the systematics of other even N precursors.

^{135}Sm: Our measured half-life for this isotope is 10(1) s, in agreement with ref. 6. The following 22 γ-rays were assigned to this decay (energies in keV, relative intensities in brackets): 49 [15], 55 [100], 77 [50], 105 [70], 116 [6], 124 [5], 127 [68], 160 [2], 182 [17], 190 [77], 237 [80], 286 [67], 313 [16], 341 [12], 351 [15], 363 [65], 418 [20], 428 [54], 543 [20], 573 [35], 755 [35], 1132 [17]. From our data we extract a delayed proton branching ratio of ~5x10^{-4} for this isotope. The 2$^+ \to$ 0$^+$ 294 keV transition in ^{134}Nd (but not the 495 keV 4$^+ \to$ 2$^+$ transition) was observed to be in coincidence with the protons. This indicates a low spin (1/2 or 3/2) for the parent ^{135}Sm.

^{135}Pm: About 40 γ-rays between 100 and 1200 keV were assigned to the decay of ^{135}Pm. The tape cycle of 40 s optimized for ^{135}Eu was too short for a reliable half-life determination. However, two distinct decay patterns of the strong γ-rays could be discerned: The 199 keV γ-ray, identified as an 11/2$^- \to$ 9/2$^-$ transition in ^{135}Nd and associated with the decay of a high spin (11/2$^-$) ^{135}Pm species[8], decays with a single component half-life of ~50 s. Most other γ-rays, such as 129, 208, 245, 263, 270 and 398 keV exhibit a complex growth-decay pattern, and are probably associated with a low spin Pm species produced in the decay of the low spin ^{135}Sm.

5. Mass 136

The reactions ^{92}Mo(^{46}Ti,pn) and ^{92}Mo(^{46}Ti,2p) at 192 MeV were used to produce

^{136}Eu and ^{136}Sm.

^{136}Eu: Twenty eight γ-rays between 100 and 1500 keV were assigned to the decay of ^{136}Eu; the lines at 255 [100], 432 [34], 535 [9] and 576 [10] represent respectively the $2^+ \to 0^+$, $4^+ \to 2^+$, $6^+ \to 4^+$ and $8^+ \to 6^+$ transitions in ^{136}Sm, indicating a high (7^+?) J^π assignment for the parent. Excess intensity of the $2^+ \to 0^+$ transition indicates however that a low spin ^{136}Eu species is also present in high abundance. This conclusion is borne out by the observation that the $4^+ \to 2^+$ transition decays with a significantly shorter half-life than the $2^+ \to 0^+$. We are thus led to postulate two ^{136}Eu species: ^{136m}Eu ($J^\pi = 7^+$?) with a half life of 3.2(5) s and ^{136g}Eu ($J^\pi = 1^+$?) with a half-life of approximately 5 s. The copious production of the low spin member in a heavy ion reaction suggests an IT path between the two species, but the relevant transitions have not been identified. Both species appear to be delayed proton precursors. Analysis of the γ-rays associated with this decay indicates a γ-vibrational band based on a second 2^+ level at 713 keV in ^{136}Sm.

^{136}Sm: The tape cycles, optimized for ^{136}Eu, were too short for a half life determination for this isotope. About 40 gamma transitions between 90 and 800 keV were assigned to its decay, most of them of low energy and roughly the same low intensity. The only exception is a strong line at 114.5 keV, about equal in intensity to the sum of all other lines. A $J^\pi = 1^+$ assignment to the 114.5 keV level in ^{136}Pm is indicated by the strong GT β^+ branch to this level. This, in turn, establishes the ground state of the ^{136}Pm species as 2^+, which is not identical with the 1.8 m 5^+ species reported in ref. 8. The existence of a $5^+/2^+$ isomeric pair is not unexpected for Pm isotopes.

6. Mass 140

^{140}Tb, ^{140}Gd, and ^{140}Eu were produced via the ^{92}Mo(^{54}Fe,xpyn) reaction at 298 MeV with x=3,4 & 5 and y=3,2 & 1. The analysis is incomplete, and no absolute normalizations are yet available. ^{140m}Eu was also produced by ^{92}Mo(^{52}Cr,3pn).

^{140m}Tb: Proton emission has been assigned to this isotope on the basis of coincidence with Gd K x-rays. The absolute proton branching ratio is $7\pm2\times10^{-3}$. γ-rays of 328.4 and 627.8 keV were observed and assigned as the $2^+ \to 0^+$ and $6^+ \to 4^+$ transitions respectively. The $4^+ \to 2^+$ transition is known to exist at 508 keV and was obscured by an intense annihilation peak. The 328.4 keV transition is twice as intense as the 627.8 keV γ-ray and no evidence of an $8^+ \to 6^+$ transition was found. We tentatively assign the isomeric decay to a 6^- isomer, predicted by systematics. The half-life of 2.4(2) s is adopted for this isotope. No evidence for a low spin ^{140g}Tb was observed.

^{140}Gd: 38 γ-rays between 160 and 1150 keV have been assigned to the decay of ^{140}Gd. A half-life of 15.8(4) s is adopted for this isotope. A partial decay scheme is shown in Fig. 3.

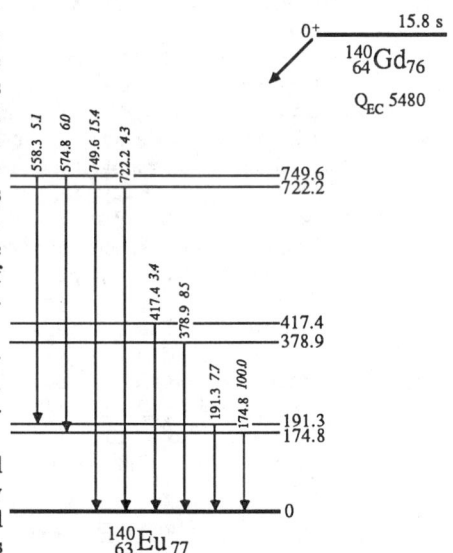

Fig. 3. Decay of ^{140}Gd.

^{140m}Eu: 174.8- and 185.2 keV γ-rays are observed with a half-life of 125(2) ms. The 174.8 keV transition is strongly populated by ^{140}Gd decay and is presumed to deexcite a 1^+ level of that energy. The 185.2 keV γ-ray is only weakly populated by ^{140g}Gd decay and is not in coincidence with the 174.8 keV γ-ray. The 185.2 keV γ-ray is intense and must deexcite to the ground state. We thus presume that both levels are populated by unobserved low-energy transitions from an isomer with a half-life of 125 ms.

^{140g}Eu: A half-life of 1.51(2) s was adopted for this isotope. 18 γ-rays between 190 and 2000 keV have been assigned to its decay and 15 have been placed in a decay scheme comprising levels in ^{140}Sm at 531, 991, 1246, 1442, 1599, 2283, 2309 and 2482 keV. The γ-rays populated in this decay are also observed in equilibrium with ^{140}Gd decay, consistent with production from a low-spin ^{140}Eu parent. However, the decay scheme is dominated by feeding to the 2^+ level at 531 keV and to excited levels which deexcite through the 2^+ but not the ground state. The 4^+ level is only weakly populated by γ-ray deexcitation and no evidence for excited 2^+ states was obtained. We cannot reconcile the ^{140g}Eu spin at this time; the most likely values are 0^-, 1^-, or 2^- with considerable unobserved decay intensity accounting for the apparent 2^+ feeding. Such a spin would be inconsistent with the systematics of the heavier Eu isotopes and might indicate a breakdown in the Z=64 semi-magic shell. No evidence could be found for a $\sim$20s ^{140m}Eu reported by Habs et al.[9], presumably due to an erroneous assignment of ^{140}Gd.

7. Mass 141

^{141}Dy, ^{141}Tb, ^{141}Gd and ^{141}Eu were produced by the ^{92}Mo(^{54}Fe,xpyn) reaction at 276 MeV, with x = 2, 3, 4 & 5 and y = 3, 2, 1 & 0, respectively. ^{141}Gd and ^{141}Eu were also produced by the ^{92}Mo(^{52}Cr,2pn) and ^{92}Mo(^{52}Cr,3p) reactions at 210 MeV.

^{141}Dy: This 0.9(2) s delayed proton precursor was reported in an earlier publication.[2] No further details of its decay features could be observed, due to the low production cross section.

^{141}Tb: The half life of this new isotope was determined as 3.5(2) s, and 35 gamma transitions between excited levels of ^{141}Gd were assigned to its decay, and fitted to a decay scheme comprising levels at 113, 198, 258, 378, 492, 514, 552, 646, 662, 753, 758, 895, 940, 990, 1100 and 1132 keV. The lower portion of this level scheme can be seen in Fig. 4. Most of the levels populated by the β^++EC decay of ^{141}Tb, decay to the $11/2^-$ isomeric state at 378 keV, and only very few transitions cross this level to the low spin structures near the ground state of ^{141}Gd. This establishes the high spin nature of the 3.5 s Tb species. No indication of a low spin ^{141}Tb could be found in the data.

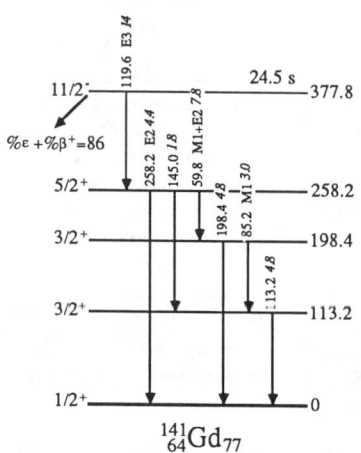

Fig. 4. IT decay of ^{141m}Gd.

^{141}Gd: Redon et al.[10] recently reported the discovery of ^{141}Gd with a half-life of 22(3) s, assigned six γ-rays to its decay and proposed a partial decay scheme. Our much more comprehensive data clearly show the presence of two distinct species: a 24.5(9) s high spin $(11/2^-)$ isomer at 378 keV, decaying by both β^+ + EC (86%) to (mostly) the $11/2^-$ 96 keV isomeric state in ^{141}Eu and by an isomeric transition (14%) a low spin $(1/2^+)$ ground state, which decays by β^+ + EC to the $5/2^+$ ground state of

^{141}Eu. The half life of ^{141g}Gd is about 20 s; a more precise value is difficult to extract from our data. The decay schemes of the two isomers are shown in Fig. 5. Thirty six of the forty γ-rays assigned to these decays have been placed with no overlap between the decay of the two species. The delayed protons that were observed in the decay of ^{141}Gd appear by half life analysis and lack of coincident γ-rays to originate from the low spin ground state, with a branching ratio of 3±1 x 10⁻⁴.

^{141}Eu: Our data on the decay of the 40 s ^{141g}Eu agree almost completely with those of Deslauriers et al.[11], and will not be elaborated here. The results regarding the 11/2⁻ isomeric state at 96.4 keV, however, differ significantly from those of Ref. 9. Our half-life value is 2.7(3) s compared to 3.3(3) s[11], but, more importantly, the β^+ + EC branching ratio is only about 7% instead of 67%, resulting in a logft value of 5.1 for the $\pi h_{11/2} \to \nu h_{11/2}$ β transition, rather than the unusually low value of 4.1 suggested by ref. 11. Since in our work ^{141}Sm is produced only by decay of ^{141}Eu, the activity ratio ^{141m}Sm/^{141g}Sm is a direct and reliable measure of the above branching ratio.

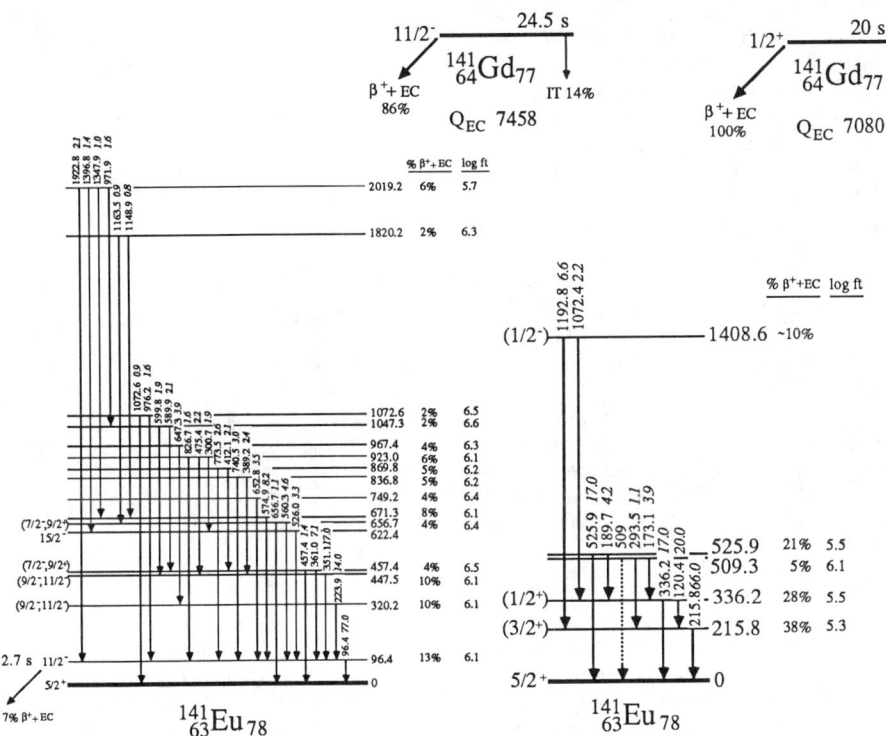

Fig. 5. Decay of ^{141g}Gd and ^{141m}Gd.

8. Mass 142

^{142}Dy, ^{142}Tb and ^{142}Gd were produced via the ^{92}Mo(^{54}Fe,xpyn) reaction at 250 MeV with x=2,3,4 and y=2,1 & 0 respectively. ^{142g}Eu was obtained from the decay of ^{142}Gd. ^{142}Tb was also procuded by ^{92}Mo(^{52}Cr,p3n). These isotopes formed a series of equilibrium pairs which together with K x-ray and annihilation data allowed the determination of absolute EC, β^+, and γ-ray intensities.

^{142}Dy: Proton emission was assigned to this decay on the basis of coincidence with Tb K x-rays[2]. The absolute proton decay branching ratio is ~8x10⁻³. A 181.9 keV γ-ray, deexciting a level of the same energy was observed with an absolute intensity of 4.3±1.2 per 100 decays. The total EC/β⁺ ratio for ¹⁴²Dy decay is 0.111±0.005. A half-life of 2.3(3) s is adopted for this isotope.

¹⁴²ᵐTb: This isotope decays by IT with a half-life of 303±7 ms. γ-rays of 181.9 (also seen in the decay of ¹⁴²Dy) and 211.6 keV are associated with this decay. The decay scheme for this isotope is given in Fig. 6.

¹⁴²ᵍTb: Proton emission was assigned to ¹⁴²ᵍTb on the basis of coincidence with Gd K x-rays. The absolute proton decay branch from this isotope is ~3x10⁻⁵. A partial decay scheme for this isotope is shown in the figure. The absolute intensity of the 515.3 keV γ-ray is 24.9±1.7 per 100 decays. The total EC/β⁺ ratio for ¹⁴²ᵍTb decay is 0.033±0.004. A half-life of 597(17) ms is adopted for this isotope.

¹⁴²Gd: 42 γ-rays between 100 and 1800 keV; were assigned to this decay and placed in a decay scheme comprising levels in ¹⁴²Eu at 178.8, 280.3, 284.4, 496.7, 503.1, 526.2, 585.8, 591.3, 614.5, 619.8, 631.7, 660.9, 935.7, 1412.7, 1438.4, 1480.8, 1485.8 and 1779.1 keV. The 178.9 keV γ-ray has an absolute intensity of 11.2±1.1 per 100 decays. The EC/β⁺ ratio for ¹⁴²Gd decay is 1.08±0.11 per 100 decays. A half-life of 70.2(6) s was adopted for this isotope.

¹⁴²ᵐEu: The previously known[12] decay scheme for this 2.4(2) s isotope was confirmed. The absolute intensity of the 768.0 keV γ-ray is 10.2±0.7 per 100 decays and the total EC/β⁺ ratio is 0.112±0.018.

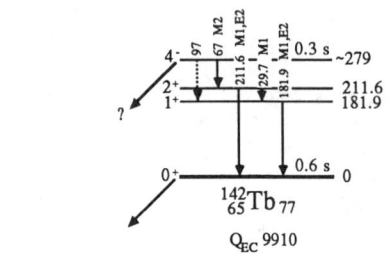

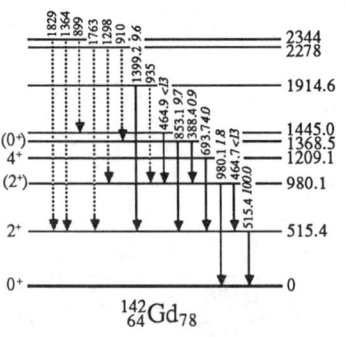

Fig. 6. Decay of ¹⁴²ᵍ⁺ᵐTb.

9. Mass 144

¹⁴⁴Ho, ¹⁴⁴Dy and ¹⁴⁴Tb were produced by the ⁹²Mo(⁵⁸Ni,xpyn) with x=3,4 & 5 and y=3,2 & 1 respectively and by the ⁹²Mo(⁵⁶Fe,xpyn) reaction with x=1,2 & 3 and y=3,2 & 1, respectively. The decays of ¹⁴⁴Ho, ¹⁴⁴Dy, and ¹⁴⁴ᵐ⁺ᵍTb were investigated. K x-ray and annihilation intensities and equilibrium information were exploited to determine absolute γ-ray intensities. Apart from minor differences, the results agree with those of refs. 10, 13 and 14

¹⁴⁴Ho: Proton emission was assigned to this isotope on the basis of coincidence with Dy K x-rays. The half-life for the protons was 0.7(1) s[2].

¹⁴⁴Dy: A half-life of 9.1(5) s was determined for ¹⁴⁴Dy. Proton emission was assigned to this isotope on the basis of coincidence with Tb K x-rays[2]. The absolute proton decay branching ratio is 6±1x10⁻³. 21 γ-rays were identified with ¹⁴⁴Dy decay populating levels at 0, 196.5, 298.7, 396.7, 469.5, 532.2, 615.9, 620.0, 774.4, 793.3, and 1237.2 keV. The absolute intensity of the 298.7 keV gamma ray is 13.7±1.1 per 100 decays.

^{144m}Tb: This isomer was determined to decay by IT (62%) and EC+β^+ (38%), with a half-life of 4.1(1) s. No evidence for proton emission was observed. 15 γ-rays were assigned to ^{144m}Tb decay. The absolute intensity of the 743 keV γ-ray is 38±3 per 100 decays of ^{144m}Tb.

^{144g}Tb: Decay to levels at 0(0$^+$), 743.0(2$^+$), 1877.2(2$^+$), and 1886.8(0$^+$)keV was observed. The 1144 keV γ-ray deexciting the 1886.8 keV level was determined to have an absolute intensity of 5.3±0.3 per 100 decays. No half-life for ^{144g}Tb could be determined because it was produced in equilibrium with both ^{144}Dy and ^{144m}Tb.

CONCLUDING REMARKS

The experiments summarized in the foregoing section have yielded a large body of new spectroscopic data on previously unknown or poorly characterized highly neutron deficient rare earth isotopes. Almost all of the odd mass and odd-odd isotopes studied exhibit isomerism due to the prominence of the $h_{11/2}$ neutron and proton shells in this mass region. The heavy ion induced reactions, which are the only practical means of producing these isotopes, exhibit a strong preference for the production of the high spin member of each isomeric pair. However, our practice of simultaneously following several members of a given isobaric decay chain, in and out of equilibrium, enabled us to study also the low spin species or, at least, infer their existence. The nuclear structure information underlying our data has not yet been fully elucidated. A pertinent problem is the persistence of the Z=64 closed shell as one moves away from N=82 shell, or the predicted onset of prolate or triaxial deformation below N=78[15-17]. The apparently anomalous J^π of the ground state of ^{140}Eu suggested by our data may indicate such an effect. In contrast, Fig. 7 shows the systematics of the $\nu h_{11/2} - \nu s_{1/2}$ isomers for N=77. Our newly established levels in ^{141}Gd follow closely the systematics of analogous levels in the other isotones[8,18]. This indicates that the Z=64 shell has little effect on the spectroscopy of these neutron levels. However, the B(E3) transition rate for the 120 keV 11/2$^-$ → 5/2$^+$ transition is only ~9x10^{-4} Weisskopf Units, i.e. about 5-10 times slower than the rate observed for the other three isotopes.

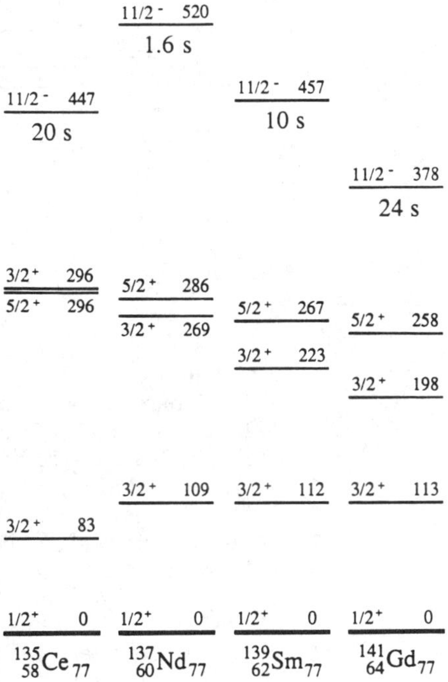

Fig. 7. Level systematics of N=77 isotones.

Fig. 8 shows the systematics of low-lying levels in even Nd and Sm isotopes. A smooth downward trend with increasing deformation is observed for both the ground state rotational band and for the γ-vibrational band. The only significant deviation is the position of the second 2$^+$ level (and the associated γ-vibrational band) in ^{132}Nd. This is in contrast with the smooth behavior of the samarium isotopes and throws a considerable doubt on the interpretation of the ^{132}Nd spectrum, recently proposed by ref. 7.

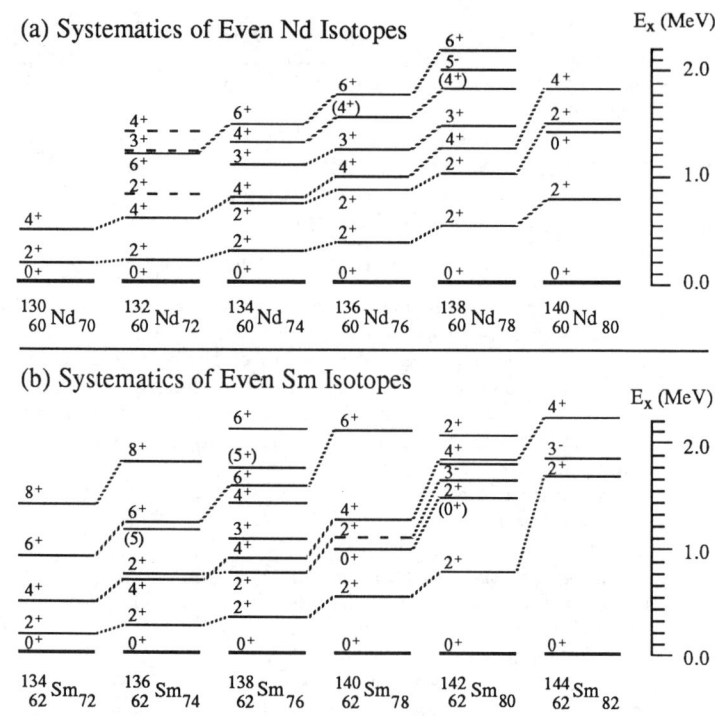

Fig. 8. Level systematics of even Nd and Sm isotopes.

REFERENCES

1. J.M. Nitschke, Nucl. Instr. and Meth. 206, 301 (1983)
2. P.A. Wilmarth et al., Z. Phys. A325, 485 (1986)
3. K.L. Ying et al., J. Phys. G: Nucl. Phys. 12, L211 (1986)
4. D.D. Bogdanov et al., Nucl. Phys. A307, 425 (1978)
5. J. Gizon et al., Z. Phys. A285, 259 (1978)
6. D.D. Bogdanov et al., Nucl. Phys. A275, 425 (1977)
7. M. Kortelahti et al., Z. Phys. A327, 231 (1987)
8. C.M. Lederer and V.S. Shirley, Table of Isotopes 7th ed., Wiley (1978)
9. D. Habs et al., Z. Phys. A250, 179 (1972)
10. N. Redon et al., Z. Phys. A325, 127 (1986)
11. J. Deslauriers et al., Z. Phys. A283, 33 (1977)
12. G.G. Kennedy et al., Phys. Rev. C12, 553 (1975)
13. D.C. Sousa et al., Phys. Rev. C25, 1012 (1982)
14. E. Nolte et al., Z. Phys. A306, 223 (1982)
15. I. Ragnarsson et al., Nucl. Phys. A233, 329 (1974)
16. G.A. Leander et al., Phys. Lett. 110B, 17 (1982)
17. J.A. Cizewski et al., Phys. Lett. 175, 11 (1986)
18. J. Deslauriers et al., Z. Phys. A325, 421 (1986)

DECAYS OF NEUTRON DEFICIENT GADOLINIUM ISOTOPES

R. Turcotte, H. Dautet, S.K. Mark and N. de Takacsy
McGill University, Montréal, Québec, Canada

E. Hagberg, V.T. Koslowsky, J.C. Hardy, H. Schmeing and X.J. Sun
Atomic Energy of Canada Ltd., Chalk River Nuclear Laboratories
Chalk River, Ontario, Canada K0J 1J0

INTRODUCTION

Nuclei in the neighbourhood of ^{146}Gd have been the subject of much experimental and theoretical interest because of their proximity to a proposed shell closure at Z=64. We report here β- and γ-decay studies of neutron deficient Gd and Eu isotopes, with the aim of determining the significance and evolution of this "shell closure" and its influence on the nuclear shape.

Our new spectroscopic data agree with known systematic trends in this region of nuclei. Calculations following the Hartree-Fock-plus-BCS method reproduce reasonably well these trends as a function of neutron and proton number. They indicate that although there is a slight increase in the proton gap as one approaches Z=64 this is by no means a magic shell. Furthermore, the nuclei studied were found to be slightly deformed and gamma soft.

EXPERIMENTAL METHOD

<u>Production</u>: The major part of this work was performed with the K=100 synchrocyclotron at the Foster Radiation Laboratory (McGill University, Montréal, Canada), the nuclei of interest being produced via the ^{144}Sm(^{3}He,xn)$^{147-x}$Gd reactions. However, due to the large number of exit channels opened with ^{3}He above 80 MeV bombarding energy, the ^{141}Eu and ^{140}Eu data (in particular the electron conversion spectra) were difficult to analyse. The study therefore was completed with ^{112}Sn(^{32}S,2pxn)$^{142-x}$Gd reactions, initiated by the MP Tandem at the Chalk River Nuclear Laboratories, which produced much cleaner sources. In both cases, the nuclear reaction products were brought in front of the detection apparatus by means of a gas jet transport system. At McGill, a mixture of 85% He, 15% C_2H_4 was utilized to carry the radioactivity, while at Chalk River, pure He was seeded with NaCℓ.

<u>Detection</u>: This work involved classical gamma-ray and electron spectroscopy. Standard gamma-ray detectors were used, ranging from small X-ray detectors to 25% HP germanium detectors. The detector for conversion electrons was Si(Li) with a 1.8 keV resolution as measured during on-line runs with gas-jet transported activity. Between it and the radioactive source, a mini-orange magnetic filter was positioned. The magnet configuration as well as the distances between source, filter and detectors were chosen to give a broad transmission curve from 50 keV to 1 MeV.

Identification of the various nuclei was performed through excitation functions, X-gamma-ray coincidences, and filiation to known activities.

EXPERIMENTAL RESULTS

Until recently, little has been known about nuclei in this region. During the last few years we have reported preliminary information on the decay of light gadolinium isotopes[1,2,3,4]). Except for ^{143}Eu, which has been studied in detail[5]), only sketchy information has been published by other groups[6,7,8]). In general, we are in agreement with these results. For reasons of brevity our results are presented here in the form of decay schemes (figs.1-4); more details will be presented in a forthcoming publication.

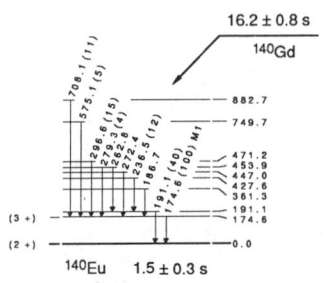

fig.1: ^{140}Gd decay scheme

fig.2: The decay of ^{141}Gd. Not shown are two isomeric states of ^{141}Tb, one with a 7.9 ± 0.6 s half-life, which decays predominantly to the 670.3 keV level in ^{141}Gd, the other with $T_{1/2}$ = 4.5 ± 0.8 s which mostly populates the 258.3 keV level.

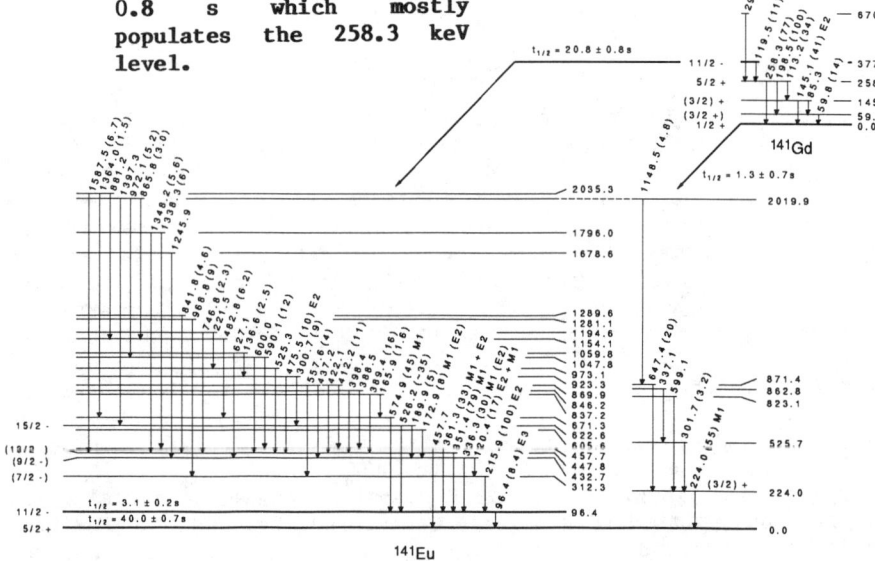

DISCUSSION

Deformed Hartree-Fock-plus-BCS[9,10]) calculations have been performed for the N=76 and N=78 isotone chains. These preliminary calculations assume an axially symmetric deformation and use semi-realistic matrix elements derived from the Tabakin[11,12]) potential. The neutrons and protons are restricted to the five orbitals between the magic numbers 50 and 82: these are $1d_{5/2}$, $0g_{7/2}$, $2s_{1/2}$, $1d_{3/2}$ and $0h_{11/2}$ orbitals. The single-

fig.3:

¹⁴²Gd decay scheme

fig.4:

¹⁴⁴Gd decay scheme

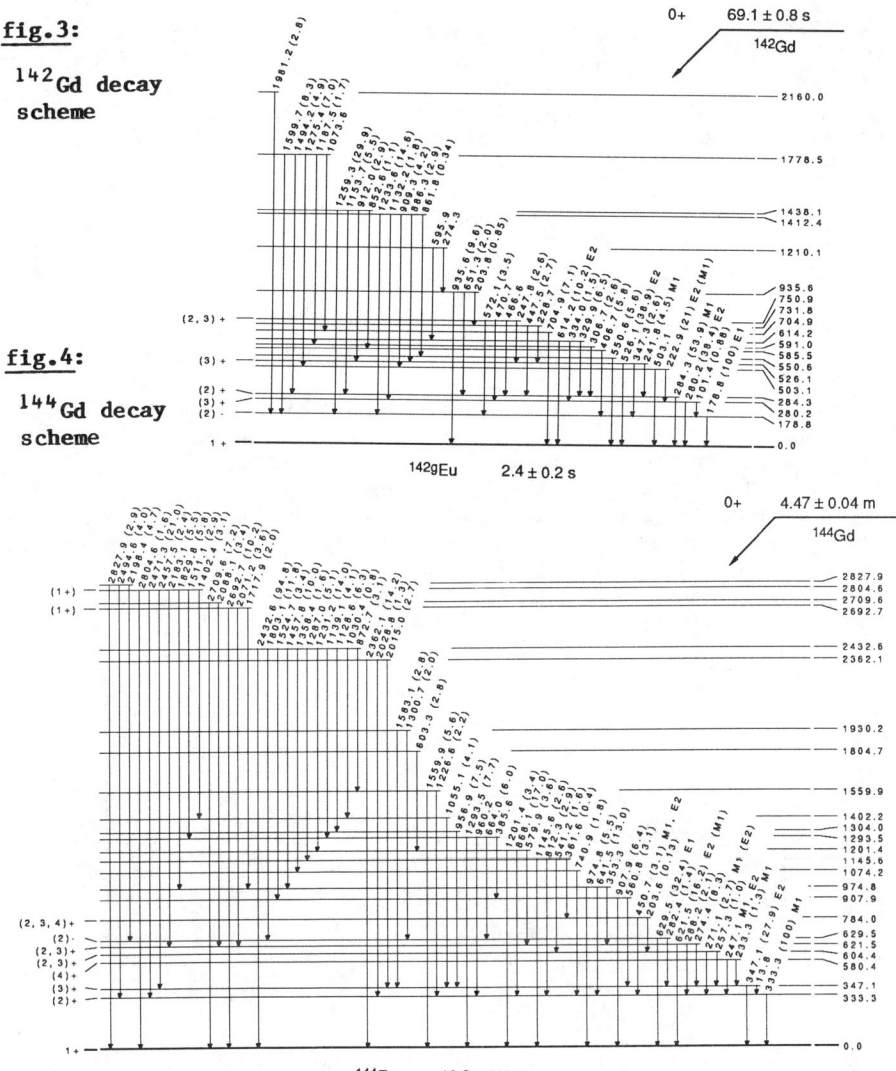

particle energies are held fixed and are obtained by fitting the odd-A, Z=50 (for neutrons) and odd-A, N=82 (for protons) known single quasi-particle states. In fitting the N=82 single-particle energies, it was found that the data could be reproduced with a modest 2.5 MeV proton energy gap, which does not qualify ¹⁴⁶Gd as a doubly magic nucleus.

The model was tested in the barium (Z=56) region where good data are available for both low-spin and high-spin states. It was found that the structure of ¹³²Ba (N=76) and ¹³⁴Ba (N=78) were fairly well reproduced and that the calculated BE(2;0⁺→2⁺) values were in good agreement with the experimental data.

Even though the model does not include triaxiality, it does

illustrate the shape evolution in this region. The N=76 and 78 isotones are found to be γ-soft. Both prolate and oblate solutions could be obtained in all cases. As can be seen in fig.5, prolate shapes seem to be favoured for 50<Z<64. This is in agreement with the observation of a decoupled band constructed on the 11/2⁻ orbital. For Z>64, the model predicts a transition to oblate shapes as the $0h_{11/2}$ proton shell-model orbital starts to be occupied. This could correspond to a change in the orientation of the angular momentum with respect to that of the symmetry axis.

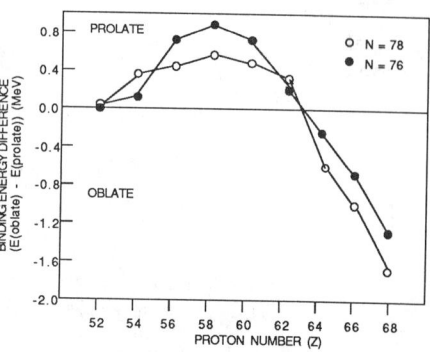

fig.5: Evolution of the preferred nuclear shape.

As an example of the results obtained with this calculation, the experimental and theoretical level spectra of ^{141}Gd and ^{143}Gd are compared in fig.6. One clearly sees that the best solution would likely be a compromise between the oblate and prolate solutions, which would thus lead to a triaxial picture.

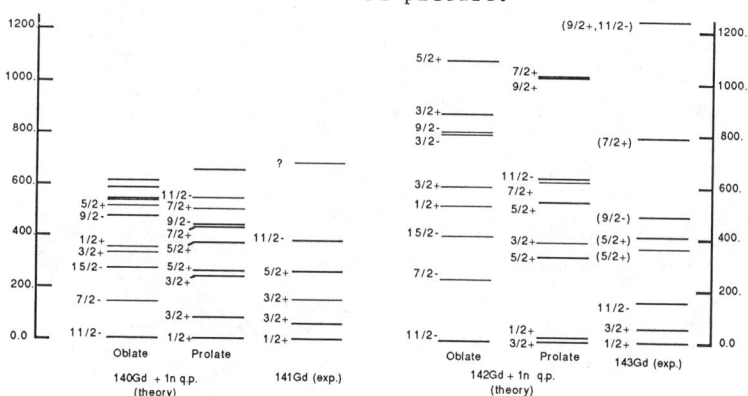

fig.6: Comparison between theory and experiment for ^{141}Gd and ^{143}Gd.

1)2)3) Progress Reports 1984-85-86, Foster Radiation Laboratory, Department of Physics, McGill University.
4) R. Turcotte et al., C.A.P. Conference, Université de Sherbrooke, June 1984, Physics in Canada **40** (1984) 69.
5) R.B. Firestone et al., Phys. Rev. **C17** (1978) 718.
6) S. Lunardi et al., Z. Phys. **A324** (1986) 433.
7) N. Redon et al., Z. Phys. **A325** (1986) 127.
8) P.A. Wilmarth et al., Z. Phys. **A325** (1986) 485.
9) S. Das Gupta and N. de Takacsy, Am. J. of Phys. **44** (1976) 47.
10) N. de Takacsy and S. Das Gupta, Phys. Rev. **C13** (1976) 399.
11) F. Tabakin, Ann. Phys. (N.Y.) **30** (1964) 51.
12) N. de Takacsy, Can. J. of Phys. **46** (1968) 2091.

Q-VALUES AND ISOMER ENERGIES FROM HIGH-RESOLUTION ALPHA-, PROTON-, AND GAMMA-RAY SPECTROSCOPY ABOVE ^{146}Gd

D. Schardt [a], R. Barden [a], R. Kirchner [a], O. Klepper [a], A. Plochocki [a], E. Roeckl [a],
P. Kleinheinz [b], M. Piiparinen [b], B. Rubio [b], K. Zuber [b], C.F. Liang [c], P. Paris [c],
A. Huck [d], G. Walter [d], G. Marguier [e], H. Gabelmann [f], J. Blomqvist [g]

[a] GSI, Postfach 110552, D-6100 Darmstadt 11, Federal Republic of Germany
[b] Institut für Kernphysik, KFA Jülich, D-5170 Jülich, Federal Republic of Germany
[c] Laboratoire Rene Bernas (CSNSM), B.P. No.1, Bat. 104, F-91406 Orsay, France
[d] CRN Strasbourg, B.P. No.20, F-67037, Strasbourg Cedex, France
[e] IPN and IN2P3, University of Lyon, F-69622 Villeurbanne Cedex, France
[f] EP Division, CERN, CH-1211 Geneva 23, Switzerland
[g] Research Institute of Physics, S-10405 Stockholm 50, Sweden

ABSTRACT

The low- and high-spin β-decay isomers and the low-lying single proton excitations in N=82 and N=84 odd-Z nuclei above ^{146}Gd were investigated by particle- and γ-ray spectroscopy following on-line mass separation. Alpha-gamma coincidence measurements establish the relative energies of the $\pi s_{1/2}$ and $\pi h_{11/2}$ isomers in both ^{151}Ho$_{84}$ and ^{147}Tb$_{82}$ with better than 1 keV accuracy. The present status of a high-resolution proton- and γ-ray study of the ^{147}Dy → ^{147}Tb β-decay, providing a link of 1.6-h^{147}Tb to the well known ^{146}Gd mass, is discussed.

INTRODUCTION

In the region of spherical nuclei around N=82 above gadolinium (Z=64) the protons fill the close-lying $s_{1/2}$, $d_{3/2}$, and $h_{11/2}$ orbitals. In odd-Z nuclei with even neutron number these single proton states and the $d_{5/2}$ and $g_{7/2}$ hole states represent the lowest-lying levels and are found well below the three-particle and collective excitations. Consequently, one expects a low-spin (1/2$^+$ or 3/2$^+$) and a high-spin (11/2$^-$) β-decay isomer in these nuclei and a number of such examples is known (Fig.1). When moving further away from the N=82 closed neutron shell, however, the $d_{5/2}$ state may come below the $h_{11/2}$ state, which then can decay by an isomeric E3 transition. Such an E3-isomer with 25 s half-live is known [1] for ^{151}Tb$_{86}$. In the odd-odd nuclei above ^{146}Gd one also expects two β-decay isomers since the coupling of an $s_{1/2}$, $d_{3/2}$, or $h_{11/2}$ proton with the $f_{7/2}$ neutron gives rise to close-lying levels with a large spin difference which cannot connect through γ-transitions. The high-spin state results from the strongly attractive $(\pi h_{11/2} \, \nu f_{7/2}) \, 9^+$ interaction, the configuration of the low-spin isomer could be either $(\pi s_{1/2} \, \nu f_{7/2}) \, 4^-$ or $(\pi d_{3/2} \, \nu f_{7/2}) \, 2^-$.

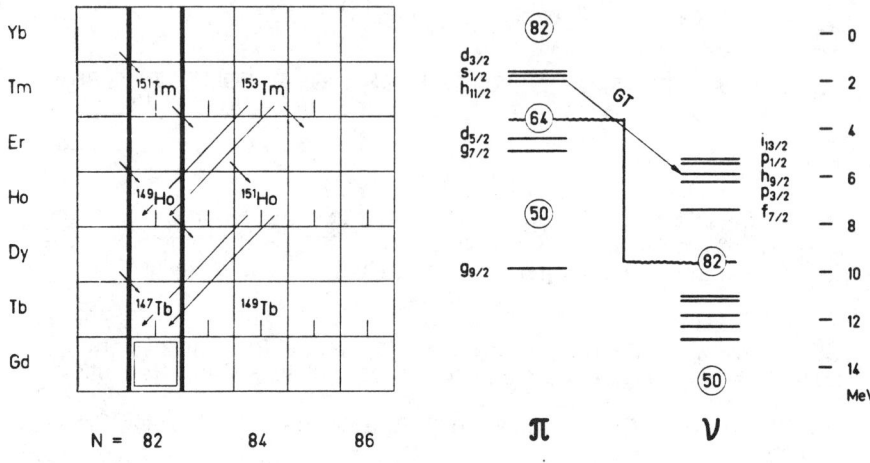

Fig. 1. Left: section of the chart of nuclides showing known low-spin and high-spin β-decay isomers in the region above ^{146}Gd. The α- and β-decays studied in the present work are indicated. Right: single particle diagram relevant for neutron-deficient rare-earths around Z = 64 and N = 82.

In the present contribution we want to discuss the systematics of the five single proton states in odd-Z N = 82 and N = 84 nuclei above gadolinium. New results have been obtained from β-decay studies of their low- and high-spin isomers, and of their respective N = 81 and N = 83 parent nuclei. We also present recent results from αγ-coincidence and high-resolution α-measurements which allow to locate the position of the 11/2⁻ isomer relative to the 1/2⁺ isomer in ^{151}Ho$_{84}$ and it's α-daughter ^{147}Tb$_{82}$, and which allow to establish the existence of similar isomers in ^{153}Tm, too. Finally, we shall discuss the present status of combined proton- and γ-ray measurements for the ^{147}Dy β-decay, aiming at precise determination of the ^{147}Tb proton binding energy which provides a link to the ^{146}Gd mass.

EXPERIMENTAL

Our measurements were performed at the GSI on-line mass separator, the ISOCELE II separator at Orsay, and the ISOLDE II facility at CERN, taking advantage of the different production mechanisms employed.
Heavy-ion induced fusion-evaporation reactions with ^{58}Ni and ^{64}Zn beams from the UNILAC and thin targets of ^{93}Nb, $^{92, 94, 96}$Mo, and ^{96}Ru were used to study the decays of ^{147}Dy, $^{149, 151}$Er, $^{151, 153}$Yb, and $^{151, 153}$Tm under very clean conditions at the GSI separator.
The ^{151}Ho αγ coincidence experiment which required a high production rate was carried out at the ISOCELE separator, using the (^{3}He, 11n) reaction with a 1.1μA beam of 240 MeV ^{3}He and a 10 g terbium target.

The very high yields [2] for neutron-deficient rare-earth isotopes, produced in spallation reactions of 600 MeV protons on a tantalum foil target offered by the ISOLDE facility, enabled the high-resolution measurements of β-delayed protons following ^{147}Dy decay. However, the strong production of isobars closer to stability restricted γ-ray spectroscopy to dysprosium and terbium isotopes.

In all these experiments the activity was collected on a tape and periodically transported into a counting position which was equipped with germanium γ- and X-ray detectors or cooled surface barrier detectors for α- and proton spectroscopy.

SINGLE PARTICLE EXCITATIONS AT N = 82 AND N = 84

The excitation energies of the five proton single-particle levels in odd-Z N = 82 isotones provide prime information for the proton single-particle energy diagram. For the isotones below gadolinium they have been identified in proton transfer reaction studies [3]. Above ^{146}Gd, where the lack of stable targets prevented such studies, information on the single-proton states has been obtained from in-beam γ-ray work as well as α- and β-decay spectroscopy.

The characteristic features of the β-decays of N = 81 to 84 odd-A nuclei above ^{146}Gd are shown in Fig.2, which may serve as a guide for the following discussion of experimental data on these decays.

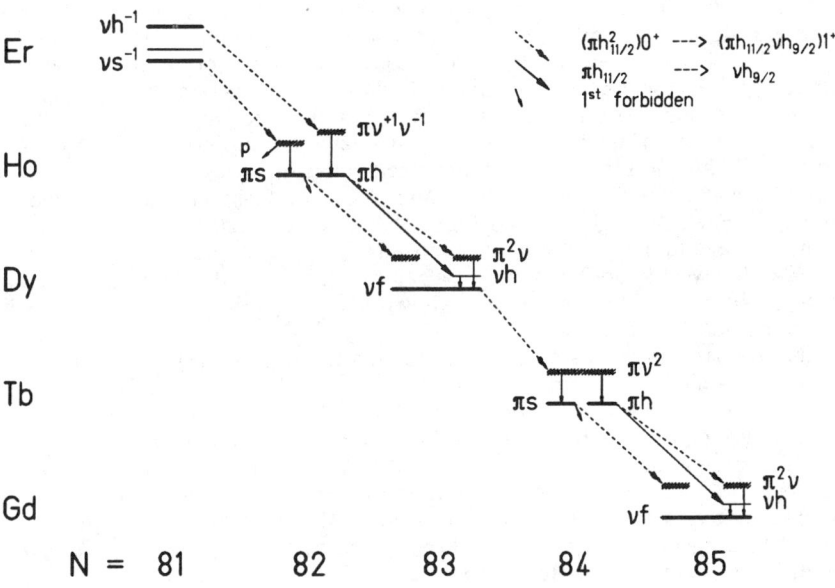

Fig. 2. Allowed β-decays of odd-A nuclei above ^{146}Gd, shown here for the A = 149 chain. Single particle orbitals are indicated, with h referring to $h_{11/2}$ or $h_{9/2}$ for respectively π and ν. The shaded areas represent three-nucleus excitations formed in decay of a $(\pi h_{11/2})0^+$ pair.

The $\pi h_{11/2}$ high-spin isomers

The β-decays shown in Fig. 2 are dominated by the $\pi h_{11/2} \to \nu h_{9/2}$ Gamow-Teller (GT) transition. In the decays of the $\pi h_{11/2}$ high-spin isomers of odd-Z N = 82 and N = 84 nuclei this leads to strong population of the low-lying $\nu h_{9/2}$ level in the N = 83 and N = 85 daughter nuclei. The observation [4-8] of such transitions in the β-decays of N = 82 ^{147}Tb, ^{149}Ho, ^{151}Tm, and N = 84 ^{149}Tb, and ^{151}Ho with log ft values between 4.2 and 3.6 firmly establishes the $\pi h_{11/2}$ assignment for the high-spin β-decay isomers of these nuclei.
In recent measurements of A = 153 decays at the GSI separator we have observed the $\pi h_{11/2} \to \nu h_{9/2}$ transition also in the previously not investigated β-decay of ^{153}Tm through detection of the known 299 keV $9/2^- \to 7/2^-$ ground-state transition [9] in ^{153}Er in coincidence with erbium KX-rays.
New results for the β-decay of ^{151}Tm($11/2^-$), where previously only the 802 keV γ-ray was identified, are shown in Fig. 5 (below). The largest fraction of the β-decay strength feeds two levels at ≅ 3 MeV excitation which are firmly characterized as $(\pi h_{11/2} \nu h_{9/2})_{1^+} \times \pi h_{11/2}$ states formed in GT decay of an $h_{11/2}$ proton out of a 0^+ pair.

The low-spin isomers and the positive-parity single-particle excitations

From the single-particle diagram (Fig. 1) the low-spin isomers in odd-Z N = 82 and N = 84 isotones above ^{146}Gd are expected to be the $\pi s_{1/2}$ or $d_{3/2}$ single proton states, while the $d_{5/2}$ and $g_{7/2}$ states occur at higher excitations as a consequence of the Z = 64 gap. In-beam γ-ray and conversion electron data [10] have established $I^\pi = 1/2^+$ for the low-spin isomer in ^{147}Tb$_{82}$ and a monotonic spin sequence for the four positive parity single proton states which are connected through consecutive M1 transitions. This result, which was confirmed by β-decay studies [11], removed the earlier incorrect $5/2^+$ assignment [12] for the 1.6-h isomer. The observation of a γ-ray triple cascade was used to identify the positive parity single proton states also in the higher-lying N = 82 isotones ^{149}Ho and ^{151}Tm from β-decay studies [13,14] of their even-Z N = 81 parent nuclei.
In the β-decay of even-Z parent nuclei an $h_{11/2}$ proton out of a 0^+ pair decays to an $h_{9/2}$ neutron, leading to the well known low-lying $(\pi h_{11/2} \nu h_{9/2})1^+$ states if the parent nucleus is doubly even.
For N = 81 parents the same decay proceeds to $(\pi h_{11/2} \nu h_{9/2})_{1^+} \times \nu^{-1}$ final states, where the spectator neutron is either the $h_{11/2}$ or the $s_{1/2}$ hole. These states lie at about 4.0 and 4.7 MeV excitation (Fig. 2). The primarily fed $9/2^-$, $11/2^-$, $13/2^-$ states can decay by the fast 4 MeV spin-flip M1 transition $\nu h_{9/2} \to \nu h_{11/2}^{-1}$ to the $\pi h_{11/2}$ β-decay isomer, whereas the γ-decay of the primarily populated $1/2^+$ and $3/2^+$ levels is expected to be more complex with frequent γ-branching. Such a decay pattern has been observed [13] for ^{147}Dy in a high-sensitivity γγ-coincidence measurement. The three M1 transitions connecting the low-lying single proton states were found to be the strongest γ-rays in this decay. In the less detailed data [13,14] for the β-decays of ^{149}Er and ^{151}Yb we observed also a cascade of three intense low-lying γ-transitions (Fig. 3). Recent conversion electron measurements [15] at the GSI separator established M1 multipolarity for the 171 keV, 344 keV, and 437 keV transitions in ^{149}Ho. Our γ-ray data for the ^{149}Er decay were confirmed by a similar study [16] at the OASIS separator.

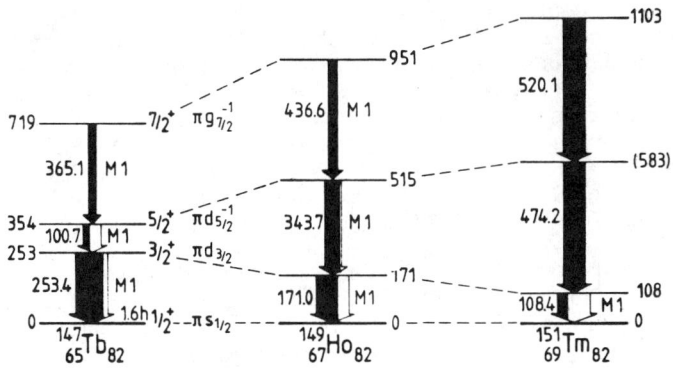

Fig. 3. Observed single proton states in odd-Z N = 82 isotones above gadolinium. Multipolarities are indicated when determined experimentally.

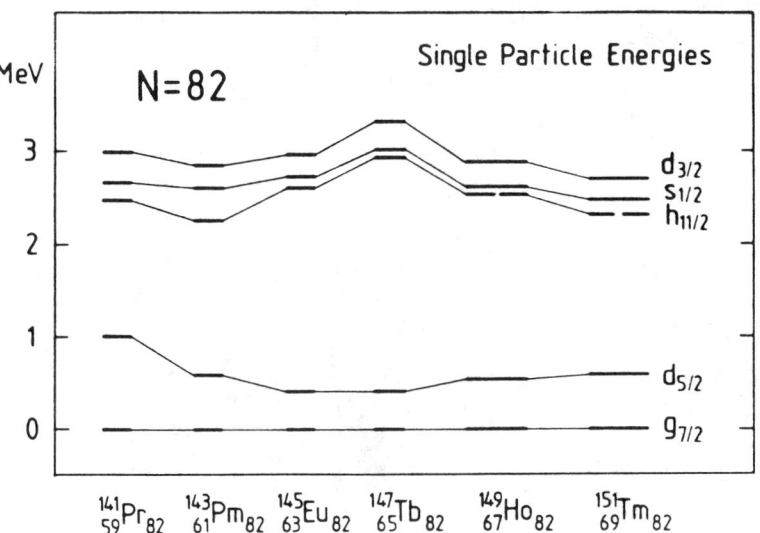

Fig. 4. Single particle energies for odd-Z N = 82 isotones resulting from a BCS analysis with G = 0.20 MeV.

The systematics of single particle states in odd-Z N = 82 isotones is shown in Fig. 4. The single particle energies obtained from a BCS calculation similar as described in Ref. 10 reproduce the observed excitation energies displayed in Fig. 3. The results for ^{149}Ho and ^{151}Tm indicate a further decrease of the Z = 64 gap when more protons are added.

The above results on the N=81→82 β-decays have identified the low-spin isomers in odd-Z N=82 isotones above ^{146}Gd as the $s_{1/2}$ single-proton state. Relatively little is known about the β-decays of these isomers. From our first results on the β-decay of ^{151}Tm (Fig. 5) we assign a 984 keV γ-line seen in coincidence with erbium X-rays but not with other ^{151}Er γ-transitions to the $\nu p_{3/2} \rightarrow \nu f_{7/2}$ ground-state transition following the decay of ^{151}Tm(1/2$^+$). The $\nu p_{3/2}$ single-particle state was proposed [13] at 1035 keV in ^{149}Dy and lies at 1153 keV [12] in ^{147}Gd. From the γ-ray timing data we obtain
$T_{1/2}(^{151}$Tm,1/2$^+) = 5.2 \pm 2.0$ s and $T_{1/2}(^{149}$Ho,1/2$^+) = 58 \pm 3$ s.

Similar as in the N=81 → 82 β-decays discussed above, the GT decay of odd-Z N=83 nuclei proceeds to high-lying three-particle states in their N=84 daughters which are here of $[(\pi h_{11/2} \nu h_{9/2})_1 + \nu f_{7/2}]$ 5/2, 7/2, 9/2$^-$ character (Fig. 2). These levels can decay either by the l-forbidden $\nu h_{9/2} \rightarrow \nu f_{7/2}$ M1 transition to the $\pi h_{11/2}$ isomer, or by forbidden E1 radiation to the $d_{3/2}, d_{5/2}$, or $g_{7/2}$ single proton states which all deexcite to the $s_{1/2}$ isomer.
In ^{149}Tb the five single proton states were identified in ^{148}Gd (τ,d) reaction studies [17]. As in the odd-Z N=82 isotones a monotonic spin sequence was found for the positive parity states and the $h_{11/2}$ isomer was located 36 keV above the $s_{1/2}$ ground-state.
A detailed investigation [18] of the highly complex decay scheme of ^{149}Dy$_{83}$ was carried out at the ISOCELE separator. The observation of γ-decay branchings from individual primarily fed levels to both β-decay isomers in ^{149}Tb enabled the determination of their relative energy with γ-spectroscopic accuracy, the result being in excellent agreement with the (τ,d) reaction studies.

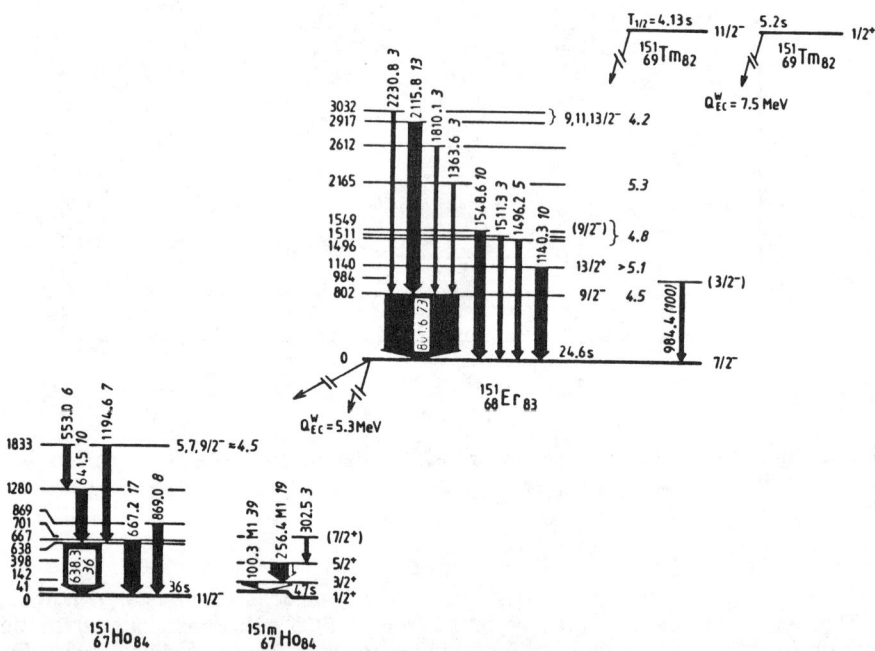

Fig. 5. Decay schemes of ^{151}Tm(11/2$^-$), ^{151}Tm(1/2$^+$), and ^{151}Er. M1 multipolarities were determined from conversion electron data.

In the N=84 isotone ^{151}Ho two isomers with half-lives of 36 s and 47 s are known [12] since long time from the observation of two strong α-transitions differing by 90 keV in energy. The $\pi h_{11/2}$ assignment for the 36-s high-spin isomer was confirmed [8] by the observation of the $\pi h_{11/2} \to \nu h_{9/2}$ GT branch through the strong 527 keV γ-transition deexciting the $\nu h_{9/2}$ state in the ^{151}Dy daughter nucleus. The low-spin isomer was proposed to be the $d_{5/2}$ single proton state in the original work [19], being based, however, on an incorrect $d_{5/2}$ assignment for both 1.6-h ^{147}Tb and 4.1-h ^{149}Tb. The two α-transitions connect the low-spin and the high-spin isomers in ^{151}Ho and its daughter ^{147}Tb, but cannot specify the isomer energies within one nucleus.

In a detailed investigation [20] of the ^{151}Ho α-decay at the ISOCELE separator we have observed weak α-transitions from both ^{151}Ho isomers in coincidence with γ-rays deexciting the 5/2$^+$ state at 354 keV in 1.6-h ^{147}Tb. These data establish the excitation energies

$E_x(^{147}\text{Tb}, 11/2^-) = 50.6 \pm 0.9$ keV and $E_x(^{151}\text{Ho}, 1/2^+) = 41.4 \pm 0.9$ keV.

The 1/2$^+$ assignment for low-spin ^{151}Ho is strongly suggested from non-observation of additional satellite lines in this decay. The latter result is in agreement with a recent study [21] of the ^{151}Er $\to$ ^{151}Ho β-decay which should proceed analogously as the ^{149}Dy decay discussed above. The low log ft value for the 1833 keV level (Fig. 5) indicates significant feeding of high-lying GT-fed states which deexcite to both ^{151}Ho isomers. The cascade of three strong low-energy γ-rays of 100, 256, and 303 keV is interpreted as a sequence of M1-transitions connecting the four positive-parity single-proton states. This assignment is based on comparison with ^{149}Tb.

A similar αγ-experiment as for ^{151}Ho should in principle also be possible for the next higher-lying N=84 isotone ^{153}Tm. However, only a single α-activity with 1.6 s half-live is known [22-24] for this nucleus. A plausible explanation might be that the two α-decays expected to proceed to the 1/2$^+$ and 11/2$^-$ isomers in ^{149}Ho$_{82}$ are very close in energy and are hence difficult to resolve experimentally. We have re-investigated the α-decay of ^{153}Tm at the GSI separator using the ^{92}Mo (^{64}Zn, 3p) reaction. In our measurements the A=153 activity was collected onto a 29 μg/cm^2 carbon foil in front of a silicon α-detector and a germanium γ-detector. The observation of both the 1035 keV and the 1091 keV γ-rays following β-decay of the two ^{153}Tm α-decay daughters ^{149}Ho (1/2$^+$) and ^{149}Ho (11/2$^-$) at mass 153 gives unambiguous evidence for the existence of a low- and high-spin isomer in ^{153}Tm (Fig. 6). This result was confirmed by a high-resolution α-measurement at ISOLDE. As can be seen from Fig. 7, only a doublet fit with the line shape taken from the ^{153}Er α-line measured in the same spectrum could reproduce the ^{153}Tm α-line satisfactorily. The energy difference of the two α-lines was found to be 7 ± 4 keV. Furthermore, the timing data for the lower and upper parts of the ^{153}Tm line give different half-lives (Fig. 7). From the doublet decomposition we obtain $T_{1/2} = 2.5(2)$ s for the low-energy component and $T_{1/2} = 1.48(1)$ s for the high-energy component of the ^{153}Tm α-activity. Based on comparison with the ^{151}Ho case, where the production at ISOLDE favours the 11/2$^-$ isomer, we assign the 5103 keV α-transition to the 11/2$^-$ high-spin isomer and the 5096 keV α-transition to the low-spin (1/2$^+$ or 3/2$^+$) isomer of ^{153}Tm.

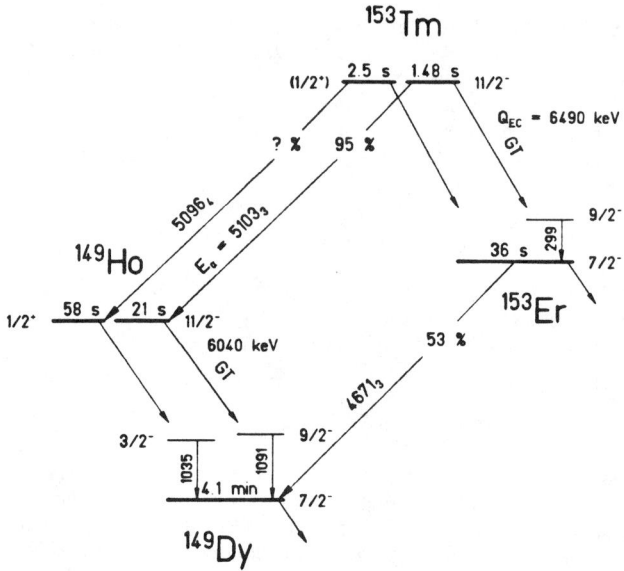

Fig. 6. Decay of the low-spin and high-spin isomer of ^{153}Tm. The α-branching ratios are taken from literature [23,24], Q_{EC}-values from Ref. 32.

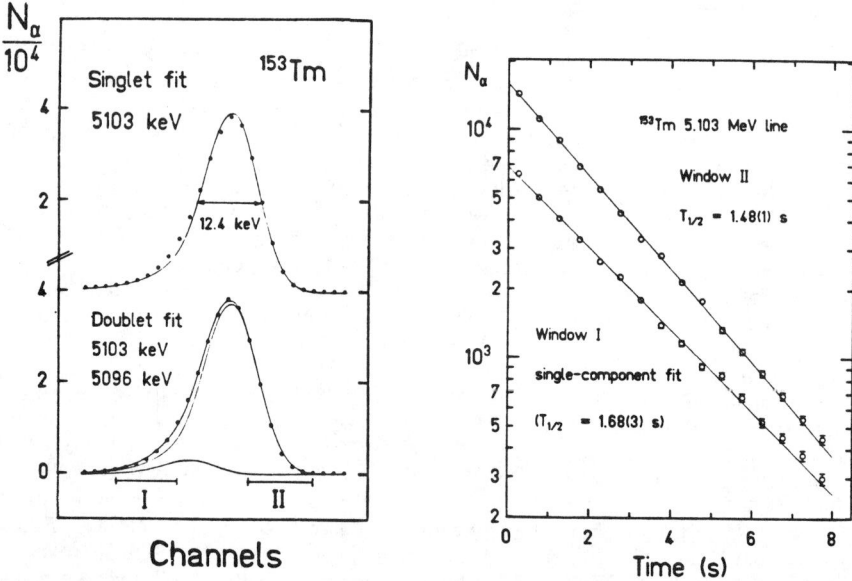

Fig. 7. Doublet structure of the ^{153}Tm α-line from high-resolution α-data. The line shape is taken from the ^{153}Er α-line measured in the same spectrum. The doublet fit reduces the χ^2 as compared to the poor singlet fit by a factor of 7. The doublet structure is confirmed by the timing data for the peak regions I and II shown to the right which give significantly different (apparent) half-lives.

BETA-DECAY OF $^{147}Dy_{81}$ AND THE PROTON SEPARATION ENERGY OF $^{147}Tb_{82}$

In the previous chapter we have presented examples of high-resolution and high-sensitivity α- and γ-ray measurements which allowed the accurate determination of mass differences such as Q_α-values and relative isomer energies. Provided a link to known absolute masses could be established, these data would give a number of precise masses in the region above ^{146}Gd. The mass surface in this region is still poorly known since Q_β-measurements are often not sufficiently accurate and transfer reactions are difficult or excluded because the appropriate target isotopes are unstable. On the other hand, these masses are of particular interest, e.g. for shell model analyses [25] of high-spin states or as base masses for the long α-decay chains [25,26] leading up to very neutron-deficient nuclei beyond the proton-drip-line.

The observation [27] of low-energy discrete proton lines in β-decay of ^{147}Dy and the identification [13] of high-lying ^{147}Tb states in γ-ray measurements open the interesting possibility of a direct determination of the ^{147}Tb proton separation energy. This would provide a link to the accurately known [28] mass of ^{146}Gd. Recent results for the ^{147}Tb mass obtained from Q_{EC}-measurements [29] and transfer reaction studies [30] have 50 to 60 keV error bars and differ by 250 keV.

The β-decay of ^{147}Dy ($1/2^+$) proceeds to high-lying particle-hole excitations of $[(\pi h_{11/2} \nu h_{9/2})_{1+} \nu^{-1}s_{1/2}]$ $1/2^+$, $3/2^+$ character in ^{147}Tb. These states lie at 4 MeV

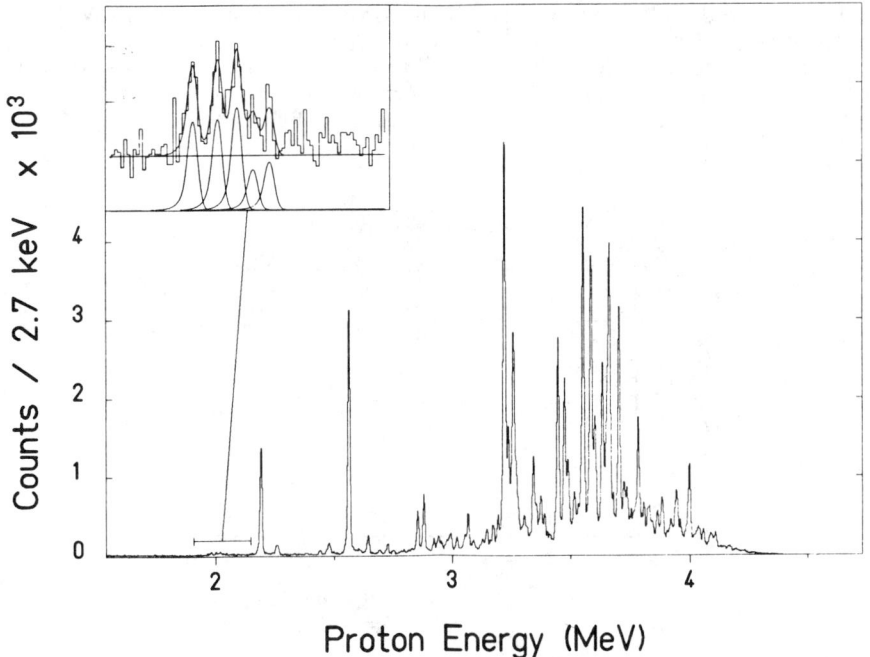

Fig. 8. Energy spectrum of protons following ^{147}Dy β-decay measured with a cooled surface barrier detector of 150 μm thickness and an active area of 50 mm^2. The activity was produced in spallation reactions with 600 MeV protons on a tantalum target at ISOLDE. The energy resolution was 11 keV FWHM.

excitation where proton emission becomes a competitive decay channel (Fig. 2). Since feeding of excited states which lie at $E_x > 1.5$ MeV in the ^{146}Gd final nucleus is negligible, the measured proton energies are directly related to excitation energies of the intermediate ^{147}Tb levels. Proton emission from high-spin levels populated in decay of ^{147}Dy(11/2$^-$) contributes very little in view of the large angular momentum hindrance.

The proton spectrum shown in Fig. 8, measured at ISOLDE with high counting statistics and 11 keV FWHM resolution, revealed a number of weak lines below 2 MeV which can be attributed to ^{147}Tb levels observed independently in the above γ-ray measurements. Fluctuation analyses based on the autocorrelation function method [31] indicate a mean level spacing of 1 to 2 keV at $<E_x> \cong 5.5$ MeV, the result depending somewhat on the smoothing function used. This supports the interpretation of the low-energy proton lines as arising from discrete ^{147}Tb levels at $E_x \cong 4$ MeV. Including the two strong proton lines at 2.2 and 2.6 MeV, seven lines were found to match corresponding ^{147}Tb levels within the experimental errors (Fig. 9). The main error contribution comes from the proton energy calibration which so far was made with α-particles from ^{148}Gd produced on-line in the same experiment. This implies a 1.4(2)% correction [33] on the slope of the calibration curve. The systematic deviation of the S_p-values (Fig.9) might indicate a small error of $\cong 0.2$% on the slope, lying within the accuracy of the above correction. The absolute proton energies, however, which are needed to determine the ^{147}Tb proton separation energy were not yet obtained with the desired precision since the above correction amounts to 30 keV at 2 MeV proton energy. Furthermore, the ^{148}Gd α-energy has to be corrected by $\cong 30$ keV for losses due to the detector dead-layer and non-ionizing processes (for 2 MeV protons this correction amounts to 5 keV only). The proton separation energy energy of 1.6-h ^{147}Tb is then determined to be 1945 keV, where in view of the

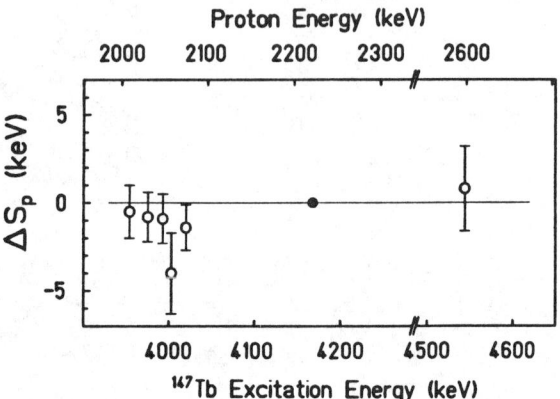

Fig. 9. Deviations in the proton separation energy S_p of ^{147}Tb(1/2$^+$) which was obtained from matching low-energy discrete proton lines observed in ^{147}Dy β-decay with corresponding high-lying ^{147}Tb levels identified independently in γ-ray measurements. The error bars include both the uncertainties in the level energies and in the proton energies relative to the 2.21 MeV proton peak.

rather large corrections estimated from tables given in the literature the uncertainty on this number is estimated as 18 keV. With this result and the ^{146}Gd mass from Ref. 28 we obtain

$$ME(^{147}Tb, 1/2^+) = -70751 \pm 19 \text{ keV}$$

for the ^{147}Tb ground-state mass. We plan, however, to improve the proton calibration in a future experiment by on-line production of ^{33}Ar which has proton lines at 1.6 and 3.2 MeV, known to 1- keV precision.

SUMMARY

We have discussed the systematics of the five single proton states $h_{11/2}$, $s_{1/2}$, $d_{3/2}$, $d_{5/2}$, and $g_{7/2}$, in odd-Z $N=82$ and $N=84$ isotones above ^{146}Gd, which in these nuclei are found well below the three-particle and collective excitations. New results have been obtained from α- and β-decay studies of their low- and high-spin isomers and of their respective $N=81$ and $N=83$ parents. Among other results, we have established that also in ^{153}Tm$_{84}$ a low-spin and high-spin isomer exist which decay predominantly through α-transitions of almost equal energy. Finally, we presented high-resolution proton and γ-ray data which can be exploited for precise determination of the ^{147}Tb mass.

1. P. Kemnitz et al., Nucl. Phys. A311, 11(1978)

2. H.L. Ravn, Proc. 11th Int. Conf. on Electromagnetic Isotope Separators and Techniques Related to their Applications, Los Alamos, NM, USA, 18-22 August, 1986, p. 72

3. B.H. Wildenthal et al., Phys. Rev. C3, 1199 (1971)

4. E. Newman et al., Phys. Rev. C9, 674 (1974)

5. K.S. Toth et al., Phys. Rev. C19, 482 (1979)

6. E. Nolte et al.,Z. Phys. A309, 33 (1982)

7. K.S. Toth et al., Phys. Rev. C11, 1370 (1975)

8. W.-D. Schmidt-Ott et al., Phys. Rev. C10, 296 (1974)

9. D. Horn et al., Phys. Rev. C23, 1047 (1981)

10. Y. Nagai et al., Phys. Rev. Lett. 47,, 1259 (1981)

11. K.S. Toth et al., Phys. Rev. C25, 667 (1982)

12. Table of isotopes, 7th ed., C.M. Lederer and V.S. Shirley (Wiley, New York, 1978), and references therein

13. D. Schardt et al., Proc. 7th Int. Conf. on Atomic Masses and Fundamental Constants (AMCO-7), 3-7 Sept. 1984, Darmstadt-Seeheim, FRG, p.222

14. P. Kleinheinz et al., Z. Phys. A323, 705 (1985)

15. R. Barden, thesis, Johannes Gutenberg-Universität Mainz, in preparation
16. K.S. Toth et al., Phys. Rev. C32 , 342 (1985)
17. D.J. Decman et al., Proc. 7th Int. Conf. on Atomic Masses and Fundamental Constants (AMCO-7), 3-7 Sept. 1984, Darmstadt-Seeheim, FRG, p. 220
18. K. Zuber et al., Annual Report 1985, Institut für Kernphysik, KFA Jülich, p. 39
19. R.D. Macfarlane and R.D. Griffioen, Phys. Rev. 130, 1491(1963)
20. C.F. Liang et al., Phys. Lett., B191 , 245(1987)
21. R. Barden et al., Beta-Decay of 27/2⁻ Isomers in N=83 Nuclei, submitted to Z. Phys.
22. R.D. Macfarlane, Phys. Rev. B136 , 941(1964)
23. E. Hagberg et al., Nucl. Phys. A293 , 1(1964)
24. S. Hofmann et al., Z. Phys. A291 , 53(1979)
25. J. Blomqvist et al., Z. Phys. A312 , 27(1983); B. Rubio et al., Z. Phys. A324 , 27(1986)
26. e.g. S. Hofmann et al., Z. Phys. A299 , 21(1981); U.J. Schrewe et al, Phys. Rev. C25 , 3091(1982); G.D. Alkhazov et al, Z. Phys. A311 , 245(1983); J.R.H. Schneider et al, Z. Phys. A312 , 21(1983); E. Runte et al, Z. Phys. A324 , 119(1986)
27. O. Klepper et al., Z. Phys. A305 , 125(1982); D. Schardt et al., , Proc. 7th Int. Conf. on Atomic Masses and Fundamental Constants (AMCO-7), 3-7 Sept. 1984, Darmstadt-Seeheim, FRG, p. 229
28. L.G. Mann et al., Phys. Rev. C34 , 729(1986)
29. U.J. Schrewe et al., Z. Phys. A320 , 595(1985)
30. R. Gyufko et al., Phys. Lett. 150B , 335(1985)
31. B. Jonson et al., CERN report 76-13, 277(1976)
32. A.H. Wapstra and G. Audi, Nucl. Phys. A432 , 1(1985)
33. W.N. Lennard et al., Nucl. Instr. Meth. A248 , 454(1986)

PROTON-NEUTRON INTERACTION
AND THE RELATIVE ISOMER MASSES IN ^{148}Tb

J. Styczen[*1], P. Kleinheinz[1], W. Starzecki[*1], B. Rubio[1],
G. de Angelis[1], H.J. Hähn, C.F. Liang[2], P. Paris[2],
R. Reinhardt[3], P. von Brentano[3], and J. Blomqvist [4]

[1] Institut f. Kernphysik, KFA Jülich, D-5170 Jülich, F.R.Germany
[2] CNRS Orsay, F-91406 Orsay, France
[3] Universität Köln, D-5000 Köln, F.R.Germany
[4] Research Institute of Physics, S-10405 Stockholm 50, Schweden

ABSTRACT

Proton-neutron states have been investigated in the odd-odd nucleus $^{148}_{65}$Tb$_{83}$ with Li-induced in-beam γ- and e^--experiments and $\alpha\gamma$-coincidence measurements of ^{152}Ho decay. The results establish the three complete $\pi^{+1}\nu f_{7/2}$ multiplets, and in addition a number of multiplet members involving a proton hole and octupole excitations. The $(\pi h_{11/2} \nu h_{9/2})$ 1^+ attraction is determined as -1.36 MeV.

The low-lying levels in the doubly-odd $^{148}_{65}$Tb$_{83}$ nucleus are due to the couplings of one valence proton with the 83rd valence neutron, and their energies provide empirical values for the proton-neutron two-body interactions. These interactions are essential in shell-model analyses for nuclei in this mass region. They are needed e.g. for calculation of multiparticle excitations or β-strength functions. One would expect that the five proton orbitals of the 50 to 82 shell will contribute to the low-lying states in ^{148}Tb since they all occur below .8 MeV in ^{147}Tb. On the other hand, the neutron should always be $f_{7/2}$ since the next higher neutron orbital, $p_{3/2}$, lies above 1 MeV in ^{147}Gd.

The nucleus ^{148}Tb has two β-decay isomers, with $I^\pi = 2^-$ and 9^+, which are interpreted as the antialigned $\pi d_{3/2} \nu f_{7/2}$ and the aligned $\pi h_{11/2} \nu f_{7/2}$ configurations. A number of excited low-spin $\pi\nu$-states with $I \leq 4$ have been identified in a recent[1] study of

^{148}Dy β-decay, but except the 9⁺ isomer no other high-spin πν-states were known, and also the relative energies of the two β-decay isomers were not well determined.

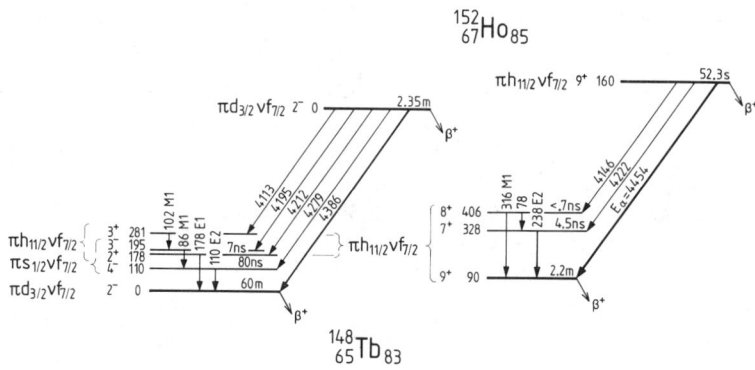

Fig. 1: α-decay of the two ^{152}Ho isomers established through αγ-coincidence measurements. All ℓ≠0 α-branches have intensities below .2 % of the respective ℓ=0 transition. Isomer excitation energies, excited state half-lives and transition multipolarities are from ^{148}Dy β-decay- and in-beam measurements.

First results on such high-spin two-body states we have obtained from a study of ^{152}Ho α-decay. This experiment was done at the Orsay ISOCELE II on-line mass separator where ^{152}Ho was produced with 1.4 μA ^{3}He-beam on 10g of Tb target. A transport tape moved the separated A = 152 activity to the measuring site with a surface-barrier α-detector and a Ge γ-ray detector in 180° close geometry where singles spectra and αγ-coincidences were recorded. The coincidence data identified six new very weak α-branches to excited states in ^{148}Tb (fig.1). Although not all satellite lines were clearly resolved in the α-singles spectrum, they are unambiguously observed in the coincidence data which firmly specify the assignment to either 2⁻ or 9⁺ ^{152}Ho decay, and also give the precise α-energies relative to the respective ℓ=0 α-transition. The four α-satellites associated with ^{152}Ho 2⁻ decay populate the four lowest-lying ^{148}Tb low-spin excited states observed in ^{148}Dy decay. The two remaining α-satellites, from ^{152}Ho 9⁺ decay, proceed to two previously unknown states, 238 and 316 keV above the ^{148}Ho 9⁺ isomer, which can be firmly assigned as the 8^+ and 7^+ couplings of the $\pi h_{11/2} \nu f_{7/2}$ multiplet, and from the systematics of two-body interactions the 316 keV state should be the 8^+ level. Only these two $\pi \nu$-states can decay through a single γ-ray to the 9^+ isomer. All other $\pi \nu$ levels expected below .6 MeV have negative parity and are limited to $I \leq 6$.

Our results also establish $I^\pi = 2^-$ and 9^+ for the two ^{152}Ho isomers. Any other than equal I^π assignments for parent and daughter nuclei would inevitably lead to a satellite intensity of ≥ 5 % for the $\ell=0$ α-transition to the respective excited state in ^{148}Tb, in clear conflict with experiment where all satellite lines are $\leq .2$ % of the high-energy α-branches.

Independently, we have investigated ^{148}Tb through in-beam γ-ray compound evaperation experiments with ^{6}Li and ^{7}Li beams on ^{144}Sm targets. Gamma-rays were measured with a Compton-suppressed multidetector array (OSIRIS) including two planar Ge-diodes for low-energy γ-detection. Transition multipolarity information derived from DCO-ratios, a singles γ-ray angular distribution measurement, and singles conversion electron spectra taken with a mini-orange.

These measurements confirmed the 7^+ and 8^+ levels found in the α-decay, and identified a significant number of other $\pi \nu$-excitations with spins down to 3. All ^{148}Tb levels below 1.2 MeV populated in these experiments are shown in fig. 2. With the I^π-values quite firmly determined, the states below 1 MeV can be assigned with some confidence to specific $\pi \nu$-multiplets. Above ~ 1 MeV members of the $\pi h_{11/2} \nu f_{7/2} \times 3^-$ octupole multiplet appear, similarly as they have recently been identified[2] in $^{146}_{63}$Eu$_{83}$ where a successful quantitative analysis of their anharmonicities could be made.

A significant result of our study is the complete identification of the $\pi h_{11/2} \nu f_{7/2}$ multiplet, which provides a connection of the ^{148}Tb 9^+ and 2^- isomers, giving

$$E_x(^{148}\text{Tb}, 9^+) = 90.1(3) \text{ keV}.$$

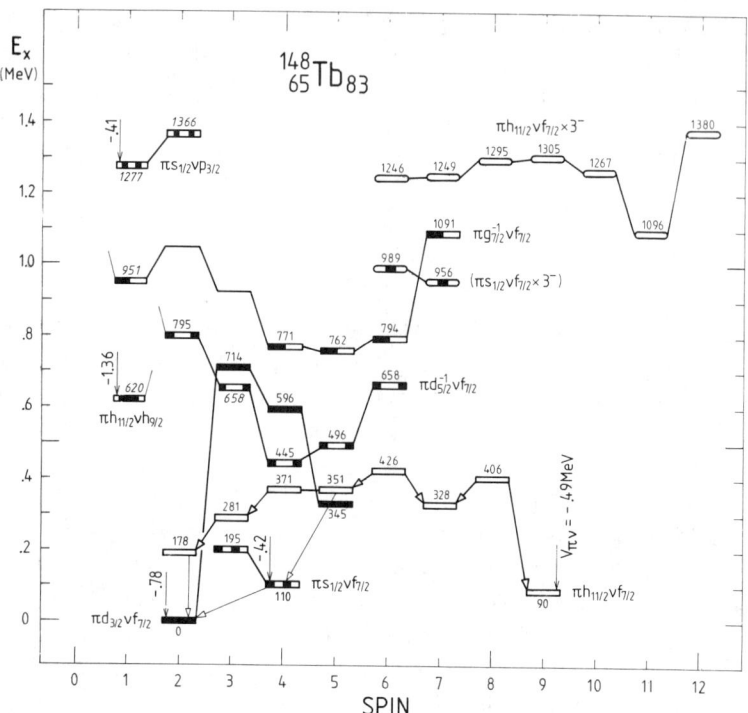

Fig. 2: πν-multiplets and octupole excitations in ^{148}Tb observed in Li-induced in-beam experiments. Selected γ-rays which connect the ^{148}Tb 2+ and 9- isomers are indicated. Residual interactions are given for the particle-particle multiplets using the masses of Rubio et al. Z. Phys. A324, 27(1986). Levels with excitation energies (in keV) given in italics were only seen in ^{148}Dy decay.

With the known α-energies the ^{152}Ho isomer excitation energy is determined as

$$E_x(^{152}\text{Ho}, 9^+) = 160(1) \text{ keV}.$$

In fig. 2 we also give the πν-interactions for the (extreme coupling of the) particle-particle multiplets which are more precisely specified with the isomer excitation energies now accurately known in both ^{147}Tb[3] and ^{148}Tb. Of these interactions e.g., the high-spin members of the $\pi h_{11/2} \nu f_{7/2}$ multiplet are vital input for the calculation of multinucleon yrast states in this region.

A particularly important result is the rather large attraction of 1.36 MeV determined for the 1^+ coupling of the $\pi h_{11/2} \nu h_{9/2}$ spin-orbit partners. Similarly large values are now also discussed[4] in nuclei around ^{96}Zr for the $\pi g_{9/2} \nu g_{7/2}$ spin-orbit pair. This 1^+ interaction is of crucial significance for the analysis of GT transition strengths[5] and of double ß-decay[6], and its large size has only recently been recognised.

*On leave from Institute of Nuclear Physics, Cracow, Poland

References

1) P. Kleinheinz et al, Phys. Rev. Lett. 55, 2664(1985)
2) A. Ercan et al, Z. Phys. A, in press
3) C.F. Liang et al, Phys. Lett B 191, 245(1987)
4) C. Sistemich, seperate contribution to this volume
5) I. Towner, Nucl. Phys. A444, 402(1985)
6) J. Engel, P. Vogel, M.R. Zirnbauer, Preprint, June 87

BETA-DECAY OF 27/2⁻ ISOMERS IN N = 83 NUCLEI

P. Kleinheinz[1], D. Schardt[2], R. Barden[2], A. Plochocki[2], B. Rubio[1],
J. Styczen[*,1], H. Güven[1], J.L. Tain[1], R. Kirchner[2], O. Klepper[2],
G. Walter[3], A. Huck[3], G. Marguier[4], J. Blomqvist[5].

[1] Institut f. Kernphysik, KFA Jülich, D-5170 Jülich, F.R. Germany
[2] GSI Darmstadt, D-6100 Darmstadt, F.R. Germany
[3] C.R.N. Strasbourg, F-67037 Strasbourg, France
[4] IPN & IN2P3, University of Lyon, Lyon, France
[5] Research Institute for Physics, S-10405 Stockholm, Sweden

ABSTRACT

In γ-ray measurements of the 27/2⁻ E3 isomer decays in $^{151}_{68}Er_{83}$ and $^{149}_{66}Dy_{83}$ at the on-line mass-separators at GSI and at ISOLDE-CERN we have identified Gamow-Teller β-decay branches proceeding to high-spin yrast states in the N=84 daughter nuclei $^{151}_{67}Ho_{84}$ and $^{149}_{65}Tb_{84}$.

INTRODUCTION

Beta-decay of high-spin multiparticle states at high excitation has been observed only in few cases. It is found in spherical nuclei near closed shells, where the combination of attractive residual two-body-interactions can significantly lower the state with fully-aligned angular momenta which then may become isomeric with a lifetime such that fast β-decay can compete. Most of these β-decays occur in light nuclei, where valence-protons and -neutrons fill the same shell, and thus proceed as Gamow-Teller(GT)-transitions between identical single particle orbitals. In heavier nuclei, GT-decay connecting spin-orbit partners may become possible whenever both the proton- and neutron Fermi-energies lie between the same spin-orbit partners. The classical 6.9-hour $(\pi g^2_{9/2} \nu d_{5/2})21/2^+$ three-particle isomer in $^{93}_{42}Mo_{51}$ discovered[1,2] in 1946 represents such a case. Its 0.12 % electron-capture branch (first observed[3] in 1977) proceeds as $\pi g_{9/2} \to \nu g_{7/2}$ to the high-lying $(\pi g_{9/2} \nu g_{7/2} d_{5/2})$ configuration in the $^{93}_{41}Nb_{52}$ daughter nucleus.

In decay studies of the analogous $(\pi h^2_{11/2} \nu f_{7/2})$ 27/2⁻ shell model isomers known[4,5] in the ^{149}Dy and ^{151}Er N=83 nuclei we have now identified the corresponding $\pi h_{11/2} \to \nu h_{9/2}$ GT β-decay branches.

EXPERIMENTAL

Our measurements for ^{151m}Er decay were performed at the GSI on-line mass separator, using 5 MeV/A beams of ^{58}Ni and ^{64}Zn on 3 mg/cm^2 metal targets of ^{92}Mo and zirconium. The ^{149}Dy isomer could be produced with very high yield at ISOLDE with 600 MeV p's on Ta-target. Dysprosium is at mass 149 close to the maximum of the yield curve, which is vital for γ-ray studies of spallation-produced activities. In both experiments a tape transport system was used, and γγ- and γ X-ray coincidences were recorded together with singles spectra in timing mode. At GSI, a mini-orange spectrometer served to measure low-energy conversion electron spectra.

THE 27/2$^-$ ISOMER IN $^{151}_{68}$Er$_{83}$

Prior to the present study, only a cascade of three γ-rays, of 289, 1100, and 1141 keV, emitted in decay of a 0.62(2)-s isomer in ^{151}Er were known[5], and from comparison with the ^{149}Dy isotone, they were assigned as a 21/2$^+$(E2) 17/2$^+$(E2) 13/2$^+$(E3) 7/2^-_g cascade emitted in decay of the $(\pi h^2_{11/2} \nu f_{7/2})27/2^-$ isomer, with the low-energy isomeric E3-transition remaining unobserved. We observed this transition in conversion electron spectra and determined its energy to be 57.7(5) keV. Electron-γ coincidence data gave the ^{151}Er isotopic assignment, and E3 multipolarity was determined from the L-subshell ratios.

The energy of the ^{151}Er 27/2$^-$ five-particle isomer can be calculated from the shell model, e.g. with the experimental excitation energies in the N=82 and 83 nuclei of Er and Dy and the appropriate six ground state masses (N = 82 and 83 Gd, Dy, Er) as

$$_{Er}E^{calc}_{27/2} = {_{Er}E_{10}} + {_{Dy}E_{27/2}} - {_{Dy}E_{10}} + S = 2531 \text{ keV},$$

within 55 keV of experiment.

β-DECAY OF THE 27/2$^-$ ISOMERS IN ^{151}Er AND ^{149}Dy

The γ-ray spectra measured for both 27/2$^-$ decays clearly reveal the principal yrast cascade transitions up to the 27/2$^-$ yrast state in the respective N=84 β-decay daughter nuclei. They decayed with the .5 sec half life of the N=83 27/2⁻ isomers, were in coincidence with each other, with annihilation radiation, and with the respective (Ho or Tb) X-rays. The ^{151m}Ho 27/2⁻ decay scheme is shown in fig. 1. From the pertinent parent- and daughter γ-ray intensities the GT-decay branches were determined as

$$\frac{I_{GT}}{I_{GT}+I_{E3}} = 4.7(5) \% \text{ for } ^{151m}\text{Er, and}$$

$$\frac{I_{GT}}{I_{GT}+I_{E3}} = 1.06(15) \% \text{ for } ^{149m}\text{Dy}$$

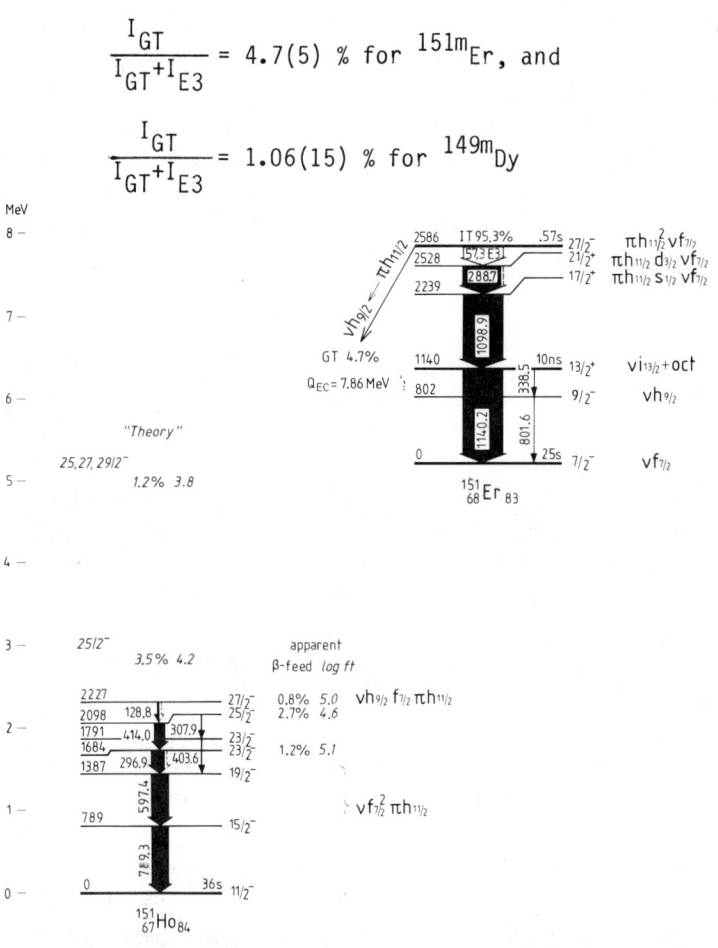

Fig. 1: Decay scheme of the 27/2⁻ isomer in ^{151}Er. The γ-intensities in ^{151}Ho are drawn ten times enhanced

With the simple, but unrealistic assumption that the entire ^{151m}Er β-branch proceeds to the 27/2⁻ level in ^{151}Ho a log-ft of 4.3 would result. But already the observed feeding (fig. 1) into the 23/2⁻ level, that is not accessible in GT-decay, suggests a more complex decay pattern. This conclusion is also supported from comparison with ^{150}Er ground state decay which is five times stronger than ^{151m}Er decay. Whereas an even-A 0^+ ground state can only decay through transformation of a proton from a 0^+ pair, leading to a single 1^+ state in the odd-odd daughter, the situation is much more complex for decay of the $(\pi h_{11/2}^4 \nu f_{7/2})$ $27/2^-$ five-particle state. Transformation of either a paired or an aligned proton leads to states with different seniority, spread over several MeV excitation in the daughter nucleus. The hatched areas in fig. 1 give a crude approximation for the expected strength function, and with the feedings as indicated the total strength becomes the same as in ^{150}Er decay.

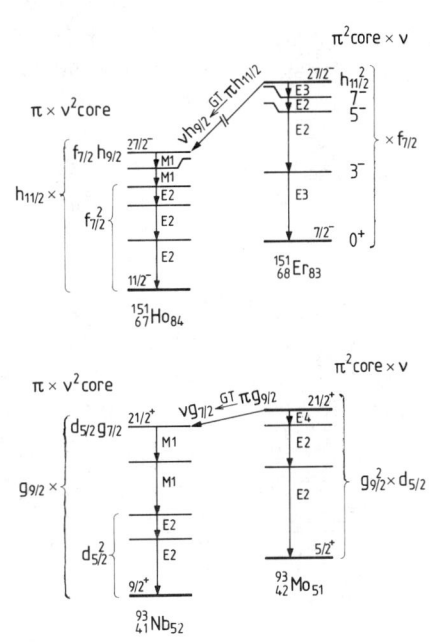

Fig. 2: Decays of the aligned $\pi^2\nu$ high-spin isomers in ^{151}Er and ^{93}Mo.

In ^{149}Dy $27/2^-$ decay the entire intensity proceeds to a $25/2^-$ level 361 keV above the $27/2^-$ $(\pi h_{11/2} \nu h_{9/2} f_{7/2})$ yrast state. This finding is in accord with theory, which predicts that 90 % of a $(\pi h_{11/2}^2 \nu f_{7/2})27/2^-$ three-particle state should decay to a $25/2^-$ $(\pi h_{11/2} \nu h_{9/2} f_{7/2})$ level. Moreover the decay strength to this state (log-ft $(25/2^-, ^{149}$Tb$)=4.07$) is within 25 % of the ^{148}Dy ground state decay (log-ft $(1^+, ^{148}$Tb$)=3.95$), in good accord with expectations.

The synopsis of fig. 2 reemphysizes the analogy of the ^{151}Er isomer decay with the ^{93}Mo case where all nucleon quantum numbers are one less. In both nuclei the high-spin isomer is formed by alignment of two valence-protons in a high-j orbit and the first valence neutron above the major shell closure. The neutron is spectator in the β-decay, but it is vital for formation of the $\pi^2\nu$ high-spin isomer through the attractive particle-particle $\pi\nu$-interaction for aligned coupling. This interaction lowers the aligned $\pi^2_{Jmax}\times\nu$ state below the $\pi^2_{Jmax-2}\times\nu$ level, thus excluding the $J_{max} \rightarrow J_{max}-2$ E2-decay occuring in the π^2 core nuclei. The different multipolarities of the isomer γ-deexcitation cascades simply reflect the more complex π^2-core yrast levels of ^{150}Er vs. the pure πj^2-sequence of ^{92}Mo. The β-decay proceeds as GT-transition where a $j_>$ valence proton decays into the empty $j_<$ neutron shell. Analogous character is also apparent for the yrast states in the two daughter nuclei, including the presently not well understood nature of the second highest state shown.

* On leave from Institute of Nuclear Physics, Cracow, Poland

REFERENCES

1. D.N. Kundu, M.L. Pool, Phys. Rev. 70, 111(1946); D.N. Kundu, et al., Phys. Rev. 77, 71(1950)
2. P. Alexander, G. Scharff-Goldhaber, Phys. Rev. 151, 964(1966)
3. R.A. Meyer, R.P. Yaffe, Phys. Rev. C15, 390(1977)
4. A.M. Stefanini, et al., Nucl. Phys. A258, 34(1976); Phys. Lett 62B, 405(1976)
5. J. Jastrzebski et al., Phys. Lett 97B, 50(1980)

SUPERDEFORMED NUCLEI AT HIGH SPIN

P.J. Twin
Science and Engineering Research Council
Daresbury Laboratory, Warrington WA4 4AD, United Kingdom
and
University of Liverpool, Liverpool L69 3BX, United Kingdom

ABSTRACT

The yrast shape of a nucleus changes with spin and in a few cases it had been predicted that a stable superdeformed shape (a prolate rotor with an axis ratio of 2:1) could exist before the nucleus fissions. A superdeformed yrast band extending from $24\hbar$ to $60\hbar$ has been observed in ^{152}Dy and the present status of the experimental data is presented. Topics covered include the spin assignments, the quadrupole moment, the moments of inertia and the decay out of the band. Recent data on the feeding of the band is reported and possible explanations are discussed of the rapid cooling at high spin which is necessary in order to see yrast states up to $60\hbar$.

INTRODUCTION

It is eighteen months since the observation[1] at Daresbury of a discrete line collective band in ^{152}Dy which extended from $24\hbar$ to $60\hbar$ and exhibited the correct moment of inertia of a superdeformed shape (a prolate rotor with an axis ratio of 2:1). Further data have been obtained on this nucleus, elucidating the properties of the band and covering the topics of moments of inertia, the quadrupole moment and the feeding of the yrast states at such a very high spin.

THE IDENTIFICATION OF THE SUPERDEFORMED BAND IN ^{152}Dy

The superdeformed band in ^{152}Dy was observed at the Daresbury tandem using TESSA3, one of the TESSA[2,3] family of instruments which have been the pioneers in the field of multi-detector high resolution arrays in gamma-ray spectroscopy. The acronyn stands for Total Energy Suppression Shield Array. The instrument combines 12 high resolution germanium detectors, each surrounded by a veto detector (suppression shield), and an inner calorimeter or total energy device. The calorimeter consists of a ball of 50 high efficiency, low resolution bismuth germanate detectors with an overall solid angle close to 4π.

The heavy ion fusion evaporation reaction, which is frequently used to populate high spin states, is not very selective in the production of a particular final nucleus. The reaction used to observe the superdeformed band involved the bombardment of a 205 MeV ^{48}Ca beam on a target of ^{108}Pd and the product nuclei ranged from ^{153}Dy(3n) to ^{150}Dy(6n). As the superdeformed band intensity is only 1% of the ^{152}Dy channel it was vital to drastically reduce the other final product nuclei. This was possible by selection on the sum

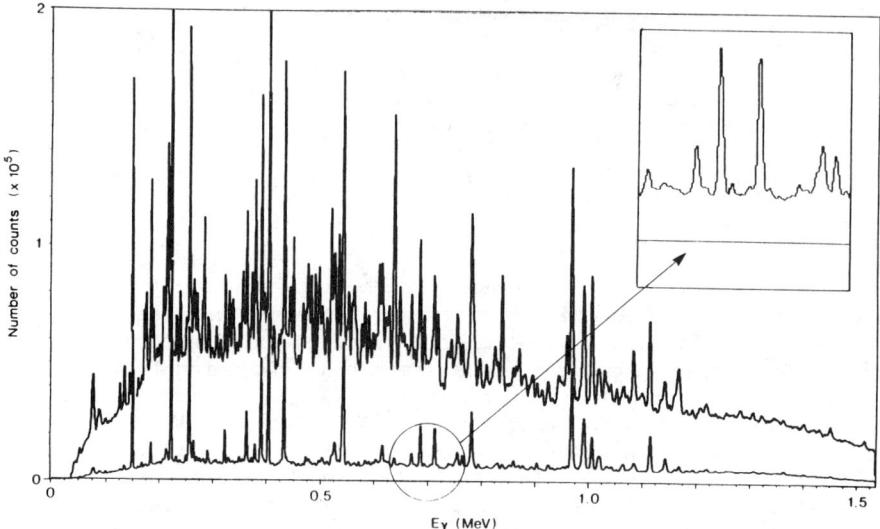

Fig.1. Spectra illustrating the improvement in signal-to-noise after using both sum-energy/fold and isomer selection to obtain a clean ^{152}Dy spectrum with 50% efficiency. The inset shows the intensity of the 692 keV γ-ray in the superdeformed band.

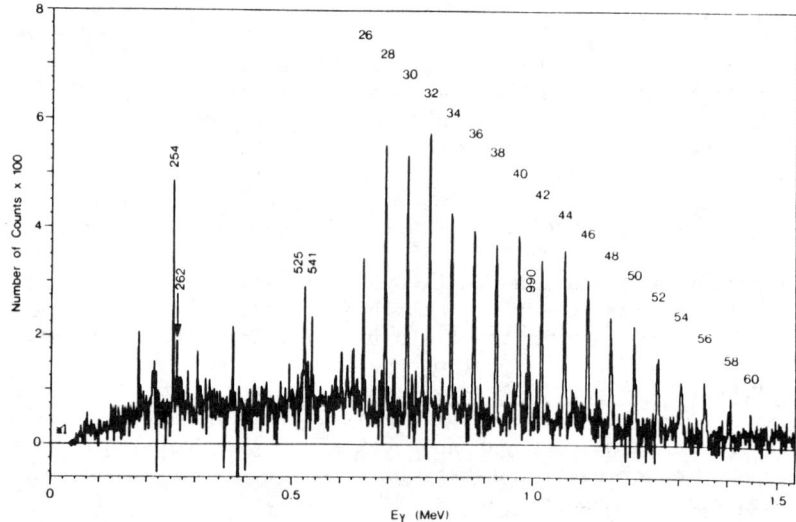

Fig.2. The spectrum of the superdeformed band in ^{152}Dy obtained by selection of the ^{152}Dy channel and by setting gates on members of the band. The background subtraction was normalised to give zero intensity in the 148 keV 29^+ - 28^+ decay between high spin oblate states. No intensity was observed remaining in the 967 keV and 221 keV γ-rays above the 25^- state but γ-ray decays between the 25^- state and 17^+ states were observed and they are indicated by their energies.

energy and fold (number of active ball elements) data from the BGO ball and also utilising it to identify delayed γ-rays from the 60ns isomer in ^{152}Dy. The channel was selected very cleanly with 50% efficiency and the resulting invaluable improvement in the signal-to-noise is seen in the spectra shown in fig.1. The spectrum of the superdeformed band itself (fig.2) was obtained by setting gates on members of the band and subtracting the underlying background under these peaks using a spectrum with a wide gate from 600 keV to 1500 keV. The normalisation was calculated using the intensities of the 148 keV and 221 keV transitions between high spin oblate states. All γ-rays above the 25⁻ state in the oblate structure were removed by the subtraction procedure but intensity remained in the 254 keV, 525 keV, 262 keV, 991 keV and 541 keV transitions below the 25⁻ state and it was assumed that this was related to the feeding from the superdeformed band. From the intensity data it was determined that the average entry spin to yrast was 21.8$\hbar$ and, therefore, the spin of the final state in the superdeformed band must be even higher. The energies of the oblate[4,5], prolate[6] and superdeformed[1] states were plotted as a function of spin (fig.3), with the assumption that the superdeformed band becomes yrast between 50$\hbar$ and 60$\hbar$. The band then must have an excitation energy several MeV above yrast at spins below 30$\hbar$, and it was proposed the γ-decay from it follows a statistical type pattern. Assuming the superdeformed band has even spin and positive parity the average spin de-exciting the band is 22.8$\hbar$, 24.8$\hbar$, 26.8$\hbar$, ... depending on the spin assignments and the spin difference of 3$\hbar$ was considered the most probable giving the spin assignments shown in fig.2.

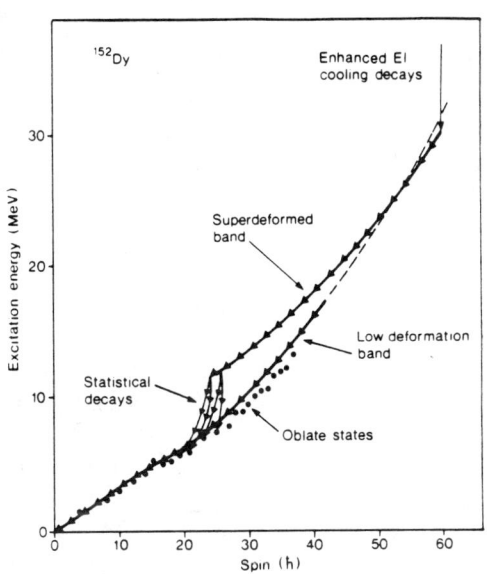

Fig.3. A schematic of the proposed decay of the superdeformed band. It was assumed it became yrast at 55$\hbar$ and it was proposed the decay out of the band is a statistical type process.

THE QUADRUPOLE MOMENT OF THE BAND

The intrinsic quadruple moment of the band is a direct measurement of the deformation. It was obtained by measuring the lifetimes of the γ-rays using the Doppler shift attenuation method with the TESSA3 spectrometer and a ^{48}Ca beam at 205 MeV bombarding a palladium target of 1.3 mg cm^{-2} on a 15 mg cm^{-2} gold foil so that the recoiling nuclei slowed up in the target and backing before finally stopping. The discrete γ-ray lines between the oblate states all have lifetimes greater than a few ps and thus showed no Doppler shift, and provided a reliable energy calibration for the detectors which were at angles of 35°, 90° and 145°. The spectra, selected for the ^{152}Dy channel and gated by members of the band, were analysed to determine the fractional Doppler shift of each superdeformed γ-ray as shown in fig.4. It is interesting to note that the measured shifts can only be explained if the feeding times of the band were very fast and of the order of a few fs. Note that as the spin of the band decreases there was a gradual decrease in the fractional shift. The calculated shifts are shown for various values of the intrinsic quadrupole moment, Q_0, with the following assumptions; the deformation and thus the quadrupole moment remained constant down the band; and the feeding into the

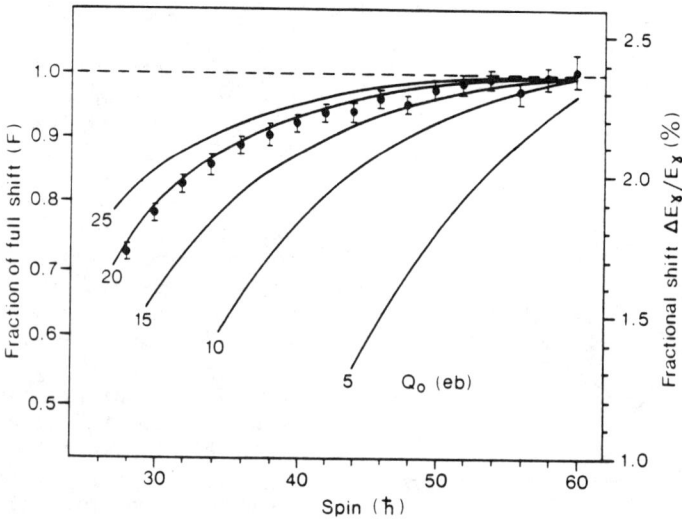

Fig.4. The measured fractional Doppler shift at θ=35° and θ=145° for transitions in the superdeformed band. The full shift is placed at the shift correponding to the centre of the target. The calculated curves are shown for values of the intrinsic quadrupole moment assuming constant deformation.

band at any point was 1 fs. The data were fitted by a Q_0 of 19±1 eb which is equivalent to a B(E2) strength of 2660 Wu. Taking into account an additional error, estimated at 15%, due to uncertainties in the slowing down process the value of Q_0 became 19±3 eb. This is very much larger than the measured values of 5 eb to 7 eb in rotational bands of normal deformation in rare earth nuclei and it is in excellent agreement with the theoretical value[10] of 18 eb. Thus the evidence is conclusive that the band is indeed associated with the superdeformed shape.

SHAPE CO-EXISTENCE IN ^{152}Dy

There have been many examples of the co-existence of different shapes over a narrow region of spin. The nucleus ^{152}Dy must, however, be the classic example as three distinct shapes co-exist over the large spin interval from 24ℏ to 40ℏ. This can be seen from the excitation energy versus spin plot shown in fig.3. The spherical or slightly oblate shape is associated with the building up of the spin by single particle configurations. The ground state sequence of stretched quadrupole transitions develops into a rotational band with a low moment of inertia which is linked with either a pure prolate deformation (ε=0.20) or a more triaxial shape (ε=0.25, γ=15°). Finally the superdeformed shape at ε=0.6 spans the spin regime at an excitation of several MeV above yrast.

THE MOMENT OF INERTIA OF THE BANDS IN ^{152}Dy

The gamma-ray energies of the bands in ^{152}Dy have been measured to high precision and thus the static moment of inertia, $\mathcal{J}^{(1)}$, and the dynamic moment of inertia, $\mathcal{J}^{(2)}$, have been accurately determined. The dynamic moment is remarkedly constant varying by less than 4% over the entire spin range of the band. The magnitude is reproduced by theoretical calculations[7,10], which also reproduce the decrease in $\mathcal{J}^{(2)}$ as the spin increases. However, values of $\mathcal{J}^{(1)}$ are predicted which are too high when pairing is completely neglected but the fit can be improved by including dynamical pairing. The moment of inertia data has been interpreted[8] to demonstrate the stiffness of the superdeformed ^{152}Dy nucleus at high spin and it is suggested the changes can be considered as a liquid-solid type transition.

FEEDING OF YRAST SUPERDEFORMED STATES IN ^{152}Dy

The intensity of the superdeformed band as a function of spin is shown in fig.5. The band is observed up to 60h and it is completely fed by 50h. This is in marked contrast to lower deformation states in rare earth nuclei where discrete states have been seen up to 50h but the total feeding into discrete states is not reached until low spin.

Various mechanisms affecting the feeding of the band have recently been investigated[9] in a model in which γ-ray decay paths are simulated between normal (triaxial with ε=0.3) states and superdeformed states. Tunnelling between the two shapes is allowed and also competition within the different decay routes of collective E2 decays (400 W.u. normal, 2500 W.u. superdeformed) statistical E1 decays. The superdeformed E1 transitions are enhanced due to both their lower level density and the large splitting of the giant dipole resonance which brings one third of the strength down to 8 MeV. The tunnelling means that, particularly at higher excitation energies, the superdeformed decays are often followed by normal decays and that the superdeformed yrast states are associated pre dominantly with higher energy E1 γ-rays which have successfully competed with the strong collective decays. The model can fit the feeding pattern but not the extremely rapid decay out of the band unless another effect, such as the reappearance of neutron pairing at a frequency of 0.3 MeV, is included.

At Daresbury, the dependence of the intensity of the discrete superdeformed band has been investigated as a function of the bombarding energy of the ^{48}Ca beam. Sufficient γ-γ coincidence data were available at each energy to produce coincidence spectra of the band following gating on the lowest six transitions. These yield accurate intensity measurements which were compared with the population of the oblate states reflecting the total intensity of the

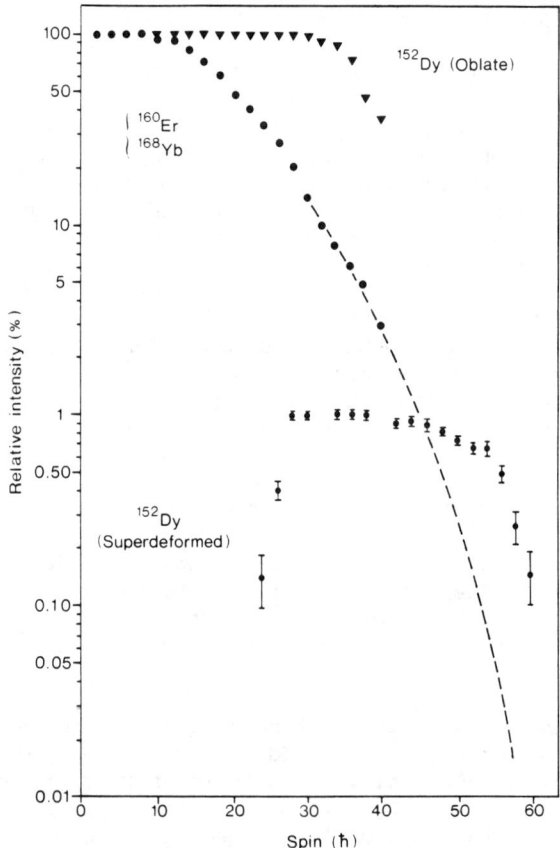

Fig.5. The total intensity measured in discrete lines for (a) well-deformed ^{160}Er and ^{168}Yb nuclei and (b) the oblate states in ^{152}Dy. The average entry spin in all cases is between 50$\hbar$ and 55$\hbar$. The intensity distribution of the superdeformed transitions in ^{152}Dy illustrates its low intensity, high spin formation and rapid de-excitation.

^{152}Dy channel. The results are shown in fig.6 and demonstrate that the fraction of the intensity of the ^{152}Dy channel populating superdeformed states increased as the bombarding energy decreased from 212 MeV to 195 MeV. This result is surprising as it was expected that the most critical factor was the angular momentum and it is at the highest energies that, on average, maximum angular momentum is observed in the ^{152}Dy channel. The reduction in intensity at 185 MeV indicates that high angular momentum is important but it is clearly not the only factor.

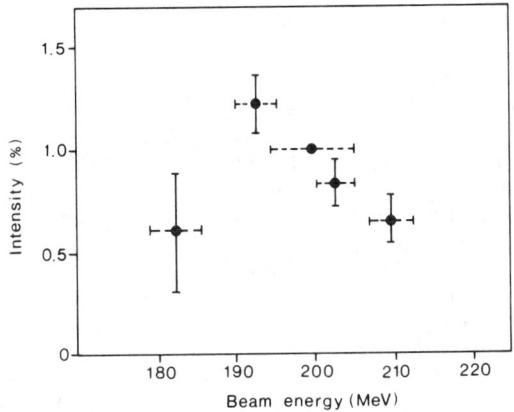

Fig.6. The intensity of the superdeformed band in ^{152}Dy relative to the total population intensity in ^{152}Dy as a function of the ^{48}Ca bombarding energy.

The explanation of this effect must be associated with lower excitation energy (or temperature) in either the compound system ^{156}Dy* or the final nucleus ^{152}Dy* or a combination of the two scenaria. The former explanation requires that the compound system ^{156}Dy* has a finite probability for formation in the superdeformed well, that this probability increases as the temperature in the system is lowered, and that there is a strong link between the yrast superdeformed states in ^{152}Dy and the superdeformed shape in ^{156}Dy*. The second explanation is that low excitation energy in the final nucleus ^{152}Dy* is a vital ingredient in the formation of the superdeformed states. The important parameters are the density of states in the superdeformed and normal structures. At the same excitation energy above yrast for each structure the density of superdeformed states is approximately an order of magnitude smaller than for the normal states. Thus only when the superdeformed states are both yrast and the excitation is low is there a large population intensity. The barrier between the two structures vanishes as the temperature increases and the tunnelling between the two structures becomes essentially unhindered with the result that due to their much larger level densities the normal states dominate at the higher temperatures.

THE ISLAND OF SUPERDEFORMATION

The search for superdeformed bands has intensified in the eighteen months since the observation in ^{152}Dy. The predicted[10,11] "island" of superdeformation has its peak around ^{152}Dy and ^{150}Gd with N=86 and Z=64,66 where the deformation has $\varepsilon=0.6$ and the axis ratio is 2:1. The stability decreases rapidly with higher N and Z but it is calculated to persist as N decreases until ^{144}Gd is reached where the deformation has decreased to $\varepsilon \simeq 0.5$. The predictions indicate that Z decreases to 58,60 and N declines further until the smaller deformation of $\varepsilon=0.4$ is reached at ^{132}Ce (Z=58, N=74) which was the first reported example[12] in this region. Extensive experimental activity by many groups has resulted in a lot of progress in the lower deformation area of cerium and neodymium but far less progress at the higher deformation close to ^{152}Dy.

The first indication in ^{152}Dy was the observation[13] of ridges with the appropriate separation in E_γ-E_γ correlation spectra. Similar weaker ridges have been observed in ^{144}Gd, ^{146}Gd and ^{150}Dy, but no discrete line band has so far been identified. Recently a second discrete line band has been found in ^{149}Gd using the Canadian 8π spectrometer[14] at Chalk River. One of the reasons for the lack of new sightings may be that experimentalists have raised the beam energy to maximise the angular momentum in the chosen final nucleus and so the temperature and excitation energy has been too high.

The band in ^{132}Ce is, to date, the only example in the cerium and neodymium nuclei for which the quadrupole moment has been measured[15] confirming that the prolate deformation has an axis ratio of 3:2 with $\varepsilon \sim 0.4$. The past year has seen the number of examples increase dramatically due to the work at Daresbury[16] and Berkeley[17] with bands in ^{133,134,135,136,137}Nd, ^{133}Pr and ^{131}Ce. The cerium and neodymium bands have several features which are different from ^{152}Dy; the larger reduction of the dynamic moment of inertia $\mathcal{J}^{(2)}$

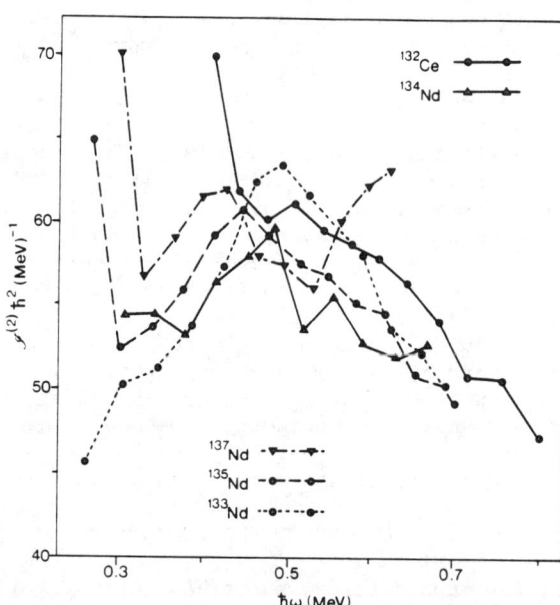

Fig. 7. The variation of the dynamic moment of inertia for the observed superdeformed bands in the cerium and neodymium nuclei as a function of frequency.

with increasing spin (fig.7); the variety of different frequencies at which the band de-excites; and a different feeding intensity into the bands.

The reduction in the moment of inertia is similar for most of the nuclei and corresponds to a drop of 30% for ^{132}Ce over the frequency range from 0.4 MeV to 1.0 MeV. It is suggested theoretically that the reduction is related to the limited angular momentum carried by the valence nucleons. However, in ^{137}Nd there is an increase of $\mathscr{J}^{(2)}$ at a frequency of 0.6 MeV and this is probably associated with a band crossing of $h_{9/2}$ protons which has been predicted to occur in calculations[18] on the neighbouring nucleus ^{135}Nd. The odd nuclei stay in the band to much lower spin (or frequency) and in the neodymium nuclei show a reduction of $\mathscr{J}^{(2)}$ as the frequency decreases below that reached in ^{132}Ce. This phenomenon is probably related to the onset of pairing and the associated influence of a band crossing. The intensity distribution curves have some interesting features. The odd neodymium nuclei exhibit very large populations of between 8% and 20% of the total intensity feeding that particular nucleus. Their fall off in intensity occurs at a lower frequency than the cerium nuclei, the half intensity point being at frequencies of 0.55 MeV in neodymium and 0.7 MeV in cerium. These data indicate that the superdeformed structure is yrast down to low spins of 30ℏ to 40ℏ and the strong population is also probably related to the larger level density at these smaller deformations. Thus a considerable body of systematic data is being built up in this mass region and detailed theoretical calculations are awaited to explain the many interesting features and particularly the role of the various proton and neutron orbitals which are involved.

CONCLUSION

This talk has surveyed the present status of the field of superdeformation in rapidly rotating nuclei. The new data on ^{152}Dy from Daresbury has confirmed the shape is indeed that of a prolate rotor with a 2:1 axis ratio, and it has yielded unexpected features in the feeding of the band. The simulation of the feeding paths is proving a useful technique to examine and test the new insights into the role of the giant dipole resonance, the level densities and the onset of pairing. The extensive new data on the cerium and neodymium nuclei are providing a wealth of systematic data to investigate various features of the bands. The past year has seen a surge of experimental activity and this seems set to increase in the coming year expanding our knowledge and understanding of superdeformation at high spin.

ACKNOWLEDGEMENTS

The work reported here is the result of hard work by many people. The major share of the ^{152}Dy analysis has been borne by M.A. Bentley and B.M. Nyakó and the experiments were carried out with the assistance of colleagues at Daresbury Laboratory, the University of Liverpool and Niels Bohr Institute.

REFERENCES

1. P.J. Twin et al, Phys. Rev. Lett., 57, 811 (1986).
2. P.J. Twin et al, Nucl. Phys., A409, 343c (1983).
3. P.J. Nolan et al, Nucl. Instr., A236, 95 (1985).
4. T.L. Khoo et al, Phys. Rev. Lett., 41, 1027 (1978).
5. J.C. Merdinger et al, Phys. Rev. Lett., 42, 23 (1979).
6. B.M. Nyakó et al, Phys. Rev. Lett., 56, 2680 (1986).
7. I. Ragnarsson and S. Åberg, Phys. Rev. Lett., 180, 191 (1986).
8. W.J. Swiatecki, Phys. Rev. Lett., 58, 1184 (1987).
9. B. Herskind et al, (to be published).
10. J. Dudek et al, (to be published).
11. R.R. Chasman, Argonne Report 4912-TH-86 (1986).
12. P.J. Nolan et al, J. Phys., G11, L17 (1985).
13. B.M. Nyakó et al, Phys. Rev. Lett., 52, 507 (1984).
14. P. Taras, private communication.
15. A.J. Kirwan et al, Phys. Rev. Lett., 58, 467 (1987).
16. R. Wadsworth et al, J. Phys. G. (in press).
17. E.M. Beck, et al, Phys. Rev. Lett., (in press).
18. T. Bengtsson, Nordita preprint, 87/14N (1987).

SPECTROSCOPY OF NEUTRON-DEFICIENT TANTALUM-TO-RHENIUM ISOTOPES [*]

F. Meissner, T. Hild, W.-D. Schmidt-Ott, E. Runte, H. Salewski, V. Freystein
University of Göttingen, 3400 Göttingen, Fed. Rep. of Germany
R. Michaelsen
Hahn-Meitner-Institut, 1000 Berlin 39, Germany

1. ABSTRACT

Alpha-decay half-lives were measured for the first time for ^{160,161}Ta of 1.2(3), and 2.7(2) s, respectively. Alpha energies and half-lives of ^{162}Ta are 4.88(1) MeV, $T_{1/2}$ =5(3) s, of ^{163}Ta are 4.63(1) MeV, $T_{1/2}$ =11(2) s, and the half-life of ^{164}Ta is 14.9(2) s. Two α-decay chains were closed and negative S_p values derived for ^{171}Ir, ^{175}Au and ^{179}Tl. For ^{166}W, β-decay was measured for the first time, a decay scheme derived and the α-branching obtained to be 0.030 (7) %. For ^{167}W, α-decay was measured for the first time with $E_α$ = 4.55(2) MeV, from γγ-coincidences a decay scheme is constructed. We have remeasured half-lives and γ-rays of ^{168}W to ^{170}W. For ^{171}W a new half-life of 2.38(4) min is reported, from γγ-coincidences a decay scheme is constructed. The decay of the new isotope ^{171}Re was measured.

2. INTRODUCTION

Scarce information was known on neutron-deficient isotopes between the magic numbers N=82 and Z=82 in the region of vanishing α-decay. The present spectroscopic investigations were started to obtain fundamental decay characteristics like α-decay energies, α-decay branchings, half-lives of new isotopes, β-delayed γ-decay, and the descriptions of excited levels. The investigations were performed on isotopes of the elements hafnium to iridium, and we are reporting on recent results on tantalum-to-rhenium isotopes. These elements are characterized by a high refractoriness, which deteriorates the use of ion-sources for mass-separation. We therefore have applied the helium-jet technique combined with a fast tape-transport system.

3. EXPERIMENTAL METHOD

Various fusion-evaporation reactions were applied by using the heavy-ion beams of the VICKSI accelerator. Typical beam intensities of 5 particle-nA and target

Table 1. Compilation of experimental parameters

Target	Beam	Energy range [MeV]	Produced isotope
^{133}CsCl	^{36}Ar	166-255	Tantalum
^{130}BaF$_2$, ^{132}BaF$_2$	^{35}Cl	156-168	Tantalum
^{136}BaF$_2$, ^{138}BaF$_2$	^{36}Ar, ^{40}Ar	177-205	Tungsten
^{136}BaF	^{36}Ar	165-194	Rhenium

[*]) The work has been funded by the German Ministry for Research and Technology (BMFT) under contract number 06 GÖ 456.

thicknesses of 1-to-2 mg/cm² were used. The experiments are compiled in table 1. The targets were evaporated on 34, and 75 µg/cm² carbon backings and foils of 0.56 mg/cm² aluminium. The compound nuclei were expelled from the target, stopped and transported with helium.

The target chamber and transport system is given in fig. 1. Tantalum degraders in front of the target were used to vary the bombarding energies as given in table 1. For identification of the element, cross bombardment and coincidence counting with X-rays were applied, the isotope assignment was performed by measurement of excitation functions. We have compared the excitation of unknown γ-rays with well known γ-rays of $^{163-166}$Ta from the ^{36}Ar + ^{133}Cs reaction.

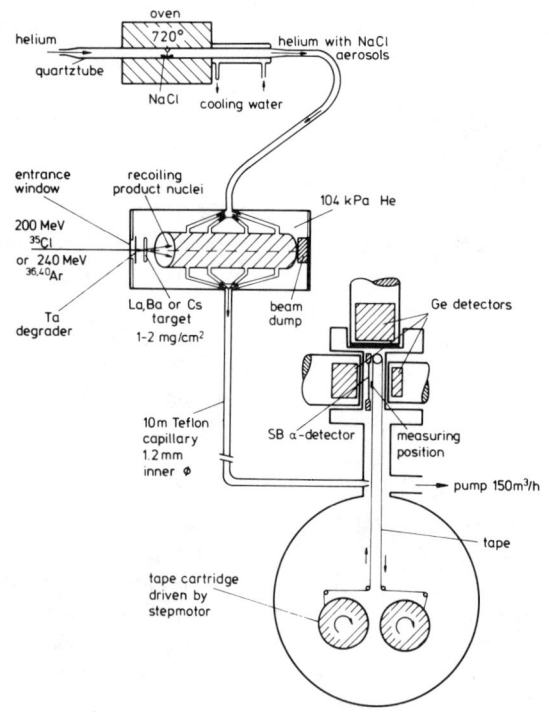

Fig. 1.
Target chamber, helium-jet transport system with aerosol oven, tape cartridge system, and detectors for α-, and γ-counting.

4. EXPERIMENTAL RESULTS

Tantalum isotopes. The new alpha half-lives of ^{160}Ta, and ^{161}Ta have recently been published [1] to be 1.2(3) s, and 2.7(2) s, respectively, while the assignment of α-energies to these isotopes was already known [2]. For ^{162}Ta, and ^{163}Ta new α-rays were found with energies 4.88(1), and 4.63(1) MeV, respectively [1]. Using these energies, and the masses of the isotopes ^{150}Ho, and ^{147}Tb [3], two α-decay chains were completed and negative proton separation energies derived for ^{171}Ir, ^{175}Au, and ^{179}Tl, characterizing these as possible proton ground-state emitters [3]. We remark that the α-assignment of 4.63 MeV to ^{163}Ta was supported by the accurate half-life measurements

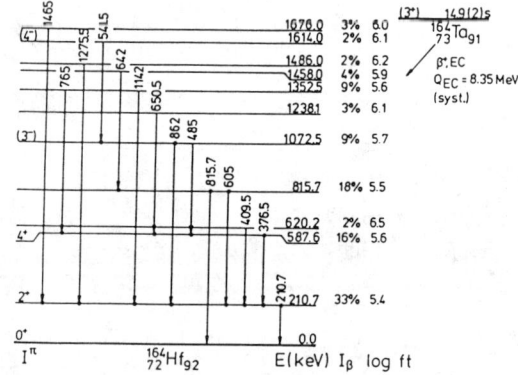

Fig. 2.
New decay scheme of ^{164}Ta. Spin values are from in-beam work [6].

of ^{163}Ta [1,4] and of ^{164}Ta, $T_{1/2}$= 14.9(2) s [1], removing an earlier assignment to ^{164}Ta [5]. In the ^{164}Ta decay besides the already known cascade between the 4^+, 2^+_1 and 0^+ states in ^{164}Hf, twelve new γ-rays were found [8] and included in the decay scheme in fig. 2.

Tungsten isotopes. The α-rays, and β-delayed γ-rays were investigated in the decays of ^{166}W, and ^{167}W. For the first, the α-decay was already known [7] with $T_{1/2}$ = 16(3) s, and $E_α$ = 4.739(5) MeV. Presently, for ^{166}W a half-life of 18.8(4) s was measured, and a complex decay scheme was derived [8], the α-branching being 0.030(7) %. The assignmnent of the strongest γ-rays of the tungsten isotopes is demonstrated by the excitation functions given in fig. 3. In ^{167}W, α-decay was measured for the first time, with $E_α$= 4.55 (2) MeV. Twelve

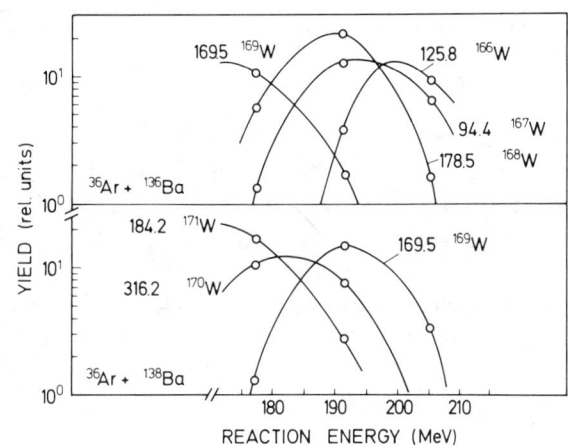

Fig. 3.
Excitation functions for $^{166-169}$W γ-rays using the ^{36}Ar beam and the ^{136}Ba target, and for $^{169-171}$W, with ^{138}Ba target.

γ-rays and their coincidence relations were measured. In the decay scheme given in fig. 4 this information is comprised.

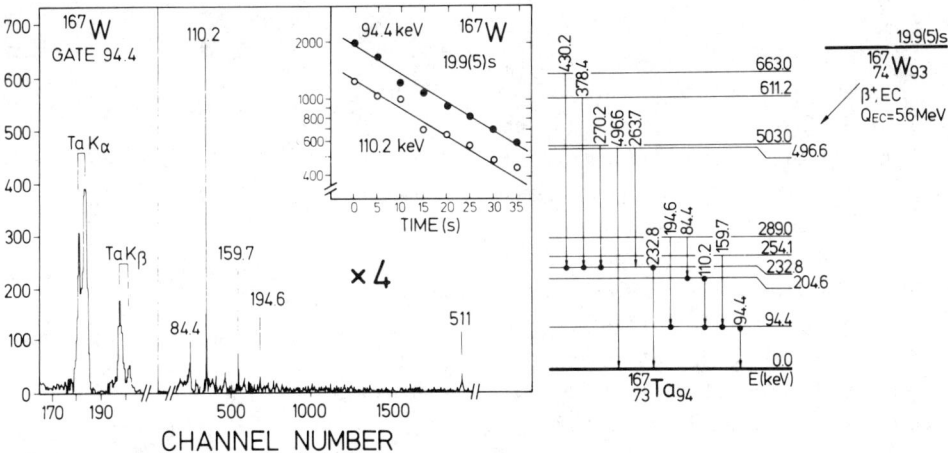

Fig. 4.
Coincidence spectrum of ^{167}W γ-rays. Inset: Decay of 94.4, and 110.2 keV γ-rays; to the right, the decay scheme, the Q_{EC}-value is measured by the EC/β$^+$-ratio.

The present data on the ^{168}W decay are comparing well with the known information [9]. We have measured the half-life of 53(2) s, and in addition to the known 178.5 keV γ-ray, two new ones of 173.8, and 352.3 keV. In the ^{169}W decay, the present half-life of 76(6) s is essentially larger than the earlier value [9]. In addition to the known γ-rays of 96.6, 136.0 and 169.5 keV, fourteen new γ-rays were identified and arranged in a level scheme [10]. For the ^{170}W decay, the earlier data [9] were essentially confirmed. Extensive coincidence data have been measured in the decay of ^{171}W. The present new half-life of this isotope is 2.38(4) min, and the decay scheme is presented in fig. 5.

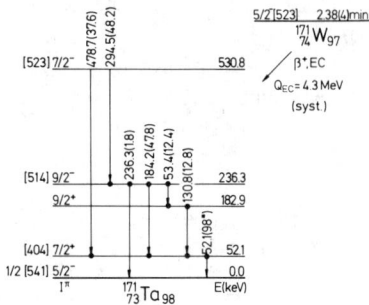

Fig. 5.
Decay scheme of ^{171}W, coincidence relations are indicated by dots. By comparison with neighbouring nuclei and with in-beam studies [11] a level assignment is suggested.

Rhenium isotopes. We have produced $^{170-172}$Re in the ^{36}Ar + ^{139}La reaction and measured the excitation functions. A new isotope, ^{171}Re, was observed with the half-life of 15.2(4) s, and a decay scheme was derived [12].

REFERENCES
[1] E. Runte, et al., Z. Phys. A324, 119 (1986)
[2] S. Hoffmann, at al., Z. Phys. A291, 53 (1979)
[3] W.-D. Schmidt-Ott, et al., HMI report 437, 23 (1987)
[4] C. F. Liang, et al., Z. Phys. A321, 695 (1985)
[5] U. J. Schrewe, et al., Z. Phys. A310, 295 (1983)
[6] K. P. Blume, et al., Nucl. Phys. A464, 445 (1987)
[7] K. S. Toth, et al., Phys. Rev. C12, 533 (1975)
[8] T. Hild, Diplomarbeit Göttingen 1987
[9] J. Estevez, et al., P6-85-778, JINC Dubna, 1985
[10] F. Meissner, et al., to be published
[11] J. C. Bacelar, et al., Nucl. Phys. A442, 547 (1985)
[12] E. Runte, et al., Z. Phys. A328 (1987)

STUDY OF TRANSITIONAL DOUBLY-ODD ^{186}Ir and ^{184}Ir.

A. Ben Braham
Faculté des Sciences de Tunis, Tunisie

C. Bourgeois, P. Kilcher, J. Oms, B. Roussière, J. Sauvage
Institut de Physique Nucléaire, 91406 Orsay, France

M.G. Porquet
C.S.N.S.M., 91406 Osay, France

A.J. Kreiner
Departamento de Fisica CNEA, 1429 Buenos Aires, Argentina

and the ISOCELE Collaboration

ABSTRACT

The transitional doubly-odd iridium nuclei with A=184 and 186 have been studied from the β^+/EC decay of the corresponding platinum isotopes using the on-line mass separator ISOCELE. Configurations can be reasonably attributed to the low-lying states of ^{184}Ir in agreement with results already known. On the other hand an E3 transition observed in ^{186}Ir suggests that the known long-lived 1.7h 2$^-$ state is located at 137.5 keV above the 16h 5$^+$ state, raising questions about structure of this latter state.

INTRODUCTION

There is currently great interest[1,2] in studying the coupling modes and residual interactions of the valence proton and neutron (Vpn) in doubly-odd nuclei, especially in the transitional region below lead where shape coexistence phenomena have been observed. In odd-proton gold nuclei, the ground state (gs) shape changes between A=187 and A=185 [refs.3,4] and shape coexistence was found in ^{186}Au [ref.5], while in odd-neutron neighboring nuclei, the shape transition is expected between ^{185}Os and ^{187}Pt [ref.6]. In the present work we report some results on ^{184,186}Ir where such phenomena could also be present.

EXPERIMENTAL PROCEDURE

Low-energy levels of ^{184}Ir and ^{186}Ir were populated in the β^+/EC decay of ^{184}Pt and ^{186}Pt produced through the decay of ^{184,186}Au. Using the ISOCELE facility numerous spectroscopic measurements have been performed from mass-separated radioactive sources. They include singles γ and e^- spectra, γ-γ-t and γ-X-t coincidences, half-life determinations, and high-resolution electron spectra obtained by means of a semi-circular magnetic spectrograph.

RESULTS AND DISCUSSION

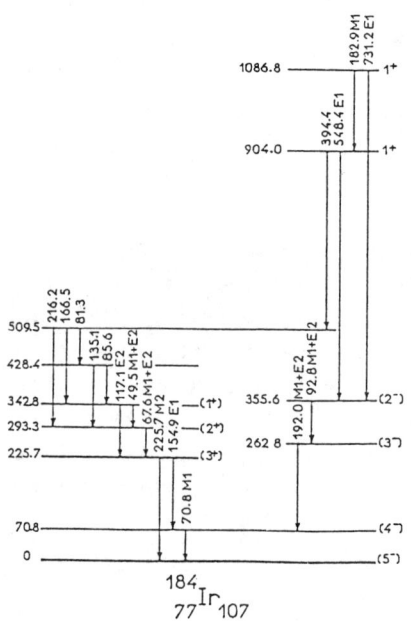

Fig.1 Partial level scheme

The spin of the gs of ^{184}Ir has very recently been determined to be 4 or 5 with K=4 or/and 5 from the log ft values deduced from a careful study of the ^{184}Ir decay towards the ^{184}Os levels[7]. This result is in agreement with the spectroscopic quadrupole moment previously obtained for the ^{184}Ir ground state[8]. A level scheme for ^{184}Ir has been established from our measurements. A $\pi 1/2^-[541]$ $\nu 9/2^+[624]$ configuration can be reasonably attributed to its gs since it gives rise to 5^- and 4^- states as lowest levels, the component proton and neutron orbitals being the gs configurations in the respective odd-mass neighbors. An isomeric state was found at 225.7 keV which decays to the gs and to a 70.8 keV level through, respectively, a 225.7 keV M2 and a 154.9 keV E1 transition. Its spin value is most likely 3^+ and it can be interpreted as the lowest level of the $\pi h9/2\nu 1/2^-[521]$ system[9]. The 4^- and 3^- states (at 70.8 and 262.8 keV) have most likely the same configuration as the ground-state whereas the 2^+ and 1^+ state (at 293.3 and 342.8 keV) possibly have the configuration of the isomeric state (fig.1).

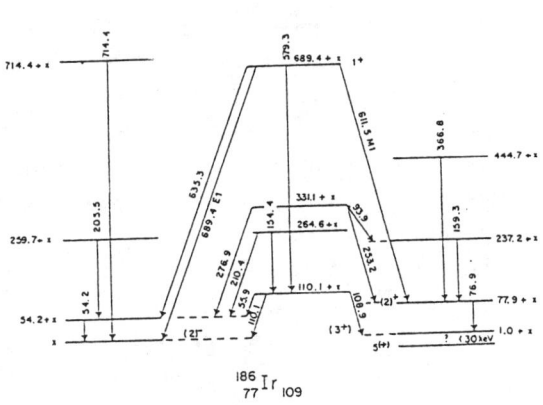

Fig.2 Partial level scheme proposed from previous data.

It is well known that two isomers exist in ^{186}Ir [ref.10] ; one has a half-life of 1.7h and its spin is believed to be 2^-, the other has a 16h half-life and its spin has been measured to be 5 with K=0 and/or 1 [ref.11]. There is strong evidence[12] suggesting positive parity for this state.

One possibility for explaining the preexisting results and our first set of data is indicated in fig.2. A search was undertaken to detect the low-energy transition which would de-excite the (3^+) level situated ∿ 1 keV above the 1.7h 2^- state, towards the 16h 5^+ state (see fig.2) but no candidate for such a transition was observed in the low-energy electron spectrum (fig.3). On the other hand, an anomalously large L conversion of the 137.1 keV $2^+ \rightarrow 0^+$

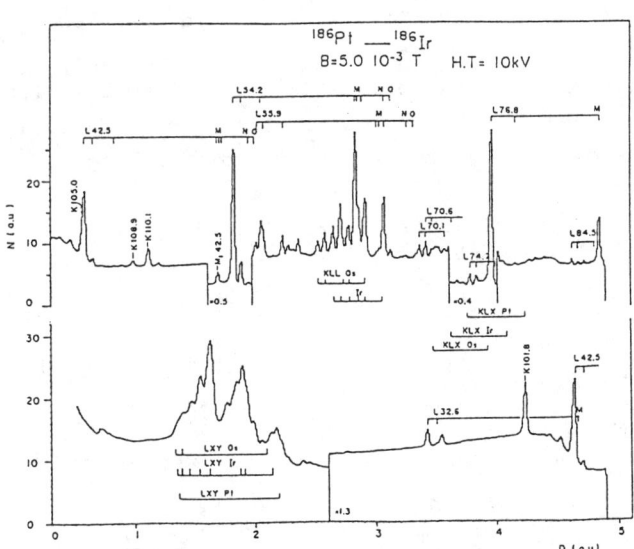

Fig. 3 Low-energy conversion-electron spectrum

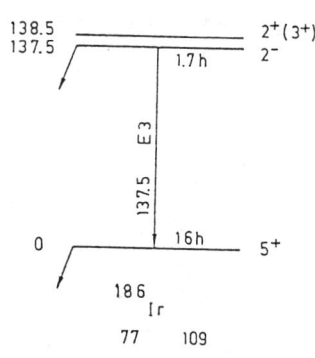

Fig. 4 Possible location of the 2⁻ state.

transition in ^{186}Os may be explained by the existence of a 137.5 keV E3 transition converting in ^{186}Ir. These results may indicate that the 1.7h 2⁻ isomeric state in ^{186}Ir is located 137.5 keV above the 5⁺ gs and partially deexcites through that isomeric E3 transition (fig.4).

The configurations which can give rise to the 5⁺ gs are $\pi 1/2^-[541]\nu 1/2^-[510]$ (admixed to the $\nu 3/2^-[512]$ and $\nu 1/2^-[521]$, see ref.12) and $\pi 1/2^-[541]$ $\nu 7/2^-[503]$, the other single-particle states being too high in energy to be considered. On the other hand, two-quasi-particle-plus-rotor model calculations[12], suggest, for both of the possibilities, the existence of other, nearby lying, closely related low-spin states ($I^\pi = 2^+, 3^+$ and 4^+). Their presence would be incompatible with the halflife of the 2⁻ isomer. Still another possibility would be to consider an oblate 5⁺ gs for ^{186}Ir, in which case candidate configurations are $\pi h11/2 \nu 3/2^-$ or $\pi 1/2^+ \nu i13/2$, where the single particle states would correspond to orbitals occurring in Pt and Au [refs.6,13].

This last possibility, however, seems unlikely in the light of the evidence obtained for ^{186}Ir, particularly from (H.I.,xn) reactions[12], which points to a predominantly prolate shape.

1. A.J. Kreiner et al., Nucl. Phys. A282, 243 (1977).
   A.J. Kreiner, Z. Phys. A288, 373 (1978).
   L. Bennour, thèse Orsay (1983).
2. A. Neskakis et al., Nucl. Phys. A390, 53 (1982).
3. C. Bourgeois et al., IPN Orsay, Rapport Annuel 19 (1986) and refs. quoted therein.
4. K. Wallmeroth et al., VIIth Int. Conf. on hyperfine interactions, Bangalore, India (1986).
5. M.G. Porquet et al., Nucl. Phys. A411, 65 (1983).
6. B. Roussière, thèse Orsay (1986), IPNO-T-86.05.
7. M.G. Porquet et al., to be published, and IPN Orsay, Rapport Annuel 28 (1986).
8. E. Hagn and E. Zech, Hyperfine Interaction, 14, 97 (1983).
9. J. Davidson et al., Z. Phys. A324, 363 (1986).
10. C.M. Lederer and V.S. Shirley, Ed., Table of isotopes (1978).
11. E. Hagn and E. Zech, Z. Phys. A297, 329 (1980).
12. A.J. Kreiner et al., Phys. Rev. C29, 1572 (1984).
    A.J. Kreiner et al., Nucl. Phys. A432, 451 (1985).
13. M.G. Porquet et al., Nucl. Phys. A451, 365 (1986).

DETECTION OF LOW-ENERGY CONVERSION ELECTRONS
AND LOCATION OF ISOMERIC STATES

P. Kilcher, J. Sauvage, C. Bourgeois, F. Le Blanc, J. Oms, B. Roussière, J. Munsch, J. Obert, A. Caruette, A. Ferro, G. Boissier, J. Fournet-Fayas, M. Ducourtieux, G. Landois, R. Sellem, D. Sznadjderman and the ISOCELE collaboration.
Institut de Physique Nucléaire, 91406 Orsay, France

A. Wojtasiewicz
Warsaw University, Poland

M.C. Abreu
Universidade de Lisboa, Portugal

A. Ben Braham
Faculté des Sciences de Tunis, Tunisia

K. Fransson
Stockholm University, Sweden

M.G. Porquet
C.S.N.S.M., 91406 Orsay, France

ABSTRACT

The detection of conversion electrons of short-lived isotopes at ISOCELE has been extended to very low-energy : a special tape-transport system has been built which allows the slowing down of the incoming radioactive ions and the acceleration of the electrons detected on photographic film in a flat magnetic spectrograph. Results have been acquired for the decay of mercury with A = 185,186,187,189,191, and of platinum with A = 184, and 186. As exemples the location of isomeric states deduced from our measurements are given to illustrate the power of high-resolution electron detection.

The studies of decay schemes of short-lived isotopes in the shape transitional region of mercury are actively pursued at the ISOCELE II on-line separator at Orsay. Of prime importance is the study of the conversion electrons of the highly-converted low-energy transition appearing frequently in odd-A or doubly-odd nuclei of this region. A flat magnetic spectrograph with photographic detection is currently used on-line at ISOCELE to obtain the electron spectra.

In order to detect the conversion electrons at very low energy (down to 1 keV) we have built a special tape transport system[1] which is able to move quickly the collected radioactive sources into the spectrograph and furthermore, must permit (i) to collect radioactive ions on the tape passing through a decelerating lens in order to keep the spectrograph good energy resolution (4/10000 for 6 keV electrons in normal operating conditions). Indeed, if the beam were not decelerated the ions would reach the collection point with a 44 keV kinetic energy and would be implanted into the tape. The electrons emitted from implanted radioactive atoms would be slowed down, making worse the electron peak outline towards low energy. (ii) to accelerate the conversion electrons at the measurement point in the spectrograph by 10 kV so the very low-energy electrons can reach the photographic film and be detected (fig.1).

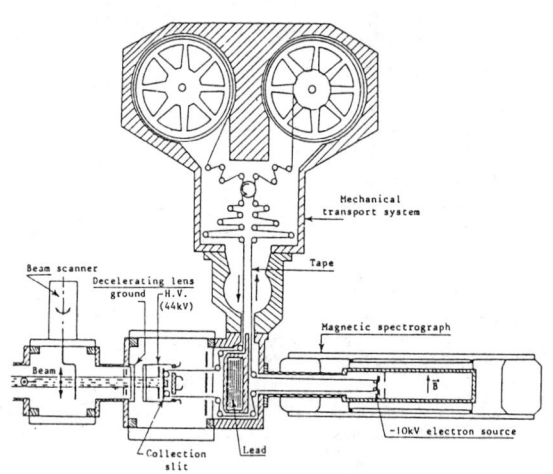

Fig.1. Lay-out of the tape transport system

The tape transport mechanics is a 16mm Debrie tape transport for cinema transformed so as the whole tape is in vacuum without damage for its own performances. The insulating tape must be put only locally at 44 kV at the collection point and at -10kV at the measurement point. Moreover the slowed-down ions must be collected on a conducting surface in order to realize a good high voltage contact and the highest collection efficiency ; therefore we chose a polyester tape on which equidistant Al-deposits have been evaporated. The difficulty was to overcome the causes of discharges developped between Al-deposits and the surrounding during the tape movement in the decelerating lens. The focusing in the decelerating lens has been studied in order to get stable and reproducible beams during long experiments. This has been obtained taking into account the two following principles : (i) The focusing will be as little as possible affected when the decelerating electric field will be established : (ii) At the lens entrance the incident beam will be precisely controlled. Then the collected ions represent about 90% of the incoming ion-

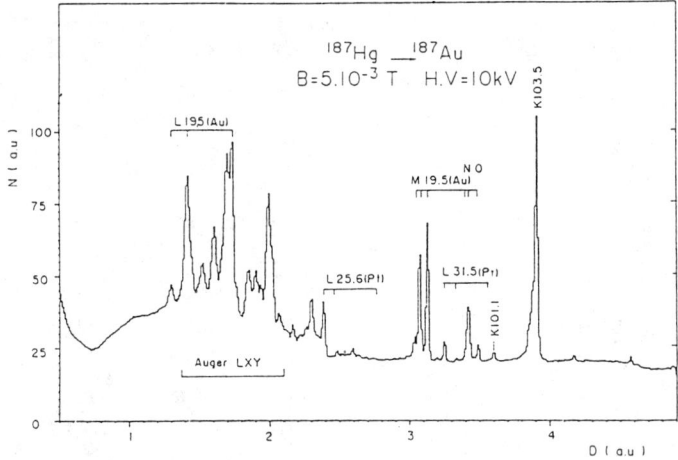

Fig.2. A low-energy conversion-electron spectrum beam. Results have been acquired for the following isotopes : ^{185}Hg, ^{185,186,187,189,191}Au and ^{184,186}Ir. One can observe in fig.2 the different L and M conversion electron lines of the 19.5 keV transition in ^{187}Au which will permit to know very precisely its mixing ratio (M1 + E2). On the same figure one can see also the line of the 101.1 keV E3 transition in this nucleus. These two transitions in cascade de-excite the isomeric state located at 120.6 keV [2] (fig.3).

The exact location of the $13/2^+$ isomeric state in ^{185}Hg was still an open question. Indeed two isomeric transitions had been observed namely an E3 65.3 keV and a (M1 + E2) 26.1 keV transition which were believed to correspond respectively to a $13/2^+ \rightarrow 7/2^-$ and $3/2^- \rightarrow 1/2^-$ transition [3]. Therefore an E2 transition with a very low-energy was expected to connect the $7/2^-$ and $3/2^-$ levels. In our low-energy conversion-electron measurements an E2 12.4 keV transition has been observed, fitting the blank of the isomeric-state de-excitation. So we can locate the $13/2^+$ state of ^{185}Hg at 103.8 keV (fig.3). From isotopic shift measurements[4] this $13/2^+$ state is known to correspond to a slightly-deformed oblate-shaped nucleus whereas the $1/2^-$ ground-state was found to correspond clearly to a well-deformed prolate-shaped one. Furthermore studies of high-spin level in ^{185}Hg have recently shown the existence of two rotational bands built on the $7/2^-$ level and on the $1/2^-$ ground state which correspond to a prolate deformation[5] since they behave like the ones observed in the prolate

^{185}Pt^{6,7}. It is worth noting that the 3/2⁻ level located at 26.1 keV could be similar to the 3/2⁻ ground-state of the heavier mercury isotopes, it would correspond then to an oblate shape. Thus in ^{185}Hg two intrinsic states corresponding to an oblate shape would exist at low-energy namely the 13/2⁺ level located at 103.8 keV and the 3/2⁻ level at 26.1 keV. The results concerning the doubly-odd Ir isotopes are discussed in another contribution in this conference. The fig.3 sums up the isomeric states in the nuclei of the transitional region around A = 186, whose energies were precisely defined thanks to measurements performed at ISOCELE.

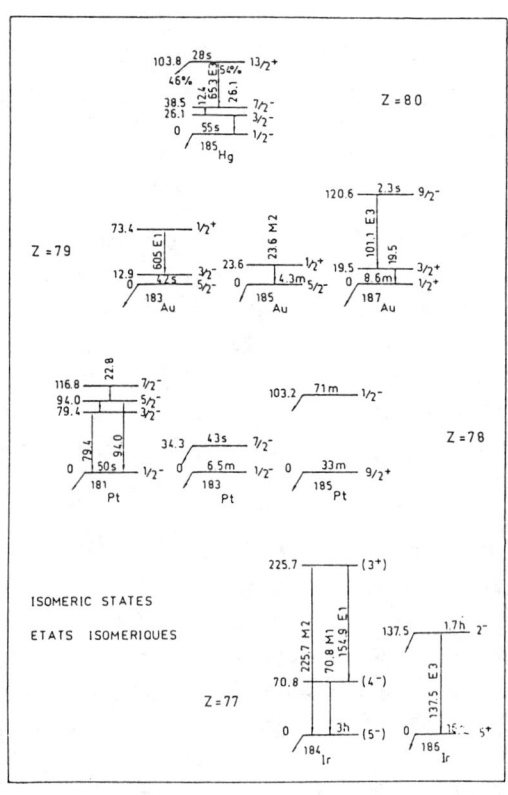

Fig.3 Recapitulation of isomeric states

1. P. Kilcher et al., submitted to Nucl. Inst. Meth.
2. C. Bourgeois et al., Nucl. Phys. A295, 424 (1979).
3. C. Bourgeois et al., Nucl. Phys. A386, 308 (1982).
4. P. Dabkiewicz et al., Phys. Lett. 82B, 199 (1979).
5. F. Hannachi, private communication.
6. S. Pilotte et al., M.M.A.L., Annual Report, 26 (1986).
7. B. Roussière, Thèse, Orsay (1986).

THE ALPHA-DECAY OF MASS-SEPARATED ^{225}Th

C.F. Liang, P. Paris, Ch. Briançon
C.S.N.S.M. Bât. 104, 91406 Campus Orsay, FRANCE

ABSTRACT

The alpha-decay of ^{225}Th ($T_{1/2}$ = 8mn) to ^{221}Ra levels was investigated. Mass-separated ^{225}Th sources were produced with the ORSAY-ISOCELE facility. The selectivity in Z was obtained using fluorination techniques. Single α and γ spectra and α-γ coincidences were simultaneously recorded.

The low-lying levels of ^{221}Ra are interpreted as two parity-coupled bands ($K^\pi = 5/2^\pm$ and $3/2^\pm$) expected for octupole deformed odd-nuclei in the Z = 88, N = 134 region.

INTRODUCTION

By involving simultaneously high-j neutron and proton systems, the light actinides provide an interesting probe of Coriolis effects. Moreover, a strong coupling between i13/2 and f7/2 proton orbitals as well as j15/2 and g9/2 neutron orbitals produces negative parity states (in even-even nuclei) of octupole nature, very low in energy (for example the 1^- ($K^\pi = 0^-$) band-head lies as low as 242 keV in ^{222}Ra). In case of reflection asymmetric shape, the parity is no more a good quantum number, this would give rise in even-even nuclei to a unique yrast band hybridized in parity ($0^+, 1^-, 2^+$...). Up to now there exists no experimental evidence of such bands at low spin. This situation appears at higher spin when the rotation plays an important role in the alignment of the octupole angular momentum along the rotational axis, which stabilizes the octupole deformation (see ref.[1] and references quoted therein).

In the odd-nuclei, one of the fingerprints of a stable octupole deformation would be the presence of bands based on states degenerated in parity. Experimentally one observes bands built on parity doublets with spin $K^{\pm}$, the splitting of the band-heads beeing related to the softness of the potential energy in respect with the axial octupole deformation. Another important characteristic is the presence of enhanced E1 transitions between the bands built on a parity-doublet. Recently theoretical calculations on odd-nuclei, implying stable octupole deformation [2,3] or based on an octupole-multiphonon model [4] agree to foresee the existence of bands based on parity-doublets.

From the experimental results it is thus difficult to say that the nuclei of this region have really a stable octupole deformation in the ground state. The study of odd nuclei is of much interest since one may expect that the odd neutron or proton can polarize the already soft potential in such a way that the octupole shape becomes more stable than in the neighbouring even-even nuclei.

Such bands based on parity doublets have been observed in other Radium isotopes : ^{225}Ra ($K^{\pi} = 1/2^{\pm}$, $K^{\pi} = 3/2^{\pm}$), ^{223}Ra ($K^{\pi} = 3/2^{\pm}$, $K^{\pi} = 1/2^{\pm}$) [5,6]. In these nuclei the $1/2^{\pm}$ bands present a decoupling parameter of the same order but with opposite signs, which is also an indication of the influence of octupole softness. A similar behaviour is also expected in ^{221}Ra.

The ^{221}Ra nucleus is essentially fed by the α-decay (> 90%) of ^{225}Th ($T_{1/2}$= 8mn). The α-spectrum has been studied by magnetic spectrography in 1961 by Ruiz [7] who identified the first low-energy levels in ^{221}Ra and observed with a NaI (Tl) detector some intense γ-transitions. Weak γ-rays of lower energies were not seen and, from these results, the existence of possible octupole configurations is not obvious. The ground state spin in ^{221}Ra was measured as 5/2 at ISOLDE by Ahmad et al. [8] by laser spectroscopy. As far as we know, since the Ruiz measurements, no other ^{225}Th-decay study has been published. This is probably due to the difficulty of processing the isotopic separation of Thorium, in addition to the short ^{225}Th half-life.

EXPERIMENTAL METHODS

The selective fluorination method, extensively used at the ISOCELE facility for the separation of refractory elements : Hf-Ta, La-Ba, Zr-Y, [9] has been extended to the homologous actinides Th-Ac-Ra. The choice of a target is limited in this region to Thorium and Uranium. We used a Th-Ce alloy (40-60% in weight) in order to decrease the target melting point. The 10 g target was bombarded by a 280 MeV-2μA ^{3}He beam. The target was used as anode in the ion-source arc-chamber [10] and was continuously fluorinated by introducing CF_4 vapour. The weak $^{225}ThF_3^+$ collected activity (at most 3000 ions/s) resulted from the slow Thorium-target evaporation. However, more than 90% of the 1 mA ion-beam intensity extracted from the source corresponded to the Cerium fluorination.

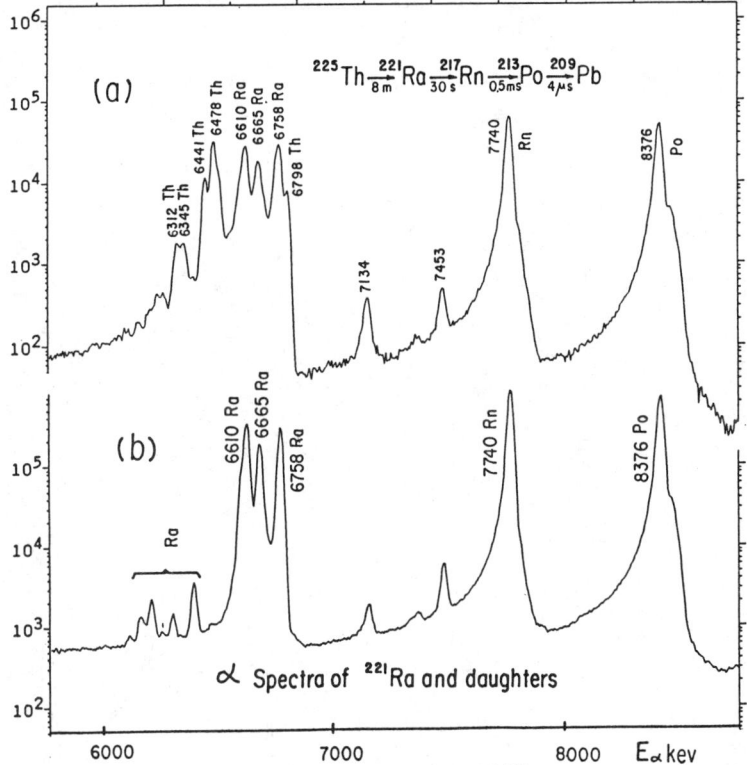

Fig. 1 - a) Direct α-spectrum, from ^{225}Th and daughters, for A=225+57 (ThF_3^+) ; b) Direct α-spectrum, from ^{221}Ra and daughters, for A = 221+19 (RaF^+).

The daughter (^{221}Ra (30 s) → ^{217}Rn → ...) is in radioactive equilibrium and the two complex α-spectra issued from ^{225}Th and ^{221}Ra overlap (fig. 1a and 1b). In order to discriminate between them, the Radium was collected, practically free of Thorium, by separating RaF$^+$. The fluorination also eliminated the otherwise overwhelming Francium. Because of the Radium volatility, the Ra production is up to two orders of magnitude higher than the Thorium one.

A tape-transport system moved the collected activities between a 2 cm^2 surface-barrier α-detector and a coaxial γ-ray detector in a 180° close geometry. Collection-measurement cycles were 13 mn (^{225}Th) and 1 mn (^{221}Ra). Single α and γ spectra and 4K-4K α-γ coincidence measurements were simultaneously recorded.

RESULTS

The Table I shows the γ-transitions observed in coincidence with the α particles issued from ^{225}Th. The total γ-spectrum coincident with α-decays is presented Fig. 2. The direct α-spectra corresponding to the ^{225}Th and ^{221}Ra decays are shown on Fig.1 (a,b).

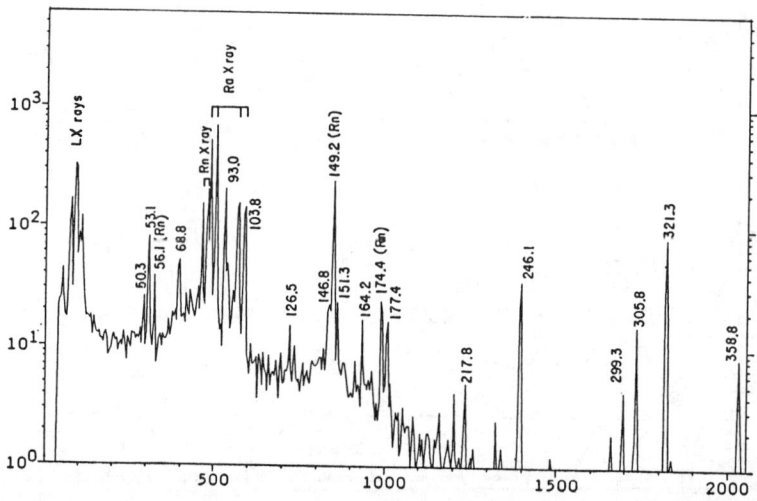

Fig. 2 : Total γ-ray spectrum coincident with α-rays of ^{225}Th and daughter products.

TABLE 1

Properties of gamma transitions in ^{221}Ra observed in this work

Energy (keV)	$I_\gamma/100\alpha$	Levels Initial → Final	Multipolarity
50.3	0.11 ± 0.02	103.4 → 53.1	
53.1	1.20 ± 0.10	53.1 → 0	M1
68.8	0.57 ± 0.05	121.9 → 53.1	M1
121.9	0.14 ± 0.03	121.9 → 0	
126.5	0.18 ± 0.04	485.3 → 358.8	(E1)
129.1	0.15 ± 0.03	450.5 → 321.3	
146.8	0.80 ± 0.10	146.8 → 0	E1
148.5	0.20 ± 0.10	201.6 → 53.1	
151.3	0.70 ± 0.10	450.5 → 299.3	E1
164.1	0.60 ± 0.10	485.3 → 321.3	E1
177.3	1.15 ± 0.10	299.3 → 121.9	
212.0	0.20 ± 0.10	358.9 → 146.8	
217.8	0.50 ± 0.08	321.3 → 103.4	(E1)
246.1	5.75 ± 0.20	299.3 → 53.1	M1
299.3	1.13 ± 0.15	299.3 → 0	M1
305.8	4.50 ± 0.30	358.9 → 53.1	M1
321.3	25.0 ± 0.5	321.3 → 0	M1
358.9	4.1 ± 0.3	358.9 → 0	M1
KX-ray	27.0 ± 1.0		

The relative intensities of the various α-groups populating the ^{221}Ra levels are indicated in Fig. 3 which shows the level scheme. The transition multipolarities are deduced from the intensity balance between α-feeding and γ-deexcitations.

The low-energy levels clearly exhibit the coupling of two bands : $K^\pi = 5/2^+$ (ground-state band) and $K^\pi = 5/2^-$, strongly related by E1 transitions, as expected by the models including octupole deformation. Such a behaviour is again more apparent on the link by several E1 transitions between a $3/2^-$ band, near 450 keV, with a $3/2^+$ one, at about 300 keV (the experimental results do not exclude the possibility for an inversion between the 3/2 and 5/2 spins at 299.3 and 321.3 keV).

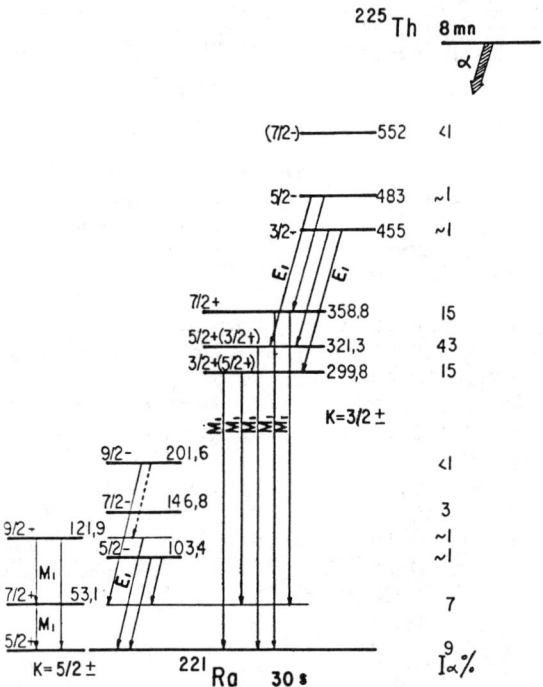

Fig. 3 : Experimental ^{221}Ra level scheme.

The weaker α-feeding to higher energy levels prevented us to observe the $K^{\pi} = 1/2^{\pm}$ coupled bands expected in this region. The Leander calculations predict a change of sign for the "decoupling invariant" "a.p", associated to the $\Omega = 1/2$ parity-mixed orbital between ^{223}Ra (+ 2.7) and ^{221}Ra (- 3.5). This reinforces the interest for improving the ^{225}Th production, in order to observe the $K^{\pi} = 1/2^{\pm}$ bands in ^{221}Ra.

CONCLUSIONS

The fluorination process seems to be as efficient in the actinide region as for the other analogous lighter elements, even if the Th-Ce alloy is probably not the best combination to maximize the Thorium yield. In that respect, other tests are needed.

The Fig. 4 summarizes the octupole configurations observed in ^{225}Ra, ^{223}Ra and, here, ^{221}Ra. The K-values implied in the coupled bands $K^\pi = 1/2^\pm$, $3/2^\pm$, $5/2^\pm$ correspond to the N = 137, 135 and 133 orbitals predicted by Leander et al., using a folded Yukawa potential [2], for an octupole deformation parameter $\varepsilon_3 \sim 0.08$ (one may notice that in these calculations the potential energies were obtained for odd nuclei by interpolation of the results of neighbouring even nuclei).

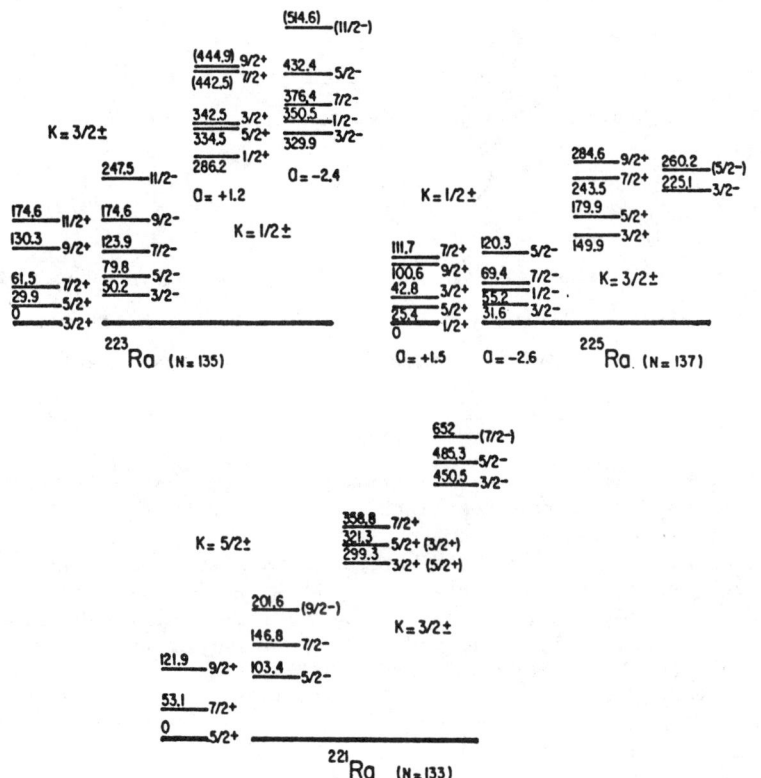

Fig. 4 : *Parity-doublet bands in ^{225}Ra, ^{223}Ra, ^{221}Ra (experimental results).*

In very recent calculations (using a Woods-Saxon potential + Strutinsky shell corrections)[11] in which the potential energy is minimized in β_2 β_3 β_4 β_5 β_6 β_7, the octupole deformation is respectively found = 0.110 (^{221}Ra), 0.108 (^{223}Ra), 0.098 (^{225}Ra) with ground-state spins respectively $5/2^+$ (^{221}Ra), $3/2^+$ (^{223}Ra), $1/2^+$ (^{225}Ra), which is in full agreement with the experiment [8]. Only by taking into account an octupole deformation one may reproduce the experimental ground-state spins, as well as the decoupling parameters.

However one must notice that the height of the barrier in the potential energy as function of β_3 is ~ 0.5 MeV-0.3 MeV respectively in ^{223}Ra and ^{225}Ra and this does not seem to be enough to get a well-localized wave-function around the minimal point in β_3. Only in the case of ^{221}Ra where the octupole barrier is found to be ~1 MeV [11] it suggests that the octupole deformation is near the limits of stability.

We are very indebted to J. Obert and J.C. Putaux for their essential collaboration in the delicate separation process during these experiments. We also gratefully acknowledge Dr. S. Cwiok for very helpful discussions and for private communication of his calculations.

REFERENCES

1. I.N. Mikhailov and Ch. Briançon, Proc. International Conf. on Nuclear Structure, Reactions and Symmetries (June 1986), Ed. World scientific, p. 263.
2. G.A. Leander, R.K. Sheline, P. Möller, P. Olanders, I. Ragnarsson and A.J. Sierk, Nucl. Phys. A388, 452 (1982) ; G.A. Leander and R.K. Sheline, Nucl. Phys. A413, 375 (1984).
3. D.M. Brink, B. Buck, R. Huby, M.A. Nagarajan and N. Rowley, J. Phys. G : Nucl. Phys. 13, 629 (1987).
4. R. Piepenbring, Z. Phys. A323, 341 (1986).
5. Ch. Briançon, G. Bastin, C.F. Liang and R.J. Walen, CSNSM, Annual Report 1983-1984.
6. R.K. Sheline et al., Phys. Lett. 133B, 13 (1983) ; R.K. Sheline, Phys. Lett. 166B, 269 (1986).
7. C.P. Ruiz, UCRL, 9511 (1961).
8. S.A. Ahmad, W. Klempt, R. Neugart, E.W. Otten, K. Wendt and C. Ekstrom, Phys. Lett. 133B, 47 (1983).
9. C.F. Liang, P. Paris, D. Bucurescu, S. Dellanegra, J. Obert and J.C. Putaux, Z. Phys. A309, 185 (1982).
10. P. Paris et al., Nucl. Instr. and Meth. 186, 321 (1981).
11. S. Cwiok et al. (private communication) to be published.

ROLE OF HIGHER-MULTIPOLARITY DEFORMATIONS IN THE PROPERTIES OF 'OCTUPOLLY' DEFORMED NUCLEI

P. Rozmej [1]
GSI, D-6100 Darmstadt, F. R. Germany

Ch. Briançon
Laboratoire de Spectrometrie Nucleaire, CSNSM, F-91406 Orsay, France

S. Ćwiok
Institute of Physics, Warsaw Technical University, PL-00-622 Warsaw, Poland

A. Sobiczewski
Institute for Nuclear Studies, Hoza 69, PL-00-681 Warsaw, Poland

ABSTRACT

Potential energy of light Th isotopes is studied in multidimensional space, in the macroscopic-microscopic approach. Deformations of higher-order multipolarities like β_5, β_6 and β_7, disregarded up to now in an analysis of this type, are found important for the shape and height of the 'octupole' barrier of these nuclei.

INTRODUCTION

One realizes for already a long time that specific properties of nuclei around radium suggest an octupole deformation for them (cf. e.g. the review paper [1]). Theoretical calculations of the potential energy for these nuclei lead to minima of this energy at the octupole deformations different from zero, $\beta_3^0 \neq 0$ (e.g. refs. [2-6]). However, the minima in the octupole degree of freedom are generally rather shalow. For example, in the calculations of ref. [5], the octupole barrier does not exceed 0.5 MeV. According to the analysis of ref. [7], this does not seem to be enough to localize the collective wave function of a nucleus around this minimal point.

The scope of the present paper is to check if a careful treatment of the deformation space, in which the energy of a nucleus is minimized, does not change the above picture. By a careful treatment, we understand here mainly two properties: (i) the deformation parameters are treated as independent variables, (ii) a sufficient number of these parameters (i.e. enough rich deformation space) is taken into account. Up to now, various combinations of deformations of different multipolarities (e.g. combinations of β_2 and β_4 or β_2, β_4 and β_6 or β_3 and β_5 etc.) have been usually used. They have been chosen in such a way as to minimize the smooth part of the energy. Thus, they did not minimize the total

---
[1] Permanent address: Institute of Physics, University MCS, PL-20-031 Lublin, Poland

energy of a nucleus. Concerning the number of deformation parameters used up to now, at most three of them ($\beta_2, \beta_3, \beta_4$) have been treated independently [3,5]. In the present study, the energy is analyzed in the 6-dimensional deformation space: β_λ, $\lambda=$ 2, 3, ..., 7 (some results are checked in a 9-dimensional space).

METHOD OF THE CALCULATIONS

The potential energy is obtained in the macroscopic-microscopic approach with the macroscopic part given by the Yukawa-plus-exponential model (with the parameters of ref. [8]) and the Strutinsky shell correction, based on the Woods-Saxon potential [9], taken for the microscopic part. The 'universal' (i.e. adjusted to experimental single-particle levels of all odd-A nuclei with A$\geq$40) parameters of the Woods-Saxon potential are taken [10]. The potential is diagonalized in the deformed-oscillator basis. All shells, from N=0 up to N=15, of the oscillator are taken into account in the diagonalization procedure. Minimization of the total energy of a nucleus in the multidimensional deformation space $\{\beta_\lambda\}$ is performed numerically by the use of the ZXMIN procedure of the IMSL library.

RESULTS AND DISCUSSION

An example of the results is given in fig.1. This is a contour map of the potential energy of ^{224}Th calculated as a function of the deformations β_2 and β_3. For each established point (β_2, β_3), the energy is minimized in the $\beta_4, \beta_5, \beta_6, \beta_7$ deformation parameters. The distance in energy between neighbouring solid lines is 0.5 MeV. The dashed lines divide this distance by two. One can see that the odd-multipolarity barrier is about 0.8 MeV and is significantly higher than in the previous study [5]. The larger barrier is due to higher-multipolarity deformations, mainly to β_5. For this reason we prefer to call it the odd-multipolarity (or reflection-asymmetric) deformation barrier rather than the octupole barrier, as used up to now.

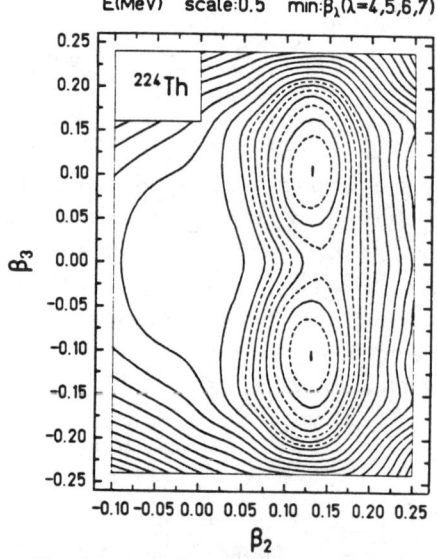

Figure 1: Contour map of the potential energy E calculated for ^{224}Th.

To illustrate the important role of β_5, let us say that minimizing the energy of ^{224}Th in the $\{\beta_2, \beta_3, \beta_4\}$ space (i.e. in 3-dimensional space, which was the space of the highest dimension considered up to now [3,5], although the effect of other deformations treated individually, e.g. $\lambda = 6$: ref. [6], has been also studied), we get the height of the odd-multipolarity barrier $E_{om}=0.36$ MeV. Performing the minimization in the enlarged $\{\beta_2, \beta_3, \beta_4, \beta_5\}$ space, we get $E_{om}=1.09$ MeV, i.e. the barrier about 3 times higher, what stresses the importance of β_5. Then the barrier is reduced by β_6, slightly increased by β_7 and again reduced (but very little) by β_8, ending at the value $E_{om}=0.79$ MeV in the $\{\beta_\lambda\}, \lambda =2, 3, ..., 8$ space. The higher multipoles β_9 and β_{10} practically do not already change this value.

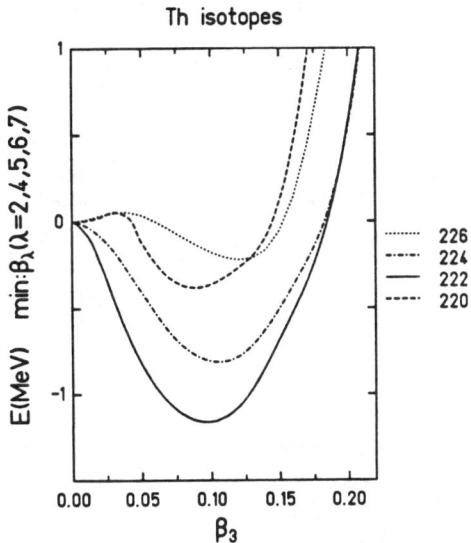

Figure 2: Shape of odd- multipolarity-deformation barrier for isotopes of Th.

Fig.2 shows the shape of the odd-multipolarity barrier for all octupolly deformed Th isotopes. The barrier is obtained by plotting the energy along the bottom of the valley which goes through the minimum. To save space, the barrier is shown only for $\beta_3 \geq 0$. For $\beta_3 < 0$, it is obtained by the mirror reflection. One can see that the highest barrier is obtained for ^{222}Th. It is rather high (about 1.1 MeV) and of a rather simple structure. More complicated structure is obtained in cases of a lower barrier (^{220,226}Th). In such cases, besides the global (deformed) minimum, a local minimum is obtained for the reflection-symmetric shape.

Table I gives the values of all deformation parameters investigated $\beta_\lambda, \lambda =2, 3, ..., 8$, at the equilibrium point for all the four isotopes of Th, which are octupolly deformed. For completeness, the values for the nearest neighbouring isotopes, ^{218}Th and ^{228}Th, are also shown. The nucleus ^{218}Th is spherical, due to the neutron number (N=128) close to the closed shell N=126. The isotope ^{228}Th appears to be rather well deformed. Its shape, however, has the reflection symmetry. Besides the equilibrium deformations β_λ^0, table I gives the odd-multipolarity-deformation barrier E_{om} and the total deformation energy E_{def}. One can see that the values of all β_λ^0 up to $\lambda = 7$ are significant and should be taken into account, at least in the study of more subtle effects (like single-particle properties of odd-A nuclei in the considered region, the odd-multipolarity-deformation barrier E_{om} and the quantities determinated by it). It seems that the higher-multipolarity deformations $\beta_\lambda, \lambda \geq 8$, may be disregarded.

Table I. Values of the equilibrium deformations β_λ^0, height of the odd-multipolarity-deformation barrier E_{om} and the total deformation energy E_{def} for Th isotopes.

N	A	β_2^0	β_3^0	β_4^0	β_5^0	β_6^0	β_7^0	β_8^0	E_{om} MeV	E_{def} MeV
128	218	0	0	0	0	0	0	0	0	0
130	220	0.085	0.098	0.056	0.048	0.016	0.031	-0.004	0.50	0.50
132	222	0.113	0.094	0.071	0.044	0.020	0.025	-0.003	1.14	1.67
134	224	0.128	0.106	0.078	0.045	0.016	0.016	-0.008	0.79	2.71
136	226	0.141	0.101	0.084	0.040	0.014	0.013	-0.010	0.20	3.56
138	228	0.176	0	0.107	0	0.032	0	-0.012	0	4.55

Two of the authors (P. R. and A. S.) would like to thank GSI-Darmstadt and one of them (S. Ć.) CSNSM-Orsay, where important part of the study has been done, for a warm hospitality.

REFERENCES

1. J. Żylicz, Proc. Int. Conf. on Nuclear Structure, Reactions and Symmetries, Dubrownik 1986, ed. R. A. Meyer and V. Paar, vol.1 (World Scientific 1986) p.79.
2. A. Gyurkovich et al., Phys. Lett. **105B**, 95 (1981).
3. G. A. Leander et al., Nucl. Phys. **A388**, 452 (1982).
4. V. V. Pashkevich, Proc. Int. School-Seminar on Heavy Ion Physics, Alushta 1983 (D7-83-644, Dubna 1983) p. 405.
5. W. Nazarewicz et al., Nucl. Phys. **A429**, 269 (1984).
6. R. R. Chasman, Phys. Lett. **175B**, 254 (1986).
7. K. Böning et al., Phys. Lett. **161B**, 231 (1985).
8. P. Möller and J.R.Nix, Nucl. Phys. **A361**, 117 (1981).
9. J. Dudek et al., J.Phys. **G5**, 1359 (1979).
10. J. Dudek et al., Phys. Rev. **C23**, 920 (1981).

SEARCH FOR STABLE OCTUPOLE DEFORMATION IN THE NUCLEUS ^{225}Fr

D.G. Burke
Dept. of Physics, McMaster Univ., Hamilton, Ontario, Canada L8S 4K1

W. Kurcewicz
Wydziat Fizyki, Uniwersytet Warszawski, 00-681 Warsaw, Poland,
and ISOLDE, CERN, 1211 Geneva 23, Switzerland

G. Løvhøiden, K. Nybø, T.F. Thorsteinsen
Fysisk Institutt, Universitetet i Bergen, N-5000 Bergen, Norway

H. Gietz, N. Kaffrell, J. Rogowski
Institut für Kernchemie, Universität Mainz, D-6500 Mainz, Germany

R.A. Naumann
Physics Dept., Princeton University, Princeton, N.J. 08540 U.S.A.

M.J.G. Borge
Instituto de Quimica-Fisica Rocasolano, CSIC, Madrid, Spain

S. Mattsson, G. Nyman
Dept. of Physics, Chalmers University of Technology,
S-41296 Göteborg, Sweden

G. L. Struble
Nuclear Chemistry Division, LLNL, Livermore, CA 94550 U.S.A.

and the ISOLDE collaboration
CERN, 1211 Geneva 23, Switzerland

ABSTRACT

The level structure of ^{225}Fr has been studied from the ^{225}Rn(β^-) decay in on-line experiments at the ISOLDE facility. A level scheme was constructed on the basis of gamma-gamma coincidence data, and the multipolarities of many transitions were established by conversion electron measurements. Levels in ^{225}Fr were also studied with the ^{226}Ra(t,α)^{225}Fr reaction at the McMaster University Accelerator Laboratory, using a target of ^{226}Ra($T_{1/2}$=1600y) and a magnetic spectrograph to analyze the alpha spectra. The first three excited states, at 28.5, 82.5 and 128.2 keV, are interpreted as rotational band members based on the ground state, which is known to have I=3/2. The (t,α) strengths to these levels indicate a 3/2$^-$[532] assignment to the ground state. No evidence for an octupole deformation in ^{225}Fr has been found so far, although analysis of data for other excited states is continuing.

© American Institute of Physics 1988

INTRODUCTION

There is now experimental evidence that suggests a number of nuclides near mass A=225 may possess stable octupole deformation[1-8]. These nuclei belong to a very limited region in the chart of the nuclides, and it is important to establish the extent of this region. It would also be of interest to find a case for which it is not only sufficient but also necessary to invoke a reflection-asymmetric shape in order to explain the nuclear properties satisfactorily. The aim of this study was to examine the nuclear structure of ^{225}Fr, for which no information about the excited states previously existed. With N=138, this nuclide is an isotone of ^{227}Ac and ^{229}Pa, which have been considered as possibly having octupole deformations[3,6]. Some properties useful as tests for possible reflection asymmetry are the single-particle level order, existence of parity doublets, enhanced B(E3) and B(E1) values, magnetic moments, decoupling parameters for K=1/2 bands, and spectroscopic strengths for single-nucleon transfer reactions.

EXPERIMENTAL DETAILS AND RESULTS

Samples of 4.65 min ^{225}Rn were obtained at the ISOLDE facility of CERN from spallation reactions induced by 600 MeV protons on a thorium carbide target. The mass-separated beam was collected on the aluminum-coated mylar tape of a tape transport system. For the gamma-gamma coincidence measurements two high-purity Ge photon detectors were placed near the source collecting point of the tape. One detector had a very thin entrance window to minimize absorption of low-energy photons, and L X-rays from the source could be readily observed. Each sample was counted during a collection time of 1 minute, and then the tape was moved to remove the old sample and permit collection on a fresh portion of tape. This procedure reduced the effects of daughter activity buildup in the samples. The singles conversion-electron spectrum was measured with a "mini-orange" spectrometer described earlier[9]. Also, a high-resolution singles gamma spectrum for the low-energy lines was obtained with a high-purity "X-ray" detector placed opposite to the electron spectrometer. For both of these measurements each ^{225}Rn source was collected for 1 minute and then transported to the measuring position for a counting time of 1 minute.

Fig. 1 shows the low-energy region of the gamma-ray spectrum recorded with the "X-ray" detector. Fig. 2 shows the internal-conversion electron spectrum. Over 300 gamma lines were observed and, mainly on the basis of the coincidence data, about 150 of them have been fitted into a level scheme with ~30 levels. The low-energy region of this scheme is shown in fig. 3. The E1 assignments were established from the internal-conversion coefficients. These, and multipolarities for other transitions, indicate that the first four levels have the same parity, and the next five levels shown in fig. 3 have opposite parity.

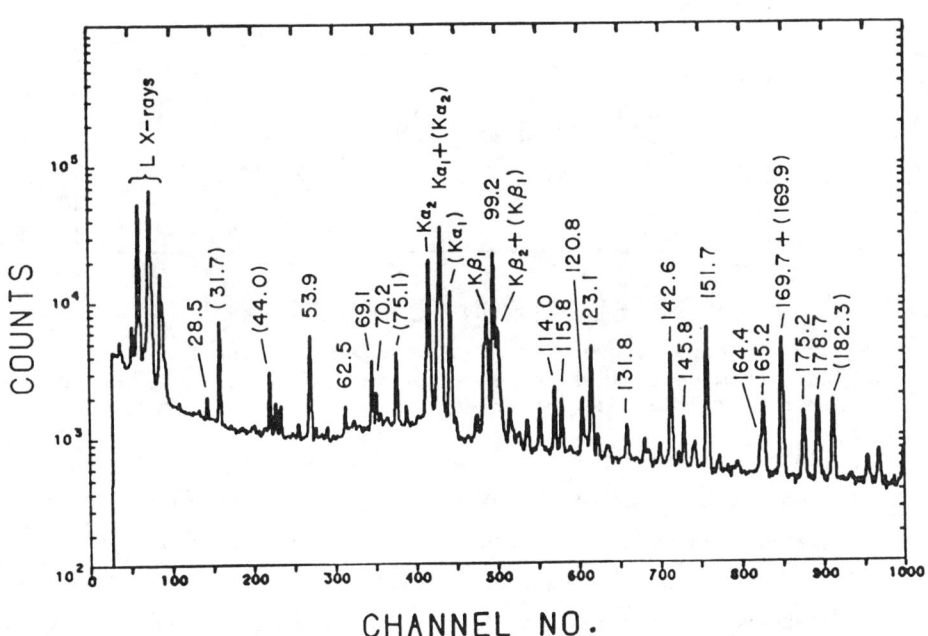

Fig. 1. Low-energy gamma spectrum from the ^{225}Rn decay. Numbers above the peaks are energies in keV. Parentheses indicate lines in ^{225}Ra due to the daughter activity.

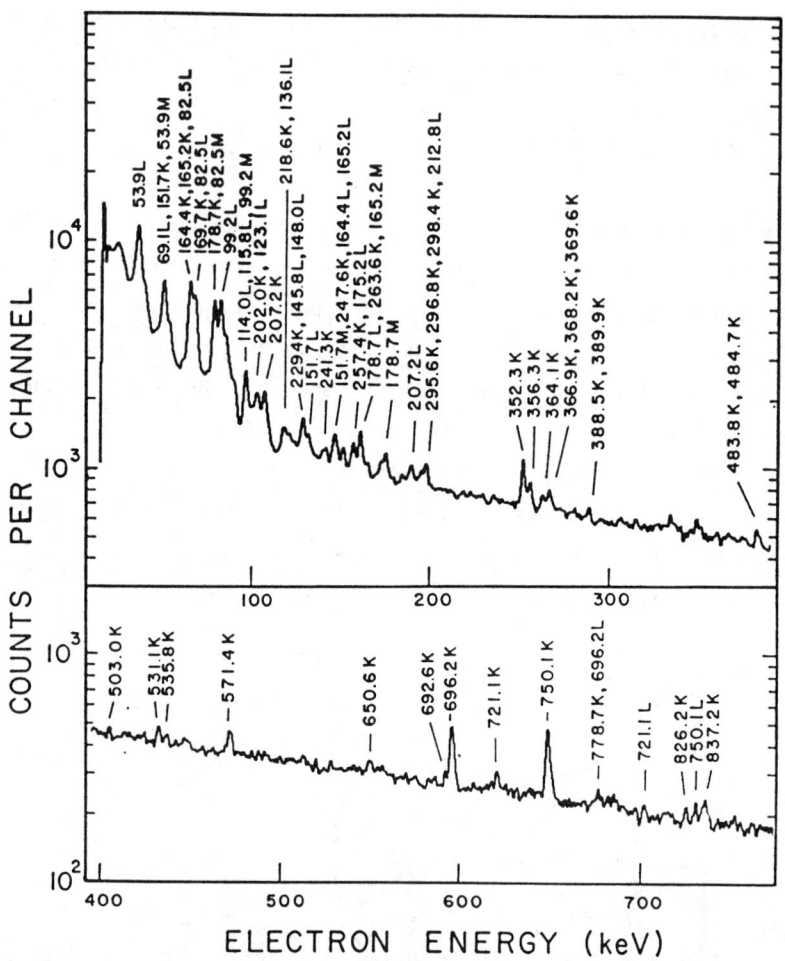

Fig. 2. Internal conversion electron spectrum from the decay of ^{225}Rn. Prominent peaks are labelled with the lines that could contribute to the observed intensity.

In a separate experiment, the ^{226}Ra$(t,\alpha)^{225}$Fr reaction was studied using a beam of 18 MeV tritons from the McMaster University Tandem Accelerator, on a 40 μg/cm^2 target of ^{226}Ra ($T_{1/2}$ = 1600 y). The reaction products were analyzed with an Enge split-pole spectrograph and the alpha particles were recorded with photographic emulsions. A typical spectrum is shown in fig. 4. The overall resolution was ~18 keV FWHM. The first four levels observed in this experiment agree very well with the first four found in the decay scheme of fig. 3.

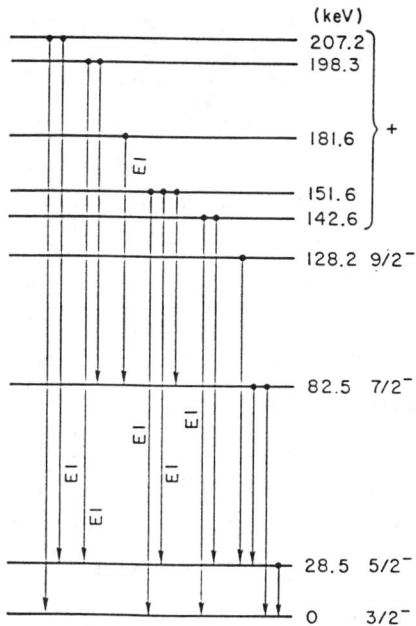

Fig. 3. Low-energy portion of the preliminary ^{225}Fr level scheme.

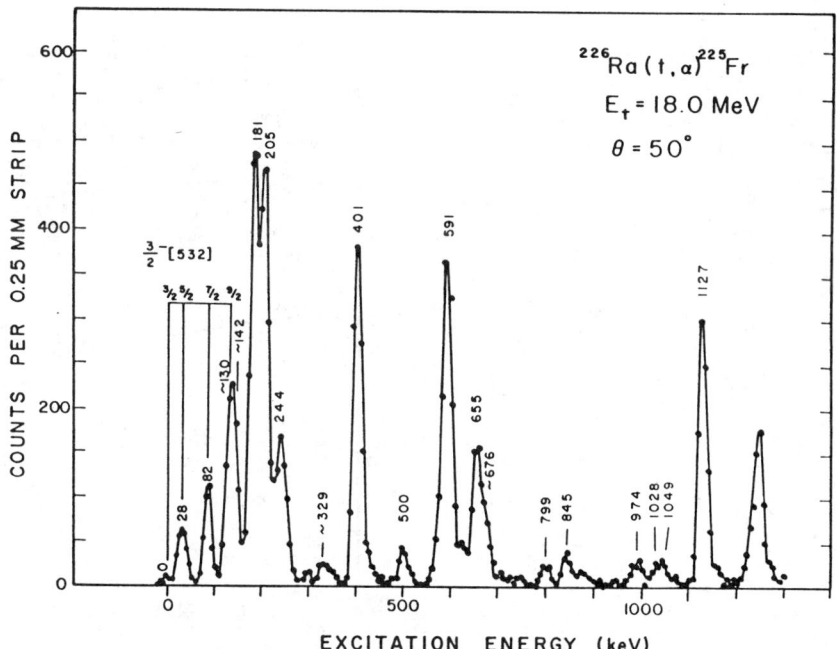

Fig. 4. Alpha spectrum from the ^{226}Ra$(t,\alpha)^{225}$Fr reaction. Prominent peaks are labelled with the excitation energy in keV.

INTERPRETATION

The ground-state spin and magnetic moment for ^{225}Fr have been measured[10] to be I=3/2 and μ =1.07±0.02 nm, respectively. Nilsson level schemes for odd protons in this region are shown in fig. 5, calculated[1] for octupole deformation parameters of ϵ_3 = 0 and ϵ_3 = 0.08. There are three K=3/2 orbitals that could be considered as candidates for the I=3/2 ground state of ^{225}Fr in the reflection-symmetric case. These include the 3/2$^-$[532] and 3/2$^+$[651] states, which form bands at 0 keV and 27 keV, respectively, in the neighbouring ^{227}Ac nucleus. In ^{227}Ac, both of these levels have magnetic moments[6,7] of 1.1 nm, and therefore the measured value of μ=1.07 nm for ^{225}Fr cannot be used to distinguish between these two possibilities.

The (t,α) data are very useful in limiting the possible assignments for the ground state band. The cross section for such a single-nucleon transfer reaction on an even-even target nucleus, to a rotational band member of spin I=j is[11]

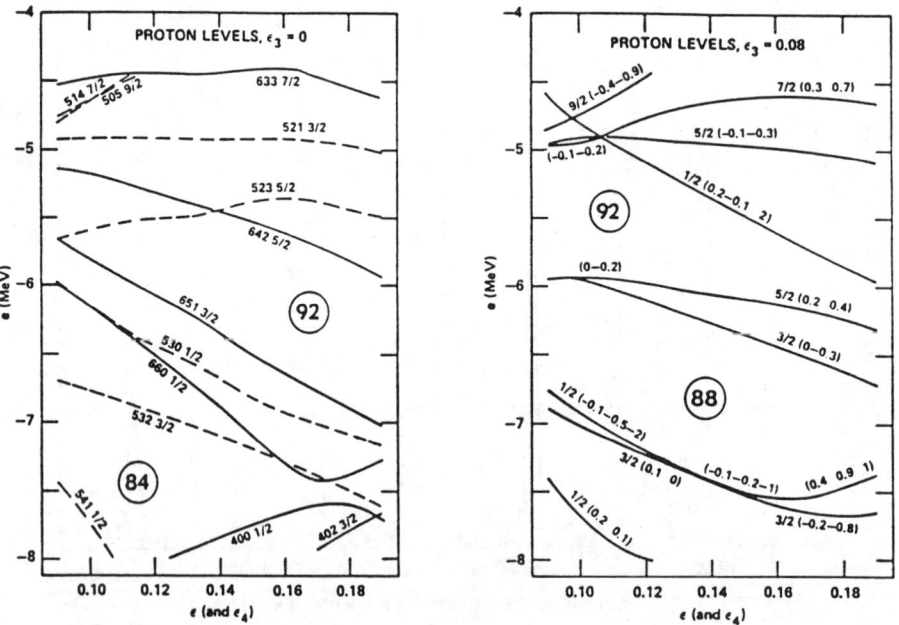

Fig. 5. Single proton orbitals calculated with and without an octupole deformation (from ref. 1).

$$\frac{d\sigma}{d\Omega} = 2N\ \sigma_{DW}\ V^2 C_{j\ell}^2$$

N is a normalization factor, and σ_{DW} is a single-particle cross section which can be obtained from a distorted-waves calculation. The $C_{j\ell}$ coefficients are the spherical amplitudes of the wave function for the Nilsson state on which the band is based, and V^2 is the probability that this orbital contains a pair of particles in the target nucleus. Thus the pattern of cross sections to the various band members depends directly on the set of $C_{j\ell}$ amplitudes and is characteristic of the Nilsson state. Each orbital has its own specific fingerprint, or signature, which can be used for identifying the orbitals involved. Table I shows predicted cross sections for the $3/2^-[532]$, $3/2^+[651]$ and $3/2^+[402]$ bands, calculated using Nilsson wavefunctions with $\kappa=0.058$, $\mu=0.646$, and $\delta=0.16$. The optical model parameters for the DWBA calculations were the same as those used for (t,α) studies[12,13] on Pb and Pt targets, with N=23. A value of $V^2=0.7$ was chosen as a reasonable estimate for an orbital in the ^{226}Ra target which is to form the ^{225}Fr ground state. The fifth column of Table 1 shows the observed (t,α) cross sections for the four levels at 0, 28.5, 82.5, and 128.2 keV, which might be considered as rotational band members with I=3/2, 5/2, 7/2, and 9/2, respectively. The experimental results are obviously most consistent with the predictions for the $3/2^-[532]$ orbital. The agreement would be further improved by including the effects of Coriolis mixing with other $h_{9/2}$ orbitals. For example, such effects would be expected to increase the calculated cross section for the $9/2^-$ member by 50% or more, depending on the excitation energies of the higher-lying bands involved in the mixing. It is therefore concluded that the $3/2^-[532]$ orbital forms the ground-state band of ^{225}Fr. It is noted that there are deviations from the I(I+1) spacing in this band very similar to those in the ground state band of ^{227}Ac, which is assigned to the same Nilsson orbital. These deviations can be ascribed in both cases to Coriolis mixing with the $1/2^-$ [541] band, which also originates from the $h_{9/2}$ shell and has a large positive decoupling parameter.

For completeness, one should also consider the possibility
that the ^{225}Fr ground state might be formed by the I=3/2 member of
one of the four K=1/2 Nilsson states shown nearby in fig. 5. The
1/2$^+$[660] and 1/2$^-$[541] orbitals both have large positive
decoupling parameters, and thus the spin 3/2 member should not be

Table I Cross sections for the ^{226}Ra(t,α) reaction

Spin	Cross sections at θ=60° in μb/sr				
	Calculated ν_3=0			Observed	Calculated ν_3=0.09
	3/2+[402]	3/2+[651]	3/2-[532]		
3/2	90	0.0	0.7	~1.5	2.2
5/2	4.1	0.08	6.2	14	8.4
7/2	1.0	0.01	3.3	20	0.0
9/2	0.04	4.8	26	~45	18
11/2		0.02	0.3		0.4
13/2			28		

below the I=1/2 bandhead. The I=3/2 members of the 1/2$^+$[400] and
1/2$^-$[530] bands have predicted (t,α) cross sections at θ=60° of 21
and 33 μb/sr, respectively, and are therefore not consistent with
the value of ~1.5 μb/sr observed for the ground state. The
Nilsson calculation described above also predicts the magnetic
moments for the I=3/2 members of these bands to be -0.3 nm and
+2.2nm, respectively, using g_s=0.6 $(g_s)_{free}$, and these values
differ significantly from the experimental value. Therefore the
3/2$^-$[532] orbital is the only reasonable assignment for the ^{225}Fr
ground state.

The (t,α) spectroscopic strength can also be used to test for
an octupole shape in the nuclear potential. The presence of such
a shape might be expected to modify the single particle
wavefunctions, which would affect the (t,α) cross sections.
Chasman has calculated wavefunctions for various values of

octupole deformations[14], and the predicted cross sections for the K=3/2 band of predominantly negative parity expected for the 87th proton at an octupole deformation of 0.09 are shown in the last column of Table I. This is the only likely K=3/2 configuration predicted by the calculations of Chasman, and in the limit of zero octupole deformation it corresponds to the $3/2^-[532]$ orbital. It is seen that the cross sections predicted with the octupole deformed wavefunctions are in slightly poorer overall agreement with experiment than those for the reflection-symmetric calculation.

CONCLUSIONS

The nuclear structure of ^{225}Fr has been studied by measurement of gamma-gamma coincidences and conversion electrons from the decay of ^{225}Rn, and also with the (t,α) reaction. Model assignments have been made for some of the lowest-lying levels but no evidence has yet been found that a stable octupole deformation exists for ^{225}Fr. In particular, since the lowest positive parity level found is at 142 keV the splitting of any possible parity doublet involving the ground state must be at least 142 keV.

The (t,α) strengths to the $3/2^-[532]$ ground state band are not extremely sensitive to the presence of an octupole deformation. In single-proton transfer studies of ^{227}Ac it was found[7] that the predicted strengths for some of the other bands (e.g., $3/2^+[651]$) were more strongly dependent upon the octupole deformation parameter. Thus there is hope that further analysis of the ^{225}Fr data may lead to identification of additional rotational bands, which may permit more definitive tests for the presence of octupole deformations.

REFERENCES

1. G.A. Leander and R.K. Sheline, Nucl. Phys. A413, 375(1984).
2. R.R. Chasman, Phys. Lett. 96B, 7(1980).
3. I. Ahmad et al., Phys. Rev. Lett. 49, 1758(1982).
4. R.K. Sheline et al., Phys. Lett. 133B, 13(1983).

5. I. Ragnarsson, Phys. Lett. 130B, 353(1983).
6. R.K. Sheline and G.A. Leander, Phys. Rev. Lett. 51, 359(1983).
7. H.E. Martz et al., Ann. Report of Nucl. Chem. Div., Lawrence Livermore National Laboratory, 1984, p. 6-109.
8. G. Løvhøiden et al., Nucl. Phys. A452, 30(1986).
9. M.J.G. Borge et al., Nucl. Phys. A464, 189(1987).
10. A. Coq et al., Phys. Lett. 163B, 66(1986).
11. B. Elbek and P.O. Tjøm, Advances in Nuclear Physics (Plenum, N.Y., 1969) Vol. 3, p. 259.
12. E.R. Flynn et al., Nucl. Phys. A279, 394(1977).
13. J.A. Cizewski et al., Phys. Rev. C27, 1040(1983).
14. R.R. Chasman, Phys. Rev. C30, 1753(1984).

NUCLEAR PHYSICS REQUIREMENTS FOR STUDIES OF
NUCLEAR ENERGY GENERATION AND NUCLEOSYNTHESIS

James W. Truran
Dept. of Astronomy, University of Illinois, Urbana, IL 61801

ABSTRACT

Processes of nuclear energy generation and nucleosynthesis in stars, novae, and supernovae are reviewed and their nuclear physics requirements are identified. Exotic, unstable, proton-rich and neutron-rich nuclear species can participate in nuclear reactions in these astrophysical environments. For nuclear processes involving lighter nuclei, critical needs include a knowledge of the cross sections at low energies for a number of specific reactions for which contributions from individual levels may dominate. Mechanisms of heavy element synthesis generally involve the participation of large numbers of stable and unstable nuclear species; here, theoretical estimates of reaction rates are necessary and experimental studies which can serve to test and guide systematics are important. An outline of the methods currently employed in statistical model calculations is presented together with a brief summary of reactions for which experimental input is of critical importance.

INTRODUCTION

Diverse astrophysical environments afford a wide range of physical conditions which naturally allow the synthesis of rare nuclear species, ranging substantially all the way from the proton drip line to the neutron drip line and from the light element region through to transuranic elements. Astrophysical processes in which stable and unstable proton-rich isotopes are produced include: hot CNO cycle hydrogen burning in thermonuclear runaways on white dwarfs leading to classical nova outbursts; high temperature hydrogen burning associated with thermonuclear outbursts on neutron stars; the formation of the iron-peak nuclei in explosive nucleosynthesis in supernovae, including ^{56}Ni, which powers the light curve; and the p-process of nucleosynthesis which forms the proton-rich isotopes of nuclei in the mass range A > 60-70. Astrophysical environments also provide settings for the formation of nuclei on the neutron-rich side of the valley of beta stability: a neutron-rich silicon burning phase accompanying core collapse can produce nuclei well off the valley of beta stability; neutron-rich equilibria characterize matter in the surface regions of neutron stars and may also be associated with the innermost layers ejected in supernova events; and the astrophysical r-process of neutron capture is characterized by flows extending deep into the neutron-rich regions, even to the neutron drip line.

The significant progress which has been achieved in nucleosynthesis theory in recent years is elaborated in recent reviews of stellar evolution[1], supernova explosions[2], and nucleosynthesis[3,4]. A brief survey of astrophysical processes of nuclear energy generation and nucleosynthesis is presented in the subsequent sections. Critical nuclear physics requirements for these studies are discussed and the sources and methods currently employed to provide this input are reviewed. A brief outline of the methods used in theoretical calculations of thermonuclear reaction rates is presented. We conclude with a summary of some astrophysical needs critical to the understanding of individual nucleosynthesis processes or of the nature of specific stellar and supernova environments.

HIGH TEMPERATURE CNO CYCLE HYDROGEN BURNING

Hydrogen burning proceeding at high temperatures and under explosive conditions can play a critical role in diverse astrophysical environments, with important implications for dynamic evolution, nuclear energy generation, and nucleosynthesis. The defining characteristics of such burning have been discussed by a number of authors[5,6,7,8]. Above a temperature of approximately 10^8 K, the rates of proton-induced reactions in the CNO cycles become comparable to or faster than the rates of beta decay. The modified CN reaction sequence at high temperatures thus becomes

$$^{12}C(p,\gamma)^{13}N(p,\gamma)^{14}O(e^+\nu)^{14}N(p,\gamma)^{15}O(e^+\nu)^{15}N(p,\alpha)^{12}C$$

where the sequence $^{13}N(p,\gamma)^{14}O(e^+\nu)^{14}N$ has replaced $^{13}N(e^+\nu)^{13}C(p,\gamma)^{14}N$. The operation of this "β-limited" CNO cycle is such that the rate of energy generation on a dynamic timescale is restricted by the slower rates of beta decay of the proton-rich oxygen isotopes, ^{14}O and ^{15}O.

This restriction plays a critical role in defining the nature of the outbursts of classical novae[9,10]. Nova outbursts are triggered by thermonuclear runaways in accreted hydrogen-rich shells on white dwarfs in close binary systems. As the runaway proceeds, the lifting of degeneracy occurs when gas pressure exceeds electron degeneracy pressure, at temperatures approaching $\sim 10^8$ K. Nuclear energy generation at high temperatures is constrained by the positron decay of ^{15}O (τ = 176 seconds). Therefore, once degeneracy is lifted, there exists a limit on the total energy available on the hydrodynamic timescale of order seconds. Specifically, on a timescale $\tau < \tau(^{15}O)$, the energy available is approximately

$$E_{nuc}/\rho \sim 2 \times 10^{15} \text{ erg g}^{-1} [n_{CNO}/n_{CNO}(SOLAR)],$$

where the ratio in brackets normalizes to the case of matter of solar composition of the CNO elements. In contrast, the binding energy per gram of the accreted envelope matter exceeds $\sim 10^{17}$ erg g^{-1}. It follows that some enrichment of the envelope in the elements carbon, nitrogen, and oxygen is necessary to explain the rapid light curve development, high ejection velocities, and super-Eddington luminosities of the fastest novae, and indeed such abundance enrichments have been observed to characterize nova ejecta[11,12,13,14]. Peak temperatures in the range $\sim 1.5 - 3 \times 10^8$ K are achieved in hydrogen burning in this environment[10,15]. Our knowledge of the operation of the CNO cycles at high temperatures thus places the nuclear physics of nova theory on a rather firm footing, although an experimental investigation of the $^{13}N(p,\gamma)^{14}O$ reaction is necessary.

The operation of the CNO cycle reactions at even higher temperatures is accompanied by breakout. Weischer et al[8] specifically find that leakage out of the hot CNO cycle ultimately occurs at a temperature 4×10^8 K. This temperature is above those typically achieved in hydrodynamic models of classical nova eruptions. Such temperatures are, however, achieved in thermonuclear runaways in accreted hydrogen-helium shells on neutron stars[6,16]. For these conditions, enhanced energy production is realized with the help of the $^{14}O(\alpha,p)^{17}F(\gamma,p)^{16}O$ reactions and breakout of the hot CNO cycle is unambiguously achieved by the $^{15}O(\alpha,\gamma)^{19}Ne(p,\gamma)^{20}Na$ reactions. At temperatures above ~ 7.5×10^8 K, this breakout can be accompanied by the buildup of heavy nuclei through the iron peak and beyond via the rapid capture of protons on seed nuclei in the "rp-process" described by Wallace and Woosley[6]. This level of breakout can also be accompanied by an energy release far in excess of that available from the β-limited CNO cycles, with implications for X-ray burst models. The rp-process conditions generally favor the production of proton-rich isotopes, with the main capture path passing through ^{56}Ni. Wallace and Woosley[6] note that, for a temperature $T = 7.5 \times 10^8$ K, the flow toward the iron peak is constrained by the rate of the slowest reactions, $^{15}O(\alpha,\gamma)^{19}Ne$ and to a lesser extent $^{14}O(\alpha,p)^{17}F$, emphasizing the importance of experimental information regarding these reactions.

The hydrogen burning environments provided by the classical novae can also contribute to galactic nucleosynthesis[17]. The less abundant isotopes of carbon, nitrogen, and oxygen, particularly ^{15}N and ^{17}O, may be formed in significant quantities. The production of the unstable species $^{22}Na(\tau_{1/2} = 2.6y)$ and $^{26}Al(\tau_{1/2} = 7.2 \times 10^5 y)$ is also predicted to occur as a consequence of the operation of modified Na-Mg and Mg-Al cycles at higher temperatures[6,8,18]. The formation of significant ^{22}Na provides an attractive explanation for the highly ^{22}Ne-enriched Ne-E anomaly identified in meteorites[19], while the existence of ^{26}Mg excesses in meteorites that correlate with Al/Mg ratios confirm the presence of live ^{26}Al in the early solar system[20]. Nuclear reaction rates of importance here include both the proton capture reactions which form these unstable nuclei, $^{21}Ne(p,\gamma)^{22}Na$ and $^{25}Mg(p,\gamma)^{26}Al$, and particularly those which destroy them, $^{22}Na(p,\gamma)^{23}Mg$ and $^{26}Al(p,\gamma)^{27}Si$.

EXPLOSIVE NUCLEOSYNTHESIS IN SUPERNOVAE

The elements from oxygen through the iron peak are now believed to have been formed collectively in the environment provided by the final stages of evolution of massive stars and the shock heating and ejection of helium-exhausted core matter in supernova explosions (see, for example, the reviews by Truran[3] and Woosley and Weaver[2]). Nucleosynthesis occurring in the cores of massive stars $M > 10\ M_\odot$ during the late stages of presupernova evolution and in the ensuing Type II supernova events themselves produces nuclei in this general mass range. It appears, however, that the iron-peak elements are underproduced in such events relative to oxygen and the neon-to-titanium elements. Theoretical studies currently suggest that nucleosynthesis contributions from Type I supernovae nicely compliment those of Type

II supernovae by forming predominantly iron-peak nuclei. Indeed, calculations of explosive nucleosynthesis associated with carbon deflagration models of Type I supernovae[21] predict that sufficient iron-peak nuclei are formed to explain both the powering of the light curves of Type I supernovae by the decays of ^{56}Ni and ^{56}Co and the observed mass fraction of iron-group nuclei in galactic matter. For purposes of illustration, a nuclear reaction network employed in early calculations of this silicon burning process is illustrated in Figure 1. Essentially all reactions involving the interactions of neutrons, protons, alpha particles, and photons on these nuclei can participate; the input nuclear physics requirements are thus substantial.

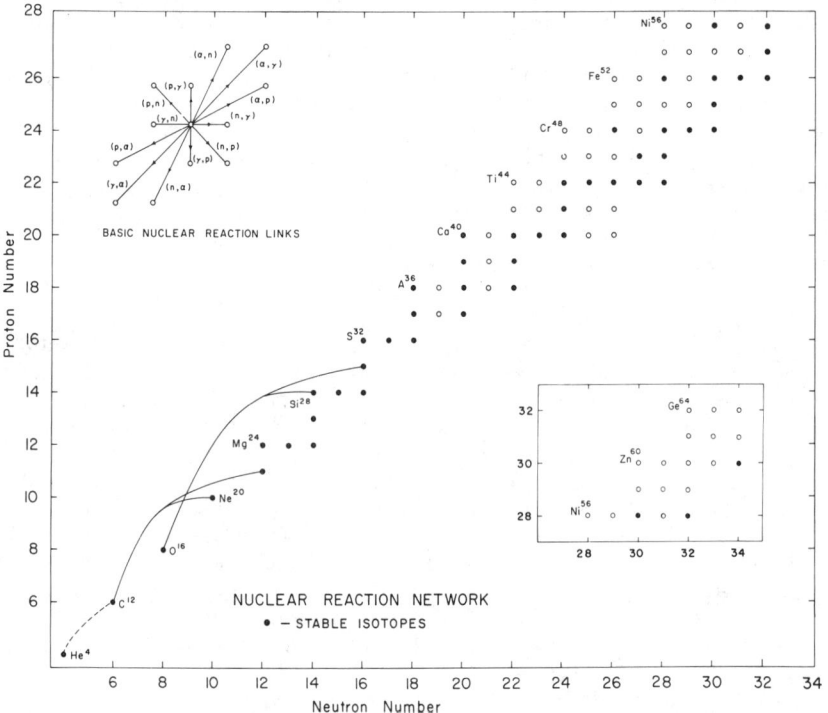

Fig. 1 Nuclear reaction network utilized in studies of carbon, oxygen, and silicon burning.

The production of mass A = 56 in situ in the form of ^{56}Ni reflects an important characteristic of the charged-particle dominated nuclear reaction sequences which form the neon-to-iron group elements. The most abundant nuclear constituents of the core during the final stages of presupernova evolution are self-conjugate nuclei (^{4}He, ^{12}C, ^{16}O, etc.). Trace amounts of odd-Z elements and of neutron-rich isotopes of even-Z elements are also present in the gas, such that the total number of neutrons exceeds that of protons by a factor of only perhaps several parts in a thousand. At the prevailing temperature and density conditions ($2 < T < 6 \times 10^9$ K and $\rho \sim 10^5\text{-}10^7$ g cm^{-3}), the timescale for significant neutronization of the matter which will ultimately be ejected in the supernova event is short compared to the timescale for thermonuclear processing. It follows that the final products of these explosive burning episodes must lie along or very near to the Z = N line. This remains true even beyond the last stable alpha-particle nucleus (^{40}Ca) and favors the production in situ of the nuclei ^{44}Ti, ^{48}Cr, ^{52}Fe, ^{56}Ni, and ^{60}Zn. The subsequent decay of these unstable nuclei and their immediate neighbors dictates the elemental and isotopic composition of the ejected matter. We note therefore that the isotopic composition of a given element in solar system matter may reflect the nuclear properties of different isotopes formed in situ as: the chlorine isotopes, ^{35}Cl and ^{37}Cl (formed as ^{37}Ar); the potassium isotopes ^{39}K and ^{41}K (^{41}Ca); the chromium isotopes ^{50}Cr, ^{52}Cr(^{52}Fe), and ^{53}Cr(^{53}Fe); the iron isotopes ^{54}Fe, ^{56}Fe(^{56}Ni), and ^{57}Fe(^{57}Ni); and the nickel isotopes ^{58}Ni, ^{60}Ni, ^{61}Ni, and ^{62}Ni (the last three formed as ^{60}Zn, ^{61}Zn, and ^{62}Zn). This general behavior and the fact that the resulting isotopic patterns are so closely in agreement with those of solar system matter strongly support the view that the iron group nuclei are formed in a relatively proton-rich environment. A knowledge of the nuclear properties of the unstable self-congugate nuclei and their immediate neighbors, from ^{44}Ti through at least ^{60}Zn, is therefore critical to studies of supernova explosions and associated nucleosynthesis.

P-PROCESS NUCLEOSYNTHESIS

There are a number of proton-rich stable isotopes of heavy nuclei (A > 60) which are shielded by stable nuclei on their respective isobars from any contributions from neutron-capture mechanisms. Cameron[22] specifically identifies 36 such "bypassed" or p-process nuclei (see Figure 2), which range in mass from ^{74}Se to ^{196}Hg and include two odd-odd nuclei, ^{138}La and ^{180}Ta. Burbidge et al[23] and Cameron[24] attributed the formation of these nuclei to a combination of (p,γ), (p,n), and (γ,n) thermonuclear reactions proceeding on preexisting heavy nuclei in stellar interiors at temperatures $T \sim 1\text{-}3 \times 10^9$ K. The fact that the abundance level of the p-process nuclei parallels that of nuclei formed by neutron capture, but is systematically lower by a factor of 100 to 1000, is consistent with the view that preexisting s-process and r-process nuclei serve as seeds. While the detailed characteristics of the astrophysical site of p-process synthesis have yet to be identified, recent numerical studies[25,26] indicate that reaction rates will be required for heavy nuclei ranging from the valley of beta stability well into the proton-rich regions, but not to the proton drip line. Here again it is clear that we must generally rely upon estimates of thermonuclear reaction rates based upon statistical model calculations.

NEUTRON CAPTURE NUCLEOSYNTHESIS

Nucleosynthesis theories attribute the production of the bulk of the heavy nuclei in the mass range A > 60 to neutron capture processes[23,24,3,4]. The abundance features in the heavy element

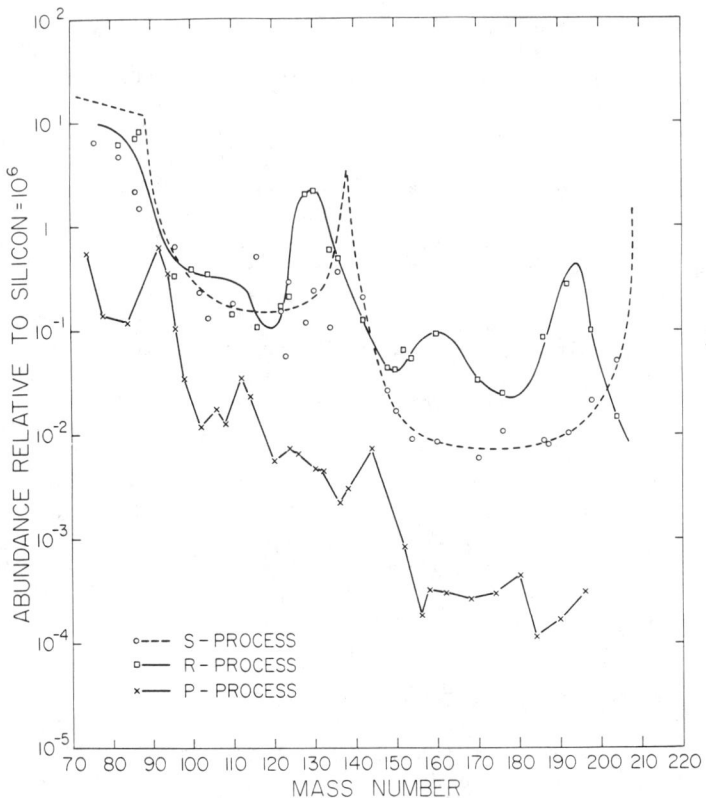

Fig. 2 Abundances of the heavy nuclei formed in the s-, r-, and p-processes[22].

region in solar system matter which are correlated with neutron shell closures strongly support this conclusion. A careful study of these abundances patterns reveals evidence for the operation of two distinct neutron capture processes characterized by two quite different neutron fluxes. These two processes can be distinguished on the basis of the relative lifetimes for neutron capture (τ_n) and beta decay (τ_β), where τ_β is a characteristic lifetime for beta unstable nuclei in the vicinity of the valley of beta stability.

The s-process of nucleosynthesis is defined by the condition $\tau_n > \tau_\beta$, which constrains the neutron capture path to remain close to the valley of beta stability. The s-process peaks in the solar system abundance distribution[22] at strontium (neutron number N = 50), barium (N = 82), and lead (N = 126) arise from the fact that cross sections for neutron captures out of the closed shells fall precipitously. Recent review discussions of the s-process mechanism have been provided by a number of authors[1,4,27,28]. Observations and theory both confirm the identification of red giant stars as the s-process site. The neutron flux is believed to be provided by either the $^{13}C(\alpha,n)^{16}O$ or the $^{22}Ne(\alpha,n)^{25}Mg$ neutron source operating in the convective helium shells of intermediate mass stars during thermal pulses. Experimental information concerning the cross sections as a function of energy for these two reactions, which will allow accurate determinations of the reaction rates as a function of temperature, is critical to our understanding of the mode of operation of the s-process. It is also essential to know the cross sections for neutron capture on ^{22}Ne, ^{25}Mg, and the immediate progeny of the ^{22}Ne(α,n)^{25}Mg reaction, which capture more than three quarters of the neutrons released by the ^{22}Ne(α,n)^{25}Mg neutron source and thus regulate the neutron exposure experienced by the heavy nuclei[29,30]. The empirical foundations of s-process theory also a knowledge of the neutron capture cross sections at 30 keV for many reactions in the vicinity of the s-process capture path along the valley of beta stability[31,32,33].

The r-process of neutron capture, defined alternatively by the condition $\tau_n < \tau_\beta$, is characterized by a high neutron density. For this condition, it follows that successive neutron captures can proceed into the neutron-rich regions far off the valley of beta stability. Following the exhaustion of the available neutrons, the capture products decay via beta decays toward the stable regions on a longer timescale. Since the neutron closed shell positions are encountered at lower mass numbers in the neutron rich regions, the abundance peaks attributable to the r-process are realized several mass numbers below the s-process peaks at strontium, barium, and lead. The double-peak features in the heavy element region in the solar system abundance pattern (Figure 2) are the signatures of these two distinct neutron-capture processes.

The astrophysical site of r-process nucleosynthesis remains to be identified. Existing theoretical models for the r-process include those involving the expansion of highly neutronized matter from high temperatures and densities[34] and those which utilize preexisting iron-peak nuclei as seeds for neutron captures leading to the production of heavier nuclei[35,36]. In either case, the character of the astrophysical r-process is dictated by nuclear properties in the neutron-rich region. The r-process capture path can lie anywhere between the valley of beta stability and the neutron drip line. Input nuclear physics requirements include nuclear masses and reaction Q values, rates of neutron capture and photodisintegration, beta decay rates, and rates of beta-delayed fission and beta-delayed neutron emission. The scope of the problem becomes evident from Figure 3, where the r-process network utilized in the calculations of Truran, Cowan, and Cameron[36], and in subsequent papers by these authors, is illustrated. This network included 6033 heavy nuclei in the region from the beta stable valley to the neutron drip line.

The approximate position of the r-process flow path through these regions can be inferred from a knowledge of the positions of the r-process peaks identified in solar system matter. Specifically, if the peaks at A = 130 and 195 are to correspond to the position at which neutron closed shells at N = 82 and 126 are encountered in the neutron-rich regions, the corresponding peak nuclei on the capture path are ^{130}Cd and ^{195}Tm. The N = 50 closed neutron shell is encountered approximately in the range ^{78}Ni to ^{80}Zn. A knowledge of the neutron binding energies and beta decay lifetimes for nuclei in these "waiting point" regions is critical to progress in our under-

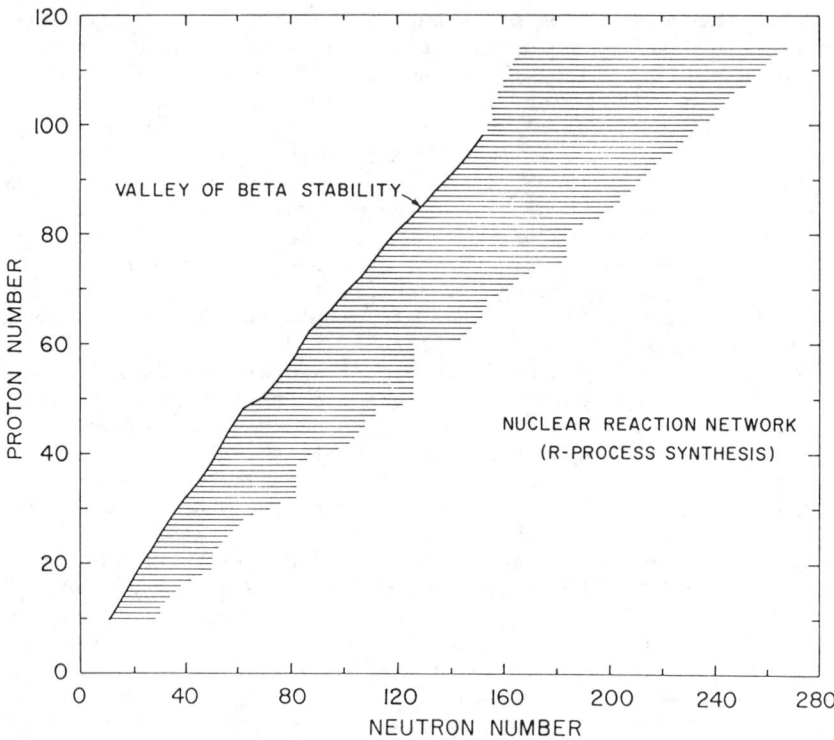

Fig. 3 Nuclear reaction network utilized in r-process studies.

standing of the r-process. Further discussions of the constraints such experimental information can impose on r-process conditions are presented in recent papers[4,37,38] and in the article by K.-L. Kratz in these proceedings.

STATISTICAL MODEL CALCULATIONS

Our brief review of astrophysical processes of nucleosynthesis has revealed that there are vast numbers of nuclear reactions involving the interactions of neutrons, protons, alpha particles, and photons with heavier nuclei that can participate in the later stages of nuclear burning in stellar interiors and in the synthesis of heavy nuclei in supernova explosions. It is clearly impractical to seek to obtain direct experimental cross section information in every case. It has thus been necessary to develop reliable theoretical methods for the determination of thermonuclear reaction rates. Substantial success has been achieved with the adoption of a Hauser-Feshbach[39] or equivalent expression for the energy-averaged cross section. Statistical model based calculations of thermonuclear reaction rates for astrophysical applications have been performed by a number of researchers[40-46].

Significant improvements over earlier work have been achieved in recent studies[47,48] with the use of: (i) realistic optical model potentials for interactions involving neutrons, protons, and alpha particles; (ii) a revised procedure for the calculation of the total photon transmission function for electric dipole radiation; and (iii) a revised procedure for the determination of nuclear level densities as a function of energy. In particular, we have solved the Schrödinger equation with the use of realistic optical model potentials for the appropriate particle-nucleus combination rather than with optical square well potentials. For protons and neutrons, we employ the optical potential derived by Jeukenne et al[49], with the correction for the imaginary part as discussed in Fantoni et al[50] and Mahaux[51]. For alpha particles, we used the phenomenological Woods-Saxon potential derived by Mann[52], based on the data of McFadden and Satchler[53]. For the gamma ray transmission coefficients, both electric and magnetic dipole transitions (E1 and M1) are included in the calculation of the total width. The more important E1 contributions were calculated on the basis of the Lorentzian representation of the Giant Dipole Resonance, utilizing the phenomenological model for the resonance width proposed by Thielemann and Arnould[54]. Finally, our treatment of nuclear level densities closely parallels that of Gilbert and Cameron[55] and combines the use of the Bethe[56] formula $\rho \propto \exp(2\sqrt{aU})$, based on the Fermi gas model, at high energies and the empirically based Ericson[57] formula $\rho(E) = T^{-1}\exp((E-E_o)/T)$ at low energies. The energy shift of the first excited state due to the necessity of breaking up a proton or neutron pair before exciting nucleus to higher states, which is not included in the independent particle Fermi gas model, is accounted for in the "back-shifted" Fermi gas formalism by reducing the energy by an amount equal to the pairing correction, $U = E-\delta$. We have similarly identified the pairing correction with the energy shift in the Ericson formula, $E_o = \delta$. The pairing energies and the shell correction energies, upon which our empirical prescription for the level density parameters a is based, are taken from the droplet model mass formula of Hilf, von Groote, and Takahashi[58].

A measure of the level of success achievable with the use of these procedures is provided by comparisons of our cross section predictions[47,48] with experiment. In Figure 4, our calculated cross sections as a function of energy for charged particle reactions are compared with experiment for representative intermediate mass nuclei which participate in stellar silicon burning. We generally find that the energy dependences and absolute values of charged particle cross sections are quite well reproduced, including specifically the magnitudes of cusps at the threshold energies of channel openings. The improvements achieved and the importance of experimental cross section information are both evident from the comparison of calculated and experimental 30 keV neutron capture cross sections presented in Figure 5. The upper figure shows the ratios of the cross sections calculated by Truran[43] to the experimental cross sections compiled by Allen, Macklin, and Gibbons[31] as a function of neutron number. The points corresponding to isotopes of identified elements are connected

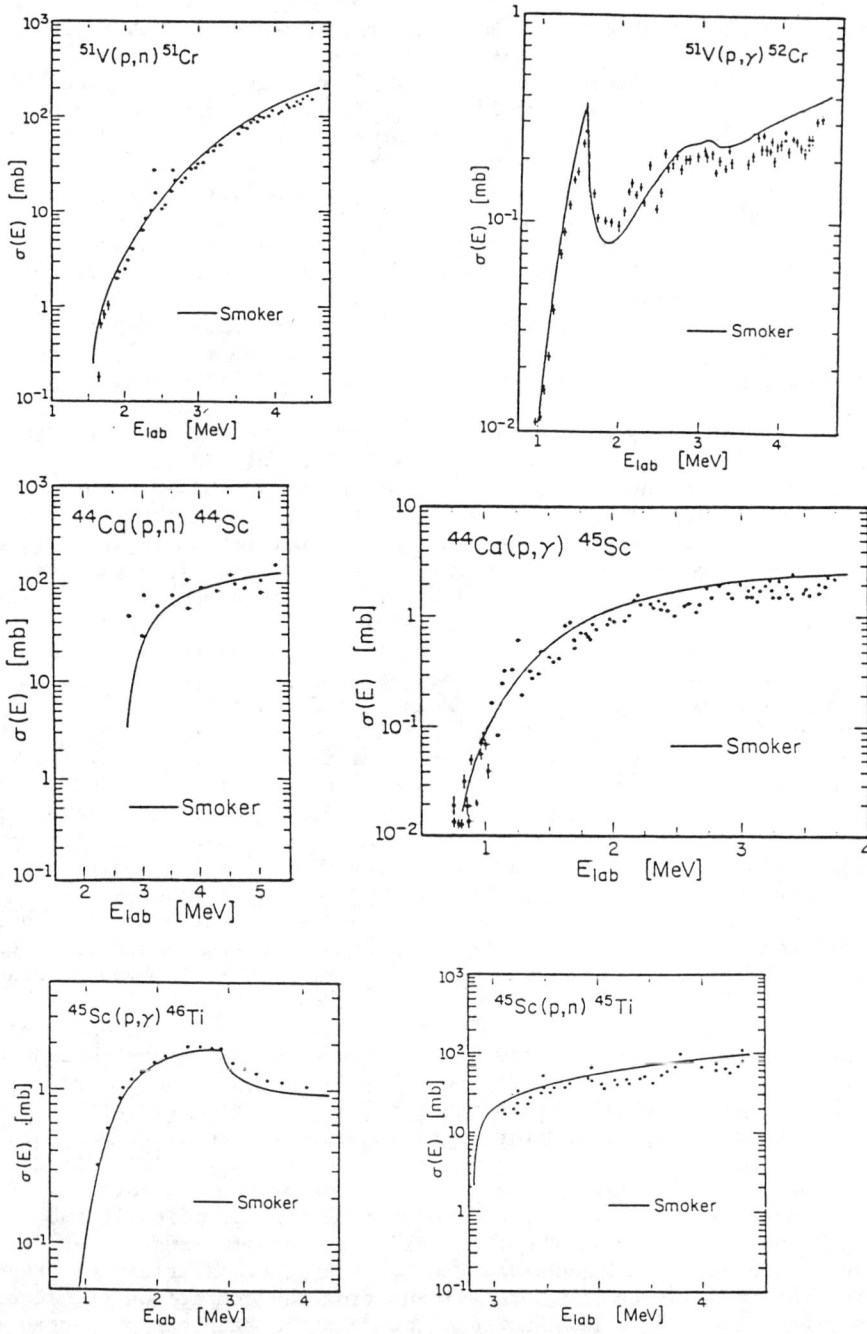

Fig. 4 Representative comparisons of theoretical and experimental cross sections for several intermediate mass nuclei.

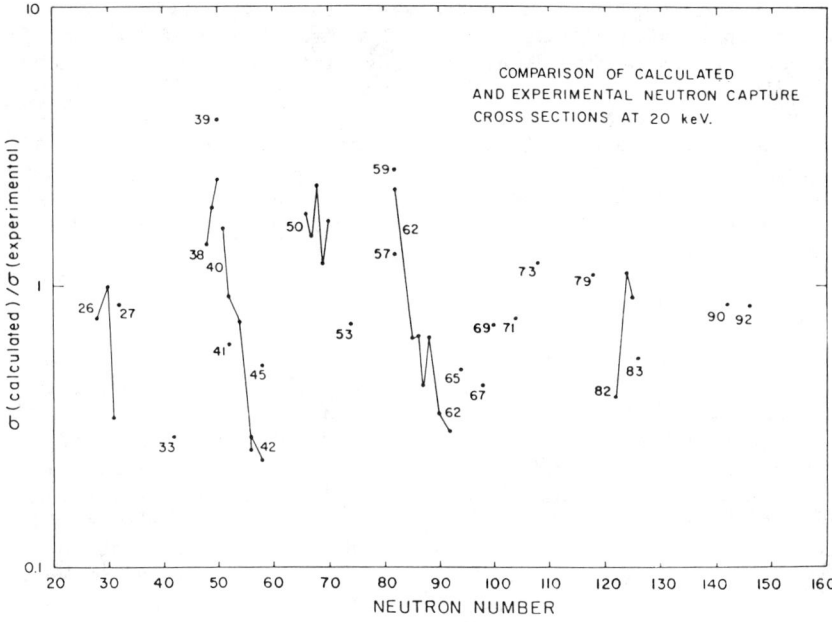

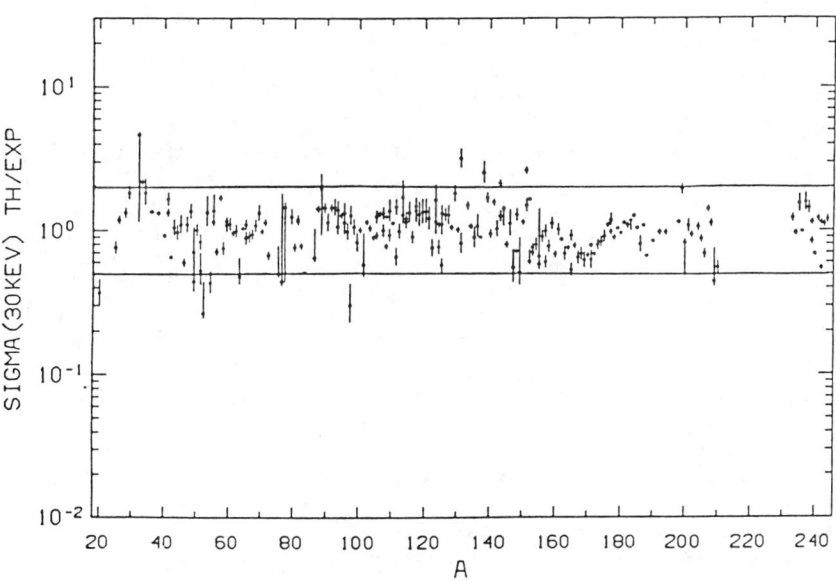

Fig. 5 Ratios of theoretical to experimental 30keV neutron capture cross sections versus A.

by solid lines, revealing systematic variations in the vicinities of closed neutron shells. These trends arise from the inadequacies of systematic treatments of nuclear level densities at low excitations, particularly in the vicinity of closed shells. The lower figure, showing a similar comparison of our recent calculations[47,48] with the experimental data compiled by Bao and Käppeler[59], reveals that the situation has been considerably improved with the use of our revised prescriptions for nuclear level densities and radiation widths. Cross section predictions obtained within this framework are expected to lie safely within a factor of 2 of experimental values.

SUMMARY AND DISCUSSION

It seems appropriate to conclude with a brief summary of some critical astrophysical needs, for input nuclear physics parameters, for various of the nuclear energy generation and nucleosynthesis processes reviewed in this paper. The nature of these needs will of course depend upon the nature and scope of the process itself. The extended phases of hydrogen and helium burning that comprise the major fraction of the active burning lifetimes of stars are dominated by reactions involving a relatively small number of nuclear species. Experimental determinations of the cross sections for many of these reactions are now available[60-62]. However, it clearly is not feasible to obtain cross-section information for each and every nuclear reaction that may be involved in the later stages of nuclear energy generation in stars or in processes of heavy element synthesis in stars and supernovae. Here statistical model calculations have proven to be an invaluable tool and nuclear physics studies which can guide these systematics are important. Some astrophysical needs for specific processes are briefly summarized below.

The empirical foundations of theories of stellar nuclear energy generation and nucleosynthesis are provided by experimentally determined cross sections for a number of critical nuclear reactions. It is often somewhat easier to single out specific reactions of interest for processes involving lighter nuclear species. An improved experimental knowledge of the rate of the $^{12}C(\alpha,\gamma)^{16}O$ reaction, for example, is clearly critical to the determination of the relative abundances of the products of stellar helium burning, ^{12}C and ^{16}O. Similarly, studies of hot CNO cycle hydrogen burning reveal important dependences upon both the rate of the $^{13}N(p,\gamma)^{14}O$ reaction, which defines the entrance to this burning sequence, and the rates of the $^{14}O(\alpha,p)^{17}F$ and $^{15}O(\alpha,\gamma)^{19}Ne$ reactions which control the breakout from the CNO cycles. The $^{22}Na(p,\gamma)^{23}Mg$ and $^{26}Al(p,\gamma)^{27}Si$ reaction rates are also of importance, since the abundances of these two unstable nuclear species hold implications for the interpretation of meteoritic anomalies and gamma ray astronomy.

Nuclear requirements for calculations of the rp-process and explosive nucleosynthesis include a knowledge of the reaction properties of nuclei along the Z=N line from the vicinity of ^{44}Ti through perhaps ^{68}Se. Individual reaction rates can also be critical to the determination of the abundances of the proton-rich progenitors of the odd-Z nuclei ^{51}V, ^{55}Mn, ^{59}Co, and ^{63}Cu formed with the iron peak nuclei in supernova explosions, as the predicted abundances of

these species are sensitive to the expansion and cooling timescale.
Processes of heavy element synthesis, as we have seen, probe the limits of stable masses on both the proton-rich and neutron-rich sides of the valley of beta stability. The vast numbers of thermonuclear reaction rates required for numerical studies of these nucleosynthesis mechanisms must generally demand the use of theoretical calculations. Nuclear masses and reaction Q values in the regions far from the valley of beta stability must be similarly provided by nuclear mass formula extrapolations. For the specific case of s-process synthesis, experimental neutron capture cross sections at 30 keV and $f\tau$ values for critical low lying excited states for nuclei along the s-process capture path can help to set constraints on the timescale, the temperature, and the neutron flux history. Experimental information concerning both the rates of neutron production as a function of temperature from the neutron sources $^{22}Ne(\alpha,n)^{25}Mg$ and $^{18}O(\alpha,n)^{21}Ne$ and the neutron capture cross sections for the $^{22}Ne(n,\gamma)^{23}Ne$ and $^{25}Mg(n,\gamma)^{26}Mg$ reactions which compete for neutrons with the heavy elements is also of great interest.

ACKNOWLEDGEMENTS

This research was supported in part by the National Science Foundation under grant AST 86-11500 at the University of Illinois.

REFERENCES

1. I. Iben, Jr. and A. Renzini, Ann. Rev. Astron. Astrophys. 21, 271 (1983).
2. S. E. Woosley and T. A. Weaver, Ann. Rev. Astron. Astrophys. 24, 205 (1986).
3. J. W. Truran, Ann. Rev. Nucl. Part. Sci. 34, 53 (1984).
4. G. J. Mathews and R. A. Ward, Rep. Prog. Phys. 48, 1371 (1985).
5. J. Audouze, J. W. Truran, and B. A. Zimmerman, Astrophys. J. 184, 493 (1973).
6. R. K. Wallace and S. E. Woosley, Astrophys. J. Suppl. 45, 389 (1981).
7. T. Hanawa, D. Sugimoto, and M. Hashimoto, Publ. Astron. Soc. Japan, 35, 491 (1983).
8. M. Weischer, J. Görres, F.-K. Thielemann, and H. Ritter, Astron. Astrophys. 160, 56 (1986).
9. J. W. Truran, in Progress in Particle and Nuclear Physics, Vol. IV, ed. D. Wilkinson (Pergamon Press, Oxford, 1981), p. 177.
10. J. W. Truran, in Essays in Nuclear Astrophysics, ed. C. A. Barnes, D. D. Clayton, and D. N. Schramm (Cambridge University Press, Cambridge, 1982), p. 467.
11. J. S. Gallagher and S. Starrfield, Ann. Rev. Astron. Astrophys. 16, 171 (1978).
12. J. W. Truran and M. Livio, Astrophys. J. 308, 721 (1987).
13. R. E. Williams, in Production and Distribution of CNO Elements, ed. I. J. Danziger (ESO, Garching, 1985), p. 225.
14. J. W. Truran, in Production and Distribution of CNO Elements, ed. I. J. Danziger, (ESO, Garching, 1985), p. 211.

15. S. Starrfield, in The Classical Novae, ed. M. F. Bode and A. Evans (Wiley, New York, 1987).
16. R. E. Taam, Ann. Rev. Nucl. Part. Sci. 35, 1 (1985).
17. J. W. Truran, in Nucleosynthesis: Challenges and New Developments, ed. W. D. Arnett and J. W. Truran (University of Chicago Press, Chicago, 1985), p. 292.
18. W. Hillebrandt and F.-K. Thielemann, Astrophys. J. 255, 617 (1982).
19. D. C. Black, Geochim. Cosmochim. Acta 36, 377 (1972).
20. T. Lee, D. A. Papanastassiou, and G. J. Wasserburg, Astrophys. J. Letters 211, L107 (1977).
21. F.-K. Thielemann, K. Nomoto, and K. Yokoi, Astron. Astrophys. 158, 17 (1986).
22. A. G. W. Cameron, in Essays in Nuclear Astrophysics, ed. C. A. Barnes, D. D. Clayton, and D. N. Schramm (Cambridge University Press, Cambridge, 1982), p. 23.
23. E. M. Burbidge, G. R. Burbidge, W. A. Fowler, and F. Hoyle, Revs. Mod. Phys. 29, 547 (1957).
24. A. G. W. Cameron, Atomic Energy of Canada Ltd., CRL-41 (1957).
25. J. Audouze and J. W. Truran, Astrophys. J. 202, 204 (1975).
26. S. E. Woosley and W. M. Howard, Astrophys. J. Suppl. 36, 285 (1978).
27. J. W. Truran, Nukleonika 25, 1463 (1980).
28. R. K. Ulrich, in Essays in Nuclear Astrophysics, ed. C. A. Barnes, D. D. Clayton, and D. N. Schramm (Cambridge University Press, Cambridge, 1982), p. 301.
29. I. Iben, Jr. Astrophys. J 196, 549 (1975).
30. J. W. Truran and I. Iben, Jr., Astrophys. J. 216, 797 (1977).
31. B. J. Allen, J. H. Gibbons, and R. L. Macklin, Adv. Nucl. Phys. 4, 205 (1971).
32. F. Käppeler, H. Beer, K. Wisshak, D. D. Clayton, R. L. Macklin, and R. A. Ward, Astrophys. J. 257, 821 (1982).
33. J. Almeida and F. Käppeler, Astrophys. J. 265, 247 (1983).
34. W. Hillebrandt, Space Sci Rev. 21, 639 (1978).
35. W. Hillebrandt and F.-K. Thielemann, Astrophys. J. 255, 617 (1978).
36. J. W. Truran, J. J. Cowan, and A. G. W. Cameron, Astrophys. J. Letters 222, L63 (1978).
37. K.-L. Kratz, H. Gabelman, W. Hillebrandt, B. Pfeiffer, K. Schlösser, and F.-K. Thielemann, Z. Phys. A-Atomic Nuclei 325, 489 (1986).
38. K.-L. Kratz, F.-K. Thielemann, W. Hillebrandt, P. Möller, V. Harms, A. Wöhr, and J. W. Truran, preprint (1987).
39. W. Hauser and H. Feshbach, Phys. Rev. 87, 366 (1952).
40. J. W. Truran, Ph.D. thesis, Yale University (1966).
41. J. W. Truran, C. J. Hansen, A. G. W. Cameron, and A. Gilbert, Can. J. Phys. 44, 151 (1966).
42. G. Michaud and W. A. Fowler, Phys. Rev. C2, 2041 (1970).
43. J. W. Truran, Astrophys. Space Sci. 18, 306 (1972).
44. M. Arnould, Astron. Astrophys. 19, 82 (1972).
45. S. E. Woosley, W. A. Fowler, J. A. Holmes, and B. A. Zimmerman, At. Data Nucl. Data Tables 22, 371 (1979).

46. F.-K. Thielemann, Thesis, Technische Hochschule Darmstadt (1980).
47. F.-K. Thielemann, M. Arnould, and J. W. Truran, in *Advances in Nuclear Astrophysics*, ed. E. Vangioni-Flam, J. Audouze, M. Casse, J.-P. Chieze, and J. Tran Thanh Van (Editions Frontieres, Gif-siu-Yvette, 1987), p. 525.
48. F.-K. Thielemann, M. Arnould, and J. W. Truran, in preparation (1987).
49. J. P. Jeukenne, A. Lejeune, and C. Mahaux, Phys. Rev. $\underline{C16}$, 80 (1977).
50. S. Fantoni, B. L. Friman, and V. R. Pandharipande, Phys. Letters $\underline{104B}$, 89 (1981).
51. C. Mahaux, Phys. Rev. $\underline{C82}$, 1848 (1982).
52. F. M. Mann, Hauser 5, A Computer Code to Calculate Nuclear Cross Sections, Hanford Engineering, HEDL-TIME 78-83 (1978).
53. L. McFadden and G. R. Satchler, Nucl. Phys. $\underline{84}$, 177 (1966).
54. F.-K. Thielemann and M. Arnould, in *Proc. Int. Conf. on Nuclear Data for Science and Technology*, ed. K. Böckhoff (Antwerpen, 1983), p. 762.
55. A. Gilbert and A. G. W. Cameron, Can. J. Phys. $\underline{43}$, 1446 (1965).
56. H. Bethe, Phys. Rev. $\underline{50}$, 352 (1936).
57. T. Ericson, Nucl. Phys. $\underline{11}$, 481 (1959).
58. E. R. Hilf, G. von Groote, and K. Takahasi, in *Proc. 3rd Int. Conf. on Nuclei Far off Stability* (CERN, Rep. 76-13, 1976), p. 142.
59. Z. Y. Bao and F. Käppeler, preprint (1986).
60. W. A. Fowler, G. R. Caughlan, and B. A. Zimmerman, Ann. Rev. Astron. Astrophys. $\underline{5}$, 525 (1967).
61. W. A. Fowler, G. R. Caughlan, and B. A. Zimmerman, Ann. Rev. Astron. Astrophys. $\underline{13}$, 69 (1975).
62. M. J. Harris, W. A. Fowler, G. R. Caughlan, and B. A. Zimmerman, Ann. Rev. Astron. Astrophys. $\underline{21}$, 165 (1983).

NUCLEAR STRUCTURE EFFECTS FAR FROM STABILITY AND THEIR CONSEQUENCES ON RAPID-NEUTRON-CAPTURE PROCESSES

K.-L. Kratz[1], P. Möller[2], W. Hillebrandt[3], W. Ziegert[1,3], V. Harms[1], A. Wöhr[1] and F.-K. Thielemann[3,4]

[1]Institut für Kernchemie, Universität Mainz, Fed. Rep. Germany
[2]Department of Mathematical Physics, Lund Inst. of Technology, Sweden
[3]MPI für Physik und Astrophysik, Garching, Fed. Rep. Germany
[4]Department of Astronomy, Harvard University, Cambridge, USA

ABSTRACT

Rapid-neutron-capture processes in explosive astrophysical environments lead to the production of nuclei far away from β-stability. It is shown at the examples of solar and meteoritic isotopic anomalies around ^{48}Ca and for the observed r-abundance peaks at A≈80 and 130, that an understanding of the nucleosynthesis in such environments is intemately related to an understanding of the nuclear-structure properties of far-unstable nuclei.

INTRODUCTION

Over the past 30 years considerable effort has been devoted to the understanding of the synthesis of heavy elements in astrophysical environments. For recent reviews see, e.g., Refs.[1-3]. At many stages in the development of ideas on neutron-capture processes the major limitation to further progress has been the degree to which the underlying microscopic nuclear physics is known. With this in mind, it is not surprising that rapid-neutron-capture processes in exotic high-temperature stellar environments, for which practically no experimental nuclear physics input data are available, still are the poorest understood phenomena in astrophysics.

Probably the most important neutron-capture process for fission product data is the classical r-process[4] which occurs very-far on the neutron-excess side of β-stability. Basically, there are two fundamental time scales which have to be determined from better nuclear physics data before the still existing r-process mystery can be solved. One is the ratio of time scales for neutron capture and β-decay. This time scale determines how high the neutron density must be to reproduce the solar r-abundance pattern[5,6]. In the limit of high neutron density (and temperature), with neutron captures being rapid compared to β-decay half-lives ($T_{1/2}$), an (n,γ)-(γ,n) equilibrium is achieved for all isotopic chains. Then the abundances will be determined by the $T_{1/2}$ of nuclei very-far from stability and by β-delayed neutron (βdn) branching (P_n-values) during the decay back to stability. For those scenarios which reproduce the observed r-abundances without reaching (n,γ)-(γ,n) equilibrium - often referred to as 'n-process' occurring in dynamical environments with lower neutron densities than the classical r-process (see e.g.Ref.[7]) - competition between β-decay and neutron capture has to be treated explicitly, and the solar abundance peaks will largely be determined by the neutron-capture cross sections (σ_n) far off β-stability[8]. The second time scale which must be determined for rapid-neutron capture processes is the dynamical time scale during which the neutron density must be high in order to build up the actinides. This quantity depends on the sum of the neutron-capture and β-decay half-lives as one moves away from β-stability. Both time scales, together with the necessary neutron densities and temperatures will place severe constraints on astrophysical processes responsible for the observed r-abundances.

In this paper, we present two examples how recent experimental data together with improved RPA shell-model predictions[9] of nuclear physics properties far from β-stability have led to a better understanding of dynamics and nucleosynthesis of (i) neutron-rich S to K isotopes as progenitors of the solar ^{46,48}Ca abundances and the corrrelated Ca-Ti-Cr anomalies in inclusions of the Allende meteorite, and (ii) N≈50 and 82 'waiting-point' nuclei as progenitors of the observed A≈80 and 130 r-abundance peaks, through rapid-neutron-capture processes.

THE SOLAR ^{48}Ca/^{46}Ca ABUNDANCE RATIO AND THE CORRELATED Ca-Ti-Cr ISOTOPIC ANOMALIES IN INCLUSIONS OF THE ALLENDE METEORITE

In the past, astrophysical models encountered severe difficulties in explaining the solar ^{46,48}Ca abundances or the correlated Ca-Ti-Cr isotopic anomalies observed in inclusions of the Allende meteorite (see e.g. Refs.[10,11]). While the lighter Ca isotopes up to ^{44}Ca, for example, are believed to be produced in explosive O and Si burning with possible s-process contributions, the rare heavy isotopes ^{46,48}Ca are probably the result of a different nucleosynthesis process. The (neutron-rich) products of this process may later have been mixed into presolar material to form the solar abundances, or - if having remained unmixed - may show up as primordial meteoritic inclusions with the original isotopic composition, which may be different from the solar one.

Among the various attempts, Sandler et al.[12] suggested the production of neutron-rich stable Ca-Ti-Cr isotopes in a special 'nβ-process' operating in a neutron-density environment of ≈10^{-7} mol/cm^3 with a neutron-exposure time of 10^3 s. Assuming the initial abundances to be solar and applying Hauser-Feshbach (HF) neutron-capture cross sections, the above authors have calculated a ^{48}Ca/^{46}Ca abundance ratio which is only a factor of 2.6 smaller than the observed solar value of 56. However, the predicted isotopic anomalies for ^{46}Ca and ^{49}Ti were too large by factors of 13 and 5, respectively, compared to those in the EK-1-4-1 inclusion of the Allende meteorite[10]. In order to reduce the abundances of these two isotopes, Sandler et al. proposed a low-lying s-wave resonance in ^{46}K(n,γ) and ^{49}Ca(n,γ), respectively. Both targets are radioactive, however, so that a direct measurement of σ_n is not possible. These resonances were assumed to enhance the HF rates by an order of magnitude. With this, the depletion of the above progenitors of ^{46}Ca and ^{49}Ti would be increased considerably.

In the case of ^{49}Ca(n,γ) we have, indeed, identified such a low-lying s-wave neutron-capture resonance in the 'inverse reaction' to neutron absorption, i.e. ßdn-emission of 740-ms ^{50}K (Ref.[13]). However, subsequent measurements of the partial decay widths of this state of Γ_n=5.4 keV and Γ_γ⩽30 eV have put constraints on the Breit-Wigner resonance neutron-capture rate. It will be smaller than the HF rate for stellar temperatures of T⩽7.5·10^8 K, but may be enhanced by up to a factor of 4 over the average continuum cross section for T⩾1.2·10^9 K. Nevertheless, this is not sufficient to support the explanation of the meteoritic ^{49}Ti abundance suggested by Sandler et al.[12].

Motivated by those difficulties and by the nuclear structure related large variations of ß-decay properties of neutron-rich isotopes near the doubly-magic nucleus ^{48}Ca, we have investigated possible implications of $T_{1/2}$ and P_n-values on the Ca-Ti-Cr isotopic abundances[14]. The idea behind our attempt is the following: If we assume the nucleosynthesis path around A=48 to lie in the K chain with considerable production of ^{49}K, its ßdn-emission of P_n≈90% will predominantly form ^{48}Ca, whereas ^{49}Ti will only be populated weakly through direct ß-decay of ^{49}K. This would enhance the ^{48}Ca abundance and reduce the abundance of ^{49}Ti. Similarly, the low ^{46}Ca abundance could be due to ßdn-emission of its A=46, 47 progenitors.

Since we intended to perform a parameter study in order to find the right conditions for both the solar-system ^{48}Ca/^{46}Ca ratio and the isotopic anomalies mentioned above, we have not used a specific astrophysical model but rather have performed complete network calculations of the abundances as functions of time for various temperatures and neutron densities. Solar-system initial abundances were always chosen. We found that the results did not vary much with temperature if T was chosen between a few times 10^8 K and 10^{9}K, but were strongly dependent on neutron densities and exposure times.

Figure 1 shows a typical example obtained for T=8·10^8 K. Such a temperature is expected, e.g., in explosive He-burning. It can be seen that the solar ^{48}Ca/^{46}Ca ratio is only reproduced if the exposure time is less than 50 ms and the neutron exposure is at least 5·10^{-5} mol·cm^{-3}s, translating into a neutron density larger than 10^{-3} mol·cm^{-3} which lies in the same range as required for the r-process. The strong time dependence seen in Fig. 1 is mainly caused by the predicted[15] short $T_{1/2}$ - 0.3s of ^{44}S. If neutron-exposure times become comparable to $T_{1/2}$(^{44}S), nuclei are fed into the Cl isotopic chain and thus the abundances of ^{46}Cl and ^{47}Cl increase which both contribute to the final yield of ^{46}Ca (Ref.[14]). If, on the other hand, the $T_{1/2}$(^{44}S) would be considerably longer, less neutrons would be needed in our model in order to explain the solar ^{48}Ca/^{46}Ca ratio. Since ^{44}S has a closed neutron shell (N=28) a longer $T_{1/2}$ is certainly possible and not out of experimental range.

A maximum of the ^{48}Ca/^{46}Ca abundance ratio exists for neutron exposures of about (5±1)·10^{-5} mol·cm^{-3}s, nearly independent of neutron density and exposure time. For fixed exposure time the maximum value is mainly determined by the P_n-values of ^{47}Cl and ^{49}K. The experimentally known large P_n≈90% of ^{49}K yields a large ^{48}Ca

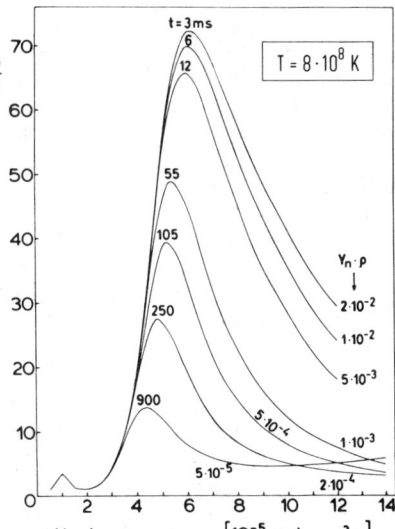

Fig. 1. $^{48}Ca/^{46}Ca$ abundance ratio as a function of neutron exposure for various combinations of neutron densities ($Y_n \cdot \varrho$) and exposure times (t). Given is the exposure time at which the maximum abundance ratio is obtained. Beta-decay half-lives from Ref.[15] have been used.

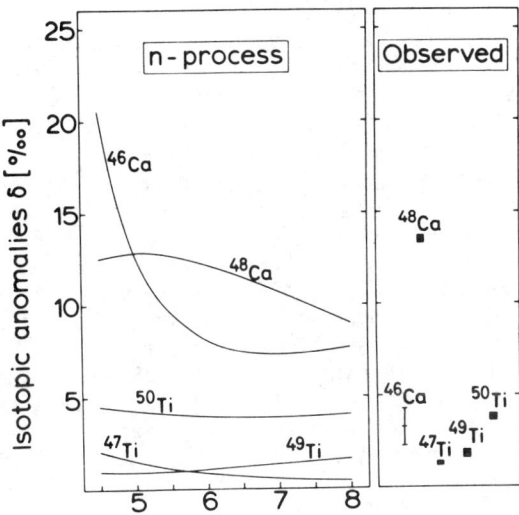

Fig. 2. Isotopic anomalies in Ca and Ti. The figure on the right shows the observed anomalies in the Allende inclusion EK-1-4-1 the figure on the left shows the computed anomalies obtained from our n-process model as a function of neutron exposure.

abundance, whereas the (so far unknown) ßdn-decay of ^{47}Cl strongly influences the ^{46}Ca production. Even if $P_n(^{47}Cl)$ would turn out to be 100%, the solar $^{48}Ca/^{46}Ca$ ratio can be reproduced for certain neutron exposures, but with the restrictions on neutron densities and exposure times shown in Fig. 1. $P_n(^{47}Cl)$-values around 30% would be more favourable. Again, this important physical quantity should be measured in order to restrict the astrophysical parameter space.

In order to derive meteoritic Ca, Ti and Cr abundances, the composition obtained from our network calculations was mixed into a reservoir with solar-system abundances. Typically a fraction of 10^{-5} of processed material gave the best fits. At a neutron-exposure of

$\approx 7 \cdot 10^{-5}$ mol·cm^{-3} s, best overall agreement with the observed Ca-Ti isotopic anomalies in EK-1-4-1 was obtained (see Fig. 2). Furthermore, our calculated abundance ratio of $[^{48}Ca]/[^{50}Ti]=3.4^{+8.2}_{-8.4}$, where square brackets denote abundances relative to their solar system values, is in excellent agreement with the measured value of 3.6±0.2. In addition, our predicted Cr isotopic anomalies of $[^{54}Cr]/[^{52}Cr]=4.5^{+3.0}_{-1.0}$ and $[^{53}Cr]/[^{52}Cr]=2.9^{+1.0}_{-0.5}$ agree reasonably well with the recent experimental values [10] of 4.8 and 1.6, respectively.

To summarize, we have demonstrated that with only slightly different neutron exposures, expected e.g. in explosive He-burning in supernovae, the <u>same</u> nucleosynthesis process may yield both the solar $^{48}Ca/^{46}Ca$ abundance ratio and the meteoritic Ca-Ti-Cr isotopic anomalies of the EK-1-4-1 inclusion. The consideration of nuclear structure effects in β-decay properties of neutron-rich S to K isotopes in high-density neutron-capture models seems to be the nuclear-physics clue to the solution of these astrophysical problems.

CONTRAINTS ON R-PROCESS CONDITIONS FROM BETA-DECAY PROPERTIES FAR OFF STABILITY AND R-ABUNDANCES

Despite years of effort, the basic astrophysical site (or sites) of the r-process is still unknown or at best vaguely defined as being associated with supernovae (SN)[1-4]. None of the detailed r-process calculations were fully consistent with the adopted astrophysical models, often pushed to the limits to fit the r-abundances.

We have followed a different approach: In order to check if the r-process operated in an equilibrium (as assumed in early attempts[4]) we relate the r-abundances of N≈82 nuclei in the r-process path which are the progenitors of the observed A≈130 r-abundance peak[5] to their $T_{1/2}$. For definition of an $(n,\gamma)-(\gamma,n)$ and a β-decay flow equilibrium, see e.g. Ref.[16]. The initial r-abundances of the progenitor isotopes ($n_{r,prog}$) can be derived from the respective observed r-abundances ($n_{n,\odot}$) by correcting them for βdn-branching which occurred at freeze-out during the decay back to stability. For consistency, we also check the relation between the recent ($n_\odot - n_s$) abundances in the A≈80 mass region as given in Ref.[6], assuming that they are predominantly of r-process origin, and the $T_{1/2}$ and P_n of their respective N≈50 r-progenitors. We then ask what information can be obtained from the $n_{r,prog} - T_{1/2}$ correlation and the knowledge of the r-process path about the necessary conditions - time scale, temperature and neutron density - under which the r-process has operated. With this information, tighter constraints may be put on the different astrophysical models.

The solar r-process abundances ($n_{r,\odot}$), expecially around the r-process peaks at A≈80, 130 and 195, can give clues to the existence of either of the above equilibria. Each abundance peak corresponds to the heaviest nucleus with one of the neutron magic numbers N=50, 82 or 126 which still lies in the r-process path, before the r-process branches off towards larger-than-magic neutron numbers. In an $(n,\gamma)-(\gamma,n)$ equilibrium the abundance distribution in one isotopic chain is usually strongly peaked at one or at most two nuclei. This follows automatically from the neutron separation energies and the

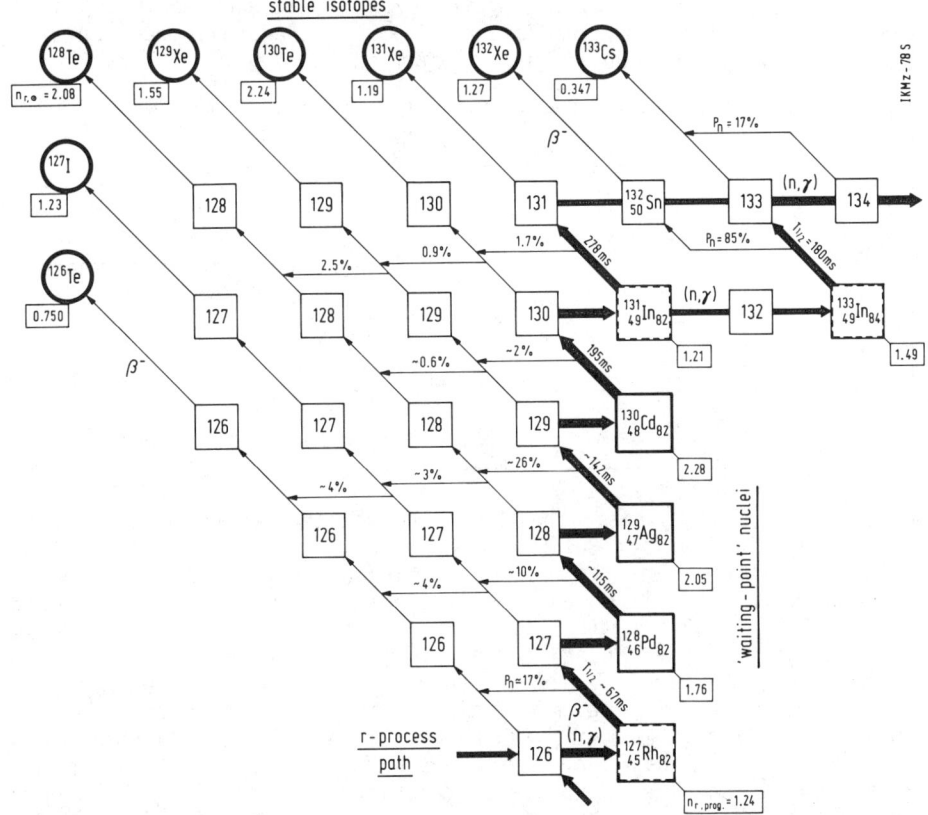

Fig. 3. Schematic view of the r-process path and the r-abundance features in the A≈130, N≈82 region. $T_{1/2}$ and P_n values are either experimental values[17,18] or RPA shell-model predictions. For details, see text.

neutron densities and temperatures of astrophysical interest.

The progenitors of the stable nuclei in the A≈130 peak, for example, are the N=82 isotones ^{127}Rh, ^{128}Pd, ^{129}Ag and ^{130}Cd. Above the heaviest waiting-point nucleus ^{130}Cd, the r-process path branches off at In, with ^{131,133}In being the most abundant nuclei in the Z=49 isotopic chain. Starting with the Cd and In isotopes in the r-process path for which the necessary β-decay properties ($T_{1/2}$ and P_n) are experimentally known[17,18], we can check whether a direct relation between r-abundances and $T_{1/2}$ is realized in nature. As can be seen from Fig. 3, for the majority of nuclei in the A≈130 mass region β-delayed neutron emission is small. In consequence, the $n_{r,prog}$ are almost identical to the observed $n_{r,\odot}$. One exception is the isotope ^{133}In with a P_n≈85%. Therefore, the main part of $n_{r,prog}(^{133}$In$)$ will be seen as $n_{r,\odot}(^{132}$Xe$)$ and only a small part will remain in the A=133 chain to form (together with the P_n-contribution from ^{134}Sn)

Tab. 1. Comparison of experimental $T_{1/2}$ of N≈82 and 50 isotopes in the r-process path with predictions from our RPA shell model and the astrophysical β-flow equilibrium prediction.

Isotope	Beta-Decay Half-Life, $T_{1/2}$ [ms]		
	Exp.	RPA-S.M.	Beta-Flow
^{133}In	180±20	123	
^{131}In	278± 3	240	
^{130}Cd	195±35	235	180
^{129}Ag		172	160
^{128}Pd		115	140
^{127}Rh		67	100
^{83}Ga	308± 4	288	
^{81}Ga	1218± 4	1070	
^{80}Zn	540±30	502	750
^{79}Cu		193	200 +)
^{78}Ni		207	185 +)
^{77}Co		6.5	≤11.5 +)

+) normalized to experimental $T_{1/2}$ of ^{80}Zn

$n_{r,\odot}(^{133}$Cs). With this, the $T_{1/2}$ prediction for ^{130}Cd from the β-flow equilibrium is (180±20)ms, where the error estimate comes from the uncertainties of the solar r-abundances[5]. The good agreement between this 'astrophysical' prediction and the recently measured $T_{1/2}(^{130}$Cd$)=(195\pm35)$ms [17] indicates that the solar r-process abundances may have been formed in a β-flow equilibrium which automatically would imply also an (n,γ)-(γ,n) equilibrium as $\tau_\beta > \tau_{n,\gamma}$.

While a single case may yield an accidental agreement between the actual β-decay $T_{1/2}$ and its β-flow prediction, additional checks - even when only based on shell-model $T_{1/2}$ - might strengthen our above conclusion. The abundance distributions in the Z=45-47 isotopic chains are dominated by the neutron closed shell nuclei ^{127}Rh to ^{129}Ag With the $n_{r,\odot}$ abundances of ^{127}I, ^{128}Te and ^{129}Xe corrected for β-delayed neutron emission (see Fig. 3), we obtain β-flow predictions for the $T_{1/2}$ of the respective N=82 r-process progenitors which are (within a 15 to 20% uncertainty limit derived from the $n_{r,\odot}$ uncertainties[5]) in agreement with our RPA shell-model predictions (see upper part of Tab.1), thus showing a consistent picture for the entire A≈130 abundance peak.

A similar check of our β-flow equilibrium assumption can also be performed for the A≈80 r-abundance peak with the N≈50 isotopes ^{77}Co to ^{80}Zn and ^{81,83}Ga. This test, however, may be less conclusive than that for the A≈130 peak because the $(n_\odot-n_s)$ abundances of Ref.[6] may not totally reflect $n_{r,\odot}$ abundances but may also contain other components. We start with the β-flow prediction of $T_{1/2}(^{80}$Zn$)$ from the $n_{r,prog}(^{81,83}$Ga$)$, and then proceed to the lighter N=50 waiting-point nuclei. As suspected, the predicted $T_{1/2}(^{80}$Zn$)$ lies with 750 ms slightly outside the experimental $T_{1/2}$ range[19]. This indicates that roughly 30% of the observed $(n_\odot-n_s)$ abundance of ^{80}Se might not be of r-process origin. When correcting for this non-r-contribution and for βdn-branching [16] we obtain reasonable agreement between the $T_{1/2}$ for ^{79}Cu and ^{78}Ni from our RPA shell model and the β-flow assumption (see lower part of Tab. 1). It is quite interesting to see that even the trend of the β-flow prediction for $T_{1/2}(^{77}Co)$ seems to be correct. The resulting $n_{r,prog}(^{77}Co) \leq 0.5$, which is compatible with zero, supports earlier conclusions[2] that it is probably inappropriate to talk about nuclei below ^{78}Ni as truly being of r-process origin.

As already mentioned, a quantity which plays an important role for the evaluation of astrophysical models is the length of time required for rapid-neutron-capture processes. Previously, with the large uncertainties in calculated $T_{1/2}$ far from stability[15,20], the r-process time scale could only be estimated to lie in the rather wide range between 0.1 and 30 s, thus preventing substantial constraints on astrophysical scenarios[2,3]. With the recent measurement of two of the most critical waiting-point nuclei[17,19] together with our improved shell-model predictions, and the evidence for the existence of a β-decay flow equilibrium, we now can give a new lower-limit-estimate for the time needed for the nucleosynthesis from Fe to the the actinides. By summing up the $T_{1/2}$ of all nuclei in the r-process path[2] we obtain a β-decay cycle time of about 3.5s.

Given the result that the r-process proceeds under the conditions of an (n,γ)-(γ,n) equilibrium, it is furthermore possible to determine the necessary neutron densities $(Y_n \rho N_A)$ and temperatures (T) which give the correct r-process path (through ^{77}Co to ^{80}Zn at N=50, and ^{127}Rh to ^{130}Cd at N=82) as well as the correct $n_{r,\odot}$ peaks. As an example $Y_n \rho N_A$ and T ranges (bracketed by the same symbol) where the N=82 waiting-point nuclei have the maximum abundance in their respective isotopic chains are shown in Fig. 4. With the requirement that the conditions for all these nuclei have to be fulfilled simultaneously, the boundary lines in the figure represent the intersection of all 'allowed' T-$Y_n \rho N_A$ regions. A further promising result is that this area for the N=82 region overlaps with the respective region for the N=50 waiting-point nuclei[16] and this

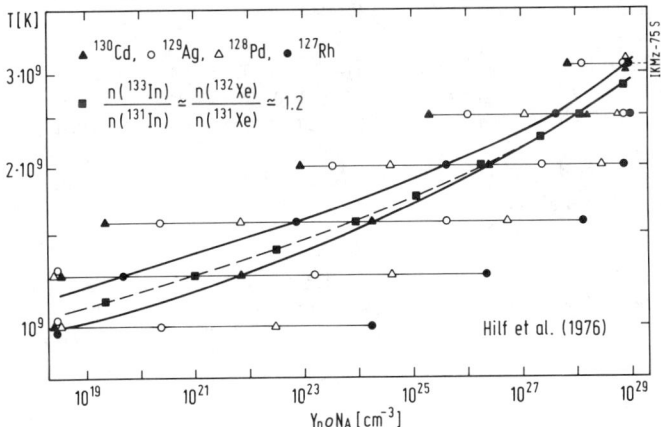

Fig. 4. Stellar temperature (T) and neutron density $(Y_n \rho N_A)$ ranges where - under the assumption of an (n,γ)-(γ,n) equilibrium - the N=82 waiting-point nuclei have the maximum r-abundance in their isotopic chains. The dashed line connecting the filled boxes indicates those conditions within the boundary lines, for which the observed $n_{r,\odot}$ ratio for $^{131,132}Xe$ (Ref.[5]) is reproduced.

not only for the (old) mass formula of Hilf et al.[21] but also for the very recent Möller-Nix mass predictions[22]. With the additional information from the $n_{r,\odot}$ abundances of ^{81}Br and ^{83}Kr which were formed by the progenitors ^{81,83}Ga, and the abundances of ^{131,132}Xe which originate from ^{131,133}In, one may give even tighter constraints for the T-$Y_n N_A$ conditions. For the latter ratio this is indicated in Fig. 4 by the dashed line connecting the boxes. It is seen that this line lies well within the boundaries of the 'allowed' region defined by the waiting-point nuclei. Moreover, as is shown in Ref.[16], the respective lines for the Ga and the In isotopic $n_{r,prog}$ ratios run rather close to each other over a wide T-$Y_n N_A$ range. Originally, there was the hope that the intersection of these lines would define a unique set of T and $Y_n N_A$ values characterizing unique conditions for the r-process. However, with the resulting band of T-$Y_n N_A$ conditions, the $n_{r,\odot}$ abundances presumably represent a last equilibrium before freeze-out due to small T and $Y_n N_A$ values. Therefore, it is probably most important that the conditions at the lower end of this area are met by astrophysical models.

The present results should help to test possible r-process models for the correct position of the r-process path at the time of freeze-out. The correct abundance pattern requires in addition time scales which guarantee that enough matter was transformed into heavy nuclei. With the assumption of a static r-process, a process duration of several seconds results, apparently in contradiction to presently favoured explosive He-burning scenarios. Such a conclusion is, however, not unconditionally necessary, as initially higher neutron densities could have caused a temporary r-process path further away from β-stability with shorter $T_{1/2}$, and the $n_{r,\odot}$ abundances correspond only to the late r-process path at freeze-out conditions.

SUMMARY

We have shown in this paper that there exists a close connection between microscopic nuclear physics and whatever constraints can be imposed to astrophysical models. Clearly, more experimental and theoretical work on the relevant nuclear physics properties far from stability is needed for a better understanding of rapid-neutron-capture process nucleosyntheses.

We acknowledge stimulating discussions with M. Arnould, F. Käppeler and J.W. Truran. This work was supported by DFG (Kr 806/1-2), BMFT (06 MZ 552), NSF (87-03 535) and the Milton Fund at Harvard University.

REFERENCES

1. W. Hillebrandt, Sp. Sci. Rev. 21, 639 (1978).
2. D.N. Schramm, in 'Essays in Nuclear Astrophysics', eds. C.A. Barnes et al. (Cambridge University Press, 1982), p.325.
3. G.J. Mathews, R.A. Ward, Rep. Prog. Phys. 48, 1371 (1985).
4. G.R. Burbidge et al., Rev. Mod. Phys. 29, 547 (1957).

5. F. Käppler et al., Ap. J. 257, 821 (1986).
6. G. Walter et al., Astr. Ap. 167, 186 (1986).
7. J.B. Blake, D.N. Schramm, Ap. J. 209, 846 (1976).
8. A.G.W. Cameron et al., Astr. Sp., Sci. 91, 335 (1983).
9. J. Krumlinde, P. Möller, Nucl. Phys. A417, 419 (1984).
10. D.A. Papanastassiou, Ap. J. 308, L30 (1986), and Refs. therein.
11. J.-L. Birck, C.J. Allegre, Geophys. Rev. Lett. 11, 943 (1984), 12, 745 (1985).
12. D.G. Sandler et al., Ap. J. 259, 908 (1982).
13. K.-L. Kratz et al., Astr. Ap. 125, 381 (1983).
14. W. Ziegert et al., Phys. Rev. Lett. 55, 1935 (1985).
15. H.V. Klapdor et al., At. Data Nucl. Data Tables 31, 81 (1984).
16. K.-L. Kratz et al., J. Phys. G, in print.
17. K.-L. Kratz et al., Z. Phys. A325, 483 (1986).
18. Proc. Specialists' Meeting on Delayed Neutron Properties (1986), Univ. of Birmingham BRC-Report, in print.
19. E. Lund et al., Proc. AMCO-7, THD-Schriftenreihe 26, 102 (1984), and Physica Scripta 34, 614 (1986).
20. K. Takahashi, M. Yamada, At. Data Nucl. Data Tables 12, 101 (1973).
21. E.R. Hilf et al., Report CERN 76-13, 142 (1976).
22. P. Möller, J.R. Nix, Report LA-UR-86-3983 (1986), and At. Data Nucl. Data Tables, in print.

NEUTRON-RICH NUCLEI AND NUCLEOSYNTHESIS

Yu.S. Lyutostansky, S.V. Malevanny
Moscow Physical-Engineering Institute, Moscow, USSR

I.V. Panov
Institute of Theoretical and Experimental Physics, Moscow, USSR

V.M. Chechetkin
Institute of Applied Mathematics, Moscow, USSR

ABSTRACT

For the description of the production of nuclei with $A \geq 80$ in the process of neutron capture nucleosynthesis a dynamical kinetic model is developed. The calculated r-process abundances were found to be strongly depended on the predictions of neutron-rich nuclei properties. The role of the β-delayed process - (β^-, kn) k=1-3 and (β^-, f) is analysed. The model dependence of the age of the Galaxy T_G is analysed in the method of actinide isotope chronometric pairs. The T_G value is calculated in the exponential model of galactic nucleosynthesis by Fowler with various conditions of the r-process proceeding and with various prognosticated nuclear data differs greatly. To coordinate numerous abundance calculations it is necessary to include a sudden spike of nucleosynthesis in the cosmological model.

The investigation of nuclei far from the stability region is concerned with a very interesting problem of nuclear astrophysics, i.e. the problem of synthesis of chemical elements in nature. In the process of rapid nucleosynthesis are involved unknown N-rich nuclei the properties of which should be predicted. In this paper in the nucleosynthesis calculations there have been used various methods for the description of N-rich nuclei, including the microscopic theory of finite Fermi systems (TFFS) by A.B. Migdal.
For the description of the production of nuclei in the process of neutron capture nucleosynthesis we developed a kinetic model[1] without considering a simplifying assumption on the equilibrium for (n, γ) and (γ, n) reactions. The system of nucleosynthesis equations (NE) describing the time dependence of nuclei concentrations n_A for the different values of A involves the rates of the following process: (n, γ) and (γ, n) on the given and neighbouring nuclides; β-decay and β-delayed multi-neutron emission - (β^-, kn)-process (k=1, 2, 3); spontaneous and β-delayed fission; and separately α-decay. The precedent papers[1,2] present a technique for the calculation of the corresponding cross sections and probabilities. We used several methods for the description of mass relations, rates of (n, γ) and (γ, n) reactions and β-decay parameters. The calculated r-process abundances were found to be strongly depended on the predictions of N-rich nuclei properties. This affects the total behaviour of the abundances with A as well as the values of

separate isotope relations, e.g. actinide isotopes for the determination the age of the Galaxy. For the calculation of β-delayed process probabilities for N-rich nuclei we used the isobaric states theory[3] defining the β-strength function developed in terms of TFFS. The β-delayed multi-neutron emission was described as earlier[4] taking into account the "tails effect" of high-lying isobaric resonances, particularly, the Gamov-Teller resonance. The calculation of β^--delayed fission, i.e. (β^-, f)-process, probability P_f was performed by using the strength function $S_\beta(E)$ calculated in terms of TFFS and mass relations[5]. The calculations indicate that the $P_{\beta-f}$ values seldom reach 40-50% and are not so large as they were assumed to be earlier[6]. So this is not opposed to the production of superheavy nuclei in nature and at the same time it enables to describe the odd-even anomalies in experiments with transuranium elements production in thermonuclear explosions (this problem is considered in details earlier[7]). The parameters in NE could be changed, if it is necessary, as a function of time according to the dependences of free neutron density $n_n(t)$, temperature $T(t)$ and $\rho(t)$. Dynamic calculations using exponential relations $\exp(-t/\tau_H)$ for ρ and T were performed with different values of the parameter.

In the system of nucleosynthesis equations (1) we used the predicted parameters of N-rich nuclei calculated in terms of various methods. As it can be seen in Fig. 1 the abundance curve n_A depends upon the calculation method using for the prognosticating N-rich nuclei properties. The influence of β-delayed neutrons substantially changes the n_A dependence when N-rich nuclei return to the region of β-stability. This leads to the n_A curve smooting and approaching the observable dependence. It should be noted that the positions of the abundance peaks are not practically depended on the prognosticating method for N-rich nuclei or the n_n, ρ and T values used in calculations.

The velocity of nucleosynthesis wave front spreading, i.e. the rate of new nuclides production depends upon the conditions under which r-process takes place. The calcualtions of N-rich nuclei half-lives make it possible to determine the minimal r-process duration time[8] t_r^{min} - the time during which nuclei with $A \geqslant 240$ are produced. This time strongly depends on temperature, e.g. when T increase only by 50% t_r^{min} value increase almost by an order of magnitude. In Fig. 2 one can see the dependence of the time of nuclei production from A value and A-intervals of the acceleration and deceleration of nucleosynthesis wave front moving. From the obtained limitation on t_r^{min} value, in such a way as in work by Schramm[9], one can get restrictions on the initial matter density of a pre-explosive star: $\rho \leqslant 10^8$ g.cm^{-3}. The calculations of half-lives were performed in this case by using TFFS.

The results of the dynamic calculations of heavy nuclei concentrations are strongly depended upon the τ_H parameter value. Nevertheless to explain the correlation among all the three peaks in the observed r-abundance it is necessary to solve the problem of initial nuclei composition as well as the problem of a source of free neut-

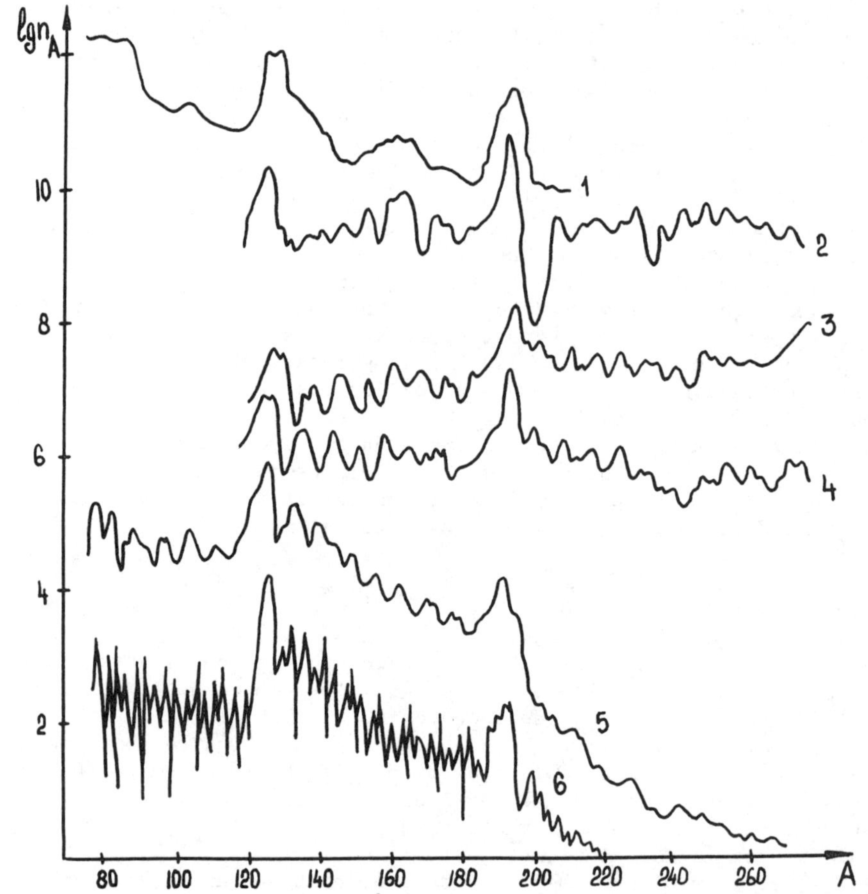

Fig. 1. The abundances of r-nuclei calculated under different external condition: constant $n_n=10^{24} cm^{-3}$, $T=10^9 K$ (curve 2-4) and exponential time decreasing with $\rho_o=2 \cdot 10^5 g/cm^3$, $T_o=2 \; 10^9 K$ (curve 5,6), 5 - is calculated including (β^-, k)-process (k=1-3), 6 without them. Shift alone the vertical axis is arbitrary with respect to curve 1 (the observed abundance). Nuclear data is taken from: curve 2) -12) - masses, 8)-$T_{1/2}$; P_n; curve 3) - 10), 8); 4) - accordingly 10), 12).

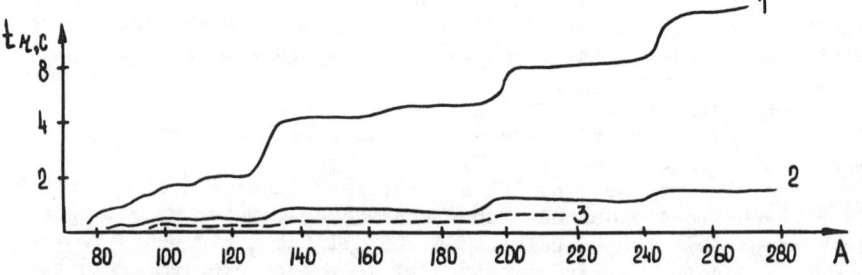

Fig. 2. The dependence of r-process duration time on mass A-value under different external conditions: curve 1) constant $n_n=10^{26} cm^{-3}$, $T=1,5 \; 10^9 K$; 2) - the same n_n, $T=10^9 K$; 3) dynamical calculations with $\rho_o=2 \; 10^5 g/cm^3$, $T_o=10^9 K$.

rons the amount of which is not enough for the production of actinides in the observed concentrations.

Within the framework of the developed kinetic model of nucleosynthesis there were calculated isotope U-Th relations by means of which the age of the Galaxy T_G was determined. We have used the Fowler model in which the galactic nucleosynthesis exponential decrease with time.

The age of Galaxy values calculated in Fowler model of galactic nucleosynthesis with various parameters and conditions of r-process proceeding, and with various prognosticated nuclear data differs greatly. Thus, they turned out to be model dependent and changed within the range 7,5 up to 40 milliard years. According to our calculations the effect of β-delayed fission doesn't influence strongly on the relative rate of U and TH isotopes production as well as on the T_G value - $\Delta T_G \approx 5 \cdot 10^8$ a. This effect certainly should not be neglected but its influence is not so great as it was assumed in[6].

By including into nucleosynthesis calculations the enrichment of a chemical composition by heavy elements due to S-spike - a supernova explosion that had taken place maybe not long ago before the formation of the solar system, we can coordinate numerous abundance calculations under different astrophysical conditions and with various description of N-rich nuclei.

Thus the nucleosynthesis calculations with various sets of nuclear data demonstrate us strong dependence both the calculated abundances and the age of the Galaxy value on the methods of prognosticating the N-rich nuclei properties.

1. Yu.S. Lyutostansky et al., Sov. Nucl. Phys., 42, 215, (1985).
2. Yu.S. Lyutostansky et al., Sov. Nucl. Phys., 44, 66, (1986).
3. Yu.V. Gaponov, Yu.S. Lyutostansky, Sov. Phys. Particles Nucl. 12, 1324, (1981).
4. Yu.S. Lyutostansky, I.V. Panov, V.K. Sirotkin, Phys. Lett. 161B, 9, (1985).
5. W.M. Howard, F. Moller, Atom. Data Nucl. Data Tables, 25, 219, (1980).
6. F.K. Thielemann, J. Metzinger, H.V. Klapdor, Zts. Physic. A309, 30, (1983).
7. R.W. Hoff, In: Weak and Electromagentic Interactions in Nuclei, eds, H.V. Klapdor, Heidelberg, 1983, p. 321.
8. Yu.S. Lyutostansky, I.V. Panov, M., Preprint ITEPh, 1986, No. 32.
9. D.N. Schramm, In: Essays in nuclear astrophysics, eds. C.A. Barnes, D.D. Clayton and D.N. Schramm (Cambridge U.P., London, 1982), p. 325.
10. W.D. Myers, W.J. Swiatecki, Ann. Phys., 55, 395, (1969).
11. K. Tasaka, Report JAERI-M 5997, Japan, 1975.
12. J. Janecle, B.P. Eynon, Atome. Data and Nucl. Data Tables, 17, 467, (1976).

THE ^{22}Na(p,γ) ^{23}Mg REACTION: PREPARATION OF A ^{22}Na TARGET

L. Buchmann, M. Dombsky and J.M. D'Auria
Simon Fraser University
Burnaby, British Columbia, Canada V5A 1S6

J.D. Vincent
TRIUMF, Vancouver, British Columbia, Canada V6T 2A3

C. Rolfs, S. Seuthe, and H.W. Becker
Institut für Kernphysik, Universität Münster, Münster,
West Germany

R. KAVANAGH
California Institute of Technology, Pasadena, CA 91125, U.S.A.

ABSTRACT

A study is in progress to measure for the first time the excitation function for proton capture on the radioactive species ^{22}Na($T_{1/2}$ = 2.6y), for incident proton energies from 0.1 to 2.0 MeV. This reaction and its strength is of considerable interest in understanding the mechanism of explosive astrophysical phenomena. This study is possible due to the recent development (C. Rolfs and R. Kavanagh, NIM A247, 507 (1986)) of a new type of gamma threshold detector, sensitive only to gamma rays with energies above 2.2 MeV. This detector, based upon the photodisintegration of deuterium followed by detection of the thermalized neutron, does not exhibit any energy resolution, and thus an isotopically pure target with only ^{22}Na is required. An isotopically pure target will be made using the TISOL facility (a new on-line isotope separator at TRIUMF) in an off-line mode. An irradiated sample of aluminum containing about 30 mCi of ^{22}Na will be used in an oven operated at about 800°C. A progress report on this study will be presented at the Conference.

INTRODUCTION

Recently[1,2] it was discovered that so-called Ne-E inclusions are present within carbonaceous chondrite meteorites. Since these inclusions are pure ^{22}Ne, within experimental limits, it has been suggested that these inclusions arose from the decay of ^{22}Na ($T_{1/2}$=2.6 a). The ^{22}Na in such a case would have been formed in explosive stellar events such as supernovae or more likely novae, ejected and later, while the stellar ejecta were cooling, condensed and incorporated into micrometer sized grains. With the formation of the solar system such grains could have been coalesced in primitive meteorites without changing their internal structure. The pure ^{22}Ne observed in the mass spectroscopy of parts of the meteorite then shows a picture integrated over many of these tiny grains.

With the observation of a βγ decay line from the long lived, ^{26}Al ($T_{1/2}$=716000 a), which is believed to be formed in similar stellar scenarios as ^{22}Na, it also seems to be feasible, given more sensitive γ-ray observing satellites, to detect the βγ decay line of ^{22}Na (1275 keV). This would prove the ongoing production of this radioisotope.

To calculate the production of ^{22}Na in stellar environments and thus to discriminate between several production scenarios, it is

© American Institute of Physics 1988

necessary to know the nuclear stellar rates for the production and destruction of ^{22}Na. Such information would also allow one to calculate the galactic abundance of ^{22}Na and stable products formed in the same process. Thus by observing ^{22}Na, conclusions could be formulated on stable isotope abundances as has been done in the case of ^{26}Al/^{27}Al.

Due to the vulnerability of ^{22}Na to the ^{22}Na(n,p)^{22}Ne reaction, which has a positive Q-Value, ^{22}Na is generally believed to be formed in proton rich, neutron free environments at temperatures moderately low, i.e. some hundred million degrees or below. One possible nuclear reaction chain leading to ^{22}Na is the NeNa cycle as displayed in Figure 1. This cycle can have some leakage into the MgAl cycle (at higher temperatures) in which ^{26}Al may be produced. The cycle starts with the stable and rather abundant isotopes of neon and propagates via proton capture and subsequent beta decay to ^{23}Na, at which point the cycle is closed by a (p,α) reaction back into ^{20}Ne. ^{22}Na is formed in this cycle by the well studied ^{21}Ne(p,γ)^{22}Na reaction[3], while even at moderate temperatures it is destroyed by proton capture (^{22}Na(p,γ)^{23}Mg) instead of the competing β$^+$ decay. Thus it is essential to know the ^{22}Na(p,γ)^{23}Mg reaction rate in order to calculate the ^{22}Na concentration in such a cycle or any proton rich environment.

Unfortunately, for any measurement of the ^{22}Na(p,γ)^{23}Mg reaction, a ^{22}Na target exhibits some nasty properties. These include extremely high βγ fluxes for any useful target (10^{17} atoms

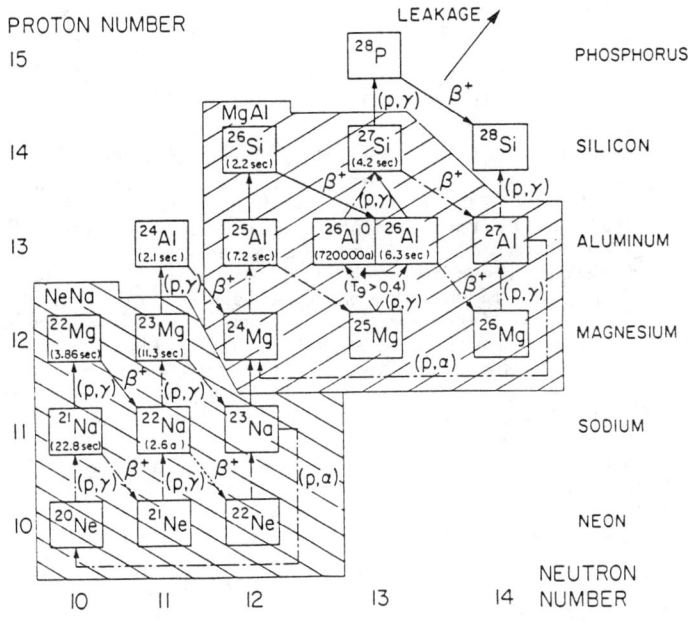

Fig. 1. The NeNa and the MgAl cycles as examples of nuclear networks in a proton rich stellar environment leading to the long-lived isotopes, ^{22}Na and ^{26}Al.

exhibit activities up to about 20mC) and reactive chemical properties for sodium which make preparation of suitable targets difficult. The approach chosen here to overcome these problems was both to develop a detector only sensitive to the high energy γ's of the proton capture and to produce a target by ion implantation which shows stability and a high degree of purity to match the experimental conditions required by the detector.

The detector is described in detail elsewhere[4]. Briefly, it is based on the photodisintegration of the deuteron, which has a threshold of 2.2 MeV. The target is placed inside the detector and is then surrounded by a tank containing about 200 ℓ of heavy water. A photon above the disintegration threshold can then split the deuteron into its constituents: a proton and a neutron. To obtain high detector efficiency the proton is not detected but rather the neutron is moderated and detected in about 20 ^{3}He counters surrounding the heavy water tank. Some polyethylene is also used outside the detector for a further moderation of the neutrons produced. Thus the entire detector is only sensitive to the number of high energy photons escaping the target while it is bombarded with protons. The total efficiency of the disintegration detector is expected to be of the order of 10^{-3}, comparable with normal Ge(Li) gamma spectrometers.

Given such a threshold detector, which does not exhibit energy resolution for the reaction gamma-rays, the target has to be of reasonably high purity. Therefore it has been decided to implant a mass separated ^{22}Na ion beam into a suitable backing. The decision to use an implanted target was also made feasible by the high ionization efficiencies demonstrated by surface ionization sources. The subsequent sections will describe the developments made towards an implanted ^{22}Na target.

THE DUAL OVEN ION SOURCE

In the implantation process to be tried here the ^{22}Na is produced by activating a thick piece of aluminum (~10 grams). Experiments with such samples have shown that radiosodium is released rapidly after the aluminum has become molten (660°C)[5]. Thus it is planned to melt such an activated piece of aluminum in a special oven which is connected to the ion source.

Given the small amount of ^{22}Na available coupled with its high activity, it was considered necessary to set up the mass analyser (as described in the subsequent section) using a stable beam (e.g. ^{39}K) prior to releasing the ^{22}Na. Thus, the separator can be operated for non-radioactive stable ion beams before the implantation of ^{22}Na is tried. To avoid any significant effects or changes on the source during switching between beams, a dual oven system connected to a surface ionization source was developed. This is shown in Figure 2 and described below.

The system has two side ovens connected to a centre oven, and each side oven may contain either the activated piece of aluminum or a mixture of sodium/potassium/rubidium compounds for the production of stable beams (1+). The ovens can be heated up alternately. To connect both of them to the source, the side ovens as well as the surface ionization source are connected to the middle/centre oven, which stays heated continuously when the source is in operation. The ion source, consisting of a rhenium ionizer, is heated continuously while the source is in operation. Thermocouples inserted in the side ovens as well as the ionizer measure the temperatures associated with the system. The ovens and the source can operate at a

DUAL OVENS AND SURFACE SOURCE FOR ION IMPLANTATION

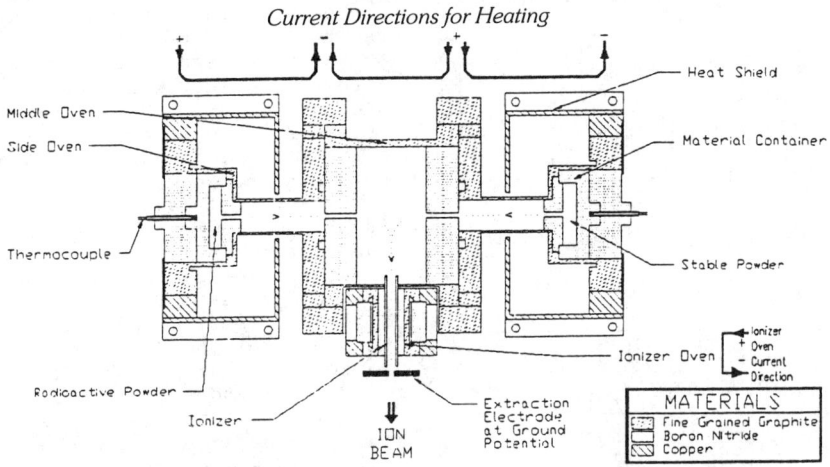

Fig. 2. The Dual Oven Ion Source to be used for the implantation of ^{22}Na. The ^{22}Na is contained in pieces of activated aluminum, which, when molten, should release it. A rhenium ionizer and subsequent extraction, produces the singly charged ion beam of ^{22}Na. For further details, see text.

high potential of 18 kV, required for extraction of the ion (1+) beam.

Given the work function of the rhenium, and relating to similar experiences with tungsten surface ionizer[6], the sodium ions diffusing through the rhenium tube (diameter about 2 mm) should be ionized with an efficiency of about 50% with the ionizer at a temperature of near 1000°C. The ions then will be extracted by a grounded 4 mm extraction electrode; its distance to the ionizer can be varied externally.

THE MASS SEPARATOR

The new TISOL mass separator facility at TRIUMF will be used to produce the implanted ^{22}Na target. This facility is designed to handle highly radioactive sources and will become contaminated as it is used in an on-line mode. The TISOL facility is described in detail elsewhere[7,8], so only a brief summary will be given here.

The entire facility is displayed in Figure 3. The Dual Oven Source is inserted into the TIS (Target Ion Source) box and connected by insulated cables to the power supplies and other services located at a Faraday cage 15 m away; all of which must be capable of operating at high tensions up to 20,000 V. The walls of the TIS box are protected from the heat radiating from the source by directly cooled copper shields. The movable extraction electrode is inserted via a side flange. A cryopump is used to obtain pressures of typically 10^{-6} Torr in the TIS box.

Following extraction, the ion beam is focussed by an einzel-lense (5 cm in diameter) at 10 cm distance from the extraction electrode to gain higher transmission through to the magnetic

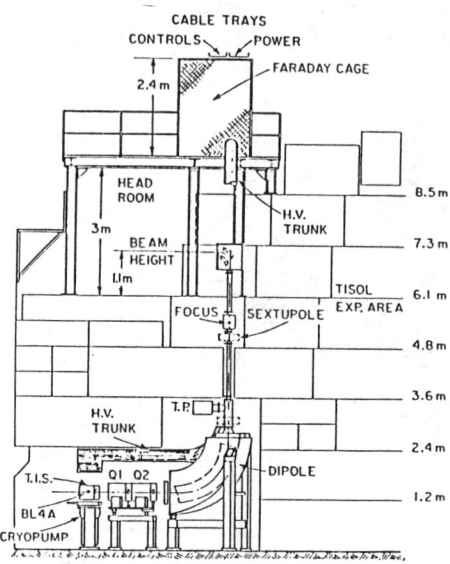

Fig. 3. A schematic, elevation view of the TISOL on-line mass separator at TRIUMF. For further details see text as well as references 7 and 8.

quadrupoles to follow. These quadrupoles, located about a metre from the ion source, focus the beam in the non-bend as well as the bend plane (in addition to the mass-separating dipole) into the collector device situated on top of the focus box. At about 3 m distance from the source, the ions enter the dipole and become mass separated with a typical mass resolution of $M/\Delta M$ of 1000 to 2000, depending on the source and the transmission. The curvature radius of the dipole is 1.25 m. The collector is situated close to the focus of the dipole, about 4m from the exit of the dipole.

Two Faraday cups, positioned after the TIS box and the magnetic quadrupoles, are available as beam analysing elements, as well as the collector system, which is described in the subsequent section. Both cups as well as the entire vacuum system, and the power supplies, are remotely controlled by an OPTUMUX/IBM clone system from the control platform located above the shielding blocks. The vacuum system for the beam line from the TIS box through the dipole contains two turbo (450 ℓ) pumps and produces pressures typically in the 10^{-6} Torr range. The three heater power supplies for the ovens described earlier are able to deliver 500 A/10 V, 500 A/10 V, 100 A/10 V, respectively, and are also controllable by the microcomputer system, even at high tension.

THE COLLECTOR DEVICE

The collector device, which contains the targetbacking and the targetholder for the subsequent proton studies, is described below.
The ion (^{22}Na) beam will be stopped in a nickel layer which is evaporated onto a 0.2 mm thick tantalum sheet. In preliminary studies nickel has proven to be the material which shows the smallest diffusion of sodium ions, when bombarded with intense beams (~200 μA) of low energy (~300 keV) protons. The tantalum sheet is directly sealed into the targetholder for the later proton bombardment. The tantalum sheet can be directly water-cooled if necessary.

To ease removal of the target and to keep the cold trap under vacuum, two gate valves separate the beam line vacuum from the target. This allows one to remove the "hot" target while maintaining the beam line vacuum. A port between the two gate valves allows for roughing and venting.

STATUS

At present the Dual Oven Ion Source and the Collection Device are being tested and modified, using stable (non-radioactive) beams.

Studies are planned using irradiated (at TRIUMF) aluminum pieces with essentially only ^{24}Na ($T_{1/2}$=15 h) to determine the overall transmission efficiency of this method, along with discovering potential "hot" spots (places where high deposits are observed). Fortunately, this radioisotope does not lead to long-term contamination of TISOL. Following a successful series of studies with ^{24}Na, irradiated pieces of aluminum, obtained from Los Alamos and with up to 100 mCi of ^{22}Na present, are then available for the preparation of a ^{22}Na target with optimally of the order of 20 mCi (or about 10^{17} g/cm^2).

REFERENCES

1. P. Eberhardt, M.H.A. Jungk, F.O. Meier and F.K. Niederer, Geochim. Cosmochim. Acta 45, 1515 (1981).
2. M. Arnould in: Proceedings of the Accelerated Radioactive Beams Workshop, ed. by L. Buchmann and J.M. D'Auria, TRI-85-1, p.29 (1985).
3. J. Goerres, H.W. Becker, L. Buchmann, C. Rolfs, P. Schmalbrock, H.P. Trautvetter and A. Vlieks, Nucl. Phys., A408, 372 (1983).
4. C. Rolfs and R.W. Kavanaugh, NIM, B26, 165 (1987).
5. H. Ravn, CERN-ISOLDE, private communication (1987).
6. H. Ravn, in reference 2, p.9 (1985).
7. K. Oxorn, J.E. Crawford, H. Dautet, J.K.P. Lee, R.B. Moore, L. Nikinnen, L. Buchmann, J. D'Auria, R. Kokke, A.J. Otter, H. Sprenger and J. Vincent, NIM, B26, 143 (1987).
8. J. D'Auria et al., to be submitted to NIM.

IDENTIFICATION AND HALF-LIFE DETERMINATION OF VERY NEUTRON-RICH COPPER ISOTOPES

E. Lund, B. Ekström, B. Fogelberg, and G. Rudstam

The Studsvik Neutron Research Laboratory
S-611 82 Nyköping, Sweden

ABSTRACT

Very neutron rich copper isotopes have been produced in the isotope separator on-line facility OSIRIS at Studsvik. The half-lives were determined for $^{74-76,78}$Cu and several gammalines identified for each of these isotopes.

INTRODUCTION

The neutron rich isotopes near the doubly closed shell at ^{78}Ni are of interest both for fundamental nuclear physics and also for astrophysical calculations because these nuclei comprise the beginning of the r-process path, the "seed region".
In this massregion the β-decay half-lives and neutron-binding energies are among those nuclear data which are needed with the highest priority for facilitating the model calculations of the r-process[1]. The neutron binding energy can be deduced from nuclear mass values which indirectly can be measured as total decay energies. This is however only possible for a few cases close to the r-process path, at the so called waiting points near the closed neutron shells. The nuclear mass values involved in r-process calculations are mainly results of mass-formula predictions, which are extremely uncertain so far from stability, and especially near closed shells.
As an example, travelling along the A=78 isobars, the predicted mass excesses are scattered within 2 MeV for ^{78}Ga where experimental Q_β-values exist. For ^{78}Cu the spread of predicted massexcess values has increased to 5.5 MeV and for ^{78}Ni to about 8 MeV.
An experimental Q_β-value in the ^{78}Ni region even with large uncertainty would be of greatest importance to improve the mass-formulas. Q_β-values and decay properties of several nuclides in this region have previously been studied at the OSIRIS facility[2,3], and, in the case of ^{80}Zn also later at TRISTAN[4]. Knowing the ground state masses of the neutron rich isotopes of Zn from these previous works the next step towards the closed shell isotopes of Ni consists of the Cu nuclei (see fig. 1).
Some decay data for neutron rich Cu isotopes was obtained already with the low temperature ion source system. Preliminary studies using the new high temperature ion source system ANUBIS[5] suggest copper yields that are 20 - 50 times better than earlier. Absolute yields have not been determined but a survey of the activities obtained in this region shows several thousand counts/h in typical gammalines of $^{74-76}$Cu.

This activity is sufficient for quite accurate Q_β-determinations. For ^{78}Cu however, an improved experimental technique is needed since the activity for this nuclide is only a few percent of that for ^{75}Cu.

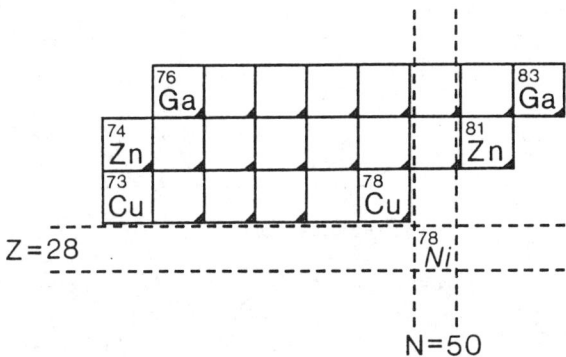

Fig. 1. Part of the nucleidic chart showing the close neighbourhood of the double closed shell nucleus ^{78}Ni. Nuclides studied at OSIRIS are indicated with a dark corner.

Hitherto the beta-decay half-lives have been determined for Cu isotopes with A=74-76 and 78, and several of their gammalines identified. The separated ion-beams contain a mixture of several isobars. The γ-rays following the decay of the Cu-isotopes were identified and the decay half-lives determined through a series of multispectrum measurements. Gamma ray singles measurements were made using a cycle comprising source collection followed by 4 to 8 consecutive recordings of γ-spectra. The collection and recording periods were chosen so that the copper isotopes were enhanced over the daughter activities. All gamma spectra were recorded using a large-volume co-axial HPGe-detector. The result of these measurements are collected in Table 1. Even if the activity of ^{78}Cu is rather low especially compared to the amount of Zn and Ga released from the ion-source the half-life was determined with an uncertainty of 30 % in a run of short duration, see fig. 2.

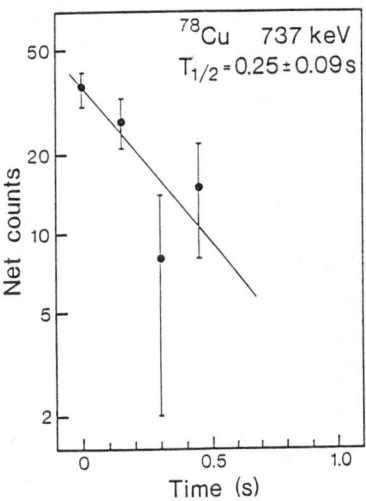

Fig. 2. The decay-curve of ^{78}Cu.

Table 1. Experimental half-life values and γ-transitions in very neutron rich copper isotopes.

Nuclide	Half-life (s) this experiment	Half-life (s) literature	γ-transitions (keV)
^{74}Cu	1.60 ± 0.15		606, 812, 1138
^{75}Cu	1.38 ± 0.08	1.3 ± 0.1 [6]	185, 421, 724
^{76}Cu	0.35 ± 0.08	0.61 ± 0.10 [7]	598, 697, 947
^{78}Cu	0.25 ± 0.09		216, 524, 737

REFERENCES

1. G. J. Mathews, "Proc. Specialists' Meeting on Yields and Decay Data of Fission Products", Brookhaven 1983, p. 485.
2. E. Lund, K. Aleklett, B. Fogelberg, and A. Sangariyavanish, Proc. of AMCO-7, Darmstadt 1984, p. 102.
3. B. Ekström, B Fogelberg, P. Hoff, E. Lund, and A. Sangariyavanish, Phys. Scripta 34, 614 (1986).
4. R. L. Gill et al., Phys. Rev. Lett. 56, 1874 (1986).
5. L. Jacobsson, B. Fogelberg, B. Ekström, and G Rudstam, Nucl. Instr. Meth., B26, 223 (1987).
6. P. L. Reeder, R. A. Warner, R. M. Liebsch, R. L. Gill, and A. Piotrowski, Phys. Rev. C31, 1029 (1985).
7. P. L. Reeder, R. A. Warner, R. L. Gill, and A. Piotrowski, Report at Specialists' Meeting on Delayed Neutrons, Birmingham, Sept. 1986.

STRANGELET MASS FORMULA AND ITS IMPLICATIONS TO THE SOLAR-NEUTRINO PROBLEM*

Kohji Takahashi and Richard N. Boyd
The Ohio State University, Columbus, Oh. 43210

ABSTRACT

We develop a mass formula for strangelets by extending the Fermi-gas formula of Berger and Jaffe with an inclusion of shell corrections in order to study reaction networks of strangelets. The results reveal a striking similarity between certain light strangelets and hypothetical "Q-nuclei", suggesting that the solar-neutrino problem (and some other problems) of the standard solar model might be solved if a tiny amount ($\sim 10^{-14}$ per normal nucleus) of low-Z strangelets existed in the Sun.

INTRODUCTION

It has been suggested by Witten[1] that matter consisting of up, down and strange quarks might be absolutely stable and might exist in nature with an obviously small but non-zero abundance. Calculations by Farhi and Jaffe[2] have shown that such "strange matter", and its finite-size lumps, "strangelets", can indeed be stable for a wide range of the values of QCD-related parameters.

We extend here the Fermi-gas mass formula of Berger and Jaffe[3] to include shell and deformation corrections, and study reaction networks of strangelets and their implications to the solar-neutrino problem. The details will be published elsewhere[4].

One of the numerous attempts to solve the solar-neutrino problem hypothesized an extra energy source denoted as "Q-nuclei"[5] which, while acting as a catalyst, would create a small convective core near the center of the Sun and thus reduce the temperature, a consequence favorable for solving the problem. We argue here that certain low-mass strangelets could be such an extra energy source. Our working assumption is that some stable strangelets could be produced, e.g. by a collision of neutron stars[1], and survive in the interstellar medium. [See Ref.[6] for the possibility of the phase transition from neutron matter to strange matter in neutron stars.]

*work supported by the NSF grant PHY-8600749 and the OSU Office of Vice-President for Research and Graduate Studies

MASS FORMULA

The liquid-drop mass formula[3] for a strangelet of N_u, N_d and N_s quarks is

$$E_{LD} = \epsilon_0 + a_S A^{2/3} + \frac{\delta_Y}{2A}(Y-Y_{min})^2 + \frac{1}{2}[\frac{\delta_Z}{A} + \frac{c_C}{A^{1/3}}](Z-Z_{min})^2, \quad (1)$$

where the mass number $A=(N_u+N_d+N_s)/3$, the hypercharge $Y=(N_u+N_d-2N_s)/3$, and the charge $Z=(2N_u-N_d-N_s)/3$, while Z_{min} and Y_{min} are the most enegetically favorable values in bulk strange matter. These and the coefficients a_S, δ_Y, δ_Z and c_C are functions[3] of two independent parameters: the energy per baryon in bulk, ϵ_0, and the mass of an s-quark, m_s. [m_u and m_d are assumed to be negligible.]

We now impose shell and deformation corrections on E_{LD} by applying the bunching procedure of Myers and Swiatecki[7] to the single-particle spectrum for each quark species

$$dN/dk = g\left[k^2 V/(2\pi^2) - kS[1 - \tan^{-1}(k/m)]/(16\pi)\right], \quad (2)$$

where $g(=6)$ is the degeneracy, and V and S are the volume and the surface area for a sphere with a radius $R=r_0 A^{1/3}$ (r_0 being the mean radius per baryon in bulk). The magic numbers are specified by making use of the spherical cavity model[8] with the linear boundary condition, as shown in Fig. 1. Our final mass formula is

$$E(A, Y, Z) = E_{LD}(\epsilon_0, m_s) + E_{sh}(c_{sh}, a/r_0), \quad (3)$$

where c_{sh} is a parameter concerning the degree of bunching and a/r_0 is an attenuation parmeter which reflects the degree of deformation. In Fig. 2, an example of the "microscopic" function, E_{sh}, is shown.

REACTION NETWORK

To be an effective source of additional energy, strangelets should form a catalytic cycle similar to those in hydrogen burning. In addition, the slowest of the reactions involved should still proceed much faster than the usual p-p chain reactions, which then would allow a very small abundance of strangelets. The above mass formula suggests that such a cycle starting from a low-A stable strangelet with $Z \lesssim 3$ exists for certain (though limited) ranges of ϵ_0 and m_s values. In order to see its efficiency in energy production in detail, the rates of various reactions[3] on strangelets have to be evaluated. In contrast to the case of a free (degenerate) quark system[9], such calculations for a finite strangelet would be difficult to do accurately. We have therefore made[4] simple estimates of the reaction rates for strangelets, analogous to those for normal nuclei, under conditions typical of those near the center of the Sun. As a result, catalytic cycles such as shown in Fig. 3 have been obtained. In this example, the cycle closes since the [A=65,Y=7,Z=3] strangelet is α-unstable.

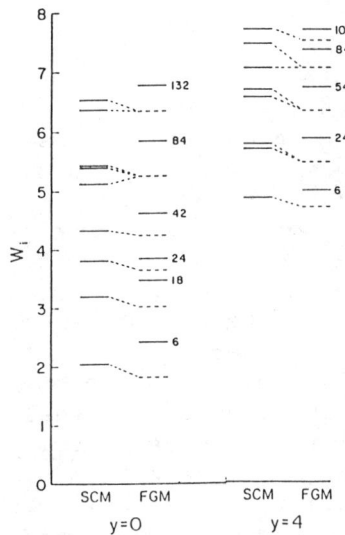
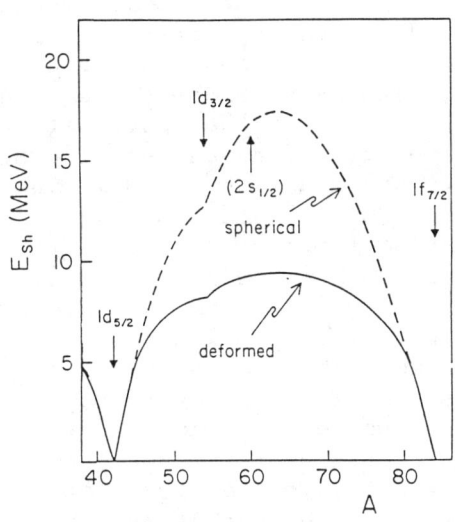

<u>Fig. 1</u> Energy spectra in <u>S</u>pherical <u>C</u>avity <u>M</u>odel and <u>F</u>ermi-<u>G</u>as <u>M</u>odel. In FGM, the bunched energies are shown by dashed lines. The scaled mass $y \equiv mR$, and W_i is also dimensionless.

<u>Fig. 2</u> Shell energies with and without deformation in the case of flavor equilibrium, $A=N_u=N_d=N_s$. $\epsilon_0=927$ MeV, $m_s=100$ MeV, $c_{sh}=0.02$ and $a/r_0=0.2$ are assumed.

$$\begin{array}{c} \text{stable} \quad p \quad \quad \beta^+\epsilon \quad \quad (p,\alpha) \quad \quad p \\ [65,5,2] \to [66,6,3] \to [66,6,2] \to [63,3,1] \to [64,4,2] \\ \beta^+\epsilon \uparrow \leftarrow\text{------------------------------------} \quad\quad\quad\quad\quad\quad\quad \downarrow p \\ \quad\quad\quad\quad\quad\quad\quad\quad\quad\quad\quad\quad\quad\quad\quad\quad\quad\quad\quad\quad\quad [65,5,3] \end{array}$$

<u>Fig. 3</u> An example of catalytic cycle with [A,Y,Z] strangelets. Parameter values are: $\epsilon_0=927$ MeV, $m_s=100$ MeV, $c_{sh}=0.02$ and $a/r_0=0.4$. The reactions involved are proton captures [radiative(p), or (p,α)] and $\beta^+\epsilon$ decays with no change of strangeness. In this specific example, the weak radiative decay process[3] is negligible.

DISCUSSION

Our analysis shows that the slowest reactions in the catalytic cycles we have found are the $\beta^+\epsilon$ decays of $Z=3$ strangelets with Q_ϵ values of typically 1.5 MeV. Such low-energy β decays of strangelets would not exacerbate the neutrino detection rate[10]. On the other hand, the β-decay rates are still expected to be about 10^{13} times larger than the p-p reaction rate at conditions near the center of the Sun. The abundance of strangelets at the level of 10^{-14} per normal nucleus could therefore produce enough energy to contribute significantly to the Sun's energy generation, and would result in a large decrease of high-energy solar-neutrino production.

Note that the actual strangelets which could exist terrestrially and affect the Sun would have $Z=2$, so would have the chemistry of He. The experimental abundance limits of anomalous nuclei in He are not nearly so stringent as those in some other nuclei[11], and do not presently preclude strangelets at an abundance level below 10^{-12}.

As was the case with Q-nuclei, the present scenario with strangelets would also solve some other problems[4,12] inherent to the standard solar model. Investigation of its success would certainly require detailed solar modelling and further studies of strangelet properties. Full confirmation of the model, though, requires detection of some unique signature of the strangelets, e.g., their observation in an accelerator mass spectroscopy experiment.

1) E. Witten, Phys. Rev. D30, 272 (1984).
2) E. Farhi and R.L. Jaffe, Phys.Rev. D30, 2379 (1984).
3) M.S. Berger and R.L. Jaffe, Phys.Rev. C35, 213 (1987).
4) K. Takahashi and R.N. Boyd, Astrophys.J., in press.
5) R.N. Boyd et al., Phys.Rev.Letters 51, 609 (1983); Astrophys.J. 289, 155 (1985).
6) P. Haensel et al., Astron.Astrophys. 160, 121 (1986); C. Alcock et al., Astrophys.J. 310, 261 (1986); H.A. Bethe et al., Nucl.Phys. A462, 791; M.A.Alpar, Phys. Rev.Letters 58, 2152.
7) W.D. Myers and W.J. Swiatecki, Nucl.Phys.81, 1 (1966).
8) C.E. DeTar and J.F. Donoghue, Ann.Rev.Part. Sci. 33, 235 (1983).
9) H. Heiselberg et al., Physica Scripta 34, 556.
10) B. Sur and R.N. Boyd, Phys.Rev.Letters 54, 485 (1985).
11) D. Nitz et al., AIP Conf. Proc., No.150, 1143 (1986).
12) C. Joseph, Nature 311, 254 (1984).

SUDBURY NEUTRINO OBSERVATORY

G.T. Ewan
Queen's University, Kingston, Ontario K7L 3N6
and the SNO Collaboration*

Abstract

A description is given of the proposed Sudbury Neutrino Observatory based on a 1000 tonne heavy water Cherenkov detector which is being designed. This detector would allow the measurement of neutrinos by the three reactions $\nu_e d \to ppe$, $\nu_x e \to \nu_x e$ and $\nu_x d \to \nu_x pn$ (where ν_x is any left-handed neutrino). The application of the detector to the resolution of the solar neutrino problem is discussed.

The only direct information on the nuclear reactions which power the sun and stars is carried by neutrinos escaping from the dense interior. These reactions are the basic step in theories of nucleosynthesis in which, for heavier elements, nuclei far from stability play an important role. This paper gives a brief description of a proposed laboratory to study neutrinos from the sun and other astrophysical sources such as stellar collapse within our galaxy.

The study of solar neutrinos can make important contributions to both astrophysics and particle physics. The pioneering experiment of Davis[1] on solar neutrinos observed a lower flux of neutrinos than that predicted by the standard solar model by a factor of ~ 4. The large discrepancy is known as the solar neutrino problem, SNP, and is widely considered as a major problem in modern physics. Two categories of solution of this problem have been extensively discussed. The first assigns some deficiency to the standard solar model and the second to some deficiency in our knowledge of neutrino properties and neutrino propagation. Much theoretical work has been stimulated by the identification by Mikheyev and Smirnov[2], following the theoretical

---

* G.T. Ewan, H.C. Evans, H.W. Lee, J.R. Leslie, J.D. MacArthur, H-.B. Mak, W. McLatchie, B.C. Robertson, P. Skensved, Queen's University. R.C. Allen, H.H. Chen, P.J. Doe, University of California at Irvine. D. Sinclair, University of Oxford. J.D. Anglin, M. Bercovitch, W.F. Davidson, C.K. Hargrove, R.S. Storey, National Research Council of Canada. E.D. Earle, Chalk River Nuclear Laboratories, P. Jagam, J.J. Simpson, University of Guelph. E.D. Hallman, Laurentian University, A.B. McDonald, Princeton University and A.L. Carter, D. Kessler, Carleton University

framework of Wolfenstein[3], of a resonance mechanism for matter enhancement of neutrino oscillations which could lead to the conversion of electron neutrinos, ν_e into other types ν_μ and ν_τ. This has become a central topic for discussion in particle physics. Weinberg[4] has suggested that the sun is a unique source offering a rare opportunity to search for such neutrino oscillations which could test unification theories beyond the electroweak sector. The detector described in this paper would measure solar ^8B neutrinos and have the potential for distinguishing among suggested solutions of the SNP.

The proposed Sudbury Neutrino Observatory is based on a large 1000 tonne heavy-water Cherenkov detector in a very low background environment deep underground to minimize effects from cosmic rays. This detector has two unique features, the use of heavy water, D_2O, and the location in a laboratory 6800 ft. underground. The proposed detector would be sensitive to the solar ^8B ν_e flux, spectrum and direction and to the total solar neutrino flux independent of neutrino flavor. It could play an important role in resolving the solar neutrino problem, testing the matter enhancement of neutrino oscillations and understanding the particle physics involved.

A conceptual design of the detector is shown in figure 1.

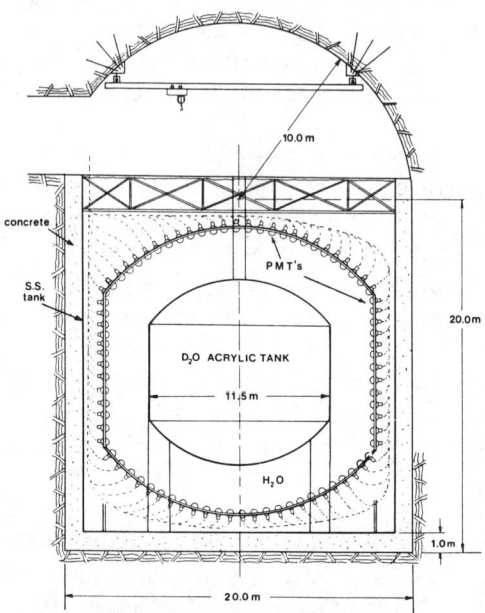

Figure 1: Conceptual design of heavy water detector

It would consist of 1000 t of 99.8% pure heavy water (D_2O) contained in a transparent tank viewed by phototubes covering 40% of the surface area. Relativistic particles are detected by the Cherenkov light they produce in the water. To address the SNP it is necessary to study electrons of energy 5-14 MeV. Monte Carlo simulations of the detector indicate that the energy of 10 MeV electrons can be measured to 15% accuracy (σ), their direction determined to 25°, and their location to 1 m. An essential consideration in the design of a solar neutrino experiments is that the backgrounds be made very low. We intend to locate the detector in the Creighton mine near Sudbury, Canada, under a flat overburden, at a depth of 6800 ft. to reduce cosmic-ray muon flux. The muon intensities and depth underground of several laboratories in shown in figure 2. The sensitive volume of the detector is shielded from radioactivity in the rock by 1 m of low activity concrete and 3 m of purified light water. All components of the detector would be made of low activity materials. A detailed assessment of the detector performance and background has been reported elsewhere[5].

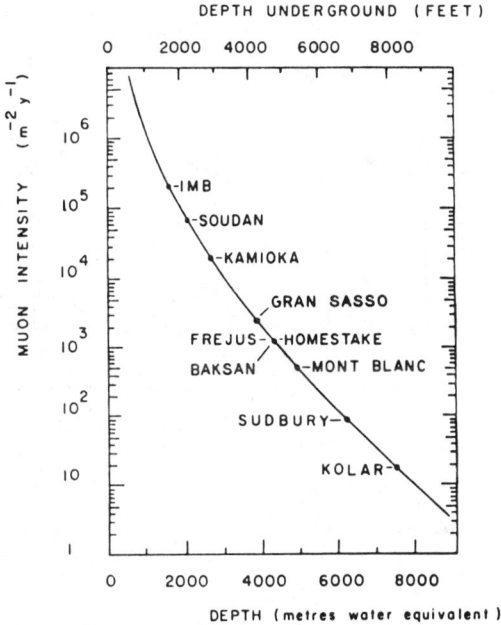

Figure 2: Cosmic ray muon intensity as function of depth

The use of D_2O enables neutrinos to be identified through three complementary reactions. Electron neutrinos ν_e, can be detected either by the inverse β-decay reaction (charged current, CC, reaction)

$$\nu_e + d \rightarrow p + p + e \tag{1}$$

or by the neutrino electron scattering (ES reaction)

$$\nu_e + e \rightarrow \nu_e + e \tag{11}$$

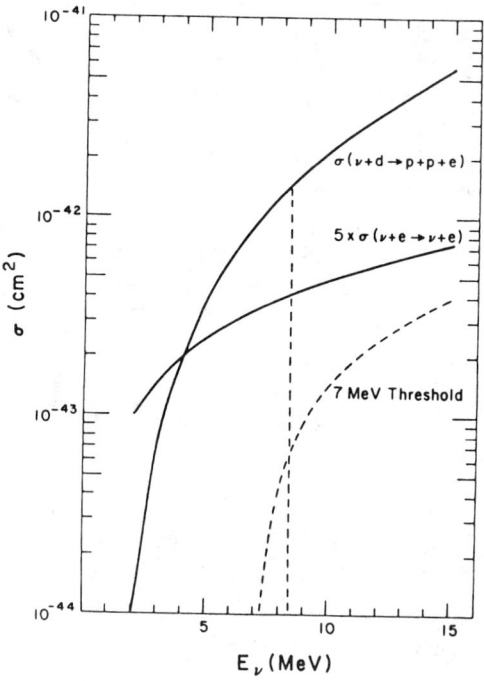

Figure 3: Cross-sections for the $(\nu_e, d) \rightarrow p + p + e$ reaction compared with the (ν, e) scattering reaction.

The cross-sections for these reactions are shown in figure 3. In both cases the signal would be the Cherenkov light from the relativistic electrons. In the inverse β-decay reaction the electron energy is uniquely given by the neutrino energy in contrast to neutrino-electron scattering where the electron may have any energy up to the neutrino energy. The effect of setting a 7 MeV threshold on the effective cross-section is shown in the figure. In light water, H_2O, electron neutrinos only interact by neutrino-electron scattering and the advantage of D_2O over H_2O in studying electron neutrinos is evident from the figure. The major problem is the availability of a large quantity of heavy water and the containment of this valuable material. Canada has a temporary surplus of heavy water and our proposal is to borrow 1000 tonnes and store it safely underground.

The total neutrino flux, independent of neutrino flavour can be measured by the neutral current reaction (NC)

$$\nu_x + d \to \nu_x + p + n \qquad (111)$$

The reaction rate would be determined by counting the free neutrons produced. The cross-section of the neutral current reaction is related to the charged current reaction ($\nu_e + d$) but independent of neutrino flavour. The neutron would be detected by the Cherenkov light from the 6.25 MeV γ-rays produced by neutron capture on the deuteron. Alternatively chlorine could be added to the detector to produce higher energy γ-rays. There is a serious background problem for this reaction from photo-disintegration of the deuteron. We hope to achieve the low background by the design of laboratory, the choice of materials and purification and monitoring of the water and acrylic vessel containing the heavy water. Details of background measurements are given in reference 5.

As indicated earlier the solution to the SNP could be due to vacuum neutrino oscillations between the sun and earth (case A) or matter enhanced neutrino oscillations in the sun (cases B and C). Another possible explanation is that the standard solar model is wrong. Table 1 lists the predicted event rates per year in a 1000 tonne D_2O Cherenkov detector with 40% PMT coverage, assuming a ^8B neutrino flux of 4 x 10^6 cm^{-2} s^{-1}. The event rates are listed for the ES, the CC, and the NC reactions. For the ES and CC reactions, the detector energy threshold is important and rates for both 5 and 9 MeV thresholds are listed. These rates also depend on the values chosen for the mixing parameters δm^2 and $\sin^2 2\theta$. Since $\nu_\mu(\nu_\tau)$ neutrinos contribute in the ES reactions, their contribution to the total event rate is noted in parenthesis. A more detailed discussion is given elsewhere[6].

In the reaction $\nu_e + d \to p + p + e$ monoenergetic neutrinos produce almost monoenergetic electrons with kinetic energies approximately E_ν - 1.44 MeV. The neutrino energy spectrum can be easily derived from the electron energies. This is much easier than from the neutrino-electron scattering where the electrons may have any energy up to the neutrino energy and in addition the cross-section is an order of magnitude lower.

the neutrino spectrum. Measurement of the spectrum shape is important in understanding matter enhanced oscillations and combined with the total neutrino flux can distinguish between possible solutions of the solar neutrino problem.

The sensitivity of the heavy water detector is much greater than the detector used by Davis[1] which was based on the inverse β-decay of $^{37}C\ell$. In addition to measuring the flux the D_2O detector also measures the energy spectrum and gives directional information. In a running period of one year the minimum observable ^8B ν_e flux allowing for expected backgrounds would be $\sim 5 \times 10^4$ cm^{-2} s^{-1} (3σ measurement). This is a factor of ~ 40 lower than the value of the ν_e flux reported by Davis. The minimum observable total neutrino flux by reaction (111) depends critically on both backgrounds and the detection techniques used for the neutrons. These are being studied with encouraging results. An ultimate sensitivity in the range $\sim 1 \times 10^5$ cm^{-2} s^{-1} is anticipated.

The detector would also be well suited to detection of neutrinos from stellar collapse in our galaxy. The detector would see ~ 10 events in the first few milliseconds from the ν_e burst and in the following seconds a burst of neutrinos of all types. Recently neutrinos from Supernova 1987a in the Large Magellanic Cloud have been observed. The observations in the large H_2O detectors at Kamioka[7] and IBM[8] were mostly of anti-neutrinos $\bar{\nu}_e$ in the second burst. The D_2O detector would be sensitive to the electron neutrinos in the first burst as well as to all neutrino types in the second burst.

The present status of the Sudbury Neutrino Observatory is that the design study has been completed including an access drift to the proposed site of the detector. A request for the major capital funding is being made to the granting agencies. The estimated time scale for the excavation of the laboratory, approximately 20 m diameter by 30 m high, and the installation of the detector is four years.

This experiment could not be contemplated without access to a large amount of heavy water and a suitable deep site. A stock heavy water exists in Canada for use in future CANDU reactors and we are grateful to AECL for agreeing to lend 1000 tonnes for this project. The encouragement and cooperation of the International Nickel Company (INCO) has been vital to this project. The support by NSERC and NRC (Canada), and NSF and DOE (US), is gratefully acknowledged.

TABLE 1. RESPONSE OF THE DETECTOR TO SOLAR NEUTRINOS

Case	ES >5 MeV	ES >9 MeV	CC >5 MeV	CC >9 MeV	a) NC
Standard solar model	730 (0)	120 (0)	6500	1530	710
A. Vacuum oscillations	320 (78)	52 (12)	2170	510	710
B. Matter oscillations E_c = 9MeV	430 (53)	30 (14)	2500	50	710
E_c = 2MeV	115 (110)	18 (17)	65	15	710
C. Non adiabatic limit in ^{8}B energy range	325 (72)	58 (10)	2500	650	710
Standard Solar Model wrong, flux 1.3×10^6 cm^{-2} s^{-1}	240 (0)	40 (0)	2170	510	240

(a) Assumes a neutron capture detection efficiency of 20%.

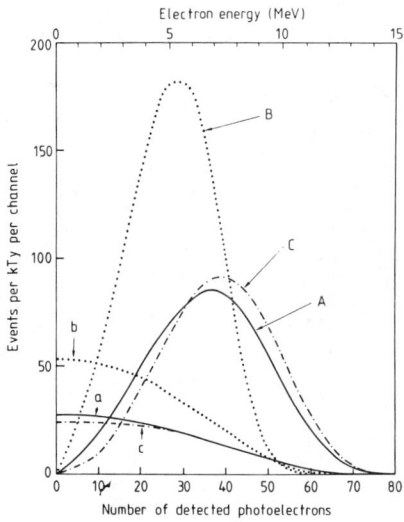

Figure 4: Calculated spectra for some possible solutions of the SNP. (see text)

Predicted neutrino spectra due to electrons for the CC reaction (capitals) and the ES reaction (lower case) are shown in figure 4 for some possible solutions of the SNP corresponding to cases of types A B and C. It is clear that the CC reaction is significantly better than the ES reaction for measuring

References

1. R. Davis Jr., D.S. Harmer and K.C. Hoffman, Phys. Rev. Lett. **20**, 1205 (1968); J.K. Rowley, B.T. Cleveland and R. Davis Jr., in: Proc. Conf. on Solar neutrinos and neutrino astronomy (Homestake, 1984), AIP Conf. Proc. 125 (AIP, New York, 1985) pl.

2. S.P. Mikheyev and A. Yu. Smirnov, Nuovo Cimento **C9**, 17 (1986).

3. L. Wolfenstein, Phys. Rev. **D17**, 2369 (1978); **D20**, 2634 (1979).

4. S. Weinberg, Proc. XXIII Intern. Conf. on High Energy Physics, Berkeley, CA, July 1986 (World Scientific, Singapore, 1986).

5. G.T. Ewan et al., Sudbury neutrino observatory, Report SNO-85-3 and SNO-86-6, Queen's University, Canada (1985, and 1986, unpublished).

6. A. Aardsma et al., Physics Letters **B194**, 321 (1987).

7. K. Hirata et al., Phys. Rev. Letters **58**, 1490 (1987).

8. R.M. Bionta et al., Phys. Rev. Letters **58**, 1494 (1987).

QUENCHING OF GAMOW-TELLER STRENGTH

I.S. Towner
Chalk River Nuclear Laboratories, Chalk River, Ontario
Canada K0J 1J0

ABSTRACT

It is becoming increasingly evident that nuclear Gamow-Teller matrix elements determined from β-decay and charge exchange reactions are significantly quenched compared to simple shell-model estimates based on one-body operators and free-nucleon coupling constants. Here we will briefly discuss the theoretical origins of this quenching giving examples from three mass regions:

(a) in light nuclei near LS-closed shells, such as ^{16}O and ^{40}Ca,

(b) in heavy nuclei near jj-closed shells, such as ^{208}Pb,

(c) in proton-rich nuclei near shell or sub-shell closures, such as ^{100}Sn and ^{146}Gd.

LS CLOSED SHELLS

Gamow-Teller matrix elements in mirror transitions in closed-LS-shell-plus (or minus)-one nuclei, ^{15}O, ^{17}F, ^{39}Ca and ^{41}Sc, have been experimentally determined[1] from β-decay data. Compared to single-particle estimates using the free-nucleon axial-vector coupling constant, $g_A \simeq 1.26$, these matrix elements are quenched by roughly 13% at the A=16 closed shell and 30% at the A=40 closed shell. Furthermore (p,n) reactions[2-4] have extended the number of data by measuring the spin-flip transitions, $j=\ell+1/2 \to j=\ell-1/2$, which again, on comparison with single-particle estimates, indicate large quenching of order 40% in the matrix element.

Theoretically these reductions must stem either from the inadequacy of the single-particle description of the nuclear state or from the inadequacy of the impulse-approximation one-body GT operator. Calculations of the first effect, frequently called core polarization, estimate through perturbation theory the admixtures of 2p-1h and 3p-2h configurations in the basically single-particle wavefunction and evaluate the impact the admixtures have on the calculated matrix element. At LS closed shells there is no contribution to the core polarization in first order. The reason is that with no spatial dependence in the one-body GT operator it cannot excite a 1p-1h state from a closed LS shell. Thus calculations must be taken to second order in perturbation theory. Computationally this leads to a time-consuming calculation as there is no selection rule to limit the intermediate-state summation and the

© American Institute of Physics 1988

convergence is slow. This is particularly true with tensor forces in the residual interaction as first stressed by Shimizu, Ichimura and Arima[5]. This propensity of the tensor force to couple strongly to high-lying excited states has been called 'tensor correlations' and the phenomenon leads to a reduction in the GT matrix element.

The other inadequacy concerns the use of one-body operators. Corrections to the impulse approximation arise because nucleons in nuclei interact through the exchange of mesons and this exchange can be perturbed by the action of the weak axial current. Since this perturbation requires at least two nucleons to be involved, corrections from this origin lead to two-body GT operators. The mathematical technique used to determine these operators follows from considering all Feynman graphs in which a meson is exchanged between two nucleons and linking the weak axial current in all possible places in the diagram. Consider first graphs involving pion exchange. These are the most important graphs because the pion, being the lightest mass meson, generates the longest-range exchange current operators. Heavy-meson exchange leads to short-range operators, which in nuclear physics are less important since the nuclear wavefunction goes rapidly to zero at short distances. The possible pion-exchange graphs can be classified into two types: Born graphs, involving only nucleons and pions; and non-Born graphs in which the axial current either excites a nucleon to an excited state, such as the isobar Δ, or converts the pion into a heavy meson, such as the ρ. The important point is that for axial currents the pion Born graphs are identically zero[6]. This is in sharp distinction to electromagnetic currents where pion Born graphs give an important contribution to such properties as magnetic moments. Thus the intimate connection between the GT operator and the spin part of the isovector M1 operator, evident in the impulse approximation, is broken when meson-exchange current (MEC) corrections are considered. For the non-Born graphs, we will distinguish between diagrams involving nucleon excitations -- referring to these as isobar graphs -- and those in which a pion is converted to a rho meson -- referring to these as $\rho\pi$ or meson-exchange-current (MEC) graphs. In general, the latter MEC graph gives a small contribution and is not an important ingredient in the quenching of the GT matrix element.

There are two further points to consider: The core polarization calculation corrects the matrix element of a one-body operator evaluated in the closed-shell-plus-one configuration for the presence of 2p-1h and 3p-2h admixtures in the single-particle wavefunction. The perturbation calculation is carried out to second order in the residual interaction (or to the fourth power in the meson-nucleon coupling constants). It is logical therefore that the matrix elements of the two-body operators should likewise be corrected for 2p-1h and 3p-2h admixtures. Since the two-body operator itself involves the meson-nucleon coupling constants to the second power, it is sufficient to estimate this correction to first order in the residual interaction. These terms have been called "crossing terms" in the work of the Tokyo group[7,8].

The second point concerns the one-body GT operator, which is obtained as a leading term in the nonrelativistic reduction of a relativistic axial-vector current. There is a few per cent correction coming from the next order terms in the nonrelativistic reduction. The GT operator, $1/2 g_A \sigma \tau_\pm$, is then modified to

$$g_A \{\sigma - \frac{1}{2} p^2/M^2 (\sigma - (\sigma \cdot \hat{p})\hat{p})\} \frac{1}{2} \tau_\pm \qquad (1)$$

where p is a nucleon momentum and M its mass. The correction depends on estimating $\langle p^2/M^2 \rangle$ for a nucleon in a nucleus.

In summary, then, corrections to lowest order shell model estimates of the GT matrix element in the impulse approximation come from: core polarization, isobar currents, MEC currents, crossing terms and relativistic corrections. All these ingredients for closed-LS-shell-plus-one nuclei have been calculated by the Tokyo group[7,8] and by Towner and Khanna[9]. We quote in Table I some results from a recent review by Towner[10], where more details can be found. It is useful to characterize the results of the calculation in terms of an equivalent effective one-body operator that contains three independent rank-one tensors

$$(GT)_{eff} = \{g_{LA,eff} L + g_{A,eff} \sigma + g_{PA,eff} [Y_2(\hat{r}) \times \sigma]^{(1)}\} \frac{1}{2} \tau_\pm \qquad (2)$$

where $g_{A,eff} = g_A + \delta g_A$ etc, with g_A the bare impulse-approximation value and δg_A the correction to it. Note that in the bare operator $g_{LA} = g_{PA} = 0$. From Table I we see the results are not good. In general, theory is underpredicting the degree of quenching in the experimental matrix element. Note, also, that the term δg_{LA} is small, while the term δg_{PA} has a rather subtle role to play. For diagonal matrix elements the quenching is proportional to $\delta g_A + K \delta g_{PA}/\sqrt{(8\pi)}$ where $K = 2\ell/(2\ell+3)$ for orbits with $j = \ell+1/2$, and $K = (2\ell+2)/(2\ell-1)$ for $j = \ell-1/2$. For the $0p_{1/2}$ orbit in particular, there is a very strong cancellation between these two terms resulting in a small predicted value for $\delta \langle GT \rangle_d / \langle GT \rangle_d$. For the spin-flip transitions, on the other hand, the quenching is proportional to $\delta g_A - 1/2 \delta g_{PA}/\sqrt{(8\pi)}$ and numerically the two terms add. Thus larger quenching is predicted for spin-flip transitions than for diagonal transitions, as experimentally observed, although the magnitude of the effect is underpredicted. In Table II we give a breakdown of the contributions to the effective operator from various sources, where it is seen the principal contribution comes from core polarization. Isobar currents give an important contribution to δg_{PA} but in this

TABLE I: Summary of all corrections to the ground-state diagonal GT matrix element, $\delta\langle GT\rangle_d$, and the off-diagonal spin-flip matrix element, $\delta\langle GT\rangle_f$, expressed as a percentage of the single-particle value for closed-shell-plus (or minus)-one configurations.

	δg_{LA}	δg_A	δg_{PA}	$\dfrac{\delta\langle GT\rangle_d}{\langle GT\rangle_d}$ calc	$\dfrac{\delta\langle GT\rangle_d}{\langle GT\rangle_d}$ expt[a]	$\dfrac{\delta\langle GT\rangle_f}{\langle GT\rangle_f}$ calc	$\dfrac{\delta\langle GT\rangle_f}{\langle GT\rangle_f}$ expt
A=16 $0p_{1/2}^{-1}$	0.013	−0.185	0.234	−1.9	−13.1±0.5	−17.0	≃−38[b,c]
A=16 $0d_{5/2}$	0.013	−0.175	0.179	−10.2	−13.8±0.3	−15.8	≃−33[b]
A=40 $0d_{3/2}^{-1}$	0.009	−0.255	0.167	−17.1	−33.7±1.0	−21.9	≃−45[b,d]
A=40 $0f_{7/2}$	0.010	−0.221	0.105	−14.2	−26.2±0.4	−18.8	

a Deduced from experimental β-decay data recorded in ref. 1

b From (p,n) measurements of Watson et al[2], where we have renormalized the (p,n) cross-section so that the deduced GT matrix element for the ground-state transition agrees with β-decay measurements. Watson et al, however, prefer to normalize their results to distorted-wave impulse approximation calculations and find for mass A=15,39 notable discrepancies between their deduced values of the ground-state GT matrix element and those deduced from β-decay.

c From Goodman et al[3].

d From Rapaport et al[4].

calculation its cancellation against δg_A leads to an overall effect that is small for diagonal matrix elements. This is somewhat controversial. Alternative theories[11] using enhanced effective interactions in the isobar-hole channels claim much larger isobar contributions. Rho[11], in particular, argues that the crossing terms should roughly cancel the tensor correlations in the core-polarization calculation. We also show the result of the Tokyo group[7,8], in Table II whose overall result is rather similar to ours.

A different approach to the determination of the effective one-body operator is that of Brown and Wildenthal[12]. The effective coupling constants are determined in a fit to a large number of data in the sd-shell using shell-model wavefunctions calculated without truncation in the complete sd-shell model space. The assumption is that the effective one-body operator is only a weakly varying function of nuclear mass; a proportionality of $A^{0.35}$ is assumed. From Table II we see that our calculated value of $\delta g_A \simeq -0.18$ is only two-thirds the empirically deduced value. If we assume that the core-polarization calculation is about right (because it is the principal correction for isoscalar magnetic moments which are well described in similar calculations[7-10]), then we can use the empirically determined value to solve for the isobar correction. The MEC, crossing term and relativistic corrections are small for GT transitions. Thus we write, in an obvious notation:

$$\delta g_A(\Delta) = \delta g_A(BW) - \delta g_A(CP) - \delta g_A(MEC) - \delta g_A(CT) - \delta g_A(Rel)$$

$$= -0.13 \pm 0.01 \qquad (3)$$

TABLE II: Contributions to the effective one-body GT operator from various sources for a 0d configuration at A=16, and a comparison with the 1983 Towner-Khanna calculation (TK), the Arima et al calculation (ASBH) and the empirical values deduced by Brown and Wildenthal (BW).

	δg_{LA}	δg_A	δg_{PA}	$\dfrac{\delta \langle GT \rangle_d}{\langle GT \rangle_d}$ % calc
Core polarization	0.011	−0.136	0.005	−9.0
Isobar currents	0.002	−0.046	0.264	−0.9
MEC, $\rho\pi$	−0.002	−0.004	−0.065	−1.2
Crossing term	0.001	0.032	0.026	3.0
Relativistic	0.000	−0.021	−0.052	−2.1
Sum	0.013	−0.175	0.179	−10.2
TK(ref. 9)	0.012	−0.191	0.103	−12.2
ASBH(ref. 8)	0.013	−0.180	0.224	−10.3
BW(ref. 12)	0.01(1)	−0.26(1)	0.09(4)	−18.7(16)

Our calculated value from Table II is $\delta g_A(\Delta) \simeq -0.046$ nearly three times smaller. This strongly suggests an enhanced effective interaction is required in the isobar-hole channels compared to the one-boson-potential used by us.

jj CLOSED SHELLS

Next we turn our attention to heavy nuclei where the shell closures separate the spin-orbit partner orbitals. In contrast to the case at the LS closed shells there are now first-order corrections from core polarization to the single-particle GT matrix element in closed-shell-plus-one nuclei. Experimentally these matrix elements are not measurable in β-decay. The Coulomb energy difference between proton and neutron states of the same orbital structure is sufficiently large that GT transitions are effectively excluded from the energy window available in β-decay. However they can be measured in charge-exchange reactions. Consider, as an example the $0^+ \to 1^+$ transitions in the ^{208}Pb (p,n) ^{208}Bi reaction. There are 12 proton particle-neutron hole 1^+ states possible with non-zero GT matrix elements, the principal ones being the non-spin-flip components $\pi h_{9/2}$-$\nu h_{9/2}^{-1}$ and $\pi i_{13/2}$-$\nu i_{13/2}^{-1}$ with unperturbed energy $\simeq 7$ MeV and spin-flip components $\pi h_{9/2}$-$\nu h_{11/2}^{-1}$ and $\pi i_{11/2}$-$\nu i_{13/2}^{-1}$ with energy $\simeq 12$ MeV. Configuration mixing among the 12 particle-hole states, which is equivalent to extending first-order core polarization to all orders in the Tamm-Dancoff approximation, concentrates most of the GT strength into one state and pushes it up in energy to $\simeq 19.2$ MeV. This is the GT giant resonance. The (p,n) reaction[13] clearly identifies the resonance but the transition strength is found to be less than that anticipated. Recall that the one-body GT operator is

$$(GT)_\pm = \frac{1}{2} g_A \sum_n \underset{\sim}{\sigma}(n) \, \tau_\pm(n) \qquad (4)$$

where, in impulse approximation, $g_A \simeq 1.26$, the axial-vector coupling constant from free-neutron decay. It is useful to define sum rules as a sum over all final states

$$S_\pm \equiv \sum_f B_\pm(GT; i \to f) = \sum_f \left| \langle i \| (GT)_\pm \| f \rangle \right|^2 \qquad (5)$$

and from the commutator algebra of isospin $[\tau_x, \tau_y] = 2i\tau_z$ obtain the Ikeda result[14]

$$S_- - S_+ = 3(N-Z) \, g_A^2 \qquad (6)$$

The difference in the sum rule for the (p,n) reaction and the (n,p) reaction on a particular target nucleus such as ^{208}Pb depends only on the neutron excess and is quite independent of any nuclear-structure model. In fact, in a heavy nucleus, S_+ is very weak so for practical purposes $3(N-Z)g_A^2$ can be viewed as a close lower bound on the summed strength S_-. Typically (p,n) reactions[13] find between 50 and 60% of this lower bound strength in the GT giant resonance using Osterfeld's RPA results[15] for background subtraction.

There have been many RPA calculations of GT strength in the literature[16-18] differing one from the other principally in the choice of effective interaction: Choices range from schematic separable interactions[16], zero-range interactions of the Skyrme or Migdal type[17], to finite-range interactions based on π- and ρ-exchange potentials[18]. Typically between 60 and 80% of the sum-rule strength, $3(N-Z)g_A^2$, is found in the strongest state, depending on the strength of the residual interaction. Experimentally[13] the observed strength peaks around 19.2 MeV, but spreads to lower energies and contains only 50-60% of the sum-rule strength. Simple RPA theory with 1p-1h states is not able to describe the spreading and fails to produce sufficient diminution of strength in the GT peak.

There are two further corrections to consider. First is the

TABLE III Percentage of sum-rule strength in the strongest state in RPA calculations in ^{208}Pb including MEC, isobar(Δ) and second-order core polarization (CP) using as residual interactions the one-boson exchange potential (OBEP) and zero-range (ZR) interaction with Migdal parameter, g'=0.6. (From ref. 10)

	RPA	RPA+MEC	RPA+MEC+Δ	RPA+MEC+Δ+CP[a]
OBEP	75	74	60	43
ZR(g'=0.6)	64	64	48	

a CP estimated from effective operator of Arima and Hyuga[7].

role played by MEC from the $\rho\pi$-graph and from isobar excitation. Second is the configuration mixing with background 2p-2h states[19], which can be estimated in second-order perturbation theory in a way exactly analogous to the core-polarization calculation discussed in the last section. Our calculational strategy is to evaluate the MEC, isobar(Δ), and core-polarization (CP) corrections for each basis particle-hole state in turn expressing the results in terms of an equivalent effective GT operator, eq. (2), and then using this effective operator in the RPA. Thus the wavefunctions are still given by a nucleons-only RPA calculation but the transition operator connecting the closed shell and 1p-1h configurations has become modified. The results of this calculation[10] are given in Table III for two choices of residual interaction: the one-boson exchange potential (OBEP) and zero-range interaction whose strength is determined by the Migdal parameter, g'.

The MEC correction from the $\rho\pi$-graph is seen to have a negligible influence on the GT resonance. On the other hand isobar currents significantly reduce the strength in the strongest state but the degree of quenching depends very much on the choice of the isobar-hole interaction. The Migdal zero-range interaction is a phenomenological one constructed for use in particle-hole channels and extended to isobar-hole channels and used only in RPA calculations. Configuration mixing with 2p-2h states are explicitly excluded, their effect being implicitly contained in the parameter, g'. By contrast the OBEP is a more fundamental interaction where one cannot single out the RPA graphs as the only important ones but all second-order and possibly higher-order processes must be considered on an equal footing. In Table III, we give an estimate of these other second-order effects and find that they too lead to a quenching of the resonance strength. In total both interactions quoted in Table III lead to roughly a 50% reduction in the resonance strength in line with experimental observation. However it is a moot point whether the quenching is dominated by isobar excitations as is sometimes claimed. It depends on the choice of residual interaction. The only reasonable statement that can be made is that both isobar mixing and 2p-2h mixing lead to a quenching of the GT resonance.

PROTON-RICH NUCLEI

Lastly, we move closer to the theme of this conference and consider proton-rich nuclei further from the valley of stable nuclei. Here the quenching of GT transitions is very pronounced. The prototype is a spin-flip transition in which a proton in a $j=\ell+1/2$ orbit decays to a neutron in an unoccupied $j'=\ell-1/2$ orbit. In such an extreme single-particle description the transition strength is

$$B_{sp}(GT; j^n \rightarrow j^{n-1} j') = n \times 4\ell/(2\ell+1) \, g_A^2 \qquad (7)$$

where n is the proton orbit occupancy in the parent nucleus. A hindrance factor, h_{exp}, is conveniently defined as the ratio of this with the experimental quantity

$$h_{exp} = B_{sp}(GT)/B_{exp}(GT) \tag{8}$$

For protons occupying the $g_{9/2}$ orbit in the region close to ^{100}Sn and protons occupying the $h_{11/2}$ orbit in the region close to ^{146}Gd, experimental data[20-21] indicate large hindrance factors, typically in the range 5 to 8. A discussion of this phenomenon has recently been given by Towner[22]. Three ingredients to the quenching are identified. The first is a consequence of a smearing out of the shell closures as described in pairing theory. The simple BCS calculations given by Towner[22] have now been superseded by more complete calculations by Conci[23] and by Dobaczewski[24]. The conclusion, however, is that with these nuclei being near to major shell closures, pairing correlations are only a small ingredient. More important is the configuration mixing induced by the strong proton-neutron forces. Towner[22] estimates the impact of this in first-order perturbation theory. This is analogous to the core

TABLE IV: Contributions from core polarization and higher-order effects to the hindrance factors for two choices of the residual proton-neutron interaction.

	h(ℓ=4)				h(ℓ=5)		
	^{94}Ru	^{96}Pd	^{98}Cd	^{100}Sn	^{148}Dy	^{150}Er	^{152}Yb
n =	4	6	8	10	2	4	6
ZR(g'=0.6)	6.4	4.6	3.5	2.7	4.1	3.6	3.1
OBEP	4.1	3.2	2.5	2.1	3.2	2.9	2.6
Experiment[20,21]	7.0	4.6			8.2	7.8	5.7

polarization discussed in the last section. Lastly is the extension of these core-polarization calculations to higher orders and the inclusion of contributions from MEC and isobar excitations. Their evaluation in heavy nuclei would be a formidable task and is not attempted here. Rather we will assume that their impact can be expressed by a simple modification of the GT operator as in eq. (2) and take $g_{A,eff} = 1.0$ rather than the free-nucleon value of 1.26.

In Table IV we give some sample calculations of hindrance factors from first-order core polarization (and multiplied by 1.6 to simulate higher-order effects, $(g_A/g_{A,eff})^2 \simeq 1.6$) for two choices of the residual proton-neutron interaction: a zero-range interaction with strength g'=0.6, and a finite-range OBEP. The calculations are close to giving the right amount of hindrance for nuclei in which the proton $g_{9/2}$ orbit is being filled, but insufficient hindrance for the $h_{11/2}$ nuclei. In the latter case, the search[21] for additional weak branches to higher excited states have failed to reveal any further GT strength. This data therefore poses a puzzle. For the same number of active protons, n, all calculations predict less hindrance in the ℓ=5 orbits around mass A≈150 than in the ℓ=4 orbits around mass A≈100 while experiment suggests the converse.

REFERENCES

1. S. Raman, C.A. Houser, T.A. Walkiewicz and I.S. Towner, Atomic Data and Nuclear Data Tables 21, 567 (1978).
2. J.W. Watson et al., Phys. Rev. Lett. 55, 1369 (1985).
3. C.D. Goodman et al., Phys. Rev. Lett. 54, 877 (1985).
4. J. Rapaport et al., Nucl. Phys. A431, 301 (1984).
5. K. Shimizu, M. Ichimura and A. Arima, Nucl. Phys. A226, 282 (1974).
6. M. Chemtob and M. Rho, Nucl. Phys. A163, 1 (1971).
7. H. Hyuga, A. Arima and K. Shimizu, Nucl. Phys. A336, 363 (1980); A. Arima and H. Hyuga, in Mesons in Nuclei, ed. D.H. Wilkinson and M. Rho (North-Holland, Amsterdam, 1979) pp. 685.
8. A. Arima, K. Shimizu, W. Bentz and H. Hyuga, "Nuclear Magnetic Properties and Gamow-Teller transitions", Adv. in Nucl. Phys. to be published.
9. I.S. Towner and F.C. Khanna, Nucl. Phys. A339, 334 (1983); I.S. Towner and F.C. Khanna, Phys. Rev. Lett. 42, 51 (1979).
10. I.S. Towner, Phys. Reports, in press.
11. E. Oset and M. Rho, Phys. Rev. Lett. 42, 47 (1979); M. Rho, Ann. Rev. Nucl. Part. Sci 34, 531 (1984).
12. B.A. Brown and B.H. Wildenthal, Phys. Rev. C28, 2397 (1983).
13. C. Gaarde et al., Nucl. Phys. A369, 258 (1981); D.E. Bainum et al., Phys. Rev. Lett. 44, 1751 (1980); D.J. Horen et al., Phys. Lett. 95B, 27 (1980).
14. K. Ikeda, S. Fujii and J.J. Fujita, Phys. Lett. 3, 271 (1963).
15. F. Osterfeld, Phys. Rev. C26, 762 (1982); F. Osterfeld and A. Schulte, Phys. Lett. 138B, 23 (1984).
16. A. Bohr and B.R. Mottelson, Phys. Lett. 100B, 10 (1981); K. Grotz et al., Phys. Lett. 126B, 417 (1983); K. Grotz, H.V. Klapdor and J. Metzinger, Phys. Lett. 132B, 22 (1983).
17. N. van Giai and H. Sagawa, Phys. Lett. 106B, 379 (1981); G.F. Bertsch, D. Cha and H. Toki, Phys. Rev. C24, 533 (1981); G.E. Brown and M. Rho, Nucl. Phys. A372, 397 (1981); H. Sagawa and N. van Giai, Phys. Lett. 113B, 119 (1982); N. Auerbach,

L. Zamick and A. Klein, Phys. Lett. 118B, 256 (1982).
18. A. Harting et al., Phys. Lett. 104B, 261 (1981); W. Knupfer, M. Dillig and A. Richter, Phys. Lett. 122B, 7 (1983); D. Cha and J. Speth, Phys. Lett. 143B, 297 (1984); F. Osterfeld, D. Cha and J. Speth, Phys. Rev. C31, 372 (1985).
19. G.F. Bertsch and I. Hamamoto, Phys. Rev. C26, (1982) 1323.
20. K. Rykaczewski et al., Z. Phys. A322, 263 (1985).
21. P. Kleinheinz et al., Phys. Rev. Lett. 55, 2664 (1985); W. Habenicht et al., in Proc. 7th Int. Conf. on At. Masses and Fundamental Constants (AMCO-7), Darmstadt-Seeheim (1984), ed. O. Klepper, p. 244.
22. I.S. Towner, Nucl. Phys. A444, 402 (1985).
23. C. Conci et al., Phys. Lett. 148B, 405 (1984).
24. J. Dobaczewski et al., in Proc. Int. Symp. on Weak and Electromagnetic Interactions in Nuclei, Heidelberg (1986), ed. H.V. Klapdor, p. 248

GAMOW–TELLER STRENGTH FUNCTIONS FROM NUCLEON SCATTERING EXPERIMENTS

O. Häusser

Simon Fraser University, Burnaby, B.C., Canada V5A 1S6
and
TRIUMF, 4004 Wesbrook Mall, Vancouver, B.C., Canada V6T 2A3

ABSTRACT

The spin–isospin structure of nuclear excitations up to ~ 50 MeV has been studied using (n,p), (p,p') and (p,n) reactions at TRIUMF. At 200–400 MeV the $L = 0$ isovector spin-transfer component of the cross section is closely related to the Gamow-Teller strength function. Results are discussed which have implications for Gamow–Teller quenching in (sd) and (fp) shell nuclei, for the Ikeda sum rule, and for the importance of 2particle-2hole correlations in nuclear wavefunctions.

INTRODUCTION

It is well known that in nuclear β decay Fermi strength (quantum numbers $L = 0$, $S = 0$, $T_i = T_f \neq 0$) is concentrated in a single transition, whereas Gamow-Teller (GT) strength ($L = 0$, $S = 1$, $T = 1$) is spread over a wide energy region as a result of the spin–dependent residual nuclear interaction. Studies of β decays in nuclei far from stability offer the opportunity to examine a large energy window and thus a large fraction of the total GT strength. Unfortunately, mixing of 2particle-2hole (2p2h) into (1p1h) states [1], or coherent isobar-hole admixtures [2] may produce a shift of GT strength to energies considerably above 10 MeV, making it undetectable in conventional β decay experiments.

Nucleon (N) scattering experiments at intermediate energies can be used to provide a substitute and extension of GT studies in beta decay. Pioneering work [3,4] using the time-of-flight setup at Indiana University (IUCF) has demonstrated that the (p,n) reaction, at $E_p = 120 - 200$ MeV and small momentum transfer, can be quantitatively related to GT strength known from analogous β decay. We discuss here results of (n,p), ($\vec{p},\vec{p}'$), and (p,n) experiments carried out at 200–400 MeV using the medium-resolution spectrometer (MRS) at TRIUMF. Recent modifications of the MRS include the installation of a dispersion-matching system, of a focal-plane polarimeter [5] to analyze the transverse spin components of scattered protons, and of a compact sweeping magnet which is part of the CHARGEX facility [6,7] for the study of (n,p) and (p,n) reactions. Although the resolution achieved is modest (typically 140 keV FWHM in (p,p') and 1 MeV in charge exchange reactions) this is compensated for by the extreme selectivity of these reactions to excite nuclear spin-isospin modes. The optimum sensitivity occurs near 300 MeV, where multistep processes are less important than at lower energies. The relationship between nucleon scattering and β decay is shown in Table 1.

Table I Nucleon scattering and beta decay

reaction	transition operator	GT strength	total strength	isospin of final state	$\sigma/B(GT)$ calibrated in
(p,n)	$\sigma\tau_-$	$B(GT_-)$	S_-	$T_0 - 1$, T_0, $T_0 + 1$	β^+ decay
($\vec{p},\vec{p}'$)	$\sigma\tau_0$			T_0, $T_0 + 1$	
(n,p)	$\sigma\tau_+$	$B(GT_+)$	S_+	$T_0 + 1$	β^- decay

© American Institute of Physics 1988

The combination of (p,n) and (n,p) reactions makes possible tests of the Ikeda sum rule [8], $S_- - S_+ = 3(N - Z)$, which might be violated if GT strength is shifted outside of the energy window over which the GT strength is summed. For targets with ground state isospin $T_0 \neq 0$ the spin response for individual isospin components (T_0-1, T_0, T_0+1) can be constructed by combining the results from all three reactions. Such a program has been carried out for the $T_3 = 1$ nucleus ^{54}Fe. The experiments were performed in collaboration with R. Abegg, P. Alford, A. Celler, D. Frekers, R. Helmer, R. Henderson, K. Hicks, P. Jackson, R. Jeppesen, A. Miller, M. Moinester, K. Raywood, R. Sawafta, M. Vetterli and S. Yen. The Rutgers group which includes C. Glashausser, R. Fergerson, and K. Jones has made substantial contributions to the ^{54}Fe$(\vec{p}, \vec{p}')$ experiment.

CONVERTING CROSS SECTIONS INTO GT STRENGTH.

The proportionality between nucleon scattering cross sections and GT strength is expected to hold at small momentum transfers in the framework of the single-step distorted wave impulse approximation (DWIA). Assuming an (n,p) reaction for which $B(GT_-)$ (and thus $B(GT_+)$) is known from corresponding β^- transitions, the cross section can be written (similar expressions are valid for (p,p') and (p,n) reactions) as

$$\frac{d\sigma_{np}}{d\Omega}(q=0) = \left(\frac{\mu}{\pi\hbar^2}\right)^2 \frac{k_i}{k_f} N_{\sigma\tau}^D |J_{\sigma\tau}|^2 B_\uparrow(GT_+)$$

$$B(GT_+) = \frac{1}{(2J_i+1)}[<J_f T_f|\Sigma\sigma\tau_+|J_i T_i>]^2$$

$$B_\uparrow(GT_+) = \frac{(2J_f-1)}{(2J_i-1)} B_\downarrow(GT_-) = \frac{(2J_f-1)}{(2J_i-1)}\left(\frac{g_A}{g_V}\right)^{-2} \frac{6163.4 \pm 3.8 sec}{ft}$$

In these expressions μ is the relativistic reduced energy divided by c^2, k_i and k_f are wave numbers, $g_A/g_V = 1.26$ is the ratio of axial vector and vector coupling constants, and $J_{\sigma\tau}$ is the volume integral of the $(\sigma\tau)$ component of the NN interaction. We do not make use of this factorized form of the cross section but employ instead a computer code which also includes spin-orbit and tensor pieces of the NN interaction. The distortion factor $N_{\sigma\tau}^D$ represents the difference between plane waves and distorted waves and reduces the cross section more and more as the target mass increases (typically by $\sim 12\%$ for $A = 12$ and a factor of three for $A = 54$). In most cases we have determined the distortions by extensive measurements of elastic scattering for the appropriate target and energy.

So far we have performed (n,p) calibration experiments [9] at 200 MeV for targets of ^{6}Li and ^{12,13}C. Figure 1 shows (n,p) spectra for targets of C (top) and CH$_2$ (bottom). Both spectra were obtained simultaneously using segmented targets separated by active wire chambers [7]. In the C spectrum (top) the large peak at Q = -12.6 MeV arises from the GT transition to the ^{12}B ground state ($B(GT_-) = 0.999 \pm 0.005$ from ^{12}B β^- decay). The small peak at Q = 0 in the top spectrum arises from hydrogen in the mylar foils and the counter gas of the segmented target chamber. We obtain $\sigma_{np}(0°, q=0)/B(GT) = 9.42 \pm 0.31$ mb/sr in good agreement, but more accurate than, previous IUCF results [4] for ^{12}C(p,n) at 200 MeV. This, and values for other pure GT transitions [9] in ^{6}Li(p,n) and ^{13}C(n,p), can be reproduced with an accuracy of 10% or better in DWIA calculations using as ingredients optical potentials derived from elastic scattering data, the effective NN interaction (essentially $J_{\sigma\tau}$) of Franey and Love [10], and transition densities from microscopic shell model calculations. The GT values quoted in the following were obtained using $\sigma/B(GT)$ calculated with the DWIA and may be subject to slight revisions as more calibration experiments will be carried out.

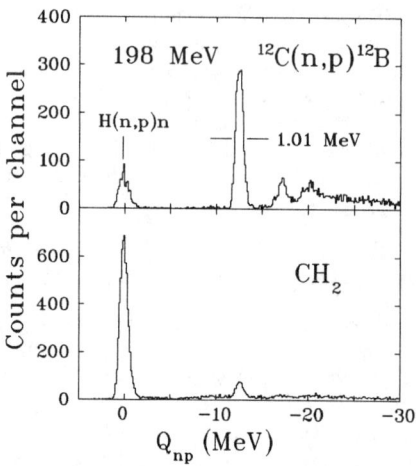

Fig. 1. (n,p) spectra obtained with a 198 MeV neutron beam and targets of carbon and CH_2.

Confirmation that the Franey–Love interaction describes well the energy-dependence of GT cross sections in the 200—400 MeV region comes from angular distribution measurements for the $^{28}Si(p,p')$ reaction leading to pure $1^+, T = 1$ states [11]. It is found that the 'quenching factors' $f(p,p')$ required to bring experimental and calculated DWIA angular distributions into agreement do not vary appreciably (to an experimental uncertainty of $\sim \pm 6\%$) at five incident proton energies between 200 and 400 MeV.

GAMOW-TELLER QUENCHING IN (sd) SHELL NUCLEI

Large-scale shell model calculations (WB [12] and WC [13]) provide comprehensive predictions of GT strength at low excitations in most (sd) shell nuclei. We have completed experiments [14] on the $^{26}Mg(n,p)^{26}Na$ reaction at 200 MeV and on inelastic proton scattering in ^{24}Mg at 250 MeV and in ^{28}Si at several energies [15, 11]. In the (p,p') reaction transitions to the $1^+, T = 1$ states are analogs of transitions excited in the charge exchange reactions. A spectrum for $^{24}Mg(\vec{p}, \vec{p}')$ and $\theta_{lab} = 2.9°$ is shown in the upper panel of Figure 2. The unwanted $S = 0$ component in the cross section can be eliminated by measurements of the spin-transfer probability S_{nn} using the focal plane polarimeter to detect a change in the proton polarization after scattering.

This is shown in the spin-transfer spectrum of Figure 2 (middle) where natural parity states with $S = 0$ are absent. The spin-transfer cross section σS_{nn} is thus a measure of the spin transfer strength in ^{24}Mg. The $L = 0$, GT components in the σS_{nn} spectra can be identified from their rapid decrease with increasing momentum transfer (e.g. compare the strong $1^+, T = 1$ peak at 10.71 MeV excitation in the middle and lower spectra). The running sum of GT strength versus excitation energy in $^{24}Mg(\vec{p}, \vec{p}')$ is compared in Figure 3 with predictions using WB [12] and WC [13] transition densities. It is apparent that both shell models give reasonably consistent estimates for the integrated GT strength but are less reliable for individual states. From the σS_{nn} data in ^{24}Mg between $E_{exc}=9-15$ MeV we deduce [15] an overall 1^+ quenching factor

$< f(p,p') > = 0.80 \pm 0.10$. The error arises from counting statistics and uncertainties in the L = 0 fraction of σS_{nn}, but excludes systematic uncertainties in the average of WC and WB transition densities and in the ratio $\sigma/B(GT)$.

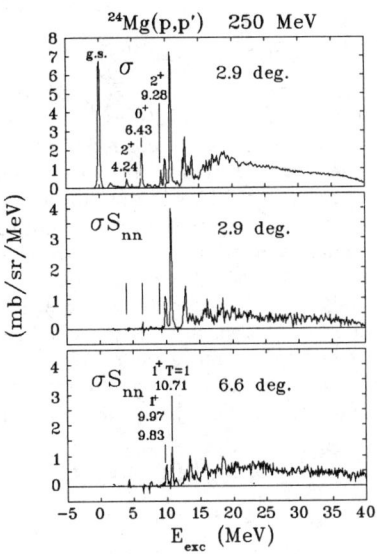

Fig. 2. Differential cross sections σ and spin-flip cross sections σS_{nn} for inelastic proton scattering from ^{24}Mg at 250 MeV.

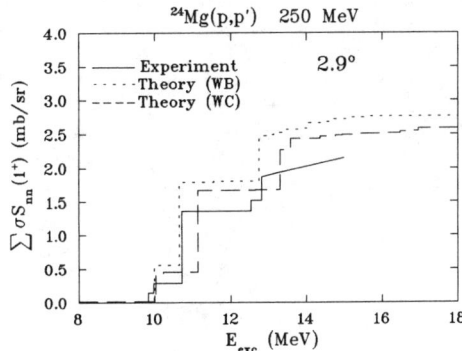

Fig. 3. Running sum of 1^+ spin-transfer strength for ^{24}Mg$(\vec{p},\vec{p}')$ at 250 MeV and $\theta = 2.9°$.

In ^{28}Si between E_{exc}=9-14 MeV we arrive [11] at an average 1^+, T = 1 quenching factor $< f(p,p') > = 0.89 \pm 0.09$. These GT quenching factors are somewhat larger than those for charge exchange reactions in neighbouring nuclei [14,16] ($f \approx 0.67$ for ^{26}Mg(n,p) and $f \approx 0.57$ for ^{26}Mg(p,n)). The variation in the individual values for f beyond experimental errors may be attributed to uncertainties in the DWIA calculations for three reactions at different energies, to different integration intervals and analysis procedures, and to systematic uncertainties in the various transition densities.

THE SPIN–ISOSPIN RESPONSE OF ^{54}Fe.

The (n,p), $(\vec{p},\vec{p}')$, and (p,n) reactions on ^{54}Fe have recently been measured at TRIUMF near 300 MeV. Analysis of these reactions [17,18] has resulted in a decomposition of the nuclear response for ^{54}Fe into spin and isospin components, and in a test of the Ikeda sum rule (see next section). Data have been obtained at 0-12° for (n,p), 3-15° for $(\vec{p},\vec{p}')$, and 0-15° for (p,n) reactions. Since in nucleon scattering different L-transfers peak at distinctly different angles, a decomposition into multipoles (L = 0, 1, 2, ...) can be carried out. This is demonstrated in Figure 4 where the $L = 0, 1, 2$ components of (n,p) angular distributions [17] are shown for different excitation energies. The GT (L = 0) component is strongly peaked at 0° and is shown in Figure 5 for the three reactions. In ^{54}Fe$(\vec{p},\vec{p}')$ at the most forward angle feasible in the experiment (3.1°) the M1 resonance at 9-15 MeV is seen much more clearly in the spin-transfer cross section σS_{nn} than in the cross section σ. For the (p,p') reaction the L = 0 component in σS_{nn} above $E_x = 20$ MeV is rather uncertain and has been omitted in the plot.

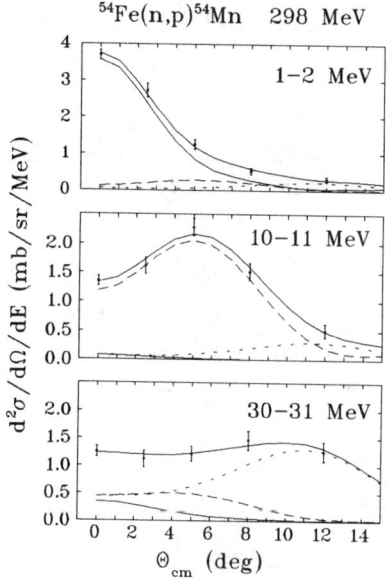

Fig. 4. Multipole decomposition of angular distributions for the ^{54}Fe(n,p)^{54}Mn reaction at three different excitation energies in ^{54}Mn. Solid lines: total and L = 0 part; dot-dashed lines: L = 1 part; short-dashed lines: L = 2 part.

The GT strength is concentrated at lower excitations, although the fitting procedure produces also a long GT tail to higher excitations. The intensity of this tail is however subject to large uncertainties since the GT fraction is small and depends on the small-angle behaviour of the L = 1, 2.. angular distributions used in the decomposition. Such a tail would be expected from admixtures of (2p2h) components in the wavefunction [1] and would provide an explanation for the quenching of GT strength at low excitations.

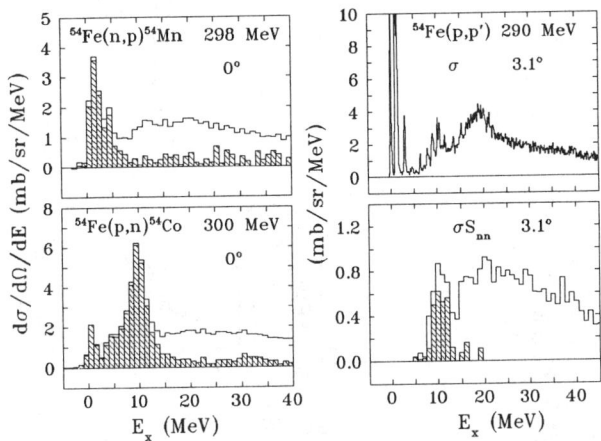

Fig. 5. Forward-angle spectra for (n,p) (top left), (p,n) (bottom left) and (p,p') (right) reactions on ^{54}Fe. The hatched areas correspond to the L = 0 (GT) components deduced from multipole decompositions.

In Table 2 we show the L = 0 component of the cross sections (or of σS_{nn}) for the three reactions, and quenching factors f based on the DWIA value of $\sigma_{np}(0°, q = 0)/B(GT) = 3.5$ mb/sr. The (p,n) result which is still preliminary is larger but statistically more accurate than that of Rapaport et al [19] obtained at 160 MeV. The quenching factors shown are for two extreme models, one in which a simple $(f_{7/2}^{14})$ ^{54}Fe ground state is assumed (f_A), and one in which RPA corrections have been estimated from a calculation in ^{60}Ni [20] (f_{RPA}). It is seen that GT quenching depends strongly on the assumptions about the Fermi surface, especially for the (n,p) reaction which tends to be Pauli blocked. In the (sd) shell where $0\hbar\omega$ configuration mixing is included fully no such asymmetry in the GT quenching factors for (p,n) and (n,p) has been observed.

Table II GT quenching factors in ^{54}Fe

reaction	E_{beam} (MeV)	measured quantity	angle (deg.)	$E_x^i \to E_x^f$ (MeV)	L = 0 strength (mb/sr)	f_A	f_{RPA}
(n,p)	298	σ	0	$0 \to 10$	12.9	0.37	0.65
$(\vec{p}, \vec{p}')$	290	σS_{nn}	3.1	$4.5 \to 14.5$	2.8	0.39	~ 0.60
(p,n)	300	σ	0	$0 \to 13.5$	33^a	0.59^a	0.77^a

^{a} preliminary value

Evidence for the importance of (2p2h) components in the wavefunctions is contained in the data at larger momentum transfers. At q> 1fm^{-1} giant resonances are no longer apparent and the nucleus can be approximated by the response of a semi-infinite slab [21, 22]. In Figure 6 we show large-angle (n,p) cross sections and σS_{nn} in (p,p') together with calculations by Smith [22] which include the free NN response (long-dashed curves), the (1p1h) RPA response (short-dashed curves) and the combined (2p2h) (1p1h) RPA reponse (solid curves). The (2p2h) correlations push spin-transfer strength towards higher energies, an effect familiar from discussions of GT quenching [1]. The (2p2h) correlations bring about reasonable agreement for (n,p), whereas the $(\vec{p}, \vec{p}')$ experiment demands even more spin-transfer strength above 30 MeV.

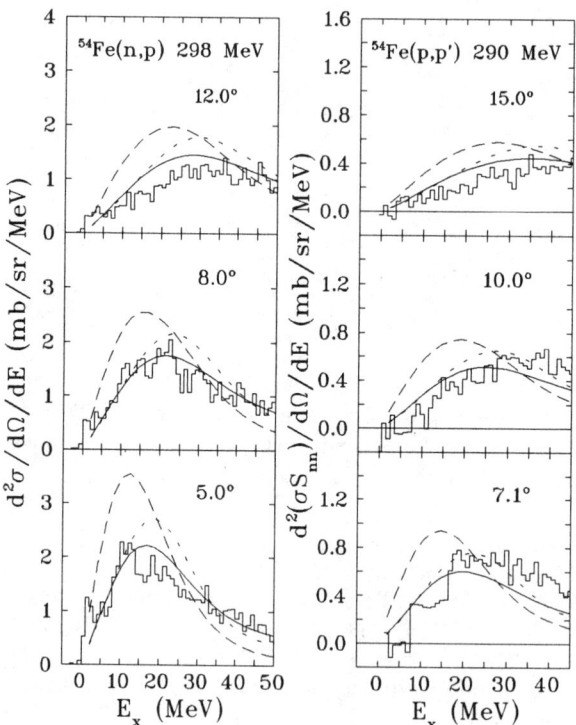

Fig. 6. Comparison between experimental (n,p) cross sections (left) and (p,p') spin-flip cross sections (right) and predictions of the RPA surface response model of Smith [22]. The curves shown correspond to the free quasielastic response (long-dashed lines), the (1p1h) RPA (short-dashed lines), and to the RPA which includes (2p2h) correlations (solid lines).

TESTS OF THE IKEDA SUM RULE

From the L = 0 strength for the (n,p) and (p,n) reactions in Table 2 we obtain for the sum rule in ^{54}Fe a (still preliminary) value of $S_- - S_+ = 5.8 \pm 0.8$, which is nearly the full value of 6.

Another test of the sum rule was carried out in ^{90}Zr where a multipole decomposition of the experimental (p,n) spectrum [23] has found most of the GT sum rule, and where the data are well reproduced by RPA calculations [24] which include (2p2h) correlations and a correction for two-step processes. However, if a significant amount of S_+ strength were to be found in the (n,p) reaction the sum rule would still be violated, making room for the isobar–hole quenching mechanism [2]. In Figure 7 we show spectra for the ^{90}Zr(n,p)^{90}Y reaction at 198 MeV obtained by Yen etal [25] at average angles of 1.8° and 6.4°.

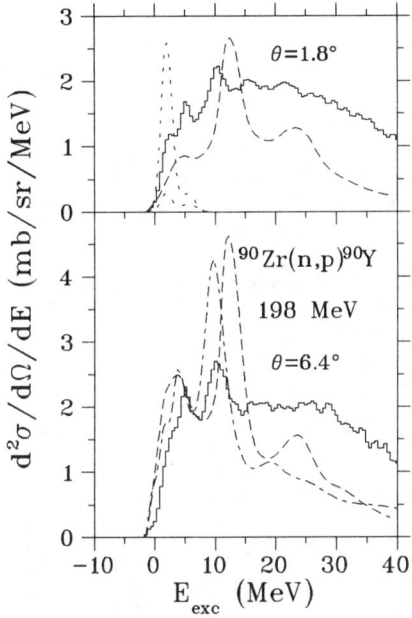

Fig. 7. Cross sections for the ^{90}Zr(n,p) reaction at 198 MeV. The theoretical curves correspond: to GT strength calculated by Bloom etal [26] (short-dashed lines in the upper panel); to the (1p1h) RPA of Klein etal [27] (long-dashed lines); to the RPA with (2p2h) correlations by Wambach etal [28] (dot-dashed lines in the lower panel).

The tall and short dotted lines in the upper spectrum represent GT strength calculated with different shell model assumptions by Bloom etal [26]. The dashed lines represent the 0° and 5° (1p1h) RPA calculation of Klein etal [27], whereas the dot-dashed curve from Wambach etal [28] includes the effects of (2p2h) correlations. There is little evidence for GT strength below 10 MeV. A safe upper limit of $S_+ \leq 3.6$ is obtained if the unrealistic assumption is made that all the cross section is L = 0; a more realistic upper limit, $S_+ \leq 1.6$, is obtained from the difference between the data and the calculation of Klein etal [27] which does not include GT stength. These upper limits are small when compared to the sum rule value of 30. Of special interest is the fact that the (2p2h) calculations produce too little damping of the spin–dipole resonance apparent in the 6.4° spectrum, and underestimate the data above $E_x \sim 18$ MeV. One may speculate that this could be remedied if correlations in the ^{90}Zr ground state (the RPA calculations assume a doubly closed shell) were included.

GAMOW–TELLER STRENGTH FUNCTIONS AND OTHER PHYSICS

GT strength functions are of special interest for other fields of physics. We mention briefly (i.e. without giving adequate references) three topics which are presently being addressed by nucleon scattering experiments at TRIUMF.

1. ASTROPHYSICS. In models of supernova explosions electron capture rates for nuclei in the iron region are important parameters. Electron capture which involves the S_+ strength function reduces the pressure of the relativistic electron gas and determines i) the rate at which gravitational collapse proceeds, ii) the mass M_i at the beginning of collapse, and iii) the mass M_h of the homologous core of maximum density. Difficulties in producing supernova explosions in the models which arise from energy absorption in remnant matter of the mantle of mass $(M_i - M_h)$ might be resolved if sufficient data on S_+ strength functions become available.

2. DOUBLE BETA DECAY. Double beta decay rates are dominated by the weak interaction process in which two neutrinos are emitted. The rates are governed by a coherent product of GT matrix elements involving 1^+ states in the intermediate odd–odd nucleus. These matrix elements can be studied independently by (p,n) and (n,p) experiments. Since S_+ strength is subject to severe Pauli blocking the (n,p) reaction may provide the key explanation why calculated lifetimes tend to be shorter than experiment, e.g. in ^{82}Se and ^{130}Te.

3. EXCHANGE CURRENTS. In $T_3 = 0$ nuclei such as ^{6}Li or ^{12}C GT matrix elements may be measured accurately by nucleon scattering if experiments to calibrate $\sigma/B(GT)$ are also carried out. Comparison of GT, and of M1 matrix elements from (e,e') work, may help to identify meson exchange contributions which are expected to be small for an axial vector (GT) current, but large for a vector (M1) current (see the article by I.S. Towner in this volume).

CONCLUSIONS

We have shown that nucleon–nucleus scattering experiments between 200 and 400 MeV provide a quantitative, optimal tool to investigate the spin and isospin structure of the nuclear response. The GT component of the response at low energies is found to be reduced compared to most nuclear models, however the amount of quenching depends sensitively on the model assumptions made about the Fermi surface, especially for the S_+ strength. Although the question regarding the quenching mechanism is not yet completely resolved, it appears now likely that (2p2h) correlations are chiefly responsible for shifting GT strength to higher energies. Supporting evidence for this view comes from the large fraction of the sum rule observed in ^{54}Fe, from the small amount of GT strength found at low excitation in the ^{90}Zr(n,p) reaction, and from the success of the slab model with (2p2h) corrections in reproducing large-angle spectra in $(\vec{p}, \vec{p}')$ and (n,p) reactions on ^{54}Fe. A puzzling aspect of the data is the excess of (spin-transfer) strength in the high-energy region for both ^{54}Fe$(\vec{p}, \vec{p}')$ and ^{90}Zr(n,p) reactions over calculations which include (2p2h) components in the RPA. We speculate that this discrepancy might be remedied when correlations in the ^{54}Fe and ^{90}Zr ground states are considered.

REFERENCES

1. G.F. Bertsch and I. Hamamoto, Phys. Rev. $C26$, 1323 (1982).
2. G.E. Brown and M. Rho, Nucl. Phys. $A372$, 397 (1981).
3. C.D. Goodman et al., Phys. Rev. Lett. 44, 1755 (1980).
4. T.N. Taddeucci et al., Nucl. Phys. $A469$, 125 (1987).
5. O. Häusser et al., Nucl. Instr. Meth. $A254$, 67 (1987).
6. R.L. Helmer, Can. J. Phys. 65 (1987), in press.
7. R.S. Henderson et al., Nucl. Instr. Meth. $A257$, 97 (1987).
8. C. Gaarde et al., Nucl. Phys. $A334$, 248 (1980).
9. K.P. Jackson et al., submitted to Phys. Lett. B.
10. M.A. Franey and W.G. Love, Phys. Rev. $C31$, 488 (1985).
11. O. Häusser et al., submitted to Phys. Rev. C.
12. B.H. Wildenthal, Progress in Particle and Nuclear Physics $Vol.11$, (Oxford: Pergamon 1984) p.5; and B.A. Brown, private communication.
13. W. Chung, PhD thesis, Michigan State University, 1976.
14. W.P. Alford et al., to be published.
15. R.S. Sawafta et al., submitted to Phys. Lett. B.
16. R.D. Madey et al., Phys. Rev. $C35$, 2011 (1987).
17. M.C. Vetterli et al., Phys. Rev. Lett. 59, 439 (1987); and to be published.
18. O. Häusser et al., to be published.
19. J.R. Rapaport et al., Nucl. Phys. $A410$, 371 (1983).
20. N. Auerbach et al., Phys. Lett. $B118$, 256 (1982).
21. H. Esbensen and G. Bertsch, Ann. Phys. 157, 255 (1984).
22. R.D. Smith, private communication.
23. M. Moinester, Can. J. Phys. 65 (1987), in press.
24. R.D. Smith and J. Wambach, preprint.
25. S. Yen et al., submitted to Phys. Lett. B.
26. S.D. Bloom, G.J. Mathews and J.A. Becker, Can. J. Phys. 65 (1987), in press.
27. A. Klein, W.G. Love and N. Auerbach, Phys. Rev. $C31$, 710 (1985).
28. J. Wambach, S. Drozdz, A. Schulte and J.A. Becker, preprint.

IMPROVEMENT OF THE GROSS THEORY OF β-DECAY

Takahiro Tachibana, Sadayuki Ohsugi and Masami Yamada
Science and Engineering Research Laboratory, Waseda University
3-4-1 Okubo, Shinjuku-ku, Tokyo 160, Japan

ABSTRACT

The gross theory of β-decay is improved in three respects. First, the effect of partial occupation of single-particle states in the pairing theory is taken into account. Second, the function to represent the Pauli exclusion principle is modified in order to observe the sum rule for the Fermi transition more accurately. The third improvement is made on the single-particle strength function. From consideration of the experimental data on Gamow-Teller giant resonances in (p,n) reactions and the results of calculations based on a many-body theory, the single-particle strength function for the Gamow-Teller transition is assumed to be a superposition of two hyperbolic secant functions, one with a width of several MeV and the other with a much wider width. The single-particle strength functions for the first forbidden transitions caused by axial-vector interaction are also modified corresponding to this modification for the Gamow-Teller transition. Results of numerical calculations are presented and discussed.

1. INTRODUCTION

The gross theory was born about twenty years ago, and was brought to the current form by Takahashi, Yamada and others in the early 1970's. A handy review of this theory can be found in the paper by Takahashi, Yamada and Kondoh[1] (referred to as TYK), in which extensive graphs of calculated half-lives are also given. Since then, this theory has been applied to various problems, but no substantial refinement of it has been made until recently as far as gross properties are concerned.* However, as we have acquired much knowledge in these years, we have undertaken to improve the gross theory. Our improvement is made within the general framework that the β-strength function is given as a certain integral of single-particle (abbreviated to SP hereafter) strength functions. Namely, the β-strength function $|M_\Omega(E)|^2$ is assumed to be given as

$$|M_\Omega(E)|^2 = \int_{\varepsilon_{min}}^{\varepsilon_{max}} D_\Omega(E,\varepsilon) W(E,\varepsilon) \frac{dn_1}{d\varepsilon} d\varepsilon \tag{1}$$

Here, Ω denotes the type of β-transition (Fermi, Gamow-Teller, etc.), E represents the energy of the final nuclear state measured from the parent nucleus, ε is the single-neutron (proton) energy for β^--decay (β^+-decay and electron capture), $dn_1/d\varepsilon$ is the neutron (proton) energy distribution, $D_\Omega(E,\varepsilon)$ is an SP strength function, and $W(E,\varepsilon)$ is a weight function to take into account the Pauli exclusion principle.

Our improvement is made in three respects: The first is related to the treatment of pairing, the second is on the function $W(E,\varepsilon)$, and the third is improvement of the SP strength functions.

---

* Apart from the gross properties, Kondoh and Yamada[2] tried to include shell effects in the formalism of gross theory.

2. EFFECT OF PAIRING UV FACTORS

In the conventional gross theory of TYK, the energy gap due to pairing is taken into account, but the effect of the so-called UV factors is not fully taken into consideration. ($U_k^2 = 1 - V_k^2$ represents the occupation probability of the k-th SP state in the BCS ground state.) Let us consider β^--decay of an odd-N nucleus, in which the odd neutron is in the k-th SP state. In the gross theory, the ground-state neucleon configuration of the parent is represented by the left part of Fig.1. However, after the β^--decay of the odd neutron, the nuclear state becomes a linear combination of zero- and two-quasi-neutron states with a mixing ratio of $U_k^2 : V_k^2$ as shown in Fig.1.

Thus, we allot U_k^2 times the SP strength to the decay of the neutron above the pairing gap, and the remaining part $V_k^2 = 1 - U_k^2$ to the decay of the neutrons just below the gap. The treatment in TYK corresponds to taking $U_k^2 = 1$ and $V_k^2 = 0$, but we keep V_k^2 in the improved treatment. A similar improvement is made for the transition to an odd hole and also for β^+-decay and electron capture.

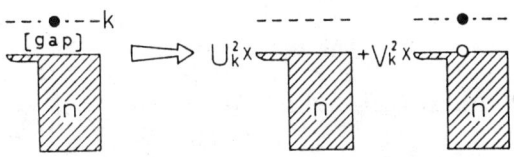

Fig.1. Change of the neutron configuration caused by β^--decay of the odd neutron in the k-th SP state. The filled circle represents a neutron and the open circle represents a neutron hole.

The values of U_k^2 and V_k^2 may be chosen for each nuclide, but for a quick survey of the whole nuclidic region, they may be taken to be a rough average value 0.5.

For more details, see our recent paper.[3]

3. MODIFICATION OF THE WEIGHT FUNCTION TO OBSERVE SUM RULES

Sum rules played an important role in constructing the gross theory. However, once Eq.(1) was taken to be the basic equation of this theory, the resulting strength functions often violate the sum rules to a considerable degree. The most conspicuous is the violation of the sum rule for the Fermi transition, which is, to a good approximation, known to be

$$\int_{-Q}^{\infty} |M_F(E)|^2 dE \approx \begin{cases} N-Z \text{ for } \beta^--\text{decay of a} \\ \qquad \text{nucleus with } N>Z \\ Z-N \text{ for } \beta^+-\text{decay of a} \\ \qquad \text{nucleus with } Z>N \\ 0 \text{ otherwise} \end{cases} \quad (2)$$

The conventional gross theory of TYK violates Eq.(2) by 30~50%. To remedy this,

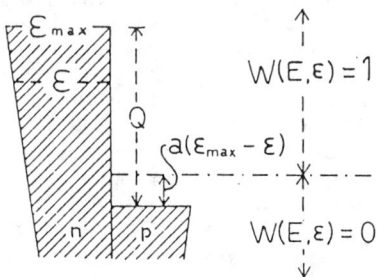

Fig.2. A simplified picture to show the modification of $W(E,\varepsilon)$. The transition from the SP state at ε can proceed only to the SP levels above the dash-dot line. In the old theory, $a=0$.

we modify the weight function $W(E,\varepsilon)$ in Eq.(1) in the manner shown in Fig.2. If we take a=0.6, the violation of the sum rule is much reduced. The modified weight function is also applied to any other transitions. For details, see our recent paper.[3]

4. IMPROVEMENT OF SINGLE-PARTICLE STRENGTH FUNCTIONS

4.1 Gamow-Teller giant resonance from (p,n) reactions

Concentration of the Gamow-Teller strength around the isobaric analog state (IAS), which had been anticipated by Ikeda, Fujii and Fujita[4] and also by Yamada in starting the gross theory, was found experimentally in (p,n) reactions.[5] The main characteristics of this Gamow-Teller giant reasonance are summarized as follows:
1. A peak with a width of several MeV is situated near IAS. The peak position measured from IAS is approximately given by[5]

$$\Delta_\tau = -30(N-Z)/A + 6.7 \text{ MeV} \qquad (3)$$

2. The sum of the strengths distributed in this giant resonance region is only about half the non-energy-weighted sum-rule value.

4.2 Sum rules for single-particle strength functions

We denote the total Hamiltonian by H and the wave function of the parent nucleus by Ψ_0. Then, the β-strength function $|M_\Omega(E)|^2$ for the transitions caused by an operator Ω satisfies the following sum rules.

$$\int_{-Q}^{\infty} |M_\Omega(E)|^2 \, dE = (\Psi_0, \Omega^\dagger \Omega \Psi_0) \qquad (4)$$

$$\int_{-Q}^{\infty} E|M_\Omega(E)|^2 \, dE = (\Psi_0, \Omega^\dagger [H,\Omega] \Psi_0) \qquad (5)$$

$$\int_{-Q}^{\infty} E^2 |M_\Omega(E)|^2 \, dE = (\Psi_0, [\Omega^\dagger, H][H,\Omega] \Psi_0) \qquad (6)$$

in which the brackets represent commutators and the dagger means Hermitian conjugate. These sum rules are qualitatively very helpful, and actually they were the motive force to construct the gross theory. It is, however, rather difficult to calculate the right-hand sides of Eqs.(4)~(6) except for a few simple cases, because exchange terms give rise to large cancellation.

In the gross theory, the above difficulty is avoided by the use of the SP strength function as an intermediary, which is supposed to be the strength function for a nucleon in the absence of the Pauli exclusion principle; the Pauli principle is taken into account in another function $W(E,\varepsilon)$ as stated in Sec.1. The SP strength function being of such a character, we can obtain sum rules related to it from Eqs.(4)~(6) by neglecting the exchange terms in the right-hand sides.

Calculations of the right-hand sides of Eqs.(4)~(6) can be made by the cluster expansion method.[6] We assume that each β-decay operator Ω is a sum of one-body operator ω_k as

$$\Omega = \sum_k \omega_k \qquad (7)$$

We also assume that the nuclear forces are of two-body type. Then, the commutator $[H,\Omega]$, which is important for the sum rules, is, in general, a sum of one-body and two-body operators. Thus, the operator $\Omega^\dagger\Omega$ in Eq.(4) is a sum of one-body and two-body operators as

$$\Omega^\dagger\Omega = \sum_k \omega_k^\dagger \omega_k + \sum_k \sum_{l(\neq k)} \omega_k^\dagger \omega_l \qquad (8)$$

The operator $\Omega^\dagger[H,\Omega]$ in Eq.(5) is a sum of one-, two-, and three-body operators, and the operator $[\Omega^\dagger,H][H,\Omega]$ is a sum of one-, two-, three-, and four-body operators. We only consider the n-body cluster in calculating the expectation value of an n-body operator, and neglect higher-order clusters.

In order to calculate the SP strength function, we further make a simplification; we neglect all the exchanges and correlations between particles unless they are explicitly connected by the nuclear-force term of $[H,\Omega]$ or $[\Omega^\dagger,H]$. It should be noted that this does not mean an insignificance of exchanges. They contribute much to the right-hand sides of Eqs.(4)~(6), but if we include them we shall get the sum rules for the strength function of the whole nucleus, not for the SP strength functions. The neglect of exchanges is a prescription to get the SP strength functions. The situation about the neglect of correlations has not yet been studied in detail; at least, it brings about a considerable simplification of calculations.

In the follwing, we concentrate on the Gamow-Teller transition. In this case, the first term in the right-hand side of Eq.(8) contributes simply $3N$ (for β^-) or $3Z$ (for β^+) to the right-hand side of Eq.(4), in which N and Z are those of the parent. These $3N$ or $3Z$ are just the sum of integrated SP strength functions. On the other hand, the second term in the right-hand side of Eq.(8) contributes nothing to the right-hand side of Eq.(4) if we neglect the exchange and correlation between the particles k and l. This result corroborates our prescription for obtaining sum rules for SP strength functions. In the case of Gamow-Teller transition, the most important part of $[H,\Omega]$ is the two-body part related to nuclear forces. We calculate the expectation value of the two-body part of $\Omega^\dagger[H,\Omega]$ as a two-body cluster integral including exchange and correlation, since the two particles are explicitly connected by the nuclear force. On the other hand, the expectation value of the three-body part of $\Omega^\dagger[H,\Omega]$ vanishes, because the third particle (in $\Omega^\dagger$) is not connected with the other two through nuclear forces and then exchanges and correlations between the first two and the third are neglected. As for the right-hand side of Eq.(6), we calculate the expectation value of the two-body and three-body parts of $[\Omega^\dagger,H][H,\Omega]$ in a similar way. The expectation value of the four-body part vanishes as long as exchanges and correlations are neglected.

Let us denote the expectation values calculated as above by $(\Psi_0,\Omega^\dagger\Omega\Psi_0)_{NX}$, $(\Psi_0,\Omega^\dagger[H,\Omega]\Psi_0)_{NX}$, and $(\Psi_0,[\Omega^\dagger,H][H,\Omega]\Psi_0)_{NX}$. Then, we can put

$$\int_{\varepsilon_{min}}^{\varepsilon_{max}} \frac{dn_1}{d\varepsilon} d\varepsilon \int_{-\infty}^{\infty} D_\Omega(E,\varepsilon)dE = (\Psi_0,\Omega^\dagger\Omega\Psi_0)_{NX} \qquad (9)$$

$$\int_{\varepsilon_{min}}^{\varepsilon_{max}} \frac{dn_1}{d\varepsilon} d\varepsilon \int_{-\infty}^{\infty} E D_\Omega(E,\varepsilon)dE = (\Psi_0,\Omega^\dagger[H,\Omega]\Psi_0)_{NX} \qquad (10)$$

$$\int_{\varepsilon_{min}}^{\varepsilon_{max}} \frac{dn_1}{d\varepsilon} d\varepsilon \int_{-\infty}^{\infty} E^2 D_\Omega(E,\varepsilon)dE = (\Psi_0,[\Omega^\dagger,H][H,\Omega]\Psi_0)_{NX} \qquad (11)$$

It is possible to arrange the forms of $(\Psi_0, \Omega^\dagger \Omega \Psi_0)_{NX}$, $(\Psi_0, \Omega^\dagger [H,\Omega]\Psi_0)_{NX}$, and $(\Psi_0, [\Omega^\dagger, H][H,\Omega]\Psi_0)_{NX}$ as sums over SP states. Therefore, from Eqs.(9)~(11), we can obtain $\int_{-\infty}^{\infty} D_\Omega(E,\varepsilon)dE$, $\int_{-\infty}^{\infty} E D_\Omega(E,\varepsilon)dE$, and $\int_{-\infty}^{\infty} E^2 D_\Omega(E,\varepsilon)dE$.

Numerical calculations of the right-hand sides of Eqs.(10) and (11) are rather laborious. So far, we have calculated only for the Gamow-Teller transition with two kinds of central potentials, OMY potential[7] with the hard-core radius $d=0.6$ fm and a large-core ($d=0.9$ fm) modified OMY potential[8] (refered to as YOY). Even for these relatively simple potentials, we have not yet obtained a reliable result for the three-body part of Eq.(11), although we can make a plausible argument that the three-body part will be considerably smaller than the two-body part. We have calculated for symmetric nuclear matter with the two-body correlation functions for even-parity ($f^+(r)$) and odd-parity ($f^-(r)$) states given by

$$f^\pm(r) = \begin{cases} 0 & (r \leq d) \\ 1 + [b_1^\pm(r-d)/d - 1]\exp[-b_2^\pm(r-d)/d] & (r > d) \end{cases} \quad (12)$$

where the parameters $b_1^\pm$ and $b_2^\pm$ are taken to be $b_1^+=1.39$, $b_2^+=2.04$, $b_1^-=0.59$, $b_2^-=1.64$, for OMY at r_0 (radius of volume per nucleon)$=1.1$ fm (unsaturated), and $b_1^+=3.72$, $b_2^+=3.74$, $b_1^-=1.20$, $b_2^-=2.65$, for YOY at $r_0=1.1$ fm (saturated). The ε dependence of the sum rules thus obtained is relatively small, and we can express, by neglecting the three-body part for the second moment, as

$$\int_{-\infty}^{\infty} D_{GT}(E,\varepsilon)dE = 3 \quad (13)$$

$$\frac{1}{3}\int_{-\infty}^{\infty}(E-\Delta_C)D_{GT}(E,\varepsilon)dE \equiv <E-\Delta_C>_D \approx \begin{cases} 29 \text{ MeV} & (OMY) \\ 67 \text{ MeV} & (YOY) \end{cases} \quad (14)$$

$$\frac{1}{3}\int_{-\infty}^{\infty}(E-\Delta_C)^2 D_{GT}(E,\varepsilon)dE \equiv <(E-\Delta_C)^2>_D \approx \begin{cases} (80 \text{ MeV})^2 & (OMY) \\ (407 \text{ MeV})^2 & (YOY) \end{cases} \quad (15)$$

where Δ_C is the Coulomb displacement energy, and the subtraction of it from E corresponds to our calculating the right-hand sides of Eqs. (14) and (15) by neglecting Coulomb forces.

Although we cannot say definitely from the above results, we think the results for OMY potential are not far from the truth (YOY potential is a very singular potential). Anyway, the largeness of the numbers appearing in the right-hand sides of Eqs.(14) and (15) should be noted; they are much larger than the usual spin-orbit splitting (several MeV).

4.3 Single-particle strength function and single-particle level density

From the description in Sec.4.1 and Sec.4.2, the SP strength functions seem to have long tails. If we assume arbitrarily long tails, we would have SP strengths even for "transitions" to the energy region below all the SP levels. In this subsection we consider a general condition on the forms of SP strength functions which forbids such unreasonable "transitions" in a natural way.

The SP strength function may be expressed as

$$D_\Omega(E,\varepsilon) = |m(E, \varepsilon + \frac{E}{2})|^2 \rho_2(E+\varepsilon) \quad (16)$$

where $\rho_2(E+\varepsilon)$ is the final SP level density, and $|m(E,\varepsilon+\frac{E}{2})|^2$ is the matrix element between the SP states at ε and $\varepsilon + E$; the second argument $\varepsilon + \frac{E}{2}$ is chosen so that the same form may be used for the reverse transition with E replaced by $-E$. Now we assume a factorized form for $|m(E,\varepsilon+\frac{E}{2})|^2$ as

$$|m(E,\varepsilon+\frac{E}{2})|^2 = C(\varepsilon)F(E)G(\varepsilon+\frac{E}{2}) \tag{17}$$

The first factor $C(\varepsilon)$ is the normalization factor, and its ε dependence is weak, so that the main behavior is due to F and G. The factor $F(E)$ is supposed to represent the most pronounced shape of the SP strength functions, and the factor G counteracts the average SP level density. We assume the form of $F(E)$ by taking account of the sum rules, but determine $G(\varepsilon+\frac{E}{2})$ from other considerations.

In the gross theory of TYK, the SP strength functions have no ε dependence. We preserve this property as far as possible. However, as it is impossible to do so exactly for arbitrary values of E, we choose $E=\Delta_C$ and make $D_\Omega(\Delta_C,\varepsilon)$ approximately independent of ε by neglecting $C(\varepsilon)$. Namely, we assume that $G(\varepsilon+\frac{\Delta_C}{2})\rho_2(\varepsilon+\Delta_C)$ is independent of ε. Thus, we get

$$G(\varepsilon+\frac{\Delta_C}{2}) = \frac{\text{const.}}{\rho_2(\varepsilon+\Delta_C)} \equiv \frac{\text{const.}}{\rho_{12}(\varepsilon+\frac{\Delta_C}{2})} \tag{18}$$

where the function ρ_{12} is defined by the second equality. This equation holds for any value of ε, so that we get

$$G(\varepsilon+\frac{E}{2}) = \frac{\text{const.}}{\rho_{12}(\varepsilon+\frac{E}{2})} \tag{19}$$

The constant numerator of Eq.(19) can be taken as unity, and we have

$$D_\Omega(E,\varepsilon) = C(\varepsilon)F(E)\frac{\rho_2(\varepsilon+E)}{\rho_{12}(\varepsilon+\frac{E}{2})} \tag{20}$$

Finally, $C(\varepsilon)$ is determined from the sum rule.

4.4 Practical forms for single-particle strength functions

In the gross theory, the SP strength function $D_\Omega(E,\varepsilon)$ which includes the pairing effects is derived from a smooth function $D_\Omega^0(E,\varepsilon)$. For the relation between $D_\Omega(E,\varepsilon)$ and $D_\Omega^0(E,\varepsilon)$, see Ref.3. Since the difference between $D_\Omega(E,\varepsilon)$ and $D_\Omega^0(E,\varepsilon)$ is small, we assume that Eq.(20) holds with $D_\Omega(E,\varepsilon)$ replaced by $D_\Omega^0(E,\varepsilon)$. In the following, we discuss the forms of $D_\Omega^0(E,\varepsilon)$.

As a basic functional form for $F(E)$ in Eq.(20), we take a hyperbolic secant type as

$$F(E) \propto \text{sech}[\pi(E-\Delta)/2\sigma] \tag{21}$$

It should be noted that σ in this equation is the root-mean-square of $E-\Delta$, i.e. $(<(E-\Delta)^2>)^{1/2}$, for this distribution, and the half-value width is approximately equal to 1.68σ.

For the Fermi transition, $\Delta=\Delta_C$, and σ is a small quantity.

For the Gamow-Teller transition, we assume a superposition of two hyperbolic secant functions, one having a width of several MeV and the other having a much larger width. The mixing ratio of these two is, after corrected for the SP level density and normalization as in Eq.(20), taken to be 6:4 considering the second property of the Gamow-Teller giant resonance mentioned in Sec.4.1. We assume two kinds of the narrower functions depending on the decaying nucleon. For the group of nucleons [1] shown in Fig.3, we assume $\Delta=\Delta_C$, while for the group [2] we assume $\Delta=\Delta_C+\Delta_\tau$. This division is necessary for explaining the first property of the Gamow-Teller giant resonance mentioned in Sec.4.1 and the relatively small β^+-strengths in some light nuclei. The wider hyperbolic secant function, for which we use $\Delta=\Delta_C$, accounts for the rather large first and second moments mentioned in Sec.4.2.

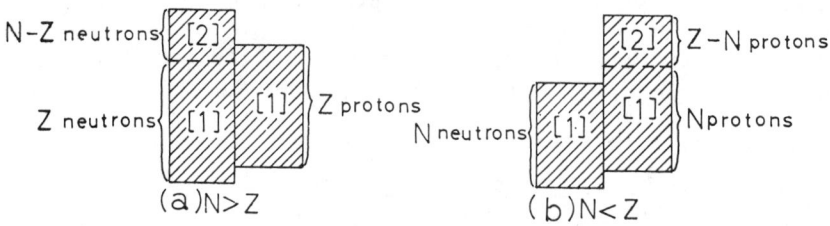

Fig.3. Division of nucleons into [1] a group having partners, and [2] a group having no partners.

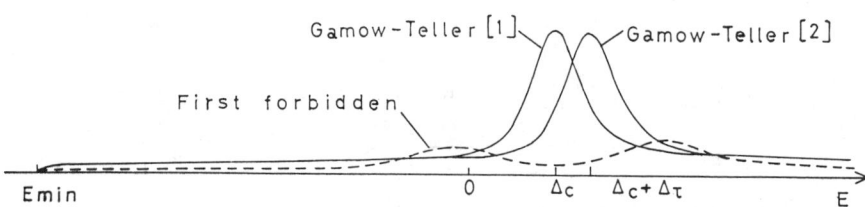

Fig.4. Schematic illustration of the Gamow-Teller and first forbidden SP strength functions. For the Gamow-Teller transition, two cases corresponding to the two groups of nucleons defined in Fig.3 are shown, but the curve for the first forbidden transition is applicable only to the transitions caused by vertor interaction and the transitions of the group [1] nucleons caused by axial-vector interaction. $E=E_{min}$ corresponds to the transition to the bottom of SP levels.

For the first forbidden transitions caused by vector interaction, we assume a superposition of two functions having peaks at
$$E=\Delta_C\pm\Delta_\gamma \qquad (22)$$
The mixing ratio is taken to be 1:1 at the stage of $F(E)$, but, as seen from Eq.(20), the effect of the SP level density makes the higher energy peak larger than the lower energy one in the stage of $D_\Omega^0(E,\varepsilon)$. For the first forbidden transitions caused by axial-vector interaction, a similar superposition of two functions is used with an addition of Δ_τ for the nucleons in group [2].

Finally, the SP level density necessary for relating $D_\Omega^0(E,\varepsilon)$ with $F(E)$ is assumed to be that of the Fermi gas. Namely, We use
$$\rho_2(\varepsilon_k)=\frac{2}{(2\pi\hbar)^3}4\pi V[2M^{*3}\varepsilon_k]^{1/2} \qquad (23)$$

where ε_k is the nucleon kinetic energy, V is the nuclear volume, and M^* is the effective nucleon mass.

In Fig.4, the shape of $D_\Omega^0(E,\varepsilon)$ is schematically illustrated for various transitions.

5. NUMERICAL CALCULATIONS, RESULTS AND DISCUSSION

In order to complete our formalism we must specify several quantities. The nuclear radius divided by $A^{1/3}$ is assumed to be

$$r_0 = 1.25 \times (1 + 0.65 A^{-2/3}) \text{ fm} \tag{24}$$

The effective nucleon mass M^* necessary for determining the SP level density is taken as

$$\frac{M^*}{M} = 0.6 + 0.4 A^{-1/3} \tag{25}$$

where M is the nucleon mass. The Coulomb displacement energy is taken from Eq.(21) of TYK. The quantity Δ_γ introduced in Eq.(22) is taken as[9]

$$\Delta_\gamma = 31.2 A^{-1/3} + 20.6 A^{-1/6} \text{ MeV} \tag{26}$$

Values of the width parameter σ (see Eq.(21)) for various transitions are taken as follows. For the Fermi transition, we use σ_C given by Eq.(22) of TYK. For the Gamow-Teller transition of a nucleon in group [1] of Fig.3, we use a superposition of two functions as stated in Sec.4.4. The values of σ for these two functions are

$$\sigma_{GT}^{narrow} = [(4 \text{ MeV})^2 + \sigma_C^2]^{1/2} \tag{27}$$

$$\sigma_{GT}^{wide} = [(135 \text{ MeV})^2 + \sigma_C^2]^{1/2} \tag{28}$$

The function $D_{GT}^0(E,\varepsilon)$ with these width parameters gives $<E-\Delta_C>_D \approx 31$ MeV for group [1] of Fig.3, $<E-\Delta_C>_D \approx (31 \text{ MeV}) + 0.6\Delta_\tau$ for group [2] of Fig.3, and $<(E-\Delta_C)^2>_D \approx (80 \text{ MeV})^2$ (see Eqs.(14) and (15)) if we replace $D_{GT}(E,\varepsilon)$ by $D_{GT}^0(E,\varepsilon)$. For the first forbidden transition caused by vector interaction, we use, for both peaks above and below IAS,

$$\sigma_{1V} = [\sigma_\gamma^2 + \sigma_C^2]^{1/2} \tag{29}$$

with $\sigma_\gamma = 3$ MeV. In the brackets on the right-hand side of Eq.(29), the first term predominates, and it gives a half-value width of about 5 MeV, which is roughly equal to the average width of E1 giant resonances.[9] Finally, for the first forbidden transitions caused by axial-vector interaction, we use

$$\sigma_{1A}^{narrow} = [\sigma_\gamma^2 + (\sigma_{GT}^{narrow})^2]^{1/2} \tag{30}$$

$$\sigma_{1A}^{wide} = [\sigma_\gamma^2 + (\sigma_{GT}^{wide})^2]^{1/2} \tag{31}$$

The function $F(E)$ (see Eq.(20)) in this case is a superposition of four hyperbolic secant functions, each of which has a center situated either above or below IAS and has a width of either Eq.(30) or Eq.(31). The mixing ratio for the former two choices is 1:1 at the stage of $F(E)$, and the mixing ratio for the latter two choices is 6:4 at the stage of $D_\Omega^0(E,\varepsilon)$ as stated in Sec.4.4.

In order to make a general survey of half-lives we approximate $D_\Omega^0(E,\varepsilon)$ to $D_\Omega^0(E,\varepsilon_{max})$. We use experimental Q-values as far as possible, but, for the nuclides for which no experimental Q-values are available, we use those obtained from the mass formula of Tachibana et al.[10] In Fig.5, we show a general tendency of agreement between calculated and experimental[11] β-decay half-lives. The β-decay half-lives depend sensitively on the orders of forbiddenness of the transitions to low-lying states. The calculated half-lives used in this figure are those corresponding to the "bottom raising" ΔQ_0 as follows:

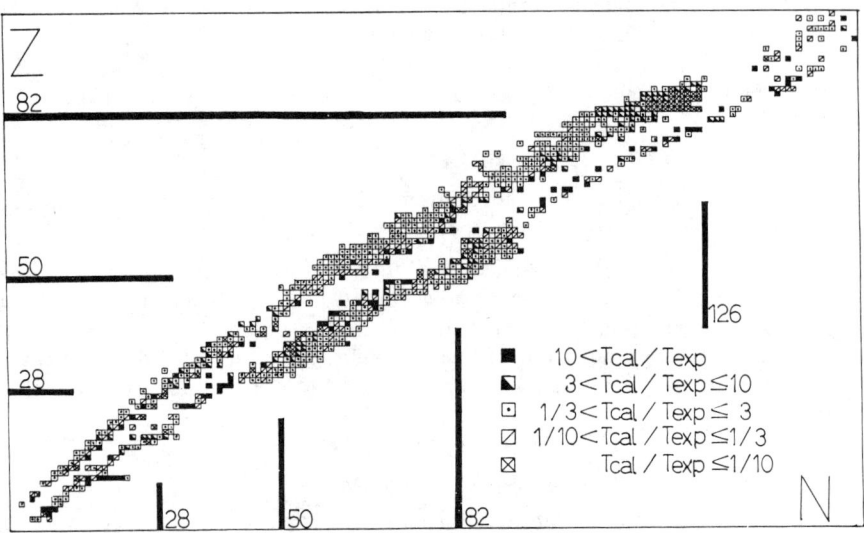

Fig.5. Comparison between calculated and experimental β-decay half-lives.

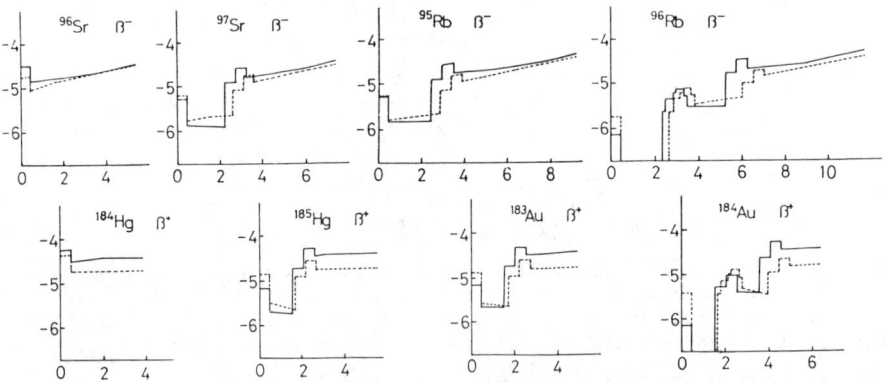

Fig.6. Total β-strength function[12] $S_\beta(E_{exc})$ from the old gross theory (dashed line) and the new gross theory (solid line). The δ-function is drawn with a width of 0.5 MeV for the ground state and 1 MeV for the other states. The abscissa is E_{exc} in MeV, and the ordinate is $\log S_\beta(E_{exc})$.

$$\Delta Q_0 = \begin{cases} 0.25 \text{ MeV} & \text{for even-even parent} \\ 1 \text{ MeV} & \text{for odd-}A \text{ parent} \\ 1.75 \text{ MeV} & \text{for odd-odd parent} \end{cases} \quad (32)$$

Namely, we assume that the transitions to the levels with excitation energies less than ΔQ_0 are highly forbidden, and pile up at $E=-Q+\Delta Q_0$ the β-strength which is originally distributed between $E=-Q$ and $E=-Q+\Delta Q_0$. The β-decays with small Q-values ($QA^{1/3}<10$ MeV) are excluded from Fig.5 as they are not suitable to be treated by the gross theory. From Fig.5, we see that the agreement between theory and experiment is fairly good and there is no large-scale systematic deviation.

Difference of the present results from those of the old gross theory should also be mentioned. As a representative of the old theory we take one in TYK with the modified-Lorentz-type SP strength functions, with the Q-values and "bottom raising" being the same as stated above. As compared with the half-lives calculated by the old theory, the new results are shorter in $β^-$-decays of very neutron-rich nuclei by a factor of two to three. On the other hand, the new results for relatively light nuclei ($A \leq 60$) are, except for very neutron-rich nuclei, considerably longer than the results of the old theory.

In Fig.6 we show, for several nuclides, the total β-strength functions[12] $S_β(E_{exc})$, which are the sums of allowed-strength equivalents of all transitions. ($E_{exc}=E+Q$). No "bottom raising" is made in this case.

More details of the results of numerical calculations will be sent upon request.

ACKNOWLEDGMENTS

The authors thanks Messrs. T.Yoshikoshi and K.Nakata for their help in calculations.

REFERENCES

1. K.Takahashi, M.Yamada and T.Kondoh, Atomic Data and Nuclear Data Tables **12**, 101 (1973).
2. T.Kondoh and M.Yamada, Prog. Theor. Phys. Supplement No.60, 136 (1976); M.Yamada, T.Kondoh and K.Yokoi, "3rd International Conference on Nuclei Far from Stability", CERN 76-13, 331 (1976).
3. T.Kondoh, T.Tachibana and M.Yamada, Prog. Theor. Phys. **74**, 708 (1985).
4. K.Ikeda, S.Fujii and J.I.Fujita, Phys. Lett. **3**, 271 (1963).
5. D.J.Horen et al., Phys. Lett. **95B**, 27 (1980); D.J.Horen et al., Phys. Lett. **99B**, 383 (1981); C.Gaarde et al., Nucl. Phys. **A369**, 258 (1981).
6. F.Iwamoto and M.Yamada, Prog. Theor. Phys. **17**, 543 (1957); **18**, 345 (1957); E.Feenberg, "Theory of Quantum Fluids", (Academic Press, New York and London, 1969); J.W.Clark, Prog. Part. Nucl. Phys. **2**, 89 (1979).
7. T.Kikuta (Ohmura), M.Morita and M.Yamada, Prog. Thor. Phys. **15**, 222 (1956).
8. T.Yoshikoshi, K.Oyamatsu and M.Yamada, Report of Science and Engineering Research Laboratory, Waseda University, No.87-4 (1987).
9. B.L.Berman and S.C.Fultz, Rev. Mod. Phys. **47**, 713 (1975).
10. T.Tachibana, M.Uno, M.Yamada and S.Yamada, to be published in Atomic Data and Nuclear Data Tables.
11. C.M.Lederer and V.S.Shirley, "Table of Isotopes, 7th ed." (J.Wiley & Sons, New York, 1978).
12. P.G.Hansen, "Advances in Nuclear Physics" Vol.7 (Plenum Press, New York-London, 1973), p.159.

GAMOW-TELLER BETA DECAY OF $^{29-31}$Na
COMPARISON WITH SHELL-MODEL ESTIMATES

P. Baumann, Ph. Dessagne, A. Huck, G. Klotz, A. Knipper
Ch. Miehé, M. Ramdane and G. Walter
Centre de Recherches Nucléaires, 67037 Strasbourg Cedex France
G. Marguier, J. Giroux
Institut de Physique Nucléaire, 69622 Villeurbanne France
C. Richard-Serre
CERN 1211 Geneva 23 Switzerland

ABSTRACT

Gamma rays and delayed-neutron processes subsequent to beta decay of the neutron-rich $^{29-31}$Na isotopes are studied in singles and coincidence mode with mass-separated sources at ISOLDE. Improved level schemes are presented for mass 29 and 30, whose features substantially agree with shell-model predictions. The deduced Gamow-Teller strengths are discussed as well as the occurrence of negative parity intruder states in the daughter nuclei.

o o o o

From the previous studies on light neutron-rich alkali isotopes several results are found in sharp contradiction to the shell-model predictions. An explanation is outlined from the inversion of the lowest fp orbits with the highest sd orbits. A goal of the present study was to delimit the mass region where the nuclear properties can be understood in terms of sd shell systematics and to locate intruder states in the beta decays. Moreover, the large Q_β values allow to test the GT operator for nuclei with neutron excesses while previous studies of the GT quenching have been carried out for mirror pairs or for GT$^+$ decay. Finally, a more comprehensive description of the ^{30}Mg level structure is a requisite to a study of the 2n disintegration mode in ^{30}Na.

Our present investigation on $^{29-31}$Na decays includes γ-ray and neutron spectra with different types of detectors. From single and coincidence spectra a systematic analysis of the B(GT) values has been made and compared with theoretical estimates in the sd shell-model space.

I. EXPERIMENTAL TECHNIQUES

The n-rich Na isotopes were produced by bombarding a uranium carbide target with a 2 µA proton beam of the 600 MeV CERN SC. The Na atoms were ionized on a tungsten surface. After mass selection in the ISOLDE separator, the isotopes are collected on a mylar tape. The activity of the short-lived Na isotopes (^{29}Na : 44.9 ms, ^{30}Na : 50 ms, ^{31}Na : 17 ms)[1] is measured at the collection point. By driving the tape, the build up of background activity from descendants (Mg,Al) and contaminants (A=116-124,4^+) is reduced.

A schematic view of the detection device is shown in Fig. 1. The ion beam is collected on the tape inside a cylindrical beta detector which gives the start of the time of flight measurement. Different types of neutron detectors were used. For efficient detection in n-γ coincidences, a neutron filter has been used. It consists of hexagonal cells (active volume : 3.75 l/cell) filled with NE 213 scintillator. For neutron spectroscopy by time of flight, a large area scintillator sheet (2880 cm^2) bent in a radius of curvature of 100 cm, was used with a time resolution of 1.1 ns (for details, see Ref.2). In order to achieve a better efficiency at low neutron energy (E_n < 1 MeV) a third detector system, similar to a device developped at Oak Ridge[3] was also used with 2.0 ns time resolution. In this system a thin slab of NE 110 scintillator is viewed by 3 phototubes, biased below the single photoelectron level and used in a majority-of-two coincidence mode. Typical spectra obtained with the two types of spectrometers are displayed in Fig.2. The high energy part of the delayed neutron spectrum of ^{29}Na appears in Fig.2b, while the low threshold of the 3 PM device makes possible to use the tof technique below 1 MeV (Fig.2a).

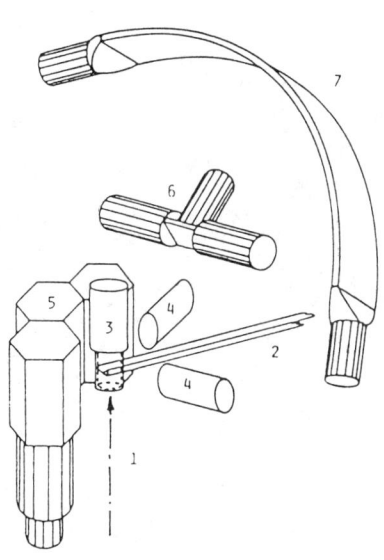

Fig.1 Experimental set-up :
1) ion beam 2) tape 3) 4πβ counter 4) γ counters 5) NE213 cells 6) 3 PM device 7) curved scintillator.

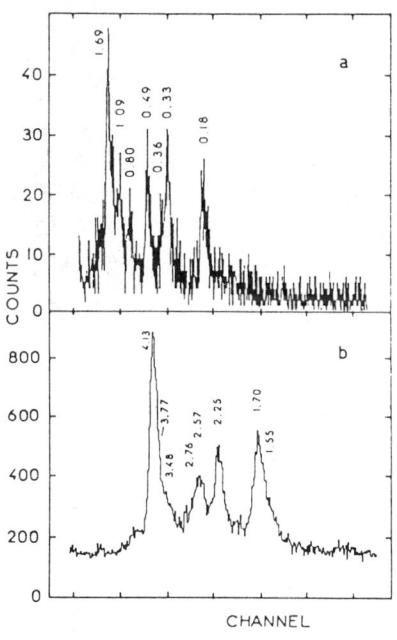

Fig.2 ^{29}Na delayed neutron spectra :
a) 3 PM device
b) curved scintillator.

II. RESULTS AND DISCUSSION

A. The ^{29}Na decay

The measurements of β-delayed γ rays and neutrons leading to a level scheme for ^{29}Mg have been recently published[4] and will therefore not be discussed. The resulting ^{29}Na (β⁻) ^{29}Mg decay scheme is presented in Fig.3.

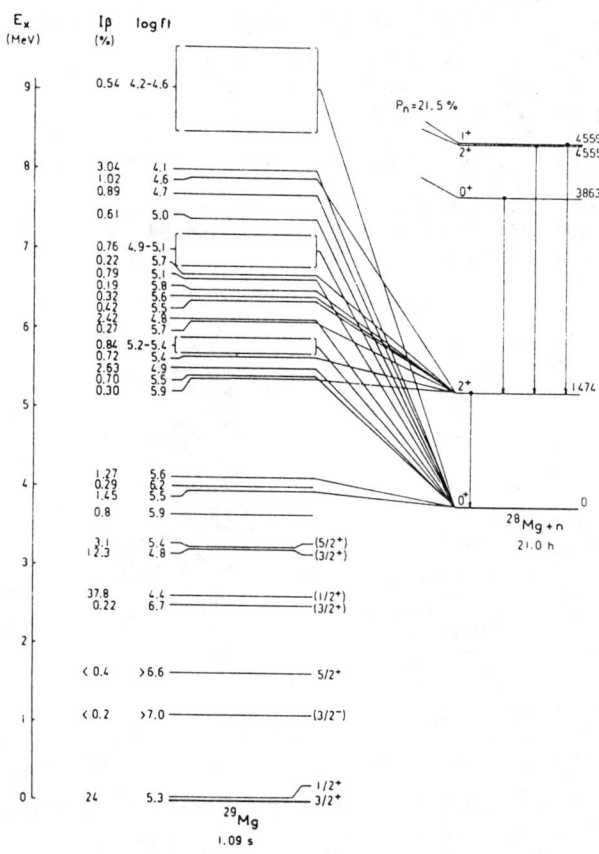

Fig.3 ^{29}Na β decay scheme.

In our experiment, the first excited state was located at 54.6 keV. From the lifetime value, (τ = 1.83 ns) deduced from γ-γ coincidences with BaF$_2$ counters, the 55 keV level has been identified as a 1/2⁺ level, decaying to the 3/2⁺ G.S. with a B (M1) = 0.11 ± 0.01 W.u. strength.

For the levels at E_x = 1638, 2500, 2615, 3224 and 3227 keV, populated through allowed beta transitions, spin-parity assignments which are proposed (Fig.3) result from the comparison of level properties with shell-model predictions (see below and Ref.4).

No beta feeding has been measured for the 1095 keV level which cannot be associated with a model state. A $3/2^-$ assignment is in accord with our results and with the analysis of Ref.5.

To establish the decay scheme to particle-unbound states in ^{29}Mg, we make use of the results of the time of flight experiment and, for the low energy part, of the data obtained by Ziegert et al.[6] with ^{3}He spectrometers. From the comparison of the two spectra, normalized to the 1.7 MeV peak, it appears that 51% of the neutron emission is observed for $E_n > 1.7$ MeV using the tof with a large scintillator whereas 49% correspond to neutrons detected with high resolution techniques[6]. The proposed decay scheme gives a satisfactory balance of the observed n and γ intensities.

B. The ^{30}Na decay

For ^{30}Na, 1n, 2n and α-delayed emissions have been reported previously with $P_{1n} = 30 \pm 4$, $P_{2n} = 1.15 \pm 0.25$ (Ref.1) and $P_\alpha = (5.5 \pm 2) 10^{-5}$

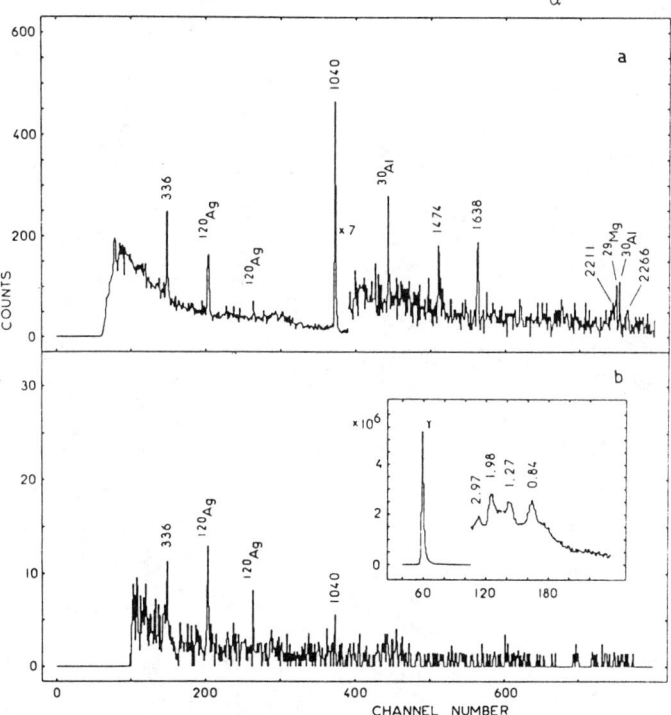

Fig.4 A = 30 γ spectra in coincidence with neutrons (see text)

(Ref.7) for 100 β decays. In order to identify unambiguously the γ-ray emitters, the n-γ coincidences have been registered in a biparametric mode (E_γ versus discrimination signal) making possible to check that a γ line, observed in coincidence with neutrons, does not correspond to accidental γ-γ coincidences resulting from insufficient rejection in the n-γ discrimination. The energy and intensity of 32 γ rays in ^{30}Na decay have been measured and the corresponding transitions belong to the level scheme of ^{30}Mg, ^{29}Mg or ^{28}Mg (see Fig.5).
In ^{30}Mg : the resulting decay scheme is in fair agreement with previous results but differences are found for γ branching ratios and therefore for β feeding and logft values.
In ^{29}Mg : from n-γ coincidences, it appears that 5 excited states of ^{29}Mg are populated after β-delayed one-neutron emission.

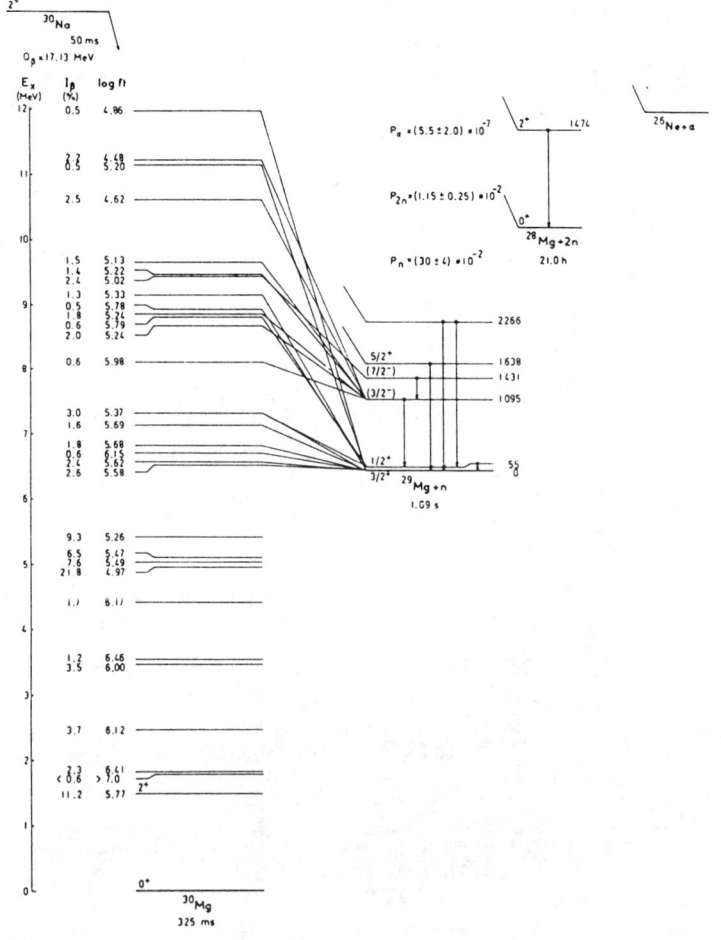

Fig.5 ^{30}Na β decay scheme.

A $J^\pi = 7/2^-$ value is proposed for the 1431 keV level which presents a single γ branch ($E_\gamma = 336$ keV) to the $J^\pi = 3/2^-$ level at 1095 keV. In Fig.4a, the γ spectrum obtained in coincidence with the efficient liquid scintillators is presented and in Fig.4b the γ spectrum obtained in coincidence with the curved spectrometer is given, with a gate on neutrons in the 1.5 MeV region. Both reveal the 336 keV ($7/2^- \to 3/2^-$) and 1040 keV ($3/2^- \to 1/2^+$) lines.

In ^{28}Mg : evidence for population of the 1474 keV ($J^\pi = 2^+$) level in ^{28}Mg is also found by the n-γ coincidences from where we conconclude that $\sim$ 25% of the 2n emission involves the ^{28}Mg 2^+ level.

The different available results have been used to establish the decay scheme (Fig.5) to particle unbound states. The high energy part of the delayed neutron spectrum is interpreted by the decay of levels around 12 MeV in ^{30}Mg, markedly above the 2n separation energy ($S_{2n} = 10.23$ MeV).

C. The ^{31}Na decay

In case of A = 31, the contribution of the contaminant activity (A = 124, 4^+) made the analysis of the β strength particularly difficult and present results do not offer useful information for a quantitative comparison with shell-model estimates. As in the previous cases, the time of flight spectrum reveals the high energy part of the delayed neutron emission.

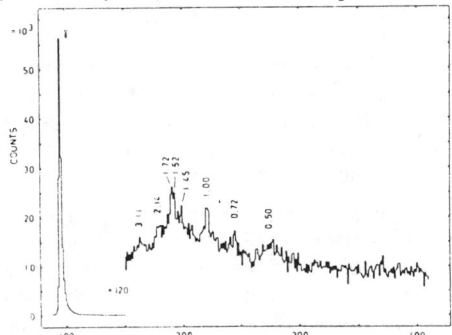

Fig.6 ^{31}Na β-delayed neutron spectrum.

III. COMPARISON WITH SHELL-MODEL PREDICTIONS

For A = 29, the observed properties of the bound levels of ^{29}Mg have been compared with predictions of USD calculations by Wildenthal[4]. The near-degeneracy of the first $1/2^+$ and $3/2^+$ levels, predicted by the calculation, has been observed ($\Delta E = 55$ keV). The measured lifetime of the $1/2^+$ level yields a B(M1) value (0.11 ± 0.01 W.u.) in good agreement with the shell-model predictions based on the free-nucleon or the effective form of the M1 operator. From a comparison of the excitation energies, branching ratios and beta decay strengths, it has been possible to associate 7 low energy levels to $3/2^+$, $1/2^+$ or $5/2^+$ model states. A remarkable agreement is found in this comparison[4].

The experimental values of B(GT), extracted from our measurements for particle bound and unbound states of ^{29}Mg are compared with the shell-model predictions based on the free-nucleon GT operator.

The ratio of the observed to the predicted GT strength is 0.5 for the whole 9.6 MeV range of the measurement and the calculation reproduces the usual quenching factor for low lying beta decay or GT$^+$ decay in the sd shell. The distribution of the strength is accurately reproduced by the calculation in the particle bound region and slightly shifted at lower excitation energy in the particle unbound region.

For A = 30, the new results obtained for the beta decay of ^{30}Na can be compared to the predictions by Wildenthal et al.[8] using complete sd-space shell-model wave functions. In Fig.7, we have reported experimental and theoretical values for the beta branching strength. The comparison is limited to experimental logft values ≤ 5.9. For the other transitions (indicated with dashed lines in Fig.7) we cannot distinguish first forbidden decays, which are not relevant to the theoretical description, from weak GT transitions.

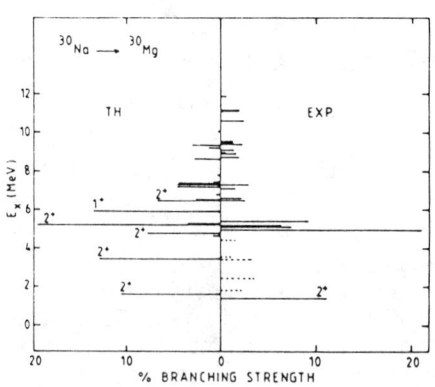

Fig.7 β branching strength for ^{30}Na decay.

The two most intense experimental β transitions (to the 1482 and 4966 keV levels) are correctly reproduced by the calculation but the strong transition predicted to the second 2$^+$ model state does not appear in the experiment. If we compare the experimental and the theoretical B(GT) strength distributions, it appears below E_x = 10 MeV an excess of calculated strength proceeding from the smaller predicted total half-life. On the other hand the upper part of the experimental distribution is not reproduced.

In summary, in the A = 30 decay, in the sd shell-model prediction, the theoretical strength function is overestimated at low excitation energy where several candidates for negative parity states can be found and underestimated above E_x = 10 MeV.

IV. NEGATIVE PARITY STATES

As stated previously, three levels below E_x = 3 MeV in ^{29}Mg cannot be related to sd model states (E_x = 1095, 1431 and 2266 keV). They have no measurable beta feeding but are strongly populated through the 1n channel. A negative parity is proposed for these levels and the measured γ branching ratios strongly support $J^{\pi} = 3/2^{-}$ (E_x = 1.09 MeV), $J^{\pi} = 7/2^{-}$ (E_x = 1.43 MeV) and $J^{\pi} = (1/2, 3/2)^{-}$ (E_x = 2.23 MeV). The 1.43 MeV level was previously observed in the (^{18}O, ^{15}O) study[5] and tentatively assigned to a $(s\,1/2)^2$ $f\,7/2$ configuration. The $3/2^{-}$ state resulting from the $(s\,1/2)^2$ $p\,3/2$ configuration would then correspond to the 1095 keV level. The excitation energy of these states is very low if we consider the systematics of the N = 17 isotones (Fig.8a) and the crossing 7/2 - 3/2 is unexpected.

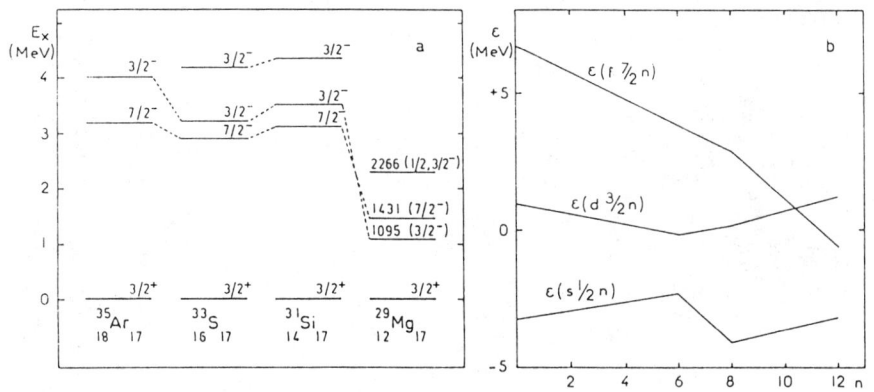

Fig.8 a) π(-) states of N = 17 isotones
       b) Effective energy of s.p. levels (Ref.9).

The lowering of the $7/2^{-}$ orbit by the large neutron excess has been discussed previously by Storm et al.[9] and we have reported in Fig.8b the effective energy of single particle levels as a function of the number n of sd-shell neutrons for the isotopes of oxygen. The cross-shell matrix elements used are those of Kuo and Brown[10]. In this simple evaluation, the $f_{7/2}$ - $d_{3/2}$ crossing occurs for n = 10. From our results in ^{29}Mg, (n = 9), a similar behaviour is suggested for 2 p3/2 single particle energies.

The selectivity of the delayed neutron emission leading to negative parity states is very apparent in the ^{30}Na decay. This selectivity can be explained by comparing the transmission coefficient, T_{ℓ}, calculated with the optical model for different ℓ values. In this mass region, ℓ = 1 waves are strongly favoured and can explain the selective population of $(3/2,7/2)^{-}$ states in ^{29}Mg from $(1,2,3)^{+}$ GT parent states in ^{30}Mg.

Very recently, Povès and Retamosa[11] have demonstrated the necessity to include the 2p3/2 orbit in the theoretical description of the N = 20 nuclei. It appears that at Z = 10,11,12 $(sd)^{-2}(fp)^2$ configurations largely dominate the ground state wave functions.

V. THE TWO-NEUTRON DECAY MODE OF ^{30}Na

To gain information on the 2n decay mode, a direct measurement of the n-n coincidences was performed. Using 6 efficient cells (A,B,C,D,E,F) placed at 33° with respect to the adjacent one, around the collection point (Fig.9), n-n coincidences have been registered and stored separately for each of the 15 different combinations (AB,BC,AC..) corresponding to 5 different relative angles (Θ = 33°,66°,99°,132° and 165°). In the same conditions, we have evaluated the rate of parasitic events by measuring in a separate experiment a ^{29}Na source (1n emitter).

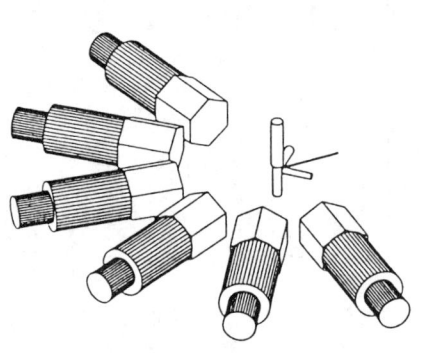

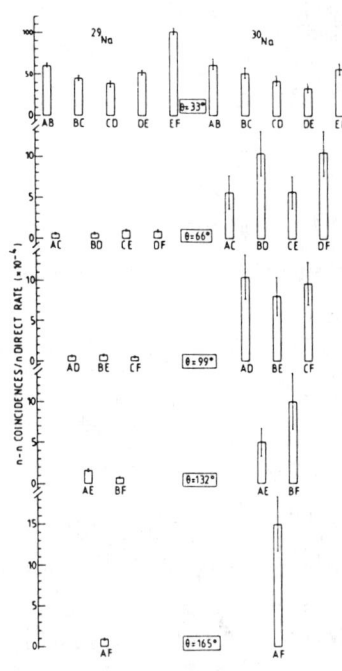

Fig.9 Set-up for n-n coincidence measurements.

Fig.10 n-n coincidence/n direct rate ratio for different counter combinations.

In Fig.10 where we have reported the ratio (coincidence/direct rate) registered with ^{29}Na (left) and ^{30}Na (right) sources.
  - For adjacent counters (Θ = 33°) the scattering from counter to counter gives the main part of the coincidence rate.
  - For non-adjacent counters, the rate is found in good agreement with the known values of P_{2n} and P_{1n}.
  - At Θ = 165°, a maximum of coincidences is observed, the origin of which has to be elucidated.

These results are in agreement with a sequential decay whereas for a dineutron emission, a small angle (Θ < 40°) is predicted between the correlated particles.

In conclusion, this work has led to a complete description of the main GT transitions between positive parity states in ^{29}Na and ^{30}Na decay. Due to the selectivity of the delayed neutron emission, information has been obtained on the level structure of negative parity states whose relevant orbits play a major role near N = 20.

It is a pleasure to acknowledge the cooperation in data taking of H. Gabelmann and K. Schlösser throughout the course of this experiment.

REFERENCES

1. D. Guillemaud-Mueller et al., Nucl. Phys. A426 37 (1984) and references therein.
2. A. Huck et al., Phys. Rev. C31 2226 (1985).
3. N.W. Hill et al., IEEE Trans. Nucl. Sci. NS-32 367 (1985).
4. P. Baumann et al. Phys. Rev. C36 765 (1987).
5. L.K. Fifield et al., Nucl. Phys. A437 141 (1985).
6. W. Ziegert et al., Proc. 4th Int. Conf. on Nuclei far from Stability, Helsingør, 1981, CERN 81-09 327 (1981).
7. C. Detraz at al., Nucl. Phys. A402 301 (1983).
8. B.H. Wildenthal et al., Phys. Rev. C28 1343 (1983).
9. M.H. Storm et al., J. Phys. G : Nucl. Phys. 9 L165 (1983).
10. T.T.S. Kuo and G.E. Brown, Nucl. Phys. 85 40 (1966).
11. A. Povès and J. Retamosa, Phys. Lett. 184B 311 (1987).

THE BETA STRENGTH IN THE PROTON-RICH ARGON ISOTOPES

K. Riisager
CERN, Geneva, Switzerland

M.J.G. Borge
Instituto 'Rocasolano', CSIC, Madrid, Spain

P.G. Hansen
Institute of Physics, University of Aarhus, Denmark

B. Jonson, S. Mattsson, and G. Nyman
Department of Physics, Chalmers University of Technology, Göteborg, Sweden

A. Richter[*]
Institut für Kernphysik, Techn. Hochschule, Darmstadt, Fed. Rep. Germany

ABSTRACT

The proton-rich isotopes ^{32}Ar–^{35}Ar represent the first sequence for which the super-allowed Gamow–Teller strength is known out to $Z - N = 4$. The results are compared with two sets of predictions from large shell-model calculations, which systematically give different quenching factors, on the average 0.54 and 0.89. Possible reasons for this are discussed.

The measured axial-vector (or Gamow–Teller) strength in nuclei is smaller than that calculated from models in which the nucleons are assumed to be point-like particles with the free-nucleon value g_A of the coupling constant. This reduction arises largely from pionic contributions and is thus one example of subnucleonic effects in nuclear physics. Very approximately one may say that part of the effect arises from virtual hole–delta pairs and part from second-order core polarization brought about by the tensor force, itself due primarily to one-pion exchange. We refer to a recent volume [1] and a forthcoming book [2] for a discussion of these phenomena and for references to the theoretical literature. We concentrate here on the Gamow–Teller strengths in the proton-rich argon isotopes, and point out that a discrepancy, previously noted for ^{34}Ar, is part of a systematic behaviour extending to ^{33}Ar and ^{32}Ar.

As a first step we attempt to relate the strength, via nuclear-model calculations, to an overall reduction factor $(g_A'/g_A)^2$, which may be thought of as accounting empirically for changes in the coupling and for corrections to the matrix element of $\sigma \cdot \tau$. The most accurate determinations of $(g_A'/g_A)^2$ have come from the analysis of magnetic moments and beta transition rates [1]. This approach was pioneered by Wilkinson [3], who used data from fast beta transitions together with wave functions based on complete diagonalization within the p-shell and lower sd-shell to calculate a reduction factor of 0.80 ± 0.06. Several analyses covering the whole sd-shell have since been performed, the most recent by Brown and Wildenthal [4–7], who, using a new set of wave functions [5], carried out a comprehensive analysis of beta decays to (mainly) low-lying final states, and found an average value

---

[*] Supported by the Bundesministerium für Forschung und Technologie, Fed. Rep. Germany.

$(g'_A/g_A)^2 = 0.58 \pm 0.04$. The difference between the results for the p-shell and for the sd-shell seems to indicate that the reduction factor decreases with increasing mass. The first results obtained [8] for proton-rich nuclei in the fp-shell indicate that the factor there also is about 0.6.

Experiments to determine the reduction factor in beta decay are made difficult by the existence of a collective spin–isospin mode, the Gamow–Teller giant resonance (GTGR). In the decay of nuclei with $N > Z$ this resonance is not energetically accessible, and the rates of the allowed beta transitions reflect mainly the screening of the interaction due to the GTGR. It is for this reason that in the following the analysis is limited to cases with $Z > N$.

It was found some years ago (see Ref. [9]) that the GTGR can be excited through (p, n) reactions at high energy on neutron-rich targets, and that there is a close, but maybe not one-to-one [10], connection between these cross-sections and the reduced transition probability B_{GT} measured in beta decay. The large energy window in this reaction permits a comparison of the measured strength directly with the sum-rule limit[*]

$$\Sigma B^-_{GT} - \Sigma B^+_{GT} = 3(N - Z),$$

where, for cases with a large neutron excess, the Pauli principle ensures that the β^+ strength is negligible. In the (p,n) experiments, one observes [9] in the resonance and lower-lying GT states a total of 50–65% of the sum-rule strength, in good agreement with the analyses of beta decays in the sd-shell. The uncertainty comes mainly from possible strength that is hidden in the high-energy tail of the resonance [9].

We have studied the super-allowed decays of ^{32}Ar [14] and ^{33}Ar [15], in which a large part of the GTGR is populated; the nuclear-model calculations can therefore be tested here in great detail. We hope that such experimental constraints can help in identifying the causes of the quenching more unambiguously (as already attempted in Ref. [7]). The decays of ^{32}Ar (99 ms) and ^{33}Ar (174 ms) populate excited levels up to 10 MeV in the daughter isotopes, so that detailed measurements are needed, including proton spectra, gamma spectra, and pγ and pβ coincidences. The element argon is an especially favourable case owing to the development at CERN's ISOLDE facility of a fast target–ion-source system [16]. It was, in fact, with a forerunner of this system that ^{32}Ar was discovered [17] as the first isotope with a proton excess of $Z - N = 4$. The details of the experiments are reported elsewhere [14, 15], but it may be useful to repeat that the measurement of a strength-function differs in some respects from normal spectroscopic studies. First of all, the strong energy dependence of the beta transition probability makes it difficult to detect the weak branches to highly excited levels; secondly, the entire proton intensity should be included, not only the part contained in resolved peaks. The results after conversion to a strength function are shown in Fig. 1 and Table I; the latter contains also data for ^{34}Ar and ^{35}Ar [18, 19].

The experimental results for the argon isotopes were compared with large shell-model calculations [4, 5]. These are based on a complete diagonalization within the space of the sd-shell, and completely define the wave functions once the basic set of interaction matrix elements has been specified. Our earlier analysis of the ^{32}Ar and ^{33}Ar experiments [14, 15] was based on predictions [20] obtained with the Chung–Wildenthal interaction [21]. The appearance of a new, presumably better, set of parameters and of calculations of

---

[*] This sum rule has been given in slightly different forms by Ikeda [11], Sorensen [12], and Gaarde et al. [13]. It can be derived from the commutation relationships of the σ and τ operators under the assumption that the nuclear wave function is fully characterized by space, spin, and isospin coordinates. Hence it is obeyed also in the large shell-model calculations that form the reference basis of the present work.

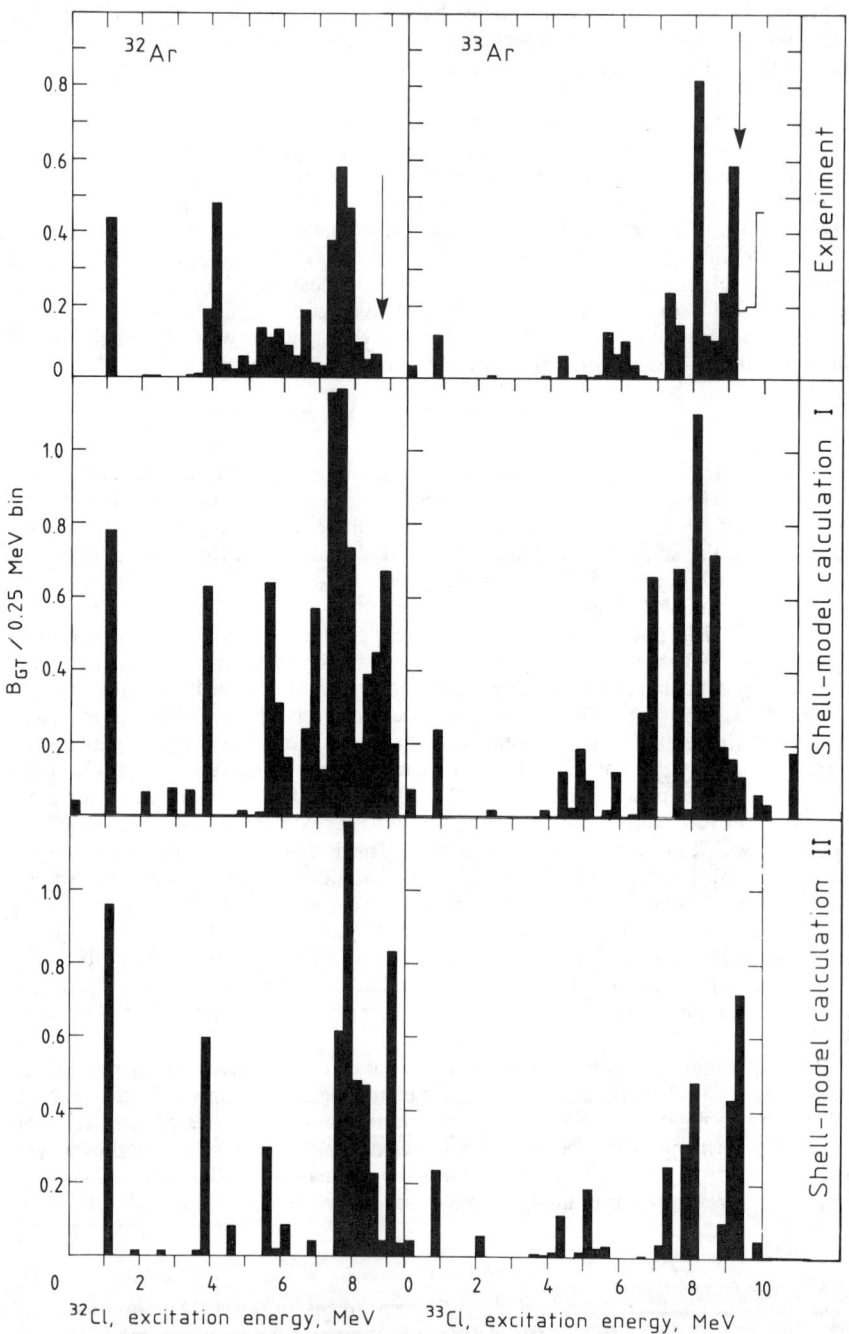

Fig. 1 The upper part of the figure shows the experimental GT strength functions [14, 15] compared with the calculations of Müller et al. [20] (middle) and of Brown and Wildenthal [4, 6] (below). The arrows indicate the energy cut-off used in the analysis.

Table I Total observed and calculated ΣB_{GT}^+
in the allowed EC,β^+ decays of the argon isotopes

Isotope, T_z	Experiment		Theory			
	Energy cut-off (MeV)	ΣB_{GT}^+	Refs. [4, 6][a]		Ref. [20]	
			ΣB_{GT}^+	$(g_A'/g_A)^2$	ΣB_{GT}^+	$(g_A'/g_A)^2$
^{37}Ar, $+1/2$	b)	0.0309	0.0723	(0.43)	0.029	(1.07)
^{35}Ar, $-1/2$	~ 4.0	0.308	0.464	0.66	0.548	0.56
^{34}Ar, -1	~ 4.0	1.73	1.82	0.95	3.30	0.52
^{33}Ar, $-3/2$	9.25	2.90	2.43	1.19	5.04	0.58
^{32}Ar, -2	8.75	3.8	5.09	0.75	7.80	0.49

a) 'Free-nucleon value'.
b) Only one transition observed. We include this case to illustrate the extremely small amount of GT strength in the configuration states as compared with the superallowed decays.

Gamow–Teller strengths [4, 6] offer an interesting possibility for using the argon data to compare the performance of the two sets (Fig. 1 and Table I). The main observation is that the large shell-model calculations have a marvellous predictive power. Figure 1 shows that, apart from an over-all scale factor, both calculations correctly predict strong transitions to states at 1 and 4 MeV in the ^{32}Ar decay and essentially no strength to low energies in the ^{33}Ar decay. The gross structure of the feeding to higher-lying levels is also well reproduced.

The two calculations differ considerably, however, for the intensity in the observed part of the GTGR. The quenching factor $(g_A'/g_A)^2$ obtained for $^{32-35}$Ar (Table I) with the Chung–Wildenthal interaction is on the average 0.54; this agrees with the overall result of 0.58 ± 0.04 obtained with the Brown–Wildenthal interaction [4, 6] for sd-shell isotopes, and also with the analysis based on the sum-rule strength for (p,n) reactions [9]. The new calculations give a completely different picture for the argon isotopes, with reduction factors ranging from 0.66 to 1.19 and with an average of 0.89. We suspect that this discrepancy arises from differences in the calculated energies for the high-lying collective states.

In their paper, Brown and Wildenthal [4] note that the most anomalous point is ^{34}Ar, and that the main strength comes from a relatively weak branch to the fourth excited 1^+ state at 3.129 MeV in ^{34}Cl. They suggest that it might be worth while to remeasure this intensity. However, the ^{34}Ar decay has been studied carefully [18], and there seems little doubt that the intensity, half-life, and Q-value are accurately known. Another explanation is suggested when one notes that the authors of Ref. [4], unlike Müller et al. [20], predict the strength to be split, with about 40% going to a level at 4.8 MeV. It might be revealing to search experimentally for GT strength in the energy range 4–5 MeV. The two calculations predict different energies for the collective states in the light argon isotopes, as may be seen from the integrated strengths: the GTGR sets in at higher energy in the calculations of Refs. [4] and [6] than in that of Ref. [20]. The displacement is roughly 1.3 MeV for ^{33}Ar and 1.5 MeV for ^{32}Ar. The existing data for other elements seem to indicate (Fig. 2) that the problem is specific to the argon isotopes, and is not one that pertains to the end of the sd-shell as a whole.

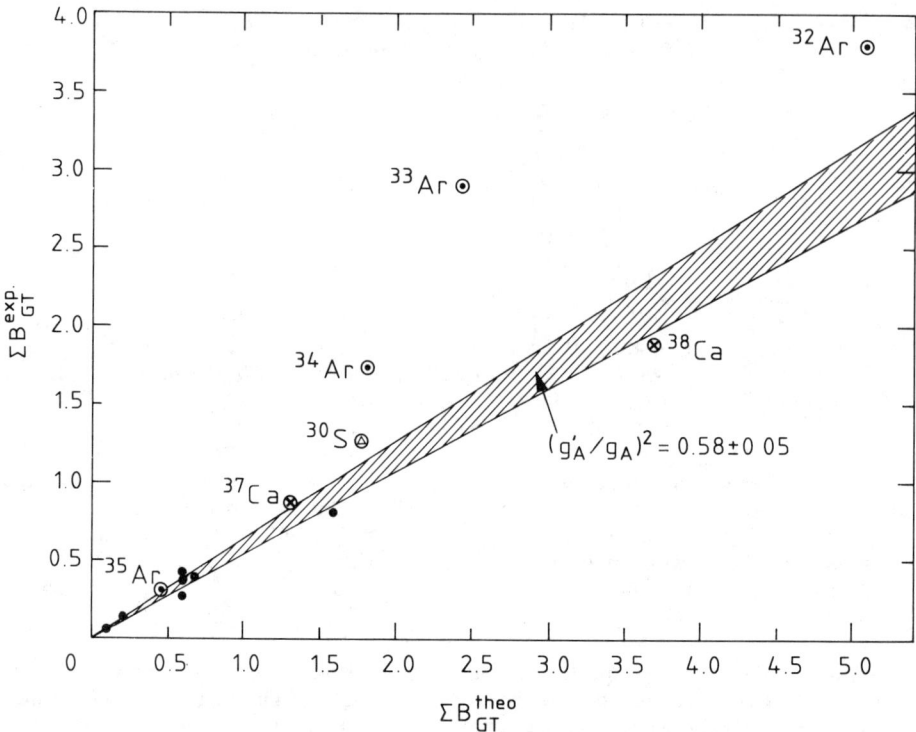

Fig. 2 The observed and calculated [4, 6] GT strengths for the proton-rich isotopes of the elements S, Cl, Ar, K, and Ca. Note that the argon isotopes seem to deviate from the general trend of all other measured cases in the sd-shell as given by the hatched area in the graph.

There is, nevertheless, also a possibility that some of the existing data for proton-rich nuclei could have a systematic bias towards underestimating the strength at high energies for purely experimental reasons. The point is that the measurement of strength functions in some ways differs from normal spectroscopic studies of complex decays. The spectroscopist, concerned with prominent transitions, is often uninterested in the distortion that could come from weak, unobserved branches. For an example of this, we refer to a study of the gamma decay of the fictitious nuclide 145-pandemonium [22].

In summary, we have pointed out a problem that may be endemic to the argon isotopes, but one that could also be a sign of a more widespread malaise in the determination of quenching factors from beta decay. The theoretical problems posed seem to be first to understand the width and energy of the GTGR, secondly to understand quantitatively the causes of the quenching. Experimentally, more, accurate data on superallowed decays with $T_z \leq -3/2$ are needed.

Acknowledgements

The authors are indebted to Drs B.A. Brown, W. Knüpfer, B.C. Metsch and B.H. Wildenthal for discussions of and assistance with this analysis.

REFERENCES

1. D. Wilkinson, ed., Proc. Int. School of Nuclear Physics: Mesons, isobars, quarks, and nuclear excitations, Erice, 1983 [Prog. Part. and Nucl. Phys. **11** (1983)].
2. T.E.O. Ericson and W. Weise, Pions in nuclei, to be published in Physics Monographs (Clarendon Press, Oxford).
3. D.H. Wilkinson, Phys. Rev. **C7**, 930 (1973); Nucl. Phys. **A209**, 470 (1973).
4. B.A. Brown and B.H. Wildenthal, At. Data and Nucl. Data Tables **33**, 347 (1985).
5. B.A. Brown and B.H. Wildenthal, Phys. Rev. **C28**, 2397 (1983).
6. B.A. Brown, personal communication.
7. B.A. Brown and B.H. Wildenthal, Michigan State Univ. preprint MSUCL-603 (1987).
8. T. Sekine et al., Nucl. Phys. **A467**, 93 (1987).
9. C. Gaarde, *in:* Nuclear Structure 1985, R. Broglia et al., eds. (North-Holland Publ. Co., Amsterdam, 1985), p. 449.
10. J.W. Watson et al., Phys. Rev. Lett. **55**, 1369 (1985).
11. K. Ikeda, Prog. Theor. Phys. **31**, 434 (1964).
12. R.A. Sorensen, Ark. Fys. **36**, 657 (1967).
13. C. Gaarde et al., Nucl. Phys. **A334**, 248 (1980).
14. T. Bjørnstad et al., Nucl.Phys **A443**, 283 (1985).
15. M.J.G. Borge et al., Phys. Scr. **36**, 218 (1987).
16. T. Bjørnstad et al., Phys. Scr. **34**, 578 (1986).
17. E. Hagberg et al., Phys. Rev. Lett. **39**, 792 (1977).
18. J.C. Hardy, H. Schmeing, J.S. Geiger and R.L. Graham, Nucl. Phys. **A223**, 157 (1974).
19. H.S. Wilson, R.W. Kavanagh and F.M. Mann, Phys. Rev. **C22**, 1696 (1980). See also E. Adelberger et al., Nucl. Phys. **A417**, 269 (1984).
20. W. Müller, B.C. Metsch, W. Knüpfer and A. Richter, Nucl. Phys **A430**, 61 (1984). W. Müller, thesis, Erlangen University (1984), unpublished.
21. W. Chung, Ph.D. thesis, Michigan State University (1976), unpublished.
22. J.C. Hardy, L.C. Carraz, B. Jonson and P.G. Hansen, Phys. Lett. **71B**, 307 (1977).

GAMOW-TELLER STRENGTH IN THE BETA DECAY OF MIRROR NUCLIDES

J. Honkanen, J. Äystö,* V. Koponen and P. Taskinen*
Department of Physics, University of Jyväskylä
SF-40100 Jyväskylä, Finland

K. Eskola
Department of Physics, University of Helsinki SF-00170 Helsinki, Finland

S. Messelt
Institute of Physics, University of Oslo, Blindern, Oslo 3, Norway

K. Ogawa†
Institute for Nuclear Study, University of Tokyo, Tokyo 188, Japan

ABSTRACT

Distribution of the Gamow-Teller strength has been studied both experimentally and theoretically in the $f_{7/2}$ shell mirror nuclides over a wide energy range. Experimental studies were performed using light ion induced reactions and the He-jet transport method or the ion-guide on-line isotope separation, IGISOL. Several transitions were observed to excited states in the decays of ^{43}Ti and ^{51}Fe and some in the decays of ^{47}Cr, ^{49}Mn, ^{53}Co and ^{55}Ni. Theoretical calculations were made by a shell model code using $f_{7/2}^n + (p_{3/2}, f_{5/2}, p_{1/2})^1$ shell space. The β-feeding has been predicted for all transitions up to about 4 MeV excitation in each daughter nucleus. The quenching of the Gamow-Teller strength has been studied by comparing the experimental strength with the calculation. The formation of the giant Gamow-Teller resonance has been studied theoretically as a function of the mass number.

INTRODUCTION

One of the objectives of the studies of the β-decay of nuclides far from stability is to determine directly the distribution and the strength of the Gamow-Teller (GT) transitions. High energy β-decay covers typically a good part of the GT strength of spin-isospin flip transitions, including a part of the so called giant GT resonance. The study of the quenching of the GT strength as compared with shell model predictions has gained an increasing interest since the observation of this phenomenon in the (p,n) reactions[1]. Although the decay energy window limits the available final states typically to energies below 10 MeV, β-decay provides good energy resolution and is free from the interpretational difficulties encountered in less selective reaction studies.

In (p,n) reactions a 50 % quenching has been observed in the β-strength except for the lightest elements[2]. In this method the (p,n) cross sections have

---

* Supported by the Research Council for Natural Sciences of Academy of Finland
† Present address: Laboratory of Physics, Kanto Gakuin University, Japan

been compared to a model independent sum rule. An analysis of a large body of GT matrix elements in the sd-shell has been carried out by Brown and Wildenthal[3]. They observed a persistent reduction of the transition matrix elements to be 0.76 ± 0.03, which corresponds to a quenching of 0.58 ± 0.05 in the β-strength. The mirror decays in the fp shell have been studied by T. Sekine et al.[4]. They observed a quenching of about 0.6 in the GT strength by using a shell model with van Hees-Glaudemans interaction. Recently a systematic analysis of the GT matrix elements has also been carried out in the $1f_{7/2}$ shell by Miyatake et al.[5]. In those calculations Kuo-Brown matrix elements were used and up to two-particle excitations from the $1f_{7/2}$ shell to the upper fp orbits were taken into account. They analysed the 52 strongest transitions in this shell and observed the quenching factors 0.35 ± 0.04, 0.59 ± 0.06 and 0.56 ± 0.04 in the β-strength for 0, 1 and 2 excited particles, respectively.

In the present paper, the distribution of the GT strength is studied in the mirror nuclides of the $f_{7/2}$ shell. The mirror nuclides in this shell possess β decay energies from 6.5 to 8.7 MeV and our interest in this work was to study the distribution of the beta strength both experimentally and theoretically over as wide of an energy range as possible. In the theoretical calculations, a similar shell model code as that used by Miyatake et al.[5] was used.

Up to now very few beta transitions to excited states were known in the decays of mirror nuclides. Recently the decay of ^{43}Ti has been studied in detail and 10 β-transitions to excited states below 3.5 MeV excitation were observed[6].

EXPERIMENTAL METHODS

Experimental studies were performed using the MC-20 cyclotron of the University of Jyväskylä and the MC-35 cyclotron of the University of Oslo. The light ion reactions (p,2n), (^{3}He,2n) and (α,n) were used. The He-jet transport method or the ion guide on-line isotope separation, IGISOL, techniques connected with a tape transport system were applied in these measurements[7]. $\beta - \gamma$ coincidence technique was used to reduce background effects.

In general, the β-branching ratios to excited states were determined by comparing the γ-ray intensity to the intensity of the short-lived component of the annihilation radiation. For ^{43}Ti the branching ratios were also deduced from mass separated samples by extracting short-lived components from the singles β and γ spectra. In the decay of ^{53g}Co a special technique had to be used, because ^{53m}Co and ^{54}Co were also present. All these activities have almost the same half-life and the relative contributions in the annihilation radiation can not be resolved. In this case the atoms released from the source by β recoil were collected on an aluminium foil. The number of preceding β-decays were obtained by measuring the daughter ^{53}Fe activity from the collection foil. The efficiency was determined by genetically linked pairs of nuclides ^{65}Ge-^{65}Ga, ^{49}Mn-^{49}Cr and ^{51}Fe-^{51}Mn.

In a precise half-life measurement, it is essential that the interfering activities with similar half-lives be eliminated and that the background level is as low as possible; therefore, we used the on-line isotope separation to measure the half-life for ^{43}Ti, ^{51}Fe and ^{55}Ni. Also in these measurements the activity was collected on a tape transport system in front of the detector and the long-lived activities were periodically transported away.

SHELL MODEL CALCULATIONS

Although full basis shell model calculations are still impossible in the fp shell, large scale calculations have now become possible. In this work, the GT matrix elements were calculated with an INS shell-model code[8], which applies ^{40}Ca as an inert core and in which one or more of the valence nucleons can be lifted from the $f_{7/2}$ orbital to the higher-lying $p_{3/2}$, $f_{5/2}$ and $p_{1/2}$ orbitals. The effective interactions were calculated using Kuo-Brown matrix elements slightly modified by Poves and Zuker[9].

Since in this paper we are discussing the β-decay to highly excited states, even above the Q_β-window, the excitation of more than one particle to the higher orbitals is not practical. Thus we use here the so called Njump=1 model i.e. a shell model under the configuration space of $f_{7/2}^n + f_{7/2}^n(p_{3/2}, f_{5/2}, p_{1/2})^1$, where the number of active particles n outside the inert core is A-40. Computational calculations were performed by using the proton-neutron coupling formalism and the Lanczos method was used for matrix diagonalization. The single-particle levels were taken from the experimental data of ^{41}Ca.

COMPARISON OF THEORETICAL AND EXPERIMENTAL RESULTS

Experimental final-state level energies, spins and GT matrix elements are given in Table I together with the values calculated with the Njump=1 shell model. The GT matrix elements have been calculated using the constants and procedures given in ref.6, where the allowed approximation and the free-neutron axial vector coupling constant $g_A/g_V = 1.2606 \pm 0.0075$[10] were used. The measured β-branching ratios are also given, but the theoretical values are not calculated because they are very sensitive to the theoretical level energies. In the following a short description of each mirror nucleus in the $f_{7/2}$ shell is given.

Decay of ^{41}Sc

^{41}Sc has only one single proton outside the ^{40}Ca core. Therefore only single particle transitions can be generated by the fp shell model. Experimentally two very weak β-transitions have been observed[11] to feed the excited states at 2.575 and 2.959 MeV.

Decay of ^{43}Ti

Recently the decay of ^{43}Ti has been studied[6] up to 5 MeV excitation both experimentally and theoretically. Among the mirror nuclides, the distribution of the GT strength is best known in this decay.

As seen in Table I the calculated β-strength is concentrated on the lowest level of each spin. Experimentally the strength is more evenly distributed. A strong transition observed to the 2288 keV $5/2^-$ state contains most of the calculated strength for the lowest $5/2^-$ state at 2001 keV. Neither a strong transition to a $9/2^-$ state nor a low lying $9/2^-$ state is observed experimentally. It is likely that its strength is distributed among the levels between 2.3 and 3.6 MeV. This is also supported by the Njump=2 calculation which brings the lowest $9/2^-$ state to 2207 keV excitation.

Table I. Theoretical and experimental exitation energies, spins and GT matrix elements.

$E_x(keV)$ theor	J^π theor	$\langle\sigma\tau\rangle$ theor	$E_x(keV)$ exp	J^π exp	BR (%) exp	$|\langle\sigma\tau\rangle|$ exp
^{41}Sc - ^{41}Ca			$T_{1/2}$=596.3(17) ms$^{a)}$		Q_{EC}=6495(1) keV$^{b)}$	
0	$7/2^-$	1.134	0	$7/2^-$	99.963(3)	0.850(6)
6500	$5/2^-$	1.309	2.575	$5/2^-$	0.0232(29)	0.075(5)
			2.959	$7/2^-$	0.0139(14)	0.080(4)
^{43}Ti - ^{43}Sc			$T_{1/2}$=509(5) ms$^{c)}$		Q_{EC}=6868(7) keV$^{b)}$	
0	$7/2^-$	0.888	0	$7/2^-$	90.2(8)	0.663(12)
1295	$9/2^-$	0.594	845.2	$5/2^-$	2.7(8)	0.257(39)
2001	$5/2^-$	0.904	1408.0	$7/2^-$	0.68(10)	0.172(13)
2258	$7/2^-$	0.110	1882.5	$(5/2, 9/2)^-$	0.20(4)	0.120(12)
2806	$5/2^-$	0.129	1962.9	$5/2^-$	0.024(10)	0.044(9)
3238	$7/2^{-*}$	0.129	2288.3	$5/2^-$	4.7(7)	0.749(56)
3246	$9/2^-$	0.360	2335.4	$(5/2, 9/2)^-$	0.39(6)	0.221(17)
3319	$5/2^{-*}$	0.189	2458.6	$(5/2, 9/2)^-$	0.92(13)	0.371(26)
3514	$5/2^-$	0.034	2760.0	$(5/2, 9/2)^-$	0.20(3)	0.209(16)
4020	$7/2^-$	0.120	3259.6	$(5/2, 9/2)^-$	0.011(3)	0.071(10)
4075	$5/2^-$	0.016	3631.5	$(5/2, 9/2)^-$	0.016(4)	0.115(14)
4367	$9/2^-$	0.064				
4468	$9/2^{-*}$	0.005				
4754	$5/2^-$	0.104				
	$\sum B_{(GT)}=$	2.200			$\sum B_{(GT)}=$	1.35(17)
^{45}V - ^{45}Ti			$T_{1/2}$=539(18) ms$^{d)}$		Q_{EC}=7132(17) keV$^{b)}$	
0	$7/2^-$	0.389	0	$7/2^-$	95.7(15)	0.505(34)
76	$5/2^-$	0.032	40	$5/2^-$	4.3(15)	0.203(36)
1169	$9/2^-$	0.181				
1509	$9/2^-$	0.276				
1928	$5/2^-$	0.217				
1962	$7/2^-$	0.071				
2238	$7/2^-$	0.434				
3077	$9/2^-$	0.315				
3193	$7/2^-$	0.089				
^{47}Cr - ^{47}V			$T_{1/2}$=508(10) ms$^{e)}$		Q_{EC}=7451(14) keV$^{b)}$	
0	$5/2^-$	0.244	0	$3/2^-$	96.3(12)	0.356(26)
196	$3/2^-$	0.405	88	$5/2^-$	3.7(12)	0.177(29)
1848	$3/2^-$	0.406				
1898	$5/2^-$	0.021				
2142	$1/2^-$	0.504				
2695	$5/2^-$	0.483				
2887	$3/2^-$	0.350				
2967	$1/2^-$	0.118				
2989	$5/2^-$	0.258				
3080	$5/2^-$	0.006				

E_x(keV) theor	J^π theor	$\langle\sigma\tau\rangle$ theor	E_x(keV) exp	J^π exp	BR (%) exp	$|\langle\sigma\tau\rangle|$ exp
^{49}Mn - ^{49}Cr			$T_{1/2}$=384(17) ms$^{f)}$		Q_{EC}=7718(24) keV$^{b)}$	
0	5/2$^-$	0.454	0	5/2$^-$	91.9(26)$^{f,g)}$	0.414(51)
167	7/2$^-$	0.248	272	7/2$^-$	5.8(26)	0.249(56)
1535	3/2$^-$	0.048	2504	7/2$^-$	2.3(9)	0.422(83)
1992	7/2$^-$	0.569				
2450	5/2$^-$	0.043				
2624	3/2$^-$	0.242				
2658	7/2$^-$	0.241				
3104	5/2$^-$	0.020				
3375	5/2$^-$*	0.182				
3394	3/2$^-$	0.041				
^{51}Fe - ^{51}Mn			$T_{1/2}$=310(5) ms$^{h)}$		Q_{EC}=8022(15) keV$^{b)}$	
0	5/2$^-$	0.601	0	5/2$^-$	93.8(13)$^{h,g)}$	0.454(21)
215	7/2$^-$	0.260	237	7/2$^-$	5.0(13)	0.230(30)
1526	3/2$^-$	0.249	1824	3/2$^-$	0.49(14)	0.134(19)
1967	7/2$^-$	0.036	2140	3/2$^-$	0.24(7)	0.109(16)
2326	3/2$^-$	0.359	2914	3/2$^-$	0.10(3)	0.104(16)
2549	5/2$^-$	0.196	3426		0.20(6)	0.201(30)
3089	7/2$^-$	0.197	3555	(1/2, 3/2)$^-$	0.16(5)	0.195(31)
3386	7/2$^-$*	0.135				
3528	5/2$^-$	0.075				
3535	7/2$^-$	0.146				
	$\sum B_{GT}=$	0.743			$\sum B_{GT}=$	0.38(13)
^{53}Co - ^{53}Fe			$T_{1/2}$=240(20) ms$^{g)}$		Q_{EC}=8304(18) keV$^{b)}$	
0	7/2$^-$	0.568	0	7/2$^-$	94.4(17)	0.534(72)
1040	9/2$^-$	0.441	1328	9/2$^-$	5.6(17)	0.376(59)
2654	7/2$^-$	0.116				
3275	5/2$^-$	0.056				
3336	7/2$^-$*	0.121				
3512	5/2$^-$	0.001				
3592	9/2$^-$	0.000				
3606	5/2$^-$*	0.144				
3754	7/2$^-$	0.045				
3835	5/2$^-$	0.028				
^{55}Ni - ^{55}Co			$T_{1/2}$=208(5) ms$^{h)}$		Q_{EC}=8696(11) keV$^{b)}$	
0	7/2$^-$	0.551	0	7/2$^-$	100	0.510(22)
3635	5/2$^-$	0.013				
3760	7/2$^-$	0.121				
3836	5/2$^-$	0.140				
3889	9/2$^-$	0.145				
4361	9/2$^-$	0.140				
4473	5/2$^-$	0.120				
4565	7/2$^-$	0.259				
4852	9/2$^-$	0.093				
4894	5/2$^-$	0.294				

a) ref. 11, b) ref. 20, c) ref. 6, d) ref.12, e) ref. 14, f) ref. 17, g) this work h) ref. 7, *T=3/

The total observed GT strength is $\Sigma B(GT)_{exp} = \Sigma \langle \sigma \tau \rangle^2_{exp} = 1.35 \pm 0.17$, whereas the Njump=1 calculation gives $\Sigma B(GT)_{theor} = 2.00$ for levels below 5 MeV. This gives a quenching factor of 0.61 ± 0.07 for the GT strength. The Njump= 2 and Njump= 3 models give slightly higher quenching factors of 0.54 ± 0.07 and 0.53 ± 0.07, respectively[6]. For the ground state and the 2288 keV $5/2^-$ state, similar values of the quenching are obtained and also these values are decreased with the number of excited particles in the calculations.

Decay of ^{45}V

In the decay of ^{45}V, one β-transition to the 40 keV excited state is observed. A branching ratio of (4.3 ± 1.5) % has been estimated[12] for this transition based on the calculated production rate of the ^{45}V nuclides. The shell model calculation predicts a very weak β feeding for this level. Instead a rather strong β branch (1-2 %) is predicted to a $7/2^-$ state at 2238 keV. A possible candidate for this level is a known[13] $(5/2^-, 7/2^-)$ 2016 keV state, which decays by a 47 % branch to the ground state and by a 53 % branch to the 40 keV level. Thus this and other transitions from higher excitations would increase the intensity of the 40 keV transition.

Decay of ^{47}Cr

Also, in the decay of ^{47}Cr, only one beta transition has been experimentally observed to the excited states[14]. However, the theoretical calculations predict rather strong transitions to several levels around 2-3 MeV. Although the shell model calculations predict the $3/2^-$ ground state and the $5/2^-$ first excited state to be in reversed order, the calculated transition matrix elements are in good agreement with the experimental values. However, the experimental energy spacing of these levels is only 88 keV.

There is an ambiguity for the half-life of ^{47}Cr. There are several half-life measurements which vary as much as between 452 ms and 508 ms, the most precisely measured values being 460.0 ± 1.5 ms[15] and 508 ± 10 ms[14]. However, a recent measurement, in which the half-life is determined from a mass separated sample, gave a value 476.4 ± 8.2 ms[16].

Decay of ^{49}Mn

The decay of ^{49}Mn was first oberved[17] by isotope separation using ^{28}Si + ^{nat}Mg reaction. A (6.4 ± 2.6) % intensity was also measured for the 272 keV γ-ray de-exciting the first excited state. In an experiment using a He-jet and the p $+^{50}$Cr reaction with 30 MeV protons, we observed weak γ-lines at 2231.5 ± 1.0 keV and 2504.8 ± 1.0 keV. The energy spacing of these lines is close to the energy of the first excited state. However, the low energy line contains a contribution of a ^{32}S 2230.2 ± 0.2 keV γ-transition, which is produced from the NaCl aerosol used as a carrier in the He-jet. If these transitions are assigned to a known 2503.4 ± 0.5 keV $7/2^-$ level[18], which has a 67.3 % branching ratio to the ground state and a 32.7 % branch to the first excited state, the contribution of the contamination can be estimated. When the intensity of the 2505 keV γ-rays are normalized to the known[17] intensity of the 272 keV line the intensities 0.6 ± 0.3 % and 1.7 ± 0.7 % are obtained for the 2232 and 2505 keV γ-lines.

Theoretically the strongest GT strength is predicted for a 1992 keV $7/2^-$ state, which can be related with the 2505 keV $7/2^-$ level. The other β-transitions, except for the ground state and the first excited state, are predicted to be rather weak. Our calculations agree with the similar ones made by Hardy et al.(ref.17) except that for the GT matrix element of the ground state transition they obtained $\langle\sigma\tau\rangle = 0.302$ while we calculated $\langle\sigma\tau\rangle = 0.454$.

Decay of ^{51}Fe

For 27 MeV ^{3}He bombardment of enriched ^{50}Cr targets, six transitions are associated with the decay of ^{51}Fe by energy and half-life considerations (see Table I). If the γ-ray intensities are normalized to the intensity of the 237 keV line, which has been earlier[7] measured to correspond to a β branch of (5.0 ± 1.3) %, the other lines will amount to a total branch of 1.2 % with the single line intensities ranging from 0.1 to 0.5 %. All the transitions match well the known level energies of ^{51}Mn[19].

Level to level correspondence with the calculated and experimental results can only be made for the three lowest levels. For them, the calculated GT matrix elements are systematically higher. If the GT strength is summed up to 4 MeV, we see that about 50 % of the calculated strength is missing.

Decay of ^{53}Co

In bombardments of ^{54}Fe targets by 30 MeV protons a weak γ-ray of 1328 keV was observed in coincidence with positrons. Both the measured energy and half-life are consistent with assigning this γ-ray with β-decay of ^{53g}Co to the 1328 keV $9/2^-$ level of ^{53}Fe. The usual technique to determine β-branchings by measuring the delayed γ-ray annihilation ratio is not possible in this case because both ^{53m}Co (247 ms) and ^{54g}Co (193 ms) are present. A method based on counting the number of daughter atoms ejected from the source by recoil due to positron emission was applied as described in the experimental section. By this method, a β-branching ratio of (5.6 ± 1.7) % was determined for this level.

In contrast to ^{43}Ti, which is the cross conjugate nucleus of ^{53}Co, the calculated low lying $9/2^-$ state and the predicted feeding of it is experimentally observed. For other levels the feeding is predicted to be small, which was also confirmed experimentally.

Decay of ^{55}Ni

The beta feeding of ^{55}Ni to excited states was also searched for by using the He-jet transport method and the ^{3}He + ^{54}Fe reaction. No γ-rays were observed and an upper limit of 0.1 % for the known $5/2^-$ state at 3302 keV was estimated from the background of the γ-ray spectrum. Instead a weak delayed proton branch was observed but because of the missing γ-rays the branching ratio could not be determined.

TOTAL GAMOW-TELLER STRENGTH DISTRIBUTION

The distribution of the total GT strength has been predicted by the Njump=1 model. In Fig. 1, the GT strength has been plotted for the transitions of ^{43}Ti, ^{53}Co and ^{55}Ni generated in this shell space. As seen in these figures, a great part of the GT strength is concentrated at high excitation energies around 9 MeV. Only 50 % of the total strength has been estimated to be below 5 MeV excitation for ^{43}Ti and 5 % for ^{53}Co and 4 % for ^{55}Ni, respectively.

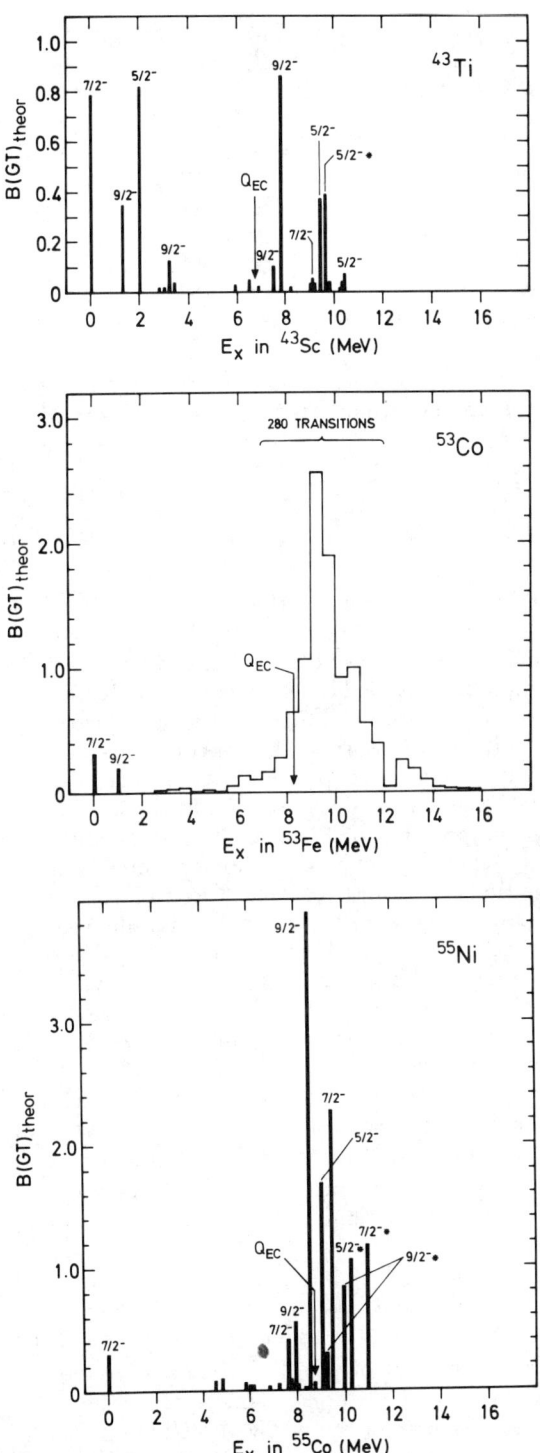

Fig. 1 also shows the formation of the GT giant resonance. In the beginning of the shell there are a few strong spin-flip $f_{7/2} - f_{5/2}$ transitions. For ^{53}Co the shell is almost full and several nucleons are taking part in the resonance formation and the resonance is well developed. For ^{55}Ni the proton shell is full and the GT strength is concentrated in a few very strong transitions, in which the dominant configuration of their wave function is $| i \otimes (7/2, 5/2)_{1+} \rangle$, where $i = 5/2^-$, $7/2^-$ or $9/2^-$. A special feature for ^{43}Ti is that there are no strong transitions to $7/2^-$ states at high excitation; instead, for ^{55}Ni all the spins $5/2^-, 7/2^-$ and $9/2^-$ are present. Also for ^{55}Ni there is another transition with the same spin but with a T = 3/2 isospin 1-1.5 MeV higher in excitation and with about half the intensity.

Total summed GT strength is given for ^{41}Sc, ^{43}Ti, ^{53}Co and ^{55}Ni in Table II. The total - strength grows rapidly with increasing mass number and depends almost linearly on the number of active particles in the shell. By using the sum rule $S_\beta^- - S_\beta^+ = 3(N-Z)$, the S_β^- strength can be estimated to be zero for ^{41}Sc but then it grows with S_β^+.

Fig.1. Theoretical Gamow-Teller strength for the decays of ^{43}Ti, ^{53}Co and ^{55}Ni. Asterisk represents $\Delta T = 1$ transitions. For ^{53}Co the strength is summed over 0.5 MeV intervals.

Table II. Total GT strength predicted by the Njump=1 shell model.

parent nucleus	number of eigenstates	S_{β^+}	S_{β^-}
^{41}Sc	2	3.000	$\leq$ 0.000
^{43}Ti	64	4.434	$\leq$ 1.434
^{53}Co	484	10.899	$\leq$ 7.899
^{55}Ni	55	12.781	$\leq$ 9.781

In the ^{54}Fe(p,n)^{54}Co reaction[21], which corresponds to the β^+ decay of ^{54}Ni, the GT giant resonance is observed at 10 MeV with the width Γ =1 MeV much alike as calculated for ^{53}Co.

CONCLUSIONS

In a theoretical study of GT matrix elements in the $f_{7/2}$ shell nuclei, the particle excitations from the $f_{7/2}$ shell have to be taken into account. It is shown in ref. 5 that the difference in the calculated values of the GT matrix elements is small between Njump=1 and Njump=2 models. The Njump=1 model predicts the lowest energy levels well except for ^{43}Sc. Also in ^{47}V the first two very close lying levels are predicted to be in reversed order. The theoretical GT matrix elements are systematically larger than the experimentally observed ones. However, rather severe discrepancies occur in the decay of ^{45}V. In this decay the experimental values are much larger, especially for the transition to the 40 keV first excited state. This might partly be explained by experimental uncertainties and feeding of this state by γ-transitions from higher levels. Due to large errors in the experimental values and because of difficulties in level assignments, it is not possible to deduce the quenching factor for other transitions than for the ground state decay. These are included in the analysis of Miyatake et al.[5]. In the decays of ^{43}Ti and ^{51}Fe several β-transitions were observed. If these are summed and compared with the sum of the theoretical values, the quenching factor of 0.61 $\pm$ 0.07 is obtained for the decay of ^{43}Ti and 0.51 $\pm$ 0.18 for ^{51}Fe. The given uncertainties include only the experimental errors. These quenching factors agree very well with the results obtained in (p,n) measurements[2] and in the systematic analysis of strong individual β-transitions[3-5]. For more detailed analysis of the GT strength in the mirror nuclides of the $f_{7/2}$ shell, the accuracy of the experiments should be improved and the decay schemes should be constructed with higher precision.

The calculations of the total GT strength in the decay of ^{43}Ti, ^{53}Co and ^{55}Ni show that the spin flip mode is well taken into account by the Njump=1 model. The position of the GT giant resonance agree well with the (p,n) measurements[21] made in this shell. The tail of the GT resonance region is in the β-decay window of the heaviest mirror nuclides. This is experimentally accessible in β-delayed proton studies as shown in the decay of ^{55}Ni and ^{59}Zn[19]. In the odd-odd T_z = -1 series, the GT giant resonance region should also be accessible by delayed protons. However, the GT giant resonance is not clearly been in the recent delayed proton measurement of ^{48}Mn[4]. According to the calculations the resonance should be better developed towards the end of the $f_{7/2}$ shell.

REFERENCES

1. C. Gaarde, J. Rapaport, T. N. Taddeucci, C. D. Goodman, C. C. Foster, D. E. Bainum, C. A. Goulding, M. B. Greenfield, D. J. Horen and E. Sugarbaker, Nucl. Phys. $\underline{A369}$, 258 (1981)
2. C. Gaarde, Nucl. Phys. $\underline{A396}$, 127c (1983)
3. B. A. Brown and B. H. Wildenthal, At. Data Nucl. Data Tables $\underline{33}$, 347 (1985)
4. T. Sekine et al. Nucl. Phys. $\underline{A467}$, 93 (1987)
5. H. Miyatake, K. Ogawa, T. Shinozuka and M. Fujioka, to be published in Nucl. Phys.
6. J. Honkanen, V. Koponen, H. Hyvönen, P. Taskinen, J. Äystö and K. Ogawa, Nucl. Phys. $\underline{A471}$, 489 (1987)
7. J. Äystö, J. Ärje, V. Koponen, P. Taskinen, H. Hyvönen, A. Hautojärvi and K. Vierinen, Phys. Lett. $\underline{138B}$, 369 (1984)
8. K. Ogawa, INS shell-model code (unpublisehed)
9. A. Poves and A. Zuker, Phys. Rept.$\underline{70}$, 235 (1981)
10. D. H. Wilkinson, Nucl. Phys. $\underline{A377}$,474 (1982)
11. H. S. Wilson, R. W. Kavanagh and F. M. Mann, Phys. Rev. $\underline{C22}$, 1696(1980)
12. P. Hornshoj, J. Kolind and N. Rud, Phys. Lett. $\underline{116B}$, 4 (1982)
13. T.W. Burrows, Nucl. Data Sheets $\underline{40}$, 149 (1983)
14. T. W. Burrows, J. W. Olness and D. E. Alburger, Phys. Rev. $\underline{C31}$, 1490 (1985)
15. M.D. Edmiston, R.A. Warner, W.C. McHarris and W.H. Kelly, Nucl. Instr. Meth.$\underline{141}$, 315 (1977)
16. H. Hama, M. Yoshii, K. Taguchi, T. Ishimatsu, T. Shinozuka, M. Fujioka, H. Miyatake and K. Ogawa, to be published
17. J. C. Hardy et al., Phys. Lett. $\underline{91B}$, 207 (1980)
18. T.W. Burrows, Nucl. Data Sheets $\underline{48}$, 569 (1986)
19. Zhou Chunmei, Zhou Enchen, LuXiane and Huo Junde, Nucl. Data Sheets $\underline{48}$, 111 (1986)
20. A. H. Wapstra and G. Audi, Nucl. Phys. $\underline{A432}$, 1 (1985)
21. J. Rapaport et al., Nucl. Phys.$\underline{A410}$, 371 (1983)
22. J. Honkanen, M. Kortelahti, K. Eskola and K. Vierinen, Nucl. Phys. $\underline{A366}$, 109 (1981)

PRECISION HALF-LIFE MEASUREMENTS OF THE MIRROR NUCLEI IN THE $f_{7/2}$-SHELL USING AN IGISOL

H. Hama*, M. Yoshii**, K. Taguchi and T. Ishimatsu
Department of Physics, Tohoku University, Sendai 980, Japan

T. Shinozuka and M. Fujioka
Cyclotron and Radioisotope Center, Tohoku University,
Sendai 980, Japan

ABSTRACT

The half-lives of $T_z=-1/2$ mirror nuclei in the $f_{7/2}$-shell region were measured with a new on-line separator, IGISOL of Tohoku University. The deduced experimental Gamow-Teller (GT) matrix elements were compared with the results of large-scale shell-model calculations. As an average quenching factor of the GT matrix elements for the mirror β-decay in the $f_{7/2}$-shell a value of 0.844±0.042 (simple average) was obtained.

INTRODUCTION

The $T_z=-1/2$ nucleus decaying to its mirror nucleus by the superallowed β-transition is playing an important role to study the quenching problem of Gamow-Teller strength. The systematic behavior of the quenching of GT matrix element, defined as the ratio of an experimental GT matrix element to a theoretical one calculated with the shell model, has been studied in the region of lighter nuclei, i.e., p-shell and sd-shell. Wildenthal derived[1] a mass-number independent quenching factor of ρ=0.76±0.03 for the sd-shell nuclei on the basis of a full sd-space shell model calculation. Recently, for the nuclei of the $f_{7/2}$-shell region a similar comparison has been made[2] on the basis of a shell model calculation considering up to two-particle jump from the $f_{7/2}$-shell, giving a similar value of ρ=0.75±0.03. The present work reports on the results of accurate measurements of the half-lives of six mirror nuclei in the $f_{7/2}$-shell region, i.e., ^{45}V, ^{47}Cr, ^{49}Mn, ^{51}Fe, ^{53}Co and ^{55}Ni, using a new technique of on-line isotope separation, IGISOL (Ion-Guide Isotope Separation On-Line), by which we focus our attention to the quenching problem of the GT matrix elements, especially of the mirror transitions in the $f_{7/2}$-shell.

EXPERIMENTAL PROCEDURE

The ion-guide method was firstly developed at University of Jyväskylä of Finland[3]. The details of our IGISOL system has been described[4]. The experimental conditions of producing the six mirror nuclei and the yields at the detector position are shown in Table 1.
The positrons from the decay of mirror nuclei except for ^{53}Co were detected by the counter telescope of a ΔE(2mm thick) and an E (2"φ×2") plastic scintillators. The detector region is shown in

Table 1. Experimental conditions

Parent Nucleus	Nuclear reaction	Projectile energy (MeV)	Yield[a] (ions/eμC)
^{45}V	^{46}V (p,2n)	30	55
^{47}Cr	^{46}V (^{3}He,2n)	27	11
^{49}Mn	^{50}Cr (p,2n)	30	15
^{51}Fe	^{50}Cr (^{3}He,2n)	27	18
^{53}Co	^{54}Fe (p,2n)	30	15
^{55}Ni	^{54}Fe (^{3}He,2n)	27	20

[a] At the tape in front of the detector.

Fig. 1. For the measurement of half-life the cyclotron beam was chopped using a pulse-type beam chopper in order to suppress the background during irradiations and the counts were accumulated in a two-dimensional list mode with energy and detection time after irradiation. The simultaneously produced ^{53m}Co is a positron emitter having a similar half-life and end-point energy as those of ^{53g}Co. Therefore, the only way to determine the half-life of ^{53g}Co is to detect the γ-ray from the 1329 keV first excited state of ^{53}Fe populated from ^{53g}Co. Although the γ-ray comes also from the decay of ^{53m}Fe populated from ^{53m}Co, the half-life of this isomeric state is so long (2.5 m) that its decay does not interfere with the decay of ^{53g}Co.

In order to check the time axis of the measurement, we measured the time spectra of positrons from the decay of ^{46}V and ^{54}Co having well established half-lives. They were produced with the ^{46}Ti(p,n) and the ^{54}Fe(p,n) reactions, respectively, and similarly mass-separated. The obtained half-lives, 421.3± 2.2 ms and 193.4±0.6 ms for ^{46}V and ^{54}Co respectively, were in good agreement with the reported values[5,6]. For the details of measurement and analysis see Ref. 9. Furthermore, we carefully checked the degree of mass identification of our IGISOL by measuring the mass spectra of β-emitters produced with different reactions; e.g., in the case of ^{47}Cr, the possibility of contamination from ^{46}V simultaneously produced by the ^{46}Ti (^{3}He,pn) reaction was excluded on the basis of a mass spectrum taken using the ^{46}Ti(p,n)^{46}V reaction which should not produce ^{47}Cr. Although the obtained average resolving power of M/ΔM ∼ 150 FWHM was considerably lower than for a conventional mass separator, disturbances from the neighboring mass were negligible.

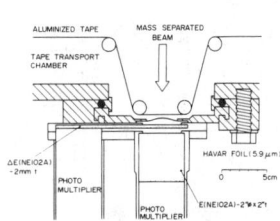

Fig. 1. The region of the tape transport and the detectors.

RESULTS AND DISCUSSION

In Table 2, the present half-life values of six mirror nuclei are shown together with the other two mirror nuclei, ^{41}Sc and ^{43}Ti. Due to sufficient statistics because of the stable operation of our IGISOL, all the experimental errors are smaller than 2% except that of ^{53}Co.

The nucleus ^{45}V was firstly mass-separated in the present experiment, and the result is in good agreement with that of Hornshøj[7]. For the half-life of ^{47}Cr, there was a considerable discrepancy among several reports as described by Burrows[8]. The present result obtained with mass-separated ^{47}Cr, however, is apparently different from that of Burrows, 508(10) ms. This discrepancy seems to be an open problem at present. In the case of ^{53}Co, the error of our result is as large as 40% but the merit of our measurement is that ^{53g}Co was definitely identified by mass separation and γ-ray detection. The previous half-life of ^{53g}Co is from Kochan[10] and Eskola[11]; the former was obtained by a β-ray measurement possibly including the positrons from ^{53m}Co and the latter by a β-γ coincidence measurement with a poor statistic similar to ours. The mirror nucleus ^{49}Mn was mass-separated by Hardy[12], and ^{51}Fe and ^{55}Ni by Äystö[13]. The present results for those nuclei are in good agreement with their data.

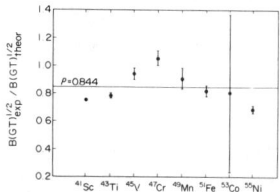

Fig. 2. The quenching factors of the mirror nuclei in the $f_{7/2}$-shell.

From the present experimental results, we obtain the reduced Gamow-Teller matrix elements[2], $<\sigma\tau>$. The method of analysis is the same as in Ref. 9. For $|g_A/g_V|$ we adopted the recent value[14] of 1.262(5). Table 2 shows the experimental GT matrix elements and the theoretical predictions from a shell-model calculation[2] using the modified Kuo-Brown interaction and permitting up to two particle excitation from the $f_{7/2}$-shell to the upper $p_{3/2}$, $f_{5/2}$ and $p_{1/2}$-shells. Figure 2 shows the quenching factor for the eight mirror nuclei as well as the simple average (i.e., with equal weight) ρ = 0.844±0.042. It is remarkable that in the middle of the $f_{7/2}$-shell the quenching factor is clearly larger than those at the beginning and end of the shell, and that the average value ρ is larger than

Table 2. Properties of the β-decay of mirror nuclei and GT matrix elements in the $f_{7/2}$-shell

Parent Nucleus	Half-life (ms)	Q_{EC} a) (keV)	Branching Ratio(%)	$<\sigma\tau>$ Exp.	Theory
^{41}Sc	596.3(17)b)	6495(1)	99.963(13)	0.850(5)	1.134
^{43}Ti	513.0(80)c)	6868(7)	100d)	0.739(15)	0.947
^{45}V	547.2(53)	7132(17)	95.7(15)	0.494(23)	0.523
^{47}Cr	472.0(63)	7452(14)	96.3(12)	0.432(23)	0.409
^{49}Mn	381.7(74)	7718(24)	93.6(26)	0.441(39)	0.487
^{51}Fe	305.0(43)	8022(15)	95.0(13)	0.483(23)	0.589
^{53}Co	267 (109)	8304(18)	(100)	0.495(359)	0.614
^{55}Ni	212.1(38)	8696(11)	100d)	0.493(20)	0.716

a) Ref. 15.
b,c) Refs. 16 and 17, respectively.
d) No branching to excited states is found.

the quenching factor[1], 0.76(3), derived for sd-shell nuclei by Wildenthal, and 0.75(3) derived for 25 GT transitions including the eight mirror ones by Miyatake[2]. The fact that the present value of ρ is larger than that derived also for the $f_{7/2}$-shell[2] is caused by; 1) we restricted ourselves to the mirror transitions, and 2) we used the present values of $T_{1/2}$; for ^{47}Cr the present value is considerably shorter than that of Burrows[8] which was used in Ref. 2. It is therefore highly required to study further the mirror β-decays in the $f_{7/2}$-shell region, especially, around the middle of the shell, both experimentally and theoretically.

We would like to thank Prof. K. Ogawa and Dr. H. Miyatake for their valuable discussions.

*Present address; Department of Physics, Tokyo Institute of Technology, Tokyo, Japan
**Present address; Accelerator Division, National Laboratory for High Energy Physics, Tsukuba, Japan

REFERENCES

1. B. H. Wildenthal, Prog. Part. Nucl. Phys. 11, 5 (1984).
2. H. Miyatake, K. Ogawa, T. Shinozuka and M. Fujioka, Nucl. Phys. A470, 328 (1987).
3. J. Ärje, J. Äystö, H. Hyvonen, P. Taskinen, V. Kopponen, J. Honkanen, K. Valli, A. Hautojärvi and K. Vierinen, Nucl. Instr. Meth. A247, 431 (1986).
4. M. Yoshii, H. Hama, K. Taguchi, T. Ishimatsu, T. Shinozuka, M. Fujioka and J. Ärje, Nucl. Instr. Meth. B26, 410 (1987).
5. G. T. A. Squier, W. E. Burcham, S. D. Hoath, J. M. Freeman, P. H. Barker and P. J. Petty, Phys. Lett. 65B, 122 (1976).
6. J. C. Hardy, H. R. Andrews, J. S. Geiger, R. L. Graham, J. A. Macdnald and H. Schmeing, Phys. Rev. Lett. 33, 1647 (1974).
7. P. Hornshøj, J. Kolind and N. Rud, Phys. Lett. 116B, 4 (1982).
8. T. W. Burrows, J. W. Olness and D. E. Alburger, Phys. Rev. C31, 1490 (1985).
9. H. Hama, M. Yoshii, K. Taguchi, T. Ishimatsu, T. Shinozuka, M. Fujioka, H. Miyatake and K. Ogawa, submitted to Phys. Rev. Lett.
10. S. Kochan, B. Rosner, I. Tserruja and R. Kalish, Nucl. Phys. A204, 185 (1973).
11. K. Eskola, A. Hautojärvi, K. Vierinen, J. Honkanen, P. Taskinen and S. Messelt, JYFL Ann. Rept. 45 (1985).
12. J. C. Hardy, H. Schmeing, E. Hagberg, W. Perry, J. Wills, E. T. H. Clifford, V. Koslowsky, I. S. Towner, J. Camplan, B. Rosenbaum, R. Kirchner and H. Evans, Phys. Lett. 91B, 207 (1980).
13. J. Äystö, J. Ärje, V. Kopponen, P. Taskinen, H. Hyvonen, A. Hautojärvi and K. Vierinen, Phys. Lett. 138B, 369 (1984).
14. P. Bopp, D. Dubbers, L. Horning E. Klemt, J. Last, H. Schutze, S. J. Fremen and O. Scharpf, Phys. Rev. Lett. 56, 919 (1986).
15. A. H. Wapstra and G. Audi, Nucl. Phys. A432, 1 (1985).
16. D. E. Alburger and D. H. Wilkinson, Phys. Rev. C8, 657 (1973).
17. V. J. Jänecke, Z. Naturtorschg, 15a, 593 (1960).

THE BETA DECAY OF ^{48}Mn: GAMOW-TELLER QUENCHING IN fp-SHELL NUCLEI

T. Sekine[a], J. Cerny[b], R. Kirchner, O. Klepper, V.T. Koslowsky[c], A. Płochocki[d], E. Roeckl, D. Schardt and B. Sherrill[e]

GSI Darmstadt, Postfach 110 552, D-6100 Darmstadt 11, Federal Republic of Germany

B.A. Brown

Michigan State University, Department of Physics and Astronomy, East Lansing, Michigan 48824, USA

A new nuclide, ^{48}Mn, has been identified and its decay has been studied, using on-line mass separation of ^{12}C (^{40}Ca, p3n) reaction products. This isotope represents the heaviest member of the A=4n, T_z=-1 family of β-delayed proton precursors known to date. The half-life of this nuclide was determined to be 150 (10) ms, and spectra of β-delayed protons and γ-rays have been obtained. A branching ratio of 2.7(12) × 10^{-3} was deduced for the emission of β-delayed protons, while an upper limit of 6 × 10^{-6} was determined for β-delayed α-particle emission. The decay is dominated, with a branching ratio of 0.598(30), by the superallowed transition to the isobaric analogue state at 5792.4(6) keV in ^{48}Cr. A partial level scheme is proposed for ^{48}Cr on the basis of γ singles and γ-γ coincidence data.

The deduced Gamow-Teller strength function is compared with predictions obtained by applying the large-basis shell-model code OXBASH [1]. The model space used in the calculation was the fp-shell with active 1f7/2, 2p3/2, 1f5/2 and 2p1/2 orbits; the two-body matrix elements and the single-particle energies were taken from the work of van Hees and Glaudemans [2]. From this comparison, a quenching factor of 0.53(17) is obtained for ^{48}Cr excitation energies up to 5.8 MeV. For a general comparison with other quenching factors from β-decay studies, mirror β-decays in the fp-shell have also been calculated in the same manner.

---

[a] Present address: Japan Atomic Energy Research Institute, Department of Radioisotopes, Tokai, Naka, Ibaraki 319-11, Japan.
[b] Humboldt awardee 1985, permanent address: University of California, Department of Chemistry and Lawrence Berkeley Laboratory, Berkeley, California 94720, USA.
[c] Present address: Atomic Energy of Canada Limited, Chalk River Nuclear Laboratories, Nuclear Physics Branch, Chalk River, Ontario K0J 1J0, Canada.
[d] Present address: Warsaw University, Institute of Experimental Physics, PL-00-681 Warsaw, Poland.
[e] On leave of absence from Michigan State University, Department of Physics and Astronomy, East Lansing, Michigan 48824, USA.

The full contents of this contribution has been published very recently [3]. In the following, therefore, we will only discuss our results in connection with experimental and theoretical information about the fp-shell that has come forth in the meantime. For example, very detailed β and γ spectroscopy of the mirror decay of ^{43}Ti by Honkanen et al. [4] yielded an improved value for the Gamow-Teller strength B(GT) = 0.440(7) of the ground-state to ground-state transition. This results in a quenching factor of 0.54(1) for the experimental strength relative to the one calculated by the OXBASH code. The quoted uncertainty relates only to the experimental error. This quenching value is close to the weighted average of 0.56 for the 8 fp-shell mirror β-decays and to the value quoted above for the ^{48}Mn β-decay.

The strengths calculated by means of the OXBASH code can now be compared with results [5,6] from another large-basis shell-model code (INS-code) [7] which employs a slightly modified Kuo-Brown interaction for the two-body interaction and the same model space as mentioned above for the OXBASH code.

For the ground-state to ground-state β-decay of ^{43}Ti strengths B(GT) of 0.79, 0.90 and 0.92 have been calculated allowing one, two or three particle-hole excitations, respectively [4,5]. These values are at most 4% higher than the ones obtained from the OXBASH code. In the case of ^{45}V both codes produce similar configuration dependences by yielding much too high strengths (higher than the experimental ones), if the basis is restricted to 1p-1h excitations. For the four heaviest nuclei ^{49}Mn, ^{51}Fe, ^{53}Co, and ^{55}Ni, in general the strengths calculated by the INS code with 2p-2h excitations and those obtained by the OXBASH code with 1p-1h configurations agree quite well. These maximum numbers of excitations (2 and 1, respectively), are presently the practical limits of the two codes for A ≥ 47. The largest discrepencies seem to be for ^{45}V and ^{47}Cr, where the INS-code gives generally 10 to 30 % lower strengths than the OXBASH code.

REFERENCES

1. A. Etchegoyen, W.D.M. Rae, N.S. Godwin, W.A. Richter, C.H. Zimmermann, B.A. Brown, W.E. Ormand, J.S. Winfield, Michigan State University Internal Report MSUCL-524 (1985).

2. A.G.M. van Hees and P.W.M. Glaudemans, Z. Phys. A 303 , (1981) 267.

3. T. Sekine, J. Cerny, R. Kirchner, O. Klepper, V.T. Koslowsky, A. Płochocki, E. Roeckl, D. Schardt, B. Sherrill, B.A. Brown, Nucl.Phys. A 467 , (1987) 93.

4. J. Honkanen, V. Koponen, H. Hyvönen, P. Taskinen, J. Äystö, K. Ogawa, "Experimental and Shell-Model Study of the Beta Decay of ^{43}Ti", Preprint JYFL 22/86 (1986), to be published.

5. H. Miyatake, K. Ogawa, T. Shinozuka, M. Fujioka, to be publ., Nucl.Phys. A470

6. K. Ogawa and H. Miyatake: "Shell Model Study of the Gamow-Teller Matrix Elements in the $0f_{7/2}$ Shell Nuclei", Abstract submitted to the Int. Symp. on Weak and Electromagnetic Interactions in Nuclei, Heidelberg, July 1986, unpublished.

7. K. Ogawa, INS-Shell-Model-Code, unpublished.

GAMOW-TELLER BETA DECAY OF EVEN NUCLEI NEAR ^{100}Sn

K. Rykaczewski [d], R. Barden [a], J. Dobaczewski [d], H. Gabelmann [b], I.S. Grant [b], L. Kalinowski [d], R. Kirchner [a], O. Klepper [a], W. Nazarewicz [e], G. Nyman [c], A. Plochocki [d], G.-E. Rathke [a], E. Roeckl [a], D. Schardt [a], J. Zylicz [a,d]

[a] GSI, Postfach 110552, D-6100 Darmstadt 11, Federal Republic of Germany

[b] EP Division, CERN, CH-1211 Geneva 23, Switzerland

[c] Department of Physics, Chálmers University of Technology, Fack, S-41296 Göteborg 5, Sweden

[d] Department of Physics, University of Warsaw, ul.Hoza 69, PL-00681 Warsaw, Poland

[e] Institute of Physics, Warsaw Institute of Technology, ul.Koszykowa 75, PL-00662 Warsaw, Poland

ABSTRACT

Using on-line separation combined with γ-ray and conversion-electron spectroscopy, decay properties of neutron-deficient isotopes were measured near ^{100}Sn. The Q_{EC} values, the excitation energies of 1^+ levels, and the strength for $0^+ \rightarrow 1^+$ Gamow-Teller β transitions, as obtained for the decay of the even nuclei ^{94}Ru, $^{96-100}$Pd, $^{100-104}$Cd, and $^{104-110}$Sn, are compared with predictions from shell-model calculations including deformation, pairing and core-polarization effects. The quenching of the Gamow-Teller strength, as obtained from this comparison, is discussed.

INTRODUCTION

The double shell closure $Z = N = 50$ at ^{100}Sn, being very far from the β stability line, has not yet been reached experimentally. The investigation of nearby nuclei, however, has made progress in recent years, and has met with considerably increased interest in studying nuclear Gamow-Teller (GT) transitions in general. This interest is related to fact that the measured GT strength tends to be retarded (quenched) in comparison to predictions from nuclear models. Both experimental and theoretical efforts aim now at obtaining more precise and more reliable quenching factors in a systematic manner (e.g., as a function of mass number, or for chains of isotopes and isotones), and at subsequently interpreting the remaining deficiency, if there is any, of the nuclear model.

As far as β decay is concerned, an average quenching factor of 0.58, defined as the ratio between measured and calculated strength, is obtained [1] for the sd-shell nuclei (^{16}O to ^{40}Ca). For this region, large-basis shell model calculations as well as an extended set of experimental data are available. Even though both requirements are not fulfilled quite as well for fp-shell nuclei (^{40}Ca to ^{56}Ni), a quenching factor similar to the one observed for the sd-shell region is suggested here, too (see e.g. ref. 2). When going to heavier nuclei, studies ought to be carried out preferentially near closed shells, since model prescriptions with a few valence nucleons are expected to be comparatively reliable.

In this paper we want to present decay studies of very neutron- deficient isotopes near ^{100}Sn. The experimental technique will be described, which consists in performing γ-ray and conversion-electron measurements of radioactive sources prepared by on-line mass separation at the GSI UNILAC and the CERN SC. Results will be exemplified by showing measured decay schemes, Q_{EC} values, excitation energies of 1^+ levels, and $0^+ \rightarrow 1^+$ β transition probabilities, as obtained for the decay of the even nuclei $^{94}_{44}$Ru, $^{96-100}_{46}$Pd (ref. 3, 4), $^{100-104}_{48}$Cd (ref. 5), $^{104-110}_{50}$Sn (ref. 6), and to discuss the data with reference to model calculations (ref. 4, 7). As these aspects can be covered only briefly here, the reader is refered to refs. 3-7 for more details on the new GSI/CERN results, on data taken from the literature, and on the model calculations. We shall conclude with a few remarks on future possibilities of decay spectroscopy near ^{100}Sn.

EXPERIMENTAL TECHNIQUES

The nuclei ^{96}Pd and $^{104-108}$Sn were produced by fusion- evaporation reactions, using ^{40}Ca and ^{58}Ni beams from the GSI UNILAC, and ^{60}Ni, ^{50}Cr and ^{58}Ni targets, respectively. In case of the ^{104}Sn measurements, the more abundantly produced isobaric contaminants were strongly reduced by applying the technique of bunched beam release [6,8] at the GSI On-line Mass Separator. This novel technique consists in cooling a trap inside the FEBIAD ion source, heating the trap subsequently and setting a time window corresponding to the element-specific release characteristic.

The decay of ^{98,100}Cd was investigated at the ISOLDE II facility [9] on-line to the CERN SC. Chemically pure mass-separated beams of cadmium isotopes were produced, using spallation reactions between a molten tin target and a beam of 600 MeV protons.

Both for the GSI and the CERN experiments, the mass-separated beams were implanted into tapes, and the resulting point-like radioactive sources were periodically transported to detector stations. Germanium detectors and a miniorange spectrometer were used to perform γ-ray and conversion-electron measurements. Singles data were taken in multispectrum mode, while coincidences were stored for γγ and γ conversion-electron events.

EXPERIMENTAL RESULTS

From the singles and coincidence data measured for the decay of the 0^+ nuclei ^{96}Pd, ^{100}Cd, and $^{104-108}$Sn, partial level schemes were constructed for the daughter nuclei. The relative β decay feeding of individual levels was obtained from the intensity balance using measured or estimated multipolarities for internal-conversion corrections. The Q_{EC} values, deduced from experimental $\beta^+/(EC+\beta^+)$ probability ratios, together with the measured β-decay halflives allowed then to determine ft values from relative branching ratios.

As an example for the experimental results, $0^+ \to 1^+$ transitions are shown in Fig. 1 as observed for the decay of ^{96}Pd, ^{100}Cd, and ^{104}Sn, these nuclei having the same isospin-projection quantum number $T_z = (N-Z)/2 = 2$. We shall return below to a discussion of the number and the energy spacing of 1^+ levels, but note here that 4 such levels were observed for the decays of ^{96}Pd and ^{104}Sn, while 7 where identified in the case of ^{100}Cd. In this context it is important to realize that the quality of the spectroscopic data for ^{100}Cd is much higher than for the other two nuclei. Leaving aside differences in chemical purity of the radioactive sources, detector geometry and counting time, this quality is connected to the relevant (maximum) intensities of mass-separated beams, which amounted to 2.8×10^4 atoms/s for ^{100}Cd, 3.5×10^3 atoms/s for ^{96}Pd, 10^3 and 200 atoms/s for ^{104}Sn in DC mode and for bunched beam release, respectively. It is thus more probable that 1^+ states may have remained unobserved for the ^{96}Pd and ^{104}Sn decay than it is for the ^{100}Cd case.

While the search for decay properties of ^{102}Sn remained unsuccessful so far, some progress has been made recently in studying ^{98}Cd, this nucleus having two protons less than ^{100}Sn. From measuring KX rays, γ-rays, and conversion electrons, 60 and 107 keV transitions in ^{98}Ag were found and the half-life of ^{98}Cd was determined to be 8.1(5) s.

NUCLEAR STRUCTURE AND GAMOW-TELLER DECAY

In this section, we compare the new experimental results with model predictions. In order to calculate the $0^+ \to 1^+$ GT transition strength, we start out with an extreme single-particle shell model (ESPSM) and add then corrections for pairing as obtained by Dobaczewski et al.[4], and for core polarization as given by Towner[7].

Before discussing GT transition probabilities in detail, we show results from the calculations of Dobaczewski et al.[4] for Q_{EC} values and 1^+ excitation energies. The macroscopic-microscopic model is based on an average deformed Woods-Saxon potential, includes a residual interaction of the monopole pairing form, and uses the Strutinsky shell-correction method with a droplet-model mass formula for calculating total binding energies[4]. The Q_{EC} values of ^{94}Ru, $^{96-100}$Pd, $^{100-104}$Cd, $^{104-110}$Sn, and the 1^+ excitation energies of the corresponding daughter nuclei were considered. Best agreement was found by including deformation effects, that is by treating quadrupole (β_2) and hexadecapole (β_4) deformation as free parameters, while spherical symmetry both for this model and for a self-consistent Hartree-Fock-Bogolyubov approach yields poorer agreement.

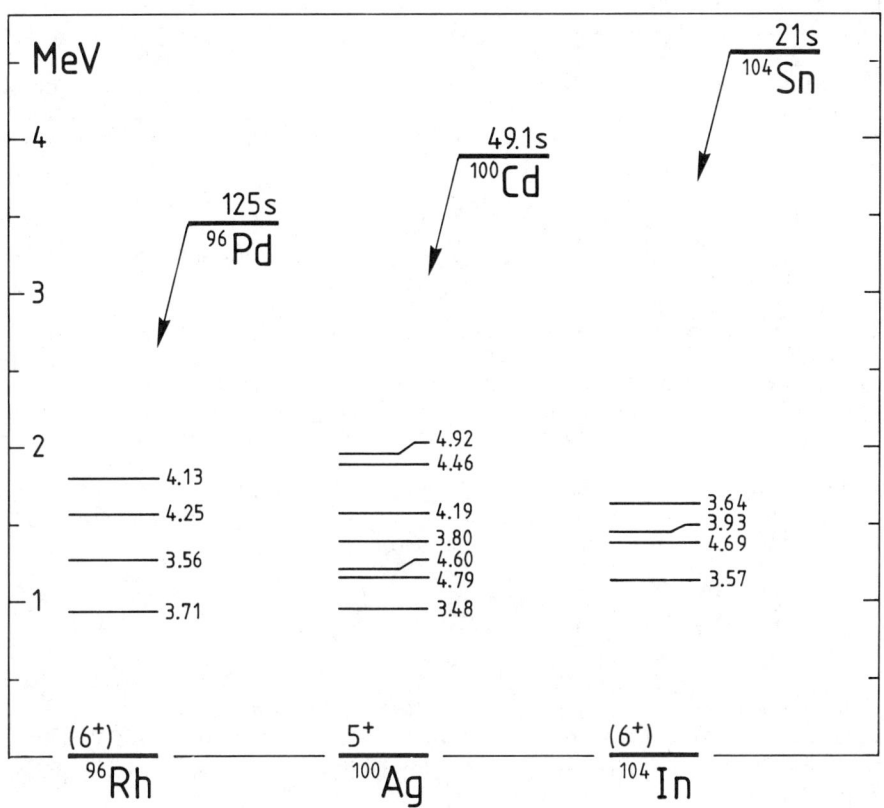

Fig. 1. Simplified level schemes for the β decay of ^{96}Pd, ^{100}Cd, and ^{104}Sn. The half-lives measured for the parent 0^+ ground-states are given. Concerning excited states in the daughter nuclei, only 1^+ levels are shown together with the measured log ft values. The 1^+ levels and the parent states are displayed on scale with reference to the respective daughter ground-state. Additional data, e.g., on excited daughter levels, and experimental uncertainties of half-lives, Q_{EC} values, level energies and log ft values, can be taken from refs. 3, 5, 6.

As an example, the results for Q_{EC} values of $^{104-110}$Sn, and for 1^+ excitation energies in $^{104-110}$In are shown in Fig. 2. The quadrupole-deformation parameters β_2, deduced from the calculation for the ground-states of odd nuclei, increase from 0.067 for ^{104}In to 0.12 for ^{110}In in fair agreement with recent results from collinear laser spectroscopy [10]. However, opposite to the ground-states, the calculations yield non-deformed shapes for the ($\pi g_{9/2}^{-1}$, $\nu g_{7/2}$) 1^+ configurations, and hence cannot explain the observed spread of the β strength over several GT transitions. Preliminary calculations indicate that this fragmentation may be related to residual proton-neutron interaction and particle-core coupling.

Turning now to a quantitive comparison between measured GT strength and model predictions, we take from experiment the reduced transition probability B_Σ (GT), summed over all identified 1^+ levels in a β-decay daughter according to

$$B_\Sigma(GT) = \frac{6160 \, s}{(g_A/g_V)^2} \sum_i \left(\frac{1}{(ft)_i}\right). \quad (1)$$

B_Σ(GT) as well as the calculated values of B(GT) to be discussed below, are given in units of $g_A^2/4\pi$ (compare ref.11). The constant in the numerator, and the free-neutron ratio of the weak coupling constants $|g_A/g_V|$ = 1.263, are taken from ref.7; we omit experimental errors of these constants in the discussion here, since the corresponding uncertainties are far below those from ft-determination.

According to the ESPSM without residual interaction the GT strength for the decay of neutron-deficient even isotopes in the ^{100}Sn region is concentrated in a single $\pi g_{9/2} \rightarrow \nu g_{7/2}$ transition and is given by

$$B(GT) = \frac{160}{9} v_\pi^2 (1 - v_\nu^2), \quad (2)$$

v_π^2 and v_ν^2 being the occupation numbers of protons and neutrons in the $\pi g_{9/2}$ and $\nu g_{7/2}$ orbits, respectively. Within the extreme model ESPSM it is assumed, that the valence protons between niobium (Z = 41) and tin (Z = 50) occupy $g_{9/2}$ orbits (v_π^2 = (Z-40)/10), and that the valence neutrons above N = 50 occupy first $d_{5/2}$ and then $g_{7/2}$ orbits (v_ν^2 = 0 for N $\leq$ 56; v_ν^2 = (N-56)/8 for 57 $\leq$ N $\leq$ 64). In order to include pairing effects, the occupation numbers in equ. 2 were taken from the macroscopic-microscopic model calculations of Dobaczewski et al.[4]. As can be seen from Fig.3, the resulting quenching factors B_Σ(GT)/B(GT) are of the order of 0.1 to 0.2. The pairing corrections shift the calculated strength values slightly towards the experimental ones except for the decay of ^{94}Ru, ^{96}Pd and ^{110}Sn where the opposite effect is observed.

An important reduction of the predicted GT strength is obtained when taking first-order core polarization [7] into account. The resulting quenching factors, displayed also in Fig.3, are of the order of 0.5 for the decays of ^{94}Ru, ^{96,98}Pd, this value being in agreement, e.g., with β-decay data obtained for sd-shell nuclei [12], and with results from (p, n) reaction studies [13]. For ^{102,104}Cd and $^{104-110}$Sn,

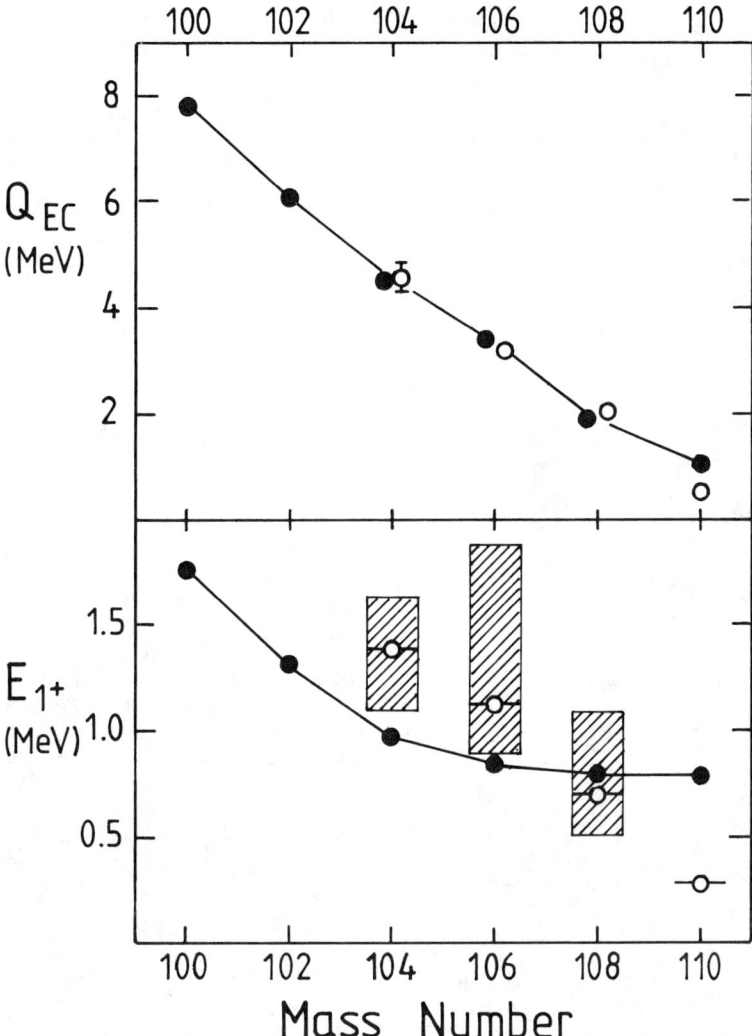

Fig. 2. Measured Q_{EC} values for $^{104-110}$Sn decays and measured average 1^+ excitation energies for $^{104-110}$In, in comparison with results obtained from a shell-correction model with a Woods-Saxon average field and monopole pairing residual interaction (model "II-def" from ref. 4). The model predictions, extending down to mass-number 100, are given as full circles connected by straight lines. Measured values are indicated as open circles. For the Q_{EC} values of $^{104-108}$Sn, the circles are drawn slightly aside from the mass-number coordinate in order to avoid overlap. For the 1^+ states of $^{104-110}$In, the hatched areas correspond to the energy ranges of experimentally identified levels, while the open circles give the average values. The latter are calculated as a sum over these levels weighted with their measured GT strenghts.

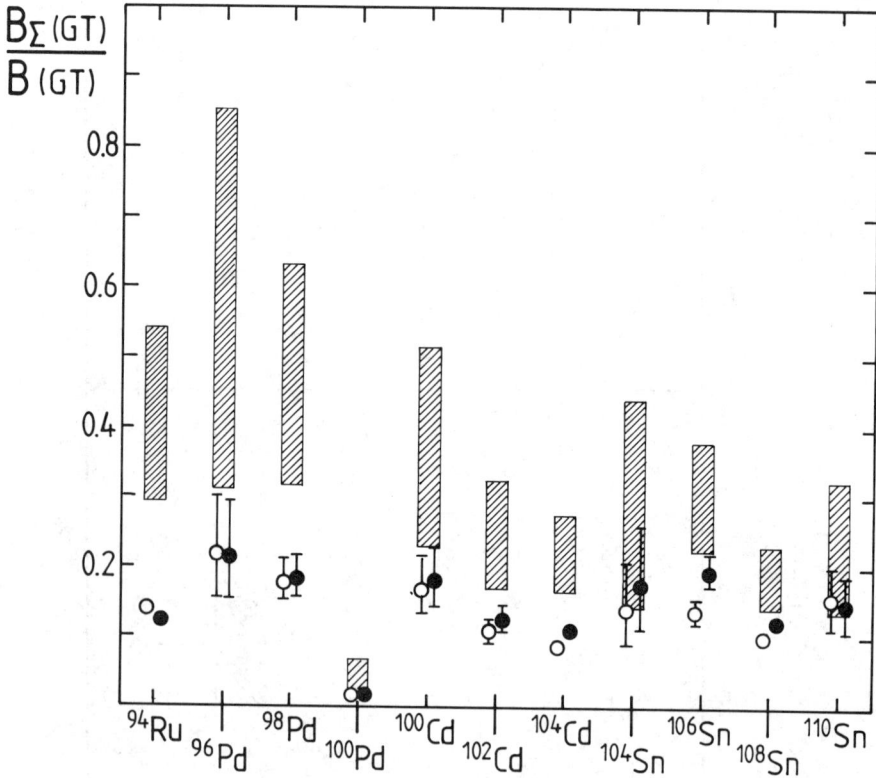

Fig. 3. GT quenching factors for $0^+ \rightarrow 1^+$ β decays of neutron-deficient ruthenium-to-tin isotopes, obtained as ratios between measured GT transition probabilities and model predictions. The experimental values $B_\Sigma(GT)$ represent summing over all observed β decay branches to 1^+ states. The predictions $B(GT)$ result from an extreme single-particle shell-model (open circles) and from a correction of this model for pairing[4] (full circles). The circles are drawn slightly aside from the isotope coordinate in order to avoid overlap, and show experimental uncertainties only. The range for quenching factors, which result from taking core-polarization corrections into account (hatched bars), is due to experimental uncertainties and to various choices of the effective interaction[7]. The calculations of ref.7 yield quenching factors for the decay of $N=50$ isotones between ^{94}Ru and ^{100}Sn. In this plot it is assumed, that for a given proton number (or isotope) these factors stay constant independent upon neutron number (or mass number).

however, the GT strength reduction appears to be stronger, the quenching factor being roughly 0.2 to 0.3.

The origin of such GT strength quenching is searched for in higher-order effects such as configuration mixing due to tensor forces or mixing with Δ-isobar nucleon-hole configurations. In this context, one should bear in mind that part of the strength may be missed experimentally in cases where the 1^+ state(s) is close or even outside the β-decay window. This represents a problem for studying the decays of ^{94}Ru, ^{100}Pd, and ^{110}Sn (see, e.g., the Q_{EC} value of ^{110}Sn and the 1^+ level energy in ^{110}In, displayed in Fig.2), and hence the quenching factors obtained for these nuclei are presumably not the most reliable ones. On the other hand, even for large Q_{EC} values weak β-decay branches to high-lying 1^+ states may remain unobserved. In this respect, higher intensities and higher purity of the radioactive sources are clearly desireable.

CONCLUDING REMARKS

We presented results from recent measurements and from literature for $0^+ \rightarrow 1^+$ GT β decays in the ^{100}Sn region, and showed that a comparison with model predictions suggests GT quenching factors of approximately 0.5 for the decays of ^{94}Ru and ^{96,98}Pd, and approximately 0.2 to 0.3 for the decays of ^{102,104}Cd and $^{104-110}$Sn.

In order to get beyond the approximate characters of these quenching factors, the model calculations ought to be improved, e.g., with respect to the residual proton-neutron interaction. For experiments, a twofold task seems to emerge. On the one hand side, the decays studied already in this area of the nuclidic chart should be re-investigated by "more complete" spectroscopic means which allow to identify also weak GT transitions to 1^+ levels at high excitations energies and/or to derive a quantitative detection limit for such transitions. In this context, spectroscopy of β-delayed protons and application of a sum spectrometer for γ-rays have to be considered. On the one hand, it is important to systematically extend these measurements to nuclei even further away from the β-stability line, where due to higher Q_{EC} values the probability of having 1^+ levels outside the observation window is smaller. The properties of such new neutron-deficient isotopes can be predicted on the basis of the results presented in this paper. For the decays of ^{98}Cd, ^{100}Sn and ^{102}Sn, for example, a log-ft value of 3.1 follows from the GT quenching systematics[6] which, together with calculated $0^+ \rightarrow 1^+$ energies (see Fig.2 and ref.4), yield half-lives of 6 s, 0.6 s, and 2 s, respectively. The measured ^{98}Cd half-live of 8.1(5) s is in good agreement with this estimate. An experiment planned at the ISOLDE facility aims at identifying $0^+ \rightarrow 1^+$ β decays of ^{98}Cd which ought to be feasible with a beam intensity of 100 atoms/s expected for this isotope. In the cases of ^{102}Sn and ^{100}Sn, perspectives are not quite as optimistic concerning the available proton or heavy-ion beams. We thus may have to wait for the fragment separator at the SIS/ESR facility[14] where, on the basis of admittedly rather tentative estimates of production cross-sections, such exotic nuclei may be produced separated in sufficient source strength.

1. B.A. Brown and B.H. Wildenthal, At. Data and Nucl. Data 33, 347 (1985).
2. T. Sekine et al., Nucl. Phys. A 467 , 93 (1987).
3. K. Rykaczewski et al., Z. Phys. A 322 , 263 (1985).
4. J. Dobaczewski et al., Structure of Nuclei near ^{100}Sn and the $\pi g_{9/2} \rightarrow \nu g_{7/2}$ Gamow-Teller Beta Decays, to be published.
5. K. Rykaczewski et al., Gamow-Teller Transitions in the Beta Decay of ^{100}Cd, to be published.
6. R. Barden et al., The Gamow-Teller Decay of Neutron-Deficient Isotopes of Tin, to be published.
7. I.S. Towner, Nucl. Phys. A 444 , 402 (1985).
8. R. Kirchner, Nucl.Instr. and Meth. in Phys. Res. B 26 , 204 (1987).
9. H.L. Ravn et al., Nucl. Instr. and Meth. 123, 131 (1975).
10. J. Eberz et al., Nucl. Phys. A 464 , 9 (1987)
11. A. Bohr and B. Mottelson, Nuclear Structure, Vol.1(Benjamin, New York, 1969), p.349.
12. B.A. Brown and B.H. Wildenthal, At. Data and Nucl. Data Tables 33, 347 (1985)
13. C. Gaarde, in Proc. Int. Conf. Weak and Elactromagnetic Interactions in Nuclei, H.V. Klapdor, ed. (Springer, Berlin-Heidelberg, 1986), p.260.
14. P. Armbruster et al., The Projectile-Fragment Separator at the Darmstadt SIS/ESR Facility, contribution to this conference.

NUCLEAR STRUCTURE EFFECTS ON α REDUCED WIDTHS

K. S. Toth, Y. A. Ellis-Akovali, H. J. Kim, and J. W. McConnell
Oak Ridge National Laboratory, Oak Ridge, TN 37831 USA

H. K. Carter
UNISOR, Oak Ridge, TN 37831 USA

D. M. Moltz
Lawrence Berkeley Laboratory, Berkeley, CA 94720 USA

ABSTRACT

A review of α widths for s-wave transitions is presented together with a discussion of the following topics: 1) a new determination of the ^{218}Ra half-life and its relation to reflection asymmetry in nuclei near N = 130, 2) a measurement of the ^{194}Pb α-decay rate and the influence of the Z = 82 gap on neutron-deficient Pb nuclei and 3) an up-date of α-decay-rate systematics for isotopes in the rare earth and medium-weight mass regions.

I. Introduction

Alpha-decay transitions between ground states of doubly-even nuclei are taken to represent unhindered decays. Reduced widths for these s-wave transitions behave in a regular fashion as a function of both neutron and atomic number. They are largest for nuclei two or four particles beyond a closed shell (with sharp minima at the shell) and they then decrease as the next closure is approached. This is shown in Fig. 1 where s-wave widths for nuclei with Z from 78 to 100 are plotted as a function of N. In our discussion, α-decay rates are considered within the theoretical framework developed by Rasmussen.[1] Here the reduced width, δ^2, is defined as; $\delta^2 = \lambda h/P$, where λ is the decay constant, h is Planck's constant, and P is the penetrability factor for the α particle to tunnel through a barrier. One sees the regularity of the reduced widths as a function of neutron number with the extremely sharp break at N = 126. This discontinuity has been shown to be a shell structure effect (see e.g. Ref. 2). A less pronounced minimum is seen at the subshell closure at N = 152.

II. Alpha-decay Widths and Nuclear Asymmetry

Interwoven bands of low-lying negative and positive parity levels, connected by enhanced E1 transitions, provide evidence that light radium and thorium isotopes have reflection asymmetry. Two explanations have been posited for the asymmetry, namely, static octupole deformation or α clustering on the nuclear surface. One set of suporting data for the second explanation consists of rather large α-decay reduced widths for nuclei in this N ⩾ 130 region.

© American Institute of Physics 1988

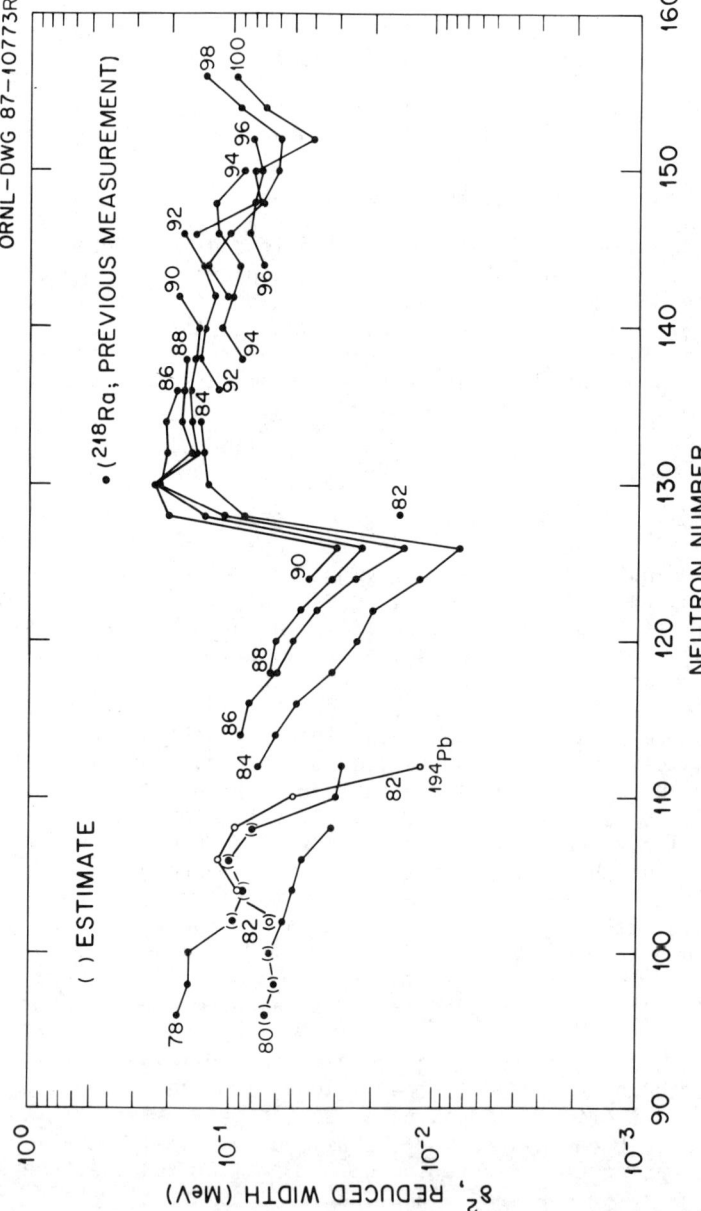

Fig. 1. Reduced widths for s-wave α transitions plotted as a function of N for even-even isotopes with Z from 78 to 100. The point at N = 130, labeled as ^{218}Ra, is the width calculated by using the data of Ref. 5. Widths for Pb nuclei are indicated by open points.

The ^{218}Ra width, in particular, exhausts about 75% of the Wigner-sum-rule limit. This is seen in Fig. 1 where the indication is that the ^{218}Ra value (open point at N = 130) is far too large with respect to other neighboring widths. It is about a factor of two greater than the δ^2 value expected from systematics.

The ^{218}Ra α-decay half-life was recently remeasured[3] at the Holifield Heavy Ion Research Facility (HHIRF) with a novel experimental technique[4] using a velocity filter. The ^{218}Ra activity was produced by bombarding a ^{208}Pb target with a ^{13}C beam from the tandem accelerator. Reaction products, including the ^{218}Ra nuclei, were implanted in a surface barrier detector after they had been separated from the direct accelerator beam. The implantation process provided a start signal for a calibrated time spectrum while subsequent α decays supplied the stop signal.

In Fig. 2 we display some of the energy and time distributions that were obtained. Parts (a) and (b) of Fig. 2 show the spectra of heavy recoils and α particles, respectively, recorded in the Si(Au) detector during the experiment. The three α groups in Fig. 2(b) are ^{218}Ra, its α-decay daughter ^{214}Rn, and a peak which encompasses events wherein both α energies are summed because of the short, 0.27-μs, half-life of ^{214}Rn. As a result of the 2-μs flight time through the velocity filter most of the short-lived ^{214}Rn nuclei have to arise from ^{218}Ra decay rather than from independent production. Since the α branch of ^{214}Rn is 100%, this is indeed reflected in the essentially equal intensities of the ^{218}Ra and ^{214}Ra α groups in Fig. 2(b).

Within uncertainties all three peaks had the same half-life, which was almost a factor of two larger than the 14-μs value previously reported[5] for ^{218}Ra. Figure 2(c) represents the time distribution for all α decays recorded in Fig. 2(b) spread over the 80-μs TAC range. The decay curve generated by setting gates only on the three peaks is shown in Fig. 2(d); the resultant half-life is 25.6 ± 1.1 μs. From this longer half-life we deduce a δ^2 value of 0.23 MeV which, in Fig. 1, is indistinguishable from the ^{216}Rn and ^{220}Th points. The result is a smooth trend of α widths from the N = 130 region to the well-deformed, prolate, Cm, Cf, and Fm nuclei, and weakens the argument for the existence of α clusters in the heavy elements. (See Ref. 3 for a more complete discussion.)

To broaden the applicability of this technique to α emitters with longer half-lives we devised a clock for time measurements in the ms regime. Here the α-decay events are tagged with a signal from a clock that is started when the evaporation residue is implanted and reset each time that an α-particle event within the expected energy range is registered. Should the α-decay event not occur within a preset time range the clock is automatically reset to await a new start pulse. This clock has been tried out successfully with half-lives in the range from 1 to 100 ms. One such test involved the investigation of light thorium isotopes produced in oxygen bombardments of lead targets. Figure 3 shows the α-particle spectrum observed in 90-MeV ^{18}O irradiations of ^{207}Pb. In that

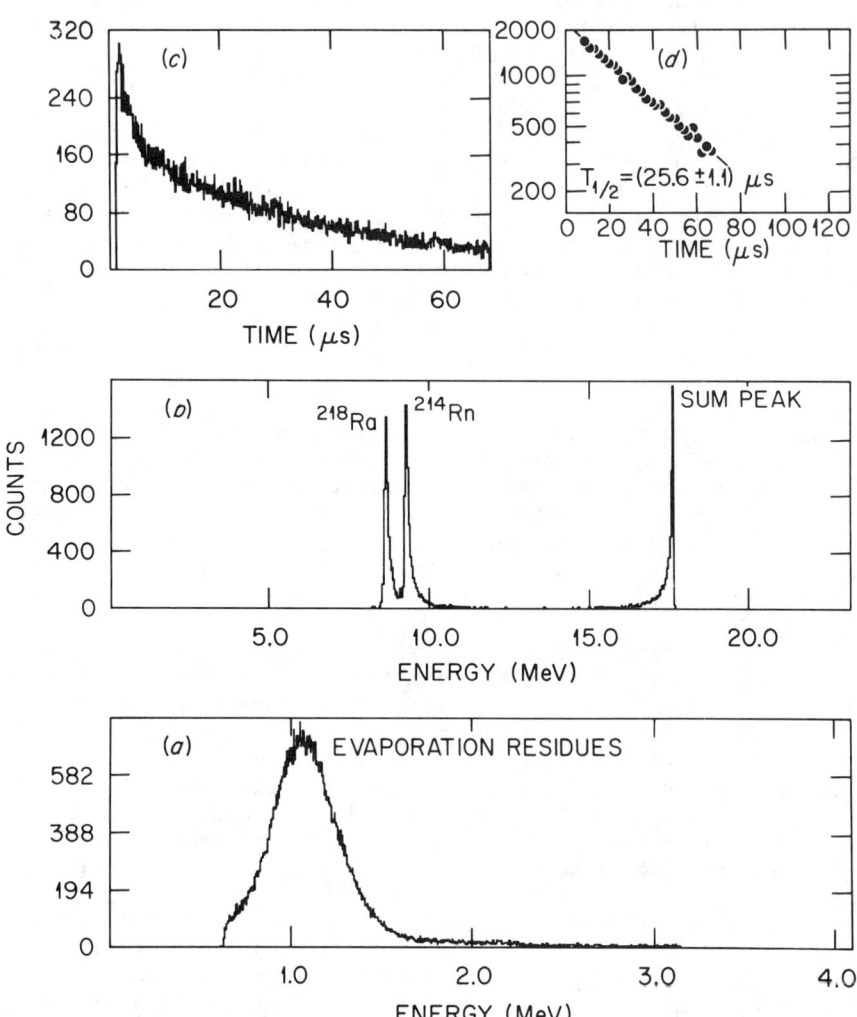

Fig. 2. Energy and time distributions measured in ^{13}C bombardments of ^{208}Pb after the product recoils had passed through the HHIRF velocity filter. Energy spectra recorded for products stopped in a Si(Au) surface barrier detector are shown in part (a), while subsequent α-particle decays registered in the same detector are shown in part (b). Part (c) is the time distribution of all recorded α decays while part (d) shows the decay curve deduced from the time distribution gated by just the three α peaks in part (b).

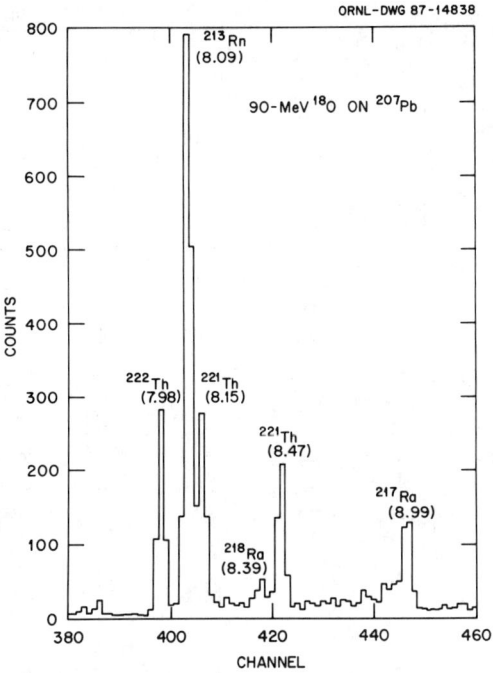

Fig. 3. Spectra recorded in a Si(Au) detector of α-particles emitted by implanted residues produced in ^{18}O bombardments of ^{207}Pb.

study the half-lives of ^{221}Th and ^{222}Th were determined to be 1.8 ± 0.1 and 2.7 ± 0.2 ms, respectively, in agreement with published values. We plan to expand our capabilities further by using a position sensitive solid state detector so that site information can be obtained to assist in establishing parent-daughter-granddaughter decay correlations.

III. Alpha-decay Rates of Neutron-deficient Lead Isotopes

We recently discussed[6] the s-wave widths of ^{186,188,190,192}Pb. These widths were found to have a dependence on N similar to that observed for other neighboring elements, i.e., they decreased in value as the N = 126 shell was approached. However, contrary to the expectation of a shell effect on α-decay rates at Z = 82, they were larger than those of mercury isotopes with the same neutron numbers. We suggested that this result could be related to a predicted[7] disappearance of the Z = 82 gap midway between N = 82 and N = 126.

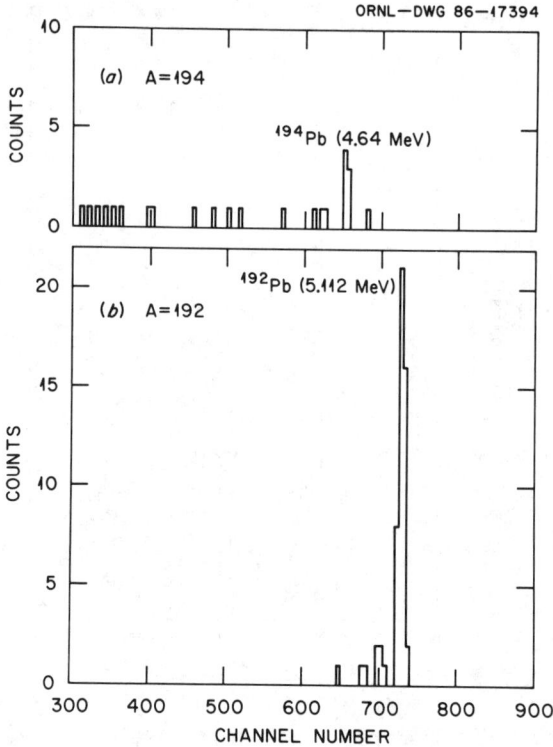

Fig. 4. Alpha-particle spectra measured for A = 194 [part (a)] and A = 192 [part (b)] radioactivities.

(A related explanation may be the varying shapes of Pb, Hg, and Pt isotopes in this mass region, i.e., α decay between the slightly oblate Pb and Hg ground states is not as hindered as it is between the oblate Hg and the prolate Pt ground states.) If correct, this suggestion implies that widths for lead nuclides with A > 192 should first come closer to, and then, for larger neutron numbers, fall below mercury values.

A major difficulty in confirming this expectation arises because α-decay energies decrease with increasing N and the concomitant α branching ratios become extremely small. In fact, the α decay of ^{194}Pb had not been observed. We decided to search for ^{194}Pb α emission and to determine the nuclide's α-branching ratio by examining its (E.C. + β^+) decay scheme in detail. Available data[8] list only a 204-keV transition in ^{194}Pb decay.

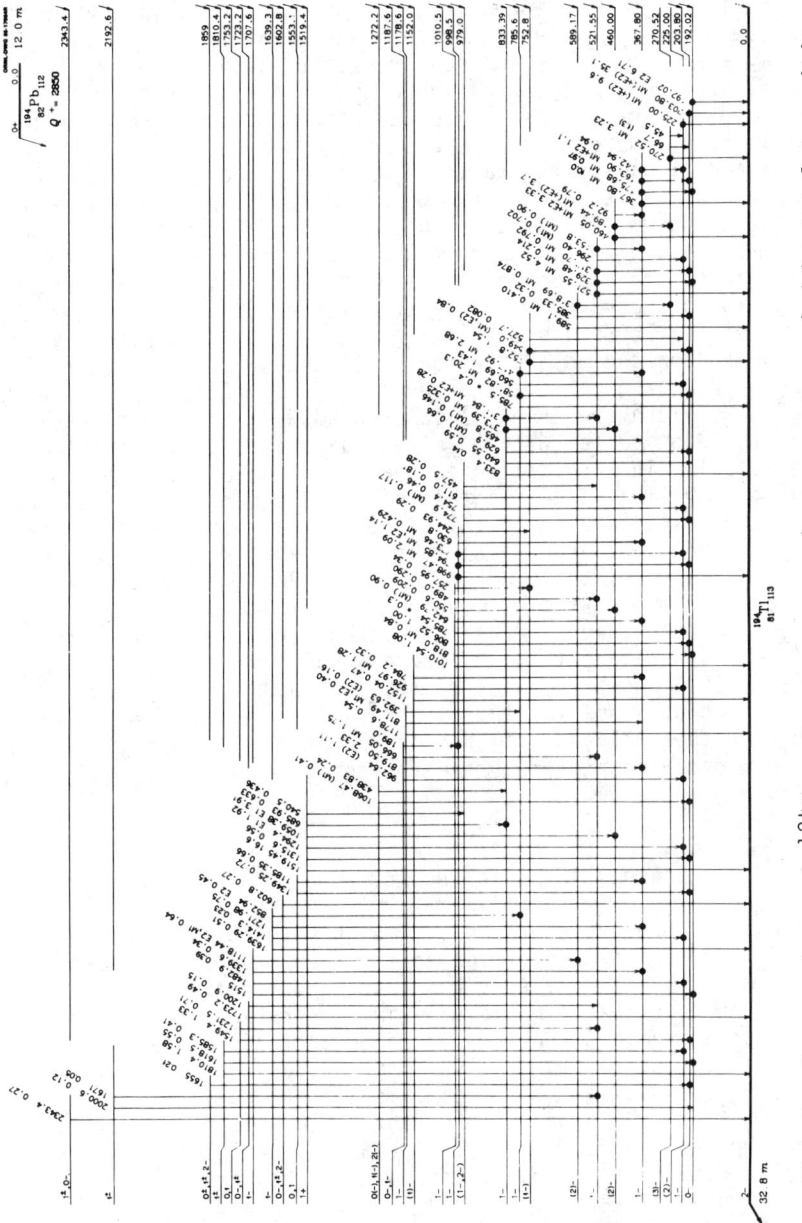

Fig. 5. Decay scheme of ^{194}Pb. Gamma-ray energies are accompanied by total transition intensities and by multipolarities where available. Coincidences are indicated by dots placed at the top and bottom of transition lines.

To this end we produced ^{194}Pb in bombardments of natural W with ^{16}O ions from the HHIRF tandem accelerator. Following mass separation at the University Isotope Separator On-line at Oak Ridge (UNISOR) facility, the isotope's decay properties were investigated with the use of γ-ray, x-ray, electron, and α-particle detectors. The α decay energy of ^{194}Pb was measured to be 4.64 ± 0.02 MeV. This is illustrated in Fig. 4(a) where the α-particle spectrum measured for A = 194 is shown. Sources of ^{240}Pu and ^{244}Cm were used as calibration standards. As a check of the geometry used and to provide an additional energy calibration point, the ^{192}Pb α-decay branching ratio[9] was also measured with the same experimental setup. The A = 192 spectrum is displayed in Fig. 4(b). A comprehensive β-decay scheme (Fig. 5) was constructed by using γ-γ and e-γ coincidence data and transition energies and intensities. As indicated in Fig. 5, it incorporates 91 γ-ray transitions into a scheme of 29 excited levels built on the 2$^-$ ^{194}Tl ground state. Based on this information and on the absolute α-particle and γ-ray intensities, the ^{194}Pb α-branching ratio was determined to be $(7.3 \pm 2.9) \times 10^{-8}$.

The ^{194}Pb width (see Fig. 1) calculated with these data is much less than those of nuclei with N ≤ 110 and is less than that of ^{190}Pt (N = 112). The indication is that the Z = 82 gap is being restored for N ≥ 112. The α widths for ^{190}Hg and ^{192}Hg need to be determined to be certain about this proposal; if correct, then the ^{192}Hg width should be larger than that of ^{194}Pb. We also show in Fig. 1 an estimated width (0.060 MeV) for ^{184}Pb based on the data of Schrewe et al.[10] and on the assumption of an α branch of 100%. [From gross β-decay theory[11] the calculated partial half-life for ^{184}Pb (β$^+$ + EC) decay is (5 - 10) seconds which yields an α branching of (89 - 96%) and a δ^2 value of (0.053 - 0.057 MeV.)] This δ^2 value is appreciably smaller than the ^{186,188,190}Pb widths and may indicate that the influence of the Z = 82 shell is also beginning to retard the α decay of extremely neutron-deficient Pb isotopes, i.e., those with N ≤ 102.

IV. Reduced Widths for Nuclei with 84 ≤ N ≤ 100

The α-decay properties of medium-weight isotopes for elements below lead have now been investigated for about 40 years, and the known number of these α emitters has increased steadily as more and more heavy-ion accelerators have gone into operation. Together with the expansion in numbers there has been an increase of reliable branching ratios so that the analysis of α-decay rates for these neutron-deficient nuclides has become more instructive.

In Fig. 6 we show s-wave widths for nuclei with 60 ≤ Z ≤ 80. The Pt and Hg data are included to provide a visual continuity with Fig. 1. The data here are not as copious as they are for the heavier elements. Nevertheless, one sees the same effect due to N = 82 as the one noted earlier at N = 126, i.e., the δ^2 values peak at about four neutrons above the shell closure. The widths then decrease as the neutron numbers begin to increase in the direction

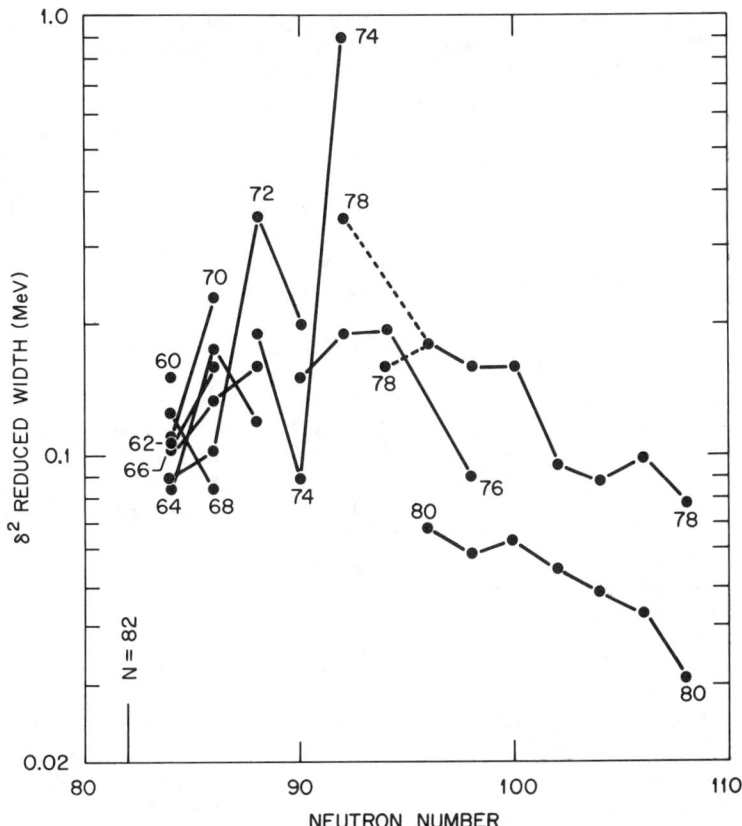

Fig. 6. Reduced widths for s-wave α transitions of even-even nuclei with 60 ≤ Z ≤ 80.

of $N = 126$. Only in the case of Er nuclei is the $N = 86$ (^{154}Er) value lower than that of the $N = 84$ (^{152}Er) nuclide. As noted in the original work[12] the α branch measured for ^{154}Er is undoubtedly low due to the presence ^{155}Er in the sources used in determining the α/total branching ratios.

No rare earth α emitters with $N < 84$ have been identified since the 82-neutron shell causes a large drop in the Q_α values for these nuclei. Also, the proximity of the proton drip line raises both Q_{EC} values and probabilities for delayed and direct proton emission. Only when one gets down to the proton-rich Te and Xe nuclides does α decay become once again observable. However, just one experimental α branch[13] for an s-wave transition is available in that mass region, namely that of ^{108}Te. The corresponding width, ~ 0.4 MeV, is consistent with values for the $N \geq 82$ nuclei shown in Fig. 6.

The existence of the $Z = 64$ subshell was established some three decades ago on the basis of Q_α systematics for $N = 84$ nuclei. More recently the closure was also noted[14] in the α-decay rates for the same set of isotones. Originally due to an incorrect determination of the ^{150}Dy α branch the experimental rates seemed to be lowest at Z = 66 rather than at Z = 64 as calculations,[15] using a Gaussian residual force in a BCS treatment, had indicated. Newer measurements[12] showed that the ^{150}Dy α branch was a factor of two greater than the previous ratio; this in turned resulted[14] in a dip for reduced widths at Z = 64 in agreement with the calculations.[15] A recent determination[16] of the ^{146}Sm α-decay energy and half-life contributes to the localization of this dip, though the availability of two rather different half-lives (see Ref. 16) for ^{148}Gd leaves the situation unclear as to just how pronounced this minimum at Z = 64 really is.

In addition to the ^{154}Er problem mentioned above, one notes in Fig. 6 several other δ^2 values that do not fit well into the general pattern. It is outside the scope of the present paper to go into details of what experimental errors may be causing these deviations; the indications are clear that better experimental data are needed.

We conclude by noting that α widths near N = 130 and N = 86 are comparable. However, while enhanced dipole transitions have been observed for light thorium and radium nuclei, similar transitions have not been shown to occur in rare earth nuclei. Large α widths therefore do not seem to be correlated with the phenomenon of reflection asymmetry.

The Oak Ridge National Laboratory is operated by Martin Marietta Energy Systems, Inc. for the U. S. Department of Energy under Contract No. DE-AC05-84OR21400. UNISOR is a consortium of Universities, State of Tennesse, Oak Ridge Associated Universities, and Oak Ridge National Laboratory and is partially supported by them and by the U. S. Department of Energy under Contract No. DE-AC05-76OR00033 with Oak Ridge Associated Universities.

REFERENCES

1. J. O. Rasmussen, Phys. Rev. 113, 1593 (1959).
2. H. J. Mang, Ann. Rev. Nucl. Sci. 14, 1 (1964).
3. K. S. Toth, et al., Phys. Rev. Lett. 56, 2360 (1986).
4. H. J. Kim, et al., Nucl. Instrum. Methods Phys. Res. A249, 386 (1986).
5. K. Valli, et al., Phys. Rev. C 1, 2115 (1970).
6. K. S. Toth, et al., Phys. Rev. Lett. 53, 1623 (1984).
7. R. A. Sorensen, Nucl. Phys. A420, 221 (1984).
8. B. Harmatz, Nucl. Data Sheets 22, 433 (1977).
9. K. S. Toth, et al., Phys. Rev. C 19, 2399 (1979).
10. U. J. Schrewe et al., Phys. Lett. 91B, 46 (1980).
11. K. Takahashi, et al., At. Data Nucl. Data Tables 12, 101 (1973).
12. K. S. Toth, et al., Phys. Rev. C 10, 2550 (1974).
13. E. Roeckl, et al., Phys. Lett. 78B, 393, (1978).
14. W.-D. Schmidt-Ott and K. S. Toth, Phys. Rev. C 13, 2574 (1976).
15. R. D. Macfarlane, et al., Phy. Rev. 134, B1196 (1964).
16. F. Meissner, et al., Z. Phys. A 327, 171 (1987).

"The submitted manuscript has been authored by a contractor of the U.S. Government under contract No. DE-AC05-84OR21400. Accordingly, the U.S. Government retains a nonexclusive, royalty-free license to publish or reproduce the published form of this contribution, or allow others to do so, for U.S. Government purposes."

ALPHA - DECAY OF LIGHT PROTACTINIUM ISOTOPES

T. Faestermann, A. Gillitzer, K. Hartel[+], W. Henning[*], P. Kienle[*]
Technische Universität München, D8046 Garching, FRG

ABSTRACT

Light protactinium isotopes have been produced with ^{204}Pb (^{19}F,xn) reactions. α-activities with E_α=9.90(5) MeV, $T_{1/2}$=53(10) ns and E_α=9.65(5) MeV, $T_{1/2}$=0.78(16) μs could be attributed to the previously unobserved nuclei ^{219}Pa and ^{220}Pa with the help of excitation functions. The peak cross sections for the 4n and 3n evaporation channels are on the order of 10 μb. The decay energies as well as the halflives fit well into the systematics of these nuclei close to the magic neutron number N=126. ^{219}Pa is the shortest lived nuclide known with directly measured halflife.

INTRODUCTION

For nuclei heavier than doubly magic ^{208}Pb the shell gap slowly decreases with increasing Z as has e.g. been demonstrated by Schmidt et al.[1]. They compared the α-decay Q-values for isotopes with Z≥84 and N=128 with those for N=126. For the protactinium isotopes however the nuclides with 128≤N≤130 are not yet known. One reason may be, that the halflives for these nuclei are expected to be shorter than 10 μs and conventional transport systems are too slow to study them. In addition the cross sections to populate these nuclei in heavy ion induced fusion evaporation reactions are rather small because the dominant channel is fission of the compound nucleus. Here we report on the investigation of light protactinium nuclides with a setup specially designed to study shortlived activities.

EXPERIMENTS

Protactinium isotopes were produced using the ^{204}Pb(^{19}F,xn) $^{223-x}$Pa reaction. The beam energy was varied between 85 MeV and 103 MeV and excitation functions were measured. In order to be sensitive to a larger range of halflives, the beam was pulsed with a gross structure of 8 μs long beam pulses every 26 μs and a fine structure with a period of 400 ns and beam bunches roughly 2 ns long. The 0.5 mg/cm² thick target was enriched to 99.7% in ^{204}Pb. Evaporation residues recoiling out of the target were stopped in an organic catcher foil 6 cm behind the target, while the primary beam passed the catcher through a small hole. The time of flight for the recoils amounts to about 30 ns. α-particles emitted in the decay of the caught recoils were detected in backward direction by an annular system of gas detectors covering 7% of 4π.

[+] Present adress: Eurosil GmbH, Eching, FRG
[*] Present adress: GSI Darmstadt, FRG

The detector system consists of a parallel plate avalanche counter (PPAC) measuring the arrival time against the pulsed heavy ion beam, and an ionization chamber (IC) with longitudinal field. The IC measures the energy of the particle and determines its Z by analyzing the Bragg curve[2]. An independent particle identification is possible by a range measurement via the drift time of the ionization track towards the anode of the IC.

An energy spectrum of α-particles taken at a beam energy of 100 MeV in the intervals between the micropulses is shown in Fig. 1. Most lines are labelled with the known α-emitters, which are produced in the reaction. An α-line at (9.90±0.05) MeV and a component of the line at (9.65±0.05) MeV are attributed to the unknown decay of the nuclei ^{219}Pa and ^{220}Pa respectively as will be explained in the following. Fig. 2a) shows the time spectrum of the 9.90 MeV line, which decays within the period between the micropulses. It is fitted by a halflife of 53±10 ns. The time spectrum of the 9.65 MeV line (Fig. 2b) shows two components. The short one is fitted by a halflife of 80±13 ns corresponding to the known[3] decay of ^{217}Ac. The slow decay is better visible between the macropulses (Fig. 3) and fitted by a halflife of 0.78±0.16 μs.

The highest yield of 12 μb for the 9.90 MeV line was obtained at a beam energy of about 102 MeV, where also the yield for the α3n evaporation channel has its maximum, supporting the assignment of the 9.90 MeV line to another 4-particle channel, the ^{204}Pb(^{19}F,4n)^{219}Pa reaction. The slowly decaying component in the 9.65 MeV line has a maximum cross section of roughly 20 μb at a beam energy around 96 MeV near the peak energy for the (α2n) evaporation channel, indicating that 3n evaporation to ^{220}Pa could be the origin of this line. A comparison of the excitation function with a calculation using the statistical model code HIVAP[4] with standard parameters shows the maxima of the 3n and 4n evaporation only 2 MeV below the observed values, supporting these assignments, but the predicted peak cross sections are a factor of 30 too large. A drastic reduction of the xn evaporation channels however has been reported earlier for Th compound nuclei by Vermeulen et al.[5] and could be explained by a decreasing influence of the N=126 shell gap at high excitation energy.

DISCUSSION

Using a simple Gamow factor we can compare the transition strength of the newly observed α-emitters with that of ^{212}Po for an L-value of 0. For ^{219}Pa the decay is a factor of two faster, while the decay of ^{220}Pa has a strength of only 0.5. These values agree with the trend observed in the lighter α-emitters above lead, that the decay of the N=128 isotones becomes increasingly faster with increasing Z, and that the N=129 isotones decay slightly slower than ^{212}Po.

For ^{219}Pa and its daughter nucleus ^{215}Ac we expect a 9/2- groundstate and the observed α-decay then connects the two ground-states with a Q_α-value of (10.08±0.05) MeV. In the case of the ^{220}Pa decay to ^{216}Ac low lying 1- and 9- states from the $(\pi h_{9/2}, \nu g_{9/2})$

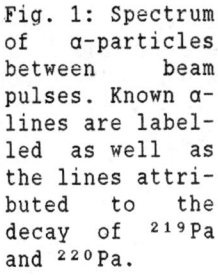
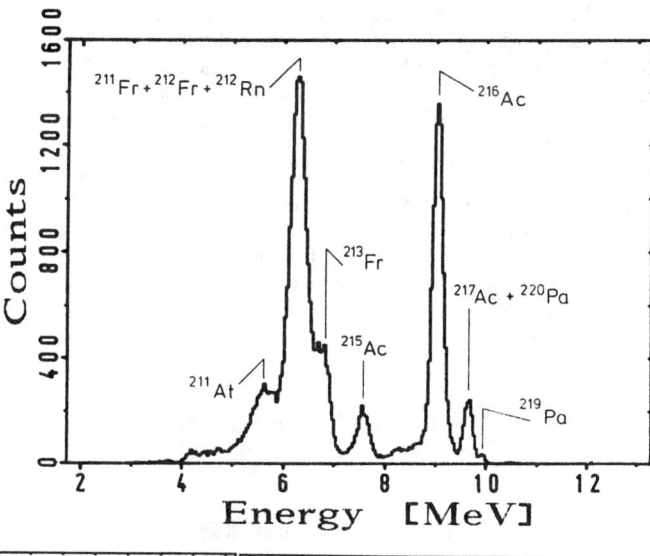

Fig. 1: Spectrum of α-particles between beam pulses. Known α-lines are labelled as well as the lines attributed to the decay of ^{219}Pa and ^{220}Pa.

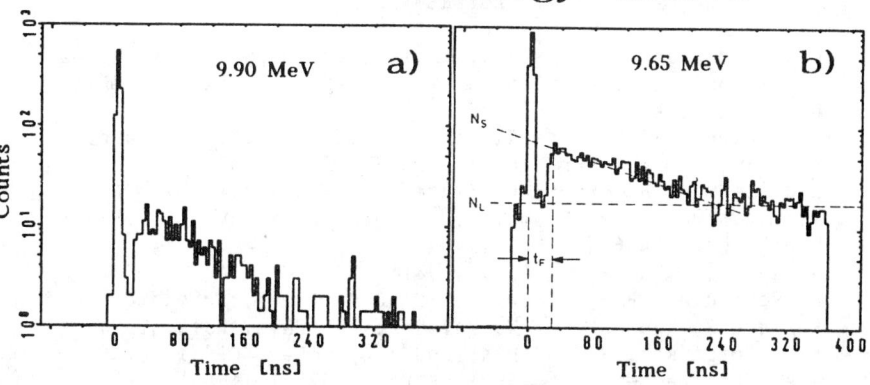

Fig. 2: Time spectra between the micropulses for a) the 9.90 MeV and b) the 9.65 MeV α-lines. In fig. 2b) the fast decaying component due to ^{217}Ac, the slow component and also the time of flight for the recoils are indicated.

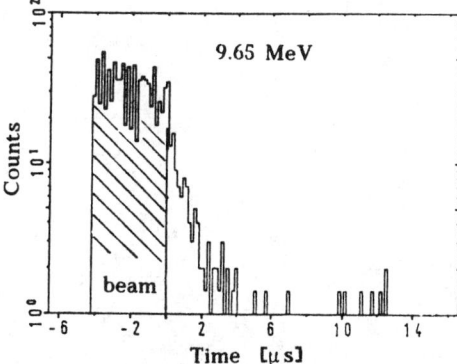

Fig. 3: Time spectrum of the 9.65 MeV α-line in the macroperiod. The decay with a half-life of 0.78 µs is attributed to ^{220}Pa.

multiplet are expected. In ^{216}Ac the 9⁻ state lies 37 keV above the 1⁻ state[6] and for ^{220}Pa the two states may be inverted, judging from the systematics of the odd-Z, N=129 nuclei. Thus we conclude, that the observed α-line from ^{220}Pa proceeds from the 9⁻ groundstate (or very low lying excited state), populated by the heavy ion reaction, to the 37 keV 9⁻ state in ^{216}Ac with a small L-value. If we allow the 9⁻ state in ^{220}Pa to lie between 0 keV and 80 keV, a Q_α-value of (9.83±0.09) MeV results. These Q_α-values compare well with the extrapolation by Wapstra and Audi[7]. As a measure of the neutron shell gap Schmidt et al.[1] have used the differences in Q_α-values for the N=128 and N=126 isotopes. Indeed for protactinium this value $\Delta Q_\alpha(Z=91)=1.59$ MeV is roughly half compared to that for polonium $\Delta Q_\alpha(Z=84)=3.55$ MeV, showing that the neutron shell gap is still substantial.

This work has been funded by the German Federal Minister for Research and Technology (BMFT) under the contract number 06-TM178/I.

REFERENCES

1. K.-H. Schmidt, W. Faust, G. Münzenberg, H.-G. Clerc, W. Lang, K. Pielenz, D. Vermeulen, H. Wohlfarth, H. Ewald, K. Güttner, Nucl. Phys. A318 (1979) 253
2. C. Schießl, W. Wagner, K. Hartel, H.J. Körner, Wa. Mayer, E. Rehm, Nucl. Instr. and Meth. 192, 291 (1982)
3. D.J. Decman, H. Grawe, H. Kluge, K.H. Maier, A. Maj, N. Roy, Y.K. Agarwal, K.P. Blume, M. Guttormsen, H. Hübel, J. Recht, Nucl. Phys. A436 (1985) 311
4. W. Reisdorf, Z. Phys. A300 (1981) 227
5. D. Vermeulen, H.-G. Clerc, C.-C. Sahm, K.-H. Schmidt, J.G. Keller, G. Münzenberg, W. Reisdorf, Z. Phys. A318 (1984) 157
6. D.F. Torgerson, R.D. Macfarlane, Phys. Rev. C2 (1970) 2309
7. A.H. Wapstra, G. Audi, Nucl. Phys. A432, 1 (1985)

ANISOTROPIC α-EMISSION OF ON-LINE SEPARATED ISOTOPES

J. Wouters, D. Vandeplassche, E. van Walle, N. Severijns,
J. Van Haverbeke and L. Vanneste,
Instituut voor Kern- en Stralingsfysika, Leuven University,
3030 Leuven, Belgium

ABSTRACT.

The technical realization of particle detection at very low temperatures (4K) has made it possible to study for the first time the anisotropic α-decay of oriented nuclei which have been produced, separated and implanted on line.
The measured α-angular distributions reveal surprising new results on nuclear aspects as well as in solid state physics. The nuclear structure information from these data questions the older α-decay theoretical interpretation and urges for a reexamination of the earliest work on anisotropic α-decay.

1. INTRODUCTION.

Alpha decay and its theoretical interpretation has gained much interest in relation with α-clustering, octupole deformations and the newly discovered exotic decay modes. Contrary to proton decay, there has been a serious discrepancy between experimental and calculated α-decay widths. The calculation of absolute reduced α-decay widths has been a problem for over 25 years already. Recently the discrepancy has been reduced to less than one order of magnitude. It has been shown for nuclei close to ^{208}Pb that the neglect or the non-sufficient description of the α-clustering at the nuclear surface is responsible for the discrepancy [1,2,3,4]. Alpha-clustering has also been applied in a more macroscopic way to explain energy spectra and reduced widths in the Ra region [5,6]. In this nuclear region, the decay by fragments heavier than α-particles has recently been discovered. These exotic decay modes lead to a reinterpretation of the α-decay phenomenon as a highly asymmetric fission process [7,8]. The α-transition probability and Q_α systematics can equally well be explained in this interpretation. Furthermore, in the octupole deformed region, the small hindrances observed between both α-branches to the members of an octupole doublet is, though not completely understood, an important phenomenological argument in the characterization of these states.
In these three examples only few observables are known, too few to test α-decay theories adequately. Studying the anisotropy of the emitted α-particles may be the way to get more decisive and specific information about the α-decay phenomenon.

Experimentally, the nuclear information from α-decay is mostly limited to reduced widths γ_L^2, deduced with the semiclassical formula $\lambda_L \sim P_L \gamma_L^2$ (where λ_L and P_L represent the transition probability and the penetrability of the spherical part of the barrier for α particles with angular momentum L), and relative intensities (particularly in the actinides). None of these data are sensitive to the phases of the α-amplitudes $a_L \sim \sqrt{P_L} \gamma_L$ and the amplitudes cannot be determined uniquely if more than one multipole L contributes to the α-transition. Moreover, as for example in the transition region with Z ⩾ 82 and N ⩽ 126, no information on L≠0 can be obtained at all from existing data: i) the reduced widths of the $I_f = I_i$ decays are completely dominated by L=0 because of the quadratic dependence, ii) the L=2 decay in even-even α-emitters (from 0^+ to excited 2^+ state) is highly hindered because of the drastic energy dependence of the barrier penetrability.

As pointed out already more than twenty years ago in α-decay calculations, the knowledge of sign and magnitude of the different α-amplitudes is very important to characterize nuclear structure and deformation and may provide a decisive answer for different theoretical approaches [9].

Among the measurable α-decay observables, only the amplitude mixing ratios are really sensitive to L≠0. This is exactly where our technique – detecting the anisotropic α-emission – is particularly powerful since only L≠0 α-particles are emitted anisotropically.

2. EXPERIMENTAL TECHNIQUE.

In our experiments, shortliving activities are produced at the Leuven Isotope Separator (LISOL) on line with the cyclotron CYCLONE at Louvain-la-Neuve. These are implanted in a magnetized iron foil soldered to the cold finger of a ^{3}He-^{4}He dilution refrigerator (KOOL). Then the angular distribution of the emitted radiation is measured as a function of temperature at 0° and 90° relative to the external magnetic field. For more details concerning the apparatus and our experimental method we refer to [10,11,12,13].

The detection of α-particles emitted by oriented nuclei requires detection at 4K in order not to deteriorate the cryogenic performance of the low temperature apparatus.

For this purpose different laboratories have – since the fifties – pursued the fabrication of solid state detectors for operation at 4K, especially at Oak Ridge, Berkeley and the National Bureau of Standards (also in view of the first parity violation experiments in β-decay). The detectors had poor stability and/or bad resolutions – some broke down at these low temperatures.

The few experiments on long lived α-emitters (^{233,235}U, ^{237}Np at Oak Ridge and ^{240}Cf, ^{253}Fs, ^{241}Am, ^{255}Fm at Berkeley) with diffused samples lead to hardly reproducible data and difficult interpretations. The development of reliable particle detection at 4K, using the present technology, was our main goal. Several types and configurations of detectors were tested and finally good results were obtained. Prototypes were constructed in cooperation with commercial suppliers. The high purity of the starting material (particularly for Ge) and the modern semiconductor contact technology (e.g. ion-implantation) were of special importance for this development. Various cryogenic problems had to be solved, e.g. the connection detector to preamplifier FET was kept as short as possible (10 cm in the present set-up).

We are now able to detect particles at 4K with energy resolutions and stabilities comparable to the ones achieved in "normal" operating conditions (18 keV for 6 MeV alphas with 300mm^2 detectors). An accumulation of α-spectra taken at different masses is shown in figure 1 and illustrates the good energy resolutions. These spectra have been recorded on-line. A drawing of the refrigerator tails as used in the particle detection mode is shown in figure 2 of a separate contribution to this conference [13]. Besides the detector functioning at 4K, also the sample preparation by (low temperature) implantation and the possibility to carry out systematic studies on line are advantages of our technique as compared to the pioneering work in the early sixties.

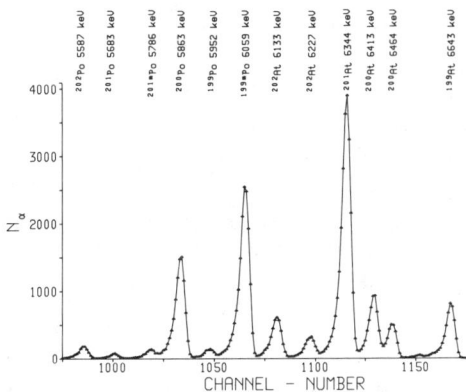

Fig. 1. An accumulation of α spectra taken at different masses with particle detectors operating at liquid helium temperature.

3. RESULTS.

In a first series of experiments we studied the α-decays of Po, At and Fr isotopes. A list of the odd-A α-emitters studied in this work is given in table 1. At present ^{205}Fr($t_{1/2}$ = 3.9 s) is the shortest lived isotope ever oriented on line after direct implantation. An α-spectrum is shown in figure 1. Alpha anisotropies, recorded for these α-decays are presented in figures 2, 3 and 4.

The α-anisotropy ε is defined as $N(0°)/N(90°)$ at temperature T normalised to the same ratio at "high" temperature with isotropic distribution (N is the number of counts in a fixed time interval, see also reference 13). Thus $\varepsilon = W(0°)/W(90°)$ where $W(\Theta)$ can be expressed as $1 + f \sum_k A_k Q_k B_k (\mu B/T) P_k(\Theta)$. For more information about the distribution formula and the extraction of parameters we refer to references 11 and 14.

The α-anisotropies have been measured with separator beam intensities ranging from 10^4 to as low as a few hundred ions per second.

As can be seen from figure 2 the α-anisotropies in the odd astatine isotopes (all $\pi h9/2 \to \pi h9/2$ favoured α-decays) show a remarkable variation: the alpha emission changes sign from preferentially parallel to preferentially perpendicular to the nuclear spin direction in going further away from the closed shell at N=126. For the odd Po isotopes, the $\nu i13/2^+$ isomers, the α-emission is highly peaked in the direction of the mother spin (see figure 3).

The α-decays of all these odd-A isotopes are favoured transitions with $I_f = I_i$ and $\pi_f \pi_i = +$. Thus, according to the selection rules, angular momenta L = 0, 2, 4, ... contribute. However, the transitions proceed mainly by L=0 α-waves: the hindrances increase rather seriously with angular momentum L because of the additional centrifugal barrier. A first analysis of the α-particle distributions reveals the admixtures with $L \neq 0$ to be very small in these decays. Still they are responsible for the observed effects, which are, moreover, described by only a second order term dependence in $W(\Theta)$. Then we can express the radiation parameters as $A_2 \approx 2\delta_{20}$ and $A_4 \approx 0$ with $\delta_{20} = a_2/a_0$ the ratio of the L=2 α-waves to the L=0 amplitude [15]. The error on δ_{20} implied by this approximation is smaller than a few percent for the α-decays under study. Thus the only free parameters in the α-particle angular distribution functions are (i) the temperature independent factor $f \delta_{20}$ and (ii) the hyperfine interaction μB. The f stands for the fraction of nuclei which experience the full hyperfine field.

The experimental anisotropies have been fitted simultaneously to both parameters. The temperature independent parameter $f \delta_{20}$ can be determined very accurately if:

i) the anisotropy shows saturation. This is the case for the At isotopes (fig. 2). The large hyperfine interaction is responsible for the saturation effect,

ii) both parameters have a small correlation (e.g. for ^{199m}Po and ^{208}Fr in figures 3 and 4).

These parameters δ_{20} and μB represent, respectively, nuclear structure information and solid state information. The deduced values from these α-data will be discussed separately.

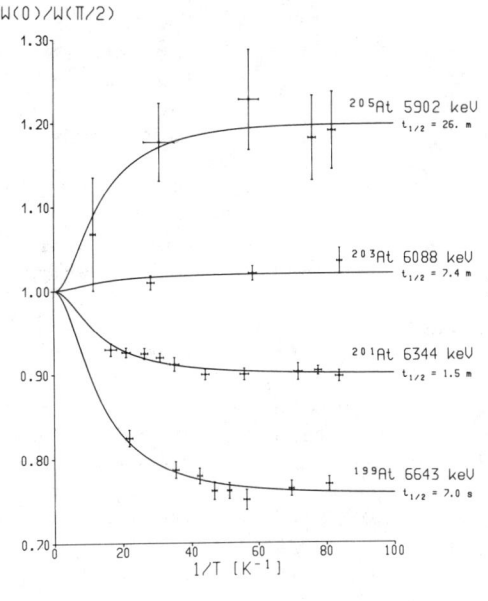

Fig. 2. The α-particle emission anisotropies as a function of 1/T of the πh9/2 → πh9/2 favoured α-decays of odd 199-205At.

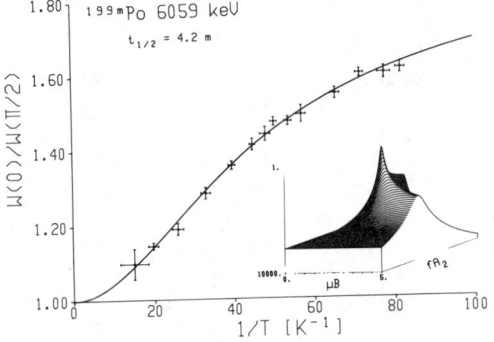

Fig. 3. The α-anisotropy of the 6059 keV transition of ^{199m}Po. The inset shows a χ^2 surface as a function of the parameters μB and fA_2 in the angular distribution

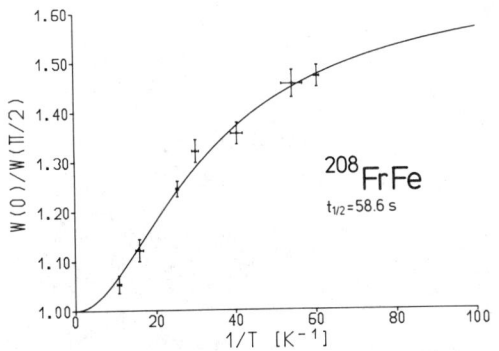

Fig. 4. The α-anisotropy as a function of 1/T of the 6636 keV transition of ^{208}Fr ($t_{1/2}$ = 58.6 s).

3.1. NUCLEAR STRUCTURE RESULTS.

The δ_{20} values extracted from the α-anisitropies are tabulated in table 1. Thus, and as described above, the changing α-anisotropy in the odd $\pi h9/2$ nuclei is caused by the variation in sign and magnitude of the mixing ratios δ_{20} (the sign of the anisotropy is determined by the sign of the δ_{20}).

Before this work no calculated α-amplitude ratios existed, except for the actinides [9,16]. We started calculations for α-decays from this work in different model approximations. We will mainly discuss the $\pi h9/2 \rightarrow \pi h9/2$ α-decays of the odd At isotopes, which show the sign change.

Starting from the spherical shell model we carried out calculations along the lines of the semiclassical framework developed by H. Mang and J.O. Rasmussen [17,18]. For relative α-amplitude ratios this should be a good approximation to the problem (not for absolute magnitudes). This model represents a quartet model: the α particle is seen as the ensemble of four particles without correlations among them. We extended this theory for nuclei further away from closed shells by incorporating pairing correlations [19]. These 2 particle correlations are very important for α-decay, as has been shown already before, but only positive δ_{20}'s result. However, negative δ_{20} values can be obtained by taking into account the quadrupole part of the p-n interaction (the monopole part only results in minor shifts of the single particle energy levels). Microscopically this means that, whereas the L=2 component of the total angular momentum of the α particle originates mainly from proton side near the N = 126 shell closure, the L=2 admixture from neutron side is clearly non-negligible, and increasing with decreasing neutron number. This is a core polarization effect by the valence neutron (holes). Moreover, this same phenomenon gives an explanation for: i) the odd-even effect in the charge radii of various chains of isotopes [20], ii) the energy variation of 0^+ intruder states in semi closed shell regions [21]. The idea of a close relationship between the quadrupole part of the p-n interaction on one hand, and deformation in one-body models on the other hand (both can equally well explain e.g. the 0^+ intruder energies [22]) are now being studied intensively [23].

Secondly, and following a suggestion of J.O. Rasmussen, we investigated a macroscopic deformed model approach in which the α-amplitudes with different angular momentum L on the surface are dependent on the deformation change from parent to daughter [24]. The alpha surface probability $|\psi_\alpha|^2$ can be written as

$|\psi_\alpha|^2 \sim \dfrac{R_i(\Theta)-R_f(\Theta)}{R_{0i} - R_{0f}}$ with $R(\Theta) = R_0[1 + \sum_k \beta_k Y_{k0}(\Theta)]$ and the α-

amplitudes with different L can be obtained by expansion in a Legendre series of even order. This simple model, with deformation parameters β_k from muonic X-ray experiments [25], can explain the α-finestructure in the actinides equally well as the older microscopic calculations [17]. For the At decays, studied in this work, deformation parameters have been calculated [26]. Although the deformation values are small ($\beta_2 \leqslant .1$), the trend of the α-amplitude calculations agrees with experiment. Exactly for the isotope with the observed isotropic α-emission, the deformation of parent and daughter (^{203}At and ^{199}Bi) are equal! In this interpretation, the α-anisotropy is a measure for the shape change from parent to daughter.

Furthermore, calculations are being done in a more modern α-decay theory: an extension of the Jackson-Rhoades Brown theory (for spherical nuclei) [4] to deformed nuclei [27]. Nevertheless, in this model, it is still rather difficult to interpret the results, esp. the selection of the right solution out of many. Yet, a sign change in the δ_{20} value would correspond to clear nuclear structure changes which may be caused by deformation effects.

In the foregoing we have mainly been discussing the structure dependence (through γ_L) in the α amplitudes a_L. For the quasi-spherical nuclei in this work the Coulomb barrier penetrability P_L can be calculated easily [9]. In the earlier works [28] and also in more recent work [29,30] on anisotropic α-emission the penetration of the asymmetric barrier is held responsible for the α-anisotropy. The arguments for this all follow the lines of the pioneering work of D.L. Hill and J.A. Wheeler [31]. We think that the structure effects are a lot more important for the anisotropy than the penetrability: the changing sign in α-anisotropy in the odd At isotopes is difficult to interpret on anisotropic penetrability arguments. For these reasons we suggest that the anisotropy work should be reexamined more carefully.

From the temperature dependence of the α-anisotropies, the hyperfine interaction strength can be deduced. In this way and in combination with several γ-anisotropies, following the EC/β$^+$ decay, the hyperfine interaction (and thus the nuclear magnetic moment) of the $\nu i13/2$ isomers ^{199m}Po and ^{201m}Po, respectively with α-branchings of 39% and 3%, have been determined with an accuracy of 3%. These measurements were performed with separator beam intensities of approximately 10^4 ions per second. For more information about these analyses we refer to another contribution to this conference [13].

3.2. SOLID STATE INFORMATION.

In addition to the nuclear structure information, we can also extract unique solid state results from α-angular distributions. The hyperfine field of Fr in Fe has been deduced [32]. For the element Fr, no stable isotope is known and halflives are in the order of minutes or shorter. Only recently it has become possible to study its atomic structure [33,34], and for the hyperfine field one had to wait for nuclear orientation facilities at on-line separators to become available. Moreover, the Fr isotopes decay mainly by α-emission, which means that the realization of α-

particle detection of oriented isotopes was a prerequisite for these hyperfine structure studies.
The hyperfine field was extracted from the ^{208}Fr ($t_{1/2}$ = 58.6 s) α-decay, which is highly anisotropic. The observed anisotropy of the 6636 keV α-transition is shown in figure 4. It is a favoured transition with $I_f = I_i = 7$ and $\pi_f \pi_i = +$, which proceeds mainly by L=0 α-waves (as in the odd-A α-emitters). A small L=2 admixture (approximately 2%) is responsible for the 50% anisotropy at the lowest attainable temperatures (see figure 4). The μB parameter has been determined in the same way as described above for the odd-A α-emitters. In the analysis a two site distribution is assumed. It is clear from an ever increasing number of isotopes of different mass numbers, sizes and solubilities that the two site approximation is quite reasonable for low dose implantations at very low temperatures. Indeed, the fractions are consistent with nearly fully substitutional positions. For experiments with the same host iron foil the same fraction f, extracted from γ anisotropies, was found for Fr, At and Po isotopes.
Then, with the magnetic moment $\mu(^{208}\text{Fr})$ = 4.75(10)μ_N, known from atomic beam laser spectroscopy combined with ABMR measurements, we deduce the hyperfine field $B(\vec{Fr}Fe)$ = (+)70(5)T.
This value, completing the alkaline systematics, represents a fundamental test case for theories of hyperfine fields. In contrast to the free atom cases and despite the simple electronic structure of the alkalines, the "ab initio" theoretical descriptions are discrepant with experiment (for Cs) and do not exist for Fr yet. With our measurement, the overall trend of the alkaline fields now clearly indicates the importance of the impurity size effects which have – up to now – mostly been neglected (except for some phenomenological models of hyperfine fields). For more information about the hyperfine field systematics and the theoretical interpretation we refer to [32] and references therein.

4. SUMMARY.

The technical realization of particle detection of on-line oriented isotopes opens new avenues for nuclear structure studies. First experiments on α-emitters in the "near above-lead" region indicate that: i) all measurements have been performed with separator beams of less than 10^4 ions per second,
    ii) an accuracy of 3% on the interaction μB can be obtained.
        iii) very small L=2 admixtures (in the order of 1% to .01%) have been detected. In the strongly deformed actinides this admixture can be as large as 20%.
This clearly illustrates the sensitivity of the technique, which is particularly important in connection with on line systems. From the α-anisotropy data we determined the hyperfine field of Fr in Fe, nuclear magnetic moments and the mixing ratios of L=2 α-parti-

cles to the main L=0 amplitude for several Po, At and Fr α-emitters. The latter intrinsically new information may yield a clear signature for changing nuclear structures, which may be caused by deformation effects. This has to be explored further experimentally as well as theoretically.

The influence of α-clustering in the L≠0 amplitudes, the dependence of the α-anisotropy on β_3 (as has been pointed out already before [35]) and the consequences of different α-decay theories for the anisotropy α-emission should also be studied further. Alpha-anisotropies have already been predicted in a fission model [36], but they are in disagreement with our experimental results.

Future experimental developments in this kind of studies may be: i) alpha anisotropies in deformed nuclei (octupole region?).
First tests show that it is not evident, even with our high efficiency, to produce and separate shortliving actinides in sufficient quantities (at least 100 ions per second).
ii) nuclear magnetic resonance on particle anisotropies (high accuracy results with low intensity separator beams?).
iii) measurements of conversion electron and X-ray anisotropies.

The authors wish to thank P. Schoovaerts for technical assistance, M. Huyse and the separator crew and the operators of CYCLONE at Louvain-la-Neuve. We are indebted to T. Berggren, K. Heyde and especially J.O. Rasmussen for interesting discussions.
We gratefully acknowledge the financial support of the "Interuniversitair Instituut voor Kernwetenschappen" and the "Nationaal Fonds voor Wetenschappelijk Onderzoek".

Isotope	$t_{1/2}$	α branching (%) (1)	E_α (keV)	Transition	δ_{20} (3)	L=2 contribution in the decay (%) (3)
^{199}At	7.0s	100	6643	$\pi h9/2 \to \pi h9/2$	−.076(4)	.6
^{201}At	1.5m	71	6344		−.029(2)	.09
^{203}At	7.4m	31	6088		+.005(3)	.003
^{205}At	26.2m	10	5902		+.059(9)	.4
^{205}Fr	3.9s	100	6916		+.040(7)(2)	.2 (2)
^{207}Fr	14.8s	93	6767		+.062(5)	.4
^{209}Fr	50.0s	89	6648		+.079(5)	.6
^{197m}Po	26s	84	6385	$\nu i13/2 \to \nu i13/2$	+.107(20)(2)	1. (2)
^{199m}Po	4.2m	39	6059		+.173(9)	3.
^{201m}Po	8.9m	3	5786		+.167(16)	3.

(1) approximate α-branching values are given
(2) this value represents only a lower limit, because relaxation effects may be present.
(3) deduced with fraction f extracted from γ-anisotropies in the EC/β+ decay of ^{199m}Po, ^{205,206}At, ^{208}Fr.

Table 1: a list of the odd A α-emitters studied in this work with the mixing ratios δ_{20} of the L=2 α-wave to the L=0 amplitude deduced from the α-anisotropies. These δ_{20}-values have been extracted i) with fraction f from γ-anisotropies, ii) assuming no relaxation effects (see other contribution to this conference [13]).

REFERENCES.

1. I. Tonozuka and A. Arima, Nucl. Phys. A323 (1979) 45
2. G. Dodig-Crnkovic et al., Nucl. Phys. A444 (1985) 419
3. S.G. Kadmensky, Z. Phys. A312 (1983) 113
4. D.F. Jackson and M. Rhoades-Brown, Ann. of Physics 105 (1977) 151
5. H. Daley and F. Iachello, Phys. Lett. 131B (1983) 281
6. M. Gai et al., Phys. Rev. Lett. 51 (1983) 646
7. A. Sandulescu et al., Sov. J. Part. Nucl. 11 (1980) 528
8. D.M. Poenaru et al., Phys. Rev. C32 (1985) 572 and references therein
9. J.O. Rasmussen, In Alpha-, Beta- and Gamma-ray Spectroscopy, Vol. 1, ed. K. Siegbahn, North-Holland, Amsterdam (1974)
10. D. Vandeplassche et al., Nucl. Instr. & Meth. 186 (1981) 211
11. Nuclear Orientation and Nuclei far from Stability, eds. B.I. Deutch and L.R. Vanneste, J.C. Baltzer AG (1985)
12. J. Wouters et al., Nucl. Instr. & Meth. B26 (1987) 463
13. D. Vandeplassche et al., this conference
14. K.S. Krane, in Nuclear Orientation, eds. N.J. Stone and H. Postma, North-Holland, Amsterdam (1986)
15. J. Wouters et al., Phys. Rev. Lett. 56 (1986) 1901
16. J.K. Poggenburg et al., Phys. Rev. 181 (1969) 1697
17. H.J. Mang and J.O. Rasmussen, Mat. Fys. Skr. Dan. Vid. Selk. 2 (1962) nr. 3
18. J.O. Rasmussen, Nucl. Phys. 44 (1963) 93
19. H.D. Zeh, Z. fur Physik 175 (1963) 490
20. I. Talmi, Nucl. Phys. A423 (1984) 189
21. K. Heyde et al., Nucl. Phys. A466 (1987) 189
22. R. Bengtsson and W. Nazarewicz, to be published in Phys. Lett.
23. K. Heyde, private communication
24. H.M.A. Radi et al., Phys. Rev. Lett. 41 (1978) 1444
25. J.D. Zumbro et al., Phys. Lett. 167B (1986) 383
26. R. Bengtsson and W. Nazarewicz, private communication
27. T. Berggren and P. Olanders, Nucl. Phys., to be published (1987)
28. J. Wouters et al., Hyp. Int. 22 (1985) 527
29. F.A. Dilmanian et al., Phys. Rev. Lett. 49 (1982) 909
30. K.H. Maier et al., in Proc. 4th Int. Conf. Nuclei far from Stability, Helsingør, eds. P.G. Hansen and O.B. Nielsen, CERN 81-09 (1981) 183
31. D.L. Hill and J.A. Wheeler, Phys. Rev. 89 (1953) 1102
32. J. Wouters et al., Phys. Lett. A, in press
33. S. Liberman et al., Phys. Rev. A 22 (1980) 2732
34. C. Ekström et al., Physica Scripta 34 (1986) 935
35. P.O. Fröman, Mat. Fys. Skr. Dan. Vid. Selsk. 1,3 (1957)
36. M.K. Basu, Proc. Int. Conf. on Nuclear Structure, Amsterdam (1982) 162

BETA DELAYED CHARGED PARTICLES DECAY FROM LIGHT NUCLEI NEAR THE DRIP LINE[*]

J.C. Jacmart, V. Borrel, F. Pougheon, A. Richard, IPN, BP 1, 91406 ORSAY, France
R. Anne. D. Bazin, H. Delagrange, C. Détraz, D. Guillemaud-Mueller,
A.C. Mueller, E. Roeckl, M.G. Saint-Laurent, GANIL, BP 5027, 14021 CAEN, France
J.P. Dufour, F. Hubert, M.S. Pravikoff, CEN Bordeaux, Le Haut Vigneau,
33170 GRADIGNAN, France

ABSTRACT

Beta delayed proton radioactivity has been observed for the $T_z=-5/2$ isotopes ^{31}Ar and ^{27}S produced by projectile fragmentation of ^{36}Ar at 85 MeV/u. Spectra of beta delayed protons have been obtained. The measured half-lives are $T_{1/2} = 15 \pm 3$ ms for ^{31}Ar and 16 ± 5 ms for ^{27}S respectively. No evidence was found for direct two-proton decay of these nuclei.

INTRODUCTION

As the proton drip line is approached, the rapid increase of mass differences leads, through β^+ decay, to a wide range of final states in the daughter nucleus, including in particular the super allowed transition to the isobaric analog state (IAS). Due to the corresponding decrease of one and two protons binding energies, most of these states decay by proton emission; the proton spectra should give informations on the position of the analog state and on the distribution of the Gamov-Teller strength.

Up to now β-delayed proton of most of the $T_z=-3/2$ and $T_z=-2$ light nuclei have been studied and only ^{35}Ca in the $T_z=-5/2$ series [1]. However the low production cross sections and the rather slow collection method (at best 25 ms) limited this procedure which had to be replaced by another production mechanism with on line recoil separation.

Light nuclei on the border of stability are produced with a rather good yield with intermediate energies heavy ions beams. Obtained by a mechanism similar to fragmentation they present a strong forward focusing which is well adapted to detection by the zero degree recoil spectrometer LISE [2]. This device has been used for identification of a large number of new isotopes far from stability [3]. As described by J.P. Dufour in the present conference [4], the use of an energy degrader in the intermediate focal plane of the spectrometer adds a range selection to the magnetic analysis so that only nuclei of a given $A^{2.5}/Z^{1.5}$ value are focused on the final focal point detector. A few nuclides among the numerous species produced by the fragmentation mechanism are selected by this technique [5] : It is thus possible to use the maximum production rate for spectroscopic studies without a too high radioactive background : maximum intensity beam of GANIL with an ECR source, thick target and full aperture of the spectrometer.

The experimental procedure and the results obtained for ^{31}Ar and ^{27}S are discussed in the following.

EXPERIMENTAL PROCEDURE

A 85 MeV/u ^{36}Ar beam bombarded a thick natural Ni target (380 mg/cm^2) with an intensity of 1 eμA (4.10^{11}pps). A 180 mg/cm^2 Al degrader was placed in the intermediate focal plane of the LISE spectrometer. The magnetic rigidities of the two dipoles have been carefully optimised on the ^{31}Ar transmission which was then collected at a rate of 1 event per second. The selected ions were implanted in a five members solid state detector (Fig. 1) lo-

---

[*]Experiment performed at the GANIL National Laboratory

cated at the final focal point of LISE. A 477 μm thick aluminium foil placed in front of the telescope slowed down the ^{31}Ar nuclei so that they stop

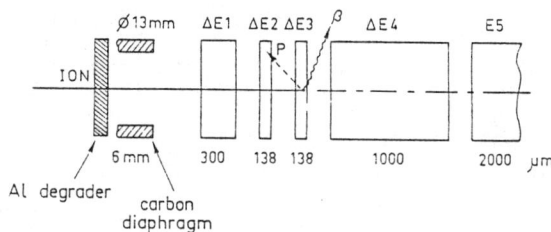

Fig. 1. Solid state telescope at the focal point of LISE.

in ΔE_3 detector.

A first data-acquisition system recorded the energy deposits of the heavy ions in the five Si detectors and the time of flight (t.o.f.) of the ions between the target and the telescope. The two dimensional plot ΔE_1 versus t.o.f provided a clear identification of the detected nuclei[15]. Two single channel analysers on the ΔE_1 and t.o.f electronic chains delivered a signal each time that an ^{31}Ar nuclei was identified : then the beam was stopped during 200 ms, the heavy-ion data acquisition system was set off line, the second data acquisition was allowed and a clock was started.

The ΔE_2, ΔE_3 and ΔE_4 detectors had to measure the energy of the emitted protons (100 keV to 10 MeV) a short time after the implantation of an energetic ^{31}Ar nucleus in the telescope. To that purpose, their charge preamplifiers fed two different electronic chains. The second one, devoted to the detection of these protons, was equipped with a high pulse rejection device[6]: after proper shaping the signal was shortened by a delay line clipping and a gate rejected all pulses corresponding to heavy ions. The amplifier saturation that would have occured after the heavy ion detection was thus eliminated with no noticeable deterioration of the signal to noise ratio. It was thus possible to detect correctly a 80 keV proton 3 μs after the implantation in the same detector of a 300 MeV heavy ion.

The proton energies and their time of arrival were recorded on the second data-acquisition system. The ions parameters and the proton parameters were stored chronologically on the same magnetic tape.

EXPERIMENTAL RESULTS

Due to the $A^{2.5}/Z^{1.5}$ selection law[4] the nuclides transmitted together with ^{31}Ar by the LISE separator are essentially the N=13 isotones. ^{30}Cl is not bound against particle emission, ^{28}P is a weak beta-delayed proton and α emitter (BR < 10^{-5}), ^{27}Si and ^{26}Al do not emit beta-delayed particles in their decay[7]. ^{29}S which is known to be a beta-delayed proton precursor with a half-life equal to 187 ms [8,9] is then the only contaminant of ^{31}Ar : indeed its strongest proton lines are clearly observed in the long lived part (100 - 200 ms) of the spectra recorded during the beam off time and trigged by ^{31}Ar detection. They yield a precise energy calibration. The difference spectrum from 1 to 40 ms obtained by substracting the ^{29}S contribution with a 187 ms half-life is shown on fig. 2. Two main peaks are located respectively at 2190 ± 10 keV (quite close to the strongest 2206 keV ^{29}S peak) and 1520 ± 40 keV. They decay with the same short half-life and have been attributed to the decay of ^{31}Ar. In addition, some smaller peaks (3700 ± 100 keV, 5450 ±100 keV and 6200 ± 100 keV and also several unresolved peaks between 2.6 and 5 MeV) were assigned to the decay of ^{31}Ar on the basis of time analysis. The half-life of the ^{31}Ar nucleus is found to be 15 ± 3 ms (fig. 3), from a two-component fit assuming the long lived component to be entirely due to the 187 ms ^{29}S.

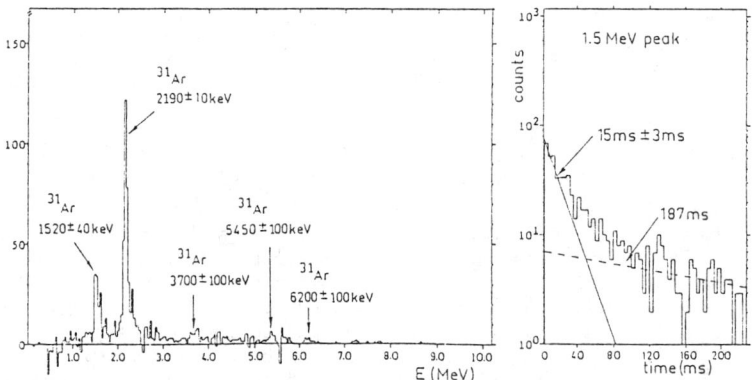

Fig. 2. ^{31}Ar spectrum from 1 to 40 ms obtained after substraction of ^{29}S contribution with 187 ms half-life.

Fig. 3. Time spectrum of the 1.5 MeV peak.

The most energetic protons are likely to leave the detector and it therefore becomes necessary to sum their energy deposits in the ΔE_3 and ΔE_2 or ΔE_4 detectors. Several high energy events (between 6.5 MeV and 10 MeV) are assigned to the ^{31}Ar decay on the basis of the observed decay pattern[15]. The efficiency for the detection of protons drops sharply when their energy increases, thus the actual relative heights of the peaks in fig. 2 do not reflect the actual relative intensities.

The procedure for the ^{27}S study was quite similar. However the beta decay of this nucleus [12] is predicted to feed mainly (BR ≃ 65 %) the first $3/2^+$ level around 1 MeV of ^{27}P leading to γ decay. Our statistics was then not sufficient to extract levels from the spectrum. We have only obtained the half-life of this nucleus which is $T_{1/2}$ (exp) = 16 ± 5 ms in agreement with the predicted value around 13 ms.

DISCUSSION ON ^{31}Ar DATA

From the present experimental results, a partial ^{31}Ar decay scheme is proposed in fig. 4. The Q_{EC} value of 18730 keV have been deduced from the mass excess measured for ^{31}Cl [10] and calculated for ^{31}Ar [11]. The excitation energy of 12490 keV for the lowest T=5/2 state in ^{31}Cl is estimated from the measured mass excess of ^{35}Ar and ^{35}Cl. If one assumes a pure Fermi decay to the isobaric analog state (B(F)=5) our experimental total half-life of 15 ms yields a branching ratio of 5 % for this transition. Due to the large Q_{EC} value most of the β strength must feed the ground state and low energy levels of ^{31}Cl.

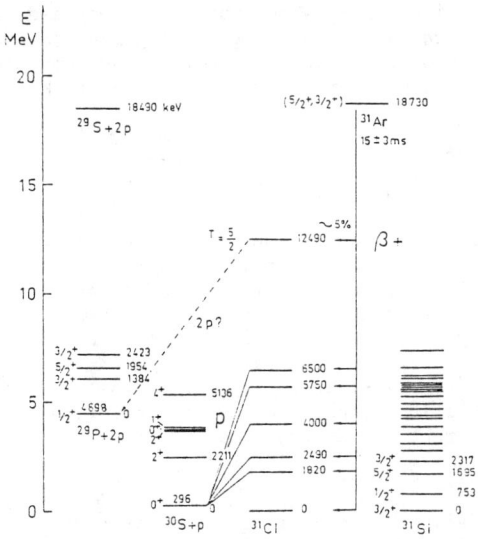

Fig. 4. Partial decay scheme of ^{31}Ar. For the ^{31}Cl nucleus, the levels are deduced from the present experiment with the exception of the T=5/2 state, for which the energy has been calculated. The energy levels of its mirror nucleus ^{31}Si are also presented.

The expected decay properties of ^{31}Ar can be predicted from the calculations of Brown and Wildenthal [12]. Their recent calculation for ^{31}Ar taking into account 50 T=3/2 levels up to 18 MeV excitation energy [13] gives branching ratios listed in table 1. The total deduced half-life $T_{1/2}$ (th) = 10.3 ± 1.7 ms (the error corresponds to ±400keV in Q value) is in good agreement whith our experimental value T(exp) = 15 ± 3 ms.

J_f	E_x(MeV)	BR %
3/2	0.000	25.6
5/2	1.606	6.9
3/2	2.295	26.7
3/2	3.825	2.0
7/2	5.632	2.0
5/2	5.764	12.2
5/2	6.074	1.4
7/2	6.506	4.3
7/2	6.555	1.1
7/2	7.009	1.2
5/2 (T=5/2)	12.061	5.0

Table 1. Calculated excitations energies and branching ratios from ref. 13 for ^{31}Ar → ^{31}Cl transition listed if BR ≥ 1 %.

The comparison with fig. 2 shows that the five main expected transitions are actually observed as proton peaks feeding the ^{30}S ground state. Our measured intensity ratios correspond roughly to the expected ones although a precise determination would require a more accurate knowledge of detection efficiencies.

The isobaric analog state in ^{31}Cl is expected to decay both by one proton and two-proton emission. There is no evidence for any proton peak with energy around 12 MeV or 9.8 MeV. This absence may be due to the combined effects of a low branching ratio and of the strong decrease of the detection efficiency as protons become more energetic. On the contrary, a group of peaks, near 7.5 MeV, decaying with the ^{31}Ar half-life, appears on the sum spectrum $\Delta E_2 + \Delta E_3 + \Delta E_4$. This group may represent the summed energies of the two protons emitted by the T=5/2 state in ^{31}Cl towards the ground state in ^{29}P, with a calculated energy of 7.8 MeV.

The measured ^{31}Ar half-life of 15 ± 3 ms, which is in good agreement with shell-model calculations, shows that ^{31}Ar does not exhibit an important branching ratio for direct two-proton decay.

CONCLUSION

The experimental method described in the present paper is very powerful for the study of beta-delayed emission of charged particles in the decay of neutron-deficient nuclei near the drip line. It should permit the investigation of direct proton-emission (of, e.g., ^{55}Cu, which was recently observed at GANIL [14] or even of new radioactive modes like two-proton emission (^{22}Si [5] and ^{39}Ti being possible candidates).

REFERENCES

1. J. Aystö, D.M. Moltz, X.J. Xu, J.E. Reiff and J. Cerny, Phys. Rev. Lett. 55,1384 (1985)
2. R. Anne, D. Bazin, A.C. Mueller, J.C. Jacmart and M. Langevin, Nucl. Instr. Methods, A 257, 215 (1987)
3. M. Langevin et al., Nucl. Phys. A455, 149 (1986) and D. Guillemaud-Mueller. Contribution to the present Conference
4. J.P. Dufour et al., Nucl. Inst. Methods A428, 267 (1986) and J.P. Dufour et al. Contribution to the present Conference
5. M.G. Saint-Laurent et al., Phys. Rev. Lett., 59, 33 (1987)
6. A. Richard, Private communication
7. P.M. Endt and C. Van der Leun, Nucl. Phys. A310, 1 (1978)
8. D.J. Vieira, R.A. Gough and J. Cerny, Phys. Rev. C19, 177 (1979)
9. Z.Y. Zhou et al., Phys. Rev. C31, 1941 (1985)
10. W. Benenson, D. Mueller, E. Kashy, H. Nann and L.W. Robinson, Phys. Rev. C15, 1187 (1977)
11. A.H. Wapstra, G. Audi and R. Hoekstra, Atomic masses from mainly experimental data, in The 1986 Atomic Mass Predictions, ed. P.E. Haustein, submitted to At. Data and Nucl. Data Tables
12. B.A. Brown and B.H. Wildenthal, At. Data and Nucl. Data Tables 33, 347 (1985)
13. B.A. Brown. Private communication
14. F. Pougheon et al., Z. Phys. A327, 17 (1987)
15. V. Borrel et al., Int. Report IPNO 87-18 and Nucl. Phys. A473, 331 (1987)

THE β^+ AND EC DECAY OF ^{69}Se
POSSIBLE SHAPE-COEXISTENCE AND SUPERDEFORMATION EFFECTS IN ^{69}As

Ph. Dessagne, Ch. Miehé, P. Baumann, A. Huck, G. Klotz
M. Ramdane, G. Walter and J. Dudeck
Centre de Recherches Nucléaires, 67037 Strasbourg Cedex France
J.M. Maison
Institut de Physique Nucléaire, 91406 Orsay France

The β^+ and EC decay of ^{69}Se to ^{69}As and the subsequent γ and proton emissions have been investigated by means of the ^{40}Ca(^{32}S,2pn)^{69}Se reaction with a 100 MeV energy beam delivered by the Strasbourg MP tandem. The He-jet technique connected to a tape system has been used.
 Direct proton spectra and proton-X rays coincidence (PXCT-Ref.1) have been registered in a closed geometry with a proton resolution of 20 keV. Our high resolution delayed proton spectrum (fig.1) reveals fine structures up to now unobserved[2]. The corresponding (Ge/As) X rays ratios, related to the lifetime of the emitting levels indicate the presence of two groups of decay times (fig.2). This feature has been interpreted in terms of two coexisting shape-configurations (moderately deformed with $|\beta_2| \sim 0.2$ and the superdeformed ones with $\beta_2 \sim 0.5$) as calculated using deformed Woods-Saxon potential and the Strutinsky method (fig.3). Gamma direct and γ-γ coincidence measurements have been performed and 13 unreported bound levels have been located in ^{69}As. The results concerning the decay of ^{69}Se to both bound and unbound levels in ^{69}As yield a GT strength of 8.6% of the sum rule, in 95% of the $Q_{\beta^+,EC}$ window. A similar intensity has been found for another $T_Z = 1/2$ nucleus, ^{65}Ge[3].

REFERENCES

1. J.C. Hardy et al., Phys. Rev. Lett., 37, 133 (1976).
2. J.A. Macdonald et al., Nucl. Phys. A288, 1 (1977).
3. K. Vierinen, Nucl. Phys. A463, 605 (1987).

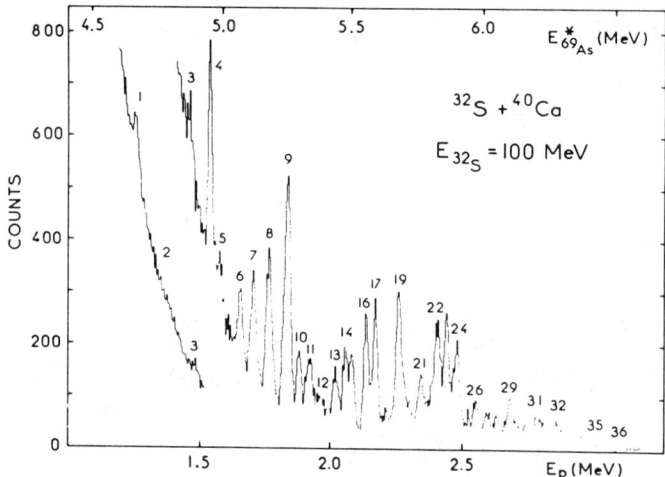

Fig. 1. Beta delayed particle spectrum.

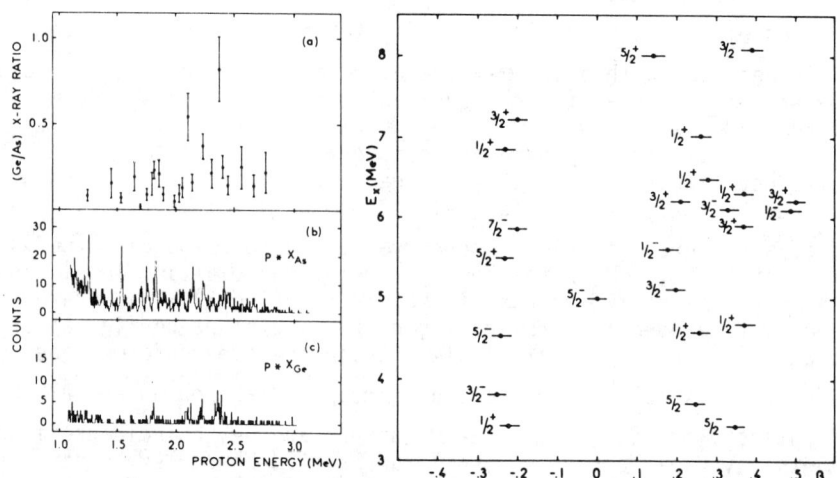

Fig. 2. Ge/As X ray ratio as a function of proton energy (a). Proton spectrum in coincidence with As X rays (b) and Ge X rays (c).

Fig. 3. Energies of single particle states in ^{69}As.

BETA-DELAYED PROTON DECAY IN THE LANTHANIDE REGION*

J.M. Nitschke, P.A. Wilmarth, J. Gilat, ** and P. Möller***
Lawrence Berkeley Laboratory, University of California, Berkeley, CA 94720

K.S. Toth
Oak Ridge National Laboratory, Oak Ridge, TN 37831

ABSTRACT

A total of 24 new β-delayed proton precursors and several new decay branches in the region of $56 < Z < 72$ and $N < 82$ have been identified with the OASIS on-line mass separator facility at the SuperHILAC in Berkeley. Proton spectra and half-lives were measured for all precursors. Additional properties determined for many of the isotopes include spins and parities of the precursors, final state feedings, proton branching ratios, (Q_{EC}–S_p) energy differences, and new levels in the proton decay daughters. Mixed β-delayed proton decay from precursor isomers and ground states has been observed. The proton decay has been compared to statistical model calculations using Gamow - Teller (GT) β-strength functions obtained from RPA calculations with Nilsson model wave functions. For the $N = 81$ precursors ^{151}Yb, ^{149}Er, and ^{147}Dy, pronounced structure in the proton spectra has been observed. An interpretation of this structure in the framework of "doorway" states is proposed.

1. INTRODUCTION

With the recent discovery of β-delayed proton emission in europium [1] this decay mode has now been observed in an unbroken sequence of elements from $Z = 52$ (Te) to $Z = 71$ (Lu), which shows, as expected, that β-delayed proton decay becomes a prominent decay mode for nuclei near the proton drip line.

Isotopes with various degrees of deformation from quasi spherical near the closed shells $Z = 50$ and $N = 82$ to highly deformed near $Z = 60$ have been studied. Nuclei in the deformed region show smooth proton spectra that are well described by a statistical model due to the close level spacings at excitation energies that are sufficiently high for proton decay. For the interpretation of our experimental results we have used a statistical model formulated originally by Hornshøj, et al.[2] with the following modifications first suggested by Hardy[3]: the original Gilbert and Cameron level density formula was replaced by a backshifted Fermi gas prescription, and the γ width calculation was modified to include the effects of the giant dipole resonance and normalized to experimental photoabsorption cross sections. We also incorporated an additional set of optical model parameters, more appropriate for the calculation of transmission coefficients below the Coulomb barrier [4]. There are several critical parameters that enter into statistical model calculations: the total decay energy Q_{EC}, the proton separation energy S_p, the level density ρ, the γ decay width Γ_γ, the optical model parameters, the spin and parity of the precursor J^π, and the β-strength function S_β. Calculated quantities that can be compared to experiment are: the proton energy spectrum, the proton branching ratios to states in the final nucleus, and the total proton/β-decay branching ratio b_p. There are large variations in the strength with which the various input parameters affect the calculated quantities. We have, for example, used the

---

* This work was supported by the Director, Office of Energy Research, Division of Nuclear Physics of the Office of High Energy and Nuclear Physics of the U.S. Department of Energy under Contract DE-AC03-76SF00098 and DE-AC05-84OR21400.
** On leave from Soreq Nuclear Research Center, Yavne 70600, Israel.
*** On leave from Lund University, Sweden.

strong coupling between J^π and the final state proton branching ratios to determine the spin and parity of several new precursors listed in Table I. In cases where proton emission occurs both from the ground and isomeric states of the precursor their intensity ratio can also be calculated from the final state proton branching ratios[5]. The most profound influence on the results of statistical model calculations is exercised by S_β. (The only parameters S_β does not affect strongly are the final state proton branching ratios.) Without detailed knowledge of S_β either from experiments such as total absorption γ-ray spectroscopy or from theory it is rarely possible to reproduce the correct centroid and shape of the proton spectrum – simultaneously with b_p, the half-life, and to some extent the final state proton branching ratios. Since no S_β measurements in our region of interest are available (except for ^{147m}Dy[6]) we have performed calculations of GT β-strength functions using deformed Nilsson model wave functions in a modified oscillator potential[7]. Pairing is treated in BCS approximation. A simple residual interaction with a coupling constant of $\chi_{GT} = 23/A$ MeV is added and is treated in the RPA approximation. The strength of this interaction has been adjusted to obtain agreement between the calculated and experimental energies of the giant GT resonance for ^{208}Pb and ^{144}Sm. The Nilsson model κ- and μ-parameters for the proton and neutron single particle potentials are: $\kappa_p = 0.0637$, $\mu_p = 0.6000$, $\kappa_n = 0.0637$, and $\mu_n = 0.4200$. These parameters have been determined from adjustment of calculated single particle levels to experimental data in the rare earth region[8]. Deformation parameters were taken from Ref. 9. We have thus far not extended these calculations to isomers.

After a brief discussion of experimental details in section 2, a comparison between experimental and calculated proton spectra is made in section 3. For precursors near the N = 82 closed shell the statistical model description breaks down, and S_β calculations have to be modified. This is discussed in section 4 for a set of three N = 81 even-odd isotones.

2. EXPERIMENTAL

The isotopes presented here were produced in compound nucleus reactions of neutron-deficient projectiles (^{36}Ar, ^{40}Ca, ^{46}Ti, ^{50,52}Cr, ^{54,56}Fe, ^{58}Ni, and ^{64}Zn) and neutron-deficient targets (^{58}Ni, ^{90}Zr, ^{92,94}Mo, and ^{96}Ru) at the LBL SuperHILAC. The evaporation residue recoils were stopped in a tantalum catcher operated close to its melting point (~3000 C) located inside a surface ionization source. After diffusing out of the catcher and being ionized, the product ions were accelerated to 50 keV and mass separated with the on-line isotope separator OASIS[10]. The separator was calibrated with stable rare earth isotopes introduced into the ion source as minute quantities of the oxides. A single isobaric chain was selected at the focal plane of the separator and transported ionoptically to a shielded, low background spectroscopy laboratory. Here the 50-keV ions were implanted in a 50-μm Mylar tape and transported within 64 ms to an array of detectors consisting of a Si ΔE–E particle telescope, a HyperPure Ge (HPGe) detector (0.7 keV FWHM resolution at 122 keV), two n-type Ge detectors with relative efficiencies of 24% (1.9 keV FWHM resolution at 1332 keV) and 52% (2.4 keV FWHM resolution at 1332 keV), and a thin plastic ΔE_β detector. A scale drawing of the detector arrangement is shown in Fig. 1. Coincidences between protons, x rays, γ rays, and positrons were recorded event-by-event on magnetic tape for subsequent off-line analysis. After the arrival of a new sample at the detectors, a quartz controlled clock was started to tag all events with a relative time signal for half-life determination. Appropriate counting time intervals were chosen, taking into account the known or expected half-lives. A new batch of collected isotopes was supplied by the computer controlled tape system at the conclusion of each counting interval. Concurrent with the event-by-event data acquisition, time resolved spectroscopy singles data were taken with the 52% γ-ray detector and the HPGe detector. Counting intervals for the singles measurements were typically divided into eight equal time bins. Singles data from the 24% detector were also recorded to correct for summing effects in the 52% detector. Energy and efficiency calibrations of the Ge detectors were carried out with standard sources; x-ray and γ-ray efficiencies were linked to the particle telescope efficiency via α particles and γ rays from ^{241}Am decay.

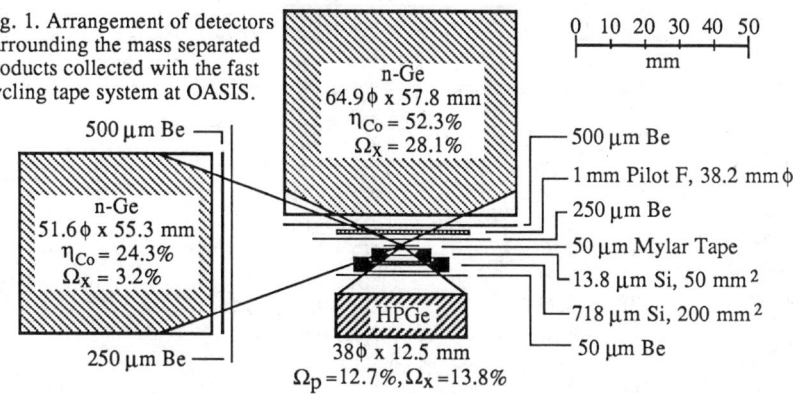

Fig. 1. Arrangement of detectors surrounding the mass separated products collected with the fast cycling tape system at OASIS.

3. RESULTS: DEFORMED REGION

Beta-delayed proton decay becomes possible when the total decay energy of a nucleus Q_{EC} exceeds the proton separation energy S_p in the β-decay daughter. The Coulomb barrier prevents proton decay from competing with γ-decay until a threshold energy Θ_p is exceeded, i.e., $(Q_{EC} - S_p - \Theta_p) > 0$. In the lanthanide region Θ_p is about 2-3 MeV. Calculations show that the observed proton emission thresholds in this region correspond to a proton decay width Γ_p of 10^{-5} eV or a proton- to γ-width ratio Γ_p/Γ_γ of $\sim 10^{-4.5}$, which yields $\Gamma_\gamma = \Gamma_p \times 10^4 = 0.1$ eV, equivalent to a γ lifetime of 7×10^{-15} s. Lifetimes of this order are associated with E1 single-particle transitions of about 0.4 MeV or M1 transitions of about 1.7 MeV, which are typical energies encountered in γ-ray cascades.

Table 1 gives a summary of all new isotopes and new decay branches observed with OASIS to date. With the exception of ^{142}Dy, ^{144}Dy, and ^{148}Er all β-delayed proton precursors, including those known previously, are odd-N nuclei. The question arises why several even-N nuclei that have a lower isospin than known β-delayed proton precursors in the same isobaric chain have no appreciable proton branches (for example $^{128}_{60}$Nd, $^{131}_{61}$Pm, and $^{141}_{65}$Tb) even though β-delayed proton decay is energetically possible. An experimental reason is that in some cases such a decay is difficult to observe because of other strong proton precursors in the same chain. Another reason supported by S_β calculations, is that most of the available β strength may go to levels at excitation energies below $(S_p + \Theta_p)$.

Rather than superficially discussing several new isotopes listed in Table 1, the state of the present experimental and interpretational capabilities will be illustrated by presenting results for one typical well deformed β-delayed proton precursor – $^{123}_{58}$Ce – in more detail. Fig. 2 shows the measured proton spectrum (histogram) and the results of several calculations. Ce-123 decays via β-delayed proton emission to the 0^+ ground state and the 2^+ and 4^+ excited states of the daughter nucleus $^{122}_{56}$Ba. The measured final state proton branching ratios are: 20 ±7% (0^+), 68 ±7% (2^+), and 12 ±3% (4^+); the half-life is 3.8 ±0.2 s. Nilsson model calculations predict the spin of ^{123}Ce to be $5/2^+$ (with a $5/2^-$ state in close proximity). Fig. 2 shows the result of an S_β

TABLE 1. New isotopes and decay branches observed at OASIS (^{119}Ba[a], ^{129}Nd[b], ^{131}Nd[c], ^{133}Sm[b], and ^{135}Sm[d] also studied but not listed below). From left to right: isotope, measured half-life, calculated half-life (see text), decay mode (βp and βg denote β-delayed proton and γ emission, respectively), mean energy and energy range of protons, number of γ rays observed, experimental spin, calculated spin from Ref. 22, and whether x rays and/or γ rays were observed in coincidence with protons.

ISOTOPE	T$_{1/2}$(exp)	T$_{1/2}$(calc)	MODE	ENERGY	N$_\gamma$	J$^\pi$(exp)	J$^\pi$(calc)	PHOTONS
^{120}La[a]	2.8(2)	1.8[e]	βp	3.7[2.1-5.6]	–	–	–	x
^{122}La[a]	8.7(7)	5.5[e]	βp	3.4[2.0-4.6]	–	–	–	x
^{123}Ce[a]	3.8(2)	2.6	βp	3.6[2.0-5.8]	–	5/2	5/2$^+$	x,γ
^{125}Ce[c]	10(1)	10	βp[f]	3.3[1.7-5.1]	25	5/2	1/2$^+$	x,γ
^{124}Pr[c]	1.2(2)	0.6[e]	βp	3.7[2.2-6.0]	1	–	–	x,γ
^{126}Pr[g]	3.2(6)	2.0[e]	βp	3.8[2.1-5.0]	–	–	–	–
^{128}Pr[b]	3.2(5)	7.7[e]	βp	3.2[1.9-5.0]	–	–	–	x
^{127}Nd[c]	1.8(4)	1.7	βp	3.7[2.2-6.0]	–	–	1/2$^+$	x,γ
^{130}Pm[b]	2.2(6)	1.7[e]	βp	3.9[2.1-5.8]	–	–	–	x
^{132}Pm[b]	5.0(8)	3.7[e]	βp[f]	3.6[2.2-5.0]	–	–	–	x
^{131}Sm[c]	1.2(2)	1.6	βp	3.7[1.8-6.6]	–	–	7/2$^-$	x,γ
^{134}Eu[d]	0.5(2)	1.3[e]	βp	3.8[1.8-6.0]	–	–	–	x
^{135}Eu[d]	1.5(2)	2.1	βγ	n.a.	1	–	1/2$^-$	n.a.
^{136}Eu[d]	3.2(5)	1.8[e]	βp	3.9[2.0-6.0]	28	–	–	x
^{137}Gd[g]	7(3)	1.2	βp	3.8[2.2-6.6]	–	–	9/2$^-$	–
^{139}Gd[g]	5(1)	4.5	βp	3.8[2.1-5.5]	–	–	7/2$^+$	–
^{140}Gd[c]	15.8(4)	15.3	βγ	n.a.	35	0$^+$	0$^+$	n.a.
^{141m}Gd[c]	24.5(9)	n.a.	βγ	n.a.	35	11/2$^-$	n.a.	n.a.
^{141g}Gd[c]	~20	16.3	βp	3.6[2.2-4.9]	10	1/2$^+$	1/2$^+$	x,γ
^{140m}Tb[c]	2.4(2)	n.a.	βp	4.2[2.2-6.6]	3	–	n.a.	x
^{141}Tb[c]	3.5(2)	3.4	βγ	n.a.	35	–	1/2$^-$	n.a.
^{142m}Tb[c]	0.3(1)	n.a.	βγ	n.a.	4	–	n.a.	n.a.
^{142g}Tb[c]	0.6(2)	3.5[e]	βp	3.9[2.5-5.2][h]	7	–	–	x,γ
^{141}Dy[c]	0.9(2)	1.4	βp	4.0[2.4-6.1]	–	–	7/2$^+$	x,γ
^{142}Dy[c]	2.3(3)	2.7	βp	3.9[2.5-5.2][h]	1	0$^+$	0$^+$	x,γ
^{143}Dy[a]	3.8(6)	2.9	βp	4.2[2.5-6.5]	–	–	3/2$^-$	x,γ
^{144}Dy[c]	9.1(5)	9.7	βp	3.2[2.6-4.5]	21	0$^+$	0$^+$	x
^{144}Ho[c]	0.7(1)	1.3[e]	βp	4.2[2.2-7.0]	–	–	–	x
^{146}Ho[c]	3.1(5)	2.9[e]	βp[f]	4.1[2.4-6.3]	9	≥6	–	x
^{148m}Ho[i,j]	9.7(3)	n.a.	βp[f]	4.1[2.2-5.4][h]	39	6$^-$	n.a.	x,γ
^{148}Er[i]	4.4(2)	n.a.	βp[f]	4.1[2.2-5.4][h]	5	0$^+$	0$^+$	x,γ
^{149}Tm[k]	0.9(?)	n.a.	βp	n.a.	7	11/2$^-$	11/2$^-$	x
^{150m}Tm[i,j]	2.2(2)	n.a.	βp[f]	4.7[2.2-7.5]	12	6$^-$	n.a.	x,γ
^{151m}Yb[l]	1.6(1)	n.a.	βp	4.0[2.5-6.0]	1	11/2$^-$	n.a.	x,γ
^{151g}Yb[l]	1.6(1)	0.9	βp	4.8[2.3-7.5]	–	1/2$^+$	11/2$^-$	x,γ
^{152m}Lu[i,j]	0.7(1)	n.a.	βp	4.6[2.3-7.9]	3	4$^-$,5$^-$,6$^-$	n.a.	x

[a]Ref. 19, [b]Ref. 20, [c]Ref. 21, [d]Ref. 1, [e]the ground state to ground state transition in odd-odd to even-even decay is not included, [f]new delayed-proton branch in known isotope, [g]Ref. 23, [h]mixture of all delayed protons in this mass chain, [i]Ref. 5, [j]relative position of low- and high-spin isomers unknown, [k]Ref. 24, [l]Ref. 12.

calculation, described in section 1, and the β intensity distribution I_β. The β strength function combined with a statistical model calculation generates four proton spectra representing transitions to the ground and the first two excited states in $^{122}_{56}$Ba and their sum, labeled 0^+, 2^+, 4^+, and "SUM." The precursor spin most consistent with the experimental results is $5/2^+$ which gives final state proton branching ratios of 17% (0^+), 73% (2^+), and 10% (4^+). Other spin/parity assumptions give proton branching ratios that are at variance with the experiment. The calculated half-life obtained by integrating I_β (Fig. 2) from 0 to Q_{EC} is 2.6 s. An empirical quenching factor of 0.5 was applied to the β strength. The other half-lives in Table 1 were obtained in the same fashion, with Q_{EC} values taken from Ref. 11. The main purpose of presenting the ^{123}Ce example is to point out that S_β acts as an "anchor" for the proton spectrum. The shape of the spectrum is sensitively dependent on S_β. In the absence of erratic variations in Γ_γ it may even be used to judge whether its shape in the energy range spanned by the proton spectrum is correct. A constant S_β for example will in general reproduce neither the correct width nor the centroid of the proton spectrum; neither will variations in the other input parameters to the statistical model.

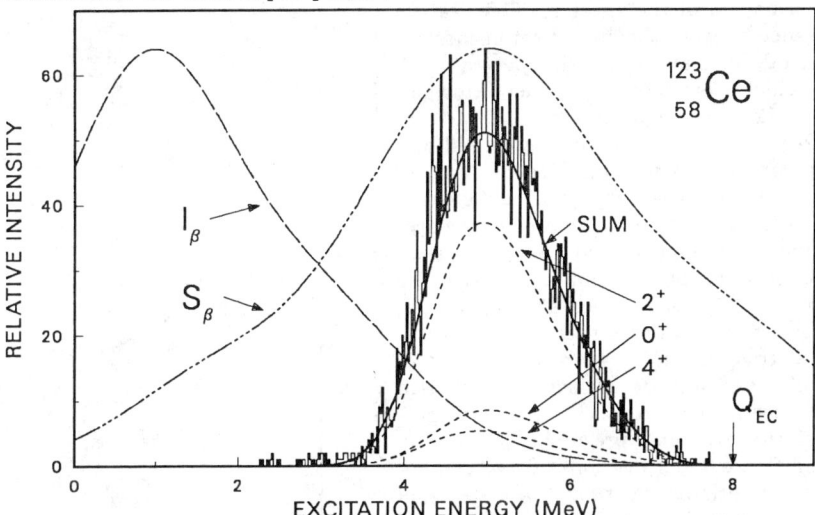

Fig. 2. Experimental proton spectrum (histogram). Proton spectra calculated with a statistical model (solid line and short dashed lines), and β- strength function S_β (dash dots). Beta intensity I_β (long dashes) as used for half-life calculation.

4. RESULTS: CLOSED SHELL REGION

In the deformed region the statistical model coupled with S_β calculations based on Nilsson model wave functions gives an adequate description of β-delayed proton decay. It is reasonable to expect that the statistical treatment will break down near shell closures because of low level densities. Another difficulty is that rapidly changing deformations will cause the β decay mother and daughter to have slightly differing basis wave functions contrary to what is assumed in the calculation of $S_\beta{}^7$. To quantify the influence of closed shells on the β-delayed proton decay we have performed a systematic study of six N = 81 proton precursors: ^{147}Dy, ^{148}Ho, ^{149}Er, ^{150}Tm, ^{151}Yb, and ^{152}Lu. The observed proton spectra as functions of excitation energy are shown in Fig. 3 (other results are summarized in Table 1).

Fig. 3. Beta-delayed proton spectra of N = 81 precursors. Proton separation energies were taken from Ref. 25 and Q_{EC} values from Ref. 11.

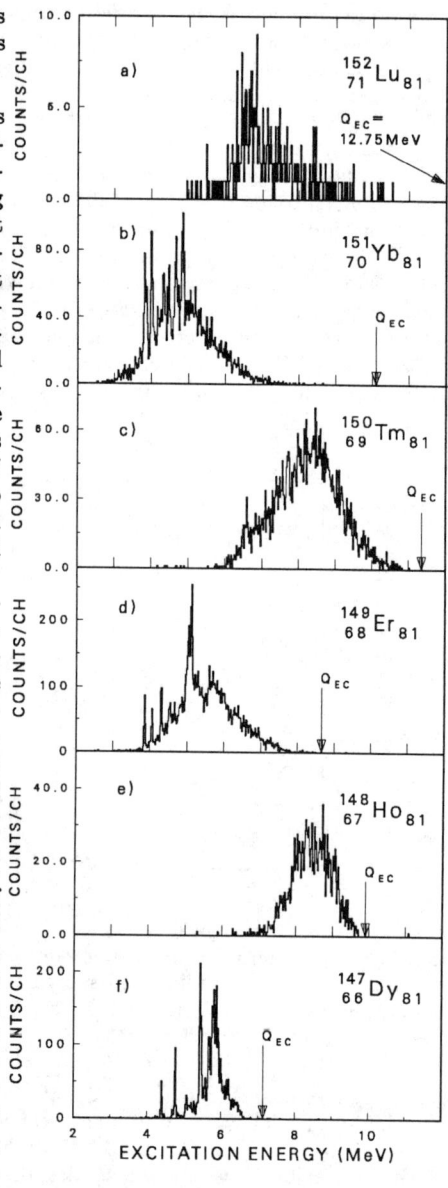

Casual inspection of Fig. 3 reveals two prominent effects: (1) an odd-even effect in the mean proton energies and (2) pronounced structure in the spectra originating from even-odd precursors. The first effect is easily explained by the fact that in odd-odd precursors GT β decay occurs without breaking a proton pair which leaves the proton emitting nucleus in a state of high excitation energy and consequently high level densities which is also the reason for the statistical nature of the spectra. The average difference between the mean proton energies of the odd-odd and even-odd spectra is 1.7 MeV which is expected since the pairing gap is $\Delta = 12/\sqrt{A}$, and thus 2Δ is ~ 2 MeV in this region. Details of the decay of odd-odd precursors are discussed in Ref. 5.

It was realized early on that the structure in the even Z, N = 81 precursor spectra may be related to the low level densities of N = 82 proton emitters. With the discovery of ^{151}Yb and our ability to separate the spectra into a structured and a statistical component[12] a more complete picture of the proton decay has emerged with the following features. The proton spectra of the even-odd precursors (Fig. 3b, d, and f) are superpositions of two components, one originating from the β decay of the $h_{11/2}$ isomeric state (unstructured), and the other associated with the decay of the $s_{1/2}$ ground state (highly structured). The decay of the isomer samples high spins, high level densities and high excitation energies in the intermediate nucleus, and the protons are emitted to final states of large angular momenta which subsequently decay by γ emission. The favored decay path of the $s_{1/2}$ precursor ground state, on the other hand, leads to low spin, low level density, and low excitation energy states in the intermediate nucleus.

Statistical model calculations show that the proton decay from these states proceeds with greater than 95% probability to the 0^+ ground state of the final nucleus. The fact that the structure in the proton spectra persists in ^{151}Yb indicates that the N = 82 rather than the Z = 64 shell closure is the relevant condition for the effect.

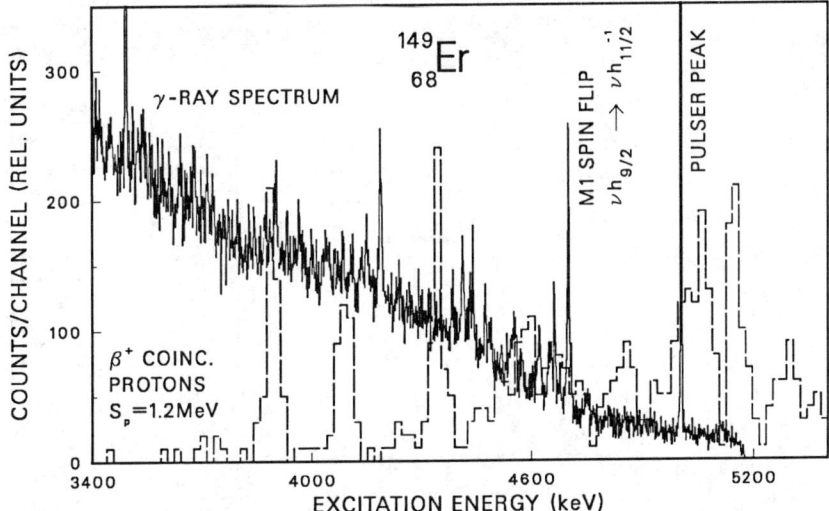

Fig. 4. Positron coincident proton spectrum and γ-ray singles spectrum following β decay of $^{149}_{68}$Er. S_p was taken from Ref. 13.

To elucidate the nature of the proton emitting states in ^{149}Ho we have measured the high energy γ-ray spectrum of ^{149}Er (Fig. 4). The spectrum ends abruptly with the $h_{9/2} \longrightarrow h_{11/2}^{-1}$ spin flip transitions, the most prominent being at 4.70 MeV, while the last strong proton "peak" occurs at 5.14 MeV. Similar observations were made for ^{147}Dy [13]. In both cases no obvious correlations between the γ-ray and proton spectra exist.

In the following an interpretation of the structured proton spectra is proposed based on the concept of doorway states (for a review on this subject see, for example, Ref. 14). The concept was originally introduced to calculate neutron resonances and strength functions[15] and later generalized for other nuclear reactions. The basic idea can best be described with the simple example of a single nucleon from the continuum interacting with a closed shell target nucleus (say, ^{146}Gd + p $\longrightarrow$ ^{147}Dy) [one particle configuration (1p0h)]. If it is assumed that the residual interaction between the incident nucleon and the target nucleons is a sum of two body potentials only two particle- one hole (2p1h) configurations can be excited. These are called "doorway" states because they exhaust most of the coupling strength to a channel in a given range of excitation energies: they are the doorways to most subsequent forms of excitations of the nucleus. The width of these states is $\Gamma = \Gamma\uparrow + \Gamma\downarrow$ where $\Gamma\uparrow$ is the escape width back to the entrance channel and $\Gamma\downarrow$ the damping or spreading width to the compound nucleus.

When treated in first order perturbation theory the probability for transitions from the doorway state with p particles and h holes to states of next higher complexity with p' particles and h' holes can be written as

$$\lambda_{ph \longrightarrow p'h'} = \frac{2\pi}{\hbar} <|M|^2> \rho_{p'h'} \qquad (1)$$

where $<|M|^2>$ is a suitably averaged matrix element and $\rho_{p'h'}$ is the density of p' particle h' hole states. For a simple uniform spacing model Ericson[16] has given a formula for the density of particle hole states:

$$\rho_{ph} = \frac{g(gU)^{p+h-1}}{p!h!(p+h-1)!} \qquad (2)$$

where U is the excitation energy of the nucleus and g the density of single particle states.

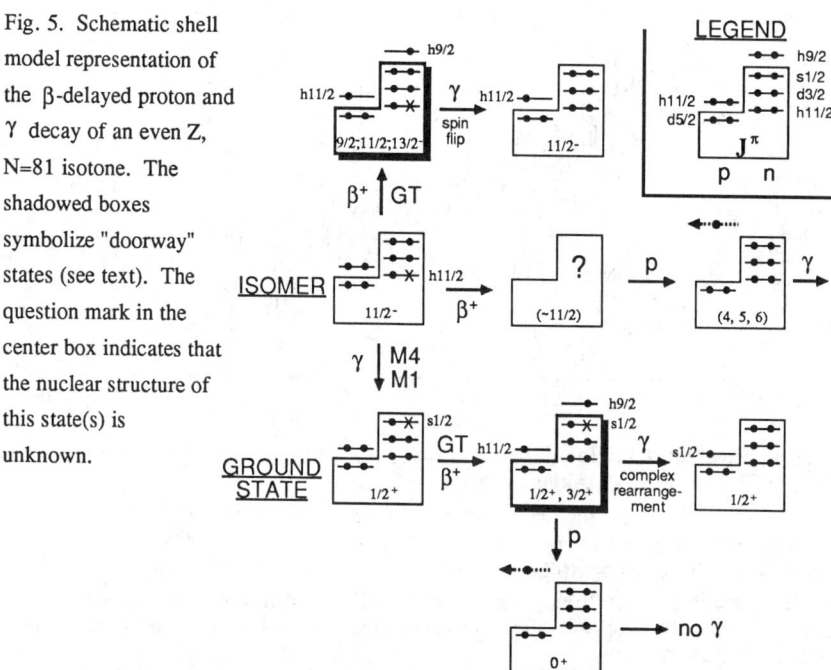

Fig. 5. Schematic shell model representation of the β-delayed proton and γ decay of an even Z, N=81 isotone. The shadowed boxes symbolize "doorway" states (see text). The question mark in the center box indicates that the nuclear structure of this state(s) is unknown.

Thus far it was assumed that the (2p1h) doorway states are excited through the residual interaction between an incident particle and the target nucleons mediated by two body potentials. It will now be argued that similar states can be excited in β decay and may under certain conditions, become manifest as structure in the subsequent proton decay. This is demonstrated by referring to a simple shell model picture (Fig. 5) which shows the major decay modes of the isomer and the ground state for a typical even-Z, N = 81 proton precursor (say, $^{147}_{66}$Dy). The decay is dominated by $\pi h_{11/2} \longrightarrow \nu h_{9/2}$ GT transitions which change the (0p1h) configurations to (2p1h) excited states. Depending on the orbital of the spectator neutron, high or low spin states are produced that have $(\pi h_{11/2} \, \nu h_{9/2}) \times \nu h_{11/2}^{-1}$ or $(\pi h_{11/2} \, \nu h_{9/2}) \times \nu s_{1/2}^{-1}$ configurations respectively. By definition both configurations are doorway states (shadowed boxes in Fig. 5) because all further excitations of the nucleus have to go through them. The channels into which the two "doorway isomers" decay are, however, distinctly different. Measurements of S_β for ^{147m}Dy[6] show a strong GT resonance at 4.7 MeV. The width of the resonance is determined by the coupling of the $h_{11/2}$ proton and $h_{9/2}$ neutron, and the coupling of this pair to different proton configurations, compatible with GT selection rules. Most of these states γ decay via a $\nu_{9/2} \longrightarrow \nu_{11/2}^{-1}$ M1 spin flip transition in $\sim 10^{-16}$ s[17] to the ground state which has been clearly observed in the spectra of ^{147}Dy[13] and ^{149}Er (Fig. 4). In contrast, proton decay to the even-even daughter nucleus from 5 MeV excitation is too weak to be observed either due to the large $(9/2^-, 11/2^-, 13/2^-) \longrightarrow 0^+$ angular momentum hindrance for decay to the ground state or the small available decay energy to the high lying high spin excited states. The unstructured β-delayed proton decay from the $h_{11/2}$ precursor in ^{149}Er and ^{151}Yb is therefore due to, as yet unknown, configurations at greater than

5 MeV excitation energy which have sufficiently small widths ($\hbar/\Gamma_\gamma \geq 10^{-14}$ s) (Fig. 5 second row). For ^{147m}Dy no significant β-delayed proton decay is expected. Statistical model calculations using S_β from Ref. 6 give a branching ratio of <2x10^{-6}, mostly due to high ℓ-wave protons. This corresponds to <0.4% of the ^{147}Dy delayed proton branch reported in Ref. 18.

In contrast to the γ decay of the high spin doorway state the γ decay from 1/2$^+$ and 3/2$^+$ (2p1h) states proceeds only very slowly, terminating in low-lying proton states which necessitates a complex rearrangement of the nucleus (Fig. 5 third row). This rearrangement requires a large number of nucleon-nucleon interactions. According to eqs. (1) and (2) the spreading width $\Gamma\downarrow$ is strongly dependent on the level density and excitation energy; for (2p1h) states it is proportional to $g(gU)^{p+h-1} = g(gU)^{3+2-1} = g^5 U^4$. The low level density of the closed shell intermediate nucleus and its low excitation energy in the region of the observed structured proton decay (~5 MeV) therefore result in a small value of $\Gamma\downarrow$ which retards the γ decay of the (2p1h) states. These "simple" states couple, however, very effectively to the open channel wave function and thus become manifest as structure in the proton spectra.

To obtain the relevant matrix elements for GT decay to (2p1h) states we have carried out Nilsson model calculations with a modified oscillator potential as described in section 1. For these calculations it was necessary to introduce "effective" ε_2 and ε_4 deformations of the nucleus, larger than the static ground state deformations. This has two effects: (1) it shifts the GT resonances to lower energies and (2) it distributes their strength over several smaller resonances. For ^{151}Yb for example the ground state deformations^9 $\varepsilon_2 = -0.091$, $\varepsilon_4 = 0.015$, and $\varepsilon_6 = 0.002$ were replaced by $\tilde{\varepsilon}_2 = -0.150$, $\tilde{\varepsilon}_4 = 0.040$, and $\tilde{\varepsilon}_6 = 0$. Since the potential mass/energy surface in this region is quite soft this corresponds to a change in ground state energy of only ~500 keV. This treatment may also be justified by the observation that the proton emission proceeds from elevated excitation energies that are in general associated with larger deformations. Fig. 6 shows the results of the S_β^{GT} calculations (c), the calculated (b), and the experimental (a) proton spectra for the three even-odd N=81 precursors. Due to limitations in computer time, only a modest effort was made to vary the ε_2 and ε_4 deformation parameters to match the experiment. The strength of the residual interaction was kept constant and ε_6 equal to zero. The calculations reproduce three basic features of the experimental results: (1) the location of the β strength, (2) the splitting into smaller resonances. and (3) their average spacing. This is taken as a strong indication that the calculated matrix elements represent the doorway states that are subsequently observed in the proton decay.

5. CONCLUSIONS AND ACKNOWLEDGMENTS

A large number of β-delayed proton emitters in the region bounded by Z > 50 and N < 82 has now been studied. Beta-delayed proton emission becomes, as expected, a prominent decay mode for all elements near the proton drip line. A significant advance in the understanding of this decay mode has been made by combining a statistical model with β-strength function calculations based on Nilsson model wave functions in a modified oscillator potential. The deformation parameters are taken from calculations that employ a folded Yukawa potential, and in the future it may be more consistent to use the same potential for the S_β calculation. Eventually the calculations will have to be extended to isomeric states since we have observed several cases of β-delayed proton decay from such states. A direct measurement of S_β – for example via total absorption γ ray spectrometry – will be necessary to gain further confidence in the calculations and to obtain quantitative information about the spreading width of the initial GT states; presently, this is an adjustable parameter (typically ~0.7 MeV in the deformed region).

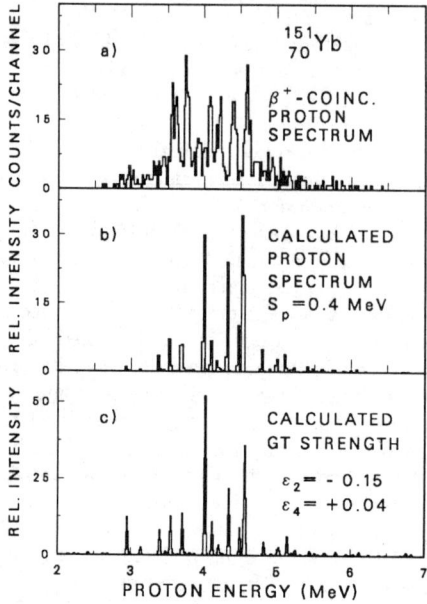
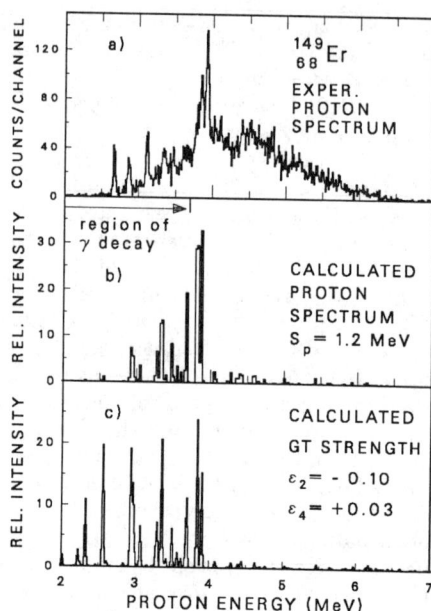
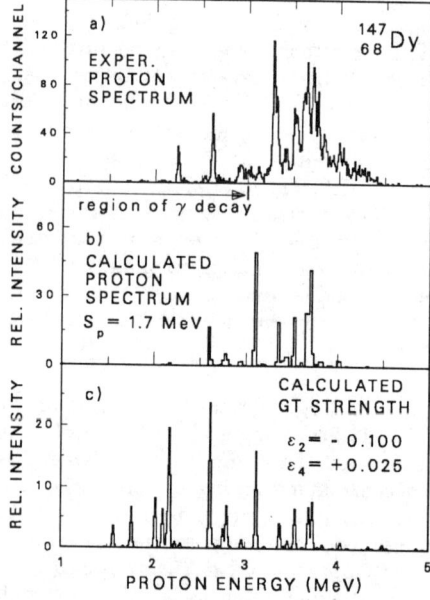

Fig. 6. (a) Beta-delayed proton spectra of even-Z, N = 81 proton precursors ; (b) calculated proton spectra using the β-strength distribution shown in (c); (c) calculated β-strength distribution (see text).

For even-Z nuclei near the N = 82 closed shell pronounced structure in the proton spectra has been observed that cannot be treated in the framework of a statistical model. An interpretation based on the concept of doorway states is proposed. Quantitative calculations of these states were carried out that reproduce the essential features of the experimental results.

The participation of P. Lemmertz in earlier experiments is appreciated. Valuable comments regarding the shell model were made by R.B. Firestone. F.T. Avignone III and Y.A. Ellis-Akovali participated in the data acquisition phase of several experiments and K. Vierinen helped in the preparation of the figures and the TEX version of the manuscript.

REFERENCES

1. K. Vierinen, et al., LBL-23221 (to be published) and this conference.
2. P. Hornshøj, et al., Nucl. Phys. A187, 609 (1972).
3. J. Hardy, Phys. Lett. 109B, 242 (1982).
4. C.H. Johnson, et al., Phys. Rev. C20, 2052 (1979).
5. J.M. Nitschke, et al., LBL-23408, submitted to Phys. Rev. C.
6. G.D. Alkhazov, et al., Nucl. Phys. A438, 482 (1985).
7. J. Krumlinde and P. Möller, Nucl. Phys. A417, 419 (1984).
8. S.G. Nilsson, et al., Nucl. Phys. A131, 1 (1969).
9. P. Möller and J.R. Nix, At. Data Nucl. Data Tables 26, 165 (1981).
10. J.M. Nitschke, Nucl. Instr. Meth. 206, 341 (1983).
11. S. Liran and N. Zeldes, At. Data Nucl. Data Tables 17, 431 (1976).
12. K.S. Toth, et al., Phys. Lett. B178, 150, (1986).
13. D. Schardt, et al., Proc. 7th Intl. Conf. Atomic Masses and Fundamental Constants (AMCO7), O. Klepper (ed.) (Darmstadt, 1984)p. 222.
14. C. Mahaux, Ann. Rev. Nucl. Sci. 23, 193 (1973).
15. B. Block and H. Feshbach, Ann. Phys. 23, 47 (1963).
16. T.E. Ericson, Adv. Phys. 9, 425 (1960).
17. J.M. Blatt and V.F. Weisskopf, "Theoretical Nuclear Physics" (John Wiley and Sons, New York, 1955) p.627.
18. K.S. Toth, et al., Phys. Rev. C30, 712 (1984).
19. J.M. Nitschke, et al., Z. Phys. A316, 249 (1984).
20. P.A. Wilmarth, et al., Z. Phys. A321, 179 (1985).
21. P.A. Wilmarth, et al., Z. Phys. A325, 485 (1986).
22. P.A. Seeger and W.M. Howard, Nucl. Phys. A238, 491 (1975).
23. J.M. Nitschke, et al., Z. Phys. A312, 256 (1983).
24. K.S. Toth, et al., Phys. Rev. C36, 826 (1987)
25. A.H. Wapstra and G. Audi, Nucl. Phys. A432, 55 (1985).

STUDY OF BETA DECAY HALF-LIVES AND BETA-DELAYED MULTI-PARTICLE FINAL STATES AT MICHIGAN STATE UNIVERSITY.

D. Mikolas, B. A. Brown, W. Benenson, Y. Chen, M. S. Curtin,
L. H. Harwood, E. Kashy, J. A. Nolen Jr., M. Samuel, B. Sherrill,
J. Stevenson, A. Vander Molen, J. S. Winfield
and Z. Q. Xie

Department of Physics and Astronomy, and
National Superconducting Cyclotron Laboratory,
Michigan State University, East Lansing, Michigan 48824-1321

R. Sherr

Department of Physics, Princeton University,
Princeton, New Jersey 08540-0708

M. Gai and Z. Zhao

Department of Physics, Wright Nuclear Structure Laboratory,
Yale University, New Haven Conneticut 06511

ABSTRACT

The first five years of operation of the Reaction Product Mass Separator (RPMS) at Michigan State University has produced information important for understanding the structure of many nuclei in the p and sd shells. We discuss the design and operation of the RPMS. The products of projectile fragmentation-like reactions leave the mass separator with a much greater range than products separated with standard on-line separators. We have taken advantage of this extended range distribution by implanting exotic species within silicon detectors, and detecting the decay products directly. Using this technique, we have measured the beta-decay half-lives of ^{14}Be, ^{15}B, ^{17}B, ^{17}C and ^{19}N. We have also measured the beta-delayed alpha emission from ^{16}N. By containing the total charged-particle decay energy of nuclear states left excited by beta decay within a single silicon detector, we have measured beta decay branching ratios of ^{9}C, which always produces a delayed three-body final state. These new measurements yield beta-decay and particle decay matrix elements that were otherwise nearly impossible to determine if only one of the three particles were detected.

© American Institute of Physics 1988

THE RPMS

The Reaction Product Mass Separator (RPMS) consists of two quadrapole doublets, an E × B velocity filter, and a dipole magnet. The first doublet establishes a parallel beam envelope of the fragments from the target. The Wein filter disperses the fragments vertically according to the difference between their velocity and the central velocity for which the RPMS is tuned. The dipole then disperses the fragments according to magnetic rigidity. The two dispersive elements are adjusted until the velocity components of their dispersion cancel. At this point fragments of a given mass to charge ratio (m/q) with a difference of up to 10% in velocity are focused by the second doublet to the same vertical position at the focal plane, and those of a different m/q to a different location.

Typically, fragments much lighter than the beam have been studied. In this case the fragments are not slowed by the target nearly as much as the beam. Most of the time we have in fact stopped the beam in the target. This is of course the most convenient place to put the beam. Up to 50 watts of ^{22}Ne beam have been deposited in a Tantallum target, and at that power the target glows so that it is visible in a TV monitor. We have exploited the large velocity acceptance of the RPMS in order to collect fragments produced not just at the front of the target, but in the first 30% of its thickness, at which point the beam has decreased by roughly 10% in velocity. We have typically used beams of 35 MeV per nucleon, and observed light neutron rich fragments with about 2/3 of the original beam velocity, or about 15 MeV per nucleon. At this energy the fragments have a flight time through the RPMS of about 250 nS. Timing off of the RF signal from the cyclotron gives between 0.5 and 1.5 nS resolution in the TOF.

By carefully matching the dispersions of the velocity filter and dipole, the isotopes ^{13}B and ^{18}N were clearly resolved. (Figure 1) They are only 0.8% different in mass to charge ratio.

In an early experiment using the RPMS, a test was made for the particle-stability of ^{10}He. While none was observed nor expected, this proved to be a useful exercise in tuning the RPMS. A rough systematic trend for all known particle stable isotopes of H, He and Li, with $T_z > 0$, demonstrated a decrease in yield of about $10^{-(3/5)}$ per excess neutron. The observation of zero counts of ^{10}He in a day of counting yields a limit four orders of magnitude below an extrapolation of the systematical trend. (figure 2))

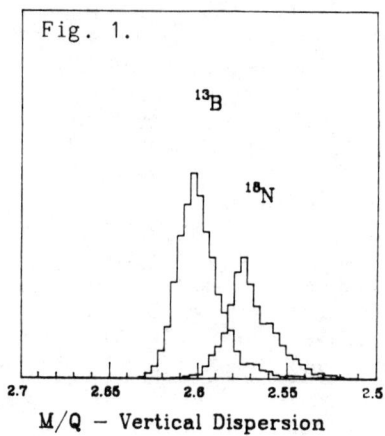
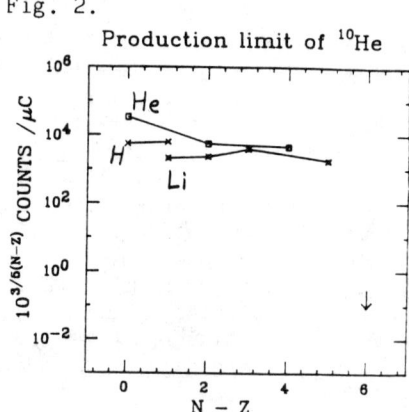

Figure 1, m/q dispersion of the RPMS as vertical displacement at the focal plane. The isotopes ^{18}N and ^{13}B are only 1% different in m/q, and provide a convenient test for tune-up of the RPMS.

Figure 2, production limit for ^{10}He. A beam of ^{18}O with an energy of 560 MeV was fragmented and stopped in a thick tin target backed by copper. Yields of H, He and Li isotopes were determined at 0° and zero counts of 10He were obtained in 12 hours. The arrow at N-Z = 6 indicates where a single count of ^{10}He would be placed if recorded.

HALF-LIFE MEASUREMENTS

We have carried out two major experiments in order to investigate half-lives of light neutron rich isotopes. In the first experimental result from the RPMS to be published[1], a beam of ^{18}O was used with a target of natural tin. The half lives of ^{14}Be and ^{17}C were determined. The isotope ^{17}C has one more neutron than the original beam, and demonstrates an in-elastic component to the reaction. In a second experiment[2], the isotopes ^{17}B, ^{18}C, and ^{19}N were produced from the fragmentation of a ^{22}Ne beam by a Ta target. The half-lives of ^{17}B and ^{19}N were measured as, 5.3 ± 0.6 ms and 235 ± 32 ms respectively, but at m/q=3.00 the count rate was overwhelmed by the production of ^{9}Li, ^{12}Be and ^{15}B. and thus prohibited a determination of a half-life for ^{18}C.

The method used to make these half-life measurements has been described elsewhere. (Refs. 1, 2) Briefly, the focal plane is equipped with m/q defining slits, a position sensitive drift chamber, and a silicon detector telescope with typically one 100μm thick element serving as a "ΔE" detector, a thick (1 to 5 mm) "E", and sometimes a third element serving as a veto. When an atom comes to rest within the "E" detector, the cyclotron beam is interrupted, the gain of the silicon detector preamplifier raised, and beta activity within the E is counted for a preset period (typically 0.1 to 1 second) before the beam resumes. The time since the arrival of the atom is recorded for all beta events.

BETA-DELAYED CHARGED PARTICLE EMISSION (^{18}N)

In addition to the half-life measurements just described, we have begun to explore ways to measure beta-delayed charged-particle emission from Nuclei Far From Stability. It is here that both the advantages and disadvantages of the use of projectile fragmentation becomes the most pronounced. Fragmentation products emerging from a target have a relatively large velocity distribution (5-10 percent), and thus a broad distribution in stopping ranges. The fact that the ranges themselves are large means that the activity can then be implanted in the environment of the experimenters choice. However, the broadness of the distribution in ranges limits the physical 'thinness' of the object in which the activity is to be implanted.

In a recent experiment performed in collaboration with Moshe Gai and Zhiping Zhao of Yale University, the absolute beta-delayed alpha emission spectrum from ^{18}N decay was measured. The ^{18}N's were produced by the fragmentation of ^{22}Ne beam by a Ta target, at a rate of about 20 per second. The activity was identified, counted, and implanted in a silicon detector telescope for one second, (approximately one mean-life of ^{18}N) so that a 'source' of ^{18}N was established. The beam was then shut off, the gain of the preamplifiers raised, and the delayed alphas were counted for another second. The process was repeated automatically, with the absolute real time and data aquisition live time continuously recorded for accurate normalization. The advantages of this technique for this application are,
1) Alpha particle energies below 1 MeV could be reliably recorded, even though the target was chosen to be arbitrarily thick.
2) No dead-layers were presented to the alpha particles.
3) The exact number of ^{18}N atoms responsible for the recorded alpha spectrum could be determined.
4) The silicon presented a solid angle of 100% of 4π Sr.

However, these advantages bring along with them some inescapable disadvantages.
1) The big range distribution means that you must present a significant amount of silicon to stop all of the activity.
2) Since the activity is within the detector itself, the beta particle invariably contributes to the recorded decay energy of the event.

These two problems present a dilemma, since their solutions are in direct competition. One would like a relatively thick silicon detector in order to collect most of the activity (and of course to minimize the capacitance and associated noise). However, the contamination of the spectrum from the energy loss of the beta particle is directly proportional to the thickness of the detector, and calls for the thinnest silicon possible.

In this experiment a compromise was chosen. Most of the ^{18}N activity was expected to be concentrated in a 300 μm thick range distribution in silicon. Three silicon detectors of 60 μm thickness tilted at 45° presented an effective thickness to the incoming ^{18}N ions of 250 μm, while behaving as 'thin' 60μm detectors as far as the beta energy loss was concerned. (Figure 3)

Figure 4 shows a delayed alpha spectrum in one of the detectors for one of the runs. Overall an energy resolution of 90 keV was obtained. The resolution was predominantly limited by the effects of the beta energy loss. This data was collected as part of an ongoing research project at Yale to measure the extremely weak alpha decay of the 0⁻ level in ^{18}O populated by the beta decay of ^{18}N. (see Figure 5) While this state is fed strongly by beta decay, only one event in perhaps 10^8 will result in the emission of an alpha particle, since the alpha decay of a 0⁻ level to 0^+ ^{14}C is parity forbidden. The analysis of the data is underway at Yale.

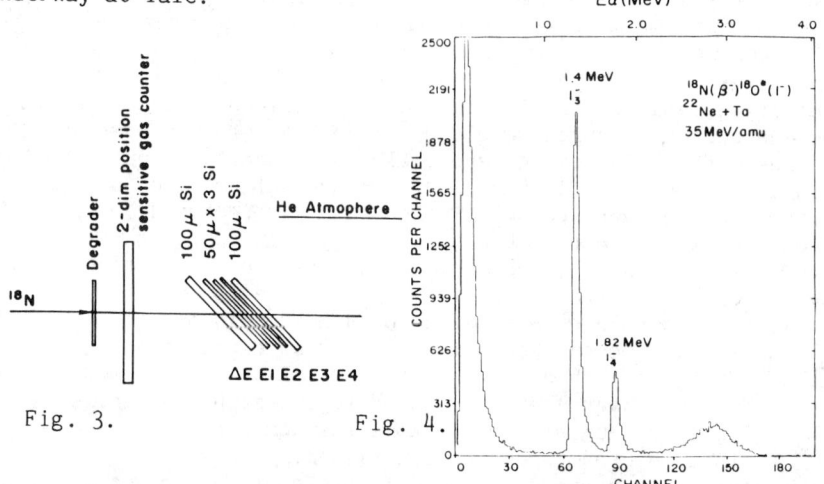

Fig. 3.   Fig. 4.

Figure 3, diagram of detector set-up at focal plane of RPMS, used for the study of beta-delayed α emission from ^{18}N. Detectors were placed in a Helium atmosphere in order to minimize the number of ^{18}N ions which came to rest in the gas between detectors.

Figure 4, delayed alpha spectrum following the decay of ^{18}N, recorded in one of the 50μm detectors.

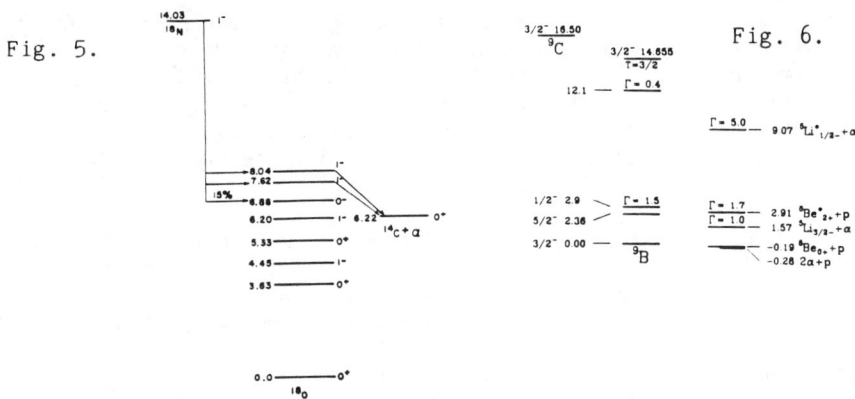

Fig. 5.

Fig. 6.

Figure 5, partial decay scheme of ^{18}N.

Figure 6, decay scheme of ^{9}C, including particle decay thresholds of ^{9}B.

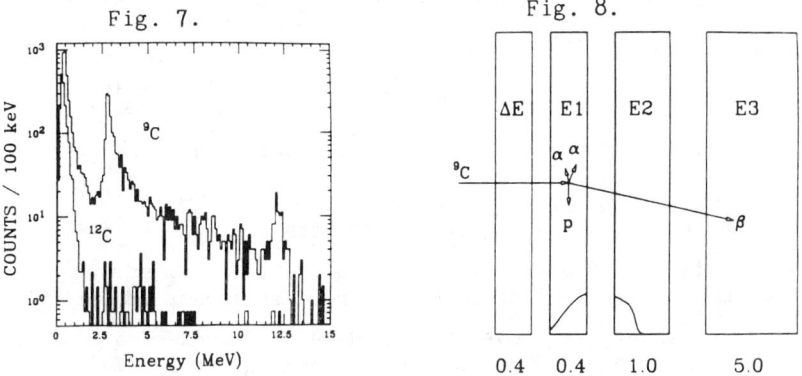

Fig. 7.

Fig. 8.

Figure 7, beta-delayed total charged particle energy from the decay of ^{9}C, recorded in a 400μm thick detector. The decay of 6000 ^{9}C ions is shown, with a background spectrum following the arrival of ^{12}C ions included.

Figure 8, detector set-up used for ^{9}C decay study.

THE DECAY OF ^{9}C

The implanting of ^{9}C deep within a silicon detector facilitated the sucess of an earlier experiment; the determination of the beta decay branching ratios of ^{9}C. From Figure 6, it is clear that the decay of ^{9}C always results in the emission of two α particles AND a proton. There are many sequential routes through which excited states in ^{9}B may decay. For a given narrow level in the daughter ^{9}B, the energy spectrum associated with any ONE of the particles will be difficult or impossible to interprate. However, when the ^{9}C decays within silicon, one records the total charged-particle energy associated with the decay, and thus measures directly the excitation spectrum in ^{9}B from the decay of ^{9}C.

Figure 7 shows the beta delayed charged particle spectrum from ^{9}C decay, as contained within a 400 μm silicon detector. (see Figure 8) The lowest peak represents the decay of ^{9}C to the ground state of ^{9}B. This level presumably decays by emission of a 185 keV (cm) proton, followed by the (90 keV) decay of ^{8}Be. The smallness of these energies would make direct observation of this branch difficult to observe in any other way. The branch to the first 5/2$^-$ of ^{9}B at 2.36 MeV is also evident.

The spectrum represents the recorded energy in a silicon detector of 400μm thickness. While the beta energy loss contamination clearly broadens the line-shapes, thinner detectors were not considered, for protons with energies up to 12 MeV (cm) energy were anticipated. Some of the continuum below the peak does indeed represent decays to the level at 12 MeV, in which the proton escaped the silicon without leaving its full energy. While a more complete account of this experiment can be found elsewhere[3], a few points of experimental interest will be included here.

In Figure 9, a simulation of the beta energy loss tail associated with the branch to the ground state and first 5/2$^-$ state in ^{9}B is subtracted from the recorded spectrum, revealing a branch to the broad 1/2$^-$ level near 3.0 MeV. In order to confirm this branch, the beta energy loss was removed event by event during analysis, by determining the dE/dX of the beta for each decay event in other elements of the silicon telescope. (see Figure 10) Thus the disadvantage of using a thick detector can be partially overcome by the addition of more active elements to detect the beta particles energy loss, and perhaps direction.

The comparison of the delayed total charged particle energy spectrum of ^{9}C with the delayed single proton spectrum of Esterl et al.[4] (Figure 11) shows some significant differences. Namely, the delayed proton spectrum contains two peaks at 9 and 12 MeV, and a huge continuum at lower energy, while the new implantation data shows only a single peak at 12 MeV, and a very modest rise at lower energy. Separately, each experiment is limited in information on the decay. However, taken together, a clear picture emerges. The level at 12 MeV is strongly populated by beta decay, but decays by the emission of protons to the ground

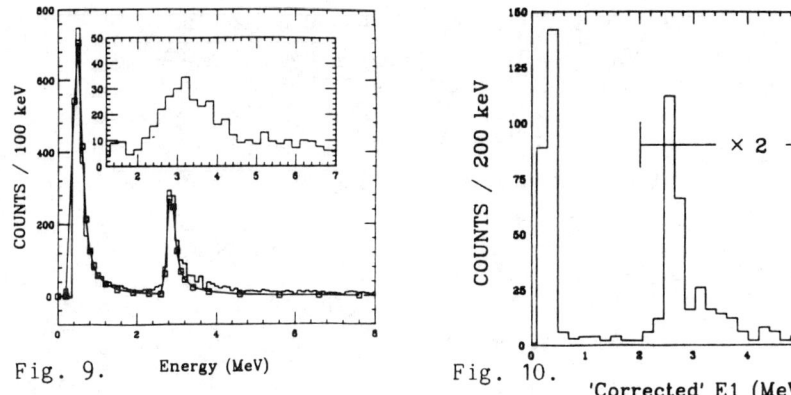

Fig. 9. Energy (MeV)

Fig. 10. 'Corrected' E1 (MeV)

Figure 9, the low energy portion of the data in Figure 7, with a simulation based on the population of the ground state and narrow first 5/2⁻ level. The difference is shown in the inset. (see text)

Figure 10, the broad 1/2⁻ level is clearly shown when the beta energy loss is removed as described in the text.

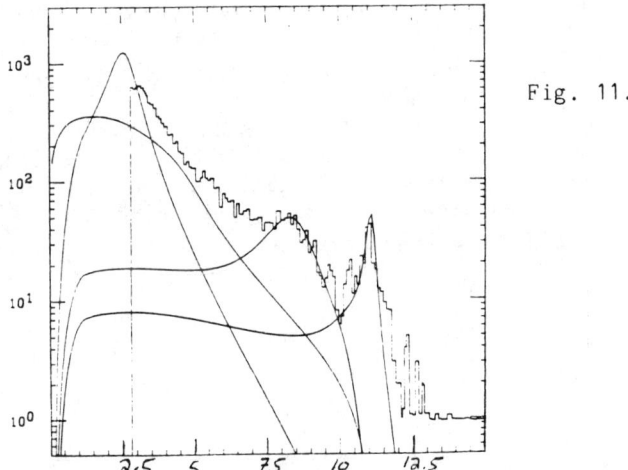

Fig. 11.

Figure 11, beta-delayed proton spectrum of Esterl et al.[4] (histogram) A simulations based on the decay of a state at 12.1 MeV in ⁹B through three channels is also shown. The highest energy peak represents the proton decay to the ground state of ⁸Be. The peak to the left represents proton decays to 2⁺ level in ⁸Be at 3 MeV. The large distribution to the left represents alpha decay to ⁵Li followed by the proton decay of the recoil. The peak at the far left represents the proton decay of the broad 1/2⁻ at 2.9 MeV.

state of ^{8}Be only weakly. Just as often the proton decay populates the broad 2^+ level in ^{8}Be near 3 MeV, and even more frequently the decay proceeds some other way, generating even lower energy protons. Studies of the decay of the mirror system ^{9}Li have suggested that a state at 12 MeV is populated, and decays by the emission of an α particle to Mass 5 (Refs.[5,6]). The solid curves in Figure 11 represent a simulation[5] of the delayed proton spectrum associated with the decay of a level in ^{9}B at 12 MeV, through the ground state and first excited states of ^{8}Be, and through the ground state of 5Li. The strong branch to the broad $1/2^-$ is also included. This match is encouraging, and has spawned further studies[7].

We have begun a series of calculations to determine the beta-delayed particle distributions from the decay of exotic nuclei, where many particle decay channels are open to a given excited state in the daughter; such as the level in ^{9}B and ^{9}Be near 12 MeV which is strongly populated by the beta decay of ^{9}C and ^{9}Li.

Spectroscopic factors are calculated in a p or p-sd shell formalism[3] and are subsequently used as input for a two or three-body (sequential decay) R-matrix calculation, similar to that of Nyman et al.[5] As an example, the relative spectroscopic factors for the nucleon decay of ^{9}B (or ^{9}Be) to states in ^{8}Be, and α decay of these states in Mass 9 to ^{5}Li (or ^{5}He) are shown in Figure 12. These levels exhibit a great amount of selectivity for the various decay modes. If this strong selectivity is characteristic of levels populated by other exotic p and p-sd nuclei, we should expect many more exotic beta-delayed decay modes to be observed in the future.

Figure 12, relative spectroscopic factors for the proton and alpha decay of levels in ^{9}B. The open rectangles have been multiplied by a factor of 10 relative to the solid rectangles. Group A includes the $3/2^-$ ground state, $5/2^-$ at 2.36 MeV, and broad $1/2^-$ at 2.9 MeV. Group B contains levels predicted by the shell model to lie between 8 and 11 MeV. Group C contains four candidates for the narrow level at 12.1 MeV in ^{9}B populated strongly by the beta decay of ^{9}C.

Fig. 12.

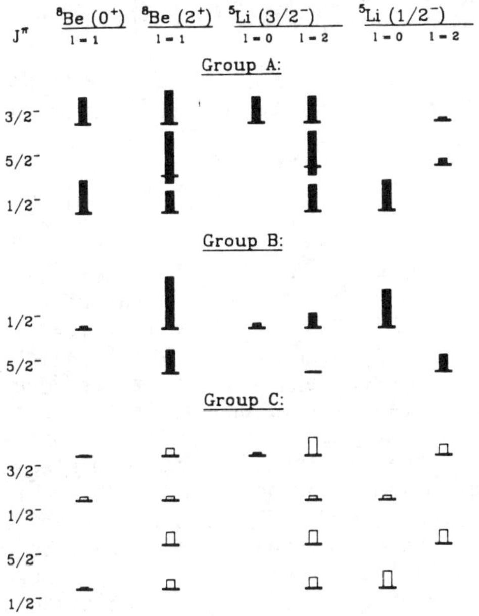

| J^π | ^{8}Be (0$^+$) $I=1$ | ^{8}Be (2$^+$) $I=1$ | ^{5}Li (3/2$^-$) $I=0$ $I=2$ | ^{5}Li (1/2$^-$) $I=0$ $I=2$ |

Group A:

3/2$^-$, 5/2$^-$, 1/2$^-$

Group B:

1/2$^-$, 5/2$^-$

Group C:

3/2$^-$, 1/2$^-$, 5/2$^-$, 1/2$^-$

REFERENCES

1) M. S. Curtin, L. H. Harwood, J. A. Nolen, B. Sherrill, Z. Q. Xie and B. A. Brown, Phys. Rev. Lett. 56, 34 (1986)

2) M. Samuel, B. A. Brown, D. Mikolas, J. Nolen, B. Sherrill, J. Stevenson, J. Winfield and Z. Q. Xie, (unpublished, 1987)

3) D. Mikolas, B. A. Brown, W. Benenson, L. H. Harwood, E. Kashy, J. A. Nolen Jr., B. Sherrill, J. Stevenson, J. S. Winfield, and Z. Q. Xie, Phys. Rev. C, (in press)

4) J. E. Esterl, David Allred, J. C. Hardy, R. G. Sextro, and Joseph Cerny, Phys. Rev. C6, 373 (1972)

5) G. Nyman, R. E. Azuma, B. Jonson, K.-L. Kratz, P. O. Larrson, S. Mattson, and W. Zieger, in Proceedings of the Fourth International Conference on Nuclei Far From Stability, Helsingør, 1981, edited by L. O. Skilen, Vol. I, p. 312.

6) M. Langevin, C. Détraz, D. Guillemaud, F. Naulin, M. Epherre, R. Klapisch, S. K. T. Mark, M. de Saint Simon, C. Thibault, and F. Touchard, Nucl. Phys. A366, 449 (1981)

7) D. Mikolas (unpublished), and Doct. Dissertation, Michigan State University, East Lansing, Mi (1987)

RADIOACTIVITIES WITH 146 ≤ A ≤ 152 INVESTIGATED AT THE OASIS FACILITY; EVIDENCE FOR ^{147}Tm β-DECAY

K. S. Toth,[a] J. M. Nitschke,[b] P. A. Wilmarth,[b] Y. A. Ellis-Akovali,[a]
D. C. Sousa,[a] K. Vierinen,[b] D. M. Moltz,[b] J. Gilat,[b] and N. M. Rao[a]

[a]Oak Ridge National Laboratory, Oak Ridge, TN 37831 USA
[b]Lawrence Berkeley Laboratory, Berkeley, CA 94720 USA

A collaborative program, involving researchers from the Lawrence Berkeley and Oak Ridge National Laboratories, has now been in progress at the SuperHILAC OASIS on-line facility[1] for more than two years. In these investigations properties of short-lived nuclei near the 82-neutron shell have been studied, their decays being assayed with particle, γ-ray, x-ray, and positron detectors. Nuclides investigated so far were produced in ^{58}Ni bombardments of ^{96}Ru, ^{94}Mo, and ^{92}Mo. They are shown in Fig. 1 where a portion of the periodic chart for N ≤ 84 and for 64 ≤ Z ≤ 72 is displayed. Note that for nuclei with less than 84 neutrons α-particle emission is not observed due to the 82-neutron shell since this closure drastically lowers Q values available for α decay. However, as shown in Fig. 1, because of the proximity of the proton drip line and the large Q_{EC} values, both β-delayed and direct proton emission are observed. Table I lists the isotopes we have investigated and summarizes some of the information obtained.

The focus of the study has been primarily two-fold: 1) to gain insight into the nature of the sharp peaks seen in the delayed-proton spectra that accompany the β decays of N = 81 even-Z precursors, and, 2) to investigate low-lying states in nuclei near N = 82, levels whose structures should be describable by shell-model analyses. Much of the accumulated data have been described in recent publications. Interested readers are referred to these papers for a discussion of the following: 1) experimental and calculational decompositions of the ^{151}Yb (Ref. 2) and of the ^{148}Ho, ^{150}Tm, and ^{152}Lu (Ref. 3) delayed-proton spectra; 2) systematics[4,5] of single-particle states in N = 81 and N = 82 isotones compared with predictions from HFB calculations; and, 3) a description of decay properties of A = 152 (Ref. 6) and A = 150 (Ref. 7) nuclides. Here we will discuss data from a recent experiment that examined A = 147 nuclides produced in ^{58}Ni bombardments of ^{92}Mo.

On the basis of proton coincidences with positrons, x rays, and γ rays, we have shown[2] that the structured and statistical components of the ^{151}Yb delayed-proton spectrum are associated with the isotope's $s_{1/2}$ ground state and $h_{11/2}$ isomeric β decays, respectively. The structured component has also been observed in the delayed-proton spectra of two other N = 81 precursors, i.e., ^{147}Dy and ^{149}Er (see e.g. Ref. 8). Because it persists for ^{151}Yb the implication is that it reflects a region of low density of 1/2 and 3/2 levels in all three β-decay daughters as a consequence of the N = 82 shell with the effect of the Z = 64 subshell being minimal.

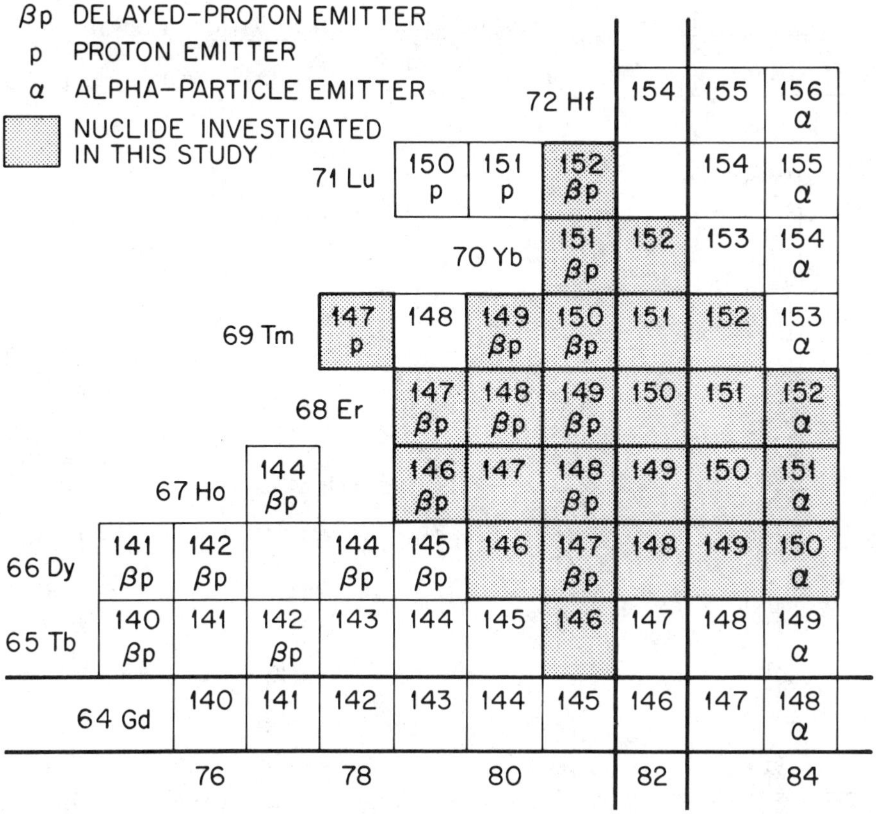

Fig. 1. Portion of nuclidic chart; investigated isotopes are indicated by shaded squares.

[Chart showing portion of nuclidic chart with the following legend:
βp DELAYED-PROTON EMITTER
p PROTON EMITTER
α ALPHA-PARTICLE EMITTER
▨ NUCLIDE INVESTIGATED IN THIS STUDY

Elements from 64 Gd to 72 Hf vs neutron numbers 76 to 84:

72 Hf: 154, 155, 156α
71 Lu: 150 p, 151 p, 152 βp, —, 154, 155α
70 Yb: 151 βp, 152, 153, 154α
69 Tm: 147 p, 148, 149 βp, 150 βp, 151, 152, 153α
68 Er: 147 βp, 148 βp, 149 βp, 150, 151, 152α
67 Ho: 144 βp, 146 βp, 147, 148 βp, 149, 150, 151α
66 Dy: 141 βp, 142 βp, —, 144 βp, 145 βp, 146, 147 βp, 148, 149, 150α
65 Tb: 140 βp, 141, 142 βp, 143, 144, 145, 146, 147, 148, 149α
64 Gd: 140, 141, 142, 143, 144, 145, 146, 147, 148α]

To provide further evidence for this suggestion we performed a decomposition of protons emitted by the N = 79 precursor ^{147}Er ($T_{1/2}$ = 2.5 s); the main difference here, vis-à-vis the N = 81 isotones, is that one is now two neutrons further removed from N = 82.

Data were taken in 4-s counting cycles to emphasize ^{147}Er activity, and in 160- and 1.28-s cycles to provide information on the contributions of ^{147}Dy and ^{147}Tm, respectively. While these results are still being analyzed, two points have emerged. First, both low- (probably $s_{1/2}$) and high-spin (probably $h_{11/2}$) precursors contribute to the ^{147}Er delayed protons, and second, neither the total spectrum (studied also by Schardt et al.[8]) nor the spectrum in coincidence with positrons (it should probe lower excitation energies in the β-decay daughter) exhibits any intense peaks.

The existence of two proton-emitting states in ^{147}Tm has been established (see e.g. Ref. 9). Cross-section measurements and barrier-penetration half-life calculations suggest that the $s_{1/2}$ state ($T_{1/2}$ = 360 μs) is a pure proton emitter but that the $h_{11/2}$ state ($T_{1/2}$ = 560 ms) has a large β-decay branch which as yet has

Table I. Isotopes investigated in this study.

Isotope	J^{π}	$T_{1/2}(s)$	Delayed protons
^{146}Ho		3.6(3)	Yes[a]
^{146}Dy	0^+	29(3)	
^{146}Tb	1^+	~8	
^{147}Tm[b]	$(11/2^-)$	0.65(7)	
^{147}Er	$(11/2^-)$	2.5(2)	Yes
^{147}Er	$(1/2^+)$		Yes
^{147}Ho	$(11/2^-)$	5.8(2)	
^{148}Er	0^+	4.4(2)	Yes[a]
^{148}Ho	(6^-)	9.7(3)	Yes[a]
^{148}Ho$_c$	(1^+)		Yes[a]
^{149}Tm	$(11/2^-)$	0.9(2)	Yes[a]
^{149}Er	$11/2^-$	9(1)	Yes
^{149}Er	$1/2^+$		Yes
^{149}Ho	$11/2^-$	21(1)	
^{149}Ho	$1/2^+$	54(5)[d]	
^{150}Tm	(6^-)	2.2(2)[d]	Yes[a]
^{150}Tm	(1^+)		Yes[a]
^{150}Er	0^+	20(1)	
^{151}Yb[c]	$11/2^-$	1.6(1)	Yes[a]
^{151}Yb[c]	$1/2^+$		Yes[a]
^{151}Tm[c]	$11/2^-$	4.3(2)	
^{151}Tm[c]	$1/2^+$	~11	
^{152}Lu[c]	(6^-)	0.7(1)	Yes[a]
^{152}Yb	0^+	3.1(2)	
^{152}Tm	(2^-)		

[a] Delayed-proton emission observed for the first time.
[b] Beta-decay branch of nuclide identified.
[c] New isotope or new isomer.
[d] New half-life.

not been observed experimentally. Despite a low production cross section (predicted to be ~0.2 mb) the 1.05-MeV proton group[9] of the $h_{11/2}$ ^{147}Tm state was clearly seen in our experiment, and, primarily in the 1.28-s data, there was evidence for this level's β-decay branch. The decay curve for the annihilation radiation had a small component of ~0.5 s, the delayed protons showed a growth and decay pattern consistent with a 2.5-s decay (^{147}Er) being fed by a 0.6-s radioactivity, and, Er K x rays were observed.

Figures 2(a) and 2(b) are portions of singles x-ray spectra taken during the first and second halves of the 1.28-s cycles; fig. 2(c) was accumulated during the first 0.5 s of the 4-s cycles. A weak Er $K\alpha_1$ peak is seen in (a), but only vestiges of it are visible in (b) and (c). Also, Er $K\alpha_1$ x rays were observed in coincidence with the annihilation radiation peak in the 1.28-s [part (d)] but not in the 4-s [part (e)] cycles. Preliminary estimates indicate that the ^{147}Tm $h_{11/2}$ level has a β-decay branch of ~90%.

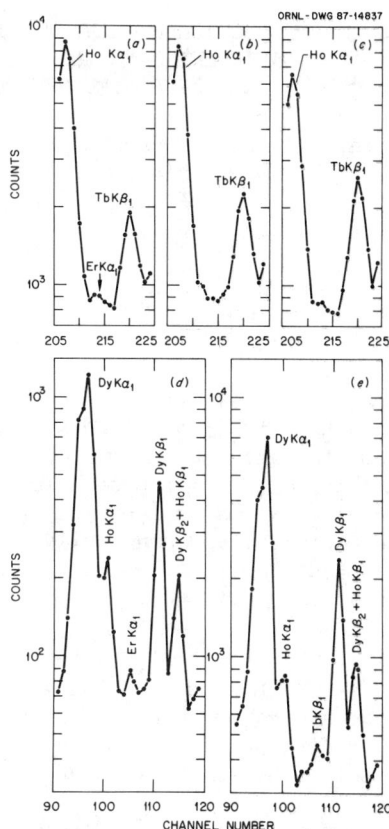

Fig. 2. Portions of K-x-ray spectra measured for A = 147 nuclides in ^{58}Ni irradiations of ^{92}Mo: (a) and (b) singles spectra taken during first and second halves of 1.28-s counting cycles, respectively; (c) singles spectrum accumulated during the first 0.5 s of 4-s cycles; (d) and (e) spectra measured in coincidence with the annihilation radiation peak in 1.28-s and 4-s counting cycles, respectively.

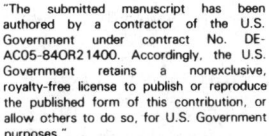

"The submitted manuscript has been authored by a contractor of the U.S. Government under contract No. DE-AC05-84OR21400. Accordingly, the U.S. Government retains a nonexclusive, royalty-free license to publish or reproduce the published form of this contribution, or allow others to do so, for U.S. Government purposes."

Oak Ridge National Laboratory is operated by Martin Marietta Energy Systems, Inc. for the U.S. Department of Energy under Contract No. DE-AC05-84OR21400. Work at the Lawrence Berkeley Laboratory is supported by the Director, Office of Energy Research, Division of Nuclear Physics of the Office of High Energy and Nuclear Physics of the U.S. Department of Energy under Contract DE-AC03-76SF00098.

REFERENCES

1. J. M. Nitschke, Nucl. Instrum. Methods 206, 341 (1983).
2. K. S. Toth et al., Phys. Lett. 178B
3. J. M. Nitschke et al., submitted to Phys. Rev. C.
4. K. S. Toth et al., Phys. Rev. C 32, 342 (1985).
5. K. S. Toth et al., Phys. Rev. C 36, 826 (1987).
6. K. S. Toth et al., Phys. Rev. C 35, 310 (1987).
7. K. S. Toth et al., Phys. Rev. C 35, 620 (1987).
8. D. Schardt et al., Proc. Seventh Intern. AMCO Conf., Sept. 3-7, 1984, Darmstadt, West Germany, p.229 (1984).
9. S. Hofmann et al., Proc. Seventh Intern. AMCO Conf., Sept. 3-7, 1984, Darmstadt, West Germany, p. 184 (1984).

ON THE PRODUCTION MECHANISMS OF NUCLEI FAR FROM STABILITY AT GANIL ENERGIES AND THEIR APPLICATION TO NEW β-DELAYED NEUTRON EMITTERS

D.Bazin, R.Anne, D.Guerreau, D.Guillemaud-Mueller, A.C.Mueller
M.G. Saint-Laurent, W.D. Schmidt-Ott
GANIL, BP 5027, 14021 CAEN Cedex, France.
V.Borrel, J.C. Jacmart, F.Pougheon, A.Richard
IPN, BP 1, 91406 ORSAY Cedex, France.

THE PRODUCTION MECHANISMS OF EXOTIC NUCLEI WITH A ^{86}Kr BEAM

For many years, it has been shown that every time a new nuclear reaction mechanism is well understood, it can be used in a certain way as a tool to produce nuclei far from stability. From this point of view, to get the maximum efficiency, one has to study the reaction mechanism of their production.

It is obvious that nuclei far from stability have the smallest binding energies, and can easily break-up or return to the valley of stability by emitting a few nucleons. Therefore, the excitation energy stored in the nuclei during the reaction is a most prominent feature.

At first approximation, the projectile fragmentation extensively used at GANIL with the LISE spectrometer[1] can be seen as a sudden break-up along two zones, one corresponding to the overlap between target and projectile (the so-called "participant zone"), and the other being the target-like and projectile-like fragments ("spectator zones").

In this picture, well described by the abrasion-ablation model of Goldhaber[2], the excitation energy of the spectators is only due to their deformation and to the "cut" in the eigenfunctions of the nuclei. That means that the reaction is fast enough to transfer *no* momentum.

However, at Fermi energies, which is the domain of GANIL, one could expect that the Goldhaber picture would no longer be valid, since collective as well as dissipative effects could be involved, giving rise to mechanisms halfway between high-energy fragmentation and dissipative reactions observed at low energy. From this idea, D. Guerreau[3] built-up a code to compute the cross sections of exotic nuclei production : starting from the hypergeometric model of abrasion, it adds up the contribution of a giant dipole resonance and calculates the excitation energy of the projectile-like fragment due to the surface energy. Then in a second step, the evaporation due to this excitation energy is computed with the LILITA code.

These calculations reproduce well, even for the most neutron-rich isotopes, the experimentally observed isotopic distributions for the fragmentation of a 44 MeV/u ^{40}Ar beam onto a ^{181}Ta target (see fig. 2 in ref. 1).

We have done the same analysis with a ^{86}Kr beam at about the same energy (43 MeV/u), measuring the yields at the image point of LISE, and introducing the spectrometer acceptance constraints in the calculations. The Fig. 1 shows the isotopic distributions of neon, calcium and zinc obtained with a ^{103}Rh target[4]. Firstly, we can see

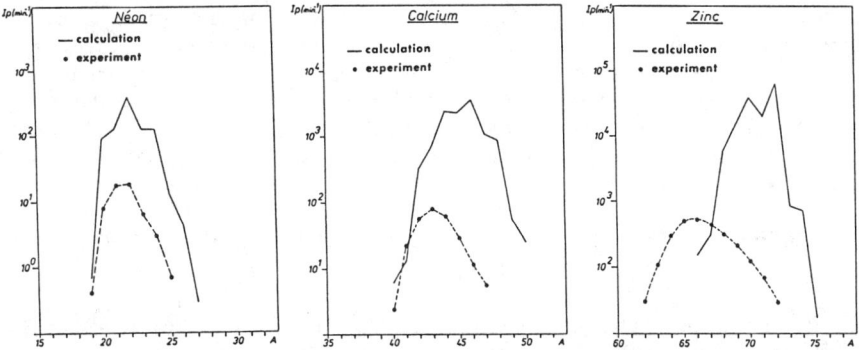

Fig. 1 : Experimental and calculated yields of fragments obtained from a 43 MeV/u ^{86}Kr beam on a ^{103}Rh target.

that the experimental distributions are shifted towards the valley of stability, and secondly, that the calculations overestimate the yields by more than one order of magnitude. These two observations point towards the same direction : they can be interpreted as the effect of a very large excitation energy of the fragments, which induces a huge evaporation of nucleons, or even a fission of the nucleus, pushing the fragments back to the valley of stability. Moreover, this excitation energy may come from very relaxed reaction processes which shift angular distributions and widen them, compared to fragmentation ones, known to be peaked at 0°[4]. Due to the limited angular acceptance of the LISE spectrometer (± 17 mrad centered at 0°), the collection of the fragments is obviously less efficient when the reaction mechanism differs from fragmentation. Furthermore, if we take a look to a reconstituted velocity distribution (Fig. 2), the typical double-humped shape of compound nucleus fission is visible (at 0°, the two bumps correspond to direct and inverse fission velocities). In this case, as already seen at 35 MeV/u[5], the observed reaction proceeds in two steps : formation of a highly-excited composite system (fusion or quasi-fusion),

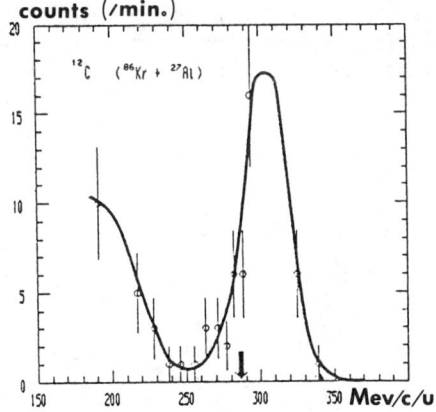

Fig. 2 : reconstituted velocity distribution of ^{12}C. The arrow corresponds to the beam velocity.

followed by its assymetric fission[5].

To summarize, we conclude that the reaction mechanisms observed with 43 MeV/u krypton has nothing to do with the fragmentation of argon observed at the same energy per nucleon. There are various probable mechanisms, characteristic of the low-energy domain, and their common point is a very large excitation energy, leading to a strong evaporation and/or a fission of the fragments. Although this excitation energy strongly modifies the production cross-sections of very exotic nuclei, the krypton beam still allows a sufficient production of nuclei for which decay properties have not been measured so far (see next chapter).

APPLICATION : STUDY OF β-DELAYED NEUTRON EMISSION OF NEUTRON-RICH NUCLEI ($5 \leq Z \leq 15$).

In order to study the β-neutron decay of neutron-rich emitters, a β-neutron coincidence detector has been built. It is composed (fig. 3) of a ΔE-E telescope for the identification and implantation of exotic nuclei, surrounded by a thin NE-102 plastic scintillator for the detection of β's, and a large volume (30 litres) NE-213 liquid scintillator for the detection of neutrons. This scintillator provides a good pulse-shape γ-neutron discrimination based on the different rise-times of the signal between the rapid response due to a γ-Compton diffusion, and the slow signal produced when an inelastic collision occurs between a neutron and a proton of this highly hydrogenized scintillator.

Because of the low production rates (down to a few nuclei per hour) of these very neutron-rich isotopes, the individual decay of every nucleus was detected : the nuclei selected and focused at the image-point of the LISE spectrometer are collected in the telescope. As soon as a potential β-delayed neutron emitter is identified (by measuring energy loss and time of flight), the primary krypton beam is instantaneously switched off to inhibit any further implantation.

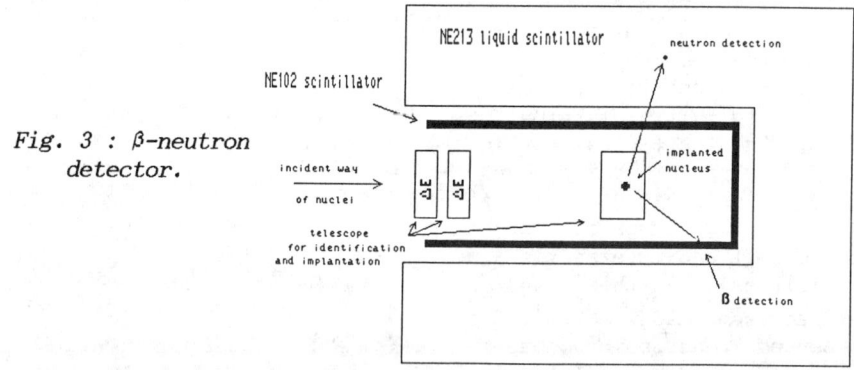

Fig. 3 : β-neutron detector.

Then, coincidences from the β-n counter are looked at during a time window matched to the expected half-life, after which the primary beam is switched on again. This method offers several advantages: the β-neutron decays are detected without background from the accelerator beam, secondly, it allows to follow different nuclei situated in the same isospin region at the same time (the spectrometer selects according to $B\rho = Av/Z$), since each β-neutron coincidence is correlated to its emitter.

Moreover, the specific detection of β-delayed neutrons garantees that a half-life measurement is not spoiled by β-γ decays of less neutron-rich isotopes (unfortunately exhibiting higher production cross sections), which are also implanted in the telescope.

A decay curve which has been obtained this way is shown in fig. 4 for the case of ^{15}B. This isotope is known to have a delayed neutron emission probability P_n close to 100% ($P_{0n} < 5\%$ and $P_{2n} < 5\%$)[6]. Assuming this value, the efficiency of our detector is found to be 21%. With this assumption, we evaluated the P_n of the other detected nuclei. All results, including first-time measurements for $T_{\frac{1}{2}}$ of ^{18}C, ^{20}N, ^{35}Al, ^{39}P, and P_n of ^{18}C, $^{19,20}N$, $^{34,35}Al$ and ^{39}P are given in the following table.

The isotopes for which the statistics were too poor did not allow for an accurate determination of their half-lives. However, they are principally even-Z isotopes (compare for instance the same isospin isotopes ^{17}C with ^{15}B or ^{19}N, and ^{37}Si with ^{35}Al or ^{39}P). This pairing effect is well reproduced by the various models we have compared to each other[4], and this underlines the strong dependence of P_n with regard to the energy window opened by the β decay. The experiment we have presented here is the first

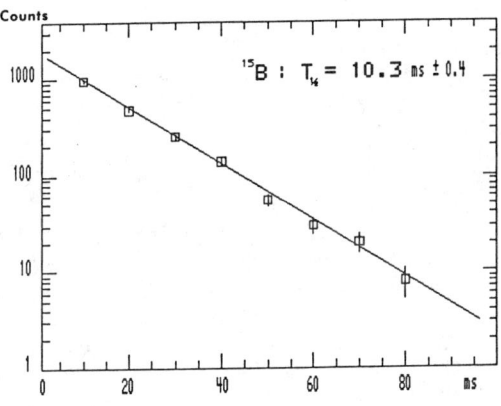

Fig. 4 : Decay curve of ^{15}B.

in a series of future measurements on very neutron-rich isotopes. We plan, for example, to take advantage of the predicted properties of a ^{48}Ca beam[1]. As underlined in first part, the reaction process will be closer to fragmentation. Consequently, we hope to see an increase of the yields of very neutron-rich isotopes by two or even three orders of magnitude. This should allow measurements of the half-lives and P_n of isotopes which are of much interest in astrophysics (^{44}S or $^{46,47}Cl$, Ar for instance[9]).

^AZ	$T_{1/2}$ (ms) (this work)	$T_{1/2}$ (ms) (previous determination)	P_n (%) (this work)
^{15}B	10.4 ± 0.3	8.8 ± 0.6[7] 11.1 ± 1[6]	(100)
^{17}C	–	–	<11
^{18}C	66^{+25}_{-15}	–	25 ± 4.5
^{19}N	210^{+200}_{-100}	320 ± 100[8]	33^{+34}_{-11}
^{20}N	100^{+30}_{-20}	–	53^{+11}_{-7}
^{34}Al	70^{+30}_{-20}	50 ± 25[8]	54 ± 12
^{35}Al	130^{+100}_{-50}	–	87^{+37}_{-25}
^{36}Si	–	–	<10
^{37}Si	–	–	<15
^{38}P	–	–	<10
^{39}P	160^{+300}_{-100}	–	41^{+32}_{-16}

REFERENCES

1. D. Guillemaud-Mueller, contribution to this conference.
2. A.S. Goldhaber, Phys. Lett. 53B, 306 (1974).
3. D. Guerreau, Journal de physique C4-8, Tome 47 (1986).
4. D. Bazin, Thèse de doctorat, GANIL T87-01 (1987).
5. F. Auger et al., Phys. Rev. C35, 190-203 (1987).
6. J.P. Dufour et al., Z. Phys. A. 319, 237-238 (1984).
7. M.S. Curtin et al., Phys. Rev. Lett. 56, 34-37 (1986).
8. J.P. Dufour et al., Z. Phys. A. 324, 487-488 (1986).
9. K.L. Kratz, contribution to this conference.

NEW REGION OF DEFORMATION FOR NEUTRON-RICH NUCLEI AND β-DELAYED NEUTRON EMISSION

Yu.S. Lyutostansky, M.V. Zverev
Moscow Physical-Engineering Institute, Moscow, USSR

I.V. Panov
Institute of Theoretical and Experimental Physics, Moscow, USSR

ABSTRACT

The existence of a new region of deformation for nuclei with $Z \sim 11$ and $N \geq 20$ results from microscopic calculations. Using the quasiparticle Lagrangean method the S_{2n} and $\langle r^2 \rangle$ values for Na isotopes were calculated. Due to single-particle levels splitting the strong deformation may change the neutron binding energy and the position of neutron drip-line. Also the ground-state parity may be changed that leads to decrease of the β-delayed neutron emission probabilities and to the sharp decrease of $T_{1/2}$ value in N-rich Na isotopes.

The problem of possible existence of a new region of deformation for neutron-rich nuclei arising at $Z \sim 11$, $N \geq 20$ has been discussed on the previous conference[1]. The existence of this region is confirmed by three experimental facts. The first one is the nonmonotonic behaviour of the two-neutron separation energy - S_{2n} for sodium isotopes with $N \geq 20$[2]. The observed irregularities occurring at A=31 couldn't be explained by calculations made by Hartree-Fock (H-F) method on the assumption of sphericity of these isotopes[3]. But the problem could be solved by taking into account the deformation in H-F calculations[4]. After a while two more experimental evidence became available on the supposed existence of a new deformation region on[5]. One of them is a sharp decrease of the excitation energy for the first 2^+ state in ^{32}Mg compared to ^{30}Mg. The other - is rapid increase of mean squares of radii $\langle r^2 \rangle$ for Na isotopes near by ^{31}Na. But at the value N=20 there is no sharp change of $\langle r^2 \rangle$, so it doesn't prove the occurrence of deformation but doesn't contradict this fact, as there exist some different explanations of abrupt change smoothing at N=20[1].

We analyze the known effects of N-rich Na isotopes possible deformation within the framework of quasi-particle Lagrangean method and suggests one more criterion sensitive to strong deformation. According to the microscopic analysis of N-rich sodium isotopes structure the strong deformation must change the β-delayed neutron emission probabilities and half-lives of these nuclei.

The quasi-particle Lagrangean method (QLM) was developed[6,7] for the description of nuclei properties: binding energy, nuclear density as well as other characteristics of nuclear ground state in terms of the quasi-particle interaction amplitude near the Fermi surface. The quasi-particle Lagrangean L_q has been constructed in such a form that the appropriate Lagrange equations should resemble the motion equations of quasi-particles near the Fermi surface.

Using the QL-method in terms of the theory of finite Fermi systems (TFFS) of Migdal we calculated two-neutron separation energy S_{2n} as well as the change of mean squares of radii $\delta \langle r^2 \rangle$ for N-even sodium isotopes. The result of the S_{2n} calculations are shown in Fig. 1. Near β-stable isotopes the theory reproduces experimental values S_{2n} rather well, but in ^{31}Na and particular in ^{33}Na it is not so. This may be explained only by deformation of these isotopes, because pairing and other possible effects for the spherical nuclei we have included in QLM calculations.

The $\delta \langle r^2 \rangle$ values are represented in Fig. 2 for N-even sodium isotopes. Not very sharp but rapid enough increase of this value near A=31 not reproducible in the calculation may confirm the deformation in this case. Really, the main difference between theory and experiment is apparently because we neglect the quadruple collective vibrations in QLM calculations. The increase of vibrations in this case confirms the sharp decrease of the first 2^+ state in ^{32}Mg. That is due to the collectivity of this state increasing together with the static deformation.

In the case of static deformation we can see the single-particle (s-p) states splitting proportional to the deformation as represented in Fig. 3. The s-p states in heavy sodium isotopes were calculated by the method[8] using deformed effective Saxon-Woods potential. We would like to emphasize the following point: a weak-unbounded nucleus with less than half occupied level $f_{7/2}^n$ can become bounded due to decreasing the energy of [330] and [321] levels at deformation, and vice versea, a weak-bounded nucleus with more than a half occupied level with $j > 1$ can accordingly become unbounded. For example, in the presence of static deformation, Na isotopes with $A \geqslant 36$ must be unbounded, though according to mass formulae N-even sodium isotopes are bounded right up to ^{39}Na. The existence of ^{36,37}Na isotopes and, therefore, the existence of static deformation in these nuclei can be tested experimentally in the nearest future.

It may be the similar situation for heavy isotopes of other elements. In the assumption on Li heavy isotopes deformation, e.g. $d_{5/2}^n$ level in a shell model approach will split and ^{13}Li nucleus must exist, while it is unbounded in terms of spherical calculations.

Thus, the assumption on the deformation of some N-rich nucleus changes the theoretical position of neutron drip-line.

Strong static deformation may as well be observed in such properties of nuclei as β-delayed neutron emission probability - P_n and half-life - $T_{1/2}$. In the heavy sodium isotopes in the case of strong deformation with $\beta_2 > 0,2$ s-p neutron levels $3^-/2$ [321] and $3^+/2$ [202] may cross (Fig. 3) that will change the ground state parity. Then β-transitions from g.s. Na to g.s. and low-lying Mg states become allowed with great intensity. This should lead to (1) a sharp decrease of $T_{1/2}$ value, and (2) to decrease of the β-delayed neutron emission probability. The last effect is connected with the redistribution of β-transitions intensities that low-lying states of the daughter Mg nucleus are mainly populated.

The P_n and $T_{1/2}$ values for heavy sodium isotopes were calcu-

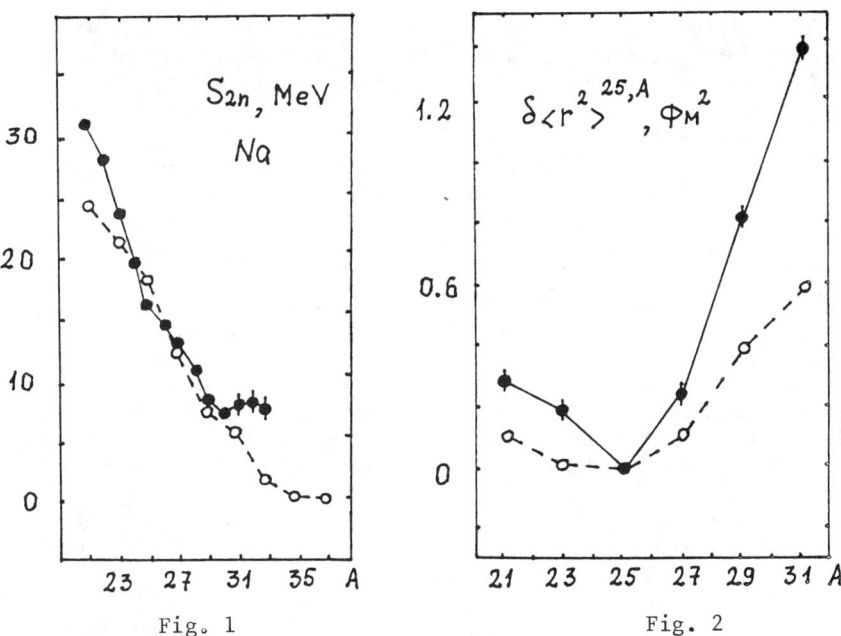

Fig. 1   Fig. 2

Two neutron separation energy S_{2n} (Fig. 1) and variations of the mean square charge radius $\langle r^2 \rangle$ for Na isotopes. Full circles - experiment, open circles - QLM calculations.

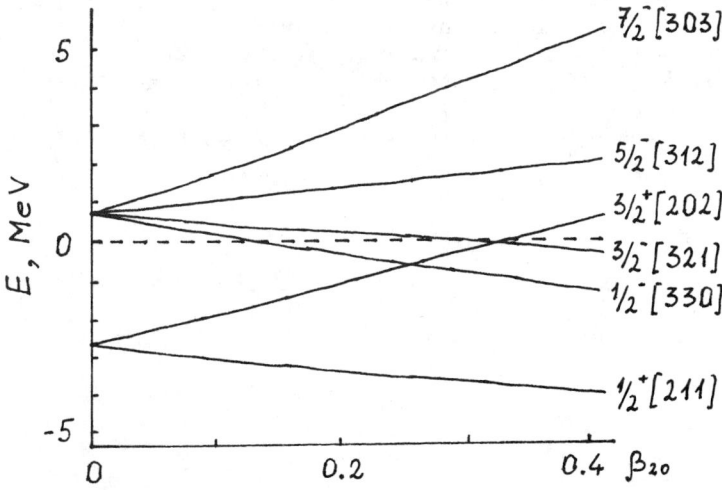

Fig. 3. Single-particle neutron levels in ^{35}Na nucleus with the Woods-Saxon deformed potential.

lated in the frame of TFFS, using theory of isobaric states (IS)[9]), i.e. charged excitations of proton-neutron hole type. The β-strength function was calculated as earlier[10]) taking into account the influence of high-lying IS of resonance type, mainly the "tail" of Gamov-Teller resonance[10]).

According to our calculations in spherical and deformed basis P_n and $T_{1/2}$ values in ^{34,35}Na decreases in the last case for 1,6-2,0 times up to the experimental data for $T_{1/2}$. But the P_n values calculated in deformed basis are smaller than experimental one in this isotopes.

Thus the existence of a new region of deformation for nuclei with $Z \sim 11$ and $N \geqslant 20$ results from microscopic calculations. The effects of deformation shall be evident in S_{2n}, $\langle r^2 \rangle$, E_2^+ and P_n, $T_{1/2}$ values and their isotopic dependence. Also the structure and neutron binding energy of heavy isotopes may be changed due to the deformation splitting of single-particle levels, that is important for the weak-bounded N-rich nuclei. The analogous effects may be observed for the Z-rich nuclei.

Acknowledgements to C. Detraz, B. Danilin, F. Gareev, V. Khodel, V. Sirotkin, E. Saperstein for useful discussions.

1. C. Detraz, Proc. 4th Int. Conf. on Nuclei far from Stability, Helsinger, CERN 01-09, 1981, p. 361.
2. C. Thibault et al., Phys. Rev. C12, 644, (1975).
3. C. Thibault, in 1), p. 47.
4. M. Beiner, R.J. Lombard and D. Mas, Nucl. Phys. A249, 11, (1975).
5. X. Campi, H. Flocard, A.K. Kerman, S. Koonin, Nucl. Phys. A251, 193 (1975).
6. E.E. Saperstein, V.A. Khodel, JETP, 81, 22, (1981).
7. V.A. Khodel, E.E. Saperstein, Phys. Report 92, 182, (1982).
8. F.A. Gareev et al., Nucl. Phys. A171, 194, (1971).
9. Yu.V. Gaponov, Yu.S. Lyutostansky, Sov. Phys. Particles Nucl. 12, 1324, (1981); Sov. Nucl. Phys. 19, 62, (1974) JETP lett. 15, 173, (1972).
10. Yu.S. Lyutostansky, V.K. Sirotkin, I.V. Panov, Phys. Lett. 161, 9, (1985).

STRUCTURE EFFECTS IN THE CALCULATION OF BETA STRENGTH FUNCTIONS AND HALF LIVES OF Rb AND Br ISOTOPES

Shaheen Rab and Adnan Shihab-Eldin
Kuwait Institute for Scientific Research
P. O. Box 24885 Safat 13109 Kuwait

ABSTRACT

The properties of beta delayed neutron emission of the precursors around mass 95 are investigated using spherical Shell Model plus pairing. The objective of this work is to find out the important parameters that effect these properties. The strength functions and half lives of both even and odd Rb and Br precursors are calculated in BCS quasi particle theory as well as in Seniority Truncated Exact Diagonalization (STED) scheme. The calculated half lives from the two schemes are in close agreement with each other. STED gave much smoother overall trend which is in better agreement with experimental values. The calculations reveal that the so called 'odd decays' which depend in an essential way on the parent ground state configuration are of crucial importance in determining both half life, $t_{1/2}$ and delayed neutron emission probability, P_n. Furthermore, it is shown that it is not possible to choose apriori global set for the single particle Shell model basis to predict the correct ground state configurations for both the odd and even nuclei. Instead, it is necessary to examine available experimental evidence in order to construct a more valid global set for single particel basis. The significant variations of the predicted half lives according to the different parent ground-state configuration choices, reveal the importance of nuclear structure effects in thes isotopes. The calculated $t_{1/2}$ and P_n of Rb and Br precursors are presented, discussed and compared with experimental values.

The beta strength function S, half life $t_{1/2}$ and β-delayed neutron emission probability P_n are calculated in spherical Shell-Model plus pairing using BCS qusi particle theory. Recently Shell Model (both spherical and deformed) theory has been used quite successfully [1,2] to describe these properties around mass 100. In our earlier work[3] the Gamow-Teller strength function in Seniority Truncated Exact Diagonalization (STED) scheme were derived and used to calculate t $_{1/2}$ and P_n of [92-102] Rb and [87-92]Br precursors. The results follow the trend of the experimental curve better than earlier BCS calculation[4]. Later, we found some discrepancies in the calculation of the transition energies[5] especially for the odd A isotopes and as a result recalculate $t_{1/2}$ and P_n for Rb and Br isotopes. It is found that the BCS and STED results agree close to each other, the STED curve being smoother to the experimental one as can be seen in figures 1-6.

The parameters that effect the β-delayed neutron emission properties are single particle energies e_j, Q values of the precursors, neutron separation energies B_n of the emitter, the force constant and above all the configuration of the mother nucleus. The sensitivity of half lives on the choice of nuclear configurations are discussed here.

The pairing force is assumed to act between like particles form level N_j to N_f, the core being $Z = 28$, $N = 50$. The treatment of pairing force in

BCS quasi particle theory are given in many references [5,6,1]. The Shell Model calculation is performed upto $1g_{7/2}$. The quasi particle energy E_j, occupation probabilities V^2, half life $t_{1/2}$ and P_n are given

$$E_j^2 = (e_j - \lambda)^2 + \Delta^2, \; j = N_f, N_{j+1}, \ldots N_f$$

$$E_j = |e_j - \lambda|, \; for \; j < N_f \tag{1}$$

$$V_j^2 = 1, \; j < N_j, \; V_j^2 = 0, \; j > N_f \tag{2}$$

$$V_j^2 = \frac{1}{2}[1 - \frac{e_j - \lambda}{[(e_j - \lambda)^2 + \Delta^2]^{1/2}}], \; j = N_j, N_{j+1} \ldots N_f$$

The half life is inversely proportional to $S_i f_i$, f being fermi function.
In the first set of calculation the same single particle energies e_j as in refernce 4 are used where $f_{5/2}$ (π) has the lowest energy i.e. assuming that the 37th proton of Rb isotopes is at $p_{3/2}$ level. The results are shown in Figure 1-3. They are compared with earlier results[a] (BCSO), STED results and with experimental value. Recent measurements of the spins' of many Rb isotopes reveal that the 37th proton is not always at $p_{3/2}$ level. For example $J\pi$,s of ^{93}Rb, ^{95}Rb, ^{96}Rb,^{97}Rb, ^{99}Rb are 5/2,5/2,2,3/2,(5/2⁻) respectively. These data show that the last proton of add A Rb isotopes should be at $f_{5/2}$ level instead of $p_{3/2}$. The spins of the even isotopes do not dictate much but gives restriction on the choice of configuration. In the second set of calculation, the single particles energies of $\pi(f_{5/2})$ and $\pi(p_{3/2})$ are flipped and the results are shown in Fig. 2. The significant difference in the two figures show structure effects on these properties especially in the lower mass region dictating the importance of the proper choice of the configuration. The most striking effect is found for ^{92}Rb for which the 37th proton should be at $p_{3/2}$. Guided by experimental evidence we flipped $\pi(f_{5/2})$ and $\pi(P_{3/2})$ single particle energies e_j for odd A_j nuclei, in the third set, to calculate the quasi particle energies and then $t_{1/2}$ of Rb isotopes and P_n of Sr Isotope. This set gives a good overall agreement with the experimental curve. The odd decays which are associated with the parent configuration are important in half life calculation as the strengths lie within the energy window (Q - B_n). The variation of half lives with configuration choices decrease with increasing mass number which can be explained from the fact that deformation starts around N = 60 i.e. ^{97}Rb and onward. The curves in figures 1,2,3 show the dependence of half lives on

733

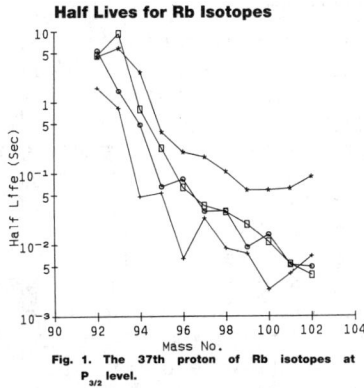

Fig. 1. The 37th proton of Rb isotopes at $P_{3/2}$ level.

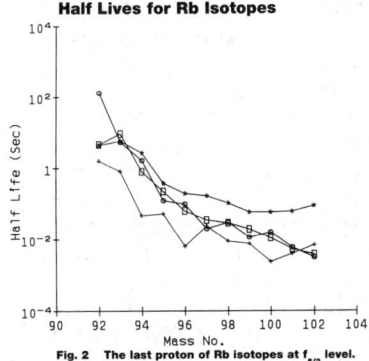

Fig. 2 The last proton of Rb isotopes at $f_{5/2}$ level.

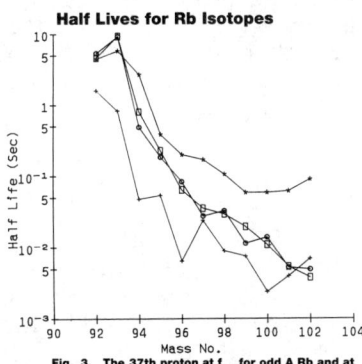

Fig. 3. The 37th proton at $f_{5/2}$ for odd A Rb and at $P_{3/2}$ for even A except for ^{101}Rb and ^{102}Rb where they are at $P_{3/2}$.

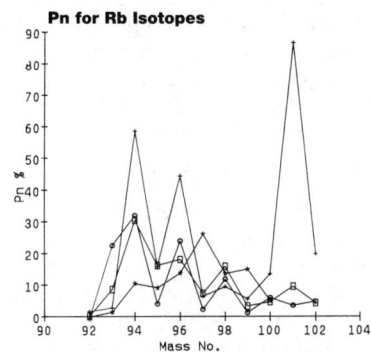

Fig. 4. The 37th proton at $f_{5/2}$ for odd A Rb and at $P_{3/2}$ for even A except for ^{101}Rb and ^{102}Rb where they are at $P_{3/2}$.

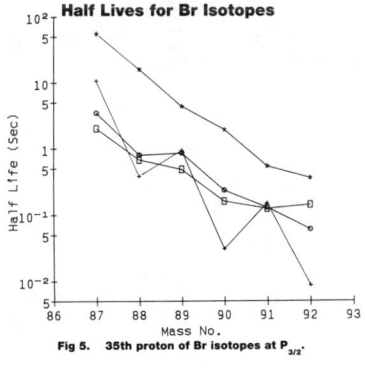

Fig 5. 35th proton of Br isotopes at $P_{3/2}$.

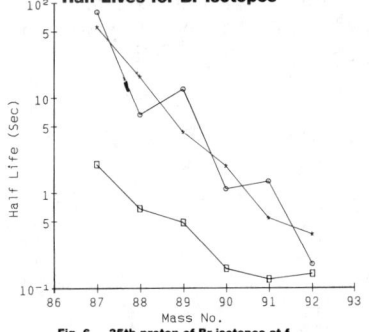

Fig. 6. 35th proton of Br isotopes at $f_{5/2}$.

○ BCS RESULTS   □ STED RESULTS   ∗ EXPERIMENTAL VALUES   + BCSO RESULTS

the spins of Rb isotopes. The striking difference between the two choices of parent ground states is observed in ^{92}Rb. The half life of ^{92}Rb with $\pi(p_{3/2})$ is close to experimental value and 25 times higher if the last proton is at $\pi(f_{5/2})$. The variation is also observed for other isotopes except for the higher ones. The measured spins of A = 93,95,97 of Rb isotopes predict the sigle particle energies of the grond states at $\pi(f_{5/2})$ which agree also with our calculation. Shell model plus pairing calculation both in STED and BCS quasi particle thoery show that the last proton of ^{92}Rb should be at $P_{3/2}$. For A = 94 and 96, the proton could be either at $P_{3/2}$ or $f_{5/2}$. The half lives of A = 98 . . . 102 are not that sensitive to the choice of single particle basis. Similarly two sets of calculation are performed for $^{87-92}$Br -> $^{87-92}$Kr decay with 35th proton at $P_{3/2}$ and $f_{5/2}$ level as shown in Figures 5 and 6. The STED calculation is done with $\pi(p_{3/2})$. The half lives increase significantly when the last proton of Br isotopes are considered to be at $f_{5/2}$ level. The significant difference between the two curves explain the importance of nuclear shell effects on half lives. Not too many data is available for Br isotopes to construct a set of single particle basis. Careful study of these results show that it is necessary to examine available experimental data in order to construct a valid global set for single particle basis. It is expected that inclusion of GT force should reduce the deviation of the calculated values from experimental results as found earlier [1,2,4,6]. Work is in progress to include GT forces in RPA to predict half lives for both even and odd precursors. Once the overall trend is reproduced, one can predict half lives for higher isotopes wich are not yet measured.

REFERENCES

1. J. Krumlinde and P. Moller. Nucl. Phys. A 417 (1984) 419.
2. K. L. Kratz, Nucl. Phys. A417 (1984) 447 and the reference in there.
3. Shaeen Rab and Adnan Shihab-Eldin, ACS Symposium Series 324, Chicago, 1985. on Nuclei Off the Line of Stability, edited by Richard A. Meyer and Daeg S. Brenner.
4. Z. M. Oliveira, Ph.D. thesus, Univ. of California, Berkeley (1980).
5. J. A. Halbleib, Sr. and R. A. Sorensen, Nucl. Phys. A98(1967) 522.
6. J. Randrup, Ph.D. thesis, University of Aarhus, Denmark, (1972).
7. Table of Radiaoactive Isotopes, E. Brown and R. B. Firestone, edited by V. S, Shirley, 1986.

DECAY OF DELAYED-NEUTRON PRECURSOR ^{93}Rb AND THE SIMPLE STATISTICAL MODEL

G.D.Alkhazov, L.H.Batist, A.A.Bykov, V.D.Wittmann, S.Yu.Orlov, V.K.Tarasov

Leningrad Nuclear Phisics Institute, Acad.Sci. USSR, 188350 Gatchina, Leningrad district, USSR.

ABSTRACT

The decay of ^{93}Rb was investigated using the total -γ-absorption technique. The intensities of β-transitions followed by γ-rays emission have been obtained. The (nγ)-competition is compared with predictions of the statistical model.

INTRODUCTION

Until recently, spectroscopy of delayed particle precursors dealt with investigations of particle spectra. Studies of β-strength functions as well as investigations of γ-rays emission competing with particle decay would enable one to gain an insight into this phenomenon. Such investigations have not yet been carried out since the routine γ-spectroscopy fails to provide sufficiently complete information on de-excitation of levels above the particle separation energy. A total γ-absorption (TGA) technique is a suitable method for this purpose. This technique was proposed in CERN [1] and subsequently it was significantly improved in LNPI [2-6]. A TGA spectrometer makes it possible now to obtain a rather accurate information on a γ-depopulation of high-excited levels without a detail knowledge of decay schemes and intensity balances. Details of this technique may be found elsewhere [2-6].

The decay of the delayed-proton precursor ^{147g}Dy was investigated by means of TGA-technique in Ref.[6]. In the present paper we report experimental data for the decay of the delayed-neutron precursor ^{93}Rb.

EXPERIMENTAL PROCEDURE AND RESULTS

The half-life, the decay energy, the neutron separation energy and the probability of neutron emission per decay for the decay of this isotope are well known from previous works: $T_{1/2}=5.85(3)$s [7], $Q_\beta=7.485(15)$MeV [8], $B_n=5.23$MeV [9], $b_n=1.40(8)\%$ [10]. As it was found in Ref.[10], 85% and 14% of neutrons are emitted to the ground and to the first excited (2^+, 0.814MeV) states of ^{93}Sr, respectively. Delayed neutron spectrum has been measured in Ref.[11]. A decay scheme of ^{93}Rb comprising 231 γ-transitions between 74 levels have been constructed in Ref.[12]. Note that the levels above B_n in this scheme are not corroborated in coincidense experiments and therefore should be regarded as doubtful. In our experiment the neutron-rich rubidium isotopes were produced by bombarding a target of UCx at 2000°C with 1GeV protons from LNPI syncrocyclotron. After mass-separation in the IRIS separator, the ^{93}Rb ions were collected on a tape and then delivered into the well of the TGA-spectrometer. A pulsed mode for operation of the accelerator and the mass-separator was employed to reduce the background. The TGA-spectra of the 290 samples were recorded. Sufficiently small source activities were used to decrease random summing of γ-rays energies. A special attention was also given to background effects due to registration of β-particles in the TGA-spectrometer. On this account, a set of control experiments was carried out using β-absorbers as of Be and of Cu. The TGA-spectrum corrected to the background effects is shown in Fig.1. This spectrum has an endpoint energy of 7.53(6)MeV in a good agreement with the Qβ value of 7.485(15) from Ref.[8]. The intensity of the β-feed to the ground state was not determined in our experiment. This intensity was found to be 42(4)% in Ref.[13] in contradiction to that of 59(3)% in Ref.[14]. We accept the averaged value of 50(10)% to correct the normalization of TGA-spectrum for unobserved transitions. The intensities of β-transitions followed by γ-rays emission,Iβγ, have been obtained by unfolding this renormalized spectra. The result is shown in Fig.2 and Tab.I. For comparision, the data of Ref.[12] are also presented. It is clear from this table that the intensity balances in Ref.[12] overestimate the β-feeds of states below 3 MeV and underestimate those of levels in the 4.5÷5.2 MeV region due to missing of many weak transitions. The problem of missing intensities was discussed in detail in Ref.[15,5]. Note that the β-feeding of levels in the (Bn-Q) -window is overestimated in Ref.[12] by a factor about 4, and we conclude that this decay scheme above Bn is erroneous.

Table I. β-intensities to the indicated energy windows for ^{93}Rb decay followed by γ-rays emission

Energy (MeV)	β-intensity (% per decay)		Number of levels in the decay scheme[12]
	our data	data of Ref.[12]	
0	(50)^	(50)^	1
0.1-3.0	4	15.0	24
3.0-3.6	6	0.4	2
3.6-4.4	26	23	16
4.4-5.2	11	4.3	9
5.2-7.5	1.5	5.6	12

^ -renormalized values (see text)

DISCUSSION

Using the results of Ref.[11] we have obtained the ratio of the averaged intensities <Iβn(En)>/<Iβγ(En+Bn°)> shown in

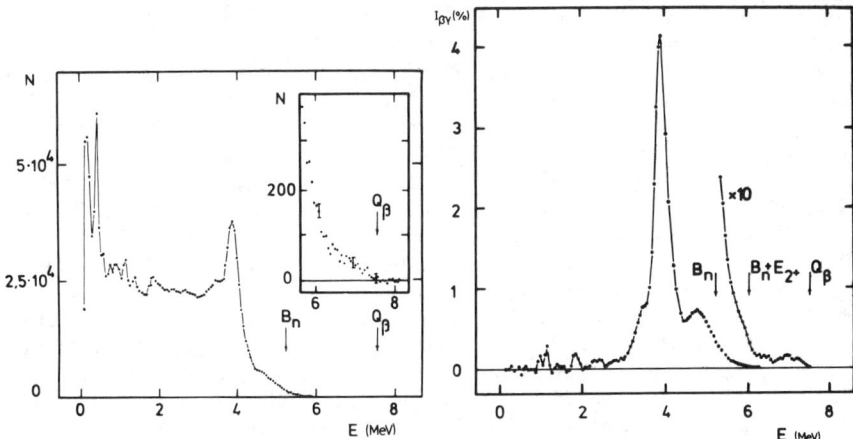

Fig.1. Total γ-absorbtion spectrum for ^{93}Rb decay. Insert: a high-energy part of this spectrum after correction to background effects.

Fig.2. Intensities of β-transitions followed by γ-rays emission for the β-decay of ^{93}Rb.

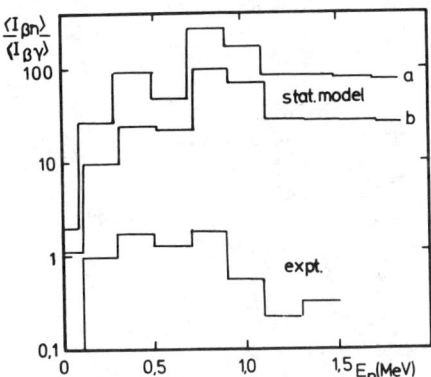

Fig.3. Ratio of averaged intensities $\langle I\beta n\rangle/\langle I\beta\gamma\rangle$ versus the neutron energies for the decay of ^{93}Rb.
expt -obtained from experiment,
stat.mod. -statistical model calculations (without the fluctuation corrections)
a) -parametrization according to Ref.[18,19],
b) -parametrization according to Ref.[20].

Fig.3. We also have evaluated this ratio in frames of the standard statistical model [16,17], the experimental values of $\langle I\beta\gamma(E+E(2^+))\rangle/\langle I\beta\gamma(E)\rangle$ was used to take account of neutron transitions to the excited 2^+ -state of ^{92}Sr. The widths were calculated with two sets of parameters. The first set is a standard one: parameters in Fermi-gas expression for the level density formula were taken to be $a=A/8$, $\Delta=12/\sqrt{A}$ (see Ref.[18]) and values of $\langle \Gamma\gamma \rangle$ were parametrized according to Ref.[19]. The fluctuation of the widths was described by Porter-Thomas law [18]. The fluctuation correction quenched the calculated ratio of the intensities by a factor about 5. The alternative choice of parameters was proposed in Ref.[20] specially for delayed neutron emitters: $a=15\text{MeV}^{-1}$ $\Delta=1.3\text{MeV}$, $\langle \Gamma\gamma \rangle=0.2\text{eV}$. In this case the fluctuation corrections are assumed to be taken into account. Both sets yield almost indistinguishable results after the correction. However, the calculations do not provide a satisfactory description of the experiment. We conclude that the simple statistical approach to the problem of delayed particle emission needs significant modification.

REFERENCES

1. B. R. Erdal, G. Rudstam, Nucl. Instr. & Meth. 104, 263(1972).
2. G.D. Alkhazov et al., Proc. of 4-th Intern. Conf on Nucl Far from Stab., CERN81-09, 238(1981).
3. G.D. Alkhazov et al., Nucl.Phys. A438, 482(1985).
4. G.D. Alkhazov et al., Yad.Phys. 42, 1313(1985).
5. G.D. Alkhazov et al., Phys.Lett. B157, 350(1985).
6. G.D. Alkhazov et al., Proc. of Inern. Symp. on Weak and Electrom. Interactions in Nuclei, Heidelberg 1986, p.239.
7. Table of isotopes, 7-th ed., edited by G. H. Lederer and V. S. Shirley, N.Y. (1987).
8. R. Decker et al., Z.Phys. A294, 35(1980).
9. A. H. Wapstra, C.Audi., Nucl.Phys. A432, 1(1985).
10. P. Hoff et al., Z.Phys. A294, 233(1980).
11. K.-L. Kratz, Nucl.Phys. A417, 447(1984).
12. C. J. Bischof, W. L. Talbert, Phys.Rev. C15, 1047(1977).
13. R. Brissot et. al., Nucl.Phys. A328, 149(1975).
14. E. Achterberg et al., Phys.Rev. C10, 2526(1974).
15. J. C. Hardy et. al., Phys.Lett. 136, 331(1984).
16. B. Jonson et al., Proc of 3rd Intern. Conf. on Nucl. Far from Stab., Corsica 1976, CERN 76-13, p.277.
17. J. M. Blatt, V. F. Weisskopf, Theoretical nuclear physics (Wiley, J. and Sons, N.Y. 1952).
18. A. Bohr and B. R. Mottelson, Nuclear structure (Benjamin Inc., N.Y., 1969), Vol. 1, p.155,187.
19. H. Malecky et al., Yad.Phys 37,284(1983).
20. P. Hoff et al., Nucl.Phys. A358, 9(1981).

PROTON RADIOACTIVITY OF MEDIUM HEAVY NUCLEI

T. Faestermann, A. Gillitzer, K. Hartel[+], P. Kienle[*], E. Nolte
Technische Universität München, D8046 Garching, FRG

ABSTRACT

For the proton radioactive nuclei ^{109}I and ^{113}Cs with decay energies of 0.83 MeV and 0.98 MeV the halflives have been measured as 109(17) μs and 33(7) μs respectively. These halflives are compared with calculations using the WKB approximation. The spectroscopic factors deduced are an order of magnitude smaller than expected for a pure $d_{5/2}$ single particle state of the emitter. For ^{105}Sb a negative proton binding energy of B_P=-0.36(20) MeV could be established, but no proton radioactivity with $T_{1/2}\geq 10$ ns and $E_P\geq 0.5$ MeV has been observed down to a cross section of 0.5 μb. In a search for proton radioactivity of ^{108}I produced with the ^{54}Fe(^{58}Ni,p3n) reaction no new proton line was found. But the cross section limit of 5 μb reached cannot definitely rule out this possibility. With the reaction ^{40}Ca(^{40}Ca,p2n)^{77}Y direct proton emission was searched for in this region of well deformed nuclei. Assuming an L=4 proton emission decay energies between 0.45 MeV and 0.90 MeV can be excluded.

INTRODUCTION

"Thus there seems a strong suspicion that H-particles are emitted from the radioactive atoms themselves, though not with uniform velocity." The possibility of proton radioactivity was for the first time mentioned in 1915 by Marsden[1]. He measured the energy loss of α-particles in air and surprisingly detected long range particles which he identified as protons. The correct explanation was then given by Rutherford[2], that - even more important - nuclear reactions had been seen for the first time, namely the ^{14}N(α,p)^{17}O reaction.

With Weizsäcker's mass formula[3] it became possible to predict the proton drip line, i.e. the line on the proton rich side of the nuclidic chart, where the Coulomb repulsion sets a limit to nuclear stability because the proton binding energy

$$B_P=BE(^AZ)-BE(^{A-1}Z-1)=-Q_P \qquad (1)$$

becomes negative and the proton decay Q-value Q_P positive. The study of direct proton emission yields information not only of binding energies , but also - as we will show - specific information on the nuclear structure very far from the valley of β-stability.

A detailed review of experiments searching for proton radioac-

[+] Present adress: Eurosil GmbH, Eching, FRG
[*] Present adress: GSI Darmstadt, FRG

tivity has been given by S. Hofmann[4] and only a few milestones shall
be mentioned here. Jackson et al.[5] and Cerny et al.[6] observed in
1970 proton emission from a spin gap isomer in ^{53}Co. In the same
year Karnaukhov et al.[7] reported first evidence for low energetic
protons after the ^{96}Ru + ^{32}S reaction. Later experiments by the same
group[8] confirmed this observation and the hypothesis was put
forward, that ground state proton radioactivity of ^{121}Pr has been
observed. But until now this hypothesis could not be corroborated.
At the last conference in this series (Helsingør 1981) Hofmann et
al.[9] presented convincing data showing, that ^{151}Lu is a ground state
proton emitter. In the meantime more cases of proton radioactivity
have been found: ^{147}Tm at the isotope separator of the GSI[10], ^{113}Cs
and ^{109}I in Munich[11,12], and ^{150}Lu and ^{147}Tm^m again at the velocity
filter SHIP of the GSI[13].

HALFLIFE CALCULATIONS

Since the discovery of the ground state proton emitter ^{151}Lu a
number of theoretical halflife calculations have been published.
Feix and Hilf[14] calculated halflives for cases between ^{103}Sb and
^{151}Lu by integrating the complex Schrödinger equation. Ogawa[15]
calculated halflives for ^{109}I and ^{113}Cs in the framework of the R-
matrix theory. In his work shell model spectroscopic factors are
included, which are not further specified and make it difficult to
compare the results. Bugrov et al.[16] constructed an integral
expression for the proton decay width. Their results vary by a
factor of three depending on the nuclear potential used but within
this range they agree with the values of Feix and Hilf. In addition
they use spectroscopic factors which generally decrease the transition strength.

Our own simple approach[17] uses the WKB approximation to
calculate the transmission through the potential barrier. As nuclear
potential the real part of an optical potential with the fitted
parameters from Perey[18] is used. A spin-orbit term is also included.
This calculation reproduces the values of Feix and Hilf within a
factor of two.

By a comparison between measured and calculated halflife the
spectroscopic factor of the decay can be deduced experimentally.
Theoretically the spectroscopic factor is identical to that of a
one-proton pickup reaction. Under the assumption of pure single
particle states, the spectroscopic factor is given[19] by

$$S_{lj}=(2j+2-n)/(2j+1) \qquad (2)$$

if one out of an odd proton number n in the subshell (lj) is emitted
in a transition to the groundstate of an even-even nucleus.

As can be expected the calculated halflife depends extremely on
the decay energy. This is illustrated in Fig. 1 for ^{109}I emitting an
L=2 proton (full drawn). Therefore it is important in the search for
proton radioactivity to be sensitive for short halflives in order to
cover a wide detection window of possible decay energies. Isotope
separators[20] have typical separation times in the order of a second

and loose three to five orders of magnitude in efficiency for a halflife of 1 ms. For partial proton decay halflives longer than the β decay halflife (expected[21] to be shorter than 1 s for medium heavy nuclei) the rapidly decreasing branching ratio limits a possible detection of a proton emitter. Better suited are recoil separators like the SHIP, which have separation times around 1 μs and combine high efficiency with good background reduction. With the setup described below we can detect decays with halflives longer than about 10 ns at the cost of a higher sensitivity for background.

The dependence of the proton decay halflife on Z and angular momentum L is also illustrated in Fig. 1 by calculations for ^{105}Sb with L=2 (dotted) and ^{109}I with L=4 (dashed).

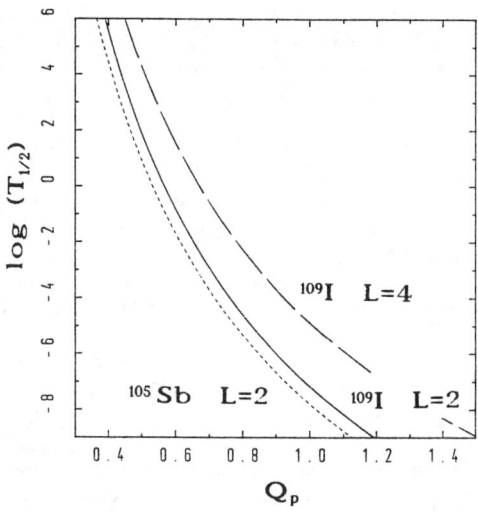

Fig. 1: Partial halflives for proton emission as a function of proton decay Q-value calculated with the WKB method for ^{109}I with L=2 (full), ^{109}I with L=4 (dashed), ^{105}Sb with L=2 (dotted).

PRODUCTION

Having seen the extreme energy dependence of the halflife it may seem hardly possible to reach a proton unstable ground state in a nuclear reaction, since usually all reactions populate the final nucleus at high excitation energy, and proton emission will be much faster than γ-decay. If however states close to the yrast line are populated, they may decay through a cascade of γ-rays to the ground state. This is illustrated in Fig. 2a. The yrast lines of mother (long dashed) and daughter (full drawn) nucleus are sketched assuming equal (rigid body) moments of inertia. States in the mother which are more than a certain energy above the yrast level in the daughter (short dashed) will decay by proton emission. Thus only states in the hatched area have a chance of feeding the ground state.

But mother and daughter nucleus do not have to have the same moment of inertia. It is well known, that an odd particle can stabilize the deformation. Therefore an odd-proton mother may have a larger deformation than the even-even daughter. This situation is sketched in Fig. 2b. The probability of feeding the ground state is increased, especially for higher spin values.

Until now we did not take into account that in even-even nuclei the ground state energy is lowered by the pairing energy. If we consider a case with the same proton decay energy as before, again the survival probability against proton evaporation will be increased, because the yrast line of the daughter lies relatively higher (Fig. 2c). Thus the same effect, that brings the proton drip line closer to stability for an odd-even proton emitter, namely the pairing energy, provides a higher cross section for the population of its ground state.

After these considerations it seems clear that preferentially a large amount of angular momentum has to be brought into a reaction feeding the ground state of a proton radioactive nucleus. That is the case with heavy ion induced fusion reactions, but may not be the case in spallation reactions induced by high energy protons.

If we tried to produce a two-proton radioactive nucleus (although there is no likely candidate among medium heavy nuclei, which is not also a single proton emitter from the ground state) the arguments are reversed as can be seen in Fig. 2d. The mother nucleus has even Z and the pairing gap now increases the energy available for single proton emission. This may be another severe handicap ever to observe a ground state two-proton radioactivity.

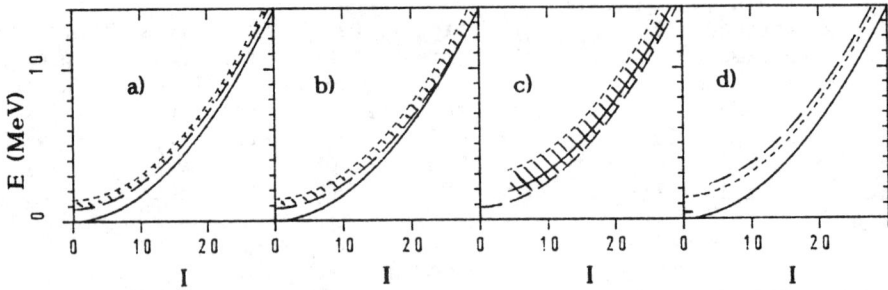

Fig. 2: Yrast line of ground state proton emitter (A=109, Q_P=0.8 MeV, long dashed) relative to that of the daughter nucleus (full drawn). Only states in the mother (shaded) within a certain energy above the daughter (short dashed) can feed mothers ground state.
a) Equal (rigid body) moments of inertia.
b) Moment of inertia of the odd-Z mother increased by 10%.
c) Equal moment of inertia, but even-even daughters ground state is lowered by the pairing energy.
d) Even-Z emitter (Q_P=0.5 MeV) whose ground state is lowered by the pairing energy. No feeding of the ground state!

EXPERIMENTAL TECHNIQUE

Candidates for proton radioactivity were produced in fusion-evaporation reactions with roughly equal target-projectile combinations. The catcher technique and the detector system are more thoroughly described by Gillitzer et al.[17]. Evaporation residues produced in a relatively thick target are recoiling out of the target and stopped in an organic catcher foil 6 cm behind the target. The primary beam passes the catcher through a small hole. Protons emitted in the decay of the caught recoils are detected in backward direction by an annular system of gas detectors covering 7%

of 4π. The catcher foil is only 0.35 mg/cm² thick to minimize the energy loss for the protons and therefore the recoil nuclei have to be slowed down to a few MeV by a target backing of variable thickness.

The detector system consists of a parallel plate avalanche counter (PPAC) measuring the arrival time against the pulsed heavy ion beam, and an ionization chamber (IC) with longitudinal field. The IC measures the energy of the particle and determines its Z by analyzing the Bragg curve[22]. An independent particle identification is possible by a range measurement via the drift time of the ionization track towards the anode of the IC. Protons are stopped in the IC for energies up to 1.5 MeV (with the gas pressure used) and can be distinguished from α-particles for energies above 0.5 MeV.

^{109}I AND ^{113}Cs

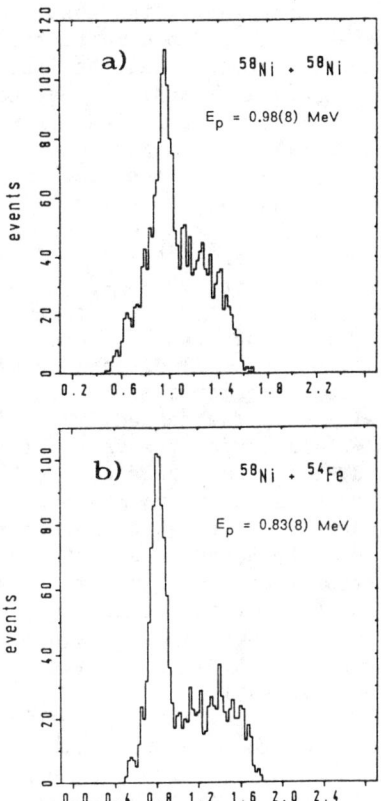

Fig. 3a and 3b show proton energy spectra from the reactions of a 250 MeV ^{58}Ni beam (from the Munich Tandem-Linac combination) with enriched ^{58}Ni and ^{54}Fe targets respectively. These spectra were taken in the beam pauses. The proton lines observed at 0.98(8) MeV and 0.83(8) MeV were confirmed later on by Hofmann et al.[13] at the SHIP and their energies measured with much better precision. The continuous background is due to β-delayed protons, which have lost energy before being detected. The intensity in the lines corresponds to cross sections of 30 μb and 40 μb with an uncertainty of a factor of two.

For the determination of the halflives the beam pulsing period had to be chosen accordingly. Since the halflives turned out to be considerably longer than anticipated, the pulsing period had to be increased from run to run. The final time spectra obtained for the proton lines from the two reactions are shown in Fig. 4a and 4b resp. These spectra are fitted by halflives of 33(7) μs and 109(17) μs. These values are much shorter than β-decay halflives and prove that the origin of the lines is proton radioactivity.

For the assignment of the proton emitters we only have to consider odd-Z nuclei because the proton bin-

Fig. 3: Energy spectra for off-beam protons. The reactions and the energy of the lines are given.

ding energy in neighbouring even-Z isotones is by twice the pairing energy (about 3 MeV) larger and no proton unstable even-Z nucleus can be reached in these reactions. Using the available information on nuclei in this region combined with arguments based on the observed cross section, we were able to assign[17] these proton radioactivities to the emitters ^{113}Cs and ^{109}I. This assignment could recently be confirmed by a direct mass determination of the two proton emitters at GSI:

Behind the SHIP a 30° bending magnet followed by two quadrupoles has been set up[23] and the hole system now can act as a mass spectrometer with a resolution of m/Δm≈130. In first experiments the two proton emitters have been produced in reactions of a ^{58}Ni beam with ^{58}Ni and ^{54}Fe targets and it could be unambiguously shown that the highest yield was observed when the system was tuned for masses A=113 and A=109 respectively.

According to the shell model we expect the odd proton to fill first the $d_{5/2}$ or $g_{7/2}$ and then the $h_{11/2}$, $s_{1/2}$ and $d_{3/2}$ orbitals beyond Z=50. In order to compare the experimental halflives with expectations we performed calculations for all known proton radioactivities with measured partial halflives using the Q_p-values from Hofmann et al.[13] and assuming the most probable single particle configurations for the emitter nucleus. The results are shown in Fig. 5 as full drawn lines for the ratio of calculated to experimental halflife. The dashed lines give the same ratio for the values from Feix and Hilf[14] extrapolated to the measured Q_p-values using the energy dependence of our WKB expression. It should be noted, that the two calculations agree within a factor of two. The shaded bars are meant to indicate the experimental uncertainties (from total halflife, branching ratio and decay energy).

These halflife ratios in Fig. 5 are just the spectroscopic factors we want to deduce from the experiments. For the ground state decays of ^{151}Lu and ^{147}Tm they are slightly smaller than unity and consistent with the shell model values 2/3 and 1/2 resp. from equation 2. For the low spin isomer in ^{147}Tm the assumption of a $d_{3/2}$ state yields a spectroscopic

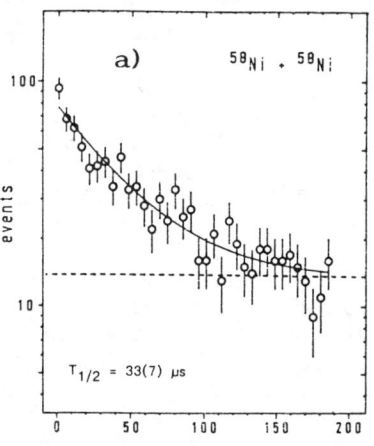

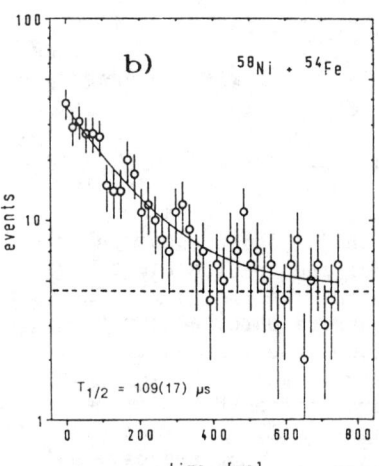

Fig. 4: Time spectra for the proton lines of Fig. 3. The reactions and the fitted halflives are given.

factor of 0.7 consistent with
the value of 1 for a single
proton in this orbital. The
spectroscopic factors for ^{109}I
and ^{113}Cs differ considerably
from unity. The assumption of
proton emission from the $g_{7/2}$
orbital signifies an enhancement of the transition. The
only effect enhancing the proton emission could be the
lowering of the Coulomb barrier by deformation. For ^{113}Cs
a substantial deformation is
possible which is certainly
larger than in ^{109}I and therefore an enhancement of the
proton emission due to deformation should be smaller in
^{109}I. This is opposite to the
observation. The more probable
assignment is that both nuclei
emit a proton from a $5/2^+$
ground state with spectroscopic factors of 0.1 instead of
$2/3$ for $(d_{5/2})^3$ in ^{109}I and
0.02 instead of $1/3$ for
$(d_{5/2})^5$ in ^{113}Cs.

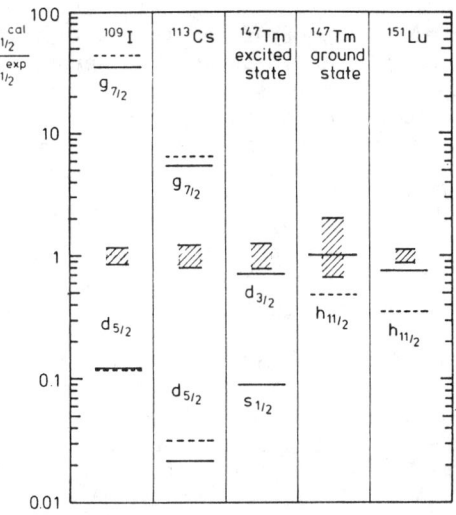

Fig. 5: Ratio of calculated (full drawn: WKB, dashed: Ref.14) to experimental halflives of known proton emitters. The orbitals assumed for the calculation are indicated as well as the experimental uncertainties (shaded).

In a recent IBFA calculation of energy levels in odd-A Cs and Xe isotopes Arias et al.[24] find the $5/2^+_1$ state below the $7/2^+_1$ state in ^{113}Cs, although in the single particle energies used as input the $d_{5/2}$ state is 0.6 MeV above the $g_{7/2}$ state. Its wavefunction in terms of ^{112}Xe core states is (only amplitudes larger than 0.05 are given):

$$|5/2^+\rangle = 0.64 |0^+ \times d_{5/2}\rangle - 0.50 |2^+ \times d_{5/2}\rangle + 0.23 |4^+ \times d_{5/2}\rangle + 0.36 |2^+ \times s_{1/2}\rangle \quad (4)$$

Only the first component takes part in the proton emission and this results in a reduction of the transition rate to 40%. But the observed reduction is not yet explained.

SEARCH FOR ^{105}Sb

In order to test the halflife calculations it would be very helpful, to find proton radioactivity of ^{105}Sb, the next lighter candidate in this $T_z=3/2$ series. With only one proton outside Z=50 it should be a good shell model nucleus and its spectroscopic factor should be unity. Mass formulae[25] predict it to have a Q_p-value similar to ^{109}I and ^{113}Cs. We irradiated an enriched ^{50}Cr target with a 250 MeV ^{58}Ni beam and searched for protons between the beam pulses which were spaced by 200 ns. No delayed protons are observed, not even β-delayed proton emitters are populated from the compound nucleus ^{108}Te. Thus we get an upper limit of 0.5 μb for the production cross section of a proton emitter if it had a halflife longer than 10 ns and an energy between 0.5 MeV and 1.5 MeV. This upper limit is nearly two orders of magnitude smaller than the cross sections measured for ^{109}I and ^{113}Cs with the same (^{58}Ni,p2n) reaction, implying that proton emission in the above mentioned range of decay energies and halflives is not the dominant decay mode of ^{105}Sb.

Fortunately this region of nuclei above ^{100}Sn contains the lightest α-emitters and is well studied far from stability. Therefore we can get a reliable estimate for the Q_P-value of ^{105}Sb since we know the binding energy difference between ^{109}I and ^{104}Sn:

$$BE(^{104}Sn) - BE(^{109}I) = Q_\alpha(^{109}I) + Q_P(^{105}Sb) = Q_P(^{109}I) + Q_\alpha(^{108}Te) \quad (5)$$

The right side adds up to 4.259(22) MeV from measured Q-values[13,26]. The α-decay Q-value for ^{109}I can confidently be extrapolated from the Q_α-values of the heavier isotopes ^{110}I to ^{113}I as 3.9(2) MeV[27]. With the same precision we then get for ^{105}Sb: $Q_P = (0.36 \pm 0.20)$ MeV.

We also have experimental evidence, that Q_α for ^{109}I is smaller than 4.1 MeV. For higher energies we expect a partial α-decay halflife of not more than 2 ms from a WKB calculation reproducing well the known α-decay halflives in this region, and an α-branch of more than 2% in the decay of ^{109}I would have been observed in our experiment. In a recent experiment[23] at the mass separator behind SHIP this limit on Q_α for ^{109}I could be lowered to $Q_\alpha \leq 3.95$ MeV, corresponding to $Q_P(^{105}Sb) \geq 0.31$ MeV.

From the limits of our search experiment the upper limit on the proton decay Q-value for ^{105}Sb is obtained. For $Q_P \leq 1.0$ MeV the halflife should be longer than 10 ns and protons should have been observed unless the partial halflife is longer than the β-decay halflife. Since our cross section limit is two orders of magnitude below the observed values for the other two reactions, the branching ratio for proton emission must be smaller than 1% if the energy is within our detection window $E_P \geq 0.5$ MeV. The β-decay halflife is of the order of 0.5 s[21]. For the emission of a 0.5 MeV proton from a $d_{5/2}$ state a 5% branch is expected and would have been observed. In the case of a $g_{7/2}$ state a branch stronger than 1% requires $Q_P \gtrsim 0.56$ MeV. We thus conclude that in ^{105}Sb the odd proton is unbound with 0.31 MeV $\lesssim Q_P \lesssim$ 0.56 MeV and a branching ratio for direct proton emission between 10^{-1} (for $Q_P = 0.5$ MeV, L=2) and 10^{-11} (for $Q_P = 0.31$ MeV, L=4).

SEARCH FOR ^{108}I AND ^{77}Y

It is expected near the proton drip line, that an odd-even nucleus and the next lighter odd-odd isotope have nearly the same proton binding energy. In a first experiment we searched for proton radioactivity of ^{108}I with our catcher technique. Since the beam energy necessary for the ^{58}Ni(^{58}Ni,p3n) reaction was not available in Munich, the UNILAC of the GSI was used to deliver a 284 MeV ^{58}Ni beam with micropulses spaced by 330 ns. Except for an indication of a line near 0.8 MeV, possibly still due to the decay of ^{109}I, no protons with energies between 0.7 MeV and 1.5 MeV and a halflife longer than 10 ns were observed in the period between beam pulses. Due to the short running time of 5 h only a detection limit of 5 μb has been reached, not sensitive enough to rule out proton radioactivity of ^{108}I in the energy and halflife region mentioned.

The proton drip line can also be crossed with reasonable yield for heavy ion reactions in the region of neutron deficient nuclei with A≈80, known for a long time as region of well deformed nuclei. It seems most interesting to study the influence of deformation on the rate of proton emission. The ^{40}Ca(^{40}Ca,p2n) reaction at a beam energy of 135 MeV was used to populate ^{77}Y. Liran and Zeldes[25] predict for ^{77}Y a proton decay Q-value of 1.06 MeV. In two runs with different pulsing periods of 0.2 μs and 3.2 μs the following detection limits were reached:
$b_p \sigma \leq 7$ μb for 0.70 MeV$\leq E_p \leq$1.5 MeV and $T_{1/2} \geq$ 10 ns
$b_p \sigma \leq 3$ μb for 0.45 MeV$\leq E_p \leq$1.5 MeV and $T_{1/2} \geq$ 1 μs
(b_p being the branching ratio for proton emission). To the ground state of the mirror nucleus ^{77}Sr a 5/2$^+$ or 7/2$^+$ assignment has been made by Lister et al.[28] the most probable possibility being the 5/2$^+$[422] Nilsson state. If we assume the same configuration for the ground state of ^{77}Y and thus an L=4 proton decay, our measurement excludes decay energies between 0.45 MeV$\leq Q_p \leq$0.90 MeV.

CONCLUSIONS

In the case of the direct proton emitters ^{109}I and ^{113}Cs it could be shown, that - just from the knowledge of decay energy and halflife - information can be obtained on the ground state configuration of nuclei beyond the limit of nuclear stability.

A search for ^{105}Sb seems still worth the effort, if an online isotope separator can make use of the intense ^{58}Ni beams available at some facilities and if the detection system is able to distinguish low energetic protons from the background of β-particles.

There are still a number of candidates for proton radioactivity, which could be produced with cross sections in excess of 1 μb. And we believe, that - though not at all easy - a continuation of the search for other cases is rewarding, because information on nuclear structure and nuclear masses can be obtained in regions, which are otherwise unaccessible.

Two of us (T.F. and A.G.) would like to thank Prof. Armbruster and the SHIPpers for the possibility of taking part in the first experiments with the separator behind SHIP.

This work has been funded by the German Federal Minister for Research and Technology (BMFT) under the contract number 06-TM178/I.

REFERENCES

1. E. Marsden, W.C. Lantsberry, Phil. Mag. 30, 240 (1915)
2. E. Rutherford, Phil. Mag. 37, 581 (1919)

3. C.F. von Weizsäcker, Z. Phys., 96, 431, (1935)
4. S. Hofmann, Preprint GSI-87-30 to be published in:
   'Particle Emission from Nuclei', M. Ivascu and D. N. Poenaru
   eds., (CRC Press, Boca Raton, USA, 1987)
5. K.P. Jackson, C.U. Cardinal, H.C. Evans, N.A. Jelley, J. Cerny
   Phys.Lett. 33B, 281 (1970)
6. J. Cerny, J.E. Esterl, R.A. Gough, R.G. Sextro,
   Phys.Lett. 33B, 284 (1970)
7. V.A. Karnaukhov, D.D. Bogdanov, L.A. Petrov, Proc. Int. Conf on
   the Properties of Nuclei far from Stability, Leysin,
   Switzerland, 1970, p.457
8. D.D. Bogdanov, V.P. Bochin, V.A. Karnaukhov, L.A. Petrov,
   Sov. J. Nucl. Phys. 16, 491 (1973)
9. S. Hofmann, W. Reisdorf, G. Münzenberg, F.P. Heßberger, J.R.H.
   Schneider, P. Armbruster, Z.Phys. A305, 111 (1982)
10. O. Klepper, T. Batsch, S. Hofmann, R. Kirchner, W. Kurcewicz,
    W. Reisdorf, E. Roeckl, D. Schardt, G. Nyman,
    Z.Phys. A305, 125 (1982)
11. T. Faestermann, A. Gillitzer, K. Hartel, P. Kienle, E. Nolte,
    Proc. Int. Conf. on Nucl. Phys., Florence 1983, Vol. 1, p. 311;
12. T. Faestermann, A. Gillitzer, K. Hartel, P. Kienle, E. Nolte,
    Phys.Lett. 137B, 23 (1984)
13. S. Hofmann, Y.K. Agarwal, P. Armbruster, F.P. Heßberger, P.O.
    Larsson, G. Münzenberg, K. Poppensieker, W. Reisdorf, J.R.H.
    Schneider, H. J. Schött, Proc. 7th Int. Conf. on Atomic Masses
    and Fundamental Constants, Darmstadt-Seeheim 1984, p. 184
14. W.F. Feix, E.R. Hilf, Phys.Lett. 120B, 14 (1983)
15. K. Ogawa, op. cit. Ref. 13, p. 530
16. V.P. Bugrov, S.G. Kadmenskii, V.I. Furman, V.G. Khlebostroev,
    Sov. J. Nucl. Phys. 41, 717 (1985)
17. A. Gillitzer, T. Faestermann, K. Hartel, P. Kienle, E. Nolte,
    Z.Phys. A326, 107 (1987)
18. F.G. Perey, Phys.Rev. 131, 745 (1963)
19. R.D. Lawson, Theory of the Nuclear Shell Model (Clarendon Press,
    Oxford,1980) pp. 155,364-368
20. O. Klepper, Nucl. Instr. and Meth. 186, 385 (1981)
21. K. Takahashi, M. Yamada, T. Kondoh,
    At. Data Nucl. Data Tables 12, 101 (1973)
22. Ch. Schießl, W. Wagner, K. Hartel, H.J. Körner, Wa. Mayer,
    E. Rehm, Nucl. Instr. and Meth. 192, 291 (1982)
23. G. Berthes, PhD Thesis, Mainz, 1987, Report GSI-87-12
24. J.M. Arias, C.E. Alonso, R. Bijker, Nucl. Phys. A445, 333 (1985)
    and priv. comm. by J.M. Arias
25. S. Liran, N. Zeldes, At. Data Nucl. Data Tables 17, 431 (1976)
26. D. Schardt, T. Batsch, R. Kirchner, O. Klepper, W. Kurcewicz, E.
    Roeckl, P. Tidemand-Petersson, Nucl.Phys. A368, 153 (1981)
27. A. H. Wapstra, G. Audi, Nucl. Phys. A432, 1 (1985)
28. C.J. Lister, B.J. Varley, D.E. Alburger, P.E.Haustein, S.K.Saha,
    J.W. Olness, H.G. Price, A.D. Irving, Phys. Rev. C28, 2127,
    (1983)

Beta-Delayed Two-Proton Emission as a Nuclear Probe

D. M. Moltz, J. E. Reiff, J. D. Robertson, T. F. Lang and Joseph Cerny
Department of Chemisty and Lawrence Berkeley Laboratory, University of California,
Berkeley, CA 94720 USA

A brief history of beta-delayed two-proton emission is given. Speculations about future experiments which would enhance our knowledge about both nuclear spectroscopy and this relatively unique decay mode are presented.

Introduction

Beta-delayed particle emission has made the study of light proton rich nuclei much easier. In fact, as experiments have proceeded further from beta stability, detection of these exotic decay modes has become the probe of choice. Studies utilizing beta-delayed alpha and proton emission have become routine. Since a recent review[1] of all proton rich light nuclei covers the advantages of using delayed particle emission in eliciting the relevant physics, we will not attempt to give these reasons here. Instead, we will concentrate on how the discovery of a new isotope via its beta-delayed proton decay branch led to the discovery of a newly predicted[2,3] decay mode, beta-delayed two-proton emission, on the mechanisms available for this decay mode, and on what new measurements might bring to light.

Extensive studies of nuclei via their strong beta-delayed proton decay branch were first made in the $A = 4n+1$, $T_z = -3/2$ series. These nuclei have been characterized in this manner from ^9C to ^{61}Ge [1,4]. Attempts to study the the $A = 4n$, $T_z = -2$ series of strong beta-delayed proton emitters were thwarted until mass separated samples were available because of the simultaneous and copious production of the $A = 4n+1$, $T_z = -3/2$ nuclei. ^{32}Ar [5], ^{20}Mg, ^{24}Si, and ^{36}Ca[6] were all studied in this manner. Mass surface systematics suggested, however, that the proton energy which would be detected following decay from the analog state in the $T_z = -1$ daughters of the $A = 4n+2$, $T_z = -2$ nuclei would be generally higher than all proton energies arising from nuclei produced in competing reactions. Thus ^{22}Al became the first member of this series to be discovered by its beta-delayed proton decay[7]. This large βp energy in ^{22}Al also provided the unique possibility that *two* protons could be emitted following beta decay to the isobaric analog state. Figure 1 shows the result of a 110 MeV ^3He + Mg bombardment [8,9] This spectrum from the decay of ^{22}Al represents the discovery of beta-delayed two-proton

© American Institute of Physics 1988

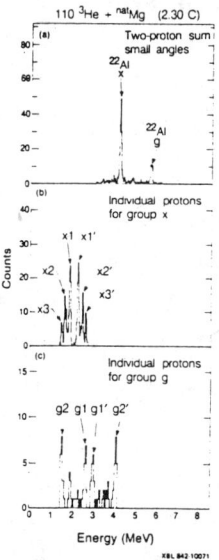

Figure 1. Proton-proton coincidence spectra obtained with a narrow angle detector system following the decay of ^{22}Al. See text and ref. 9. a) Two-proton summed energy spectrum. Groups x and g correspond to transitions involving the ^{20}Ne first excited state and ground state, respectively. b) Individual proton energy spectrum for protons forming group x in part a). Primed peaks probably correspond to the first emitted proton. c) Individual proton energy spectrum for protons forming group g in part a). In this case, g1 and g2' correspond to the first emitted protons.

radioactivity. (The importance of the narrow angle designation in Fig. 1 will become obvious later.) The 'g' and 'x' designations refer to the ground and first excited state in the 2p daughter, ^{20}Ne. The individual proton spectra of these two sum peaks are also given in Fig. 1. It is in these single proton spectra in which the decay mechanism is most evident. There exist three distinct mechanism possibilities: sequential, correlated simultaneous, and uncorrelated simultaneous emission. These three mechanisms would manifest themselves in the individual spectra in very different ways. Sequential emission for this decay would be nearly isotropic with the second proton energy kinematically dependent upon the relative observation angle. Correlated simultaneous emission in the $^1S_0$ state (^{2}He) would yield a strong angular dependence peaked at ~40° and essentially no yield observed beyond ~65° in addition to a continuum of single proton energies. Uncorrelated simultaneous emission would exhibit this same continuum of single proton energies, but would instead be essentially isotropic. Utilizing data obtained from a wide angle setup centered at 120°[9] and from an angular distribution measurement[10], it was possible to ascertain that although a 15% admixture of ^{2}He emission cannot be excluded, the β2p decay of ^{22}Al proceeds predominantly by sequential emission. When statistics are sufficient, it is possible to determine which proton is emitted first by comparing data at different angles. This technique has been used with the proton groups shown for ^{22}Al in Fig. 1 to construct the decay scheme given in Fig. 2[9]. The beta-delayed single proton groups observed in ref.7 have also been incorporated into Fig. 2.

Subsequent experiments discovered that the next member of the A = 4n+2, T_z = -2 series, ^{26}P, is also a beta-delayed two-proton emitter[11]. Unfortunately, ^{2}He emission from the isobaric analog state in ^{26}Si to the ^{24}Mg ground state is spin/parity forbidden. Thus this decay must be sequential. The following section will discuss not only the potential for observing ^{2}He emission, but how beta-delayed two-proton emission can be used as a spectroscopic probe.

New Studies

Once the initial characterization of beta-delayed two-proton emission was complete, the uniqueness of the decay mode could be exploited to search for heretofore experimentally inaccessible nuclei. The β2p spectrum arising from the 135 MeV ^{3}He + ^{40}Ca reaction shown in Fig. 3 represents the discovery of the first $T_z = -5/2$ nucleus, ^{35}Ca [12], and the first time that beta-delayed two-proton decay had been used to discover a new nuclide. The evidence supporting the assignment of this β2p activity to ^{35}Ca has been summarized elsewhere [12]. Subsequently, the existence of four $T_z = -5/2$ nuclei, ^{23}Si, ^{27}S, ^{31}Ar, and ^{35}Ca [13], was established in a fragmentation reaction study. Although the beta-delayed proton decay of ^{31}Ar has been determined[14], the predicted βp and β2p decay branches for the rest of these nuclei have yet to be reported.

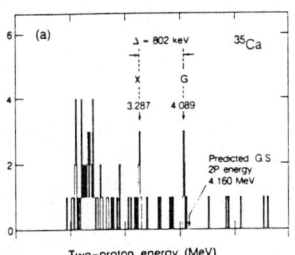

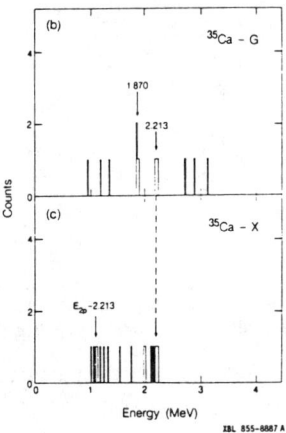

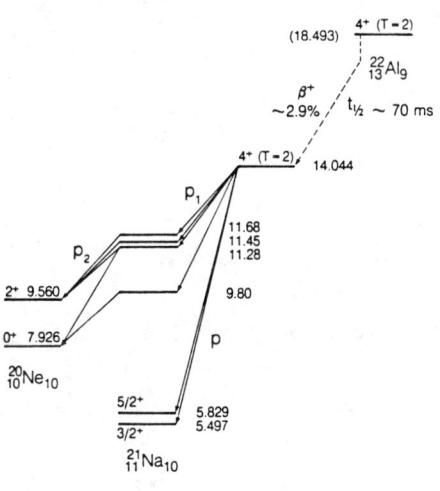

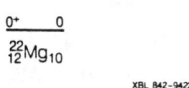

Figure 2. Proposed partial decay scheme for ^{22}Al.

Figure 3. a) Beta-delayed two-proton sum spectrum of ^{35}Ca. Groups labeled by G and X are related to the two-proton transitions to the ground and first excited states in the daughter nucleus ^{33}Cl. Individual proton energy spectra are shown in b) and c) corresponding to the ^{35}Ca G and X 2p peaks, respectively.

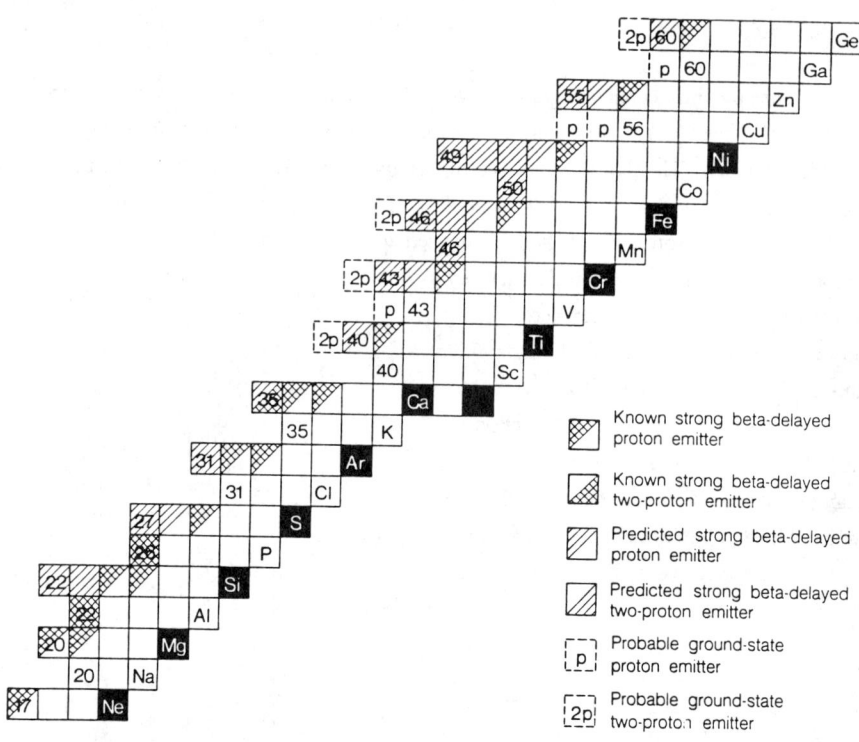

Figure 4. Proton rich portion of the chart of the nuclides from Z = 10-32 depicting known and predicted beta-delayed and direct particle decaying nuclides.

Although only three examples of this decay mode have so far been demonstrated, many more could in principle exist. Figure 4 shows the light proton-rich section of the chart of the nuclides where those nuclei which could undergo β2p decay are depicted. There are ~12 nuclei in the Z = 10-32 region alone which would possibly exhibit this decay mode. The notation "strong" in Fig. 4. refers to cases where the isobaric analog state is open to the given decay mode. Searches for new beta-delayed two-proton emitters are, of course, hampered by the same problems which affect all studies as one proceeds further from beta stability: lower production rates and competing reactions which exhibit the same decay mode. For example, any compound nucleus reaction which would produce ^{27}S or ^{31}Ar would also produce ^{22}Al as background at some yield. Recoil spectrometers can in principle overcome this background problem, but detection of the relevant decays can be difficult. Although medium energy fragmentation reactions have in general larger cross sections than similar compound nuclear reactions for such studies, the unique proton/neutron ratio of ^{3}He permits comparable cross section compound nuclear reactions. (Unfortunately, this is not true above Z > 20 due to the absence of stable Z = N targets.) Recently, we attempted to observe the beta-delayed two-proton decays of ^{46}Mn and ^{50}Co in a ^{14}N + ^{40}Ca bombardment. Not only was the predicted cross section more than a factor of ten smaller than an "equivalent" ^{3}He induced reaction, the predicted β2p decay energies are < 2 MeV, requiring the observation of protons with energies less than 1 MeV.

Utilizing beta-delayed two-proton emission to search for new nuclei is only one interesting aspect of this nuclear probe. There are many unanswered questions about the decay itself. Three possible decay mechanisms have already been given. Since simultaneous uncorrelated two-proton emission appears least likely, we will only consider the other two mechanisms. Figure 5 gives the result of a Monte Carlo simulation [9] of what would be observed for both sequential and ^{2}He emission using a detector system with ~40° relative angular range. Sequential emission would give a fixed energy first proton(s) but because the second proton is

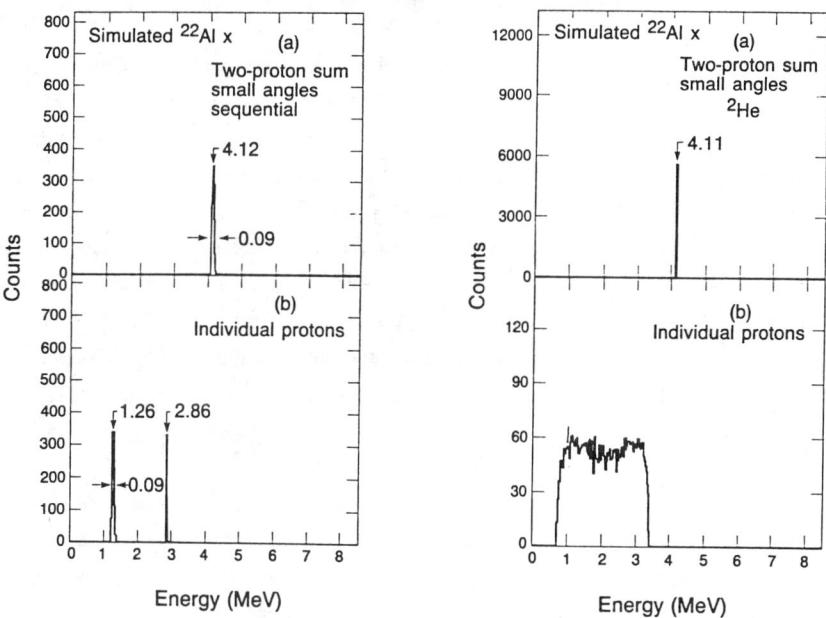

Figure 5. Monte Carlo simulation of A) sequential and B) ^{2}He emission of two protons with E^{cm} = 4.48 MeV (a predicted ^{22}Al decay energy) as would be observed with a detector system capable of detecting the two protons with relative anagular range of 0-70°. a) Two-proton summed energy spectra. b) Individual proton energy spectra.

emitted from a moving source, the measured energy of this proton is dependent upon the relative emission angle. The width of the second proton peak and therefore the summed two-proton peak is dependent upon the solid angle subtended by the second detector. For instance, the summed peak width difference shown for ^{22}Al decay detected at narrow and wide angles in Ref. 9 is dependent upon the twofold difference in total solid angle subtended by the detector telescopes. The emission of ^{2}He would have a totally different signature. Such a decay would give a monoenergetic two-proton peak, independent of angle. The individual protons would comprise, however, part of an energy continuum centered at $E_{p1} = E_{p2}$. The relative angle of the two observed protons would peak at ~40°; essentially no yield would be observed beyond 70°. Although evidence for this decay mechanism has not been found, it remains a very interesting possibility that deserves more experimental attention. Unfortunately, most of the remaining undiscovered β2p emitters will necessarily have higher proton numbers, where the Coulomb barrier for ^{2}He emission might favor sequential decay.

Another facet of beta-delayed two-proton emission that has to date remained unexplored involves the observation of low-energy single protons (100-1000 keV). This stems from a lack of suitable detectors, which in some cases (e.g., ^{50}Co) may have prevented the observation of the β2p decay branch. Since all known β2p groups have been assigned to decays from the isobaric analog state, detection of low energy protons might also permit the observation of β2p groups associated with strongly populated Gamow-Teller beta decays. Thus, the development of suitable low-energy proton detectors would significantly further our knowledge of beta-delayed two-proton emission.

Experiments designed to study the β2p decay mechanism and to observe new nuclei exhibiting β2p emission will be difficult. To fully appreciate the magnitude of the experimental difficulties which might be encountered, one can consider two examples. Figure 6 depicts a typical detector setup

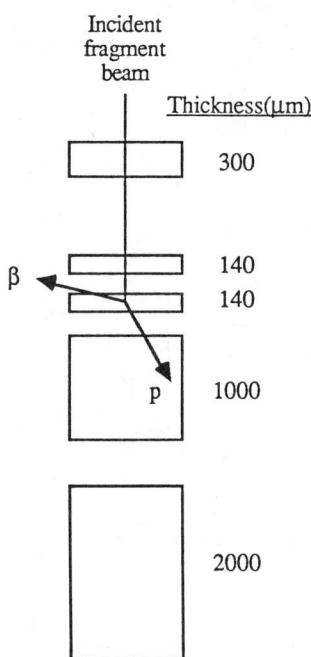

Figure 6. Typical detector setup used with the LISE spectrometer for identifying proton rich nuclei and subsequently observing a beta-delayed proton event. See ref. 14 for additional details.

used with the LISE recoil spectrometer [15] to identify separated reaction products. To study the subsequent radioactive decay of the collected sample (one atom at a time), the identification electronics is rapidly switched to utilize the silicon counters as proton detectors. If one wished to detect and properly identify two protons from the decay of an implanted atom, one would be faced with an extremely difficult proposition. In a recent experiment which observed the βp decay of ^{31}Ar [14], it was suggested that some of the events were potential β2p counts. Unfortunately, without proper identification of both protons, these assignments may remain inconclusive.

All initial beta-delayed two-proton studies were conducted utilizing the helium-jet technique. The short half-lives which are encountered near the proton drip line, however, soon renders the 25 ms helium-jet transit time as a severe liability. Faster collection techniques must be used. One example of such a technique is the fast rotating wheel shown schematically in Fig. 7. This wheel is capable of studying nuclides down to <100 μs half-life range. However, in order to obtain this short lifetime capability, the detectors must be operated very near the primary beam. This intense radiation field forces the acquisition of data only during the beam-off phase of a pulsed accelerator beam. These experimental difficulties and low production cross sections can often be overcome because it is possible to observe this decay mode in a very high background: $1:10^5$ protons or $1:10^9$ betas. Finally, beta-delayed two-proton decay can provide unique insights into the general mechanisms in which nuclei in excited states lose their energy.

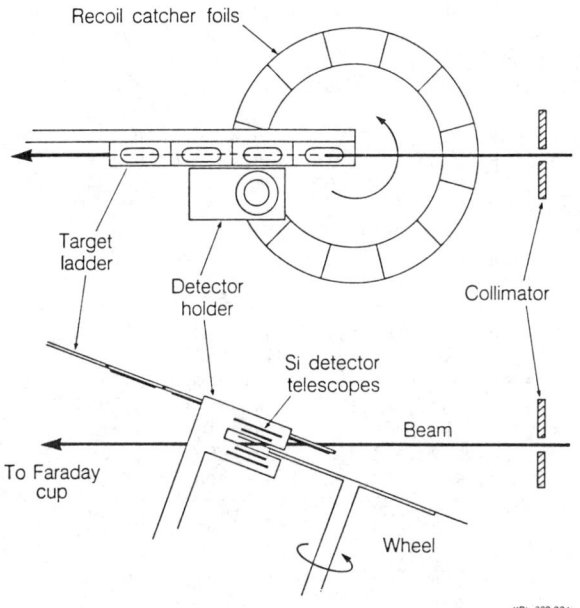

Figure 7. Schematic diagram of the fast rotating wheel setup for the study of short-lived radioactivities.

Summary

Beta-delayed two-proton emission is an interesting decay mode about which much more is to be learned. A more complete review of the current state of this decay mode may be found elsewhere [16]. Many experiments remain to be performed which should give exciting new insights into the use of beta-delayed two-proton emission as a nuclear probe.

References

1. Äystö, J. and J. Cerny, to be published in v. 8 of Treatise on Heavy Ion Science, D. A. Bromley, ed., (1987).
2. Gol'danskii, V. I., Pis'ma Zh. Eksp. Theor. Fiz. 32, 572 (1980), (Sov. Phys. -JETP Lett. 32, 554 (1980).)
3. Gol'danskii, V. I., Usp. Fiz. Nauk. 141, 715 (1983), (Sov. Phys. Usp. 26, 1112 (1985).)
4. Cerny, J. and J. C. Hardy, Ann. Rev. Nucl. Sci. 27, 333 (1977).
5. Hagberg, E., P. G. Hansen, J. C. Hardy, A. Huck, B. Jonson, S. Mattson, H. L. Ravn, P. Tidemand-Petersson, and G. Walter, Phys. Rev. Lett. 39, 792 (1977).
6. Äystö, J., M. D. Cable, R. F. Parry, J. M. Wouters, D. M. Moltz, and J. Cerny, Phys. Rev. C 23, 879 (1981).
7. Cable, M. D., J. Honkanen, R. F. Parry, H. M. Thierens, J. M. Wouters, Z. Y. Zhou, and J. Cerny, Phys. Rev. C 26, 1778 (1982).
8. Cable, M. D., J. Honkanen, R. F. Parry, S. H. Zhou, Z. Y. Zhou, and J. Cerny, Phys. Rev. Lett. 50, 404 (1983).
9. Cable, M. D., J. Honkanen, E. C. Schloemer, M. Ahmed, J. E. Reiff, Z. Y. Zhou, and J. Cerny, Phys. Rev. C 30, 1276 (1984).
10. Jahn, R., R. L. McGrath, D. M. Moltz, J. E. Reiff, X.J. Xu, J. Äystö and J. Cerny, Phys. Rev. C 31, 1576 (1985).
11. Honkanan, J., M. D. Cable, R. F. Parry, S. H. Zhou, Z. Y., Zhou, and J. Cerny, Phys. Lett. 133B, 146 (1983).
12. Äystö, J., D. M. Moltz, X. J. Xu, J. E. Reiff, and J. Cerny, Phys. Rev. Lett. 55, 1384 (1985).
13. Langevin, M., A. C. Mueller, D. Guillemaud-Mueller, M. G. Saint-Laurent, R. Anne, M. Bernas, J. Galin, D. Guerreau, J.C. Jacmart, S. D. Hoath, F. Naulin, F. Pougheon, E. Quinious, and C. Detraz, Nucl. Phys. A455, 149 (1986).
14. Borrel, V. et al., Orsay Report Number IPNODRE 87-18.
15. Anne, R., D. Bazin, A. C. Mueller, J. C. Jacmart and M. Langevin, submitted to Nucl. Instr. Meth., Report Number GANIL 86-23.
16. Moltz, D. M. and J. Cerny, to be published in "Charged Particle Emission from Nuclei", Vol. III, M. Ivascu and D. Poenaru, eds., (1987).

PRODUCTION AND IDENTIFICATION OF NEW NUCLEI AT THE NEUTRON AND PROTON DRIP-LINES FOR LIGHT ISOTOPES

D. Guillemaud-Mueller, R. Anne, D. Bazin, C. Détraz, J. Galin,
D. Guerreau, A.C. Mueller, E. Roeckl, M.G. Saint-Laurent
GANIL, BP 5027, 14021 CAEN-CEDEX, France

M. Bernas, V. Borrel, J.C. Jacmart, M. Langevin, F. Naulin,
F. Pougheon, E. Quiniou
IPN, BP 1, 91406 ORSAY-CEDEX, France

ABSTRACT

By combining the effectiveness of the projectile-fragmentation mechanism at intermediate energy, the relatively high beam intensities available at GANIL and the performances of the doubly achromatic spectrometer LISE[1], a breakthrough has been achieved in reaching the neutron and proton drip-lines. The achromatic refocusing of the products by the spectrometer allows identification by a small high-resolution solid-state detector telescope. Combined with the time of flight measurement of the nuclei, a redundant determination of A and Z is ensured. By fragmentation of various projectiles such as ^{40}Ar, ^{36}Ar, ^{86}Kr, ^{40}Ca, ^{58}Ni, it has been possible to produce new light neutron- rich or neutron-deficient nuclei.

INTRODUCTION

At the last conference on nuclei far from stability in 1981, the effectiveness of high-energy projectile fragmentation in producing new light exotic nuclei has been demonstrated by T.J.M. Symons[1]. In the meantime, the availability of intermediate energy heavy-ion beams at GANIL with high intensities has allowed to pursue the production of new light isotopes at both the neutron and the proton drip-lines. Indeed the projectile fragmentation still remains the predominant reaction mechanism at GANIL energy for light to medium projectiles[2]. By combining the effectiveness of this mechanism, the relatively high beam intensities available and the performances of the doubly achromatic spectrometer LISE it has been possible to reach the proton and neutron drip-lines for light isotopes.

The following chapters will dwell on some essential aspects of the production reaction, of the recoil spectrometer and report on experiments performed so far.

PROPERTIES OF THE REACTION MECHANISM

Important features of the projectile fragmentation are :

small momentum transfer between target and projectile and an angular distribution peaked around zero degree. Therefore the projectile-like fragments are emitted with a velocity close to the one of the incident beam in a narrow cone. This allows for an efficient magnetic separation of the products.

In such a process one can expect the isotopic distributions to be dominated by the isospin asymmetry of the projectile ; a more neutron-rich projectile will produce more neutron-rich fragments. A calculation has been made by D. Guerreau[3] to obtain the isotopic distribution of the fragments. There, a simple geometrical model of abrasion-ablation determined by the overlap between the two interacting nuclei for the primary fragment distribution and a stage of desexcitation for the observable final distribution are used. The result, for silicon isotopes, produced with three different projectiles is shown on fig. 1.

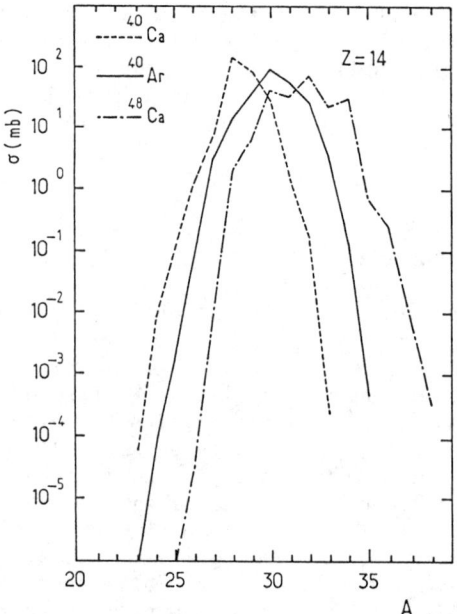

Fig. 1 : Calculated final distributions for $Z = 14$ isotopes produced with three different projectiles (E_{inc} = 44 MeV/u, Ta target) (taken from ref.[3]).

It clearly demonstrates the influence of the N/Z ratio of the projectile on the production cross-section changing by several orders of magnitude. Calculated final isotopic distributions for the reaction ^{40}Ar (44 MeV/u) + Ta are compared to experimental results on figure 2 which have been obtained for stable nuclei in a scattering chamber experiment[2] and for nuclei far off stability in experiments[4] with LISE. The agreement is quite satisfactory considering the simplicity of the model.

There is however a difference observed between fragmentation at relativistic energy as compared to intermediate energy. The N/Z ratio of the target contributes due to the nucleon exchange between projectile and target in the intermediate energy domain[2,5].

In order to obtain the maximum production yields, in addition to the considerations above, it is important to take into account

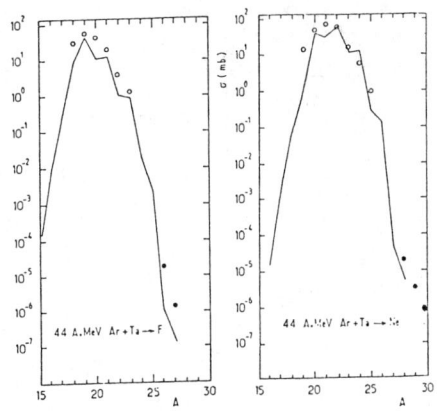

Fig. 2 : Calculated final isotopic distributions for the reaction ^{40}Ar (44 MeV/u)+Ta (solid lines). Comparison is made with results of ref.[2] (open circles) and of ref.[4] (solid circles) (taken from ref.[3]).

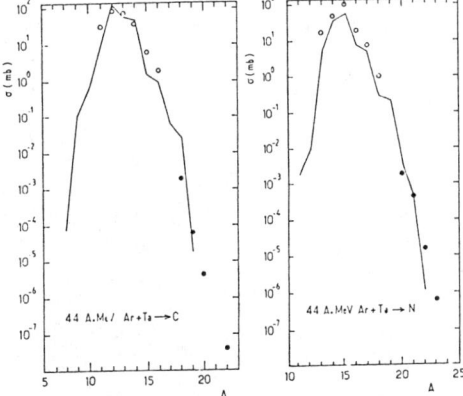

the efficiency of the recoil spectrometer. As shown in figure 3 this efficiency decreases with increasing distance of the fragments from the projectile. This is due to the increase of the widths of the angular and velocity distributions which exceed the limited acceptance of the spectrometer. Therefore the choice of the projectile is important to give access to a definite region. For projectiles as heavy as krypton, however, the mechanism described above seems to change at an energy of 44 MeV/u (see contribution of D. Bazin et al in this conference).
In order to show the actual possibilities of the GANIL in Table I are summurized the beams available with their energies and intensities.

THE LISE SPECTROMETER

LISE (Fig. 4) is a 0° magnetic analyser with an acceptance of 1 msr[6]. It is essentially composed of two identical dipole magnets (45°, ρ = 2 m, B = 1.6T). The incoming beam is directed and focused onto the target location by means of four magnetic quadrupole lenses (q1 to q4). The target, in order to accept high beam intensity, of 1.5 μAe for ^{40}Ar (44 MeV/u) for example, is water-cooled. Ten different targets are mounted on a wheel and

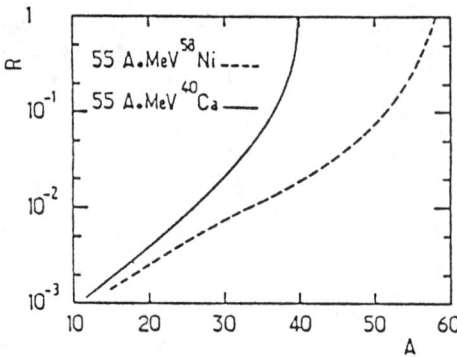

Fig. 3 The calculated efficiency of the LISE spectrometer for both projectiles on a 100 mg/cm² Ni target (taken from ref.[3])

Table I : Beams currently available at Ganil

ion	energy MeV/u	Intensity pps x 10^{11}
^{12}C	70, 40	10.4, 10.4
^{13}C	50	6.2
^{14}N	60	7.1
^{16}O	50, 95	10.7, 11.7
^{17}O, ^{18}O	84, 65	5.4, 15.6
^{20}Ne	40, 60	4.9, 2.8
^{22}Ne	60	3.5
^{36}Ar	85	3.5
^{40}Ar	44, 60	3.9, 3.4
^{40}Ca	60, 77	≤0.35, ≤0.35
^{58}Ni	55	0.13
^{84}, ^{86}Kr	33, 45	0.4, 0.3

so, without breaking the vacuum system, are available. The target changes are automatically accomplished by an on-line computer system which has also access to all the values of the quadrupole lenses, magnetic dipoles, setting in and setting out the multiwire proportional chambers monitoring the beam profile, in fact of every device which controls the beam. The first dipole analyses the outgoing beam in A/Z with a resolution of 1.2×10^{-3}. Four quadrupole lenses Q1 to Q4 adjacent to the magnet provide an adjustment of the beam size in the first analyser and of the resulting dispersion in the image plane. At this place, a variable aperture slit controls the Bρ acceptance up to a maximum of ± 2.5%. It is also possible to either install stripper foils for additional stripping or energy degraders to slow down the beam. The second dipole compensates the dispersion of the first one

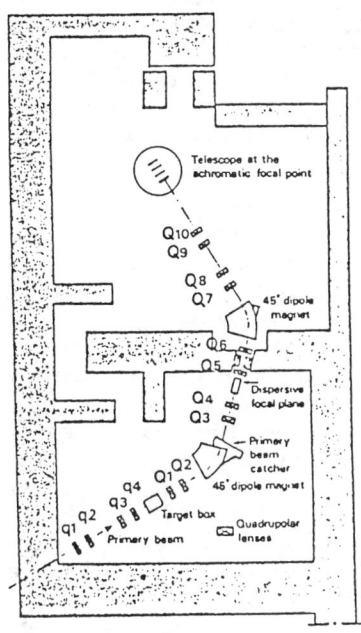

Fig.4 : Schematic lay-out of the LISE spectrometer

which provides a system doubly achromatic in angle as well as in position. Then, the four last quadrupole lenses refocuse the beam and adapt its size in function of the experiments.

Moreover LISE has been designed to maintain its optical properties even after the installation of an energy degrader in the intermediate focal plane.

This degrader according to the Bethe's formula provides an additional selection according to the different slowing down of the nuclei. With that criterion the fragments are selected according to a $A^{2.5}/Z^{1.5}$ law. Operating in such a way LISE acts as an isotope separator (see the contribution of J.P. Dufour in this conference). Therefore only some nuclides are collected at the final focal point of LISE instead of a few tens. This additional selection is powerful in different configurations. For the identification of nuclei it allows to improve the sensitivity of the experiment by reducing the number of transmitted nuclei out of interest (described below for the case of ^{22}Si[7]). It was essential, in order to avoid the contamination of these isotopes for γ-decay[8] studies of neutron-rich isotopes and experiments on the β-delayed proton emission[9].

The double achromatism of the LISE line ensures a practically constant flight-path length ($\frac{\Delta l}{l} \approx 10^{-4}$, l = 18 m) between the initial and final focal points of the line i.e. the production target and the detector position independent of angle of entry into the spectrometer. Thus, the time of flight of the fragments becomes an important parameter for particle identification techniques. This feature played an essential rôle for the experiments which led to the production of new nuclei at both drip-lines. For the identification of the fragment, a four stage solid-state detector telescope (ΔE_1, ΔE_2, E, $\bar{E}$) has been mounted at the final focal point of LISE. The informations of the telescope combined to the time of flight of the product and the magnetic rigidity setting of the spectrometer ensures an overdetermined identification in A and Z. The time of flight is obtained from the time signal of a ΔE detector and the radio-frequency signal of the last GANIL cyclotron which is correlated with the beam impact on the target. Depending on the beam, the time resolution is ranging between 0.2 and 1%. A resolution in Z of 1% is easily achieved.

TOWARDS THE NEUTRON DRIP-LINE

In the fragmentation of an argon beam at 44 MeV/u [4,15] it has been possible to prove the existence of the following new nuclei : ^{22}C, ^{23}N, ^{29}Ne, ^{30}Ne and the particle unstability of ^{18}B, ^{21}C, ^{25}O. Figure 5 shows a typical bidimensionnal diagram Z versus time of flight. This representation shows the quality of the spectra within respect to background, and the confirmation at a level of much improved statistics of the existence of ^{19}B[10].

Comparing these results to the current mass formulae[11,12,13], agreement is found for stability and non stability of the above nuclei.

The search of particle stability of nuclei provides a stringent test for mass formulae. This is illustrated by the case of ^{29}Ne. This nucleus was predicted unstable by the old Möller and Nix calculations[14]. With their improved calculations[13] especially in the treatment of the pairing energy, this nucleus is now particle-stable which corresponds to the experimental result. Neon isotopes whose even nuclei are predicted stable up to ^{38}Ne by the Uno and Yamada prescriptions are interesting since they cross the N = 20 region found to be a new region of deformation[16].

The predicted one, and two neutron separation energies S_{1n}, S_{2n} calculated with the formulae of Garvey. Kelson[12] and Uno and Yamada[11] are given on fig. 6. It predicts the last bound isotopes to be ^{19}B, ^{22}C and ^{23}N which means that the predicted drip-line is now mapped up to Z = 7.

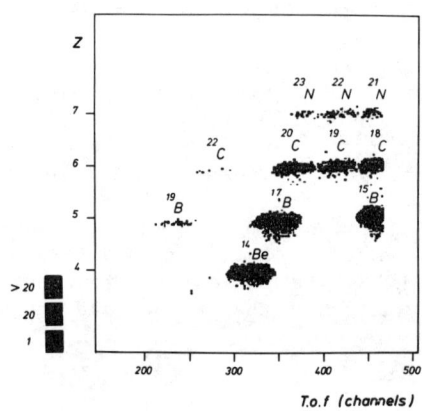

Fig. 5 : Two dimensional plot of events in the diagram Z versus time of flight in the reaction $^{40}Ar(44MeV/u)$ + Ta which shows the new isotope ^{22}C.

The potential of the use of an argon beam is certainly exhausted. The observation of ^{26}O and ^{32}Ne unfruitful with argon might however be within the reach with an heavier neutron-rich projectile such as ^{48}Ca.

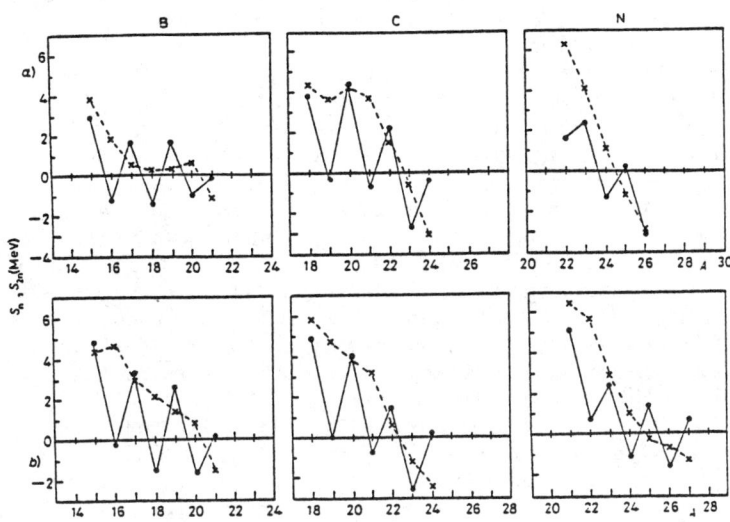

Fig. 6 : Predicted one and two-neutron separation energies S_{1n}, S_{2n} calculated with the formulae of Garvey-Kelson[11a) and Uno-Yamada[11]b).

TOWARDS THE PROTON DRIP-LINE

The proton drip-lines has been investigated using three different projectile : ^{40}Ca (77 MeV/u)[17], ^{36}Ar (84 MeV/u)[7], ^{58}Ni (55 MeV/u)[18] on natural nickel targets.

In the first experiment, the series of the four $T_z = -5/2$ isotopes ^{23}Si, ^{27}S, ^{31}Ar, ^{35}Ca which was predicted stable or nearly stable against two-proton emission has been identified. The last one ^{35}Ca was yet observed through its β-delayed two proton emission[19]. All the nuclei with $T_z = -2, -5/2, -3$ which was not observed in this reaction was predicted to be unstable against the emission of one or two protons by applying the charge symmetry formula of Kelson-Garvey[20] with the exception of the ^{22}Si bound against 2p emission by 16 ± 300 KeV. The search of the stability of this nucleus was the goal of the experiment of fragmentation of the ^{36}Ar beam. In order to make use of all the possibilities of the machine (intensity of the beam) and of the spectrometer (momentum acceptance) a degrader has been installed in the intermediate focal plane of LISE, which reduced the counting rate of the telescope fragments avoiding pile up due to fragments out of interest. Thus only three N = 8 isotones of ^{22}Si were transmitted simultaneously with the choosen nucleus : ^{17}F, ^{18}Ne, ^{20}Mg as shown on the bidimensionnal Z versus time of flight (fig. 7). The unbound character of ^{19}Na (ref. 21) and ^{21}Al which is predicted to be proton-unbound by 1.09 MeV[12,20] is confirmed. ^{22}Si is the first $T_z = -3$ nucleus ever evidenced.

The fragmentation of ^{58}Ni has allowed to produce twelve new proton-rich nuclei : ^{43}V, ^{44}Cr, ^{46,47}Mn, ^{48}Fe, ^{50,51,52}Co, ^{52,51}Ni and ^{55,56}Cu. In table II are given the 1p and 2p separation energies for proton rich nuclei with $23 \leq Z \leq 29$ and $-5/2 \leq T_z \leq -1$. This is done by applying the charge symmetry formula of Kelson-Garvey[20] and using the most recent experimental masses[22]. All the nuclei which have been produced are predicted to be bound or slightly bound against one or two-proton emission with the exception of ^{55}Cu which is predicted unbound against one-proton emission by 165KeV. Two nuclei in the $T_z = -2$ series which are predicted to be unbound against one-proton emission by some 300 KeV and 400 KeV respectively, ^{54}Cu and ^{42}V have not been observed. This might indicate that their proton radioactive half lives are shorter than 200 ns.

In the $T_z = -5/2$ series only ^{51}Ni predicted bound against proton emision is observed. ^{43}Cr and ^{47}Fe which are in the same range of separation energy are not observed (only 1 count of ^{47}Fe is observed and it is not possible to conclude).

Concerning the proton drip-line, it has probably been reached for Z = 23, 25, 29. For cobalt isotopes it can not be concluded since only two counts are observed for ^{49}Co which is predicted to be unbound by $S_{1p} = -940$ KeV.

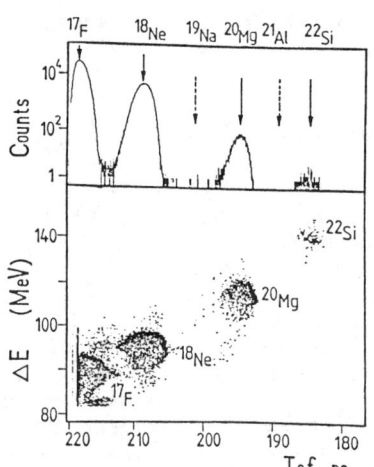

Fig. 7 : Experimental features of ^{22}Si and neighbouring nuclei observed in the reaction ^{36}Ar (85 MeV/u) + Ni. Lower part : the two-dimensional display of fragments identified in the focal plane of LISE. Upper part : the corresponding projected time-of-flight spectrum.

Table II (see text)

ISOTOPE	T_z	S_{1p} (keV)	S_{2p} (keV)
^{44}V	-1	1733 ± 38	6269 ± 33
^{46}Cr		4886 ±37	6500 ± 21
^{48}Mn		2038 ± 59	6806 ± 46
^{50}Fe		4148 ± 64	6230 ± 60
^{52}Co		1449 ± 54	6330 ± 41
^{54}Ni		3859 ± 50	5457 ± 50
^{56}Cu		578 ± 65	5188 ± 55
^{43}V	-3/2	81 ± 39	3849 ± 34
^{45}Cr		2933 ± 181	4717 ± 157
^{47}Mn		447 ± 83	5334 ± 80
^{49}Fe		2455 ± 205	4494 ± 174
^{51}Co		224 ± 125	4372 ± 89
^{53}Ni		2320 ± 219	3770 ± 195
^{55}Cu		-165 ± 123	3693 ± 91
^{42}V	-2	-413 ± 76	2048 ± 40
^{44}Cr		2594 ± 213	2676 ± 186
^{46}Mn		419 ± 222	3353 ± 103
^{48}Fe		2822 ± 142	3270 ± 99
^{50}Co		544 ± 240	3000 ± 125
^{52}Ni		2559 ± 142	2784 ± 137
^{54}Cu		-310 ± 270	2010 ± 129
^{41}V	-5/2	-1881 ± 50	403 ± 51
^{43}Cr		1047 ± 97	993 ± 100
^{45}Mn		-1881 ± 50	403 ± 51
^{47}Fe		1839 ± 162	2259 ± 239
^{49}Co		-943 ± 179	1878 ± 162
^{51}Ni		1573 ± 173	2118 ± 253
^{53}Cu		-1565 ± 194	994 ± 181

For even Z isotopes the drip-line is not yet mapped since isotopes of ^{46}Fe and ^{50}Ni should be bound by local mass relations[20].

CONCLUSION

It may be noted that also several new neutron-rich isotopes with Z ⩾ 18 have been found in the fragmentation of Kr at intermediate energy albeit still far away from the drip-line[23,24]. This underlines the potential of the fragmentation also for heavier beams of GANIL especially in view of the upgrading of the accelerator, now under the way, concerning its maximum energy as well as its intensity. It will be very interesting to note at the next conference which progress may have been obtained due to the use of strong relativistic heavy-ion beam for the future SIS facility at Darmstadt[25].

REFERENCES

1. T.J.M. Symons, Proceedings of the 4th International Conference on Nuclei far from stability, Helsingor, Denmark, 1981, CERN-81-09, p. 668
2. D. Guerreau, V. Borrel, D. Jacquet, J. Galin, B. Gatty, X. Tarrago, Phys. Lett. 131B, 293 (1983)
3. D. Guerreau, Conference on Heavy Ion Nuclear Collisions in the Fermi Energy Domain, Caen, France, 1986, J. de physique C4-8, p. 205
4. M. Langevin, E. Quiniou, M. Bernas, J. Galin, J.C. Jacmart, F.Naulin, F. Pougheon, R. Anne, C. Détraz, D. Guerreau, D. Guillemaud-Mueller and A.C. Mueller, Phys. Lett. 105B, 71 (1985)
5. V. Borrel, B. Gatty, D. Guerreau, J. Galin, D. Jacquet Z. Phys. A324, 205 (1986)
6. R. Anne, D. Bazin, A.C. Mueller, J.C. Jacmart, M. Langevin NIM. A257, 215 (1987)
7. M.G. Saint-Laurent, J.P. Dufour, R. Anne, D. Bazin, V. Borrel, H. Delagrange, C. Détraz, D. Guillemaud-Mueller, F. Hubert, J.C. Jacmart, A.C. Mueller, F. Pougheon, M.S. Pravikoff, E. Roeckl, Phys. Rev. Lett 59, 33 (1987)
8. J.P. Dufour, Contribution to this conference and references therein
9. J.C. Jacmart, Contribution to this conference and references therein
10. J.A. Musser and J.D. Stevenson, Phys. Rev. Lett. 53, 2544 (1984)
11. M. Uno and M. Yamada, INS Report NUMA 40 (1982)
12. J. Jänecke, in At. Data and Nucl. Data Tables, S. Maripuu Ed. 17, 445 (1976)
13. P. Möller and J.R. Nix, Submitted to At. Data and Nucl. Data Tables and preprint LA-UR-86-3983
14. P. Möller and J.R. Nix, At. Data and Nucl. Data Tables 26, 165 (1981)
15. F. Pougheon, D. Guillemaud-Mueller, E. Quiniou, M.G. Saint-Laurent, R. Anne, D. Bazin, M. Bernas, D. Guerreau, J.C. Jacmart, S.D. Hoath, A.C. Mueller, C. Détraz, Europhys. Lett. 2, 505 (1986)
16. D. Guillemaud-Mueller, C. Détraz, M. Langevin, F. Naulin, M. de Saint-Simon, C. Thibault, F. Touchard, M. Epherre Nucl. Phys. A426, 37 (1984)
17. M. Langevin, A.C. Mueller, D. Guillemaud-Mueller, M.G. Saint-Laurent, R. Anne, M. Bernas, J. Galin, D. Guerreau J.C. Jacmart, S.D. Hoath, F. Naulin, F. Pougheon, E. Quiniou, C. Détraz, Nucl. Phys. A455, 149 (1986)

18. F. Pougheon, J.C. Jacmart, E. Quiniou, R. Anne, D. Bazin,
    V. Borrel, J. Galin, D. Guerreau, D. Guillemaud-Mueller,
    A.C. Mueller, E. Roeckl, M.G. Saint-Laurent, C. Détraz
    Z. Phys. A327, 17 (1987)
19. J. Aysto, D.M. Moltz, X.J. Xu, J.E. Reiff, J. Cerny,
    Phys. Rev. Lett. 55, 1384 (1985)
20. I. Kelson, G. Garvey Phys. Lett. 23, 689 (1966)
21. J. Cerny, R.A. Mendelson, Jr., G.J. Wozniak, J.E. Esterl,
    J.C. Hardy, Phys. Rev. Lett. 22, 612 (1969)
22. A.H. Wapstra, G. Audi, Nucl. Phys. A432, 15 (1985)
23. D. Guillemaud-Mueller, A.C. Mueller, D. Guerreau,
    F. Pougheon, R. Anne, M. Bernas, J. Galin, J.C. Jacmart,
    M. Langevin, F. Naulin, E. Quiniou, C. Détraz, Z. Phys. A322,
    415 (1985).
24. W. Mittig, private communication.
25. Proceeding of the Workshop on experiments and experimental
    facilities at SIS-ESR March 30-April 1, 1987, GSI Report
    87-7.

IDENTIFICATION OF THE NEW NEUTRON-RICH ISOTOPES
$^{70-74}$Ni and $^{74-77}$Cu IN THERMAL NEUTRON FISSION OF ^{235}U.

M. BERNAS[1], P. ARMBRUSTER[2], J.P. BOCQUET, R. BRISSOT[3], H.R. FAUST and P. ROUSSEL[1].

Institut Laue-Langevin, Grenoble, France

(1) Institut de Physique Nucleaire, Orsay, France
(2) Ges. Für Schwerioneuforschung, Darmstadt, F.R. Germany
(3) DRF/PN, CEN Grenoble, France

ABSTRACT

The unknown isotopes $^{70-74}$Ni and $^{74-77}$Cu were identified in thermal neutron fission of ^{235}U, using the recoil separator Lohengrin and ΔE-E detection techniques. Isotopic yields for preselected values of ionic charge and kinetic energy of the fission fragments in the mass range A = 70-80 were measured down to 10^{-9} fragment/fission (f/f).

INTRODUCTION

The area covered by known nuclei on the map of the stability valley is constantly increasing due to the improvements of both production methods - intense, energetic or exotic ion beams, cooled, mass separated targets and new productive reactions - and rejection power of separation techniques - magnetic-electrostatic, or physico-chemical filters.

In this work, the fifty year old neutron induced fission has been used as a source reaction. The recent improvement of the detection system following the recoil mass spectrometer Lohengrin at the high flux reactor of the ILL in Grenoble, has permitted identification of new neutron rich isotopes at the frontier of the very small production yields.

The observation of new neutron rich isotopes, in the region lying from Fe (Z = 26) to Zn (Z = 30), even undoubtely particle stable, is of interest for nuclear spectroscopy and astrophysics.

With the closed proton shell (Z = 28) the Ni isotopes provide good probes for nuclear model calculations[1]. The nickel isotopes, especially the doubly magic isotope of ^{78}Ni, when discoved and studied, will show how shell and pairing effects act apart from stability.

Besides, these neutron rich nuclei are involved in the astrophysical r-process[2]. Since rapid neutron capture and β decay determine the build-up of heavier elements, the mass excess and β decay parameters of these nuclei are needed to calculate natural mass abundances in order to better understand element synthesis[3]. In this respect, the N = 50 closed shell-nuclei and Z = 28 Ni isotopes are key ingredients.

The heaviest Ni isotope reported until now is ^{69}Ni. Its mass excess and half-life were measured in Orsay[4] and GSI respectively[5]. With 40 neutrons, ^{68}Ni having a 0^+ level as first excited states, shows shape isomerism[1]. On the N = 50 line the lightest known isotope is ^{80}Zn which was produced in the thermal neutron fission with low yield. Its half-life, β decay scheme and mass excess were measured in 1983[6] and confirmed in 1985[7].

In the following, the experimental set-up of the present experiment will be described after some comments on fission kinematics. The results of the measurements will be given and preliminary fission yields indicated. The results and associated production methods will be discussed in comparison with other recent works concerning isotopes of the same region.

1. KINEMATICS OF FISSION

In order to identify the very rare fission channels which would lead to neutron rich Cu and Ni, the high flux reactor of the ILL facility (10^{15} n/cm²/sec) with the recoil mass spectrometer Lohengrin, provides one of the most appropriate experimental systems. It has been extensively used for measurement of fission mass yields and spectroscopy of fission products.

The Z and A dependance of fission production yields is illustrated by Fig.1 which summarizes the results obtained for

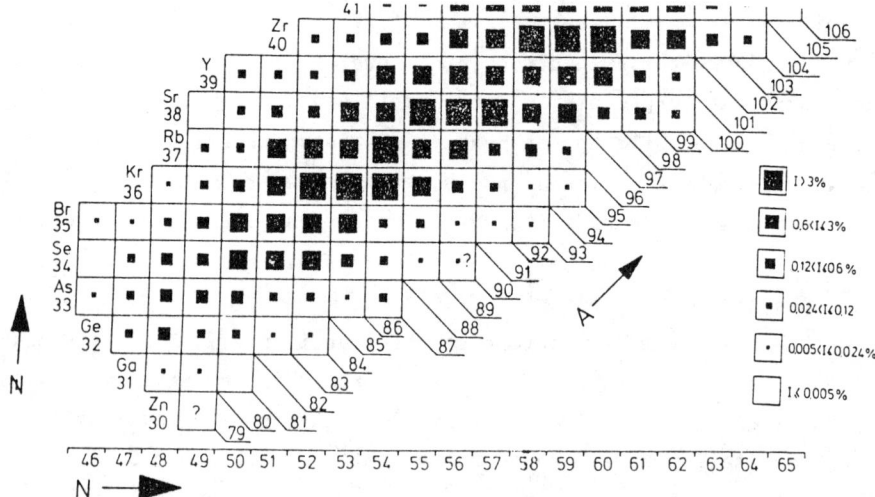

Fig. 1 : Schematic view of ^{233}U n-induced fission rates.

neutron induced fission of ^{233}U from ref. (8). For each mass number A, a most probable Z value ($\bar{Z}$) is found with a rapid fall of the yields on each side. But, at lower A the extrapolation, even when performed with adjusted gaussian shapes, are not accurate enough to predict isotopic yields better than within an order of magnitude.

The more probable fissions are, at first order, the ones delivering the larger Q-value. In Fig. 2a is plotted the sum of the mass excess of the pair of fragments produced in the fission of $^{235}U+n$, a member of which is the required isotope (horizontal scale). From there, a most probable Z value is indicated and a possible odd even effect can be observed. However no neutron shell closure effect for N = 50 can be expected from this curve.

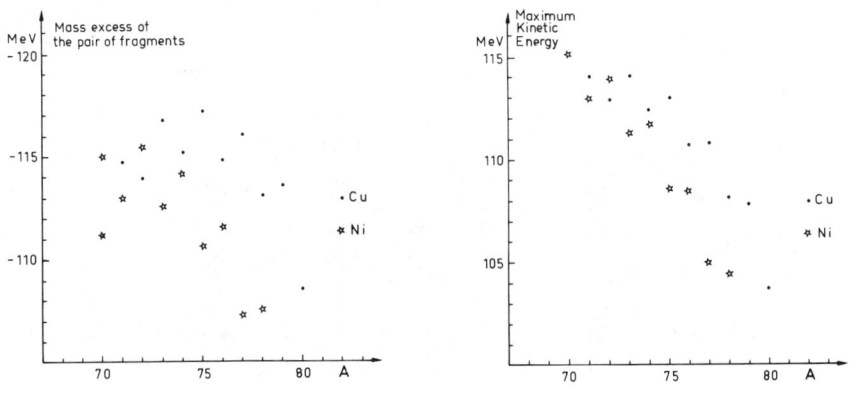

Fig. 2 : a) Q-values of ^{235}U fission as a function of outgoing channel b) Maximum kinetic energy of the light fragment.

In the extreme case of cold fission, the mass energy is totally converted into kinetic energy. Applying conservation laws, the maximum energy of the light fragment is calculated as a function of the light partner of the pair (Fig. 2b). The lighter this fragment, the higher its kinetic energy. When evolving towards more symetric fission the kinetic energy of the light fragment stays the same

Fig. 3 : Schematic and perspective view of the arrangement of the fields at Lohengrin.

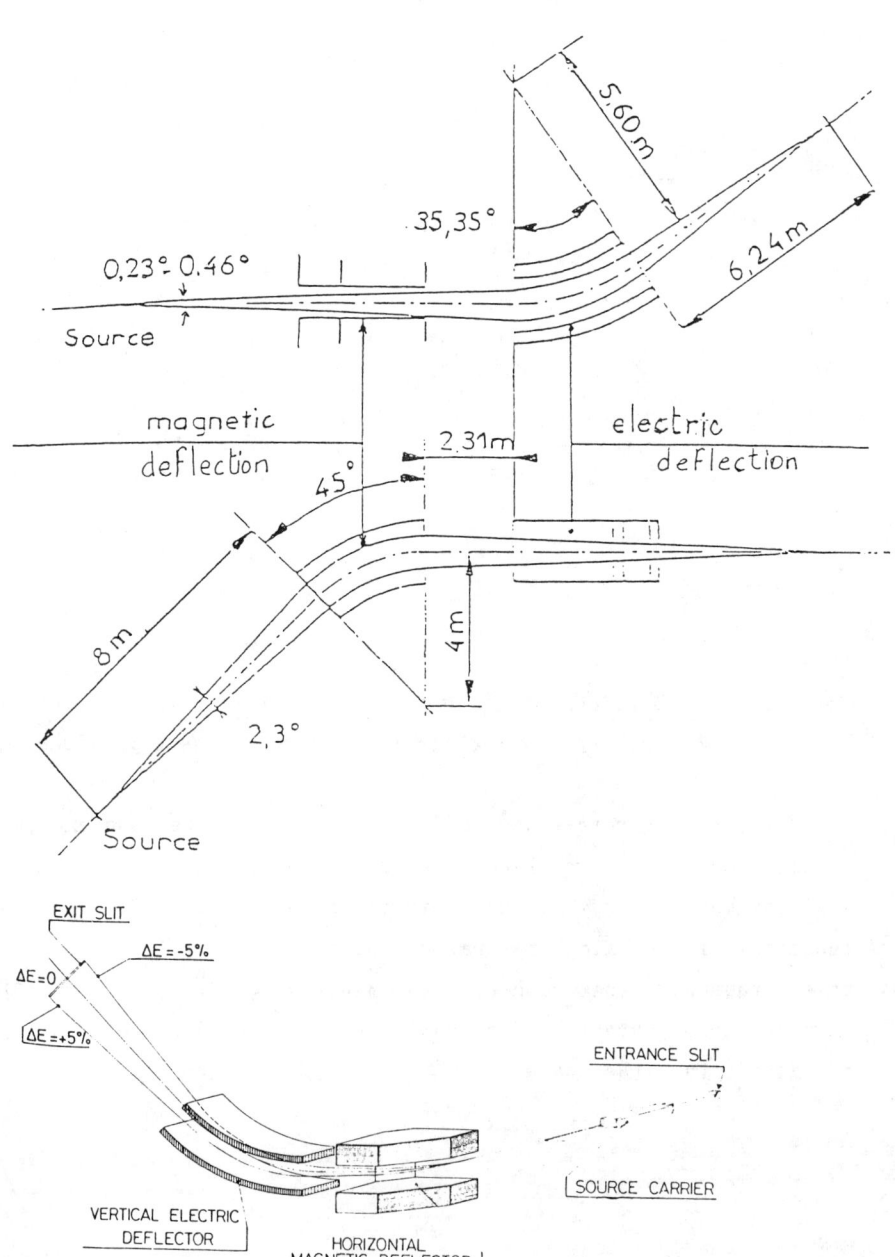

2. EXPERIMENTAL SET-UP

The "source" of fission fragments is made of 150 μgr/cm² ^{235}U covered by a 220 μgr/cm² Ni foil to avoid the burn-up of ^{235}U. The 5 10^{11} fission fragments produced per second, are analysed through the Lohengrin spectrometer[9] within a solid angle of the order of 10^{-5}sr. The spectrometer combines an electric and a magnetic deflection perpendicular one to the other (Fig. 3). The fission products are thus separated into spectra of A/q lines (A = mass number, q = ionic charge) with a mass dispersion, perpendicular to the parabola lines, of 3.24 cm for a 1 % mass relative difference. Because of the requirement of a high energy resolution in the detector, only a tenth of the focal length (70 cm) was selected, corresponding to 1 % in energy aperture (~ 1 MeV).

The selection in A/q is insured by setting a narrow entrance slit (3 mm) before the detector. The reported data were taken partly with this slit, and with no slit. The detector must identify ions of almost similar A/q and E/q values.

Ionization chambers were developped to measure E and thus provide a splitting of q and consequently A values. An ultimate version of the detector, a two section Frish grid ionization chamber (Fig. 4) has been recently optimized[9] in order to resolve, in addition, elements up to Z = 40 at the price of selecting only 7 cm of the focal line

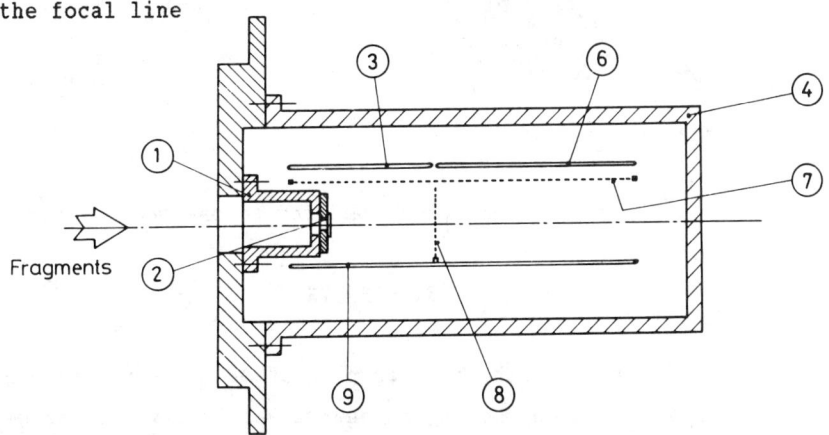

Fig. 4 : Scheme of the ΔE-E ionization chamber.

Such a performance requires special care since the fragments are slow (E ≲ 1 MeV/A) The chamber is filled with a circulating 50 mb pressure of isobutane. The entrance window is made of a thin (40 μg/cm^2) extended polypropylene film, sustained by a metallic grid. The ΔE and E electric volumes are isolated by a grounded grid like the cathode.

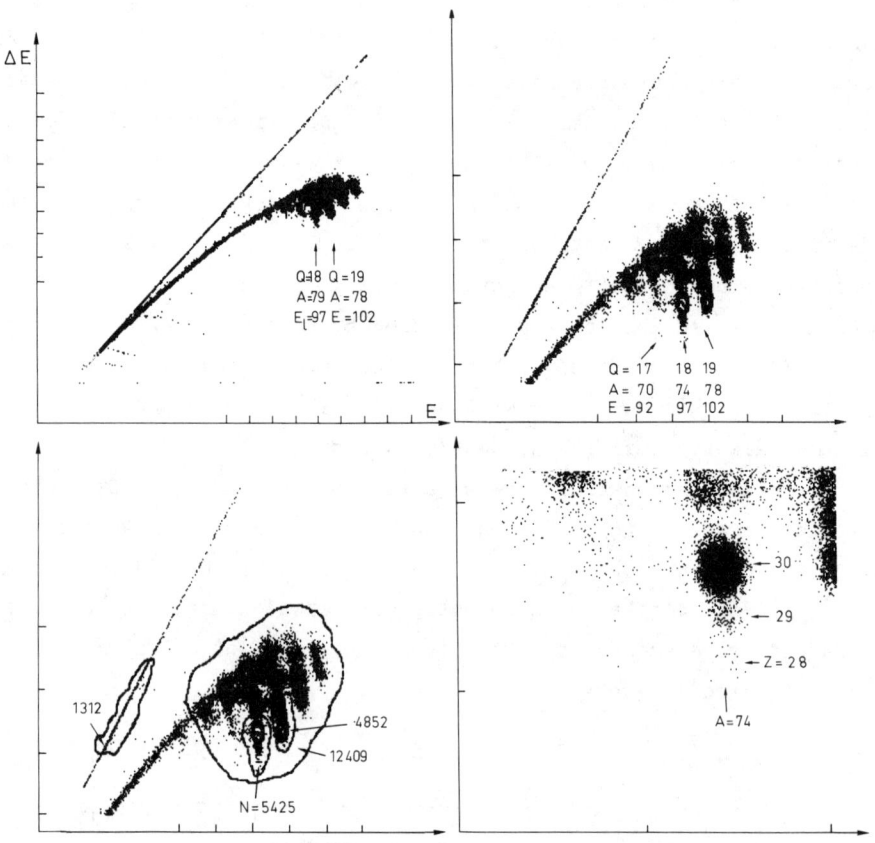

Fig. 5 : Scatter plot ΔE-E corresponding to one measurement (A=74).

3. RESULTS

Fig. 5 reports the scatter plot of ΔE-E events associated to A = 74. The choosen Lohengrin energy was 97 MeV and charge state q=18. The counting rates of relevant events were not too sensitive to these parameters. Within 3 units the value of q could be chosen so as to increase the distance with other fragments which could have similar or even exactly same A/q (for example A/q = 4 or 5 are excluded since many different ions would exactly overlap). The choice of the E value is guided by the fission kinematics (in this case it is slighly higher than the most probable E value). When the energy was decreased by 10 MeV the background yield due to very abundant fission products of larger Z and similar energy was enhanced, while the counting rate of "good" events remained stable.

From frame a to d, the E and ΔE scales are increased in order to show for one value of A the series of "clouds" associated to successive elements. On frame a and b the events, cut off because of the intermediate grid, are shown (10 % of relevant events). Further, the window grid intercepts ions which appear as energy degraded.

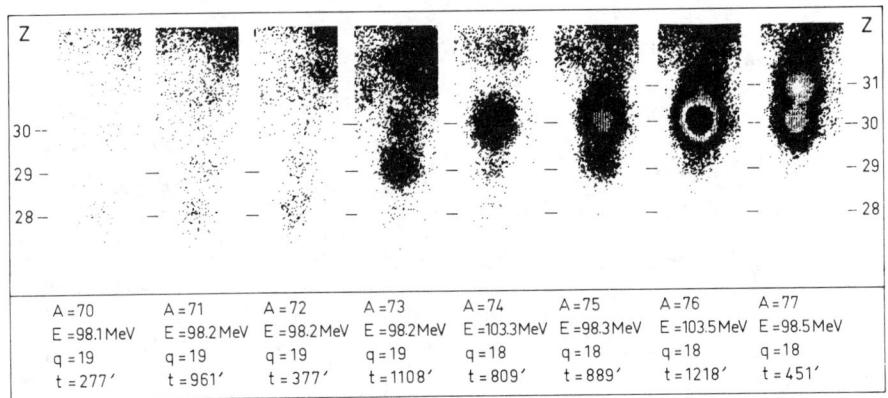

Fig. 6 : Series of mass scatter plots showing the new isotopes.

A series of successive measurements was performed changing the selected values of A and eventually E and q. The significant part of each scattered plot is shown on Fig. 6 where the accumulation times are indicated in minutes. The ΔE/Z calibration and the normalization were deduced from comparing our results on mass 81 and 82 with known elementary yields[10].

In addition to new isotopes, Fig. 6 provides a didactive view of known features of fission. The overall yield strongly decreases with A, while the Z distribution shifts towards smaller Z values. Events associated with different elements are well enough separated to show unambiguously the occurence of $^{70-74}$Ni and $^{74-77}$Cu at the border of other isotopes already known. At masses A = 76 and 77 there are no background events in the region of Ni isotopes. For all masses, not a single event is seen in the region of Co isotopes (Z = 27). The observed yields in the mass range 70-80 are displayed on Fig. 7 for the set values of ionic charge and chosen energy

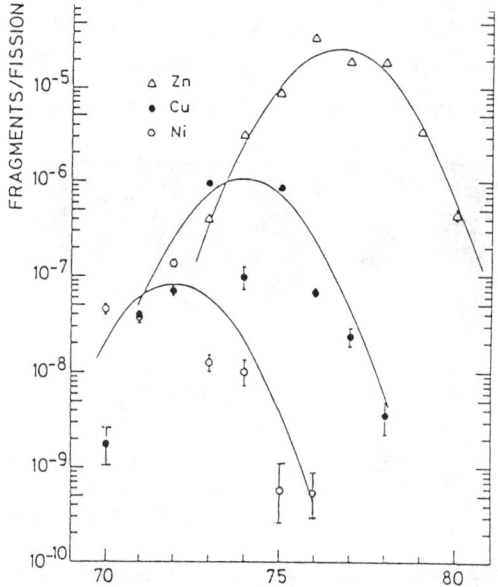

Fig. 7 : Fission rates for selected A/q and E values.

window. Although not integrated over kinetic energies and charge states, the rates measured for the present values of q and E are normalized to the already known independant yields of ^{79,80}Ga and ^{79,80}Ge. For ^{75,76}Ni and ^{78}Cu, yields are also reported, although those isotopes are not claimed to be discovered since only one event is observed in each case, which does not presently suffice. The smaller yield of ^{74}Cu may be due to an internal conversion process shifting the ionic charge distribution of this isotope toward higher value. Keeping in mind the above restrictions, the sums over masses at constant Z (elementary yields) are 2.5 10^{-7}, 2 10^{-6} and 8 10^{-5} for Ni, Cu, and Zn respectively. The fission yields of masses 70 and 71 can also be estimated for the first time : 9 10^{-6}% and 1.5 10^{-5}%. These values, as well as the other cumulative mass yields, agree within a factor of 2 with the extrapolated or known yields for thermal fission of ^{235}U (10).

To a certain extent the curves of production yields are related to those of available energy in the fission (Fig. 2). An indication for an odd-even effect is not excluded. Further measurements over different q and E values are to be performed. Other fissile sources will be tested in order to try to optimize production rates. But it is already remarkable that one fission product could be unambiguously isolated over 10^8 (ref. 11). Even if counting rates are low, the cleanliness of the experiment gives a chance for half life measurements.

4. COMPARISON WITH OTHER EXPERIMENTS SEEKING FOR NEUTRON RICH ISOTOPES

In the recent experimental studies devoted to neutron rich isotopes in the region of nickel, different methods have been utilised. The overall counting rates in term of number of atom/sec are reported on Fig. 8. The challenge of the production yields is only one parameter but still significant in view of prospective. Production reactions, coupled with analysing system can also be examined as a function of the spectroscopy achieved or of any other criteria.

If we consider how far the reaction kinematics is exploited, the two body transfer reaction has been very performant. From conservation of momentum and energy, mass excesses and excitation energies were deduced and the 0^+ spin was attributed to the ^{68}Ni first excited level with the (^{14}C,^{16}O) reaction on mass separated ^{70}Zn target (a) $^{(12)}$.

The deep inelastic fragmentation reaction, utilised in Ganil works on the kinematic properties of fragments since most of them occur in a small solid angle and a narrow energy gap. The efficiency of making light new isotopes is presently the best but it decreases rapidly with the excess of neutrons, because production probabilities reduce and the neutron evaporation probability increases. In Fig. 7 the isotopic production rates of Ni measured with a ^{86}Kr beam of 33 MeV/A and 1nA/p on a 50 mgr/cm^2 Au target are reported. The calculations of D. Guerreau

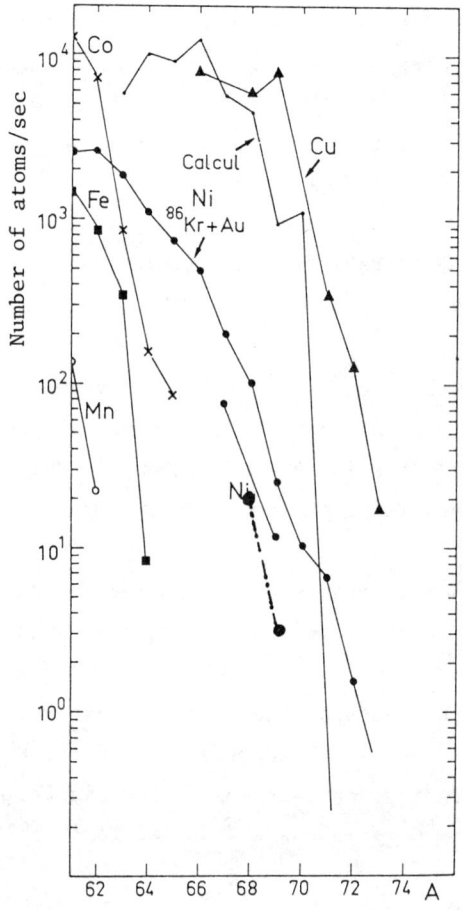

Fig. 8 : Counting rates obtained with different experimental set-up.
- *Full points stands for Ganil measurements.*
- *Overloaded circles for IPN results.*
- *The other data are from GSI on line mass separator.*

in which a fragmentation process is assumed with neutron evaporation, overestimate the slope of the isotopic distribution. It indicates that the reaction process is rather a mixture of processes. The more neutron rich isotopes of Co and Fe, ^{70}Co and ^{68}Fe were identified at L.I.S.E.$^{(13)}$.

The experiment reported in this paper, takes some benefit from the kinematics since the similar energy of the fragments helps for their separation (Fig. 5). Concerning the counting rates, this study cannot compete efficiently due mainly to the very low angular efficiency (10^{-5} sr) of Lohengrin. Nevertheless fission is the only case for which isotopes with 10 neutrons in excess are produced with the largest yields.

When the spectroscopic results are regarded, the on-line mass separator techniques, systematically applied in GSI, provide the most accurate results on daughter nuclei but wiping all records of production reactions, except the isotopic rates. In Fig. 8, counting rates for Mn, Fe, Co, Ni and Cu have been plotted, resulting from multi-transfer reactions from a ^{82}Se beam at 11 MeV/A. on a W-U target$^{(14)}$. The isotopic slopes are rather similar to the one measured at Ganil for Ni. However the extraction of Ni and Co ions from the mass separator ion sources is inhibited due to their physico-chemical properties.

5. OUTLOOK

The observation of new isotopes is a first step necessary to look at forthcoming spectroscopic studies. First, the β decay half lives should be measured but two major difficulties have to be overcome : a) the low counting rates ; for Ni they range from 0.02 event/min (^{74}Ni) to 0.3 event/min (^{72}Ni). They reach 2 events/min for ^{75}Cu. b) The long half-lives expected for the new isotopes ; for Ni, they range from 90 (or 8 sec) for ^{70}Ni to 3 (or 0.9 sec) for ^{73}Ni and for copper from 4 (or 0.6 sec) for ^{75}Cu to 0.5 (or 0.2 sec) for ^{78}Cu, according to predictions of Takahashy and Klapdor, respectively.

In order to test the feasibility of such a study, the required β detector efficiency, ϵ_β, and the acceptable background rate n_β have been calculated.

During a measurement time T, the number of delayed coincidences would be $N_t = n_1 \epsilon_\beta T/2$ (when a correlation time τ is choosen), while the number of random would be $N_f = n_1 n_\beta \tau T$.

To be significant, the measurement must be so that $N_t > 3\sqrt{N_f}$ which means $n_1 T > 36 \tau n_\beta/\epsilon_\beta^2$, with $n_1 T$ = number of events of the selected isotope as shown in fig.5. An optimum β counting efficiency and a low β background rate are required ; with the actual experimental conditions $n_\beta/\epsilon_\beta^2 < 1$ is needed.

This goal could be achieved by a multi-cell solid state detector where fragments could be implanted and β decay correlated.

A β background comes from decaying fission fragments stopped in the ionization chamber (0.3 f.f./sec on Fig. 5) which could be correlated randomly with the selected fragment. Other β particles are uncorrelated. Their measured rate (few counts/min/10 cm³) should be reduced by improving the shielding from the reactor.

CONCLUSION

The highly selective mass separator Lohengrin, combined with the newly designed ionization chamber made possible a direct identification of nine neutron rich isotopes of Ni and Cu. This work will be continued along two lines : first to perform a systematic exploration of the experimental conditions to optimize production rates and eventually to identify further new species. Second, given the low background, to settle β-decay or β-γ correlation techniques in order to measure half-lives (the values of which are estimated to lie between 10 ms and 10 s). For the ultimate goal of ^{78}Ni however, it is clear that better production reactions with much higher yields must be found while keeping separation efficient.

REFERENCES

1) M. Girod, Ph. Dessagne, M. Bernas, M. Langevin, F. Pougheon, P. Roussel. -To be published.
2) G.R. Burbridge, E.M. Burbridge, W.A. Fowler, F. Hoyle
    -Rev. Mod. Phys. 29 (1957) 547.
3) K.L. Kratz, H. Gabelman, W. Hillebrandt, B. Pfeiffer, K. Schlösser, F.K. Thielemann   -Z. Phys. A325 (1986) 489.
4) Ph. Dessagne, M. Bernas, M. Langevin, G.C. Morrison, G.J. Payet, F. Pougheon, P. Roussel
    -Nucl. Phys. A426 (1984) 399.
5) E. Lundt, K. Aleklett, B. Fogelberg, A. Sangariyavanish Proc. Int. Conf. AMCO7, p. 102,   -T.H. Darmstadt 1984.
6) R.L. Gill, R.F. Casten, D.D. Warner, A. Pietrowski, H. Mach, J.C. Hill, F.K. Wohn, J.A. Winger, R. Moreh
    -Phys. Rev. Lett. 56 (1986) 1874.
7) U. Quade   -PhD. Thesis, University of Munich - 1984 -
8) E. Moll, H. Schrader, G. Siegert et al. -NIM 123 (1975) 615.
9) J.P. Bocquet, R. Brissot and H.R. Faust
    -To be published in NIM.
10) A.C. Wahl-data tables   -To be published.
11) P. Armbruster, M. Bernas, J.P.Bocquet, R. Brissot,H.R.Faust and P. Roussel   -Europhysics Letters in press.
12) M. Bernas, Ph. Dessagne, M. Langevin, J. Payet, F. Pougheon and P. Roussel   -Phys. Lett. 113B (1983) 279.
13) D. Guillemaud-Mueller, A.C. Mueller, D. Guerreau, F. Pougheon et al.   -Z Phys. A322 (1985) 415.
14) E. Runte, K.L. Gippert, W.D. Schmidt-Ott et al.
    - Nucl. Phys. A441 (1985) 237.

IDENTIFICATION OF NEW NEUTRON-RICH RARE-EARTH ISOTOPES PRODUCED IN ^{252}Cf SPONTANEOUS FISSION

R. C. Greenwood, R. A. Anderl, J. D. Cole and M. A. Lee
Idaho National Engineering Laboratory
EG&G Idaho Inc., Idaho Falls, ID 83415, USA

H. Willmes
University of Idaho, Moscow, ID 83843, USA

ABSTRACT

New neutron-rich isotopes of several rare-earth elements have been identified using the ^{252}Cf-based ISOL facility at the INEL. New isotopes observed to date, with values of their half-lives in parentheses, are ^{153}Pr (4.28 ± 0.11 s), ^{155}Nd (8.9 ± 0.2 s), ^{156}Nd (5.47 ± 0.11 s), ^{157}Pm (10.50 ± 0.12 s), ^{158}Pm (4.8 ± 0.5 s), ^{162}Eu (10.6 ± 1.0 s) and ^{164}Gd (45 ± 3 s). Also, new half-life values have been measured for ^{151}Pr (18.90 ± 0.07 s), ^{152}Pr (3.7 ± 0.2 s), ^{153}Nd (28.9 ± 0.4 s), ^{154}Nd (25.9 ± 0.2 s), ^{153}Pm (315 ± 1 s), ^{155}Pm (41.5 ± 0.2 s), ^{156}Pm (26.70 ± 0.10 s), ^{157}Sm (483 ± 2 s), ^{158}Sm (318 ± 2 s), ^{159}Sm (11.37 ± 0.15 s), ^{160}Sm (9.6 ± 0.3 s) and ^{161}Eu (24 ± 4 s). The half-life values are compared to the results of theoretical predictions.

A unique feature of the ISOL facility at the INEL is its use of spontaneous fission of ^{252}Cf as the source of fission isotopes. As noted in Ref. 1, fission yields of nuclides with 105<A<125 and A>150 are significently enhanced in spontaneous fission of ^{252}Cf over those obtained from thermal-neutron fission of ^{235}U. Thus, the INEL ISOL facility is particularly well suited for nuclear-structure studies of neutron-rich nuclides within these mass ranges. In a continuation of our earlier effort to identify new neutron-rich isotopes produced in spontaneous fission of ^{252}Cf using fast radiochemical separation techniques (e.g. Refs. 1 - 3), we have recently reported on the identification of six new neutron-rich rare-earth isotopes using the INEL ISOL facility.[4] In recent work, we have extended that effort to include another new isotope, ^{164}Gd, refined the half-life values of ^{153}Pr and ^{157}Pm based on new measurements, and made other measurement to confirm, and improve the half-life values for, a number of other neutron-rich rare-earth nuclides. These results represent the initial ones of a program to systematically study nuclear properties (i.e., half-lives, excitation modes, beta-decay energies, and beta-strength functions) of nuclides with A>150 and 105<A<125.

A description of the INEL ISOL facility is given in Ref. 5. Briefly, the facility consists of two ∼200-µg electrodeposits of ^{252}Cf (Ref. 6) mounted in a pressurized chamber in a hot cell. After thermalization, and attachment to NaCl aerosols seeded in the He-gas flowing through the chamber, the fission fragments are transported with high efficiency by a gas-jet transport technique, through a skimmer chamber, into the ion source of the mass separator. Ion sources used in these experiments were of the electron-bombardment-heated hollow cathode design[5] and were constructed from either Ta or W. The principal difference between the operation of these ion sources is that the rare-earth oxide-to-elemental ion intensity ratios are significantly reduced using the Ta ion source. Thus, the Ta ion source was of particular use for observation of elemental mass fractions. For certain special cases, however, e.g., ^{153}Pr and ^{164}Gd, there were advantages in sensitivity to be gained from observing the monoxide mass fraction using the W ion source.

Following mass separation, the selected mass fraction was collected on a tape, which was moved after a preset collection time to place the collected mass fraction in front of the x-ray and γ-ray detectors. To obtain values for the half-lives corresponding to each of the K x-ray and γ-ray peaks observed with these detectors, three to eight time-multiscaled spectra were recorded within the collection cycle time. Typically, this collection time was chosen to be two to four times the isotopic half-life. Two detectors were used in the present experiments. The K x-rays and low-energy γ-rays were observed with a 200-mm² x 10-mm planar Ge detector whose energy resolution was adequate to allow the Kα x-ray peaks for adjacent rare-elements to be resolved. With this detector then, identification of the K x-rays, produced in the daughter nucleus due to internal conversion, combined with the mass selection provides a unique identification of the parent nucleus. Higher energy γ-rays were observed with a 50 cm³ Ge(Li) detector.

The half-life values obtained for both the new isotopes and for the other isotopes studied in the present work are summarized in Table I. Specific comments on the identification and on the half-life values have been given in Ref. 4 for many of these isotopes. Additional comments on selected isotopes are as follows:[7]

^{151}Pr. The earliest report of ^{151}Pr as having a 4 s half-life is, in retrospect, most probably a missassignment of ^{152}Pr. In the present work, we confirm the decay modes for this isotope reported in Ref. 8, with our half-life value being intermediate between their value and that reported in Ref. 15.

^{152}Pr. We confirm the decay modes reported in Refs. 10 and 11. However, we see no evidence for Pr K x rays in the multiscaled spectra which might be associated with the decay of ^{152}Ce having a comparable half-life to that of ^{152}Pr. Thus, the tentative report of the observation of ^{152}Ce in Ref. 10 is not supported.

^{164}Gd. The assignment of this activity to ^{164}Gd is based upon the observation of Tb K x-ray lines in spectra measured from the separated mass-164 and mass-180 (as ^{164}GdO) fractions.

In view of the importance of gross nuclear properties, such as half-lives, of neutron rich nuclei off the line of β⁻ stability to astrophysical calculations involving the rapid neutron-capture process (r-process), it is instructive to compare the measured half-life values in Table I to the predictions of existing models. At the present time, there exist two sets of half-life

TABLE I. Half-life values measured for the new neutron rich rare-earth isotopes and their comparison with the results of two theoretical predictions.

Isotope	Experimental		Theoretical		R = (observed/predicted)	
	Present work (sec)	Other work (sec)	Takahashi et al. (Ref.18)	Klapdor et al. (Ref.19)	Takahashi et al. (Ref.18)	Klapdor et al. (Ref.19)
^{151}Pr	18.90±0.07	4.0±0.7 (Ref.7) 22.4±1.5 (Ref.8) 14 (Ref.9)	34		0.56	
^{152}Pr	3.7±0.2	3.24±0.19(Ref.10) 3.8±0.2 (Ref.11)	13	3.93	0.28	0.94
^{153}Pr	4.28±0.11[a]		13	2.81	0.33	1.52
^{153}Nd	28.9±0.4	32±4 (Ref.8)	58		0.50	
^{154}Nd	25.9±0.2	26±2 (Ref.12)	77		0.34	
^{155}Nd	8.9±0.2		18	15.3	0.49	0.58
^{156}Nd	5.47±0.11		21	8.36	0.26	0.65
^{153}Pm	315±1	330±12 (Ref.13) 318±18 (Ref.14)	229		1.38	
^{155}Pm	41.5±0.2	48± 4 (Ref.1)	74	21.2	0.56	1.96
^{156}Pm	26.70±0.10	28.2±1.1 (Ref.15)	25	8.39	1.07	3.18
^{157}Pm	10.50±0.12[a]		25	4.68	0.42	2.24
^{158}Pm	4.8±0.5		10	2.31	0.48	2.08
^{157}Sm	483±2	480±30 (Ref.16) 480±60 (Ref.17) 402±24 (Ref.15)	110		5.3	
^{158}Sm	318±2	331±5 (Ref. 2) 312±12 (Ref.15)	390		0.82	
^{159}Sm	11.37±0.15	15±2 (Ref. 15)	37	10.1	0.31	1.13
^{160}Sm	9.6±0.3	8.7±1.4 (Ref.15)	74	6.06	0.13	1.58
^{161}Eu	24±4	27±3 (Ref.15)	64	24.8	0.38	0.97
^{162}Eu	10.6±1.0		20	9.35	0.53	1.13
^{163}Gd		68±3 (Ref.3)	110	66.3	0.62	1.03
^{164}Gd	45±3		229	31.8	0.20	1.42

[a] These new half-life values are based upon more recent measurements and consequently supersede the values given in Ref. 4.

predictions, one by Takahashi et al.[18] based on the gross theory of beta decay and another by Klapdor et al.[19] based on a microscopic model employing a description of the distribution of beta strength. Such a comparison for each of the measured isotopes is shown in Table I. The results of this comparison follow the general trend noted by Kratz [20,21] and confirmed in our earlier study[4], namely that the model of Takahashi et al.[18] for nuclei far from stability systematically overestimates and the model of Klapdor et al.[19] generally underestimates the measured half-lives. Specifically though, for the neutron-rich isotopes of Pr through Gd for which we have obtained half-life values we conclude, (1) that as one goes further away from the line of beta stability, the ratios R computed using the model of Takahashi et al.[18] tend towards values which are <0.4, and (2) that the values of R computed using the model of Klapdor et al.[19] are systematically too low for the Nd and too high for the Pm isotopes, respectively. Furthermore, the sawtooth pattern in R values proposed in Ref. 15 using the predictions of Klapdor et al.,[19] with turning points at Z = 59 and 65, is not supported by the present data. Rather, we note that it is the abnormally large values of R for the Pm data alone which might lead one to suggest such a sawtooth pattern.

We are grateful to M. A. Oates for operation of the INEL ISOL facility. This research was supported by the U. S. Department of Energy under Contract NO. DE-AC07-76ID01570 with EG&G Idaho Inc.

REFERENCES

1. R. C. Greenwood et al., Radiochimica Acta 30, 57 (1982).
2. J. D. Baker et al., J. Inorg. Nucl. Chem. 42, 1547 (1980).
3. R. J. Gehrke et al., Radiochimica Acta 31, 1 (1982).
4. R. C. Greenwood et al., Phys. Rev. C 35, 1965 (1987).
5. R. A. Anderl et al., Nucl. Instrum. Methods B 26, 333 (1987).
6. ^{252}Cf made available through the National Transplutonium Element Production Program, ORNL, Oak Ridge, TN 37830.
7. J. B. Wilhelmy et al., UCRL-19530, 178 (1970).
8. J. A. Pinston et al., Atomic Masses and Fundamental Constants 6, ed. J. A. Nolan, W. Benenson (Plenum, New York, 1980), p. 493.
9. H. Mach et al., Bull. Am. Phys. Soc. 32, 1018 (1987).
10. J. C. Hill et al., Phys. Rev. C 27, 2857 (1983).
11. M. Brügger et al., Nucl. Instrum. Methods A 234, 218 (1985).
12. T. Karlewski et al., Z. Phys. A 322, 177 (1985).
13. K. Kotajima, Nucl. Phys. 39, 89 (1962).
14. R. K. Smither et al., Phys. Rev. 187, 1632 (1969).
15. H. Mach et al., Phys. Rev. Lett. 56, 1547 (1969).
16. N. Kaffrell Phys. Rev. C 8, 414 (1973).
17. J. M. D'Auria et al., Can. J. Phys. 51, 686 (1973).
18. K. Takahashi et al., At. Data Nucl. Data Tables 12, 101 (1973).
19. H. V. Klapdor et al., At. Data Nucl. Data Tables 31, 81 (1984).
20. K. -L. Kratz, Z. Phys. A 312, 236 (1983).
21. K. -L. Kratz, Nucl. Phys. A 417, 447 (1984).

PRODUCTION OF THE HEAVIEST ELEMENTS ON THE BORDERLINE OF NUCLEAR STABILITY BY COLD FUSION REACTIONS

F.P. Heßberger, G. Münzenberg, S. Hofmann, H. Folger, K. Poppensieker
K.-H. Schmidt, W. Reisdorf, P. Armbruster, M.E. Leino[1]

Gesellschaft für Schwerionenforschung mbH, Darmstadt, FRG
[1] University of Helsinki, Finland

ABSTRACT

In cold fusion reactions leading to compound nuclei with $Z > 103$, isotopes of the new elements 107, 108, 109 were identified unambigouosly for the first time. In these experiments we observed α decay of 14 isotopes with $Z \geq 103$, which were either completely unknown or of which only an sf branch had been reported before. Measured α energies were used to determine ground state masses and to derive experimental shell correction energies and fission barriers. The existence of a peninsula of purely shell stabilized deformed nuclei could be proved experimentally. Measured excitation functions showed a hindrance to complete fusion at energies close to the classical barrier. We also report on our experiments to synthesize element 110.

INTRODUCTION

Since the discovery of Neptunium ($Z = 93$) by E. McMillan and P.E. Abelson [1] in 1940, 17 new transuranium elements have been identified until today. While the elements $Z = 93$ to 106 were synthesized in nuclear reactions using 'light' projectiles (nucleons or nuclei with $Z < 10$ and $A < 20$) and actinide targets, the production of heavier elements by this method failed.

A new, promising method to produce transfermium elements was introduced by Oganessian [2]. Cold fusion of targets around the doubly magic ^{208}Pb with neutron rich, medium heavy projectiles ($Z > 18$, $A > 40$) profits from favourable Q-values. So, compound nuclei (CN) can be produced with excitation energies of $E^* \simeq 20$ MeV. Drawbacks of the Dubna experiments [2] were the lack of selectivity to products from complete fusion and the sensitivity to spontaneous fission (sf) alone. Since an identification of an isotope by observing only sf is difficult, a new, powerful method to identify evaporation residues (ER) from these kind of reactions unambiguously was demanded.

© American Institute of Physics 1988

The velocity filter SHIP [3] at GSI, Darmstadt, has turned out to meet these requirements. A fast and efficient separation of the ER from the projectile beam and products from other reactions is achieved. The implantation of the separated nuclei into silicon detectors allows to identify α and sf activities by genetic relationship with production cross sections down to 10 pb and halflives from some μs to several tenth of seconds [4].

Subject of this contribution will be a report on the experiments to produce isotopes with Z $\geq$ 103 performed at SHIP in the period 1980 - 1987, and a discussion about the consequences of our results concerning the stability of the heaviest elements and the possibility to produce them in cold fusion reactions. For better orientation, an overview of the produced isotopes and the used reactions is presented in Fig. 1.

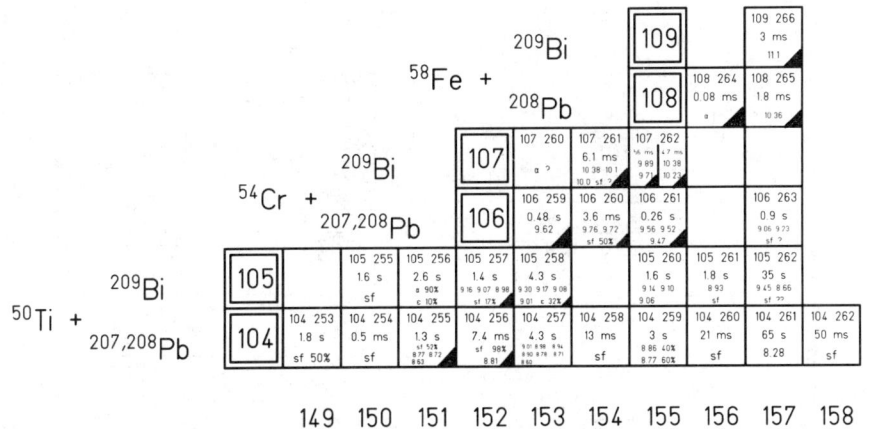

Figure 1: Excerpt from the chart of nuclei for Z $\geq$ 104. The new isotopes are marked by black triangles.

IDENTIFICATION OF THE NEW ELEMENTS 107, 109 AND 108

Already in 1976, the observation of an (1-2) ms sf activity, which was produced in reactions of ^{209}Bi with ^{54}Cr and assigned to ^{261}107, was reported by Oganessian et al. [5]. Due to the difficulties connected with the assignment of unknown sf activities, it seemed necessary to identify an isotope of element 107 unambigouously by correlated α decay sequences. For this purpose we bombarded ^{209}Bi with ^{54}Cr at beam energies of 4.85, 4.95 MeV/u, to produce the odd-odd isotope ^{262}107, which was assumed to be an α emitter, in the reaction ^{209}Bi(^{54}Cr,1n)^{262}107. In an experiment of four days we observed six α decay sequences, which could be assigned to ^{262}107. More detailed information on the decay properties of this isotope could be obtained in following experiments, in which also a second α emitter was found that could be assigned to ^{261}107.

The positive results in the synthesis of element 107 encouraged us to produce heavier elements. From considerations concerning the nuclear stability of the heaviest elements it seemed promising to synthesize an odd-odd isotope of element 109, since we knew from element 107, that for odd-odd isotopes α decay could compete with sf, while doubly even

Table 1: Decay properties of the new α-emitters. The error for the decay energies is ± 30 keV for Z ≥ 106 and ± 20 keV for Z < 106 if not indicated otherwise, the numbers in brackets give the relative intensities.
*Probably α-decay from groundstate and isomeric state observed.

Isotope	E_α/MeV	$T_{1/2}$	Decay
^{266}109	11.0 ± 0.04	3.5 ms	α
^{265}108	10.36	$1.8^{+2.2}_{-0.7}$ ms	α
^{264}108		76^{+364}_{-36} μs	α
^{262}107*	10.35(∼ 0.7), 10.23(∼ 0.3)	$4.7^{+2.3}_{-1.6}$ ms	α
	9.89(∼ 0.5), 9.71(∼ 0.5)	56^{+53}_{-18} ms	α
^{261}107	10.38(∼ 0.3), 10.10(∼ 0.5)	$6.1^{+4.3}_{-1.8}$ ms	α
	10.00(∼ 0.2)		
^{261}106	9.56(0.6), 9.52(0.3)	260^{+110}_{-60} ms	α
	9.47(0.1)		
^{260}106	9.77(0.9), 9.72(0.1)	$3.6^{+0.9}_{-0.6}$ ms	α(0.5), sf(0.5)
^{259}106	9.62	480^{+280}_{-130} ms	α
^{258}105	9.299(0.08), 9.172(0.59)	$4.4^{+0.9}_{-0.6}$ s	α(0.67), EC(0.33)
^{257}105	9.160(0.3), 9.071(0.3)	$1.4^{+0.6}_{-0.3}$ s	α(≥ 0.83)
	8.970(0.4)		sf(≤ 0.17)
^{256}104	8.812	$7.4^{+0.9}_{-0.7}$ ms	α(∼ 0.02)
			sf(∼ 0.98)
^{255}104	8.766(0.2), 8.715(0.7)	$1.4^{+0.2}_{-0.2}$ s	α(0.48)
	8.625(0.1)		sf(0.52)
^{254}Lr	8.460(0.69), 8.408(0.36)	13^{+3}_{-2} s	α(0.78)
			EC(0.22)
^{253}Lr	8.800(0.56), 8.722(0.44)	$1.3^{+0.6}_{-0.3}$ s	α

(ee) isotopes were expected to decay predominantly by sf [6]. In a 13 days experiment ^{209}Bi was irradiated with ^{58}Fe at beam energies of 4.95, 5.05, and 5.15 MeV/u. A total projectile dose of 0.7 × 10^{18} particles was collected. We observed one decay sequence that could be attributed to a heavy nucleus, corresponding to a production cross section of 15^{+35}_{-12} pb. This decay chain was assigned to ^{266}109 [7], produced in the reaction ^{209}Bi(^{58}Fe,n)^{266}109. From a statistical analysis, the probability that this sequence belongs to an isotope of another element was obtained to be ≤ 0.0038 [7].

Despite of the expectations of very short sf halflives [6], the investigation of element 108 isotopes remained a challenging task. An investigation of neutron deficient element 106 isotopes turned out to be necessary before attempting to produce element 108. In bombardments of ^{207}Pb, ^{208}Pb with ^{54}Cr we observed α decay of ^{261}106, ^{260}106 and ^{259}106. The observation of a 50 % α-decay branch of ^{260}106 was a surprise, since it proved that the sf halflife did not change from ^{258}104 to the neighbouring ee-isotope ^{260}106 in contrast to the expectations from systematics and from predictions [6].

In a subsequent experiment we aimed to synthesize element 108. In an experiment of 14 days we bombarded ^{208}Pb targets with a beam of ^{58}Fe at an energy of 5.04 MeV/u. A projectile dose of 0.6×10^{18} particles was collected. We observed three α-decay chains that could be attributed to ^{265}108, produced in the reaction ^{208}Pb(^{58}Fe,n)^{265}108. The formation cross section corresponded to 19^{+18}_{-11} pb.

SPECTROSCOPIC RESULTS

The decay properties of the new α emitters observed in our experiments are listed in table 1. The trend of the α energies fit quite well into the systematics of heavy isotopes already known from literature (Fig. 2). The local minimum around N = 152, which reflects a gap in the neutron levels is observed up to element 106.

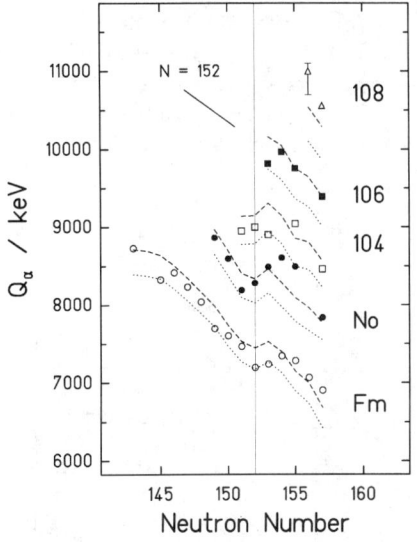

Figure 2: Q_α-systematics for even Z isotopes with $Z \geq 100$.

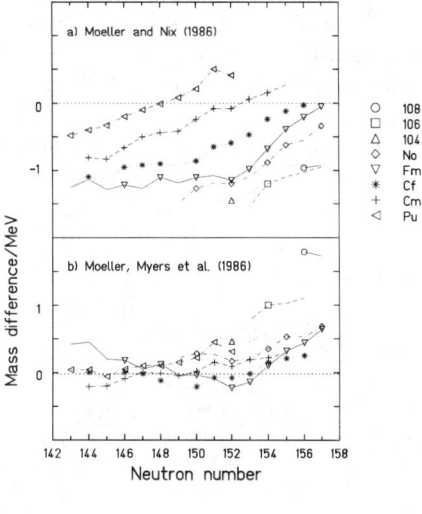

Figure 3: Mass differences between experimental data and predictions of a) Ref. [8], and b) Ref. [9] (figure taken from [10]).

In Fig. 2 we compare the experimental Q_α-values with the results derived from mass predictions published recently by Møller and Nix [8] and by Møller, Myers et al. [9]. The main difference between these two publications is a different treatment of the macroscopic part of the mass evaluation. While in Ref. 8 a Yukawa plus exponential potential is taken to calculate the macroscopic masses, in Ref. 9 the finite range droplet model is used. The shell correction energies are the same in both papers.

The experimental data are, in general, better reproduced by the predictions of Ref. 8 than by those of Ref. 9. Especially an increasing underestimation of the data for $Z >$ 104 is noticed for the predictions of Ref. 9. These deviations are obviously due to the different treatment of the macroscopic masses, a result which is of special importance for the discussions of evaluations of experimental shell effects and fission barriers in the next sections.

ALPHA DECAY OF EVEN-EVEN ISOTOPES WITH Z > 102

The study of α decay of ee-nuclei is of special interest, since ground state transitions are assumed to be the dominating α-decay mode. So mass excesses can be derived from the Q_α-values, if the mass excesses of the daughter products are known. This is the case for the isotopes characterized by N-Z = 48. Mass excesses for ^{256}104 and ^{260}106 could be obtained from the measured α energies.

For ^{264}108, the heaviest ee-nucleus known at present, the situation is more complicated: In an experiment of 14 days, where a total projectile dose of 1.7×10^{18} particles was collected, we produced this isotope in the reaction ^{207}Pb(^{58}Fe,n)^{264}108. Only one decay sequence that could be assigned to ^{264}108 - from probability arguments based on known or expected decay properties of the heaviest isotopes - was observed. The formation cross section was $3.2^{+6.1}_{-2.1}$ pb. The decay α particles were not registered with their full energy. So we used the measured time difference of 110 μs to determine the Q_α-value. From the formula of Rasmussen [11] and $w_\alpha = 1$ (which is valid for ^{260}106), we obtained $Q_\alpha = (11.0^{+0.1}_{-0.3})$ MeV.

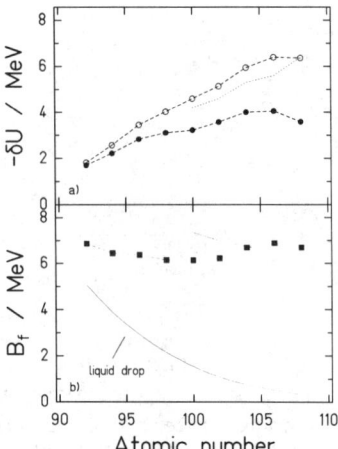

Figure 4: a) Comparison between 'experimental' and theoretical shell effects;
b) Comparison between 'experimental' and theoretical fission barriers.

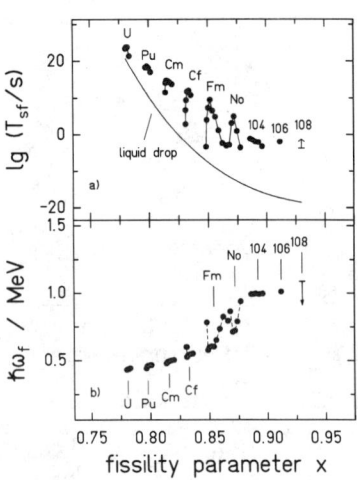

Figure 5: a) sf halflives of even-even nuclei
b) barrier curvature energies $\hbar\omega_f$ for even-even nuclei.

In Figs. 3a,b we compare the mass predictions [8,9] with the experimental data, which are in general better reproduced by Ref. 9 than by Ref. 8. For the heaviest isotopes, however, significant deviations occur. For this reason (and also due to the deviations in the Q_α-values) we prefer in the following the predictions of Ref. 8 for the heaviest isotopes. We determined experimental shell correction energies from the difference between our experimental groundstate masses and 'macroscopic' masses from Ref. 8 (open circles) and Ref. 9 (dots), Fig. 4a). The theoretical shell correction energies are represented by the full line. The dotted line refers to calculations of Cwiok et al. [12] which are typically $\simeq 0.8$ MeV higher than the values of Refs. 8, 9. The general trend of increasing shell effects from U to element 108, predicted by the theory, is qualitatively reproduced by the experimental data. The agreement of the experimental shell effects obtained with the macroscopic masses of Ref. 9 is better than 0.5 MeV for $Z \leq 104$, a serious disagreement is obtained for $Z = 106$ and $Z = 108$, while the shell effects obtained with those of Ref. 8 overestimate the theoretical values by $\simeq 1.5$ MeV in general.

The fission barriers derived from our experimental shell effects using the masses [8] are shown in Fig. 4b. They are expressed as the sum of the liquid drop fission barriers calculated according to [13] and the shell correction of the groundstate and saddlepoint masses. The latter were expected to be a small contribution and thus neglected. A fair agreement with theoretical predictions of Sobiczewski [14] (full line) is obtained.

Moreover one can see that the fission barriers remain high due to large ground state shell corrections, despite of the drastic drop of the liquid drop fission barriers, which are expected to be lower than the zero point motion of 0.5 MeV for $Z > 106$. Thus isotopes $Z > 106$ might be regarded as purely shell stabilized nuclei ('deformed superheavy nuclei' [15]). The consequences concerning the nuclear stability can be seen from fig.5a, the stabilisation against sf is more than 14 orders of magnitude for the heaviest isotopes. The analysis of the fission barrier in terms of the barrier curvature energy $\hbar\omega_f$ exhibits the increasing importance of the shell stabilisation towards the transactinide region. While $\hbar\omega_f$ is rather small ($\simeq 0.4$ MeV) for the light actinides, which indicates a broad ('liquid drop-like') barrier, the high values of $\hbar\omega_f \simeq 1$ MeV for the heaviest nuclei indicate narrow ('shell effect') barriers (Fig. 5b).

CROSS SECTIONS AND HINDRANCE TO COMPLETE FUSION AT ENERGIES CLOSE TO THE BARRIER

Our cross sections measured for production of nuclei with $Z \geq 100$ are summarized in Fig. 6. The figure shows, that the maxima occur always close to the Bass model barrier [16]. Although the excitation energies are around $E^* \simeq 20$ MeV, the drop of the production rates with increasing Z is drastic (see also Fig. 8a), despite of nearly constant fission barriers. Although the ground state shell stabilisation will be washed out with increasing excitation energy, an increasing instability of the CN against prompt fission is — in our opinion — not the only reason for the drop of the cross sections, as all nuclei are produced with nearly the same excitation energy of $\simeq 20$ MeV. A serious hindrance to complete fusion at energies close to the Bass model fusion barrier has to be assumed.

This situation is demonstrated in Fig. 7. Here we compare the excitation function for ^{50}Ti + ^{208}Pb with calculations using the evaporation code HIVAP [17], which was adjusted to reproduce the 4n cross section for the light system ^{12}C + ^{249}Cf [18], leading to a closely neighboured CN. The Ti + Pb system is hindered by a factor of $\simeq 30$ compared to C + Cf, which can be explained by a shift of the fusion barrier ('extra push') of $\Delta B = 21$ MeV, and a barrier fluctuation of 11 MeV. The hindrance to complete fusion is expected to increase with increasing projectile mass, for ^{266}109 it is about 10^4.

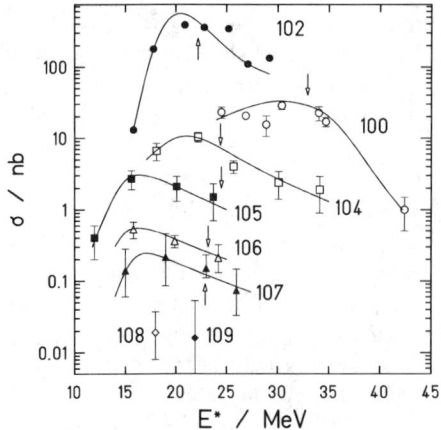

Figure 6: Production cross-sections for elements $Z \geq 100$, measured at SHIP. The Bass-model barrier is marked by an arrow.

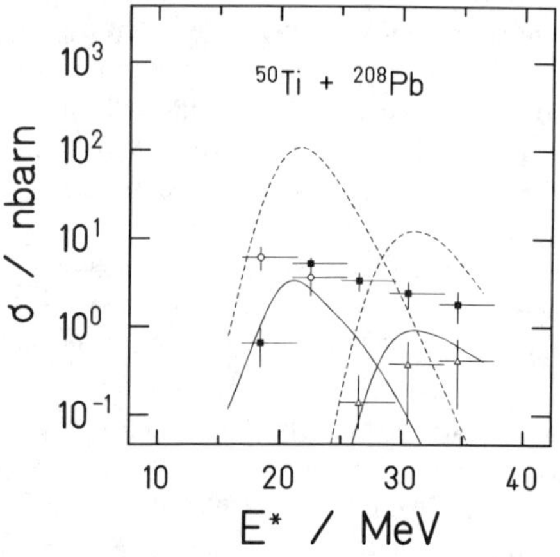

Figure 7: Comparison of ER production cross-sections for ^{50}Ti + ^{208}Pb with HIVAP calculations.
dashed line : $\Delta B = 0$
full line : $\Delta B = 21$ MeV, $\sigma B = 6$ %.

ATTEMPTS TO SYNTHESIZE ELEMENT 110 AT SHIP

Up to now we have carried out two experiments at SHIP to produce an isotope of element 110: The reactions ^{208}Pb + ^{64}Ni and ^{235}U + ^{40}Ar.

The first reaction was a straight forward continuation of the cold fusion method, which was used successfully to synthesize elements up to 109. The cross section was expected to be (1-10) pb from simple extrapolations (Fig. 8a). The choice of the second reaction was more delicious: Due to a strong increase of the hindrance to complete fusion for the 'heavy' cold fusion systems it is tempting to investigate reactions where liquid drop dynamics predict a lower hindrance, i.e. more asymmetric systems. In a recent model Blocki et al. [19] scaled the 'extra push' by the parameter $x_a = 1/3 \cdot x^{UCD} + 2/3 \cdot x_{CN}$. x^{UCD} and x_{CN} denote the fissility parameters in the entrance channel and of the CN, respectively. 'Extra push' is expected for $x_a > 0.723$. Concerning both, hindrance and Q-value, the most asymmetric systems ^{23}Na + ^{254}Es and ^{22}Ne + ^{257}Fm would be the best choices, but are not technically feasible presently, while for ^{40}Ar + ^{235}U an 'extra push' of 132 MeV is predicted [19]. Some experimental information about the hindrance of actinide based systems could be obtained from a comparison of different reactions leading to the compound nucleus ^{264}104. The system ^{26}Mg + ^{238}U ($x_a = 0.745$) was found to be hindered by a factor of 10 with respect to the undercritical system ^{16}O + ^{248}Cm ($x_a = 0.709$), which is in agreement with the expectations from the Blocki model. So, presently, no evidence from a deviation from the x_a scaling is observed for actinide based systems.

Extrapolations of the general trends of the production cross section lead into the pbarn region for the best asymmetric reactions and into the femtobarn region for the U

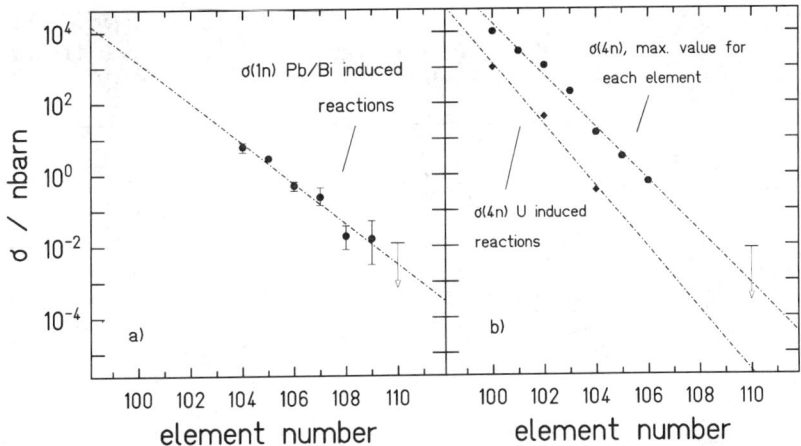

Figure 8: a) Trend of 1n-cross sections for cold fusion reactions; b) Trend of 4n-cross sections actinide based reactions.

Table 2: Upper limits for production of element 110

Projectile	Target	Irr. Time/h	Beam Dose	E^*/MeV	σ/pb
^{64}Ni	^{208}Pb	250	1×10^{18}	15 – 21	< 12
^{40}Ar	^{235}U	120	1.7×10^{18}	33 – 37	< 8

induced reactions (Fig. 8b). Therefore the synthesis of element 110 in the reaction Ar + U is a gamble. One only can hope for a reduced 'extra push' value accompanied by large barrier fluctuations. Large fluctuations have been observed for symmetric systems [15].

Both experiments carried out at SHIP had negative results: No α decay or sf, that could be attributed to an isotope of element 110 was observed in any experiment. Our upper limits a summarized in table 2 and are shown, together with the systematic trends in Fig. 8a, 8b. It is seen that our experiments might not have been sensitive enough, since our upper limits are lying just at the upper limits of the extrapolations, but they represent the lower limit of sensitivity available with the presently used technique. So, in our opinion, technical effort to improve the experimental sensitivity by at least one, or better, two orders of magnitude has to be done before making a new, promising attempt to synthesize element 110.

CONCLUSIONS

In the course of our experiments we have identified the three heaviest elements 107, 108, 109 by observing α decay of at least one isotope of each element unambigouosly for the first time. Altogether 14 new α emitters were found. The measured α energies allowed to deduce ground-state masses, shell correction energies and fission barriers. Theoretical trends were fully established, predicted values reproduced within ± 1.5 MeV.

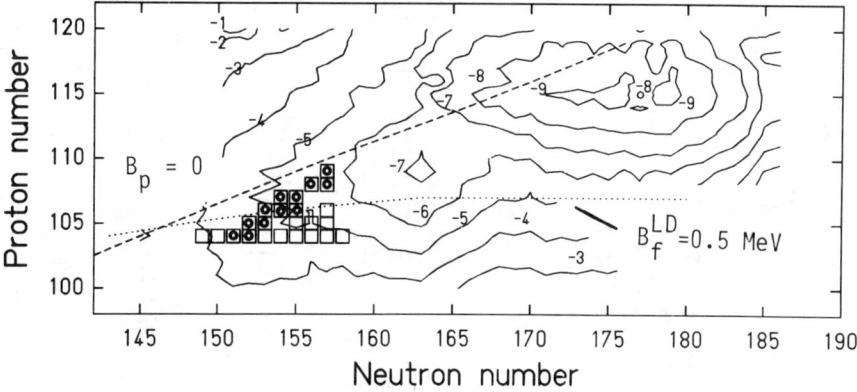

Figure 9: Landscape of the 'superheavy region'; the heaviest nuclei observed at SHIP are marked by dots. Data taken from Ref. 8.

The fission barriers turned out to be single humped and narrow for the heaviest isotopes, while the liquid drop fission barriers are negligible. So far, we have reached the peninsula of 'deformed' super heavy elements, which is located 'southwest' of the island of the 'spherical' super heavy elements (Fig. 9). The possibilities to reach the latter island are not very promising, due to a strong hindrance to complete fusion close to the barrier.

Perhaps cold fusion reactions are the best way to make elements with Z > 106, since actinide based systems may suffer not only from the hindrance but also from high excitation energies of the CN. So, investigations of the reaction mechanisms, to get a better understanding of the nature of the hindrance, is the most import task in the near future.

References

[1] E. McMillan, P.H. Abelson, Phys. Rev. $\underline{57}$ (12), 1185 (1940)

[2] Yu.Ts. Oganessian, Classical and Quantum Mechanical Aspects of Heavy Ion Collisions. In: Lecture Notes in Physics. H.L.Harney, P.Braun-Munzinger, C.K.Gelbke (eds), Vol.33 (Springer-Verlag, Berlin, New York, 1974), p.221

[3] G. Münzenberg, W. Faust, S. Hofmann, P. Armbruster, K. Güttner, H. Ewald, Nucl. Instrum. Methods $\underline{161}$,65 (1979)

[4] S. Hofmann, G. Münzenberg, F.P. Heßberger, H.J. Schött, Nucl. Instrum. Methods $\underline{223}$, 312 (1984);
G. Münzenberg, S. Hofmann, F.P. Heßberger, W. Reisdorf, K.-H. Schmidt, W.

Faust, P. Armbruster, K. Güttner, B. Thuma, D. Vermeulen, C.-C.Sahm, Proc. 4th International Conference on Nuclei far off Stability, Helsingør 1981, CERN-Report 81-09, 1981

[5] Yu.Ts. Oganessian, A.G. Demin, N.A. Danilov, G.N. Flerov, M.P. Ivanov, A.S. Iljinov, N.N. Kolesnikov, B.N. Markov, V.M. Plotko, S.P.Tretyakova, Nucl. Phys. A273, 505 (1976)

[6] J. Randrup, S.E. Larsson, P. Møller, S.G. Nilsson, K. Pomorski, A. Sobiczewski, Phys. Rev. C13, 1, 229 (1976)

[7] G. Münzenberg, W. Reisdorf, S. Hofmann, Y.K. Agarwal, F.P. Heßberger, K. Poppensieker, J.R.H. Schneider, W.F.W. Schneider, K.-H. Schmidt, H.-J. Schött, P. Armbruster, C.-C. Sahm, D. Vermeulen, Z. Phys. A315, 145 (1984)

[8] P.Møller, J.R. Nix, Preprint LA-UR-86-3266, Los Alamos National Laboratory, 1986 (subm. to Atomic and Nuclear Data Tables)

[9] P.Møller, W.D. Myers, W.J. Swiatecki, J. Treiner, Preprint LBL-22686, 1986

[10] G. Münzenberg, P. Armbruster, G. Berthes, H. Folger, F.P. Heßberger, S. Hofmann, J. Keller, K. Poppensieker, A.B. Quint, W. Reisdorf, K.-H. Schmidt, H.-J. Schött, K. Sümmerer, I. Zychor, M.E. Leino, R. Hingmann, U. Gollerthan, E. Hanelt, Z. Phys. A, in press, 1987

[11] J.O. Rasmussen, Phys. Rev. 113, 6, 1593 (1959)

[12] S. Cwiok, V.V. Pashkevich, J. Dudek, W. Nazarewicz, Nucl. Phys. A410, 154 (1983)

[13] M. Dahlinger, D. Vermeulen, K.-H. Schmidt, Nucl.Phys. A376, 94 (1982)

[14] A. Sobiczewski, Z. Patyk, S. Cwiok, Phys. Lett. B186 (1), 6 (1987); A. Sobiczewski, priv. cummunication, 1987

[15] P. Armbruster, Lectures given at the International School on Heavy Ion Physics, Erice, Italy, Oct. 12 - 22, 1986, GSI Preprint GSI-86-56

[16] R. Bass, Proc. Conf. Deep Inelastic and Fusion Reactions with Heavy Ions, Berlin 1979, Springer Verlag, Berlin, New York, Heidelberg, 1980

[17] W. Reisdorf, Z.Phys. A300, 227 (1981)

[18] P. Eskola, K. Eskola, M. Nurmia, A. Ghiorso, Phys. Rev. C2 (3), 1058 (1970)

[19] J. Blocki, H. Feldmeier, W.J. Swiatecki, Nucl.Phys. A459, 145 (1986)

STUDY OF SECONDARY REACTION EXPERIMENTS IN THE SUPERHEAVY AND ACTINIDE REGIONS

A. Marinov and S. Eshhar
Racah Institute of Physics, The Hebrew University, Jerusalem, Israel.

D. Kolb
Department of Physics, Kassel University, 35 Kassel, West Germany.

Secondary reactions[1,2] as a possible mean for production of superheavy elements have several advantages over direct heavy-ion reactions, namely: (a) Large variety of isotopes, both stable and unstable are available as "projectiles". (b) The second step of the reaction takes place within about 5×10^{-14} sec after the fragment has been formed. During this short time the fragment is still at high excitation energy and quite deformed. This may affect strongly the fusion cross-section[3,4].

Recently a consistent interpretation has been presented of the spontaneous fission events previously observed in Hg sources separated from CERN W targets[5]. This interpretation was based on the assumption that the production of neutron-deficient superheavy isotopes is not impossible. In Table I the measured masses of the fissioning nuclei are interpreted in terms of various possible molecules of element 112 which is the chemical homolog of Hg. The molecules chosen are to be common molecules found with Hg and may be formed by combination of element 112 with various impurity molecular ions present in the ion source of the mass separator[5]. It is seen that five different molecules can be related to the isotopes of element 112 with 160-161 neutrons.

TABLE I. Results of mass separator measurements on the Hg source. Number of fission tracks are given in parentheses for each mass. The masses are arranged according to various possible molecules of element 112 (see text).

A^+	$A^{16}O^+$	$A^{35}Cl^+$	$A^{12}C^{14}N^{16}O^+$, $A^{14}N_3^+$	$A^{14}N^{16}O_2^+$	N
269(1)					157
272(1)	288(1)	308(3)ᵃ	315(2)	317-318(4)	160-161
276(1)	292(1)	311(1)ᵇ			164

ᵃMass 308 may also be interpreted as $^{276}AO_2^+$.
ᵇMass 311 may also be interpreted as $^{269}AN_3^+$ or $^{269}A^{12}C^{14}N^{16}O^+$.

Fig. 1 shows a two-dimensional spectrum of correlated energies of fission fragments from the Hg source [2,5]. Evidence for three groups of coincidences which are outside the region where most of the ^{252}Cf fragments would have been found, is observed. These groups are: 69-MeV fragments in the front detector in coincidence with 62 MeV fragments in the back detector, 100 MeV in coincidence with 67 MeV and 83 MeV in coincidence with 116 MeV. It is difficult[5] to understand these data, in particular the coincidences between two frag-

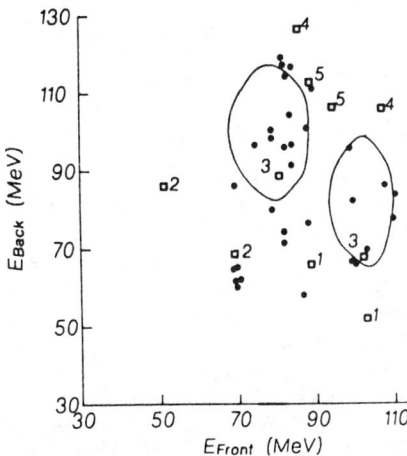

FIG. 1. Correlated energies of fission fragments from the Hg source. The pairs of numbered points show the alternative interpretations when the analysis of the events is ambiguous (Ref. 2). The contours enclose the region in which 90% of the ^{252}Cf fragments would be found and are asymmetric because of energy loss in the source backing.

ments of about 65 MeV each, in terms of binary fission. Total kinetic energy of about 130 MeV is expected from isotopes of Hg or nearby elements. However, spontaneous fission in the Hg region is very unlikely[6,7]. A consistent interpretation has been given[5] in terms of four-particle break-up where the two 65-MeV fragments are connected to the others around 108 MeV (average of 100 and 116 MeV) and 83 MeV. Under this assumption the total measured kinetic energy of the fragments would be about 320 MeV, and the value of $\bar{\nu}$ was estimated to be between 3 and 5 in accord with the experimental result[2].

From the point of view of reaction mechanism it is easier to understand the production of neutron-deficient superheavy nuclei with N value around 160. In particular it was shown[5] that the radiative capture process is possible. The Q values for reactions like

$$^{88}Sr + ^{184}W \rightarrow ^{272}_{112}X_{160} \text{ or } ^{86}Sr + ^{186}W \rightarrow ^{272}_{112}X_{160}$$

are -282.4 and -275.8 MeV respectively, while the Coulomb barrier between the targets and the projectiles is around 285 MeV. It was mentioned[5] that under these conditions, the positive experimental results are not in contradiction[5] with the results of Katcoff and Perlman[6] nor with other experiments which used U or Th targets[8]. The radiative capture process is not possible with these targets, and the cross sections for reactions like $^{238}U(^{39}Ca,5n)^{272}112$ is probably quite low, because of the small expected value of $(\Gamma_n/\Gamma_f)^5$.

Recently large fission barriers have been predicted[9,10] for nuclei around Z=110 and N=162. However, the predicted lifetimes[10] against alpha-particle emission are in the msec region. Our assumption that the production of neutron-deficient superheavy elements is not impossible was based on evidence[11] for production of long-lived isomeric states in neutron-deficient actinide nuclei. Alpha particle spectra were measured[11,12] with actinide sources which were separated from the same W target. A pronounced group at (5.76±0.40) MeV has been seen with the Am and Bk sources. Fig. 2 shows the intensities (normalized to 30 days) of this group in the Am and Bk sources as a function of time. It is seen that in both cases the intensity

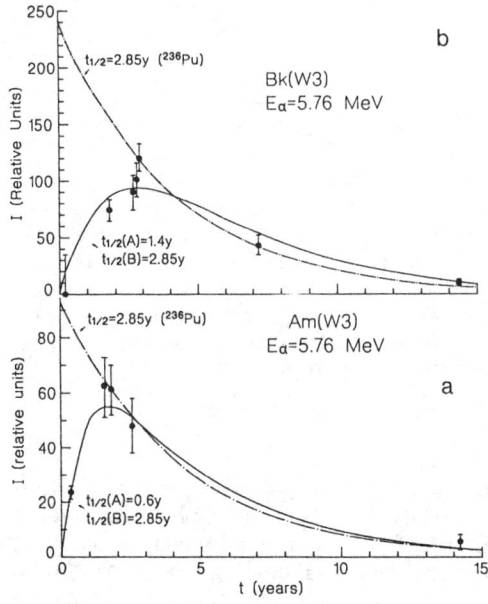

FIG. 2. (a) Decay curve of the 5.76 MeV alpha-particle group in Am. In between the first and second measurements the source was evaporated. The relative intensities were obtained by using the 3.18 MeV group of ^{148}Gd as a normalization (see Ref. 11 for details). (b) Decay curve of the 5.76 MeV alpha-particle group in Bk (see Ref. 11).

first grew with time and then decayed by more than a factor of ten. On the basis of energy, lifetime and mass of the recoiling nuclei[11,12] the 5.76-MeV group both in Am and Bk was identified as due to ^{236}Pu. However, it could not be ^{236}Pu from the beginning in the sources, since, among other reasons[11], first the intensity grew with time in strong contradiction to the dot-dashed lines in Fig. 2. It was argued[11] that these data can consistently be interpreted if one assumes that long-lived isomeric states have been produced in previously unknown ^{236}Am and ^{236}Bk nuclei, and decayed by electron capture or β^+ to ^{236}Pu. The lifetimes of the isomeric states are about 10^4 to 10^5 longer as expected for the normal ground states. In principle one may consider two kinds of isomeric states, high spin states or a new kind of shape isomeric states. It should be mentioned, however, that high spin states have not been observed before in very heavy neutron-deficient actinides far from closed-shells.

A theoretical support for the existence of a new kind of shape isomeric states comes from Hartree-Fock calculations[13,14] on ^{236}Cm. In addition to the ground state and to the fission isomeric state at $\beta_p=0.45$, two oblate isomeric states have been predicted. Of particular interest is the state with $\beta_p=-0.6$ which has about 30 MeV lower Coulomb energy as compared to the ground state. On the basis of the cluster model the predicted density structure of this state[13,14] may be interpreted by one ^{48}Ca and four ^{47}K (which is one proton hole in ^{48}Ca) clustering on a ring. It was argued[14] that the predicted properties of such states may offer explanation to the experimental results both in the superheavy and in the actinide regions. They will be more stable against spontaneous fission decay because of their considerable lower Coulomb energy as compared to the other configurations, and it is reasonable[14] that they will undergo fission to more than two fragments, since the normal binary fission requires first a transition from oblate to prolate shape. The mass parabolas of the shape isomers are predicted[14] to be shallower as compared to normal mass parabolas. The long-lived isomeric state of ^{236}Am and ^{236}Bk mentioned above may be due to such shape isomers, since the lifetime of the EC process depends strongly on the mass difference between the two isobaric nuclei.

The shape isomeric states may decay by abnormal alpha particle groups[14]. The Q values for transition between such states will be different as compared to normal transitions. Also, because of their special shape the barrier against alpha particle emission will be reduced, and the normal relationships between energies and lifetimes will not hold for the decays of such states. The observed[11,12,14,15] unidentified alpha particles groups with abnormal lifetimes may be due to transitions between such shape isomeric states.

According to the extra-push theory[16,17] the results in the superheavy element region cannot be understood. However, the value of the extra-push energy will be reduced considerably[17] for formation of a deformed nucleus, like for instance the ring clustered shape isomer predicted by the Hartree-Fock calculations. In addition, sub-barrier fusion may be very effective[3,4] for interaction between the deformed W target nuclei and the deformed excited projectiles.

REFERENCES

1. A. Marinov et al., Nature (London) 229, 464 (1971).
2. A. Marinov et al., Nature (London) 234, 212 (1971).
3. W. Reisdorf, in Proc. Int. Nucl. Phys. Conf. Harrogate, U.K., 1986, Vol. 2, p. 205.
4. A.B. Quint et al., GSI Annual Report GSI-87-1, 34 (1987).
5. A. Marinov, S. Eshhar, J.L. Weil and D. Kolb, Phys. Rev. Lett. 52, 2209 (1984): 53, 1120 (1984) (E), and references therein.
6. S. Katcoff and M.L. Perlman, Nature (London) 231, 522 (1971).
7. W.D. Myers, Droplet Model of Atomic Nuclei (Plenum, New York, 1977).
8. A complete list of references of these publications is given by G. Herrmann, Radiochemistry (London) 8, 221 (1975).
9. S. Čwiok, V.V. Pashkevich, J. Dubek and W. Nazarewicz, Nucl. Phys. A410, 254 (1983).
10. P. Möller, G.A. Leander and J.R. Nix, Z. Phys. A323, 41 (1986).
11. A. Marinov, S. Eshhar and D. Kolb, Phys. Lett. 191B, 36 (1987).
12. A. Marinov, S. Eshhar and J.L. Weil, in Proc. Inter. Symp. on Superheavy Elements, Lubbock, Texas, 1978, ed. M.H.K. Lodhi, (Pergamon, New York, 1978) p. 72.
13. D. Kolb and A. Marinov, in Proc. Inter. Conf. Nucl. Phys., Florence, 1983, Vol. 1, p. 92.
14. D. Kolb and A. Marinov, preprint, submitted for publication in Phys. Rev. Lett.
15. A. Marinov, S. Eshhar and D. Kolb, Fizika 19, (Supp. 1), 67 (1987).
16. S. Bjørnholm and W.J. Swiatecki, Nucl. Phys. A391, 471 (1982).
17. J.P. Blocki, H. Feldmeier and W.J. Swiatecki, Nucl. Phys. A459, 145(1986).

Heavy Fragment Radioactivities

P. B. Price
Physics Department, University of California, Berkeley, CA 94720

ABSTRACT

This recently discovered mode of radioactive decay, like alpha decay and spontaneous fission, is believed to involve tunneling through the deformation-energy barrier between a very heavy nucleus and two separated fragments the sum of whose masses is less than the mass of the parent nucleus. In all known cases the heavier of the two fragments is close to doubly magic ^{208}Pb, and the lighter fragment has even Z. Four isotopes of Ra are known to emit ^{14}C nuclei; several isotopes of U as well as ^{230}Th and ^{231}Pa emit Ne nuclei; and ^{234}U exhibits four hadronic decay modes -- alpha decay, spontaneous fission, Ne decay and Mg decay.

INTRODUCTION

The possibility that heavy nuclei might decay by monoenergetic heavy fragment emission was predicted[1] in 1980. The experimental study of heavy fragment radioactivity did not begin until four years later, when Rose and Jones[2] discovered that ^{223}Ra spontaneously emits 30 MeV ^{14}C ions with a branching ratio $\approx 10^{-9}$ relative to alpha decay. This new decay mode was quickly confirmed by several other groups[3,4,5,6] and followed by discoveries of ^{14}C decay of ^{222}Ra, ^{224}Ra and ^{226}Ra; Ne decay of ^{230}Th, ^{231}Pa, ^{232}U, ^{233}U and ^{234}U; and Mg decay of ^{234}U. All but one[7] of these results have been obtained by us[5,8,9,10] and by a Dubna group[11,12,13] using solid state nuclear track detectors. Current activity is concentrating on searches for emission of neutron-rich isotopes of Mg and heavier fragments at branching ratios $B(X/\alpha) < 10^{-15}$, some six orders of magnitude lower in B than originally seen by Rose and Jones. The only technique capable of rejecting alpha particles at such low branching ratios is that of solid state nuclear track detectors with sensitivity optimized for the fragment being sought. An attempt[14] to detect gamma-ray line emission from ^{24}Na, following β-decay of ^{24}Ne produced in a 10 gm ^{233}U source, succeeded only in setting an upper limit of $B \approx 10^{-12}$, somewhat higher than the actual value measured with plastic track detectors[11].

SOLID STATE NUCLEAR TRACK DETECTORS

Ref. 15 discusses techniques for particle identification by measurements of the dimensions of etched tracks, from which charge and range of a heavily ionizing particle can be determined provided the alpha particle background is below a limiting density at which radiation damage causes the general etch rate to increase and the sensitivity to decrease. Three types of track detectors have been used in the study of heavy ion radioactivity: polycarbonate film such as Rodyne or Lexan, which records nuclei down to $Z \approx 5$ at the energies of interest, ≈ 2 MeV/N, and can tolerate alpha particle fluences up to $\approx 3 \times 10^{10}/cm^2$; polyester film such as Cronar, which records nuclei down to $Z \approx 8$ at α fluences up to $\approx 10^{12}/cm^2$; and phosphate glasses[10,16], with charge thresholds in the range $Z \approx 8$ to 10, and maximum tolerable α fluences $\approx 10^{14}/cm^2$. Polycarbonate film is satisfactory for the study of ^{14}C decay. Both polyester and phosphate glass have been used in the study of Ne decay, whereas only phosphate glass has been used to study Mg decay. My opinion is that, even though it is more expensive than polyester film, phosphate glass is preferable because it is much less sensitive to α-particle radiation damage, its charge resolution in the region $9 < Z < 15$ is much better than that of polyester, and the signal to noise ratio is high enough for automated scanning techniques to work. My colleagues and I are working on highly sensitive new phosphate glasses that may be able to detect ^{14}C nuclei in a background of alpha fluences much higher than is possible with polycarbonate film. This would make possible the study of carbon emission by a number of nuclides in the Rn to Ra region with branching ratios less than 10^{-12}.

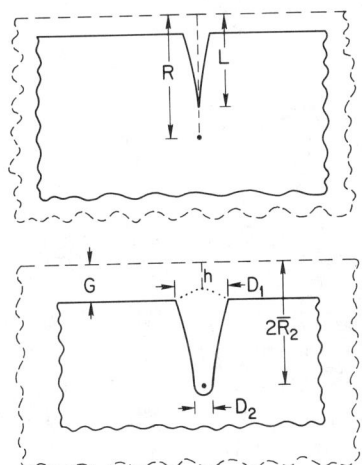

Fig.1. Identification of a heavy ion with range R. After a short etch time, length L is a measure of charge Z if the range is known. After longer etch time, D_1 and D_2 and depth give Z and R.

Figure 1 illustrates the geometry of the etched track of a normally incident particle with range R insufficient to penetrate a single sheet of a detector. To identify a heavy ion one relates two quantities, s and R, to the charge Z and energy E of the ion. We define $s = v_T/v_G = 1/\sin \varphi$ as the ratio of etch rate along the track to general etch rate (φ is the half-angle of the conical etchpit); by means of accelerator calibrations one establishes the relation between s and dE/dx for a given detector and etching reagent. I shall present a sample of such data in the subsection on Mg decay. For a track etched to the end of its range as in Fig. 1 b, the quantities D_1, D_2, and R_2 (and a knowledge of G) provide a pair of measurements of s and R that determine Z and E rather accurately. The average value of s_1 is inferred from the mouth diameter, D_1, through the geometric relation

$$s_1 \cos \theta = [1 + r^2]/[1 - r^2] \qquad (1)$$

where θ is the track zenith angle, $G = v_G t$ is the amount of material removed during the etching time t, and $r \equiv D_1/2G$. As explained in ref. 15, a least time analysis shows that the value of D_1 is determined by etchant traveling at a rate v_T for a distance h given by

$$h = G s_1 /(1 + s_1 \cos \theta) \qquad (2)$$

followed by etching at a rate v_G along the direction indicated in Fig. 1 until the mouth is reached. The average value of the range, R_1, at which s is represented by its average value is given by

$$R_1 = R - h/2 \qquad (3)$$

This method of identification works best if the sensitivity of the detector is chosen so that s < 6, corresponding to an etchpit half-cone-angle < 10^0. Measurement of the diameter of the rounded tip of the etched track gives an independent identification of the particle. The larger is s, the sooner the etchant will reach the tip and the larger its diameter will become. Once the etchant reaches the tip, its diameter grows at a rate v_G. The average rate s_2 at the average range $R_2 = R/2$ is given by

$$s_2 = R/(G - D_2/2) \qquad (4)$$

and the pair $s_2(D_2)$, R_2 gives a second way of identifying the particle that is independent of s_1, R_1. Detailed discussion of calibrations and techniques such as the use of plastic replicas of etched tracks in glass can be found in refs. 5, 8-11.

STUDIES OF C, Ne, AND Mg DECAY MODES

The ^{14}C decay mode. Rose and Jones[2] used a semiconductor detector telescope to identify carbon ions emitted by ^{223}Ra daughters in a ^{227}Th source. Hourani et al.[7] used a magnetic spectrometer to concentrate the ^{14}C ions from a ^{226}Ra source and a semiconductor telescope to identify them. My colleagues and I[5,9] used the ISOLDE on-line isotope separator at CERN to produce beams of 60 keV ions of ^{221}Fr, ^{221}Ra, ^{222}Ra, ^{223}Ra, ^{224}Ra, ^{225}Fr, and ^{225}Ra. (The latter two, with half-lives of 3.9 min and 14.6 days, beta-decayed in the collector to produce a ^{225}Ac source.) Each beam was collected on a different foil installed in the vacuum line and surrounded by its own array of Rodyne polycarbonate detecting films. We detected ^{14}C emission from ^{222}Ra, ^{223}Ra, ^{224}Ra and ^{226}Ra, and set stringent upper limits on the branching ratios for ^{14}C emission from the other nuclides. (Our result for ^{226}Ra confirmed the result obtained by Hourani et al.[7] and improved the statistics.) Table 1 and Figs. 4 and 5 give a compilation of these and other results. Other nuclides in the vicinity of Ra are predicted to have very small branching ratios for ^{14}C (or ^{12}C) emission, and no attempt has been made to study them.

The Ne decay mode. ISOLDE does not produce useful quantities of any nuclides expected to decay by Ne emission. Both my group[8,10,16,17] and the Dubna group[11,12,13] have obtained sources of various long-lived nuclides ranging from ^{230}Th up to ^{241}Am. In some cases we have placed polyester films or plates of phosphate glass in contact with the sources; in other cases, to spread the α-particle radiation damage over a large area, we placed the source at the center of a hemispherical shell lined with detectors at a radial distance of 5 cm or more and placed the entire assembly inside a vacuum system.

The results are summarized in Table 1 and in Figs. 4 and 5. Measured values of B(Ne/α), in those experiments that led to positive results, range from $10^{-11.2}$ for ^{231}Pa down to $\approx 10^{-12.2}$ for ^{230}Th and ^{234}U. After a 10-month exposure of an 80 cm^2 sheet of Cronar in contact with a thick sheet of ^{232}Th, we obtained a null result, corresponding to an upper limit of $B < 10^{-10.3}$(90% C.L.) for Ne emission[18]. Because of the low specific activity of ^{232}Th, it is difficult to go much lower.

The Mg decay mode. A number of uranium and transuranic isotopes are predicted to emit Mg ions, at branching ratios that vary by orders of magnitude from one model to another. Some nuclides, e.g., ^{234}U, are predicted to emit both Ne and Mg isotopes. My colleagues and I at Berkeley have collaborated with K. Moody and K. Hulet at Livermore in a successful search for Mg emission from ^{234}U. It is crucial to have a source with a very low level of ^{232}U present as a contaminant. Our source contained 5% ^{232}U by activity, which proved to be acceptably low. Our calibrations at the Berkeley Superhilac, displayed in Fig. 2, showed that PSK-50 phosphate glass (Schott) has excellent charge resolution in the region of Ne and Mg, considerably better than the resolution of plastic detectors. The lines in Fig. 2 are fits to data for ^{28}Si, ^{24}Mg, and ^{20}Ne ions. The points are our data for heavy ions emitted from thin sources of ^{234}U of total mass 135 mg during a collection time of 100 days. The data for s vs R comprise three populations -- due to

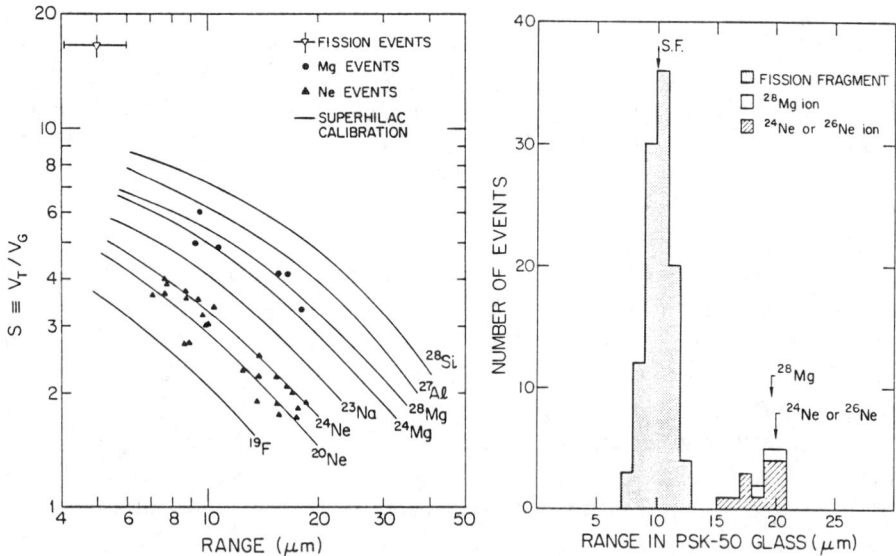

Fig.2. Identification of fission fragments, Ne ions, and Mg ions emitted from ^{234}U source. Curves result from calibrations with ion beams. Fig.3. Range distributions for ions emitted by ^{234}U.

spontaneous fission, to emission of a Mg isotope, and to emission of one or more Ne isotopes. Figure 3 shows the measurements of range. The vertical arrows indicate the expected ranges for several predicted modes of decay, taking into account self-absorption in the ^{234}U source. The agreement with the observed ranges is quite good. The resolution of our technique is not quite adequate to identify isotopes. To obtain quantitative values for branching ratios, we had to determine detection efficiency as a function of angle of entry. This was done both by calibrations at the Superhilac and by looking at the angular distribution of spontaneously emitted ions. We detected 590 fissions, 14 Ne decays, and 3 Mg decays. (Each of the Ne and Mg events is represented in Fig. 2 by two data points.) Note that, because of different efficiencies for the three ions and because we did not scan the same areas in searching for the three types of ions, the branching ratios are not exactly proportional to the number of events seen. From previous work on ^{232}U decay, we concluded that the contributions of ^{232}U to the observed events were < 2 fissions, ≈2 Ne ions, and < 0.1 Mg ion.

The nucleus ^{234}U is the first and only example to date of a parent for which all three categories of hadronic decay modes -- alpha decay, heavy ion radioactivity, and spontaneous fission -- have been detected. The relative rates for the different decay modes are predicted surprisingly well by the most recent version of the unified model of Poenaru et al.[19] For the two heavy ion decay modes, the observed ratio of partial half-lives, τ_{Mg}/τ_{Ne} = 3.1 ± 2, is reasonably consistent with the value 0.8 predicted by the same authors.

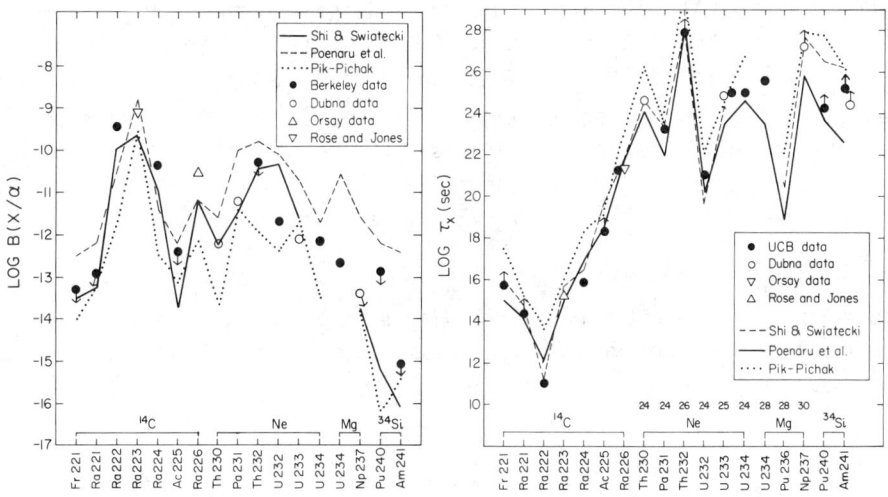

Fig. 4. Comparison of measurements of branching ratios for heavy ion emission with the models in refs. 23, 24, 28, 32. Fig. 5. Comparison of measured lifetimes for heavy ion emission with models.

The Si decay mode. ^{241}Am would appear to be a favorable nuclide to examine for possible Si emission. The unified models predict branching ratios for emission of 80.6 MeV ^{34}Si ions ranging from $10^{-12.4}$ to $10^{-15.4}$, which makes a determined search feasible.

Furthermore, its branching ratio for spontaneous fission relative to alpha decay is only $\approx 10^{-11.4}$ which is at worst only 10^4 times higher than the lowest branching ratio predicted for Si emission. Several groups[7,13,20,21] have obtained negative results. Recently our colleagues at Livermore and we[22] made a rigorous search using LG750 phosphate glass detectors with response optimized at Si ions and capable of withstanding a background dose of at least 10^{14} alphas/cm^2 without shift of sensitivity. The characteristics of the spontaneous fission tracks recorded during a six-month exposure to 8 mg of ^{241}Am showed that tracks of Si ions would have been correctly identified. Our null result sets an upper limit (90% C.L.) of $10^{-15.1}$ on the branching ratio for Si emission, which is not inconsistent with the ratio predicted by Shi and Swiatecki[23] and Pik-Pichak[24], but is about a factor six lower than predicted by Poenaru et al.[19]

To continue the search for the Si decay mode, we are preparing strong sources of ^{237}Np and ^{238}Pu to be examined with LG750 glass detectors. In the latter case we intend to range out the intense spontaneous fission background with an Al foil. Si (or Mg) ions should penetrate the Al with a residual range large enough to detect. Predictions for the branching ratio for Si emission from ^{238}Pu range from $10^{-15.5}$ to 10^{-17} and predictions for the branching ratio for Mg emission from ^{237}Np range from $10^{-13.3}$ to $10^{-13.8}$. The low specific activity of Np^{237} makes its study as difficult as that of ^{238}Pu.

ODD-EVEN EFFECTS

Figure 6 displays the available data on heavy ion radioactivities in a form that enables odd-even effects to be seen. The abscissa is proportional to the Gamow penetration factor for a square-well plus Coulomb potential.

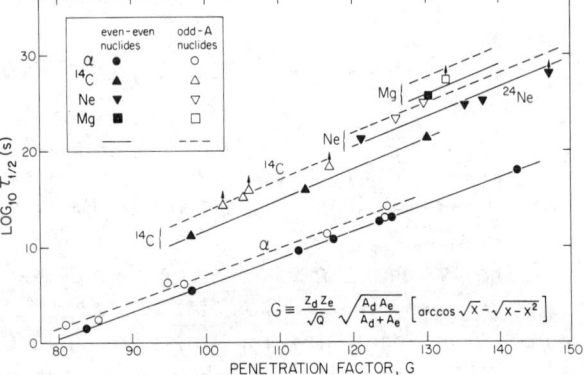

Fig.6 Odd-even effects exhibited on a plot of log τ vs penetration factor.

Rates of decay of an odd-A nuclide by emission of a particular ion are hindered by a factor of $\approx 10^2$ relative to the same type of decay by an even-even nuclide. For alpha decay the odd-even effect is only a factor 5 or 10.

COLD FISSION

Sandulescu[25] and Poenaru et al.[26] have pointed out that cold fission, in which essentially all of the Q value goes into the kinetic energy of two fission fragments, is a special case of heavy ion radioactivity. Here the mass distribution is very narrow and centered around an isotope such as Sn, with a closed or nearly closed proton or neutron shell, and the path to scission is believed to lead through two compact shapes -- the incipient fragments. Hulet et al.[27] have given beautiful experimental examples of spontaneous fission of certain transuranic nuclides such as ^{264}Fm by two different modes -- normal fission and cold fission or "Sn-decay".

COMPARISON OF MODELS

Four groups have developed unified models that differ only in quantitative details. All four groups included in the potential energy expression not only the Coulomb barrier that appears after creation of the daughter nuclei but also the barrier before scission, caused by the joint action of the Coulomb and nuclear forces. To compensate for the artificially sharp cusp in their barrier at the point of scission of the two fragments, Poenaru et al.[19,28] raised the decay energy Q by an adjustable amount E_V, which scales with Q and depends on the odd-even nature of the parent nucleus and on the mass of the emitted fragment. Shi and Swiatecki[23,29] represented the nuclear potential by an attractive "proximity potential", which was added to the repulsive Coulomb potential at points beyond scission. At distances between scission and the parent configuration they represented the barrier by a smooth power-law interpolation to zero. In their second paper they included nuclear ground-state deformations and the attenuation of fragment shell effects by the interaction of the fragments. Pik-Pichak[24] took into account nuclear deformations, used a Lagrangian coordinate for the distance between the centers of gravity of the fragments, described the parent nucleus and heavy fragment by intersecting ellipsoids, and evaluated the mass coefficient by a hydrodynamical model. Barranco et al.[32] let the inertial mass depend on the superfluid pairing gap of the nucleus.

Poenaru et al.[19] have prepared a detailed table of predicted half-lives and branching ratios for the most probable heavy ions, with light fragment charge less than 29, and for cold fission. This is valuable to the experimentalist, but one should note that it tends to be overly optimistic in its prediction of branching ratio for ions heavier than Ne.

Cluster models[30,31] regard the heavy ion as having a certain probability of forming out of a cluster of nucleons inside or at the surface of the parent nucleus and of assaulting the barrier with a characteristic frequency. The calculated preformation probability decreases from a value ≈ 0.03 for an alpha particle to $\approx 5 \times 10^{-8}$ for a Mg ion.

None of the models predicts branching ratios that agree with data as well as does a simple square well (with $r_0 = 0.98$ fm) plus Coulomb potential.[5] Experimental studies of heavy fragment radioactivities over a wider range of fragment charges are needed.

This research was supported in part by DOE. I have benefitted from stimulating collaborations with S. W. Barwick, J. D. Stevenson, H. L. Ravn, K. J. Moody, E. K. Hulet, and Wang Shicheng.

Table 1. Nuclei with Measured Branching Ratios for Heavy Ion Decay

Decay	E_k (MeV)	Theoretical Predictions of $-\log B$ (refs.)				Measured $-\log B$ (refs.)	$\log(\tau_{1/2}\,\mathrm{exp})$ (seconds)
		(28)	(23)	(24)	(32)		
^{221}Fr→^{14}C	29.28	12.5	13.6	14.0	-	>13.3 (9)	>15.77
^{221}Ra→^{14}C	30.34	11.9	13.3	13.2	10.4	>12.9 (9)	>14.35
^{222}Ra→^{14}C	30.97	11.0	9.7	11.7	9.1	9.43±0.06 (5)	11.02±0.06
^{223}Ra→^{14}C	29.85	8.7	9.7	9.6	7.3	9.21±0.05 (2-6)	15.2±0.05
^{224}Ra→^{14}C	28.63	11.9	11.0	12.5	11.1	10.37±0.12 (5)	15.9±0.12
^{225}Ac→^{14}C	28.57	12.2	13.7	13.1	-	>12.4 (9)	>18.34
^{226}Ra→^{14}C	26.46	11.7	10.2	12.2	11.8	10.6±0.2 (7,9)	21.33±0.2
^{230}Th→^{24}Ne	51.75	12.5	12.2	13.7	-	12.25±0.07 (13)	24.64±0.07
^{232}Th→^{26}Ne	55.37	10.8	10.4	12.0	-	>10.3 (18)	>27.94
^{231}Pa→^{24}Ne	54.14	10.0	11.4	11.4	12.5	11.22 (12)	23.23
^{232}U→^{24}Ne	55.86	10.9	10.4	12.4	13.3	11.7±0.1 (7)	21.06±0.1
^{233}U→^{24}Ne	54.27	10.3	11.7	11.7	12.6	12.12±0.15 (11)	24.82±0.15
^{234}U→^{24}Ne	52.81	11.9	-	13.5	16.1	12.18±0.12 (10)	25.07±0.12
^{234}U→^{28}Mg	65.26	10.6	-	-	21.0	12.66±0.25 (10)	25.55±0.25
^{237}Np→^{30}Mg	61.16	11.6	13.7	13.9	-	>13.4 (13)	>27.27
^{240}Pu→^{34}Si	78.07	13.3	15.2	16.2	-	>12.88 (18)	>24.25
^{241}Am→^{34}Si	80.60	12.4	16.1	15.4	13.1	>15.1(22);>14.1(13)	>25.3 (22); >24.2 (13)

REFERENCES

1. A. Sandulescu et al., Sov. J. Part. Nucl. 11, 528 (1980).
2. H. J. Rose and G. A. Jones, Nature 307, 245 (1984).
3. D. V. Aleksandrov et al., JETP Lett. 40, 909 (1984).
4. S. Gales et al., Phys. Rev. Lett. 53, 759 (1984).
5. P. B. Price et al., Phys. Rev. Lett. 54, 297 (1985).
6. W. Kutschera et al., Phys. Rev. C 32, 2036 (1985).
7. E. Hourani et al., Phys. Lett. 160B, 375 (1985).
8. S. W. Barwick et al., Phys. Rev. C 31, 1984 (1985).
9. S. W. Barwick et al., Phys. Rev. C 34, 362 (1986).
10. Wang Shicheng, P. B. Price, S. W. Barwick, K. J. Moody, and E. K. Hulet, submitted to Phys. Rev. Lett. (1987).
11. S. P. Tretyakova et al., JINR Rapid Comm. 7, 23 (1985).
12. A. Sandulescu et al., JINR Rapid Comm. 5, 5 (1984).
13. S. P. Tretyakova et al., JINR Rapid Comm. 13, 34 (1985).
14. A. Ya. Balysh et al., Sov. Phys. JETP 64, 21 (1986).
15. R. L. Fleischer, P. B. Price, and R. M. Walker, Nuclear Tracks in Solids, California Press, Berkeley (1975).
16. P. B. Price et al., Nature 325, 137 (1987).
17. S. W. Barwick, Ph.D. thesis, Department of Physics, University of California, Berkeley (1986).
18. P. B. Price and S. W. Barwick, Vol. 2 of Charged Particle Emission from Nuclei, ed. D. N. Poenaru and M. Ivascu (CRC Press, Boca Raton, Florida, 1987).
19. D. N. Poenaru et al., preprint NP-54-86, Bucharest (1986).
20. P. B. Price, Phys. Bull. 36, 589 (1985).
21. M. Paul et al., Phys. Rev. C 34, 1980 (1986).
22. K. J. Moody, E. K. Hulet, Shicheng Wang, P. B. Price, and S. W. Barwick, submitted to Phys. Rev. C Rapid Comm. (1987).
23. Y.-J. Shi and W. J. Swiatecki, Nucl. Phys. A464, 205 (1987).
24. G. A. Pik-Pichak, Sov. J. Nucl. Phys. 44, 923 (1987).
25. A. Sandulescu, Proc. Inter. School on Nuclear Structure, ed. V. G. Soloviev and Yu. P. Popov (Alushta, October 14-22, 1985).
26. D. N. Poenaru et al., University of Frankfurt preprint UFTP 201, 1987.
27. E. K. Hulet et al., Phys. Rev. Lett. 56, 313 (1986).
28. D. N. Poenaru et al., At. Data Nucl. Data Tables 34, 423 (1986).
29. Y.-J. Shi and W. J. Swiatecki, Nucl. Phys. A438, 450 (1985).
30. R. Blendowske et al., Nucl. Phys. A464, 75 (1987).
31. M. Iriondo et al., Nucl. Phys. A454, 252 (1986).
32. F. Barranco, R. A. Broglia, and G. F. Bertsch, to be published.

EVIDENCE FOR BIMODAL FISSION IN THE HEAVIEST ELEMENTS

E. K. Hulet*
University of California, Lawrence Livermore National Laboratory,
Livermore, CA 94550

ABSTRACT

We have measured the mass and kinetic-energy partitioning in the spontaneous fission of five heavy nuclides: ^{258}Fm, ^{259}Md, ^{260}Md, ^{258}No, and ^{260}[104]. Each was produced by heavy-ion reactions with either ^{248}Cm, ^{249}Bk, or ^{254}Es targets. Energies of correlated fragments from the isotopes with millisecond half lives, ^{258}No and ^{260}[104], were measured on-line by a special rotating-wheel instrument, while the others were determined off-line after mass separation. All fissioned with mass distributions that were symmetric. Total-kinetic-energy distributions peaked near either 200 or 235 MeV. Surprisingly, because only a single Gaussian energy distribution had been observed previously in actinide fission, these energy distributions were skewed upward or downward from the peak in each case, except for ^{260}[104], indicating a composite of two energy distributions. We were able to fit accurately two Gaussian curves to the gross energy distributions from the four remaining nuclides. From the multiple TKE distributions and the shapes of the mass distributions, we conclude that there is a low-energy fission component with liquid-drop characteristics which is admixed with a much higher-energy component due to closed fragment shells. We now have further evidence for this conclusion from measurements of the neutron multiplicity in the spontaneous fission of ^{260}Md.

INTRODUCTION

A central feature of low-energy and spontaneous fission of the heavy elements is the division into fragments of unequal mass. Mass-symmetric fission is a rarer mode, occurring only in two restricted regions of the chart of the nuclides: the Tl to Ac region preceding the actinides, and near the termination of the actinide series of elements. The causes advanced for mass symmetry and asymmetry differ markedly for each region of nuclides. The liquid-drop model describes the broadly symmetric mass distributions and moderate kinetic energies found for the low-energy induced fission of nuclei between Tl and Ac.[1] In contrast, the highly symmetric mass division found in the heaviest Fm isotopes has been ascribed to the strong shell effects emerging near scission from fragments approaching the doubly-magic ^{132}Sn nucleus.[2,3] Because these fission product nuclei are spherical and stiff toward deformation, the compact configuration at scission

---

* Coauthors and collaborators: J. F. Wild, R. W. Lougheed, R. J. Dougan, J. H. Landrum, A. D. Dougan, M. Schädel, R. L. Hahn, P. A. Baisden, C. M. Henderson, R. J. Dupzyk, K. Sümmerer, G. R. Bethune, J. v. Aarle, W. Westmeier, R. Brandt, and P. Patzelt.

results in a high Coulomb energy which is translated into unusually large kinetic energies for the fragments. For these nuclides, fragment energies closely approach and some are equal to the Q value for the fission reaction. We have called this "fragment-shell directed" symmetric fission. Previously, these two causes for mass symmetry were thought to be mutually exclusive because they depended on the very different balance between macroscopic forces and single-particle couplings in separate regions of nuclides.

We have gathered evidence from fragment energy measurements in the spontaneous fission (sf) of the heaviest nuclides that strongly suggests symmetric mass division in this region may arise as much from the liquid-drop process as from the influence of emerging fragment shells.[4] We found the total kinetic energy (TKE) distributions of the fragments strongly deviated from the Gaussian distributions which are observed in the fission of lighter actinides. In four of five nuclides studied, the anomalous TKE distribution was skewed sufficiently that it could be decomposed into two Gaussian distributions. From the multiple TKEs and the shapes of the mass distributions, we conclude that there is a lower energy fission component with liquid-drop characteristics which is admixed with a much higher-energy component due to fragment shells. These observations have provided new insights into the fission process which have inspired significant theoretical advances.

EXPERIMENTAL

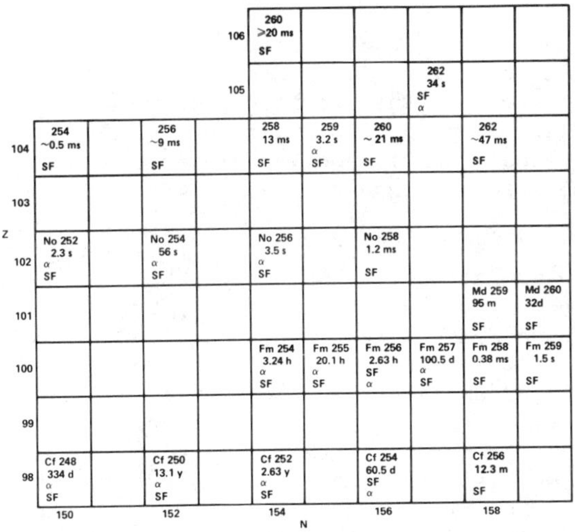

We have measured the energies of coincident fragments arising from the sf of five nuclides with Z≥100 and N≥156.[4] A portion of the nuclide chart showing these isotopes and surrounding ones decaying by sf is given in Fig. 1. The very short-lived nuclides, 1.2-ms ^{258}No (Ref. 5) and 20-ms ^{260}[104] (Ref. 6), were produced in fusion reactions of 68-MeV ^{13}C ions with ^{248}Cm and of 81-MeV ^{15}N ions with ^{249}Bk, respectively. Fragment-correlated energy data for these were collected with a new instrument that provides a continuous on-line method for producing short-lived isotopes and measuring the energy of the fragments emitted in sf (Fig. 2). Recoil products emerging from the target were stopped in a band of thin Al foils mounted on the rim of a 30-cm diam wheel that spins up to 5000 rpm. These foils were rotated past opposing banks of trapezoidal-shaped surface-barrier

Fig. 1 Portion of the Nuclide Chart showing isotopes known to decay by sf.

detectors, which measured the energies deposited by coincident fragments. With event rates averaging only 5-8 coincident fissions per hour, we obtained 382 ^{258}No sf events together with 59 ^{256}Fm sf events.[7] For ^{260}[104], 300 events were recorded along with 41 events from ^{256}Fm produced by transfer reactions.[8]

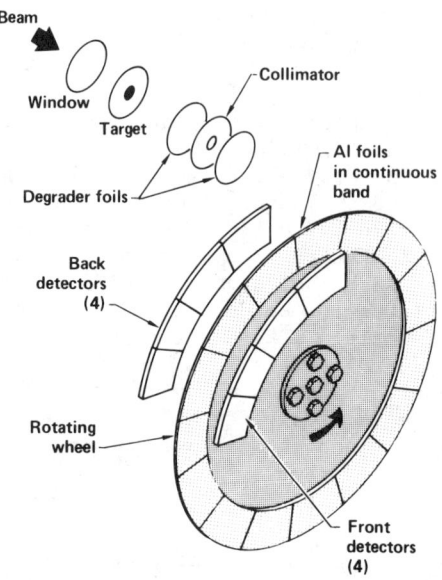

Fig. 2. Schematic diagram of the SWAMI instrument used for measuring coincident fragment energies arising from the sf of millisecond nuclides. A portion of the recoil nuclei produced in the nuclear reactions are stopped in 100 μg·cm^{-2} Al foils surrounding the rotating wheel, where they are moved between opposing pairs of surface-barrier detectors.

Half-lives of ^{258m}Md, ^{259}Md, and ^{260}Md are sufficiently long to allow time for off-line mass separation for the preparation of isotopically pure sources. These Md isotopes were produced by transfer reactions in bombardments of an ^{254}Es target with beams of 105-MeV ^{18}O and 126-MeV ^{22}Ne from the 88-in cyclotron at the Lawrence Berkeley Laboratory. Recoil products from the bombardments were trapped in Ta foils which were then flown by helicopter from the cyclotron to the Lawrence Livermore National Laboratory for mass separation. Because ^{256}Fm (157-min sf) is one of the most abundant products of these synthesizing reactions, mass separation was a key technique in freeing $^{258-260}$Md from the massive interference caused by the sf of ^{256}Fm. The background correction due to this isotope ranged from 8.6% for ^{259}Md to zero for ^{260}Md.

The desired mass fraction was collected on an Al foil (50 μg·cm^{-2}) which was subsequently placed between two surface-barrier detectors for the measurement of the correlated fission-fragment energies. The elapsed time between the end of bombardment and the start of counting was 1 h. This short interval was an important factor which allowed us to study the sf of ^{259}Md and 370-μs ^{258}Fm, which is produced by the electron-capture (E.C.) decay of 60-m ^{258m}Md.[9] Fragment energies were determined from calibrations of the detectors with ^{252}Cf for which we used 181.03 MeV for the average TKE.[10] Fragments that passed through the Al supporting foil were corrected for energy losses amounting to an average of 3.2 MeV. For all of our results, fragment masses were calculated on the basis of the conservation of mass and linear momentum in the fission process.

The sf properties of ^{258}Fm (Ref. 9,11) had been roughly determined in an earlier study.[3] Previous fission studies had also been made on 95-min ^{259}Md, which decays directly by sf.[12] In comparison with our older work,[3,12] these re-measurements on mass-separated samples of ^{258}Fm and ^{259}Md resulted in more events, much lower contributions from the sf of ^{256}Fm, and better energy resolution. Because of these improvements, skewing of the TKE distributions became apparent for the first time.

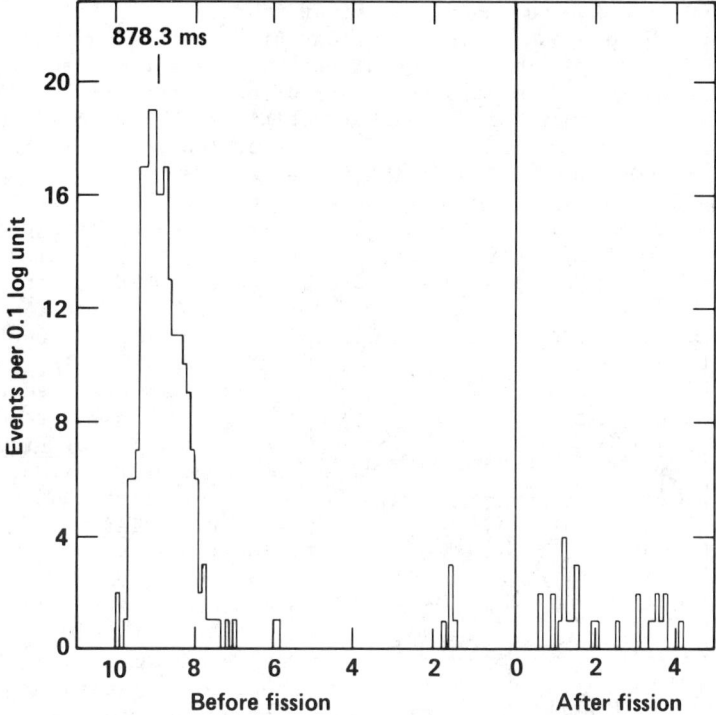

Fig. 3. Logarithmic time distributions for the last energy-windowed photon detected before the sf of ^{260}Md. The energy window corresponds to the K x-ray region of Fm (112-145 keV).

The longest-lived nuclide, ^{260}Md with a half-life of 32 d, had been discovered in the A=260 mass fraction after off-line mass separation.[13] Chemical methods have identified a Md isotope as the source of this activity; thus, the Z and A are certain. Although we attribute the decay period to ^{260}Md, this sf activity could, in principle, result from the decay of its possible daughters, ^{260}No or ^{260}Fm. These nuclides are expected to have subsecond sf half-lives and, therefore, could possibly be in secular equilibrium with the parent ^{260}Md. We have eliminated β^- decay to ^{260}No as a possible source of the observed fissions by measuring the time intervals between 0.04- to 1-MeV betas and subsequent fissions. We found that this time distribution was random and the same as that between any two successive beta particles. Measurements of the time correlation between Fm K or L x-rays and fission events were performed using the same methods and apparatus described in Ref. 9. As shown in Fig. 3 for the spectrum of time intervals between K x-rays and fissions, there are very few photons, except for the random background peak at 878 ms, in the Fm K x-ray energy range occurring before fission. L x-ray time-spectra were similar and demonstrated that branching by E. C. decay to ^{260}Fm was at most 20%, providing the mean lifetime of ^{260}Fm is 100 ms or less. Therefore, we ascribe our sf results for the A=260 fraction to the direct sf of ^{260}Md.

For the purpose of verifying our bimodal-fission interpretation of the TKE distributions, we have recently measured the multiplicity of prompt neutrons emitted in the sf of ^{260}Md (Ref. 14). Similar to observations from fission in other actinides, we expected an inverse relation between the TKE and the number of neutrons emitted in each event. Because our TKE distribution appeared to be a composite of two distributions, so should the neutron-multiplicity distribution. A small number of neutrons (1 or less) might be expected from fissions with TKEs approaching the Q value of the reaction, whereas the lower-energy TKE peak near 200 MeV should provide a distribution comparable with ^{252}Cf (an average of 3.8 neutrons per fission). Preliminary results from measurements on a single sample are shown in the histogram of Fig. 4. The average neutron multiplicity for this distribution is 2.4 neutrons/fission, which is considerably less than values ranging from 3.5 to 4.15 measured for the heavier actinides. Over 32% of the fissions from ^{260}Md emit 1 neutron or less; by contrast, only about 2.8% of the fissions from ^{252}Cf emit 1 neutron or less. We intend to repeat this experiment with the added measurement of fragment energies correlated with the number of neutrons emitted in each fission.

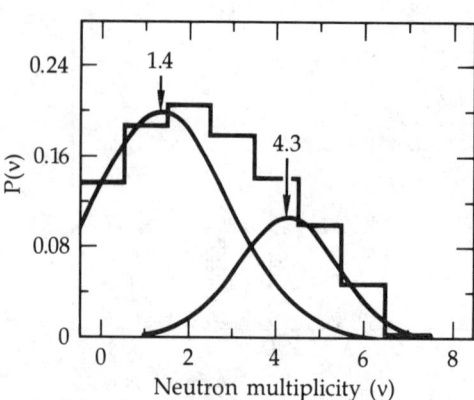

Fig. 4. Neutron multiplicity distribution from the sf of ^{260}Md showing a decomposition into two Gaussian distributions.

RESULTS AND DISCUSSION

Figure 5 shows the mass distributions we obtained for the five nuclides. All are symmetric but some more sharply than others. The narrowest, from ^{258}Fm and ^{260}Md, have full-widths at half-maximum of 7.5 u while the broadest, that for ^{260}[104], is 36 u. In a majority of these nuclides, there are wings extending far outward in mass from the central peak, a feature not so clearly obvious from earlier studies. The fraction of events in these wings is lowest in ^{258}Fm and ^{260}Md and increases with Z. A finding common to all these five and to other very heavy nuclides[15,16] is that events with masses residing in these wings are associated with low TKEs while events with TKEs ≥220 MeV give an exceedingly sharp mass distribution around symmetry. When sf events with TKEs less than 200 MeV are chosen, the resulting mass distributions are very broad and flat. Some might even be characterized as asymmetric if our statistical samples were larger. Thus, we find a low-energy form of fission with very broad mass distributions and a high-energy form associated with sharply symmetric mass division.

In four of the nuclides, the TKE distributions deviated substantially from Gaussian distributions, as seen in Fig. 6. This is a phenomenon not previously observed in the sf of actinide nuclei; for example, the TKE distribution for ^{240}Pu is given in Fig. 7.[17] Fig. 6 makes it clear that asymmetric tailing from the peak energy can occur toward either higher or lower energies. Furthermore, we note that the peak in each of the TKE curves falls in one of two distinct positions, either near 200 MeV or 235 MeV. Skewing of the TKE curves results in distributing an appreciable portion of the events into each of these two main energy locations. Least-mean-squares fitting of two Gaussian distributions to the TKE curves (e.g., see Fig. 8) gave the following centroids: ^{258}Fm - 205 and 232 MeV; ^{259}Md - 201 and 235 MeV; ^{258}No - 203 and 235 MeV; ^{260}Md - 195 and 234 MeV. These values were calculated by setting the width of the lower energy Gaussian to the value obtained for ^{260}[104].

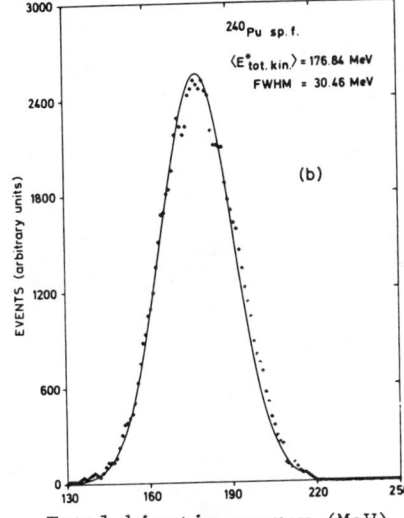

Fig. 7. Gaussian distribution fitted to TKE points for ^{240}Pu (Ref. 17) A small deviation is seen on the high-energy side of the distribution.

The sf of ^{260}[104], in which the low-energy mode dominates, provides an exception to the largely asymmetric TKE distributions noted above for the other nuclides. Because of the low average TKE and the broad TKE and mass distributions obtained for this nuclide, we have interpreted these properties as being characteristic of a liquid-drop

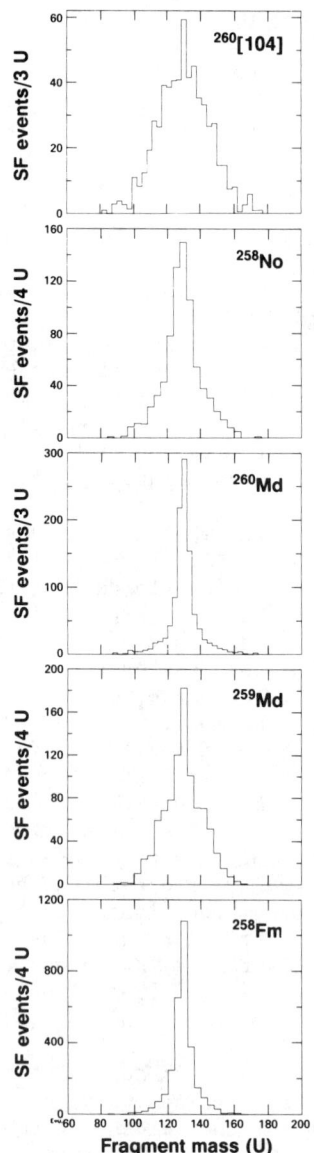

Figure 5. Provisional mass distributions obtained from correlated fragment energies. A small contribution from ^{256}Fm has been subtracted from most. The mass bins have been chosen to be slightly different for each nuclide.

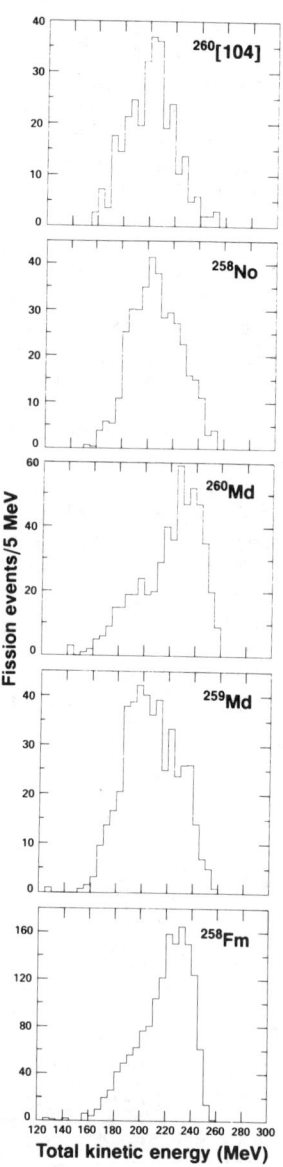

Figure 6. Provisional total-kinetic-energy distributions. A small contribution equivalent to the known amount of ^{256}Fm has been subtracted from all but the ^{260}Md distribution.

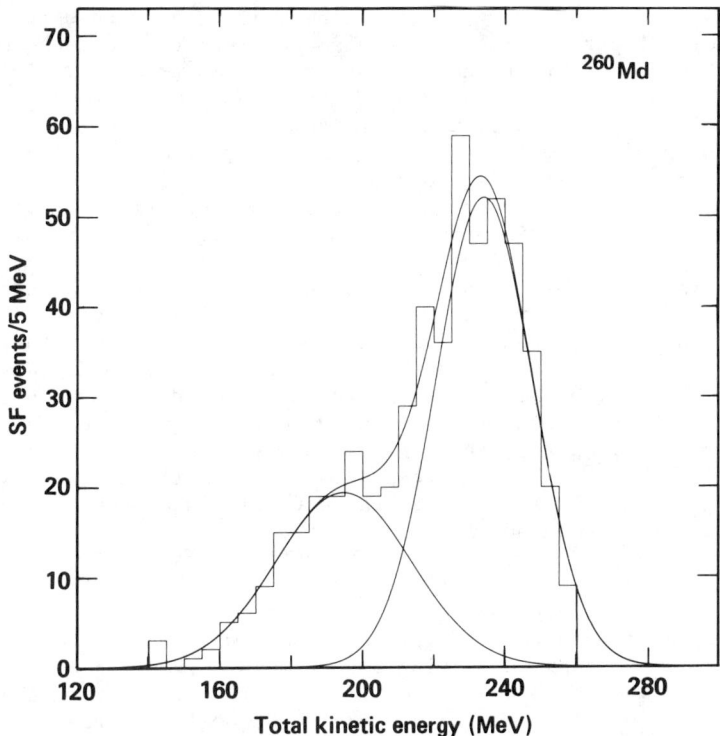

Fig. 8. Decomposition into two components of the asymmetrical TKE distribution measured for ^{260}Md. A low- and high-energy mode was obtained by fitting with two Gaussian distributions.

mode of fission.[8] The outer asymmetric fission barrier, which may be responsible for asymmetric mass distributions in the fission of all but the heaviest actinides,[18] is predicted to have disappeared below the ground state in ^{260}[104].[19] Passage through the remaining inner barrier, which contains liquid-drop and shell components, should yield symmetric mass distributions and average TKEs that conform to the estimates of Viola et al.[20] Mass symmetry is expected because only reflection-symmetric shapes are allowed by the underlying liquid-drop fission barrier.

Because there are unmistakable high- and low-energy fission distributions occurring concurrently in the same nuclide, we conclude these nuclides sustain two strikingly different modes of fission. We have suggested that the liquid-drop and the "fragment shell" processes are separately responsible for the distinct portions of the TKE distributions.[4] The high-TKE mode depends on the influence of fragment shells which are emerging between the saddle and scission point.[2,3] Fragment shells near the doubly-magic ^{132}Sn lower the potential-energy path and, thus, guide the mass division toward Sn isotopes near the 82-neutron closed shell. Only the two fission models noted can account for the large (≈35 MeV) difference between the low- and high-energy modes; the symmetric mass distributions, with their disparate widths, are also consistent with these models.

These new observations indicate that symmetric mass division and high TKEs are no longer unique to the heavy Fm isotopes, ^{258}Fm and ^{259}Fm. It should be expected that symmetric mass division caused by liquid-drop contributions to the first barrier most likely will extend beyond element-104 toward the region of the "superheavy elements". Even in this unexplored area, exceptionally large TKEs may not be unusual since four of the five nuclides we studied have a significant high-TKE component. In nuclides with neutron numbers approaching 164, high TKEs will be a possibility because symmetric fission leads to spherical fragments containing a closed 82-neutron shell. But as N decreases below 158 neutrons, and Z of the fissioning species increases beyond 100, the opportunity to divide into two Sn fragments diminishes. Within the ranges of Z and N reported here, we observe a trend away from sf characterized by unusually high TKEs and sharply symmetric mass splits and toward the liquid-drop mode represented by ^{260}[104]. Irrespective of this trend, one of the most striking features is the extraordinarily sharp change in fission properties with the addition of a single nucleon. For instance, adding a proton to ^{258}Fm or a neutron to ^{259}Md results in an abrupt inversion in the population of the two fission modes. Such rapid changes are beyond theoretical explanation because the underlying physical parameters tend to vary only smoothly with nucleon numbers.

Earlier,[15] based on having seen a mass-asymmetric mode together with a distinct high-energy, mass-symmetric component in the neutron induced fission of ^{255}Fm, we had argued strongly for the two-mode hypothesis first proposed by Turkevich and Niday.[21] However, this was not accepted because theoretical calculations of potential-energy surfaces failed to show a hint of two distinct fission paths along such surfaces. Now that our experimental evidence is nearly conclusive, we have suggested there, indeed, must be separate, competitive valleys in the potential-energy surfaces.[4] As a consequence, Brosa et al.,[22] Möller et at.,[23] Pashkevich and Sandulescu[24], and Depta et al.[25] have each reported finding several separated paths in the region of the scission point from revised calculations of the potential-energy surface of ^{258}Fm. One illustration of these newly calculated surfaces is provided in Fig. 9 (Ref. 23), which shows two minimum-energy paths after the first barrier, one leading to a very compact configuration at scission and the other to an elongated, liquid-drop shape. Our experimental TKEs require two different charge separation distances and, therefore, these scission configurations, to reproduce the desired Coulomb energies. Still, all is not well with this picture because there should be nearly equal probabilities of passing through either valley in order to conform to our measurements of the relative amounts of the low- and high-energy modes. The surfaces derived from quasi-static deformations don't provide useful guidance on this point and we feel it will be necessary to introduce dynamical aspects (inertial mass, etc.) to resolve this question.

We thank the staff and operating crew of the 88-in cyclotron for the irradiations. This work was supported by the U. S. Department of Energy under contract No. W-7405-Eng-48.

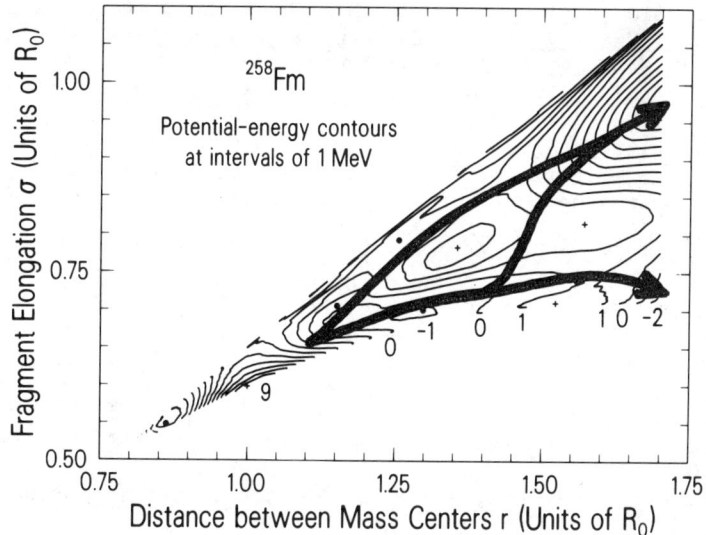

Fig. 9. Potential-energy map of ^{258}Fm as a function of deformation space (Ref. 23). Two minimum-energy exit channels are indicated.

REFERENCES

1. M. G. Itkis et al., Z. Phys. A320, 443 (1985); V. V. Pashkevich, Nucl. Phys. A169, 275 (1971); C. Gustafsson, P. Möller, and S. G. Nilsson, Phys. Lett. 34B, 349 (1971); U. Mösel, Phys. Rev. C 6, 971 (1972).
2. U. Mösel and H. W. Schmitt, Phys. Rev. C 4, 2185 (1971); M. G. Mustafa, Phys. Rev. C 11, 1059 (1975).
3. D. C. Hoffman et al., Phys. Rev. C 21, 1972 (1980).
4. E. K. Hulet et al., Phys. Rev. Lett. 56, 313 (1986).
5. M. Nurmia, K. Eskola, P. Eskola, and A. Ghiorso, in Lawrence Berkeley Laboratory Report, Berkeley, CA 94720, UCRL-18667 (1969) p. 63 (unpublished).
6. J. M. Nitschke et al., Nucl. Phys. A352, 138 (1981); L. P. Somerville et al., Phys. Rev. C 31, 1801 (1985).
7. J. F. Wild et al., Nucl. Chem. Div. Ann. Report, Lawrence Livermore National Laboratory, Livermore, CA 94550, UCAR 10062-84/1 (1984), p. 6-23 (unpublished).
8. E. K. Hulet, in Proceeding of the International School-Seminar on Heavy Ion Physics, Alushta, 1983 (Joint Institute for Nuclear Research Report D7-83-644, Dubna, USSR, 1983) p. 431.
9. E. K. Hulet et al., Phys. Rev. C 34, 1394 (1986).
10. E. Weissenberger, P. Geltenbort, A. Oed, F. Gönnenwein, and H. Faust, Nucl. Instrum. & Methods A248, 506 (1986).
11. E. K. Hulet et al., Phys. Rev. Lett. 26, 523 (1971).
12. J. F. Wild et al., Phys. Rev. C 26, 1531 (1982).
13. R. W. Lougheed et al., J. Less Common Metals, 122, 411 (1986).

14. J. F. Wild et al., paper 109, Div. Nucl. Chem. & Tech., 194th National Meeting of the American Chemical Society, New Orleans, Aug. 30-Sept. 4, 1987 (unpublished).
15. R. C. Ragaini, E. K. Hulet, R. W. Lougheed, and J. F. Wild, Phys. Rev. C 9, 399 (1974).
16. H. C. Britt et al., Phys. Rev. C 30, 559 (1984).
17. See, e.g., Fig. 1, A. J. Deruytter and G. Wegener-Penning, in Physics and Chemistry of Fission-1973 (Proc. Symp. Rochester, 1973), IAEA, Vienna (1974), p. 55.
18. P. Möller and S. G. Nilsson, Phys. Lett. 31B, 283 (1970); H. C. Pauli, T. Ledergerber, and M. Brack, Phys. Lett. 34B, 264 (1971); J. Maruhn and W. Greiner, Phys. Rev. Lett. 32, 548 (1974).
19. J. Randrup et al., Phys. Rev. C 13, 229 (1976).
20. V. E. Viola, K. Kwiatkowski, and M. Walker, Phys. Rev. C 31, 1550 (1985).
21. A. Turkevich and J. B. Niday, Phys. Rev 84, 52 (1951).
22. U. Brosa, S. Grossmann, and A. Müller, Z. Phys. A 325, 242 (1986).
23. P. Möller, J. R. Nix, and W. J. Swiatecki, submitted to Nucl. Phys. A (1986).
24. V. V. Pashkevich and A. Sandulescu, Rapid Communications No.16-86, Joint Institute for Nuclear Research, Dubna, USSR, p 19-23 (1986).
25. K. Depta et al., Mod. Phys. Lett. A 1, 377 (1986).

ON THE FISSION OF THE HEAVIEST FERMIUM ISOTOPES

S. Ćwiok
Institute of Physics, Warsaw Technical University, PL-00-622 Warsaw, Poland

P. Rozmej [1]
GSI, D-6100 Darmstadt, F. R. Germany

A. Sobiczewski
Institute for Nuclear Studies, Hoza 69, PL-00-681 Warsaw, Poland

ABSTRACT

Potential energy of ^{258}Fm is calculated in multidimensional deformation space as a function of both (reflection-) symmetric and asymmetric shapes. The inclusion of the asymmetric shapes is found important as it "opens a pass" between the two fission valleys: one coresponding to compact and the other to elongated shapes of the nucleus. Thus, it makes both valleys accessible to the fissioning nucleus.

INTRODUCTION

Recent experiments on the mass and total-kinetic-energy (TKE) distributions for the spontaneous-fission fragments of the heaviest Fm isotopes and nuclides close to them has increased an interest in the shapes involved in the fission process of these nuclei. An observation of two distinct components in the TKE distribution (for some of the nuclides): one, traditional, at low energy (about 200 MeV) and the other, new, at high energy (about 235 MeV) suggests[1] the existence of two fission valleys: one, traditional, corresponding to elongated shapes at the scission point and the other (usually not considered in description of the fission process), corresponding to more compact shapes. In accordance with this suggestion, the compact-shape valley has been obtained in a number of recent theoretical calculations [2-5] (cf. also the older papers [6-8]). In particular, an extensive and detailed calculations of the potential-energy surface for nuclei around heavy Fm isotopes have been performed in ref.[5] A specific (three-quadratic-surface) parametrization has been used. Only the reflection-symmetric shapes have been considered.

---

[1]Permanent address: Institute of Physics, University MCS, PL-20-031 Lublin, Poland

The objective of the present paper is to study the potential energy in another deformation space. The main scope is to examine the effect of asymmetric deformations on this energy. However, also the symmetric case, alone, is considered as it allows us to study the sensitivity of the energy to changes of the deformation of a nucleus, by comparing our results with those of ref.[5] It is because our shapes differ from the shapes of that reference.

We concentrate on the nucleus ^{258}Fm, for which both groups of low and high TKE-fission fragments are observed [1].

DESCRIPTION OF THE CALCULATIONS

The potential energy is calculated by the macroscopic-microscopic method with the macroscopic part given by the Yukawa-plus-exponential model with the parameters of ref.[9] The Strutinski shell correction, based on the Woods-Saxon potential [10], is taken for the microscopic part. The "universal" (i.e. adjusted to experimental single-particle levels of all odd-A nuclei with $A \geq 40$) parameters of the Woods-Saxon potential are used [11]. The deformation of a nucleus is described by usual deformation parameters β_λ, i.e the coefficients in the expansion of the radius into spherical harmonics.

Minimization of the total energy of a nucleus in the multidimensional deformation space $\{\beta_\lambda\}$ is performed numerically by use of the ZXMIN procedure of the IMSL library. For the symmetric-shape case, the 3-dimensional space $\{\beta_\lambda\}$, $\lambda=2,4,6$, is taken, while for the analysis of the asymmetric shapes, the 6-dimensional space $\{\beta_\lambda\}$, $\lambda=2,3,...,7$, is used.

RESULTS AND DISCUSSION

Fig.1 gives the result of an analysis of the potential energy of ^{258}Fm for reflection-symmetric shapes. The energy is calculated in 3-dimensional deformation space $\{\beta_\lambda\}$, $\lambda=2,4,6$. It is presented in fig.1 as a function of β_2 and β_4, but for each point (β_2,β_4), the energy is minimized in β_6. One can see that two valleys of the energy are obtained. One is the valley situated in the lower part of the figure. It corresponds to compact shapes (as will be seen in fig.4). The other is placed in the upper part of the figure and is connected with the elongated shapes. The fission trajectory L of the nucleus starts in the first minimum I (obtained at $\beta_2 \approx 0.24$, $\beta_4 \approx$ -0.01), passes through the saddle point A (at $\beta_2 \approx 0.46$, $\beta_4 \approx 0.08$) and goes to the compact-shape (CS) valley. The first dashed contour line, neighbouring the solid line, denotes energy which is by 1 MeV lower than the energy of the ground state (i.e the energy at the first minimum I). One can see that there is only one hump in the fission barrier, in distinction to ref.[5] where two humps are obtained. To get into the elongated-shape (ES) valley, one should cross a rather high ridge separating the two valleys. For example, to follow the trajectory L_2 (departing from the main trajectory L_1 at the point D, before the scission point expected somewhat farther on L_1) one should overcome an additional barrier of about 6 or 5 MeV (depending on the assumption for the zero-point energy E_0 in the ground state). This makes the trajectory L_2 (and thus the observation of low-energy fragments) rather unprobable.

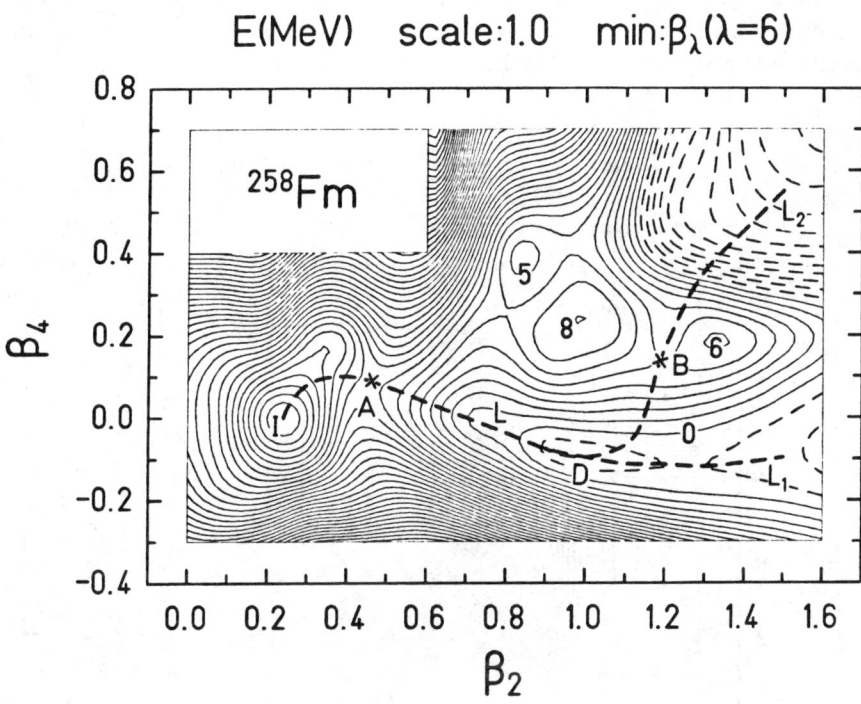

Figure 1: Contour map of the potential energy of ^{258}Fm for symmetric shapes.

Fig.2 shows the potential energy when asymmetric shapes are also included. Here, for each point (β_2,β_4), the energy is minimized in β_3, β_5 and β_6 degrees of freedom. One can see that now the ridge between the valleys is significantly reduced. One can pass from the CS valley to the ES one with no additional barrier, e.g. along the trajectory L_2.

The behaviour of the potential energy along both the trajectories L_1 and L_2 is explicitly shown in fig.3. The quantity, which parametrizes here the position of the nucleus on the trajectory, is the distance d between the mass centers of its fission fragments. One can see that the fissioning nucleus experiences the same barrier along both the trajectories. The barrier extends from the deformation d_1 up to d_2. Thus, the half-life for both fission modes is the same. The energy at the point D, where the two trajectories split is already much (by about 2 MeV) lower than that of the ground-state. As the energy surface around the point D, i.e. already in some distance from the exit point off the barrier, is much flat and does not give any significant barrier between the two valleys, the choice of the valley by the fissioning nucleus will be the result of a rather subtle play of the dynamics (cf. ref.[12]) and not of the potential energy. But this will be rather the dynamics on the way from the exit point off the barrier to the scission point than the dynamics inside the barrier, as expected earlier.

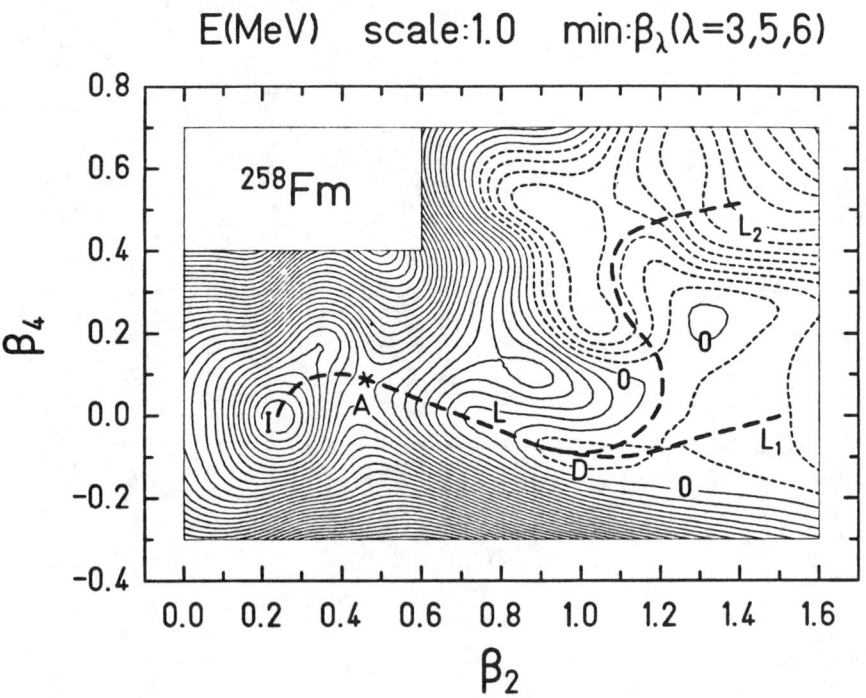

Figure 2: Contour map of the potential energy of ^{258}Fm after inclusion of asymmetric shapes.

Fig.4 illustrates the shapes of the considered nucleus ^{258}Fm, corresponding to the energy presented in fig.2. Thus, for each point (β_2,β_4), for which the shape is given, the deformations β_3, β_5 and β_6 are those which give minimal energy. The figure explains the names for both the compact-shape (along the trajectory L_1) and elongated shape (along the trajectory L_2) valleys.

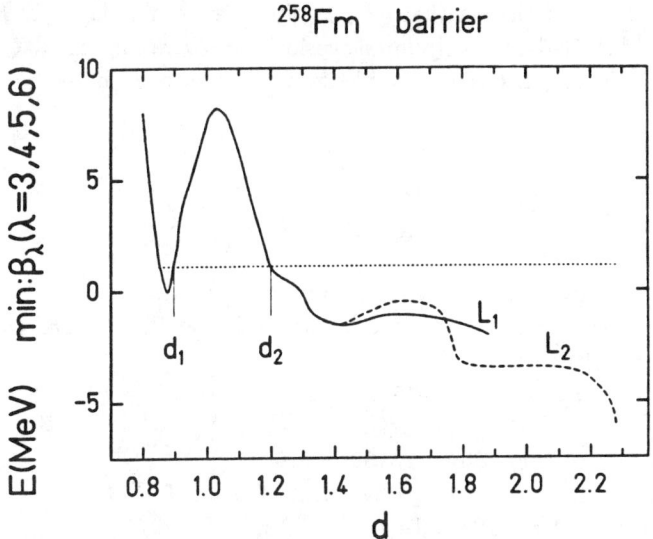

Figure 3: Fission barriers of ^{258}Fm along compact- (L_1) and elongated-shape (L_2) trajectories.

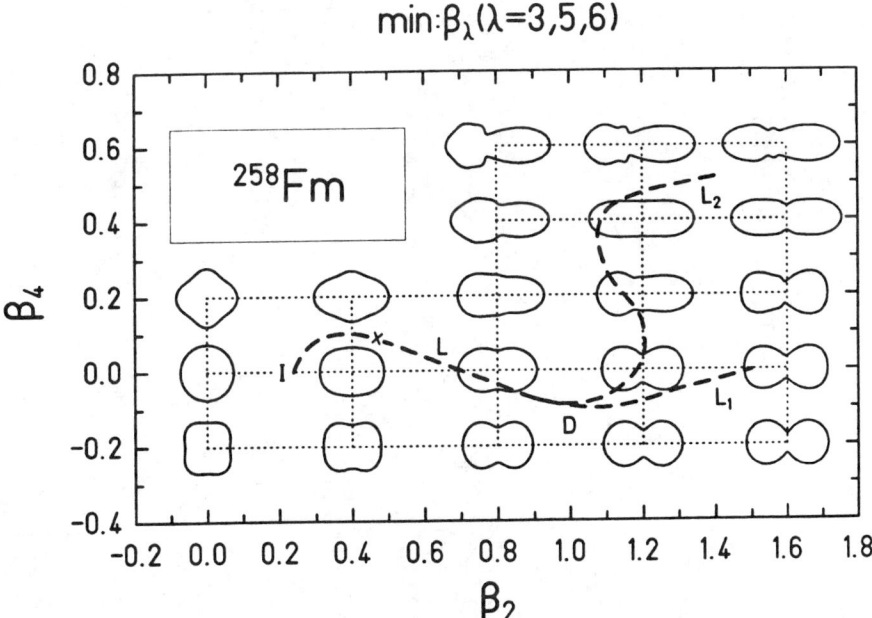

Figure 4: Shapes of the ^{258}Fm nucleus corresponding to the energies given in fig.2.

The authors would like to thank P. Armbruster, R. W. Hasse, E. K. Hulet, P. Möller and M. Schädel for helpful discussions. They would also like to thank GSI-Darmstadt, where a large part of the study has been done, for a warm hospitality.

REFERENCES

1. E. K. Hulet et al., Phys. Rev. Lett. **56**, 313 (1986).
2. V. V. Pashkevich and A. Sandulescu, JINR Rapid Comm. no. 16-86 (Dubna 1986) p.19.
3. U. Brosa et al., Z. Phys. **A325**, 241 (1986).
4. K. Depta et al., Mod. Phys. Lett. **A1**, 377 (1986).
5. P. Möller et al., Nucl. Phys. **A469**, 1 (1987).
6. U. Mosel and H. W. Schmitt, Phys. Rev. **C4**, 2185 (1971).
7. M. G. Mustafa et al., Phys. Rev. Lett. **28**, 1536 (1972); Phys. Rev. **C7**, 1519 (1973).
8. M. G. Mustafa, Phys. Rev. **C11**, 1059 (1975).
9. P. Möller and J. R. Nix, Nucl. Phys. **A361**, 117 (1981).
10. J. Dudek et al., J. Phys. **G5**, 1359 (1979).
11. J. Dudek et al., Phys. Rev. **C23**, 920 (1981).
12. V. Schneider et al., GSI Scientific Report (GSI 87-1) p.106.

FROM SYMMETRIC COLD FISSION FRAGMENT MASS DISTRIBUTIONS TO EXTREMELY ASYMMETRIC ALPHA DECAY

D.N.Poenaru, M.Ivaşcu, J.A.Maruhn* and W.Greiner*
Central Institute of Physics, R-76900 Bucharest,Romania
*Institut für Theoretische Physik der Universität,
D-6000 Frankfurt am Main, F.R.Germany

Abstract: The analytical superasymmetric fission model, successful in the study of extremely asymmetric decay modes like α - decay and heavy ion radioactivities, is applied to cold fission phenomena. The three groups of processes are described in a unified manner, showing that cold fission could be considered heavy cluster emission. For ^{234}U all groups have been detected. The highest symmetry of the fragment mass distributions should be observed for the neutron rich nucleus ^{264}Fm, leading to doubly magic products ^{132}Sn. The most probable light fragments from cold fission of ^{234,236}U, ^{239}Np and ^{240}Pu are ^{100}Zr, ^{104,106,106}Mo respectively, in good agreement with experimental data.

^{28}Mg emission and two heavy ion radioactivities (Ne and Mg) of ^{234}U have been observed recently[1]). In this way all three groups of decay modes (α, heavy ions and cold fission) of this nucleus are experimentally determined.

There are two different fission mechanisms. 1)Usually the fragments are deformed and excited; they reach final states by neutron emission and gamma decay. Their most probable total kinetic energy (TKE) are by 20-45 MeV lower than the Q-value). 2) Cold fission (TKE≈Q) meaning no (or very low) excitation and compact shapes, was observed in two regions of heavy nuclei: a) U, Np, and Pu isotopes, where it is a very weak phenomenon (small fraction of fission events)[2-4]; b) Fm, Md, No and the element 104, where it is comparable with the usual mechanism, leading to the bimodal fission[5]).

Theoretical works on cold fission have been performed by various groups[6-12]).

The purpose of this contribution is to present some results obtained in the framework of the last version of the analytical superasymmetric fission model (ASAFM). Details on the model and references of experimental works concerning heavy ion radioactivities are given elsewhere[13,14]). Estimated masses[15]) are used in the region beyond the measured ones[16]) to calculate the Q-values. ASAFM was extended to study transitions toward excited final states and the Coulomb induced cluster emission[17]). Another fission model for heavy ion radioactivities was also improved[18]).

From a systematic study[11]) of the cold fission process one can see a smooth variation of nuclear properties when Z or N of the parent nucleus is increased by four units - a consequence of the odd-even effects in both light and heavy fragments. The cold fission properties of transuranium nuclei are dominated by the interplay between the magic number of neutrons N=82 and protons Z=50 in one or both fragments. Hence symmetric cold fission appear at Z=100 (see Fig.1) and N=164, leading to identic doubly magic fragments ^{132}Sn. The peak-to-valley ratio (P/V) of the half-life spectra in the cold fission region is very large for Z<100 and N<164, where the closed shells are expected in the heavy fragment. Then, for Z>100 and N>164, where these magic numbers are found in the most probable light fragments, smaller asymmetry is observed. The transition from asymmetry to symmetry is very sharp and that from symmetry to asymmetry is much smoother.

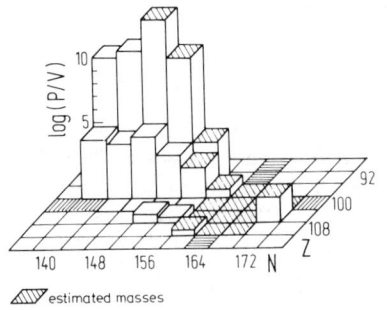

Fig.1.- The peak-to-valley ratio of the half-life spectra for the cold fission of transuranium nuclei with Z and N multiple of four.

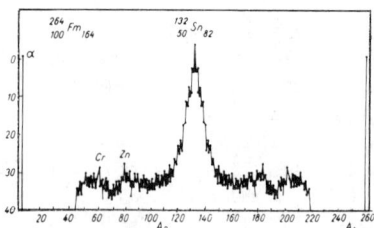

Fig.2.- Half-life spectrum for spontaneous emissions from ^{264}Fm.

The highest symmetry should be observed for the neutron rich ^{264}Fm (see Fig. 2) for which cold fission is expected to be the most probable decay mode (much stronger than any heavy ion emission and stronger even than α - decay).

A typical half-life spectrum for the region of U, Np, Pu nuclei, is plotted in Fig. 3, illustrating the idea of a unified treatment of different decay modes in a wide range of the asymmetry parameter $\eta = (A_1-A_2)/A$, where A is the mass number of the parent, and A_1, A_2- that of the daughter (heavy fragment) and of the emitted (light fragment) nuclei.

Three groups, already observed for this special nucleus, are clearly seen both in the light and heavy fragments, in good agreement with experimental results; a) α - decay; b) heavy ion radioactivities (^{24,26}Ne and ^{28}Mg light fragments and the corresponding ^{210,208}Pb and ^{208}Hg heavy ones) and c) cold fission (the most probable split in ^{100}Zr and ^{134}Te having a closed shell neutron number).

Similar spectra of ^{236}U, ^{239}Np and ^{240}Pu are showing the most probable light fragments ^{104}Mo, ^{106}Mo and ^{106}Mo respectively, in agreement with the experimental data.

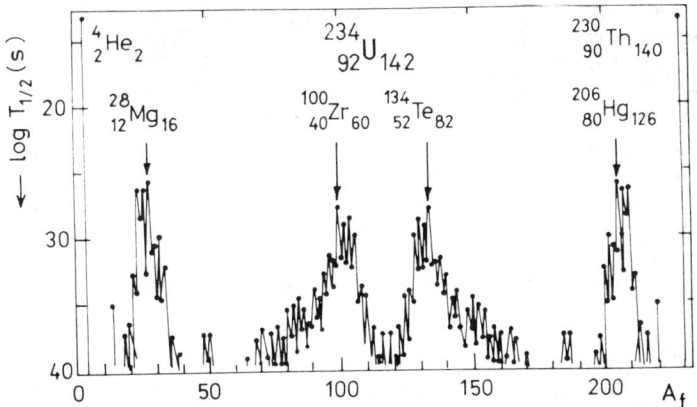

Fig.3.- Three groups of ^{234}U decay modes. All are experimentally observed.

In conclusion, even at low mass asymmetry, ASAFM can be used. It allows to obtain results reproducing the experimental data and describes in a unified way α - decay, heavy ion radioactivities and cold fission.

Acknowledgements. *We are grateful to P.B.Price, F.Gönnenwein, P.Armbruster, K.Depta, R.Gherghescu, Martin Greiner, E.Hourani, K.Hulet, D.Mazilu, M.Schädel, W.Scheid, C.Signarbieux and K.Sümmerer for fruitful discussions. One of us (D.N.P.) would like to thank to the Internationales Bureau KfK Karlsruhe for support in the framework of the German-Romanian Programme for scientific research and technological development.*

REFERENCES

1. S.Wang, P.B.Price, S.W.Barwick, K.J.Moody, E.K.Hulet, to be published.
2. C.Signarbieux, G.Simon, J.Trochon, F.Brissard, J.Phys.Lettres 46 (1985) L 1095
3. H.G.Clerc, W.Lang, M.Mutterer, C.Schmitt, J.P.Theobald, U.Quade, E.Rudolph, P.Armbruster, F.Gönnenwein, H.Schrader, E.Engelhardt, Nucl.Phys. A452 (1986) 277.
4. F.Gönnenwein, in Proc.Seminar on Fission (C.Wagemans Ed.), Report BLG 586, SCK/CEN Mol, 1986, pp. 106.
5. E.K.Hulet, J.F.Wild, R.J.Dougan, R.W.Lougheed, J.H.Landrum, A.D.Dougan, M.Schädel, R.L.Hahn, P.A.Baisden, C.M.Henderson, R.J.Dupzyk, K.Sümmerer, G.R.Bethune, Phys.Rev.Lett. 56 (1986)313
6. J.F.Berger, M.Girod, D.Gogny, Nucl.Phys. A428 (1984) 230.
7. K.Depta, R.Herrmann, J.A.Maruhn, W.Greiner, in Proc.Int.Symp. on Collective Phenomena in Nuclear and Subnuclear Long Range Interactions, Bad Honnef, May 4-7, 1987.

8. P.Möller, J.R.Nix, W.J.Swiatecki, Nucl.Phys., to be published.
9. U.Brosa, S.Grossmann, A.Müller, Z.Phys. A325 (1986) 241.
10. D.N.Poenaru, M.Ivaşcu, W.Greiner, Nucl.Tracks 12 (1986) 313
11. D.N.Poenaru, M.Ivaşcu, D.Mazilu, R.Gherghescu, K.Depta, W.Greiner, Central Institute of Physics Bucharest, Report NP-54-86, 1986.
12. V.V.Pashkevich, A.Sandulescu, JINR Rapid Communic. 16 (1986) 19.
13. D.N.Poenaru, W.Greiner, M.Ivaşcu, D.Mazilu, I.H.Plonski, Z.Phys. A325 (1986) 435.
14. S.W.Barwick, P.B.Price, H.L.Ravn, E.Hourani, M.Hussonnois, Phys.Rev. C34 (1986) 362.
15. P.Möller, J.R.Nix, Atomic Data and Nucl.Data Tables 26 (1981) 165.
16. A.H.Wapstra, G.Audi, Nucl. Phys. A432 (1985) 1.
17. M.Greiner, W.Scheid, J.Phys.G: Nucl.Phys. 12 (1986) L229.
    M.Greiner, W.Scheid, V.Oberacker, Phys.Rev.D, to be published.
18. Y.J.Shi, W.J.Swiatecki, Nucl.Phys. A464 (1987) 205.

A RESIDUE IMPLANTATION DETECTION SYSTEM ON THE DARESBURY RECOIL SEPARATOR

P.J. Woods, S.J. Bennett, B.R. Fulton and R.D. Page,
Department of Physics, University of Birmingham,
Birmingham B15 2TT, UK.

K.A. Connell, J. Groves and J. Simpson,
Daresbury Laboratory, Warrington WA4 4AD, UK.

A.N. James,
Oliver Lodge Laboratory, Liverpool University, Liverpool L69 3BX, UK.

W.D.M. Rae,
Nuclear Physics Laboratory, Oxford University, Oxford OX1 3RH, UK.

ABSTRACT

A two-dimensionally position sensitive silicon surface barrier detector has been positioned behind the focal plane of the Daresbury Recoil Separator. Proton-rich residue products formed in heavy-ion fusion-evaporation reactions are implanted into the detector after being velocity and momentum analysed by the Separator. Subsequent causally related alpha and proton decays are then precisely correlated using the position information to obtain clean decay time spectra.

INTRODUCTION

The combination of a velocity filter and an implantation detection system has proved an immensely powerful tool for studying the alpha and proton radioactivity of evaporation residues formed in fusion reactions. In particular the first case of ground-state proton radioactivity was discovered at GSI using the velocity filter SHIP[1]. The same technique has also been used to great effect to map out alpha-decay systematics close to the proton drip line and as a means of identifying new heavy elements up to Z=109[2,3].

In the present work we have sought to utilise the A/q resolution (1 part in 300) of the Daresbury Recoil Separator[4] (RS) to select and identify evaporation residues. The main advantages of such a device over a simple velocity filter are that mass information is directly obtained without having recourse to excitation functions and that the background from unwanted residues and scattered beam particles is greatly reduced. This increases the maximum beam current that can be tolerated without saturating the detectors. Both these factors make the RS particularly suitable for studying specific low cross-section reaction channels. To complement these features a two-dimensionally position sensitive

silicon surface barrier detector has been positioned behind the
focal plane of the RS in order to optimally correlate alpha and
proton decays from implanted evaporation residues. Recent
publications indicate that a similar combination of sector magnet
and 2D P.S.D. is also now being employed on SHIP[5,6]. In the
following sections a description of the design and performance of
the Residue Implantation Detection System (RIDS) will be presented.

THE RESIDUE IMPLANTATION DETECTION SYSTEM

Figure 1 shows a schematic diagram of RIDS. The carbon foil
and microchannel plate assemblies have been used in all RS
experiments to date, often in conjunction with a gas ionisation
chamber. The foil lies at an angle of 30° w.r.t. the beam direction
which approximately corresponds to the focal plane of the RS.
Signals generated by secondary electrons drifted and cascaded onto
the microchannel plates are used to obtain measurements of ion
velocity, A/q and also a crude energy loss value (through the foil).
This latter capability is particularly useful when implantations are
performed since this provides the only available Z discrimination
between beam and residue ions.

Ions passing through the carbon foil are implanted normally
into the 300 μm thick 2D P.S.D. so as to ensure maximum penetration
into the body of silicon. The implantation depth is typically 20
μm which compares with a range of 30 μm for 5 MeV alpha-particles
in silicon. Some residues (about 20%) therefore decay without
depositing all their energy in the detector.

Horizontal (X) and vertical (Y) position signals are generated
by charge division across two 100 Angstrom thick Au layers situated
on the front and rear faces of the detector. Residue implantation
and decay energies are obtained by adding either the X or Y position
signals together.

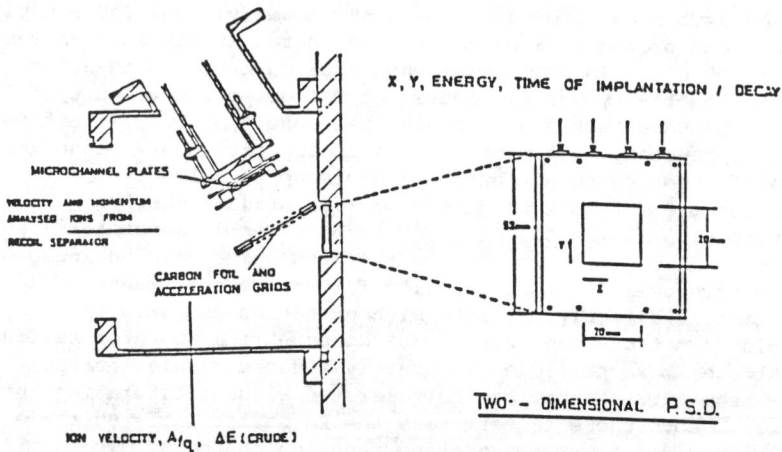

Fig. 1. Schematic diagram of the Residue
Implantation Detection System (RIDS).

RIDS PERFORMANCE TESTS

The following fusion reaction:

$^{58}Ni + {}^{110}Cd \rightarrow {}^{168}Os$ Ex = 40 MeV   E(beam) = 260 MeV

was used to produce alpha-decaying evaporation residues. The RS was tuned to analyse residue ions having A = 163 - 166 and q = 26. Figure 2 shows the alpha-decay energy spectrum measured by the silicon detector using an amplification dynamic range of 0-10 MeV, where the energy resolution FWHM = 80 keV. Decays were assigned on the basis of their energy and the centroid of the X co-ordinate which is proportional to the mass of the originally implanted parent nucleus. Figure 3 illustrates how the different decay lines separate according to mass (X). The X FWHM is equivalent to 1 part in 180 in A/q. This mass resolution value is inferior to the intrinsic RS value because the silicon detector is situated approximately 5 cm downstream of the centre of the focal plane. However, the resolution is quite adequate for mass assignment purposes and has the benefit of creating a larger horizontal image which improves correlation performance.

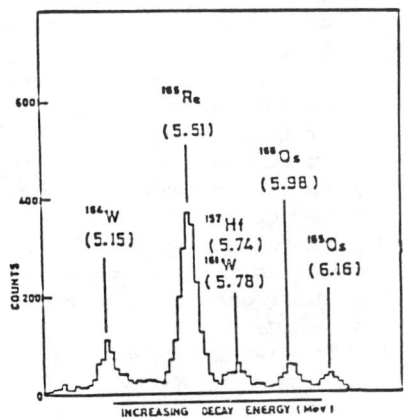
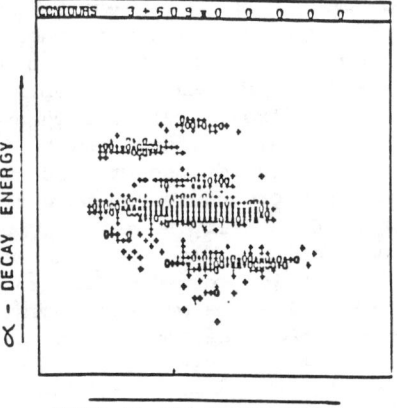

Figure 2. Alpha particle energy spectrum.

Figure 3. Alpha particle energies shown as a function of the implanted ion mass.

An example of how the position information is used to correlate successive decays is shown in Figures 4a) and b) for the sequential alpha decays of ^{165}Os and ^{161}W. Gating around the peaks corresponding to small (1mm<) relative displacements one can obtain decay time spectra almost entirely free of background. An identical approach has been used to correlate implantations and decays though these are more difficult to perform due to the high implantation rate. However, by using the X and Y signals one can effectively grid the detector into approx. 100 segments. This means that 100x longer lifetimes can be measured than would be possible without position information. This factor is subject to variation due to the variable implantation rate across the face of the detector in a given experiment.

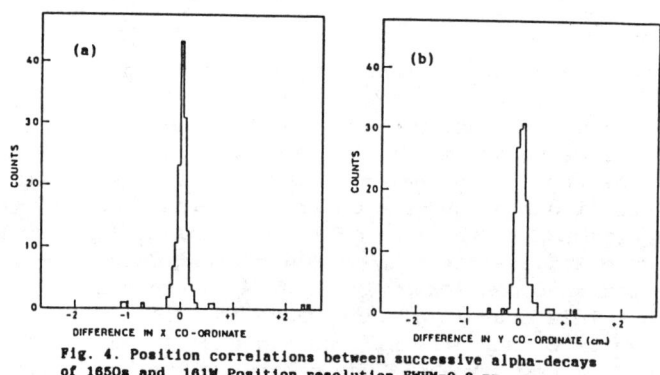

Fig. 4. Position correlations between successive alpha-decays of 165Os and 161W. Position resolution FWHM=0.8 mm.

DISCUSSION AND CONCLUSION

RIDS has been successfully commissioned. In conjunction with the Daresbury RS RIDS represents an exceptionally versatile tool for studying the radioactivity of nuclei lying in the proximity of the proton drip-line. The ability to select and identify residue ions at the focal plane has great implications for the study of low cross-section fusion reaction channels. In future experiments we intend to use alpha-decays emitted from first generation daughter nuclei to obtain retrospective correlations w.r.t. preceding proton decays. This procedure can be used to circumvent the problem of low energy β activity submerging potential proton activity. The signature for this process is an alpha line at an X co-ordinate centroid position corresponding to one mass unit greater than would be the case if alpha emission occurred initially.

Future modifications to the present system will include the addition of a LEPS detector for coincidence gamma and X-ray measurements and an extra silicon detector element to identify β decays. RIDS may also be used in future as a tagging device in prompt gamma ray studies involving alpha-decaying nuclei. This may prove to be particularly useful in the study of high Z(>50) nuclei where the Z resolution of a gas ionisation detector may be inadequate.

REFERENCES

1. S. Hofmann et al., Z. Phys. A305, 111, (1982)
2. S. Hofmann et al., Z. Phys. A291, 53, (1979)
3. G. Munzenberg et al., Z. Phys. A309, 89, (1982)
4. A.N. James et al., submitted to NIMS
5. G. Munzenberg, NIMS B26, 294, (1987)
6. W. Morawek et al., NIMS A258, 82, (1987)

PROPOSED RECOIL MASS SPECTROMETER FOR HEAVY ION REACTIONS

J. D. Cole
INEL, P. O. Box 1625, Idaho Falls, ID 83401, USA

T. M. Cormier
NSRL, University of Rochester, Rochester, NY 14627, USA

J. H. Hamilton and A. V. Ramayya
Physics Department, Vanderbilt University, Nashville, TN 37235, USA

ABSTRACT

A proposed recoil mass spectrometer of high rigidity, large solid angle acceptance, and good mass resolution for use with the HHIRF accelerators at ORNL is described.

The recoil mass spectrometer (RMS) described is designed to analyze heavy nuclear products from a heavy-ion induced reactions. Several papers[1-3] give reviews of spectrometers used in nuclear physics, including recoil mass spectrometers. The rigidity and dispersion desired has been obtained in many spectrometers, but the high beam rejection at 0° and large solid angle acceptance are problems that cannot be overcome together in traditional magnetic spectrometers. Although mass separators can be made to have large acceptance, they have problems not found in a RMS: different ionization efficiency for different species, the loss of correlation between information at the target location and the separator focal plane, and the long hold-up time of the species in the ion-source.

We are proposing a RMS that is to be used to carry out a broad research program in heavy ion science. The proposed RMS is to facilitate the study of otherwise inaccessible exotic nuclei. Careful attention was given to match the RMS to all the beams available from HHIRF accelerators, including those with the highest energy and massive particles for use in inverse reactions. In coupled operation or with heavier beams in inverse reactions, the energy per charge will rise thus, requiring the high rigidity. In addition, large solid angle and energy dispersion, small reaction product beam size for different masses at the focal plane, and ample space at the target area were considered crucial to the design.

The direct identification of the reaction products is limited by two problems. First, the reaction products and the primary beam particles usually both lie within a few degrees of one another in the forward direction. The problem is to reject the beam and accept the recoils. The second problem is that for detectors alone with the more massive recoils ($A \geq 100$), the energy resolution of the detectors is insufficient to determine mass using time-of-flight and direct energy measurements. This is the problem of analysis of the fragments. In recent years there has been an effort to combine the beam rejection function and the analysis function into one "recoil spectrometer".

Table I: SPECTROMETER DESIGN PARAMETERS		
Dipole & Deflectors	MD	ED1, ED2
Radius (ρ)	2 m	6 m
Deflection angle	43.6°	15°
Effective Field Boundary Shim Angles	9°	0°
Effective Field Boundary Radii	0.5 m	0 m
Element Gaps	100 mm	100 mm
Maximum Fields	10 kG	40 kV/cm
Maximum Rigidity	$B\rho$=20 kG-m =2 T-m ≡193.0 MeV-amu/q²	$E\rho$=24 MV ≡12 MeV/q
Quadrupoles & Sextupoles	Q1,Q2 Q3,Q4,Q5	S1,S2
Aperture Diameter	100 mm 150 mm	200 mm
Length	280 mm, 200 mm 150 mm	100 mm
Maximum Field at Pole Tip	8 kG 8 kG	0.6 kG
m/q range	±5%	
ΔE range	±10%	
Ω (solid angle)	8.3 msr	
m/Δm	≥300	
mass dispersion	≈10 mm/%	
primary beam rejection	≥10¹³	
time dispersion	≈0.05%	
Target to Q1 distance	600 mm	
Overall Length	≈11 m	

Figure 1 schematically shows the physical layout of the proposed spectrometer, and Table I summarizes its parameters. The design uses a split cylindrical electrostatic deflector. The magnetic dipole separates the two parts of the electric deflector, which results in a high separation of the reaction products and the elastically scattered beam par-

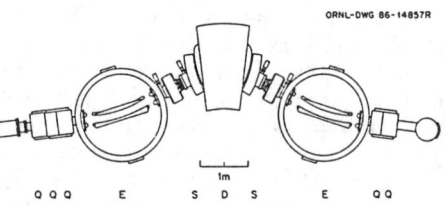

Figure 1

ticles. This is because the beam particles and the fusion reaction products have very different energy E/q, and the electric deflector disperses based upon (E/q). Also, the magnetic dipole acts as a lens and focuses particles of the same mass, but different energies, to the same point. This results in scattered particles not making it to the focal plane because they are focused onto the second electric deflector. For these particles to reach the detector at the focal plane, further scattering must occur to pass through the final quadrupoles. The net result is that the beam particles must undergo at least three scatterings to reach the focal plane. This is not impossible, but it does make the beam rejection of such a design very high. The fact that the primary beam strikes the first positive potential deflection plate in the system is not considered a problem, as the Rochester RMS has been working in this manner for several years.

The general view of an experiment with a RMS is that the RMS is but one element in the experiment. Detectors in single or various array configurations will be used with the spectrometer. A variety of different detectors may be used at the focal plane for different purposes. However, detectors around the target itself are essential to most experiments, and so the target location is important. The physical distance from the target position to the first spectrometer element (in this case a quadrupole) needs to be as large as possible to accommodate many detectors (both in number and kind). Our experiments[4,5] with the Rochester RMS have made us aware of how crucial this distance is. Other considerations affecting the target location are the solid angle subtended by the spectrometer and the magnification of the instrument.

The target distance chosen in the present case is 60 cm. This gives good space for both gamma-ray and neutron detectors. This could include large parts of the HHIRF Spin Spectrometer or several of the BGO Compton suppressed germanium detectors. This distance can be crucial when multiple particle detectors are needed in inverse reactions. Moreover, this distance with quadrupoles that have an aperture with a diameter of 10 cm still gives a large solid angle of 8.3 msr. The angular acceptance is not symmetrical, as the dispersive plane has an angular acceptance that is one-third that of the vertical plane.

Velocity dispersion matching constraints, beam rigidity, and beam rejection requirements fix the layout of the dispersive elements of the spectrometer. Therefore, the main design problems are the layout of the focusing elements and the correction for major aberrations. Effort was made to minimize the second order aberrations, but without the loss in spatial focusing. A small vertical beam size at the focal plane will allow several experiments (e.g. radioactive ion beams, isotope separation, tilted foils, laser gas cells) to be performed.

The system has been limited to two active sextupoles which yield an incompletely corrected focus with respect to angle variation in the dispersive plane (θ) and energy (δ_E) when used to: rotate the focal plane perpendicular to the optic axis, and aid in the vertical focus. In particular, a modest x/θ^2 and a small x/δ_E^2 aberration remains. The velocity chromatic terms that must be reduced by the sextupoles are best minimized by placing the sextupoles where x/δ_v (fractional velocity dispersion) is large. The velocity dispersion is not its largest at the entrance quadrupoles, but at the entrance and exit of the magnetic dipole. Thus, the magnetic dipole is preceded and followed by a sextupole. By using separate sextupoles one has greater control over the correction for chromatic aberration while keeping the magnetic dipole somewhat simpler and cheaper. Even if the effective pole face curvature is not exactly as planned on the dipole, it can be compensated with the sextupoles. Other aberrations including $x/\theta\delta$ are negligible.

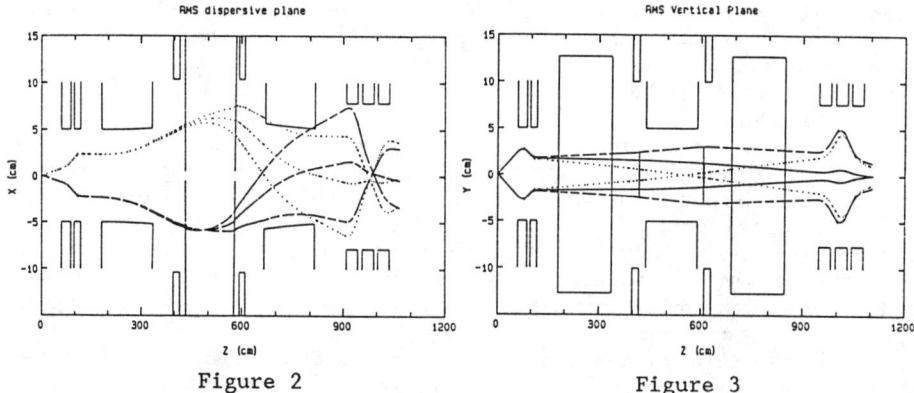

Figure 2   Figure 3

Figure 2 shows the θ acceptance and focus of the spectrometer for the maximum values of θ, three masses of A=95, 100, and 105 amu and a $\Delta E=\pm 5\%$ in the dispersive plane. Figure 3 shows the vertical plane for mass A=100. The mass resolution at full solid angle is m/Δm$\approx$300. Above 5 msr the resolution is approximately constant but rises to m/Δm$\approx$400 at 1 msr and m/Δm$\approx$700 at 0.25 msr.

The first elements are a pair of magnetic quadrupoles. The first lens is diverging in the dispersive plane, and the second lens is converging. This is basically a point-to-parallel and point-to-point focusing in the dispersive and vertical planes, respectively. This is the same as in the Rochester device, but in that case a triplet set of quadrupoles is used. The electric deflectors (157 cm by 25 cm) are planned to have initially a maximum field of 40 kV/cm, which would yield an electric rigidity, Eρ=24 MV or 12 MeV/q. By adding the quadrupole triplet, a vertical focus can be obtained and the mass dispersion increased by changing the magnification. These improvements have their price in other aberrations becoming larger, but some of the worst of these can be corrected in the focal plane detector.

The design proposed is new in the sense that a configuration was selected and calculations were performed to match the spectrometer to the accelerators at HHIRF so that a broad experimental program could be performed. This RMS is a device matched to the goal of performing frontier research in heavy ion physics.

This work was supported in part by the U.S. Department of Energy under contracts No. DE-AC07-76ID01570, No. DE-AS05-76ER05034, and NSF under contract No. PHY8515908.

REFERENCES

1. H.A. Enge, Nucl. Instrum. Methods 186, 413 (1981).
2. H.A. Enge, Nucl. Instrum. Methods 162, 161 (1979).
3. K. Sistemich, Nucl. Instrum. Methods 139 203 (1976).
4. S. Wen et al., J. Phys. G:Nucl. Phys. 11, 173 (1985); S.J. Robinson et al., BAPS 30, 726 (1985).
5. X. Zhao et al., BAPS 31, 772 (1986).

THE PROJECTILE-FRAGMENT SEPARATOR AT THE DARMSTADT SIS/ESR FACILITY

P. Armbruster, H.-G. Clerc[1], J.P. Dufour[2], B. Franczak, H. Geissel,
E. Hanelt[1], O. Klepper, B. Langenbeck, G. Münzenberg, F. Nickel,
M.S. Pravikoff[2], E. Roeckl, D. Schardt, K.-H. Schmidt, D. Schüll,
T. Schwab[4], B. Sherrill[3], K. Sümmerer and H. Wollnik[4]

GSI Darmstadt, [1]TH Darmstadt, [2]C.E.N Bordeaux, [3]MSU, [4]Univ. Giessen

ABSTRACT

The projectile-fragment separator (FRS) is presented, which is under construction at the Darmstadt SIS/ESR facility.

At GSI heavy-ion research will be extended to relativistic energies with the new SIS/ESR facility [1]. All ions up to uranium can be accelerated with the synchrotron SIS up to specific energies of 1-2 GeV/u. Secondary beams of radioactive isotopes can be produced with a high rate and efficiently separated by the projectile-fragment separator FRS [2] (see Fig. 1), presently under construction.

The separated fragments can be studied and used directly at the final focal plane of the FRS, see indicated experimental area 2 in Fig. 1, or they will be injected into the storage- and cooler-ring ESR for experiments in the ring. Using the cooling facilities of the ESR the phase-space density of the radioactive beam is highly increased, i.e., the momentum spread is reduced to approximately 10^{-5} allowing to decelerate down to the Coulomb barrier. This specially prepared beam can then be extraced from the ESR and delivered to the different experimental facilities in the main experimental area. In the following the production rates for projectile fragmentation at relativistic energies are outlined, the method of in-flight separation is presented, and first experiments with the FRS are exemplified.

PRODUCTION OF RELATIVISTIC RADIOACTIVE BEAMS

At relativistic energies projectile fragmentation is a well suited process for producing and separating new exotic nuclei [3]. For radioactive beams near stability [4] intensities of up to 5×10^8/s can be achieved using target thicknesses of $\cong 1 g/cm^2$ and projectile intensities of 5×10^{11}/s. The separation of light fragments ($A_f < 40$) at energies in the order of 50 MeV/u has been sucessfully demonstrated with the separator LISE at GANIL [5,6].

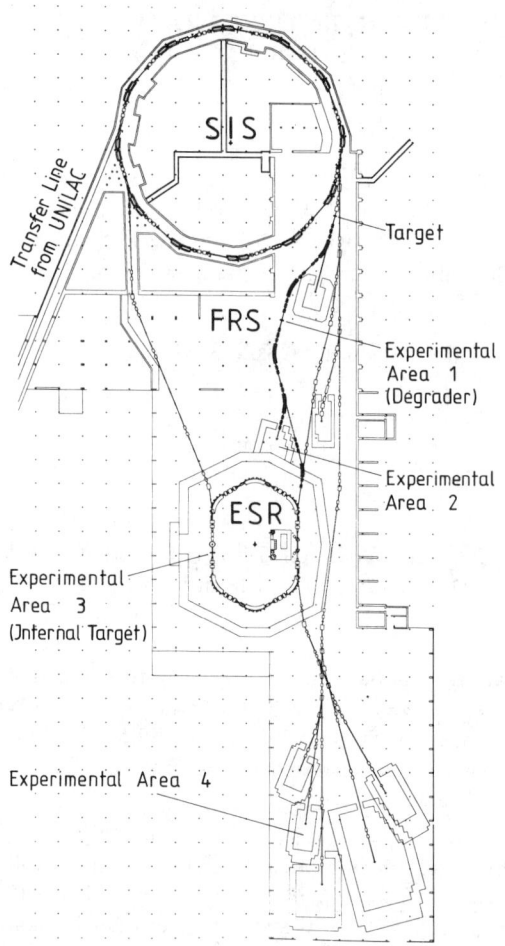

Figure 1: Layout of the synchrotron SIS, the fragment separator FRS, the cooler and storage ring ESR, and the experimental areas for investigations with radioactive beams.

The kinematic properties of the projectile fragments are determined by the nuclear reaction mechanisms [7] and the slowing down processes in the production target. The corresponding velocity spreads are the basic limitation for achieving a high resolving power in A / Z using only magnetic sector fields. Angular and energy-loss straggling are of minor importance at relativistic energies even when the transversed material thickness is a significant portion of the range of the selected fragments.

IN-FLIGHT SEPARATION OF SELECTED RADIOACTIVE ISOTOPES

The main design goals of the FRS are to separate the nuclear reaction products from the primary beam and additionally to perform an efficient separation for the projectile fragments with respect to A and Z, this means, to separate a given isotope. FRS will separate in-flight isotopes up to uranium at specific energies in the range of (0.1 - 1) GeV/u. These optimum operation energies are determined by atomic charge-exchange processes at the lower energy side and by the increasing influence of secondary reactions at the higher energies.

The principle of the isotopic separation is based on a combination of magnetic analysis and electronic energy loss of the fragments in matter [8]. The separation method is schematically presented in Fig. 2,

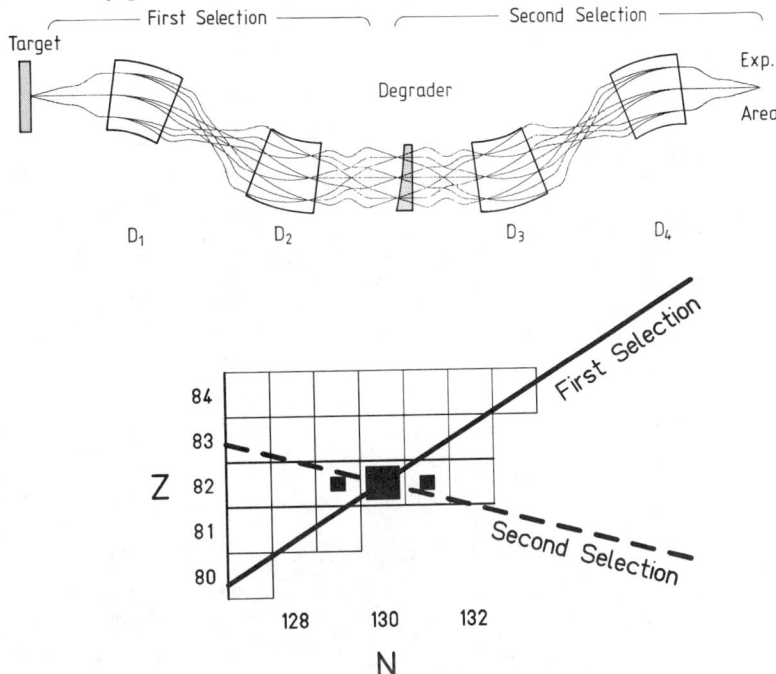

Figure 2: The separation principle [8] of the FRS for a difficult test case at 1 GeV/u. The two selection lines on the chart of nuclei are calculated for the separation of ^{212}Pb produced via uranium fragmentation. The squares represent linearly the transmisson values of the selected fragment and the contaminants [9].

where the different dipole sections are indicated by D_1-D_4. The keys for the separation are an achromatic system characterized by a high resolving power independent of the velocity spread of the fragments ($\Delta v/v \cong 1$ %) and a profiled degrader at the dispersive focal plane providing the separation in A and Z. Due to the reaction kinematics the first separation selects only a chosen A/Z ratio. All fragments with the same magnetic rigidity are focused on the degrader. The electronic energy loss of the fragments penetrating the degrader provides the additional selection needed for the separation of a single nuclide. The shape of the degrader is chosen to preserve the velocity achromatism of the system. For a difficult test case the two selection criteria are demonstrated for the separation of ^{212}Pb produced by 1 GeV/u ^{238}U fragmentation in a 0.5 g cm^{-2} Be target. The degrader was chosen to be 5.3 g cm^{-2} Pb in this sample calculation.

The transmission values of the selected ^{212}Pb and the contaminants are shown by the corresponding sizes of the squares on a chart of nuclides. Detailed considerations about the resolving power and separation characteristics are presented in reference [9].

The FRS consists of 4 magnetic dipole stages, 20 quadrupoles, and 8 sextupoles, see Fig. 3. The production target is positioned at the

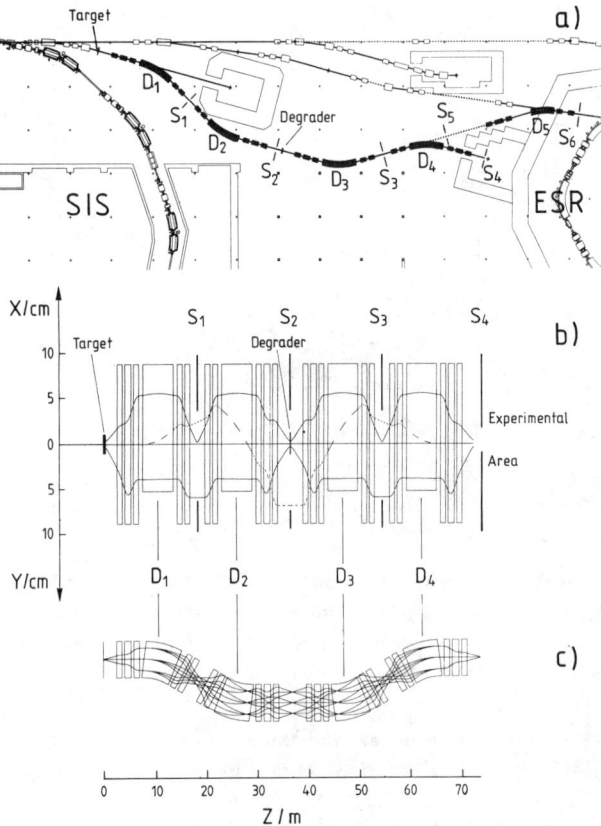

Figure 3: The ion-optical elements of FRS. The 30^0-dipoles are indicated by D_1-D_4. The different focal planes and the corresponding slit positions are marked by S_1-S_4. The full lines in part b) represent the envelopes for a 18 Tm-beam with an emittance of 20 πmm mr. The dashed line is the corresponding dispersion curve for $\Delta p/p = 1$ %. In part c) the dipole and quadrupole magnets are shown in the dispersion plane for three different angles and three momenta.

entrance of the FRS. The primary beam has to be separated from the fragments at the focal plane of the first dipole section (slit S_1 in Fig. 3). The separation quality at the final focal plane S_4 is strongly dependent on the ion-optical resolution of the separator, the thickness and the perfection of the degrader and the energy of the fragments. The momentum resolution at the degrader position is 1500 for an emittance of 20 πmm mrad. Separation in A and Z of all fragments up to mass 240 will be possible with low background.

EXPERIMENTS WITH THE PROJECTILE FRAGMENT SEPARATOR

The scientific motivation for nuclear physics experiments with exotic nuclei using the new separator is a straightforward continuation at SIS-ESR of the successful programme of the investigations of exotic nuclei during the past ten years of heavy ion research at the UNILAC [10].

New exotic nuclei can be produced at relativistic energies by projectile fragmentation using the high intensities planned for SIS. In Fig. 4 the chart of nuclei is

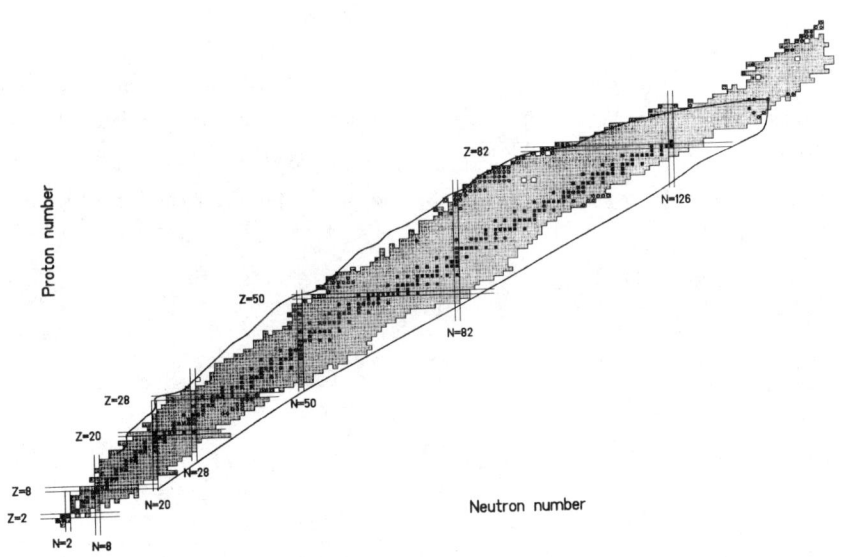

Figure 4: Chart of nuclides with estimated regions of new isotopes, which can be produced via projectile fragmentation with a rate larger than 1 / s.

shown with the nuclides which are expected to be produced and separated with FRS. The estimated boarder lines for production rates of 1 s^{-1} are indicated. Production rates of 5×10^8/s can be achieved for nuclei near stability. An extensive region of new neutron-rich heavy nuclei can be discovered, e.g., ^{100}Sn and ^{78}Ni are within reach.

Although FRS will be optimized to separate efficiently projectile fragments with respect to A and Z, the different dipole sections may ideally be used to prepare beams with well-defined charge states. Experiments which can be performed directly at the separator are: decay spectroscopy of nuclei far from stability, studies of ground state properties of exotic nuclei, spectroscopy of few-electron systems, investigations of atomic collisions with few-electron ions, measurements on the slowing-down and charge exchange of relativistic ions in matter and applications in other fields of science [11].

The magnetic system can also be operated as a high-resolution energy-loss spectrometer, which might offer advantages for reaction studies with a target positioned at the intermediate focal plane. Independent of the relatively large momentum spread of the incident projectile beams, high resolution measurements of the energy transfer in nuclear reactions become possible.

A new generation of experiments will be possible with the combination of SIS, the fragment separator FRS, and the storage ring ESR. The separated fragments can be stored, cooled and investigated in the ESR. For example, accumulated fragments of high phase-space density allow to study their nuclear structure in inelastic collisions with light internal target atoms (H, D, ^{3}He) [1]. Measurements of β-decay into bound final electron states can be performed for the first time [1]. FRS in combination with the ESR will be a unique tool for direct mass measurements of exotic nuclei up to $Z = 92$.

References

[1] Proceeding of the Workshop on Experiments and Experimental Facilities at SIS/ESR, GSI Report 87-7 (1987)

[2] H. Geissel et al., Projectile-Fragment Separator, Proposal for the SIS-ESR Experimental Programme (1987)

[3] T. J. Symons et al. PRL 42, 40, (1979)
G. D. Westfall et al. PRL 43, 1859 (1979)

[4] J. Alonso et al. Trans. on Nucl. Sci. NS-26, 3003 (1979)

[5] R. Anne et al., GANIL Report RA/NJ 278 (1982)

[6] J.P. Dufour et al., Z. f. Phys. A234, 487 (1986)

[7] J. Hüfner, Phys. Report 125 No.4 (1985) and references therein

[8] J.P. Dufour et al., Nucl Instr. Meth. A248, 487 (1986)

[9] K.H. Schmidt et al., GSI Report 1987, to be published in Nucl. Instr. and Meth. A

[10] N. Angert and P. Kienle eds., "10 Years Uranium Beam at the UNILAC", GSI 86-19 (1986)

[11] H. Geissel et al., in Abstracts of the Third Workshop of Heavy Charged Particles in Biology and Medicine GSI-87 11 (1987)

TESTS OF A LARGE AIR-CORE SUPERCONDUCTING SOLENOID AS A NUCLEAR-REACTION-PRODUCT SPECTROMETER

R. L. Stern, F. D. Becchetti, T. Casey, J. W. Jänecke, P. M. Lister
and W. Z. Liu
University of Michigan, Ann Arbor, Mi. 48109

D. G. Kovar, R. V. F. Janssens, M. F. Vineyard and W. R. Phillips
Argonne National Laboratory, Argonne, Il. 60439

J. J. Kolata
University of Notre Dame, South Bend, In. 46556

ABSTRACT

An air-core superconducting solenoid, with a diameter of 0.2 m and a length of 0.4 m, has been configured for use as a heavy-ion reaction-product spectrometer (E/A ≤ 5 MeV/u) near θ = 0° (10 to 35 msr). The performance of the spectrometer was established using α-particle sources and nuclear-reaction products from (^{18}O, ^{18}Ne), (^{18}O, ^{20}Ne) and (^{18}O, ^{14}O) and masses determined for ^{30}Mg, ^{108}Ru and ^{109}Rh. A system suitable for production of radioactive beams has been constructed, and in-beam tests are in progress at the University of Notre Dame. Large air-core solenoids with $d\Omega$ ≥ 20 msr and capable of focusing ions with E/A ≥ 30 MeV/u appear feasible.

The ideal magnetic spectrometer for study of heavy-ion reactions would i) operate near 0° (0° to 30°); ii) focus heavy ions up to 50 or 100 MeV/u; iii) have a very large solid angle (> 20 msr); iv) have variable energy dispersion and magnification; v) have both long and short TOF flight path capability; vi) utilize high-resolution solid-state [ΔE, t, E, xy] detectors at the focal plane; and vii) have simple optics. A device which can satisfy most of these needs as a nuclear reaction product collector and filter is the superconducting-solenoid spectrometer.[1,2] A solenoid, when aligned at $\theta \approx 0°$, acts as a large-aperture, broad-range lens ($d\Omega$ > 20 msr) with nearly isochronous, variable flight paths (2 m ≤ ℓ ≤ 8 m).

A 0.2 m bore, 0.4 m long, 3.5 T air-core superconducting solenoid magnet has been configured as an ion spectrometer as shown in Fig. 1, with the characteristics given in Fig. 2. Recent in-beam tests at ANL-ATLAS have utilized a (ΔE - t), (E - xy) solid-state telescope consisting of a large area (450 mm^2), planar, 28 μm thick silicon detector for ΔE and timing backed by a special 25 × 25 mm^2 two-dimensional position-sensitive silicon surface-barrier E detector 200 μm thick. This position-sensitive detector was developed at LBL[3] and utilizes a resistive anode for xy position readout. It has a position resolution of ≤ 0.2 mm and 90 to 140 keV energy resolution for 8.78 MeV alpha particles. Position (xy) and E - TOF spectra, the latter gated via xy to limit the Bρ range, are displayed in Figs. 3

and 4.

The mass resolution of the spectrometer is limited by the non-isochronism versus θ, the kinematic shift versus θ, and the timing resolution of the accelerator and the detector. Like a dipole spectrometer, one can correct for kinematic shifts in energy (Fig. 5). With Δt = 0.6 ns typical of the silicon-detector system, one has $\Delta M/M \approx 1/50$ for a 2 m flight path.

Energy spectra for $^{26}Mg(^{18}O,^{20}Ne)^{24}Ne$ and $^{26}Mg(^{18}O,^{14}O)^{30}Mg$ are shown in Fig. 6 and 7. The energy accuracy (±150 keV) and resolution, typically ~1 MeV FWHM, were limited primarily by properties of the accelerator beam and somewhat by the solid-state telescope. Spectra for $^{110}Pd(^{18}O,^{19}F)^{109}Rh$ and $^{110}Pd(^{18}O,^{20}Ne)^{108}Ru$ were also obtained, the latter being one of the first direct observations of ^{108}Ru and establishing a mass for this neutron-rich nuclei. The mass excesses obtained are ΔM (^{30}Mg) = -10.32 ± 0.16 MeV, ΔM (^{108}Ru) = -83.95 ± 0.15 MeV and ΔM (^{109}Rh) = -84.24 ± 0.15. While these values are not particularly accurate, they were obtained during relatively short runs (1-2 hours) and illustrate the potential of this type of magnet for mass measurements.

A solenoid similar to the one used at ANL has been set up at the University of Notre Dame Tandem Van de Graaff facility to produce and focus secondary, radioactive beams at the focal plane. This spectrometer will be run in an asymmetric mode, with $z_{tgt} \approx 50$ cm and $d\Omega \approx 100$ msr (Fig. 2). With a production cross section of ~ 10 mb/sr and $d\Omega \approx 100$ msr we expect a ^8Li production rate of 10^5 to 10^6 ions/sec at the secondary focus. Preliminary tests are in progress.

The existing magnet is limited to E/A < 6 MeV/u. We propose construction of a large-bore, high-field, air-core superconducting solenoid with a 0.4 m bore, ca. 1 m long, and B $\geq$ 5 T. Our design work indicates that such a magnet would be a very cost-effective system capable of focusing ions up to E/A = 50 MeV/u with solid angles of $d\Omega$ = 20 to 800 msr. Alternately, one could use the magnet with its large bore as a very-long-flight-path TOF spectrometer for high-mass ions, e.g. fusion products. The very large solid angle possible with a 0.4 m bore magnet makes this magnet well suited as a secondary and radioactive beam collector. This work has been supported by NSF, DoE grants PHY-83-08072, PHY-86-05907, and W-31-109-ENG-38.

REFERENCES

[1] J. P. Schapira, S. Gales, and H. Laurent, Orsay Report IPNO-PhN-7921 (1979); J. P Schapira, et al., Nucl. Inst. Meth. 224, 337 (1984); S. Gales, et al., Phys. Rev. Lett. 53, 759 (1984).

[2] R. L. Stern, et al. Proc. Conf. on Instrumentation for Heavy-Ion Nuclear Research, ed. D. Schapira (ORNL Publ. Conf.-841005 -Absts., 1984), p. 95; R. L. Stern, et al., Rev. Sci. Instr. 58, 1682 (1987). R. L. Stern, Ph.D. thesis (University of Michigan, 1987) unpublished.

[3] J. Walton, LBL detector group

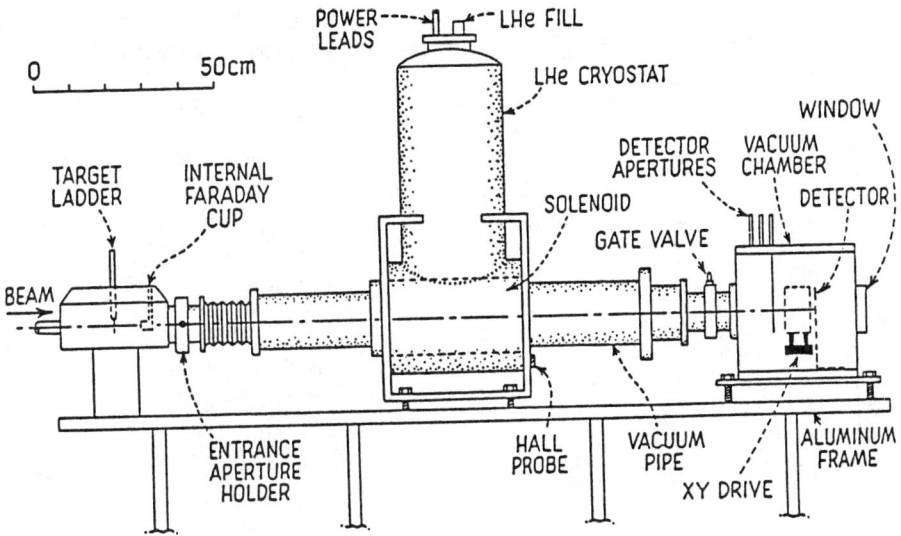

Fig. 1. Schematic diagram of the University of Michigan 0.2 m bore superconducting solenoid spectrometer.

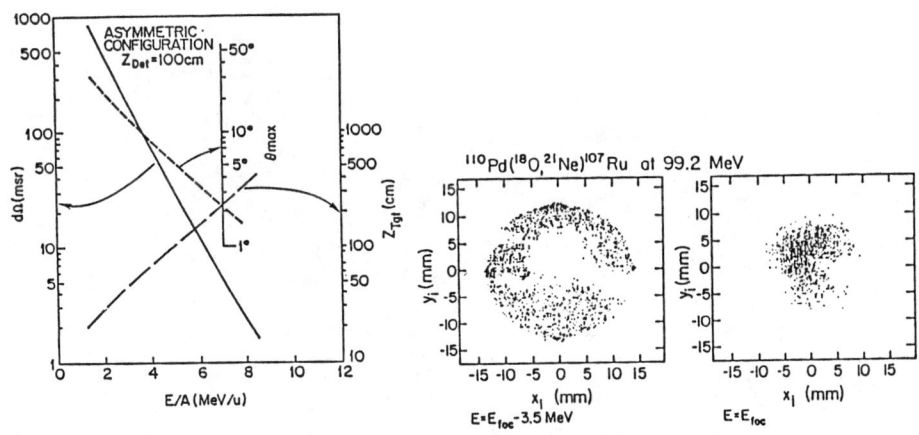

Fig. 2. Calculated solid angle vs. image and object distance and E/A for asymmetric configurations of the spectrometer. The energies are for non-relativistic ions with $q = Z = A/2$.

Fig. 3. Images of ^{21}Ne ions ^{18}O + ^{110}Pd. The ^{21}Ne ions are gated by windows 1 MeV wide in energy: a) $E = E_{foc} - 3.5$ MeV; b) $E = E_{foc}$.

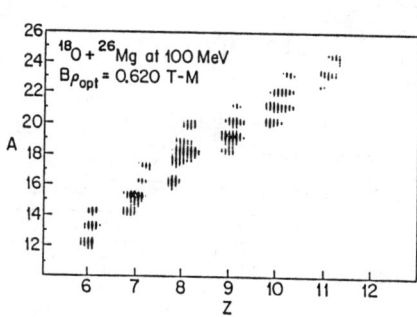

Fig. 4. An ion identification spectrum for 99.2 MeV ^{18}O + ^{26}Mg obtained with the silicon solid-state detector telescope.

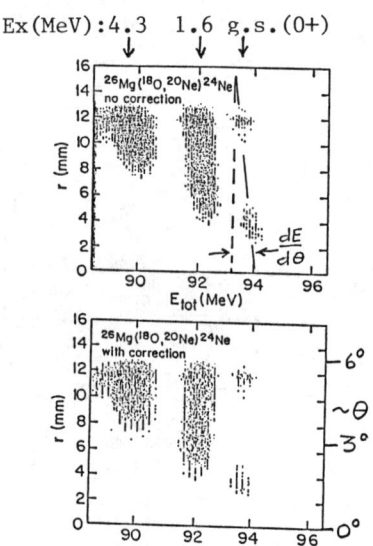

Fig. 5. E_{tot} vs. r for $^{26}Mg(^{18}O,^{20}Ne)^{24}Ne$ at E_{beam} = 99.2 MeV before (top) and after (bottom) kinematic corrections.

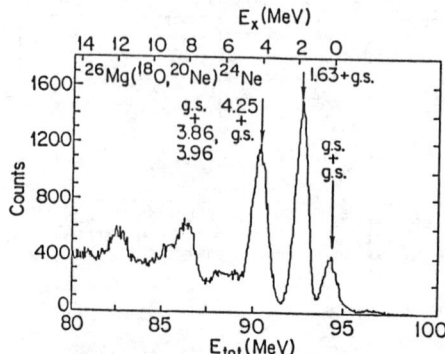

Fig. 6. Energy spectrum for $^{26}Mg(^{18}O,^{20}Ne)^{24}Ne$ at E_{beam} = 99.2 MeV.

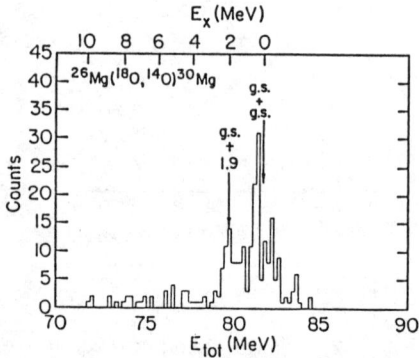

Fig. 7. Energy spectrum for $^{26}Mg(^{18}O,^{14}O)^{30}Mg$ at E_{beam} = 99.2 MeV.

UNISOR ON-LINE NUCLEAR ORIENTATION FACILITY

I. C. Girit
Vanderbilt University, Nashville, TN 37235 and
Joint Institute for Heavy-Ion Research, Oak Ridge, TN 37830

G. D. Alton
Oak Ridge National Laboratory, Oak Ridge, TN 37830

C. R. Bingham
University of Tennessee, Knoxville, TN 37996

H. K. Carter and M. L. Simpson
UNISOR, Oak Ridge Associated Universities, Oak Ridge, TN 37831

J. D. Cole
INEL Laboratory, Idaho Falls, ID 83415

J. H. Hamilton
Vanderbilt University, Nashville, TN 37235

B. D. Kern
University of Kentucky, Lexington, KY 40506

K. S. Krane
Oregon State University, Corvallis, OR 97331

E. F. Zganjar
Louisiana State University, Baton Rouge, LA 70803

ABSTRACT

The UNISOR on-line nuclear orientation facility consists of a He_3-He_4 dilution refrigerator on line to the isotope separator. Nuclei are implanted directly onto a target foil soldered to the bottom accessed cold finger of the refrigerator. A 1.5 T superconducting magnet polarizes the ferromagnetic target foils and determines the axis of symmetry. Up to eight gamma detectors can be positioned around the refrigerator, each 9 cm from the target. A unique feature of this system is that the k=4 term in the directional distribution function can be deduced so that a single solution for the mixing ratio can be found.

INTRODUCTION

Low temperature nuclear orientation has been a very productive experimental technique for studying nuclear structure from radioactive decays. The combination of an on-line isotope separator and dilution refrigerator has increased the applicability of low temperature nuclear orientation to a wide range of short-lived nuclei far from stability since about 1980.[1,2] The UNISOR system described in this paper is now added to the list of fully operational systems located at Daresbury, England and Leuven, Belgium.

The UNISOR system features a unique design in that the beam access to the target is from the bottom of the refrigerator. This configuration allows up to eight Ge detectors to view the target. The superconducting magnet is designed so that four of the eight detectors can be positioned at 45° angles, enabling the deduction of k=4 terms in the directional distribution function, and thus removing the ambiguity in the determination of the multipole mixing ratios.

The UNISOR facility is now completely installed. The initial performance tests are being done. In this paper, the design features, and some preliminary performance results of the system will be presented.

EXPERIMENTAL EQUIPMENT

A schematic view of the UNISOR on-line NO facility is shown in Fig. 1. The Holifield Heavy-Ion Research Facility (HHIRF) 20 MV Tandem accelerator (TM) provides the primary beam. This beam is focused on a target in a modified FEBIAD ion source (IS). The ionized reaction products are accelerated (typically 50 kV) and separated by a 90° bending magnet (MS). The chosen mass beam then is focused by several Einzel lenses in the central beam line and is bent by an electrostatic deflector by 90° and is transported to the cold finger of the refrigerator. The vertical beam is focused by a quadrupole triplet lens before entering the cold beam tube (CBT). The foil (T) on which the activity is implanted is soldered to the cold finger of the He_3-He_4 dilution refrigerator, which should maintain the temperature at about 12 mK. The magnetic domains in the ferromagnetic target foils (usually Fe) are polarized by a 1.5 T superconducting magnet (SM).

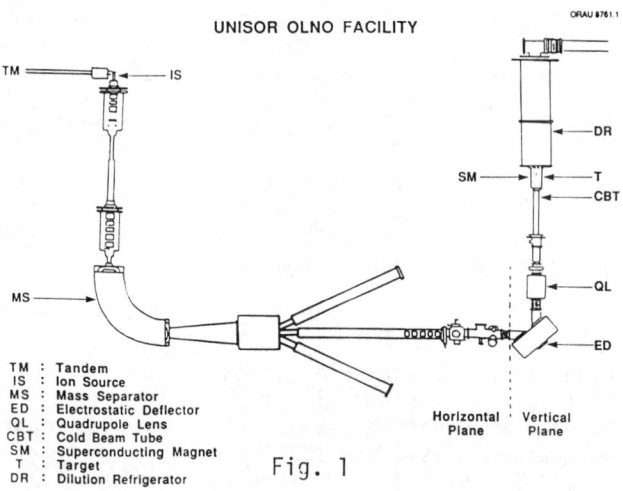

TM : Tandem
IS : Ion Source
MS : Mass Separator
ED : Electrostatic Deflector
QL : Quadrupole Lens
CBT : Cold Beam Tube
SM : Superconducting Magnet
T : Target
DR : Dilution Refrigerator

Fig. 1

THE He_3-He_4 REFRIGERATOR

The He_3-He_4 refrigerator and the bottom access tubes were manufactured by the Oxford Instruments Co., Ltd. Both are standard designs, but in order to allow bottom access, the tail section and magnet are modified. The tails are cylindrical, the outer one is 18 cm in diameter, so that the closest source-to-detector distance is 9 cm. At this distance, for detectors 5 cm in diameter, the magnet allows 1.65, 1.29, and 1.07% solid angle for 0°, 45° and 90° detectors respectively. There is also a 2.2 cm diameter opening at the bottom of the magnet for beam entrance. The refrigerator is equipped with a "top-load" facility to change target material while the refrigerator is operational. The cooling power of the refrigerator as a function of temperature at a circulation rate of 500 µmol/s is given by $Q = 0.017(T^2-44.9)$ µW.

The target foils and thermometer sources are soldered on the face of a 9 cm diameter cylindrical, copper sample holder which is screwed into the mixing chamber of the dilution unit by the top-loading rod. This sample holder lies at the center of a 1.5 T Helmholtz type superconducting magnet, which can also be run in persistent mode. The field direction is horizontal and determines the axis of symmetry (0 degrees). The beam direction is perpendicular to this field. Since the beam comes from the bottom of the refrigerator, a perfect cylindrical symmetry is preserved so that two 0°, two 90°, and four 45° detectors can be positioned around the tail section.

The refrigerator cryostat and all the detectors are accommodated by a 6 m high platform. This platform is surrounded by another platform for detection electronics, storage dewar, experimentalists and for cryogenic liquid transfer (see Fig. 2 for an overall view).

THE BEAM LINE

The 50 kV beam from the separator is controlled and monitored by several lenses and deflector assemblies. At first, the beam is focused in the dispersion chamber of the separator. Its shape is checked by a beam profile monitor, and the intensity of the beam is measured by a Faraday cup. This beam is focused by two sets of Einzel lenses in the horizontal beam line, and transmission through the line can be checked with a wire beam scanner and a Faraday cup before it is bent up. A 90° electrostatic deflector made out of 7 equally spaced parallel plates deflects the beam. A quadrupole triplet lens working at ~ 18 kV potential provides the final focusing before the beam enters the cold beam tube. It is also possible to steer this beam just after the lens through a set of x-y deflectors. The beam monitoring at this stage is done by a modified radiation baffle which a Faraday cup with 6 mm entrance aperature is attached. An iris in the cold beam line about 70 cm away from the target allows control of the beam size. The transmission from the separator to the target was measured to be 96%.

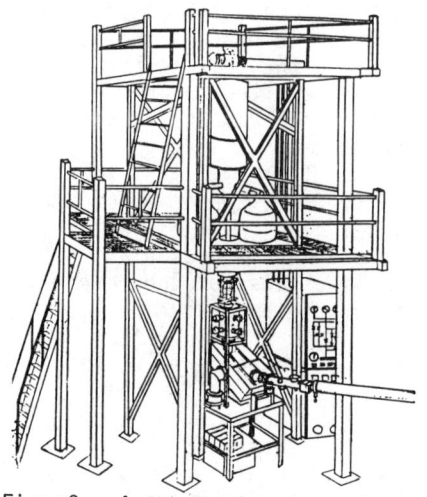

Fig. 2. A schematic view of the UNISOR-OLNO facility.

DATA ANALYSIS

In UNISOR nuclear orientation experiments, the anisotropy or the effect is measured experimentally as

$$A(\theta) = 1 - \frac{N(\theta,C)}{N(\theta,W)} \bigg/ \frac{N(0°,C)}{N(0°,W)}$$

where $N(\theta,C)$, and $N(\theta,W)$ refer to areas of the peaks taken when the implanted nuclei are oriented (cold) and are at 1K (warm), respectively. Since in the UNISOR system it is possible to have up to 8 detectors positioned around the target at 45° intervals, this definition produces 12 anisotropy values for every peak in the spectrum, each 45° and 90° detector data normalized to 0° and 180° detectors. For decay scheme studies, where no temperature dependence of the anisotropy is measured, the anisotropy data for each detector are averaged for every detector separately within a limited temperature range, ($\Delta T=0.5mK$). In the final analysis, if the orientation coefficients B_k are known, one obtains 16 $a_k=U_k A_k$ values for every peak which then can be averaged for further error reduction. The 0° averaging corrects for errors due to the movement of the beam spot or the cryostat in the polar direction, and 90° averaging in the equatorial direction.

SUMMARY

A helium dilution refrigerator on-line to the UNISOR mass separator is installed. Initial performance tests of the beam line, refrigerator, and data acquisition systems have been carried out. The three on-line experiments, the study of ^{189}Hg, ^{201}Po decays and Ge-Se region are scheduled to take place next year.

UNISOR is a consortium of universities, State of Tenn., ORAU, ORNL, and is partially supported by them and by the U.S. DOE under Contract DE-AC05-76OR00033. The work at Vanderbilt Univ., Univ. of TN, INEL Lab., Oregon State Univ., and Louisiana State Univ. is supported by U.S. DOE Contracts DE-AS05-76ER05034, DE-AS05-76ER04936, DE-AC07-76ID01570, DE-FG06-87ER40345, and DE-FG05-84ER40159.

REFERENCES

1. D. Vandeplasche, et al., NIM 186 (1981) 211
2. "Low Temperature Nuclear Orientation," Ed. N. J. Stone and H. Postma, North Holland, 1986

ISOL - FACILITY YASNAPP-2

K.Ya.Gromov, V.G.Kalinnikov, V.M.Tsupko-Sitnikov
Joint Institute for Nuclear Research,
Laboratory of Nuclear Problems, Dubna, USSR 141980

ABSTRACT

The nuclei far from stability have been investigated on the synchrocyclotron beam in the Laboratory of Nuclear Problems about 30 years. The investigations have been developing from the first off-line works (1956), which showed that 660 MeV protons were effective for production of neutron-deficient nuclides, through the semi-on-line system YASNAPP-1, and to the recently-built ISOL facility YASNAPP-2. The facility and the spectrometric devices prepared for it are described.

Over several years the Laboratory of Nuclear Problems in the Joint Institute for Nuclear Research has been studying properties of nuclei far from stability. In 1956-67 a series of investigations was performed with neutron-deficient nuclides produced by irradiation of various targets with 660 MeV protons /1/. This technique was shown to be most efficient and universal for production of neutron-deficient nuclei. These works were a stimulus to the creation of the facility ISOLDE in Geneva. The papers of 1956-67 identified over 100 new isotopes, and much new information was obtained about properties of excited states of neutron-deficient nuclei.

The main stages in the development of spectroscopy of nuclei far from stability in Dubna were as follows:
- Development of a mass-separator (1969) to separate large quantities of radioactive isotopes in the off-line mode. Development of a highly efficient tube-like ion source with surface ionisation. Spectroscopy of Rn, At, Tl, rare earth and other isotopes (1969-1975).
- Development of a semi-on-line isotope separator YASNAPP-1 /2/ comprising of: a target for E_p=660 MeV, a rabbit post, a mass-separator. Nuclides with $T_{1/2} \approx 1$ min (1970-1979) were then investigated.
- Participation in the construction of the ISOL facility IRIS (1975) in the Leningrad Institute of Nuclear Physics. The α-, β-, γ-spectroscopy of rare-earth nuclides with $T_{1/2}$ up to 0.5 sec. The limits of nuclear stability against the radioactive p-decay were established for a wide range of nuclei (from Tm to Au). A considerable increase in pair energies was found in nuclei with the excess of protons.

Now an ISOL facility, YASNAPP-2 has been constructed at the proton beam of the JINR phasotron in Dubna. The "target - ion source" device of the mass-separator (I, Fig.1) was installed on the extracted proton beam with E_p=680 MeV, $I_p \approx$ 1-2 μA. The semi-automatic system for replacing ion sources is also there. The "collector" hall houses the dispersion chamber and the beam splitting system (II) with the begining of four ion supply lines (2).

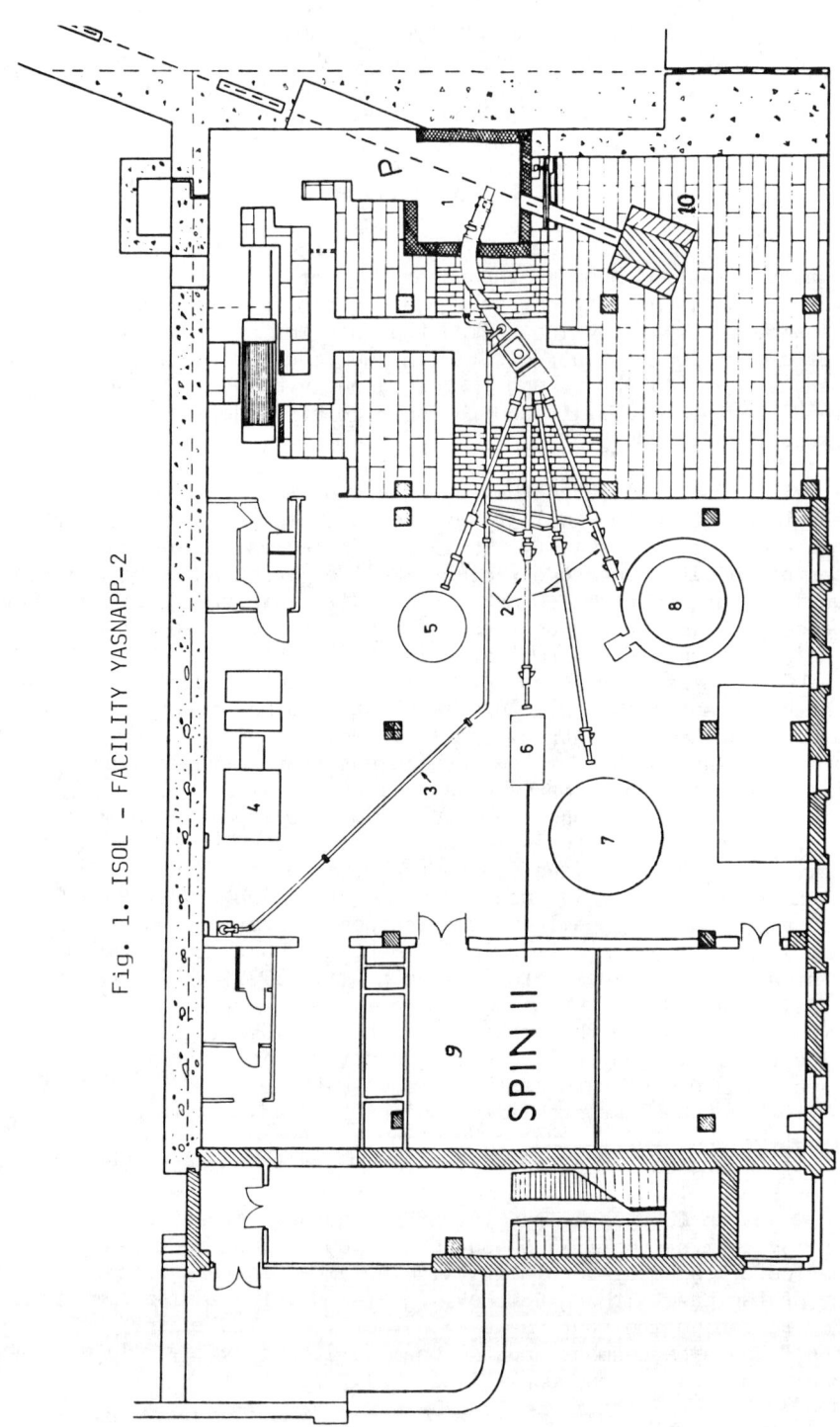

Fig. 1. ISOL - FACILITY YASNAPP-2

The main parameters of the mass-separator /3/ are as follows: the mean trajectory deviation angle is 55°, the mean trajectory radius in the magnet is 1.5 m, dispersion is 15 mm, the resolution is up to 2000, and the range of masses simultaneously injected in the collector chamber is ±15%. An on-line source /3/ has been made; it has thermal surface ionisation with direct glow of Ta, W, Re ionisers used as targets. To ensure long-time (~100 hours) irradiation of the ion source by a proton beam, it has four successively and automatically replaced ionisers. A plasma ion source with a hollow cathode /4/ is being developed to produce radionuclides of elements with an ionisation potential > 7 eV, first of all - Tl, In, Ga. Melted Pb, Sn or Ge are also supposed to be good targets /5/.

At the ion beams (2) of the mass-separator there are spectrometers:

- The multidetector unit ELGA /11/ (7) is designed for the measurement of the spectra of γ-rays, R-ray radiation, internal conversion electrons, positrons, α-particles and of instant and delayed $\gamma\gamma$-and eγ-coincidences. Its measurement chamber is directly installed on the ion supply line and is attached to a belt transport system. There are two SCDs and a beta-spectrometer of the "mini-orange" type in the chamber. Without using the system of radioactive source transportation, one can investigate nuclei with half-lifes up to 0.1 s.

- The multidetector correlation unit MUK /7/ (5) consists of 7 Ge(Li) detectors, an electronic system and a belt transport device to transport radioactive sources. The detectors are rigidly fixed to the γ-radiation collimators in the lead protection block. This unit allows the study of functions of the space-correlation of cascade γ-rays and, naturally, simple γ-spectra, spectra of $\gamma\gamma$-coincidences and life-times of nuclear states in the nanosecond range.

- The magnetic-lens spectrometer MLS (6) is the spectrometer for eγ-coincidences /8/ and is based on a fast magnetic-lens β-spectrometer with a triangular magnetic field /9/ which ensures high isochronism of the trajectories of the electrons. The e-detector is a NF-104 type plastic scintillator. A resolution of $\Delta B\rho/B\rho = 0.4\%$ has been achieved at the luminosity T=0.6% and $\Delta B\rho/B\rho = 0.8\%$ at T=2%. Using a plastic scintillator in the γ-spectrometer, a time resolution has been obtained of $2\tau = 410$ ps at the slope of the curve of instant coincidences $T_{1/2}=50$ ps, and $E_\gamma > 900$ keV at $E_\beta = 250$ keV (^{60}Co). The influence of the charge collection time upon the time resolution of SCDs has been studied /10/. Choosing pulses with the maximum rise rate allows one to improve the resolution of the time spectrometer in the mode of ($\beta_{photom.} - \gamma_{SCD}$)- coincidences from $2\tau = 5.50$ ns to $2\tau = 1.65$ ns for $\gamma 1332$ keV ^{60}Co.

- The magnetic analyser of heavy particles /12/ (8) is a device based on a large precise alpha-spectrometer with double focussing. In the focal plane of the device there is a detector of charged particles consisting of two coordinate-sensitive proportional chambers placed one behind the other. The size of the inlet window is 500x10 mm². One-coordinate read-out is performed with the distributed delay line ($\tau \sim 1.1 \mu$s). A resolution of < 2 mm was obtained for alpha-particles. The connection of the chambers in coincidence reduces detection of background events.

- A laser spectrometric unit (4) is being built on the basis of both pulse and continuous lasers with variable wavelength. This unit is intended to measure superfine structures and isotope shifts of optical lines of short-lived nuclides and to investigate angular correlations of particles during the beta-decay of atomic nuclei /6/. In future it is planned to increase the sensitivity of the technique by bunching and moderating the atomic beams.

- The SPIN-2 facility (9) is designed for the investigation of spin effects during the decay of oriented nuclei (ON) and the nuclear magnetic resonance of oriented radioactive nuclei (NMR/ON). To orient nuclei in the on-line mode the ion beam is directly connected with the ^{3}He-^{4}He dilution refrigerator. One will be able to orient all nuclei with spin $J > 1/2$ and with the half-life more than the time of spin-lattice relaxation in ferromagnetic matrices (10-100 s at a temperature of 10 mK). To orient short-lived nuclei ($T_{1/2} > 1$ ns) it is planned to employ the technique of laser-induced nuclear orientation.

Experiments at all spectrometric units of the facility YASNAPP-2 are automated by the use of several microcomputers. Besides conventional systems and libraries for analysis of γ-spectra and information from the multidimensional measurements a diagram-plotting package /13/ and a system for construction of decay modes /14,15/ are also used.

REFERENCES

1. K.Ya.Gromov, B.S.Dzhelepov. Atomnaya Energiya, 1969, 26, 362-369.
2. R.Arlt et al. JINR, B3-6-7258, Dubna, 1973.
3. V.P.Afanasiev et al. Preprint LINP No 532, Leningrad, 1979.
4. A.Latuszynski et al. Proceed. of the EMIS Conf. Los Alamos, 1986, p.136.
5. A.F.Novgorodov et al. JINR, P6-86-379, Dubna, 1986.
6. V.E.Egorov, A.A.Solnyshkin. JINR, P15-85-862, Dubna, 1985, p.27.
7. V.N.Abrosimov et al. JINR, P6-86-320, Dubna, 1986.
8. V.N.Abrosimov et al. JINR, P6-86-675, Dubna, 1986.
9. V.M.Abazov. JINR, P6-85-238, Dubna, 1985.
10. V.G.Zinov et al. Preprint JINR, 13-84-542, Dubna, 1984.
11. Z.Arvai et al. Preprint JINR, P13-85-774, Dubna, 1985.
12. A.V.Tokmakov et al. JINR, 13-86-488, Dubna, 1986.
13. V.M.Gorozhankin et al. JINR, P10-85-217, Dubna, 1985.
14. V.M.Gorozhankin et al. JINR, P10-85-528, Dubna, 1985.
15. V.B.Brudanin et al. P10-85-116, Dubna, 1985.

PULSED LASER SPECTROSCOPY OF BUNCHED BEAMS

V.Egorov
Joint Institute for Nuclear Research,
Laboratory of Nuclear Problems, Dubna, USSR   141980

ABSTRACT

A new laser-spectroscopic method for investigation of short-lived nuclides (PLSBB) is described. The method consists in the synchronous irradiating of ion or atom bunches with laser pulses. The method is developed in JINR, Dubna, for ISOL-facility ᵧASNAPP-2.

There is a large variety of methods for investigation of short-lived nuclides with laser light beams. The most famous and widely exploited one is the Collinear Laser Spectroscopy of Fast Beams (CLSFB) /1/. The only disadvantage of CLSFB is its relatively low sensitivity in the sense that it needs mass-separated beams with intensity over $10^5...10^6$ particles per second /2/. On the other hend, the most interesting information about nuclear processes is being now obtained from investigation of very short-lived nuclei far from stability but close to a proton- or to a neutron-drip line. The cross-sections of nuclear reactions producing these nuclei are extremely low, and therefore yield of nuclei under investigation is low too (in the case of rare-earth isotopes with halflives less then 10s this yield is usually not more than 1...100 particles per second). Under these conditions more sensitive methods must be developed. Some of them were proposed by the ISOLDE-group of CERN, Geneva /3/, the IRIS-group of LINP, Leningrad /4/, etc.

In this article one more method proposed in JINR, Dubna is described, that is the Pulsed Laser Spectroscopy of Bunched Beams (PLSBB) /5/. The aim of this method is to make all short-lived nuclides in a beam (atoms or ions) interact with a laser beam.

Since the increase of the average laser power to values over 1...2 Watts is connected with a lot of technical difficulties, we have decided to use pulsed dye lasers with a high repeatition rate. Although the average power of such laser is not large, its intensity is quite sufficient to saturate optical transitions of all nuclides inside the irradiated zone during the laser pulse. To avoid the loss of nuclides crossing this zone between laser pulses it is suggested to transform a continuous atomic or ion beam from ISOL-facility into a pulsed one by bunching it.

It must be mentioned that something like that has been already realized in Leningrad University /6/, when a pulsed atomic beam was produced in the off-line facility by evaporating a solid sample with the pulsed laser beam. But in the case of the ISOL-facility such evaporation is not effective because it takes place only on the sur-face of the target and percentage of nuclides under investigation is just negligible. The most suitable solution is the electrostatic ion beam bunching by means of an immersion lens fed with a pulsed voltage of special shape.

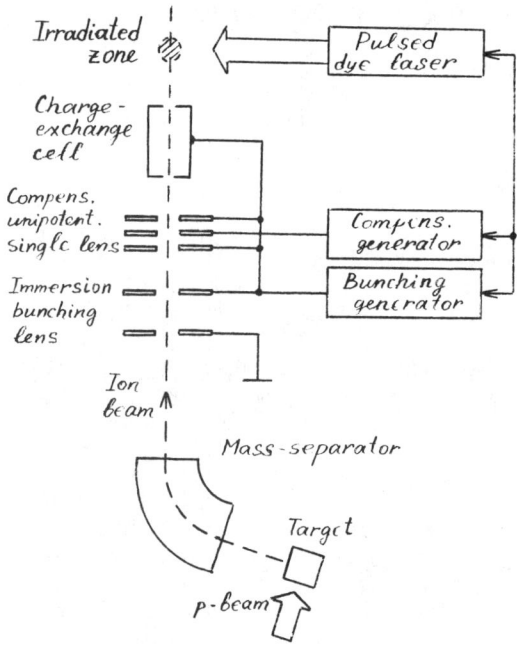

Fig. 1

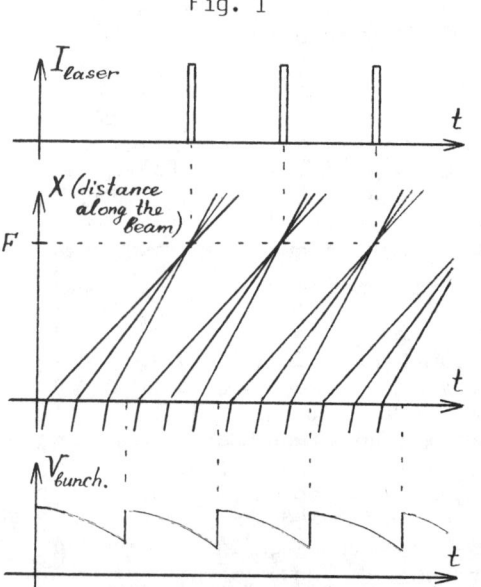

Fig. 2

Fig. 1 shows a simplified lay-out of the ISOL-facility using the PLSBB, and fig. 2 presents diagrams of laser pulses, ion moving along the beam, and bunching voltage.

When passing through the lens, ions are slowed down by a contrary directed electric field to the energy depending on the phase of the feeding pulse. At the beginning of the pulse ions become slower than at the end of the pulse, so that the earlier ions are being caught by the later ones and at some distance from the lens an ion bunch appears. After some time the ions drift apart and the bunch disappears.

The above process is illustrated in fig. 3, showing an ion beam linear density distribution along its axis, calculated for different successive moments after ions' passing through the bunching lens (the lens' thickness is neglected). In the case of ideal shape of voltage pulses (fig. 3a) the bunching would be quite complete, i.e. in a certain moment 100% of ions having passed through the lens would form a point-like bunch. It is difficult to electronically achieve such an ideal shape of the pulse, but even in the case of a simple quasi-harmonic pulse shape (fig. 3b) about 50% of ions can be gathered to a 1cm-bunch.

Irradiating ion bunches by laser pulses

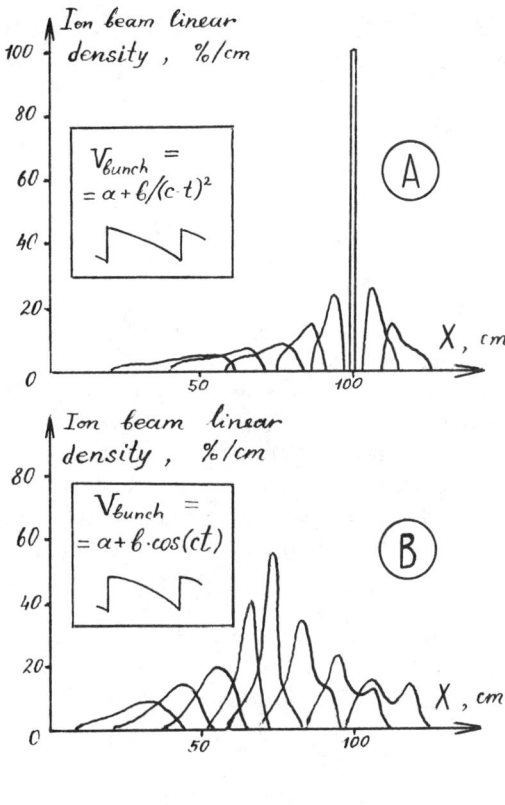

Fig. 3

strictly synchronized with the phase of lens feeding pulses, one can achieve an extremely high efficiency of ion interaction with laser light. To deal with neutral atoms and not with ions, the charge-exchange cell can be put between the last bunching electrode and the bunch formation point: the ions move inertially there and therefore it would not interfer in the bunching process.

Of course, this way has its own difficulties. For example, it is necessary to avoid the radial defocusing of the ion beam, which is a side effect produced by the immersion lens and leading to a large Doppler broadening of resonance lines. This harmful effect can be compensated by an additional unipotential lens with a synchronously varying focus distance, as it is usually done in television engineering.

Naturally, the described method (PLSBB) does not have such a good resolution as CLSFB, but thanks to its large efficiency it can be used to measure isotope shifts of most short-lived nuclides, to separate isobaric beams by the atomic number Z, to polarize separated beams, etc.

The laser-spectroscopic unit utilizing PLSBB is now being installed at the ISOL-facility YASNAPP-2, JINR, Dubna.

REFERENCES

1. S.L.Kaufmann, Opt.Commun. 1976, v.17, 309.
2. E.Roeckl, D.Schardt, Proc. of Int. Symp. on In-Beam Nucl. Spectroscopy, Debrecen, Hungary, May 14-18, 1984.
3. ISOLDE experiments IS81, IS82, IS150, IS01-9, IS01-16; see, for ex., ISOLDE Users'Guide, CERN 86-05.
4. V.N.Fedoseev et al. Opt.Commun. 1984, v.52, 24; G.D.Alchazov et al. Letters to JETP, 1984, v.40, 95.
5. V.Egorov, A.Solnyshkin, Avt.svid. SU-1208499A, Bull. OIPOTZ, 1986, N 4.
6. I.N.Izosimov et al. Izv. AN SSSR, Ser.fiz., 1981, v.45, 2035.

"A View of the Future"

P. Armbruster

Gesellschaft für Schwerionenforschung mbH, Darmstadt, FRG

I am not a prophet, even if I have a beard you could swear by. An honest prophet makes you sad. He tells you the truth, which will stay behind your expectations. A false prophet is a liar. He tells you what makes you feel good. I will define my task differently, and to speak on evident trends, everybody observing the conferences "Far off Stability" could have told you as well. As there are only 4 at the meeting who attended all conferences since the first one 1966 at Lysekil (W. Talbert, N. Zeldes, J. Żylicz, and myself), the choice of having given the so defined task to me even seems to have some sound foundation. The first conference had the title "Why and How Should we Study Nuclei Far Off Stability" or short the "Why and How" conference. It was organised by Ingmar Bergström in his hometown Lysekil in Southern Sweden.

One evident trend: The question of "Why" is not put any more, the question of "How" rarely. We feel established by now two decades of success. We just do it. Anyhow, let me bring back the question "Why and How". It stays important, otherwise we may drop to a "business as usual" level.

"Why" has still two very convincing general answers.

1. In this onion-like world, one layer are the nuclei. We want to study them, the stable ones the world is made out of, the radioactive ones we still find in nature which bring the message that this planet has been created once, the radioactive ones which we make artificially since more than fifty years and which are being made now somewhere far in the universe. *Understanding the nuclear layer of our world and how it is created by astrophysical processes is our primary goal.* This research contributes to a deeper understanding of the world, our cultural mission. Besides its applications, bombs and reactors, there is more in nuclear physics, as we learnt yesterday by the message Monique Bernas wanted to transmit by her film on exotic nuclei. We should be aware of this cultural mission and encouraged to spread it.

2. Why do we like to make and study nuclei ? Why is our field expanding ? About 600 names are listed in the author index of our book of abstracts. *Making things is fun and more than pure understanding.* Having made something and looking at it is deeply rewarding. In the making we experience the freedom of artists, in the understanding of what we made we experience intellectual satisfaction. However, looking at nature's work of art and admiring the beauty of cosmic order we know, we are small and still we feel reassured by this order. We are a priviledged group of people. Our work not only feeds our daily needs, but, moreover may give the part necessary beyond the daily bread.

To compose a new nucleus and to understand it is our task. We are happy with protons and neutrons as building blocks and admire their dance in the nucleus. From the film yesterday we learnt the dance is our message, not the dancers, knowing well that the dancer itself may be worth studying, as we do in β-decay, a transmutation of a dancer. Let the intestines of our nucleons be studied by others which we kindly ask to respect that looking for new dances in the thing we do, and blowing our ballets into pieces and dissecting our dancers in another game, which they may like, but many of us just do not. Working in the layer underlying the nucleons is an important field, but not our central interest. We are connected to this layer via our studies on β-decay, as we depend on the upper level, atomic physics, via our studies using laser techniques. But, we must defend our field as a part of nuclear physics, and the two "Whys" I defined may give you some munition for the next 5 years.

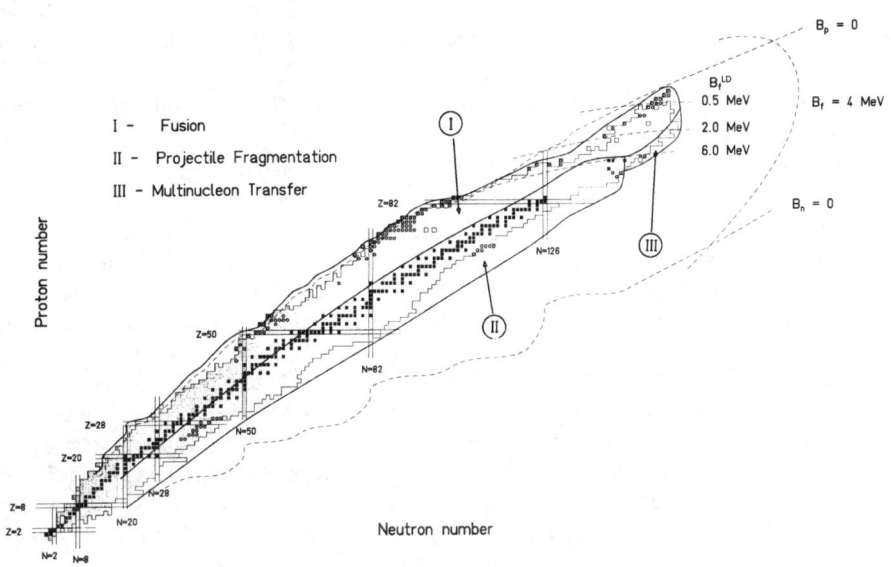

Figure 1: The figure shows the known nuclides and the the boundaries of nuclear instability (hatched lines). The 3 reactions indicated allow the production of further nuclides within the region given by thick lines. Most of the still accessible nuclei are neutron-rich.

Our field is sound, as it was 1966 at Lysekil. Let me show the perspectives of making further nuclides, Fig. 1. The nuclear instabilities make the number of possible nuclei finite. The proton- and the neutron-dripline and the spontaeneous fission boundary, ($T_{1/2} > 10^{-6}$ s) surround a continent of about 6000 nuclides. They all may exist somewhere in the universe but humans live on this wonderful, well tempered, cool planet and not on hot stars with 10^9 K. We found about 265 stable and about 60 radioactive nuclei on earth. Out of these we have to cook other radioactive species. We made about 2000 of them since the discovery of artificial radioactivity in 1934 by F. Joliot and I. Curie. Since then they are coming in with a rate of 40/year. The making of isotopes will limit the number of nuclei, which we may still reach to about 600, mainly on the neutron-rich side. So all in all, we will arrive at about 3000 nuclei, about 50 % of those which would live longer than 10^{-6} s. Finding these nuclei at the same rate of 40/year will guarantee the next 15 years. Finding and showing their existence is only the first step. Studying and understanding the nuclei the real task. It takes about 20 man-years to study the ground state properties and the decay of a nucleus. The community of 600 persons cited in the book of abstracts has to work for 20 years to do the job, in case society is willing to pay further 2×10^6 \$ for a nucleus, that is 1.2×10^9 \$ in total. Do not be afraid of the numbers. The yearly costs of our field amount to 1/hour of society's military spending. Another 5 conferences certainly are necessary to cover the work to come. The field following this extrapolation slowly could vanish to the occupation of a small minority beyond 2010, unless new production methods are discovered. Thus, a baby now borne may have difficulties to find still unknown species. Perhaps new tools allow him to restudy what we were doing and to find new aspects of our game, or physics with exotic beams may have become the most important branch. This is enough, I am sliding into prophecy.

The second part of this summary deals with the two "Hows": *How do we produce and how investigate new nuclei ?*

There are four main reactions: Fission, Spallation or Projectile Fragmentation, Multinucleontransfer, and Fusion. Another trend from Lysekil to Rosseau Lake: Nobody asks for production mechanisms. I am sure we cannot continue this trend. In this conference program the "How" has degenerated to this summary talk. Certainly many speakers mentioned aspects of reaction mechanisms and they indicated that we do not fully understand the production reactions. Especially the complex cold rearrangement processes governed by nuclear structure are poorly studied. There are surprises everywhere, in 50 year old fission, ^{258}Fm (H2) and the heavy Ni-isotopes (M4), in spallation the N/Z-dependence and mean field effects in production rates (M3), in multi-nucleontransfer the recent discovery of new heavy isotopes as ^{261}103 and ^{262}103 and the multitransfer at the Coulomb barrier producing cold heavy neutron-rich nuclei with N = 126, in fusion the production of the heaviest elements by using closed shell Pb- and Bi-targets (I1), and the use of ^{48}Ca to produce superdeformation in ^{152}Dy (L1). To turn the trend I propose for the next conference a session on reaction mechanisms.

The experimental developments since the 1981 Helsingor-Conference are impressing. I will close with a survey of the most important ones.

Direct mass measurements became possible by new instruments. The TOFI-spectrome (A1) allows measurements for recoils with A < 20. The SPEG at GANIL (A2) opened direct mass measurements for recoils with A ≤ 40. With an ion trap and cyclotron frequency resonance techniques masses with A ~ 100 have been measured with the until now unachieved resolution of 300 000 (A7). The small efficiency of 5×10^{-6} is still a problem, but here certainly large improvements can be expected.

There was large progress in the *measurements of moments and radii*. Collinear laser spectroscopy (C2, C3, B1) and resonance ion spectroscopy (C1) has given a wealth of information on isotopic chains for many elements throughout the periodic table. In low temperature orientation techniques nuclear orientation has been complemented by nuclear resonance techniques (B12). Reaction polarization of recoils has been reported and the g-factor of ^{15}C has been measured (B3). Here evidently the reaction mechanism has to be understood better in order to proceed further.

Improvements of existing separation techniques and new spectrometers have been reported.

There are about 30 ISOL-systems working and 2/3 of all contributions to the conference have been obtained by this working horse. A pulsed ion source finally allowed to measure the halflife of ^{68}Ni (E2), a technique which will be useful very generally. He-jet coupled ISOL-systems are working at different places, and we heard a report on the system at SARA (G8). He-jet ion-guide systems using singly charged recoils after slowing down in He have been reported. Systems are working at Jyväskylä (J3, G18) and at Leuven (F3, K13). They separate all products, even the refractive elements within 0.1 ms.

Recoil mass separators and velocity filters are advancing, 8 are in operation and 5 are in construction or planned. We heard reports from Daresbury (L2, N6, N8) and SHIP at GSI (I1). A project for ORNL has been presented as a poster (N9). Correlation techniques together with recoil separators promise further applications in in-beam-spectroscopy.

The technique of combining atomic processes with ion optical devices has a new brother, the Momentum Loss Achromat. For fast projectile fragments it solves the problem of separating recoils of equal velocity and A/Z-ratio, which principally cannot be done by electric and magnetic field combinations. The technique successfully is applied with LISE at GANIL (D1, M3) and will be used at MSU (K3) and the new accelerator combination SIS/ESR at GSI (N10).

In the *study of decay properties* there are interesting innovations. The β-decay studies aiming at the β strength function are complemented by (n,p)-(p,n) and $(\vec{p},\vec{p})$-reaction studies (J2) and work at electron accelerators. These links have been important in the past and should not be neglected in future. α-spectroscopy has discovered the $(\alpha-\gamma)$ coincidence technique and it was applied convincingly for nuclei above ^{146}Gd (G1) and for ^{225}Th (G13). The new BGO-shields for Ge-detectors allow to build high resolution crystalballs for γ-spectroscopy. A highlight of this technique is the discovery of another island of superdeformation besides the fission isomers near ^{152}Dy (L1). This technique combined with the recoil mass spectrometer (L2) as done at Daresbury seems to me a break through in spectroscopy. The discovery of ^{80}Zr is a first success and a wide application can be expected in the future. Neutron spectroscopy will become more important as we go to neutron-rich nuclei. We have seen a TOF-system applied to the study of ^{29}Na (J7). Both, TOF and ^{3}He-spectrometers need further improvements.

A new branch of the studies with nuclei far off stability has been introduced recently, *studies with radioactive beams*. Secondary beams have been used to measure total reaction cross sections of projectile fragments (L4). It was found that ^{11}Li has a radius nearly as large as ^{28}Si. The study of nuclei at the edge of the neutron drip line for the light elements up to Ne is of primary importance (G6). Secondary beams can be produced as beams of fast recoils or via a slowing-down and reionisation process using ISOL-systems and post-acceleration. The latter technique will be applied at Leuven (N15) and has been discussed for TRIUMPH. It is complementary to the fast recoil beams and certainly should be developed further. Projectile fragmentation beams will be delivered by the new GSI-facility (N10). Secondary beam physics will be one of the major programs starting in 1989 / 1990.

Summarizing, ISOL-systems will remain the working horse, but more innovations were presented for recoil techniques. More complex systems will come up. Not every laboratory can afford the expensive instruments. Collaborations in large groups at external facilities will become necessary demanding new ways of organisation. The division of experimentalists into spectrometer operators and users of a facility should be blocked by adequate means. The majority of us are experimentalists, we pass our time mainly with our instruments and the evaluation of data. We should not be afraid to celebrate the next time an instrumental session and invite our collegues from theoretical physics to participate. The wide spectrum from building instruments to theory should be covered in the program, as it was once at Lysekil. Hiding in a position in the center excluding instruments and advanced theory, is a danger for a field, it creates isolation and favours "GURUism", which stand for a closed system, what we hopefully are still far from being or becoming.

Let me close by thanking John Hardy and his team in the name of all of us for organising this interesting conference, choosing this marvellous site and arranging the outing to Honey Island. I hope to see you again in good health, enthusiastic, with new results and achievements, 1992 at the 6th conference to be held in Europe.

(Instead of references the nomenclature given in the book of abstracts has been chosen.)

CONFERENCE PROGRAM

Monday, September 14

OPENING SESSION

8.20   J.C. Hardy (Chalk River)
        WELCOME

A. MASSES (Chair: A.H. Wapstra)

8.30   D.J. Vieira (Los Alamos)
        DIRECT MASS MEASUREMENTS OF LIGHT NEUTRON-RICH NUCLEI USING
        FAST RECOIL SPECTROMETERS

9.05   W. Mittig (GANIL)
        MASS MEASUREMENTS FAR FROM STABILITY OF NEUTRON-RICH LIGHT
        NUCLEI

9.40   P.E. Haustein (Brookhaven)
        THE 1986-87 ATOMIC MASS PREDICTIONS

10.05   D.H. Feng (Drexel)
        THE FERMION DYNAMICAL SYMMETRY MODEL AND COLLECTIVE NUCLEAR
        STRUCTURE PHYSICS

10.50   E. Hagberg (Chalk River)
        MASS MEASUREMENTS OF THE NUCLIDES IN THE ^{162,163}Ta
        ALPHA-DECAY CHAINS

11.15   J.M. Pearson (Montreal)
        NUCLEAR MASSES, POTENTIAL SURFACES AND FISSION BARRIERS FAR
        FROM STABILITY IN THE ETFSI MODEL

11.40   F. Münnich (Braunschweig)
        BETA-DECAY ENERGIES AND SYSTEMATICS OF NUCLEAR STRUCTURE
        EFFECTS

12.05   H.J. Kluge (Mainz and CERN)
        ABSOLUTE MASS MEASUREMENTS OF SHORT-LIVED ISOTOPES USING A
        PENNING TRAP

B. MOMENTS (Chair: C. Détraz)

16.30   G. Huber (Mainz)
        STATIC MAGNETIC AND ELECTRIC MOMENTS NEAR THE CLOSED SHELL
        Z=50 FROM LASER SPECTROSCOPY

17.15   R.L. Gill (Brookhaven)
        MAGNETIC MOMENTS IN NEUTRON-RICH NUCLEI

17.40   K. Asahi (RIKEN)
        PRODUCTION OF SPIN POLARISED ^{15}C IN A HEAVY-ION REACTION AND
        MEASUREMENT OF THE G-FACTOR FOR THE $1/2^+$ GROUND STATE

18.25   D. Vandeplassche (Leuven)
        STUDIES OF ORIENTED NUCLEI AT THE KOOL FACILITY

19.00   W.B. Walters (Maryland and Oxford)
        DEFORMATION OF LIGHT BROMINE NUCLIDES

Tuesday, September 15

C. RADII (Chair: W.D. Myers)

8.30 G.D. Alkhazov (Leningrad)
INVESTIGATION OF NUCLEAR ELECTROMAGNETIC MOMENTS AND CHARGE RADII OF RARE-EARTH ELEMENTS BY RESONANCE PHOTOIONIZATION SPECTROSCOPY

9.05 R. Neugart (Mainz)
NEW TECHNIQUES AND RESULTS OF COLLINEAR LASER SPECTROSCOPY -- Xe, Rn AND Ho ISOTOPES

9.30 I. Tanihata (RIKEN)
NUCLEAR STRUCTURE STUDIES USING BEAMS OF RADIOACTIVE NUCLEI

10.05 B. Jonson (Chalmers)
THE NEUTRON HALO OF EXTREMELY NEUTRON-RICH NUCLEI

D. SPECTROSCOPY, EXOTIC NUCLEI (Chair: J.H. Äystö)

10.50 J.P. Dufour (CEN Bordeaux)
SPECTROSCOPIC MEASUREMENTS WITH A NEW METHOD: PROJECTILE-FRAGMENTS ISOTOPIC SEPARATION

11.35 N.A.F.M. Poppelier (Utrecht)
A SHELL-MODEL STUDY OF LIGHT EXOTIC NUCLEI

12.00 K.K. Seth (Northwestern)
EXCITED STATES OF THE EXOTIC NUCLEUS HELIUM-9

12.25 J.C. Hill (Iowa)
IS THE REGION ABOVE ^{78}Ni DOUBLY MAGIC?

16.00 POSTER SESSION

E. ASTROPHYSICS (Chair: E. Roeckl)

17.00 J. Truran (Illinois)
EXOTIC NUCLEI IN ASTROPHYSICAL ENVIRONMENTS

17.45 K.L. Kratz (Mainz)
NUCLEAR STRUCTURE EFFECTS FAR FROM STABILITY AND THEIR CONSEQUENCES ON RAPID NEUTRON-CAPTURE PROCESSES

18.30 W.D. Schmidt-Ott (Göttingen)
NEUTRON-RICH NUCLEI PRODUCED BY MULTI-NUCLEON TRANSFER REACTIONS

18.55 G.T. Ewan (Queen's)
SUDBURY NEUTRINO OBSERVATORY

Wednesday, September 16

F. <u>SHAPE COEXISTENCE, INTRUDERS</u> (Chair: M. Kirson)

8.30   J.L. Wood (Georgia)
NUCLEAR SHAPE COEXISTENCE AND THE STUDY OF NUCLEI FAR FROM STABILITY

9.15   E.F. Zganjar (Louisiana)
LOW-ENERGY E0 TRANSITIONS IN ODD-MASS NUCLEI OF THE NEUTRON-DEFICIENT $180<A<200$ REGION

9.40   M. Huyse (Leuven)
SHAPE COEXISTENCE IN THE REGION AROUND $Z=82$

10.25   K. Heyde (Gent)
A SHELL-MODEL DESCRIPTION OF INTRUDER STATES

11.10   B. Fogelberg (Studsvik)
TRANSITION PROBABILITIES BETWEEN INTRUDER LEVELS IN HEAVY Ag ISOTOPES

11.35   N. Kaffrell (Mainz)
EVIDENCE FOR SHAPE COEXISTENCE IN NEUTRON-RICH Rh AND Ag NUCLEI

Thursday, September 17

G. <u>SPECTROSCOPY</u> (Chair: S.K. Mark)

8.30   D. Schardt (GSI)
Q-VALUES AND ISOMER ENERGIES FROM HIGH-RESOLUTION ALPHA-, PROTON- AND GAMMA-RAY SPECTROSCOPY ABOVE ^{146}Gd

9.05   P. Kleinheinz (Jülich)
NUCLEAR STRUCTURE OF NUCLEI ABOVE ^{146}Gd FROM IN-BEAM AND GAMOW-TELLER DECAY STUDIES

9.40   J. Gilat (Soreq and LBL Berkeley)
DECAY STUDIES OF NEUTRON-DEFICIENT RARE-EARTH ISOTOPES WITH OASIS

10.05   R. Beraud (Lyon)
IDENTIFICATION AND STRUCTURE OF P-RICH RARE-EARTH NUCLEI INVESTIGATED USING A He-JET FED ON-LINE MASS SEPARATOR

10.50   K. Sistemich (Jülich)
THE PROTON G9/2 - NEUTRON G7/2 INTERACTION IN NEUTRON-RICH NUCLEI AT MASSES AROUND 100

11.15   C.A. Stone (Maryland)
ISOMERISM AND MULTIPLET STRUCTURE IN Sb ISOTOPES

11.40   D.G. Burke (McMaster)
SEARCH FOR STABLE OCTUPOLE DEFORMATION IN THE NUCLEUS ^{225}Fr

12.05   C.F. Liang (Orsay)
THE ALPHA DECAY OF MASS-SEPARATED ^{225}Th

H. FISSION (Chair: R. Sherr)

16.00 P.B. Price (Berkeley)
HEAVY-ION RADIOACTIVITIES

16.35 E.K. Hulet (LLL Livermore)
EVIDENCE FOR BIMODAL FISSION IN THE HEAVIEST ELEMENTS

17.10 P. Rozmej (GSI and Lublin)
ON THE FISSION OF THE HEAVIEST FERMIUM ISOTOPES

I. HEAVY ELEMENT SYNTHESIS (Chair: M. Fujioka)

17.55 F.P. Hessberger (GSI)
PRODUCTION OF THE HEAVIEST ELEMENTS ON THE BORDERLINE OF NUCLEAR STABILITY BY COLD FUSION REACTIONS

Friday, September 18

J. GAMOW-TELLER STRENGTH (Chair: J. Zylicz)

8.30 I.S. Towner (Chalk River)
QUENCHING OF GAMOW-TELLER STRENGTH

9.15 O. Häusser (SFU and TRIUMF)
GAMOW-TELLER STRENGTH FUNCTIONS FROM NUCLEON SCATTERING EXPERIMENTS

10.00 J. Honkanen (Jyväskylä)
GAMOW-TELLER STRENGTH IN THE BETA DECAY OF MIRROR NUCLIDES

10.45 K. Rykaczewski (Warsaw)
GAMOW-TELLER BETA DECAY OF EVEN NUCLEI NEAR ^{100}Sn

11.10 K. Riisager (CERN)
THE BETA STRENGTH IN PROTON-RICH NUCLEI

11.35 G. Walter (Strasbourg)
GAMOW-TELLER BETA DECAY OF $^{29-31}$Na: COMPARISON WITH SHELL MODEL ESTIMATES

12.00 T. Tachibana (Waseda)
IMPROVEMENT OF THE GROSS THEORY OF BETA DECAY

K. PARTICLE DECAY (Chair: Z.Y. Zhou)

16.30 D.M. Moltz (LBL Berkeley)
BETA-DELAYED TWO-PROTON EMISSION: NEW DECAY MECHANISM?

17.05 J.M. Nitschke (LBL Berkeley)
BETA-DELAYED PROTON DECAY IN THE LANTHANIDE REGION

17.30 D. Mikolas (NSCL)
STUDY OF BETA DECAY HALF-LIVES AND BETA-DELAYED MULTI-PARTICLE FINAL STATES AT MICHIGAN STATE UNIVERSITY

18.15   T. Faestermann (München)
          PROTON RADIOACTIVITY OF MEDIUM-HEAVY NUCLEI

18.40   L. Vanneste (Leuven)
          ANISOTROPIC ALPHA-EMISSION FROM ON-LINE SEPARATED ISOTOPES

19.05   K.S. Toth (Oak Ridge)
          NUCLEAR STRUCTURE EFFECTS ON ALPHA REDUCED WIDTHS

Saturday, September 19

L. GAMMA-RAY SPECTROSCOPY (Chair: S.P. Pandya)

8.30   P.J. Twin (Daresbury)
          SUPERDEFORMED NUCLEI AT HIGH SPIN

9.15   C.J. Lister (Manchester and Yale)
          SPECTROSCOPY OF THE N=Z NUCLEI FROM GERMANIUM TOWARDS TIN

9.40   J.H. Hamilton (Vanderbilt)
          COMPETITION OF OBLATE AND PROLATE BANDS IN THE MASS-70 REGION

M. EXISTENCE (Chair: N. Zeldes)

10.25   D. Guillemaud-Mueller (GANIL)
          PRODUCTION AND IDENTIFICATION OF NEW NUCLEI AT THE NEUTRON AND
          PROTON DRIP-LINES FOR LIGHT ISOTOPES

10.50   M. Bernas (Orsay)
          IDENTIFICATION OF THE NEW NEUTRON-RICH ISOTOPES $^{70-74}$Ni and
          $^{74-77}$Cu IN THERMAL NEUTRON FISSION OF ^{235}U

N. CLOSING SESSION

11.15   P. Armbruster (GSI)
          A VIEW OF THE FUTURE

LIST OF PARTICIPANTS

G. Alkhazov
Leningrad Inst. of Nuclear Physics
Gattchina, Leningrad
U.S.S.R 188350

Brian Allardyce
PS Division
CERN
Geneva
SWITZERLAND

Peter Armbruster
GSI Darmstadt
61 Darmstadt
Max Planck Str. 1
Fed. Rep. of GERMANY

K. Asahi
Radiation Laboratory
The Institute of Physical
   and Chemical Research
2-1 Hirosawa, Wako-shi, Saitama 351
JAPAN

Georges Audi
CSNSM
Batiment 108, BP no.1
91406 Orsay Cedex
FRANCE

J.H. Äystö
Department of Physics
University of Jyväskylä
Seminaarinkatu 15
SF–40100 Jyväskylä
FINLAND

Daniel Bazin
GANIL
BP No. 5027
14021 Caen Cedex
FRANCE

Fred D. Becchetti
Dept. of Physics
Randall Lab
University of Michigan
Ann Arbor, MI 48109
U.S.A.

Z. Berant
Physics Department
Nuclear Research Centre–Negev
P.O.B. 9001
84190 Beer Sheva
ISRAEL

R. Beraud
Institut de Physique Nucléaire
Université Lyon–1
43, Bd. du 11 Nov. 1918
F 69622 Villeurbanne Cedex
FRANCE

Istvan Berkes
Institut de Physique Nucléaire
Université Lyon–1
43, Bd. du 11 Nov. 1918
F 69622 Villeurbanne Cedex
FRANCE

Monique Bernas
IPN d'ORSAY
B.P. no.1
91406 Orsay
FRANCE

Carrol Bingham
Dept. of Physics
University of Tennessee
Knoxville, TN 37916
U.S.A.

Jochen Bonn
Institut für Physik
Johannes Gutenberg-Universität
Staudinger Weg 7
6500 Mainz
Fed. Rep. of GERMANY

M.J.G. Borge
Instituto "Rocasolano"
C.S.I.C.
Serrano 119
E–28006 Madrid
SPAIN

D.S. Brenner
Physics Department–510A
Brookhaven National Laboratory
Upton, NY 11973
U.S.A.

F. Buchinger
McGill University
3610 University St.
Montreal, PQ H3A 2B2
CANADA

Merle E. Bunker
Los Alamos National Laboratory
Group INC–5, MS–G776
P.O. Box 1663
Los Alamos, NM 87545
U.S.A.

Dennis G. Burke
Tandem Accelerator Laboratory,
   GSB105
McMaster University
1280 Main St. West
Hamilton, ON L8S 4K1
CANADA

A. Bykov
Leningrad Inst. of Nuclear Physics
Gattchina, Leningrad
U.S.S.R. 188350

J.D. Cole
Idaho National Engineering Laboratory
P.O. Box 1625
Idaho Falls, ID 83415
U.S.A.

John E. Crawford
Foster Radiation Laboratory
McGill University
3610 University Street
Montréal, PQ H3A 2B2
CANADA

J.M. D'Auria
Chemistry Department
Simon Fraser University
Burnaby, BC V5A 1S6
CANADA

Henri Dautet
Foster Radiation Laboratory
McGill University
3610 University 57
Montreal, PQ H3A 2B2
CANADA

Cary N. Davids
Argonne National Laboratory
PHY–203, 9700 S. Cass Ave.
Argonne, IL 60439
U.S.A.

Denis De Frenne
Laboratorium voor Kernfysica
Proeftuinstraat 86
B–9000 Gent
BELGIUM

Ph. Dessagne
Centre de Recherches Nucléaires
Physique Nucléaire Expérimentale
B.P. 20
67037 Strasbourg Cedex
FRANCE

Claude Detraz
GANIL
B.P. 5027
14021 Caen Cedex
FRANCE

Udo Dinger
Institut für Physik, Universität Mainz
Staudinger Weg 7
D–6500 Mainz
Fed. Rep. of GERMANY

Roger Duffait
Institut de Physique Nucléaire de Lyon
43, Bd. du 11 Nov. 1918
F 69622 Villeurbanne Cédex
FRANCE

Jean-Pierre Dufour
C.E.N. Bordeaux-Gradignan
Domaine du Haut Vigneau
33170 Gradignan
FRANCE

H.C. Duong
Lab Aimé Cotton
Faculté des Sciences d'Orsay, Bat.505
91405 Orsay
FRANCE

Derek Eastham
Daresbury Laboratory
Daresbury
Warrington WA4 4AD
Cheshire
U.K.

V.G. Egorov
Joint Institute for Nuclear Research
Head Post Office, P.O. Box 79
101000 Moscow
U.S.S.R.

Birgitta Ekström
The Studsvik Neutron Research
  Laboratory
S–61182 Nyköping
SWEDEN

Kari Eskola
Dept. of Physics
University of Helsinki
Siltavuorenpenger 20D
SF–00170 Helsinki
FINLAND

H.C. Evans
Physics Department
Queen's University
Kingston, ON K7L 3N6
CANADA

G.T. Ewan
Physics Department
Queen's University
Kingston, ON K7L 3N6
CANADA

Thomas Faestermann
Physik Department E12
Technische Universität München
James-Franck-Strasse
D–8046 Garching
Fed. Rep. of GERMANY

D.H. Feng
Dept. of Physics and Atmospheric Science
Drexel University College of Science
Philadelphia, PA 19104
U.S.A.

A. Fleury
C.E.N. Bordeaux-Gradignan
Domaine du Haut Vigneau
33170 Gradignan
FRANCE

B. Fogelberg
The Studsvik Science Research
  Laboratories
S–61182 Nyköping
SWEDEN

Antonino Foti
Dipartimento di Fisica dell Universita
Corso Italia, 57
I–95129 Catania
ITALY

M. Fujioka
Cyclotron and Radioisotope Center
Tohoku University
Aramaki-Aoba
Sendai 980
JAPAN

Janine Genevey
Institut des Sciences Nucléaires
53, Av. des Martyrs
F–38026 Grenoble Cedex
FRANCE

Jacob Gilat
Soreq Nuclear Research Center
Yavne, 70600
ISRAEL

R.L. Gill
Department of Physics
Bldg. 510A
Brookhaven National Laboratory
Upton, NY 11973
U.S.A.

Alain Gillibert
GANIL
BP 5027
14021 Caen Cedex
FRANCE

Albrecht Gillitzer
Technical University of Munich
James Franck Str.
D–8046 Garching
Fed. Rep. of GERMANY

I.C. Girit
ORNL/UNISOR
P.O. Box X, Bldg 6008
Oak Ridge, TN
U.S.A.

Andrée Gizon
Institut des Sciences Nucléaires
53, Avenue des Martyrs
38026 Grenoble Cedex
FRANCE

Reginald C. Greenwood
Idaho National Engineering Laboratory
EG&G Idaho Inc.
P.O. Box 1625
Idaho Falls, ID 83415
U.S.A.

D. Guillemaud-Mueller
GANIL
B.P. 5027
14021 Caen Cedex
FRANCE

Heinz Haas
EP Division
CERN
CH–1211 Geneva
SWITZERLAND

Erik Hagberg
Chalk River Nuclear Laboratories
Chalk River, ON K0J 1J0
CANADA

J.H. Hamilton
Department of Physics and Astronomy
Vanderbilt University
Nashville, TN 37235
U.S.A.

John C. Hardy
Chalk River Nuclear Laboratories
Chalk River, ON K0J 1J0
CANADA

O. Häusser
Department of Physics
Simon Fraser University
Burnaby, BC V5A 1S6
CANADA

Peter E. Haustein
Department of Chemistry
Bldg. 555A
Brookhaven National Laboratory
Upton, New York 11973
U.S.A.

Fritz Hessberger
Gesellschaft für
 Schwerionenforschung mbH
Postfach 11 05 52
D–61 Darmstadt 11
Fed. Rep. of GERMANY

K. Heyde
Institute for Nuclear Physics
Proeftuinstraat, 86
B–9000 Gent
BELGIUM

John C. Hill
Dept. of Physics
Iowa State University
Ames, IA 50011
U.S.A.

Jorma Honkanen
Univ. of Jyväskylä
Dept. of Physics
Seminaarinkatu 15
SF–40100 Jyväskylä
FINLAND

Gerhard Huber
Institut F. Physik
Staudinger Weg 7
65 Mainz
Fed. Rep. of GERMANY

F. Hubert
C.E.N. Bordeaux-Gradignan
Domaine du Haut Vigneau
33170 Gradignan
FRANCE

E.K. Hulet
Lawrence Livermore National
  Laboratory
Box 808, L232
Livermore, CA 94550
U.S.A.

Mark Huyse
LISOL, K.U. Leuven
Celestijnenlaan 200 D
3030 Leuven
BELGIUM

J.C. Jacmart
Institut de Physique Nucléaire
B.P. No.1
91406 Orsay
FRANCE

Arthur N. James
Oliver Lodge Laboratory
University of Liverpool
P.O. Box 147
Merseyside L48 8AY
U.K.

Joachim Jänecke
Dept. of Physics
Univ. of Michigan
Ann Arbor, MI 48109
U.S.A.

Aksel S. Jensen
Institute of Physics
University of Aarhus
DK–8000 Aarhus C
DENMARK

Björn Jonson
Department of Physics
Chalmers University of Technology
S 41296 Göteborg
SWEDEN

N. Kaffrell
Institute für Kernchemie
Universität Mainz
Postfach 3980
D–6500 Mainz
Fed. Rep. of GERMANY

Bernard D. Kern
Dept. of Physics and Astronomy 00552
Univ. of Kentucky
Lexington, KY 40502-0055
U.S.A.

Uwe Keyser
Institut F. Metallphysik und Nukleare
Festkoerperphysik Techn.-Universitaet
D3300 Braunschweig
Mendelssohnstr.3
Fed. Rep. of GERMANY

Pierrette Kilcher
Institut de Physique Nucléaire
Batiment 104–B.P. no.1
91406 Orsay
FRANCE

Michael Kirson
Dept. of Nuclear Physics
Weizmann Institute of Science
76100 Rehovot
ISRAEL

Peter Kleinheinz
Kernforschungsanlage Jülich GmbH
Institut für Kernphysik
Postfach 1913
D–5170 Jülich 1
Fed Rep. of GERMANY

Otto Klepper
Gesellschaft für Schwerionenforschung
(GSI)
Postfach 11 0552
D–6100 Darmstadt 11
Fed. Rep. of GERMANY

Hans Jürgen Kluge
Institut für Physik
Postbox 3980
D–65 Mainz
Fed. Rep. of GERMANY

Jan Kormicki
Physics Department
Vanderbilt University
P.O. Box 1807
Nashville, TN 37235
U.S.A.

V. Koslowsky
Chalk River Nuclear Laboratories
Chalk River, ON K0J 1J0
CANADA

Karl-Ludwig Kratz
Inst. F. Kernchemie
Universität Mainz
Fritz-Strassmann-weg 2
D–6500 Mainz
Fed. Rep. of GERMANY

Wiktor Kurcewicz
CERN/EP
CH–1211 Geneve 23
SWITZERLAND

Thomas Lang
Bldg. 88
Lawrence Berkeley Laboratory
1 Cyclotron Rd.
Berkeley, CA 94720
U.S.A.

J.K.P. Lee
Foster Radiation Laboratory
McGill University
3610 University St.
Montreal, PQ H3A 2B2
CANADA

C.F. Liang
CSNSM BP no.1
91406 Orsay
FRANCE

Peter Lievens
Instituut voor Kern- en Stralingsfysika
K.U. Leuven Celestijnenlaan 200 D
B–3030 Leuven
BELGIUM

Kim Lister
Department of Physics
Yale University
New Haven, CT 06520
U.S.A.

G.K.E. Löbner
Sektion Physik
Universitat München
Am Coulombwall 1
8046 Garching
Fed. Rep. of GERMANY

L. LoMonaco
Centres de Recherches Nucléaires
Laboratoire de Physique Nucléaire
    Théorique
BP 20
67037 Strasbourg Cedex
FRANCE

Henryk Mach
510A, Physics Department
Brookhaven National Laboratory
Upton, NY 11973
U.S.A.

Paul F. Mantica
Dept. of Chemistry
University of Maryland
College Park, MD 20781
U.S.A.

S.K. Mark
Rutherford Physics Bldg.
McGill University
3610 University Street
Montreal, PQ H3A 2B2
CANADA

David Mikolas
NSCL/Cylotron Laboratory
Michigan State University
East Lansing, MI 48824-1321
U.S.A.

W. Mittig
GANIL
B.P. 5027
14021 Caen Cedex
FRANCE

D.M. Moltz
Lawrence Berkeley Laboratory
1 Cyclotron Road, Bldg. 88
Berkeley, CA 94720
U.S.A.

Alex C. Mueller
GANIL
B.P. No. 5027
F–14021 Caen Cedex
FRANCE

Fritz Münnich
Institut für Metallphysik und Nukleare
Tech. Universität Braunschweig
D–33 Braunschweig
Mendelssohnstr. 3
Fed. Rep. of GERMANY

William D. Myers
Lawrence Berkeley Laboratory
University of California
Bldg. 70a, 3307
Berkeley, CA 94720
U.S.A

Rainer Neugart
Institut für Physik
Universität Mainz
Staudinger-Weg7
Postfach 3980
D–6500 Mainz
Fed. Rep. GERMANY

J.M. Nitschke
Lawrence Berkeley Laboratory
Bldg. 71–259
1 Cyclotron Blvd.
Berkeley, CA 94720
U.S.A.

Göran Nyman
Department of Physics
Chalmers University of Technology
S 41296 Göteborg
SWEDEN

S.P. Pandya
Physical Research Laboratory
Navrangpura
Ahmedabad–380 009
INDIA

P. Paris
Laboratoire Rene Bernas (CSNSM)
Bat. 104
F–91406 Orsay
FRANCE

J.M. Pearson
Departement de Physique
Université de Montréal
Case Postale 6128
Succursale "A"
Montréal, PQ H3C 3J7
CANADA

B. Pfeiffer
Universität Mainz
Institut für Kernchemie
Fritz-Stassmann-Weg 2
D–6500 Mainz
Fed. Rep. of GERMANY

J. Pinard
Lab Aimé Cotton
CNRS II BC 505
91405 Orsay
FRANCE

Domitian Popescu
McMaster University
Hamilton, ON L8S 4K1
CANADA

N.A.F.M. Poppelier
Vakgroep Kernfysica
Fysisch Laboratorium
Rijksuniversiteit Utrecht
postbus 80.000
THE NETHERLANDS

F. Pougheon
I.P.N.
BP No. 1
91406 Orsay
FRANCE

P. B. Price
1056 Overlook Road
Berkeley, CA 94708
U.S.A.

Shaheen Rab
Nuclear Data Project
Kuwait Institute for Scientific Research
P.O. Box 24885
13109 Safat
KUWAIT

Paul L. Reeder
Battelle, Pacific Northwest Laboratory
P.O. Box 999
Richland, WA 99352
U.S.A.

Jay Reiff
Bldg. 88
Lawrence Berkeley Laboratory
Berkeley, CA 94720
U.S.A.

Karsten Riisager
EP–Division
CERN
CH–1211 Genève
SWITZERLAND

J. David Robertson
1 Cyclotron Road, Bldg. 88
Lawrence Berkeley Laboratory
Berkeley, CA 94720
U.S.A.

Ernst Roeckl
Gesellschaft für Schwerionenforschung
Postfach 110552
D–6100 Darmstadt 11
Fed. Rep. of GERMANY

P. Rozmej
GSI Planck St. 1
Postfach 110552
6100 Darmstadt 11
Fed. Rep. of GERMANY

Krzysztof Rykaczewski
Institute of Experimental Physics
University of Warsaw
Ul. Hoza 69
P1-00-681 Warszawa
POLAND

M.G. Saint-Laurent
GANIL
B.P. No. 5027
F–14021 Caen Cedex
FRANCE

Lakshmidhar Satpathy
Institute of Physics
Sachivalayamarg
Bhubaneswar 751005
INDIA

Dieter Schardt
GSI Darmstadt
Postfach 11 05 41
D–6100 Darmstadt 11
Fed. Rep. of GERMANY

H. Schmeing
Chalk River Nuclear Laboratories
Chalk River, ON K0J 1J0
CANADA

Wolf-Dieter Schmidt-Ott
Bunsenstrasse 7-9
II. Physikalisches Institut
Universität Göttingen
D–3400 Göttingen
Fed. Rep. of GERMANY

Johannes Schwarzenberg
School of Chemistry
Georgia Inst. of Technology
Atlanta, GA 30332
U.S.A.

Kamal K. Seth
Physics Department
Northwestern University
Evanston, IL 60201
U.S.A.

Nathal Severijns
I.K.S.
Celestijnenlaan, 200 D
3030 Heverlee
BELGIUM

K.S. Sharma
Dept. of Physics
University of Manitoba
Winnipeg, MB R3T 2N2
CANADA

T.M. Shaw
The Clarendon Lab
Parks Road
Oxford OX1 3PU
U.K.

R. Sherr
Princeton University
Dept. of Physics
Jadwin Hall
Princeton, NJ 08540
U.S.A.

Brad Sherrill
National Superconducting Cyclotron Lab.
Michigan State University
East Lansing, MI 48824
U.S.A.

Tadashi Shimoda
Insitute of Physics
College of General Education
Osaka University
Toyonaka, Osaka 560
JAPAN

T. Shinozuka
Cyclotron and Radioisotope Center
Tohoku University
Aramaki-Aoba, Sendai 980
JAPAN

Kornelius Sistemich
KFA Jülich GmbH
Institut für Kernphysik
Postfach 1913
5170 Jülich 1
Fed. Rep. of GERMANY

Gunnar Skarnemark
Chalmers University of Technology
Department of Nuclear Chemistry
S–412 96 Göteborg
SWEDEN

Francesca Soramel
Dipartimento di Fisica "Galileo Galilei"
Universita Degli Studi di Padova
Via F.Marzolo 8
I–35131 Padova
ITALY

Paola Spolaore
I.N.F.N., Laboratori Naz. di Legnaro
Via Romea, 4
35020 Legnaro (PD)
ITALY

D.W.L. Sprung
Physics Department SS348
McMaster University
Hamilton, ON L8S 4M1
CANADA

Craig A. Stone
Department of Chemistry
University of Maryland
College Park, MD 20742
U.S.A.

X.J. Sun
Chalk River Nuclear Laboratories
Chalk River, ON K0J 1J0
CANADA

T. Tachibana
Science and Engineering Research Lab.
Waseda University
3–4–1 Okubo, Shinjuku-Ku
Tokyo 160
JAPAN

W.L. Talbert
Group INC–11, MS–J514
Los Alamos National Laboratory
P.O. Box 1663
Los Alamos, NM 87545
U.S.A.

Isao Tanihata
RIKEN Accelerator
2-1 Hirosawa
Wako-shi
Saitama 351-01
JAPAN

Olof Tengblad
Dept. of Subatomic Physics
Chalmers University of Technology
S-412 96 Göteborg
SWEDEN

K.S. Toth
Oak Ridge National Laboratory
Physics Division, Bldg 6000, MS371
P.O. Box X
Oak Ridge, TN 37831-6371
U.S.A.

Ian S. Towner
Chalk River Nuclear Laboratories
Chalk River, ON K0J 1J0
CANADA

James W. Truran
Department of Astronomy
University of Illinois
1011 West Springfield Avenue
Urbana, IL 61801
U.S.A.

V.M. Tsupko-Sitnikov
Joint Institute for Nuclear Research
Head Post Office, P.O. Box 79
101000 Moscow
U.S.S.R.

P.J. Twin
Daresbury Laboratory
Science and Engineering Research
   Council
Warrington WA4 4AD
Cheshire
U.K.

Dirk Vandeplassche
Laboratorium voor Kern-en
   Stralingsfysika
Celestijnenlaan 200D
B-3030 Leuven
BELGIUM

Ludo Vanneste
I.K.S.
University of Leuven
Celestijnenlaan, 200 D
3030 Leuven
BELGIUM

Brian James Varley
Schuster Lab.
University of Manchester
Manchester M13 9PL
U.K.

D.J. Vieira
Group INC-11, MS-H824
Los Alamos National Laboratory
P.O. Box 1663
Los Alamos, NM 87545
U.S.A.

Guy Walter
Centre de Recherches Nucléaires
23, rue du Loess
67037 Strasbourg Cedex
FRANCE

William B. Walters
Dept. of Chemistry
University of Maryland
College Park, MD 20742
U.S.A.

Aaldert H. Wapstra
NIKHEF-K
P.O. Box 41882
100gDB Amsterdam
THE NETHERLANDS

Ray A. Warner
Batelle, Pacific Northwest Laboratory
P.O. Box 999
Richland, WA 99352
U.S.A.

Henry Willmes
Physics Department
University of Idaho
Moscow, ID 83843
U.S.A.

Jeff Winger
Physics Dept., Bldg. 510A
Brookhaven National Laboratory
Upton, NY 11973
U.S.A.

F.K. Wohn
Department of Physics
Iowa State University
Ames, IA 50011
U.S.A.

John L. Wood
School of Physics
Georgia Inst. of Technology
Atlanta, GA 30332
U.S.A.

P.J. Woods
Department of Physics
The University of Birmingham
P.O. Box 363
Birmingham B15 2TT
U.K.

Jan Wouters
INC–11, MS H824/LAMPF
Los Alamos National Laboratory
Los Alamos, NM 87545
U.S.A.

Masami Yamada
Science and Engineering Research Lab.
Waseda University
3–4–1 Okubo, Shinjuku-Ku
Tokyo 160
JAPAN

Nissan Zeldes
Racah Institute of Physics
Hebrew University of Jerusalem
Jerusalem 91904
ISRAEL

Edward Zganjar
Dept. of Physics
Louisiana State Univ.
Baton Rouge, LA 70803
U.S.A

Z.Y. Zhou
Department of Physics
Nanjing University
Nanjing
People's Republic of CHINA

Jan Zylicz
Institute of Experimental Physics
University of Warsaw
Ul. Hoza 69
Pl-00-681 Warszawa
POLAND

Author Index

A

Ahalpara, D. P, 278
Alkhazov, G. D., 115, 735
Alstad, J., 286
Alton, G. D., 849
Anderl, R. A., 782
Anne, R., 690, 722, 757
Armbruster, P., 768, 786, 839, 860
Arnold, E., 126, 161, 197
Asahi, K., 165
Ashworth, C. J., 184
Audi, G., 22
Aysto, J. H., 411, 640

B

Barden, R., 477, 494, 656
Barzakh, A. E., 115
Batist, L. N., 735
Baum, E. M., 455, 459
Baumann, P., 51, 624, 695
Bazin, D., 690, 722, 757
Beau, M., 344
Becchetti, F. D., 845
Becker, H. W., 572
Becker, St., 201
Ben Braham, A., 513
Benenson, W., 708
Bennett, S. J., 831
Beraud, R., 419, 445
Berdichevsky, D., 197, 237
Berkes, I., 193
Bernas, M., 757, 768
Bianchi, L., 11, 233
Billowes, J., 136
Bingham, C. R., 849
Blomqvist, J., 477, 489, 494
Bocquet, J. P., 768
Boissier, G., 517
Bollen, G., 22
Bonn, J., 161
Borchers, W., 126
Borge, M. J. G., 223, 533, 634
Borrel, V., 690, 722, 757
Bosch, U., 365
Bourgeois, C., 513, 517
Boyd, R. N., 581

Braga, R. A., 441
Brant, S., 393
Brenner, D. S., 84, 174
Briancon, Ch., 521, 529
Brissot, R., 768
Brown, B. A., 654, 708
Buchinger, F., 197
Buchmann, L., 572
Burke, D. G., 533
Busch, J., 268
Buyanov, N. B., 115
Bykov, A. A., 735

C

Carter, H. K., 441, 665, 849
Caruette, A., 517
Casey, T., 845
Casten, R. F., 84, 174, 282
Cerny, J., 654, 749
Charvet, A., 419, 445
Chaturvedi, L., 268
Chechetkin, V. M., 568
Chen, Y., 708
Chen, Z.-M., 268
Chishti, A. A., 354
Chouvel, J. M., 233
Chubukov, I. Ya., 115
Clerc, H.-G., 839
Coenen, E., 305
Cole, J. D., 268, 782, 835, 849
Connell, K. A., 831
Cormier, T. M., 268, 835
Crawford, J. E., 205
Cristancho, F., 385
Cunsolo, A., 11, 233
Curtin, M. S., 708
Cwiok, S., 529, 821

D

D'Auria, J. M., 572
Dammrich, U., 184
Daniels, W. R., 286
Dautet, H., 473
Davids, C. N., 411
de Angelis, G., 489
De Frenne, D., 286

de Takacsy, N., 473
de Vries, J. H., 334
Del Moral, R., 344
Delagrange, H., 344, 690
Deneffe, K., 305
Denisov, V. P., 115
Dessagne, Ph., 51, 624, 695
Detraz, C., 690, 757
Dinger, U., 209
Dobaczewski, J., 656
Dombsky, M., 572
Ducourtieux, M., 517
Dudeck, J., 695
Duffait, R., 419, 445
Dufour, J. P., 344, 690, 839
Dumont, H., 11, 233
Duong, H. T., 205
Dutta, A. K., 66

E

Eastham, D. A., 136
Eberth, J., 268, 385
Eberz, J., 209
Egelhof, P., 22
Egorov, V., 857
Ekstrom, B., 296, 578
El Hajjaji, O., 193
Ellis-Akovali, Y. A., 665, 718
Emsallem, A., 419, 445
Eshhar, S., 796
Eskola, K., 640
Evans, D. E., 136
Ewan, G. T., 585

F

Faestermann, T., 675, 739
Fahad, M., 193
Faller, S. H., 429, 459
Faust, H. R., 768
Fawcett, M. J., 136
Fedoseev, V. N., 115
Feng, D. H., 89
Fernandez, B., 11, 233
Ferro, A., 517
Finger, M., 157
Fink, R. W., 441
Firestone, R. B., 463
Fleming, R., 197, 237
Fleury, A., 344
Fogelberg, B., 296, 578

Folger, H., 365, 786
Foti, A., 11, 233
Fournet-Fayas, J., 517
Franczak, B., 839
Frehaut, J., 344
Freystein, V., 509
Fujioka, M., 650
Fukuda, T., 165
Fulton, B. R., 831

G

Gabelmann, H., 477, 656
Gai, M., 708
Galin, J., 757
Gastebois, J., 11, 233
Geissel, H., 839
Gelletly, W., 354
Genevey, J., 419, 445
Gerber, M., 201
Gietz, H., 533
Gilat, J., 463, 697, 718
Gill, R. L., 174, 282, 375, 407, 459
Gillibert, A., 11, 233
Gillitzer, A., 675, 739
Giraudet, G., 344
Girit, I. C., 849
Giroux, J., 193, 624
Gizon, A., 415, 419, 445
Glaudemans, P. W. M., 334
Goulden, J. D., 407
Graefenstedt, M., 30
Grant, I. S., 136, 184, 656
Green, V. R., 193
Greenwood, R. C., 782
Gregoire, Ch. 11, 233
Greiner, W., 827
Griffith, J. A. R., 136
Griffiths, A. G., 184
Gromov, K. Ya., 853
Groves, J., 136, 831
Guerreau, D., 722, 757
Guillemaud-Mueller, D., 690, 722, 757
Guven, H., 494

H

Hagberg, E., 41, 473
Hahn, H. J., 489
Hama, H., 650
Hamilton, J. H., 268, 835, 849
Hanakawa, K., 165

Hanelt, E., 344, 839
Hansen, P. G., 223, 634
Harakeh, M. N., 286
Hardy, J. C., 41, 473
Harms, V., 403, 558
Hartel, K., 675, 739
Harwood, L. H., 708
Hassani, R., 193
Hausser, O., 604
Haustein, P. E., 53, 84
Heese, J., 385
Henning, W., 675
Herath-Banda, M. A., 268
Herzog, P., 184
Hessberger, F. P., 786
Heyde, K., 255, 286
Hilberath, Th., 22, 201
Hild, T., 509
Hill, J. C., 174, 375, 407
Hillebrandt, W., 558
Hofmann, S., 786
Honkanen, J. A., 411, 640
Huber, G., 105, 209
Hubert, F., 344, 690
Huck, A., 51, 477, 494, 624, 695
Hulet, E. K., 810
Huyse, M., 305

I

Idrissi, N., 419
Ikeda, N., 165
Ishihara, M., 165
Ishimatsu, T., 650
Itahashi, T., 165
Ivanov, V. S., 115
Ivascu, M., 827

J

Jacmart, J. C., 690, 722, 757
Jacobs, E., 286
James, A. N., 354, 425, 831
Janecke, J., 62, 845
Janssens, R. V. F., 845
Jauho, P., 411
Jean, D., 344
Jensen, A., 101, 241
Ji, X., 375
Jonson, B., 223, 634

K

Kaffrell, N., 286, 533
Kalinnikov, V. G., 853
Kalinowski, L., 656
Kalinowsky, H., 22
Kashy, E., 708
Katori, K., 165
Kavanagh, R., 572
Kern, B. D., 441, 849
Kern, F., 22
Keyser, U., 30
Kienle, P., 675, 739
Kilcher, P., 205, 513, 517
Kim, H. J., 665
Kirchner, R., 209, 365, 477, 494, 654, 656
Kleinheinz, P., 477, 489, 494
Klepper, O., 209, 365, 477, 494, 654, 656, 839
Klotz, G., 51, 624, 695
Kluge, H.-J., 22, 201
Knipper, A., 51, 624
Kolata, J. J., 845
Kolb, D., 796
Koponen, V., 411, 640
Kormicki, J., 268
Kortelahti, M. O., 268, 313, 441
Koschel, P., 365
Koslowsky, V. T., 41, 473, 654
Kovar, D. G., 845
Kracikova, T. I., 157
Krane, K. S., 849
Kratz, K.-L., 403, 558
Kreiner, A. J., 513
Kronert, U., 201
Kuhl, T., 209
Kunz, K. 22
Kurcewicz, W., 533
Kusnezov, D. F., 455

L

Landois, G., 517
Lang, T. F., 749
Langenbeck, B., 839
Langevin, M., 757
Le Blanc, F., 205, 517
Leander, G. A., 268, 441
Lee, J. K. P., 205
Lee, M. A., 782
Leino, M. E., 786

Lenz, U., 389
Letokhov, V. S., 115
Lhersonneau, G., 393
Liang, C. F., 415, 477, 489, 521
Liberman, S., 205
Lieb, K. P., 385
Lievens, P., 197
Lister, C. J., 354
Lister, P. M., 845
Liu, W. Z., 845
Lobner, K. E. G., 389
Long, Z. W., 11
Lovhoiden, G., 533
Lund, E., 296, 578
Lyutostansky, Yu. S., 568, 727

M

Ma, W.-C., 268
Mach, H., 174, 282, 407
Maguire, C. F., 268
Maison, J. M., 695
Malevanny, S. V., 568
Mantica, P. F., 455, 459
Marest, G., 193
Marguier, G., 51, 193, 477, 494, 624
Marinov, A., 796
Mark, S. K., 473
Maruhn, J. A., 827
Marx, D., 209
Masson, P. J., 62
Mattsson, S., 223, 533, 634
McConnell, J. W., 665
Meissner, F., 365, 509
Menges, R., 209
Messelt, S., 640
Meyer, M., 419, 445
Meyer, R. A., 455
Michaelsen, R., 509
Miehe, Ch., 51, 624, 695
Mikolas, D., 708
Minamisono, T., 165
Miranda, A., 101, 241
Mishin, V. I., 115
Mittig, W., 11, 233
Mlekodaj, R. L., 441
Moller, P., 403, 558, 697
Molnar, G., 282
Moltz, D. M., 665, 718, 749
Monnand, E., 403
Moore, R. B., 22
Morjean, M., 11, 233

Morrison, T., 354
Mueller, A. C., 344, 690, 722, 757
Munnich, F., 30
Munsch, J., 517
Munzenberg, G., 786, 839
Mylaeus, Th., 268, 385

N

Nakamura, A., 165
Naulin, F., 757
Naumann, R. A., 533
Nayak, R. C., 80
Nazarewicz, W., 268, 441, 656
Neu, W., 126, 161, 197
Neugart, R., 126, 161, 197
Nickel, F., 839
Nitschke, J. M., 463, 697, 718
Nojiri, Y., 165
Nolen Jr., J. A., 708
Nolte, E., 739
Nybo, K., 533
Nyman, G., 223, 533, 634, 656

O

Obert, J., 205, 517
Ogawa, K., 640
Ohm, H., 393
Ohsugi, S., 614
Ollivier, T., 419
Oms, J., 205, 513, 517
Orlov, S. Yu., 735
Osipowicz, Th., 385
Otten, E. W., 161

P

Page, R. D., 831
Pandya, S. P., 278
Panov, I. V., 568, 727
Papanicolopulos, C. D., 313
Paris, P., 415, 477, 489, 521
Pavlov, V. N., 157
Pearson, J. M., 66
Penttila, H., 411
Perez, C. P., 441
Petry, R. F., 407
Pfeiffer, B., 403
Phillips, W. R., 845
Piercey, R. B., 268
Piiparinen, M., 477
Pinard, J., 205

Piotrowski, A., 375, 407, 459
Plochocki, A., 477, 494, 654, 656
Poenaru, D. N., 827
Poppelier, N. A. F. M., 334
Poppensieker, K., 786
Porquet, M. G., 513
Pougheon, F., 690, 722, 757
Pranal, Y., 11, 233
Pravikoff, M. S., 344, 690, 839
Price, H. G., 354
Price, P. B., 800
Putaux, J. C., 205

Q

Quade, U., 389
Quiniou, E., 757

R

Rab, S., 731
Rae, W. D. M., 831
Rainhard, R., 489
Ramayya, A. V., 268, 835
Ramdane, M., 624, 695
Ramsay, E. B., 197
Rao, M. N., 718
Rathke, G.-E., 656
Redon, N., 419, 445
Reiff, J. E., 749
Reisdorf, W., 786
Reynolds, M. J., 184
Richard, A., 690, 722
Richard-Serre, C., 51, 624
Richter, A., 634
Riisager, K., 223, 634
Rikovska, J., 184, 193
Robertson, J. D., 429, 455, 459, 749
Roeckl, E., 365, 477, 654, 656, 690, 757, 839
Rogowski, J., 286, 533
Rolando-Eugio, D., 445
Rolfs, C., 572
Roussel, P., 768
Roussiere, B., 205, 513, 517
Rozmej, P., 529, 821
Rubio, B., 477, 489, 494
Rudolph, K., 389
Rudstam, G., 578
Runte, E., 365, 509
Rykaczewski, K., 365, 656

S

Saint-Laurent, M. G., 690, 722, 757
Salewski, H., 509
Samuel, M., 708
Satpathy, L., 76, 80
Satteson, M., 268
Sauvage, J., 205, 513, 517
Savard, G., 205
Schardt, D., 365, 477, 494, 654, 656, 839
Schippers, J. M., 286
Schloesser, K., 184
Schmal, N., 268
Schmeing, H., 41, 473
Schmidt, K.-H., 344, 786, 839
Schmidt-Ott, W.-D., 365, 509, 722
Schomburg, W., 389
Schreiber, F., 30
Schroder, S., 209
Schull, D., 839
Schutz, Y., 11, 233
Schwab, T., 839
Schweikhard, L., 22
Sekatsky, S. K., 115
Sekine, T., 654
Sellem, R., 517
Semmes, P. B., 441
Seth, K., 324
Seuthe, S., 572
Severijns, N., 140, 679
Sherr, R., 708
Sherrill, B., 654, 708, 839
Shihab-Eldin, A., 731
Shimoda, T., 165
Shimoura, S., 165
Shinozuka, T., 650
Silverans, R. E., 197
Simpson, J., 354, 831
Simpson, M. L., 849
Sistemich, K., 393
Skarnemark, G., 286
Skeppstedt, O., 354
Skoda, S., 268
Skorka, S. J., 389
Smith, J. R. H., 136
Sobiczewski, A., 529, 821
Sousa, D. C., 718
Sprouse, G. D., 209
Sprung, D. W. L., 197, 237
Starzecki, W., 489
Steinmayer, M., 389

Stephan, Cl., 11, 233
Stern, R. L., 845
Stevenson, J., 708
Stolzenberg, H., 22
Stone, C. A., 429, 455, 459
Stone, N. J., 184, 193
Struble, G. L., 533
Styczen, J., 489, 494
Summerer, K., 344, 839
Sun, X. J., 41, 473
Sznadjderman, D., 517

T

Tachibana, T., 97, 614
Taguchi, K., 650
Tain, J. L., 494
Takahashi, K., 581
Takahashi, N., 165
Tanihata, I., 213
Tarasov, V. K., 735
Taskinen, P., 411, 640
Teichert, W., 268
Tengblad, O., 223
Tetzlaff, H., 286
Thekkadath, G., 205
Thielemann, F.-K., 558
Thorsteinsen, T. F., 533
Tolfree, D. W. L., 136
Tondeur, F., 66
Toth, K. S., 665, 697, 718
Towner, I. S., 593
Trautmann, N., 286
Truran, J. W., 543
Tsupko-Sitnikov, V. M., 157, 853
Turcotte, R., 473
Twin, P. J., 499

U

Ulbig, S., 385
Ulm, G., 126, 161, 197
Uno, M., 97

V

Van Duppen, P., 305
Van Haverbeke, J., 140, 679
van Walle, E., 140, 679
van der Werf, S. Y., 286
Vandeplassche, D., 140, 679
Vander Molen, A., 708

Vanneste, L., 140, 679
Varley, B. J., 354
Vieira, D. J., 1
Vierinen, K., 463, 718
Vincent, J., 572
Vineyard, M. F., 845
Von Brentano, P., 268, 489

W

Walker, P. M., 136, 184
Wallmeroth, K., 201
Walter, G., 51, 477, 494, 624, 695
Walters, W. B., 184, 429, 455, 459
Warner, D. D.
Watson, D. C. B., 425
Weiss, B., 415, 419
Wells, S. A., 136
Wen, S., 268
Wendt, K., 126, 161, 197
White, J. P., 184
Wildenthal, B. H., 375
Willmes, H., 782
Wilmarth, P. A., 463, 697, 718
Winfield, J. S., 708
Winger, J. A., 174, 282, 375, 407
Wiosna, E., 268, 385
Wittmann, V. D., 735
Wohn, F. K., 174, 375, 407
Wohr, A., 558
Wolf, A., 174, 282
Wolfsberg, K., 286
Wollnik, H., 839
Wolters, A. A., 334
Wood, J. L., 245, 313
Woods, P. J., 831
Wormann, B., 385
Wouters, J., 140, 679
Wouters, J. M., 1

X

Xie, Z. Q., 708

Y

Yamada, M., 97, 614
Yamada, S., 97
Yoshii, M., 411, 650

Z

Zganjar, E. F., 268, 313, 441, 849
Zhan, W. L., 233
Zhao, X., 268
Zhao, Z., 708
Ziegert, W., 558
Zongyuan, Ye., 296
Zuber, K., 477
Zverev, M. V., 727
Zylicz, J., 411, 656

AIP Conference Proceedings

		L.C. Number	ISBN
No. 1	Feedback and Dynamic Control of Plasmas – 1970	70-141596	0-88318-100-2
No. 2	Particles and Fields – 1971 (Rochester)	71-184662	0-88318-101-0
No. 3	Thermal Expansion – 1971 (Corning)	72-76970	0-88318-102-9
No. 4	Superconductivity in d- and f-Band Metals (Rochester, 1971)	74-18879	0-88318-103-7
No. 5	Magnetism and Magnetic Materials – 1971 (2 parts) (Chicago)	59-2468	0-88318-104-5
No. 6	Particle Physics (Irvine, 1971)	72-81239	0-88318-105-3
No. 7	Exploring the History of Nuclear Physics – 1972	72-81883	0-88318-106-1
No. 8	Experimental Meson Spectroscopy –1972	72-88226	0-88318-107-X
No. 9	Cyclotrons – 1972 (Vancouver)	72-92798	0-88318-108-8
No. 10	Magnetism and Magnetic Materials – 1972	72-623469	0-88318-109-6
No. 11	Transport Phenomena – 1973 (Brown University Conference)	73-80682	0-88318-110-X
No. 12	Experiments on High Energy Particle Collisions – 1973 (Vanderbilt Conference)	73-81705	0-88318-111-8
No. 13	π-π Scattering – 1973 (Tallahassee Conference)	73-81704	0-88318-112-6
No. 14	Particles and Fields – 1973 (APS/DPF Berkeley)	73-91923	0-88318-113-4
No. 15	High Energy Collisions – 1973 (Stony Brook)	73-92324	0-88318-114-2
No. 16	Causality and Physical Theories (Wayne State University, 1973)	73-93420	0-88318-115-0
No. 17	Thermal Expansion – 1973 (Lake of the Ozarks)	73-94415	0-88318-116-9
No. 18	Magnetism and Magnetic Materials – 1973 (2 parts) (Boston)	59-2468	0-88318-117-7
No. 19	Physics and the Energy Problem – 1974 (APS Chicago)	73-94416	0-88318-118-5
No. 20	Tetrahedrally Bonded Amorphous Semiconductors (Yorktown Heights, 1974)	74-80145	0-88318-119-3
No. 21	Experimental Meson Spectroscopy – 1974 (Boston)	74-82628	0-88318-120-7
No. 22	Neutrinos – 1974 (Philadelphia)	74-82413	0-88318-121-5
No. 23	Particles and Fields – 1974 (APS/DPF Williamsburg)	74-27575	0-88318-122-3
No. 24	Magnetism and Magnetic Materials – 1974 (20th Annual Conference, San Francisco)	75-2647	0-88318-123-1
No. 25	Efficient Use of Energy (The APS Studies on the Technical Aspects of the More Efficient Use of Energy)	75-18227	0-88318-124-X
No. 26	High-Energy Physics and Nuclear Structure – 1975 (Santa Fe and Los Alamos)	75-26411	0-88318-125-8

No. 27	Topics in Statistical Mechanics and Biophysics: A Memorial to Julius L. Jackson (Wayne State University, 1975)	75-36309	0-88318-126-6
No. 28	Physics and Our World: A Symposium in Honor of Victor F. Weisskopf (M.I.T., 1974)	76-7207	0-88318-127-4
No. 29	Magnetism and Magnetic Materials – 1975 (21st Annual Conference, Philadelphia)	76-10931	0-88318-128-2
No. 30	Particle Searches and Discoveries – 1976 (Vanderbilt Conference)	76-19949	0-88318-129-0
No. 31	Structure and Excitations of Amorphous Solids (Williamsburg, VA, 1976)	76-22279	0-88318-130-4
No. 32	Materials Technology – 1976 (APS New York Meeting)	76-27967	0-88318-131-2
No. 33	Meson-Nuclear Physics – 1976 (Carnegie-Mellon Conference)	76-26811	0-88318-132-0
No. 34	Magnetism and Magnetic Materials – 1976 (Joint MMM-Intermag Conference, Pittsburgh)	76-47106	0-88318-133-9
No. 35	High Energy Physics with Polarized Beams and Targets (Argonne, 1976)	76-50181	0-88318-134-7
No. 36	Momentum Wave Functions – 1976 (Indiana University)	77-82145	0-88318-135-5
No. 37	Weak Interaction Physics – 1977 (Indiana University)	77-83344	0-88318-136-3
No. 38	Workshop on New Directions in Mossbauer Spectroscopy (Argonne, 1977)	77-90635	0-88318-137-1
No. 39	Physics Careers, Employment and Education (Penn State, 1977)	77-94053	0-88318-138-X
No. 40	Electrical Transport and Optical Properties of Inhomogeneous Media (Ohio State University, 1977)	78-54319	0-88318-139-8
No. 41	Nucleon-Nucleon Interactions – 1977 (Vancouver)	78-54249	0-88318-140-1
No. 42	Higher Energy Polarized Proton Beams (Ann Arbor, 1977)	78-55682	0-88318-141-X
No. 43	Particles and Fields – 1977 (APS/DPF, Argonne)	78-55683	0-88318-142-8
No. 44	Future Trends in Superconductive Electronics (Charlottesville, 1978)	77-9240	0-88318-143-6
No. 45	New Results in High Energy Physics – 1978 (Vanderbilt Conference)	78-67196	0-88318-144-4
No. 46	Topics in Nonlinear Dynamics (La Jolla Institute)	78-57870	0-88318-145-2
No. 47	Clustering Aspects of Nuclear Structure and Nuclear Reactions (Winnepeg, 1978)	78-64942	0-88318-146-0
No. 48	Current Trends in the Theory of Fields (Tallahassee, 1978)	78-72948	0-88318-147-9
No. 49	Cosmic Rays and Particle Physics – 1978 (Bartol Conference)	79-50489	0-88318-148-7
No. 50	Laser-Solid Interactions and Laser Processing – 1978 (Boston)	79-51564	0-88318-149-5

No. 51	High Energy Physics with Polarized Beams and Polarized Targets (Argonne, 1978)	79-64565	0-88318-150-9
No. 52	Long-Distance Neutrino Detection – 1978 (C.L. Cowan Memorial Symposium)	79-52078	0-88318-151-7
No. 53	Modulated Structures – 1979 (Kailua Kona, Hawaii)	79-53846	0-88318-152-5
No. 54	Meson-Nuclear Physics – 1979 (Houston)	79-53978	0-88318-153-3
No. 55	Quantum Chromodynamics (La Jolla, 1978)	79-54969	0-88318-154-1
No. 56	Particle Acceleration Mechanisms in Astrophysics (La Jolla, 1979)	79-55844	0-88318-155-X
No. 57	Nonlinear Dynamics and the Beam-Beam Interaction (Brookhaven, 1979)	79-57341	0-88318-156-8
No. 58	Inhomogeneous Superconductors – 1979 (Berkeley Springs, W.V.)	79-57620	0-88318-157-6
No. 59	Particles and Fields – 1979 (APS/DPF Montreal)	80-66631	0-88318-158-4
No. 60	History of the ZGS (Argonne, 1979)	80-67694	0-88318-159-2
No. 61	Aspects of the Kinetics and Dynamics of Surface Reactions (La Jolla Institute, 1979)	80-68004	0-88318-160-6
No. 62	High Energy e^+e^- Interactions (Vanderbilt, 1980)	80-53377	0-88318-161-4
No. 63	Supernovae Spectra (La Jolla, 1980)	80-70019	0-88318-162-2
No. 64	Laboratory EXAFS Facilities – 1980 (Univ. of Washington)	80-70579	0-88318-163-0
No. 65	Optics in Four Dimensions – 1980 (ICO, Ensenada)	80-70771	0-88318-164-9
No. 66	Physics in the Automotive Industry – 1980 (APS/AAPT Topical Conference)	80-70987	0-88318-165-7
No. 67	Experimental Meson Spectroscopy – 1980 (Sixth International Conference, Brookhaven)	80-71123	0-88318-166-5
No. 68	High Energy Physics – 1980 (XX International Conference, Madison)	81-65032	0-88318-167-3
No. 69	Polarization Phenomena in Nuclear Physics – 1980 (Fifth International Symposium, Santa Fe)	81-65107	0-88318-168-1
No. 70	Chemistry and Physics of Coal Utilization – 1980 (APS, Morgantown)	81-65106	0-88318-169-X
No. 71	Group Theory and its Applications in Physics – 1980 (Latin American School of Physics, Mexico City)	81-66132	0-88318-170-3
No. 72	Weak Interactions as a Probe of Unification (Virginia Polytechnic Institute – 1980)	81-67184	0-88318-171-1
No. 73	Tetrahedrally Bonded Amorphous Semiconductors (Carefree, Arizona, 1981)	81-67419	0-88318-172-X
No. 74	Perturbative Quantum Chromodynamics (Tallahassee, 1981)	81-70372	0-88318-173-8
No. 75	Low Energy X-Ray Diagnostics – 1981 (Monterey)	81-69841	0-88318-174-6
No. 76	Nonlinear Properties of Internal Waves (La Jolla Institute, 1981)	81-71062	0-88318-175-4

No. 77	Gamma Ray Transients and Related Astrophysical Phenomena (La Jolla Institute, 1981)	81-71543	0-88318-176-2
No. 78	Shock Waves in Condensed Matter – 1981 (Menlo Park)	82-70014	0-88318-177-0
No. 79	Pion Production and Absorption in Nuclei – 1981 (Indiana University Cyclotron Facility)	82-70678	0-88318-178-9
No. 80	Polarized Proton Ion Sources (Ann Arbor, 1981)	82-71025	0-88318-179-7
No. 81	Particles and Fields –1981: Testing the Standard Model (APS/DPF, Santa Cruz)	82-71156	0-88318-180-0
No. 82	Interpretation of Climate and Photochemical Models, Ozone and Temperature Measurements (La Jolla Institute, 1981)	82-71345	0-88318-181-9
No. 83	The Galactic Center (Cal. Inst. of Tech., 1982)	82-71635	0-88318-182-7
No. 84	Physics in the Steel Industry (APS/AISI, Lehigh University, 1981)	82-72033	0-88318-183-5
No. 85	Proton-Antiproton Collider Physics –1981 (Madison, Wisconsin)	82-72141	0-88318-184-3
No. 86	Momentum Wave Functions – 1982 (Adelaide, Australia)	82-72375	0-88318-185-1
No. 87	Physics of High Energy Particle Accelerators (Fermilab Summer School, 1981)	82-72421	0-88318-186-X
No. 88	Mathematical Methods in Hydrodynamics and Integrability in Dynamical Systems (La Jolla Institute, 1981)	82-72462	0-88318-187-8
No. 89	Neutron Scattering – 1981 (Argonne National Laboratory)	82-73094	0-88318-188-6
No. 90	Laser Techniques for Extreme Ultraviolt Spectroscopy (Boulder, 1982)	82-73205	0-88318-189-4
No. 91	Laser Acceleration of Particles (Los Alamos, 1982)	82-73361	0-88318-190-8
No. 92	The State of Particle Accelerators and High Energy Physics (Fermilab, 1981)	82-73861	0-88318-191-6
No. 93	Novel Results in Particle Physics (Vanderbilt, 1982)	82-73954	0-88318-192-4
No. 94	X-Ray and Atomic Inner-Shell Physics – 1982 (International Conference, U. of Oregon)	82-74075	0-88318-193-2
No. 95	High Energy Spin Physics – 1982 (Brookhaven National Laboratory)	83-70154	0-88318-194-0
No. 96	Science Underground (Los Alamos, 1982)	83-70377	0-88318-195-9
No. 97	The Interaction Between Medium Energy Nucleons in Nuclei – 1982 (Indiana University)	83-70649	0-88318-196-7
No. 98	Particles and Fields – 1982 (APS/DPF University of Maryland)	83-70807	0-88318-197-5
No. 99	Neutrino Mass and Gauge Structure of Weak Interactions (Telemark, 1982)	83-71072	0-88318-198-3
No. 100	Excimer Lasers – 1983 (OSA, Lake Tahoe, Nevada)	83-71437	0-88318-199-1

No. 101	Positron-Electron Pairs in Astrophysics (Goddard Space Flight Center, 1983)	83-71926	0-88318-200-9
No. 102	Intense Medium Energy Sources of Strangeness (UC-Sant Cruz, 1983)	83-72261	0-88318-201-7
No. 103	Quantum Fluids and Solids – 1983 (Sanibel Island, Florida)	83-72440	0-88318-202-5
No. 104	Physics, Technology and the Nuclear Arms Race (APS Baltimore –1983)	83-72533	0-88318-203-3
No. 105	Physics of High Energy Particle Accelerators (SLAC Summer School, 1982)	83-72986	0-88318-304-8
No. 106	Predictability of Fluid Motions (La Jolla Institute, 1983)	83-73641	0-88318-305-6
No. 107	Physics and Chemistry of Porous Media (Schlumberger-Doll Research, 1983)	83-73640	0-88318-306-4
No. 108	The Time Projection Chamber (TRIUMF, Vancouver, 1983)	83-83445	0-88318-307-2
No. 109	Random Walks and Their Applications in the Physical and Biological Sciences (NBS/La Jolla Institute, 1982)	84-70208	0-88318-308-0
No. 110	Hadron Substructure in Nuclear Physics (Indiana University, 1983)	84-70165	0-88318-309-9
No. 111	Production and Neutralization of Negative Ions and Beams (3rd Int'l Symposium, Brookhaven, 1983)	84-70379	0-88318-310-2
No. 112	Particles and Fields – 1983 (APS/DPF, Blacksburg, VA)	84-70378	0-88318-311-0
No. 113	Experimental Meson Spectroscopy – 1983 (Seventh International Conference, Brookhaven)	84-70910	0-88318-312-9
No. 114	Low Energy Tests of Conservation Laws in Particle Physics (Blacksburg, VA, 1983)	84-71157	0-88318-313-7
No. 115	High Energy Transients in Astrophysics (Santa Cruz, CA, 1983)	84-71205	0-88318-314-5
No. 116	Problems in Unification and Supergravity (La Jolla Institute, 1983)	84-71246	0-88318-315-3
No. 117	Polarized Proton Ion Sources (TRIUMF, Vancouver, 1983)	84-71235	0-88318-316-1
No. 118	Free Electron Generation of Extreme Ultraviolet Coherent Radiation (Brookhaven/OSA, 1983)	84-71539	0-88318-317-X
No. 119	Laser Techniques in the Extreme Ultraviolet (OSA, Boulder, Colorado, 1984)	84-72128	0-88318-318-8
No. 120	Optical Effects in Amorphous Semiconductors (Snowbird, Utah, 1984)	84-72419	0-88318-319-6
No. 121	High Energy e e Interactions (Vanderbilt, 1984)	84-72632	0-88318-320-X
No. 122	The Physics of VLSI (Xerox, Palo Alto, 1984)	84-72729	0-88318-321-8
No. 123	Intersections Between Particle and Nuclear Physics (Steamboat Springs, 1984)	84-72790	0-88318-322-6

No. 124	Neutron-Nucleus Collisions – A Probe of Nuclear Structure (Burr Oak State Park - 1984)	84-73216	0-88318-323-4
No. 125	Capture Gamma-Ray Spectroscopy and Related Topics – 1984 (Internat. Symposium, Knoxville)	84-73303	0-88318-324-2
No. 126	Solar Neutrinos and Neutrino Astronomy (Homestake, 1984)	84-63143	0-88318-325-0
No. 127	Physics of High Energy Particle Accelerators (BNL/SUNY Summer School, 1983)	85-70057	0-88318-326-9
No. 128	Nuclear Physics with Stored, Cooled Beams (McCormick's Creek State Park, Indiana, 1984)	85-71167	0-88318-327-7
No. 129	Radiofrequency Plasma Heating (Sixth Topical Conference, Callaway Gardens, GA, 1985)	85-48027	0-88318-328-5
No. 130	Laser Acceleration of Particles (Malibu, California, 1985)	85-48028	0-88318-329-3
No. 131	Workshop on Polarized ^{3}He Beams and Targets (Princeton, New Jersey, 1984)	85-48026	0-88318-330-7
No. 132	Hadron Spectroscopy–1985 (International Conference, Univ. of Maryland)	85-72537	0-88318-331-5
No. 133	Hadronic Probes and Nuclear Interactions (Arizona State University, 1985)	85-72638	0-88318-332-3
No. 134	The State of High Energy Physics (BNL/SUNY Summer School, 1983)	85-73170	0-88318-333-1
No. 135	Energy Sources: Conservation and Renewables (APS, Washington, DC, 1985)	85-73019	0-88318-334-X
No. 136	Atomic Theory Workshop on Relativistic and QED Effects in Heavy Atoms	85-73790	0-88318-335-8
No. 137	Polymer-Flow Interaction (La Jolla Institute, 1985)	85-73915	0-88318-336-6
No. 138	Frontiers in Electronic Materials and Processing (Houston, TX, 1985)	86-70108	0-88318-337-4
No. 139	High-Current, High-Brightness, and High-Duty Factor Ion Injectors (La Jolla Institute, 1985)	86-70245	0-88318-338-2
No. 140	Boron-Rich Solids (Albuquerque, NM, 1985)	86-70246	0-88318-339-0
No. 141	Gamma-Ray Bursts (Stanford, CA, 1984)	86-70761	0-88318-340-4
No. 142	Nuclear Structure at High Spin, Excitation, and Momentum Transfer (Indiana University, 1985)	86-70837	0-88318-341-2
No. 143	Mexican School of Particles and Fields (Oaxtepec, México, 1984)	86-81187	0-88318-342-0
No. 144	Magnetospheric Phenomena in Astrophysics (Los Alamos, 1984)	86-71149	0-88318-343-9
No. 145	Polarized Beams at SSC & Polarized Antiprotons (Ann Arbor, MI & Bodega Bay, CA 1985)	86-71343	0-88318-344-7

No. 146	Advances in Laser Science–I (Dallas, TX, 1985)	86-71536	0-88318-345-5
No. 147	Short Wavelength Coherent Radiation: Generation and Applications (Monterey, CA, 1986)	86-71674	0-88318-346-3
No. 148	Space Colonization: Technology and The Liberal Arts (Geneva, NY, 1985)	86-71675	0-88318-347-1
No. 149	Physics and Chemistry of Protective Coatings (Universal City, CA, 1985)	86-72019	0-88318-348-X
No. 150	Intersections Between Particle and Nuclear Physics (Lake Louise, Canada, 1986)	86-72018	0-88318-349-8
No. 151	Neural Networks for Computing (Snowbird, UT, 1986)	86-72481	0-88318-351-X
No. 152	Heavy Ion Inertial Fusion (Washington, DC, 1986)	86-73185	0-88318-352-8
No. 153	Physics of Particle Accelerators (SLAC Summer School, 1985) (Fermilab Summer School, 1984)	87-70103	0-88318-353-6
No. 154	Physics and Chemistry of Porous Media—II (Ridge Field, CT, 1986)	83-73640	0-88318-354-4
No. 155	The Galactic Center: Proceedings of the Symposium Honoring C. H. Townes (Berkeley, CA, 1986)	86-73186	0-88318-355-2
No. 156	Advanced Accelerator Concepts (Madison, WI, 1986)	87-70635	0-88318-358-0
No. 157	Stability of Amorphous Silicon Alloy Materials and Devices (Palo Alto, CA, 1987)	87-70990	0-88318-359-9
No. 158	Production and Neutralization of Negative Ions and Beams (Brookhaven, NY, 1986)	87-71695	0-88318-358-7
No. 159	Applications of Radio-Frequency Power to Plasma: Seventh Topical Conference (Kissimmee, FL, 1987)	87-71812	0-88318-359-5
No. 160	Advances in Laser Science–II (Seattle, WA, 1986)	87-71962	0-88318-360-9
No. 161	Electron Scattering in Nuclear and Particle Science: In Commemoration of the 35th Anniversary of the Lyman-Hanson-Scott Experiment (Urbana, IL, 1986)	87-72403	0-88318-361-7
No. 162	Few-Body Systems and Multiparticle Dynamics	87-72594	0-88318-362-5
No. 163	Pion–Nucleus Physics: Future Directions and New Facilities at LAMPF	87-72961	0-88318-363-3

RAYMOND H. FOGLER LIBRARY
DATE DUE

BOOKS ARE SUBJECT TO
RECALL AFTER TWO WEEKS

MAY 27 1988